# Transgenic Plants and Crops

# TRANSGENIC PLANTS AND CROPS

EDITED BY

## George G. Khachatourians
*University of Saskatchewan*
*Saskatoon, Saskatchewan, Canada*

## Alan McHughen
*University of Saskatchewan*
*Saskatoon, Saskatchewan, Canada*

## Ralph Scorza
*U.S. Department of Agriculture*
*Kearneysville, West Virginia*

## Wai-Kit Nip
*University of Hawaii at Manoa*
*Honolulu, Hawaii*

## Y. H. Hui
*Science Technology System*
*West Sacramento, California*

MARCEL DEKKER, INC.   NEW YORK · BASEL

**Library of Congress Cataloging-in-Publication Data**

Transgenic plants and crops / edited by George G. Khachatourians . . . [et al.].
   p. cm.
   ISBN 0-8247-0545-9 (alk. paper)
   1. Transgenic plants  I. Khachatourians, George G.

SB123.57 .T7315 2002
631.5′233—dc21                                                                  2001055252

This book is printed on acid-free paper.

**Headquarters**
Marcel Dekker, Inc.
270 Madison Avenue, New York, NY 10016
tel: 212-696-9000; fax: 212-685-4540

**Eastern Hemisphere Distribution**
Marcel Dekker AG
Hutgasse 4, Postfach 812, CH-4001 Basel, Switzerland
tel: 41-61-261-8482; fax: 41-61-261-8896

**World Wide Web**
http://www.dekker.com

The publisher offers discounts on this book when ordered in bulk quantities. For more information, write to Special Sales/Professional Marketing at the headquarters address above.

**Copyright © 2002 by Marcel Dekker, Inc. All Rights Reserved.**

Neither this book nor any part may be reproduced or transmitted in any form or by any means, electronic or mechanical, including photocopying, microfilming, and recording, or by any information storage and retrieval system, without permission in writing from the publisher.

Current printing (last digit):
10 9 8 7 6 5 4 3 2 1

**PRINTED IN THE UNITED STATES OF AMERICA**

# Preface

The world's population has increased steadily through the centuries and is forecast to reach 7.5 billion by the year 2020. Unless it stabilizes at 8.5 to 10 billion people within the next 50 years, all the available arable land will be dedicated to producing food of plant origin. Even at the current levels of agricultural land use, there are significant concerns about water usage. Of all water on Earth, 70% is available for use. Of this available water, 65% is used for agriculture, 22% is used for industries, and 7% is used for households and municipalities. Increasing human population will constrain future agricultural water use. The production of food crops is further held back by plant pests and diseases, unpredictable changes in the weather, and increased pressure on agricultural land in use for increased productivity, diversification, and encroaching urbanization. A new paradigm in the supply of crops and the demand created by the world's population is required, necessitating better management of natural resources, by knowledge-based innovations in new technologies, including genetic engineering, as well as by regional and global institutional policies and regulations.

In the last century, efforts through traditional plant breeding techniques have increased many agronomic aspects and nutritional values of food plants. The main drawback of traditional plant breeding is that it relies on the use of germplasm of the same or closely related species, which is sometimes a serious limiting factor. In addition, progress is time-consuming and relies on the extensive use of natural resources. Food crop biotechnology is an alternative option for plant improvement.

Early developments arising from transgenic techniques in plants in the mid-1970s produced tomatoes, corn, soybean, and canola with modified traits. Since the 1980s, these techniques have matured, become more target specific, and generated transgenic plants of most major food crops. However, these techniques still need to become more precise to avoid the inadvertent introduction of undesirable genes, such as those carrying allergenicity or those that can cause weediness and endanger natural ecosystems. Political, economic, social, regulatory, and legal issues also must be considered as society debates the deployment of these new food production technologies. The challenge lies in resolving these societal issues while at the same time meeting the demands of an increasing world population. Meeting these needs requires the intellectual and technical resources of agricultural sciences, knowledge and manipulation of genetics, the tools of molecular biology, and open communication among and between scientists and society at large.

We have brought together in this volume essential information related to the production of select transgenic food plants. Our 130 contributors present both general and specific examples of recent achievements in transgenic plants and crops. This book is divided into four parts with a total of 55 chapters: "Principles and Applications," which covers social, political, regulatory, and legal issues; "Fruits"; "Vegetables"; and "Grains and Other Seeds." We hope our readers share our excitement about genetics and what may lie ahead.

The completion of this task could not have been achieved without the cooperation of the chapter contributors and reviewers, the support of Peter McCann at AgWest Biotech (Saskatoon), and the editors at Marcel Dekker, Inc., especially Theresa Dominick.

*George G. Khachatourians*
*Alan McHughen*
*Ralph Scorza*
*Wai-Kit Nip*
*Y. H. Hui*

# Reviewers

To ensure accuracy of information, each chapter was reviewed by experts in industry, government, and academia. Although most chapters had one or two reviewers, some had as many as five or six.

The authors of the chapters also served as reviewers for chapters other than their own. They are not all included in this list.

This list is incomplete for a variety of other reasons as well. The review process was spread over three years and some names have been misplaced, some reviewers wish to remain anonymous, and so on. A note of appreciation and/or apology is extended to those reviewers not included in this list.

**Roger N. Beachy**  Division of Plant Biology, The Scripps Research Institute, La Jolla, California

**Robert L. Brown**  Southern Regional Research Center, Agricultural Research Service, U.S. Department of Agriculture, New Orleans, Louisiana

**Ravindra N. Chibbar**  Plant Biotechnology Institute, National Research Council, Saskatoon, Saskatchewan, Canada

**H. A. Collin**  Department of Biology, University of Liverpool, Liverpool, England

**Glenn B. Collins**  Department of Agronomy, University of Kentucky, Lexington, Kentucky

**Raju Datla**  Plant Biotechnology Institute, National Research Council, Saskatoon, Saskatchewan, Canada

**Christopher D'Hulst**  University of Lille, Lille, France

**Alison M. R. Ferrie**  Plant Biotechnology Institute, National Research Council, Saskatoon, Saskatchewan, Canada

**Tatsuhito Fujimura**  Institute of Agricultural and Forest Engineering, University of Tsukuba, Tsukuba, Ibaraki, Japan

**Fawzy Gerges**  Plant Biotechnology Institute, National Research Council, Saskatoon, Saskatchewan, Canada

**Steve Gleddie**  Eastern Cereal and Oilseed Research Centre, Agriculture and Agri-Food Canada, Ottawa, Ontario, Canada

**Dennis Gonsalves**  Department of Plant Pathology, Cornell University, Geneva, New York

**Anju Gulati**  Crop Development Centre, Department of Plant Sciences, College of Agriculture, University of Saskatchewan, Saskatoon, Saskatchewan, Canada

**Timothy C. Hall**  Institute of Developmental and Molecular Biology, Department of Biology, Texas A&M University, College Station, Texas

**Abdelali Hannouffa**  Research Center, Agriculture and Agri-Food Canada, Saskatoon, Saskatchewan, Canada

**M. Hasegawa**  Horticulture Department, Purdue University, West Lafayette, Indiana

**Maarten A. Jongsma**  Department of Molecular Biology, DLO-Centre for Plant Breeding and Reproduction Research (CPRO-DLO), Wageningen, The Netherlands

**John Kemp**  Plant Genetic Engineering Laboratory, New Mexico State University, Las Cruces, New Mexico

**George G. Khachatourians**  Department of Applied Microbiology and Food Science, College of Agriculture, University of Saskatchewan, Saskatoon, Saskatchewan, Canada

**L. M. Khachatourians**  Prairie Region Health Promotion Research Center, Department of Community Health and Epidemiology, College of Medicine, Royal University Hospital, Saskatoon, Saskatchewan, Canada

**Danny J. Llewellyn**  CSIRO Plant Industry, Canberra, Australian Capital Territory, Australia

**John Mahon**  Plant Biotechnology Institute, National Research Council, Saskatoon, Saskatchewan, Canada

**Alan McHughen**  Crop Development Centre, Department of Plant Sciences, College of Agriculture, University of Saskatchewan, Saskatoon, Saskatchewan, Canada

**Steve Millam**  Scottish Crop Research Institute, Dundee, Scotland

**Norimoto Murai**  Department of Plant Pathology and Crop Physiology, Louisiana State University, Baton Rouge, Louisiana

**Tatsuro Murata**  Faculty of Agriculture, Kyushu Tokai University, Kumamoto, Japan

**Constantine E. Palmer**  Department of Plant Science, University of Manitoba, Winnipeg, Manitoba, Canada

**Larry Pelcher**  Plant Biotechnology Institute, National Research Council, Saskatoon, Saskatchewan, Canada

**Reviewers**

**Joseph F. Petolino**   Biotechnology and Plant Genetics, Dow AgroSciences, Indianapolis, Indiana

**Peter W. B. Phillips**   Department of Agriculural Economics, College of Agriculture, University of Saskatchewan, Saskatoon, Saskatchewan, Canada

**Patricia Polowick**   Plant Biotechnology Institute, National Research Council, Saskatoon, Saskatchewan, Canada

**William M. Roca**   Biotechnology Unit, CIAT, Cali, Colombia

**William L. Rooney**   Soil and Crop Sciences Department, College of Agriculture and Life Sciences, Texas A&M University, College Station, Texas

**Kevin Rozwadowski**   Saskatoon Research Centre, Agriculture and Agri-Food Canada, Saskatoon, Saskatchewan, Canada

**C. Schmidt**   Laboratoire de Virologie, Institut National de la Recherche Agronomique, Colmar, France

**Andrew Sharp**   Plant Biotechnology Institute, National Research Council, Saskatoon, Saskatchewan, Canada

**Ko Shimamoto**   Plant Molecular Genetics, Department of Molecular Biology, Nara Institute of Science and Technology, Nara, Japan

**K. R. Shivana**   Department of Botany, University of Delhi, Delhi, India

**Phil Simon**   Department of Horticulture, University of Wisconsin–Madison, Madison, Wisconsin

**Taylor Steeves**   Department of Biology, University of Saskatchewan, Saskatoon, Saskatchewan, Canada

**Daryl Summers**   Plant Biotechnology Institute, National Research Council, Saskatoon, Saskatchewan, Canada

**R. T. Tyler**   Department of Applied Microbiology and Food Science, College of Agriculture, University of Saskatchewan, Saskatoon, Saskatchewan, Canada

**Richard G. F. Visser**   Department of Plant Breeding, Agricultural University of Wageningen, Wageningen, The Netherlands

**Hung-Mei Wang**   Aventis Crop Science Canada Co., Saskatoon, Saskatchewan, Canada

**A. Woytowich**   Department of Applied Microbiology and Food Science, College of Agriculture, University of Saskatchewan, Saskatoon, Saskatchewan, Canada

**Shyi-Dong Yeh**   Department of Plant Pathology, National Chung Hsing University, Taichung, Taiwan

**L. M. Yepes**   Laboratoire de Virologie, Institut National de la Recherche Agronomique, Colmar, France

**Su-May Yu**   Institute of Molecular Science, Academia Sinica, Nankang, Taipei, Taiwan

# Contents

*Preface*   *iii*
*Reviewers*   *v*
*Contributors*   *xv*

## I. Principles and Applications

1. Agricultural and Food Crops: Development, Science, and Society   1
   *George G. Khachatourians*

2. The Dynamics of Plant Genome Organization   29
   *Isobel A. Parkin and Derek J. Lydiate*

3. Embryogenesis   43
   *Alison M. R. Ferrie*

4. Shoot Regeneration and Proliferation   69
   *Seedhabadee Ganeshan, Karen L. Caswell, Kutty K. Kartha, and Ravindra N. Chibbar*

5. Techniques for Gene Marking, Transferring, and Tagging   85
   *Albert Abbott*

6. Pollen Biotechnology   99
   *Vipen K. Sawhney*

7. Parent-of-Origin Effects and Seed Development: Genetics and Epigenetics   109
   *Charles Spillane, Jean-Philippe Vielle-Calzada, and Ueli Grossniklaus*

8. Direct DNA Delivery Into Intact Cells and Tissues   137
   *Joseph F. Petolino*

| | | |
|---|---|---|
| 9. | Electroporation and Cell Energy Factor<br>*Paul F. Lurquin, Guangyu Chen, and Anthony Conner* | 145 |
| 10. | Cell Culture and Regeneration of Plant Tissues<br>*Wei Wen Su* | 151 |
| 11. | Genetic Engineering for Modified Starch Structure in Cereals<br>*Ming Gao, Monica Båga, and Ravindra N. Chibbar* | 169 |
| 12. | Improving Crop Performance Through Transgenic Modification of Flowering<br>*Pierre Fobert* | 183 |
| 13. | Genetic Technology in Peas for Improved Field Performance and Enhanced Grain Quality<br>*Roger Leslie Morton, Stephanie Gollasch, Hart E. Schroeder, Kaye S. Bateman, and Thomas J. Higgins* | 197 |
| 14. | Genetic Engineering for Levels of Select Phytonutrients Affecting Human Health<br>*George G. Khachatourians* | 207 |
| 15. | Genetic Engineering and Resistance to Viruses<br>*Marc Fuchs and Dennis Gonsalves* | 217 |
| 16. | Genetic Engineering for Resistance to Nematodes<br>*Daguang Cai, Urs Wyss, Christian Jung, and Michael Kleine* | 233 |
| 17. | Genetic Engineering and Resistance to Insects<br>*Dwayne D. Hegedus, Margaret Y. Gruber, Lorraine Braun, and George G. Khachatourians* | 249 |
| 18. | Intellectual Property Protection for Transgenic Plants<br>*Brian G. Kingwell and Joy D. Morrow* | 279 |
| 19. | Public Perceptions of Transgenic Plants<br>*Thomas Jefferson Hoban* | 293 |
| 20. | Industry Perspectives<br>*Katherine A. Means* | 305 |
| 21. | Political and Economic Consequences<br>*Peter W. B. Phillips and George G. Khachatourians* | 311 |

## II. Fruits

| | | |
|---|---|---|
| 22. | Introduction and Expression of Transgenes in Apples<br>*Abhaya M. Dandekar* | 327 |
| 23. | Genetic Transformation of Avocado<br>*Richard E. Litz and Witjaksono* | 345 |

## Contents

| | | |
|---|---|---|
| 24. | Production of Transgenic Banana (*Musa* species)<br>László Sági, Serge Remy, Juan Bernardo Pérez Hernández, and Rony Swennen | 359 |
| 25. | Production of Transgenic Melon<br>Ekaterini Papadopoulou and Rebecca Grumet | 371 |
| 26. | Cranberry Transformation and Regeneration<br>James J. Polashock and Nicholi Vorsa | 383 |
| 27. | Transgenic Grapevines<br>Dennis J. Gray, Subramanian Jayasankar, Zhijian Li, John Cordts, Ralph Scorza, and C. Srinivasan | 397 |
| 28. | Genetic Transformation of Kiwifruit (*Actinidia* species)<br>M. Margarida Oliveira and M. Helena Raquel | 407 |
| 29. | Genetic Transformation of Mango (*Mangifera indica* L.)<br>Richard E. Litz and Miguel A. Gomez-Lim | 421 |
| 30. | Transgenic Papayas in Hawaii—A Useful Tool for New Cultivar Development and Clonal Propagation<br>Maureen M. M. Fitch | 437 |
| 31. | Genetic Engineering of Strawberries and Raspberries<br>Robert R. Martin | 449 |

### III. Vegetables

| | | |
|---|---|---|
| 32. | Progress in Asparagus Biotechnology<br>Amnon Levi and Kenneth C. Sink | 465 |
| 33. | Generation of Transgenic Bean (*Phaseolus* species) Plants for Improvement of Nutritional Quality<br>Jae-Whune Kim and Chee H. Harn | 475 |
| 34. | Genetic Engineering of Beet and the Concept of the Plant as a Factory<br>Robert Sévenier, Andries J. Koops, and Robert D. Hall | 485 |
| 35. | Transgenic Carrots with Enhanced Tolerance to Fungal Pathogens<br>Zamir K. Punja | 503 |
| 36. | Transgenic Cassava for Food Security and Economic Development<br>Nigel J. Taylor, M. V. Masona, Claude M. Fauquet, and Christian Schöpke | 523 |
| 37. | Transgenic Cauliflower with Insect Resistance<br>K. Chengalrayan, Yih-Ming Chen, Kai-Wun Yeh, and Po-Jen Wang | 547 |

| | | |
|---|---|---|
| 38. | Virus-Resistant Chili Pepper Produced by *Agrobacterium* Species–Mediated Transformation<br>Wen-qi Cai, Rong-Xiang Fang, Feng-li Zhang, Jiu-chun Zhang, Xiaoying Chen, Gui-ling Wang, Ke-qiang Mang, Hong-sheng Shang, Xu Wang, and Yue-ren Li | 563 |
| 39. | Transgenic Cucumber with Resistance to Cucumber Mosaic Virus<br>Soryu Nishibayashi | 579 |
| 40. | Transgenic Parthenocarpic and Insect-Resistant Eggplant<br>Giuseppe Leonardo Rotino, Francesco Sunseri, Nazareno Acciarri, Salvatore Arpaia, Giuseppe Mennella, Angelo Spena, and Michela Zottini | 587 |
| 41. | Transgenic Cowpea, Lentil, and Chickpea with Reporter and Agronomically Relevant Genes<br>Paul F. Lurquin, Edgardo Filippone, and Gabriella Colucci | 603 |
| 42. | Genetic Manipulation of Lettuce<br>Michael Raymond Davey, Matthew Sean McCabe, Umaballava Mohapatra, and J. Brian Power | 613 |
| 43. | Maize Food and Feed: A Current Perspective and Consideration of Future Possibilities<br>Bruce R. Hamaker and Brian A. Larkins | 637 |
| 44. | The Transformation of Onions and Related Alliums<br>Colin C. Eady | 655 |
| 45. | Potato Transformation Produces Value-Added Traits<br>Lawrence M. Kawchuk | 673 |
| 46. | Transgenic Sweet Potato with Agronomically Important Genes<br>Motoyasu Otani and Takiko Shimada | 699 |

## IV. Grains and Other Seeds

| | | |
|---|---|---|
| 47. | Genetic Enrichment of Barley: Opportunities and Challenges<br>Seedhabadee Ganeshan, Monica Båga, and Ravindra N. Chibbar | 717 |
| 48. | Transgenic Coffee<br>Maria Filomena Carneiro and John I. Stiles | 737 |
| 49. | Transgenic Linseed Flax<br>Alan McHughen | 747 |
| 50. | Antimicrobial Peptides from Macadamia Nuts: Potential Source of Novel Resistance in Transgenic Crops<br>Kemal Kazan, John P. Marcus, Ken C. Goulter, and John M. Manners | 763 |

**Contents**

| | | |
|---|---|---|
| 51. | Transgenic Oilseed Brassicas<br>*Constantine E. Palmer and Wilf A. Keller* | 773 |
| 52. | Studies on Genetic Engineering of Rice<br>*Honghong Zheng, Xiaotian Ming, Yi Li, Hongya Gu, and Zhangliang Chen* | 793 |
| 53. | Transgenic Sorghum with Improved Nutritional Quality<br>*Yohannes Tadesse and Michel Jacobs* | 801 |
| 54. | Sunflower Seed<br>*Günther Hahne* | 813 |
| 55. | Transformation of Wheat<br>*Mark C. Jordan, Fredy Altpeter, and Javed A. Qureshi* | 835 |

*Index*     *849*

# Contributors

**Albert Abbott**  Department of Biological Sciences, Clemson University, Clemson, South Carolina

**Nazareno Acciarri**  Research Institute for Vegetable Crops, Monsampolo del Tronto, Italy

**Fredy Altpeter**  Department of Molecular Cell Biology, Institut für Pflanzengenetik und Kulturpflanzenforschung Gatersleben, Gatersleben, Germany

**Salvatore Arpaia**  Unit of Agronomy and Crop Protection, Metapontum Agrobios, Metaponto, Italy

**Monica Båga**  Plant Biotechnology Institute, National Research Council, Saskatoon, Saskatchewan, Canada

**Kaye S. Bateman**  CSIRO Plant Industry, Canberra, Australian Capital Territory, Australia

**Lorraine Braun**  Department of Ecological Crop Protection, Agriculture and Agri-Food Canada, Saskatoon, Saskatchewan, Canada

**Daguang Cai**  Institute of Crop Science and Plant Breeding, University of Kiel, Kiel, Germany

**Wen-qi Cai**  Laboratory of Plant Biotechnology, Institute of Microbiology, Chinese Academy of Sciences, Beijing, China

**Maria Filomena Carneiro**  Centro de Investigação das Ferrugens do Cafeeiro, Instituto de Investigação Científica Tropical, Oeiras, Portugal

**Karen L. Caswell**  Plant Biotechnology Institute, National Research Council, Saskatoon, Saskatchewan, Canada

**Guangyu Chen**  Biotechnology Center, Jiangxi Academy of Agricultural Sciences, Nanchang, China

**Xiaoying Chen**  Laboratory of Plant Biotechnology, Institute of Microbiology, Chinese Academy of Sciences, Beijing, China

**Yih-Ming Chen**  Center for Biotechnology, Department of Botany, National Taiwan University, Taipei, Taiwan

**Zhangliang Chen**  National Laboratory of Protein Engineering and Plant Genetics Engineering, College of Life Sciences, Peking University, Beijing, China

**K. Chengalrayan**  Department of Botany, National Taiwan University, Taipei, Taiwan

**Ravindra N. Chibbar**  Plant Biotechnology Institute, National Research Council, Saskatoon, Saskatchewan, Canada

**Gabriella Colucci**  Department of Agronomy and Plant Genetics, University of Naples, Naples, Italy

**Anthony Conner**  Soil Plant and Ecological Sciences Division, Lincoln University, Canterbury, New Zealand

**John Cordts**  Profigen, Inc., Paterson, Washington

**Abhaya M. Dandekar**  Department of Pomology, University of California, Davis, California

**Michael Raymond Davey**  Plant Science Division, School of Biosciences, University of Nottingham, Nottingham, England

**Colin C. Eady**  New Zealand Institute for Crop & Food Research Ltd., Christchurch, New Zealand

**Rong-Xiang Fang**  Laboratory of Plant Biotechnology, Institute of Microbiology, Chinese Academy of Sciences, Beijing, China

**Claude M. Fauquet**  International Laboratory for Tropical Agricultural Biotechnology, ILTAB/Donald Danforth Plant Science Center, St. Louis, Missouri

**Alison M. R. Ferrie**  Plant Biotechnology Institute, National Research Council, Saskatoon, Saskatchewan, Canada

**Edgardo Filippone**  Department of Agronomy and Plant Genetics, University of Naples, Naples, Italy

**Maureen M. M. Fitch**  Pacific Basin Agricultural Center, Agricultural Research Service, U.S. Department of Agriculture, Aiea, Hawaii

**Pierre Fobert**  Plant Biotechnology Institute, National Research Council, Saskatoon, Saskatchewan, Canada

**Marc Fuchs**  Laboratoire de Virologie, Unité de Recherche Vigne et Vin, Institut National de la Recherche Agronomique, Colmar, France

**Contributors**

**Seedhabadee Ganeshan**  Plant Biotechnology Institute, National Research Council, Saskatoon, Saskatchewan, Canada

**Ming Gao**  Promoter and Gene Discovery Group, Plant Biotechnology Institute, National Research Council, Saskatoon, Saskatchewan, Canada

**Stephanie Gollasch**  CSIRO Plant Industry, Canberra, Australian Capital Territory, Australia

**Miguel A. Gomez-Lim**  CINVESTAV Unidad Irapuato, Irapuato, Mexico

**Dennis Gonsalves**  Department of Plant Pathology, Cornell University, Geneva, New York

**Ken C. Goulter**  Cooperative Research Centre for Tropical Plant Protection, The University of Queensland, Brisbane, Queensland, Australia

**Dennis J. Gray**  Mid-Florida Research and Education Center, Institute of Food and Agricultural Sciences, University of Florida, Apopka, Florida

**Ueli Grossniklaus**  Institute of Plant Biology, University of Zurich, Zurich, Switzerland

**Margaret Y. Gruber**  Department of Molecular Genetics, Agriculture and Agri-Food Canada, Saskatoon, Saskatchewan, Canada

**Rebecca Grumet**  Department of Horticulture, Michigan State University, East Lansing, Michigan

**Hongya Gu**  Biotechnology Department, College of Life Sciences, Peking University, Beijing, China

**Günther Hahne**  Institut de Biologie Moléculaire des Plantes, CNRS and Louis Pasteur University, Strasbourg, France

**Robert D. Hall**  Cell Cybernetics, Plant Research International, Wageningen, The Netherlands

**Bruce R. Hamaker**  Department of Food Science, Purdue University, West Lafayette, Indiana

**Chee H. Harn**  Biotechnology Center, Nong Woo Bio Co., Ltd., Kyonggi-do, Korea

**Dwayne D. Hegedus**  Department of Molecular Genetics, Agriculture and Agri-Food Canada, Saskatoon, Saskatchewan, Canada

**Juan Bernardo Pérez Hernández**  Laboratory of Tropical Crop Improvement, Catholic University of Leuven, Leuven, Belgium

**Thomas J. Higgins**  CSIRO Plant Industry, Canberra, Australian Capital Territory, Australia

**Thomas Jefferson Hoban**  Department of Sociology and Food Science, North Carolina State University, Raleigh, North Carolina

**Michel Jacobs**   Laboratory of Plant Genetics, Department of Biology, Free University of Brussels, Sint-Genesius-Rode, Belgium

**Subramanian Jayasankar**   Mid-Florida Research and Education Center, Institute of Food and Agricultural Sciences, University of Florida, Apopka, Florida

**Mark C. Jordan**   Cereal Research Centre, Agriculture and Agri-Food Canada, Winnipeg, Manitoba, Canada

**Christian Jung**   Institute of Crop Science and Plant Breeding, University of Kiel, Kiel, Germany

**Kutty K. Kartha**   Plant Biotechnology Institute, National Research Council, Saskatoon, Saskatchewan, Canada

**Lawrence M. Kawchuk**   Lethbridge Research Centre, Agriculture and Agri-Food Canada, Lethbridge, Alberta, Canada

**Kemal Kazan**   Cooperative Research Centre for Tropical Plant Protection, The University of Queensland, Brisbane, Queensland, Australia

**Wilf A. Keller**   Plant Biotechnology Institute, National Research Council, Saskatoon, Saskatchewan, Canada

**George G. Khachatourians**   Department of Applied Microbiology and Food Science, College of Agriculture, University of Saskatchewan, Saskatoon, Saskatchewan, Canada

**Jae-Whune Kim**   Plant Cell Culture Research Unit, Microplants Co., Ltd., Chollabukdo, Korea

**Brian G. Kingwell**   Smart and Biggar/Fetherstonhaugh and Co., Vancouver, British Columbia, Canada

**Michael Kleine**   Institute of Crop Science and Plant Breeding, University of Kiel, Kiel, Germany

**Andries J. Koops**   Cell Cybernetics, Plant Research International, Wageningen, The Netherlands

**Brian A. Larkins**   Department of Plant Sciences, The University of Arizona, Tucson, Arizona

**Amnon Levi**   U.S. Vegetable Laboratory, Agricultural Research Service, U.S. Department of Agriculture, Charleston, South Carolina

**Yi Li**   National Laboratory of Protein Engineering and Plant Genetics Engineering, College of Life Sciences, Peking University, Beijing, China

**Yue-ren Li**   College of Plant Protection, Northwest Sci-Tech University of Agriculture and Forestry, Yangling, Shaanxi, China

**Contributors**  xix

**Zhijian Li**  Mid-Florida Research and Education Center, Institute of Food and Agricultural Sciences, University of Florida, Apopka, Florida

**Richard E. Litz**  Tropical Research and Education Center, University of Florida, Homestead, Florida

**Paul F. Lurquin**  School of Molecular Biosciences, Washington State University, Pullman, Washington

**Derek J. Lydiate**  Molecular Genetics Section, Saskatoon Research Centre, Agriculture and Agri-Food Canada, Saskatoon, Saskatchewan, Canada

**Ke-qiang Mang**  Laboratory of Plant Biotechnology, Institute of Microbiology, Chinese Academy of Sciences, Beijing, China

**John M. Manners**  Cooperative Research Centre for Tropical Plant Protection, The University of Queensland, Brisbane, Queensland, Australia

**John P. Marcus**  Cooperative Research Centre for Tropical Plant Protection, The University of Queensland, Brisbane, Queensland, Australia

**Robert R. Martin**  Agricultural Research Service, U.S. Department of Agriculture, Corvallis, Oregon

**M. V. Masona**  International Laboratory for Tropical Agricultural Biotechnology, ILTAB/Donald Danforth Plant Science Center, St. Louis, Missouri

**Matthew Sean McCabe**  Plant Science Division, School of Biosciences, University of Nottingham, Nottingham, England

**Alan McHughen**  Crop Development Centre, Department of Plant Sciences, College of Agriculture, University of Saskatchewan, Saskatoon, Saskatchewan, Canada

**Katherine A. Means**  Produce Marketing Association, Newark, Delaware

**Giuseppe Mennella**  Research Institute for Vegetable Crops, Pontecagnano, Italy

**Xiaotian Ming**  National Laboratory of Protein Engineering and Plant Genetics Engineering, College of Life Sciences, Peking University, Beijing, China

**Umaballava Mohapatra**  Plant Science Division, School of Biosciences, University of Nottingham, Nottingham, England

**Joy D. Morrow**  Smart and Biggar/Fetherstonhaugh and Co., Vancouver, British Columbia, Canada

**Roger Leslie Morton**  Department of Genetic Engineering for Plant Improvement, CSIRO Plant Industry, Canberra, Australian Capital Territory, Australia

**Soryu Nishibayashi**   Bioinformatics Project Department, Mitsubishi Space Software Co., Ltd., Amagasaki City, Japan

**M. Margarida Oliveira**   Biologia Vegetal, Faculdade de Ciências de Lisboa, Lisbon, and Instituto de Biologia Experimental e Technológica, Oeiras, Portugal

**Motoyasu Otani**   Research Institute of Agricultural Resources, Ishikawa Agricultural College, Nonoichi, Ishikawa, Japan

**Constantine E. Palmer**   Department of Plant Science, University of Manitoba, Winnipeg, Manitoba, Canada

**Ekaterini Papadopoulou**   Department of Horticulture, Michigan State University, East Lansing, Michigan

**Isobel A. Parkin**   Molecular Genetics Section, Saskatoon Research Centre, Agriculture and Agri-Food Canada, Saskatoon, Saskatchewan, Canada

**Joseph F. Petolino**   Biotechnology and Plant Genetics, Dow AgroSciences, Indianapolis, Indiana

**Peter W. B. Phillips**   Department of Agricultural Economics, College of Agriculture, University of Saskatchewan, Saskatoon, Saskatchewan, Canada

**James J. Polashock**   P. E. Marucci Center, Department of Plant Science, Rutgers University, Chatsworth, New Jersey

**J. Brian Power**   Plant Science Division, School of Biosciences, University of Nottingham, Nottingham, England

**Zamir K. Punja**   Centre for Environmental Biology, Department of Biological Sciences, Simon Fraser University, Burnaby, British Columbia, Canada

**Javed A. Qureshi**   Novartis Agrbusiness Biotechnology Research, Inc., Research Triangle Park, North Carolina

**M. Helena Raquel**   Instituto de Biologia Experimental e Technológica, Oeiras, Portugal

**Serge Remy**   Laboratory of Tropical Crop Improvement, Catholic University of Leuven, Leuven, Belgium

**Giuseppe Leonardo Rotino**   Research Institute for Vegetable Crops, Montanaso Lombardo, Italy

**László Sági**   Laboratory of Tropical Crop Improvement, Catholic University of Leuven, Leuven, Belgium

**Vipen K. Sawhney**   Department of Biology, University of Saskatchewan, Saskatoon, Saskatchewan, Canada

**Christian Schöpke**  Plant and Industrial Products Division, ValiGen, San Diego, California

**Hart E. Schroeder**  CSIRO Plant Industry, Canberra, Australian Capital Territory, Australia

**Ralph Scorza**  Appalachian Fruit Research Station, Agricultural Research Service, U.S. Department of Agriculture, Kearneysville, West Virginia

**Robert Sévenier**  Cell Cybernetics, Plant Research International, Wageningen, The Netherlands

**Hong-sheng Shang**  College of Plant Protection, Northwest Sci-Tech University of Agriculture and Forestry, Yangling, Shaanxi, China

**Takiko Shimada**  Research Institute of Agricultural Resources, Ishikawa Agricultural College, Nonoichi, Ishikawa, Japan

**Kenneth C. Sink**  Department of Horticulture, Michigan State University, East Lansing, Michigan

**Angelo Spena**  Faculty of Science, University of Verona, Verona, Italy

**Charles Spillane**  Institute of Plant Biology, University of Zurich, Zurich, Switzerland

**C. Srinivasan**  Appalachian Fruit Research Station, Agricultural Research Service, U.S. Department of Agriculture, Kearneysville, West Virginia

**John I. Stiles**  Integrated Coffee Technologies, Inc., Honolulu, Hawaii

**Wei Wen Su**  Department of Molecular Biosciences and Biosystems Engineering, University of Hawaii at Manoa, Honolulu, Hawaii

**Francesco Sunseri**  Department of Biology, University of Basilicata, Potenza, Italy

**Rony Swennen**  Laboratory of Tropical Crop Improvement, Catholic University of Leuven, Leuven, Belgium

**Yohannes Tadesse**  Laboratory of Plant Genetics, Institute of Molecular Biology and Biotechnology, Free University of Brussels, Sint-Genesius-Rode, Belgium

**Nigel J. Taylor**  International Laboratory for Tropical Agricultural Biotechnology, ILTAB/Donald Danforth Plant Science Center, St. Louis, Missouri

**Jean-Philippe Vielle-Calzada**  Cold Spring Harbor Laboratory, Cold Spring Harbor, New York

**Nicholi Vorsa**  P. E. Marucci Center, Department of Plant Science, Rutgers University, Chatsworth, New Jersey

**Gui-ling Wang**  Laboratory of Plant Biotechnology, Institute of Microbiology, Chinese Academy of Sciences, Beijing, China

**Po-Jen Wang** Department of Molecular Biology, National Chung Hsing University, Taichung, Taiwan

**Xu Wang** College of Plant Protection, Northwest Sci-Tech University of Agriculture and Forestry, Yangling, Shaanxi, China

**Witjaksono** Tropical Research and Education Center, University of Florida, Homestead, Florida

**Urs Wyss** Institute of Phytopathology, University of Kiel, Kiel, Germany

**Kai-Wun Yeh** Department of Botany, National Taiwan University, Taipei, Taiwan

**Feng-li Zhang** Laboratory of Plant Biotechnology, Institute of Microbiology, Chinese Academy of Sciences, Beijing, China

**Jiu-chun Zhang** Laboratory of Plant Biotechnology, Institute of Microbiology, Chinese Academy of Sciences, Beijing, China

**Honghong Zheng** National Laboratory of Protein Engineering and Plant Genetics Engineering, College of Life Sciences, Peking University, Beijing, China

**Michela Zottini** Faculty of Science, University of Verona, Verona, Italy

# Transgenic Plants and Crops

# 1
# Agriculture and Food Crops: Development, Science, and Society

**George G. Khachatourians**
*University of Saskatchewan, Saskatoon, Saskatchewan, Canada*

| | | |
|---|---|---|
| I. | INTRODUCTION | 1 |
| II. | AGRICULTURE AND EMERGENCE OF FOOD CROPS | 2 |
| III. | TECHNOLOGICAL DEVELOPMENTS IN INTENSIFIED FOOD CROP PRODUCTION | 3 |
| IV. | THE ULTIMATE EQUATION: CROP PRODUCTION AND CONSUMPTION | 6 |
| | A. World Supply of Grains | 7 |
| | B. Complexities in Feeding the World | 7 |
| V. | IMPACT OF GENETICALLY MODIFIED CROPS ON FOOD | 10 |
| | A. Crops and Nutrition | 10 |
| | B. Crops and Food Security | 11 |
| VI. | TRANSGENIC PLANTS AND TRADE | 11 |
| | A. Transgenic Crops: New Trade Rules | 12 |
| | B. Transgenic Crops: Biodiversity and Germplasms | 13 |
| VII. | TRANSGENIC CROPS AND THEIR GENOMICS | 14 |
| VIII. | TWO EXAMPLES OF RESEARCH IN TRANSGENIC CROPS | 16 |
| | A. Mycopathogens and Mycotoxins | 16 |
| | B. Ingredients for Food Production or Processing | 17 |
| IX. | THE TRANSGENIC CROPS: A COMPLEX PARADIGM | 17 |
| | A. Intellectual Property, Technology Transfer, and Consumers | 18 |
| | B. The Interface Between Social and Technological Issues | 18 |
| | C. Issues of Ethics and Safety | 20 |
| | D. Transgenic Crops, Food Security and Policy | 22 |
| X. | CONCLUSIONS | 23 |
| | REFERENCES | 24 |

## I. INTRODUCTION

Edible plant products are the major component of our food. During the next 25 years, the food demand will triple, the world population will increase by at least 40%, and cultivated land area will increase by perhaps 10%. The connections among land area, agricultural practices, small to very large farms, farmers, cropping, plant breeding for production, processing, and adding of value will become very obvious. Sadly enough, all of these issues will place great demands on each step of world agricultural productivity. Part of the knowledge needed to enhance productivity will arise from newer applications of the sciences of biotechnology (and not solely genetic engineering) and informatics to agriculture and food production. The other factor important to agricultural production climate and environmental change, despite much improved weather forecasting and reporting, will remain immutable and nonchangeable by humans. Political and economic policy considerations and consumer confidence and acceptance will further be constrained by the global epidemic of malnutrition, whether people are underfed or overfed. Progress against a shrinking timeline should make this period in the life of humanity difficult. These challenges are indeed the necessary impetus for human ingenuity and inventiveness once again to rise to the occasion. In this chapter, the above issues are examined from developmental, scientific and societal perspectives.

## II. AGRICULTURE AND EMERGENCE OF FOOD CROPS

Although it is difficult to pinpoint where or how our relationship with food crops and agriculture began, certainly it has been a long and enduring one. Knowledge and use of plant and animal diversity help sustain human life. In terms of abundance, 0.25 to 0.75 million plants constitute the third largest category in terms of species and diversity after fungi (1.5 million species) and insects (6 to 10 million species). The most important scene in the evolutionary drama is the manner in which diverse plants connect the community of organisms and food webs.

The importance of plants relates to photosynthesis and the production of food, fiber, fuel, and structural material. Humans, through experience and understanding of the earliest edible plants as food crops, have organized community growth around centers of diverse plants and productivity. In Diamond's (1) survey of the chain of causation in the broadcast of humans, horses, the earliest technologies of steel, and the development and oceangoing ships were factors dependent on the domestication of plant and animal species. With the development of food crop productivity additional concerns, such as the need for the storage of surplus food, arose. Collectively these changes transformed nomadic people into an organization of large, dense, sedentary, and stratified societies. Independently of origins, the domestication of plants and animals occurred between 8500 B.C. and 2500 B.C. (Table 1). The earliest crops and food production spread both to and from other centers. At this time, the major food crops were obtained from cereals and

**Table 1** Domestication of Food Crops

| World region | Location | Date (B.C.) |
|---|---|---|
| Eurasia | Fertile Crescent | 8500 |
| | China | 7500 |
| | England | 3500 |
| Native Americas | Andes | 3000 |
| | Argentina | 3000 |
| | Mesoamerica | 3000 |
| | Eastern America | 2500 |

other grasses (wheat, millet, rice, corn, sugarcane); pulses (pea, lentil, chickpea, bean); fiber plants (flax, hemp, yucca, agave, cotton); and roots and tubers (yams, jicama, potato, sweet potato, taro, Jerusalem artichoke). The relative influence of some other food plants, such as melons, squash, and bananas, arose in the areas where these plants were adapted and abundant (e.g., the Fertile Crescent, Mesoamerica, the Andes, and West Africa).

Through changes in the types of society, economy, religion, government, and membership to the patterns of human settlement, decision making and leadership, modes of conflict settlement, labor, food production, control of lands, and societal acquisition, exchange, and/or organized theft (kleptocracy) occurred (1). Whereas bands had no need for intensification of food production, tribes and chiefdoms moved in this direction. As a result, organized economies and states have adopted intensive agriculture and food production. Prehistoric agriculture in tropical highlands (2), through vegiculture or the cultivation of starchy tubers and rhizomes (root crops) and seed cultures or selection and propagation of seed-bearing plants in South America, occurred between 3000 and 6000 B.C. Archaeological and historical records indicate that by 500 B.C. these choices were leading to change (2), as a result of manipulation of ecosystems or their breakdown. Technological ascendancy in the New World tropics aided the maize-bean agriculture in the highlands, and the root-tree crops in Atlantic sector, all of which were cultivated by the same ancient people.

Early people's experience and knowledge of food crop production given, the abundance of land were at the mercy of serious pest problems and unpredictable and variable water supply. Irrigation and control over the culturing of plants must have been key developments in intensification of food production. Some societies became well positioned to lead the new technology of agriculture and food production. Yet this knowledge even then must have been trailing so far as its use in the ever-growing human population and its need for foods were concerned. During the ensuing centuries, the realization that there is an interlocking of population, land use, environment, and plant-based food products required new understanding. Ultimately, by the early part of the 20th century, enhanced understanding led to greater inputs of fertilizers, pesticides, capital, cooperation, and trade. On a global scale we now need a different paradigm and set of relationships to feed the world.

With the geometric doubling of population and marginally arithmetic doubling of food production, which was and still is at the mercy of natural disasters, the ideas of Robert Malthus proposed in 1798 remain as controversial as when they first appeared. Malthus's hypothesis remained an important unanswered question at the bicentennial of Malthus's paper. Certainly, the conventional practice of agriculture has not doubled and cannot double the production of plant-based foods. One major intervening force has been the emergence of new and reemerging infectious diseases, which in the absence of global war have had an equivalent effect in terms of human suffering and death worldwide. Ironically these epidemic events are reoccurring in spite of advanced medical technologies and accelerated methods of health care delivery and immunization.

## III. TECHNOLOGICAL DEVELOPMENTS IN INTENSIFIED FOOD CROP PRODUCTION

In part we create the future from our experiences. Human history, through experience, experimentation, knowledge, and wisdom, has aided us in understanding our dependency on plants for food, fiber, and fuel. The paradigm of obtaining our food from the land and its plants incrementally and possibly deliberately must have expanded over millennia. The dimensions of hunger and exploration of edible foods must have forced humans to accept their dependence on plants and animals instinctively. Over the centuries, people came to know about the need for improved agronomic practices, enhanced food production, and preservation and storage schemes. With generational experience and records of correlation and causality the mastery of early agrarian society must have become sophisticated. During the period from 8000 to 2000 B.C. domestication of several plants—foxtail, einkorn, emmer, lentil pea, millet, squash, gourds, and others—in parts of the Old and New Worlds, Middle East, China, Americas, and North Africa was taking place (see Table 1). Later civilizations with the advanced knowledge of the day and in the development of scientific experimentation were changing planting and cropping practices. At the same time, substantial improvements in harvesting from a given area of farmland through better agronomic practices, improved crops through breeding, and amplification of agricultural production occurred. Hopper (3) illustrates the events with rice yields in Asian agriculture (Table 2). Whereas an increase from 0.8 to slightly over 2.5 metric tons per hectare of land was accomplished from 600 to 1900 A.D., a yield increase to 6 metric tons per hectare was achieved just 30 years ago. The task of harvesting an acre in the 1830s took 2400 person-minutes (0–40hrs). This was reduced to 240 person minutes or 4 hours in 1890, and 1 hour in 1925 and further reduced to 10 minutes or 0.17 hour in 1965 (Fig. 1). Advances in harvest technology made these changes possible. With better understanding of applied microbiology and food science new dimensions in food processing, preservation, canning, prevention of spoilage, and avoidance of pathogenic microbes and food refrigeration had positive impacts on food safety, quality of life, and economic prosperity of nations.

After the 1940s significant developments in other fields of human inquiry made an impact on the paradigm of agriculture and food production (Fig. 2). Much of the unprecedented increase in developing countries' food production was due to chemical input and advances in agricultural engineering–based technologies. However, these successes were achieved with favorable environmental conditions, availability of irrigational water, and economic resources (4). During the 1960s, problems of soil, lack of essential nutrients, buildup of salts, iron or aluminum excess, and high acidity were critical constraints on food crop production in the developing countries (4).

Disciplinary crossovers of genetics, microbiology, nutritional sciences, and engineering set the stage for reconsidering the paradigm of agriculture from traditional breeding for food plants. The strongest impact on agriculture in this area occurred after the discovery of in vitro genetic engineering and the use of transgenic plants. This new revolutionary era of biotechnology was 25

**Table 2** Intensification of Production of Rice

| Cultivation era | Location | Year (A.D.) | Rice yield (MT/Ha) |
| --- | --- | --- | --- |
| Primitive farming | NA | 600–700 | 0.8 |
| Irrigated cultivation | Laos | 900 | 1.3 |
| | Cambodia | 970 | 1.4 |
| | India, Philippines, Thailand | 1350–1500 | 1.7–1.9 |
| | Burma, North Vietnam, Bangladesh | 1560 | 2.1 |
| | Sri Lanka, Pakistan | 1685 | 2.2 |
| | South Vietnam | 1800 | 2.5 |
| Technical innovation | Indonesia | 1910s | 2.7 |
| | Malaysia | 1920s | 3.2 |
| | China | 1940s | 3.5 |
| | Taiwan | 1950s | 4.3 |
| Structural reforms | Korea | 1960s | 5.3 |
| | Japan | 1970s | 5.9 |
| Biotechnology | Many | 1990s | NA |

# Agriculture and Food Crops: Development, Science, and Society

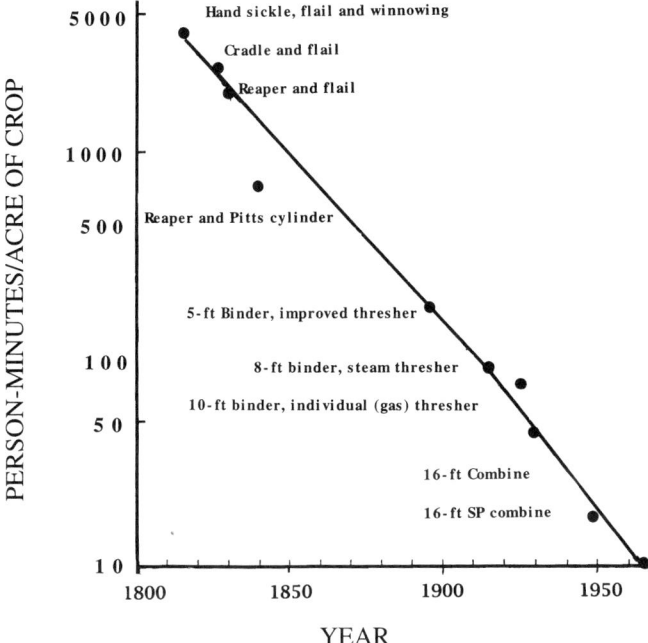

**Figure 1** Technological developments and innovations in wheat harvest. The rate of harvest (acre of crop per person per minute) between 1820 to 1963 is shown. Advances in harvest technology are labeled with each datum. (Data from Dr. L. Katz).

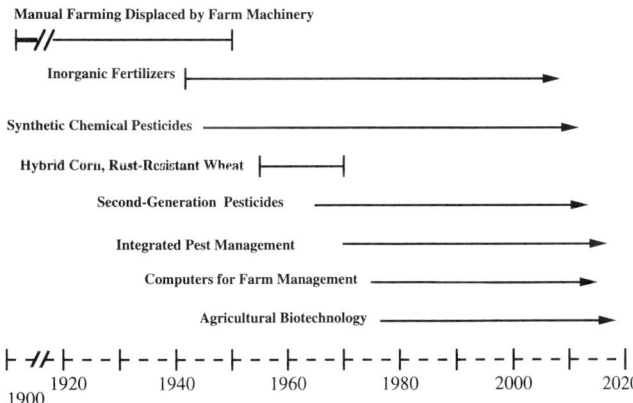

**Figure 2** Major technological or developmental changes in agriculture and food production paradigm. Input products or processes that have contributed to agricultural innovation and increased food production from an application point are indicated with the beginning of heavy lines. Continuation of use is shown with arrows and end of practice with diagonal lines for each technology or practice.

years in development before it could demonstrate positive impacts in production agriculture and new food crops.

Today we are confident that judicious and timely applications of modern genetics to the understanding of plant science will be an important driver of world agriculture. Given the rise in population, we must also understand that abundance of food through biological technologies does not necessarily translate into abundant supply for people. To feed the world population we must strive overcome global deficiencies in food transportation and storage in many countries as well as affordability. Certainly, agricultural biotechnology can provide part of the answer, but global sociopolitical factors, including ethics of farming, farmers, corporate agribusiness, world trade organizations and states, and international treaties and enforcement agencies, will also be influential. Plants can do their share in the production of foods; another variable of the equation is clearly the role of humans.

## IV. THE ULTIMATE EQUATION: CROP PRODUCTION AND CONSUMPTION

The production and consumption of food are locked in a delicate balance. Twenty-eight plants whose production exceeded 10 million metric tons (MMT) in 1975 (Table 3) made the most substantial contribution to global food production (5), which by 2000 had increased notably. Today Food and Agriculture Organization (FAO) data indicates over 135 plants are major food crops. In 2000 45 food crops produced more than 10 MMT, 11 more than 100 MMT, and five more than

**Table 3** Production of Food Crops with Over 10 MMT 1975[a]

| Crop Product | MMT | Rank |
|---|---|---|
| Apple | 31.1 | 18 |
| Banana | 31.6 | 16 |
| Barley | 135.4 | 9 |
| Beans (broad + dry + green) | 26.5 | 21 |
| Cabbage | 31.4 | 17 |
| Cassava | 110.3 | 10 |
| Coconut | 30.7 | 19 |
| Grape | 58.4 | 13 |
| Maize | 341.6 | 4 |
| Millet | 27.4 | 20 |
| Oats | 25.9 | 22 |
| Onion (dry + green + shallots) | 18.7 | 26 |
| Orange | 32.5 | 15 |
| Peanut | 20.0 | 25 |
| Peas (dry + green) | 13.1 | 27 |
| Potato | 270.3 | 5 |
| Rice | 356.9 | 2 |
| Rye | 24.1 | 23 |
| Sorghum | 61.9 | 12 |
| Soybean | 64.2 | 11 |
| Sugar beet | 251.3 | 6 |
| Sugarcane | 655.6 | 1 |
| Sunflower seed | 9.8 | 28 |
| Sweet potato | 141.6 | 8 |
| Tomato | 47.0 | 14 |
| Watermelon | 23.6 | 24 |
| Wheat | 355.8 | 3 |
| Yam | 214.6 | 7 |

[a]MMT, million metric tons.
*Source*: Food and Agriculture Organization of the United Nations, http://www.fao.org/.

500 MMT (Table 4). By rank, sugarcane, wheat, rice, maize, and potato have been the top five crops, with production of over 300 and 500 MMT in 1975 and 2000 respectively. Production and demand Food and Agriculture organization (FAO) data ever increasing population made these changes possible. More significantly, overall demand for both varieties and volume of many food crop products, tubers, fruits, vegetables, and seeds has increased during the past 1975 years. In a significant manner breeding of agronomic, quality, and yield value through conventional genetics has been part of the success story of agriculture. Modern genetics and construction of transgenic traits, as shown in this book, are the trends for building production capacity in the next 25 years.

Plant-based food production is also advancing more rapidly than in past decades. We have significant new understandings of the processes involved in production and postproduction agriculture. We have the traditional and biotechnological means for combating pathogenic microorganisms and their toxic metabolites. We have also learned to add value after production agriculture to foods and new products by microorganisms (6). These trends by all criteria will continue because of human ingenuity and should help to maintain an equilibrium of food production and consumption (7).

## A. World Supply of Grains

Historically applications of agronomic practices and free availability of plants for food production have been exploited by society. Many countries rely on importation of food grains, fruits, and vegetables. However, many nations have shortfalls of food crops that are aggravated by unpredictable global climate change. The world production of grains per person has remained at about 300 to 340 kg since the 1970 yr. To explain the world supply of grains, Borlaug (8) is said to have said, "Picture the whole world's grain harvest as a highway—circling the earth at the equator. In 1971 it would have been 55 feet wide, six feet deep and 25,000 miles long. That was the greatest harvest in history. And we ate up all of it the following year. With the world population increasing each year, we have to do more than maintain the highway of grain. We have to add 625 miles each year."

However, this picture has changed. Of the three—wheat, rice, and corn—the world wheat carryover stock was at 78 days in 1999. This is the third lowest supply on record since the Green Revolution in terms of the food security threshold of 70 days. The supply of rice stocks was down to 42 days of consumption; consumption had been on the rise for 26 consecutive years since 1973 (9). Climate change and population increase are changing food security. If we add the role of water in irrigation, sustainability of production, and its scarcity, the outlook deteriorates dramatically with the future forecast pointing at worsening of events. A total of 1000 tons of water is required to produce 1 ton of grain. Aquifer depletion or contamination is an ignored but real threat to food grain production.

## B. Complexities in Feeding the World

Foods in the developed countries greatly depend on an animal-based protein economy, which has five sources—beef, pork, poultry, fish, and soybeans. Improved economic conditions in China and India have created greater demand for better and more varied food products, including meats. Although some ethnic food groups satisfy part of the need, whether by agriculture or ocean fisheries, increases in production have limits and efficiency can have only a small deciding role. With annual growth rates of 11.6% for aquaculture and 1.3% in oceanic fish catch and 5.2%, 3.4 yr and 0.3% for poultry, pork, and beef, the demand on plant-based feed products is putting added constraints on foods (9); already 35% of world grain is used as livestock feed. Finally, grain harvest area has largely remained unchanged since the 1960s. To translate these figures, the decline in

**Table 4** Production of Food Crops with Over 1.2 MT in 2000[a]

| Rank | MT | Crop product |
|---|---|---|
| 1 | 1,281,767,380 | Sugarcane |
| 2 | 597,154,664 | Rice, paddy |
| 3 | 589,355,356 | Maize |
| 4 | 580,014,595 | Wheat |
| 5 | 308,216,588 | Potato |
| 6 | 249,888,712 | Sugar beet |
| 7 | 171,517,343 | Cassava |
| 8 | 161,042,126 | Soybean |
| 9 | 141,069,941 | Sweet potato |
| 10 | 132,896,783 | Barley |
| 11 | 100,761,543 | Tomato |
| 12 | 97,761,398 | Oil palm fruit |
| 13 | 66,054,079 | Orange |
| 14 | 63,185,776 | Watermelon |
| 15 | 62,312,769 | Grape |
| 16 | 59,963,060 | Apple |
| 17 | 58,764,543 | Sorghum |
| 18 | 58,687,214 | Banana |
| 19 | 52,351,974 | Cabbage |
| 20 | 50,777,191 | Cottonseed |
| 21 | 48,374,677 | Coconut |
| 22 | 47,781,146 | Dry onions |
| 23 | 40,242,599 | Rapeseed |
| 24 | 37,772,511 | Yam |
| 25 | 34,522,077 | Groundnut in shell |
| 26 | 30,583,162 | Plantain |
| 27 | 29,963,141 | Cucumber and gherkin |
| 28 | 27,491,475 | Millet |
| 29 | 27,482,145 | Sunflower seed |
| 30 | 25,994,503 | Oat |
| 31 | 21,987,062 | Eggplant |
| 32 | 19,971,448 | Rye |
| 33 | 19,558,989 | Carrot |
| 34 | 19,468,796 | Cantaloupe and other melon |
| 35 | 18,825,653 | Dry beans |
| 36 | 18,636,610 | Tangerines |
| 37 | 18,614,569 | Pepper-chili and green |
| 38 | 16,926,929 | Lettuce |
| 39 | 16,626,473 | Pear |
| 40 | 15,415,747 | Pumpkin, squash and gourd |
| 41 | 14,240,402 | Cauliflower |
| 42 | 13,741,161 | Olive |
| 43 | 13,456,924 | Peach and nectarine |
| 44 | 13,455,362 | Pineapple |
| 45 | 10,928,256 | Dry pea |
| 46 | 9,974,584 | Garlic |
| 47 | 9,972,298 | Lemon and lime |
| 48 | 9,319,616 | Triticale |
| 49 | 8,821,530 | Taro (coco yam) |
| 50 | 8,801,590 | Chickpea |

**Table 4** (*Continued*)

| Rank | MT | Crop product |
|---|---|---|
| 51 | 8,222,955 | Plum |
| 52 | 7,755,161 | Spinach |
| 53 | 7,058,028 | Coffee, green |
| 54 | 7,001,343 | Green pea |
| 55 | 5,363,167 | Papaya |
| 56 | 5,312,000 | Vanilla |
| 57 | 5,237,941 | Date |
| 58 | 5,088,450 | Grapefruit and pomelo |
| 59 | 4,963,959 | Cauliflower |
| 60 | 4,571,955 | Green bean |
| 61 | 4,167,960 | Asparagus |
| 62 | 4,109,099 | Dry onion, shallot and green |
| 63 | 3,874,099 | Okra |
| 64 | 3,303,590 | Dry cowpea |
| 65 | 3,222,329 | Pigeon pea |
| 66 | 3,183,427 | Dry broad bean |
| 67 | 3,172,531 | Lentil |
| 68 | 3,117,405 | Cocoa bean |
| 69 | 3,110,186 | Strawberry |
| 70 | 2,880,122 | Tea |
| 71 | 2,742,100 | Apricot |
| 72 | 2,383,710 | Mushroom |
| 73 | 2,336,765 | Avocado |
| 74 | 2,143,942 | Persimmon |
| 75 | 1,934,071 | Pimento |
| 76 | 1,769,197 | Cherry |
| 77 | 1,452,465 | Almond |
| 78 | 1,289,997 | Artichoke |
| 79 | 1,201,558 | Cashew nut |

[a]MT, metric tons.
*Source*: Food and Agriculture Organization of the United Nations, http://www.fao.org/.

grain area harvested per person (0.12 hectare for the period 1992–1999) represents half of the level of 1950 (9). With shrinkage in the amount of arable land, which at this time comprises only about 3% of the Earth's surface, and deterioration of topsoil quality and quantity there will be further decreases in the global per capita arable land from the current 0.26 hectare to half that in just 50 years (10). Excluded in any future calculation are the roles of drought, pollution, soil erosion, floods, insect attacks, warming trends, and lack of irrigation water (11). Separately the outcomes of these predictors are well known; however, when they are combined, the particular synergisms and antagonisms that the elements create are not. As indicated in State of the World 2000 (12), human pressure on Earth's natural systems has reached a point at which it is more and more likely to engender unanticipated problems. Further, without new and comprehensive principles, we fathom neither the complexity of nature nor its homeostasis. The contributions of new research in agricultural biotechnology, creation of new cooperatives, and implementation of sustainable agriculture should be a strong consideration. New solutions are urgently needed, as nature gives away nothing for free and has no reset button.

## V. IMPACT OF GENETICALLY MODIFIED CROPS ON FOOD

The options provided by genitically modified (GM) and transgenic (TG) plants, the latter involving the incorporation of genes from other species than the specie in question, therefore, trans, by genetic techniques and food biotechnology can have a significant impact on human nutrition. Augmentation of foods to yield nutritionally balanced and adequate micronutrient content is being achieved through innovation in food science and technology. With the advances in biotechnology and genetic engineering, value can be added to foods by physiological, biochemical, and genetic techniques. The goal of nourishing the world's people, 1.2 billion of them unfed, 2 billion having an unbalanced diet, and 1.2 billion overfed, has faced calamities. Norman Borlaug devoted his life to the oldest struggle of human life, the battle to grow food and avoid starvation (8). What Borlaug created came to be known as the *Green Revolution,* which provided a means for meeting the demand for food production for the next few decades. The battle, however, was far from over. Although Borlaug's work was recognized by the Nobel Peace Prize, certainly an exceptional honor for an agronomist, it did not nullify the Malthusian theory. Deficiencies in daily food are the major challenge in agriculture and food production. Shortage of the essential daily required proteins and oils and the search for life-sustaining vitamins and minerals, whether associated with hunger and malnutrition in the developing countries or overconsumption of food and excessive calories in the developed world, do not have simple cause-effect relationships. Although the conventional wisdom that income growth results in improved health held true for a time, recent studies on health-led development show that the converse also holds true (13). Indeed, economic analysis indicates that health status, as measured by life expectancy, is a significant predictor of subsequent economic growth (13,14). In this regard, the nature of food crops from the nutrition and health perspectives becomes especially important. Although much of plant-based food contains an adequate supply of minerals, trace elements, vitamins, and phytochemicals, inadequacies can be met through fortification or transgenic techniques (see chapter 14). Countries with health-led development, that is, healthier and nutritious foods, safe water, and good quality health services, tend to generate better economic growth and wealth.

### A. Crops and Nutrition

The well-known facts on population growth and economics are that 800 million people are malnourished today, most of them in Africa. About 1.3 billion people live on less than U.S. $1 per day and another 3 billion on less than U.S. $2 per day. In addition to these figures, 1.3 billion people do not have access to clean water and 2 billion live without any sanitation. These are all concerns for several countries, the World Health Organization (WHO), and the World Bank. Other global themes of urgent consideration are meeting basic human needs: the ever-increasing global demand for food, energy, and sustainable development (15). In conjunction with the rapid rate of climate change, temporal patterns of precipitation, and high temperatures have effects on agricultural systems and the quantity and quality of food plants. Some of these changes have profound effects on the emergence and spread of past and new infectious diseases: malaria, cholera, dysentery, and so on (15,16). If changes due to climate continue, then crop yields, quality, and storage and delivery of foods will be severely affected. Serious concerns are raised by yield forecasts of a 30% decrease in overall agricultural productivity doubling in production of carbon dioxide (15). Any further increase in the temperature of the tropics and subtropics, especially in arid or semiarid areas, where some food crops are at their maximal temperature tolerance, should aggravate food production.

Some of the richer countries in the Middle East and Africa are equally vulnerable to the elements of nature. The dependency of these areas on water for irrigation and drinking is extremely high. There are 19 countries in these regions that have national water scarcity. Conservation and sharing of water as a limited resource will be harder if water continues to have its value as a public rather than an economic good. Should water be looked at as an economic good, other ethical issues will emerge. It is paradoxical that 90% of the composition of many crops and fruits is water, yet dependence on water cannot be included in the food equation. In the view of the World Bank, the scientific efforts directed to this end should also help in plant and animal-based food production and its sustainability. Biotechnology and particularly TG drought-tolerant crops have been touted as an important aid in this regard, but their outcomes are still unsettled. It remains to be seen whether scientific innovations and R&D will have an impact on water, particularly in places where it could make a difference.

## B. Crops and Food Security

In spite of the positive prospects for feeding the hungry world through new agriculture, there is no assurance of success. Of some 250,000 to 500,000 plant species, only 7000 can be cultivated and possibly 2000 to 5000 are edible and nontoxic to humans. Of these the 99% top food crops for which FAO keeps production statistics (Tables 3 and 4) constitute only a small proportion of the much greater diversity of plants of which current agricultural production cannot make effective use. The European Union and 174 national governments are signatories to the 1992 Convention on Biological Diversity (CBD) treaty. A group of scientists generated the information on the State of the World's Plant Genetic Resources for Food and Agriculture and an accompanying plan of action for conservation and sustainable utilization of it. Today what are at stake are over 3 million crop accessions held in germplasm collections throughout the world (17,18). Plant genetic resources are tremendously valuable, in terms of market opportunity, some U.S.$500 to U.S.$800 billion worth of market products is derived from these plant resources (19). Global crop germplasm storage capacity will require a significan development fund (U.S. $130 to U.S. $304 million). Opportunities that are being eroded and lost are novel foods, phytonutrients, and other medicinal products that can be built into TG food crops for socioeconomic returns (20).

However, there is a disconnect in the new paradigm of food production through biotechnology even with the great power of one of its ingredient technologies, genetic engineering. Since the late 1980s low-input sustainable agriculture (LISA), in spite of wide advocacy has fallen short of major subscription. Management of low-input farming would require high levels of integration of multidisciplinary knowledge. This in turn would depend on the education of farmers in natural and engineering sciences and the social, economic, and political sciences. Individuals or groups of traditional and new cooperative farmers must have the crosscutting knowledge base for keeping their agriculture and income sustainable. Three clusters of technology—information, precision farming, and biotechnology—will promote sustainability of agriculture. The addition of specific features to food crops provides the opportunity for an important social experiment in the agriculture and food production continuum. In order for LISA and other community-supported agriculture to work, institutional innovation and reform are urgently needed (21,22). Presumably, we will see whether our generation's stewardship of the land will be characterized by the phrase, "We did not just inherit the land from our grandparents, but are borrowing from our next generation."

## VI. TRANSGENIC PLANTS AND TRADE

Food production, population growth, and the environment are interconnected so far as humanity's success in feeding itself is concerned (23). Economic and ecological operating systems are linked. They will determine the proportion of people whose access to nutritional requirements is secure and whose global food production is sustainable. The link between economic and ecological operations works primarily at the local level, but at a higher level, policy and intervention are the regulators of the link (23). Changes in trade policies as enacted throughout the globe have an impact on the linkage and trends for TG food crop development, production, and trade.

The period since the 1970s has seen a remarkable reversal of positions between the developed and developing countries in world trade negotiations. In the 1970s, the developed countries started extending, reintroducing, and inventing nontariff barriers to trade, which since the 1950s had been recognized as contrary to the principles of the General Agreement on Tariffs and Trade (GATT). In agriculture it was accepted that this was an undesirable relic, perhaps needed until a sector or population could fully adapt to international trade. In the 1980s, the reversal went further, with the European Community (EC) and the United States. In contrast, in the 1970s and 1980s developing countries were recognizing that import substitution, usually behind tariff and nontariff barriers, was neither the only nor a sufficient path to opening their economies to imports and emphasizing exports. At the same time, the newly industrialized countries (NICs) gave examples of an alternative road, which by the early 1990s led to participation in multilateral negotiations. As remarked by Page (24), the Uruguay Round of trade negotiations attracted much attention in the developing countries by promising to weaken moves toward bilateral trade agreements and to break open trade in temperate agricultural products. As such these actions should remove the last vestiges of protection against tropical products under awkward but escalating tariffs. Overall, the main gains for the developing world will probably accrue to exporters of temperate products, above all those in Latin America that have captured the resources, strategies, and technological advantages (25). For some of the poorer developing countries, losses occur as shortage of technologically trained human resources, weak research infrastructure, and lack of appreciation of local resources for plant productivity erode their access to markets.

### A. Transgenic Crops: New Trade Rules

Current and evolving agricultural trade policy has widened to include the necessity to protect public health through monitoring the movements of pests, pathogens, and contaminants associated with plants and plant-based commodities. There is a greater understanding of origins of material or agents that threaten public health across international borders. In part this is associated with lack of specific sanitary measures for food crops. Internationally, these threats impede the free movement of goods. The World Trade Organization (WTO) at a meeting in 1994 adopted the Agreement on Sanitary and Phytosanitary (SPS) measures (26). This and the Food and Agriculture Organization's International Plant Protection Convention (IPPC) agreement have placed an increased emphasis on science-based phytosanitary regulations. As Roberts (26) points out, the SPS measures and process create a set of multilateral trade rules for their legitimate use for the protection of environment and human health while disallowing its use for mercantilist regulatory protectionism.

A major issue in SPS is the effect of animal and plant pathogenic fungi on cut flowers, horticultural and food crop plants, and edible crop products in a host of diseases and mycotoxicoses. Foodstuff contamination can occur in the field before and after crop harvest, in storage, and during food preparation. Since the discovery of aflatoxins in the early 1960s, trichothecenes (deoxynivalenol and T-2 toxin) in 1970's, ochratoxins in 1980, and fuminosins in the 1990s, it has been estimated that one-quarter of the world's foods and feeds are contaminated annually. Mycotoxins' adverse health effects include cancer (aflatoxins, ochratoxins, strigmatocystin, fuminosins), mutations, teratogenicity or induced birth defects, immunosuppression, dermotoxicity, neurotoxicity, and changes in estrogenic activities. Target plants widely susceptible to mycotoxins that produce fungi are maize, peanuts, oil seeds, nuts (almond, brazil, hazel, pecans, pistachio, walnut), and fruit-producing plant spices.

Until TG plants resistant to attack by mycotoxin-producing fungi are developed, we have to rely on SPS measures. What is clear is that as we move through the 21st century, the prospects for grain storage, transportation, and processing will change as a result of the same forces that have an impact on all other facets of the economy (27). In addition to government regulation, in the public and private sectors social, environmental, and economic changes must occur. The subscription and implementation of SPS measures will be responsibilities for everyone. Applied mycology and biotechnology approaches can ensure that many aspects of agriculture-based commodities are free of mycotoxins, mold allergens, and other problems of quality loss during storage.

Many vegetables, fruits, and seeds lose their nutritive and other qualitative values through loss of moisture, infection with spoilage microorganisms, and senescence. This wastage occurs during transport, handling, and redistribution. Loss of shelf life alone, e.g., due to lack of refrigeration, is a major contributor to limited market expansion of foods. Saprophytic and pathogenic fungi are major determinants of fruit and vegetable freshness and safety. Application of antifungal peptides and antimicrobial peptides (28,29) could significantly change this situation. For example, gene regulators that will cause expression of plant protectants at the desired time, control of growth and development of plants of ethnobotanical importance, and alteration of the composition of the harvested product will provide major opportunities in fungal biotechnology for application in trade.

## B. Transgenic Crops: Biodiversity and Germplasms

Intensification of agriculture in itself has had a serious negative effect on biodiversity since the 1970s. In part this has manifested itself in terms of control or loss of weeds, insects, and other animal and plant species due to use of herbicides and insecticides. Monoculture of plants with certain agronomic values and indiscriminate use of pesticides along with the TGs raise concerns about further erosion of biodiversity. Additional criticism relates to the monopolistic practices permitted through intellectual property rights (IPRs) and protection by patents (7, also see Chapter 18).

Since the 1980s, globalization of the world economy has had an effect on the use of patents, namely, special rewards and benefits for the intellectual property rights of owners and licensees. It is argued that in the context of agriculture and food use of patents further exacerbates its commodification. It is in this context that IPRs and biodiversity, community, indigenous people, compensation, and other issues collide (7, 30, 31–32 and chapter 21).

Transgenic food crops provoke two topics of debate—knowledge and food—that give a new shape to intellectual property legislation. The reaffirmation of state sovereignty over genetic resources (Convention on Biological Diversity and FAO) calls for the protection of plant varieties by either patents, effective *sui generis* systems, or a combination thereof (GATT agreement on Trade-Related Aspects of Intellectual Property Rights 1994). Modifications of plant breeders' rights extend rights on protected materials (International Convention for the Protection of New Varieties of Plants [UPOV] Amendment 1991). These treaties signal a shift in the aim of property protection systems, particularly for plant genetic resources. Key social objectives of intellectual property legislation are promoting and rewarding innovation while ensuring access of the public to useful information. Increasingly, however, greater emphasis is being placed on intellectual property as a tool to secure the exclusivity of information and to maximize the rights of innovators to profits (7,30).

The drive to monopolize benefits from the research and development of new products has led to vast increases in the number of patent applications and the increasing employment of intellectual property systems by outside users of indigenous resources. Faced with this situation, indigenous and local communities experience growing pressure to develop their own legal protection systems (31). The protection of plants and genetic resources is needed for food security. At the same time the recognition of the rights of people, indigenous or otherwise, is also needed. If these issues are not resolved, the entire area of TG food plants and ownership, production, and distribution of improved seed, especially in the developing world, will be a problem. As well, in many African countries the future of agriculture and food production depends on intensified land use rather than cropping area to solve major local or poor rural population's problems (32). In such an event, countries such as Egypt, Kenya, and Zimbabwe, where seed production industries are developed, and other countries, where they are being developed (Malawi and Zambia) or progress has been very limited in spite of investment (Cameroon, Ethiopia, Ghana, Tanzania, and Uganda) will benefit from genetic engineered food crops and biotechnological agriculture accordingly. The main outcome of course will be a continuation of disparity in Africa and similar other parts of the world. Iwu (30) calls for establishment of meaningful and just collaborations, cooperation, and functional partnerships as a paralegal requirement if these issues are to be resolved and implemented. The developing world has a great deal of advantage in TG food crops and food development that is not recognized by the developed world. As indicated in Table 5, for the developing countries, plant diversity and genetic resources are only two aspects of the many advantages and opportunities in the utilization of TG food production.

## VII. TRANSGENIC CROPS AND THEIR GENOMICS

The history of plant sciences and most importantly of plant genetics stands on the shoulders of giants. The discovery of plant genetics, from Gregory Mendel, to Barbara McClintock, to the current generation of molecular geneticists, has paved the road for a much easier entry into theoret-

**Table 5** Advantages of the Developing Countries for Use of Transgenic Crops[a]

| Item | Advantage | Organizations and year commenced |
|---|---|---|
| Plant diversity | Native plants | IARC |
| Germplasms | Existing seed/gene banks | IBPGR, 1973 |
| Infrastructure | IARC research institutes fermentation facilities | IRRI, 1960; IITA, 1965; CIAT, 1968, CIP, 1972; ICRISAT, 1972; ICARDA, 1976 |
| Human resources | Trained collaborators Highly productive workers Low labor costs | IRRI, 1960; IITA, 1965; CIAT, 1968, CIP, 1972; CRISAT, 1972; ICARDA, 1976; local centers/institutions |
| Collaborations | International and cross-cultural | Many |
| Knowledge base | Fermented foods/beverages | National level |
| Production agriculture | Established in many areas | Local and regional levels |

[a]IARC, International Agricultural Research Center; IBPGR, International Board for Plant Genetic Resources; CGIAR, Consultative Group on International Agricultural Research; IRRI, International Rice Research Institute; IITA, International Institute of Tropical Agriculture; CIAT, Centro International de Agricultura Tropical; CIP, International Potato Center; ICRISAT International Crop Research Institute for the Semi-Arid Tropics; ICARDA, International Center for Agricultural Research in Dry Areas.

ical, quantitative, and applied genetics of many food plants, most notably maize. Few scientists have had the impact on our understanding of genetic process like Barbara McClintock, who discovered the transposable elements in the maize. Today we have a better comprehension of these elements and the occurrence of retrotransposons, which make up huge intergenic segments of deoxyribonucleic acid (DNA) in maize, rice, and sorghum. Although the retrotransposon insertion into chromosomes and the long terminal repeats were evolved within the past 5 million years, the idea that these transposable elements in maize have undergone massive amplification has also been supported (33).

Whether under natural environmental selection or the plant breeder's activity, understanding of the even simplest trait and its alleles determining its location and its inheritance has been difficult. Through molecular markers and electrophoretic analysis of whole chromosomes, such as pulsed-field gel electophoresis (PFGE), contour clamped homogeneous electric field (CHEF), chromosome length polymorphism (CLP), restriction fragment length polymorphism (RELP), and randomly amplified polymorphic DNA (RAPD), we are able to map and dissect the control of complex plant traits into its elements (see Chapter 2). Further, the availability of *Arabidopsis* spp. for dicot and rice for monocot plant models combined with their genome sequence data is making it easier to study others.

Introduced genes must be expressed at the appropriate time to be effective (34). Understanding of the regulation of gene expression is critical. Recent work has identified genetic elements involved in regulation of plant gene expression. Beyond the curiosity of dissection, we are beginning to have the genetic and mechanistic details of how epigenetic factors interact with and control plant development (see Chapter 7). In addition, studies of plant embryogenesis (see Chapter 3), development (35–37), photosynthesis (38), pollination (see also Chapter 39, 40), and cell culture (see also Chapter 4), and regeneration of plant tissues, indicate that we can dissect and manipulate those genetic processes that control root proliferation and interaction with microorganisms (42), plant height, leaf size, numbers, flower timing, color, size and shape (43); plant defenses against microbial and insect pests and pathogens, (29, 44–48 and Chapter 17); environmental stress (49,50); size and shape of seeds and seed contents (51); metabolic pathways and engineering of such pathways; and organelles within the cell. Certainly a substantial number of achievements have occurred in the past 50 years, a testimony to science in its efforts to furthering the frontiers to new limits.

The evolutionary trend in genetic research has moved from genes to genomics, the science of studying the genome of organisms. Plants have broad classes of dispensable metabolic pathways involved in catabolism of low-molecular-mass nutrients and natural product synthesis. These gene clusters can have as many as two dozen genes and occupy 60kb of DNA to contribute to survival from fungal infections and ecological stress through other features (52). Best studied in this context are the shared clusters of genes in *Arabidopsis* sp. and various *Brassica* sp. plants (36,60). What began with the height reduction genetics of rice, which significantly aided the Green Revolution, can now be done with many plants. However, in the era of TG plants and biotechnology the rules of the road differ from those of the open access to improved varieties during the Green Revolution (53).

There are three major elements of genomics: structure, evolution, and functionality. Today the genomes of *Arabidopsis thaliana,* barley, corn, cotton, foxtail millet, legumes, maize, oats, pearl millet, rice, sorghum, sugarcane, tomato, triticale, and wheat are being sequenced (54,55). The information on plants' genomics is accumulating and can be updated from various institutions. Further, genomics has direct and substantial economic ties with many industries.

Research in the area of functional genomics include, molecular and structural biology, bioinformatics, combinatorial chemistry, proteomics, high-throughput technologies, model plants, transgenics and differential gene expression (56,57 and Chapter 2, 5, 7 and 16). Both researchers and companies are using functional genomics to determine gene function and transfer of genetic information to particular dimensions of products and processes.

The technology push from the other microbial and animal genome projects has had an impact on the development of a generic technology for genetic analysis (57,58). New generations of analytical instruments and systems that speed up gene sequencing and biochip technology allow 100 to 200 analysis to be performed in a day whereas in the early 1990s only 1 to 5 such analyses could have been performed by using conventional technology. Single-nucleotide polymorphism (SNP), which is a mutation due to a single base pair, can be detected in amplicons ranging from 70 to 700 base pairs in size. Such measurements are possible with better than 90% sensitivity by, e.g., Varian Inc. (Palo Alto, CA) advance technologies. Biochip technologies, such as those of Gene Logic (Gaithersburg, MD), have developed a porous glass chip with 1 million microchannels of 10-micrometer size running in three-dimensions to analyze complementary ribonucleic acid (cRNA) or cDNA and immunoassays.

## VIII. TWO EXAMPLES OF RESEARCH IN TRANSGENIC CROPS

### A. Mycopathogens and Mycotoxins

Historical records on consequences of drought, pests, and shortcomings in food crops are numerous. As shown in Table 6, since the 1600s, several records indicate the negative social impact of food crop microbial pathogens in reducing or seriously threatening the availability and safety of food. It is estimated that over 400 fungi can be considered potentially toxigenic of which about 20 are confirmed producers of mycotoxins (59,62). Food crops and their products and feedstuff contaminated with single or multiple toxigenic fungi are well known today.

Human and animal exposure to these fungal metabolites results in well-known toxopathological manifestations and death. Compared to singular toxins, in the environment there are multiple and often structurally different mycotoxins, which at subthreshold levels by their interactions with multiple sites and targets often produce devastating synergistic effects on living cells and whole animals (60,66). Because of the polygenic nature of some of these plant-fungus interactions, the rational choice is genetic engineering of mycotoxin resistance in food crop plants (63,64). So far, however, breeding of corn and cereal grain plants for resistance has been attempted but remained unsuccessful. This is possibly due to multiple modes of action and polygenic nature of resistance (24–26). Perhaps with the isolation of target specific genes a better and fuller resistance could be achieved.

**Table 6** History of Major Plant Diseases Since 1600 A.D.

| Year(s) | Location | Food plant | Disease | Social impact |
|---|---|---|---|---|
| 1840–1846 | Ireland | Potato | Blight | Famine |
| 1851 | Europe | Vineyards | Powdery mildew | Reduced wine production |
| 1878 | Europe | Vineyards | Downy mildew | Reduced wine production |
| 1870s | Ceylon | Coffee | Rust | Beginning of coffee cultivation in South America |
| 1904 | United States | Chestnut | Fungal blight | Near eradication of U.S. chestnuts |
| 1958 | Central America | Banana | Microbial | Economic losses |
| 1600–1816 | Europe | Rye | Ergot | Poisoning and death of populations |

The plant-based food threat to public health in this regard arise, from the fungal alkaloids the lead to ergotism. The fungus *Claviceps purpurea* is prevalent in the cool climates where rye is grown, and *Clavicepes africana* in the last few years has spread through sorghum from Brazil to the United States, Australia, and Africa (65). Sorghum is the fifth most important cereal crop in the world. It would be desirable to engineer *Claviceps* resistant sorghum.

A 1999 publication (66) draws to our attention a large number of plant pathogenic and toxigenic *Fusarium* species that were isolated from human blood; autopsies or biopsies of organs, cerebrospinal, bronachioalveolar lavage, and peritoneal fluid, and wounds of patients in Japan. Further, all 18 of the *Fusarium solani* specimens from this collection were cyclosporin A (an immunosuppressant) producers. Vigilance in management of fungal spread among edible plant products and environments is needed. At the same time plant breeding and genetic engineering options for combating these threats to public health are urgently needed. Implementation of present-day knowledge in the production, formulation, and application of mycoinsecticides and mycoherbicides will be necessary (29,67).

## B.  Ingredients for Food Production or Processing

Research and development work on food products and processes is less advanced than that in the plant and animal area. For the most part, research to decrease process costs is just beginning. Undoubtedly, biotechnology will allow improvement in important consumer and health-associated aspects of food (14). These may include longer shelf life, improved appearance, improved flavor, and increased perceived healthfulness of the food. "Light beer" is an example of the few completed products or processes in the food and beverage industries. New technology development in fermentation can be through genetics or epigenetics; either of these is valid and powerful in its contribution to the agrifood industries. In the production of certain foods, food additives, phytochemicals, nutraceuticals, and ingredients, another facet of biotechnology, fermentation technology, enters the picture.

## IX.  THE TRANSGENIC CROPS: A COMPLEX PARADIGM

The new research in agriculture and food has become even more complex than the previous paradigm, in which food production was reasonably in par with its consumption, most likely for regional and national needs and then some exports. Today the challenges to agriculture and food production are many: to name a few, production-versus value-added agriculture and food, environmental constraints, ecological expectations, distortions from national to transnational and international policymakers, disparity of poor versus wealthy nations, unpredictable weather patterns, large corporate bodies ownership of intensive livestock operations, and vertical integration of the food production chain. The new technologies have created multiple shifts in the agricultural and food research and development paradigms. As shown in Fig. 3, four levels of complex social factors lead this phenomenon: multidisciplinary, multidimensional (spatial, temporal, scale, quantitative, and qualitative aspects), multisectoral, and multiperspective. Furthermore, the promise of genetic engineering of plants and the associated intellectual property rights and ownership of patented seeds, genes, and input products have precipitated unprecedented societal reaction. Current uneasiness about potential pitfalls, concern about ethics of science, and other questions indicate unusual global public reaction. These issues have been subject of many books, forums, and protests. From my perspective, there are four immediate issues that face the transgenic crops that will be addressed here.

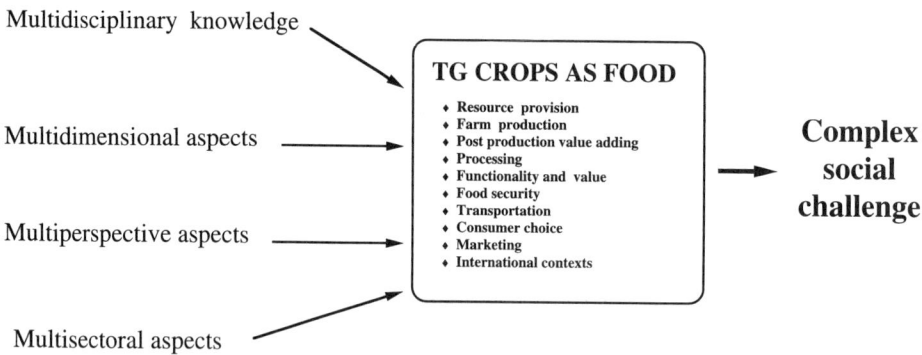

**Figure 3** Relationship between transgenic crops as food and multiple levels of scientific and social issues. Multiple inputs (left) converge on all contextual aspects of transgenic crops as food (box at right). Individually and collectively these outcomes provide a complex social challenge.

### A. Intellectual Property, Technology Transfer, and Consumers

The new inputs derived from biotechnology, especially those from the private sector, are finding wide utility in agriculture. This is made evident by the number of products reaching fields and markets in industrialized countries and the amount of ongoing cross-licensing of biotechnology products by commercial agricultural research organizations. Most of these inputs are protected through some form of IPR (7 and Chapter 18). However, it is not only the commercial sector that is using and developing materials for which intellectual property protection is being sought (68). The development and use of protected materials are also occurring among public, national, and international agricultural research organizations working for and with developing countries.

### B. The Interface Between Social and Technological Issues

As with other technologies the social and ethical issues of genetically modified crops have been the subject of journalistic, academic, and scholarly debate and writing. The context for risk and benefit assessment is well established and indeed the subject of many universities degree programs. Fig. 4 depicts the elements, tool kit, and operational requirements for a general risk/benefit assessment and decision making. The issues are many, the Nuffield Council on Bioethics Consultation (69) considers (1) that genetically modified (GM) crops can greatly contribute to human well-being; (2) that there are human health, environmental safety, and proprietary concerns; (3) that with a proper regulatory system, benefits are likely to outweigh concerns; (4) that there should be complete transparency as to the presence of GM foods in the human food chain; (5) that it is not unethical to favor some crop genes, but there are, for some, ethical concerns as to the size of the species gap across which genes are transferred; and (6) that continued GM research is required to maximize benefit and minimize risk.

Spires (70) presents in some detail the interface of technology and society, in particular the examination of animal and plant cell biotechnology. Historically, the components of the technology-society interface have been the societal needs for new products, services, or product opportunities to drive the universities, government research establishments (GREs), and industry research establishments (IREs) to generate new and beneficial goods, services, or technology. This is transferred to end users directly, as in the case of information or knowledge (say, through ex-

# Agriculture and Food Crops: Development, Science, and Society

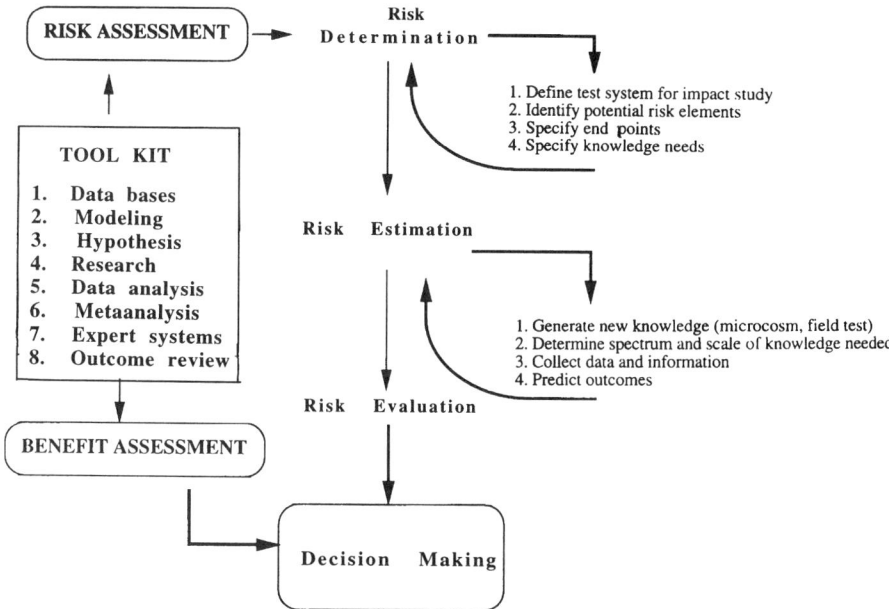

**Figure 4** A generalized model for risk-benefit assessment in transgenic crops. The elements of the tool kit (left) needed for risk- and benefit-assessment and processes required in risk determination, risk estimation and risk evaluation (right) lead to the process of decision making.

tension division or public domain materials). Alternately, through transfer of the technology or prototype industry, R&D generate agricultural technologies and agrifood products for consumption. Social acceptance or "buying in" demonstrates itself at the cash register, to begin with, and through "Darwinian selection" for the long-term survival and success in the marketplace. In most classic technologies the new goods must be socially beneficial and market-adaptable. In the case of plant biotechnology and more specifically TG plants, newer constraints of ethics and medicine feed into social acceptability both before and after the maturation of research and development for a new technology or product.

To respond to all players, governments intervene at the regulatory level through national and international agencies. Spire (70) argues that there is a continual thrusting up of new issues and opportunities to improve the "human condition" that, in turn, leads to the requirement to review the way we behave. This is particularly acute when we survey the way in which the products of animal and plant cell technology impinge on the public domain. Areas in which ethical issues have been raised include the culture of the universities' interface and alliance with industry. Furthermore we see that the regulatory agencies, the media, and ethical society become influential agencies in the determination of which products enter the marketplace. Part of the decision-making process is dependent on the ethical views held by the regulators (71). Between extreme cautiousness, to the point of inaction, and quick action, to a point of irresponsibility, lies the balance point where answers to questions of fairness, safety, nature, and purity will sit (72, also see Chapter 19).

Scientific and public perceptions are the two distinctive aspects of public attitudes about risk due to GM plants, i.e., the belief that the process is risky and is morally wrong. There is also a disparity between expert and lay perceptions of recombinant (rDNA) technology and its applications to food crops. This makes providing public information a difficult task. Ruibal-Mendiet and Lints, and Hoban (73 and see Chapter 19) present these conflicting points of view and make recommendations pertaining to public information in Europe. It appears that consensus conferences might be a good approach to stimulate public knowledge and public debate. The consequence of public reaction lies in the ways that perception of risk is modified by the so-called outrage reaction. The components of outrage reaction are many for GE plants, including issues of morality, risk, unfamiliarity, partial or total lack of knowledge, unnaturalness, artificiality, disrespect for life, fairness, and control (see Chapter 20). Polls show that although knowledge about GE is lacking, outrage plays a major role in public perception and its effects. Boulter suggests (74) that scientists should be prepared to discuss their work, its nature, and the way in which generalizations are made and to participate in nonscientific forums. The overall objectives should be to communicate and provide opportunity for public debate to consider the risk and/or benefit trade-offs, including ethical and social issues.

Consideration of safety of food, whether from plant crops or animal sources, has always been required and expected in the history of food production, whether traditional or commercial. Currently, the methods for assessment of hazard, that is, the result of risk times exposure, determination of consequences, and reduction or elimination are well known and practiced (60–63). A variety of national and international agencies are responsible for setting guidelines and standards for such hazard assessment and remediation measures. Risk to humans can be assessed for food crops–based on traditional plant breeding practices and those based on genetic engineering. Kappeli and Auberson (75) of the Agency for Biosafety Research and Assessment of Technology Impacts of the Swiss Priority Programme Biotechnology contrast the detailed national guidelines for safety and hazards of TG crops to those for traditional plant breeding. Hazards, real or otherwise, are perceived to originate from TG plants and necessitate accountability for at least three broad issues: (a) the congruence at genotypes, phenotypes, and epigenetic relationships; (b) the role of pleiotropic events; and (c) genomic plasticity. *Genome plasticity* is the capacity of the genome to reorganize itself; it can occur through external or internal cues and can result in mutations, transpositions, translocations, recombinations, selfing, and other primary and secondary chromosomal effects.

## C. Issues of Ethics and Safety

Whereas the past practices of plant breeding of food crops had routine ways of testing for biochemicals, phytochemicals, plant-borne mutagens, phytotoxins, and sensory analysis, the perceived hazards of TG crops will require many more and different types of considerations. For example, if a TG plant–engineered gene for an improved agronomic phenotype has been deemed allergenic, the TG plant's source of the introduced gene must be checked for allergenicity, including testing for amino acid similarity with known allergens and whether or not such a protein will be susceptible to digestive degradation and stability to heat and acid pH. In this case safety considerations regarding TG plants will be scientifically quantitative and measurable. With TG plants, three potential hazards, that posed by the TG crop itself, that posed by introgression and gene flow, and that posed by horizontal gene transfer, have to be considered. In the case of an unexpected or unintended phenotype that arises long after the TG construction and is due to genomic plasticity, there are solutions. One approach is to employ safety considerations based on substantial equivalence with the unmodified plants, while remaining aware of hazard and monitoring and containment.

The wider and longer-term acceptance of TG food crops should be secured after the intended or unintended effects are known. The collection of such information would be incremental and the knowledge required for its comprehension still developmental. The critical approach and solution will require long-term (5 to 15-year) collection of experimental data, baseline information from reports of adverse consumer health reactions or interaction with other medicinals and foods, trait stability and environmental or biotic community interaction, cross reaction, and evolutionary fate. The ultimate biosafety and environmental safety protocols will be derived in part from regulatory guidelines but also from long and painstaking trickling of individual case reports. This system is beginning to work well for pharmaceuticals and is worth considering for TG food plants. As with the Food and Drug Administrations (FDA's) approval of drugs, multiphase preclinical and clinical trials do not ensure absence of side effects or drug interactions. At times, newer research does force the recall of a drug; why not a TG food? Do we not consume both alike? DeKathen (76) suggests that a balance must be found with respect to TG plants. This should be done irrespective of viewpoints on the assessment of impact or the reluctance to make such an assessment. Such an outlook could end fundamentalism and reductionism by weighing potential risks against potential benefits and causes of and solutions to problems.

Democratic societies of the current era have operated and ideally will continue to operate in the best interest of the citizens. From a crossover view point, facets of environment and agriculture, economics and employment, and production of foods, industrial goods, and human services, we balance our actions and reactions (see Fig. 5). The regulators of our deeds are ethics, law, and, depending on the day, the scrutiny of legal statutes or media. Our societies function and do it well because of our desire to have a civil society, governed by freedoms, respect for democracy, education, law, and security. With respect to provision of food and related industrial products, subsets of these items interact and sustain our society. The principal institutions, universities and colleges, industry, GREs and IREs, media establishments and agencies of governance, through sets of laws, statutes, and conventions, create and maintain the interface of today's society and the flow of its relationships.

There are few TG plants that have been cultivated in fields for over a decade and more that are under construction (19). Collectively, the first wave of aggregate technical, social and ethical consequences and outcomes will not be known until 2010 or late. This process must be decidedly

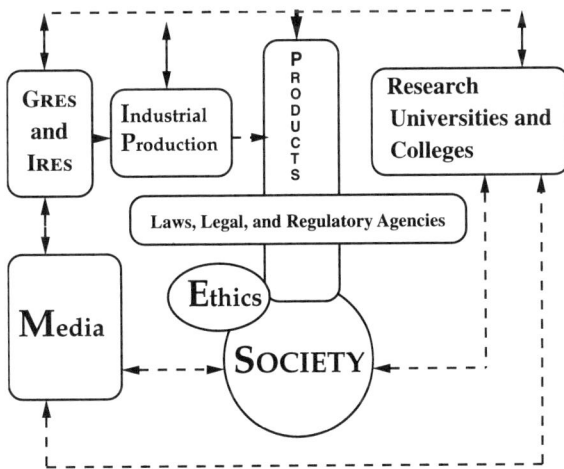

**Figure 5** A conceptual model for transgenic agri-food products research and production and their linkages with larger social phenomena. The transgenic crops or products are governed by legal and regulatory frameworks and receive perspectives of ethics, media and society. Arrowheads show directionality of interactions and impacts. GREs and IREs represent government and industry research establishments respectively. (Adapted from Ref. 70, with modifications.)

open, interactive, constructive, and inclusive. Consumer demand for information on TG product labels is one outcome of such a process that is already unfolding (77). The process indicated should be in force with an intended resolution, which might be to label. If so, other aspects of TG foods must follow. For example, there could be a language used in labeling in which the limited amount of space (depending on the size of the label) should be used for information, warnings, and brand advertisements or selling points (see Chapter 19). Alternatives such as a food-line telephone service or on-line food safety information site should be considered, too. In any event, accessible technical language, not emotive language (78), would better serve the intended purpose of informing the public and offering the essential facts for decision making. In communication words convey meaning but also have a psychological effect on the listener.

It would also help were the biotechnologists to agree to become professionals like doctors, lawyers, and architects, who have a formal contract with society. Because the issues and developments under debate and consideration are momentous, we have to resist the temptation to issue blanket declarations, e.g., to ban all TG development. Present-day communication media handle the TG and biotechnology issues in a manner that may comply with the realities of the situation. A cautious, open, and inviting approach is what citizens require. As scientists and technologists we can no longer remain in a back room and foist on an unprepared public whatever we have concocted for instruction. This is an unprecedented era of intellectual and technical advancement, and it is best if issues of TG food crops discussion demonstrate conjoint effort in a caring, deliberate, and well-considered manner.

## D. Transgenic Crops, Food Security, and Policy

There are beliefs that future agricultural policy lies in reorienting emphasis from maximal production and toward sustainability. Such a shift could focus the food production system on nourishment, food security, and sustainability. Certainly, chapters in this volume discussing strategies for TG crops with increased agronomic value and enhanced qualitative or quantitative nutritional attributes are a part of such a system. To create congruence, focus is needed on a new policy-making system and building of integrative themes compatible with the responsibilities and activities of feeding the world. Emphasis on agriculture and food policy development must become more transdisciplinary and proximal to the diverse groups affected needing the policy resolutions for daily food.

To the extent that the Green Revolution dominated the last 50 years, plant biotechnology has the power to lead the next 50 years (9). However, it will not be the magic bullet that will solve all of the problems. It is argued that the food security system in a community should be considered in a manner that relinks production and consumption and seriously restructures research trends and policy issues (7,76,78–80). Arguably, community food security cannot be a substitute for governmental and international means for food security, but it should be an important addition to it (90). In this regard, Rod MacRac and the Toronto Food Policy Council (81) emphasize a new impetus for rethinking about food and technology. Agriculture, although promising to be more productive to feed the hungry world, suffers from fragmentation of issues, knowledge, and responsibilities. The hidden costs of planning in the food sector by governments reside in environmental, social, and health factors that converge in sociopolitical issues. In a government some issues of agriculture and food are considered in ministries with histories unrelated to agricultural policy. The current system is preoccupied with traditional views of competitiveness and efficiency. Policies, programs, and regulations are organized to support specific commodities, not farming and food systems. Responsibilities are extremely fragmented and frequently uncoordinated. In this environment, the focus on nourishment, food security, and environmental sustainability has become subordinate to economic issues. Associating issues of TG food plants and food products with agricultural economics, demographics, and food security remains nothing less than a Herculean task.

Altieri (82) argues that unexpected results that follow TG releases have potential impacts. For these TG plants the environmental risks ought to be evaluated in the context of agroecological goals of making agriculture more socially just, economically viable, and ecologically sound. The author believes that public funding of research on TG crops should be diverted to ecological sustainability, alternative low-input technologies, needs of small farmers, and human health and nutrition rather than biotechnology. On the other hand, Ismail Serageldin, chairman of CGIAR (83), argues that biotechnology can contribute to future security with one proviso, that is, if the research outputs of agricultural biotechnology benefit the sustainability of small-farm agriculture in developing countries. In such a case a double shift in the agricultural research paradigm is necessary: first, the integration of crop-specific research with larger multisectoral, multiperspective, multidimensional aspects of the complex and changing dimensions of the agriculture-forestry-food complex, second, harnessing of genetics for the benefit of poor people and the environment.

As in most policy-making, formal and informal players are involved in setting agricultural policy. Agricultural policy is based on explicit and implicit values and assumptions and both public and private sector influences. The formal system is that of federal and provincial governments. Central governmental responsibilities lie mostly with trade and national standard setting for food safety, grading, and labeling. Provincial or regional responsibilities focus on extension, land use, environment, and internal movement of goods. Finally, development of policy related to funding human resource training in agriculture, education, and public sector R&D could occur within either national or local governments and might include input from the private sector.

## X. CONCLUSIONS

Biotechnology as an umbrella of concepts, methodologies, and tools under which genetic engineering of food crops rests has created a new foundation for our foods. Transgenic food plants and their products should make agriculture-based food production more successful and productive. Now biotechnology must create a framework for agriculture and food sustainability and food security. However, fragmentation of issues, knowledge, and responsibilities could hide the costs incurred in the success of TG crops. These are mainly environmental, social, and health costs, assigned to various levels of government and world governance. At the national level, each country has had its own history of lack of connections between policy, population growth, economics, food security, and cost. The disappearance of traditional farming and the remaining 1–2% that continue farming to feed the world will pose new challenges, difficulties, and opportunities. Creating TG crops, as described in this volume, is only one side of the food equation. The presence and adequacies or deficiencies of the ancillary technologies in food transportation, refrigeration, public health, and safety will also have significant impact.

The complex problems facing people involved in the chain of food and agriculture systems today may become even greater in the future. It is desirable for biotechnology to actualize its significant potential by continued transformation of plant-based food production. The economic base of the global agrifood industry has been fundamentally changed by the introduction of TG plant research. How the agricultural sciences governing food and nutrition will be perceived in a few decades from now will greatly depend on our activities today. To continue with this chapter of human history, the cooperation of citizens at all levels will be required. Instead of repeating

Dickens's "It was the best of times and it was the worst of times," I hope that future researchers will characterize the period that lies ahead as unequivacally the best of times.

## ACKNOWLEDGMENTS

I am grateful to Dr. Leon Katz, who has been an inspiration to me and some time back introduced to me to the progress in harvest technology depicted in Fig 1. I am also thankful to Lorraine M. Khachatourians; Drs. Holmes Tiessen, and Peter Phillips, for input, discussions, and review.

## REFERENCES

1. JM Diamond. Guns, Germs, and Steel: The Fates of Human Societies. New York: WW Norton, 1997.
2. NC Brady. Chemistry and world food supplies. Science 218: 847–853, 1982.
3. WD Hopper. The development of agriculture in developing countries. Sci Am 235: 196–205, 1976.
4. OF Linares, PD Sheets, EJ Rosenthal. Prehistoric agriculture in tropical highlands. Science 184: 137–145, 1975.
5. JR Harlan. The plants and animals that nourish man. Sci Am 235: 89–97, 1976.
6. Y-H Hui, GG Khachatourians, eds. Food Biotechnology: Microorganisms. New York: VCH, 1995.
7. PWB Phillips, GG Khachatourians. The Biotechnology Revolution in Global Agriculture: Innovation, Invention and Innvestment in the Canola Industry, Oxon, United Kingdom: CABI, 2001.
8. L Bickel. Facing Starvation: Norman Borlaug and the Fight Against Hunger. New York: Reader's Digest Press, 1974.
9. LR Brown, M Renner, B Halweil. Vital Signs, The Environmental Trends That Are Shaping the Future. New York: WW Norton, 1999.
10. IK Vasil. Biotechnology and food security for the 21st century: A real world perspective. Nature Biotechnol 16: 399–400, 1998.
11. AK Biswas. Population-Resources-Environment-Development: A systems view. Int Soc Ecol Model J 6: 11–24, 1984.
12. LR Brown, C Flavin, H French, JN Abramovitz, C Bright, S Dunn, B Halweil, G. Gardner, A Mattoon, A Platt McGinn, M O'Mcara, S Postel, M Renner, L Starke. State of the World 2000. New York: WW Norton, 2000.
13. DE Bloom, D Canning. The health and wealth of nations. Science 287: 1207–1209, 2000.
14. World Health Organization. World Health Report 1999: Making a Difference. Geneva: WHO, 1999.
15. R Watson. Common themes for ecologists in global issues. J Appl Ecol 36: 1–10, 1999.
16. J Lederberg. Emerging infections: An evolutionary perspective. Emerging Infect Dis 4:366–371, 1998.
17. DL Plucknett, NJH Smith, JT Williams, N Murthi Anishetty. Gene Banks and the World's Food. Princeton NJ: Princeton University Press, 1987.
18. National Research Council, Board on Agriculture. Managing Global Genetic Resources: Agricultural Crop issues and Policies. Washington, DC: National Academy Press, 1993, pp 1–28.
19. B Hitz. Economic aspects of transgenic crops which produce novel products. Curr Opin Plant Biol 2: 135–138, 1999.
20. K van Kate, SA Laird. The business behind biodiversity. Seed Trade News 12: 28–30, 1999.
21. VW Ruttan. Constraints on the design of sustainable systems of agricultural production. Ecol Econ 10: 209–219, 1994.
22. LB DeLind. Close encounters with a CSA: The reflections of a bruised and somewhat wiser anthropologist. Agric Hum Values 16: 3–9, 1999.
23. G Daily, P Dasgupta, B Bolin, P Crosson, J du Guerny, P Ehrlich, C Folke, AM Jansson, N Kautsky, A Kinzig, S Levin, K-G Maler, P Pinstrup-Andersen, D Siniscalco, B Walker. Food production, population growth and the environment. Science 281: 1291–1292, 1998.

24. S Page. The impact of changes in trade policy on developing country agriculture. J Agric Econ 42: 171–176, 1994.
25. GG Khachatourians, E Valencia. Integrated pest management and entomopathogenic fungal biotechnology in the Latin Americas: II. Key research and development prerequisites. Rev Acad Colomb Cien Exact Fisic Natur 23: 489–496, 1999.
26. D Roberts. Preliminary assessment of the effects of the WTO agreement on sanitary and phytosanitary trade regulations. J Int Econ Law 1: 377–405, 1998.
27. BR Champ. Prospects for grain storage technology in the 21st century. Postharvest Newsl 47: 8–13, 1998.
28. T Girbes, JM Ferreras. Ribosome inactivating proteins from plants. Rec Res Dev Agric Biol Chem 2: 1, 1–16, 1998.
29. AE Woytowich, GG Khachatourians. Plant antifungal peptides and their use in transgenic food crops. In: GG Khachatourians, DK Arora, eds. Applied Mycology and Biotechnology. Amsterdam: Elsevier Science 2001, pp. 145–164.
30. MM Iwu. Implementing the Biodiversity Treaty: How to make international co-operative agreement work. Trends Biotechnol 14: 78–82, 1996.
31. AB King, PB Eyzaguirre. Intellectual property rights and agricultural biodiversity: Literature addressing the suitability of IRP for the protection of indigenous resources. Agric Hum Values 16: 41–49, 1999.
32. S Lanteri, L Quagliotti. Problems related to seed production in the African region. Euphytica 96:173–183, 1997.
33. S Wessler, K Dawe, M Scanlon. Standing on the shoulders of giants. Trend Plant Sci 3:246–248, 1998.
34. P Meyer. Understanding and controlling transgene expression. Trends Biotechnol 13:332–337, 1995.
35. RKM Hay, RP Ellis. The control of flowering in wheat and barely: What recent advances in molecular genetics can reveal. Ann Bot 82:541–554, 1998.
36. SJ Lolle, RE Pruitt. Epidermal cell interaction: A case for local talk. Trend Plant Sci 4:14–19, 1999.
37. RE Pruitt, J-P Vielle-Calzada, SE Ploense, U Grossniklaus, SJ Lolle. FIDDLEHEAD, a gene required to suppress epidermal cell interactions in *Arabidopsis,* encodes a putative lipid biosynthetic enzyme. Proc Natl Acad Sci 97:1311–1316, 2000.
38. CC Mann. Genetic engineers aim to soup up crop photosynthesis. Science 283:314–316, 1999.
39. DT Luu, D Marty-Mazars, C Dumas, P Heizmann. Pollen stigma adhesion in *Brassica* involves SLG and SLRI glycoproteins. Plant Cell 11:251–262, 1999.
40. JE Wilkinson, K Lindsey, D Twell. Antisense-mediated suppression of transgene expression targeted specifically to pollen. J Exp Bot 49:1481–1490, 1998.
41. OJM Gopdijn, J Pen. Plants as bioreactors. Trends Biotechnol 13:379–387, 1995.
42. C Baron, PC Zambryski. The plant response in pathogenesis, symbiosis, and wounding: Variations on a common theme. Annu Rev Gent 29:107–129, 1995.
43. J Mol, E Grotewald, R Kotes. How genes paint flowers and seeds. Trends Plant Sci 3:212–217, 1995.
44. F Mourgues, MN Brisste, E Chevreau. Strategies to improve plant resistance to bacterial diseases though genetic engineering. Trends Biotechnol 16:203–210, 1998.
45. KJ Brunke, RL Meeusen. Insect control with genetically engineered crops. Trends Biotechnol 9:197, 1991.
46. WR Bushnell, DA Sommers, RW Giroux, LJ Szabo, RJ Zeyen. Genetic engineering of disease resistance in cereals. Can J Plant Pathol 20:137–220, 1998.
47. RA de Maagd, D Bosch, W Stiekema. *Bacillus thuringiensis* toxin-mediated insect resistance in plants. Trend Plant Sci 4:9–13, 1999.
48. TH Schuler, GM Poppy, BR Kerry, I Denholm. Insect-resistant transgenic plants. Trends Biotechnol 16:168–174, 1998.
49. J-K Zhu, PM Hasegawa, RA Bressan. Molecular aspects of osmotic stress. Crit Rev Plant Sci 16:253–277, 1997.
50. M Kasuga, Q Liu, K Yamaguchi-Shinozaki, K Shinozaki. Improving plant drought, salt and freezing tolerance by gene transfer of a single stress-inducible transcription factor. Nat Biotchnol 17:287–291, 1999.

51. AM Chaudhury, L Ming, C Miller, S Craig, ES Dennis. Fertilization-independent seed development in *Arabidopsis thaliana*. Proc Natl Acad Sci USA 94:4223–4228, 1997.
52. DW Meinke, JM Cherry, C Dean, SD Rounsley, M Koornneef. *Arabidopsis thaliana:* A model plant for genome analysis. Science 282:662–682, 1998.
53. CL Ives, BM Bedford, eds. Agricultural Biotechnology in International Development. Wallington, CABI 1998.
54. MD Gale, KM Devos. Plant comparative genetics after 10 years. Science 282:656–659, 1998.
55. V Walbot. Genes, genomes, genomics: What can plant biologists expect from the 1998 National Science Foundation Plant Genome Research Program? Plant Physiol 119:1151–1155, 1999.
56. R Datla, J William, G Selvaraj. Plant promoters for transgene expression. Biotechnol Annu Rev 3:269–296, 1997.
57. C Somerville, S Somerville. Plant functional genomics. Science 285:380–383, 1999.
58. MR Ponce, P Robles, JL Micol. High throughput genetic mapping in *Arabidopsis thaliana*. Mol Gen Genet 261:408–415, 1999.
59. W De Koe. In: JRN Taylor, PR Randall, VH Viljoen, eds. Cereal Science and Technology: Impact on Changing Africa. Pretoria: Sci Indust Rsch, 1993, pp. 807–822.
60. HA Koshinsky, GG Khachatourians. Trichothecene synergism, additivity and antagonism: The significance of the maximally quiescent ratio. Nat Toxins 1:38–47, 1992.
61. HA Koshinsky, GG Khachatourians. In: YH Hui, ed. in chief. Handbook of Foodborne Diseases, Vol. II. New York: Marcel Dekker, 1994, pp. 463–520.
62. HA Koshinsky, AL Woytowich, GG Khachatourians. Plant Toxicants. In: Foodborne Disease Handbook, 2nd Ed. Vol. 3. YH Hui, RA Smith, DG Spoerke Jr., eds. New York: Marcel Dekker, 2000, pp 627–652.
63. JD Miller, MA Ewen. Toxic effects of deoxynivalenon on ribosomes and tissues of the spring wheat cultivars frotana and casavant. Nat Toxins 5:23–237, 1997.
64. AE Woytowich, HA Koshinsky, GG Khachatourians. Identification of *Saccharomyces cerevisiae* gene which confers resistance to T-2 toxin. Plant Physiol 114:176, 1997.
65. R Bandyopadhyay, DE Fredrickson, NW McLaren, GN Odvody, MJ Ryley. Ergot a new disease threat to sorghum in the Americas and Australia. Plant Dis 82:356–367, 1998.
66. Y Sugiura, JR Barr, DB Barr, JW Brock, CM Eie, Y Ueno, DG Patterson Jr, ME Potter, E Reiss. Physiological characteristics and mycotoxins of human clinical isolates of *Fusarium* species. Mycol Res 103:1462–1468, 1999.
67. DG Panaccione, SL Annis. Use of fungal peptides and peptide synthetases in agri-food industry. In: Applied Mycology and Biotechnology. GG Khachatourians, DK Arora, eds. Amsterdam: Elsevier Science, 2001, pp. 114–143.
68. JI Cohen, C Falconi, J Komen, M Blakeney. Proprietary biotechnology inputs and international agricultural research. Int Serv Nat Agric Res Brief, Paper #39, 1998.
69. United Kingdom Institute of Biology. Genetically modified crops: the social and ethical issues. J Sustain Dev World Ecol 6:79–88, 1999.
70. RE Spire. Animal and plant cell technology: A critical evaluation of the technology/society interface. J Biotechnol 65:111–125, 1998.
71. BR Stillings. Regulatory environment: Incentive or impediment to developments in food science and technology. Crit Rev Food Sci Nutr 34:223–227, 1994.
72. R Straughan. Social and ethical issues surrounding biotechnological advance. Outlook Agric 20:89–94, 1991.
73. NL Ruibal-Mendieta, FA Lints, Novel and transgenic food crops: overview of scientific versus public perception. Transgenic Res 7:379–386, 1998.
74. D Boulter. Scientific and public perception of plant genetic manipulation—a critical review. Crit Rev Plant Sci 16:231–251, 1997.
75. O Kappeli, L Auberson. How safe is safe enough in plant genetic engineering? Trends Plant Sci 7:276–281, 1998.
76. A deKathen. The debate on risks from plant biotechnology: The end of reductionism? Plant Tissue Cult Biotechnol 4:136–147, 1998.

77. A McHughen. Pandora's Picnic Basket: The potential and hazards of genetically modified foods. Oxford (England: Oxford University Press, 2000.
78. P Allen. Reweaving the food security safety net: Mediating entitlement and enterprenurship. Agric Hum Values 16:117–129, 1999.
79. D Pelletier, V Kraak, C McCullum, U Uusitalo R Rich. Community food security: Salience and participation at community level. Agric Hum Values 16:401–419, 1999.
80. M Koc, KA Dahlberg. The restructuring of food system: Trends, research, and policy issues. Agric Hum Values 16:109–116, 1999.
81. R MacRae and the Toronto Food Policy Council. Not just what, but how: Creating agricultural sustainability and food security by changing Canada's agricultural policy making process. Agric Hum Values 16:187–201, 1999.
82. MA Altieri. The environmental risks of transgenic crops: an agroecological assessment. AgBiotech News Inform 10:405N–410N, 1998.
83. I Serageldin. Biotechnology and food security in the 21st century. Science 285:387–389, 1999.

# 2
# The Dynamics of Plant Genome Organization

**Isobel A. Parkin and Derek J. Lydiate**
*Agriculture and Agri-Food Canada, Saskatoon, Saskatchewan, Canada*

| | | |
|---|---|---|
| I. | WHAT IS THE GENOME? | 29 |
| II. | HISTORICAL PERSPECTIVES | 30 |
| | A. Cytogenetics | 30 |
| | B. Genome Mapping | 31 |
| | C. Genome Sequencing | 32 |
| III. | FACTORS DRIVING THE EVOLUTION OF GENOME STRUCTURE | 33 |
| | A. Polyploidy | 34 |
| | B. Chromosomal Rearrangements | 34 |
| | C. Transposable Elements | 35 |
| IV. | IMPACT OF GENOME DYNAMICS ON CROP IMPROVEMENT | 35 |
| | A. Comparative Mapping | 35 |
| | B. Intergenomic Gene Transfer | 37 |
| | C. Candidate Gene Analysis | 37 |
| V. | FUTURE PERSPECTIVES | 38 |
| | REFERENCES | 38 |

## I. WHAT IS THE GENOME?

The nuclear genome of a plant includes all the deoxyribonucleic acid (DNA) found in the nucleus of each cell and all the genes encoded by this DNA. The genome also comprises the cytological and biochemical structures that regulate the expression of these genes and ensure the stable inheritance of the DNA, in the form of chromosomes, from generation to generation. Plants also have two organeller genomes, one in the mitochondrion and one in the chloroplast, which are believed to be the degenerate genomes of ancient prokaryotic symbiotes. It appears that genes have migrated from the organellar genomes into the nuclear genome during millions of years of evolution, leaving the organellar genomes completely incapable of supporting a free living prokaryotic organism. Much of the DNA of all three genomes is organized into genes that encode either proteins or biochemically active ribonucleic acid (RNA) molecules such as transfer RNA and ribosomal RNA. The protein encoding regions of genes are flanked by regulatory elements that control the temporal and spatial expression of these genes. The programming that defines the architecture and the biochemical and physiological processes of a plant is contained within these genes and their associated regulatory elements, but the expression of this programming is subject to modification as a response to environmental factors.

The nuclear DNA is arranged in a small number of incredibly long DNA molecules. Each molecule along with its associated proteins can be visualised as a distinct chromosome. The DNA molecule at the core of each chromosome is composed of two complementary strands, with each strand carrying essentially all the information required to replicate an identical copy of its sister strand. This duplication of DNA molecules occurs during the replicative phase of each cell cycle, when two sister chromatids are formed from each chromosome. During mitosis, or division, one chromatid of each chromosome is inherited by each of the new cells. During meiosis, the sexual cell division that forms gametes, homologous chromosomes pair and recombine to form hybrid DNA molecules, and one of the four chromatids from each homologous pair of chromosomes is inherited by each gamete at the second meiotic division. This recombination process allows sequence variations present on different DNA molecules representing the same chromosome and of benefit to the organism to be combined on a single superior DNA molecule.

## II. HISTORICAL PERSPECTIVES

The study of plant genomes has changed dramatically with advances in technology. Cytogenetics, the microscopic examination of chromosome behavior, charted the associations of related chromosomes at meiosis and the patterns of inheritance of normal and aberrant chromosomes. The advent of molecular markers gave new impetus to studies of the genome organization and provides insights into genome evolution through comparative mapping. DNA sequencing promises to provide a considerably more detailed picture of the genome, and the DNA sequence of all five chromosomes of the model species *Arabidopsis thaliana* has been completed.

### A. Cytogenetics

In the early 1900s the two disciplines cytology, which is the study of the structure and life history of the cell, and genetics, which is the study of heredity, united with the realization that most aspects of genetic behavior could be explained by the recombination and reassortment of chromosomes. Cytogenetics revealed the organization of chromosomes for a number of the major crop species, contributing to the study of plant evolution and facilitating the development of cytogenetic stocks that have been used for both crop improvement and for positioning of genes on chromosome segments. Evolutionary relationships could be inferred from the study of chromosome pairing at meiosis by observing the number of bivalents formed and the presence of any abnormal pairing structures (Fig. 1). Chromosome doubled synthetic hybrids can be formed from the interspecific hybridization of diploid and tetraploid species related to more complex crop species. On crossing these resynthesized plants with major crops such as wheat and canola the evolutionary origins and genome organizations of these crops could be inferred from the cytological characteristics of the resulting hybrids. For example, three diploid *Brassica* species (*B. oleracea*, *B. rapa*, and *B. nigra*) have fused in each pairwise combination to generate three amphidiploid *Brassica* species (*B. napus*, *B. juncea*, and *B. carinata*) (1). Cereals have proved to be extremely amenable to cytogenetic techniques because their relatively large genomes are distributed over comparatively small numbers of chromosomes, allowing easy visualization of individual chromosomes under the light microscope. Wheat has a hexaploid genome with six related copies of each of seven basic types of chromosome that was probably formed from three distinct parental species each containing seven pairs of chromosomes (2). A set of 21 monosomic lines have been

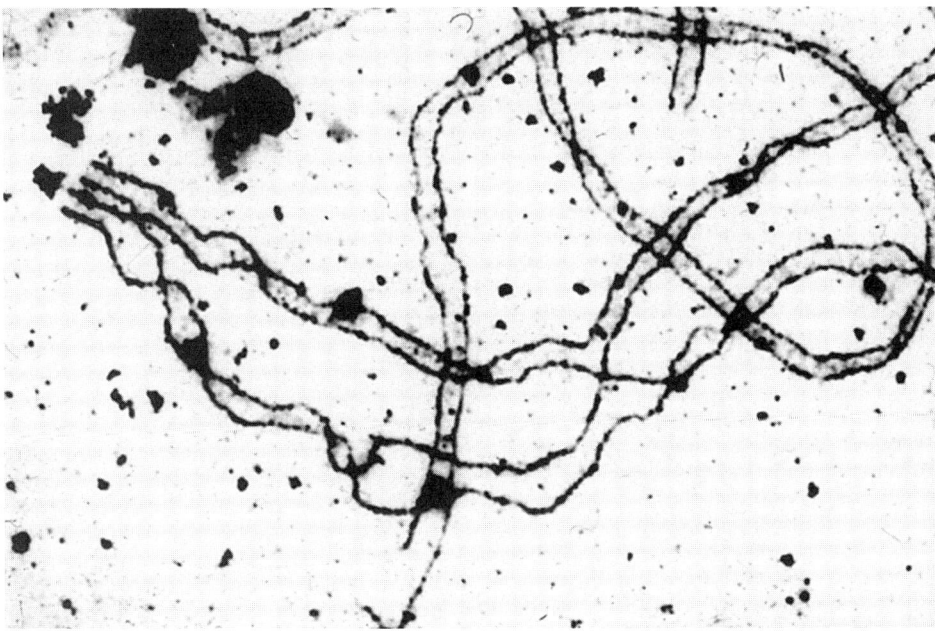

**Figure 1** Synaptonemal complex showing meiotic association of two pairs of homoeologous chromosomes (quadrivalent) common to all $F_1$ lines derived from a cross between a spring and a winter variety of canola. (From DJ Lydiate, AG Sharpe, and JS Parker, unpublished data.)

generated, each missing a chromosome from one of the 21 chromosome pairs of hexaploid wheat (3). Such lines have proved very useful for positioning genes and in interspecific hybridizations. For example, variation at the Ph locus on chromosome 5B controls whether the chromosomes of wheat associate and recombine in 21 pairs of strict homologues or as seven sets each with six homoeologous chromosomes (4,5), and lines nullisomic for 5B allow enhanced transfer of agronomically important genes from alien chromosomes to the wheat genome (6). Lines derived from interspecies crosses tend to inherit deleterious genes along with the gene of interest, and the removal of these unwanted characteristics involves years of breeding. The use of molecular markers to tag the genes of interest has revolutionized the introgression of genes from diverse genotypes into established cultivars. The molecular makers allow not only selection of the region of the genome containing the gene of interest but concurrent elimination of unwanted regions of the foreign genotype (7).

## B. Genome Mapping

Molecular biology techniques have furnished an almost unlimited number of phenotypically neutral and highly reproducible markers for genetic analysis. The first class of molecular markers were restriction fragment length polymorphisms (RFLPs) (8). RFLPs are based on size differences in DNA fragments generated from equivalent regions of homologous chromosomes after digestion with restriction enzymes (Fig. 2). The molecular basis of these polymorphisms is sequence variation at endonuclease cleavage sites and the insertion and excision of mobile genetic elements. RFLP markers were quickly applied to a number of crop plants because they are codominant (allowing every possible genotype to be determined), because they are immune to the

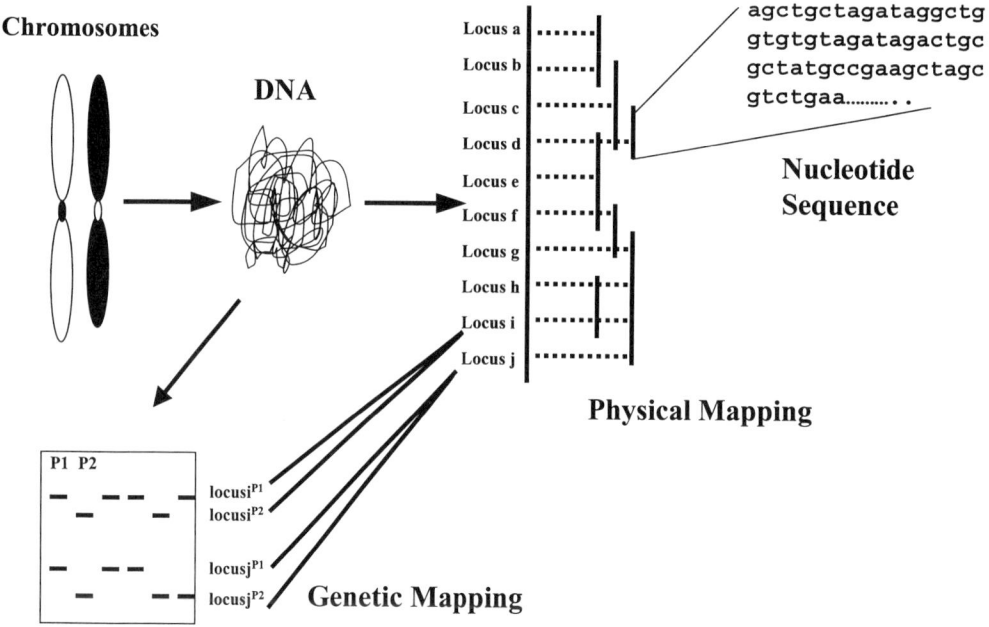

**Figure 2** Schematic diagram showing the pathway from visualization of chromosome structure to the final nucleotide sequence of the DNA, through molecular marker analysis of DNA (genetic mapping) and the use of those markers to identify overlapping clones containing large inserts of DNA (physical mapping).

effects of the environment and epistasis, and because a number of naturally occurring alleles are commonly found in crop germplasm (9). A number of PCR-based marker systems have been developed, including random amplified polymorphic DNAs (RAPDs) (10) and amplified fragment length polymorphisms (AFLPs) (11). Such markers allow considerable numbers of loci to be positioned on the genome relatively easily and quickly; however, these markers are of limited use for genome analysis and in particular comparative mapping, because the different loci detected by a particular marker are evolutionarily unrelated. A number of genetic maps of crop plants have now been established by using all three marker systems; such maps show the linear arrangement of genetic loci along linkage groups (12–15). In some crops these linkage groups have been assigned to distinct chromosomes by integration with the original cytogenetic maps using morphological markers as anchor points. The low copy RFLP probes can also be used as probes for in situ experiments in which the homologous region on the genome is directly detected on extended chromosome fibers (16). In *A. thaliana* and rice the genetic map has now been aligned with a physical map of the genome, in which the chromosomes have been cut into large fragments (ranging from 90 to 600 kb); each fragment is maintained individually (as a bacterial or yeast artificial chromosome, BAC or YAC, respectively), and the order of the fragments along the chromosome is determined (17,18) (Fig. 2). Physical maps underpin genome sequencing programs that promise to uncover the organization of the chromosomes down to the DNA nucleotide level and also serve as resources for map-based gene cloning projects.

## C. Genome Sequencing

The Maxim-Gilbert method of chemical cleavage made it possible to determine the exact nucleotide sequence of DNA. DNA sequencing was simplified and accelerated with the advent of

Sanger sequencing, which involves the controlled termination of DNA replication and which, with the addition of flourescent labels, led to the development of automated high-throughput DNA sequencers. These machines paved the way for projects to derive the exact nucleotide sequence of entire genomes. *Arabidopsis thaliana* (which is closely related to *Brassica* crop species) and *Ozyra sativa* (rice) are the first two plant subjects for genome sequencing projects. This will reinforce the position of *Arabidopsis* as the model dicotyledonous plant and rice as the model monocot. *Arabidopsis* and rice have been chosen due to their relatively small genome size (Table 1) (19–21), low proportions of repetitive DNA, and the extensive genomics tools (such as BACs, YACs, and markers) (22,23). The sequence of the entire genome of *Arabidopsis thaliana* was completed at the end of 2000 (24). The data for the Arabidopsis genome has shown an average gene density of approximately one gene every 5 kilobases, it has uncovered the organization of disease resistance loci, has revealed the distribution of retroelements, and has identified an unexpectedly high number of duplicated genes (25). The usefulness of all the sequencing data is dictated by the efficacy of the bioinformatics tools that are used to analyze the data, and caution should be exercised when evaluating the exact sequences of genes predicted by existing algorithms. Of the predicted genes, 56% have been found to have a high level of similarity to partially sequenced cDNAs (expressed sequence tags [ESTs]), confirming that the predicted genes are indeed transcribed. However, in most cases the functions of these genes can only be inferred from homology to known genes. The large and often highly duplicated genomes of most crop species have made them technically unattractive subjects for whole genome sequencing. EST sequencing has provided a relatively cheap and quick method for obtaining data about the genes that define a species, and this technology has been adopted for a number of crops including maize, soybean, and rice. This type of data has proved to be extremely useful for phylogenetic studies, for comparative genome studies, as anchor points in physical maps, for exon prediction in large expanses of genomic DNA, and most recently for expression studies as targets on microarrays (26).

## III. FACTORS DRIVING THE EVOLUTION OF GENOME STRUCTURE

Normal genetic analysis assumes that the linear order of genes along the chromosomes is constant. This inferred stability underpins the construction of genetic and physical maps. However,

Table 1 Haploid Genome Size and the Basic Chromosome Number for Various Plant Species

| Plant species | Family | Chromosome Number | Nuclear DNA Content (~Mbp) |
|---|---|---|---|
| *Arabidopsis thaliana* | Cruciferae | $x = 5$ | 100,[a] 145,[b] 190[c] |
| *Oryza sativa* (rice) | Gramineae | $x = 12$ | 430,[b] 580[c] |
| *Lycopersicon esculentum* (tomato) | Solanaceae | $x = 12$ | 950,[b] 965[c] |
| *Brassica napus* (canola) | Cruciferae | $2x = 19$ | 1200,[b] 1500[c] |
| *Zea mays* (corn) | Gramineae | $x = 10$ | 2500,[b] 2300[c] |
| *Hordeum vulgare* (barley) | Gramineae | $x = 7$ | 4900,[b] 5300[c] |
| *Triticum aestivum* (wheat) | Gramineae | $3x = 21$ | 16,000,[b] 16,700[c] |

[a]See Ref. 19.
[b]See Ref. 20.
[c]See Ref. 21.

the processes of polyploidy and chromosomal rearrangement and the activities of transposable elements introduce fluidity into the organization of plant genomes (described later).

### A. Polyploidy

It has been estimated that up to 70% of modern angiosperms could be polyploids (27), indicating the pervasive and recurrent nature of genome fusion as a factor in plant evolution. Allopolyploid and amphidiploid genomes are formed from the fusion of two or more distinct but closely related genomes (28,29). They probably arise from chromosome doubled interspecies hybrids and contain the full chromosome complement of each parent. Autopolyploids, which are rarer than allopolyploids, contain three or more haploid copies of the same genome (28,29).

The increase in genome size resulting from polyploidy causes a related increase in cell size and this normally translates into an increase in plant size. The duplicate sets of parental genes expressed in allopolyploids also produce an effect akin to heterosis. Duplicate genes protect the plant from naturally occurring recessive mutations that are masked by functional gene copies. As a result polyploid plants tend to be larger and exhibit more vigorous vegetative growth than their parental species and hence have the potential to dominate habitats. In the long term, polyploidy probably also plays an adaptive role, since the duplicate copies of genes can be seconded to perform distinct functions. However, in the short term, polyploidy reduces the rate of adaptation, because recessive mutations are masked by dominant wild-type alleles.

In new polyploids and particularly autopolyploids, the pairing of multiple copies of related chromosomes at meiosis can be irregular and lead to unbalanced gametes and/or chromosomal rearrangements. In most cases, the resulting aneuploidy causes reduced fertility and reduced fitness. Successful allopolyploids have generally evolved a genetic mechanism to control chromosome pairing, ensuring the faithful alignment of homologous chromosomes. However, seemingly well adapted amphidiploids such as *B. napus* still produce chromosomal rearrangements at meiosis at a detectable frequency as a result of pairing between related but nonhomologous (homoeologous) chromosomes (30) (Fig. 1).

### B. Chromosomal Rearrangements

The structure of individual chromosomes evolves through centric fusions, centric fissions, inversions, and translocations, both reciprocal and nonreciprocal. In the Solanaceae, comparative mapping revealed that the chromosomes of potato and tomato were differentiated by chromosomal inversions, all derived from single breakpoints mapped at or near to the centromeres and resulting in the inversion of entire chromosome arms (31). In the Brassiceae, comparative mapping has similarly positioned centromeres at the breakpoints of collinearity between the ancestral A and C genomes of *B. napus*, again suggesting that centric fission and fusion have been strong driving forces in the evolution of the *B. rapa* (the A genome) and *B. oleracea* (the C genome) as they diverged from a common ancestor (32, IAP Parkin and DJ Lydiate, unpublished). The intragenomic and intergenomic translocations that differentiate the genomes of wheat and rye have been identified and the breakpoints defined (33). From the preceding it is clear that some (rare) chromosomal rearrangements have been fixed during evolution and speciation; however, rearrangements occur continuously in nature and the vast majority probably cause reduced fitness and are eliminated. This is illustrated by *B. napus*, in which the amphidiploid genome structure has been maintained for several hundred years, presumably as a result of continual selection for fitness, although 1 in 15 gametes carry nonreciprocal translocations at each generation (30, DJ Lydiate, unpublished).

## C. Transposable Elements

With the discovery of transposons, or "jumping genes," by Barbara McClintock in the 1940s (34) it became apparent that the linear organization of genomic DNA is far from rigid, but the extent to which mobile elements can constitute a high proportion of genomic DNA only became apparent with the advent of Southern hybridization and DNA sequencing (35,36). *Triticum aestivum* (wheat) and *Zea mays* (corn) appear to be fairly extreme examples of genomes carrying heavy loads of dispersed repetitive elements (37,38). In the case of *Z. mays*, these elements are composed of numerous fairly well characterized families of transposons, retrotransposons, and related defective elements (39,40). A common feature of plant transposable elements is their ability to excise from one chromosomal location and randomly reinsert into another location. Excision events are normally imprecise and leave potentially mutagenic "footprints," which are most commonly insertions of a few nucleotides. Retrotransposons and retroposons are replicated through the reverse transcription of RNA intermediates, and new copies are again inserted randomly into the genome. The combined activity of mobile elements in corn is such that they are likely to have produced a high proportion of the allelic variation observed in corn germplasm. The genetic load of continually increasing numbers of mobile elements will erode the success of well-adapted organisms in stable environmental conditions where the ability to adapt further is not at a premium and it is likely that the processes of DNA methylation and the high mutation rate of methylated cytosine residues have evolved at least in part in order to curb the activity of wayward mobile elements.

## IV. IMPACT OF GENOME DYNAMICS ON CROP IMPROVEMENT

### A. Comparative Mapping

Comparative mapping involves the charting of regions of orthology between the genomes of different plant species, that is, regions that have related structures by virtue of having evolved from a common ancestral structure. Such studies should eventually allow researchers to deduce the genome organizations of the ancestral progenitors of modern-day plants. Typically, the relative positions of RFLP-defined loci detected by a common set of probes in a number of related species have revealed chromosomal regions with conserved gene content and indeed conserved gene order. For example, all crop species of the Gramineae, which after 60 million years of evolution have markedly different genome sizes and chromosome numbers, exhibit striking genome collinearity (41). More high-resolution experiments using physical information based on cloned contigs and DNA sequence have confirmed these patterns of collinearity (42,43). Analysis of the Gramineae has revealed some chromosomal rearrangements that characterize taxonomic groups and others that have occurred more recently, during or even after speciation (41).

Moore (44) presented a holistic model of cereal genome evolution (Fig. 3) in which all genomes, including those of corn, wheat, and rice, can be broken down into 19 linkage blocks. Similar, but less extensive, comparative mapping studies have been carried out in the Solanaceae, involving tomato, pepper, and potato (45,31); between a number of legume species (46,47); and in the Crucifereae between *B. napus*, the three diploid *Brassica* crop species (48,49), and *Arabidopsis*, the model dicot (50–52). Large regions of collinearity have been observed in each of these families, but areas of the genome that appear to be hot spots for rearrangements have also been identified (53,54). These regions with clustered disruptions in the collinearity appear to be associated with centromeric and telomeric repeats, suggesting that such regions are common sites for chromosome breakage and fusion in plant evolution (32, 54, DJ Lydiate and IAP Parkin, unpublished).

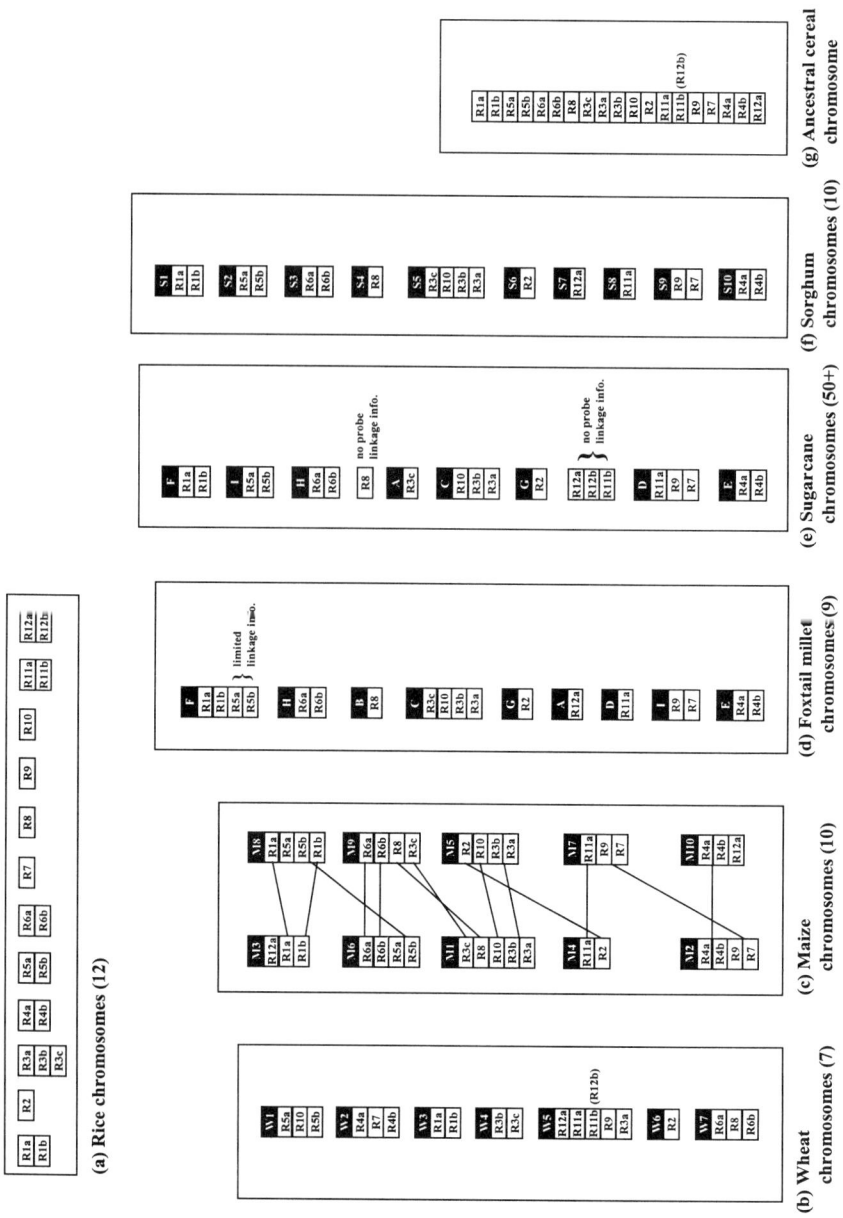

**Figure 3** Comparisons of cereal genome evolution based on rice linkage segments. (a) Rice chromosomes dissected into linkage blocks, (b) wheat, (c) maize, (d) foxtail millet, (e) sugarcane, and (f) sorghum chromosomes represented as rice blocks on the basis of homology and/or conservation of gene order. Connecting lines indicate duplicated segments within the maize chromosomes. (g) An ancestral "single chromosome" reconstructed on the basis of these linkage blocks. (Redrawn from Ref. 44.)

Comparative mapping, which is identifying regions of collinearity between complex crop genomes and the model genomes of rice and *Arabidopsis,* will allow researchers working with crop species to exploit the genomic resources and data being developed for the model species.

## B. Intergenomic Gene Transfer

The wild relatives of crop species often contain unique genetic variation of potential benefit to crop improvement, such as genes for disease and insect resistance (55,56). Similarly, related crop species often contain complementary sets of useful genetic variation. Many crops can form fertile interspecies hybrids with related species, and these hybrids can function as bridges for the interspecies transfer of useful genetic variation. Two examples of this process are the interspecies transfer of resistance to club root in *Brassica* (57) and the transfer of resistance to leaf rust, stripe rust, and powdery mildew from rye to wheat (58,59). Successful intergenomic gene transfer is a vital component of systems where the interspecies transfer of useful genes have long-term utility in plant breeding. In instances in which whole donor chromosomes or large segments of donor chromosome containing the gene of interest are transferred into a crop species but in which these large pieces of donor genome are unable to recombine efficiently with the recipient genome, the resulting linkage drag normally prevents successful crop development. The association of poor quality (high seed glucosinolates) with the restorer of cytoplasmic male sterility transferred from *Raphanus* to *Brassica* (60) and the association of poor protein profiles with the disease resistance loci transferred from rye to wheat (61) are examples of this problem. The development of effective genetic marker systems and accurate comparative maps will assist in the future selection of high-precision intergenomic gene transfer events, in which hybrid chromosomes with structures that should allow them to pair efficiently with recipient chromosomes can be recognized in the early generations of interspecies gene transfer programs. In the future it might even be possible to develop variants of crop species in which recombination between homoeologous chromosomes is more frequent.

## C. Candidate Gene Analysis

The conservation of gene content and gene order between crop species and related model species offers the enticing prospect of identifying the molecularly well characterized genes in model genomes that correspond to genes in related crops that control major agronomic or quality traits but have only been characterized genetically. The imminent availability of the entire genome sequence for *Arabidopsis* will allow the prediction of many genes that might have counterparts that contribute to agronomically important traits in crop species. This candidate gene prediction should be feasible in crucifers, in which *Brassica napus* and *Arabidopsis* share on average 85% sequence identity between orthologous genes (52) and wild-type genes cloned from *B. napus* can complement the corresponding mutant phenotype in *Arabidopsis* (62). Candidate genes that have been identified in *Arabidopsis* appear to control similar traits in *B. napus*; fatty acid elongase (FAE1) maps coincidentally with a locus controlling variation in erucic acid content (63); similarly curly leaf (CLF) has been shown to map at the loci controlling variation for an apetalous phenotype (64). In other plant families candidate genes have been less forthcoming; however, detailed comparative mapping between wheat and the model genome rice has identified a rice YAC contig that corresponds to the region of wheat containing the *Ph1* locus and still might yield possible candidate genes (65). However, the high density of genes and the relative imprecision of genetic mapping make it likely that candidate gene analysis will sometimes result in spurious associations between attractive candidates in model species with homologues that are closely linked to genes controlling important traits in crop species (66).

## V. FUTURE PERSPECTIVES

We are presently entering a new stage in genomics research with large organizations worldwide investing significantly in plant genomics and particularly functional genomics. Over the next 10 years this effort will yield an enormous reservoir of information generated through high-capacity methods of research such as the analysis of gene expression patterns using DNA chips (67), map-based gene cloning using ordered BAC and YAC libraries (68), genome and EST sequencing, and finally functional genomics using populations of insertion mutagenized lines (69). Currently only approximately 50% of the genes predicted from the Arabidopsis sequence can be assigned putative functions, indicating the vast wealth of genes for which functions still have to be identified (25). On a cautionary note, with huge volumes of data comes the potential for huge numbers of errors and an absolute requirement for automated data analysis. Efficient bioinformatics tempered with informed biological insight will be vital if genomics data are to be applied successfully to crop improvement.

## REFERENCES

1. N U. Genome analysis in *Brassica* with special reference to the experimental formation of *B. napus* and peculiar mode of fertilisation. Jpn J Bot 7:389–452, 1935.
2. ER Sears. Homoeologous chromosomes in *Triticum aestivum*. Genetics 37:624, 1952.
3. ER Sears. The aneuploids of common wheat. University of Missouri Reseach Bulletin 572, pp 1–58, 1954.
4. M Okamoto. Asynaptic effect of chromosome V. Wheat Inform Serv 5:6–7, 1958.
5. R Riley, V Chapman. Genetic control of the cytologically diploid behaviour of hexaploid wheat. Nature 182:713–715, 1958.
6. AKMR Islam, KW Shepherd. In: PK Gupta, T Tsuchiya, eds. Alien Genetic Variation in Wheat Improvement. In: Chromosome Engineering in Plants: Genetics, Breeding, Evolution. Part A. Amsterdam: Elsevier Science, 1991, pp 291–312.
7. ND Young, SD Tanksley. RFLP analysis of the size of chromosomal segments retained around the *Tm-2* locus of tomato during backcross breeding. Theor Appl Genet 77:353–359, 1989.
8. D Botstein, RL White, M Skolnick, RW Davies. Construction of a genetic linkage map in man using restriction fragment length polymorphisms. Am J Hum Genet 32:314–331, 1980.
9. AH Paterson, SD Tanksley, ME Sorrels. DNA markers in plant improvement. Adv Agron 46:39–90, 1991.
10. JGK Williams, AR Kubelik, KJ Livak, JA Rafalski, SV Tingey. DNA polymorphisms amplified by arbitrary primers are useful as genetic markers. Nucleic Acids Res 18:6531–6535, 1990.
11. P Vos, R Hogers, M Bleeker, M Reijans, T van de Lee, M Hornes, A Frijters, J Pot, J Peleman, M Kuiper, M Zabeau. AFLP: A new technique for DNA fingerprinting. Nucleic Acids Res 23:4407–4414, 1995.
12. T Helentjaris. A genetic linkage map for maize based on RFLPs. Trends Genet 3:217–221, 1987.
13. GB Kiss, G Csanadi, K Kalman, P Kalo, L Okresz. Construction of a basic genetic map for alfafa using RFLP, RAPD, isozyme and morphological markers. Mol Gen Genet 238:129–137, 1993.
14. IAP Parkin, AG Sharpe, DJ Keith, DJ Lydiate. Identification of the A and C genomes of amphidiploid *Brassica napus* (oilseed rape). Genome 38:1122–1131, 1995.
15. W Spielmeyer, AG Green, D Bittisnich, N Mendham, ES Lagudah. Identification of quantitative trait loci contributing to Fusarium wilt resistance on an AFLP linkage map of flax (*Linum usitatissimum*). Theor Appl Genet 97:633–641, 1998.
16. P Fransz, S Armstrong, C Alonso-Blanco, TC Fischer, RA Torres-Ruiz, G Jones. Cytogenetics for the model system *Arabidopsis thaliana*. Plant J 13:867–876, 1996.
17. BA Antonio, M Emoto, J Wu, I Ashikawa, Y Umehara, N Kurata, T Sasaki. Physical mapping of rice chromosomes 8 and 9 with YAC clones. DNA Res 3:393–400, 1996.
18. R Schmidt, J West, G Cnops, K Love, A Balestrazzi, C Dean. Detailed description of four YAC con-

tigs representing 17 Mb of chromosome 4 of *Arabidopsis* thaliana ecotype Columbia. Plant J 9:755–765, 1996.
19. I Hwang, T Kohchi, BM Hauge, H Goodman, R Schmidt, G Cnops, C Dean, S Gibson, K Iba, B Lemieux. Identification and map position of YAC clones comprising one-third of the *Arabidopsis* genome. Plant J 1:367–374, 1991.
20. K Arumuganathan, ED Earle. Nuclear DNA content of some important plant species. Plant Mol Biol Rep 9:208–218, 1991.
21. MD Bennet, JB Smith. Nuclear DNA amounts in angiosperms. Philos Trans R Soc London B 274:227–274, 1976.
22. DW Meinke, JM Cherry, C Dean, SD Rounsley, M Koorneef. *Arabidopsis thaliana*: A model plant for genome analysis. Science 282:662–682, 1998.
23. SA Goff. Rice as a model for cereal genomics. Curr Opin Plant Biol 2:86–89, 1999.
24. The Arabidopsis Genome Initiative. Analysis of the genome sequence of the flowering plant *Arabidopsis thaliana*. Nature 408:796–815, 2000.
25. M Bevan, I Bancroft, E Bent, K Love, H Goodman, C Dean, R Bergkamp, W Dirkse, M Van Staveren, W Stiekema, L Drost, P Ridley, SA Hudson, K Patel, G Murphy, P Piffanelli, H Wedler, E Wedler, R Wanbutt, T Weitzenegger, TM Pohl, N Terryn, J Gielen, R Villarroel, R De Clerck, M Van Montagu, A Lecharny, S Auborg, I Gy, M Kreis, N Lao, T Kavanagh, S Hempel, P Kotter, KD Entian, M Rieger, M Schaeffer, B Funk, S Muller-Auer, M Silvey, R James, A Montfort, A Pons, P Puigdomenech, A Douka, E Voukelatou, D Millioni, P Hatzopoulos, E Piravandi, B Obermaier, H Hilbert, A Düsterhöft, T Moores, JDG Jones, T Eneva, K Palme, V Benes, S Rechman, W Ansorge, R Cooke, C Berger, M Delseny, M Voet, G Volckaert, HW Mewes, S Klosterman, C Shueller, N Chalwatzis. Analysis of 1.9Mb of contiguous sequence from chromosome 4 of *Arabidopsis thaliana*. Nature 391:485–488, 1998.
26. D Bouchez, H Höfte. Functional genomics in plants. Plant Physiol 118:725–732, 1998.
27. J Masterson. Stomatal size in fossil plants: Evidence for polyploidy in majority of angiosperms. Science 264:421–424, 1994.
28. GL Stebbins. Polyploidy. I. Occurrence and nature of polyploid types. In: Variation and Evolution in Plants. New York: Columbia University Press, 1950, pp 298–341.
29. V Grant. Plant Speciation. New York: Columbia University Press, 1971, pp 219–261.
30. AG Sharpe, IAP Parkin, DJ Keith, DJ Lydiate. Frequent nonreciprocal translocations in the amphidiploid genome of oilseed rape (*Brassica napus*). Genome 38:1112–1121, 1995.
31. SD Tanksley, MW Ganal, JP Prince, MC de Vicente, MW Bonierbale, P Broun, TM Fulton, JJ Giovannoni, S Grandillo, GB Martin, R Messeguer, JC Miller, L Miller, A Paterson, O Pineda, MS Röder, RA Wing, W Wu, ND Young. High density molecular linkage maps of the tomato and potato genomes. Genetics 132:1141–1160, 1992.
32. A Kelly. The genetic basis of petal number and pod orientation in oilseed rape (*B. napus*). PhD dissertation, University of Newcastle-upon-Tyne, England, 1996.
33. KD Devos, MD Atkinson, CD Chinoy, HA Francis, RL Harcourt, RMD Koebner, CJ Liu, P Masojé, DX Xie, MD Gale. Chromosomal rearrangements in the rye genome relative to that of wheat. Theor Appl Genet 85:673–680, 1993.
34. B McClintock. Chromosomal organisation and genic expression. Cold Spring Harbor Symp Quan Biol 16:13–47, 1951.
35. R Flavell. The molecular characterisation and organization of plant chromosomal DNA sequences. Annu Rev Plant Physiol 31:569–596, 1980.
36. P SanMiguel, A Tikhonov, YK Jin, N Motchoulskaia, D Zakharov, A Melake-Berhan, PS Springer, KJ Edwards, M Lee, Z Avramova, JL Bennetzen. Nested retrotransposons in the intergenic regions of the maize genome. Science 274:765–768, 1996.
37. RB Flavell, MD Bennet, JB Smith, DB Smith. Genome size and the proportion of repeated nucleotide sequence in DNA in plants. Biochem Genet 12:257–269, 1974.
38. JL Bennetzen, P SanMiguel, M Chen, A Tikhonov, M Francki, Z Avramova. Grass genomes. Proc Natl Acad Sci USA 95:1975–1978, 1998.
39. PA Peterson. Transposable elements in maize: Their role in creating plant genetic variability. Adv Agronom 51:79–123, 1993.

40. JL Bennetzen. The contributions of retroelements to plant genome organization, function and evolution. Trends Micro 4:347–353, 1996.
41. KM Devos, MD Gale. Comparative genetics in grasses. Plant Mol Biol 35:3–15, 1997.
42. A Killian, J Chen, F Han, B Steffenson, A Kleinhofs. Towards map-based cloning of the barley stem rust resistance genes *Rpg1* and *rpg4* using rice as as intergenomic cloning vehicle. Plant Mol Biol 35: 187–195, 1997.
43. M Chen, P SanMiguel, JL Bennetzen. Sequence organisation and conservation in *sh2/a1*-homologous regions of sorghum and rice. Genetics 148:435–443, 1998.
44. G Moore. Cereal genome evolution: pastoral pursuits with 'Lego' genomes. Curr Opin Genet Dev 5: 717–724, 1995.
45. SD Tanksley, R Bernatzky, NL Lapitan, JP Prince. Conservation of gene repetoire but not gene order in pepper and tomato. Proc Natl Acad Sci USA 85:6419–6423, 1988.
46. NF Weeden, FJ Muehlbauer, G Ladizinsky. Extensive conservation of linkage relationships between pea and lentil genetic maps. Heredity 83:123–129, 1992.
47. D Menancio-Hautea, CA Fatokun, L Kumar, D Danesh, ND Young. Comparative genome analysis of mungbean (*Vigna radiata* L. *Wilczek*) and cowpea (*V. Unguiculata* L. *Walpers*) using RFLP mapping data. Theor Appl Genet 86:797–810, 1993.
48. EJ Bohuon, DJ Keith, IAP Parkin, AG Sharpe, DJ Lydiate. Alignment of the conserved C genomes of *Brassica oleracea* and *Brassica napus*. Theor Appl Genet 93:833–839, 1996.
49. U Lagercrantz, DJ Lydiate. Comparative genome mapping in *Brassica*. Genetics 144:1903–1910, 1996.
50. U Lagercrantz, J Putterill, G Coupland, D Lydiate. Comparative mapping in *Arabidopsis* and *Brassica*, fine scale genome collinearity and congruence of genes controlling flowering time. Plant J 9:13–20, 1996.
51. JA Scheffler, AG Sharpe, H Schmidt, P Sperling, IAP Parkin, W Luhs, DJ Lydiate, E Heinz. Destaurase multigene families of *Brassica napus* arose through genome duplication. Theor Appl Genet 94:583–591, 1997.
52. AC Cavell, DJ Lydiate, IA Parkin, C Dean, M Trick. Collinearity between a 30-centimorgan segment of *Arabidopsis thaliana* chromosome 4 and duplicated regions within the *Brassica napus* genome. Genome 41:62–69, 1998.
53. TC Osborn, C Kole, IA Parkin, AG Sharpe, M Kuiper, DJ Lydiate, M Trick. Comparison of flowering time genes in *Brassica rapa*, *B. napus* and *Arabidopsis thaliana*. Genetics 146:1123–1129, 1997.
54. G Moore, M Roberts, L Aragon-Alcaide, T Foote. Centromeric sites and cereal chromosome evolution. Chromosoma 105:321–232, 1997.
55. P Moreau, P Thoquet, J Olivier, H Laterrot, N Grimsley. Genetic mapping of *Ph-2*, a single locus controlling resistance to *Phytophera infestans* in tomato. Mol Plant Microbe Interact 11:259–269, 1998.
56. MD Romero, MJ Montes, E Sin, I Lopez-Brana, A Duce, JA Martin-SAnchez, ME Andres, A Delibes. A cereal cyst nematode (*Heterodera avenae* Woll) resistance gene transferred from *Aegilops triuncialis* to hexaploid wheat. Theor Appl Genet 96:1135–1140, 1998.
57. E Diederichesen, B Wagenblatt, V Schallehn, U Deppe, MD Sacristan. Transfer of clubroot resistance from resynthesised *Brassica napus* into oilseed rape: Identification of race-specific interactions with *Plasmodiophora brassicae*. Acta Hortic 407:423–429, 1996.
58. FJ Zeller, SLK Hsam. Broadening the genetic variability of cultivated wheat by utilizing rye chromatin. Proceedings of the 6th International Wheat Genetics Symposium, Kyoto, Japan, 1983, pp 161–173.
59. B Friebe, M Heun, N Tuleen, FJ Zeller, BS Gill. Cytogenetically monitored transfer of powdery mildew resistance from rye to wheat. Crop Sci 34:621–625, 1994.
60. R Delourme, N Foisset, R Horvais, P Barret, G Champagne, WY Cheung, BS Landry, M Renard. Characterisation of the radish introgression carrying the *Rfo* restorer gene for the Ogu-INRA cytoplasmic male sterility in rapeseed (*Brassica napus* L.). Theor Appl Genet 97:129–134, 1998.
61. FJ Zeller, G Günzel, G Fischbeck, P Gerstenkorn, P Weipert. Veränderung der backeigenschaften der weizen-roggen-chromosomen-translokation 1B/1R. Getreide Mehl Brot 36:141–143, 1982.
62. LS Robert, F Robson, A Sharpe, D Lydiate, G Coupland. Conserved structure and function of the *Arabidopsis* flowering time gene *CONSTANS* in *Brassica napus*. Plant Mol Biol 37:763–772, 1998.

63. P Barret, R Delourme, M Renard, F Domergue, R Lessire, M Delseny, TJ Roscoe. A rapeseed *FAE1* gene is linked to the E1 locus associated with variation in the content of erucic acid. Theor Appl Genet 96:177–186, 1998.
64. MJ Fray, P Puangsomlee, J Goodrich, G Coupland, EJ Evans, AE Arthur, DJ Lydiate. The genetics of stamenoid petal production in oilseed rape (*Brassica napus*) and equivalent variation in *Arabidopsis thaliana*. Theor Appl Genet 94:731–736, 1997.
65. T Foote, M Roberts, N Kurata, T Sasaki, G Moore. Detailed comparative mapping of cereal chromosome regions corresponding to the *Ph1* locus in wheat. Genetics 147:801–807, 1997.
66. I Parkin, D Lydiate. Can we use Arabidopsis to understand the control of flowering time in *Brassica* crops. PBI Bulletin, May, 1998, pp. 12–13.
67. B Lemieux, A Aharoni, M Schena. Overview of DNA chip technology. Mol Breed 4:277–289, 1998.
68. SD Tanksley, MW Ganal, GB Martin. Chromosome landing: A paradigm for map-based gene cloning in plants with large genomes. Trends Genet 11:63–68, 1995.
69. RA Martienssen. Functional genomics: Probing plant gene function and expression with transposons. Proc Natl Acad Sci USA 95:2021–2026, 1998.

# 3
# Embryogenesis

**Alison M. R. Ferrie**
*National Research Council, Saskatoon, Saskatchewan, Canada*

|      |                                                      |    |
|------|------------------------------------------------------|----|
| I.   | INTRODUCTION                                         | 43 |
| II.  | ANDROGENESIS                                         | 44 |
|      | A. Factors Influencing Androgenesis                  | 45 |
| III. | GYNOGENESIS                                          | 50 |
|      | A. Factors Influencing Gynogenesis                   | 50 |
| IV.  | HAPLOID OR DOUBLED HAPLOID PLANT PRODUCTION          | 52 |
|      | A. Regeneration                                      | 52 |
|      | B. Chromosome Doubling                               | 52 |
| V.   | DEVELOPMENTAL ASPECTS                                | 53 |
| VI.  | APPLICATION OF HAPLOID AND DOUBLED HAPLOID PLANTS    | 54 |
|      | A. Varietal Development                              | 54 |
|      | B. Mutation Breeding                                 | 55 |
|      | C. Gene Transfer                                     | 55 |
|      | D. Biochemical and Physiological Studies             | 56 |
| VII. | CONCLUSION                                           | 56 |
|      | REFERENCES                                           | 56 |

## I. INTRODUCTION

Embryogenesis is a process that can occur in a fertilized egg cell, reproductive cells, or somatic tissue. This chapter reviews the published literature on the induction of embryogenesis from the male and female gametophyte with emphasis on advances since 1995.

Androgenesis is the process of embryo development from male gametophytic cells. When given appropriate conditions, microspores switch from a normal gametophytic developmental process to a sequence of events that lead to the formation of embryos. A number of factors influence this type of embryogenesis, and these are discussed in this chapter.

Gynogenesis is the process of embryo development from the culture of unfertilized ovaries or ovules. It is extremely difficult to isolate and culture the egg cell or other haploid cells, and therefore a group of cells, i.e., embryo sac, is cultured at one time. The embryo sac undergoes division and develops into an embryo without fertilization when given the appropriate culture conditions. A number of factors similar to those in androgenesis influence gynogenesis; they are discussed in this chapter. Gynogenic protocols have been used in a number of species to develop haploid plants, although the culture of anthers or microspores is the preferred method of haploid plant production for many species. There are fewer reports on development of haploid embryos via gynogenesis, partly because microspores are easier to handle and gynogenesis is very labor-intensive. Microspores are also more abundant and uniform in size and developmental stage. However, gynogenesis is an important technology for species that do not respond to androgenic techniques.

Haploid cells, embryos, plants, and doubled haploid plants are useful for both fundamental research and practical applications. Doubled haploid technology has been used for varietal development, mutagenesis, and gene transfer. The developmental pathway for in vitro haploid embryogenesis is similar to that of in vitro somatic embryogenesis and in vivo zygotic embryogenesis. Because of this, microspore-derived embryos have been used in biochemical and physiological studies. The uses, both applied and basic, have been well documented and are reviewed briefly in this chapter.

## II. ANDROGENESIS

In 1964, Guha and Maheshwari first demonstrated development of haploid plants from cultured anthers (1), although haploids had been reported several decades before (2). The early work on anther and isolated microspore culture was mostly with the Solanaceae species, e.g., *Datura innoxia* Mill. (1) and tobacco (*Nicotiana tabacum*) (3,4). Since the 1960s, haploid calli, embryos, or plants have been produced in over 250 species, including many economically important crops. In 1996, Maheshwari estimated that there were over 2000 publications on androgenesis and its applications (5). Table 1 shows a listing of some species in which plants have been regenerated via anther or microspore culture since 1995 (6–20). For a listing of other species producing haploids, refer to reviews by Maheshwari et al. (21), Dunwell (22), and Ferrie et al. (23).

A number of factors influence haploid embryogenesis from male gametophytic tissue. Donor plant genotype, donor plant growing conditions, developmental stage of the pollen grain, pretreatment of the floral organs, media composition, and culture conditions are factors that have

**Table 1** Haploid or Doubled Haploid Plant Production via Androgenesis Since 1995

| Species | Reference |
|---|---|
| *Aesculus hippocastranum* | 6 |
| *Apium graveolens* | 7 |
| *Atriplex glauca* | 8 |
| *Avena sativa* | 9,10 |
| *Avena sterilis* | 11 |
| *Bupleurum falcatum* | 12 |
| *Cajanus cajan* | 13 |
| *Cichorium intybus* var. *foliosum* | 14 |
| *Cicer arietinum* | 15 |
| *Guizotia abyssinica* | 16 |
| *Helianthus annuus* | 17 |
| *Oryza* species | 18 |
| *Quercus suber* | 19 |
| *Solanum melongena* | 20 |

been studied in many species in order to develop efficient protocols for generating haploid or doubled haploid plants.

## A. Factors Influencing Androgenesis

### 1. Donor Plant Genotype

As with many tissue culture systems, genotypic differences are reflected in androgenic response. Genotype screening studies have shown that there are differences among genotypes and even individual plants within the same genotype for both embryo induction and plantlet regeneration (24, 25). Genetic control of androgenesis has been reported in a number of species. In wheat (*Triticum aestivum* L.), additive and dominant gene action affected callus induction frequency, embryo induction frequency, and green plant regeneration (26). Cytoplasmic effects have also been reported in wheat (27,28). Reciprocal substitution analysis has shown that chromosomes of genomes A, B, and D influenced anther culture response, whereas green plant production was influenced by chromosomes of A and D (29). Genes on chromosomes of the B and D genomes affected albino plant regeneration (29). For a cross-pollinating cereal like rye (*Secale cereale* L.), regeneration via androgenesis is limited to genotypes derived from crosses between *S. cereale* and *S. vavilovii* Grossh (30,31). Crosses with this wild rye ancestor resulted in a line (SC35) capable of high-frequency embryogenesis (32). Further crosses with *S. cereale* indicated that it was possible to transfer genes for embryogenic ability to low-responding types (33). This response has also been shown in pepper (*Capsicum annuum* L.) (34). The highest frequency of embryo formation occurred in large-fruited pepper genotypes. Small-fruit genotypes gave poor response, and crosses between large and small-fruited genotypes gave an intermediate response (34). In rice (*Oryza sativa* L.), japonica types were found to be more responsive in terms of embryogenic capability in comparison to indica types (35). When comparing wild *Oryza* species, it has been suggested that those species possessing the A genome have a higher regeneration capability than those with the B, C, E, BC, or CD genomes (18). In tomato (*Lycopersicon esculentum*), the gene ms $10^{35}$ controlling male sterility was found to play a role in callus induction from anther culture (36). Callus formation was greater from genotypes that had the gene in the homozygous state than in genotypes carrying the gene in the heterozygous state (36).

### 2. Donor Plant Conditions

The environmental conditions in which donor plants are grown can influence embryogenic response. Healthy, vigorous donor plants are essential for successful anther/microspore culture. In vitro culture studies have shown that embryo yield can be affected by temperature, photoperiod, light intensity, time of year, fertilizer regime, age of the plant, and location where the plant is grown (i.e., greenhouse, field, or growth chamber) (Table 2). For many species, especially the *Brassica* species, growth of the donor plants at a low temperature (10°/5°C) is beneficial (Table 2) (37–39). It may not always be necessary to grow the donor plants under low temperatures, as a cold pretreatment of buds can have the equivalent effect (13,17,47,48).

### 3. Developmental Stage of the Pollen Grain

The optimal pollen developmental stage for in vitro culture can vary, depending on the species, genotype, donor plant conditions, and technique used (i.e., isolated microspore culture or anther culture) (Table 3). For microspore culture of most species, this developmental stage is usually the mid-uninucleate to early binucleate stage. However, recent studies have shown that the "narrow

**Table 2** Donor Plant Conditions Influencing In Vitro Embryogenesis

| Factor | Factor | Species | Reference |
|---|---|---|---|
| Temperature | 10°/5°C | *Brassica* species | 37,38,39 |
| | 14°/8°C | *Linum usitatissimum* | 40 |
| | 25°/15°C | *Triticum aestivum* | 41 |
| | 25°/18°C | *Capsicum annuum* | 42 |
| Photoperiod | 14 hr | *Oryza sativa* | 43 |
| | 18–24 h | *Triticum aestivum* | 44 |
| Light intensity | 300–350 µmol m$^{-2}$s$^{-1}$ | *Triticum aestivum* | 44 |
| Growth environment | Field vs. greenhouse | *S. chacoense* | 45 |
| | | *Lycopersicon esculentum* | 46 |
| Time of year | Spring/early summer | *Atriplex glauca* | 8 |

**Table 3** The Developmental Stage of the Pollen Grain Most Responsive to Embryogenesis

| Developmental stage | Technique | Species | Reference |
|---|---|---|---|
| Tetrad | Anther culture | *Apium graveolens* | 7 |
| | Anther culture | *Daucus carota* | 49 |
| Uninucleate | Anther culture | *Lilium longiflorum* | 50 |
| | Microspore culture | *Gingko bilboa* | 51 |
| | Microspore culture | *Cichorium intybus* | 14 |
| | Microspore culture | *Cajanus cajan* | 13 |
| | Microspore culture | *Apium graveolens* | 7 |
| | Microspore culture | *Daucus carota* | 52 |
| | Microspore culture | *Brassica* species | 53,54 |
| Binucleate—trinucleate | Microspore culture | *B. oleracea* | 55 |

window" for microspore embryogenesis induction can be manipulated to allow embryo development from a number of developmental stages (56–58). Embryos can develop from tobacco microspores at the early uninucleate ($G_1$ phase) to mid-binucleate stage (56,57). In *B. napus*, a severe heat shock (41°C) induced bicellular pollen grains to develop into embryos (58). These microspores had previously been considered unresponsive as starch had already formed within the cells. Starch has been used as an indicator of nonembryogenic cells (59,60).

In rye, embryo induction frequency increased as the microspores developed from uninucleate to binucleate, being highest at the binucleate stage (61). However, plant regeneration decreased with an increase in microspore development, with the highest regeneration efficiency at the early- to mid-uninucleate stage. Studies have shown that callus induction from anther culture involves several chromosome regions, which are different from those involved in plant regeneration (62). Other studies have shown that microspores earlier or later than the optimal stage for embryogenesis may produce toxic substances, which could lead to a reduction in embryogenesis and the production of abnormal embryos (53,54).

The developmental stage of the pollen may also have an effect on the ploidy of the resulting embryo/plant. This has been observed in *Atropa* and *Datura* spp. (63–65). Haploid embryos were obtained from microspores cultured at the tetrad stage, whereas higher-ploidy embryos developed from binucleate pollen grains.

# Embryogenesis

**Table 4** Pretreatment Factors Influencing Microspore Embryogenesis

| Factor | Pretreated plant organ | Species | Reference |
|---|---|---|---|
| Low temperature | | | |
| 4°C, 24 h | Buds | *Morus indica* | 47 |
| 4°C, 3–7 days | Buds | *Cajanus cajan* | 13 |
| 4°C, 4 days | Buds | *Helianthus annuus* | 17 |
| 4°C, 7 days | Buds | *Triticum aestivum* | 66 |
| 10°C, 21 or 42 days | Buds | *Oryza sativa* | 48 |
| Elevated temperature | | | |
| 32°C, 2.5 h | Buds | *Oryza sativa* | 35 |
| Slow desiccation | Buds | *Oryza sativa* | 35 |
| Reduced atmospheric pressure | Buds | *Sinapis alba* | 67 |
| Anaerobic conditions | Anthers | *Nicotiana tabacum* | 68 |
| Gamma irradiation | Microspores | *B. napus* | 69 |
| | Donor plants | *B. napus* | 70 |
| | Buds | *Malus × domestica* | 71 |
| Gamma irradiation and low temperature | Buds | *Lycopersicon esculentum* | 46 |
| Colchicine | Anthers, microspores | *B. napus* | 72,73 |
| Ethanol stress | Buds | *B. napus* | 70 |
| Chemical mutagens | Anthers | *Nicotiana tabacum* | 74 |
| Chemical hybridizing agents | Donor plants | *Triticum durum* | 75 |

## 4. Pretreatments

Both physical and chemical pretreatments have been used to enhance microspore embryogenesis (Table 4). Cold treatment of plants or floral organs prior to in vitro culture has been successfully used (13,17,47,48,66). Both temperature and duration of pretreatment are important and in many species, the duration of the cold pretreatment is genotype-dependent. The role of a cold pretreatment in inducing embryogenesis is unclear, and a number of theories exist. It has been speculated that the cold pretreatment delays the first haploid mitosis and hence delays pollen development or increases the viability of the microspore and increases the permeability of the pollen wall (76). A delay in the senescence of the anther wall allowed a supply of nutrients to the developing embryos (77). Some have also suggested that a cold treatment increases the number of microspores that divide into two identical nuclei, possibly by destroying the microtubules (78–80). Most species seem to respond to cold pretreatments, however, elevated temperatures have also enhanced embryo induction (Table 4).

## 5. Media Constituents

The composition of the media plays a major role in embryogenesis. There are numerous studies evaluating the different components of media and their role in anther/microspore culture (Table 5). Critical factors are carbon source and concentration, macro- and micronutrients, growth regulators, pH, physical factors (e.g., gelling agents, microspore density, and anther orientation), and other beneficial additives.

One of the essential media components is carbon. The source and the concentration are important in both microspore and anther culture. Sucrose is one of the most common carbohydrates used in androgenesis as both a nutrient and an osmoticum. Elevated levels of sucrose (>8%) have been used in microspore culture of *Brassica* spp. (37,81,82). It has also been observed that microspore-derived embryos take up excess sucrose when it is provided in the media, leading to

**Table 5** Media Constituents Positively Influencing Embryogenesis from the Male Gametophyte

| Factor studied | Species | Reference |
|---|---|---|
| Carbon source | | |
|   Sucrose | *Brassica* species | 37,81,82 |
| | *Zea mays* | 83 |
| | *Cichorium intybus* | 14 |
| | *Cucurbita pepo* | 84 |
| | *Asparagus officinalis* | 85 |
|   Sucrose starvation | *Nicotiana* species | 86,87 |
| | *Triticum aestivum* | 88,89 |
| | *Quercus suber* | 19 |
| | *Oryza sativa* | 48 |
| | *Solanum melongena* | 20 |
|   Maltose | *Hordeum vulgare* | 90 |
| | *Dactylis glomerata* | 91 |
| | *Fagopyrum esculentum* | 92 |
| | *Helianthus annuus* | 93 |
| | *Triticum aestivum* | 94 |
| | *Oryza sativa* | 95,96 |
| | *Avena* species | 9 |
|   Glucose | *Fragaria* × *ananassa* Duch. | 97 |
| | *Camellia japonica* | 98 |
|   Lactose | *Solanum tuberosum* | 99 |
| Growth regulators | | |
|   *N*-1-Naphthylphthalamic acid (NPA) | *Avena* species | 10 |
|   Polyvinylpyrrolidone (PVP) | *Helianthus annuus* | 100 |
|   Silver nitrate (AgNO$_3$) | *Oryza sativa* | 96 |
|   Phenylacetic Acid (PAA) | *Triticum aestivum* | 94 |
|   Elevated levels of 2,4-D | *Avena sterilis* | 10 |
| | *Avena sativa* | 10 |
|   2,3,5-Triiodobenzoic acid (TIBA) | *Zea mays* | 83 |
|   Ancymidol | *Avena sativa* | 10 |
| | *Triticum durum* | 29 |
| | *Asparagus officinalis* | 101 |
| Other medium components | | |
|   W14 salts | *Avena* species | 10 |
|   Ficoll | *Hordeum vulgare* | 102 |
|   PEG | *Brassica napus* | 103 |

abundant starch accumulation (104,105). This has an effect on the morphological characteristics and subsequent germination of the embryos. Studies have tried to separate the nutritional and the osmotic effects of sucrose. A novel system was reported in which embryos were induced and developed in very low levels of sucrose (103). Very low levels of sucrose were used along with polyethylene glycol (PEG), an osmoticum. PEG is a neutral polymer, which is highly soluble in water. Frequency of embryogenesis was similar using either this novel system or sucrose as the main medium osmoticum (103). However, embryo quality differed; those embryos induced in PEG more closely resembled zygotic embryos. Embryos induced in sucrose differed in size, morphological features, and color of the cotyledons (103). Mannitol and sorbitol have also been used to induce osmotic stress in *Brassica* sp. but were detrimental to embryo induction (103,106).

Carbohydrates other than sucrose have also been used successfully in androgenesis (Table 5). Replacing sucrose with maltose has resulted in an increase in embryo induction and green plant regeneration in cereals (90,94,95). Glucose and lactose have also been used to induce embryogenesis in some species (97–99).

The type and concentration of growth regulators can influence microspore embryogenesis in many species. The beneficial growth regulators differ among species and genotypes, and therefore conflicting results are reported in the literature. For the *Brassica* species, growth regulators are generally not required. The auxin 2,4-dichlorophenoxyacetic acid (2,4-D) was reported beneficial in oat (*Avena sativa, A. sterilis* (10), rice (95), wheat (107), chicory (*Cichorium intybus*) (14), rye (108), celery (*Apium graveolens*) (7), soybean (*Glycine max*) (109), and maize (*Zea mays*) (83). Cytokinins have also been used. Zeatin was found beneficial in chicory (14). Kinetin, in combination with 2,4-D, was beneficial for *Camellia japonica*, but kinetin caused severe browning in *Avena* cultures (10). Benzyladenine (BA) improved embryo induction in barley (*Hordeum vulgare*) (110), soybean (109), and celery (7).

## 6. Culture Conditions

In vitro embryogenesis can be influenced by culture conditions (Table 6). Much research has focused on culture temperature and duration, which vary among species. Culture temperature usually ranges from 24°C to 27°C. For some species (e.g., *Brassica, Triticum* spp.), an elevated temperature (30°C to 35°C) for 12–72 hours is required for embryo induction (37,111–113). It has been shown that colchicine can also induce embryogenesis in *Brassica* sp. (73,117). Different developmental stages are affected by the culture treatments. Heat induction was beneficial for microspores that were more advanced than those induced by colchicine. The microtubules of the early uninucleate microspores were more sensitive to colchicine than the binucleate microspores (117). Pollen development is disrupted by colchicine by depolymerizing microspore microtubules. This reorganization of the cytoskeleton results in a loss of cell asymmetry and normal pollen development. In addition, a combination of heat shock and colchicine treatment was reported beneficial especially in poorly responding *Brassica* sp. genotypes (121,122). Besides improving embryo induction, the colchicine treatments induced high rates of chromosome doubling.

Table 6 Culture Conditions Affecting Microspore Embryogenesis

| Factor studied | Species | Reference |
| --- | --- | --- |
| Heat shock | | |
| 32°C, 2–3 days | *Brassica* species | 24,37,111,112 |
| 32°C, 5 days | *Avena sativa, A. sterilis* | 9,10 |
| 32°C, 6 days | *Triticum aestivum* | 113 |
| 33°C, 5 days | *Quercus suber* | 19 |
| 35°C, 12 h | *Solanum tuberosum* | 114 |
| 35°C, 1 day | *Linum usitatissimum* | 115 |
| 35°C, 8 days | *Capsicum annuum* | 116 |
| Colchicine treatment | *B. napus* | 73,117 |
| Low light intensity (80-μmol m$^{-2}$s$^{-1}$) | *Triticum aestivum* | 44 |
| Plating density | *Hordeum vulgare* | 118 |
| | *Ginkgo biloba* | 51 |
| Coculture with other tissue | *Triticum aestivum* | 119,120 |

## III. GYNOGENESIS

The same year as Guha and Maheshwari (1964) demonstrated haploid plant production from anthers, Tulecke (123) obtained haploid callus from the female gametophyte of *Ginkgo biloba*. However, because of the success of anther culture compared to ovary culture, much of the focus remained on haploid plant production via androgenesis. It was not until 1976 that the first plants derived from in vitro gynogenesis were reported (124). Since then, haploid plant production from ovule or ovary culture has been reported in a number of species (124–128). In some species, in which both gynogenesis and androgenesis are possible, androgenesis is the preferred technique. Practical application of gynogenesis is limited to only a few species (e.g., *Beta* and *Allium* spp.).

Haploid embryos have been produced from the female gametophytic cells, including unfertilized ovaries, ovules, or flower buds. Compared to that of androgenesis, the embryo yield is lower because there are fewer cells to work with and they are more difficult to manipulate physically. However, gynogenesis is most useful where there is defective pollen (e.g., male sterile lines) or where pollen is unresponsive to culture (e.g., *Beta*, *Allium*, and *Gerbera* spp.). It has also been shown that plants derived via gynogenesis are of better quality than those derived via androgenesis. Albinism is usually a problem in cereal haploid production and may be alleviated by using ovule or ovary culture. For example, green plants were derived via gynogenesis of *Hordeum* sp. (124,129), *Oryza sativa* (130), and *Triticum aestivum* (131). Spontaneous chromosome doubling in some species is higher from plants derived via gynogenesis than via androgenesis (124, 125,129,132).

Similar to those in androgenesis, genotype, donor plant conditions, developmental stage, pretreatments, media composition, and culture environment influence the induction of embryogenesis from the female gametophyte.

### A.  Factors Influencing Gynogenesis

#### 1.  Donor Plant Genotype

Genotypic differences exist for embryo induction and plantlet regeneration in most species evaluated for gynogenic response. This has been shown in durum wheat (133), *Allium* spp. (134–137), and *Oryza sativa* (138). As with androgenesis, japonica rice genotypes were more responsive than indica rice genotypes (138,139). Selection among genotypes for high embryogenic response can lead to lines with greater ability to produce haploid embryos and plants.

#### 2.  Donor Plant Conditions

Donor plants for ovule/ovary culture have been grown in the field, greenhouse, or growth cabinet. Embryogenic frequency differs, depending on where the plants are grown. A higher embryo yield was observed when *Beta vulgaris* (sugar beet) donor plants were grown in the greenhouse or growth cabinet than under field conditions (140). The time of year also appears to influence embryogenesis in *Beta* and *Gerbera* spp. The summer months (May–September) were favorable for embryo production in *Beta* sp., whereas callus induction of *Gerbera* sp. was highest during the autumn months (140–142).

In *Beta vulgaris*, the location and age of the florets on the plant are important. Florets from the lateral branches gave a higher ovule response in terms of embryo induction, when compared to florets from the stem apex. In addition, florets from the lateral branches that formed first gave a higher response than florets at the top of the plant or that formed later (143).

## 3. Developmental Stage of the Female Gametophytic Cell

The female gametophyte, i.e., embryo sac, generally consists of seven cells (egg cell, two synergids, three antipodals, and central cell). The embryo sac can be cultured and potentially develop into a haploid embryo. Histological studies in *Helianthus*, *Beta*, and *Nicotiana* spp. and *Salvia sclarea* have shown that the egg cell develops into the embryo (144–147); however, antipodal or synergid cells may also develop into embryos (148). Similar to that in androgenesis, the developmental stage of the female gametophyte influences embryogenesis (Table 7). There are a limited number of studies in this area as it is difficult to determine the developmental stage of the embryo sac. For some species, there is a correlation between the development of the embryo sac and the development of the pollen grain; hence, this correlation has been used to estimate the optimal stage of culture (Table 7). San and Demarley (152) found that barley, wheat, maize, sugar beet, and lettuce (*Lactuca sativa*) cells were most responsive in culture when the embryo sac was mature or nearly mature. At this stage, all cells had the capability of dividing and producing calli or embryos. This is in contrast to androgenesis, in which microspores were immature, usually at the mid-uninucleate to early-binucleate stage, for optimal embryogenic response.

## 4. Pretreatments

A cold pretreatment has proved beneficial for gynogenesis in a number of species. The temperature and duration of the cold pretreatment vary among different species. In rice, a treatment of 8°C for 6–14 days enhanced embryo production (138). A pretreatment temperature of 4°C was beneficial for both *Helianthus* and *Beta* spp. However, a short duration of 24–48 hours was required for *Helianthus* sp. (126) but 4–5 days was beneficial for *Beta vulgaris* (140). Cold pretreatments were also beneficial for durum wheat (133) and *Salvia sclarea* (147). For *Picea sitchensis*, a cold treatment was ineffective, whereas a heat treatment of 33°C for 2–4 days was beneficial for callus induction (153).

## 5. Media Constituents

Similar to the culture of male gametophytic cells, the composition of the media plays a major role in influencing haploid embryogenesis via the female gametophyte. The induction of calli occurs most frequently on solid media although liquid media has been used for wheat and rice cultures (154,155). Basal media like Murashige and Skoog (MS) (156), B5 (Gamborg B5), (157) and N6 (158,159) and modifications of these have been used in ovule/ovary cultures with success (129, 133,138,160).

Carbohydrate source and concentration are also important. Sucrose has been used most widely for ovule/ovary culture with concentrations ranging from 2% to 10% (129,133,138,161).

Table 7 Developmental Stage of the Female Gametophyte Most Responsive to Gynogenesis

| Female gametophytic stage | Corresponding male gametophytic stage | Species | Reference |
|---|---|---|---|
| Uninucleate to tetranucleate embryo sac | Late uninucleate to early binucleate | *Oryza sativa* | 149 |
| Mature embryo sac | Uninucleate | *Solanum tuberosum* | 150,151 |
| | Trinucleate | *Oryza sativa* | 138 |
| | Trinucleate | *Hordeum vulgare* | 124,129 |
| | Bicellular | *Triticum durum* | 133 |

Higher levels of sucrose may inhibit growth and development. Other carbohydrates such as maltose have also been used (92).

Growth regulators have been used to culture the female gametophyte and their effects differ among species. Beneficial effects of auxins and cytokinins have been reported either alone or in combination with other growth regulators. The auxins 2,4-D (52,133,134,136), 4-chloro-2-methylphenoxyacetic acid (MCPA) (126,129), indole-3-acetic acid (IAA) (134), and napthaleneacetic acid (NAA) (138) have been used. Cytokinins having beneficial effects include BA (92,129) and $N^6$-2-isopentenyladenine (2iP) (92). Other additives used successfully in the culture of the female gametophyte include thiadiazorun (TDZ) (134,161), coconut water (129), and dimethyl sulfoxide (DMSO) (138).

## 6. Culture Conditions

Very few studies have looked at the culture conditions required for continued development of the female gametophyte. Usually callus induction or embryo development takes place in the dark; however, light is required for plant regeneration. A dark period of 2–5 weeks was beneficial for durum wheat (133). Culture temperatures usually range from 25°C to 28°C (127). A heat treatment of 30°C led to a reduction in regeneration in *Allium* spp. (162). Other environmental factors may also be beneficial. For example, up to 90% of the female gametophytes in a responsive *Picea sitchensis* could be induced to form calli when given a low oxygen treatment for 7 weeks (153).

## IV. HAPLOID OR DOUBLED HAPLOID PLANT PRODUCTION

### A. Regeneration

Plants can be regenerated from cultures via direct embryogenesis or indirectly via a callus phase. Genotype, media, and culture conditions can influence frequency of plantlet regeneration as well as plantlet quality. For tobacco, regeneration takes place on anther culture medium, but for most other species, a specific embryo culture medium is required. Usually embryo culture medium has a lower level of carbohydrate and different growth regulators when compared to the induction medium. Light is required for embryo germination and the temperature is usually 24°C to 26°C.

Poor plantlet regeneration and quality may limit utilization of the plants produced. Tissue culture manipulations are required to improve embryo germination, plantlet regeneration, and normality of the plantlets (163). An example of abnormality is that of albinism, which is a problem in cereals especially those derived from the male gametophytic cells. However, recent improvements to the protocol have resulted in an increase in the frequency of green plants regenerated (102,164–166). Studies have shown that alterations or deletions in plastid deoxyribonucleic acid (DNA) may be the cause of albinism in cereals (167–169). At the ribonucleic acid (RNA) level, transcription does occur in plastids; however, the pattern of plastid transcription of albino plants is different from that of green plants (169). A large reduction of ribosomal RNA was also observed in all albino plants.

### B. Chromosome Doubling

Doubling the chromosome number of haploid plants is necessary for the production of doubled haploid plants, which can then be used in breeding, genetic studies, mutagenesis, or gene transfer. For some species, a chemical treatment is required because spontaneous doubling is very low. Traditionally colchicine has been used to double the chromosome number. This compound is classified as a carcinogen, and therefore other compounds that are less toxic have been evaluated as chromosome doubling agents. Oryzalin, trifluralin, amiprophos-methyl, and pronamide are antimicrotubule agents that can be used for this purpose. Both colchicine and antimicrotubule agents have been used successfully to double the chromosome number of haploids generated via androgenesis or gynogenesis (170–173).

Early chromosome doubling techniques involved colchicine application to the plants by root immersion or application of colchicine to the buds. These techniques were time-consuming and inefficient. More recently, in vitro chromosome doubling protocols have been developed. The addition of colchicine or antimicrotubule agents to the microspore induction medium has resulted in doubling frequencies similar to or greater than those achieved with doubling at the plant stage (174–176). One additional feature of in vitro colchicine application to microspores has been an improvement in the frequency of embryogenesis, which has been shown in *Brassica* spp. (122), *Zea mays* (176), and *Triticum aestivum* (175). However, other studies have shown that colchicine causes a reduction in embryogenesis (177).

## V. DEVELOPMENTAL ASPECTS

Many of the molecular, biochemical, and physiological aspects of haploid embryogenesis are studied in *Nicotiana*, *Triticum*, and *Brassica* spp. These studies were feasible after the development of microspore culture systems that provided high efficiency, uniform, synchronized embryogenesis, as well as direct embryogenesis from the microspore with no intervening callus phase. Pollen development follows a precise sequence of events. To change the developmental pathway from the gametophytic to the embryogenic requires specific inductive conditions. It has been shown that stress triggers the embryogenic pathway. These stress triggers include nitrogen or sucrose starvation (56,88,89,178,179), thermal shock (cold or heat) (88,89,180–184), and colchicine (185). Different species respond to different stress triggers. In *B. napus*, a heat shock of 32°C for 8 hours results in embryo development, whereas microspores kept at 18°C develop into mature pollen grains (184). Cultures maintained at 25°C contain both gametophytic cells and embryogenic cells. Similarly, in *Nicotiana* sp. and wheat, sucrose starvation treatment in combination with heat shock was required for embryo induction (88).

Prior to culture, microspores have a large central vacuole, thin tonoplast, parietal cytoplasm, and peripheral nucleus. During stress the microspores swell, the cytoplasm reorganizes, and the nucleus moves into the central position. Techniques have been developed so that the development of an individual microspore can be followed from single cell to embryo (186). Embryogenic microspores are difficult to identify, although methods have been reported. Sangwan and Camefort (187) identified a cytological marker. They observed that within 12 hours, embryogenic microspores developed a uniform coating, which consisted of tannins. Telmer et al. (60) used fluorescence microscopy to identify embryogenic pollen grains. Flow cytometric techniques have also been used to isolate embryogenic cells (188–190).

After 8 hours of culture, *B. napus* cells are irreversibly committed to embryogenesis. When comparing heat-stressed cells (32°C) and nonstressed (18°C) cells, there was a twofold increase in protein synthesis (192). Twenty-five proteins were differentially synthesized during the first 8 h of microspore embryogenesis (193). Seventeen of the proteins belonged to the class of heat shock proteins (HSPs). Of the 25 proteins, 4 have been identified as HSP17, HSP68, HSP70, and one that was not a HSP, which could be used as a marker for embryo induction (193). Further studies have reported strong correlation among embryo induction, HSP70 synthesis, and location of the nucleus (192). The expression of HSP68 and HSP70 under nonembryogenic conditions was similar to in vivo pollen development (192,194). Smykal and Pechan (195) observed the expression of HSP17 corresponded to the frequency of embryogenesis. Expression of HSP17 was lower in plants that had poor embryogenic response than those plants with high embryogenic response. There was also an association with HSP17 expression and embryogenesis induced by colchicine treatment (195). Boutilier et al. (196) observed that high levels of napin seed storage protein gene expression was correlated with induction of embryogenesis. No expression of the napin gene was observed in microspores following normal pollen development. Napin gene expression was induced by elevated temperature (196). However, gene expression did not resemble heat shock response, as expression remained high for a long time.

In *Nicotiana* sp. there was a decrease in overall synthesis of RNA and protein during the 7-day starvation period. The degradation of protein or suppression of protein synthesis was necessary to induce a switch from pollen to embryo development. In *Brassica* sp. there was an increase in protein synthesis during the inductive period (first 48 hours) and then a gradual decrease in synthesis. It has been postulated that there is gamete-specific gene expression that starts after the first pollen mitosis when the microspores divide to produce a small generative cell and a large vegetative cell (192). In *Brassica* sp. where mid to late-uninucleate microspores are used, there is a fast switch to embryogenesis (8 hours), whereas in *Nicotiana* sp., where mid-bicellular stage is used, a longer stress is required (7 days).

Several approaches are being undertaken to isolate and characterize genes expressed during induction of microspore embryogenesis. These include differential display of reverse transcribed mRNAs, suppression subtractive hybridization (SSH), and mutagenesis. Several cDNA fragments have been identified, of which some show homology to known genes (186,197). Zárský et al. (198) first characterized a gene that was transcriptionally activated during pollen embryogenesis. A genetic approach to identify quantitative trait loci (QTLs) is also being used (199–201). Comparison has been made between responding and nonresponding *Zea mays* lines (199–201). QTLs for anther culture response have been identified, however, the location of these markers did not overlap and seemed to be genotype-specific.

## VI. APPLICATION OF HAPLOID AND DOUBLED HAPLOID PLANTS

In vitro haploidy techniques aimed to produce haploid and doubled haploid plants were initially developed for plant breeding. In some species, these techniques are well established and routinely used by breeding groups to develop advanced breeding lines and cultivars (202,203). Cultivars, derived via doubled haploidy, have recently been released and include Quantum and Q2 canola (204) and McKenzie (R Graf, personal communication) and GK Délibáb (205) wheat. Haploids and doubled haploids also have other practical and fundamental applications.

### A. Varietal Development

The use of doubled haploids has a number of advantages in plant breeding. The main advantage is the ability to produce a homozygous line in one generation rather than after several cycles of selfing and backcrossing. This saves 3–4 years in the development of a cultivar compared with a conventional plant-breeding program (206). The canola cultivar Quantum was developed in 6 years (204) instead of the usual 9–10 years required for a conventionally derived cultivar. With haploidy technology, traits are immediately fixed in the homozygous condition, allowing a greater efficiency of selection since there is no masking of recessive genes by dominant genes. The homozygous doubled haploid plants are also an advantage in hybrid breeding in that they allow immediate testing of combining ability. The use of doubled haploids also allows the plant breeders to use smaller population sizes when screening for desirable genotypes (207). Doubled haploids in combination with molecular markers can enhance the transfer of genes or chromosomal segments between lines or species. These advantages result in savings of time, space, and effort.

# Embryogenesis

Anther/microspore culture techniques have been used to develop disease-resistant pepper cultivars (116,208,209), higher-yielding eggplant (*Solanum melongene*) lines with better fruiting ability (210), "supermale" asparagus (211), and high linolenic flax (*Linum usitassimum*) (212). In addition, haploids or doubled haploids are used in genetic studies in detecting linkage and in calculating recombination values between linked genes (28,213) as well as genome mapping for major genes or QTLs (214–216). Genetic linkage and mapping studies are additional tools developed to enhance plant breeding.

## B. Mutation Breeding

Conventional mutagenesis using seeds has been used to develop varieties in both ornamental and crop species. The main disadvantage of using conventional methods of mutagenesis is that seed is multicellular and therefore the resulting plant could potentially be chimeric. Also there is a limited chance of the mutant cells being part of the germ line. Mutagenesis of single cells, such as microspores, which have a high regenerative potential, is desirable and has been used in mutation breeding and selection studies. Because the microspore is a single cell, any genetic variation induced by the mutagen is expressed in all cells of the regenerated plant and its progeny, therefore eliminating chimeras and segregation. Both recessive and dominant traits are expressed and are easily selectable in culture. Since selection can be made at the haploid or doubled haploid level, undesirable traits can be readily discarded instead of being carried for generations in the heterozygote. Apart from the high regenerative potential of microspores, there are other advantages for their use in mutagenesis. A large number of uniform cells can be exposed to chemical or physical mutagens in a relatively small space. Appropriate selection pressure can be applied to the culture to select mutants. In the case of herbicide-tolerant mutants, the chemical can be incorporated directly into the culture medium and potential mutants can be selected. Screening thousands of microspores in a petri plate can save resources compared to growing thousands of plants in the greenhouse or field.

Although such advantages are evident, there are few reports of the successful use of microspores in mutation breeding. One limitation is the small number of species and genotypes in which microspore embryogenesis can be reliably achieved in numbers sufficient for a mutagenesis approach. There may be negative effects of mutagens on regeneration capacity of the microspores. Microspore mutagenesis has been used in *B. napus* and *B. rapa* to develop lines resistant to herbicide (217,218) or possessing alterations in the fatty acid profile (219–221). Isolated *B. napus* microspores subjected to ultraviolet (UV) irradiation resulted in two lines showing increased resistance to *Alternaria brassicicola* (222).

## C. Gene Transfer

The haploid microspore system is advantageous for genetic transformation. This is due to the availability of large numbers of uniform single cells, from which haploid or doubled haploid embryos can be easily developed and regenerated to doubled haploid plants homozygous for the introduced gene. Other advantages to transforming microspores are that the resulting transformants are nonchimeric and as the male gametophyte is less tolerant to genetic aberration, there is a natural selection for normality. In addition, in systems where direct embryogenesis occurs, undesirable somaclonal variation is largely eliminated, as no callus phase is present. However, this technique relies on efficient microspore culture protocols, and since both species and genotype play a major role in embryogenic response, this method of gene transfer cannot yet be applied to all species and genotypes. An efficient screening procedure is another requirement. Green fluorescent protein (GFP) is a nontoxic, nondestructive marker, which has been used successfully in screening for transgenic pollen in male germ line transformation (223,224) although this study did not involve in vitro haploidy techniques. Other genes like β-glucuronidase (GUS) and firefly luciferase (Luc) have also been used as markers to identify transformed microspores and embryos (225–230).

Microspores or microspore-derived embryos have been the subject of transformation studies. Several methods have been investigated, including microinjection (231,232), *Agrobacterium* mediation (233–235), *Agrobacterium* and particle bombardment (236), electroporation and PEG delivery (225,227,237), and particle bombardment (226,228–230,238–240). Most of this work has been conducted in barley, tobacco, and *B. napus* since in vitro haploidy techniques are well established. However, these studies generally report transient expression and low efficiency of transformation. This topic has been reviewed by Harwood et al. (241) and Morikawa and Nishihara (242).

### D. Biochemical and Physiological Studies

The developmental sequences of many biochemical pathways of a zygotic embryo and a microspore-derived embryo are similar (243), and therefore microspore-derived embryos can be used in fundamental biochemical and physiological studies. Much of this work has been done in the *Brassica* species. This includes lipid storage (244,245), oil quality (246), glucosinolate metabolism (247), chlorophyll metabolism (248), freezing tolerance (249,250), and embryo maturation studies (251).

## VII. CONCLUSION

Androgenesis and gynogenesis are powerful tools in crop improvement. Substantial advances have been made in the area of in vitro haploidy techniques since the original experiments in the 1960s. The focus of the early experiments was to develop a protocol for generating embryos/plants from gametophytic cells (i.e., microspores or ovaries/ovules). Constant improvements in media and culture conditions have allowed us to produce embryos/plants reliably and in sufficient numbers to allow the use of the system for practical application. For some species, in vitro haploidy techniques are well established and are being routinely used to develop breeding lines and cultivars. However, improvements are still required to widen the range of responding species and genotypes and to improve quantity and quality of doubled haploid plants. Many species such as legumes, specialty crops, woody plants, and medicinal plants are still considered recalcitrant to haploid induction. Fewer species are amenable to gynogenic procedures. The numerous advantages of in vitro haploidy technology are only starting to be exploited, and exciting opportunities lie ahead to develop these technologies. Since reliable in vitro haploidy technologies are now available in some species, there are a number of questions to be answered: What is the underlying mechanism of gametophytic embryogenesis? What genes control embryogenesis? What is the potential of transferring this ability to other genotypes? New technologies such as DNA arrays and DNA chip technologies may have application in the area of gene expression. In vitro haploidy technology would also greatly promote advances in the understanding of the basic biochemical and physiological processes in nature.

## REFERENCES

1. S Guha, SC Maheshwari. In vitro production of embryos from anthers of *Datura*. Nature 204:497, 1964.
2. AF Blakeslee, J Belling, ME Farnham, AD Bergner. A haploid mutant in the Jimson weed, *Datura stramonium*. Science 55:646–647, 1922.
3. JP Nitsch, C Nitsch. Haploid plants from pollen grains. Science 163:85–87, 1969.

4. C Nitsch, B Norreel. La cultures de pollen isolé sur milieu synthétique. C R Acad Sci Paris 278D: 1031–1034, 1974.
5. SC Maheshwari. The discovery of anther culture technique for the production of haploids. In: SM Jain, SK Sopory, RE Veilleux, eds. In Vitro Haploid Production in Higher Plants. Vol 1. Dordrecht: Kluwer Academic, 1996, 1–10.
6. L Radojevic, N Marinkovic, S Jevremovic, D Calic. Regeneration of androgenic *Aesculus hippocastanum* L. plantlets through uninuclear microspore suspension cultures. In IX International Association of Plant Tissue Culture, Jerusalem, 1998, p 170.
7. N Dohya, S Matsubara, K Murakami. Callus formation and regeneration of adventitious embryos from celery microspores by anther and isolated microspore cultures. J Jpn Soc Hortic Sci 65:747–752, 1997.
8. L Kenny, PDS Caligari. Androgenesis of the salt tolerant shrub *Atriplex glauca*. Plant Cell Rep 15: 829–832, 1996.
9. EM Kiviharju, E Pehu. The effect of cold and heat pretreatments on anther culture response of *Avena sativa* and *A. sterilis*. Plant Cell Tissue Org Cult 54:97–104, 1998.
10. EM Kiviharju, AA Tauriainen. 2,4-Dichlorophenoxyacetic acid and kinetin in anther culture of cultivated and wild oats and their interspecific crosses: plant regeneration from *A. sativa* L. Plant Cell Rep 18:582–588, 1999.
11. EM Kiviharju, M Puolimatka, E Pehu. Regeneration of anther-derived plants of *Avena sterilis*. Plant Cell Tissue Org Cult 48:147–152, 1997.
12. TK Shon, T Yoshida. Induction of haploid plantlets by anther culture of *Bupleurum falcatum* L. Jpn J Crop Sci 66:137–138, 1997.
13. P Kaur, JK Bhalla. Regeneration of haploid plants from microspore culture of pigeonpea (*Cajanus cajan* L.). Indian J Exp Biol 36:736–738, 1998.
14. R Theiler-Hedtrich, CS Hunter. Regeneration of dihaploid chicory (*Cichorium intybus* L. var. *foliosum hegi*) via microspore culture. Plant Breed 114:18–23, 1995.
15. JS Croser, JB Brouwer, PWJ Taylor, ECK Pang. Somatic embryogenesis from anthers of Australian Kabuli chickpea cultivars. Proceedings of the 3rd International Int Food Legume Research Conference, Adelaide, 1997.
16. A Sarvesh, TP Reddy, PB Kavi Kishor. In vitro anther culture and flowering in *Guizotia abyssinica* Cass. Indian J Exp Biol 34:565–568, 1996.
17. KV Saji, M Sujatha. Embryogenesis and plant regeneration in anther culture of sunflower (*Helianthus annuus* L.). Euphytica 103:1–7, 1998.
18. K Tang, X Sun, Y He, Z Zhang. Anther culture response of wild *Oryza* species. Plant Breed 117:443–446, 1998.
19. MA Bueno, A Gomez, M Boscaiu, JA Manzanera, O Vicente. Stress-induced formation of haploid plants through anther culture in cork oak (*Quercus suber*). Physiologia Plantarum 99:335–341, 1997.
20. K Miyoshi. Callus induction and plantlet formation through culture of isolated microspores of eggplant (*Solanum melongena* L.). Plant Cell Rep 15:391–395, 1996.
21. SC Maheshwari, A Rashid, TK Tyagi. Haploids from pollen grains—retrospects and prospects. Am J Bot 69:865–879, 1982.
22. JM Dunwell. Pollen, ovule and embryo culture as tools in plant breeding. In: LA Withers, PG Alderson, eds. Plant Tissue Culture and Its Agricultural Applications. London: Butterworths, 1986, pp 375–404.
23. AMR Ferrie, CE Palmer, WA Keller. Haploid embryogenesis. In: TA Thorpe, ed. In Vitro Embryogenesis in Plants. Dordrecht: Kluwer Academic, 1995, pp 309–344.
24. M Hiramatsu, K Odahara, Y Matsue. A survey of microspore embryogenesis in leaf mustard (*Brassica juncea*). Acta Hortic 392:139–145, 1995.
25. C Phippen, DJ Ockendon. Genotype, plant, bud size and media factors affecting anther culture of cauliflowers (*Brassica oleracea* var *botrytis*). Theor Appl Genet 79:33–38, 1990.
26. MBM Bruins, CHA Snijders. Inheritance of anther culture derived green plantlet regeneration in wheat (*Triticum aestivum* L.). Plant Cell Tissue Org Cult 43:13–19, 1995.
27. L Sági, B Barnabás. Evidence for cytoplasmic control of in vitro microspore embryogenesis in the anther culture of wheat (*Triticum aestivum* L.) Theor Appl Genet 78:867–872, 1989.

28. H Ekiz, CF Konzak. Preliminary diallel analysis of anther culture response in wheat (*Triticum aestivum* L.). Plant Breed 113:47–52, 1994.
29. M Ghaemi, A Sarrafi, R Morris. Reciprocal substitutions analysis of embryo induction and plant regeneration from anther culture in wheat (*Triticum aestivum* L.) Genome 38:158–165, 1995.
30. G Wenzel, F Hoffman, E. Thomas. Increased induction and chromosome doubling of androgenetic haploid rye. Theor Appl Genet 51:81–86, 1977.
31. G Daniel. Anther culture in rye: Improved plant regeneration using modified MS-media. Plant Breed 110:259–261, 1993.
32. T Flehinghaus, S Deimling, HH Geiger. Methodical improvements in rye anther culture. Plant Cell Rep 10:397–400, 1991.
33. T Flehinghaus-Roux, S Deimling, HH Geiger. Anther-culture ability in *Secale cereale* L. Plant Breed 114:259–261, 1995.
34. J Mitykó, A Andrásfalvy, G Csilléry, M Fári. Anther-culture response in different genotypes and $F_1$ hybrids of pepper (*Capsicum annuum* L.). Plant Breed 114:78–80, 1995.
35. P Sathish, OL Gamborg, MW Nabors. Rice anther culture: Callus initiation and androclonal variation in progenies of regenerated plants. Plant Cell Rep 14:432–436, 1995.
36. NA Zagorska, A Shtereva, BD Dimitrov, MM Kruleva. Induced androgenesis in tomato (*Lycopersicon esculentum* Mill.) I. Influence of genotype on androgenetic ability. Plant Cell Rep 17:968–973, 1988.
37. AMR Baillie, DJ Epp, D Hutcheson, WA Keller. In vitro culture of isolated microspores and regeneration of plants in *Brassica campestris*. Plant Cell Rep 11:234–237, 1992.
38. Y Takahata, WA Keller. High frequency embryogenesis and plant regeneration in isolated microspore culture of *Brassica oleracea* L. Plant Sci 74:235–242, 1991.
39. WA Keller, PG Arnison, BK Cardy. Haploids from gametophytic cells: Recent developments and future prospects. In: CE Green, DA Somers, WP Hackett, DD Biesboer eds. Plant Tissue and Cell Culture, New York: Allan R. Liss, 1987, pp 233–241.
40. K Nichterlein, H Umbach, W Friedt. Genotypic and exogenous factors affecting shoot regeneration from anther callus of linseed (*Linum usitatissimum* L.). Euphytica 58:157–164, 1991.
41. BR Orshinsky, RS Sadasivaiah. Effect of plant growth conditions, plating density, and genotype on the anther culture response of soft white spring wheat hybrids. Plant Cell Rep 16:758–762, 1997.
42. IP Munyon, JF Hubstenberger, GC Phillips. Origin of plantlets and callus obtained from chile pepper anther cultures. In Vitro Cell Dev Biol 25:293–296, 1989.
43. ZX Sun, HM Si, XY Zhan, SH Chen. The effect of thermo-photoperiod for donor plant growth on anther culture of Indica rice. Curr Plant Sci Biotechnol Agric 15:361–364, 1992.
44. H Ekiz, CF Konzak. Effect of light regimes on anther culture response in bread wheat. Plant Cell Tissue Org Cult 50:7–12, 1997.
45. M Cappadocia, DSK Cheng, R Ludlum-Simonette. Plant regeneration from in vitro culture of anthers of *Solanum chacoense* Bitt. and interspecific diploid hybrids *S. tuberosum* L. x *S. chacoense* Bitt. Theor Appl Genet 69:139–143, 1984.
46. LA Shtereva, NA Zagorska, BD Dimitrov, MM Kruleva, HK Oanh. Induced androgenesis in tomato (*Lycopersicon esculentum* Mill.). II. Factors affecting induction of androgenesis. Plant Cell Rep 18: 312–317, 1998.
47. AK Jain, A Sarkar, RK Datta. Induction of haploid callus and embryogenesis in in vitro cultured anthers of mulberry (*Morus indica*). Plant Cell Tissue Org Cult 44:143–147, 1996.
48. T Ogawa, H Fukuoka, Y Ohkawa. Plant regeneration through direct culture of isolated pollen grains in rice. Breed Sci 45:301–307, 1995.
49. S Matsubara, N Dohya, K Murakami. Callus formation and regeneration of adventitious embryos from carrot, fennel and mitsuba microspores by anther and isolated microspore cultures. Acta Hortic 392:129–137, 1995.
50. AM Arzate-Fernández, T Nakazaki, H Yamagata, T Tanisaka. Production of doubled-haploid plants from *Lilium longiflorum* Thunb. anther culture. Plant Sci 123:179–187, 1997.
51. D Laurain, J Trémouillaux-Guiller, J Chénieux. Embryogenesis from microspores of *Ginkgo biloba* L., a medicinal woody species. Plant Cell Rep 12:501–505, 1993.

52. GB Tjukavin, NA Shmykova. Anther culture and unpollinated ovules of carrot. IX International Association of Plant Tissue Culture, Jerusalem, 1998 p 115.
53. Z Fan, L Holbrook, WA Keller. Isolation and enrichment of embryogenic microspores in *Brassica napus* L. by fractionation using percoll density gradients. Proc. 7th Int. Rapeseed Cong. Paznan, Poland, 1988, pp 92–96.
54. LS Kott, L Polonsi, B Ellis, WD Beversdorf. Autotoxicity in isolated microspore cultures of *Brassica napus*. Can J Bot 66:1665–1670, 1988.
55. MQ Cao, F Charlot, C Doré. Embrogenèse et régénération de plantes de chou à choucroute (*Brassica oleracea* L ssp. *capitata*) par culture in vitro de microspores isolées, C R Acad Sci III 310:203–209, 1990.
56. M Kyo, H Harada. Control of the developmental pathway of tobacco pollen in vitro. Planta 168:427–432, 1986.
57. C Stauffer, RMB Moreno, E Heberle-Bors. In situ pollination with in vitro matured pollen of *Triticum aestivum*. Theor Appl Genet 81:576–580, 1991.
58. P Binarova, G Hause, V Cenklová, JHG Cordewener, MM van Lookeren Campagne. A short severe heat shock is required to induce embryogenesis in late bicellular pollen of *Brassica napus* L. Sex Plant Reprod 10:200–208, 1997.
59. RS Sangwan, BS Sangwan-Norreel. Ultrastructural cytology of plastids in pollen grains of certain androgenic and non-androgenic plants. Protoplasma 138:11–22, 1987.
60. CA Telmer, DH Simmonds, W Newcomb. Determination of developmental stage to obtain high frequencies of embryogenic microspores in *Brassica napus*. Physiol Plant 84:417–424, 1992.
61. S Immonen, H Anttila. Impact of microspore developmental stage on induction and plant regeneration in rye anther culture. Plant Sci 139:213–222, 1998.
62. BA Grosse, S Deimling, HH Geiger. Mapping of genes for anther culture ability in rye by molecular markers. EUCARPIA: International Symposium on Rye Breeding and Genetics, 1996. Vortr Pflanzenzuchtg 35:282–283, 1996.
63. S Narayanaswamy, LP Chandy. In vitro induction of haploid, diploid androgenic embryoids and plantlets in *Datura metel* L. Ann Bot 35:535–542, 1971.
64. KC Engvild, I Linde-Laursen, A Lundqvist. Anther cultures of *Datura innoxia* flower bud stage and embryoid level of ploidy. Hereditas 72:331–332, 1972.
65. B Norreel. Etude physiologique, cytochimique et ultrastructurale de l'embryogenèse somatique chez le *Daucus carota* et de l'androgenèse chez le *Nicotiana tabacum* L. et le *Datura innoxia* Mill. Thèse de Doctorat d'Etat, Université P. et M. Curie, Paris, 1975.
66. A Stober, D Hess. Spike pretreatments, anther culture conditions and anther culture response of 17 German varieties of spring wheat (*Triticum aestivum* L.). Plant Breed 116:443–447, 1997.
67. K Klimaszewska, WA Keller. The production of haploids from *Brassica hirta* Moench (*Sinapis alba* L.) anther cultures. Z Pflanzenphysiol 109:235–241, 1983.
68. J Imamura, H Harada. Stimulation of tobacco pollen embryogenesis by anaerobic treatments. Z Pflanzenphysiol 103:259–263, 1981.
69. MV MacDonald, MA Hadwiger, FN Aslam, DS Ingram. The enhancement of anther culture efficiency in *Brassica napus* ssp. *oleifera* Metzg. (Sinsk.) using low doses of gamma irradiation. New Phytol 110:101–107.
70. PM Pechan, WA Keller. Induction of microspore embryogenesis in *Brassica napus* L. by gamma irradiation and ethanol stress. In Vitro Cell Dev Biol 25:1073–1074, 1989.
71. YX Zhang, L Bouvier, Y Lespinasse. Microspore embryogenesis induced by low gamma dose irradiation in apple. Plant Breed 108:173–176, 1992.
72. MAM Zaki, HG Dickinson. Microspore-derived embryos in *Brassica*: The significance of division symmetry in pollen mitosis I to embryogenic development. Sex Plant Reprod 4:48–55, 1991.
73. M Zaki, H Dickinson. Modification of cell development in vitro: The effect of colchicine on anther and isolated microspore culture in *Brassica napus*. Plant Cell Tissue Org Cult 40:255–270, 1995.
74. H Medrano, E Primo-Millo, J Guerri. Ethyl-methane-sulfonate effects on anther cultures of *Nicotiana tabacum*. Euphytica 35:161–168, 1986.

75. N Saidi, S Cherkaoui, A Chylah, H Chlyah. Embryo formation and regeneration in *Triticum turgidum* ssp. *durum* anther culture. Plant Cell Tissue Org Cult 51:27–33, 1997.
76. YPS Bajaj. Regeneration of haploid tobacco plants from isolated pollen grown in drop culture. Indian J Exp Biol 16:407–409, 1978.
77. P Dieu, JM Dunwell. Anther culture with different genotypes of opium poppy (*Papover somniferum* L.): Effect of cold treatment. Plant Cell Tissue Org Cult 12:263, 1988.
78. C Nitsch. Pollen culture—a new technique for mass production of haploids and homozygous plants. In: KJ Kasha ed. Haploids in Higher Plants: Advances and Potential. Guelph: Univeristy of Guelph, 1974, pp 123–125.
79. EJ Duncan, E Heberle. Effect of temperature shock on nuclear phenomena in microspores of *Nicotiana tabacum* and consequently of plantlet production. Protoplasma 90:173–177, 1976.
80. BS Sangwan-Norreel. Androgenic stimulating factors in the anther and isolated pollen grain culture of *Datura innoxia*. Mill. J Exp Bot 28:843–852, 1977.
81. Z Chen, Z Chen. High frequency induction of pollen-derived embryoids from anther cultures of rape (*Brassica napus*). Kexue Tongbao 28:1690–1694, 1983.
82. JM Dunwell, N Thurling. Role of sucrose in microspore embryo production in *Brassica napus* ssp. *oleifera*. J Exp Bot 36:1478–1491, 1985.
83. B Büter. In vitro haploid production in maize. In: SM Jain, SK Sopory, RE Veilleux, eds. In Vitro Haploid Production in Higher Plants, Vol 4. Dordrecht: Kluwer Academic, 1997, pp 37–71.
84. EI Metwally, SA Moustafa, EI El-Sawy, TA Shalaby. Haploid plantlets derived via anther culture of *Cucurbita pepo*. Plant Cell Tissue Org Cult 32.171–176, 1998.
85. XR Feng, DJ Wolyn. High frequency production of haploid embryos in asparagus anther culture. Plant Cell Rep 10:574–578, 1991.
86. A Touraev, M Pfosser, O Vicente, E Heberle-Bors. Stress as the major signal controlling the developmental fate of tobacco microspores: Towards a unified model of induction of microspore/pollen embryogenesis. Planta 200:144–152, 1996.
87. A Touraev, A Ilham, O Vicente, E Heberle-Bors. Stress induced microspore embryogenesis from tobacco microspores: An optimized system for molecular studies. Plant Cell Rep 15:561–565, 1996.
88. A Touraev, A Indrianto, I Wratschko, O Vincente, E Heberle-Bors. Starvation and heat-shock induced in vitro microspore embryogenesis in wheat (*Triticum aestivum* L.). Sex Plant Reprod 9:209–215, 1996.
89. A Touraev, A Indrianto, I Wratschko, O Vicente, E Heberle-Bors. Efficient microspore embryogenesis in wheat (*Triticum aestivum* L.) induced by starvation at high temperatures. Sex Plant Reprod 9:209–215, 1996.
90. U Kuhlmann, B Foroughi-Wehr. Production of double haploid lines in frequencies sufficient for barley breeding programs. Plant Cell Rep 8:78–81, 1989.
91. JR Christensen, E Borrino, A Olesen, SB Andersen. Diploid, tetraploid, and octoploid plants from anther culture of tetraploid orchard grass, *Dactylis glomerata* L. Plant Breed 116:267–270, 1997.
92. B Bohanec. Haploid induction in buckwheat (*Fagopyrum esculentum* Moench). In: SM Jain, SK Sopory, RE Veilleux, eds. In Vitro Haploid Production in Higher Plants. Vol. 4. Dordrecht: Kluwer Academic Press, 1997, 163–170.
93. M Coumans, D Zhong. Doubled haploid sunflower (*Helianthus annuus*) plant production by androgenesis: Fact or artifact? Part 2. In vitro isolated microspore culture. Plant Cell Tissue Org Cult 41: 203–209, 1995.
94. TC Hu, A Ziauddin, E Simion, KJ Kasha. Isolated microspore culture of wheat (*Triticum aestivum* L.) in a defined media. I. Effects of pretreatment, isolation methods, and hormones. In Vitro Cell Dev Biol 31:79–81, 1995.
95. J Xie, M Gao, Q Cai, X Cheng, Y Shen, Z Liang. Improved isolated microspore culture efficiency in medium with maltose and optimized growth regulator combination in japonica rice (*Oryza sativa*). Plant Cell Tissue Org Cult 42:245–250, 1995.
96. Z Lentini, P Reyes, CP Martínez, WM Roca. Androgenesis of highly recalcitrant rice genotypes with maltose and silver nitrate. Plant Sci 110:127–138, 1995.
97. HR Owen, AR Miller. Haploid plant regeneration from anther cultures of three North American cultivars of strawberry (*Fragaria* x *ananassa* Duch.). Plant Cell Rep 15:905–909, 1996.

98. MC Pedroso, MS Pais. Induction of microspore embryogenesis in *Camellia japonica* cv. Elegans. Plant Cell Tissue Org Cult 37:129–136, 1994.
99. L Ríhová, J Tupý. Influence of 2,4-D and lactose on pollen embryogenesis in anther culture of potato. Plant Cell Tissue Org Cult 45:269–272, 1996.
100. D Zhong, N Michaux-Ferriére, M Coumans. Assay for doubled haploid sunflower (*Helianthus annuus*) plant production by androgenesis: fact or artifact? Part 1. In vitro anther culture. Plant Cell, Tissue Org Cult 41:91–97, 1995.
101. XR Feng, DJ Wolyn. Development of haploid asparagus embryos from liquid cultures of anther-derived calli is enhanced by ancymidol. Plant Cell Rep 12:281–285, 1993.
102. KN Kao, M Saleem, S Abrams, M Pedras, D Horn, C Mallard. Culture conditions for induction of green plants from barley microspores by anther culture methods. Plant Cell Rep 9:595–601, 1991.
103. K Ilic-Grubor, SM Attree, LC Fowke. Induction of microspore-derived embryos of *Brassica napus* L. with polyethylene glycol (PEG) as osmoticum in a low sucrose medium. Plant Cell Rep 17:329–333, 1998.
104. MH Rahman. Microspore-derived embryos of *Brassica napus* L.: Stress tolerance and embryo development. Ph.D. Thesis, University of Calgary, Calgary, Canada, 1993.
105. CE Yeung, MH Rahman, TA Thorpe. Comparative development of zygotic and microspore-derived embryos in *Brassica napus* L. cv. Topas. I. Histodifferentiation. Int J Plant Sci 157:27–39, 1996.
106. Y Hamaoka, Y Fujita, S Iwati. Effects of temperature on the mode of pollen development in anther culture of *Brassica campestris*. Physiol Plant 82:67–72, 1991.
107. H Hu. In vitro induced haploids in wheat. In: SM Jain, SK Sopory, RE Veilleux, eds. In Vitro Haploid Production in Higher Plants. Vol. 3. Dordrecht: Kluwer Academic, 1997, 377–395.
108. M Rakoczy-Trojanowska, M Smiech, S Malepszy. The influence of genotype and medium on rye (*Secale cereale* L.) anther culture. Plant Cell Tissue Org Cult 48:15–21, 1997.
109. E Kaltchuk-Santos, JE Mariath, E Mundstock, CY Hu, MH Bodanese-Zanettini. Cytological analysis of early microspore divisions and embryo formation in cultured soybean anthers. Plant Cell Tissue Org Cult 49:107–115, 1997.
110. L Cistué, A Ziauddin, E Simion, KJ Kasha. Effects of culture conditions on isolated microspore response of barley cultivar Igri. Plant Cell Tissue Org Cult 42:163–169, 1995.
111. N Roulund, L Hansted, SB Anderson, B Farestveit. Effect of genotype, environment and carbohydrate on anther culture response in head cabbage (*Brassica oleracea* L. convar *capitata* (L.) Alef). Euphytica 49:237–242, 1990.
112. WA Keller, KC Armstrong. Stimulation of embryogenesis and haploid production in *Brassica campestris* anther cultures by elevated temperature treatments. Theor Appl Genet 55:65–67, 1979.
113. H Li, JA Qureshi, KK Kartha. The influence of different temperature treatments on anther culture response of spring wheat (*Triticum aestivum* L.). Plant Sci 57:55–61, 1988.
114. LY Shen, RE Veilleux. Effect of temperature shock and elevated incubation temperature on androgenic embryo yield of diploid potato. Plant Cell Tissue Org Cult 43:29–35, 1995.
115. Y Chen, EO Kenaschuk, JD Procunier. Plant regeneration from anther culture in Canadian cultivars of flax (*Linum usitatissimum* L.). Euphytica 102:183–189, 1998.
116. JK Hwang, KY Paek, DH Cho. Breeding of resistant pepper lines (*Capsicum annuum* L.) to bacterial spot (*Xanthomonas campestris* Pv. *Vesicatoria*) through anther culture. Acta Hortic 461:301–307, 1998.
117. J Zhao, DH Simmonds, W Newcomb. High frequency production of doubled haploid plants of *Brassica napus* cv. Topas derived from colchicine-induced microspore embryogenesis without heat shock. Plant Cell Rep 15:668–671, 1996.
118. PA Davies, S Morton. A comparison of barley isolated microspore and anther culture and the influence of cell culture density. Plant Cell Rep 17:206–210, 1998.
119. M Puolimatka, J Pauk. Impact of explant type, duration and initiation time on the co-culture effect in isolated microspore culture of wheat (*Triticum aestivum* L.). J Plant Physiol 154:367–373, 1999.
120. T Hu, K Kasha. Improvement of isolated microspore culture of wheat (*Triticum aestivum* L.) through ovary co-culture. Plant Cell Rep 16:520–525, 1997.
121. ZZ Chen, S Snyder, ZG Fan, WH Loh. Efficient production of doubled haploid plants through chromosome doubling of isolated microspores of *Brassica napus*. Plant Breed 113:217–221, 1994.

122. MCM Iqbal, C Möllers, G Röbbelen. Increased embryogenesis after colchicine treatment of microspore cultures of *Brassica napus* L. J Plant Physiol 143:222–226, 1994.
123. W Tulecke. A haploid tissue culture from the female gametophyte of *Ginkgo biloba*. Nature 203:94–95, 1964.
124. LH San Noeum. Haploïdes d'*Hordeum vulgare* L. par culture in vitro d'ovaires no fecondés. Ann Amelior Plantes 26:751–754, 1976.
125. LH San Noeum, P Gelebart. Production of gynogenetic haploids. In: IK Vasil, ed. Cell Culture and Somatic Cell Genetics of Plants. Oxford, NY:Academic Press, 1986, pp 305–322.
126. HY Yang, C Zhou. In Vitro gynogenesis. In: SS Bhojwani, ed. Plant Tissue Culture Applications and Limitations. Oxford, NY: Elsevier, 1990, pp 242–258.
127. ERJ Keller, L Korzun. Ovary and ovule culture for haploid production. In: SM Jain, SK Sopory, RE Veilleux, eds. In Vitro Haploid Production in Higher Plants. Vol. 1. Dordrecht: Kluwer Academic, 1996, pp 217–235.
128. G Lakshmi Sita. Gynogenic haploids in vitro. In: SM Jain, SK Sopory, RE Veilleux, eds. In Vitro Haploid Production in Higher Plants. Vol. 5. Dordrecht: Kluwer Academic Publishers, 1997, pp 175–193.
129. AM Castillo, L Cistué. Production of gynogenic haploid of *Hordeum vulgare* L. Plant Cell Rep 12: 139–143, 1993.
130. M Asselin de Beauville. Obtention d'haploïdes in vitro à partir d'ovaires no fecondés de riz, *Oryza sativa* L. C R Acad Sci III 290:489–492, 1980.
131. ZC Zhu, HS Wu, QK An, ZY Liu. Induction of haploid plantlets from unpollinated ovaries of *Triticum aestivum* cultured in vitro. Acta Genet Sin 8:386–390, 1981.
132. GJ Speckmann, JPC van Geyt, M Jacobs. The induction of haploids of sugarbeet (*Beta vulgaris* L.) using anther and free pollen culture or ovule and ovary culture. In: W Horn, CJ Jensen, W Odenbach, O Schieder eds. Proceedings of the International Symposium on Genetic Manipulation in Plant Breeding, EUCARPIA, Berlin, 1985, pp 351–353.
133. M Mdarhri-Alaoui, N Saidi, A Chlyah, H Chlyah. Green haploid plant formation in durum wheat through in vitro gynogenesis. C R Acad Sci III 321:25–30, 1998.
134. B Bohanec, M Jakše, A Ihan, B Javornik. Studies of gynogenesis in onion (*Allium cepa* L.): induction procedures and genetic analysis of regenerants. Plant Sci 104:215–224, 1995.
135. J Cohat. Obtention chez l'échalote (*Allium cepa* L. var. *aggregatum*) de plantes gynogénétiques par culture in vitro de boutons floraux. Agronomie 14:299–304, 1994.
136. M Jakše, B Bohanec, A Ihan. Effect of media components on the gynogenic regeneration of onion (*Allium cepa* L.) cultivars and analysis of regenerants. Plant Cell Rep 15:934–938, 1996.
137. B Bohanec, M Jakše. Variations in gynogenic response among long-day onion (*Allium cepa* L.) accessions. Plant Cell Rep 18:737–742, 1999.
138. L Rongbai, MP Pandey, GK Garg, SK Pandey, DK Dwivedi, Ashima. Development of a technique for in vitro unpollinated ovary culture in rice, *Oryza sativa* L. Euphytica 104:159–166, 1998.
139. C Zhou, HY Yang. In vitro embryogenesis in unfertilized embryo sacs of *Oryza sativa* L. Acta Bot Sin 23:176–179, 1981.
140. H Lux, L Herrmann, C Wetzel. Production of haploid sugar beet (*Beta vulgaris* L.) by culturing unpollinated ovules. Plant Breed 104:177–183, 1990.
141. M Doctrinal, RS Sangwan, BS Sangwan-Norreel. In vitro gynogenesis in *Beta vulgaris* L. Effect of plant growth regulators, temperature, genotypes and season. Plant Cell Tissue Org Cult 17:1–12, 1989.
142. M Cappadocia, L Chrétien, G Laublin. Production of haploids in *Gerbera jamesonii* via ovule culture: Influence of fall versus spring sampling on callus formation and shoot regeneration. Can J Bot 66:1107–1110, 1988.
143. K D'Halluin, B Keimer. Production of haploid sugar beets (*Beta vulgaris* L.) by ovule culture. In: W Horn, CJ Jensen, W Odenbach, O Schieder, eds. Genetic manipulation in Plant Breed, Berlin: De Gruyter, 1986, pp 307–309.
144. H Yan, HY Yang, WA Jensen. An electron microscope study on in vitro parthenogenesis in sunflower. Sex Plant Reprod 2:154–166, 1989.

145. CH Bornman. Haploidization of sugar beet (*Beta vulgaris*) via gynogenesis. In Vitro 1:3, 1985.
146. BJ Wu, KC Cheng. Cytological and embryological studies on haploid plant production from cultured unpollinated ovaries of *Nicotiana tabacum*. Acta Bot Sin 2:125–129, 1982.
147. AM Bugara, LV Rusina. Haploid callus formation in the culture of unfertilized ovules of clary sage. Fiziol Biokh Kul't rast 20:554–560, 1989.
148. HQ Tian, HY Yang. Haploid embryogeny and plant regeneration in unpollinated ovary culture of *Allium tuberosum*. Acta Biol Exp Sin 22:139–147, 1989.
149. C Zhou, HY Yang. Induction of haploid rice plantlets by ovary culture. Plant Sci Lett 20:231–237, 1981.
150. ZR Tao, MS Liu, ZC Zhu. In vitro production of haploid plantlets from unpollinated ovaries of potato. Hereditas 7:5, 1985.
151. ZR Tao, MS Liu, ZC Zhu. Induction of dihaploid plantlets from unpollinated ovaries of potato in vitro. Acta Genet Sin 15:329–334, 1988.
152. LH San, Y Demarly. Gynogenesis in vitro and biometric studies of doubled haploids obtained by three techniques in *Hordeum vulgare* L. In: W Lange, AC Zeven, NG Hogenboom eds. Efficiency in Plant Breeding. Proc. 10th Congress EUCARPIA, Wageningen, 1983, p 347.
153. S Baldursson. Promotion of haploid callus induction in cultured mega-gametophytes of *Picea sitchensis*. Proceedings 5th Int. IUFRO Work. Party S2.04–07 "Biotechnology of Trees", Balsain, Spain, 1993.
154. C Zhou, H Yang, H Tian, Z Liu, H Yan. In vitro culture of unpollinated ovaries in *Oryza sativa* L. In: H Hu, H Yang, eds. Haploids of Higher Plants In Vitro. Berlin: Springer-Verlag, 1986, pp 165–181.
155. MA Gusakovskaya, MA Nadzhar. Gynogenesis in the culture of unfertilized wheat ovaries and ovules. Bot Zurn 79:70–79, 1994.
156. T Murashige, F Skoog. A revised medium for rapid growth and bioassays with tobacco tissue cultures. Physiol Plant 15:473–497, 1962.
157. OL Gamborg, RA Miller, K Ojima. Nutrient requirements of suspension cultures of soybean root cells. Exp Cell Res 50:151–158, 1968.
158. CC Chu. The N6 medium and its application to anther culture of cereal crops. Proceedings on the Symposium on Plant Tissue Culture, Beijing, 1978, pp 43–50.
159. T Eriksson. Studies on the growth requirements and growth measurements of cell culture of *Haplopappus gracilis*. Physiol Plant 18:976–993, 1965.
160. MW Galatowitch, GA Smith. Regeneration from unfertilized ovule callus of sugarbeet (*Beta vulgaris* L.). Can J Plant Sci 70:83–89, 1990.
161. G Juhász, G Venczel, P Balogh. Haploid plant induction in zucchini (*Cucurbia pepo* L. convar. giromontiina Duch) and in cucumber (*Cucumis sativus* L.) lines through in vitro gynogenesis. Acta Hortic 447:623–624, 1997.
162. A Schum, L Mattiesch, E-M Timmann, K Hofmann. Regeneration of dihaploids via gynogenesis in *Allium porrum* L. Gartenbauwissenschaft 58:227–232, 1993.
163. S Yoshida, Y Kasai, K Watanabe, M Fujino. Prolines stimulates albino regeneration from anther-and seed-derived rice callus under high osmosis. J Plant Physiol 155:107–109, 1999.
164. KN Kao. Embryogenesis and plant regeneration from cereal microspores. 4th Canadian Plant Tissue Culture and Genetic Engineering Conference, Saskatoon, Canada 1996, pp 17.
165. S Hoekstra, MH van Zijderveld, F Heidekamp, F van der Mark. Microspore culture of *Hordeum vulgare* L.: The influence of density and osmolality. Plant Cell Rep 12:661–665, 1993.
166. M Salmenkallio-Marttila, U Kurtén, V Kauppinen. Culture conditions for efficient induction of green plants from isolated microspores of barley. Plant Cell Tissue Organ Cult 43:79–81, 1995.
167. A Day, THN Ellis. Deleted forms of plastid DNA in albino plants from cereal anther culture. Curr Genet 9:671–678, 1985.
168. A Elisabeth, E Heberle-Bors, M Pfosser. A specific plastid DNA deletion discovered in wheat microspores. In Vitro Cell Dev Biol 35:169, 1999.
169. B Hofinger, E Heberle-Bors, M Pfosser. Distinct alterations of transcript levels are correlated with albinism in microspore- and anther culture-derived wheat plants. In Vitro Cell Dev Biol 35:172, 1999.

170. A Tosca, R Pandolfi, S Citterio, A Fasoli, S Sgorbati. Determination by flow cytometry of the chromosome doubling capacity of colchicine and oryzalin in gynogenetic haploids of *Gerbera*. Plant Cell Rep 14:455–458, 1995.
171. J Zhao, DH Simmonds. Application of trifluralin to embryogenic microspore cultures to generate doubled haploid plants in *Brassica napus*. Physiol Plant 95:304–309, 1995.
172. NJP Hansen, SB Andersen. In vitro chromosome doubling potential of colchicine, oryzalin, trifluralin, and APM in *Brassica napus* microspore culture. Euphytica 88:159–164, 1996.
173. AL Hansen, A Gertz, M Joersbo, SB Andersen. Antimicrotubule herbicides for in vitro chromosome doubling in *Beta vulgaris* L. ovule culture. Euphytica 101:231–237, 1998.
174. B Barnabás, PL Pfahler, G Kovács. Direct effect of colchicine on the microspore embryogenesis to produce dihaploid plants in wheat (*Triticum aestivum* L.). Theor Appl Genet 81:675–678, 1991.
175. B Barnabás, B Obert, G Kovács. Colchicine, an efficient genome-doubling agent for maize (*Zea mays* L.) microspores cultured in anthers. Plant Cell Rep 18:858–862, 1999.
176. S Antoine-Michard, M Beckert. Spontaneous versus colchicine-induced chromosome doubling in maize anther culture. Plant Cell Tissue Org Cult 48:203–207, 1997.
177. A Redha, T Attia, B Büter, S Saisingtong, P Stamp, JE Schmid. Improved production of doubled haploids by colchicine application to wheat (*Triticum aestivum* L.) anther culture. Plant Cell Rep 17: 974–979, 1998.
178. T Ogawa, H Fukuoka, Y Ohkawa. Induction of cell division of isolated pollen grains by sugar starvation in rice. Breed Sci 44:75–77, 1994.
179. S Hoekstra, MH van Zijderveld, JD Louwerse, F Heidekamp, F van der Mark. Anther and microspore culture of *Hordeum vulgare* L. cv. Igri. Plant Sci 86:89–96, 1992.
180. A Gaillard, P Vergne, M Beckert. Optimization of maize isolation and conditions for reliable plant regeneration. Plant Cell Rep 10:55–58, 1991.
181. VD Gustafson, PS Baenziger, MS Wright, WW Stroup, Y Yen. Isolated wheat microspore culture. Plant Cell Tissue Org Cult 42:207–213, 1995.
182. B Huang, N Sunderland. Temperature stress pretreatment in barley anther culture. Ann Bot 49:77–88, 1982.
183. MS Cho, FJ Zapata. Callus formation and plant regeneration in isolated pollen culture of rice (*Oryza sativa* L., cv. Taipei 309). Plant Sci 58:239–244, 1988.
184. JBM Custers, JHG Cordenwener, Y Nöllen, HJM Dons, MM Van Lookern Campagne. Temperature controls both gametophytic and sporophytic development in microspore cultures of *Brassica napus*. Plant Cell Rep 13:267–271, 1994.
185. J Zhao, DH Simmonds, W Newcomb. Induction of embryogenesis with colchicine instead of heat in microspores of *Brassica napus* L. cv. Topas. Planta 198:433–439, 1996.
186. A Touraev, A Indrianto, A Tashpulatov, C Haller, H Katholnigg, E Heberle-Bors. Molecular biology and physiology of microspore embryogenesis. In Vitro Cell Dev Biol 35:178, 1999.
187. RS Sangwan, H Camefort. The tonoplast, a specific marker of embryogenic microspores of *Datura* cultured in vitro. Histochemistry 78:473–480, 1983.
188. K-H Lo, KP Pauls. Plant growth environment effects on rapeseed microspore development and culture. Plant Physiol 99:468–472, 1992.
189. PM Pechan, WA Keller, F Mandy, M Bergeron. Selection of *Brassica napus* L. embryogenic microspores by flow sorting. Plant Cell Rep 7:396–398, 1988.
190. C Deslauriers, AD Powell, K Fuchs, KP Pauls. Flow cytometric characterization and sorting of cultured *Brassica napus* microspores. Biochim Biophys Acta 1091:165–172, 1991.
191. D Schulze, KP Pauls. Flow cytometric characterization of embryogenic and gametophytic development in *Brassica napus* microspore cultures. Plant Cell Physiol 39:226–234, 1998.
192. JHG Cordewener, G Hause, E Görgen, R Busink, B Hause, HJM Dons, AAM Van Lammeren, MM Van Lookeren Campagne, P Pechan. Changes in synthesis and localization of the 70 kDa class of heat shock proteins accompany the induction of embryogenesis in *Brassica napus* microspores. Planta 196:747–755, 1995.
193. JBM Custers, JHG Cordewener, HJM Dons, MM van Lookeren Campagne. Regulation of the inductive phase of microspore embryogenesis in *Brassica napus*. Acta Hortic 407:209–217, 1996.
194. AAM Van Lammeren, T Havlicky, P Binarova, K Straatman, B Hause, G Hause. Expression of cy-

toskeletal and heat shock proteins in embryogenic microspore culture of *Brassica napus*. Visualized by immunocytochemistry. In: E Heberle-Bors, M Hess, O Vicente eds. Frontiers in sexual plant reproduction research. Vienna: University of Vienna, 1994, p 13.

195. P Smykal, PM Pechan. Expression of Hsp17 during induction of pollen embryogenesis in *B. napus* L. Hamburg: Plant Embryogenesis Workshop, 1996, pp 68.
196. K Boutilier, M-J Ginés, JM DeMoor, B Huang, CL Baszcynski, VN Iyer, BL Miki. Expression of the BnmNAP subfamily of napin genes coincides with the induction of *Brassica* microspore embryogenesis. Plant Mol Biol 26:1711–1723, 1994.
197. HJ Jansen, JHG Cordewener, HJM Dons, MM van Lookeren Campagne. Cloning of differentially expressed mRNAs during the induction of microspore embryogenesis in *Brassica napus*. Hamburg: Plant Embryogenesis Workshop, 1996, pp 36.
198. V Zárský, N Eller, D Garrido, J Tupý, F Schoffl, O Vicente E Heberle-Bors. Activation of a small heat shock gene during induction of tobacco pollen embryogenesis by starvation. In: E Heberle-Bors, M Hesse, O Vicente eds. Proceedings of the 13th International Congress on Sexual Plant Reproduction. Vienna: University of Vienna, 1994, pp 15.
199. A Murigneux, S Bentolila, T Hardy, S Baud, C Guitton, H Jullien, S Ben Tahar, G Freyssinet, M Beckert. Genotypic variation of quantitative trait loci controlling in vitro androgenesis in maize. Genome 37:970–976, 1994.
200. Y Wan, TR Rocheford, JM Widholm. RFLP analysis to identify putative chromosomal regions involved in the anther culture response and callus formation of maize. Theor Appl Genet 85:360–365, 1992.
201. NM Cowen, CD Johnson, K Armstrong, M Miller, A Woosley, S Pescitelli, M Skokut, S Belmar, JF Petolino. Mapping genes conditioning in vitro androgenesis in maize using RFLP analysis. Theor Appl Genet 84:720–724, 1992.
202. GS Khush, SS Virmani. Haploids in plant breeding. In: SM Jain, SK Sopory, RE Veilleux, eds. In Vitro Haploid Production in Higher Plants. Vol 1. Dordrecht: Kluwer Academic, 1996, 11–33.
203. G Pelletier. Use of haplo-diploidisation in plant breeding. In: Y Chupeau, M Caboche, Y Henry, eds. Androgenesis and Haploid Plants. Berlin: Springer-Verlag, 1998, pp 104–111.
204. GR Stringam, VK Bansal, MR Thiagarajah, DF Degenhardt, JP Tewari. Development of an agronomically superior blackleg resistant canola cultivar in *Brassica napus* L. using doubled haploidy. Can J Plant Sci 75:437–439, 1995.
205. J Pauk, Z Kertész, B Beke, L Bóna, M Csosz, J Matuz. New winter wheat variety: 'GK Délibáb' developed via combining conventional breeding and in vitro androgenesis. Cereal Res Commun 23:251–256, 1995.
206. A Ulrich, WH Furtan, RK Downey. Biotechnology and rapeseed breeding: Some economic considerations. Science Council of Canada Report, Ottawa, 1984.
207. T Rajhathy. Haploid flax revisited. Z Pflanzenzuecht 76:1–10, 1976.
208. E Pochard, A Palloix, AM Daubeze. The use of androgenetic autodiploid lines for the analysis of complex resistance systems in the pepper. 6th Eucarpia Meeting on Genetics and Breeding on Capsicum and Eggplant, Saragoza, 1986, pp 105–109.
209. AM Daubeze, E Pochard, A Paloix. Inheritance of resistance to *Leveillula taurica* and relation to other phenotypic characters in the haploid progeny issued from an African pepper line. 7th Eucarpia Genetics and Breeding *Capsicum* and Eggplant, Kragujevac,, 1989, pp 299–232.
210. C Doré. Asparagus anther culture and field trials of dihaploids and F1 hybrids. In: YPS Bajaj, ed. Biotechnology in Agriculture and Forestry, Heidelberg; Springer-Verlag, 1990, 12:322–345.
211. L Corriols, C Doré, C Rameau. Commercial release in France of Andreas, the first asparagus all male F1 hybrid. Acta Hortic 271:249–252, 1990.
212. W Friedt, C Bickert, H Schaub. In vitro breeding of high-linolenic, doubled-haploid lines of linseed (*Linum usitassimum* L) via androgenesis. Plant Breed 114:322–326, 1995.
213. P Lashermes, E Couturon, A Charrier. Combining ability of doubled haploids in *Coffea canephora* P. Plant Breed 112:330–337, 1994.
214. M Zivy, P Devaux, J Blaisonneaux, R. Jean, H Threllment. Segregation distortion and linkage studies in microspore-derived doubled haploid lines of *Hordeum vulgare* L. Theor Appl Genet 83:919–924, 1992.

215. N Huang, S McCrouch, T Mew, A Parco, E Guiderdoni. Development of an RFLP map from a doubled haploid population in rice. Rice Genet Newsl 11:134–137, 1994.
216. PK Tanhuanpää, HJ Vilkki, JP Vilkki, SK Pulli. Segregation distortion of DNA markers in a microspore derived population and identification of markers linked to a locus affecting linolenic acid in oilseed rape (*Brassica napus* L.). VIII Crucifer Genetics Workshop, Saskatoon, Canada, 1993, p 61.
217. EB Swanson, MJ Herrgesell, M Arnoldo. Microspore mutagenesis and selection: Canola plants with field tolerance to the imidazolinones. Theor Appl Genet 78:525–530, 1989.
218. WD Beversdorf, LS Kott. An in vitro mutagenesis/selection system for *Brassica napus*. Iowa State J Res 61:435–443, 1987.
219. J Turner, D Facciotti. High oleic acid *Brassica napus* from mutagenized microspores. Proceeding of the 6th Crucifer Genetics Workshop, Ithaca, NY, 1990, p 24.
220. B Huang, EB Swanson, CL Baszczynski, WD Macrae, E Barbour, V Armavil, L Woke, M Arnoldo, S Rozakis, M Westecott, RF Keats, R Kemble. Application of microspore culture to canola improvement. Proceedings of the 8th International Rapeseed Congress, Saskatoon, 1991, pp 298–303.
221. AMR Ferrie. Combining microspores and mutagenesis. PBI Bulletin, January 1999.
222. I Ahmad, JP Day, MV MacDonald, DS Ingram. Haploid culture and UV mutagenesis in rapid-cycling *Brassica napus* for the generation of resistance to chlorsulfuron and *Alternaria brassicicola*. Ann Bot 67:521–525, 1991.
223. A Touraev, E Stöger, V Voronin, I Ottenschlager, M Dahl, I Barinova, E Heberle-Bors. Plant germ line transformation (MAGELITR). In Vitro Cell Biol Dev 35:179, 1999.
224. J Wang, HZ Shi, C Zhou, HY Yang, XL Zhang, RD Zhang. β-glucuronidase gene and green fluorescent protein gene expression in de-exined pollen of *Nicotiana tabacum* by microprojectile bombardment. Sex Plant Reprod 11:159–162, 1998.
225. MF Jardinaud, A Souvré, G Alibert. Transient GUS gene expression in *Brassica napus* electroporated microspores. Plant Sci 93:177–184, 1993.
226. MF Jardinaud, A Souvré, G Alibert, M Beckert. UidA gene transfer and expression in maize microspores using the biolistic method. Protoplasma 187:138–143, 1995.
227. U Kuhlmann, B Foroughi-Wehr, A Graner, G Wenzel. Improved culture system for microspores of barley to become target for DNA uptake. Plant Breed 107:165–168, 1991.
228. M Nishihara, M Seki, M Kyo, K Irifune, H Morikawa. Transgenic haploid plants of *Nicotiana rustica* produced by bombardment-mediated transformation of pollen. Transgenic Res 4:341–348, 1995.
229. E Stöger, C Fink, M Pfosser, E Heberle-Bors. Plant transformation by particle bombardment of embryogenic pollen. Plant Cell Rep 14:273–278, 1995.
230. H Fukuoka, T Ogawa, M Matsuoka, Y Ohkawa, H Yano. Direct gene delivery into isolated microspores of rapeseed (*Brassica napus* L.) and the production of fertile transgenic plants. Plant Cell Rep 17:323–328.
231. A Gaillard, E Matthysrochon, C Dumas. Selection of microspore derived embryogenic structures in maize related to transformation potential by microinjection. Bot Acta 105:313–318, 1992.
232. G Neuhaus, G Spangenberg, O Mittelsten Scheid, H-G Schweiger. Transgenic rapeseed plants obtained by the microinjection of DNA into microspore-derived embryoids. Theor Appl Genet 75:30–36, 1987.
233. EB Swanson, LR Erickson. Haploid transformation in *Brassica napus* using an octopine-producing strain of *Agrobacterium tumefaciens*. Theor Appl Genet 78:831–835, 1989.
234. M Dormann, H-M Wang, N Datla, AMR Ferrie, WA Keller, MM Oelck. Transformation of freshly isolated *Brassica* microspores and regeneration to fertile homozygous plants. 9th International Rapeseed Congress, Vol. 3, Cambridge, England, 1995, 816–818.
235. M Dormann, N Datla, A Hayden, D Puttick, J Quandt. Non-destructive screening of haploid embryos for glufosinate ammonium resistance four weeks after microspore transformation in *Brassica*. Acta Hortic 459:191–197, 1998.
236. G Creissen, C Smith, R Francis, H Reynolds, P Mullineaux. *Agrobacterium*- and microprojectile-mediated viral DNA delivery into barley microspore-derived cultures. Plant Cell Rep 8:680–683, 1990.

237. A Fennell, R Hauptmann. Electroporation and PEG delivery of DNA into maize microspores. Plant Cell Rep 11:567–570, 1992.
238. A Jähne, D Becker, R Brettschneider, H Lörz. Regeneration of transgenic microspore-derived, fertile barley. Theor Appl Genet 89:525–533, 1994.
239. JL Chen, WD Beversdorf. A combined use of microprojectile bombardment and DNA imbibition enhances transformation frequency of canola (*Brassica napus* L.). Theor Appl Genet 88:187–192, 1994.
240. QA Yao, E Simion, M William, J Krochko, KJ Kasha. Biolistic transformation of haploid isolated microspores of barley (*Hordeum vulgare* L.). Genome 40:570–581, 1997.
241. WA Harwood, DF Chen, GP Creissen. Transformation of pollen and microspores. In: SM Jain, SK Sopory, RE Veilleux, eds. In Vitro Haploid Production in Higher Plants. Vol. 2. Dordrecht: Kluwer Academic, 1996, pp 53–71.
242. H Morikawa, M Nishihara. Use of pollen in gene transfer. In: KR Shivanna, VK Sawhney, eds. Pollen Biotechnology for Crop Production and Improvement. New York: Cambridge University Press, 1997, pp 423–437.
243. ML Crouch, Nonzygotic embryos of *Brassica napus* L. contain embryo-specific storage proteins. Planta 156:520–524, 1982.
244. E Wiberg, I Rahlen, M Hellman, E Tillberg, K Glimelius, S Stymne. The microspore-derived embryo of *Brassica napus* L. as a tool for studying embryo-specific lipid biogenesis and regulation of oil quality. Theor Appl Genet 82:515–520, 1991.
245. LA Holbrook, GJH Van Rooijen, RW Wilen, MM Moloney. 1991. Oil-body proteins in microspore-derived embryos of *Brassica napus*: Hormonal, osmotic and developmental regulation of synthesis. Plant Physiol 97:1051–1058, 1991.
246. DC Taylor, DL Barton, KP Roux, SI Mackenzie, DW Reed, EW Underhill, MK Pomeroy, N Weber. Biosynthesis of acyllipids containing very long chain fatty acids in microspore-derived and zygotic embryos of *Brassica napus* L. cv. Reston. Plant Physiol 99:1609–1618, 1992.
247. D McClellan, L Kott, W Beversdorf, BE Ellis. Glucosinolate metabolism in zygotic and microspore-derived embryos of *Brassica napus* L. J Plant Physiol 141:153–159, 1993.
248. AM Johnson-Flanagan, J Singh. A method to study seed degreening using haploid embryos of *Brassica napus* cv. Topas. J Plant Physiol 141:487–493, 1993.
249. W Orr, AM Johnson-Flanagan, WA Keller, J Singh. Induction of freezing tolerance in microspore-derived embryos of winter *Brassica napus*. Plant Cell Rep 8:579–581, 1990.
250. S Cloutier. In vitro selection for freezing tolerance using *Brassica napus* microspore culture. MS thesis, University of Guelph, Ontario, 1990.
251. AM Johnson-Flanagan, Z Huiwey, X-M Geng, DCW Brown, CL Nykiforuk, J Singh. Frost, abscisic acid and desiccation hasten embryo development in *Brassica napus* L. Plant Physiol 94:700–706, 1992.

# 4
# Shoot Regeneration and Proliferation

**Seedhabadee Ganeshan, Karen L. Caswell, Kutty K. Kartha, and Ravindra N. Chibbar**
*National Research Council, Saskatoon, Saskatchewan, Canada*

| | | |
|---|---|---|
| I. | INTRODUCTION | 69 |
| II. | STRATEGIES FOR IN VITRO SHOOT PRODUCTION | 70 |
| III. | IMPLICATIONS OF IN VITRO SHOOT PROLIFERATION IN TRANSGENE TECHNOLOGY | 71 |
| | A. Axillary Shoot Formation | 71 |
| | B. Direct Adventitious Shoot Formation | 71 |
| | C. Callus-Mediated Production of Adventitious Shoots | 74 |
| IV. | FACTORS AFFECTING IN VITRO SHOOT PRODUCTION AND PROLIFERATION | 74 |
| | A. Biotic Factors | 75 |
| | B. Abiotic Factors | 76 |
| V. | CONCLUDING REMARKS AND FUTURE PROSPECTS | 78 |
| | REFERENCES | 80 |

## I. INTRODUCTION

The ability of plant cells cultured in vitro, under suitable conditions, to form completely normal plantlets efficiently and reproducibly is the cornerstone for the production of transgenic plants. The retention of totipotency in differentiated plant cells provides the theoretical basis for this unique phenomenon in plants. Thus, plant tissues cultured in vitro can differentiate to form de novo organs such as shoots, roots, flowers, and embryos. Although the concept of plant cell totipotency dates back to the late 1800s, it was first demonstrated in 1939 when Gautheret (1), Nobécourt (2), and White (3) independently reported that continuously growing callus cultures were derived from meristematic tissues. However, Skoog (4) and Skoog and Tsui (5) are credited with the induction of callus and adventitious shoots from isolated mature and differentiated cells. They also implicated plant growth substances in the induction of shoot and/or root initiation from

NRCC No. 43799

tobacco callus, thereby providing the principle on which all micropropagation depends. Since these pioneering discoveries related to de novo organ formation from plant tissue cultures, this research area has developed into a scientific discipline that has become critical to the production of transgenic plants. Although the major impediments encountered in in vitro culture of most plant species have been overcome, the process still requires careful optimization in order to develop an efficient transgenic technology for plant improvement. In addition to development of high-frequency transformation efficiencies, the transgene must be stably integrated into the genome, inherited in a predictable manner, and expressed with fidelity. The foundation for fulfilling these requirements lies in the efficient regeneration of plants from a single cell that has received the gene of interest.

The in vitro production of shoots has been studied for a number of years with different objectives in mind. Initial experiments focused on the effect of plant growth regulators on such processes as callogenesis (callus formation), embryogenesis (somatic embryo formation), caulogenesis (shoot formation), and rhizogenesis (root formation) under controlled conditions, all of which, with the exception of rhizogenesis, culminate in the production of shoots in vitro. In the mid-1970s, in vitro shoot proliferation was used for clonal propagation of elite germplasm of economically important plant species. Optimized protocols based on a number of parameters such as the physiological status of donor plants, explant size and type, culture medium composition, and environmental conditions during incubation of cultures were developed. In the mid 1980s, with the advent of transgenic technology, the interaction of optimized regeneration protocols with deoxyribonucleic acid (DNA) delivery methods and selective agents became the focus of many reports. The decade of the 1990s saw the extension of in vitro shoot regeneration systems to a wide range of plant species, including many that were previously considered recalcitrant. It is an onerous task to present an extensive commentary on the vast volume of literature devoted to in vitro shoot proliferation. However, we will attempt to highlight critical issues that pertain to in vitro shoot proliferation and to the application of such a system to produce transgenic plants.

## II. STRATEGIES FOR IN VITRO SHOOT PRODUCTION

The developmental fates of cultured cells are dictated by a complex number of stimuli, which trigger a cascade of events at the molecular level, hitherto only superficially understood. The results of such interactions lead to in vitro shoot production according to the following developmental pathways: (a) elongation of dormant meristems, (b) adventitious shoot formation, (c) organogenesis from callus or cell cultures, and (d) somatic embryogenesis. The elongation of preformed meristems is essentially the proliferation in culture of axillary meristems. This form of in vitro shoot production is widely used in commercial micropropagation. Its use in the production of transgenic plants is unclear at present, as discussed in the following paragraphs. The remaining three approaches for in vitro shoot production are of vital importance to transgenic plant production. In vitro regeneration through somatic embryogenesis has been discussed in an accompanying chapter, and therefore is not discussed here. The focus of this review is on the in vitro production and proliferation of shoots from meristems or meristemlike tissues, depending on the starting explant type.

However, before proceeding further, the pertinent terminologies used in this chapter are defined. The terms *adventitious shoot formation* and *organogenesis* are often cited interchangeably in tissue culture literature, and a very fine line of distinction between these two processes is generally perceived. *Organogenesis* refers to the formation of unipolar structures such as shoot or root primordia from cells or tissues in culture (6). *Adventitious shoot formation* refers to the de novo development of a shoot or shoots from points of origin other than the axils of leaves or apices (7). In nature, the occurrence of adventitious shoots on leaves, modified leaves, stems, or roots is common (8). By extrapolation, shoots produced directly from calli can also be referred to as *adventitious*. Moreover, these processes of shoot formation, from callus or other plant tissues, are initiated from meristemlike layers of cells often referred to as *shoot meristemoids* (6), as an analogy to shoot meristems in planta. Meristemoids were described by Torrey (9) as meristematic layers of cells organized within calli that have the potential to differentiate into shoots, roots, or embryos. Throughout this chapter the term *adventitious shoot* is used to encompass in vitro shoot production from either plant tissues or calli.

## III. IMPLICATIONS OF IN VITRO SHOOT PROLIFERATION IN TRANSGENE TECHNOLOGY

### A. Axillary Shoot Formation

As mentioned earlier, elongation of dormant meristems is widely used for clonal micropropagation. For some species that are recalcitrant to classical vegetative propagation methods, in vitro shoot development has proved to be of great value (10). Some advantages of axillary shoot multiplication (refer to Ref. 10) are rapid production of a large number of plants, expeditious international exchange of germplasm, and production of disease-free stocks (11). A shoot proliferation system such as this would be a prime target for transformation. Although feasible, it raises the problem of production of chimeras (12), because it is highly unlikely that all the cells of a meristematic dome could have received the gene. However, there are reports claiming transformation of shoot apices. Successful transformation of shoot apices of petunia using *Agrobacterium* sp. was achieved for the first time by Ulian et al. (13). Hussey et al. (14) reported on the transformation of meristematic cells within shoot apices and young primordia of pea shoots by *Agrobacterium* sp. Although tumor formation was reported, no shoot regeneration was mentioned. Gould et al. (12) transformed shoot apical meristems of *Zea mays* using *Agrobacterium tumefaciens* and recovered transgenic, as well as chimeric, plants from preexisting shoot apical meristems. A similar observation was reported by Schrammeijer et al. (15), in an attempt to transform shoot meristems of sunflower.

Considering the advantages of shoot meristems for in vitro shoot production, a modification of the microprojectile bombardment technique for DNA transfer to meristematic cells was suggested (16). In order to target the gene to the few cells in the meristematic region that would ultimately give rise to a plant, Sautter et al. (16) developed a system referred to as *biolistic micro targeting*. Using this method, transient expression of the GUS and anthocyanin genes was reported in wheat meristem cells (17).

### B. Direct Adventitious Shoot Formation

There are numerous reports on the production of adventitious shoots directly from plant tissues, and no attempt is made here to present an exhaustive list. There are several advantages to this method, both for regeneration of plants, as well as for their genetic transformation (18). It has been suggested that direct initiation of shoots would alleviate the problems associated with somaclonal variation (19), which is generally attributed to the passage of plant tissues in culture via a callus-mediated phase (20). Furthermore, the time from initiation of culture to production of plants is significantly reduced. It is a particularly appealing system for production of transgenic plants. Table 1 shows selected examples of transformation strategies using adventitious shoot initiation. The explants used include stem segments, leaf disks, cotyledons, hypocotyls, embryonic axes, shoot apices, and buds. Both *Agrobacterium* sp. and particle bombardment have been used

**Table 1**
Examples of Transformed Plants Produced Through Organogenesis

Agrobacterium tumefaciens–mediated transformation

| Species | Common Name | Explant | Regeneration | Ref. |
|---|---|---|---|---|
| *Arachis hypogaea* L. | Peanut | Epicotyl or leaf | Multiple shoots | 80 |
| *Brassica napus* | Rapeseed | Hypocotyl sections | Shoots from callus | 81 |
| *Brassica oleracea* var. *botrytis* L. | Taiwan cauliflower | Hypocotyl | Adventitious bud regeneration | 82 |
| *Brassica oleracea* var. *capitata* | Cabbage | Hypocotyl or petiole | Shoots from explant or callus | 83 |
| *Brassica oleracea* var. *italica* | Broccoli | Peduncle, hypocotyl or petiole | Shoots from explant or callus | 83 |
| *Cicer arietinum* L. | Chickpea | Embryo axis from seed | Direct mutiple shoot production | 22 |
| *Cichorium intybus* L. var. *sativum* | Chicory | Shoot buds | Multiple shoots produced per bud | 84 |
| *Citrus aurantifolia* Swing. | Lime | Intermodal stem segments | Adventitious shoots | 85 |
| *Citrus sinensis* L. Osbeck | Sweet orange | Intermodal stem segments | Multiple shoots and in vitro grafting | 86 |
| *Citrus sinensis* L. Osbeck × *Poncirus trifoliata* L. Raf. | Carrizo citrange | Intermodal stem segments | Little callus formed before shoots | 87 |
| *Dendranthema indicum* L. Des Moul | Chrysanthemum | Leaf pieces | Shoots directly from leaf pieces | 88 |
| *Dendranthema* × *Grandiflorum* | Florists' chrysanthemum | Leaf pieces | Organogenesis with or without callus | 89 |
| *Fragaria* × *ananassa* Dush. | Strawberry | Leaf disks | Shoots from meristematic regions | 90 |
| *Glycine max* L. Merr. | Soybean | Cotyledonary node | Multiple shoots | 91 |
| *Glycine max* L. Merr. | Soybean | Cotyledons | De novo adventitious shoots | 92 |
| *Lycopersicon chilense* Dun. | Wild tomato | Leaf disks | Direct or indirect organogenesis | 93 |
| *Lycopersicon esculentum* | Tomato | Leaf sections | Shoots from callus | 94 |
| *Malus* × *domestica* Borkh. | Apple | Leaves | Shoots from callus | 95 |
| *Medicago truncatula* | Barrel medic | Cotyledons plus split embryonic axis | Direct organogenesis | 96 |
| *Nicotiana alata* Link & Otto | Flowering tobacco | Hypocotyls | Organogenesis, no callus | 97 |
| *Oryza sativa* L. | Rice | Shoot meristems | Elongation of preformed meristems | 98 |
| *Pelargonium* × *domesticum* Dubonnet | Regal pelargonium | Leaf-lamina explants | Adventitious shoots from callus | 99 |
| *Pisum sativum* L. | Pea | Epicotyl segments or nodes | Organogenesis | 100 |
| *Pisum sativum* L. | Pea | Seed embryonic axis | Organogenesis | 101 |
| *Poncirus trifoliata* Raf. | Trifoliate orange | Epicotyl segments | Adventitious shoots | 102 |
| *Populus alba* × *P. grandidentata* | Hybrid poplar | Suspension cultures | Shoots from callus | 103 |

# Shoot Regeneration and Proliferation

| | | | | |
|---|---|---|---|---|
| *Rubus* | Blackberry, raspberry | Internodes | Organogenesis | 104 |
| *Solanum melongena* L. | Eggplant | Cotyledonary leaves | Organogenesis or embryogenesis | 105 |
| *Solanum tuberosum* L. | Potato | Leaf disks | Adventitious shoots with or without callus | 106 |
| *Solanum tuberosum* L. | Potato | Leaf strips | Shoots from callus | 107 |
| *Solanum tuberosum* L. | Potato | Stem internode sections | Shoots from callus | 108 |
| *Trifolium repens* L. | White clover | Cotyledons and apical shoot | Direct shoot organogenesis | 109 |
| *Trifolium subterraneum* L. | Subterranean clover | Hypocotyl segments | Adventitious shoots, organogenesis | 110 |
| *Ulmus procera* | English elm | Shoots or internodes | Shoots from tumors | 111 |
| *Vigna radiata* | Mung bean | Cotyledons | Organogenesis | 112 |

### *Agrobacterium rhizogenes*–Mediated Transformation

| | | | | |
|---|---|---|---|---|
| *Astragalus sinicus* | Chinese milk vetch | Seedlings | Callus, roots, shoots from roots | 113 |
| *Lotus angustissimus* L. | Slender bird's-foot-trefoil | Plantlets | Shoots from adventitious hairy root calli | 114 |

### Microprojectile–Mediated Transformation

| | | | | |
|---|---|---|---|---|
| *Avena sativa* L. | Oat | Shoot meristematic culture | Adventitious shoot meristems | 18 |
| *Catharanthus roseus* | Periwinkle | Nodal explants with axillary buds | Adventitious organogenesis | 21 |
| *Eucalyptus globulus* | Eucalyptus | Zygotic embryos | Shoots from callus | 23 |
| *Hordeum vulgare* L. | Barley | Shoot meristematic culture | Adventitious shoot meristems | 18 |
| *Phaseolus vulgaris* L. | Common bean | Seed meristems | De novo shoot formation | 115 |
| *Phaseolus vulgaris* L. | Bean | Shoot apices | Multiple shoot formation | 116 |
| *Zea mays* L. | Maize | Shoot apices | Shoot tip multiplication | 117 |

to deliver DNA to the target tissues. Particle bombardment was used to transform nodal explants of *Catharanthus roseus* with GFP or GUS reporter genes, and adventitious bud induction was achieved on medium containing 1 mg/l benzyladenine (BA) (21). A 98% survival rate of rooted shoots was also reported upon transfer to soil. The embryo axis of chickpea, devoid of the root meristem and shoot apex, was transformed with *Agrobacterium tumefaciens* and cultured on medium containing 3.0 mg/l BA and 0.004 mg/l naphthaleneacetic acid (NAA), wherein multiple shoots were induced (22). Zhang et al. (18) reported on the transformation of commercial cultivars of oat and barley from in vitro shoot meristematic cultures by using particle bombardment. Transformed plants for both cultivars were obtained.

### C. Callus-Mediated Production of Adventitious Shoots

The induction of adventitious shoots directly from calli is also a very useful approach for the in vitro proliferation of shoots. Since production of adventitious shoots directly from tissues is not always possible for many species, a callus-mediated phase is beneficial. Although the phenomenon of somaclonal variation has been attributed to callus-mediated cultures, the frequency of occurrence of such genetic instability varies from genotype to genotype (19). Nonetheless, callus or transformed tissue derived callus has been used for shoot induction for a number of plant species (Table 1). Serrano et al. (23) transformed zygotic embryos of *Eucalyptus globulus* and produced calli, which led to shoot formation by organogenesis. Internodes from in vitro grown plants of *Forsythia × intermedia* were transformed using *Agrobacterium* sp., and after an intermediary callus phase, shoots were induced (24). Studies have also been conducted to determine the effect of selection agents on shoot induction. Casas et al. (25) observed that callus induced from immature inflorescences transformed by particle bombardment underwent two pathways for shoot production upon selection with bialaphos for the bar gene. In the absence of bialaphos embryogenesis occurred preferentially, whereas the presence of bialaphos led to shoot induction mainly by organogenesis.

## IV. FACTORS AFFECTING IN VITRO SHOOT PRODUCTION AND PROLIFERATION

Although the production and proliferation of shoots in vitro are possible for many species, there are critical parameters to be followed in order to alter the developmental pathway of cultured cells and to regenerate plants efficiently. Interaction of explant tissues with components of culture medium alone is not sufficient to achieve this. A plethora of other factors such as explant type, growth conditions of donor plants, light, temperature, and humidity affect the culture process at every stage of in vitro regeneration. Furthermore, many studies have implicated the genotype effect on regeneration in vitro (6); the consequence is that a specific optimal set of conditions is required for each genotype. However, the time-consuming and labor-intensive nature of this optimization process is not practical. Instead, a consensus is reached on a set of culture medium compositions and conditions for use with genotypes of the same species, sometimes with minor modifications.

The process of in vitro shoot production can be broadly divided into three stages as follows (6): (a) shoot induction, (b) shoot development and proliferation, and (c) rhizogenesis from developed shoots. At each stage different sets of conditions may be required. For some genotypes or explants, all three stages can be effectively completed with one set of conditions. It is therefore imperative to set preliminary experiments to fine tune the whole process a priori, then proceed with actual large-scale experiments. The factors involved in the in vitro shoot proliferation process can, therefore, be divided into biotic and abiotic. Whereas biotic factors are associated with the genotype per se, abiotic factors comprise all the physical and chemical environments sustaining the in vitro culture process. Essentially, this is analogous to the genotype × environment interaction encountered by plant breeders and quantitative geneticists to account for the performance of a particular cultivar under different environments, albeit at a microenvironmental level in tissue culture.

## A. Biotic Factors

Although the concept of totipotency iterates that every cell in a plant is capable of regenerating a whole new plant, choice of the initial starting material is important. Explants for culture have been obtained from a number of sources such as leaves, stems, shoots, roots, flowers, or immature tissues. However, for a specific genotype one explant may be more responsive in culture than others. This is because the extent to which a differentiated cell can dedifferentiate depends on the cytological and physiological states it has attained (26), in conjunction with the genotypic background of the donor plant material (27,28). The physiological state can be due to the growth conditions of the donor plant and the origin, size, and relative maturity of the explant. Many reports have alluded to the impact of the physiological state of explants and/or whole donor plants on in vitro response. However, most of the studies have focused on factors affecting androgenesis and somatic embryogenesis. Nonetheless, those factors influence in vitro shoot development in a similar manner. For example, plants grown in growth chambers and greenhouses, as opposed to field-grown plants, have been found to be more responsive to in vitro regeneration. The fluctuation in temperature, incident radiation, humidity, exposure to pests and diseases, and nutrient levels would very easily affect the physiological state of the plant. Using immature embryos derived from four cultivars of *Zea mays* grown for three consecutive years in the field, Santos and Torné (29) showed variation in the production of totipotent callus within and among the cultivars over the three years.

### 1. Explant

The choice of the explant is crucial in that it may affect the success of the whole process leading to shoot production. Reference to the explant in terms of position on the donor plant, size, and maturity implicates the explant's physiological and developmental state. Generally, embryonic, meristematic, and reproductive tissues have been found to be more amenable to culture (30). A number of studies have been conducted to determine the efficiency of different explants for organogenesis. For example, in a comparative study of *Panax ginseng* explants obtained from leaves, petioles, flower stalks, and roots of in vitro grown plants, Lim et al. (31) found the petioles to be more suited for callus induction. Petiole-derived callus was subsequently used for adventitious shoot induction. The morphogenetic pathways of explants from *Citrus grandis* were found to vary with the type of the explant and medium composition (32). Multiple shoots were formed de novo from epicotyl and root segments, when cultured on MS medium supplemented with BA, although a lower level of BA was required for root segments. However, cotyledonary and leaf explants produced calli, which subsequently regenerated shoots. High frequencies of shoots were also obtained from shoot tips and nodal explants. Therefore, judicious selection of explants, in combination with specific culture medium types, would determine the pathway and efficiency of shoot production. Orientation and polarity of the explant on culture medium can also affect the developmental pathway of cultured cells. Garcia et al. (33) showed that epicotyl segments of Troyer citrange produced adventitious shoots directly from the apical end when placed vertically, with the basal end in the culture medium. However, when the explant was placed horizontally, callus induction occurred at both ends of the epicotyl and adventitious shoots developed from the calli when the culture medium was supplemented with BA.

## 2. Genotype

The process of shoot regeneration is further compounded by the genotypic influence inherent in the donor plant, the so-called genotype dependency. In 1975, Steward et al. (34) observed differences in the abundance and form of somatic embryos derived from four cultivars of carrot. Green and Phillips (35) found that of four cultivars of maize they studied, three produced shoots by organogenesis and one produced somatic embryos. Other studies have demonstrated varying frequencies of callus induction, somatic embryogenesis, androgenesis, and regeneration among different genotypes. Most of the studies have focused on genotypic effects with regard to androgenesis and somatic embryogenesis; reports on the genotypic influence on organogenesis are less widespread. Banerjee et al. (36) found variation in the rate of shoot bud proliferation from meristem tip cultures of eight triploid cultivars of *Musa* spp. The variation was suggested to be due to the presence of one or two B genomes. Higher frequency of bud proliferation tended to correlate with ABB or AAB genomic composition. The genetics and inheritance of organogenic potential have been studied in only few species (e.g., 37, 38). For example, in melon, the organogenic response was suggested to be under the control of two genes, which were partially dominant and segregated independently (37).

## B. Abiotic Factors

The influence of the physicochemical environment of the explant during culture is as important as that of the biotic factors. The composition of the culture medium and physical factors such as light, temperature, and humidity can greatly affect the developmental pathway of the cultured cells and preclude attainment of desired morphogenesis and morphogenic efficiency.

## 1. Culture Medium Composition

The basic composition of a culture medium includes the following: macro- and microelements, vitamins, plant growth substances, carbon source, and sometimes miscellaneous compounds. The macro- and microelements and vitamin constituents have remained fairly constant, although different formulations have been developed over the years to suit particular culture systems. The Murashige and Skoog (39) medium has been one of the most commonly used culture media since its inception. Prior to that, medium formulations devised by White (40) and Heller (41) were widely used. Many of the medium formulations, such as Eriksson (42), B5 (43), and Schenk and Hildebrandt (44), have evolved from the MS medium composition. One or more of these medium types are often tried in preliminary experiments to assess the efficiency of each.

## 2. Plant Growth Regulators

The role of plant growth regulators in in vitro cultures has been documented extensively. In fact, the pioneering work of Skoog and Miller (45), implicating the interaction between auxins and cytokinins in the initiation of roots and shoots, has made possible the in vitro regeneration of a wide array of plant species. It is now common knowledge that the concentrations and ratios of these two classes of compounds influence callus production, organogenesis, and somatic embryogenesis. For callus induction high levels of auxins are usually required, whereas shoot production generally requires higher levels of cytokinins. However, there are exceptions. The most commonly used auxins are 2,4-dichlorophenoxyacetic acid (2,4-D), naphthaleneacetic acid (NAA), and indole acetic acid (IAA). The commonly used cytokinins are kinetin and 6-benzyladenine (BA). Several other types of these plant growth substances, or analogues thereof, are also used. Compounds that are unrelated to auxins and cytokinins but produce similar effects have also been used.

The in vitro proliferation of shoots requires careful optimization of the levels of auxins and cytokinins at each stage during the culture process. Four potential pathways exist when an explant is inoculated onto a culture medium, viz., callogenesis, caulogenesis, rhizogenesis, and somatic embryogenesis; the latter three occur either directly or via callus mediation. Plant growth regulators play an important role in determining the outcome of these processes. A detailed account of plant growth regulator combinations and levels for in vitro proliferation of shoots is not given here, since these have been the subject of a number of research publications and reviews. In general, higher levels of cytokinins and lower levels of auxins have been found to induce multiple shoots. Other plant growth regulators such as gibberellins, abscisic acid, and ethylene have been used in conjunction with auxins and cytokinins for improvement of morphogenesis in vitro. The ability of a novel plant growth substance, thidiazuron (TDZ), to induce multiple shoots is now well documented (46,47). It has been claimed to have both auxinlike and cytokininlike activities (46), but its mode of action has not been fully elucidated. TDZ, alone or in combination with auxins or cytokinins, has been used in diverse species of plants for in vitro shoot proliferation. A combination of TDZ and BA (at 2 mg/l of each) was used for multiple shoot production from cotyledonary nodes of *Vicia faba* (48). When TDZ or BA was used alone, the frequency of shoot production was lower. Cotyledons and hypocotyl segments of *Glycine max* cultured on medium containing 2 mg/l TDZ produced more shoots than those cultured on medium containing 1.15 mg/l BA (49). TDZ has also been found to switch developmental pathways in cultures of *Cicer arietinum*, by supplementing the medium with an amino acid (50). MS medium supplemented with TDZ induced multiple shoots directly from cotyledonary notches of seedlings. When the MS medium was supplemented with L-proline, somatic embryos were mainly induced. TDZ has also been used to induce multiple shoots from a graminaceous species. Gupta and Conger (51) were able to induce multiple shoots from seedlings of switchgrass, *Panicum virgatum*. Another plant growth substance, brassin, which is a synthetic analogue of brassinolide (52), has also been implicated in in vitro shoot proliferation, albeit indirectly. Ponsamuel et al. (53) demonstrated that although caulogenesis was induced from plumular explants of *Arachis hypogaea*, most of the shoot buds were dormant. Conversion of these dormant shoots occurred upon transfer to a medium containing brassin, BA, and naphthoxyacetic acid. Polyamines have been demonstrated to play a role in morphogenesis, either directly or indirectly (for a review, see Refs. 54 and 55). Chi et al. (56) reported on the enhancement of de novo shoot formation from cotyledonary explants of *Brassica campestris* spp. *pekinensis*, in response to polyamines such as spermidine, spermine, and putrescine.

## 3. Carbon Source in Culture Media

Sucrose has been one of the most commonly used carbon sources in tissue cultures. Glucose has also been used to some extent. However, in recent years studies have demonstrated that maltose is a very effective carbohydrate source for improved response of plant tissues in culture (57,58). Besides providing the required energy source to the cultured tissues, carbohydrates have also been shown to contribute to the osmoticum of the medium (59).

## 4. Environmental Factors

Physical requirements such as light, temperature, and humidity are equally important in the process of in vitro growth and development. Culture response can be affected by duration of exposure to light and its intensity and quality. Although photosynthesis may not necessarily occur in cultured tissues, light may be required for other photomorphogenetic processes (60). Light requirements tend to be overlooked but can affect morphogenesis (61,62). For callus induction, cultures are generally incubated in the dark, wherein prolific callus development occurs. Low light intensity favors shoot-bud induction in tobacco (4,63). Organogenesis is also affected by exposure to continuous light (64), and generally cultures are incubated under a photoperiod of 16/8-hour light/dark cycle. The influence of light quality on shoot production has been studied to some extent with varying results. Although fluorescent lamps (cool white) are the most commonly used, sometimes better results have been obtained when combining incandescent lamps with cool white lamps. Schneider-Moldrickx (65) studied the effect of light from six different types of fluorescent lamps on adventitious shoot production from leaf explants of *Kalanchoë* sp. It was found that lamps that emitted mostly orange-red light increased the frequency of adventitious shoot formation, whereas lamps that principally emitted in the ultraviolet (UV) or near-UV range inhibited adventitious shoot formation.

Growth and differentiation of cultures in vitro are expected to proceed at altered rates at different temperatures. Cultures are usually incubated at 24°C–26°C. Ideally, for each genotype an optimal temperature for in vitro growth and development is required (62,66) and is generally 3°C–4°C higher than in vivo (10). Better shoot bud initiation was obtained when hypocotyl segments of *Linum usitatissimum* were incubated at 30°C (67). Skoog (4) studied the response of tobacco callus growth and differentiation at temperatures ranging from 5°C to 33°C. Callus growth increased with increasing temperatures, but no shoot buds were initiated at 33°C. The optimal temperature for shoot bud initiation was found to be 18°C. Higher temperatures, which may produce optimal explant differentiation, may potentially encourage growth of contaminants in cultures. In some cases, alternating temperatures may be required to induce growth and differentiation, much as in a photoperiod. This was demonstrated by Capite (68); best callus growth was obtained when cultures of *Helianthus tuberosus*, *Parthenocissus* sp., and carrot were incubated at 26°C during the day and at 20°C at night.

Humidity has not been given much attention with regard to in vitro shoot development. Culture vessels are sealed and are, therefore, expected to maintain a relative humidity of close to 100%. High relative humidity can lead to growth of contaminants. Besides humidity inside a culture vessel, gaseous components hovering over cultures can influence growth and development (62). A number of gases such as ethylene, ethanol, and acetaldehyde, in addition to oxygen and carbon dioxide, can be found in the culture vessels. Ethylene buildup inside culture vessels can either adversely or positively affect morphogenetic potential (69). Gonzalez et al. (70) showed that use of aminoethoxyvinylglycine (AVG) (an inhibitor of ethylene synthesis) in culture medium inhibited organogenesis from nodal explants of *Populus tremula*. Medium containing either 1-aminocyclopropane-1-carboxylic acid (ACC), a precursor of ethylene, or ethepon (ethrel, 2-chloroethyl phosphonic acid [CEPA]), which decomposes to release ethylene, induced organogenesis. These two compounds are involved in ethylene production and therefore indicate a positive effect of ethylene for shoot induction. A study by Chraibi et al. (71) demonstrated that addition of silver nitrate (an inhibitor of the physiological action of ethylene) to culture medium enhanced shoot production from cotyledonary cultures of *Helianthus annuus*. In the same study, cobalt chloride (an inhibitor of ethylene biosynthesis) stimulated shoot production.

## V. CONCLUDING REMARKS AND FUTURE PROSPECTS

The production of shoots in vitro is the culmination of a series of complex events triggered by physical and chemical stimuli perceived by receptors within the explant. Although the combination of these stimuli catalyzes the reprogramming of cells for dedifferentiation, only a judicious approach can determine the proper developmental fates of those cells via callogenesis, caulogenesis, rhizogenesis, or somatic embyogenesis. These events can occur independently or concurrently, or callogenesis may lead to the other three processes. In this review, several issues related to the production of shoots in vitro have been discussed. Besides the selection of the explants and cultural aspects for efficient shoot production, the genotypic constitution of the donor plants was implicated. Unfortunately, the latter cannot be obviated currently. To compound matters further, very often the most desirable genotypes of commercial value tend to be the least responsive in culture. The elucidation of this genotypic dependence has been the subject of many studies, particularly in anther culture responsiveness. Nonetheless, results from those studies may possibly be correlated to organogenic response. In fact, a report by Veronneau et al. (72) established a correlation between anther culture response and leaf disk culture response in *Solanum chacoense*. Two anther culture responsive clones and eight of their reciprocal $F_1$ hybrids were analyzed for anther and leaf disk culture response. Genetic analysis of the reciprocal hybrids revealed a significant correlation ($r = 0.82$) between callus induction from anthers and shoot induction from leaf disks. Therefore, under those specific culture conditions, it was suggested that the genetic control mechanism for these two types of cultures might be similar. Furthermore, the study also indicated estimates of broad-sense heritability to be 83% for leaf disk culture, indicating that the gene(s) for culture responsiveness may be transferred to nonresponsive genotypes with relative ease. More research on the genetic analysis of caulogenetic responsiveness needs to be conducted, so as to maximize the efficiency of this simple culture system.

Characterization of a number of genes involved in meristem-related expression has been achieved (73). The role of two of these genes is illustrated to highlight the potential of molecular tools and transformation technologies in dissecting developmental pathways to contribute to the understanding of genetic switches involved during in vitro shoot production. The gene cdc2Zm encodes a cyclin-dependent kinase involved in cell division (74) and the gene knotted1 (KN1) (75) encodes a protein associated with shoot meristem formation (76). Zhang et al. (77) studied the expression of these two genes in maize and their cross-reacting proteins in barley during in vitro axillary shoot meristem proliferation and adventitious shoot formation. Expression of CDC2Zm approximately corresponded with in vitro cell proliferation. Also, in meristematic domes its expression was initiated during in vitro proliferation. Expression of KN1, or its homologue, was localized in meristematic cells during proliferation of axillary shoots in vitro. Cells in the proliferating meristematic domes expressing KN1, or its homologue, seemed to form multiple adventitious shoot meristems. In transgenic maize, leaves overexpressing KN1 did not lead to initiation of adventitious shoot meristem on their surfaces. However, ectopic expression of KN1 was observed in leaves of *Arabidopsis* sp. (78) and tobacco (79). It was inferred that KN1 alone was not responsible for adventitious shoot meristem formation from in vitro proliferating axillary shoot meristems in maize.

For the transgene technology to be more effective, the advantages offered by in vitro shoot proliferation systems would be of great value. The transfer of genes to explants from a number of species used for in vitro adventitious shoot induction is now possible by particle bombardment and *Agrobacterium* sp.–mediated techniques. Refinement of the methodologies and assessment of possible genotypic instability due to insertion of the gene and/or culture process must be undertaken.

## ACKNOWLEDGMENTS

Professor C.E. (Don) Palmer (University of Manitoba) and Dr. Patricia Polowick are gratefully acknowledged for the review of this manuscript.

## REFERENCES

1. RJ Gautheret. Sur la possibilité de réaliser la culture indéfinie des tissus de tubercules de carotte. C R Acad Sci, 208:118–120, 1939.
2. P Nobécourt. Sur les radicelles naissant des cultures de tissus du tubercule de carotte. C R Séances Soc Biol Ses Fil 130:1271–1272, 1939.
3. PR White. Potentially unlimited growth of excised plant callus in an artificial medium. Am J Bot 26:59–64, 1939.
4. F Skoog. Growth and organ formation in tobacco tissue cultures. Am J Bot 31:9–24, 1944.
5. F Skoog, C Tsui. Growth substances and the formation of buds in plant tissues. In: F Skoog, ed. Plant Growth Substances. Madison: University of Wisconsin Press, 1951, pp 263–285.
6. TA Thorpe. Morphogenesis and regeneration. In: IK Vasil, TA Thorpe, eds. Plant Cell and Tissue Culture. Dordrecht: Kluwer Academic, 1994, pp 17–36.
7. SS Bhojwani, MK Razdan, eds. Plant Tissue Culture: Theory and Practice, Developments in Crop Science, Vol. 5. New York: Elsevier, 1983.
8. RA Kerstetter, S Hake. Shoot meristem formation in vegetative development: The Pl. Cell 9:1001–1010, 1997.
9. JG Torrey. The initiation of organized development in plants. In: M Abercrombie, J Brachet, eds. Advances in Morphogenesis. Vol. 5. New York: Academic Press, 1966, pp. 39–91.
10. RLM Pierik. In vitro culture of higher plants. Dordrecht: Martinus Nijhoff, 1987, p 183.
11. CY Hu, PJ Wang. Meristem, shoot tip and bud cultures. In: DA Evans, WR Sharp, PV Ammirato, Y Yamada, eds. Handbook of Plant Cell Culture. Vol. 1. Techniques for Propagation and Breeding, New York: MacMillan, 1983, pp 177.
12. J Gould, M Devey, O Hasegawa, EC Ulian, G Peterson, RH Smith. Transformation of Zea Mays L. using Agrobacterium tumefaciens and the shoot apex. Plant Physiol 95:426–434, 1991.
13. EC Ulian, RH Smith, JH Gould, TD McKnight. Transformation of plants via the shoot apex. In Vitro Cell Dev Biol Plant 24:951–954, 1988.
14. G Hussey, RD Johnson, S Warren. Transformation of meristematic cells in the shoot apex of cultured pea shoots by *Agrobacterium tumefaciens* and *A. rhizogenes*. Protoplasma 148:101–105, 1989.
15. B Schrammeijer, PC Sijmons, PJM van den Elzen, A Hoekema. Meristem transformation of sunflower via *Agrobacterium*. Plant Cell Rep 9:55–60, 1990.
16. C Sautter, H Waldner, G Neuhaus-Url, A Galli, G Neuhaus, I Potrykus. Microtargeting: High efficiency gene transfer using a novel approach for the acceleration of micro-projectiles. Biotechnology 9:1080–1085, 1991.
17. VA Iglesias, A Gisel, R Bilang, N Leduc, I Potrykus, C Sautter. Transient expression of visible marker genes in meristem cells of wheat embryos after ballistic microtargeting. Planta 192:84–91, 1994.
18. S Zhang, M-J Cho, T Koprek, R Yun, P Bregitzer, PG Lemaux. Genetic transformation of commercial cultivars of oat (*Avena sativa* L.) and barley (*Hordeum vulgare* L.) using in vitro shoot meristematic cultures derived from germinated seedlings. Plant Cell Rep 18:959–966, 1999.
19. A Karp. Origins, causes and uses of variation in plant tissue cultures. In: IK Vasil, TA Thorpe, eds. Plant Cell and Tissue Culture. Dordrecht: Kluwer Acad Publ, 1994, pp 139–151.
20. PJ Larkin, WR Scowcroft. Somaclonal variation—a novel source of variability from cell cultures for plant improvement. Theor Appl Genet 60:197–214, 1981.
21. R Zárate, J Memelink, R van der Heijden, R Verpoorte. Genetic transformation via particle bombardment of *Catharanthus roseus* plants through adventitious organogenesis of buds. Biotech Lett 21:997–1002, 1999.
22. S Kar, TM Johnson, P Nayak, SK Sen. Efficient transgenic plant regeneration through *Agrobacterium*-mediated transformation of chickpea (*Cicer arietinum* L.). Plant Cell Rep 16:32–37, 1996.
23. L Serrano, F Rochange, JP Semblat, C Marque, C Teulières, A-M Boudet. Genetic transformation of *Eucalyptus globulus* through biolistics: complementary development of procedures for organogenesis from zygotic embryos and stable transformation of corresponding proliferating tissue. J Exp Bot 47:285–290, 1996.

## Shoot Regeneration and Proliferation

24. C Rosati, A Cadic, JP Renou, M Duron. Regeneration and *Agrobacterium*-mediated transformation of *Forsythia* × *intermedia* "Spring Glory." Plant Cell Rep 16:114–117, 1996.
25. AM Casas, AK Kononowicz, TG Haan, LY Zhang, DT Tomes, RA Bressan, PM Hasegawa. Transgenic sorghum plants obtained after microprojectile bombardment of immature inflorescences. In Vitro Cell Dev Biol Plant 33:92–100, 1997.
26. RJ Gautheret. Factors affecting differentiation of plant tissues grown in vitro. In: W Beerman, ed. Cell Differentiation and Morphogenesis. Amsterdam: North-Holland, 1966, pp 55–95.
27. JM Dunwell. Influence of genotype and environment on growth of barley embryos in vitro. Ann Bot 48:535–542, 1981.
28. C Raquin. Genetic control of embryo production and embryo quality in anther culture of Petunia. Theor Appl Genet 63:151–154, 1982.
29. MA Santos, JM Torné. A comparative analysis between totipotency and growth environment conditions of the donor plants in tissue culture of *Zea mays* L. J Plant Physiol 123:299–305, 1986.
30. PV Ammirato. Control and expression of morphogenesis in culture. In: LA Withers, PG Alderson, eds. Plant Tissue Culture and Its Agricultural Applications. London: Butterworths, 1986, pp 23–45.
31. H-T Lim, H-S Lee, T Eriksson. Regeneration of Panax ginseng C.A. Meyer by organogenesis and nuclear DNA analysis of regenerants. Plant Cell Tissue Org Cult 49:179–187, 1997.
32. CJ Goh, GE Sim, CL Morales, CS Loh. Plantlet regeneration through different morphogenic pathways in pommelo tissue culture. Plant Cell Tissue Org Cult 43:301–303, 1995.
33. LA Garcia, Y Bordon, JM Moreira-Dias, RV Molina, JL Guardiola. Explant orientation and polarity determine the morphogenic response of epicotyl segments of Troyer citrange. Ann Bot 84:715–723, 1999.
34. HE Steward, HW Israel, RL Mott, HJ Wilson, AD Krikorian. Observations on growth and morphogenesis in cultured cells of carrot (*Daucus carota* L.). Philos Trans R Soc 273:33–53, 1975.
35. CE Green, RL Phillips. Plant regeneration from tissue cultures of maize. Crop Sci 15:417–421, 1975.
36. N Banerjee, D Vuylsteke, EAL de Langhe. Meristem tip culture of *Musa*: Histomorphological studies of shoot bud proliferation. In: LA Withers, PG Alderson, eds. Plant Tissue Culture and Its Agricultural Applications, University of Nottingham School of Agriculture, London: Butterworths, 1986, pp 139–147.
37. RV Molina, F Nuez. The inheritance of organogenic response in melon. Plant Cell Tissue Org Cult 46:251–256, 1996.
38. A Sarrafi, JP Roustan, J Fallot, G Alibert. Genetic analysis of organogenesis in the cotyledons of zygotic embryos of sunflower (*Helianthus annuus* L.). Theor Appl Genet 92:225–229, 1996.
39. T Murashige, F Skoog. A revised medium for rapid growth and bioassays with tobacco tissue cultures. Physiol Plant 15:473–497, 1962.
40. PR White. Nutrient deficiency studies and an improved inorganic nutrient for cultivation of excised tomato roots. Growth 7:53–65, 1943.
41. R Heller. Recherches sur la nutrition minérale des tissus végétaux cultivés in vitro. Ann Sci Natl Biol Veg 14:1–223, 1953.
42. T Eriksson. Studies on growth requirements and growth measurements of cell cultures of *Haplopappus gracilis*. Physiol Plant 18:976–993, 1965.
43. OL Gamborg, RA Miller, K Ojima. Nutrient requirements of suspension cultures of soybean root cells. Exp Cell Res 50:151–158, 1968.
44. RU Schenk, AC Hildebrandt. Medium and techniques for induction and growth of monocotyledonous and dicotyledonous plant cell cultures. Can J Bot 50:199–204, 1972.
45. F Skoog, CO Miller. Chemical regulation of growth and organ formation in plant tissues cultured in vitro. Symp Soc Exp Biol 11:118–131, 1957.
46. BNS Murthy, SJ Murch, PK Saxena. Thidiazuron: A potent regulator of in vitro plant morphogenesis. In Vitro Cell Dev Biol Plant 34:267–275, 1998.
47. CA Huetteman, JE Preece. Thidiazuron: A potent cytokinin for woody plant tissue culture. Plant Cell Tissue Org Cult 33:105–119, 1993.
48. MM Khalafalla, K Hattori. A combination of thidiazuron and benzyladenine promotes multiple

shoot production from cotyledonary node explants of faba bean (*Vicia faba* L.). Plant Growth Reg 27:145–148, 1999.
49. Y Kaneda, Y Tabei, S Nishimura, K Harada, T Akihama, K Kitamura. Combination of thidiazuron and basal media with low salt concentrations increases the frequency of shoot organogenesis in soybeans [*Glycine max* (L.) Merr.]. Plant Cell Rep 17:8–12, 1997.
50. BNS Murthy, J Victor, RP Singh, RA Fletcher, PK Saxena. In vitro regeneration of chickpea (*Cicer arietinum* L.): Stimulation of direct organogenesis and somatic embryogenesis by thidiazuron. Plant Growth Reg 19:233–240, 1996.
51. SD Gupta, BV Conger. In vitro differentiation of multiple shoot clumps from intact seedlings of switchgrass. In Vitro Cell Dev Biol Plant 34:196–202, 1998.
52. TH Maugh. New chemicals promise larger crops. Science 212:33–34, 1981.
53. J Ponsamuel, DV Huhman, BG Cassidy, D Post-Beittenmiller. In vitro regeneration via caulogenesis and brassin-induced shoot conversion of dormant buds from plumular explants of peanut (*Arachis hypogaea* L. cv 'Okrun'). Plant Cell Rep 17:373–378, 1998.
54. N Bagni, P Torrigiani. Polyamines: A new class of growth substances. Curr Plant Sci Biotechnol Agric 13:264–275, 1992.
55. N Bagni, MM Altamura, S Biondi, M Mengoli, P Torrigiani. Polyamines and morphogenesis in normal and transgenic plant cultures. In: KA Roubelakis-Angelakis, K Tran Thanh Van, eds. Morphogenesis in Plants. New York: Plenum Press, 1993, pp 89–111.
56. GL Chi, WS Lin, JEE Lee, EC Pua. Role of polyamines on de novo shoot morphogenesis from cotyledons of *Brassica campestris* ssp. *pekinensis* (Lour) Olsson in vitro. Plant Cell Rep 13:323–329, 1994.
57. N Batty, JM Dunwell. Effect of maltose on the response of potato anthers in culture. Plant Cell Tissue Org Cult 18:221–226, 1989.
58. BR Orshinsky, LJ McGregor, GEI Johnson, P Hucl, KK Kartha. Improved embryoid induction and green shoot regeneration from wheat anthers cultured in medium with maltose. Plant Cell Rep 9:365–369, 1990.
59. DCW Brown, DWM Leung, TA Thorpe. Osmotic requirement for shoot formation in tobacco callus. Physiol Plant 46:36–41, 1979.
60. T Murashige. Plant propagation through tissue culture. Annu Rev Plant Physiol 25:135–166, 1974.
61. TA Thorpe. Organogenesis in vitro: Structural, physiological and biochemical aspects. Int Rev Cytol Suppl 11A:71–111, 1980.
62. KW Hughes. In vitro ecology: Exogenous factors affecting growth and morphogenesis in plant tissue cultures. Environ Exp Bot 21:281–288, 1981.
63. TA Thorpe, T Murashige. Some histochemical changes underlying shoot initiation in tobacco callus cultures. Can J Bot 48:277–285, 1970.
64. SK Pillai. Alternating light and dark periods in the differentiation of geranium callus cultures. In: Seminar on Plant Morphogenesis. Delhi: University of Delhi, 1968, pp 66–68.
65. RC Schneider-Moldrickx. The influence of light quality and light intensity on regeneration of *Kalanchoë blossfeldiana* hybrids in vitro. Acta Hortic 131:163–170, 1983.
66. V Chalupa. Temperature. In: JM Bonga, DJ Durzan, eds. Cell and tissue culture in forestry. Vol. 1. Dordrecht: Martinus Nijhoff, 1987, pp 142–151.
67. BE Murray, RJ Handyside, WA Keller. In vitro regeneration of shoots on stem explants of haploid and diploid flax (*Linum usitatissimum*). Can J Genet Cytol 19:177–186, 1977.
68. LD Capite. Action of light and temperature on growth of plant tissue cultures in vitro. Am J Bot 42:869–873, 1955.
69. PP Kumar, P Lakshmanan, TA Thorpe. Regulation of morphogenesis in plant tissue culture by ethylene. In Vitro Cell Dev Biol Plant 34:94–103, 1998.
70. A Gonzalez, L Arigita, J Majada, R Sanchez-Tames. Ethylene involvement in in vitro organogenesis and plant growth of *Populus tremula* L. Plant Growth Reg 22:1–6, 1997.
71. BKM Chraibi, A Latche, JP Roustan, J Fallot. Stimulation of shoot regeneration from cotyledons of *Helianthus annus* by the ethylene inhibitors, silver and cobalt. Plant Cell Rep 10:204–207, 1991.
72. H Veronneau, G Lavoie, M Cappadocia. Genetic analysis of anther and leaf disc culture in two clones of *Solanum chacoense* Bitt. and their reciprocal hybrids. Plant Cell Tissue Org Cult 30:199–209, 1992.

73. JI Medford. Vegetative and apical meristems. Plant Cell 4:1029–1039, 1992.
74. J Colosanti, SO Cho, S Wick, V Sundaresan. Localization of the functional p34cdc2 homologue of maize in root tip and stomatal complex cells: association with predicted division sites. Plant Cell 5:1101–1111, 1993.
75. S Hake, E Vollbrecht, M Freeling. Cloning Knotted, the dominant morphological mutant in maize using Ds2 as a transposon tag. EMBO J 8:15–22, 1989.
76. LG Smith, D Jackson, S Hake. Expression of Knotted1 marks shoot meristem formation during maize embryogenesis. Dev Genet 16:344–348, 1995.
77. S Zhang, R Williams-Carrier, D Jackson, PG Lemaux. Expression of CDC2Zm and KNOTTED1 during in-vitro axillary shoot meristem proliferation and adventitious shoot meristem formation in maize (*Zea Mays* L.) and barley (*Hordeum vulgare* L.). Planta 204:542–549, 1998.
78. G Chuck, C Lincoln, S Hake. KNAT1 induces lobed leaves with ectopic meristems when overexpressed in *Arabidopsis*. Plant Cell 8:1277–1289, 1996.
79. N Sinha, R Williams, S Hake. Overexpression of the maize homeobox gene, Knotted1, causes a switch from determinate to indeterminate cell fates. Genes Dev 7:787–795, 1993.
80. M Egnin, A Mora, CS Prakash. Factors enhancing Agrobacterium tumefaciens-mediated gene transfer in peanut (*Arachis hypogaea* L.). In Vitro Cell Dev Biol Plant 34:310–318, 1998.
81. I Stefanov, S Fekete, L Bögre, J Pauk, A Fehér, D Dudits. Differential activity of the mannopine synthase and the CAMV 35S promoters during development of transgenic rapeseed plants. Plant Sci 95:175–186, 1994.
82. L-C Ding, C-Y Hu, K-W Yeh, P-J Wang. Development of insect-resistant transgenic cauliflower plants expressing the trypsin inhibitor gene isolated from local sweet potato. Plant Cell Rep 17:854–860, 1998.
83. TD Metz, R Dixit, ED Earle. Agrobacterium tumefaciens-mediated transformation of broccoli (*Brassica oleracea* var. *italica*) and cabbage (*B. oleracea* var. *capitata*). Plant Cell Rep 15:287–292.
84. F Frulleux, G Weyens, M Jacobs. Agrobacterium tumefaciens-mediated transformation of shoot-buds of chicory. Plant Cell Tissue Org Cult 50:107–112, 1997.
85. L Peña, M Cervera, J Juárez, A Navarro, JA Pina, L Navarro. Genetic transformation of lime (*Citrus aurantifolia* Swing.): factors affecting transformation and regeneration. Plant Cell Rep 16:731–737, 1997.
86. L Peña, M Cervera, J Juárez, A Navarro, JA Pina, N Durán-Vila, L Navarro. Agrobacterium-mediated transformation of sweet orange and regeneration of transgenic plants. Plant Cell Rep 14:616–619, 1995.
87. L Peña, M Cervera, J Juárez, C Ortega, JA Pina, N Durán-Vila, L Navarro. High efficiency *Agrobacterium*-mediated transformation and regeneration of citrus. Plant Sci 104:183–191, 1995.
88. SE Ledger, SC Deroles, NK Given. Regeneration and *Agrobacterium*-mediated transformation of chrysanthemum. Plant Cell Rep 10:195–199, 1991.
89. MR Boase, JM Bradley, NK Borst. Genetic transformation mediated by Agrobacterium tumefaciens of florists' chrysanthemum (*Dendranthema* × *grandiflorum*) cultivar 'Peach Margaret'. In Vitro Cell Dev Biol Plant 34:46–51, 1998.
90. NS Nehra, RN Chibbar, KK Kartha, RSS Datla, WL Crosby, C Stushnoff. Genetic transformation of strawberry by Agrobacterium tumefaciens using a leaf disk regeneration system. Plant Cell Rep 9:293–298, 1990.
91. CA Meurer, RD Dinkins, GB Collins. Factors affecting soybean cotyledonary node transformation. Plant Cell Rep 18:180–186, 1998.
92. MAW Hinchee, DV Connor-Ward, CA Newell, RE McDonnell, SJ Sato, CS Gasser, DA Fischhoff, DB Re, RT Fraley, RB Horsch. Production of transgenic soybean plants using *Agrobacterium*-mediated DNA transfer. Biotechnology 6:915–922, 1988.
93. Z Agharbaoui, AF Greer, Z Tabaeizadeh. Transformation of the wild tomato Lycopersicon chilense Dun. by Agrobacterium tumefaciens. Plant Cell Rep 15:102–105, 1995.
94. S McCormick, J Niedermeyer, J Fry, A Barnason, R Horsch, R Fraley. Leaf disc transformation of cultivated tomato (*L. esculentum*) using *Agrobacterium tumefaciens*. Plant Cell Rep 5:81–84, 1986.
95. A De Bondt, K Eggermont, I Pennickx, I Goderis, WF Broekaert. *Agrobacterium*-mediated transformation of apple (*Malus* × *domestica* Borkh.): an assessment of factors affecting regeneration of transgenic plants. Plant Cell Rep 15:549–554, 1996.

96. AT Trieu, MJ Harrison. Rapid transformation of Medicago truncatula: Regeneration via shoot organogenesis. Plant Cell Rep 16:6–11, 1996.
97. PR Ebert, AE Clarke. Transformation and regeneration of the self-incompatible species *Nicotiana alata* Link & Otto. Plant Mol Biol 14:815–824, 1990.
98. SH Park, SRM Pinson, RH Smith. T-DNA integration into genomic DNA of rice following *Agrobacterium* inoculation of isolated shoot apices. Plant Mol Biol 14:815–824, 1996.
99. MR Boase, JM Bradley, NK Borst. An improved method for transformation of regal pelargonium (*Pelargonium* × *domesticum* Dubbonnet) by *Agrobacterium tumefaciens*. Plant Sci 139:59–69, 1998.
100. A De Kathen, H-J Jacobsen. *Agrobacterium* tumefaciens-mediated transformation of *Pisum sativum* L. using binary and cointegrate vectors. Plant Cell Rep 9:276–279, 1990.
101. HE Schroeder, AH Schotz, T Wardley-Richardson, D Spencer, TJV Higgins. Transformation and regeneration of two cultivars of pea (*Pisum sativum* L.). Plant Physiol 101:751–757, 1993.
102. J Kaneyoshi (Hiramatsu), S Kobayashi, Y Nakamura, N Shigemoto, Y Doi. A Simple and efficient gene transfer system of trifoliate orange (*Poncirus trifoliata* Raf.) Plant Cell Rep 13:541–545, 1994.
103. GT Howe, B Goldfarb, SH Strauss. Agrobacterium-mediated transformation of hybrid poplar suspension cultures and regeneration of transformed plants. Plant Cell Tissue Org Cult 36:59–71, 1994.
104. MA Hassan, HJ Swartz, G Inamine, P Mullineaux. Agrobacterium tumefaciens-mediated transformation of several *Rubus* genotypes and recovery of transformed plants. Plant Cell Tissue Org Cult 33:9–17, 1993.
105. M Fári, I Nagy, M Csányi, J Mitykó, A Andrásfalvy. Agrobacterium mediated genetic transformation and plant regeneration via organogenesis and somatic embryogenesis from cotyledon leaves in eggplant (*Solanum melongena* L. cv. 'Kecskeméti lila'). Plant Cell Rep 15:82–86, 1995.
106. R Tavazza, M Tavazza, RJ Ordas, G Ancora, E Benvenuto. Genetic transformation of potato (*Solanum tuberosum*): an efficient method to obtain transgenic plants. Plant Sci 59:175–181, 1988.
107. H Wenzler, G Mignery, G May, W Park. A Rapid and efficient transformation method for the production of large numbers of transgenic potato plants. Plant Sci 63:79–85, 1989.
108. CA Newell, R Rozman, MA Hinchee, EC Lawson, L Haley, P Sanders, W Kaniewski, NE Tumer, RB Horsch, RT Fraley. Agrobacterium-mediated transformation of Solanum tuberosum L. cv. 'Russet Burbank'. Plant Cell Rep 10:30–34, 1991.
109. CR Voisey, DWR White, B Dudas, RD Appleby, PM Ealing, AG Scott. Agrobacterium-mediated transformation of white clover using direct shoot organogenesis. Plant Cell Rep 13:309–314, 1994.
110. MRI Khan, LM Tabe, LC Heath, D Spencer, TJV Higgins. *Agrobacterium*-mediated transformation of subterranean clover (*Trifolium subterraneum* L.). Plant Physiol 105:81–88, 1994.
111. TM Fenning, SS Tymens, JS Gartland, CM Brasier, KMA Gartland. Transformation and regeneration of English elm using wild-type *Agrobacterium tumefaciens*. Plant Sci 116:37–46, 1996.
112. M Pal, U Ghosh, M Chandra, A Pal, BB Biswas. Transformation and regeneration of mung bean (*Vigna radiata*). Indian J Biochem Biophys 28:449–455, 1991.
113. H-J Cho, JM Widholm, N Tanaka, Y Nakanishi, Y Murooka. *Agrobacterium* rhizogenes-mediated transformation and regeneration of the legume *Astragalus sinicus* (Chinese milk vetch). Plant Sci 138:53–65, 1998.
114. E Nenz, F Pupilli, F Paolocci, F Damiani F, CA Cenci, S Arcioni. Plant regeneration and genetic transformation of *Lotus angustissimus*. Plant Cell Tissue Org Cult 45:145–152, 1996.
115. DR Russell, KM Wallace, JH Bathe, BJ Martinell, DE McCabe. Stable transformation of *Phaseolus vulgaris* via electric-discharge mediated particle acceleration. Plant Cell Rep 12:165–169, 1993.
116. JL Aragao-Francisco, L Rech-Elibio. Morphological factors influencing recovery of transgenic bean plants (*Phaseolus vulgaris* L.) of a *Carioca* cultivar. Int J Plant Sci 158(2):157–163, 1997.
117. H Zhong, B Sun, D Warkentin, S Zhang, R Wu, T Wu, MB Sticklen. The competence of maize shoot meristems for integrative transformation and inherited expression of transgenes. Plant Physiol 110:1097–1107, 1996.

# 5
# Techniques for Gene Marking, Transferring, and Tagging

**Albert Abbott**
*Clemson University, Clemson, South Carolina*

| | | |
|---|---|---|
| I. | MOLECULAR MARKER SYSTEMS | 85 |
| II. | PROTEIN MARKER SYSTEMS | 86 |
| III. | DNA MARKER SYSTEMS | 86 |
| | A. Hybridization-Based Markers | 86 |
| | B. DNA Amplification Fingerprinting (DAF) | 88 |
| IV. | TAGGING SIMPLE TRAITS | 90 |
| V. | TAGGING COMPLEX TRAITS | 90 |
| | A. Quantitative Trait Loci Mapping | 90 |
| VI. | COMPARATIVE GENOMICS | 91 |
| | REFERENCES | 91 |

## I. MOLECULAR MARKER SYSTEMS

The discovery that allelic forms of enzymes (isozymes) could be separated electrophoretically on gels and detected with histochemical activity stains heralded the introduction of molecular marker technology into the field of genetics (Smithies, 1955; Hunter and Markert, 1957). With these technologies, it was no longer necessary to have a visible change in the phenotype of the organism to identify a marker locus. This significantly increased the number of markers identifiable in genetic material and made possible the production of highly saturated genetic maps for use in marker-assisted breeding, gene transfer, and genetic manipulation of crop species. These marker systems have now become the major tools for genetic analysis. Depending on the molecular technology employed, the markers can be highly abundant, phenotypically neutral, and detectable at early stages of growth and can show no environmental effects on detectability. These markers have been employed for deoxyribonucleic acid (DNA) fingerprinting, for construction of genetic linkage maps, for tagging of genes controlling certain traits, and as molecular landmarks for map-based cloning of genes.

## II. PROTEIN MARKER SYSTEMS

One of the earliest molecular marker technologies was the application of gel electrophoresis and activity staining techniques for the visualization of differently charged enzymatic forms of particular enzymes (Markert and Moller, 1959). Isoenzymatic forms of particular enzymes (isozymes) are derived from genetic mutations that change the primary structure of the individual peptides of the protein, thereby producing allelic variants that can be distinguished by protein gel electrophoresis technologies. Isozyme markers have been developed in many species such as *Phaseolus vulgaris* (Vallejos et al., 1992), barley (Zhang et al., 1993), maize (Gardiner et al., 1993), grape (Lodhi et al., 1995), peach (Messeguer et al., 1987; Chaparro et al., 1994; Foolad et al., 1995), apple (Lawson et al., 1995), almond (Vezvaei et al., 1995), and sweet cherry (Granger, 1996). Isozyme markers have been used for genetic applications (Causse et al., 1994; Benito et al., 1994; Freyre and Douches, 1994; Ragot et al., 1995) and for assessment of genetic diversity among different species (Simonsen and Heneen, 1995; Maas and Klaas, 1995; Sonnante et al., 1994; Stalker et al., 1994). However; the paucity of isozyme loci and the fact that they are subject to posttranslational modifications often restricts their utility (Staub et al., 1996).

## III. DNA MARKER SYSTEMS

The utilization of DNA-based genetic markers has signaled a new era in genome analysis. DNA polymorphisms are more abundant than conventional phenotypic and biochemical markers, enabling saturated maps to be developed in a single segregating population. This abundance of molecular markers throughout the genome greatly facilitates the development of highly saturated molecular marker maps that in turn allow the tagging of quantitative trait loci (QTL), as well as those controlled by single genes. Molecular marker-based genetic maps facilitate the development and use of indirect selection schemes for germplasm improvement. Such strategies strive to increase precision and efficiency in the manipulation of both qualitative and quantitative traits.

There are two major technologies that have been employed to detect DNA polymorphism: (a) molecular hybridization, which employs the use of specific probes to detect polymorphic restriction enzyme fragments; (b) the polymerase chain reaction (PCR) where oligonucleotide primers are used to amplify polymorphic DNA fragments. In either case, polymorphic fragments are subsequently linked to traits of interest and serve as molecular markers for tagging and transfer of genes in conventional breeding schemes. Each technology has its merit and limitations in specific applications.

### A. Hybridization-Based Markers

#### 1. Restriction Fragment Length Polymorphism Analysis

Historically restriction fragment length polymorphism (RFLP) analysis paved the way for the implementation of DNA marker-based systems for genetic mapping and gene tagging purposes. This technology couples the use of restriction enzyme digestion of genomic DNA, southern transfer, and molecular hybridization for the detection of specific regions of genomic DNA that have undergone sequence change through mutation. The markers revealed are highly reproducible in different laboratories, generally codominant, easily visualized, and theoretically abundant. However, the process requires large amounts of genomic DNA and restriction enzymes, requires several days for display and detection of polymorphism, and, depending on the degree of genetic divergence of individuals under study, may require numerous hybridizations to detect a limited number of polymorphic loci. RFLP based markers are highly portable within species and, depending on the nature of the RFLP probe (gene encoding or random genomic sequences), may be portable across much greater taxonomic distances.

Mapping of chromosomes in a number of species belonging to a wide variety of plant families has been accomplished through the implementation of RFLP marker systems [e.g., Brassicaceae (Chang et al., 1988; Song et al., 1988a, 1988b; Ferreira et al., 1995; van Denzye et al., 1995), Fabaceae (Apuya et al., 1988; Keim et al., 1990; Kochert et al., 1991; McCoy et al., 1991), Poaceae (Helentjaris et al., 1986; McCouch et al., 1988) Ragab et al., 1994; Causse et al., 1994; Yu et al., 1995; Galiba et al., 1995), Pinaceae (Neale and Williams, 1991), Solanaceae (Bonierbale et al., 1988; Tanksley et al., 1988; Gebhardt et al., 1989; McLean et al., 1990; Messegeur et al., 1991), Rosaceae (Nybom and Schall, 1990; Eldredge et al., 1992; Foolad et al., 1995; Rajapakse et al. 1995; Lawson et al., 1995), Malvaceae (Reinisch et al., 1994), Liliaceae (Restivo et al., 1995), and Chenopodiaceae (Pillen et al., 1992). In many of these maps, economically important genes have been tagged, e.g., downy mildew resistance genes in lettuce (Paran and Michelmore, 1993) and a gene for resistance to tobacco mosaic virus in tomato (Young et al., 1988).

Clearly this technology has not only been instrumental for gene tagging and marker mapping, but has been fundamental to the development of our current understanding of plant genome structure and evolution.

## 2. Single Nucleotide Polymorphisms

As sequence databases become more complete for model organisms, it becomes possible to utilize DNA chip hybridization technologies to detect single nucleotide polymorphisms with unprecedented efficiency (Chee et al., 1996; Lashkari et al., 1997). This approach has already been demonstrated to be of great value in yeast and human genomics (Chee et al., 1996; Lashkari et al., 1997; Wang et al., 1998; Winzeler et al., 1998) and holds great promise for application in mutation analysis and gene discovery in other systems.

## 3. DNA Amplification-Based Markers

With the development of polymerase chain reaction (PCR) technologies (Mullis et al., 1986) alternative approaches to hybridization-based detection of DNA polymorphism were possible. In general, these approaches utilize randomly or specifically derived primer sequences to amplify regions of the genome that may display polymorphic differences among individuals. Polymorphic amplification products are then used as markers for linkage mapping and tagging of traits of interest. PCR-based technologies have the advantages of being easily automated, capable of high throughput, suited to easy transfer and application of marker sequences from laboratory to laboratory, and, depending on the marker system, relatively inexpensive in cost, time, and effort. However, they can present problems of reproducibility, contaminate amplification, and are characterized by low level of transferability between different crosses; in most cases markers are dominant in genetic nature.

## 4. Random Amplified Polymorphic DNA

One of the first applications of PCR methods to detection of polymorphic regions of the genome utilized a single short (e.g., ≤10 nt) oligonucleotide primer of arbitrary sequence to amplify regions of genomic DNA that by chance have this sequence and its inverse, separated by a sufficient number of nucleotides to produce a visible PCR product. The presence of this sequence and its inverse within amplifiable distance of each other usually occurs within the genome in a number of places, producing a primer/DNA template-specific amplification pattern of fragments. Polymorphism is detected by the presence and absence of amplified products among individual DNA samples. This method, referred to *as random amplified polymorphic DNA* (RAPD) *analysis,* was developed independently in two laboratories (Welsh and McClelland, 1990; Williams et al., 1990). This approach has many advantages. It requires very little DNA, has a high throughput, is easily automated, and utilizes relatively simple gel detection systems. However, in most cases the markers are dominant markers, may not transfer well between crosses, and may suffer from low reproducibility in different laboratories. Some of these limitations have been overcome by cloning and sequencing polymorphic RAPD fragments to develop sequence-specific primers for amplification of specific marker regions (Paran and Michelmore, 1993). These sequence characterized sites (SCARS) greatly simplify the process of marker detection by other laboratories, are more reproducible from experiment to experiment than are RAPDs, and in many cases are codominant markers.

RAPD-based marker mapping has been used in various species such as *Arabidopsis thaliana* (Reiter et al., 1992), alfalfa (Echt et al., 1993), oat (Penner et al., 1993), lettuce (Kesseli et al., 1994), tomato (Klein-Lankhorts et al., 1991; Williamson et al., 1994), peach (Chaparro et al., 1994; Rajapakse et al., 1995; Warburton et al., 1996), and sweet cherry (Stockinger et al., 1996). They have also been used for cultivar identification (Gregor et al., 1994; Myneni et al., 1995; Salimath et al., 1995), for obtaining of markers closely linked to specific genes (Paran and Michelmore., 1993; Dickinson et al., 1993; Barua et al., 1993), for saturation of genomic regions in marker assisted cloning studies (Martin et al., 1991; Michelmore et al., 1991), and for plant population genetic studies (Yeh et al., 1995; Bonnin et al., 1996; Dawson et al., 1996; Yan et al., 1997).

## B. DNA Amplification Fingerprinting (DAF)

DNA amplification fingerprinting is a PCR-based method related in principal to the RAPD technology that involves the use of shorter primers, thereby increasing the frequency of primer-template annealings. This yields a significant increase in the number of randomly amplified fragments of the genome. These amplification products are visualized on silver stained polyacrylamide gels (Caetano-Anolles et al., 1991; Bassam et al., 1991). DAF produces relatively complex DNA profiles that have utility for genetic mapping analysis (Prabhu and Greshoff, 1994; Jianq and Greshoff, 1997) and for assessing genetic relatedness (He et al., 1995). The procedure has many of the same advantages as RAPD analysis for polymorphism detection. However, amplification patterns are more complex and the polymorphic amplification products are dominant markers.

### 1. Simple Sequence Repeats

The very short simple sequence repeats (SSR) of extremely high polymorphic content, the so-called microsatellites (Litt and Luty, 1989), are arguably the best molecular marker system available for the molecular analysis of plant genomes, as they are with other organisms (Dietrich et al., 1992; Weissenbach et al., 1992).

Microsatellite loci are regions of the genome containing simple sequence repeats of varying complexity and repeat number. The repeat units are usually less than six bases in length, and individual loci may contain several to many tandemly arranged repeats. The frequency of occurrence of these loci varies, depending on the complexity of the repeat sequence. For example, from our studies in *Prunus persica* (peach), (CT) repeats occur at least once in every 78 kb in peach, compared to once in every 120 kb in apple (Guilford et al., 1997) and 225 kb in rice (Wu and Tanksley, 1993). In concurrence with previous observations, (CA) repeats were less frequent, occurring once in every 156 kb in peach. In apple and rice, (CA) repeats occur every 190 kb and 480 kb, respectively. Of the microsatellite motifs we have examined in peach, the (AGG) repeat motif was found to be the least common, occurring approximately once in every 700 kb. Low frequencies of occurrence of trinucleotide repeats have also been reported: in apple, trinucleotide repeats occur every 3 Mb (Guilford et al., 1997); in wheat, trinucleotides are as much as 10 times less frequent than dinucleotide repeats (Ma et al., 1996). Studies on the utility of microsatellites reveal that many of the most informative repeats are dinucleotides [e.g., (CA), (AT) or (AG)] (Weber and May, 1989). Once microsatellite containing loci are identified, cloned, and sequenced, unique primers flanking the tandem repeats are synthesized and used to search for codominant, simple sequence length polymorphisms (SSLPs) (Tautz, 1989), usually by size fractionating labeled amplification products on denaturing polyacrylamide gels. This class of markers is in essence analyzed as sequence tagged sites (STS) (Olson et al., 1989) or sequence tagged microsatellites (STMS) (Beckmann and Soller, 1990).

A number of reports have appeared demonstrating the utility of these markers for molecular genetic mapping and phylogenetic analysis of plant species (Condit and Hubbel, 1991; Akkaya et al, 1995; Zhao and Kochert, 1993; Morgante and Olivieri, 1993; Bell and Becker, 1994; Akagi et al., 1996; Guilford et al., 1997; Provan et al., 1996).

When compared to other codominant markers, such as RFLPs, microsatellite loci exhibit more variability at a given locus. This was observed in rice (Wu and Tanksley, 1993), wheat (Roder et al., 1995), and other species. For example, in barley, a predominantly inbred species, as many as 33 alleles were observed at a single locus (Saghai Maroof et al., 1994). Cregan et al. (1994) found 23 alleles in soybean. We have observed in Rosaceae species that the number of alleles in peach appeared to be relatively low (1–4) when compared to that of other species such as apple (1–9). However, the polymorphism level in peach germplasm is still quite satisfactory (average heterozygosity = 0.5) for most genetic studies. Microsatellites are thus highly informative even in species in which low variability exists. Since the markers generated from microsatellite sequences identify significant levels of polymorphism, are highly transportable, and occur in reasonable abundance, it is evident that microsatellites have significant potential for genetic mapping, map merging, and cultivar identification.

Microsatellites can also be used in an oligonucleotide "fingerprinting approach." In this method, oligonucleotide probes complementary to simple tandem repeats are used as hybridization probes on southern transfers of restriction digested, electrophoretically separated DNA, and the products visualized by autoradiography (Sharma et al., 1995). An alternative method of generating microsatellite based DNA fingerprints is by hybridizing microsatellite containing oligonucleotides to southern blotted RAPD products (Richardson et al., 1995).

## 2. Single Primer Amplification Reactions

Single primer amplification reaction (SPAR) markers are derived from the use of microsatellites as primers to amplify intermicrosatellite DNA sequences (Gupta et al., 1994). This system can produce multiple markers per assay, but these markers are usually dominant in genetic nature.

## 3. Amplified Fragment Length Polymorphism

One of the most recent molecular marker technologies implemented in plant genome analysis is amplified fragment length polymorphism (AFLP) analysis (Vos et al., 1995). AFLP analysis couples the technologies of restriction enzyme digestion of genomic DNA with PCR amplification of select digestion products. For this analysis, genomic DNA is first digested with restriction enzymes (typically Eco RI and Mse I), then specific adapters are ligated onto the resulting fragments. The adapter sequences, typically with two or third additional specific bases at the 3′ end, are then used as primers for PCR amplification of a specific subset of the adapted restriction fragments. The products of the PCR step are visualized by polyacrylamide gel electrophoresis.

This marker technique allows the inspection of numerous RFLP-derived polymorphisms simultaneously, making AFLP a powerful tool for genome analysis. The markers identified by AFLP are typically dominant, requiring conversion to STSs before they are useful for positional cloning and marker assisted selections. AFLP has been used for DNA fingerprinting (Vos et al., 1995) and for the construction of linkage maps in crops such as barley (Becker and Heun, 1995), potato (Meksem et al., 1995), tomato (Thomas et al., 1995), and peach (Lu et al., 1998, Dirlewanger et al. 1998). They have also been used to assess gene-pool similarities of populations (Folkerstma et al., 1996) and for messenger RNA (mRNA) fingerprinting (Money et al., 1996).

Powell et al. (1995) evaluated the relative utility of various molecular marker systems in terms of their Marker Index (MI). This is a numerical product of the marker's expected heterozygosity and the multiplex ratio (number of polymorphic products per reaction). They determined that dinucleotide SSRs had the highest expected heterozygosity, almost twice that of RFLPs and RAPDs. AFLPs, on the other hand, had the highest multiplex ratios (more than an order of magnitude greater than that of SSRs). They determined MI values in soybean for AFLPs, SSRs, RAPDs, and RFLPs as 6.14, 0.60, 0.48, and 0.10, respectively. The same relative order of MI values is likely to be found in other crop plants.

We believe that AFLPs, RFLPs, and SSRs can play complementary roles in strategies to quickly obtain marker-saturated linkage maps. AFLPs produce more polymorphisms per unit effort and provide the largest number of markers on the saturated map. However, codominant loci such as RFLPs and SSRs are needed to tie these loci together into a single map and to integrate information from different maps. Due to their high polymorphism level and codominant expression, SSRs provide the best avenue for the integration of map information from different crosses, both within and among species.

## IV. TAGGING SIMPLE TRAITS

The utility of molecular markers for the tagging and manipulation of simple gene controlled traits is well established. With the earliest reports of molecular map construction, it was evident that these technologies were capable of producing tightly linked markers that could be used to streamline the breeding process. Once a trait is closely tagged, the processes of transferring the trait to other cultivars is greatly simplified, because large numbers of progeny can be scored easily and cost-effectively for the presence of the tags. Highly saturated maps can significantly shorten the time required for introgression of traits by allowing the breeder to identify and maintain important genomic components during advanced introgressive crossing (Tanksley et al., 1992).

## V. TAGGING COMPLEX TRAITS

### A. Quantitative Trait Loci Mapping

Mapping of quantitative trait loci (QTL) is important in plant breeding because most traits of agricultural importance (e.g., yield, hardiness, fruit flavor) are influenced by several loci. QTL analysis can be used to identify most of these loci, facilitating marker-assisted introgression, thereby enriching the genetic base and accelerating the rate of crop improvement (Tanksley, 1995).

An early example of this approach is provided by the development and application of genetic linkage maps for tomato (Helentjaris et al., 1986; Tanksley et al., 1987). These maps were employed to analyze and localize quantitative trait loci for such characters as insect resistance

(Nienhuis et al., 1987), water use efficiency (Martin et al., 1989), fruit mass, pH, yield, and soluble solids (Paterson et al., 1990). These pioneering efforts provided the first molecular maps of quantitative traits and highlighted the potential benefits of this analytical method for plant breeding. Subsequently, researchers working on many different crop species have demonstrated the utility of QTL mapping using molecular marker systems, as a valuable tool for the identification of genes controlling complex traits and for their manipulation by marker assisted breeding methods (Kowalski et al., 1994; Groover et al., 1994; Eshed and Zamir, 1995; Toroser et al., 1995; Lee, 1995; Lin et al., 1996; Tanksley et al., 1996; Bernacchi and Tanksley, 1997; Yano and Sasaki 1997; McCouch and Doerge, 1995; Meyer et al., 1998; Mitchell-Olds and Pederson, 1998; Lubberstedt et al. 1998; Austin and Lee, 1998).

Another benefit resulting from the molecular genetic localization of QTL is the opportunity to clone them by using map-based approaches. In a few cases, for genes with relatively large effects, the map location has been precise enough to allow "chromosomal landing" rather than "walking" (Tanksley, 1995).

## VI. COMPARATIVE GENOMICS

As the expressed sequence tag (EST) and genomic marker databases of model species are developed in each of the major plant families, cross correlation of genomic regions among maps will facilitate marker identification in regions where important genes controlling simple and complex traits are located. This approach to tagging traits will be particularly simplified if it can be demonstrated that there is significant genome microsynteny among the model species. Microsyntenic studies have revealed that among closely related species, the degree of preservation of genomic organization can be quite high (Dunford et al., 1995; Lagercrantz et al., 1996; Bennetzen et al., 1996; Chen et al., 1997; Kilian et al., 1997; Avramova et al., 1998). Thus, by using model genomic physical maps, EST, and genomic markers, in particular SSRs, and large insert clone library resources, close tagging of traits in related species will be significantly facilitated.

Indeed, from the coupling of QTL analysis and comparative molecular marker mapping, it appears that there may be significant functional conservation of QTL intervals in different plant species (Paterson et al., 1991; Fatokun et al., 1992, Pireira and Lee, 1995; Lin et al., 1995). Observations from these reports suggest that QTL information from one species may be broadly applicable to many species (Paterson, 1996). However, this needs to be tested rigorously in different plant families, and particularly in those such as the Rosaceae, in which long generation time and juvenility hamper traditional breeding approaches. In this regard, we and others are working closely with a number of laboratories worldwide to integrate the markers from various maps and obtain comparative map information for a number of plant species. Conservation of QTL-containing genomic regions suggests that markers that tag significant QTL intervals in one species may be broadly applicable for crop improvement strategies in other related species in the same family. These observations further underscore the importance of establishing genetic model species for each family of major agricultural importance.

## REFERENCES

Akagi, H., Yokozeki, Y. and Fujimura, T. 1996. Microsatellite DNA markers for rice chromosomes. Theor. Appl. Genet. 93: 1071–1077.

Akkaya, M.S., Shoemaker, R.C., Specht, J.E., Bhagwat, A.A. and Cregan, P.B. 1995. Integration of simple sequence repeat DNA markers into a soybean linkage map. Crop Sci 35: 1439–1445.

Apuya, N., Frizzier, B., Keim, P., Roth, J. and Lark, K. 1988. Restriction fragment length polymorphisms as genetic markers in soybean, *Glycine max* (L.) Merrill. Theor. Appl. Genet. 75: 889–901.

Arondel, V.B., Lemieux, B., Huang, I., Gibson, S., Goodman, H. and Sommerville, C. 1992. Map-based cloning of a gene controlling omega-3 fatty acid desaturase in arabidopsis. Science 258: 1353–1355.

Austin, D.F. and Lee, M. 1998. Detection of Quantitative Trait Loci for grain yield and yield components in maize across generations in stress and nonstress environments. Crop Sci 38(5): 1296–1308.

Avramova, Z, Tikhonov, A., Chen, M. and Bennetzen, J.L. 1998. Matrix attachment regions and structural colinearity in the genomes of two grass species. Nucleic Acids Res. 26(3): 761–767.

Barua, U.M., Chalmers, K.J., Hackett, C.A., Thomas, W.T., Powell, W. and Waugh, R. 1993. Identification of RAPD markers linked to a *Rhynchosporium secalis* resistance locus in barley using near-isogenic lines and bulked segregant analysis. Heredity 71: 177–184.

Bassam, B., Caetano-Annoles, G., and Gresshoff, P. 1991. A fast and sensitive silver-staining for DNA in polyacrylamide gels. Anal. Biochem. 1996: 80–83.

Becker, J. and Heun, M. 1995. Barley microsatellites: Allele variation and mapping. Plant Mol. Biol. 27: 835–845.

Beckmann, J. and Soller, M. 1991. Biotechnol. 8: 930–932.

Bell, C.L. and Becker, J.R. 1994. Assignment of 30 microsatellite loci to the linkage map of *Arabidopsis*. Genomics 19: 137–144.

Benito, C., Llorente, F., Henriques, G.N., Gallego, F.J., Zaragoza, C., Delibes, A. and Figueiras, A.M. 1994. A map of rye chromosome 4R with cytological and isozyme markers. Theor. Appl. Genet. 87: 941–946.

Bennetzen, J.L., SanMiguel, P., Liu, C.N., Chen, M., Tikhonov, A., Costa de Oliveira, A., Jin, Y.K., Avramova, Z., Woo, S.S., Zhang, H. and Wing, R.A. 1996. The Hybaid Lecture: Microcollinearity and segmental duplication in the evolution of grass nuclear genomes. Symp Soc Exp Biol. 50: 1–3.

Bernacchi, D. and Tanksley, S. 1997. An interspecific backcross of *Lycopersicon esculentum* × *L. hirsutum*: linkage analysis and a QTL study of sexual compatibility factors and floral traits. Genetics 147: 861–877.

Besse, P., Seguin, M., Chevallier, M.H., Nicholas, D. and Lanaud, C. 1994. Genetic diversity among wild and cultivated populations of *Hevea brasiliensis* assessed by nuclear RFLP analysis. Theor. Appl. Genet. 88: 199–207.

Bonierbale, M.W., Plaisted, R.L. and Tanklsey, S.D. 1988. RFLP maps based on a common set of clones reveal modes of chromosomal evolution in potato and tomato. Genetics 120: 1095–1103.

Bonnin, I., Prosperi, J.M., Olivieri, I. 1996. Genetic markers and quantitative genetic variation in *Medicago truncatula* (Leguminosae): a comparative analysis of population structure. Genetics 143(4): 1795–1805.

Caetano-Annoles, G., Bassam, B. and Gresshof, P. 1991. DNA amplification fingerprinting using very short arbitrary oligonucleotide primers. Biotechnology 9: 619–623.

Causse, M.A., Fulton, T.M., Cho, Y.G., Ahn, S.N., Chunwongse, J., Wu, K., Xiao, J., Yu, Z., Ronald, P.C., Harrington, S.E., Second, G., McCouch, S. and Tanksley, S.D. 1994. Saturated molecular map of the rice genome based on an interspecific backcross population. Genetics 138: 1251–1274.

Chang, C., Bowman, J., DeJohn, A., Lander, E. and Meyerowitz, E. 1988. Restriction fragment length polymorphism linkage map for *Arabidopsis thaliana*. Proc. Natl. Acad. Sci USA 85: 6856–6860.

Chapparo, J.X., Werner, D.J., O'Malley, D. and Sederoff, R.R. 1994. Targeted mapping and linkage analysis of morphological, isoenzyme and RAPD markers in peach. Theor. Appl. Genet. 87: 805–815.

Chee, M., Yang, R., Hubbell, E., Berno, A., Huang, X.C., Stern, D., Winkler, J., Lockhart, D.J., Morris, M.S., Fodor, S.P. 1996. Accessing genetic information with high-density DNA arrays. Science 274(5287): 610–614.

Chen, M, SanMiguel, P., de Oliveira, A.C., Woo, S.S., Zhang, H., Wing, R.A. and Bennetzen, J.L. 1997. Microcolinearity in *sh2*-homologous regions of the maize, rice, and sorghum genomes. Proc Natl Acad Sci USA 94: 3431–3435.

Condit, R. and Hubbel, S.P. 1991. Abundance and DNA sequence of two-base repeat regions in tropical tree genomes. Genome 34: 66–71.

Cregan, P.B., Akkaya, M.S., Bhagwat, A.A., Lavi, U. and Rongwen, J. 1994. Length polymorphisms of simple sequence repeat (SSR) DNA as molecular markers in plants. In: Plant Genome Analysis. CRC Press, pp 47–56.

Dietrich, W., Katz, H., Lincoln, S., Shin, H.-S., Friedman, J., Dracopoli, N., and Lander, E. 1992. A genetic map of the mouse suitable for typing intraspecific crossovers. Genetics 131: 423–447.

Dickinson, M.J., Jones, D.A. and Jones, J.D. 1993. Close linkage between the *Cf-2/Cf-5* and *Mi* resistance loci in tomato. Mol. Plant Microbe Interact. 6: 341–347.

Dirlwanger, E., Moing, A., Pronier, V., Svanella, L., Guye, A., Monet, R., Rothan, C. 1998. Detection of QTLs controlling peach fruit acidity and sweetness. Acta Hortic. 465: 89–98.

Dunford, R.P., Kurata, N., Laurie, D.A., Money, T.A., Minobe, Y. and Moore, G. 1995. Conservation of fine-scale DNA marker order in the genomes of rice and the Triticeae. Nucleic Acids Res. 23: 2724–2728.

Echt, C.S., Kidwell, K.K., Knapp, S.J., Osborn, T.C. and McCoy, T.J. 1993. Linkage mapping in diploid alfalfa (*Medicago sativa*). Genome 37: 61–71.

Eldredge, L., Ballard, R.E., Baird, W.V., Abbott, A.G., Morgens, P., Callahan, A., Scorza, R. and Monet, R. 1992. Application of RFLP analysis to genetic linkage mapping in peaches. HortScience 27: 160–164.

Eshed, Y. and Zamir, D. 1995. An introgression line population of *Lycopersicon pennellii* in the culivated tomato enables the identification and fine mapping of yield-associated QTL. Genetics 141: 1147–1162.

Fatokun, C.A., Menancio-Hautea, D.I., Danesh, D. and Young, N.D. 1992. Evidence for orthologous seed weight genes in cowpea and mung bean based on RFLP mapping. Genetics 132: 841–846.

Ferreira, M.E., Rimmer, S.R., Williams, P.H. and Osborn, T.C. 1995. Mapping loci controlling *Brassica napus* resistance to *Leptosphaeria maculans* under different screening conditions. Phytopathology 85: 213–217.

Folkerstma, R.T., Rouppe van der Voort, J.N., de Groot, K.E., van Zandvoort, P.M., Schots, A., Gommers, F.J., Helder, J. and Bakker, J. 1996. Gene pool similarities of potato cyst nematode populatins assessed by AFLP analysis. Mol. Plant Microbe Interact. 9: 47–54.

Foolad, M.R., Arulsekar, S., Beccera, V. and Bliss, F.A. 1995. A genetic map of *Prunus* based on an interspecific cross between peach and almond. Theor. Appl. Genet. 91: 262–269.

Freyre, R. and Douches, D.S. 1994. Isoenzymatic identification of quantitative traits in crosses between heterozygous parents: mapping tuber traits in diploid potato (*Solanum* spp.). Theor. Appl. Genet. 87: 764–772.

Galiba, G., Quarrie, S.A., Sutka, J., Moruonov, A. and Snape, J.W. 1995. RFLP mapping of the vernalization (*Vrn 1*) and frost resistance (*Fr1*) genes on chromosome 5A of wheat. Theor. Appl. Genet. 90: 1174–1179.

Gardiner, J.M., Coe, E.H., Melia-Hancock, S., Hoisington, D.A. and Chao, S. 1993. Development of a core RFLP map in maize using an immortalized F2 population. Genetics 134: 917–930.

Gebhardt, C., Ritter, E., Debener, T., Schachtschabel, U., Walkemeier, B., Uhrig, H. and Salamini, F., 1989. RFLP analysis and linkage mapping in *Solanum tuberosum*. Theor. Appl. Genet. 78: 65–75.

Granger, A.R. 1996. Inheritance and linkage of isozymes in sweet cherry (*Prunus avium* L.) Theor. Appl. Genet. 93: 426–430.

Gregor, D., Hartmann, W. and Stosser, R. 1994. Cultivar identification in *Prunus domestica* using random amplified polymorphic DNA markers. Acta Hortic. 359: 33–40.

Groover, A., Devey, M., Fiddler, T., Lee, J., Megraw, R., Mitchel-Olds, T., Sherman, B., Vujcic, S., Williams, C. and D. Neale. 1994. Identification of quantitative trait loci infuencing wood specific gravity in an outbred pedigree of loblolly pine. Genetics 138: 1293–1300.

Guilford, P., Prakash, S., Zhu, J.M., Rikkerink, E., Gardiner, S., Bassett, H. and Forster, R. 1997. Microsatellites in *Malus × Domestica* (apple): abundance, polymorphism and cultivar identification. Theor. Appl. Genet. 94: 249–254.

Gupta, M., Chyi, Y.S., Romero-Severson, J. and Owen, J.L. 1994. Amplification of DNA markers from evolutionarily diverse genomes usong single primers of simple-sequence repeats. Theor. Appl. Genet. 89: 998–1006.

He, G., Prakash, C.S., Jarret, R.L. 1995. Analysis of genetic diversity in a sweetpotato (*Ipomoea batatas*) germplasm collection using DNA amplification fingerprinting. Genome 38(5): 938–945.

Helentjaris, T., Slocum, M., Wright, S., Schaefer, A. and Nienhuis, J. 1986. Construction of genetic linkage maps in maize and tomato using restriction fragment length polymorphisms. Theor. Appl. Genet. 72: 761–769.

Hunter, R.L., and Markert, C.L. 1957. Histochemical demonstration of enzymes separated by zone electrophoresis in starch gels. Science 125: 1294–1295.

Jiang, Q., Greshoff, P.M. 1997. Classical and molecular genetics of the model legume *Lotus japonicus*. Mol Plant Microbe Interact 10(1): 59–68.

Keim, P., Diers, B., Olson T. and Shoemaker, R. 1990. RFLP Mapping in soybean: Association between marker loci and variation in quantitative traits. Genetics 126: 735–742.

Kesseli, R.V., Paran, I. and Michelmore, R.W. 1994. Analysis of a detailed genetic linkage map of *Lactuca sativa* (lettuce) constructed from RFLP and RAPD markers. Genetics 136: 1435–1446.

Kilian, A., Chen, J., Han, F., Steffenson, B. and Kleinhofs, A. 1997. Towards map-based cloning of the barley stem rust resistance genes *Rpg1* and *rpg4* using rice as an intergenomic cloning vehicle. Plant Mol. Biol. 35: 187–195.

Klein-Lankhorts, R.M., Vermunt, A., Weide, R., Liharska, T. and Zabel, P. 1991. Isolation of molecular markers for tomato (*L. esculentum*) using random amplified polymorphic DNA (RAPD). Theor. Appl. Genet. 83: 108–114.

Kochert, G., Halward, T., Branch, W. and Simpson, C. 1991. RFLP variability in peanut (*Arachis hypogea* L.) cultivars and wild species. Theor. Appl. Genet. 81: 565–570.

Kowalski, S.P., Lan, T.H., Feldmann, K.A. and Paterson, A.H. 1994. QTL mapping of naturally-occuring variation in flowering time of *Arabidopsis thaliana*. Mol. Gen. Genet. 245: 548–555.

Lagercrantz, U., Puterill, J., Coupland, G. and D. Lydiate. 1996. Comparative mapping in *Arabidopsis* and *Brassica*, fine scale collinearity and congruence of genes controlling flowering time. Plant J. 9: 13–20.

Lashkari, D.A., DeRisi, J.L., McCusker, J.H., Namath, A.F., Gentile, C., Hwang, S.Y., Brown, P.O., Davis, R.W. 1997. Yeast microarrays for genome wide parallel genetic and gene expression analysis. Proc Natl Acad Sci USA 94(24): 13057–13062.

Lawson, D.M., Hemmat, M. and Weeden, N.F. 1995. The use of molecular markes to analyze the inheritance of morphological and developmental traits in apple. J. Am. Soc. Hortic. Sci. 120: 532–537.

Lee, M. 1995. DNA markers and plant breeding programs. Adv. Agron. 55: 265–345.

Lin, H.X., Quian, H.R., Zhuang, J.Y., Lu, J., Min, S.K., Xiong, Z.M., Huang, N. and Zheng, K.L. 1996. RFLP mapping of QTL for yield and related characters in rice (*Oryza sativa* L.) Theor. Appl. Genet. 92: 920–927.

Lin, Y.R., Schertz, K.F. and Paterson, A.H. 1995. Comparative analysis of QTLs affecting plant height and maturity across the Poaceae, in reference to an interspecific sorghum population. Genetics 141: 391–411.

Litt, M. and Luty, J. 1989. A hypervariable microsatellite revealed by in vitro amplification of a dinucleotide repeat within the cardiac muscle actin gene. Amer. J. Hum. Genet. 44: 397–401.

Lodhi, M.A., Daly, M.J., Ye, G.N., Weeden, N.F. and Reisch, B.I. 1995. A molecular marker based linkage map of *Vitis*. Genome 38: 786–794.

Lu, Z.X., Sosinski, B., Reighard, G.L., Baird, W.V., Abbott, A.G. 1998. Construction of a genetic linkage map and identification of AFLP markers for resistance to root-knot nematodes in peach rootstocks. Genome 41: 199–207.

Lubberstedt, T., Melchinger, A.E., Fahr, S., Klein, S., Dally, A. and Westhoff, P. 1998. QTL mapping in testcrosses of flint lines of maize. III. Comparison across populations for forage traits. Crop Sci. 38: 1278–1289.

Ma, Z.Q., Roder, M. and Sorrells, M.E. 1996. Frequencies and sequence characteristics of di-, tri-, and tetranucleotide microsatellites in wheat. Genome 39: 123–130.

Maas, H.I. and Klaas, M. 1995. Infraspecific differentiation of garlic (*Allium sativum* L.) by isozyme and RAPD markers. Theor. Appl. Genet. 91: 89–97.

Markert, C.L. and Moller, F. 1959. Multiple forms of enzymes: tissues, ontogenetic and species specific patterns. Proc. Natl. Acad. Sci. USA 45: 753–763.

Martin, B., Nienhuis, J., King G. and Schaefer, A. 1989. Restriction fragment length polymorphisms associated with water use efficiency in tomato. Science 243: 1725–1728.

Martin, G.B., Williams, J. and Tanksley, S. 1991. Rapid identification of markers linked to a *Pseudomonas* resistance gene in tomato by using random primers and near-isogenic lines. Proc. Natl. Acad. Sci. USA 88: 2336–2340.

McCouch, S.R. and Doerge, R.W. 1995. QTL mapping in rice. Trends Genet. 11: 482–487.

McCouch, S., Kochert, G., Yu, Z., Wang, Z., Khush, G., Coffman, W. and Tanksley, S. 1988. Molecular mapping of rice chromosomes. Theor. Appl. Genet. 76: 815–829.

McCoy, T., Echt, C. and Mancino, L. 1991: Segregation of molecular markers supports an allotetraploid structure for *Medicago sativa* × *M. papillosa* interspecific hybrid. Genome 34: 574–578.

McLean, M., Gerats, A., Baird, V. and Meagher, R. 1990. Six actin gene subfamilies map to five chromosomes of *Petunia hybrida*. J. Hered. 81: 341–346.

Meksem, K., Leister, D., Peleman, J., Zabeau, M., Salamini, F. and Gebhardt, C. 1995. A high resolution map of the vicinity of the R1 locus on chromosome V of potato based on RFLP and AFLP markers. Mol. Gen. Genet. 249: 74–81.

Messegeur, R., Arus, P. and Carrera, M. 1987. Identification of peach cultivars with pollen isozymes. Sci. Hortic. 31: 107–117.

Messegeur, R., Ganal, M., De-Vincente, M.C., Young, N.D., Bolkan, H. and Tanksley, S.D. 1991. A high resolution RFLP map around the root knot nematode resistance gene (*Mi*) in tomato. Theor. Appl. Genet. 82: 529–539.

Meyer, R.C., Milbourne, D., Hackett, C.A., Bradshaw, J.E., McNichol, J.W. and Waugh, R. 1998. Linkage analysis in tetraploid potato and association with quantitative resistance to late blight (*Phytophthora infestans*). Mol. Gen. Genet. 259: 150–160.

Michelmore, R.W., Paran, I. and Kesseli, R.V. 1991. Identification of markers linked to disease resistance genes by bulked segregant analysis: a rapid method to detect markers in specific genomic regions using segregating populations. Proc. Natl. Acad. Sci. USA 88: 9828–9832.

Mitchell-Olds, T. and Pederson, D. 1998. The molecular basis of quantitative variation in central and secondary metabolism in *Arabidopsis*. Genetics 149: 739–747.

Money, T., Reader, S., Qu, T.J., Dunford, R.P. and Moore, G. 1996. AFLP-based mRNA fingerprinting. Nucleic Acids Res. 24: 2616–2617.

Morgante, M. and Olivieri, A.M. 1993. PCR-amplified microsatellites as markers in plant genetics. Plant J. 3: 175–182.

Mullis, K.B., Faloona, F., Scharf, S., Saiki, R., Horn, G., and Erlich, H. 1986. Specific enzymatic amplification of DNA in vitro: the polymerase chain reaction. Cold Spring Harbor Symp. Quant. Biol. 51: 263–273.

Myneni, A., Austin, M.E. and Ozias-Akins, P. 1995. Randomly amplified polymorphic DNA fingerprinting for identifying rabbiteye blueberry (*Vaccinium ashei* Reade) cultivars. J Am. Soc. Hortic. Sci. 120: 710–713.

Neale, D. and Williams, C. 1991. Restriction fragment length polymorphism mapping in conifers and application to forest genetics and tree improvement. Can. J. For. Res. 21: 545–554.

Nienhuis, J., Helentjaris, T., Slocum, M., Ruggero B. and Schaefer, A. 1987. Restriction fragment length polymorphism analysis of loci associated with insect resistance in tomato. Crop Sci. 27: 797–803.

Nybom, H., and Schall, B.A. 1990. DNA 'fingerprints' applied to paternity analysis in apples (*Malus* × *domestica*). Theor. Appl. Genet. 79: 763–768.

Olson, M., Hood, L., Cantor, C., Botstein, D. 1989. A common language for physical mapping of the human genome. Science 245: 1434–1435.

Paran, I. and Michelmore, R.W. 1993. Development of reliable PCR-based markers linked to downy mildew resistance genes in lettuce. Theor. Appl. Genet. 85: 985–993.

Paterson, A.H. 1996. Comparative mapping sows seeds of progress. Newsl USDA Plant Genome Res Prog. 6: 1–3.

Paterson, A.H., Damon, S., Hewitt, J.D., Zamir, D., Rabinowitch, H.D., Lincoln, S.E., Lander, E.S. and Tanksley, S. 1991. Mendelian factors underlying quantitative traits in tomato: Comparison across species, generations, and environments. Genetics 127: 181–197.

Paterson, A., DeVerna, J., Lanini, B. and Tanksley, S. 1990. Fine mapping of quantitative trait loci using selected overlapping recombinant chromosomes in an interspecies cross of tomato. Genetics 124: 735–742.

Penner, C.A., Chong, J., Levesque-Lemay, M., Molnar, S.J. and Fedak, G. 1993. Identification of a RAPD marker linked to oat stem rust gene *Pg3*. Theor. Appl. Genet. 85: 702–705.

Pillen, K., Streinruchen, G., Wricke, G., Herrman, R.G. and Jung, C. 1992. A linkage map of sugar beet (*Beta vulgaris* L.). Theor. Appl. Genet. 84: 129–135.

Pireira, M.G. and Lee, M. 1995. Identification of genomic regions affecting plant height in sorghum and maize. Theor. Appl. Genet. 90: 380–388.

Powell, W., Morgante, M., Andre, C., Hanafey, M., Vogel, J., Tingey, S., Rafalski, A. 1996. The comparison of RFLP, RAPD, AFLP and SSR (microsatellite) markers for germplasm analysis. Mol Breed, 2(3): 225–238.

Prabhu, R.R., Greshoff, P.M. 1994. Inheritance of polymorphic markers generated by DNA amplification fingerprinting and their use as genetic markers in soybean. Plant Mol Biol 26(1): 105–116.

Provan, J., Powell, W. and Waugh, R. 1996. Microsatellite analysis of relationships within cultivated potato (*Solanum tuberosum*). Theor. Appl. Genet. 92: 1078–1084.

Ragab, R.A., Dronavalli, S., Saghai Maroof, M.A. and Yu, Y.G. 1994. Construction of a sorghum RFLP linkage map using sorghum and maize DNA probes. Genome 37: 590–594.

Ragot, M., Sisco, P.H., Hoisington, D.A., Stuber, C.W. 1995. Molecular-marker-mediated characterization of favorable exotic alleles at quantitative trait loci in maize. Crop Sci v. 35(5): 1306–1315.

Rajapakse, S., Belthoff, L.E., He, G., Estager, A.E., Scorza, R., Verde, I., Ballard, R.E., Baird, W.V., Callahan, A., Monet, R. and Abbott, A.G. 1995. Genetic linkage mapping in peach using morphological, RFLP and RAPD markers. Theor. Appl. Genet. 90: 503–510.

Reinisch, A.J., Dong, J., Brubaker, C.L., Stelly, D.M., Wendel, J.F. and Paterson, A.H. 1994. A detailed RFLP map of cotton, *Gossypium hirsutum* × *Gossypium barbadense*: chromosome organization and evolution in a disomic polyploid genome. Genetics 138: 829–847.

Reiter, R.S., Williams, J.G.K., Feldman, K.A., Rafalski, A., Tingey, S.V. and Scolnik, P.A. 1992. Global and local genome mapping in *Arabidopsis thaliana* by using recombinant inbred lines and random amplified polymorphic DNAs. Proc. Natl. Acad. Sci. USA 89: 1477–1481.

Restivo, F.M., Tassi, F., Biffi, R., Falavigna, A., Caporali, E., Carboni, A., Doldi, M.L., Spada, A. and Marziani, G.P. 1995. Linkage arrangement of RFLP loci in progenies from crosses between double haploid *Asparagus officinalis* L. clones. Theor. Appl. Genet. 90: 124–128.

Richardson, T., Cato, S., Ramser, J., Kahl, G. and Weising, K. 1995. Hybridization of microsatellites to RAPD: a new source of polymorphic markers. Nucleic Acids Res. 23: 3798–3799.

Roder, M.S., Plaschke, J., Konig, S.U., Borner, A., Sorrells, M.E., Tanksley, S.D. and Ganal, M. 1995. Abundance, variability and chromosomal location of microsatellites in wheat. Mol. Gen. Genet. 246: 327–333.

Saghai Maroof, M.A., Biyashev, R.M., Yang, G.P., Zhang, Q. and Allard, R.W. 1994. Extraordinarily polymorphic microsatellite DNA in barley: Species diversity, chromosomal locations, and population dynamics. Proc. Natl. Acad. Sci. USA 91: 5466–5470.

Salimath, S.S., de Oliveira, A.C., Godwin, I.D. and Bennetzen, J.L. 1995. Assessment of genome origins and genetic diversity in the genus *Eleusine* with DNA markers. Genome 38: 757–763.

Sharma, P.C., Huttel, B., Winter, P., Kahl, G., Gardner, R.C. and Weising, K. 1995. The potential for microsatellites for hybridization- and polymerase chain reaction-based DNA fingerprinting of chickpes (*Cicer arietinum* L.) and related species. Electrophoresis 16: 1755–1761.

Simonsen, V. and Heneen, W.K. 1995. Inheritance of isozymes in *Brassica campestris* L. and genetic divergence among different species of Brassiceae. Theor. Appl. Genet. 91: 353–360.

Smithies, O. 1955. Zone electrophoresis in starch gels. Biochem. J. 61: 629.

Song, K., Osborn, T. and Williams, P. 1988a. *Brassica* taxonomy based on nuclear restriction fragment length polymorphisms (RFLPs). 1. Genome evolution of diploid and amphidiploid species. Theor. Appl. Genet. 75: 784–794.

Song, K., Osborn, T. and Williams, P. 1988b. *Brassica* taxonomy based on nuclear restriction fragment length polymorphisms (RFLPs). 2. Preliminary analysis of subspecies within *B. rapa* and *B. oleracea*. Theor. Appl. Genet. 75: 784–794.

Sonnante, G., Stockton, T., Nodari, R.O., Beccera-Velasquez, V.L. and Gepts, P. 1994. Evolution of genetic diversity during the domestication of common-bean (*Phaseolus vulgaris* L.). Theor. Appl. Genet. 89: 629–635.

Stalker, H.T., Phillips, T.D., Murphy, J.P. and Jones, T.M. 1994. Variation of isozyme patterns among *Arachis* species. Theor. Appl. Genet. 87: 746–755.

Staub, J.E., Serquen, F.C. and Gupta, M. 1996. Genetic markers, map construction, and their application in plant breeding. HortScience 31: 729–738.

Stockinger, E.J., Mulinix, C.A., Long, C.M., Brettin, T.S. and Iezzoni, A.F. 1996. A linkage map of sweet cherry based on RAPD analysis of a microspore derived callus culture population. J. Hered. 87: 214–218.

Tanksley, S.D. 1995. Impact of genome research on plant breeding. PG III Abstracts. P13.

Tanksley, S.D., Ganal, M.W., Prince, J.P., di Vicente, M.C., Bonierbale, M.W., Broun, P., Fulton, T.M., Giovannoni, J.J., Grandillo, S. and Martin, G.B. 1992. High density molecular linkage maps of the tomato and potato genomes. Genetics. 132(4): 1141–1160.

Tanksley, S.D., Grandillo, S., Fulton, T.M., Zamir, D., Eshed, Y., Petiard, V., Lopez, J., Beck Bunn, T. 1996. Advanced backcross qTL analysis in a cross between an elite processing line of tomato and its wild relative *L. pimpinellifolium*. Theor. Appl. Genet. 92: 213–224.

Tanksley, S., Mutschler, M. and Rick, C. 1987. Linkage map of the tomato (*Lycopersicon esculentum*) (2n = 24). pp 655–669, In Genetic Maps 1987: A compilation of linkage and restriction maps of genetically studied organisms. S. O'Brien, ed. Cold Spring Harbor Laboratory Press, Cold Spring Harbor, NY.

Tautz, D. 1989. Hypervariability of simple sequences as a general source for polymorphic markers. Nucl. Acids Res. 17: 6463–6471.

Thomas, C.M., Vos, P., Zabeau, M., Jones, D.A., Norcott, K.A., Chadwick, B.P. and Jones, J.D. 1995. Identification of amplified restriction fragment length polymorphism (AFLP) markers tightly linked to the tomato Cf-9 gene for resistance to *Cladosporium fulvum*. Plant J. 8: 785–794.

Toroser, D., Thormann, C.E., Osborn, T.C. and Mithen, R. 1995. RFLP mapping of quantitative trait loci controlling seed aliphaticglucosinolate content in oilseed rape (*Brassica napus* L.). Theor. Appl. Genet. 91: 802–808.

Vallejos, C.E., Sakiyama, N.S. and Chase, C.D. 1992. A molecular marker-based linkage map of *Phaseolus vulgaris* L. Genetics 131: 733–740.

van Denzye, A.E., Landry, B.S. and Pauls, K.P. 1995. The identification of restriction fragment length polymorphisms linked to seed color genes in *Brassica napus*. Genome 38: 534–542.

van Deynze, A.E., Nelson, J.C., O'Donoughue, L.S., Ahn, S.N., Siripoonwiwat, W., Harringtion, S.E., Yglesias, E.S., Braga, D.P., McCouch, S.R., and M.E. Sorrells. 1995. Comparative mapping in grasses. Oat relationships. Mol. Gen. Genet. 249: 349–356.

van Deynze, A.E., Nelson, J.C., Yglesias, E.S., Harrington, S.E., Braga, D.P., McCouch, S.R., and Sorrells, M.E. 1995. Comparative mapping in grasses: Wheat relationships. Mol. Gen. Genet. 248: 744–754.

Vezvaei, A., Hancock, T.W., Giles, L.C., Clarke, G.R. and Jackson, J.F. 1995. Inheritance and linkage of isozyme loci in almond. Theor. Appl. Genet. 91: 432–438.

Vos, P., Hagers, R., Bleeler, M., Reijans, M., van de Lee, T., Hornes, M., Frijters, A., Pot, J., Peleman, J., Kuiper, M. and Zabeau, M. 1995. AFLP: A new technique for DNA fingerprinting. Nucleic Acids Res. 23: 4407–4414.

Wang, D.G., Fan, J.B., Siao, C.J., Berno, A., Young, P., Sapolsky, R., Ghandour, G., Perkins, N., Winchester, E., Spencer, J., Kruglyak, L., Stein, L., Hsie, L., Topalogou, T., Hubbell, E., Robinson, E., Mittmann, M., Morris, M.S., Shen, N., Kilburn, D., Rioux, J., Nusbaum, C., Rozen, S., Hudson, T.J., Lander, E.S. 1998. Large-scale identification, mapping, and genotyping of single-nucleotidepolymorphisms in the human genome. Science 280 (5366): 1077–1082.

Warburton, J.L., Becerra-Velasquez, V.L., Goffreda, J.C. and Bliss, F.A. 1996. Utility of RAPD markers identifying genetic linkages to genes of economic interest in peach. Theor. Appl. Genet. 93: 920–925.

Weber, J.L. and May, P.E. 1989. Abundant class of human DNA polymorphisms which can be typed using the polymerase chain reaction. Am. J. Hum. Genet. 44: 388–396.

Weissenbach, J., Gyapay, G., Dib, C., Vignal, A., Morissette, J., Millasseau, P., Vaysseix, G., and Lathrop, M. 1992. A second-generation linkage map of the human genome. Nature 359: 794–801.

Welsh, J. and McClelland, M. 1990. Fingerprinting genomes using PCR with arbitrary primers. Nucleic Acids Res. 18: 7213–7218.

Williams, J.G.K., Kubelik, A.R., Livak, K.J., Rafalski, J.A. and Tingey, S.V. 1990. DNA polymorphisms amplified by arbitrary primers are useful as genetic markers. Nucleic Acids Res. 18: 6531–6535.

Williamson, V.M., Ho, J.-Y., Wu, F.F., Miller, N. and Kaloshian, I. 1994. A PCR-based marker tightly linked to the nematode resistance gene, *Mi*, in tomato. Theor. Appl. Genet. 87: 757–763.

Winzeler, E.A., Richards, D.R., Conway, A.R., Goldstein, A.L., Kalman, S., McCullough, M.J., McCusker, J.H., Stevens, D.A., Wodicka, L., Lockhart, D.J., Davis, R.W. 1998. Direct allelic variation scanning of the yeast genome. Science 281(5380): 1194–1197.

Wu, K.S. and Tanksley, S.D. 1993. Abundance, polymorphism and genetic mapping of microsatellites in rice. Mol. Gen. Genet. 241: 225–235.

Yan, H.J., Dai, S.L., Wu, N.H. 1997. RAPD analysis of natural populations of *Acanthopanax brachypus*. Cell Res 7(1): 99–106.

Yano, M. and Sasaki, T. 1997. Genetic and molecular dissection of quantitative traits in rice. Plant Mol. Biol. 35: 145–153.

Yeh, F.C., Chong, D.K., Yang, R.C. 1995. RAPD variation within and among natural populations of trembling aspen (*Populus tremuloides* Michx.) from Alberta. J Hered 86(6): 454–460.

Young, N., Zamir, D., Ganal, M., and Tanksley, S. 1988. Use of isogenic lines and simultaneous probing to identify DNA markers tightly linked to the *Tm-2a* gene in tomato. Genetics 120: 579–585.

Yu, Y.G., Saghai Maroof, M.A., Buss, G.R., Maughan, P.J. and S.A. Tolin. 1994. RFLP and microsatellite mapping of a gene for soybean mosaic virus resistance. Phytopathology 84: 60–64.

Zhang, Q., Saghai Maroof, M.A. and Kleinhofs, A. 1993. Comparative diversity analysis of RFLPs and isozymes within and among populations of *Hordeum vulgare* ssp. *spontaneum*. Genetics 134: 909–916.

Zhao, X. and Kochert, G. 1993. Phylogenetic distribution and genetic mapping of a (GGC)n microsatellite from rice (*Oryza sativa* L.) Plant Mol. Biol. 21: 607–614.

# 6
# Pollen Biotechnology

**Vipen K. Sawhney**
*University of Saskatchewan, Saskatoon, Saskatchewan, Canada*

I. INTRODUCTION 99
II. GENETIC CONTROL OF POLLEN DEVELOPMENT 100
   A. Male Sterile Mutants 100
   B. Chemical Hybridizing Agents 102
   C. Genetic Engineering 102
III. GENETIC TRANSFORMATION OF POLLEN 104
   REFERENCES 105

## I. INTRODUCTION

Pollen grain, the male gametophyte, in flowering plants is a microscopic two- or three-celled structure, and its primary function is to deliver male gametes (sperm cells) to the female reproductive organ, the carpel. Pollen grains are highly desiccated structures when mature and contain either two cells, a vegetative cell and a generative cell; the latter divides to form two sperm cells before germination, or three cells: a vegetative cell and two sperm cells. Pollen are produced in large numbers in the anther of a stamen and may be carried long distance by wind, insects, or other animals for fertilization. Pollen grains have a thick outer protective wall, the exine (1,2), which contains sporopollenin, a complex polymer that makes the pollen resistant to decay, a useful feature for pollen storage and function. The exine also forms a distinct sculpturing pattern on the pollen surface that is characteristic of a plant species (2). Since plants are sedentary organisms, they are entirely dependent on pollen for the transport of sperm cells either to the carpel of the same flower in which they are produced (self-pollination) or to a flower of another plant (cross-pollination) of the same species. Thus, pollen development is crucial for successful sexual reproduction in angiosperms, and for subsequent fruit and seed development. In addition, pollen grains are of direct, or indirect, importance in several areas of human interest, e.g., in honey production, in pharmaceutical products, in pollen allergens, and as food supplement (see e.g., 3–5).

Over the years, there has been a tremendous interest in understanding the genetic, physiological, and environmental control of pollen development. The major reason for this has been the ultimate ability to control pollination and fertilization in plants, especially in crops. The control of pollination, which in essence involves the control of pollen development or pollen function, is of significance in the production of hybrid seed. Since the discovery of heterosis (hybrid vigor) in maize plants in the later part of the 19th century, most cultivated cereals and vegetable crops are now grown from hybrid seed, instead of seed from inbred lines. Indeed, the hybrid seed technology has revolutionized crop production throughout the world.

*Pollen biotechnology* is the "manipulation of pollen development and/or function with the objective(s) of increased production and improvement of crops, and other pollen products" (6). There are several steps during pollen development, and later during the delivery of male gametes, that are known to be regulated by genetic, chemical, and environmental factors. These include premeiotic and meiotic events in the microsporogenous tissue, the development of microspores and pollen in the anther, pollen maturation and pollen release, pollen dispersal, attachment of pollen to the stigma of carpel, pollen germination and tube growth, and release of sperm cells into the female gametophyte in the ovule of an ovary. A disruption in any one or more of these events can lead to failure of fertilization and the lack of seed and fruit development. This chapter is focused on discussing the genetic control and manipulation of pollen development and the genetic transformation of pollen grains with the objectives of crop improvement. Discussions on related areas, e.g., pollen-stigma interactions and self-incompatibility, can be found elsewhere (e.g. 7–10).

## II. GENETIC CONTROL OF POLLEN DEVELOPMENT

There are a large number of genes controlling different steps in pollen development, starting from the differentiation of the microsporogenous tissue through to the maturation and release of pollen from the anther. Genes controlling pollen development reside both in the nucleus and in the cytoplasm, mainly mitochondria. A number of nuclear-encoded or genic male sterile (GMS) and cytoplasmic male sterile (CMS) mutants are known in every major crop (see, e.g., Ref. 11). For example, in maize over 100, in soybean approximately 20, and in tomato 50 GMS mutants have been reported. In 1999 by using T-DNA and EMS mutagenesis, Sanders et al. (12) reported over 800 male sterile mutants in *Arabidopsis thaliana*. This exemplifies the complexity of the genetic control of pollen development.

It is estimated that there are approximately 24,000 different mRNA transcripts in maize pollen. Whereas the majority of these represent housekeeping genes, or the genes expressed in vegetative tissues, about 355 genes are specifically expressed in the pollen (13). These estimates are in line with other species examined. In general, pollen genes are categorized into two classes, "early" and "late" genes (14). The early class genes are associated with the development of the microspores from the sporogenous tissue, and the later genes are considered to have a role in pollen maturation, germination, and pollen tube growth on the stigma (14–16). The majority of pollen genes are the late genes and the expression patterns of many of these show homologies to wall degrading enzymes, e.g., pectate lyase (17,18), and to proteins involved in cytoskeleton (19,20). The wall degrading enzymes are probably required for the growth of the pollen tube through the stigma and style of the female reproductive organ, and the cytoskeleton proteins, e.g., actin, α- and β-tubulin, and profilin, have a role in cytoplasmic streaming and the growth of pollen tubes (15). A list of some of the pollen-specific genes can be found in Hamilton and Mascarenhas (16).

The primary objective of manipulating pollen development has been to develop "pollination control systems" for use in hybrid seed production. There have been several approaches used to manipulate male fertility/sterility in angiosperms, especially in crop plants. These include the use of spontaneous and induced male sterile (ms) mutants; application of chemical hybridizing agents (CHAs), also called *gametocides,* for the induction of male sterility; and genetic engineering.

## A. Male Sterile Mutants

Considerable research has gone into the characterization of various CMS and GMS systems at the genetic, cytological, hormonal, and, more recently, molecular levels. As well, the role of environmental factors, e.g., temperature, photoperiod, and drought, in male sterility has been examined. The latter are particularly important if ms systems are to be used under field conditions for commercial scale production of hybrid seed.

CMS is maternally inherited and generally represents an alteration in the mitochondrial genome (21–23). Molecular analysis of CMS systems in a number of species has shown that genes concerned with male sterility are novel open reading frames (ORFs) linked to 5' or 3' ends of the normal genes. ORFs may be formed by different mechanisms, including duplication, recombination, insertion, or deletion of the mitochondrial genes (see, e.g., Refs. 24,25). The interesting feature of CMS is that although the ORFs are expressed in different parts of the plant, the phenotypic changes are only observed in the anther. Research has shown that the expression of sterility genes in the cytoplasm is influenced by fertility restoration (RF) genes in the nucleus. However, the mechanisms by which RF genes regulate pollen fertility but have no influence on other tissues are not well understood.

The expression of CMS can be affected by environmental factors, e.g., temperature and light. In *Brassica napus*, for example, in the *nap* and *ogu* CMS systems, high temperatures restore partial to complete fertility in ms plants (26,27). Similarly, the expression of some of the CMS systems can be altered by plant hormones (reviewed in Ref. 28), and CMS mutants may also contain altered levels of endogenous hormones, particularly cytokinins (29). Thus, it appears that both the physiological and molecular factors are responsible for the expression of CMS. Although CMS has been successfully used in some crops, e.g., sunflower, sorghum, sugarbeet, corn, and recently rapeseed or canola (30,31), there can be some limitations to its use. These include negative pleiotropic effects of CMS, increased disease susceptibility, instability of male sterility and fertility restoration due to environmental factors, and absence of restorer lines (11,31).

GMS is more commonly known than CMS in most plant species examined. The majority of GMS mutants are recessive, although some dominant mutants also exist (11). At the cytological level, lesions in nuclear genes controlling male fertility may affect pollen development at any stage, i.e., from premeiosis to pollen release. In the anther, the tapetum, the tissue layer that surrounds the developing pollen grain, is known to provide many essential precursors and metabolites for pollen development (2,32,33). In many of the GMS, and some of the CMS systems, breakdown in pollen development is associated with aberrations in tapetum development (e.g. 31, 34,35). Indeed, many of the genes controlling pollen development are believed to reside in the tapetum. This has led to the development of molecular strategies that selectively destroy tapetum, thereby inducing male sterility in crops (discussed later).

The expression of GMS can also be regulated by environmental and hormonal factors. For example, in male sterile lines in *Brassica oleracea* (36) and tomato (37) low temperatures restore fertility. Photoperiod can also affect pollen development in some of the GMS mutants, e.g., rice (38) and tomato (35). Similarly, hormones, e.g., gibberellins (GAs), are known to restore fertility in some of the male sterile systems and this is correlated with reduction in the level of GAs, but increase in the concentration of indole acetic acid (IAA) and abscisic acid (ABA) in ms tissues (reviewed in Ref. 28). This ability to manipulate male sterility/fertility in GMS lines by environmental or hormonal factors is especially useful in hybrid seed programs.

Since GMS is generally expressed in the recessive condition, the mutant lines are maintained by backcrossing with the heterozygotes. The progeny produced are one half fertile and one half male sterile. Thus, in the field male fertile plants have to be manually removed (rogued) so that ms plants alone are used as female parents. This poses an impediment for the use of GMS systems in large-scale hybrid seed production. Different strategies have been proposed to overcome this limitation, including the linking of ms genes to marker genes (reviewed in Ref. 35). The approach of restoring fertility by chemical or environmental factors permits the production of 100% pure male sterile seed, which can be directly used in the field as female parents for hybrid seed production (35).

## B. Chemical Hybridizing Agents

Another approach to controlling pollen development is to treat normal wild-type plants with growth regulating chemicals that would result in abnormal pollen structure or function. This approach has been used with a number of crops, but most notably with wheat (39–41). Various CHAs, or male gametocides, have been used, including known plant hormones, e.g., auxins, gibberellins, ABA, and ethrel, and a number of synthetic compounds. The effects of these compounds have, in general, not been specific on pollen development. For example, in addition to inducing pollen sterility, the effects of an auxin, naphthalene acetic acid (NAA), on lentil plants included increased branching, inhibition of plant growth, and flower abnormalities (42). Similarly gibberellic acid, i.e., $GA_3$, affects pollen development in *Capsicum annuum*, but it also causes feminization of flowers and affects female fertility (43). Abscisic acid is known to cause male sterility in wheat (44) and tomato (45), but ABA is also known to affect plant growth, including growth of floral organs.

There have been a few promising synthetic compounds that affect pollen development, including Fenridazon (46), phenylcinnoline carboxylates (47), and aztidine-3-carboxylates. These compounds, along with some trade name substances (reviewed in Ref. 41), have largely been tested on wheat and have had variable success in inducing complete male sterility in the plant and ultimate use in hybrid seed production. The mode of action of many of these compounds is, however, not well understood.

## C. Genetic Engineering

In recent years transgenic approaches, which involve the linking of an anther-specific promoter with a gene encoding an enzyme that selectively affects certain aspects of pollen development, have been used effectively for inducing male sterility. The first such success was obtained by Mariani et al. (48), who used a construct of a tapetum-specific promoter TA29 and a ribonuclease gene (*barnase*) from the bacterium *Bacillus amyloliqueafaciens* and expressed it in tobacco and rapeseed plants. The expression of this chimeric gene resulted in the destruction of tapetum in the anther and thus failure of pollen development or induction of male sterility (Fig. 1). However, the male sterility induced was dominant, and, as in the case of recessive GMS systems (described previously), when ms plants were crossed with the heterozygotes, 50% of the resulting plants were male fertile. This problem was overcome by linking the male sterile gene to a gene that is resistant to the herbicide glufosinate ammonium. Thus, treating plants of the ms and heterozygote cross with the herbicide at an early stage eliminated the fertile plants, i.e., with sensitivity to the herbicide, and 100% male sterile population could be obtained (49,50).

Since male sterility in the genetically transformed plants is dominant, one half of the $F_1$ hybrids produced by such a system would be male sterile. Thus, a restorer gene is required to obtain full fertility in the hybrids. Mariani et al. (51) used another gene, *barstar* also from *B. amyloliqueafaciens*, which is a specific inhibitor of barnase, and linked it to the same tapetum-specific promoter, TA29. The pollen donating plant contains the barstar, which inactivates the barnase in $F_1$, and the hybrids are 100% fertile (Fig. 1). This genetic engineering approach has now been successfully used to produce hybrid seed in canola (oilseed rape) and in corn (49,51).

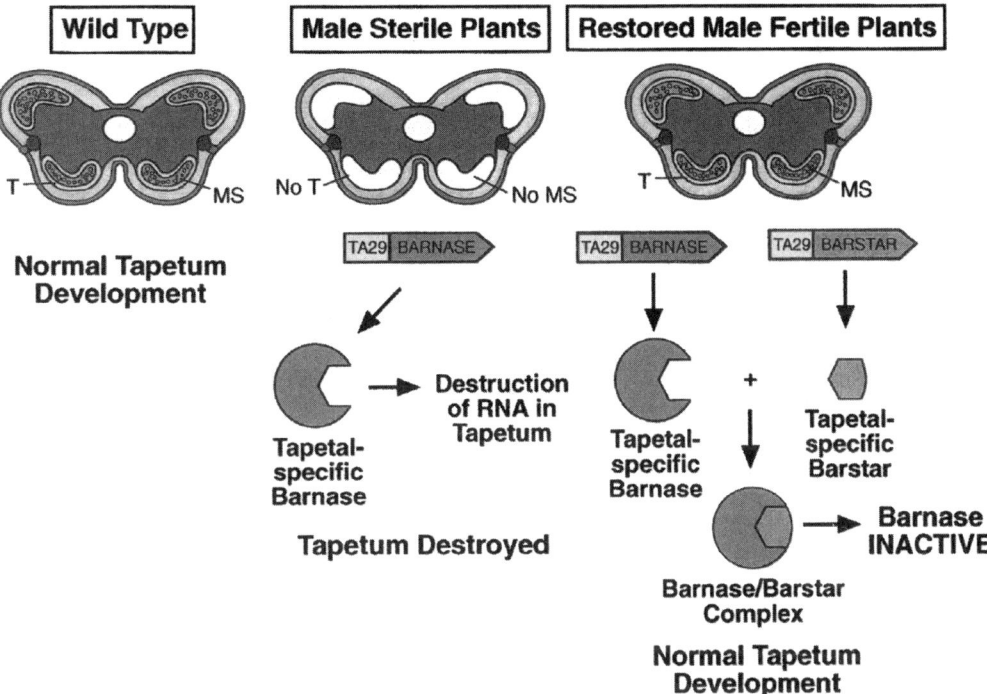

**Figure 1** Genetic engineering of male sterility/fertility using the barnase and barstar system. MS, microspores; T, Tapetum. (From Ref. 48a. Reproduced with permission.)

An alternative approach for the restoration of male fertility in transgenic plants is to treat the male sterile plants with a chemical to induce pollen development. In such a system, called *reversible male sterility* (RMS), Greenland et al. (52) fused the *barstar* gene with a promoter *GST27*, isolated from corn, that is up-regulated by chemicals called *safeners;* treatment with safeners causes herbicide tolerance in plants. Thus, transgenic male sterile plants were treated with safeners and normal pollen development was obtained. The restored plants were selfed and the resulting population was 100% male sterile.

Male sterility has also been induced by developing constructs containing a tapetum-specific promoter and the β-1,3-glucanase (callase) gene (53,54). In the anther, callase breaks down callose (β-1,3-glucan), which is deposited around the tetrads of microspores during meiosis, thereby releasing microspores in the anther locule. The timing of both the callase release and the dissolution of callose is critical, and the early release of callase in the transformants resulted in abnormal pollen development (53,54). Other approaches for inducing male sterility include inhibiting the expression of chalcone synthase (CHS), an enzyme required for flavonoid synthesis. Flavonoids are critical for pollen maturation, and in their absence functional male sterility results (55,56). Male sterility was induced by expressing an antisense CHS cDNA fused with a *35S CaMV* promoter in the tapetum tissue. In transgenic male sterile plants produced by this method, fertility was restored by spraying plants with flavonoids during pollination (57).

In an interesting, opposite approach, Kriete et al. (58) developed a system in which the tapetum-specific promoter *TA29* was fused with the *argE* gene from the bacterium *Escherichia coli*.

The argE product codes for an enzyme that causes deacetylation of the compound $N$-acetyl-L-phosphinothricin (N-ac-Pt). N-ac-Pt is otherwise nontoxic to plant tissues, but the deacetylated compound has cytotoxic effects on anther tissues. Thus, when transgenic plants containing the *TA 29/ argE* construct were sprayed with N-ac-Pt, it was deacetylated and male sterility resulted: i.e., anthers were devoid of pollen. In the absence of spray treatment, plants were male fertile, and therefore, fertility restoration in $F_1$ generation was not required.

## III. GENETIC TRANSFORMATION OF POLLEN

Several attempts have been made to transform pollen by gene delivery with two objectives: (a) to produce genetically modified seed by pollinating plants with transformed pollen and (b) to produce haploid plants from transformed pollen and microspores. The following is a brief summary of the efforts made, and successes obtained, in this area.

Different approaches have been used to transform pollen genetically. These include *Agrobacterium*- and polyethylene glycol– (PEG-) mediated transformation, electroporation, and particle bombardment (for review see Ref. 59). Each of these techniques has limitations, partly because of the tough nature of the pollen wall, the exine. However, significant successes have been obtained with one or the other approach. For example, by using particle bombardment, transient expression of foreign genes has been observed in pollen of a number of plant species e.g., tomato (60), tobacco (61), corn (62), and lily (63). Anther-specific promoters *LAT52* and, *LAT 56* from tomato and *ZM13* from corn were used to drive the expression of *gus* gene in the transformed pollen. A number of factors influenced the expression of foreign genes, including the stage of microspore/pollen development, preculture of pollen in a growth medium, and type of the medium, and type and size of the particle used in particle bombardment. Various levels of expression were observed in transformed pollen, depending upon the species and the promoter used. Although there are no published reports on the production of transgenic seed using transformed pollen, the successes obtained so far are significant and will form the ground work for future research in this direction.

The production of haploid plants from transformed microspores has been achieved in *Nicotiana rustica* (59) and in *Brassica napus* and *B. rapa* (64). In the *Brassica* species, young microspores were transformed by *Agrobacterium* containing the *GUS* or *PAT* gene. The modified microspores were cultured in vitro and more than 100 putative transformants were obtained that developed into haploid embryos. After colchicine treatment, doubled haploid plants were produced that were fully fertile and later set seed. In 1998, by using a nondestructive marker, the *luciferase* (*Luc*) gene, Fukuoka et al. (65) also transformed *B. napus* microspores and obtained haploid embryos. These embryos were diploidized by colchicine, and the progeny produced from mature fertile transgenic plants showed *Luc* activity, indicating that the introduced gene was fixed in the $T_1$ generation. This study documents that it is possible to obtain double haploid transgenic plants by combining the genetic transformation of pollen with the techniques of production of pollen embryos.

## ACKNOWLEDGMENTS

The author expresses his gratitude to Prof. K. R. Shivanna for his critical review of the manuscript. The author's research reported here was supported by the Natural Sciences and Engineering Research Council of Canada.

## REFERENCES

1. Y Iwanami, T Sasakuma, Y Yamada. Pollen: Illustrations and Scanning Electron Micrographs. Tokyo: Kodansha & Springer, 1988.
2. M Cresti, S Blackmore, JL van Went. Atlas of Sexual Reproduction in Flowering Plants. Berlin: Springer-Verlag, 1992.
3. RG Stanley, HF Linskens. Pollen: Biology, Biochemistry and Management. Berlin: Springer-Verlag, 1972.
4. JO Schmidt, Sl Buchman. Other products of the hive. In: JM Graham, ed. The Hive and the Honey Bee. Hamilton IL: Dadant & Sons, 1992, pp 927–982.
5. SS Mohapatra, RB Knox. Pollen Biotechnology: Gene Expression and Allergen Characterization. New York: Chapman & Hall, 1996.
6. KR Shivanna, VK Sawhney. Pollen Biotechnology for Crop Production and Improvement. New York: Cambridge University Press, 1997.
7. JB Nasrallah, JC Stein, MK Kandasamy, ME Nasrallah. Signalling the arrest of pollen tube development in self-incompatible plants. Science 266:1505–1508, 1994.
8. DP Mattan, N Nass, AE Clarke, E Newbigin. Self-incompatibility: How plants avoid illegitimate offsprings. Proc Natl Acad Sci USA 91:1992–1997, 1994.
9. FCH Franklin, MJ Lawrence, VE Frankin-Tong. Cell and molecular biology of self-incompatibility in flowering plants. Int Rev Cytol 158:1–64, 1995.
10. A McCubbin, HG Dickinson. Self-incompatibility. In: KR Shivanna, VK Sawhney, eds. Pollen Biotechnology for Crop Production and Improvement. New York: Cambridge University Press, 1997, pp 199–217.
11. MLH Kaul. Male sterility in higher plants. Berlin: Springer-Verlag, 1988.
12. PM Sanders. AQ Bui, K Wetterings, KN McIntire, Y-C Hsu, PY Lee, MT Truong, TP Beals, RB Goldberg. Anther developmental defects in *Arabidopsis thaliana* male sterile mutants. Sex Plant Reprod 11: 297–322, 1999.
13. RP Willing, D Basche, JP Mascarenhas. An analysis of the quantity and diversity of messenger RNAs from pollen and shoots of *Zea mays*. Theor Appl Genet 75:751–753, 1988.
14. JP Mascarenhas. Gene activity during pollen development. Annu Rev Plant Physiol Plant Mol Biol 41: 317–338, 1990.
15. JP Mascarenhas. Molecular mechanisms of pollen tube growth and differentiation. Plant Cell 5:1303–1314, 1993.
16. DA Hamilton, JP Mascarenhas. Gene expression during pollen development. In: KR Shivanna, VK Sawhney, eds. Pollen Biotechnology for Crop Production and Improvement. New York: Cambridge University Press, 1997, pp 40–58.
17. S McCormick. Male gametophyte development. Plant Cell 5:1265–1275, 1993.
18. MP Turcich, DA Hamilton, JP Mascarenhas. Isolation and characterization of pollen-specific maize genes with sequence homology to ragweed allergens and pectate lyases. Plant Mol Biol 23:1061–1065, 1993.
19. JL Carpenter, SE Ploense, DP Snustad, C Silflow. Preferential expression of an $\alpha$-tubulin gene of *Arabidopsis* in pollen. Plant Cell 4:557–571, 1992.
20. M Thangavelu, D Belostotsky, MW Bevan, RB Flavell, HJ Rogers, DM Lonsdale. Partial characterization of the *Nicotiana tabacum* actin gene family: Evidence for pollen-specific expression of one of the gene family members. Mol Gen Genet 240:290–295, 1993.
21. MR Hanson. Plant mitochondrial mutations and male sterility. Annu Rev Genet. 25:461–486, 1991.
22. CS Levings. Thoughts on cytoplasmic male sterility in CMS-T maize. Plant Cell 5:1285–1290, 1993.
23. F Kempken, D Pring. Plant breeding: Male sterility in higher plants—fundamentals and applications. Prog Bot 60:139–166, 1999.
24. M Singh, G Brown. Suppression of cytoplasmic male sterility by nuclear genes alters expression of a novel mitochondrial gene region. Plant Cell 3:1349–1362, 1991.
25. S Krishnasamy, CA Makaroff. Characterization of the radish mitochondrial *orfB* locus: Possible relationship with male sterility in Ogura radish. Curr Genet 24:156–163, 1993.

26. Z Fan, BR Stefansson. Influence of temperature on sterility of two cytoplasmic male sterile systems in rape (*Brassica napus* L.). Can J Plant Sci 66:221–227, 1986.
27. PL Polowick, VK Sawhney. High temperature induced male and female sterility in canola (*Brassica napus* L.). Ann Bot 62:83–86, 1988.
28. VK Sawhney, A Shukla. Male sterility in flowering plants: Are plant growth substances involved? Am J Bot 81:1640–1647, 1994.
29. S Singh, VK Sawhney. Cytokinins in a normal and the ogura (*ogu*) cytoplasmic male sterile line of rapeseed (*Brassica napus*). Plant Sci 86:147–154, 1992.
30. WR Feistritzer, AF Kelly. Hybrid Seed Production of Selected Cereal, Oil and Vegetable Crops. Rome: FAO, 1987.
31. PBE McVetty. Cytoplasmic male sterility. In: KR Shivanna, VK Sawhney, eds. Pollen Biotechnology for Crop Production and Improvement. New York: Cambridge University Press, 1997, pp 155–182.
32. KR Shivanna, BM Johri. The angiosperm pollen: Structure and function. New Delhi: Wiley Eastern, 1985.
33. E Pacini, GG Franchi, M Hesse. The tapetum: Its form, function and possible phylogeny in embryophyta. Plant Syst Evol 149:155–185, 1985.
34. P Bedinger. The remarkable biology of pollen. Plant Cell 4:879–887, 1992.
35. VK Sawhney. Geneic male sterility. In: KR Shivanna, VK Sawhney, eds. Pollen Biotechnology for Crop Production and Improvement. New York: Cambridge University Press, 1997, pp 183–198.
36. MH Dickson. A temperature-sensitive male sterile gene in brocolli, *Brassica oleracea* L. var. *italica*. J Am Soc Hortic Sci 95:13–14, 1970.
37. VK Sawhney. Temperature control of male sterility in a tomato mutant. J Hered 74:51–54, 1983.
38. MS Shi. The discovery, determination and utilization of the Hubei photosensitive genic male sterile rice (*Oryza sativa* ssp. *Japonica*). Acta Genet Sinica 13:107–112, 1986.
39. DH McRae. Advances in chemical hybridization. Plant Breed Rev 3:169–191, 1985.
40. JW Cross, JAR Ladyman. Chemical agents that inhibit pollen development: Tools for research. Sex Plant Reprod 4:235–243, 1991.
41. JW Cross, PJ Schulz. Chemical induction of male sterility. In KR Shivanna, VK Sawhney, eds. Pollen Biotechnology for Crop Production and Improvement. New York: Cambridge University Press, 1997, pp 218–236.
42. NNC Awasthi, DK Dubey. Pollen abortion in chemically-induced male sterile lentil (*Lens culinaris*). Lens Newsletter 12:12–16, 1985.
43. VK Sawhney. Abnormalities in pepper (*Capsicum annuum*) flowers induced by gibberellic acid. Can J Bot 59:8–16, 1981.
44. HS Saini, D Aspinall. Sterility in wheat (Triticum aestivum L.) induced by water deficit or high temperature: possible mediation by abscisic acid. Aust J Plant Physiol 9:529–537, 1982.
45. R Rastogi, VK Sawhney. Suppression of stamen development by CCC and ABA in tomato floral buds cultured in vitro. J Plant Physiol 133:620–624, 1988.
46. GA El-Ghazaly, WA Jensen. Development of wheat (*Triticum aestivum*) pollen wall before and after the effect of a gametocide. Can J Bot 68:2509–2516, 1990.
47. WJ Guilford, TG Patterson, RO Vega, L Fang, Y Liang, HA Lewis, JN Labovitz. Synthesis and pollen suppressant activity of phenylcinnoline-3-carboxylic acids. J Agric Food Chem 40:2026–2032, 1992.
48. C Mariani, C De Beuckeleer, J Truettener, J Leemans, RB Goldberg. Induction of male sterility by a chimaeric ribonuclease gene. Nature 347:737–741, 1990.
48a. RB Goldberg, TP Beals PM Sanders. Plant Cell 5:1217–1229, 1993.
49. ME Williams. Genetic engineering for pollination control. Trends Biotech 13:344–349, 1995.
50. ME Williams. Male sterility through recombinant DNA technology. In: KR Shivanna, VK Sawhney, eds. Pollen Biotechnology for Crop Production and Improvement. New York: Cambridge University Press, 1997, pp 237–257.
51. C Mariani, K D'Halluin, C Dickburt, C De Beuckeleer, M De Block, RB Goldberg, W De Greef, J Leemans. A chimaeric ribonuclease-inhibitor gene restores fertility to male sterile plants. Nature 357:384–387, 1992.
52. A Greenland, P Bell, C Hart, I Jepson, T Nevshermal, J Register, S Wright. Reversible male sterility:

A novel system for the production of hybrid corn. In AJ Greenland, EM Meyerowitz, M Steer, eds. Proceedings of the Symposium on Control of Plant Development: Genes and Signals. Cambridge, 1998, pp 141–147.
53. D Worrall, DL Hird, R Hodge, W Paul, J Draper, R Scott. Premature dissolution of the microsporocyte callose wall causes male sterility in transgenic tobacco. Plant Cell 4:759–771, 1992.
54. T Tsuchiya, K Toriyama, M Yoshikawa, S Ejhiri, K Hinata. Tapetum-specific expression of the gene for an endo-beta 1,3-glucanase causes male sterility in transgenic tobacco. Plant Cell Physiol 36:487–494, 1995.
55. IM van der Meer, ME Stam, AJ van Tunen, JNM Mol, AR Stuitje. Antisense inhibition of flavonoid biosynthesis in petunia anthers results in male sterility. Plant Cell 4:253–263, 1992.
56. LP Taylor, R Jorgensen. Conditional male fertility in chalcone synthase-deficient petunia. J Hered 83: 11–17, 1992.
57. AJ van Tunen, IM van der Meer, JNBM Mol. Flavonoids and genetic modification of male fertility. In: EG Williams, AE Clarke, RB Knox, eds. Genetic control of self-incompatibility and reproductive development in flowering plants. Dordrecht: Kluwer, 1994, pp 423–442.
58. G. Kriete, K Niehaus, AM Perlick, A Puhler, I Broer. Male sterility in transgenic tobacco plants induced by tapetum-specific deacetylation of the externally applied non-toxic compound N-acetyl-L-phosphinothricin. Plant J 9:809–818, 1996.
59. H Morikawa, M Nishihara. Use of pollen in gene transfer. In: KR Shivanna, VK Sawhney, eds. Pollen Biotechnology for Crop Production and Improvement. New York: Cambridge University Press, 1997, pp 423–437.
60. D Twell, R Wing, J Yamaguchi, S. McCormick. Transient expression of chimeric genes delivered into pollen by microprojectile bombardment. Plant Physiol 91:1270–1274, 1989.
61. M Nishihara, M Ito, I Tanaka, M Kyo, K Ono, K Irifune, H Morikawa. Expression of the β-glucuronidase gene in pollen of lily (*Lilium longiflorum*), tobacco (*Nicotiana tabacum*), *N. rustica* and peony (*Paeonia lactiflora*) by particle bombardment. Plant Physiol 102:357–361, 1993.
62. DA Hamilton, M Roy, J Rueda, RK Sindhu, J Sanford, JP Mascarenhas. Dissection of a pollen-specific promoter from maize by transient transformation assays. Plant Mol Biol 18:211–218, 1992.
63. LM Plegt, BCE Ven, RJ Bino, TPM Salm, AJ van Tunen. Introduction and differential use of various promoters in pollen grains of *Nicotiana glutinosa* and *Lilium longiflorum*. Plant Cell Rep 11:20–24, 1992.
64. M. Dormann, H-M Wang, N Datla, AMR Ferrie, WA Keller, MM Oelck. Transformation of freshly isolated *Brassica* microspores and regeneration to fertile homozygous plants. Procedings of the 9th International Rapeseed Congress, Cambridge, 1995, pp 816–818.
65. H Fukuoka, T Ogawa, M Matsuoka, Yohkawa, H Yuno. Direct gene delivery into isolated microspores of rapeseed (*Brassica napus* L.) and the production of fertile transgenic plants. Plant Cell Rep 17:323–328, 1998.

# 7
# Parent-of-Origin Effects and Seed Development: Genetics and Epigenetics

**Charles Spillane and Ueli Grossniklaus**
*University of Zurich, Zurich, Switzerland*

**Jean-Philippe Vielle-Calzada**
*Cold Spring Harbor Laboratory, Cold Spring Harbor, New York*

| | | |
|---|---|---|
| I. | INTRODUCTION | 109 |
| II. | PARENT-OF-ORIGIN EFFECTS AT THE GENE LEVEL | 111 |
| III. | GENOMIC IMPRINTING IN HIGHER PLANTS | 113 |
| IV. | PARENT-OF-ORIGIN EFFECTS IN PLANT BREEDING | 114 |
| | A. Parent-of-Origin Effects for Crossability | 115 |
| | B. Endosperm Genome Dosage Ratios and Imprinting Effects in Interploidy Crosses | 115 |
| | C. Pseudogamous Apomixis and Genomic Imprinting | 118 |
| | D. Epigenetic Regulation of Seed Quality Traits | 120 |
| | E. Parent-of-Origin Effects for Combining Ability and Heterosis | 120 |
| V. | THE *medea* MUTANT: A GATEWAY FOR THE DISSECTION OF GENOMIC IMPRINTING MECHANISMS | 122 |
| VI. | EPIGENETIC ENGINEERING OF SEED TRAITS | 124 |
| VII. | CONCLUSIONS | 126 |
| | REFERENCES | 126 |

## I. INTRODUCTION

Seeds play a crucial role in the evolution of both higher plants and human civilization. The domestication of the major crop plants about 13,000 years ago was contingent upon altering key seed characteristics, especially within the Gramineae and Leguminosae (1,2). The seeds of just three crops (rice, wheat, and maize) provide more than half of the global plant-derived energy intake (3). The persistent endosperm of wheat, rice, maize, sorghum, millet, barley, and oats contributes much of the world's food supply (4). The world market for crop seed is approximately U.S. $45 billion, which can be roughly divided into three equal categories—commercial seed, farm-saved seed, and seed provided by government institutions (5). The importance of crop seeds has provided much impetus for research to understand seed ontogeny and develop improved seed characteristics (e.g., improved starch, protein, and oil profiles) (6).

Understanding seed formation and ontogeny requires due consideration of the characteristic life strategy of plants (7). The plant life cycle alternates between a diploid and a haploid generation, the sporophyte (spore-producing organism) and the gametophyte (gamete-producing organism). In the sexual organs of the sporophyte, mega- and microspores are produced: the haploid products of meiosis that mark the beginning of the gametophytic phase. The fusion of the gametes at fertilization concludes the gametophytic phase to reconstitute the diploid sporophyte by forming the zygote. After fertilization the ovule bearing the female gametophyte develops into a seed, in a complex process that depends on interactions among various tissues of zygotic and maternal origin. The seed is the vegetative propagule for the plant embryo.

Seed development in all angiosperms depends on double fertilization, involving the fusion of two pairs of gametic cells. One of the sperm cells delivered by the pollen tube fuses with the egg cell to form the diploid zygote; the second fuses with the binucleate central cell to form the triploid primary endosperm nucleus (8,9). The angiosperm seed is usually comprised of (a) the embryo; (b) the endosperm; (c) the perisperm, derived from the nucellar tissue of the ovule; and (d) the testa or seed coat, formed from one or both of the integuments of the ovule (Fig. 1). Although all mature seeds contain an embryo, and many are surrounded by a seed coat, the extent to which the endosperm or perisperm persists varies between species. The embryo consists of the embryo proper and a suspensor, which is linked to the sporophyte and may play a nutritional role. Endosperm is a nutritive tissue found only in the angiosperms (4). The endosperm may have a differentiated epidermal cell layer, the aleurone. Where it is persistent, endosperm typically accumulates storage proteins, lipids, and starch used later during seed germination. Current research on endosperm development mainly focuses on cereals and more recently on the model dicot *Arabidopsis thaliana* (hereafter referred to as *Arabidopsis*) (10).

Viable seed formation depends on the coordinated development of the embryo, endosperm, and maternal seed coat. The interactions between these cells and tissues remain an unresolved and complex aspect of seed development and it is not clear to what extent one tissue influences the development of the other. Studies of maize (11) and rice (12) mutants indicate that the develop-

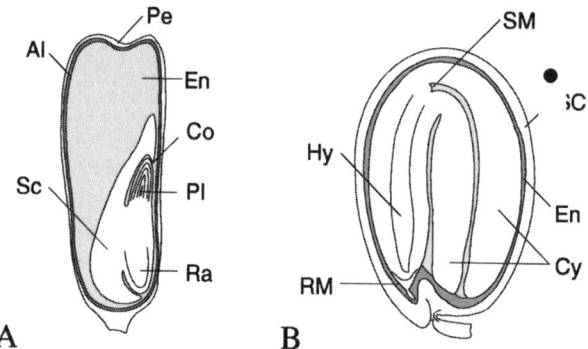

**Figure 1** Mature seeds of *Zea mays* and *Arabidopsis thaliana*. (A) Maize kernel showing endosperm (En), aleurone (Al), pericarp (Pe), scutellum (Sc), radicle (Ra), coleoptile (Co), and plumule (Pl); (B) *Arabidopsis* seed showing endosperm (En), cotyledons (Cy), hypocotyl (Hy), root meristem (RM), shoot meristem (SM), and seed coat (SC). Not drawn to scale: maize seed 8–9 mm long, *Arabidopsis* seed approximately 0.5 mm long. (Reprinted with permission, copyright 1999, Springer Verlag. From Ref. 27.)

ment of embryo and endosperm are interrelated. However, mutations affecting embryogenesis but apparently not the endosperm have also been described in rice and maize (13,14). Hence, normal embryogenesis appears to not be strictly required for endosperm development. Although a failure of endosperm development in maize often results in embryo abortion (15), the embryo can undergo normal morphogenesis in some *dek* mutants with very little endosperm (16). Similar observations have been made for endospermless rice grains (17).

Because of double fertilization and the clonal origin of both male and female gametes in most angiosperm species, it is technically difficult to determine whether a mutant primarily affects the embryo or the endosperm. Chang and Neuffer used B-A translocations with a range of maize *dek* mutants to generate kernels that had a genetically normal embryo in the presence of a genetically mutant endosperm, or vice versa (18). Their studies suggest that although interactions between endosperm and embryo may play a role with respect to nutrition, embryo and endosperm determine their independent morphogeneses, and neither tissue requires a normal counterpart for its own normal development (16).

Different types of parent-of-origin effects can arise during seed development because of complex interactions (a) between maternally provided factors and zygotically expressed genes, (b) between sporophytic and zygotic tissues, and (c) between the embryo and endosperm. Successful seed development may not only depend on the specialized cytoplasm of the female gametes, but also require genomes from both parents.

Parent-of-origin effects can have a genetic or an epigenetic basis. Epigenetics generally relates to mitotically and/or meiotically heritable changes in gene function that do not involve any changes in deoxyribonucleic acid (DNA) sequence (19). An epimutation at any locus can phenotypically behave as a genetic mutation but does not result from a change in the DNA sequence. Epimutations may result from differential effects of chromatin packing and act in *cis* (20). Whereas epigenetic inheritance has been observed in many species, the underlying molecular mechanisms remain largely unknown. Epigenetic effects have been observed for many seemingly unrelated phenomena in plants, including paramutation (21), dioecious sex determination (22), nucleolar dominance (23), transposition (24,25), genomic imprinting (26–28), meiotic drive (29), and transgene silencing (30).

In this chapter we review the evidence for parent-of-origin effects and epigenetic regulation of seed development and plant reproduction and outline some molecular and genetic approaches underway to dissect the underlying genetic and epigenetic mechanisms.

## II. PARENT-OF-ORIGIN EFFECTS AT THE GENE LEVEL

Parent-of-origin effects are observed if reciprocal crosses confer differing phenotypes on the F1 (maternal or paternal effect) or F2 progeny (grandparent-of-origin effect). Parent-of-origin effects are seen in many organisms (31,32). Maternal effect genes have been extensively studied in *Drosophila melanogaster* (hereafter referred to as *Drosophila*) and *Caenorhabditis elegans* (hereafter referred to as *Ceanorhabditis*), where most affect a cytoplasmic factor stored in the egg cell.

In plants, maternal effects can be more complex because double fertilization allows for maternal control over both embryo and endosperm development (33,34). Maternal control can be exerted via gametophytic (i.e., the female gametophyte) and/or sporophytic tissues (i.e., maternal ovule tissues surrounding the female gametophyte or developing embryo and endosperm) (Fig. 2) (9,35). On the basis of mutational analyses it is known that seed morphogenesis requires maternal gene activity in the gametophytic (34,36,37) as well as in the sporophytic tissues of the developing ovule (38–40).

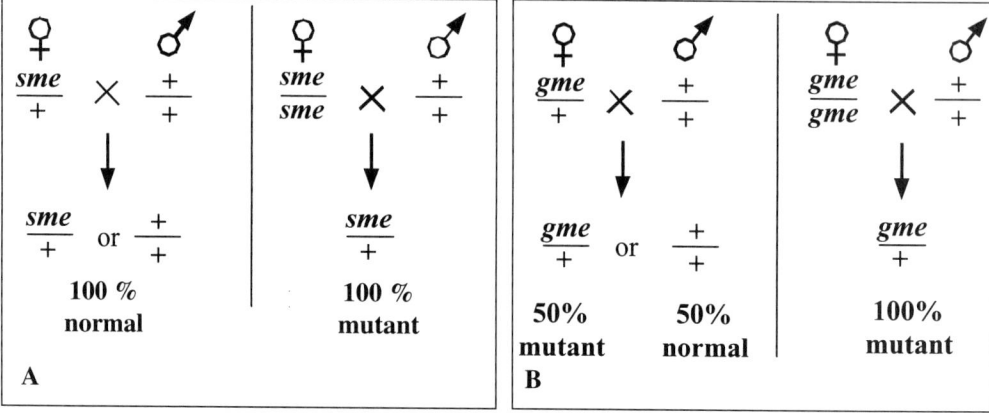

**Figure 2** Genetic behavior of maternal effect mutants in plants. (A) Results of test crosses that identify a sporophytic maternal effect (*sme*), such as *sin1* in *Arabidopsis*. (B) Results of test crosses that identify a gametophytic maternal effect (*gme*), such as the *mea*, *fie*, and *fis2* mutants in *Arabidopsis*.

Early events in embryogenesis depend on a complex interplay between maternally provided factors and zygotically expressed genes. In many animals, maternal products (messenger ribonucleic acid [mRNA] or protein), deposited in the egg, can support the embryo for a limited number of division cycles, prior to the activation of zygotic transcription. Although at least 50 maternal effect genes affecting *Drosophila* embryogenesis have been identified, it is unknown to what extent maternally provided components direct early embryogenesis in plants. The polarity of the egg and early embryo is suggestive of a maternal influence on axis determination (9,41,42). Whereas considerable information on gene expression is available for later stages of embryogenesis (43,44), very little is known about the molecular events occurring in the early morphogenetic phase.

In plants, a gametophytic maternal effect mutation can be caused by (a) a mutation in a cytoplasmic factor deposited into the egg and/or central cell, (b) disruption of a dosage-sensitive gene, or (c) disruption of an imprinted gene (28,34). In angiosperms, gametophytic maternal effects can be of egg cell or central cell (endosperm effect) origin (35). Only a few gametophytic maternal effect mutations that influence embryogenesis have been identified (9,35,45). The gametophytic maternal effect mutants known to date in *Arabidopsis* are the *medea (mea), fertilization independent seed2 (fis2),* and *fertilization independent endosperm (fie)* genes. In these mutants, seeds that inherit a mutant allele from the mother abort (34,36,37,46).

Maternal sporophytic tissues play important roles in embryo and/or endosperm formation; mutations that disrupt such roles can display a sporophytic maternal effect. These tissues include (a) the inner and outer integuments that are the precursors of the seed coat, (b) the nucellus that forms the perisperm, and (c) the ovary wall, which can differentiate into the pericarp or fruit coat. The first maternal effect embryo-defective mutation (*short integument1*) discovered is essential for normal embryo development in *Arabidopsis* (39). There are more identified sporophytic maternal effect mutants affecting the endosperm: at least six maize mutants display a defective aleurone pigmentation phenotype (*Dap*) (47), and in *Petunia hybrida*, sporophytic expression of two MADS-box transcription factors (*FBP7* and *FBP11*) is required for normal endosperm development (40). A number of barley *shrunken endosperm* mutants that exhibit sporophytic maternal effects on endosperm formation have been identified (38,48).

Parent-of-origin effects have also been observed for quantitative trait loci (QTL). Reciprocal crosses between several plant species have shown strong maternal effects on seed size (33). Maternal effect QTL responsible for differences in seed size and number between two *Arabidopsis* ecotypes (Cvi, L*er*) have been mapped (49). The Cape Verde Islands (Cvi) ecotype yields on average about 40% fewer seeds than Landsberg *erecta* (L*er*), but Cvi seeds are almost twice as heavy. The seed size differences between the two ecotypes resulted from changes in the final cell number and cell size of the seed coat and the embryo. Cell number variation was controlled mainly by maternal factors, whereas nonmaternal effect allelic variation mostly affected cell size. Five of the seed size QTL colocated with QTL for other traits, suggesting that they control seed size via trade-offs with maternal components affecting ovule number, carpel or ovule development, or reproductive resource allocation in the mother plant. It is possible that the large size observed in Cvi × L*er* hybrid seeds and the slightly smaller than L*er* mean size of the reciprocal L*er* × Cvi hybrid seeds could result from genomic imprinting at loci controlling seed size (49).

## III. GENOMIC IMPRINTING IN HIGHER PLANTS

Genomic imprinting is a parent-of-origin-dependent epigenetic phenomenon. Imprinted genes are differentially expressed depending on their parental origin, resulting in non-Mendelian segregation and parent-of-origin-specific functional hemizygosity at the imprinted locus (50). Differential imprinting can be specific to a developmental stage and/or tissue.

Two types of genomic imprinting have been identified. Imprinting was first shown for parent-of-origin-dependent inheritance of chromosomes in coccoid insects (*Sciara* and allies), in which the entire paternal genome is inactivated or eliminated (51–53). Parent-of-origin-specific chromosome elimination has also been observed in interspecific barley crosses (54,55). An analogous form of imprinting that affects most if not all of the paternal genome may also occur during early stages of seed development in *Arabidopsis*, in which paternally inherited genes are silenced until the mid-globular stage of embryogenesis (55a).

Gene-specific genomic imprinting refers to the differential marking of maternally or paternally inherited alleles of certain autosomal genes during gametogenesis, whereby the expression level of an allele depends upon its parental origin (56). Such gene-specific genomic imprinting was first described for the maize *red color (r)* locus (57) and for mammals in the early 1980s (58–60). Because genomic imprinting leads to functional hemizygousity, successful embryogenesis in mammals (50) and seed development in many plant species (26,61), respectively, requires both a paternal and a maternal genome in addition to the specialized cytoplasm of the female gametes. It is estimated that there are approximately 100 imprinted genes in the mammalian genome (62), of which over 25 have been identified (50). In contrast, very few imprinted genes have been studied in plants (27). For the remainder of this chapter, we focus on gene-specific genomic imprinting.

In mammals, both androgenesis and gynogenesis are highly restricted as a result of genomic imprinting. The failure of mammalian androgenotes and gynogenotes is probably due to an imbalance in the dosage of imprinted genes (50). However, the requirements for genomic imprinting in plants may be less strict than in mammals and the situation is more complex because of interactions between tissues of different genetic composition and developmental origin. In many angiosperms there is no strict requirement for both paternal and maternal genomes for successful embryogenesis (26,27,63). For instance, plant embryos can form independently of fertilization in apomictic plants (64), and from somatic cells or microspores under appropriate conditions (65). In addition, both maternal (gynogenesis) and paternal (androgenesis) haploid embryos produce viable seedlings in many flowering plants (66–68).

A role of imprinted genes for seed formation has been inferred from interploidy crosses, in which entire parental genomes or individual chromosomes are manipulated (15,26). In maize, changes of the parental genome ratio lead to endosperm abortion but have little effect on embryo development, suggesting that imprinting is specific to the endosperm in this species (69). In *Arabidopsis*, interploidy crosses have an effect on both endosperm and embryo proliferation (70). The recently identified *MEA* gene provides the first instance of an imprinted gene that affects seed development in plants (28,34,45,71). Previously, gene-specific genomic imprinting in plants had only been described for three genes, all of which are expressed in the endosperm of maize but do not affect seed development (27). These are the *r* gene involved in the regulation of anthocyanin biosynthesis (57), the seed storage protein regulatory gene *dzr1* (72), and some α-*tubulin* genes (73). In all four cases (*r, dzr1, α-tubulin, mea*) only maternally inherited alleles are expressed. For the *r, dzr1,* and α-*tubulin* loci only specific alleles are subject to epigenetic regulation by imprinting. In contrast, imprinting in mammals, and at the *mea* locus in *Arabidopsis* is generally locus-specific and all alleles are subject to imprinting (28).

With a few possible exceptions (74,75) gene-specific imprinting has only been observed in mammals and plants (27,50). In both, reproduction is characterized by a placental habit whereby the embryo receives most or all of its postfertilization nutrients from the mother. Haig and Westoby (76) proposed a model whereby genomic imprinting evolved as a consequence of a conflict between paternal and maternal genomes over the allocation of nutrients from the mother to the offspring. The model predicts that some parentally controlled loci should influence the growth rate of the embryo, with paternally expressed genes promoting growth and maternally expressed ones tending to reduce growth. However, a parental conflict is only expected if a female carries offspring from more than one male over the span of her lifetime. Importantly, even a very limited partner exchange in a species is sufficient to induce a parent-offspring conflict (77). In mammals, many imprinted genes are involved in fetal growth and the vast majority are expressed and imprinted in the placenta, although the precise function and role of many imprinted genes are unknown (50,62). In plants, larger seeds with higher food reserves produce more vigorous seedlings. According to Haig and Westoby's (76) evolutionary theory, genes expressed in the offspring that influence seed size are selected to promote greater nutrient flow from the mother and, thus, larger seed size when the genes are of paternal origin, than when the same genes are of maternal origin. This selection pressure ultimately leads to differential expression of maternal and paternal alleles.

Evidence directly supporting Haig and Westoby's (76) model has come from both studies on gynogenetic and androgenetic mouse embryos and imprinted loci in mammals (50,78,79), as well as from interploidy and interspecific crosses in flowering plants (61,70,80). Although not all cases of imprinting can be easily explained by Haig and Westoby's model (50,81), it has been extremely successful in accommodating many of the functional observations on imprinted genes.

## IV. PARENT-OF-ORIGIN EFFECTS IN PLANT BREEDING

Parent-of-origin effects are widespread in plant reproduction and breeding at the whole genome level (82). Many empirically developed plant breeding methodologies require directional crosses at various stages (82,83). A number of plant breeding models based on quantitative genetics have been proposed for the analysis of maternal effects on endosperm and embryo (84–86). The molecular and genetic mechanisms underlying parent-of-origin effects are generally unknown, although they are often interpreted as resulting from gene dosage imbalances caused by genomic imprinting. The following sections outline four important areas of plant breeding where parent-of-origin phenomena are encountered, yet poorly understood at the molecular genetic level.

## A. Parent-of-Origin Effects for Crossability

In higher plants parent-of-origin effects are prevalent in many interspecies reciprocal crosses, and in some intraspecies crosses (61). The phenotype of the F1 generation in such crosses can differ substantially, depending on which of the two parents was the pollen donor or recipient. In many cases, crosses are successful in one direction but not in the other even though the same parents are used. Such effects are particularly common in crosses between crops and their wild relatives.

Haig and Westoby (61) have reviewed the early evidence for parent-of-origin contingent incompatibility in crosses between species. Interspecies "crossability" of different lines between the primary (GP1) to tertiary (GP3) gene pools of crops is of major interest to plant breeders in terms of transferring useful traits into elite germplasm (87–89). The definition of primary to tertiary genepools is empirically based on crossability polygons (90), and many crop genepools are classified into different sections on the basis of their crossability (91,92). For instance, over 1500 Asian rice cultivars that were classified into six different phyletic groups by molecular analysis exhibit various levels of sterility in intergroup crosses but not in intragroup crosses (93,94).

The success of wide hybridization efforts depends on overcoming pre- and postzygotic crossing barriers, the achievement of high fertility, and stable expression of desirable traits in F1 progeny and backcross derivatives. Wide crossing techniques such as hybrid embryo culture (95) and the development of novel crossing strategies (e.g., ploidy changes and bridge crosses) are often necessary to overcome interspecies or interploidy crossability barriers (96). Although such wide crossing techniques are an integral element of plant breeding, the underlying molecular genetic factors conditioning whether a wide cross will work or not are largely unknown (97,98). Prezygotic barriers to wide crossability such as pollen-pistil incompatibility may be responsible for some crossing failures. Syntenic differences between chromosomes causing irregular meiotic pairing and chromosome segregation are thought to be a major cause of hybrid sterility in wide hybridization efforts. Some postzygotic barriers may be due to gene dosage imbalances caused by imprinting that lead to seed abortion (61). However, the contribution of epigenetic barriers to inter- or intraspecies crossability has not been explored to any significant level.

Parent-of-origin effects on crossability have been observed, for instance, in potato (88,89), *Alstroemeria* sp. (99), and intergeneric *Brassica/Crambe* hybrids (100). The cowpea wild relative *Vigna rhomboidea* can only be crossed with the cultivated *V. unguiculata* in a unidirectional manner, with *V. rhomboidea* as the pollen donor (101). There are also instances of segregation distortion in progeny of wide crosses that are suggestive of parent-of-origin effects. For instance, crosses between chickpea and its wild relative *Cicer echinospermum* are reported to produce F1 hybrids with 50% fertility (92).

Studies of wheat-rye hybrids suggest a role for genomic imprinting/gene silencing via the methylation of DNA of rye origin (102). In some inbred polyploid hybrids, such as Triticale (wheat × rye), treatment with the demethylation agent 5-azacytidine releases "hidden" phenotypic variation, suggesting that some aspects of gene expression in such hybrids are regulated by methylation (102). Methylation-defective plants often exhibit developmental abnormalities including the ectopic expression of meristem identity and homeotic genes leading to changes in flowering time and transformations of floral organs (103,104).

## B. Endosperm Genome Dosage Ratios and Imprinting Effects in Interploidy Crosses

Failure of endosperm development can be a cause of seed abortion in both intra- and interspecies hybrids (105). Successful seed development in some species depends on a particular ratio of maternal to paternal genomes in the endosperm. In maize and potato this ratio is two maternal to one paternal genome (2m:1p) (106,107). Deviations from a 2m:1p genome dosage ratio (GDR) typically lead to endosperm abortion and failure of seed development. Seed abortion has been interpreted as resulting from gene dosage imbalances of imprinted genes (26,106). Deviations from the 2m:1p GDR occur in interploidy crosses, and, indeed, many crosses between diploids and autotetraploids fail. If a diploid is used as the pollen parent, the endosperm has a 4m:1p GDR, whereas in the reciprocal cross the resultant endosperm has a 2m:2p GDR. Imprinting effects are also likely to be responsible for the breakdown of endosperm, which is often seen in wide crosses between related plants of different ploidies. However, interploidy crosses provide better evidence for genomic imprinting since there is less allelic variation between the parents potentially contributing to the parent-of-origin effect than in interspecies crosses (76).

Genome instability in plants can be triggered by a change in chromosome number arising from genome duplications (polyploidy) or loss/gain of individual chromosomes (108). Evidently some interploidy crosses do not give rise to viable seeds. Elucidating the causes of instability of some polyploids, aneuploids, and trisomics is of major importance to understanding crop evolution and to devising improved means of plant breeding involving polyploids. It will be interesting to determine whether there are any links (e.g., in terms of GDR) between the genetic mechanisms controlling genome instability and the mechanisms controlling genomic imprinting.

Ploidy manipulations are often necessary to effect gene flow between the secondary genepool and the primary cultivated genepool of many crops. For instance, ploidy-based barriers to wide crossing are observed in most cultivated plants, including cassava (109), pearl millet (110), alfalfa (111), potato (88,89), sweet potato (112), *Musa* spp. (113), and groundnut (114,115). Parent-of-origin-specific effects in diploid × autotetraploid crosses have also been demonstrated for a wide range of crops such as maize (116), barley (117), rye (118), and oilseed rape (119). Similar results have been obtained for wild plants such as *Primula* spp. (120) and *Lycopersicon pimpinellifolium* (121). Evidence for genomic imprinting in reciprocal diploid × tetraploid crosses was strengthened by work on *Oenothera hookeri* (122,123). *Oenothera* sp. is unusual because its endosperm is normally diploid, produced by the fertilization of a single polar nucleus. Therefore, endosperm is triploid in both 4n × 2n and 2n × 4n crosses, such that reciprocal differences could be attributed to parental origin rather than ploidy per se (122,123).

Many models were initially proposed to account for the behavior of reciprocal interploidy crosses (15,61,76). However, not until genetic studies in maize was it conclusively demonstrated that normal maize endosperm development required a 2m:1p GDR (69,106). Although any plant species require a 2m:1p GDR in the endosperm, this requirement is not axiomatic. Many species, including *Arabidopsis* can tolerate deviations from the 2m:1p GDR, as long as the imbalance is not too extreme (70,124).

The embryological evidence for inter-ploidy crosses supports Haig and Westoby's predictions (61,76). An excess of maternal genome dosage in the endosperm often leads to slower nuclear division in the endosperm, earlier cellularization, smaller seeds, and high germination levels. Conversely, excesses of paternal genome dosage in the endosperm are often associated with normal-sized but shrivelled seeds, poor germination, delayed cellularization, and faster nuclear division. Such effects are often seen in seeds resulting from incompatible reciprocal crosses, especially for species with nuclear endosperm. These generalizations are based on supporting evidence from reciprocal crosses in *Avena, Triticum, Zea mays, Hordeum vulgare, Lolium/Festuca, Brassica* spp., *Primula, Oenothera,* and *Citrus* spp. (61,76). More recent genetic evidence from maize (69) and *Arabidopsis* (70) confirms the reciprocal associations between seed size and an excess of either maternal or paternal genome dosage in the endosperm. We briefly review the evidence from interploidy crosses in (a) *Arabidopsis* (b) potato, and (c) maize.

## 1. *Arabidopsis* Interploidy Crosses

In *Arabidopsis*, reciprocal crosses between diploids and tetraploids produce viable seeds containing triploid embryos, indicating that deviations from the 2m:1p GDR can be tolerated (70, 124). *Arabidopsis* is not totally intolerant of GDR imbalances because reciprocal crosses between diploids and hexaploids invariably produce seeds that abort.

Viable seeds produced from the diploid × tetraploid reciprocal crosses exhibit abnormal endosperm development, with maternal and paternal genomic excess producing complementary phenotypes. When the maternal genome dosage in the endosperm is doubled, endosperm development is reduced and a smaller embryo is produced. When the paternal genome dosage in the endosperm is doubled, endosperm and embryo growth is promoted. The alteration of the parental GDR in the endosperm affects seed size at maturity, with larger seed size correlating with increased paternal genome dosage in the endosperm, and vice versa, for increased maternal dosage as expected from Haig and Westoby's model (61).

Changes in the parental genome dosage ratio have effects on seed viability, seed weight, rate of mitosis in the endosperm, and timing of endosperm cellularization. Because interploidy crosses change the genome balance of the embryo and endosperm simultaneously, Scott et al. (70) were not able to determine whether the effects seen were due to imbalance in the embryo, the endosperm, or both. However, they concluded that genomic imprinting is the most likely explanation for the differential phenotypes.

## 2. Potato Interploidy Crosses

In some genera certain interploidy crosses may be more successful than same-ploidy crosses. Within *Solanum* species all intraspecies interploidy crosses conform to the 2m:1p GDR hypothesis (125,126). The failure of a particular species to produce triploids in 4x × 2x intraspecific crosses is commonly called "triploid block" (55). However, this generalization regarding the 2m:1p GDR does not hold at the interspecies level, where some crosses between *Solanum* species of the same ploidy are incompatible, whereas other crosses between species of different ploidies are compatible (88,89). Consequently, Johnston et al. (125) proposed that each species be assigned a specific endosperm value, the endosperm balance number (EBN), which describes its effective ploidy in crosses with an arbitrarily chosen tester species. As a basis for the EBN designation, it was hypothesized that the genes involved in the development of the hybrid endosperm may have different EBN values or "strength" among species, regardless of ploidies. The EBN has been determined for most *Solanum* species by crossing each with standard species of known EBN (127). Through this empirical testing species were identified that were 2x (1EBN), 2x (2EBN), 4x (2EBN), 4x (4EBN), and 6x (4EBN).

According to the EBN hypothesis, normal endosperm development requires a 2:1 ratio of the sum of the EBNs of the polar nuclei to the EBN of the sperm nuclei. For instance, diploid (2n = 2x = 24) *Solanum* species with EBN = 1 are sexually isolated from diploid 2EBN species and both tetraploid (2n = 4x = 48, 4EBN) and dihaploid (2n = 2x = 24, 2EBN) *S. tuberosum* group Tuberosum. Carputo et al. (96) have manipulated EBN levels by scaling ploidy levels both up and down to overcome crossing barriers between the diploid species *S. commersonii* (1 EBN) and the 4x (2EBN) gene pool of cultivated potato.

In a complete diallel cross between wild potato species of different ploidies and EBNs, Masuelli and Camadro (88) observed that in incompatible EBN combinations, more than 85% of the seeds were not well developed or were shrunken. The inviable seeds had poorly developed or collapsed endosperms and thick endothelia, whereas the viable seeds have large endosperms and thin endothelia. In the same study, parent-of-origin effects on crossability were observed for both the self-compatible *S. acaule* and the self-incompatible *S. gourlayi,* which had better crossabilities as female or male parent, respectively. Jackson and Hanneman (89) noted better success in crosses with more unrelated potato species if they were used as the male.

However, not all inter-EBN crosses conform to the EBN hypothesis. Jackson and Hanneman (89) performed reciprocal crosses between cultivated potato (2n = 4x = 48; 4EBN) and over 400 accessions of 134 wild potato species and found that a few crosses were successful despite predicted failure due to EBN differences. In some inter-EBN crosses where the EBN ratio deviates from 2:1, normal or plump seeds are occasionally formed (128,129). Interestingly, some plump seeds are formed in both intra- and inter-EBN crosses, where the 2:1 EBN requirement seems leaky (130).

The EBN seems to be under oligogenic control in potato (131) and models based on two or three unlinked loci that control endosperm development in a threshold-dependent manner have been proposed (55,132,133). The same loci may control endosperm incompatibility both between and within *Solanum* species. However, nothing is yet known about the genes that are responsible for the EBN phenomenon.

### 3. Maize Endosperm Genome Dosage Ratios

In maize, it has been conclusively shown that any deviation from the normal 2m:1p GDR in the endosperm results in seed abortion or the production of small kernels in the case of a 3m:1p GDR (15,69,134). The evidence for the 2m:1p GDR requirement in maize has been extensively reviewed (15,26,27) and is briefly summarized here.

That normal endosperm development in maize requires a 2m:1p GDR was elegantly shown by a series of crosses, using 2n or 4n pollen parents, and 2n seed parents (69), that contributed differing numbers of polar nuclei to the endosperm because of the *indeterminate gametophyte (ig)* mutation (134). Hence, it was possible to generate a range of kernels that had normal 2n sporophytic tissues and either 2n or 3n embryos, but crucially had a complete range of maternal:paternal GDRs in the endosperm. The results demonstrated that it is the parental GDR in the endosperm, rather than the ploidy per se, that is critical, whereas the relative ploidy ratios between the endosperm and embryo or sporophytic tissue are not important. The detection of parental effects in maize endosperm but not in the embryo suggests that imprinting is specific to the endosperm in this species.

It has been proposed that imprinting may play a role in the epigenetic differentiation of egg and central cell during megagametogenesis (26,27). Although genes playing a role in the differentiation of cells in the megagametophyte may not be subject to selection resulting from a parental conflict, the endosperm phenotypes observed in interploidy crosses are in agreement with Haig and Westoby's model (80). Indeed, several regions of the maize genome are suspected of carrying imprinted genes involved in endosperm growth (e.g., chromosome arms 1L, 4S, and 10L) (106,135). For instance, the lack of a paternal copy of any of the endosperm factors (Ef1–4) on chromosome arm 10L leads to the production of small kernels. Kernel size cannot be restored by adding extra maternally derived alleles, suggesting that the Efs are only active if inherited paternally. Thus, two distinct classes of imprinted genes may exist in plants, one involved in the differentiation of the cells in the megagametophyte, and a second that shows functional similarities to imprinted genes in mammals and controls cell proliferation and growth during seed development (27).

### C. Pseudogamous Apomixis and Genomic Imprinting

Apomixis is a naturally occurring process that allows asexual reproduction through seeds, resulting in offspring that are genetically identical to the mother plant (64,136). In theory, apomixis could be used to generate true-breeding F1 hybrids that maintain the benefits of heterosis (137–141). The introduction of apomixis into crop species is expected to revolutionize plant breeding, seed production, and crop improvement and is perceived as one of the most important scientific challenges faced by modern agriculture (138,140–142). It has been estimated that the benefit of apomixis technology to hybrid rice alone could amount to more than U.S.$2.5 billion per annum (143).

Apomictic wild relatives have only been identified for a few agriculturally important grain crops such as pearl millet, wheat, and maize (137). For other major crops such as rice, no apomictic wild relatives have been identified (94). Although apomixis has been introduced from wild relatives into pearl millet (144–146) and maize (147–150), only apomictic plants with a high degree of seed abortion and additional chromosomes have been recovered. Introgression of apomixis through backcrossing programs is dependent on the occurrence of apomixis in wild relatives and is often impeded by breeding barriers between the cultivated and wild species (138).

Imprinting phenomena may account for the high degree of sterility observed in some sexual × apomictic hybrids resulting from attempts to introgress apomixis into sexual crops, e.g., *Tripsacum*/maize hybrids (149,151); *Pennisetum* spp./pearl millet hybrids (144,152,153), and *Poa longifolia/P. pratensis* hybrids (154). A better understanding of genomic imprinting in plants will be essential for introgression or de novo engineering of apomixis in sexual crops with persistent endosperm, e.g., cereals (141).

Apomicts can be classified as either pseudogamous or autonomous. In autonomous apomicts, both the embryo and endosperm develop without fertilization. However, the majority of apomicts require fertilization of the central cell (pseudogamy) to ensure endosperm formation if viable apomictic seeds are to be obtained. In contrast to the formation of unreduced megagametophytes, apomicts usually produce pollen with a reduced chromosome number (64). Fertilization of an unreduced central cell with reduced sperm results in a GDR imbalance that may result in endosperm abortion (141). To overcome this epigenetic constraint apomicts use two strategies: (a) They are insensitive to genomic imbalances in the endosperm, or (b) they modify gametogenesis or fertilization in a manner that results in viable endosperm with the correct GDR (61,141). Such modifications include the production of (a) unreduced megagametophytes that are 4-nucleate (*Panicum* type), with usually only one rather than two polar nuclei (64,155); (b) unreduced male gametes (156,157); (c) fertilization of the two polar nuclei with both sperm cells delivered by the pollen tube (158,159); or (d) karyogamy of a single unfused polar nucleus with one reduced sperm nucleus (160).

In *Tripsacum dactyloides* (hereafter referred to as *Tripsacum*) and *Paspalum notatum* there are no GDR constraints for endosperm development. Grimanelli et al. (161) have investigated dosage effects in the endosperm of diplosporous apomictic *Tripsacum*. They found that endosperm develops normally over a wide range of maternal to paternal GDRs in both apomicitc and sexual accessions. Thus, a specific GDR is not required for normal endosperm development in *Tripascum* Quarin (162) analyzed the effect of different sources and ploidy levels of pollen donors on endosperm and seed development in the aposporous tetraploid apomict *P. notatum*. Although a GDR of 2m:1p is required for endosperm and seed formation in sexual *P. notatum* plants, pseudogamous apomictic *P. notatum* exhibit an insensitivity to such requirements. This insensitivity seems to be effective when the maternal genomic input exceeds the 4x level, possibly as a result of imprinting changes. Whether this is also the case in *Tripsacum* is not clear since no controlled pollinations were made to address whether diploid sexual *Tripsacum* is sensitive or insensitive to imbalances in the parental GDR. However, tetraploid sexual *Tripsacum* accessions were also found to be insensitive to dosage effects due to imprinting (161).

In any species in which genomic imprinting is strictly essential to embryogenesis and/or endosperm development, it constitutes an obstacle to the introduction of apomixis (141,161). The successful engineering of apomixis in cereals depends upon the production of functional, high-quality endosperm for consumption, in combination with the production of a viable embryo that can be clonally propagated. Although autonomous apomixis is highly desirable (139), genomic imprinting constraints on endosperm and/or embryo development may make the engineering of autonomous apomixis difficult, especially in sexual crops such as cereals, which are especially sensitive to GDR imbalances in the endosperm (140,141).

### D. Epigenetic Regulation of Seed Quality Traits

Physiologically seeds behave as sinks for assimilates from maternal sources. During assimilate partitioning, seed growth and development are controlled by coordinated interactions among the embryo, endosperm, seed coat, and other maternal tissues. As a result, many seed quality traits exhibit a complex inheritance because they are controlled by several genetic systems (e.g., nuclear genes of maternal plant, nuclear genes of endosperm and embryo, and organellar genes). Cooking quality characters in cereals are largely defined by the constitution of the endosperm. A number of studies have demonstrated that seed quality traits in crops such as cotton, rice, and barley can be controlled by gene effects of embryo, endosperm, and/or maternal tissues (163–166).

Development of cereal grains can be divided into two stages: grain enlargement and grain filling. Grain enlargement is the result of cell division followed by cell expansion. Grain filling occurs as the reserves, starch and proteins, are deposited within the persistent endosperm. Grain filling is a highly important agronomic trait in cereals, where a number of cell layers in the endosperm play a critical role. The endosperm's strength as a nutrient sink during grain filling is primarily determined by the number of endosperm cells, and this number is stabilized before grain filling begins (167–169). Endosperm with more cells accumulates more storage materials. The development of endosperm depends on both sink capacity and assimilates supplied by sporophytic maternal tissues, and hence the maternal genotype.

In angiosperm seeds the embryo sac is symplastically isolated from the maternal tissues. The embryo and endosperm also share no symplastic continuity. Specialized transfer cell layers are thought to play important roles in nutrient transfer. The aleurone cell layer surrounds the reserve cells of the endosperm and can consist of one layer (maize, rice, wheat) or up to four layers (barley). The transfer cell layer is composed of modified aleurone cells and is considered to have a role in nutrient transfer from maternal to gametophytically derived tissues.

Charlton et al. (170) performed crosses in maize between diploid female plants and autotetraploid pollen donors to generate seeds that contained an unbalanced GDR (2m:2p) in the endosperm. Such seeds abort but undergo apparently normal development until 10–12 days after pollination (DAP). Structural comparisons of aberrant and normal endosperm indicate that the formation of the transfer cell layer is almost completely suppressed when paternal genome dosage is in excess. Both triploid and tetraploid endosperms have a similar ontogeny until 6 DAP, when in the normal triploid endosperm early signs of transfer cell layer development are first observed. In the tetraploid endosperms, transfer cell layer development fails to be initiated.

Imprinted genes may play crucial roles in the regulation of endosperm cell number, endoreduplication, and differentiation of the transfer cell layers and thus grain size, grain filling, and other seed quality traits. Thus, it should be possible to improve seed quality traits through targeted manipulation of the (epi)genetic regulation of endosperm cell number and transfer cell layer development.

### E. Parent-of-Origin Effects for Combining Ability and Heterosis

Heterosis or hybrid vigor is usually described in terms of the superiority of F1 hybrid performance over some measure (midparent or high-parent heterosis) of its parents (82). Yet, despite decades of theoretical and empirical studies the genetic basis of heterosis and the factors causing heterotic hybrid genesis and breakdown remain unclear. Most studies propose a role for dominance (masking of deleterious recessives), overdominance (single-locus heterosis), or epistasis (traits derived from two different lines that give superior performance in combination). In rice, both dominance and epistasis have been suggested to be the genetic basis of heterosis (171,172).

Many plant breeding methodologies are proposed on the basis of different theoretical models for heterosis: recurrent selection for general combining ability, and inbred per se selection (additive effects), recurrent selection for specific combining ability (dominance effects), and reciprocal recurrent selection (both additive and dominance effects) (173). Prediction of hybrid performance is of major interest to hybrid crop breeding programs. Although some studies have shown positive correlations between genetic distance and heterosis (174), in general such prediction has proved difficult as the molecular genetic basis of heterosis remains unknown. In both plant and animal breeding, the relative level of heterosis observed in F1 and F2 generations can be affected by maternal effects. Plant breeders routinely conduct polycrosses, top-crosses, or diallel crosses with other (tester) lines to determine empirically which lines are likely to be the best candidates for use in further breeding.

Elucidation of the molecular or biochemical basis of heterosis remains a challenging and elusive task (175–177). Rood et al. (178) have positively correlated higher gibberellin levels with heterosis in maize. Single-gene (overdominance) heterosis is rare but has been demonstrated in a number of cases (179–182). There may be links between paramutation and single-gene heterosis. Paramutation is the meiotically heritable alteration of one allele after exposure to another allele in particular heterozygous combinations (21,183). For instance, two alleles of a gene may interact such that one of the alleles is epigenetically silenced and the silenced state is genetically transmissible for one or more generations. Paramutation-like phenomena are being found in a wide variety of organisms (21,184). Paramutation at the multigenic *R-r* locus in maize has been correlated with increases in its level of cytosine methylation (185). Hollick and Chandler (182) revealed interesting parallels between paramutation and heterosis. They found that a paramutable maize locus (*Pl-mah*) could exhibit single-locus heterosis, or overdominance, whereby the heterozygote displays a gene activity that is greater than in either homozygote.

It is still an open question whether there is some epigenetic basis to heterosis (177,186–188). Although there is certainly no strong evidence for epigenetic components of heterosis, there are disparate reports that warrant further testing. Chakraborty (189) proposed that genomic imprinting may mimic observations that are often construed to be due to hybrid vigor and/or inbreeding depression. Interestingly, tissue-specific heterosis may also be observed for endosperm qualities (55,190,191).

The methylation status of parental lines and heterotic F1 hybrid progeny has only been investigated in a few instances (177,192). In a number of studies it was found that hybrids were less methylated than inbreds and that improved inbred lines were less methylated than older low-yielding lines (177). An investigation of the heterochromatin and euchromatin composition in maize F1 hybrids and their parental inbreds found that in some F1 hybrids there was an increased level of heterochromatin (192). It was proposed that chromatin structure changes within hybrid nuclei appeared to be necessary for proper organization of the F1 hybrid genome (192). As with most methylation studies, it is not known whether the changed levels of methylation seen in some F1 hybrids are a cause or an effect of heterosis.

Heterosis is considered as the converse to inbreeding depression (82). Highly inbred parental lines are developed for use in F1 hybrid seed production through successive generations of self-pollination with selection for desired attributes in the F1 progeny. Specific combining abilities are considered to be positively correlated with the level of inbreeding that line has undergone. Interestingly, Brink and Cooper (190) showed that seed failure due to the inbreeding of alfalfa was often the result of the collapse of the endosperm, suggesting that inbreeding in some species is detrimental to endosperm function and development (55).

Very little is known about the genetic basis of inbreeding depression in plants. Because the observed heterosis from inbred line combinations may be due to the recovery of inbreeding depression, light may be shed on the genetics of heterosis by studying the genetics of inbreeding depression. Pray and Goodnight (193) found evidence for nonlinearity in inbreeding depression that suggested that epistasis may be important for some traits. They also concluded that the genetic variation present for inbreeding depression may suggest that inbreeding depression is a heritable trait.

## V. THE *medea* MUTANT: A GATEWAY FOR THE DISSECTION OF GENOMIC IMPRINTING MECHANISMS

The elucidation of the genetic basis of genomic imprinting may have important implications for the development of novel or more efficient crop improvement strategies, wide or interploidy crosses, development of apomictic crops, improvement of endosperm quality, and heterosis breeding. An applied perspective on the predictions of Haig and Westoby's intergenomic conflict theories (61,76) suggests that it may be possible to deliberately manipulate imprinted genes in order to change seed size and other seed traits. Model systems for genomic imprinting in plants will greatly facilitate the identification of genes involved in genomic imprinting, and possibly other parent-of-origin effects.

The amenability of *Arabidopsis* to genetic and molecular analysis makes it an ideal system for the identification and molecular isolation of genes and mutants controlling sexual plant reproduction (9,41,194). The imprinted *MEA* gene in *Arabidopsis* provides the first example of an imprinted gene in a model dicot plant species (28,34) and a powerful tool for the dissection of genomic imprinting in higher plants. Grossniklaus and coworkers (34) used a transposon mutagenesis approach to isolate *mea*, a gametophytic maternal effect embryo-lethal mutant that exhibits aberrant seed development. The *MEA* gene regulates cell proliferation by exerting gametophytic maternal control during seed development, producing a phenotype that is in agreement with Haig and Westoby's theory (34,61,76). The *mea* locus is regulated by genomic imprinting and defines a new class of imprinted genes in higher plants (28).

Self-fertilization of plants heterozygous for the *mea* mutation produce 50% aborted seeds that collapse, accumulate anthocyanin, and do not germinate (Fig. 3). Reciprocal crosses between *mea* mutants and the wild type have demonstrated that embryo lethality is under strict maternal (gametophytic) control. Seeds derived from embryo sacs carrying a mutant *mea* allele abort after delayed morphogenesis with excessive cell proliferation in the embryo and reduced free nuclear divisions in the endosperm. Embryos derived from *mea* eggs grow to a giant size, suggesting that the wild-type function of *MEA* is to restrict cell proliferation and embryo size. Morphogenetic progression of *mea* embryos appears normal, but these embryos eventually die during seed desiccation as a result of delayed progression through embryogenesis (34).

*MEA* encodes a SET domain protein with homology to members of the *Polycomb* (Pc-G) and *trithorax* group (trx-G) (34). The 130 amino acid SET domain is present in *Drosophila* proteins, e.g., **Su**(*var*)*3–9*, *Enhancer of zeste (E(z))*, **trithorax**, which are best known for regulating homeotic genes but also play crucial roles in the regulation of cell proliferation (195). *MEA* belongs to the *E(z)* SET domain subfamily (34), of which *CURLY LEAF (CLF)*, a regulator of the floral homeotic gene *AGAMOUS*, was the first plant member to be identified (196). Pc-G and trx-G proteins are thought to regulate gene expression via the control of higher-order chromatin structure, possibly via mechanisms with similarity to imprinting (195). Therefore, *MEA* may have two links to imprinting: (a) its expression is regulated by genomic imprinting, and (b) it may regulate target genes by an imprinting-like mechanism that involves chromatin remodeling.

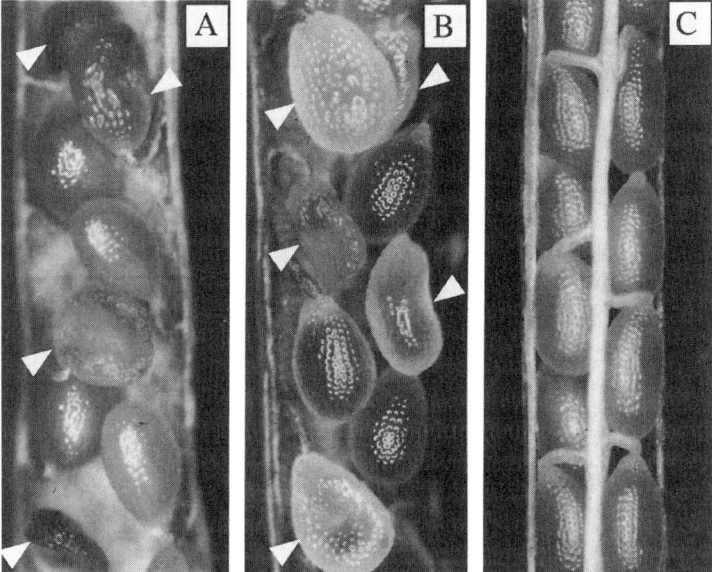

**Figure 3** The *medea* (*mea*) mutant displays a gametophytic maternal effect. (A) In a heteozygous *mea/MEA* plant, half of the seeds abort after self-fertilization. (B) In an outcross with a wild-type male, 50% of the seeds also abort. (C) All seeds are normal if the *mea/MEA* plant is used as a male in a cross to a wild-type female. (B and C reprinted with permission from Ref. 34. Copyright 1998 American Association for the Advancement of Science.)

Five alleles of *mea* have been described (34,37,46,197,198), all of which are likely to be recessive loss-of-function mutations, although this has only been demonstrated for three of them (34,198). Given that truncations of *E(z)* can create antimorphic alleles (203,204), the situation, however, may be more complex. Ovules carrying a mutant *mea* megagametophyte are able to initiate endosperm development and seed coat differentiation, and to induce silique (fruit) maturation in the absence of fertilization at a low frequency (45,197,198). Hence, in addition to the gametophytic maternal effect phenotype, all of the five known *mea* alleles display a fertilization-independent endosperm phenotype, a feature of autonomous apomixis. The relationship between the two phenotypes is unknown.

The *fie* and *fis2* mutants also show autonomous endosperm development and a gametophytic maternal effect on seed formation (36,37). *FIE* is most similar to *Extra-sex-combs (Esc)*, another member of the *Drosophila* Pc-G (36), *FIS2* encodes a protein with a TFIIIA-like Zn-finger motif (197). Although it is now known that *MEA* is an imprinted gene (28), the nature of the maternal effect on seed development in *fie* and *fis2* mutants has not been elucidated yet. Either they may be regulated by imprinting themselves, or their activity may depend on interactions with a factor that is regulated by imprinting such as *MEA*.

The interactions between *E(z)* and *Esc* or their homologues have been well characterized in *Drosophila* (201,202) and mammals (205,206). Similar interactions and modes of action for the regulation of higher-order chromatin structure might be inferred for the homologous plant components (e.g., *MEA, CLF, FIE*) of Pc-G and trx-G regulatory complexes. It is possible that the fertilization-independent endosperm phenotype is a manifestation of a disruption in protein complexes that are involved in epigenetic regulation of endosperm proliferation or seed development (197,198).

Molecular and genetic characterization of the *mea* mutant continues with the goal of elucidating the regulation of imprinting and its function in seed development. Because large-scale mutagenesis approaches are not facile in mammalian models, the *mea* mutant provides a unique entry point for identifying genes involved in genomic imprinting, including potential "imprintor" genes. Using a variety of strategies we aim to identify both modifiers and target genes of *MEA*.

## VI. EPIGENETIC ENGINEERING OF SEED TRAITS

Deliberate modification of higher-order chromatin structure to generate useful phenotypic effects may be considered a form of "epigenetic engineering." Chromatin serves as the structural organizer of DNA. Advances are being made in understanding the molecular factors required to propagate or maintain gene activation or gene silencing, through modification of higher-order chromatin domains (205). A number of multiprotein complexes that affect local domains of chromatin structure provide a cellular memory system for epigenetic inheritance of repressed or activated chromosome regions (206).

Genes of the Pc-G and trx-G are part of an evolutionarily conserved cellular memory system that, by modulating higher-order chromatin structure, maintains inactive and active states of homeotic gene expression (207). Related Pc-G and trx-G loci in mammals encode components of multiprotein complexes that regulate transcriptional activation, repression, and aspects of chromatin structure, including the regulation of *HOX* genes (208). Both Pc-G and trx-G protein complexes alter the accessibility of DNA to factors required for transcription by remodeling chromatin. Once the respective chromatin states are established early in embryogenesis, they are faithfully inherited during development. As there is no DNA methylation in *Drosophila*, such epigenetic regulation cannot be methylation-dependent.

Pc-G and trx-G protein complexes are likely to have different compositions at different target genes (209). In *Drosophila,* formation of these multiprotein complexes appears to depend on particular DNA elements, the overlapping Pc-G response elements (PREs), and trx-G response elements (TREs) (206). PREs, TREs, and cis-regulatory imprinting control elements found in mammalian imprinted genes may be functionally or structurally similar, even through they recruit different trans-acting factors (206).

A number of multiprotein complexes in yeast, *Drosophila caenorhabditis,* and humans are involved in the regulation of gene expression through remodeling of chromatin (205,210). For instance, the evolutionarily conserved *SWI/SNF* multiprotein complex functions to open chromatin, permitting access to transcription factors (211,212). In comparison to those in animal systems, little is yet known about the corresponding multiprotein complexes in plants, which may regulate higher-order chromatin structure, or any phenotypic effects resulting from the perturbation of such complexes.

The ability to control key genes that affect epigenetic control of developmental or agronomic traits may open new avenues for crop improvement and facilitate existing plant breeding approaches. Jaenisch (79) suggested that imprinting may be dispensable under conditions in which the imprints on both parental genomes are erased. If such erasure can be induced in a developmental or tissue-specific manner, then it may be possible to reduce genomic imprinting barriers to crop improvement. Both transgene silencing and dominant-negative approaches to the perturbation of such epigenetic regulatory systems may be possible once the underlying genes and protein interactions controlling epigenetic phenomena are elucidated. The deliberate alteration of chromatin structure at key stages of crop development or in a tissue-specific manner may allow for transgenic approaches to the modulation of epigenetic agronomic traits or a form of epigenetic engineering.

Epigenetic engineering of seed size may be possible. In 1999 Vielle Calzada et al. (28) used a "candidate modifier" approach to deliberately perturb genomic imprinting at the *mea* locus. Potential modifiers of *mea* may exist among mutants known to affect DNA methylation or gene silencing such as the *decreased in DNA methylation1 (ddm1)* mutant in which genomic DNA methylation is reduced by 70% (213–215). *DDM1* was shown to encode a chromatin remodeling factor of the *SWI2/SNF2* family of DNA-dependent adenosine triphosphatases (ATPases) (215).

Mutations in *DDM1* can rescue *mea* seeds by activating the paternally inherited *MEA* wildtype allele later during seed development (28). Our results indicate that *DDM1* is required for the maintenance but not the establishment of the imprint at the *mea* locus. Remarkably, rescue of aborting *mea* embryos through a lack of zygotic *DDM1* activity led to the formation of giant seeds (Fig. 4). The *mea* seeds rescued by *ddm1* show overgrowth of the embryo and some persistent endosperm that may or may not be cellularized. The phenotype of the enlarged rescued seeds suggests that during early seed development there is no *MEA* activity in $mea^m/MEA^p$; $ddm1/ddm1$ seeds, leading to delayed embryogenesis and larger embryos as in the *mea* mutant. Later in seed development, *MEA* activity is provided from the paternally inherited allele, which is reactivated as a result of a lack of *DDM1* activity and allows embryogenesis to resume and form viable seeds of giant size (28). Thus, seed size can be manipulated by inhibiting *MEA* activity early in seed development but providing *MEA* activity later.

Seed size is a key adaptive trait for all plants, and an important agronomic trait for cultivated plants. Selection for larger seeds in wheat and maize through conventional breeding, although successful, has proved difficult as a result of inverse correlations observed with other desirable traits such as number of seeds, fruits, and/or inflorescences (216–218). A number of major QTL that affect seed size have been identified in crops such as sorghum (219), rice (220,221), pea (222), tomato (223), soybean (224), and other legumes (225).

Haig and Westoby's parental conflict theory (61,76) predicts that deliberate perturbations of genomic imprinting could lead to increases in embryo or endosperm size. The phenotype of

**Figure 4** Rescue of $mea^m/MEA^p$ seeds in a $ddm1/ddm1$ background. The rescued $mea^m/MEA^p$; $ddm1/ddm1$ seeds grow to a giant size (top row), in comparison to normal siblings that are $MEA^m/mea^p$ and segregate for *ddm1* (bottom row). (With permission from Ref. 28, copyright 1999 Cold Spring Harbor Press.)

the *mea* mutant suggests that resultant increases in embryo size may lead to inviable seeds, unless it is possible to rescue seed abortion by reactivation of *MEA,* for instance, through a second site mutation such as *ddm1.* Unlike most of the cereals, Brassicaceae such as *Arabidopsis* do not have a persistent endosperm of economic importance. Nevertheless, it may be possible to engineer increased seed (especially endosperm) size in cereals by an appropriate manipulation of *MEA* activity (or the activity of its functional homologues) because *MEA* also affects cell proliferation in the endosperm (34). It should be possible to increase embryo size in dicot crops such as cotton and oilseed rape in which the main part of the seed is of embryonic origin (165), by either controlling *MEA* expression stage-specifically in the seed or through creating second site modifier mutations. Finally, abolishing *MEA* activity should lead to an overall decrease of seed size in species in which seedlessness is a desirable trait (226).

## VII. CONCLUSIONS

A greater knowledge of genomic imprinting in plants could prove important for future transgenic and conventional breeding approaches to crop improvement. It is possible that the gametic imprints (marks) for many imprinted genes lie in their promoter regions (50,81). Given that minor changes in the promoter activity of key regulatory genes can lead to major phenotypic changes (227,228), alterations in regulatory regions of key developmental genes could be major determinants of reproductive compatibility and embryogenesis.

Whole genome duplication (polyploidy) is an important source of evolutionary novelty in many eukaryotic organisms (229,230). If imprinting and other dosage-related gene silencing phenomena are ubiquitous in polyploids of higher plants, then such phenomena probably play a major role in crop evolution and plant breeding (230–232).

It is likely that epigenetic regulation such as genomic imprinting has significant effects both on reproductive biological characteristics (e.g., crossability barriers) and seed morphogenesis (e.g., embryo or endosperm size, grain filling). The use of genes that regulate epigenetic phenomena and are controlled by transgenic inducer/repressor systems may make it possible to engineer conditional epigenetic effects, such as the perturbation of genomic imprinting at key stages of plant breeding programs or crop production. Such manipulations may open new avenues for biotechnology through epigenetic engineering.

## REFERENCES

1. WR Lush, LT Evans. The seedcoats of cowpeas and other grain legumes: structure in relation to function. Field Crops Res 3:267–286, 1980.
2. AH Paterson, Y-R Lin, Z Li, KFM Schertz, JF Doebley, SRM Pinson, S-C Liu, JW Stansel, JE Irvine. Convergent domestication of cereal crops by independent mutations at corresponding genetic loci. Science 269:1714–1718, 1995.
3. R Prescott-Allen, C Prescott-Allen. How many plants feed the world? Conservation Biol 4:365–374, 1990.
4. MA Lopes, BA Larkins. Endosperm origin, development and function. Plant Cell 5:1383–1399,
5. Rabobank. The World Seed Market: Developments and Strategy. Rabobank: Netherlands, 1994.
6. JD Bewley, M Black. Seeds: Physiology of Development and Germination. New York: Plenum Press, 1994, p 445.
7. V Walbot. Sources and consequences of phenotypic and genotypic plasticity in flowering plants. Trends Plant Sci 1:27–32, 1996.
8. SD Russell. The egg cell: Development and role in fertilization and early embryogenesis. Plant Cell 5:1349–1359, 1993.

9. U Grossniklaus, K Schneitz. The molecular and genetic basis of ovule and megagametophyte development. Semin Cell Dev Biol 9:227–238, 1998.
10. F Berger. Endosperm development. Curr Opin Plant Biol 2:28–32, 1999.
11. MG Neuffer, WF Sheridan. Defective kernel mutants of maize. I. Genetic and lethality studies. Genetics 95:929–944, 1980.
12. SK Hong, H Kitano, H Satoh, Y Nagato. How is embryo size genetically regulated in rice? Development 122:2051–2058, 1996.
13. SK Hong, T Aoki, H Kitano, H Satoh, Y Nagato. Phenotypic diversity of 188 rice embryo mutants. Dev Genet 16:298–310, 1995.
14. JK Clark, WF Sheridan. Isolation and characterisation of 51 embryo-specific mutations of maize. Plant Cell 3:935–951, 1991.
15. JA Birchler. Dosage analysis of maize endosperm development. Annu Rev Genet 27:181–204, 1993.
16. WF Sheridan, JK Clark, MG Neuffer. Embryo-endosperm interactions in early seed development. Abstract t6 presented at Conference on Harnessing Apomixis, College Station, TX, September 25–27, 1995.
17. Y Kageyama, H Fukuoka, K Yamamoto, G Takeda. The rice plant bearing endospermless grains: A novel mutant induced by gamma-irradiation of tetraploid rice (*Oryza sativa* L.) Japan J Breed 41:341–345, 1999.
18. MT Chang, MG Neuffer. Endosperm-embryo interactions in maize. Maydica 39:9–1, 1994.
19. AP Wolffe, MA Matzke. Epigenetic regulation through repression. Science 286:481–6, 1999.
20. BD Hendrich, HF Willard. Epigenetic regulation of gene expression: the effect of altered chromatin structure from yeast to mammals. Hum Mol Genet 4:1765–1777, 1995.
21. JB Hollick, JE Dorweiler, VL Chandler. Paramutation and related allelic interactions. Trends Genet 13:302–8, 1997.
22. A Lardon, S Georgiev, A Aghmir, G Le Merrer, I Negrutiu. Sexual dimorphism in white campion: Complex control of carpel number is revealed by Y chromosome deletions. Genetics 151:1173–1185, 1999.
23. ZJ Chen, CS Pikaard. Epigenetic silencing of RNA polymerase I transcription: A role for DNA methylation and histone modification in nucleolar dominance. Genes & Dev 11:2124–2136, 1997.
24. NV Federoff. Transposable elements as a molecular evolutionary force. Ann NY Acad Sci 18:251–264, 1999.
25. NV Federoff. The suppressor-mutator element and the evolutionary riddle of transposons. Genes Cells 4:11–19, 1999.
26. JL Kermicle, M Alleman. Gametic imprinting in maize in relation to the angiosperm life cycle. Development Suppl. 1:9–14, 1990.
27. J Messing, U Grossniklaus. Genomic imprinting in plants. In: Ohlsson R, ed. Genomic Imprinting: Results and Problems in Cell Differentiation. Berlin-Heidelberg, Germany: Springer Verlag, 1999, pp 23–40.
28. JP Vielle-Calzada, J Thomas, C Spillane, A Coluccio, MA Hoeppner, U Grossniklaus. Maintenance of genomic imprinting at the *Arabidopsis medea* locus requires zygotic *DDM1* activity. Genes Dev 13:2971–2982, 1999.
29. ES Buckler IV, TL Phelps-Durr, CS Buckler, RK Dawe, JF Doebley, TP Holtsford. Meiotic drive of chromosomal knobs reshaped the maize genome. Genetics 153:415–426, 1999.
30. H Vaucheret, C Beclin, T Elmayan, F Feuerbach, C Godon, JB Morel, P Mourrain, JC Palauqui, S Vernhettes. Transgene-induced gene silencing in plants. Plant J 16:651–659,
31. RM John, MA Surani. Imprinted genes and regulation of gene expression by epigenetic inheritance. Curr Opin Cell Biol 8:348–353, 1996.
32. IM Morison, AE Reeve. A catalogue of imprinted genes and parent-of-origin effects in humans and animals. Hum Mol Genet 7:1599–1609, 1998.
33. DA Roach, RD Wulff. Maternal effects in plants. Annu Rev Ecol Syst 18:209–235, 1987.
34. U Grossniklaus, J-P Vielle-Calzada, MA Hoeppner, WB Gagliano. Maternal control of embryogenesis by *MEDEA*, a *Polycomb* group gene in *Arabidopsis*. Science 280:446–450, 1998.
35. A Ray. New paradigms in plant embryogenesis: Maternal control comes in different flavours. Trends Plant Sci 3:325–327, 1998.

36. N Ohad, L Margossian, Y-C Hsu, C Williams, P Repetti, RL Fischer. A mutation that allows endosperm development without fertilization. Proc Natl Acad Sci USA 93:5319–5324, 1996.
37. AM Chaudhury, L Ming, C Miller, S Craig, ES Dennis, WJ Peacock. Fertilization-independent seed development in *Arabidopsis thaliana*. Proc Natl Acad Sci USA 94:4223–4228, 1997.
38. FC Felkner, DM Peterson, OE Nelson. Anatomy of immature grains of eight maternal effect shrunken endosperm barley mutants. Am J Bot 72:248–256, 1985.
39. S Ray, T Golden, A Ray. Maternal effects of the *short integument* mutation on embryo development in *Arabidopsis*. Dev Biol 180:365–369, 1996.
40. L Colombo, J Franken, AR van der Krol, PE Wittich, HJ Dons, GC Angenent. Downregulation of ovule specific MADS box genes from petunia results in maternally controlled defects in seed development. Plant Cell 9:703–715, 1999.
41. AM Chaudhury, S Craig, ES Dennis, WJ Peacock. Ovule and embryo development, apomixis and fertilization. Curr Opin Plant Biol 1:26–31, 1998.
42. JJ Harada. Signaling in plant embryogenesis. Curr Opin Plant Biol 2:23–27, 1999.
43. RB Goldberg, SJ Barker, L Perez-Grau. Regulation of gene expression during plant embryogenesis. Cell 56:149–160, 1989.
44. TL Thomas. Gene expression during plant embryogenesis and germination: An overview. Plant Cell 5:1401–1410, 1993.
45. U Grossniklaus, J-P Vielle-Calzada. Response: parental conflict and infanticide during embryogenesis. Trends Plant Sci 3:328, 1998.
46. LA Castle, D Errampalli, TL Atherton, LH Franzmann, ES Yoon, DW Meinke. Genetic and molecular characterization of embryonic mutants identified following seed transformation in *Arabidopsis*. Mol Gen Genet 241:504–514, 1993.
47. G Gavazzi, S Dolfini, D Allegra, P Castiglioni, G Todesco, M Hoxha. *Dap (Defective aleurone pigmentation)* mutations affect maize aleurone development. Mol Gen Genet 256:223–230, 1997.
48. AJ Jarvi. Shrunken endosperm mutants in barley. Crop Sci 15:363–366, 1975.
49. C Alonso-Blanco, H Blankestijn-de Vries, CJ Hanhart, M Koornneef. Natural allelic variation at seed size loci in relation to other life history traits of *Arabidopsis thaliana*. Proc Natl Acad Sci USA 96:4710–4717, 1999.
50. SM Tilghman. The sins of the fathers and mothers: Genomic imprinting in mammalian development. Cell 96:185–193, 1999.
51. HV Crouse. The controlling element in sex chromosome behaviour in *Sciara*. Genetics 45:1429–1443, 1960.
52. KR Fitch, GK Yasuda, KN Owens, BT Wakimoto. Paternal effects in *Drosophila:* Implications for mechanisms of early development. Curr Top Dev Biol 38:1–34, 1998.
53. L Sanchez, ALP Perondini. Sex determination in sciarid flies: A model for the control of differential X-chromosome elimination. J Theor Biol 197:247–259, 1999.
54. NC Subrahmanyam, KJ Kasha. Selective chromosomal elimination during haploid formation in barley following inter-specific hybridization. Chromosoma 42:111–125, 1973.
55. MK Ehlenfeldt, R Ortiz. On the origins of endosperm dosage requirements in *Solanum* and other angiosperm genera. Sex Plant Reprod 8:189–196, 1995.
55a. J-P Vielle-Calzada, R Baskar, U Grossniklaus. Delayed activation of the paternal genome. Nature 404:91–94, 2000.
56. MA Surani. Genomic imprinting: Developmental significance and molecular mechanism. Curr Opin Genet Dev 1:241–246, 1991.
57. JL Kermicle. Dependence of the *R*-mottled aleurone phenotype in maize on the mode of sexual transmission. Genetics 66:69–85, 1970.
58. J McGrath, D Solter. Completion of mouse embryogenesis requires both the maternal and paternal genomes. Cell 37:179–183, 1984.
59. SC Barton, MAH Surani, ML Norris. Role of paternal and maternal genomes in mouse development. Nature 311:374–376, 1984.
60. MAH Surani, SC Barton, ML Norris. Development of reconstituted mouse eggs suggests imprinting of the genome during gametogenesis. Nature 308:548–550, 1984.

61. D Haig, M Westoby. Genomic imprinting in endosperm: Its effect on seed development in crosses between species, and between different ploidies of the same species, and its implications for the evolution of apomixis. Philos Trans R Soc Lond 333:1–13, 1991.
62. DP Barlow. Gametic imprinting in mammals. Science 270:1610–1613, 1995.
63. R Martienssen. Chromosomal imprinting in plants. Curr Opin Genet Dev 8:240–244, 1998.
64. GA Nogler. Gametophytic apomixis. In: BM Johri, ed. Embryology of Angiosperms. Berlin: Springer-Verlag, 1984, pp 475–518.
65. AP Mordhurst, MAJ Toonen, SC de Vries. Plant embryogenesis. Crit Rev Plant Sci 16:535–576, 1997.
66. G Kimber, R Riley. Haploid angiosperms. Bot Rev 29:490–531, 1963.
67. KR Sarkar, EH Coe Jr. A genetic analysis of the origin of maternal haploids in maize. Genetics 54:453–464, 1966.
68. JL Kermicle. Androgenesis conditioned by a mutation in maize. Science 166:1422–1424, 1969.
69. B-Y Lin. Ploidy barrier to endosperm development in maize. Genetics 107:103–115, 1984.
70. R Scott, M Spielman, J Bailey, HG Dickinson. Parent-of-origin effects on seed development in *Arabidopsis thaliana*. Development 125:3329–3341, 1998.
71. T Kinoshita, R Yadegari, JJ Harada, RB Goldberg, RL Fischer. Imprinting of the *MEDEA Polycomb* gene in the *Arabidopsis* endosperm. Plant Cell 11:1945–1952, 1999.
72. S Chaudhuri, J Messing. Allele-specific parental imprinting of *dzr1*, a posttranscriptional regulator of zein accumulation. Proc Natl Acad Sci USA 91:4867–4871, 1994.
73. G Lund, J Messing, A Viotti. Endosperm-specific demethylation and activation of specific alleles of α-tubulin genes of *Zea mays* L. Mol Gen Genet 246:716–722, 1995.
74. VK Lloyd, DA Sinclair, TA Grigliatti. Genomic imprinting and position-effect variegation in *Drosophila melanogaster* Genetics 151:1503–1516, 1999.
75. KG Golic, MM Golic, S Pimpinelli. Imprinted control of gene activity in *Drosophila*. Curr Biol 8:1273–1276, 1998.
76. D Haig, M Westoby. Parent-specific gene expression and the triploid endosperm. Am Nat 134:147–155, 1989.
77. A Mochizuki, Y Takeda, Y Iwasa. The evolution of genomic imprinting. Genetics 144:1283–1295, 1996.
78. D Haig, C Graham. Genomic imprinting and the strange case of the insulin-like growth factor II receptor. Cell 64:1045–1046, 1991.
79. R Jaenisch. DNA methylation and imprinting; Why bother? Trends Genet 13:323–329, 1997.
80. T Moore, D Haig. Genomic imprinting in mammalian development: A parental tug-of-war. Trends Genet 7:45–49, 1991.
81. LD Hurst, GT McVean. Do we understand the evolution of genomic imprinting? Curr Opin Genet Devel 8:701–708, 1998.
82. DS Falconer, TFC Mackay. Introduction to Quantitative Genetics. Essex, England: Longman, 1996.
83. NW Simmonds. Principles of Crop Improvement. London: Longman, 1979, p 408.
84. HS Pooni, I Kumar, GS Khush. A comprehensive model for disomically inherited metrical traits expressed in triploid tissues. Heredity 69:166–174, 1992.
85. MR Foolad, RA Jones. Models to estimate maternally controlled variation in quantitative seed characters. Theor Appl Genet 83:360–366, 1992.
86. J Zhu, BS Weir. Analysis of cytoplasmic and maternal effects. II. Genetic models for triploid endosperms. Theor Appl Genet 89:160–166, 1994.
87. M Rui, DS Zheng, L Fan. The crossability percentages of 96 bread wheat landraces and cultivars from Japan with rye. Euphytica 92:301–306, 1996.
88. RW Masuelli, EL Camadro. Crossability relationships between wild potato species with different ploidies and endosperm balance numbers (EBN). Euphytica 94:227–235, 1997.
89. SA Jackson, RE Hanneman Jr. Crossability between cultivated and wild tuber- and non-tuber-bearing solanums. Euphytica 109:51–67, 1999.
90. JR Harlan, JMJ de Wet. Toward a rational clasification of cultivated plants. Taxon 20:509–517, 1971.
91. JG Hawkes. The Potato: Evolution, Biodiversity and Genetic Resources. Washington, DC: Smithsonian Institution Press, 1990, p 259.

92. KB Singh, B Ocampo. Exploitation of wild *Cicer* species for yield improvement in chickpea. Theor Appl Genet 95:418–423, 1997.
93. JC Glaszmann. Isozymes and classification of Asian rice varieties. Theor Appl Genet 74:21–30, 1987.
94. GS Khush. Origin, dispersal, cultivation and variation of rice. Plant Mol Biol 35:25–34, 1997.
95. D Sharma, R Kaur, K Kumar. Embryo rescue in plants. Euphytica 89:325–337, 1996.
96. D Carputo, A Barone, T Cardi, A Sebastiano, L Frusciante, SJ Peloquin. Endosperm balance number manipulation for direct *in vivo* germplasm introgression to potato from a sexually isolated relative (*Solanum commersonii* Dun.). Proc Natl Acad Sci USA 94:12013–12017, 1997.
97. A Barone, A Del Giudice, NQ Ng. Barriers to interspecific hybridization between *Vigna unguiculata* and *Vigna vexillata*. Sex Plant Reprod 5:195–200, 1992.
98. HC Sharma. How wide can a wide cross be? Euphytica 82:43–64, 1995.
99. MJ de Jeu, E Jacobsen. Early postfertilization ovule culture in *Alstroemeria* L. and barriers to interspecific hybridization. Euphytica 86:15–23, 1995.
100. W Youping, L Peng. Intergeneric hybridization between *Brassica* species and *Crambe abyssinica*. Euphytica 101:1–7, 1998.
101. NQ Ng, BB Singh. Cowpea. In: D Fuccillo, L Sears, P Stapleton, eds. Biodiversity in Trust: Conservation and Use of Plant Genetic Resources in CGIAR Centres. Cambridge: Cambridge University Press, 1997, pp 82–99.
102. JS Heslop-Harrison. Gene expression and parental dominance in hybrid plants. Dev Suppl 1:21–28, 1990.
103. JA Yoder, TH Bestor. Genetic analysis of genomic methylation patterns in plants and mammals. Biol Chem 377:605–610, 1996.
104. EJ Richards. DNA methylation and plant development. Trends Genet 13:319–323, 1997.
105. RA Brink, DC Cooper. The endosperm in seed development. Bot Rev 13:488–494, 1947.
106. B-Y Lin. Association of endosperm reduction with parental imprinting in maize. Genetics 100:475–486, 1982.
107. S Johnston, R Hanneman. Manipulations of endosperm balance number overcome crossing barriers between diploid *Solanum* species. Science 217:446–448, 1982.
108. MA Matzke, OM Scheid, AJ Matzke. Rapid structural and epigenetic changes in polyploid and aneuploid genomes. Bioessays 21:761–767, 1999.
109. SK Hahn, KV Bai, R Asiedu. Tetraploids, triploids and 2n pollen from diploid interspecific crosses with cassava. Theor Appl Genet 79:433–439, 1990.
110. L Marchais, S Tostain. Analysis of reproductive isolation between pearl millet (*Pennisetum glaucum* (L.) R.Br.) and *P. ramosum, P. schweinfurthii, P. squamulatum, Cenchrus ciliaris*. Euphytica 93:97–105, 1997.
111. G Barcaccia, D Rosellini, M Falcinelli, F Veronesi. Reproductive behaviour of tetraploid alfalfa plants obtained by unilateral and bilateral sexual polyploidisation. Euphytica 99:199–203, 1998.
112. G Orjeda. Ploidy manipulations for sweet potato breeding and genetic studies. PhD Thesis, Faculty of Science, University of Birmingham, England, 1995.
113. R Ortiz, D Vuylsteke. Factors affecting seed set in triploid *Musa* spp. L. Ann Bot 75:151–155, 1995.
114. AK Singh. Utilization of wild relatives in genetic improvement of *Arachis hypogaea* L. 8. Synthetic amphidiploids and their importance in interspecific breeding. Theor Appl Genet 72:433–439, 1986.
115. KB Singh, B Ocampo. Interspecific hybridization in annual *Cicer* species. J Genet Breed 47:199–204, 1993.
116. KR Sarkar, EH Coe. Anomalous fertilization in diploid-tetraploid crosses in maize. Crop Sci 11:539–542, 1971.
117. A Hakansson. Endosperm formation after 2x, 4x crosses in certain cereals, especially in *Hordeum vulgare*. Hereditas 39:57–64, 1953.
118. A Hakansson, S Ellerstrom. Seed development after reciprocal crosses between diploid and tetraploid rye. Hereditas 36:256–296, 1950.
119. A Hakansson. Seed development of *Brassica oleracea* and *B. rapa* after certain reciprocal pollinations. Hereditas 42:373–396, 1956.
120. SRJ Woodell, DH Valentine. Studies in British primulas. IX. Seed incompatibility in diploid autotetraploid crosses. New Phytol 60:282–294, 1961.

121. DC Cooper, RA Brink. Seed collapse following matings between diploid and tetraploid strains of *Lycopersicon pimpinellifolium*. Genetics 30:376–401, 1945.
122. KH Von Wangenheim. Zur Ursache der Abortion von Samenanlagen in diploid-polyploid-Kreuzungen. I. Die Chromosomenzahlen von Mütterlichem Gewebe, Endosperm and Embryo. Z Pflanzenzüchtung 46:13–19, 1962.
123. KH Von Wangenheim. Entwicklungsphysiologische Untersuchungen über die Beteiligung nuklearer Erbträger an der Phänogenese. Ber Dtsch Bot Ges 80:228–236, 1967.
124. GP Redei. Crossing experiences with polyploids. Arabidopsis Inf Serv 1:13, 1964.
125. SA Johnston, TPM den Nijs, SJ Peloquin, RE Hanneman. The significance of genic balance to endosperm development in inter-specific crosses. Theor Appl Genet 57:5–9, 1980.
126. S Johnston, R Hanneman. Manipulations of endosperm balance number overcome crossing barriers between diploid *Solanum* species. Science 217:446–448, 1982.
127. RE Hanneman. Assignment of endosperm balance number to the tuber bearing solanums and their close non-tuber bearing solanums. Euphytica 74:19–25, 1994.
128. TR Tarn, JG Hawkes. Cytogenetic studies and the occurrence of triploidy in the wild potato species *Solanum commersonii* Dun. Euphytica 35:293–302, 1986.
129. MT Jackson, PR Rowe, JG Hawkes. Crossability relationships of Andean potato varieties of three ploidy levels. Euphytica 27:541–551, 1978.
130. SA Johnston, RE Hanneman. The genetics of triploid formation and its relationship to endosperm balance number in potato. Genome 38:60–67, 1995.
131. SA Johnston, RE Hanneman. Genetic control of endosperm balance number (EBN) in the Solanaceae based on trisomic and mutation analysis. Genome 39:314–321, 1996.
132. MK Ehlenfeldt, RE Hanneman. Genetic control of endosperm balance number (EBN): Three additive loci in a threshold-like system. Theor Appl Genet 75:825–832, 1988.
133. EL Camadro, RW Masuelli. A genetic model for endosperm balance number (EBN) in the wild potato *S. acaule* Bitt. and two related diploid species. Sex Plant Reprod 8:283–288, 1995.
134. JL Kermicle. Pleiotropic effects on seed development of the *indeterminate gametophyte* gene in maize. Am J Bot 58:1–7, 1971.
135. JA Birchler, JR Hart. Interaction of endosperm size factors in maize. Genetics 117:309–317, 1987.
136. SE Asker, L Jerling. Apomixis in plants. London: CRC Press, 1992.
137. EC Bashaw, WW Hanna. Apomictic reproduction. In: GP Chapman ed. Reproductive versatility in the grasses. Cambridge: Cambridge University Press, 1990, pp 100–130.
138. AM Koltunow, RA Bicknell, AM Chaudhury. Apomixis: Molecular strategies for the generation of genetically identical seeds without fertilization. Plant Physiol 108:1345–1352, 1995.
139. RA Jefferson, R. Bicknell. The potential impacts of apomixis: A molecular genetics approach. In: BWS Sobral, ed. The Impact of Plant Molecular Genetics. Boston: Birkhäuser, 1996, pp 87–101.
140. U Grossniklaus, A Koltunow, M van Lookeren Campagne. A bright future for apomixis. Trend Plant Sci 3:415–416, 1998.
141. U Grossniklaus, JM Moore, WB Gagliano. Molecular and genetic approaches to understanding and engineering apomixis: *Arabidopsis* as a powerful tool. In: SS Virmani, EA Siddiq, K Muralidharan, eds. Advances in Hybrid Rice Technology. Proceedings of the 3$^{rd}$ International Symposium on Hybrid Rice 1996, Hyderabad, India. Los Banos, Philippines: IRRI, 1998, p 187.
142. J-P Vielle-Calzada, CF Crane, DM Stelly. Apomixis: The asexual revolution. Science 274:1322–1323, 1996.
143. S McMeniman, G Lubulwa. Project Development Assessment: An Economic Evaluation of the Potential Benefits of Integrating Apomixis in Hybrid Rice. Canberra: Australian Centre for International Agricultural Research, 1997.
144. M Dujardin, WW Hanna. Developing apomictic pearl millet: Characterization of a BC3 plant. J Genet Breed 43:145–151, 1989.
145. P Ozias-Akins, EL Lubbers, WW Hanna, JW McNay. Transmission of the apomictic mode of reproduction in *Pennisetum:* Co-inheritance of the trait and molecular markers. Theor Appl Genet 85:632–638, 1993.
146. WW Hanna, D Roche, P Ozias-Akins. Use of apomixis in crop improvement. In: SS Virmani, EA

Siddiq, K Muralidharan, eds. Advances in Hybrid Rice Technology. Los Banos, Philippines: IRRI, 1998, pp 283–296.

147. D Grimanelli, O Leblanc, D Gonzales de Leon, Y Savidan. Apomixis expression in maize-*Tripsacum* hybrid derivatives and the implications regarding its control and potential for manipulation. Apomixis Newsl 8:35–37, 1995.

148. D Grimanelli, O Leblanc, E Espinosa, E Perotti, D Gonzalez de Leon, Y Savidan. Non-Mendelian transmission of apomixis in maize-*Tripsacum* hybrids caused by a transmission ratio distortion. Heredity 80:40–47, 1998.

149. BF Yudin, VA Sokolow. Towards regular apomixis in maize, achieved by experiment. Genet Manip Plants 5:36–40, 1989.

150. B Kindiger, V Sokolov, IV Khatypova. Evaluation of apomictic reproduction in a set of 39 chromosome maize-*Tripascum* backcross hybrids. Crop Sci 36:1108–1113, 1996.

151. Y Savidan. Les promesses de l'apomomixie. ORSTOM Actualities 47:2–7, 1995.

152. M Dujardin, WW Hanna. Crossability of pearl millet with wild *Pennisetum* species. Crop Sci 29:77–80, 1989.

153. WW Hanna, M Dujardin, P Ozias-Akins, E Lubbers, L Arthur. Reproduction, cytology and fertility of pearl millet × *Pennisetum squamulatum* BC4 plants. J Hered 84:213–216, 1993.

154. CJ Williamson, PJ Watson. Production and description of interspecific hybrids between *P. pratensis* and *P. longifolia*. Euphytica 29:715–725, 1980.

155. JP Vielle-Calzada, ML Nuccio, MA Budiman, TL Thomas, BL Burson, MA Hussey, RA Wing. Comparative gene expression in sexual and apomictic ovaries of *Pennisetum ciliare* (L.) Link. Plant Mol Biol 32:1085–1092, 1996.

156. LA Snyder. Apomixis in *Paspalum secans*. Am J Bot 44:318–324, 1957.

157. CY Chao. Autonomous development of the embryo in *Paspalum conjugatum* berg. Bot Notiser 133:215–222, 1980.

158. A Rutishauser. Die Entwicklungserregung des Endosperms bei pseudogamen *Ranunculus*-Arten. Mitt Naturforsch Ges Schaffhausen 25:1–45, 1954.

159. PS Reddy, R d'Cruz. Mechanism of apomixis in *Dichanthium annulatum* (Forssk). Stapf Bot Gaz 139:71–79, 1969.

160. Y Savidan. Chromosomal and embryological analyses in sexual × apoictic hybrids of *Panicum maximum* Jacq. Theor Appl Genet 57:153–156, 1980.

161. D Grimanelli, M Hernandez, E Perotti, Y Savidan. Dosage effects in the endosperm of diplosporous apomictic *Tripsacum* (Poaceae). Sex Plant Reprod 10:279–282, 1997.

162. CL Quarin. Effect of pollen source and pollen ploidy on endosperm formation and seed set In pseudogamous apomictic *Paspalum notatum*. Sex Plant Reprod 11:331–335, 1999.

163. JX Wu, GJ Wang, J Zhu, FH Xu, DF Ji. Genetic analysis on direct and maternal effects of seed traits in upland cotton (*Gossypium hirsutum L.*). Acta Agronomica Sinica 21:559–664, 1995.

164. CH Shi, J Zhu, RC Zang, GL Chen. Genetic and heterosis analysis for cooking quality traits of Indica rice in different environments. Theor Appl Genet 95:294–300, 1997.

165. XF Yan, J Zhu, SY Xu, TH Xu. Genetic effects of embryo and endosperm for four malting quality traits of barley. Euphytica 106:27–34, 1999.

166. J Chen, J Zhu. Genetic effects and genotype × environment interactions for cooking quality traits in Indica-Japonica crosses of rice (*Oryza sativa* L.). Euphytica 109:9–15, 1999.

167. PA Brocklehurst. Factors controlling grain weight in wheat. Nature 266:348–349, 1977.

168. AJS Chojecki, MW Bayliss, MD Gale. Cell production and DNA accumulation in the wheat endosperm, and their association with grain weight. Ann Bot 58:702–708, 1986.

169. S Ouattar, RJ Jones, RK Crookston. Effect of water deficit during grain filling on the pattern of maize kernel growth and development. Crop. Sci 27:726–730, 1987.

170. WL Charlton, CL Keen, C Merriman, P Lynch, AJ Greenland, HG Dickinson. Endosperm development in *Zea mays*: Implications of gametic imprinting and paternal excess in regulation of transfer layer development. Development 121:3089–3097, 1995.

171. JH Xiao, JM Li, LP Yuan, SD Tanksley. Dominance is the major genetic basis of heterosis in rice as revealed by QTL analysis using molecular markers. Genetics 140:745–754, 1995.

172. SB Yu, JX Li, CG Xu, YF Tan, YJ Gao, XH Li, Q Zhang, MA Saghai-Maroof. Importance of epistasis as the genetic basis of heterosis in an elite rice hybrid. Proc Natl Acad Sci USA 94:9226–9231, 1997.
173. CW Stuber. Heterosis in plant breeding. Plant Breed Rev. 12:227–251, 1994.
174. RM Li, CG Xu, ZY Yang, XK Wang. The extent of parental genotypic divergence determines maximal heterosis by increasing fertility in inter-subspecific hybrids of rice (*Oryza sativa L.*) Mol Breed 4:205–214, 1998.
175. A Leonardi, C Damerval, Y Herbert, A Gallais, D de Vienne. Association of protein amount polymorphism (PAP) among maize lines with performances of their hybrids. Theor Appl Genet 82:552–560, 1991.
176. AS Tsaftaris. Molecular aspects of heterosis in plants. Physiol Plant 94:362–370, 1995.
177. AS Tsaftaris, M Kafka. Mechanisms of heterosis in crop plants. J Crop Prod 1:95–111, 1998.
178. SB Rood, RI Buzzell, DJ Major, RP Pharis. Gibberellins and heterosis in maize: Quantitative relationships. Crop Sci 30:281–286, 1990.
179. D Schwartz, WJ Laughner. A molecular basis for heterosis. Science 166:626–627, 1969.
180. AJ Pryor. Allelic glutamic dehydrogenase isozymes in maize—a single hybrid isozyme in heterozygotes. Heredity 32:397–401, 1974.
181. JG Hall, C Wills. Conditional overdominance at an alcohol dehydrogenase locus in yeast. Genetics 117:421–427, 1987.
182. JB Hollick, VL Chandler. Epigenetic allelic states of a maize transcriptional regulatory locus exhibit overdominant gene action. Genetics 150:891–897, 1998.
183. RA Brink. Paramutation. Annu Rev Genet 7:129–152, 1973.
184. R Martienssen. Epigenetic phenomena: Paramutation and gene silencing in plants. Curr Biol 6:810–813, 1996.
185. EL Walker. Paramutation of the *r1* locus of maize is associated with increased cytosine methylation. Genetics 148:1973–1981, 1998.
186. DC Rasmussen, RL Phillips. Plant breeding progress and genetic diversity from *de novo* variation and elevated epistasis. Crop Sci 37:303–310, 1997.
187. S Palmer, V Ulrich. Nuclease accessibility of chromatin from a heterotic hybrid and from parental inbreds. Biol Plant 34:361–366, 1992.
188. LM McMurphy, AL Rayburn. Cytological evidence for nucleolar competition in a maize hybrid. J Hered 85:407–410, 1994.
189. R Chakraborty. Can molecular imprinting explain heterozygote deficiency and hybrid vigour? Genetics 122:713–717, 1989.
190. RA Brink, DC Cooper. Double fertilisation and development of the seeds in angiosperms. Bot Gaz 102:1–25, 1940.
191. GL Stebbins. Flowering Plants: Evolution above the Species Level. Cambridge, MA: Harvard University Press, 1974, 1997.
192. AL Rayburn. Chromatin composition in maize F1 hybrids. Maydica 42:393–399, .
193. LA Pray, CJ Goodnight. Genetic variation in inbreeding depression in the red flour beetle *Tribolium castaneum*. Evolution 49:176–188, 1995.
194. GN Drews, D Lee, CA Christensen. Genetic analysis of female gametophyte development and function. Plant Cell 10:5–17, 1998.
195. T Jenuwein, G Laible, R Dorn, G Reuter. SET domain proteins modulate chromatin domains in eu- and hetero-chromatin. Cell Mol Life Sci 54:80–93, 1998.
196. J Goodrich, P Puangsomlee, M Martin, D Long, EM Meyerowitz, G Coupland. A *Polycomb*-group gene regulates homeotic gene expression in *Arabidopsis*. Nature 386:44–51, 1997.
197. L Ming, P Bilodeau, A Koltunow, ES Dennis, WJ Peacock, AM Chaudhury. Genes controlling fertilization-independent seed development in *Arabidopsis* thaliana. Proc Natl Acad Sci USA 96:296–301, 1999.
198. T Kiyosue, N Ohad, R Yadegari, M Hannon, J Dinneny, D Wells, A Katz, L Margossian, JJ Harada, RB Goldberg, RL Fischer. Control of fertilization-independent endosperm development by the *MEDEA* polycomb gene in *Arabidopsis*. Proc Natl Acad Sci USA 96:4186–4191, 1999.

199. RS Jones, WM Gelbart. The *Drosophila Polycomb*-group gene *Enhancer of zeste* contains a region with sequence similarity to *trithorax*. Mol Cell Biol 13:6357–6366, 1993.
200. EA Carrington, RS Jones. The *Drosophila Enhancer of zeste* gene encodes a chromosomal protein: Examination of wild-type and mutant protein distribution Development 122:4073–4083, 1996.
201. CA Jones, J Ng, AJ Peterson, K Morgan, J Simon, RS Jones. The *Drosophila esc* and *E(z)* proteins are direct partners in *Polycomb* Group-mediated repression. Mol Cell Biol 18:2825–2834, 1998.
202. T Tie, T Furuyama, PJ Harte. The *Drosophila Polycomb* Group proteins *ESC* and *E(Z)* bind directly to each other and co-localise at multiple chromosomal sites. Development 125:3483–3496, 1998.
203. RGAB Sewalt, J van der Vlag, MJ Gunster, KM Hamer, JL den Blaauwen, DPE Satijn, T Hendrix, R van Driel, AP Otte. Characterisation of interactions between the mammalian *Polycomb*-Group proteins *Enx1/EZH2* and *EED* suggests the existence of different mammalian *Polycomb*-Group protein complexes. Mol Cell Biol 18:3586–3595, 1998.
204. M Van Lohuizen, M Tijms, JW Vonckens, A Schumacher, T Magnuson, E Wientjens. Interaction of mouse *Polycomb*-Group (Pc-G) proteins *Enx1* and *Enx2* with *Eed:* Indication for separate Pc-G complexes. Mol Cell Biol 18:3572–3579, 1998.
205. K Hagstrom, P Schedl. Remembrance of things past: Maintaining gene expression patterns with altered chromatin. Curr Opin Genet Dev 7:814–821, 1997.
206. F Lyko, R Paro. Chromosomal elements conferring epigenetic inheritance. Bioessays 21:824–832, 1999.
207. V Pirrotta. Pc-G complexes and chromatin silencing. Curr Opin Genet Dev 7:249–258, 1997.
208. S Bel, N Core, MDjabali, K Kieboom, N Van der Lugt, MJ Alkema, M Van Lohuizen. Genetic interactions and dosage effects of *Polycomb* group genes in mice. Development 125:3543–3551, 1998.
209. H Strutt, R Paro. The *Polycomb* group protein complex of *Drosophila melanogaster* has different compositions at different target genes. Mol Cell Biol 17:6773–6783, 1997.
210. RD Kornberg, Y Lorch. Chromatin modifying and remodelling complexes. Curr Opin Genet Dev 9:148–151, 1999.
211. S Bjorklund, G Almouzni, I Davidson, KP Nightingale, J Weiss. Global transcription regulators of eukaryotes. Cell 96:759–767, 1999.
212. G Schnitzler, S Sif, RE Kingston. Human SWI/SNF interconverts a nucleosome between its base state and a stable remodeled state. Cell 94:17–27, 1998.
213. A Vongs, T Kakutani, RA Martienssen, EJ Richards, *Arabidopsis thaliana* DNA methylation mutants. Science 260:1926–1928, 1993.
214. T Kakutani, K Munakata, EJ Richards, H Hirochika. Meiotically and mitotically stable inheritance of DNA hypomethylation induced by *ddm1* mutation of *Arabidopsis thaliana.* Genetics 151:831–838, 1999.
215. JA Jeddeloh, TL Stokes, EJ Richards. Maintenance of genomic methylation requires a SWI2/SNF2-like protein. Nat Genet 22:94–97, 1999.
216. RH Busch, K Kofoid. Recurrent selection for kernel weight in spring wheat. Crop Sci 22:568–572, 1982.
217. MO Odhiambo, WA Compton. Twenty cycles of divergent mass selection for seed size in corn. Crop Sci 27:1113–1116, 1987.
218. P Martiniello, G Delogu, M Odoardi, G Boggini, AM Stanca. Breeding progress in grain yield and selected agronomic characters of winter barley (*Hordeum vulgare* L.) over the last quarter of a century. Plant Breed 99:289–294, 1987.
219. J-F Rami, P Dufour, G Trouche, G Fliedel, C Mestres, F Davrieux, P Blanchard P Hamon. Quantitative trait loci for grain quality, productivity, morphological and agronomical traits in sorghum (*Sorghum bicolor* L. Moench). Theor Appl Genet 97:605–616, 1998.
220. J Xiao, J Li, L Yuan, SD Tanksley. Identification of QTLs affecting traits of agronomic importance in a recombinant inbred population derived from a subspecific rice cross. Theor Appl Genet 92:230–244, 1996.
221. C Lu, L Shen, Z Tan, Y Xu, P He, Y Chen, L Zhu. Comparative mapping of QTLs for agronomic traits of rice across environments by using a doubled-haploid population. Theor Appl Genet 94:145–150, 1997.

222. GM Timmerman-Vaughan, JA McCallum, TJ Frew, NF Weeden, C Russel. Linkage mapping of quantitative trait loci controlling seed weight in pea (*Pisum sativum* L.). Theor Appl Genet 93:431–439, 1996.
223. S Grandillo, SD Tanksley. QTL analysis of horticultural traits differentiating the cultivated tomato from the closely related species *Lycopersicon pimpinellifolium*. Theor Appl Genet 92:935–951, 1996.
224. MAR Mian, MA Bailey, JP Tamulonis, ER Shioe, TE Carter, WA Parrott, DA Ashley, RS Hussey, HR Boerma. Molecular markers associated with seed weight in two soybean populations. Theor Appl Genet 93:1011–1016, 1996.
225. PJ Maughan, MA Saghai Maroof, GR Buss. Molecular-marker analysis of seed weight: Genomic locations, gene action, and evidence for orthologous evolution among three legume species. Theor Appl Genet 93:574–579, 1996.
226. AM Koltunow, P Brenna, JE Bond, SJ Barker. Evaluation of genes to reduce seed size in *Arabidopsis* and tobacco and their application to Citrus. Mol Breed 4:235–251, 1998.
227. J Doebley, L Lukens. Transcriptional regulators and the evolution of plant form. Plant Cell 10:1075–1082, 1998.
228. RL Wang, A Stec, J Hey, L Lukens, J Doebley. The limits of selection during maize domestication. Nature 398:236–239, 1999.
229. LB Skrabanek, KH Wolfe. Eukaryote genome duplication—where's the evidence? Curr Opin Genet Dev 8:694–700, 1998.
230. AM Matzke, AJM Matzke. Polyploidy and transposons. Trends Ecol Evol 13:241, 1998.
231. A Bird. Gene number, noise reduction and biological complexity. Trends Genet 11:94–100, 1995.
232. RB Flavell. Inactivation of gene expression in plants as a consequence of specific sequence duplication. Proc Natl Acad Sci USA 91:3490–3496, 1994.

# 8
# Direct DNA Delivery Into Intact Cells and Tissues

**Joseph F. Petolino**
*Dow AgroSciences, Indianapolis, Indiana*

| | | |
|---|---|---|
| I. | INTRODUCTION | 137 |
| II. | VALIDATED METHODS OF DIRECT DNA DELIVERY INTO INTACT CELLS AND TISSUES | 138 |
| | A. Microparticle Bombardment | 138 |
| | B. Tissue Electroporation | 139 |
| | C. Whiskers | 139 |
| III. | TISSUE CULTURE TARGETS | 140 |
| | A. Target Tissue Characteristics Necessary for Direct DNA Delivery | 140 |
| | B. Embryogenic Cell and Tissue Culture | 140 |
| IV. | CONCLUSION | 141 |
| | REFERENCES | 141 |

## I. INTRODUCTION

Transgenic plant production today seems a far cry from the mid-1980s, when recalcitrance to in vitro manipulation in the major cereal and legume crops appeared to be the major limitation to the advancement of agricultural biotechnology. These days, no species should be considered a priori to fall outside the range of those amenable to transformation. The term *recalcitrant species* has largely disappeared from our vocabulary in recent years. Experience has taught that, given sufficient effort, any plant species can be transformed. Maize is a case in point. Because maize was seemingly nontransformable by virtue of its resistance to Agrobacterium infection (1) and its relatively recalcitrant to in vitro manipulation (2), not until the invention of a novel deoxyribonucleic acid (DNA) delivery means (3), the development of a new selectable marker system (4), the cloning of monocot expression elements (5), and major advances in tissue culture (6) did reliable transgenic maize production become a reality (7,8).

Over the last decade, a whole generation of technology to transform cells and tissues has revolutionized the field of gene transfer into plants, with the result that virtually all of the major crop species have been transformed. A veritable "toolbox" of direct DNA delivery methods has cropped up over the last several years. Although very different from each other at first glance, most of these methods share a common paradigm: since all of these procedures involve intact cells and tissues as targets, there is a need to breach the cell wall. This is accomplished by causing some degree of cellular injury such that DNA enters cells. Perhaps during some sort of wound response, the cells become competent for dedifferentiation and ultimately DNA integration. This chapter explores some of the critical issues relative to transgenic production via direct DNA delivery into intact cells and tissues. The focus is on the interaction between the delivery mechanism and the recipient cells.

## II. VALIDATED METHODS OF DIRECT DNA DELIVERY INTO INTACT CELLS AND TISSUES

### A. Microparticle Bombardment

United States Patent 4945050 defines *microparticle bombardment* as follows:

> A method for introducing particles into cells comprising accelerating particles having a diameter sufficiently small to penetrate and be retained in a preselected cell without killing the cell, and propelling said particles at said cells whereby said particles penetrate the surface of said cells and become incorporated into the interior of said cells. (9)

The definition outlines the basic principle of microparticle bombardment as the acceleration of high-density, micrometer-sized particles to penetrating velocities such that materials can be delivered to living cells. Numerous devices capable of accelerating microparticles to velocities greater than 300 m/s have been developed for accurate and reliable delivery into intact cells and tissues (10). The main differences between the various microparticle propulsion devices relate to the means by which the particles are accelerated, i.e., macrocarriers and stopping plates (3), continuous gas flow (11), helium blasting (12), and so forth.

For gene transfer into plant tissues, DNA-coated gold or tungsten microparticles (1–2 micrometers) are typically accelerated at targets comprising in vitro–cultured cells or meristematic regions. The challenge is to deliver DNA effectively into a large number of target cells without causing too much damage, thereby reducing survival. Transient expression of a reporter gene, such as uidA or GUS (13), after bombardment has been used for determining DNA delivery efficiency. Although the identification of physical and biological parameters associated with optimal DNA delivery must be empirically determined for each particular system, general principles are emerging (14).

After microparticle bombardment, transiently expressing tissues were found to correspond to those cells in which a particle had hit a cell's nucleus (15). Most, if not all, transient GUS expressing tissues were observed to contain at least one cell, usually in the central region of the expressing tissue, that contained an intranuclear particle. Under optimal conditions, DNA delivery, as measured by transient expression of a reporter gene such as GUS, was as high as several hundred units per target (100–500 mg of tissue). However, stable transformation was typically orders of magnitude less (0.1–5 transgenic colonies per target). A kinetic study of cell survival after microparticle bombardment showed that most cells (99%) that received a particle did not survive 48 hours after bombardment (16). This might explain some of the discrepancy between the number of cells receiving DNA (transient expression) and those integratively transformed (stable colonies).

Nonetheless, microparticle bombardment has become one of the most broadly applicable DNA delivery methods in current use. Reliable transgenic plant production in agronomically important and historically recalcitrant crops such as maize (17,18), wheat (19), barley (20), and soybean (21) is testimony to the significance of this technology to applied agriculture. In addition, microparticle bombardment, in combination with selection for antibiotic resistance, has contributed to the demonstration of plastid transformation (22,23), which, once adapted to the major crop species, will have far-reaching repercussions relative to transgene expression.

## B. Tissue Electroporation

Canadian patent 1208146 defines *tissue electroporation* as follows:

> The method of transferring genes into cells which comprises the step of subjecting a mixture of said genes and said cells to an electric field (24).

The exposure of cells to a high-voltage electric field pulse results in the formation of temporary pores in the plasma membrane whereby they become transiently permeable to large molecules, such as DNA (25). Although not completely understood, DNA uptake via electroporation appears to involve a two-step process whereby a prepulse adsorption to the membrane surface is followed by an electric field–mediated endocytosis-like internalization (26). How the DNA reaches the nucleus in not known, but many involve vesicular entrapment (27).

Although historically used to transfer genes to isolated plant protoplasts (28), electroporation was shown to mediate DNA delivery into intact plant cells (29) as well as organized tissues (30). Clearly, the cell wall must not represent a totally impenetrable barrier to DNA internalization (25). Apparently, DNA diffusion– and/or electric field–mediated electrophoresis through cell wall interstices allows cellular uptake via subsequent transient membrane pore formation. The first report of stable transformation via electroporation of intact cells involved tobacco suspension cultures (31). Compared to that for isolated protoplasts, a somewhat longer pulse length was required for DNA uptake into intact cells, perhaps supporting the electrophoresis concept. Subsequently, this method has been adapted for use with embryogenic callus (32,33) and isolated embryos (34). As with protoplasts, specific parameters for optimal delivery need to be determined for each particular tissue type. Although not nearly as well studied as microparticle bombardment, electroporation appears to be a generally applicable means of delivering DNA into various types of intact cells and tissues.

## C. Whiskers

U.S. Patent 5302523 defines *whiskers* as follows:

> A method of introducing a nucleic acid into plant cells comprising providing in a liquid medium i) plant cells suspended therein, ii) a multiplicity of metal or ceramic whisker bodies and iii) a nucleic acid, and subjecting said liquid medium containing the said suspended cells, the said metal or ceramic whisker bodies and said nucleic acid to physical motion so as to create collisions between said metal or ceramic whiskers and said plant cells whereby said nucleic acid is introduced into said plant cells (35).

Silicon carbide whiskers are microfibers 10–80 µm long and 0.6 µm in diameter. The vigorous agitation of intact cells in the presence of whiskers and DNA is yet another validated method for plant cell transformation (36). The mechanism by which whisker-mediated transformation occurs is not completely understood, although evidence suggests that it is a largely physical process (37). Silicon carbide is a hard ceramic substance that fractures readily, resulting in sharp cutting edges. Scanning electron microscopy of whisker-treated cells clearly shows that the fine fibers are capable of cell wall penetration (38). It is not known whether or not the whiskers actually carry the

DNA into the cells, although precipitation is not a requirement for uptake (37). Nonetheless, collisions between the whiskers and the plant cells probably result in sublethal damage, i.e., micrometer-sized holes in the cell walls, thereby allowing DNA entry via either active or passive means.

DNA delivery after whisker treatment, as measured by transient reporter gene expression, has been reported in several plant species (36–41). Fertile transgenic maize plants have been recovered after whisker-mediated transformation of embryogenic suspension cultures (36). In addition, transgenic plants have been regenerated from suspension cultures of *Lolium multiflorum*, *Lolium perenne*, *Festuca arundinacea*, and *Agrostis stolonifera* after whisker treatment (39). The observation of transient GUS expression after whisker treatment of suspension cells of rice (40) and isolated wheat embryos (41) suggests that this method may be of general applicability.

## III. TISSUE CULTURE TARGETS

### A. Target Tissue Characteristics Necessary for Direct DNA Delivery

In addition to an effective means of delivering DNA, successful transgenic plant production requires appropriate target cells and the ability to manipulate them effectively. Although tissue culture–free transformation systems may one day be available for all crop species, efficient target tissue manipulation, including in vitro culture and plant regeneration, is currently an absolute requirement in all but a few systems. In spite of claims being made for stable transformation via direct DNA delivery to pollen (42) or reproductive meristems (43), it is still in vitro cell and tissue cultures upon which virtually all direct DNA delivery systems depend. With regard to gene transfer via direct DNA delivery, there are three basic requirements: (a) reliable production and maintenance of tissue cultures of particular morphological characteristics, (b) efficient delivery of DNA to the appropriate cells without irreversibly inhibiting their capacity to contribute to the future of the culture, and (c) effective isolation and recovery of rare integration events.

The following factors should be considered in relation to selecting an appropriate target tissue for direct DNA delivery. Microparticle bombardment, electroporation, and whiskers all deliver DNA most effectively to surface cell layers (12,30,41). Conditions that allow deeper penetration are usually not conducive to cell survival (16,25). Thus, surface cells must be competent for DNA uptake and integration, including resistance to delivery-induced stress. Since integrative transformation is a relatively rare event, transformed cells must be competent to proliferate in the presence of a selection agent that inhibits the growth of surrounding "wild-type" tissue such that de novo meristem formation, preferably after repetitive cycles of dedifferentiation, leads to transgenic, morphogenically competent colony isolation.

### B. Embryogenic Cell and Tissue Cultures

The availability of target cells with the necessary features for direct DNA delivery is an absolute prerequisite for transgenic production via any of the methods mentioned. The initiation and maintenance of embryogenic cultures, derived from immature tissue explants isolated at defined stages of development, have been paramount to success in this area (44). Immature embryos, or callus and suspension cultures derived from them, have been the most successful targets for direct DNA delivery into intact cells and tissues to date. This is due to the fact that these cultures fulfill the basic requirements of a target tissue: they contain cells that are accessible, selectable, and ultimately totipotent.

Most of the early reports of successful transgenic production via direct DNA delivery into intact cells and tissues involved embryogenic suspensions as target cultures (17,31,36). Once an appropriate suspension was established, transgenic production with these cultures was quite efficient; as long as the cell line was maintained, i.e., using cryopreservation (45). Although embryogenic suspensions display all of the characteristics required of a target tissue, their use is severely limited by the difficulties associated with establishing these types of cultures in all but a few specific genotypes. Later studies identified conditions such that embryogenic callus (19,32) and immature embryos (44) could be used directly as targets. Although the target cells in these tissues are not always as accessible and the proliferative capacity not nearly as high as that seen in suspensions, the use of these more organized tissues reduces the time and effort required to establish the target cells and increases the range of genotypes available.

## IV. CONCLUSION

There has been exceptionally rapid progress in recent years in extending gene transfer capabilities to include economically important crop species that fall outside the natural host range of *Agrobacterium* sp. Methods that allow direct DNA delivery into intact cells and tissues, such as microparticle bombardment, electroporation, and whiskers, have contributed significantly to this expanded capability, primarily by circumventing the need to regenerate plants from isolated protoplasts. In principle, no plant species should be considered to fall outside the range of those amenable to transformation. Indeed, experience has taught that, given sufficient effort, any plant species can be transformed.

In current practice, the availability of regenerable target tissue displaying the necessary features, such as immature embryos and embryogenic callus and suspension cultures, is an essential component of all gene transfer systems involving direct DNA delivery. This requirement is unlikely to change in the near future for most major crop species. Since little is actually known about the control of in vitro morphogenesis, the development and optimization of tissue culture and regeneration protocols tend to be rather empirical activities and, as such, require sustained commitment over several years to acquire the necessary expertise. In practice, the establishment and maintenance of cultures capable of exhibiting particular in vitro behavior are usually highly genotype-dependent. Thus, once beyond a few model species or genotypes, transgenic production is a highly specialized and resource-intensive pursuit. Even in those species in which transgenic production is currently being successfully performed, the systems are far from optimized.

Future work should include the identification of conditions that result in effective DNA delivery into intact cells without causing significant cellular or tissue damage as well as the development of improved in vitro culture systems to afford greater accessibility and selectability of totipotent cells across a broad spectrum of agronomically useful germplasm. This will require a strengthening of basic understanding of the physical and biological events leading to DNA internalization and integration as well as increased knowledge of the factors associated with cell proliferation and in vitro morphogenesis.

## REFERENCES

1. B Hohn, Z Koukolinkova-Nicola, G Bakkeren, N Grimsley. *Agrobacterium*-mediated gene transfer to monocots and dicots. Genome 31:987–993, 1989.
2. Potrykus. Gene transfer to cereals: An assessment. Biotechnology 8:535–542, 1990.
3. J Sanford. The biolistic process. Trends Biotechnol 6:299–302, 1988.
4. M De Block, J Botterman, M Vandewiele, J Dockx, C Thoen, V Gossele, N Rao Movva, C Thompson, M Van Montagu, J Leemans. Engineering herbicide resistance in plants by expression of a detoxifying enzyme. EMBO J 6:2513–2518, 1987.

5. AH Christensen, RA Sharrock, PH Quail. Maize polyubiquitin genes: Thermal perturbation of expression and transcript splicing, and promoter activity following transfer to protoplasts by electroporation. Plant Mol Biol 18:675–689, 1992.
6. V Vasil, IK Vasil. Induction and maintenance of embryogenic callus cultures of Gramineae. In: IK Vasil, ed. Cell Culture and Somatic Cell Genetics of Plants. Orlando, FL: Academic Press, 1984, pp 36–42.
7. MG Koziel, GL Beland, C Bowman, NB Carozzi, R Crenshaw, L Crossland, J Dawson, N Desai, M Hill, S Kadwell, K Launis, K Lewis, D Maddox, K McPherson, MR Meghji, E Merlin, R Rhodes, GW Warren, M Wright, SV Evola. Field performance of elite transgenic maize plants expressing an insecticidal protein derived from *Bacillus thuringiensis*. Biotechnology 11:194–200, 1993.
8. AR Gould, NM Cowen, DJ Merlo, JF Petolino, SA Thompson, TA Walsh. Insect control via transgenic hybrid maize. Proceedings of the 48th Annual Corn and Sorghum Industry Research Conference, 1993, pp 63–75.
9. J Sanford, E Wolf, N Allen. Method for transporting substances into living cells and tissues and apparatus therefor. United States Patent #4945050, Issued July 30, 1990.
10. DJ Gray, JJ Finer. Development and operation of five particle guns for introduction of DNA into plant cells. Plant Cell Tissue Organ Cult 33:219, 1993.
11. C Sautter, H Waldner, G Neuhaus-Url, A Galli, G Neuhaus, I Potrykus. Micro-targeting: High efficiency gene transfer using a novel approach for the acceleration of micro-particles. Biotechnology 9:1080–1085, 1991.
12. D Pareddy, J Petolino, T Skokut, N Hopkins, M Miller, M Welter, K Smith, D Clayton, S Pescitelli, A Gould. Maize transformation via helium blasting. Maydica 42:143–154, 1997.
13. RA Jefferson. Assaying chimeric genes in plants: The GUS gene fusion system. Plant Mol Biol Rep 5:387–405, 1987.
14. R Birch, R Bower. Principles of gene transfer using particle bombardment. In: NS Yang, P Christou, ed. Particle Bombardment Technology for Gene Transfer. Oxford: Oxford University Press, pp 3–37.
15. T Yamashita, A Iida, H Morikawa. Evidence that more than 90% of β-glucuronidase expressing cells after particle bombardment directly receive the foreign gene in their nucleus. Plant Physiol 97:829–831, 1991.
16. R Hunold, R Bronner, G Hahne. Early events in microparticle bombardment: Cell viability and particle location. Plant J 5:593–604, 1994.
17. WJ Gordon-Kamm, TM Spenser, ML Mangano, TR Adams, RJ Daines, WG Start, JV O'Brien, SA Chambers, WR Adams, NG Willetts, TB Rice, CJ Mackey, RW Krueger, AP Kausch, PG Lemaux. Transformation of maize cells and regeneration of fertile transgenic plants. Plant Cell 2:603–618, 1990.
18. ME Fromm, F Morrish, C Armstrong, R Thomas, TM Klein. Inheritance and expression of chimeric genes in the progeny of transgenic maize plants. Biotechnology 8:833–839, 1990.
19. V Vasil, AM Castillo, ME Fromm, IK Vasil. Herbicide resistant fertile transgenic wheat plants obtained by microprojectile bombardment of regenerable embryogenic callus. Biotechnology 10:667–674, 1992.
20. Y Wan, PG Lemaux. Generation of large numbers of independently transformed fertile barley plants. Plant Physiol 104:37–38, 1994.
21. DE McCabe, WF Swain, BJ Martinelli, P Christou. Stable transformation of soybean (*Glycine max*) by particle acceleration. Biotechnology 6:923–926, 1988.
22. Z Svab, P Hajdukiewicz, P Maliga. Stable transformation of plastids in higher plants. Proc Natl Acad Sci USA 87:8526–8530, 1990.
23. H Daniell. Foreign gene expression in chloroplasts of higher plants mediated by tungsten particle bombardment. Methods Enzymol 217:536–556, 1993.
24. TK Wong. Method of transferring genes into cells. Canadian Patent #1208146, Issued July 22, 1986.
25. PF Lurquin. Gene transfer by electroporation. Mol Biotechnol 7:5–35, 1997.
26. N Eynard, MP Rols, V Ganeva, B Galutzov, N Sabri, J Teissie. Electrotransformation pathways of procaryotic and eucaryotic cells: Recent developments. Bioelectrochem Bioenerg 44:103–110, 1997.
27. LV Chernomordik, AV Sokolov, VG Budker. Electrostimulated uptake of DNA by liposomes. Biochim Biophys Acta 1024:179–183, 1990.

28. ME Fromm, LP Taylor, V Walbot. Expression of genes transferred into monocot and dicot plant cells by electroporation. Proc Natl Acad Sci USA 82:5824–5828, 1985.
29. H Morikawa, A Iida, C Matui, M Ikegami, Y Yamada. Gene transfer into intact plant cells by electroinjection through cell walls and membranes. Gene 41:121–124, 1986.
30. RA Dekeyser, B Claes, RMU De Rycke, ME Habets, MC Van Montegu, AB Caplan. Transient gene expression in intact and organized rice tissues. Plant Cell 2:591–602, 1990.
31. JS Lee, JW Suh, KW Lee. Gene transfer into intact cells of tobacco by electroporation. Korean J Genet 11(2):65–72, 1989.
32. K D'Halluin, K Bonne, M Bossut, M De Beuckeleer, J Leemans. Transgenic maize plants by tissue electroporation. Plant Cell 4:1495–1505, 1992.
33. SM Pescitelli, K Sukhapinda. Stable transformation via electroporation into maize type II callus and regeneration of fertile transgenic plants. Plant Cell Rep 14:712–716, 1995.
34. X Xu, B Li. Fertile transgenic Indica rice plants obtained by electroporation of the seed embryos cells. Plant Cell Rep 13:237–242, 1994.
35. R Coffee, JM Dunwell. Transformation of plant cells. United States Patent 5302523, Issued April 12, 1994.
36. BR Frame, PR Drayton, SV Bagnall, CJ Lewnau, WP Bullock, HM Wilson, JM Dunwell, JA Thompson, K Wang. Production of fertile transgenic maize plants by silicon carbide whisker-mediated transformation. Plant J 6:941–948, 1994.
37. JA Thompson, PR Drayton, BR Frame, K Wang, JM Dunwell. Maize transformation utilizing silicon carbide whiskers: A review. Euphytica 85:75–80, 1995.
38. HF Kaeppler, W Gu, DA Somers, HW Rines, AF Cockburn. Silicon carbide fiber-mediated DNA delivery into plant cells. Plant Cell Rep 8:415–418, 1990.
39. SJ Dalton, AJE Bettany, E Timms, P Morris. Transgenic plants of *Lolium multiflorum, Lolium perenne, Festuca arundinacea,* and *Agrostis stolonifera* by silicon carbide fibre-mediated transformation of cell suspension cultures. Plant Sci 132:31–43, 1998.
40. N Nagatani, H Honda, T Shimada, T Kobayashi. DNA delivery into rice cells and transformation using silicon carbide whiskers. Biotechnol Tech 11:471–473, 1997.
41. O Serik, I Ainur, K Murat, M Tetsuo, I Masaki. Silicon carbide fiber-mediated DNA delivery into cells of wheat (*Triticum aestivum* L.) mature embryos. Plant Cell Rep 16:133–136, 1996.
42. CR Smith, JA Saunders, S Van Wert, J Cheng, BF Matthews. Expression of GUS and CAT activities using electrotransformed pollen. Plant Sci 104:49–58, 1994.
43. A De la Pena, H Lorz, J Schell. Transgenic rye plants obtained by injecting DNA into young floral tillers. Nature 235:274–276, 1987.
44. IK Vasil. Molecular improvement of cereals. Plant Mol Biol Bio 25:925–937, 1994.
45. AP Kausch, TR Adams, M Mangano, SJ Zachwieja, W Gordon-Kamm, R Daines, NG Willetts, SA Chambers, W Adams, A Anderson, G Williams, G Haines. Effects of microprojectile bombardment on embryogenic suspension cell cultures of maize (*Zea mays* L.) used for genetic transformation. Planta 196:501–509, 1995.

# 9
# Electroporation and Cell Energy Factor

**Paul F. Lurquin**
*Washington State University, Pullman, Washington*

**Guangyu Chen**
*Jiangxi Academy of Agricultural Sciences, Nanchang, China*

**Anthony Conner**
*Lincoln University, Canterbury, New Zealand*

| | | |
|---|---|---|
| I. | INTRODUCTION | 145 |
| II. | THEORETICAL BACKGROUND | 146 |
| III. | LITERATURE SURVEY | 146 |
| IV. | ADDITIONAL EMPIRICAL EVIDENCE | 147 |
| V. | CONCLUSION | 148 |

## I. INTRODUCTION

Since its inception in 1982, electroporation has possibly become the most widespread and versatile technique used to introduce nucleic acids into living cells. Basically, cells are subjected to short electric pulses, delivered through capacitor discharge (exponential decay) or fast switching (square wave or pulse), in the presence of deoxyribonucleic acid (DNA) or ribonucleic acid (RNA). So far, it has generally been assumed that $E$ (electric field strength between the electrodes) and $RC$ (also designated $\tau$, the time taken by a capacitor to release 63% of its charge), or pulse time, in the case of a square wave, are the most critical parameters in electroporation success: "E and $\tau$ are the most important electrical variables affecting electroporation" (1). We wish to replace this statement with a new paradigm that unifies all electroporation parameters: "Once the cell membrane breakdown voltage has been reached, the most important parameter affecting electroporation is energy dissipation in the system."

This new concept came about as a result of efforts to reconcile seemingly disparate electroporation conditions (exponential decay vs. square wave, high voltage and low capacitance vs. low voltage and high capacitance) reported by many authors. Further, electrical parameters were systematically investigated in attempts to optimize electroporation of *Asparagus officinalis* protoplasts. Similar conclusions were drawn independently in our two laboratories. Thus, both these theoretical considerations and empirical results demonstrate that it is the release of electrical energy in the electroporation cell that determines cellular permeabilization (2–4). The examples used in the following discussion to illustrate this concept all deal with plant protoplasts (cells devoid of cell wall) but are also applicable to mammalian cells (4).

## II. THEORETICAL BACKGROUND

Plant protoplasts are ideally suited for the exploration of electroporation parameters, given their perfect spherical shape. In this case, the integrated Laplace equation, where $V$ (in volts) is the membrane breakdown voltage, $r$ (in centimeters) is the radius of the sphere, and $E$ (in volts/centimeter) is the electric field at the poles of the protoplasts

$$V = 1.5 r E \tag{1}$$

is directly applicable. Since the radius of plant protoplasts is not less than 10 μm, and since pulses longer than 20 μs can lower the membrane breakdown voltage to 0.5 V (5), permeabilization of these cells will occur at field strengths as low as about 300 V/cm or less for larger protoplasts.

The theory of electrical circuits shows that the energy ($\varepsilon$, in joules) released by a capacitor is

$$\varepsilon = 0.5 C V^2 \tag{2}$$

where $C$ is the capacitance (in microfarads) and $V$ (in volts) is the voltage applied to the poles of the capacitor. In the case of a square wave, the equation becomes

$$\varepsilon = t V^2 / R = t V I = t I^2 R \tag{3}$$

where $V$ is the set voltage (in volts), $R$ the resistance of the system (in ohms), $t$ the discharge time (in seconds), and $I$ the current (in amperes).

## III. LITERATURE SURVEY

Electroporation conditions published by a series of authors have allowed us to calculate energy dissipation values in their systems. The first two examples given in Table 1 show that successful electroporation of plant protoplasts in the presence of RNA only occurs at high enough energy dissipation per unit volume (EDV) value, regardless of the values attributed to voltage (which must be at breakdown level), and directly depends on the capacitance (which must be high enough to reach a proper EDV value). It can be seen that *Nicotiana tabacum* protoplasts were not electroporated at 1.00 kV/cm (a value well above breakdown level) using a capacitance of 1 μF but were permeabilized at the much lower field strength of 0.50 kV/cm when the capacitance was 790μF (6). The EDV was 0.08 J/ml in the first case and 15.8 J/ml in the second. Table 1 also shows several other examples of successful protoplast electroporation by exponential decay (this time with DNA), with EDV values all exceeding 14 J/ml, and using widely different values for $C$ (and hence $RC$) and $E/d$ (field strength). *A. officinalis* protoplasts could not be electrotransformed at 8 J/ml, but they expressed a donor chimeric transgene at 16 J/ml. In cases in which single discharges were delivered at the low EDV value of 0.72 J/ml, 20 consecutive pulses (for a total EDV of 14.4 J/ml) were necessary to observe optimal expression of the donor transgene in *N. tabacum* protoplasts (7). The same authors observed that the number of pulses could be reduced to 10 when the value of the capacitance was doubled. This is in full agreement with our model, since doubling the capacitance while keeping the voltage constant simply doubles the EDV. At an EDV level of 34.4 J/ml per discharge, only single pulses sufficed to achieve electrotransformation of *Glycine max* protoplasts (8). Similarly, and using this time a square pulse, *Daucus carota* protoplasts were not electrotransformed at 7.2 J/ml (with $E/d$ = 0.35 kV/cm, above the breakdown

## Electroporation and Energy

**Table 1** Electroporation of Plant Protoplasts in the Presence of RNA or DNA

| Origin | Exponential decay (capacitor discharge) | | | | |
|---|---|---|---|---|---|
| | $C$ ($\mu$F) | $E/d$ (kV/cm) | $V$ (kV) | Total EDV (J/ml) | Electroporation |
| N. tabacum (Ref. 6) | 1 | 1.00 | 0.40 | 0.08[a] | No |
| N. tabacum (Ref. 6) | 790 | 0.50 | 0.20 | 15.8[a] | Yes |
| N. tabacum (Ref. 7) | 16 | 0.30 | 0.30 | 14.4[b] | Yes |
| Glycine max (Ref. 8) | 490 | 0.375 | 0.375 | 34.4[a] | Yes |
| A. officinalis (Ref. 2) | 500 | 0.28 | 0.113 | 8[a] | No |
| A. officinalis (Ref. 2) | 500 | 0.28 | 0.113 | 16[a] | Yes |
| | Square wave pulse | | | | |
| D. carota (Ref. 9) | | 0.54 | 0.405 | 17.2[c] | Yes |
| D. carota (Ref. 9) | | 0.35 | 0.262 | 7.2[c] | No |

All energy levels were normalized to joules per milliliter (J/ml). RNA or DNA concentrations were optimized and all electroporation parameters (including volume) were defined in the examples. This table shows that electrotransformation (or electrotransfection) does not occur below a certain energy value even when the breakdown voltage is reached or exceeded. A. officinalis protoplasts were electroporated at the same capacitance, voltage, and energy (3.2 J) in both cases; what differed was the volume, 0.4 ml in the first case and 0.2 ml in the second. Yes, presence of RNA or DNA expression in the electroporated protoplasts; no, such expression not detected.
[a]Single pulse.
[b]20 Pulses of 0.72 J/ml each.
[c]6 Pulses of 2.87 J/ml and 1.20 J/ml each.

threshold) but were electrotransformed at an EDV of 17.2 J/ml. In this case, 6 consecutive square wave pulses at 2.87 J/ml (for a total of 17.2 J/ml) were needed to achieve electrotransformation (9). Interestingly, Joersbo et al. (10) were able to derive a general empirical equation correlating square wave electroporation parameters of the form

$$t E^q = K \tag{4}$$

where $t$ is pulse time, $q$ is a constant, and $K$ is a value that varies with the size of the protoplasts and their origin. The average value of $q$ calculated from these authors' data is 1.89, indicating that electroporation efficiency varied with the first power of time and, within 5.5%, with the square of the electric field. Thus, it can be seen that $t$ and $E$ in Eq. (4) assume the same exponents as they do in Eq. (3), which describes energy dissipation in a system containing no capacitor, as was the case in their experiments.

## IV. ADDITIONAL EMPIRICAL EVIDENCE

Experimental evidence further demonstrated that the electroporation efficiency (electropermeabilization) of A. officinalis protoplasts in the presence of methylene blue (which is not subject to biological variables, contrary to transgene expression) increased linearly with EDV from 0 to 95 J/ml (3). Figure 1 shows that the percentage of stained protoplasts was positively correlated with energy dissipation, whereas survival rate was negatively correlated. The $RC$ constant, at three different energy levels (6.6, 12.1, and 24.2 J/ml), had a minor effect on electroporation efficiency between 10 and 100 ms; for example, longer pulses increased electroporation efficiency from 40% to 47% at 24.2 J/ml (3).

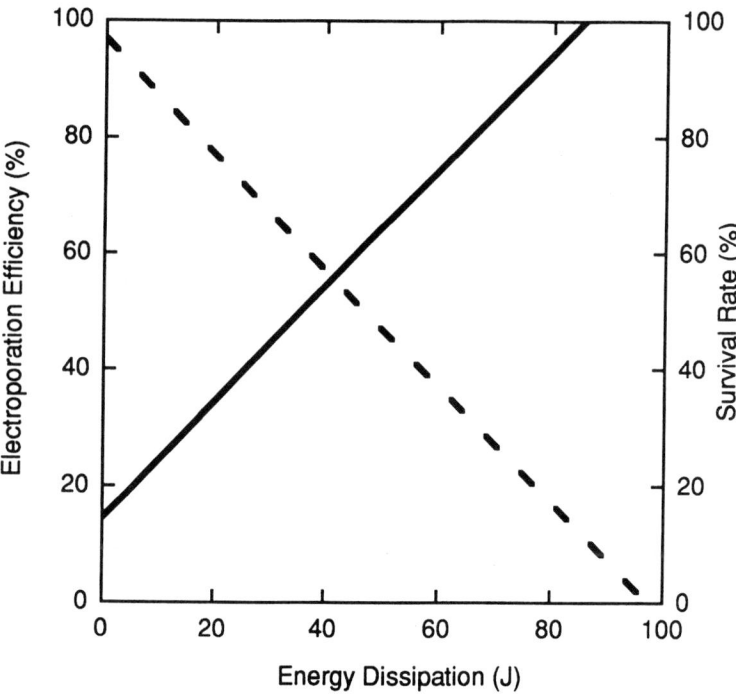

**Figure 1** Regression plots of asparagus protoplast electropermeabilization efficiency (solid line) and survival (dashed line). The regression lines were calculated from four replicate points at 16 energy values as published in Ref. 3. Electroporation volume was 0.2 ml. Equations are $Y = 14.4 + (1 \times \text{energy})$ for electroporation efficiency and $Y = 96.9 - (0.995 \times \text{energy})$ for survival.

## V. CONCLUSION

Taken together, theoretical and empirical observations show that EDV, and not the $RC$ constant, is the major factor in protoplast electroporation. Similarly, the concept of EDV unifies results obtained with square and exponential waveforms, making the electroporation phenomenon equipment-independent. Optimization of electroporation conditions must take into account protoplast survival as a function of energy level. Since electroporation efficiency increases linearly with energy dissipation whereas survival is negatively correlated, the optimal point is at the intersection of both lines (Fig. 1). The existence of this linear relationship completely eliminates the guesswork in establishing electroporation parameters and will greatly simplify the optimization of electrical parameters during electroporation.

Although our conclusions are based here on the response of higher plant protoplasts, we believe that mammalian and microbial cells obey the same general principles (as reviewed in Ref. 4) since cell electroporation is based on the laws of physics. Of course, optimal EDV values must be determined in individual cases, since electroporation efficiency directly depends on cell size, as mandated by the integrated Laplace equation. Thus, bacterial cells require high EDV values since their breakdown voltage is necessarily high, whereas larger cells by definition necessitate lower EDV values.

## ACKNOWLEDGMENT

We thank Dr. Charlotte Omoto (Washington State University) for help with computer graphics.

## REFERENCES

1. K Shikegawa, WJ Dower. Electroporation of prokaryotes and eukaryotes: A general approach to the introduction of macromolecules into cells. Biotechniques 6:742–751, 1988.
2. GY Chen, AJ Conner, AG Fautrier, RJ Field. Transient β-glucuronidase expression in asparagus protoplasts following electroporation. Proceedings of the Ninth International Asparagus Symposium. Acta Horticulturae 479:339–345, 1999.
3. GY Chen, AJ Conner, J Wang, AG Fautrier, RJ Field. Energy dissipation as a key factor for electroporation of protoplasts. Mol Biotechnol 10:209–216, 1998.
4. PF Lurquin. Gene transfer by electroporation. Mol Biotechnol 7:5–35, 1997.
5. U Zimmermann, J Vienken. Electric field–induced cell-to-cell fusion. J Membrane Biol 67:165–182, 1982.
6. K Okada, T Nagata, I Takebe. Introduction of functional RNA into plant protoplasts by electroporation. Plant Cell Physiol 27:619–626, 1986.
7. P Guerche, C Bellini, J-M Le Moullec, M Caboche. Use of a transient expression assay for the optimization of direct gene transfer into tobacco mesophyll protoplasts by electroporation. Biochimie 69:621–628, 1987.
8. P Christou, JE Murphy, WF Swain. Stable transformation of soybean by electroporation and root formation from transformed callus. Proc Natl Acad Sci USA 84:3962–3966, 1987.
9. R Bower, RG Birch. Competence for gene transfer by electroporation in a sub-population of protoplasts from uniform carrot cell suspension cultures. Plant Cell Rep 9:386–389, 1990.
10. M Joersbo, J Brunstedt, F Floto. Quantitative relationship between parameters of electroporation. J Plant Physiol 137:169–174, 1990.

# 10
# Cell Culture and Regeneration of Plant Tissues

**Wei Wen Su**
*University of Hawaii at Manoa, Honolulu, Hawaii*

| | | |
|---|---|---|
| I. | INTRODUCTION | 151 |
| II. | BASIC CELL PROPERTIES AND DEVELOPMENTAL PROCESSES FOR IN VITRO PLANT REGENERATION | 152 |
| III. | PATHWAYS FOR IN VITRO PLANT REGENERATION | 155 |
| | A.  Organogenesis | 156 |
| | B.  Embryogenesis | 157 |
| IV. | MOLECULAR AND BIOCHEMICAL MARKERS OF IN VITRO REGENERATION | 158 |
| V. | FACTORS AFFECTING IN VITRO CULTURE | 159 |
| | A.  Species and Cultivars | 159 |
| | B.  Explant Source | 160 |
| | C.  Culture Conditions | 161 |
| | D.  Transformation Methods and Selection Agents | 161 |
| | REFERENCES | 163 |

## I. INTRODUCTION

To obtain a transgenic plant, a common genetic transformation scheme involves the introduction of foreign genes into a plant explant, followed by selection and regeneration. With the need to convert the explants into whole plants, tissue culture techniques are indispensable in this process. Development of transgenic plants encompasses an integration of a genetic transformation technique, a functional plant regeneration system, and a selectable or screenable marker gene system. As a result of requirements for the genetic transformation process, the insertion of the foreign gene(s) and the selection marker, as well as the selection procedures employed, there exist unique challenges in developing tissue culture and regeneration systems for transgenic plants. Regeneration systems are generally species- and often cultivar-specific (1). Not all transformation techniques are compatible with all regeneration systems. Furthermore, not all selectable markers work well with all species. As such, tissue culture and regeneration methodology needs to be optimized for individual plant systems. Nonetheless, certain existing guidelines and general techniques from the basis for developing the more specific regeneration protocols.

There have been several excellent reviews concerning in vitro regeneration of plants. These reviews have covered general experimental protocols as well as developmental biological characteristics of in vitro plant regeneration or morphogenesis (1,2). There also exists a large body of literature published in the past few years that details specific experimental manipulations required for transformation and regeneration of transgenic plants from many commercially important species. A survey of recent literature published in 1998 on the transformation and regeneration of a variety of plant species is presented in Table 1. The scope of this review is to cover the general principles of in vitro plant tissue cultures and the recent advances of in vitro plant regeneration. Special attention has been directed, wherever applicable, to issues specific to transgenic plants as opposed to nontransgenic plant systems. This review does not detail the species- and cultivar-specific manipulations for plant regeneration. Specific protocols on several commercially important crops can also be found in a number of recently published books and manuals on transgenic plants (3–5).

## II. BASIC CELL PROPERTIES AND DEVELOPMENTAL PROCESSES FOR IN VITRO PLANT REGENERATION

An important requirement for in vitro plant regeneration is that cultured somatic cells remain totipotent and competent. The capability of cultured cells or protoplasts to proliferate and organize into tissues and eventually develop into a whole plant is termed *totipotency*. The term *competence* refers to the capability of a cell or cell clusters to respond to an inductive stimulus for a developmental process. The competent states represent unique genetic, epigenetic, and physiological characteristics of the responding cells in particular developmental processes. It is possible for a cell to be totipotent, but not competent to express totipotency under a particular set of experimental inductive conditions. Identification of competent cells has been achieved to some degree morphologically and cytologically, although the biochemical and molecular basis for competency is still poorly understood (1). Note that not all individual cells in an in vitro culture are capable of expressing totipotency. After cultures have been maintained at a dedifferentiated state for a long period, chromosomal aberration (e.g., polyploidization) often occurs, rendering the cultured cells incapable of expressing totipotency although they may be capable of proliferation. The main in vitro developmental pathways consist of three analogous phases. Using the terms of de Klerk et al. (6), these three phases ar: (a) dedifferentiation or acquisition of competence (during which the tissue becomes competent to respond to the organogenesis or embryogenesis stimuli), (b) induction (during which cells become determined to form either a root, a shoot, or an embryo), and (c) realization (outgrowth to an organ or an embryo). The general sequence of these developmental phases during regeneration is depicted in Fig. 1. This general scheme has been observed in several systems in which a certain period of time is found necessary before a tissue is competent for morphogenic induction. After the attainment of competence, a certain additional amount of exposure time to induction medium is required for the tissue to become determined for the developmental pathway (6,7). It is well known now that not only the medium composition is important in morphogenesis (8); the amount of time in which the tissue is exposed to a particular medium is also critical. Further, the amount of time required for dedifferentiation and for attainment of competence for induction is surprisingly short in many instances but varies considerably with different genotypes. It has been noted that competence for induction is transient. There is a window of competence in which induction for morphogenesis is possible; on either side of this window, only callus proliferation occurs (1). As is discussed further in Sec. IV,

Table 1  A Survey of 1998 Literature on Plant Transformation and Regeneration

| Plant | Transformation | Explant | Selection agent | Regeneration method | Reference |
|---|---|---|---|---|---|
| Astragalus sinicus (Chinese milk vetch) | A. rhizogenes | Seeding | — | Shoot organogenesis | 15 |
| Manihot esculenta (cassava plant) | A. tumefaciens | Embryogenic suspension culture | Paromomycin | Somatic embryogenesis | 66 |
| Vitis vinifera L. | A. tumefaciens | Embryogenic callus | Kanamycin | Somatic embryogenesis | 67 |
| Brassica oleracea (cauliflower) | A. tumefaciens | Hypocotyl segments | Kanamycin | Shoot organogenesis | 68 |
| Gerbera hybrida | A. tumefaciens | Petiole, leaf, and shoot tip explants from micropropagated shoots | Kanamycin | Indirect shoot organogenesis | 69 |
| Rosa hybrida L. (rose) | Particle bombardment | Embryogenic callus | Kanamycin | Somatic embryogenesis | 70 |
| Saccharum officinarum L. (sugarcane) | A. tumefaciens | Meristem | Phosphinothricine | Micropropagation | 14 |
| Pinus sylvestris L. (Scots pine) | Particle bombardment | Cotyledons and embryogenic cultures | Geneticin | Shoot organogenesis and somatic embryogenesis | 71 |
| Lycopersicon esculentum (tomato) | A. tumefaciens | Leaf discs | Kanamycin | Shoot organogenesis | 72 |
| Manihot esculenta (cassava) | Particle bombardment | Embryogenic callus | Phosphinothricin | Somatic embryogenesis | 73 |
| Nicotiana tabacum | A. tumefaciens | Leaf discs | Kanamycin and methotrexate | Shoot organogenesis | 74 |
| Arachis hypogaea L. (peanut) | Particle bombardment | Somatic embryos | Hygromycin | Somatic embryogenesis | 75 |
| Eucalyptus camaldulensis | A. tumefaciens | Hypocotyl segments | Kanamycin | Indirect shoot organogenesis | 11 |
| Citrus aurantifolia (Mexican lime) | A. rhizogenes | Internodal stem segments | — | Shoot organogenesis | 76 |
| Apple | A. tumefaciens | Leaf explants | Kanamycin | Indirect shoot organogenesis | 12 |

**Table 1** (Continued)

| Plant | Transformation | Explant | Selection agent | Regeneration method | Reference |
|---|---|---|---|---|---|
| *Rubus ideaus* L. (Red raspberry) | *A. tumefaciens* | Leaf discs | Kanamycin | Shoot organogenesis | 77 |
| *Populus nigra* (poplar) | *A. tumefaciens* | Leaf explants | Kanamycin | Shoot organogenesis | 78 |
| *Daucus carota* L. (carrot) | *A. tumefaciens* | Hypocotyl segments | Kanamycin | Indirect somatic embryogenesis | 79 |
| Avocado | *A. tumefaciens* | Somatic embryos | Kanamycin | Somatic embryogenesis | 80 |
| *Prunus avium* (cherry) | *A. rhizogenes* | In vitro grown shoots | — | Shoot organogenesis | 81 |
| *Diospyros kaki* (Japanese persimmon) | *A. tumefaciens* | Hypocotyl segment | Kanamycin | Shoot organogenesis | 82 |
| *Pinus radiata* | Particle bombardment | Embryogenic tissue | Kanamycin | Somatic embryogenesis | 83 |
| *Arabidopsis thaliana* | *A. tumefaciens* | Root explants | Kanamycin | Shoot organogenesis | 84 |
| *Zoysia japonica* (Japanese lawngrass) | Polyethylene glycol-mediated gene transfer | Protoplasts | Hygromycin | Indirect Shoot organogenesis | 85 |
| *Medicago truncatula* and *Medicago sativa* | *A. tumefaciens* | Leaf explants | Kanamycin | Somatic embryogenesis | 86 |
| Rice | Particle bombardment | Immature embryos | Hygromycin | Embryogenesis | 87 |
| *Lilium longiflorum* (lily) | Particle bombardment | Morphogenic calli derived from bulblet scales | Bialaphos | Shoot organogenesis | 88 |
| *Antirrhinum majus* L. | *A. rhizogenes* | Seedlings | — | Shoot organogenesis | 89 |
| Sugar beet | *A. tumefaciens* | Cotyledons | D-mannose | Shoot organogenesis | 10 |
| *Capsicum annuum* L. (Chilli) | *A. tumefaciens* | Cotyledonary tissues | Kanamycin | Shoot organogenesis | 90 |
| *Santalum album* L. (sandalwood) | *A. tumefaciens* | Somatic embryos | Kanamycin | Somatic embryogenesis | 91 |
| *Dendranthema grandiflora* | *A. tumefaciens* | Leaf explants | Kanamycin | Shoot organogenesis | 92 |
| *Brassica carinata* | *A. tumefaciens* | Cotyledonary petioles and hypocotyls | Kanamycin | Shoot organogenesis | 93 |
| *Lotus corniculatus* (bird's-foot trefoil) | *A. tumefaciens* | Cotyledon segments | Kanamycin | Embryogenesis | 94 |
| *Rosa hybrida* L. (rose) | Particle bombardment | Embryogenic callus | Kanamycin | Somatic embryogenesis | 95 |

# Cell Culture and Regeneration

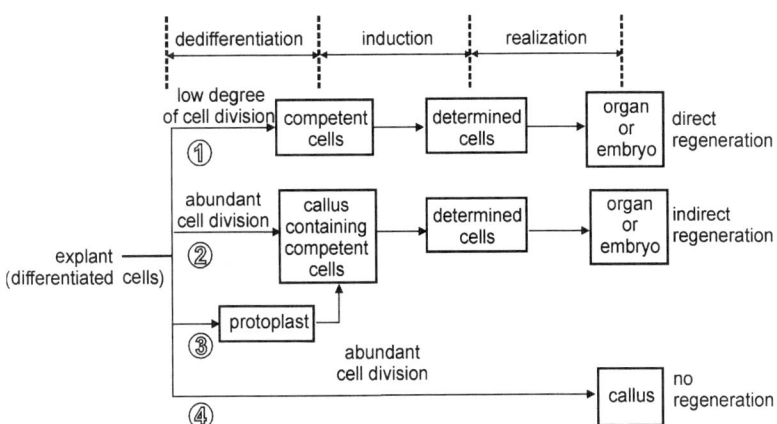

**Figure 1** General sequence of developmental phases during in vitro regeneration. (Modified from Ref. 6.)

recent studies on the molecular and biochemical basis of the competent states have shed new light on this complex phenomenon.

## III. PATHWAYS FOR IN VITRO PLANT REGENERATION

Depending on the type of explants used for transformation and the transformation method employed, various tissue culture strategies can be utilized to establish transgenic plants. The main pathways for in vitro plant regeneration involve formation of shoots and roots via organogenesis, and of somatic embryos via embryogenesis. These two main pathways are discussed separately, and some of the most common tissue culture strategies are depicted schematically in Fig. 2.

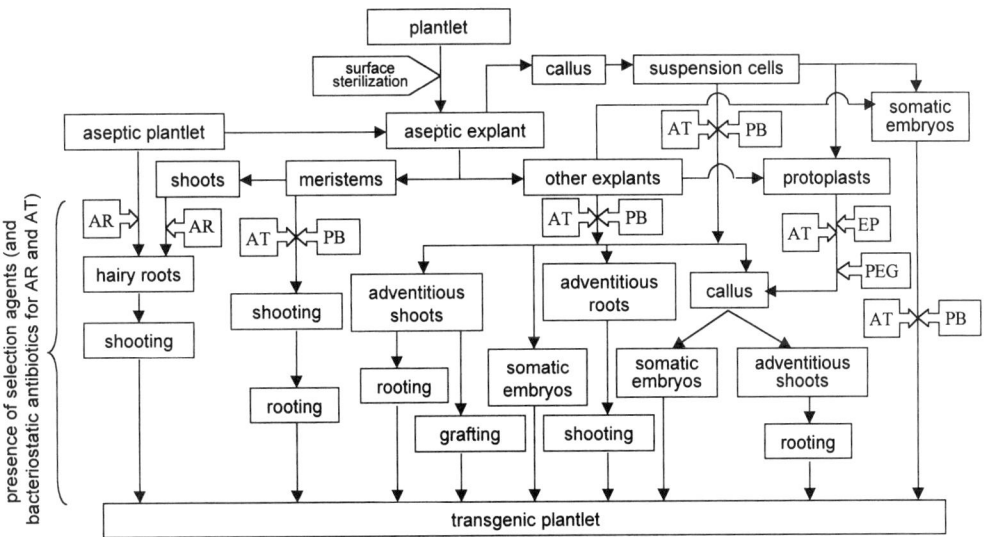

**Figure 2** Pathways for in vitro plant regeneration. AR, *Agrobacterium rhizogenes;* AT, *Agrobacterium tumefaciens;* PB, particle bombardment; EP, electroporation; PEG, PEG-mediated transformation.

Newly regenerated plant tissues lack fully functional cuticle. There is usually a low level of wax found in the cuticles of regenerated plant tissues. It is necessary, therefore, to acclimate newly regenerated plants slowly to the normal growth conditions, during which time there is a buildup of cuticular wax. Acclimatization can be achieved by transferring the plantlets to a growth environment that has a lower relative humidity and a higher light level. It can also be achieved by covering the potted-out plants with polyethylene bags and by punching an increasing number of holes in them (2). The hardened plantlets can then be grown in a greenhouse or growth chamber.

## A. Organogenesis

Organogenesis is a developmental pathway in which shoots or roots have been induced to differentiate from a cell or cell clusters. In vitro plant regeneration by organogenesis usually involves induction and development of shoots from the explant tissue (shoot organogenesis), followed by transfer to a different medium to induce root formation and development (Fig. 2). If the shoot or root is induced and develops directly from the explant without undergoing an initial callus phase, this is termed *direct* or *adventitious organogenesis*. An example of direct in vitro organogenesis is found with tobacco leaf disks (9) or cotyledonary tissues of sugar beet (10). Indirect organogenesis involves an initial phase of callus proliferation and growth, followed by shoot or root induction and development from this proliferated callus tissue that contains competent cells (Fig. 2). An example of indirect in vitro organogenesis is found with *Agrobacterium tumefaciens* transformed *Eucalyptus* hypocotyl segments (11) and apple leaf explants (12). As a result of potential problems with somaclonal variation in callus cultures, it is more desirable to regenerate transgenic plants by direct rather than indirect organogenesis, or at least to minimize callus proliferation before regeneration. In addition to adventitious shoot regeneration, an important type of direct organogenesis, termed *micropropagation*, involves regeneration via existing meristems. Micropropagation of apical meristems has been used as the regeneration system for transformation of sunflower (13) and sugarcane (14). Regeneration by micropropagation has the advantage that plants are regenerated directly from an organized tissue without an intervening callus stage. This not only saves time, but also eliminates undesirable somaclonal variation associated with long callus culture period. Given that when regeneration is done via micropropagation, a stringent selection procedure should be followed to minimize nontransformed plants, as is discussed in Sec. V.2. Besides shoot regeneration, it is also possible to generate roots from the explants, followed by induction of shoot formation. This is commonly practiced when *A. rhizogenes* is used to transform the plants; hairy roots are formed from the wound sites of the seedlings or explants as a result of the insertion of the *Ri* plasmid (15). Regeneration from adventitious or hairy roots usually is more difficult to achieve than regeneration from shoots (6).

Organogenesis has been chosen for in vitro plant regeneration from a variety of transformed explants, including protoplasts (Fig. 2; Table 1). As seen in Table 1 and current literature, *Agrobacterium* transformation (with *A. tumefaciens* or *A. rhizogenes*) and particle bombardment (biolistic; gene gun) are the most widely used techniques for plant transformation. For both techniques, organogenesis has been used successfully in regenerating transgenic plants (Table 1). The first breakthrough on control of organogesis was reported by Skoog and Miller (8), who showed that alterations of the auxin and cytokinin ratios were sufficient to control morphogenesis in tobacco. High cytokinin-to-auxin ratios were found to produce shoots, low cytokinin-to-auxin ratios produced roots, and more equal concentrations of these phytohormones were found to cause callus proliferation. Since this initial report, media formulations for callus, shoot, and root induction have been devised for many plant species (2,9). In general, these media formulations have been derived empirically as a result of lack of understanding of plant development and mechanisms of hormone action, and inability to apply plant regeneration protocols from one species or

# Cell Culture and Regeneration

cultivar within a species to other species or cultivars successfully. Considering the sequential phases leading to organogenesis (Fig. 1), different media are usually needed to cause dedifferentiation, attainment of competence, induction for the organogenic pathway, and determination for the pathway, and not to interfere with the morphogenic expression of the developmental pathway. For the regeneration system used in the *Agrobacterium*-mediated leaf disk transformation system (16), however, a single consensus medium was used.

## B. Embryogenesis

In somatic embryogenesis, somatic cells develop to form complete embryos analogous to zygotic embryos. The bipolar structure of the somatic embryo contains both shoot and root meristems. As with organogenesis, somatic embryogenesis can occur from cells of the explant tissue with or without an intervening callus phase (Fig. 2). For direct somatic embryogenesis, the immature zygotic embryo is most often used as the explant and the response of the embryo depends largely on the developmental stage of the explant (9). The indirect embryogenesis pathway, whereby somatic embryos are induced and develop from proliferated callus or suspension cells, is generally more common (1). Embryogenic cells appear very similar to meristematic cells in that they are small and densely cytoplasmic, have a large nucleus and prominent nucleoli, and contain many small vacuoles, lipid droplets, and starch grains (17). Embryogenic cells in suspension culture commonly form small compact clumps that have been termed *proembryonal complexes* or *proembryogenic masses* (PEMs) (17). Nonembryogenic cells, on the other hand, are usually highly vacuolated and have variable shapes. Embryogenic cultures are highly heterogeneous; they may contain a mix of organogenic and embryogenic structures, as well as nonembryogenic cells (18).

Analogous to organogenesis, embryogenesis can be dissected into a series of successive phases (Fig. 1). For newly initiated suspension or callus cultures, a certain period is usually required for the cells to dedifferentiate and attain competence for the embryogenic pathway (19). It is generally believed that the embryogenic pathway is induced and becomes determined very early in embryogenic cultures, and this clearly seems to be the case in the model species carrot (19). For species other than carrot, embryogenic cultures probably also comprise determined cells in which some level of embryo development is maintained in culture. The acquisition of competence and induction of somatic embryogenesis depend upon auxin (usually 2,4-dichlorophenoxyacetic acid) (6). In established embryogenic cultures, the exogenously supplied auxin maintains cellular proliferation but to some degree represses morphogenic expression of embryogenicity. The degree of morphogenic repression depends primarily on the auxin concentration or cultural practices that affect the auxin concentration (e.g., extended time between subculturing) but may also depend on the cell density for suspension cultures. It is hypothesized that the auxin represses morphogenesis by disrupting important cell-to-cell interaction. This allows some embryogenic cells in the clump to develop autonomously and break away from the clump, rather than remain as part of the developmentally integrated primordial complex. This fragmentation of single cells or small groups of cells from the clump is believed to be the mechanism for proliferation of embryogenic suspension cultures (17). In order to allow the formation of matured embryos from the embryogenic cell culture during the "realization" phase of the developmental process (Fig. 1), the auxin usually has to be removed. After its removal, the amount of embryonic development that can occur in liquid suspension varies from complete somatic embryo development in carrot suspensions to blockage at the late globular or early scutellar stage for most cereals and grasses (1). Several species have been transformed, by using embryogenic suspension cultures, including carrot, corn, rice, cotton (1), and more recently rose, cassava, and peanut (Table 1), as starting material. Embryogenic tissues or cells are most often transformed with *A. tumefaciens* or particle bombardment.

## IV. MOLECULAR AND BIOCHEMICAL MARKERS OF IN VITRO REGENERATION

The identification of molecular markers involved in early phases of in vitro morphogenesis not only contributes to our understanding of the processes underlying growth regulator–controlled determination, but also provides a useful tool for evaluating the regeneration potential of the target explant tissues. It has been found that in the carrot system, significant changes in the organization and modification of DNA occur during the callus phase (20,21). These are possibly related to developmental events of the regeneration process. A progressive reduction in DNA was also demonstrated by Giorgetti et al. (22) during the generation of an embryogenic cell culture from carrot hypocotyls. Deumling and Clermont (23) reported a complex pattern of chromosome diminution during cell culture and plant regeneration of *Scilla siberica*. They concluded that an excessive and specific chromatin loss is a prerequisite for plant regeneration. Many studies have shown that the state of dedifferentiation is related to DNA methylation (24,25). During the linear growth phase of carrot callus, the extent of genome methylation increases; it then decreases during the stationary phase. These changes are believed to be related to concurrent developmental events in the callus. Auxin increases DNA methylation reversibly, whereas kinetin tends to block changes in DNA methylation (20). An increase in genome methylation was suggested to be necessary to disorganize an ongoing cell program (6). Fractionation of cell types of an embryogenic carrot cell line indicated a characteristic low genome methylation level of a fraction enriched in precursor cells of somatic embryos (24). Although the causal relationships concerning DNA methylation are still obscure, the effect of methylated cytosine residues in the DNA is believed to be mainly due to interference with DNA-protein binding, influencing the regulation of genome activities (transcription, replication, rearrangements) (6).

Typical changes of gene expression during the early stages of differentiation have been reviewed by Sterck and de Vries (26). Using mitochrondrial ribonucleic acids (mRNAs) isolated from polysomal fractions, Zimmermann et al. (27) by subtractive hybridization isolated 30 carrot (cDNA) clones that are enhanced in globular embryos. By adapting a mRNA differential display technique to the comparative analysis of a model system of tomato cotyledons that can be driven selectively toward either shoot or callus formation by means of previously determined growth regulator supplementation, Torelli et al. (28) were able to monitor changes in gene expression accompanying in vitro regeneration and identified two potential morphogenetic marker genes. In barley, Stirn et al. (29) showed that expression of two embryo-specific genes was limited to embryogenic cell cultures but not to nonembryogenic cultures. Nevertheless, cell aggregates that were embryogenic but no longer able to regenerate plants expressed both genes. Comparing the protein pattern of embryogenic/regenerable and embryogenic/nonregenerable suspension cultures of barley, Stirn et al. (29) identified a 85-kDa polypeptide (pI 5.8) that accumulates only in nonregenerable cultures. Moreover, after the electrophoretic pattern of secreted proteins, two glycoproteins correlating with the embryogenic capacity (46 kDa, pI 6.1) and the loss of regenerative potential (17.4 kDa), respectively, were found. These data indicate that at least two levels of control exist in somatic embryogenesis. One is the induction of bipolar embryos, and the second the germination of somatic embryos into plantlets. In addition to the work of Stirn et al. (29), correlation of excreted proteins, in particular glycoproteins, with differentiation of somatic embryos has been shown in several reports (30,31). Various intracellular soluble proteins have also been identified as potential markers for embryogeneic capability (32,33).

Another potential biochemical marker for regeneration is based upon differences of enzyme activities in recalcitrant vs. easy-to-regenerate explants. For instance, peroxidase activity has been correlated with rooting (34) and somatic embryogenesis (35), and esterase activity with somatic embryogenesis (36). Polyphenoloxidase activity in *Euphorbia pulcherrima* was used for the characterization of the embryogenic status of cell suspension cultures (37). Differences in enzyme activity may also be detected by the occurrence of the end products of the enzymatic reaction. Short-lived starch grains occur in the very early stages of shoot and embryo regeneration (38). In slices cut from apple microshoots and treated with auxin, short-lived starch grains appear during the dedifferentiation phase in a ring consisting of cells of the vascular bundles and primary rays. These cells enter division and from those in the primary rays, root primordia may develop (39). When auxin is not supplied, starch grains are formed at a much later time. In *Cucumis sativus*, differential uptake of carbon sources from the medium of suspension cultures has been used as a biochemical marker for embryogenic cultures (40).

Various other biochemical and physiological features have been correlated with regeneration. Variation in endogenous hormone (e.g., indole-3-acetic acid [IAA]) distribution was shown to accompany morphological polarity in the process of somatic embryogenesis in *Freesia refracta* (41). Enhancement of the in vitro organogenesis ability of rubber-tree clones after somatic embryogenesis or repeated grafting onto juvenile rootstocks was accompanied by an increase of zeatin riboside levels in shoots used as starting material for in vitro micropropagation. Furthermore, the zeatin level in in vitro shoots capable of organogenesis was higher than in their nonorganogenic counterparts. Perrin et al. (42) then concluded that the endogenous zeatinlike cytokinin level (free and ribosylated forms) could be considered as a reliable marker for the recovery of in vitro shoot and root organogenesis after rejuvenating treatments in rubber-tree clones. The concentrations of polyamines, especially of putrescine and spermidine, are higher in embryogenic than in nonembryogenic cells and media of suspension cultures (43). Inhibition of polyamine synthesis reduces the number of embryos, whereas addition of polyamines to inhibitor-supplemented cultures restores embryo formation at the original level. Accumulation of ethylene is less in embryogenic suspensions than in nonembryogenic cultures (44), as well as the amount of glutathion. The redox status of cells, characterized by the ability to reduce $Fe^{3+}$, is far higher in nonembryogenic cells. Phenolics are a very heterogeneous group of substances, interacting with intra- and intercellular processes, e.g., with auxin metabolism. The phenolic content is used in woody plants to differentiate between juvenile and adult phases and thus serves as a marker for the capacity for root formation (45).

A major challenge in the study of regeneration markers is the need for very sensitive analytical techniques. Besides the classical approaches involving mutant isolation and differential display, recent advances in cDNA microarrays (gene chips) and proteomics should significantly improve our ability to isolate genes that are involved in the early phases of regeneration.

## V. FACTORS AFFECTING IN VITRO CULTURE

The main factors affecting in vitro transgenic plant regeneration are genotype, explant source, culture conditions, and transformation/selection methods.

### A. Species and Cultivars

Regeneration capacity differs between plant species and cultivars. It is well known that some families and genera, such as Solanaceae, Cruciferae, Gesneriaceae, Compositae, and Liliaceae, have higher regeneration ability (46). Herbaceous plants regenerate far more readily than trees and shrubs. With its relative ease and repeatability of culture initiation and plant regeneration, tobacco (*Nicotiana tabacum*) has been the model species for organogenic studies and carrot (*Daucus carota*) has been the model species for the study of embryogenesis (1). Other species within the same family as tobacco (Solanaceae) and within the same family as carrot (Umbelliferae) have also been readily amenable to in vitro culture and regenerability, whereas other families (e.g., Gramineae) have been less amenable (2,47). Different cultivars of the same species can also exhibit very different regeneration capacity. For instance, Machii et al. (48) screened 107 wheat genotypes for callus induction and regeneration capability from anther and immature embryo cultures. For anther cultures, only 9 genotypes produced normal plants. For immature embryo cultures, 74 genotypes regenerated plants. Apparently, the genetic component is highly influential on success in in vitro culture and plant regeneration.

## B. Explant Source

To achieve effective in vitro plant regeneration, the major requirements for explant tissues are high cell division potential and morphogenic plasticity. These criteria are usually satisfied by immature, rapidly growing tissues. The use of young tissue is especially important for cereal monocots such as maize in which parenchyma cells in vivo quickly lose the ability to redifferentiate as they mature (49). Although plants have been regenerated in vitro from many different tissues of monocots (9,47), the immature organs or meristematic and undifferentiated tissues are generally the most responsive and reliable explant sources (9,47). The reason for this stage-specific response may be genetic, epigenetic, or physiological changes that occur in mature cells (47). For dicot species, successful in vitro regeneration has been achieved by using leaf pieces (disks or strips), leaf petiole segments, cotyledonary petioles, cotyledonary pieces, hypocotyl segments, root segments, stem segments, various floral and inflorescence structures, storage root and tuber pieces, embryos, and immature embryos (47).

Before explant tissues can be cultured, contaminating microbes must be destroyed. Although most microorganisms are confined to the surface of plant tissue and can be destroyed by surface sterilization, using sodium hypochloride, for example, some microbes may invade the plant vascular tissues and hence be difficult to remove. The use of aseptically grown material is a convenient way to circumvent sterility problems. During the initial culture of the explant tissues, common problems are browning and eventually the death of the tissue, which is due to the excessive production of polyphenolics. These problems can sometimes be alleviated by incorporating adsorbants such as charcoal or polyvinylpyrrolidone, or an antioxidant such as ascorbic acid. The inclusion of adsorbants must be carefully controlled, however, because they can adsorb medium components as well.

When using *Agrobacterium* sp. as a transformation vehicle it is essential that a reasonable proportion of the intact cells at the wound sites of explanted tissues undergo dedifferentiation and cell division. Direct organogenesis of adventitious organs and continued growth from preexisting meristems have to be kept to a minimum as there is no evidence that cells undergoing this type of development can be transformed (50). If these latter two modes of development are not controlled, laborious screening techniques may have to be employed to identify transformed organs among a much larger population of organs derived from nontransformed tissues.

Leaf disks from in vitro propagated or greenhouse-grown plants have proved useful for transformation. Leaf lamina slices do not contain quiescent buds or preformed organ primordia, and therefore the majority of adventitious organs forming at the wound sites of explanted tissues originate from dedifferentiated cells susceptible to transformation by *Agrobacterium* sp. (50). Many other types of explants contain quiescent meristems or buds and may preferentially produce shoot primordia, from cambial meristems or other cell layers deep within the explant, which are not accessible to *Agrobacterium* sp. This is the case with hypocotyl explants. For such explants, it is desirable to include an intervening callus stage in the regeneration process (i.e., indirect organogenesis or embryogenesis). It should also be stressed that the leaf-disk method is not applicable to all species as it depends entirely on the ability to regenerate shoots efficiently from dedifferentiated cells at the wound site. For instance, it has proved rather difficult to adapt this technique for potato (again a solanaceous species) (50).

## Cell Culture and Regeneration

Initiation of in vitro cultures almost invariably requires some injury to the explanted tissue; wound response is hence an unpreventable consequence during the initial culture. Many dicot species produce a callus in response to the injury at the wound site. Some dicot explants (e.g., potato tuber slices) exhibit wound response in which cellular proliferation and wound healing are less random because of the establishment of a wound periderm (1). Monocots typically show a poor wound response by increasing cell wall lignification of cells adjacent to the wound and by showing little or no proliferating wound callus in the wound site. Additionally, monocot cells may lose their ability to dedifferentiate and resume meristematic activity upon maturity (1). As discussed success with *Agrobacterium*-mediated transformation is dependent in part on dedifferentiation and cell division in cells adjacent to the wound (50). For a species such as a cereal monocot that shows a poor wound response (i.e., a small amount of cell division near the wound site by only a few cells), the transformation efficiency with *Agrobacterium* sp. is expected to be reduced.

Suspension cell/callus culture may be considered as a unique "explant." It is an important component of indirect organogenesis and embryogenesis (Fig. 2). For monocots, suspension cells and suspension protoplasts have been the primary target explants for plant transformation. Suspension cells have been used to obtain transgenic maize via particle bombardment (51). In addition, suspension protoplasts combined with electroporation or polyethylene glycol- (PEG-) mediated transformation techniques have been utilized to obtain transgenic rice, corn, and orchard grass (1). Numerous species have regenerated completely to plants after protoplast culture. The versatility of protoplasts is illustrated by the fact that they have been utilized in transformation systems using *Agrobaeterium* sp., PEG, and electroporation (Fig. 2), and in protoplast fusion and microinjection (9).

### C. Culture Conditions

Medium composition is usually the most important culture condition to consider for in vitro plant regeneration. Of the various medium components, growth regulator plays a central role in culture initiation and morphogenesis. Other compounds such as reduced nitrogen and sugar may also affect morphogenesis (52). In addition, the type of medium gelling agent has been shown to affect plant regeneration (53). Despite the complexity of the cell-growth regulator interaction, growth regulator effects on differentiation do show a degree of consistency that make them useful guiding principles, even if they do not hold universally. These principles are as follows:

1. High auxin concentrations suppress organized growth and promote formation of meristemlike cells.
2. Auxin/cytokinin ratio influences the balance between root and shoot formation. As a rule of thumb, high auxin/cytokinin ratio favors root formation and the converse situation favors shoot formation.
3. High cytokinin concentrations inhibit root formation.
4. High auxin concentrations induce somatic embryogenesis but suppress further embryo development and maturation.

Aspects of various culture conditions are discussed in greater detail by Vasil and Thorpe (47).

### D. Transformation Methods and Selection Agents

Selectable genes encode proteins that render the transformed plants resistant to phytotoxic agents (negative selection) or confering capability to outgrow nontransformed plants (positive selection). For either negative or positive selections, specific selection agents need to be added to the selection media to select the transformed plants. In negative selection, chemicals such as antibiotics and herbicides are used to kill off nontransformed plants. The selection agent is usually applied early in the plant regeneration program to allow more efficient elimination of the nontransgenic cells. In some cases, exposure to the selection agent may have to continue throughout the regeneration process to reduce the number of escapes. The concentration of the selection agents has to be determined empirically for different plant species by generating killing curves. The resistance of the transgenic plant tissues ultimately depends on the strength and tissue specificity of the promoter driving the selectable marker gene. One frequently used selectable marker is the neomycin phosphotransferase gene, which confers resistance to aminoglycosides such as kanamycin. This system is very efficient for the selection of transgenic shoots of Solanaceae but is less suited for some members of, e.g., the Fabaceae because of high levels of inherent tolerance to the selective agent (10). New selection systems based on "positive selection" have been developed recently. One such system has been established using inactive cytokinin glucuronides as selective agents and a β-glucuronidase (GUS) gene as selectable gene, releasing active cytokinin in the transgenic cells, which, in turn, stimulates growth and regeneration (54). Another system is based on mannose as selective agent and a phosphomannose isomerase (PMI) gene as selectable gene, which has been shown to be superior to kanamycin in transforming sugar beet (10,55). The working principle of this selection system is quite different from that of the kanamycin selection system. First, the toxicity of mannose is not mediated by the compound per se but is considered to be a consequence of its phosphorylation to mannose-6-phosphate by hexokinase, whereby the nontransgenic PMI-negative cells are starved for phosphate and adenosine triphosphate (ATP). Second, the transgenic PMI-positive cells convert the selective agent to readily metabolized compound, fructose-6-phosphate, thus improving the energy status of the transgenic cells and preventing accumulation of derivatized selective agent. The mannose concentration is increased stepwise during selection. Several parameters affect the efficiency of the mannose selection system, and hence it is not as straightforward to apply as the classical resistance-based selection systems. Some of these parameters are the interaction between the sugar and mannose, phosphate concentration, and light intensity (10,55). Transgenic plants can also be selected or screened on the basis of a unique phenotype. For instance, in *A. rhizogenes* transformation, transformants exhibit the hairy root phenotype and hence no chemical selection agent is necessary. The resulting plants, however, exhibit the "hairy root" syndrome such as small thin leaves and short internodes. Plants have also been cotransformed by using both *A. rhizogenes* oncogenic Ri plasmid and a separate *A. tumefaciens* binary vector carrying the gene of interest (56,57).

In *Agrobacterium*-mediated transformation, supplementation by bacteriostatic antibiotics, such as carbenicillin or cefotaxime, of the selection medium is necessary to eliminate the *agrobacteria* from the plant material. The antibiotic solution is filter-sterilized and added to the medium after autoclaving and cooling. Carbenicillin usually is used at concentrations of 300–500 mg/l, whereas cefotaxime has been used most commonly at concentrations around 250 mg/l. At these concentrations, these antibiotics (by themselves or in combination with kanamycin) may cause inhibition on growth or regeneration in some plant species. The effect of carbenicillin and cefotaxime on regeneration of various plant species has been surveyed by Otani et al. (58). A number of alternative antibiotics may be used for plant species that are sensitive to carbenicillin and cefotaxime. Timentin, a mixture of ticarcillin (a penicillin derivative) and clavulanic acid, has been shown to be superior to carbenicillin and cefotaxime in tobacco (58,59) and tomato (60). This antibiotic mixture is commonly used at 150 mg/l. At this concentration, Ling and coworkers (60) have shown that timentin is not toxic to tomato tissues and promotes callus formation and shoot regeneration. A similar promoting effect of timentin was also observed by Nauerby and associates (59) in the tobacco system. For tomato, the transformation frequency was raised more than 40% in comparison to that of cefotaxime. In this case, cefotaxime itself did not inhibit callus growth in culture medium, but it decreased shoot differentiation. Together with kanamycin, cefotaxime strongly reduced callus growth, shoot regeneration, and transformation efficiency. For pears, Chevreau and colleagues al. (53) reported that cefotaxime (200 mg/l) plus ticarcillin/clavulanic acid (100 mg/l) could be used in the culture medium without affecting the frequency of bud regeneration. Another antibiotic, amoxicillin trihydrate (Augmentin) at 300 mg/l, was able to eliminate *A. tumefaciens* and enhance shoot proliferation of eggplant (*Solanum melongena* L.) (61). In a study of apple transformation and regeneration, Hammerschlag et al. (62) demonstrated that the incidence of *A. tumefaciens* contamination could be reduced to 28% without negatively impacting shoot regeneration by using a 1-hour vacuum infiltration with an acidified medium, an 18-hour vacuum infiltration with cefotaxime (5000 mg/l), and 52-day incubation of regeneration and elongation media containing 100 mg/l each of cefoxitin and carbenicillin.

Depending on the property and level of the foreign gene product expressed, it may have positive or negative effects on plant regeneration. When a tomato antisense ACC synthase gene was expressed in tobacco, shoot proliferation during plant regeneration was significantly enhanced, indicating the regulatory role of ethylene in shoot formation (63). On the other hand, when expressing a toxic protein constitutively (e.g., ribosome-inactivating protein (64) or when an inert protein is expressed at an exceedingly high level (e.g., a modified green fluorescent protein (65), plant regeneration may be hampered.

## ACKNOWLEDGMENT

The author would like to thank Dr. Kung-Ta Lee of the Development Center for Biotechnology (Taiwan) for helpful discussion.

## REFERENCES

1. SW Ritchie, TK Hodges. Cell culture and regeneration of transgenic plants. In: S-D Kung, R Wu, eds. Transgenic Plants. San Diego: Academic Press, 1993, pp 147–180.
2. G Warren. The regeneration of plants from cultured cells and tissues. In: A Stafford, G Warren, eds. Plant Cell Tissue Culture. Milton Keynes: Open University Press, 1991, pp 82–102.
3. C Cunningham, AJR Porter, eds. Recombinant Proteins from Plants: Methods in Biotechnology. Vol. 3. Totowa, NJ: Humana Press, 1998.
4. GB Collins, RJ Shepherd, eds. Engineering plants for commercial products and applications. Ann NY Acad Sci 792, 1996.
5. MRL Owen, J Pen, eds. Transgenic Plants: A Production System for Industrial and Pharmaceutical Proteins. Chichester, England: Wiley, 1996.
6. GJ de Klerk, B Arnholdt-Schmitt, R Lieberei, KH Neumann. Regeneration of roots, shoots and embryos: Physiological, biochemical and molecular aspects. Biol Planta 39:53–66, 1997.
7. ML Christianson, DA Warnick. Competence and determination in the process of in vitro shoot organogenesis. Dev Biol 95:288–293, 1983.
8. F Skoog, CO Miller. Chemical regulation of growth and organ formation in plant tissues cultured in vitro. Soc Exp Biol Symp 11:118–131, 1957.
9. OL Gamborg, GC Phillips, eds. Plant Cell, Tissue and Organ Culture: Fundamental Methods. Berlin: Springer, 1995.
10. M Joersbo, I Donaldson, J Kreiberg, SG Petersen. Analysis of mannose selection used for transformation of sugar beet. Mol Breed 4:111–117, 1998.
11. CK Ho, SH Chang, JY Tsay, CJ Tsai, VL Chiang, ZZ Chen. *Agrobacterium tumefaciens*–mediated transformation of *Eucalyptus camaldulensis* and production of transgenic plants. Plant Cell Rep 17: 675–680, 1998.
12. SN Maximova, AM Dandekar, MJ Guiltinan. Investigation of *Agrobacterium*-medium transformation

of apple using green fluorescent protein: High transient expression and low stable transformation suggest that factors other than T-DNA transfer are rate-limiting. Plant Mol Biol 37:549–559, 1998.
13. B Schrammeijer, PC Sijmons, PJM Van Den Elzen, A Hoekema. Meristem transformation of sunflower via agrobacterium. Plant Cell Rep 9:55–60, 1990.
14. GA Enriquez Obregon, RI Vazquez Padron, DL Prieto Samsonov, GAD Riva, G Selman Housein. Herbicide-resistant sugarcane (*Saccharum officinarum* L.) plants by *Agrobacterium*-mediated transformation. Planta 206:20–27, 1998.
15. HJ Cho, JM Widholm, N Tanaka, Y Nakanishi, Y Murooka. *Agrobacterium rhizogenes*–mediated transformation and regeneration of the legume *Astragalus sinicus* (Chinese milk vetch). Plant Sci 138: 53–65, 1998.
16. RB Horsch, JF Fry, NL Hoffman, D Eichholtz, SG Rogers, RT Fraley. A simple and general method for transferring genes into plants. Science 227:226–228, 1985.
17. G Maheswaran, EG Williams. Primary and secondary direct somatic embryogenesis from immature zygotic embryos of *Brassica campestris*. J Plant Physiol 124:455–464, 1986.
18. A Komamine, R Kawahara, M Matsumoto, S Sunobori, T Toya, A Fujiwara, M Tsukahara, J Smith, M Ito, H Fukuda, K Nomura, T Fujimura. Mechanisms of somatic embryogenesis in cell cultures: Physiology, biochemistry and molecular biology. In Vitro Cell Dev Biol 28:11–14, 1992.
19. SC De Vries, H Booij, P Meyerink, G Huisman, HD Wilde, TL Thomas, A Van Kammen. Acquisition of embryogenic potential in carrot cell–suspension cultures. Planta 176:196–204, 1988.
20. B Arnholdt-Schmitt, S Herterich, KH Neumann. Physiological aspects of genome variability in tissue culture. I. Growth phase–dependent differential DNA methylation of the carrot genome (*Daucus carota* L.) during primary culture. Theor Appl Genet 91:809–815, 1995.
21. B Arnholdt-Schmitt. Physiological aspects of genome variability in tissue culture. II. Growth phase–dependent quantitative variability of repetitive BstNI fragments of primary cultures of *Daucus carota* L. Theor Appl Genet 91:816–823, 1995.
22. L Giorgetti, MR Vergara, M Evangelista, F Loschiavo, M Terzi, VN Ronchi. On the occurrence of somatic meiosis in embryogenic carrot cell cultures. Mol Gen Genet 246:657–662, 1995.
23. B Deumling, L Clermont. Changes in DNA content and chromosomal size during cell culture and plant regeneration of *Scilla siberica* selective chromatin diminution in response to environmental conditions. Chromosoma 97:439–448, 1989.
24. G Palmgren, O Mattsson, FT Okkels. Specific levels of DNA methylation in various tissues, cell lines and cell types of *Daucus carota*. Plant Physiol 95:174–178, 1991.
25. CD Riggs, MJ Chrispeels. The expression of phytohemagglutinin genes in *Phaseolus vulgaris* is associated with organ-specific DNA methylation patterns. Plant Mol Biol 14:629–632, 1990.
26. P Sterck, SC de Vries. Molecular markers for plant embryos. In: K Redenbaugh, ed. Synseeds: Applications of Synthetic Seeds to Crop Improvement. Boca Raton, FL: CRC Press, 1993, pp 115–132.
27. JL Zimmermann, X Lin, GH Hwang. The molecular basis for somatic embryo development in carrot. Acta Hortic 336:217–224, 1993.
28. A Torelli, E Soragni, A Bolchi, S Petrucco, S Ottonello, C Branca. New potential markers of in vitro tomato morphogenesis identified by mRNA differential display. Plant Mol Biol 32:891–900, 1996.
29. S Stirn, AP Mordhorst, S Fuchs, H Loerz. Molecular and biochemical markers for embryogenic potential and regenerative capacity of barley (*Hordeum vulgare* L.) cell cultures. Plant Sci 106:195–206, 1995.
30. J-M Domon, B Dumas, E Laine, Y Meyer, A David, H David. Three glycosylated polypeptides secreted by several embryogenic cell cultures of pine show highly specific serological affinity to antibodies directed against the wheat Germin apoprotein monomer. Plant Physiol 108:141–148, 1995.
31. J-M Domon. Extracellular (glyco) proteins in embryogenic and non-embryogenic cell lines of Caribbean pine: Comparison between phenotypes of stage one somatic embryos. Plant Physiol Biochem 32:137–147, 1994.
32. MD Blanco, N Nieves, M Sanchez, CG Borroto, R Castillo, JL Gonzalez, M Escalona, E Baez, Z Hernandez. Protein changes associated with plant regeneration in embryogenic calli of sugarcane (*Saccharum* sp.). Plant Cell Tissue Org Cult 51:153–158, 1997.
33. JP Fellers, AC Guenzi, DR Porter. Marker proteins associated with somatic embryogenesis of wheat callus cultures. J Plant Physiol 151:201–208, 1997.

34. JD Choi, KW Kim. Peroxidase activity as a biochemical marker for organogenesis during *Gladiolus* callus culture. J Korean Soc Hortic Sci 38:581–587, 1997.
35. S Wochok, G Burleson. Isoperoxidase activity and induction in cultured tissue of wild carrot. Physiol Plant 31:73–75, 1974.
36. L Coppens, D Dewitte. Esterase and peroxidase zymograms from barley callus as a biochemical marker system of embryogenesis and organogenesis. Plant Sci 67:97–105, 1990.
37. C Grotkass, R Lieberei, W Preil. Polyphenoloxidase-activity and -activation in embryogenic and non-embryogenic suspension cultures of *Euphorbia pulcherrima*. Plant Cell Rep 14:428–431, 1995.
38. BS Mangat, MK Pelekis, AC Cassells. Changes in the starch content during organogenesis in in-vitro cultured *Begonia rex* stem explants. Physiol Planta 79:267–274, 1990.
39. J Jasik, GJ De Klerk. Anatomical and ultrastructural examination of adventitious root formation in stem slices of apple. Biol Planta 39:79–90, 1997.
40. A Callebaut, JC Motte, W De Cat. Substrate utilization by embryogenic and non-embryogenic cell suspension cultures of *Cucumis sativus* L. J Plant Physiol 127:271–280, 1987.
41. L Wang, M Bao Xiao, Q Huang Bai, S Hao. Somatic embryogenic potential determined by the morphological polarity of the explant in tissue cultures of *Freesia refracta*. Acta Botanica Sinica 40:138–143, 1998.
42. Y Perrin, P Doumas, L Lardet, MP Carron. Endogenous cytokinins as biochemical markers of rubbertree (*Hevea brasiliensis*) clone rejuvenation. Plant Cell Tissue Org Cult 47:1996–1997, 1996.
43. M Danin, SJ Upfold, N Levin, BL Nadel, A Altman, J Van Staden. Polyamines and cytokinins in celery embryogenic cell cultures. Plant Growth Regul 12:245–254, 1993.
44. T Hatanaka, E Sawabe, T Azuma, N Uchida, T Yasuda. The role of ethylene in somatic embryogenesis from leaf discs of *Coffea canephora*. Plant Sci 107:199–204, 1995.
45. C Jay Allemand, D Cornu, JJ Macheix. Biochemical attributes associated with rejuvenation of walnut tree. Plant Physiol Biochem 26:139–144, 1988.
46. RLM Pierik. In Vitro Culture of Higher Plants. Boston: Martinus Nijhoff, 1987.
47. IK Vasil, TA Thorpe, eds. Plant Cell and Tissue Culture. New York: Kluwer Academic, 1994.
48. H Machii, H Mizuno, T Hirabayashi, H Li, T Hagio. Screening wheat genotypes for high callus induction and regeneration capability from anther and immature embryo cultures. Plant Cell Tissue Org Cult 53:67–74, 1998.
49. R Aloni, T Plotkin. Wound-induced and naturally occurring regenerative differentiation of xylem in *Zea mays*. Planta 163:126–132, 1985.
50. J Draper, R Scott, P Armitage, R Walden, eds. Plant Genetic Transformation and Gene Expression: A Laboratory Manual. Oxford: Blackwell Scientific, 1989.
51. ME Fromm, F Morrish, C Armstrong, R Williams, J Thomas, TM Klein. Inheritance and expression of chimeric genes in the progeny of transgenic maize plants. Biotechnology 8:833–839, 1990.
52. WW Su, WI Hwang, SY Kim, Y Sagawa. Induction of somatic embryogenesis in *Azadirachta indica*. Plant Cell Tissue Org Cult 50:91–95, 1997.
53. E Chevreau, F Mourgues, M Neveu, M Chevalier. Effect of gelling agents and antibiotics on adventitious bud regeneration from in vitro leaves of pear. In Vitro Cell Dev Biol Plant 33:173–179, 1997.
54. M Joersbo, FT Okkels. A novel principle for selection of transgenic plant cells: Positive selection. Plant Cell Rep 16:219–221, 1996.
55. M Joersbo, S Guldager Petersen, FT Okkels. Parameters interacting with mannose selection employed for the production of transgenic sugar beet. Physiol Planta 105:109–115, 1999.
56. EA Shahin, K Sukhapinda, RB Simpson, R Spivey. Transformation of cultivated tomato *Lycopersicon esculentum* by a binary vector in *Agrobacterium rhizogenes* transgenic plants with normal phenotypes harbor binary vector transferred DNA but no Ri-plasmid transferred DNA. Theor Appl Genet 72:770–777, 1986.
57. L Torregrosa, A Bouquet. *Agrobacterium rhizogenes* and *A. tumefaciens* co-transformation to obtain grapevine hairy roots producing the coat protein of grapevine chrome mosaic nepovirus. Plant Cell Tissue Org Cult 49:53–62, 1997.
58. M Otani, T Shimada, H Kamada, H Teruya, M Mii. Fertile transgenic plants of *Ipomoea trichocarpa* Ell. induced by different strains of *Agrobacterium rhizogenes*. Plant Sci 116:169–175, 1996.
59. B Nauerby, K Billing, R Wyndaele. Influence of the antibiotic timentin on plant regeneration compared

to carbenicillin and cefotaxime in concentrations suitable for elimination of *Agrobacterium tumefaciens*. Plant Sci 123:169–177, 1997.
60. HQ Ling, D Krieseleit, MW Ganal. Effect of ticarcillin/potassium clavulanate on callus growth and shoot regeneration in *Agrobacterium*-mediated transformation of tomato (*Lycopersicon esculentum* Mill.). Plant Cell Rep 17:843–847, 1998.
61. S Billings, G Jelenkovic, CK Chin, J Eberhardt. The effect of growth regulators and antibiotics on eggplant transformation. J Am Soc Hortic Sci 122:158–162, 1997.
62. FA Hammerschlag, RH Zimmerman, UL Yadava, S Hunsucker, P Gercheva. Effect of antibiotics and exposure to an acidified medium on the elimination of *Agrobacterium tumefaciens* from apple leaf explants and on shoot regeneration. J Am Soc Hortic Sci 122:758–763, 1997.
63. QH Ma, YR Song. Expression of tomato antisense ACC synthase gene in transgenic tobacco and its role in shoot formation. Acta Botanica Sinica 39:1047–1052, 1997.
64. Y Hong, K Saunders, R Hartley Martin, J Stanley. Resistance to geminivirus infection by virus-induced expression of dianthin in transgenic plants. Virology 220:119–127, 1996.
65. J Haseloff, R Siemering Kirby, C Prasher Douglas, S Hodge. Removal of a cryptic intron and subcellular localization of green fluorescent protein are required to mark transgenic *Arabidopsis* plants brightly. Proc Natl Acad Sci USA 94:2122–2127, 1997.
66. AE Gonzalez, C Shopke, NJ Taylor, RN Beachy, CM Fauquet. Regeneration of transgenic cassava plants (*Manihot esculenta* Crantz) through *Agrobacterium*-mediated transformation of embryogenic suspension cultures. Plant Cell Rep 17:827–831, 1998.
67. T Franks, DG He, M Thomas. Regeneration of transgenic *Vitis vinifera* L. Sultana plants: Genotypic and phenotypic analysis. Mol Breed 4:321–333, 1998.
68. LC Ding, CY Hu, KW Yeh, PJ Wang. Development of insect-resistant transgenic cauliflower plants expressing the trypsin inhibitor gene isolated from local sweet potato. Plant Cell Rep 17:854–860, 1998.
69. V Nagaraju, GSL Srinivas, GL Sita. *Agrobacterium*-mediated genetic transformation in *Gerbera* hybrida. Curr Sci 74:630–634, 1997.
70. R Marchant, MR Davey, JA Lucas, CJ Lamb, RA Dixon, JB Power. Expression of a chitinase transgene in rose (*Rosa hybrida* L.) reduces development of blackspot disease (*Diplocarpon rosae* Wolf). Mol Breed 4:187–194, 1998.
71. H Haggman, T Aronen. Transgene expression in regenerating cotyledons and embryogenic cultures of Scots pine. J Exp Bot 49:1147–1156, 1998.
72. G Honee, J Buitink, T Jabs, J De Kloe, F Sijbolts, M Apotheker, R Weide, T Sijen, M Stuiver, PJGM De Wit. Induction of defense-related responses in Cf9 tomato cells by the AVR9 elicitor peptide of *Cladosporium fulvum* is developmentally regulated. Plant Physiol 117:809–820, 1998.
73. TRI Munyikwa, KCJM Raemakers, M Schreuder, R Kok, M Schippers, E Jacobsen, RGF Visser. Pinpointing towards improved transformation and regeneration of cassava (*Manihot esculenta* Crantz). Plant Sci 135:87–101, 1998.
74. T Irdani, P Bogani, A Mengoni, G Mastromei, M Buiatti. Construction of a new vector conferring methotrexate resistance in *Nicotiana tabacum* plants. Plant Mol Biol 37:1079–1084, 1998.
75. H Yang, C Singsit, A Wang, D Gonsalves, P Ozias Akins. Transgenic peanut plants containing a nucleocapsid protein gene of tomato spotted wilt virus show divergent levels of gene expression. Plant Cell Rep 17:693–699, 1998.
76. E Perez Molphe Balch, N Ochoa Alejo. Regeneration of transgenic plants of Mexican lime from *Agrobacterium rhizogenes*–transformed tissues. Plant Cell Rep 17:591–596, 1998.
77. MJSSD Faria, DJ Donnelly, JC Cousineau. Adventitious shoot regeneration and agrobacterium-mediated transformation of red raspberry. Arq Biol Technol 40:518–529, 1997.
78. M Confalonieri, G Allegro, A Balestrazzi, C Fogher, M Delledone. Regeneration of *Populus nigra* transgenic plants expressing a Kunitz proteinase inhibitor (KT1-3) gene. Mol Breed 4:137–145, 1998.
79. M Hardegger, A Sturm. Transformation and regeneration of carrot (*Daucus carota* L.). Mol Breed 4: 119–127, 1998.
80. A Cruz Hernandez, Witjaksono, RE Litz, M Gomez Lim. *Agrobacterium tumefaciens* mediated transformation of embryogenic avocado cultures and regeneration of somatic embryos. Plant Cell Rep 17: 497–503, 1998.
81. P Gutierrez Pescep, K Taylor, R Muleo, E Rugini. Somatic embryogenesis and shoot regeneration from

transgenic roots of the cherry rootstock colt (*Prunus avium* × *P pseudocerasus*) mediated by pRi 1855 T-DNA *Agrobacterium rhizogenes*. Plant Cell Rep 17:574–580, 1998.
82. Y Nakamura, S Kobayashi, I Nakajima. *Agrobacterium*-mediated transformation and plant regeneration from hypocotyl segment of Japanese persimmon (*Diospyros kaki* Thunb). Plant Cell Rep 17:435–440, 1998.
83. C Walter, LJ Grace, A Wagner, DWR White, AR Walden, SS Donaldson. Stable transformation and regeneration of transgenic plants of *Pinus radiata* D. Don. Plant Cell Rep 17:460–468, 1998.
84. J Isaksson, S Karim, A Mandal. Extracellular enzymes of *Erwinia carotovora* eliminate the need for azacytidine treatment for high frequency transformation of *Arabidopsis thaliana*. In Vitro Cell Dev Biol Plant 34:41–45, 1998.
85. C Inokuma, K Sugiura, N Imaizumi, C Cho. Transgenic Japanese lawngrass (*Zoysia japonica* Steud.) plants regenerated from protoplasts. Plant Cell Rep 17:334–338, 1998.
86. TH Trinh, P Ratet, E Kondorosi, P Durand, K Kamate, P Bauer, A Kondorosi. Rapid and efficient transformation of diploid *Medicago truncatula* and *Medicago sativa* ssp. *falcata* lines improved in somatic embryogenesis. Plant Cell Rep 17:345–355, 1998.
87. P Vain, B Worland, MC Clarke, G Richard, M Beavis, H Liu, A Kohli, M Leech, J Snape, P Christou, H Atkinson. Expression of an engineered cysteine proteinase inhibitor (Oryzacystatin-I-DELTA-D86) for nematode resistance in transgenic rice plants. Theor Appl Genet 96:266–271, 1998.
88. AA Watad, DJ Yun, T Matsumoto, X Niu, Y Wu, AK Kononowicz, RA Bressan, PM Hasegawa. Microprojectile bombardment–mediated transformation of *Lilium longiflorum*. Plant Cell Rep 17:262–267, 1998.
89. Y Hoshino, M Mii. *Bialaphos* stimulates shoot regeneration from hairy roots of snapdragon (*Antirrhinum majus* L.) transformed by *Agrobacterium rhizogenes*. Plant Cell Rep 17:256–261, 1998.
90. M Manoharan, CSS Vidya, GL Sita. *Agrobacterium*-mediated genetic transformation in hot chilli (*Capsicum annuum* L. var. Pusa jwala). Plant Sci 131:77–83, 1998.
91. V Shiri, KS Rao. Introduction and expression of marker genes in sandalwood (*Santalum album* L.) following *Agrobacterium*-mediated transformation. Plant Sci 131:53–63, 1998.
92. JM Sherman, JW Moyer, ME Daub. A regeneration and *Agrobacterium*-mediated transformation system for genetically diverse *Chrysanthemum* cultivars. J Am Soc Hortic Sci 123:189–194, 1998.
93. V Babic, RS Datla, GJ Scoles, WA Keller. Development of an efficient *Agrobacterium*-mediated transformation system for *Brassica carinata*. Plant Cell Rep 17:83–188, 1998.
94. R Akashi, T Uchiyama, A Sakamoto, O Kawamura, F Hoffmann. High-frequency embryogenesis from cotyledons of bird's-foot trefoil (*Lotus corniculatus*) and its effective utilization in *Agrobacterium tumefaciens*–mediated transformation. J Plant Physiol 152:84–91, 1998.
95. R Marchant, JB Power, JA Lucas, MR Davey. Biolistic transformation of rose (*Rosa hybrida* L.). Ann Bot 81:109–114, 1998.

# 11
# Genetic Engineering for Modified Starch Structure in Cereals

**Ming Gao, Monica Båga, and Ravindra N. Chibbar**
*National Research Council, Saskatoon, Saskatchewan, Canada*

I. INTRODUCTION 169
II. STARCH AND ITS GENETIC MODIFICATION 170
    A. Starch Granules in Cereal Grains and Potential Modifications 170
    B. Amylose and Its Modification 171
    C. Amylopectin and Its Modification 173
    REFERENCES 178

## I. INTRODUCTION

Cereal grains, particularly those from wheat, maize, rice and barley, provide staple foods and important industrial raw materials for human beings. Starch is the major component of cereal grains, accounting for 60% to 75% of the grain weight. It provides 70% to 80% of the calories consumed by humans worldwide. Cereal grains also contain an important amount of proteins, and a very small amount of lipids. Thus, the two major components, starch and protein, mostly determine the quality of cereal grains.

    Genetic engineering of starch and proteins could potentially be a very powerful means of improving cereal grain quality for various dietary and industrial applications. Although the process is still difficult and time-consuming, major cereal species, including maize, wheat, and rice, can now be successfully transformed by either biolistic or agrobacteria-mediated methods. These genetic transformation technologies make it feasible to manipulate enzymes involved in starch biosynthesis selectively for the production of starch with desired or novel functional properties and to modify storage proteins directly for desired functional properties in cereal grains. In recent years, great progress has been made in our understanding of the enzymatic machinery for starch biosynthesis and of the structural and functional property relationships of starch and storage proteins. This ever-increasing body of knowledge could lead to genetically modified starch and proteins in cereal grains with novel or significantly improved functional and nutritional properties.

    Cereal grains will undoubtedly continue to be a staple food and to provide essential indus-

NRCC No. 43798

trial raw materials. Genetic engineering of starch and proteins has a great potential for the quality improvement of cereal grains to meet the requirements of many established dietary and industrial applications, and for the development of new dietary and industrial products. In this chapter, we attempt to identify those structural properties of starch amenable to genetic modification and discuss characteristics of the enzymatic machinery for starch biosynthesis in cereal grains, which pertain to their genetic modification.

## II. STARCH AND ITS GENETIC MODIFICATION

*Starch* refers to two major homoglucans in starch granules, amylose and amylopectin. Amylose is a mostly linear polymer of α-1,4 linked α-D-glucopyranosyl units. Some amylose molecules may contain several branches of α-1,6 linkages per molecule. Amylopectin is a highly branched glucan with α-1,4 forming the main chain and α-1,6 at branching points (reviewed in Refs. 1–4). In addition to two major polysaccharides, cereal starch granules contain a small amount of proteins on the surface or as integral components, and a small amount of free fatty acids and lysophospholipids.

Starch biosynthesis in cereal endosperm involves at least four groups of enzymes: adenosine diphosphate (ADP)-glucose pyrophosphorylase (AGPase), starch synthases (SSs), starch branching enzymes (SBEs), and starch debranching enzymes (SDBEs). The AGPase synthesizes ADP-glucose from glucose-1-phosphate derived from photosynthesis assimilate. Most of ADP-glucose for starch synthesis in cereal endosperm is synthesized by a cytosolic AGPase isoform (5,6), and is translocated into amyloplasts through an adenylate translocator (7,8). Starch synthases catalyze the formation of α-1,4 linkages in amylose and amylopectin, by adding a glucose moiety from ADP-glucose to a nonreducing end of elongating glucans. SBEs are responsible for the formation of α-1,6 linkages at branch points in amylopectin. SDBEs are also involved in the starch biosynthesis, although their specific roles remain uncertain (4,9).

Our understanding of enzymatic machinery for starch synthesis is far from complete. Aimed modifications of particular structural elements of cereal starch are presently not achievable. However, recent progress in molecular characterization of genes encoding these four groups of enzymes in cereal species enable us to manipulate these enzymes genetically for the modification of cereal starch.

### A. Starch Granules in Cereal Grains and Potential Modifications

The size and shape of plant storage starch granules are species-specific. Starch granules in cereal endosperm are generally small and polyhedric and are heterogeneous in size, e.g., 3 to 26 μm in diameter in maize and 1 to 40 μm in wheat. Starch granules in endosperm of maize, wheat, and barely are simple granules, i.e., one granule per amyloplast. Those from oats and rice are compound.

The size and shape of starch granules have an important impact on many dietary and industrial uses of granular starch (10). For example, the size and shape of starch granules are critical in the production of biodegradable plastic thin film (11). Relatively uniform and small starch granules are highly desirable for some applications, such as face and talcum powders. On the other hand, larger starch granules may be useful in other applications such as absorbent materials. Thus, the size and shape of cereal starch granules are potential targets for genetic modification to suit various applications.

Wheat, barley, and rye produce two types of starch granules, the large lenticular and the small spherical granules (A and B types). A-type granules are developed from granules initiated at an early developmental stage and increase in size, but not in number in developing endosperm

(12). At a later developmental stage, small B-type granules are produced within the evaginations of the amyloplasts already containing developing A-type granules. The B-type granules are later separated from the original amyloplasts. The number of B-type granules, but not the size, increases throughout grain development (12). The two types of granules seem to differ in their amylose concentration and gelatinization properties (13).

Very little is known about the genetic and biochemical bases for the formation of these two populations of granules. Some starch synthetic enzymes and/or a particular type of amylopectin may play an important role in their formation. Starch granules from the mature endosperm of barley *shx* mutant, with greatly reduced SSI activity showed an altered size distribution, comprising normal B-type granules and small-size A-type granules (14). Therefore, starch granules in cereal grains, especially those from wheat, could be potentially modified to give a narrower and unimodal size distribution by genetic modification of starch synthetic enzymes, such as starch synthases. However, any targeted modification of the size distribution or structures of cereal starch granules will have to await relatively complete knowledge of the genetic and biochemical mechanism for the formation of the granule structure and the size distribution.

Starch granules in cereal grains have a very complex semicrystalline structure comprising concentric rings of alternating amorphous and semicrystalline composition. The semicrystalline growth ring contains stacks of amorphous and crystalline lamellae. Amylopectin is mostly responsible for the granule crystallinity. Short side chains of amylopectin are believed to form double helices that associate into clusters. These clusters pack together to form a structure of alternating amorphous and crystalline lamellae, with most of the branch points in the amorphous lamellae and with double helices forming the crystalline lamellae. The size of a layer of amorphous and crystalline lamella is 9 nm, as it is throughout the plant kingdom.

Despite recent progress in our understanding of starch biosynthesis (reviewed in Refs. 1–4), it remains unknown how this complex granule structure is formed from its components and what specific contribution amylose and amylopectin make to the granule structure. Amylose and amylopectin are most likely synthesized in the amyloplast stroma and deposited on growing starch granules (15), although it cannot be completely ruled out that some polysaccharides may be synthesized on the surface of, or within, growing starch granules. The formation of granule structure could be a complete crystallization process or an enzyme-mediated biochemical process. If the former is true, amylopectin should essentially determine the size and morphological characteristics of starch granules, in addition to their structure. Modification of the granule size and morphological features may, thus, be achieved simply by changing amylopectin structure. Some evidence indeed suggests that amylopectin could have an essential impact on the granule size and morphological characteristics and that amylose may not have much influence on it. The size and morphological traits of starch granules were significantly altered in a number of maize, pea, and *Chlamydomonas* sp. mutants deficient in various starch synthases and starch branching enzymes involved in amylopectin synthesis (15,16). On the other hand, almost complete elimination of amylose in starch granules from waxy mutants of maize and wheat did not result in any significant changes in the size, morphological features, and structure of these starch granules (15,17). Amylose also seems to disrupt the structural order within the amylopectin crystallites (18). If the granule formation is an enzyme-mediated process, the size, morphological features, and structures of starch granules may be modified through manipulating yet to be identified enzymes, without changing amylopectin.

## B. Amylose and Its Modification

The amylose concentration of nonmutant cereal starch varies among species and cultivars and depends on the growing conditions and on the developmental stages at which starch is isolated. It can vary from 20% to 36% of starch among 399 maize cultivars, and 17% to 29% among 167 wheat cultivars (15). Amylose, isolated from fractionation of cereal starch, has a rather broad size distribution, e.g., 100 to 10,000 glucose residues in maize (19). In addition, the amylose fraction comprises a mixture of linear and loosely branched chains (20). However, branches may not significantly alter the solution behavior of the amylose chains (4). Although the conformation and location of amylose in starch granules remain unclear, most amylose molecules may form a complex with lipids present in starch granules (2,21).

The amylose concentration or the amylose/amylopectin ratio has a very important impact on many physicochemical properties, and on many dietary and industrial applications of cereal starches. High amylose concentration has a negative effect on the structural order within the amylopectin crystallites (18) and may be thus responsible for the conversion of the X-ray diffraction pattern from the A pattern for the normal maize starch to the B pattern for the high-amylose maize starch (22). High amylose concentration in a cereal starch diet can yield many health benefits, largely because of the increased amount of resistant starch (23–25). On the other hand, high-amylose concentrations of wheat flour seem to have a negative effect on the quality of certain types of noodles (26–29). The amylose concentration of cereal starch also influences paste properties (30,31), gelatinization and retrogradation (32–37), the texture of cooked rice (38), and the properties of starch-based films, foams, and plastics (39–41).

Amylose concentrations of cereal starches have been successfully modified by using mutants deficient in starch synthases and branching enzymes, using classical plant breeding techniques. Cereal endosperm waxy mutants lacking the granule-bound starch synthase I (GBSSI) almost completely eliminate all amylose in cereal grains (15,17). Starches from *waxy* maize, rice (glutinous), and barley have been utilized for many dietary and industrial applications for the past few decades. Wheat mutants with one or two null *waxy* alleles have been recently identified and shown to have reduced amylose concentration (42). Amylose-free *waxy* wheat has been recently developed by combining three homoeologous null *waxy* alleles using traditional breeding methods (reviewed in Ref. 17).

Cereal starches with high amylose concentration have also been developed from endosperm mutants. Maize endosperm mutant amylose-extender (*ae*) deficient in one of three starch branching enzyme isoforms, SBEIIb, accumulated more apparent amylose, but less total starch (15), and is mostly responsible for up to 70% of amylose content in commercial high-amylose corn cultivars (43,44). Similar *ae* endosperm mutants have been also identified in rice and barley. It's extremely difficult to identify null *ae* alleles in allohexaploid wheat. Therefore, we have transformed wheat with starch branching enzyme antisense gene constructs to regulate the activity of starch branching enzyme. Preliminary results show wheat grain starch with altered amylose concentration in transgenic plants (45). Maize *dull* mutation with reduced activities of the biochemically defined soluble starch synthase II (or SSIII in this chapter) and SBEIIa, and *su2* mutation with unknown genetic lesion also elevate amylose concentration in certain genetic backgrounds (15). These two mutants may also be used for breeding high-amylose corn cultivars.

Relative to the amylose concentration, the GBSSI activity seems not to be a limiting factor in the triploid endosperm of cereal crops such as maize, rice, and wheat. Although the normal *Wx* allele is not completely dominant to the mutant *Wx* allele relative to the amylose content in triploid maize endosperm (15), the amylose content does not increase substantially with increased dosages of the normal *Wx* allele in maize endosperm (21). A similar conclusion was also drawn from the dosage study of the wheat *waxy* proteins on the amylose concentration in wheat endosperm (46). Moreover, in a potato GBSSI gene dosage population, a certain level of GBSSI activity led to a maximal amount of amylose although the GBSSI activity increased almost linearly with the increase of the number of wild-type GBSSI alleles (47). Thus, the GBSSI activity is most likely not limiting relative to the amylose synthesis in the normal triploid endosperm of cereal species.

The limiting factors for the amylose content are likely to be the availability of substrates for the GBSSI, either ADP-glucose or glucan primers, or both. The increase of amylose concentration in maize *dull* and *su2* mutants in some genetic backgrounds might have resulted from shifting more ADP-glucose to the amylose synthesis by GBSSI. The maize GBSSI, in its intact granule-bound form, has an affinity for ADP-glucose 10-fold lower than that of other starch synthases (48). This suggests that the amylose content in normal cereal endosperm may be restricted by the affinity of GBSSI for limited ADP-glucose in vivo. Thus, transformation of cereal crops with heterologous or modified homeologous genes coding for GBSSI with higher affinity for ADP-glucose could potentially elevate substantially the amylose concentrations in cereal endosperm starch, especially when mutations such as *dull* and *su2* are incorporated in the genetic background of transfomants. A substantial increase in starch production has been reported in a revertant of the maize sh2 mutant with increased AGPase activity (49). Therefore, genetic combination of a GBSSI with high affinity for ADP-glucose, a AGPase capable of producing a higher amount of ADP-glucose, and controlled allocation of more ADP-glucose for amylose synthesis may be achieved in the future by using genetic transformation and traditional breeding for the production of high-amylose cereal starch.

The nature of the primers required for GBSSI to synthesize amylose remains unclear. Two types of primers have been proposed, small maltooligosaccharides (50) and short linear oligosaccharides derived from trimming of branches of preamylopectin by SDBE (51,51a). If the latter is indeed the primer in cereal endosperm, and if the full GBSSI activity for amylose synthesis requires the granule structure, amylose should be a side product of amylopectin synthesis. Thus, there would exist an upper limit on the amylose concentration in cereal starches.

Another challenge or interesting area for genetic modification of amylose is its size distribution. The amylose fraction of cereal starch is heterogeneous and has a wide size distribution that depends on species or cultivar. With the progress in our understanding of the mechanism of amylose synthesis, we may, in the future, alter and narrow the size distribution of amylose in cereal starch, through the expression of a foreign or modified GBSSI gene.

## C. Amylopectin and Its Modification

Amylopectin is the major polysaccharide in cereal endosperm starch, ranging from 72% to 82% (w/w). It determines the crystalline structure of starch granule and the physicochemical properties of starch. Although amylopectin and glycogen share the same glucose building blocks and the same chemical linkages ($\alpha$-1,4 and $\alpha$-1,6), amylopectin has far more organized structures that are mostly determined by the clustered $\alpha$-1,6 branch structure and unique branch chain profiles (reviewed in Refs. 1, 4). In fact, the fine structure of amylopectin still remains unclear, mostly because of the lack of methods to resolve the heterogeneous amylopectin population and to determine locations of $\alpha$-1,6 branch linkages in amylopectin molecules precisely. Complicating further our understanding of the amylopectin structure are the secondary crystalline structure of amylopectin molecules and its interactions with amylose, lipid, and proteins in a starch granule of very complex structure itself (reviewed in Ref. 2). Direct correlation between physicochemical properties of starch and amylopectin structural features is currently not possible. It is not currently possible to modify amylopectin and produce predictable structural changes for targeted physicochemical properties.

However, now it is practical and achievable to manipulate enzymes involved in amylopectin synthesis genetically, with a certain degree of predictability for the production of novel or modified starches. Novel dietary and industrial applications may be subsequently developed from these modified starches. In cereal endosperm, the assembly of amylopectin molecules involves three groups of enzymes, SS, SBE, and SDBE. Many mutants with deficiency in these enzyme activities have been identified in cereal species, particularly in maize (reviewed in Ref. 15), and have been utilized for production of specialty starches. Genetic transformation technology could further extend the spectrum of the manipulation of these enzymes. Currently, the following approaches may be employed individually, or in combination, to manipulate these enzymes for the production of novel or modified starch:

1. Use of different combinations of mutations in double or triple mutants to reduce or eliminate various enzyme activities. Many mutants deficient in starch synthetic enzymes are available or can be selected with relatively simple techniques in maize, rice, and barley. The drawback of this approach is the reduction of starch yield in these mutants. This may be alleviated in the future by overexpression of a modified AGPase with higher enzymatic activity for the production of more ADP-glucose, such as the revertant of the maize *sh2* mutant (49).

2. Antisense technology to reduce the enzyme activities involved in amylopectin synthesis. This approach has been successfully used in potato to down-regulate the activity of SSII and III (52,53) and SBE A and B (54,55). It may be particularly useful in bread wheat because mutations are extremely difficult to identify because of the allohexaploid nature of wheat.

3. Expression of foreign or modified genes for enzymes involved in, or nonrelated to, amylopectin synthesis. The diversity of plant starch synthesis, as reflected in the diversity of granule structure and morphological characteristics, amylose/amylopectin ratio, and structures of amylose and amylopectin in various plant starches, may be exploited for modification of cereal starches. With the progress in our understanding of biochemical properties of starch synthetic enzymes, modification of these enzymes for desired properties at the DNA level could be very useful in the future for engineering amylopectin synthesis. Enzymes that could modify side groups of glucose residues in amylopectin, such as the potato R1 enzyme (56), may be used to modify cereal starch for some desired properties, e.g., elevated phosphate content.

Our knowledge about the enzymatic machinery for amylopectin synthesis is far from complete. Recent progress has been elegantly summarized in a series of reviews (4,16,57–59). Important features of three types of enzymes involved in amylopectin synthesis are discussed here from the perspective of genetic modification of amylopectin:

1. Multiple SS and SBE isoforms and their enzymatic activities in the triploid cereal endosperm
2. The clustered branching patterns and side chain profiles of amylopectin, which are most likely determined by the combined action of starch synthase, starch branching enzyme, and starch debranching enzyme
3. Multienzyme complexes containing three groups of enzymes for amylopectin synthesis that may be present in cereal endosperm

These features may profoundly impact the strategies for genetically modifying amylopectin synthesis in cereal grains.

Multiple isoforms of starch synthases have been identified in the same tissue or different tissues in many plants, including maize, rice and wheat, pea, and potato (60–67). On the basis of their sequence relatedness, these isoforms can be classified into four basic groups, arbitrarily named GBSSI, SSI, SSII, and SSIII (Fig. 1). Endosperm of cereal species, as represented by maize (reviewed in Ref. 48), may contain all four types of starch synthases. These isoforms have different degrees of association with starch granules, which may reflect their unique contributions to the synthesis of amylose and/or amylopectin and the formation of starch granule structure.

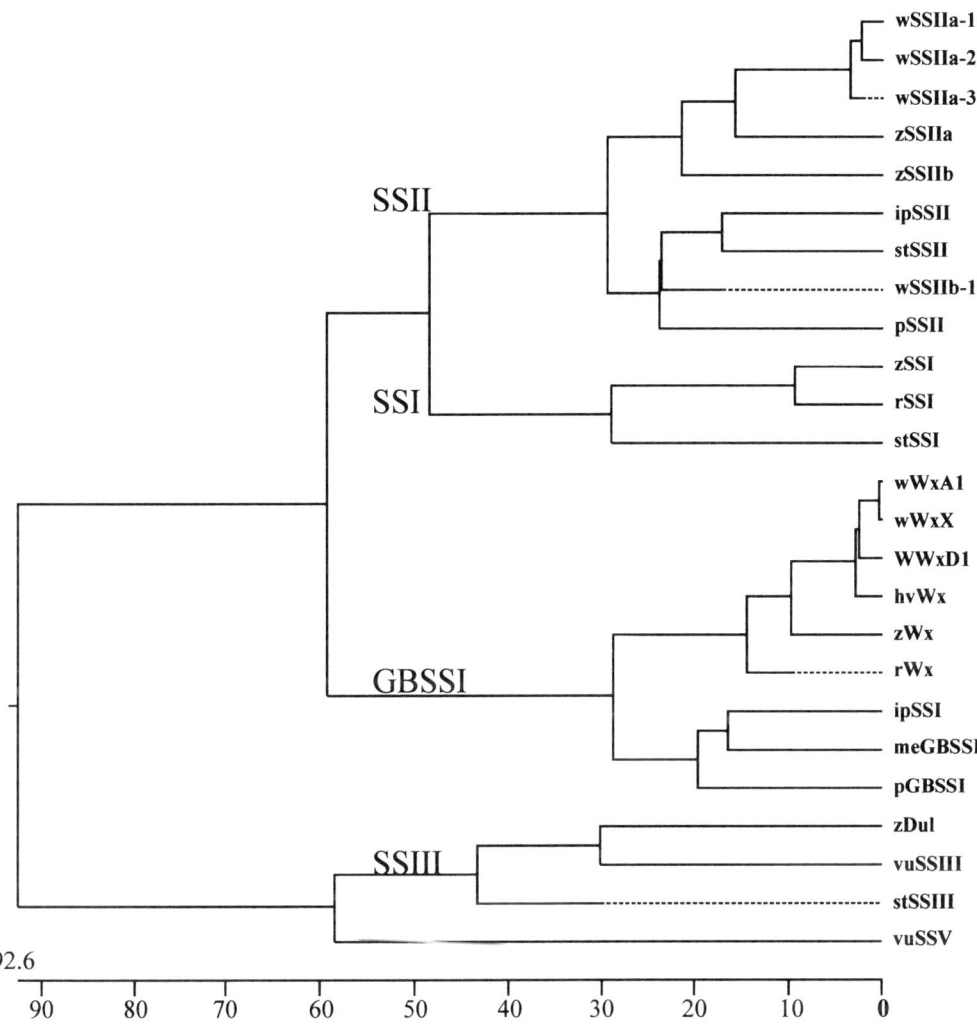

**Figure 1** Phylogenetic relationships of starch synthase isoforms from higher plants. Amino acid sequences of various starch synthases were aligned by using the Clustal method with PAM250 residue weight table. The dendrogram was constructed by using the MegAlign program. The starch synthases used in the alignment and their Genbank accession numbers are the maize zSSIIa (AF019296), zSSIIb (AF019297), zSSI(AF036891), waxy protein (zWx, X03935), and Dull protein (zDul, AF023159); the rice waxy protein (rWx, X62134.1) and SSI (rSSI, D16202); the pea GBSSI (pGBSSI, X88789) and SSII (pSSII, X88790); the potato SSI (stSSI, Y10416), SSII (st SSII, X87988), and SSIII (stSSIII, X94400); the cowpea SSIII (vuSSIII, AJ225088) and SSV (vuSSV AJ006752); the sweet potato SSI (ipSSI, U44126) and SSII (ipSSII, AF068834); the cassava GBSSI (meGBSSI, X74160); the wheat waxy proteins wWxA1b (AF113843), wWxD1b (AF113844) and WWx (X57233); the barley Waxy protein (hvWx, X07932); the wheat wSSIIb-1 (Gao and Chibbar, unpublished) and wSSIIa-1 (Gao and Chibbar, unpublished), wSSIIa-2 (Gao and Chibbar, unpublished), and wSSIIa-3 (Gao and Chibbar, unpublished).

The four groups of starch synthase isoforms seem to make distinct contributions to amylopectin synthesis and to the formation of starch granules. The exclusively granule-bound GBSSI or Waxy protein may contribute to amylopectin synthesis to some degree, in addition to playing an essential role in amylose synthesis (50). The SSI isoforms, including the maize SSI (68,69), rice SSI (70), and potato SSI (71), exist both in starch granules and in stroma of amyloplasts but account for a small portion of total soluble starch synthase activity. The antisense inhibition of the potato SSI expression resulted in accumulation of starch granules with changed morphological features and contained amylopectin with altered chain length distribution (71). Moreover, the barley *shx* mutants with reduced activity of SSI resulted in A-type granules with reduced size (14). The starch synthase II (SSII), as represented by those in pea embryo and potato tubers, also occurs in both stroma of amyloplasts and starch granules (71,72). Deficiency in the SSII in the pea *rug5* mutant embryo (73) and the antisense inhibition of the SSII expression in transgenic potato tubers (52,53) also resulted in starch granule morphological alterations and in changes of the chain length profiles of amylopectin. The maize Dull starch synthase (or SSIII in this chapter) (68, 74,75) and the potato SSIII (67,76) may be representatives of the major starch synthase that are exclusively present in stroma of amyloplasts in cereal species. Analyses of starches from the dull mutant endosperm (77,78) and from tubers of transgenic potato plants containing antisense SsII cDNA constructs suggest that this group of isoforms make unique contributions to the synthesis of amylopectin. Distinct impacts of various starch synthase isoforms on the chain length profile of amylopectin have been well documented in a series of studies of mutants in higher plants and Chlamydamonas sp. (reviewed in Refs. 4,15,16,59).

Three possible modes for multiple SS isoforms to synthesize amylopectin in cereal endosperm can be speculated. First, they may act individually so that each of them synthesizes a particular type of amylopectin molecule that differs slightly from others in its degree of polymerization, the clustered branching pattern, and the profile of branch chain length. By this mode, elimination or reduction of one isoform by mutations or antisense inhibition could result in the loss or reduction of a particular type of amylopectin but would not affect other types of amylopectin. Addition of a foreign or modified isoform through genetic transformation may add a new type of amylopectin but would not affect the synthesis of the original types of amylopectin. Second, they may synthesize different parts of the same type of amylopectin molecules at various biosynthetic stages, so that each isoform may act only on certain type of intermediates. By this mode, the reduction or elimination of an isoform through antisense inhibition or mutations would affect the synthesis of all types of amylopectin and result in accumulation of a particular type of intermediate. The introduction of a foreign or modified isoform may, or may not, have an impact on amylopectin synthesis. Both operating modes for different SS isoforms could be present at the same time in cereal endosperm. Some SSs may act independently, others may act together with other SSs for amylopectin synthesis. Finally, a mixture of both modes for a particular type of SS could be present, too. Until we understand individual roles of each isoform in amylopectin synthesis, genetic manipulation of SS for novel starch can be only a trial and error process.

The activity of at least some starch synthase isoforms may be redundant in the triploid cereal endosperm such as those of maize and rice, and more so in wheat because of the presence of three homoeologous alleles coding for the same type of SS isoforms. Gene dosage studies of maize *dull* mutants indicate that the normal *du* allele is completely dominant relative to the content and structure of amylopectin (15) and thus suggest the redundancy of the activity of the Dull starch synthase in normal triploid maize endosperm. In other words, the activity of the Dull starch synthase is not a limiting factor for amylopectin synthesis. Therefore, it may not be necessary to overexpress any endogenous starch synthase isoform in cereal endosperm for modification of the content and structure of amylopectin.

At least three SBE isoforms belonging to two major groups, SBEI, SBEIIa, and SBEIIb, may be present in endosperm of all cereal species as suggested by their presence in endosperm of maize (reviewed in Refs. 48,57) and barley (79). Each of the three diploid donor genomes in hexaploid wheat may code for counterparts of all three types of SBE isoforms ((80) and unpublished data). In other words, nine SBE isoforms encoded by three groups of homoeologous alleles may be present in endosperm of hexaploid wheat. The deduced amino acid sequences of SBEIIa and SBEIIb in maize (81), barley (82), and wheat (unpublished data) are very similar, except for those at their N termini, but are quite divergent from that of SBEI isoforms.

Individual roles of SBE isoforms in the formation of the clustered branch structure of amylopectin are not well understood. The dramatic elevation of amylose content and the accumulation of loosely branched amylopectin in maize *ae* mutant strongly suggest that each SBE isoform may have a unique role in the formation of the clustered branching pattern in amylopectin. Moreover, their preference for different in vitro substrates also points to their unique roles in amylopectin synthesis. The SBEI isoform prefers amylopectin as an in vitro substrate and tends to transfer longer chains, whereas SBEIIa and IIb prefer long glucan chains as in vitro substrates and are inclined to transfer shorter chains (83,84). The two types of SBE isoforms also require different minimal size of glucans as in vitro substrates (85). If this preference for different substrates holds true for these SBE isoforms in vivo, the formation of branches of a particular type on amylopectin molecules may involve all three SBE isoforms. A foreign or modified SBE introduced into cereal endosperm through genetic transformation may alter the structure of all types of amylopectin molecules or may produce a new type of amylopectin.

The SBE isoforms are not exclusively responsible for the clustered branching structure. Deficiency in a debranching enzyme in maize and rice *su* mutant endosperm and in the *Chlamydamonas* sp. mutant *sta7* resulted in accumulation of a highly branched polysaccharide, phytoglycogen (reviewed in Refs. 4,15,51,86). This indicates that the debranching enzyme may play an important role in the formation of the clustered branching structure.

The specific role of the debranching enzyme in the formation of the clustered branching structure remains unclear. It has been proposed that the debranching enzyme may trim random branches formed by SBE to form an ordered branch cluster on part of a growing amylopectin molecule (51). However, it is unclear how the specificity of debranching action could be determined if the debranching enzyme is solely responsible for the clustered branching structure. The random branches on restricted parts of a growing amylopectin molecule should not have much structural difference to distinguish one from another for specific debranching. Moreover, the recombinant maize debranching enzyme did not show a very high intrinsic specificity in debranching in vitro substrates (87). Nonetheless, the debranching enzyme may play an essential role in the formation of the clustered branch structure of amylopectin in maize and rice endosperm. On the whole, amylopectin synthesis is a highly integrated process that may require coordinated action of three types of enzymes.

Some evidence suggests that the three groups of enzymes, SS, SBE, and SDBE, may function as a multienzyme complex for the assembly of amylopectin molecules. The genetic lesion at the maize Du locus encoding SSIII (74,75) also reduced the SBEIIa activity (60). This finding strongly suggests that this particular SS may be physically associated with SBEIIa in vivo. Consistent with this hypothesis is the observation that peak activities of this starch synthase and SBEIIa, and SBEIIb and SSI coincide in the same anion exchange column fraction (60,88). Similar tight association of peak activities of SS and SBE has been also observed in wheat (62).

The genetic interaction between the maize *du* mutation and various *su* alleles also suggests that the soluble starch synthase may also interact physically with the debranching enzyme. Four major mutant alleles at the maize *su* locus have been identified. However, three *su* alleles, *su-am*, *su-st*, and *su-bn2*, give rise to near-normal kernel phenotype in a single mutant but result in more obvious and synthetic phenotypes in combination with the *du* or *su2* mutant alleles (15). The unique synthetic mutant phenotype that resulted from the combination of the maize *du* and *su-am* mutations implies that the starch synthase encoded at the Du locus may interact with the debranching enzyme encoded at the Su locus in direct physical association. The occurrence of three *su* alleles may be best explained as mutations that may have interrupted interactions between the SDBE and various SS or SBE isoforms. If multienzyme complexes containing all three types of enzymes are indeed present for the amylopectin synthesis, that could post an obstacle for modification of amylopectin in cereal endosperm using foreign or modified enzymes. Some foreign enzymes introduced by transformation may not have any impact on the amylopectin synthesis, because they may not be able to interact with those endogenous SDBE and SBE isoforms.

In summary, genetic manipulation of starch synthetic enzymes for modification of cereal starch is still a trial and error process. With the progress in our understanding of enzymatic mechanisms of starch biosynthesis and functional and structural property relationships of cereal starch, we will be able to produce starches, not only with novel functional properties, but also with desired structural modifications for specific functions. As a renewable and biodegradable natural polymer, starch can yield many new and important dietary and industrial applications, which will continue to be developed. Genetically modified starches with novel functionality could further expand the spectrum of starch utilization.

## ACKNOWLEDGMENTS

Drs. Fawzy Georges (Plant Biotechnology Institute) and Christopher D'Hulst (University of Lille, France) are gratefully acknowledged for their review of this manuscript.

## REFERENCES

1. DJ Manners. Review paper: Recent developments in our understanding of amylopectin structure. Carbohydr Polym 11:87–112, 1989.
2. IAM Appelqvist, MRM Debet. Starch-biopolymer interactions—a review. Food Rev Int 13:163–224, 1997.
3. SG Ball, MHBJ van de Wal, RGF Visser. Progress in understanding the biosynthesis of amylose. Trends Plant Sci 3:462–467, 1998.
4. A Buleon, P Colonna, V Planchot, S Ball. Starch granules: Structure and biosynthesis. Int J Biol Macromol 23:85–112, 1998.
5. K Denyer, F Dunlap, T Thorbjornsen, P Keeling, AM Smith. The major form of ADP-glucose pyrophosphorylase in maize endosperm is extra-plastidial. Plant Physiol 112:779–785, 1996.
6. JC Shannon, F-M Pien, H Cao, K-C Liu. Brittle-1, an adenylate translocator, facilitates transfer of extraplastidial synthesized ADP-glucose into amyloplasts of maize endosperm. Plant Physiol 117:1235–1252, 1998.
7. H Cao, TD Sullivan, CD Boyer, JC Shannon. Bt1, a structural gene for the major 39–44 kDa amyloplast membrane polypeptides. Physiol Planta 95:176–186, 1995.
8. TD Sullivan, Y Kaneko. The maize brittle1 gene encodes amyloplast membrane polypeptides. Planta 196:477–484, 1995.
9. MG James, DS Robertson, AM Myers. Characterization of the maize gene sugary1, a determinant of starch composition in kernels. Plant Cell 7:417–429, 1995.
10. RP Ellis, MP Cochrane, MFB Dale, CM Duffus, A Lynn, IM Morrison, RDM Prentice, JS Swanston, SA Tiller. Starch production and industrial use. J Sci Food Agric 77:289–311, 1998.

11. S Limm, J Jane, S Rajagopalan, P Seib. Effect of starch granule size on physical properties of starch filled polyethylene film. Biotechnol Prog 8:51–57, 1992.
12. WR Morrison, H Gadan. The amylose and lipid contents of starch granules in developing wheat endosperm. J Cereal Sci 5:263–275, 1986.
13. M Peng, M Gao, ESM Abdel-Aal, P Hucl, RN Chibbar. Separation and characterization of A- and B-type starch granules in wheat endosperm. Cereal Chem 76:375–379, 1999.
14. AH Schulman, RF Tester, H Ahokas, WR Morrison. The effect of the shrunken endosperm mutation *shx* on starch granule development in barley seeds. J Cereal Sci 19:49–55, 1994.
15. JC Shannon, DL Garwood. Genetics and physiology of starch development. In: RL Wishtler, JN Bemiller, EF Paschall, eds. Starch: Chemistry and Technology. Orlando, FL: Academic Press, 1984, pp 25–86.
16. A Buleon, DJ Gallant, B Bouchet, G Mouille, C D'Hulst, J Kossmann, S Ball. Starches from A to C: *Chlamydomonas reinhardtii* as a model microbial system to investigate the biosynthesis of the plant amylopectin crystal. Plant Physiol 115:949–957, 1997.
17. RA Graybosch. Waxy wheats: Origin, properties and prospects. Trends Food Sci Tech 9:135–142, 1998.
18. PJ Jenkins, AM Donald. The influence of amylose on starch granule structure. Int J Biol Macromol 17:315–321, 1995.
19. ML Fishman, P Cooke, B White, W Damert. Size distributions of amylose and amylopectin solubilized from corn starch granules. Carbohydr Polym 26:245–253, 1995.
20. Y Takeda, S Hizukuri, C Takeda, A Suzuki. Structures of branched molecules of amylose of various origins, and molar fractions of branched and unbranched molecules. Carbohydr Res 165:139–145, 1987.
21. JB South, WR Morrison, OE Nelson. A relationship between the amylose and lipid contents of starches from various mutants for amylose content in maize. J Cereal Sci 14:267–278, 1991.
22. Y-C Shi, T Capitani, P Trzasko, R Jeffcoat. Molecular structure of a low amylopectin starch and other high-amylose maize starches. J Cereal Sci 27:289–299, 1998.
23. MS Goddard, G Young, R Marcus. The effect of amylose content on insulin and glucose responses to ingested rice. Am J Clin Nutr 39:388–392, 1984.
24. JG Muir, GP Young, K O'Dea. Resistant starch—implications for health. Proc Nutr Soc Aust 18:23–32, 1994.
25. PA Baghurst, KI Baghurst, SJ Record. Dietary fibre, non-starch polysaccharides and resistant starch. Food Aust (Suppl) 48:S3–S35, 1996.
26. CM Konik, LM Mikkelsen, PW Gras. Contribution of starch and non-starch parameters to the eating quality of Japanese white salted noodles. J Sci Food Agric 58:403, 1992.
27. CM Konik, LM Mikkelsen, R Moss, PJ Gore. Relationships between physical starch properties and yellow alkaline noodle quality. Starch 46:292, 1994.
28. L Wang, PA Seib. Australian salt-noodle flours and their starches compared to U.S. wheat flours and their starches. Cereal Chem 73:167–175, 1996.
29. XC Zhao, IL Batey, PJ Sharp, G Crosbie, I Barclay, R Wilson, MK Morell, R Appels. A single genetic locus associated with starch granule properties and noodle quality in wheat. J Cereal Sci 27:7–13, 1998.
30. J-L Jane, J-F Chen. Effect of amylose molecular size and amylopectin branch chain length on paste properties of starch. Cereal Chem 69:60–65, 1992.
31. C Kiribuchi-Otobe, T Yanagisawa, I Yamaguchi, H Yoshida. Wheat mutant with waxy starch showing stable hot paste viscosity. Cereal Chem 75(5):671–672, 1998.
32. RF Tester, WR Morrison. Swelling and gelatinization of cereal starches. I. Effects of amylopectin, amylose, and lipids. Cereal Chem 67:551–557, 1990.
33. R Hoover. Starch retrogradation. Food Rev Int 11:331–346, 1995.
34. C-Y Lii, M-L Tsai, K-H Tseng. Effect of amylose content on the rheological property of rice starch. Cereal Chem 73:415–420, 1996.
35. T Yasui, J Matsuki, T Sasaki, M Yamamori. Amylose and lipid contents, amylopectin structure, and gelatinisation properties of waxy wheat (*Triticum aestivum*) starch. J Cereal Sci 24:131–137, 1996.

36. JP Mua, DS Jackson. Retrogradation and gel textural attributes of corn starch amylose and amylopectin fractions. J Cereal Sci 27:157–166, 1998.
37. H Fredriksson, J Silverio, R Andersson, AC Eliasson, P Aman. The influence of amylose and amylopectin characteristics on gelatinization and retrogradation properties of different starches. Carbohydr Polym 35:119–134, 1998.
38. MH Ong, JMV Blanshard. Texture determinants in cooked, parboiled rice. I. Rice starch amylose and the fine structure of amylopectin. J Cereal Sci 21:251–260, 1995.
39. JM Mayer, DL Kaplan. Biodegradable materials: Balancing degradability and performance. Trends Polym Sci 2:227, 1994.
40. D Lourdin, GD Valle, P Colonna. Influence of amylose content on starch films and foams. Carbohydr Polym 27:261–270, 1995.
41. S Simmons, EL Thomas. Structural characteristics of biodegradable thermoplastic starch/poly(ethylene-vinyl alcohol) blends. J Appl Polym Sci 68:2250–2285, 1995.
42. M Yamamori, T Nakamura, TR Endo, T Nagmine. Waxy protein deficiency and chromosomal location of coding genes in common wheat. Theor Appl Genet 89:179–184, 1994.
43. V Fergason. High amylose and waxy corns. In: AR Hallaver, ed. Specialty Corns. Boca Raton, FL: CRC Press, 1994, pp 55–77.
44. C Sidebottom, M Kirkland, B Strongitharm, R Jeffcoat. Characterization of the difference of starch branching enzyme activities in normal and low-amylopectin maize during kernel development. J Cereal Sci 27:279–287, 1998.
45. A Repellin, M Baga, T Demeke, K Caswell, M Gao, ESM Abdel-Aal, P Hucl, RN Chibbar. Development of transgenic wheat with modified carbohydrate composition. Proceedings ICC and AACC Symposium: Genetic Engineering of Cereals, Vienna, 1998, pp 32–45.
46. H Miura, A Sugawara. Dosage effects of the three Wx genes on amylose synthesis in wheat endosperm. Theor Appl Genet 93:1066–1070, 1996.
47. E Flipse, CJAM Keetels, E Jacobsen. The dosage effect of the wildtype GBSS allele is linear for GBSS activity but not for amylose content: Absence of amylose has a distinct influence on the physico-chemical properties of starch. Theor Appl Genet 92:121–127, 1996.
48. J Preiss. Biology and molecular biology of starch synthesis and its regulation. In: BJ Miflin, HF Miflin, eds. Oxford Surveys of Plant Molecular and Cell Biology. Oxford: Oxford University Press, 1991, pp 59–114.
49. MJ Giroux, J Shaw, G Barry, BG Cobb, T Green, T Okita, CL Hannah. A single gene mutation that increases maize seed weight. Proc Natl Acad Sci USA 93:5824–5829, 1996.
50. K Denyer, B Clarke, C Hylton, H Tatge, AM Smith. The elongation of amylose and amylopectin chains in isolated starch granules. Plant J 10:1135–1143, 1996.
51. S Ball, HP Guan, M James, A Myers, P Keeling, G Mouille, A Buleon, P Colonna, J Preiss. From glycogen to amylopectin: A model for the biogenesis of the plant starch granule. Cell 86:349–352, 1996.
51a. M van de Wall, C D'Hulst, J-P Vincken, A Buleon, R Visser, S Ball. Amylose is synthesized in vitro by extension of and cleavage from amylopectin. J Biol Chem 273:22232–22240, 1998.
52. A Edwards, DC Fulton, CM Hylton, SA Jobling, M Gidley, U Rossner, C Martin, AM Smith. A combined reduction in activity of starch synthases II and III of potato has novel effects on the starch of tubers. Plant J 17:251–261, 1999.
53. J Lloyd, R, V Landschutze, J Kossmann. Simultaneous antisense inhibition of two starch-synthase isoforms in potato tubers leads to accumulation of grossly modified amylopectin. Biochem J 338:515–521, 1999.
54. R Safford, SA Jobling, C Sidebottom, M, RJ Westcott, D Cooke, K Tober, J, BH Strongitharm, AL Russel, MJ Gidley. Consequences of antisense RNA inhibition of starch branching enzyme activity on properties of potato starch. Carbohydr Polym 35:155–168, 1998.
55. SA Jobling, GP Schwall, RJ Westcott, CM Sidebottom, M Debet, MJ Gidley, R Jeffcoat, R Safford. A minor form of starch branching enzyme in potato (*Solanum tuberosum* L.) tubers has a major effect on starch structure: Cloning and characterization of multiple forms of SBE A. Plant J 18:163–171, 1999.
56. R Lorberth, G Ritte, L Willmitzer, J Kossmann. Inhibition of a starch-granule-bound protein leads to modified starch and repression of cold sweetening. Nat Biotechnol 16:473–477, 1998.

57. C Martin, AM Smith. Starch biosynthesis. Plant Cell 7:971–985, 1995.
58. O Nelson, D Pan. Starch synthesis in maize endosperms. Annu Rev Plant Physiol 46:475–496, 1995.
59. AM Smith, K Denyer, C Martin. The synthesis of the starch granule. Annu Rev Plant Physiol 48:67–87, 1997.
60. C Boyer, J Preiss. Evidence for independent genetic control of the multiple forms of maize endosperm branching enzymes and starch syntheses. Plant Physiol 67:1141–1145, 1981.
61. T Baba, M Nishihara, K Mizuno, T Kawasaki, H Shimada, E Kobayahsi, S Ohnishi, K-I Tanaka, Y Arai. Identification, cDNA cloning, and gene expression of soluble starch synthase in rice (*Oryza sativa* L.) immature seeds. Plant Physiol 103:565–573, 1993.
62. K Denyer, CM Hylton, CF Jenner, AM Smith. Identification of multiple isoforms of soluble and granule-bound starch synthase in developing wheat endosperm. Planta 196:256–265, 1995.
63. S Rahman, B Kosar-Hashemi, M Samuel, A Hill, DC Abbott, JH Skerritt, J Preiss, R Appels, MK Morell. The major proteins of wheat endosperm starch granules. Aust J Plant Physiol 22:793–803, 1995.
64. M Takaoka, S Watanabe, H Sassa, M Yamamori, T Nakamura, T Sasakuma, H Hisashi. Structural characterization of high molecular weight starch granule-bound proteins in wheat (*Triticum aestivum* L.). J Agric Food Chem 45:2929–2934, 1997.
65. I Dry, A Smith, A Edwards, M Bhattacharyya, D Paul, C Martin. Characterization of cDNAs encoding two isoforms of granule-bound starch synthase which show differential expression in developing storage organs of pea and potato. Plant J 2:193–202, 1992.
66. K Tomlinson, J Craig, AM Smith. Major differences in isoform composition of starch synthase between leaves and embryos of pea (*Pisum sativum* L.). Planta 204:86–92, 1998.
67. J Marshall, C Sidebottom, M Debet, C Martin, AM Smith. Identification of the major starch synthase in the soluble fraction of potato tubers. Plant Cell 8:1121–1135, 1996.
68. CD Boyer, J Preiss. Multiple forms of (1-4)-α-D-glucan, (1-4)-α-D-glucan-6-glucosyl transferase from developing *Zea mays* L. kernels. Carbohydr Res 61(321–334):321–334, 1978.
69. C Mu, C Harn, Y-T Ko, GW Singletary, PL Keeling, BP Wasserman. Association of a 76 kDa polypeptide with soluble starch synthase I activity in maize (cv B73) endosperm. Plant J 6(2):151–159, 1994.
70. K-I Tanaka, S Ohnishi, N Kishimoto, T Kawasaki, T Baba. Structure, organization, and chromosomal location of the gene encoding a form of rice soluble starch synthase. Plant Physiol 108:677–683, 1995.
71. A Edwards, J Marshall, C Sidebottom, RGF Visser, AM Smith, C Martin. Biochemical and molecular characterization of a novel starch synthase from potato tubers. Plant J 8:283–294, 1995.
72. A Edwards, J Marshall, K Denyer, C Sidebottom, RGF Visser, C Martin, AM Smith. Evidence that a 77-kilodalton protein from the starch of pea embryos is an isoform of starch synthase that is both soluble and granule bound. Plant Physiol 112:89–97, 1996.
73. J Craig, JR Lloyd, K Tomlinson, L Barber, A Edwards, T Wang, L, C Martin, CL Hedley, AM Smith. Mutations in the gene encoding starch synthase II profoundly alter amylopectin structure in pea embryos. Plant Cell 10:413–426, 1998.
74. M Gao, J Wanat, PS Stinard, M James, G, A Myers, M Characterization of *dull* 1, a maize gene coding for a novel starch synthase. Plant Cell 10:399–412, 1998.
75. H Cao, J Imparl-Radosevich, H Guan, PL Keeling, MG James, AM Myers. Identification of the soluble starch synthase activities of maize endosperm. Plant Physiol 120:205–215, 1999.
76. GJW Abel, F Springer, L Willmitzer, J Kossmann. Cloning and functional analysis of a cDNA encoding a novel 139 kDa starch synthase from potato (*Solanum tuberosum* L.). Plant J 10:981–991, 1996.
77. YJ Wang, P White, L Pollak, JL Jane. Characterization of starch structures of 17 maize endosperm mutant genotypes with Oh43 inbred line background. Cereal Chem 70:171–179, 1993.
78. YJ Wang, P White, L Pollak, JL Jane. Amylopectin and intermediate materials in starches from mutant genotypes of the Oh43 inbred lines. Cereal Chem 70:521–525, 1993.
79. C Sun, P Sathish, S Ahlandsberg, A Deiber, C Jansson. Identification of four starch branching enzymes in barley endosperm: Partial purification of forms I, IIa and IIb. New Phytol 137:215–222, 1997.
80. MK Morell, A Blennow, B Kosar-Hashemi, MS Samuel. Differential expression and properties of starch branching enzyme isoforms in developing wheat endosperm. Plant Physiol 113:201–208, 1997.
81. M Gao, D Fisher, K-N Kim, JC Shannon, MJ Guiltinan. Independent genetic control of maize starch-branching enzymes IIa and IIb. Plant Physiol 114:69–78, 1997.

82. C Sun, P Sathish, S Ahlandsberg, C Jansson. The two genes encoding starch-branching enzymes IIa and IIb are differentially expressed in barley. Plant Physiol 118:37–49, 1998.
83. Y Takeda, H Guan, J Preiss. Branching of amylose by the branching isozymes of maize endosperm. Carbohydr Res 240:253–263, 1993.
84. H Guan, J Preiss. Differentiation of the properties of the branching enzymes from maize (*Zea mays*). Plant Physiol 102:1269–1273, 1993.
85. H Guan, J Imarl-Rasedovitch, J Preiss, P Keeling. Comparing the properties of *Escherichia coli* branching enzyme and maize branching enzyme. Arch Biochem Biophys 342:92–98, 1997.
86. Y Nakamura. Some properties of starch debranching enzymes and their possible role in amylopectin biosynthesis. Plant Sci 121:1–18, 1996.
87. A Rahman, K-S Wong, J-L Jane, AM Myers, MG James. Characterization of SU1 isoamylase, a determinant of storage starch structure in maize. Plant Physiol 117:425–435, 1998.
88. PL Dang, CD Boyer. Maize leaf and kernel starch synthases and starch branching enzymes. Phytochem 27:1255–1259, 1988.

# 12
# Improving Crop Performance Through Transgenic Modification of Flowering

**Pierre Fobert**
*National Research Council, Saskatoon, Saskatchewan, Canada*

|       |                                                           |     |
|-------|-----------------------------------------------------------|-----|
| I.    | INTRODUCTION                                              | 183 |
| II.   | STRATEGIES FOR ISOLATING AND STUDYING GENES THAT CONTROL FLOWERING | 184 |
| III.  | FLOWERS ON DEMAND                                         | 185 |
|       | A. Genes That Control Flowering Time                      | 185 |
|       | B. Genes That Control Floral Meristem Identity            | 186 |
|       | C. Possible Biotechnology Applications                    | 187 |
| IV.   | REDESIGNING PLANT ARCHITECTURE                            | 188 |
| V.    | MAKING DESIGNER FLOWERS                                   | 189 |
|       | A. Floral Organ Identity Genes                            | 189 |
|       | B. Creating New Arrangements of Floral Organs             | 189 |
|       | C. Producing Apterous Canola                              | 190 |
| VI.   | NEW SYSTEMS FOR SEED AND FRUIT PRODUCTION                 | 191 |
|       | A. Hybrid Seed Production                                 | 191 |
|       | B. Seeds Without Fertilization                            | 191 |
| VII.  | PERSPECTIVES                                              | 192 |
|       | REFERENCES                                                | 193 |

## I. INTRODUCTION

The goal of most breeding programs is ultimately to increase crop yield. Given the wide range of intrinsic (i.e., genetic) and extrinsic (i.e., environmental) factors that influence "yield," it is not surprising that many aspects of plant growth, development, and interaction with the environment have been targeted for modification in an effort to achieve that goal. In particular, altering plant size and shape has played an important role in the development of our current crop cultivars. One only has to compare a crop with its wild relatives to witness the often dramatic changes achieved by years of selection by humans. Manipulation of plant development continues to be an important target for crop improvement in current breeding programs. For example, field peas are being selected for better standing abilities to resist lodging (1), and *Brassica* cultivars with flatter, upright pods are being sought for more efficient interception of available light (2).

This chapter examines how transgenic technology may be used to modify plant development for the purpose of enhancing crop performance, and hence yield. Unlike conventional breeding, transgenic approaches require that the gene(s) controlling traits of interest first be isolated as defined segments of deoxyribonucleic acid (DNA) (i.e., cloned). Identification and isolation of genes controlling plant development represent huge challenges to molecular biologists and are currently areas of intensive study. Excellent progress is being made in several areas, with flower development representing the best understood process at the present. Consequently, this chapter focuses specifically on the manipulation of various aspects of flowering. Each section first highlights what is known about the genes that regulate the relevant phases of flower development, emphasizing information obtained from transgenic studies. Examples of how the information has been, or may be, applied to produce novel transgenic crops is then considered. A common theme emerging from current research is that transgenic technology not only provides a means to produce genetically engineered crops, but also represents an important research tool for the identification and study of genes controlling plant development.

## II. STRATEGIES FOR ISOLATING AND STUDYING GENES THAT CONTROL FLOWERING

As stated, the availability of cloned genes is absolutely required for transgenic projects. One approach that has proved particularly effective for identifying and cloning genes that control flowering is molecular genetics (3–6). First, relevant genes are *identified* through the analysis of mutants (induced or naturally occurring). Characterization of mutant phenotypes provides important information about the possible function of the affected genes. Because most mutations lead to gene inactivation, those in key flowering genes may result in plants that produce abnormal flowers or produce normal-looking flowers but at inappropriate times or places. Combining different mutations in the same genetic background permits the study of gene interactions and allows the relative positioning of different genes acting in the same pathways. Second, molecular genetics allows *cloning* of the affected gene, simply based on the presence of the mutation, using molecular tagging or positional cloning approaches (3). These methods are particularly powerful because no additional information about the biochemical function, structure, or expression of the gene or gene product is required.

Although mutations affecting flower development have been known for centuries, strategies for cloning the affected genes have only been developed recently. Two species that are well suited for cloning genes solely in terms of the presence of mutations are the garden snapdragon (*Antirrhinum majus*) and a small cruciferous weed, *Arabidopsis thaliana* (5–7). Consequently, these species have emerged as model systems for studying the molecular genetics of flowering. This chapter focuses on results obtained in *Arabidopsis* because most transgenic research has been performed in this species or with genes isolated from this species. However, it is noteworthy that several of the *Arabidopsis* flowering genes were isolated by using sequence similarity to that of previously identified genes from Antirrhinum.

Transgenic technology also provides a powerful tool for studying genes that control flowering. In many ways, this technology complements that of molecular genetics. For example, transgenics allow the introduction of genes isolated by molecular genetics into mutant plants to test whether they can rescue the mutant phenotype, thus providing conclusive proof that the desired gene has been cloned (for example, see Ref. 8). Transgenic approaches also permit certain types of targeted gene manipulations that are more difficult, or impossible, to accomplish through traditional mutagenesis approaches. Studies with so-called gain-of-function transgenes, which are expressed at higher levels or at times or places where the genes are not normally expressed, have been particularly effective in cases in which gene function is difficult to ascertain, including situations in which several genes share common or related functions or genes function as part of a complex pathway. Another powerful research application of transgenic technology is the testing of putative flowering genes in species in which the creation or isolation of mutants is difficult. It is possible to suppress the expression of genes in transgenic plants using antisense or cosuppression methods (9), thereby simulating the effect of mutations, and allowing the assignment of gene function. Because transgenic approaches allow the transfer of genes between species, they also provide a means of testing evolutionary relationships between genes. For example, flowering genes from conifers have been introduced into *Arabidopsis* and found to induce developmental changes very similar to those of the related genes from *Arabidopsis* (10,11). This finding suggests that the function of these genes has been well conserved between angiosperms and conifers, even though the former produce true flowers and the latter produce cones. Finally, transgenic technology has permitted the transfer of well-characterized transposable elements, such as those from maize, into heterologous plants in which transposons are uncharacterized (12,13). This has made possible the generation and eventual cloning of insertional mutants in the heterologous hosts, using the transposons as molecular tags (13). In some cases, the *Agrobacterium* transfer DNA (T-DNA) itself has served as an insertional mutagen and molecular tag for gene isolation (14).

## III. FLOWERS ON DEMAND

The time at which a crop initiates flowering can have a major impact on yield. Controlling when in the growing season plants flower ensures that seeds develop and mature during the most favorable environmental conditions, while avoiding unfavorable risk factors such as drought and frost. It follows that the optimal time to flower varies considerably, depending on geographic region and local environmental conditions. By manipulating the window of time during which plants flower, cultivars adapted to different geographic regions have been developed for many crops. Transgenic manipulation of two classes of flowering genes appears to have application in this area and is discussed in the following sections.

### A. Genes That Control Flowering Time

Genes controlling flowering time have been identified in many plant species. For example, approximately 80 loci affecting flowering time have been identified in *Arabidopsis* (4). Mutational analyses indicate that these act in at least four pathways to control flowering time. Interactions between different genes and pathways may be complex, as some genes have redundant functions. More than a quarter of the *Arabidopsis* genes known to affect flowering time has now been cloned (4). Transgenic plants in which expression of these genes has been changed all display altered flowering time (15–17). The *CONSTANS* (*CO*) gene has been particularly well studied and is used here as an example.

In wild-type *Arabidopsis* exposure to long days (16 hours light) promotes flowering (8). Mutations in *CO* delay flowering under long days but have no effect under short days (10 hours light). The CO protein contains zinc fingers and probably functions as a transcription factor, regulating the expression of other genes that promote flowering in response to long days (8). On the basis of the observation that the *CO* messenger ribonucleic acid (mRNA) is more abundant in leaves of plants grown under long days (8), Coupland and coworkers decided to test whether levels of CO were important for regulating flowering time. To this end, they produced transgenic *Arabidopsis* plants containing a chemically inducible version of CO (CO:GR). In this instance, the chemical inducer was a mammalian steroid hormone not normally found in plants. In the absence of the hormone, CO:GR protein does not enter the nucleus and therefore cannot regulate the expression of its target genes. Addition of the hormone causes a large amount of the modified CO rapidly to enter the nucleus, where it is active. Coupland's group observed that transgenic plants containing CO:GR flowered rapidly after addition of the steroid hormone (15). The earlier the hormone was applied, the earlier the plants flowered. In some cases, plants flowered before the time normally required by the wild-type plants. Furthermore, flowering was no longer regulated by day length. Expression of high levels of CO:GR also caused the apical inflorescence meristem to develop into a terminal flower (15). In untransformed *Arabidopsis* plants, the inflorescence meristem does not develop into a terminal flower but grows indeterminately until it senesces (18). Consequently, hormone-induced CO:GR plants produced fewer flowers than non-induced transgenic plants or untransformed plants (15). Taken together, these results suggest that although several genes normally interact to determine flowering time in *Arabidopsis* a sufficient increase in the level of a key genetic regulator such as CO can "override" the normal control systems and trigger flowering at any point during development.

It is not known how well conserved CO function is among different plant species. At least four *CO*-related genes have been cloned in *Brassica napus* (19). These map to two regions of the *B. napus* genome, where quantitative trait loci (QTLs) affecting flowering time are localized. At least one of the *B. napus* genes can complement a *co* mutation when transferred into *Arabidopsis* (19), indicating that CO function is conserved in *B. napus*. It will be interesting to determine the role of CO homologues in other plant species, especially those in which flower induction is distinct from the bolting strategy found in *Arabidopsis* and *Brassica* spp.

### B. Genes That Control Floral Meristem Identity

Once a plant is induced to flower, a group of genes that specify the floral program is activated in incipient meristematic cells (20,21). Loss-of-function mutations in these genes may not alter flowering time but cause plants to produce structures characteristic of shoots instead of flowers. For example, plants with mutations in both the *LEAFY* (*LFY*) and *APETALA1* (*AP1*) genes completely fail to form flowers (22). In CO:GR plants, both *LFY* and *AP1* are activated after hormone application, although *LFY* activation is much more rapid (15). This suggests that *LFY* may be a direct target of CO. Like *CO*, *LFY* and *AP1* also encode putative transcription factors, although the three proteins are structurally unrelated (23,24).

Plants containing gain-of-function transgenes of *LFY* and *AP1* have been produced. In these cases, the transgenes were not inducible but simply fused to a strong constitutive promoter (the cauliflower mosaic virus 35S promoter, or simply 35S) (25). These fusion genes were expressed at very high levels, and much earlier in the life cycle than the native genes, which are normally activated at high levels only in floral meristems (23,24,26). As observed with CO:GR plants, transgenic plants expressing 35S:*LFY* or 35S:*AP1* flowered precociously (27,28). However, 35S:*LFY* plants did not flower as early as CO:GR plants. Specifically, CO:GR was able to induce flowering in very young plants grown under short days, whereas 35S:*LFY* was not (15,28). Differences were also observed between 35S:*LFY* and 35S:*AP1* plants: the latter developed the terminal flower before all lateral flowers had developed, the former developed all axillary flowers before the terminal flower (29).

Encouraged by results obtained in *Arabidopsis* Weigel and Nilsson (28) transferred the

35S:*LFY* gene construct into an evolutionarily distant plant, aspen. Results obtained were dramatic. Although aspen trees do not normally begin to flower until they are 8–20 years old, some transgenic lines expressing 35S:*LFY* initiated flowering after only 7 months. Tobacco plants containing the 35S:*LFY* gene also flowered very early (29). In addition, putative LFY homologues from *Pinus radiata* (10) and *Eucalyptus* sp. (30) were able to induce precocious flowering and terminal flowers when fused to the 35S promoter and introduced into *Arabidopsis*. Taken together, these results demonstrate that *LFY*, when expressed at high levels, is capable of triggering precocious flowering in distantly related plant species and that this function may be conserved among *LFY* homologues. There is at least one exception to this hypothesis. A putative *LFY* homologue from rice (*RFL*) expressed from the 35S promoter did not induce early flowering in transgenic *Arabidopsis* (31). This could suggest either that the ability to induce early flowering may not be conserved in *LFY* homologues from monocots or alternatively that rice contains multiple *LFY*-like genes, and the *RFL* is not the true *LFY* homologue from this species.

In the case of *AP1*, it has been shown that related genes from *Eucalyptus* sp. fused to the 35S promoter can trigger flowering in *Arabidopsis* (32). However, the 35S:*AP1* gene from *Arabidopsis* does not appear to be able to induce early flowering when introduced into aspen (29). Therefore, *AP1* and *LFY* may differ in their ability to trigger early flowering in distantly related plant species.

## C. Possible Biotechnology Applications

### 1. Reducing Flowering Time

The precocious flowering of transgenic plants expressing high levels of different floral regulatory genes suggests that a similar strategy could be exploited to reduce flowering time in crops. In particular, the short growing season found in Canada and other northern climates requires cultivars that flower earlier than those grown in warmer climates. Adequate natural variation in flowering time exists for many plant species, but there remain some crops for which additional reduction in flowering time would be beneficial (e.g., *Brassica carinata*). Transgenic lines expressing different levels of a floral regulatory transgene would be expected to flower at different times, and therefore it should be possible to select lines having the appropriate flowering time for various geographical locations. Even in crops in which earliness is not a problem, transgenic approaches may be able to provide better control of when plants flower. One possibility would be to fuse genes such as *CO*, *LFY*, or *AP1* to a suitable chemically inducible promoter, thus permitting the grower to trigger flowering in his crops by spraying fields with chemical inducers. The decision of when or whether to spray would be based on environmental conditions and crop performance during the growing season. In certain circumstances reducing flowering time by as little as a day may have a significant impact on yield.

Reducing time to flowering may also be beneficial in plant species that have a prolonged juvenile phase, such as trees. This would allow breeders to begin crossing material much earlier than currently possible, thus accelerating their programs. In the event that early flowering is ultimately an undesirable trait, the transgenes could be removed by site-directed recombination (33) before the material enters commercial production.

### 2. Preventing Flower Production

In some situations, it may be desirable to eliminate flower production. For example, flowers are aesthetically undesirable in turfgrass, and they reduce digestibility of forage grasses (34). In the case of transgenic forest trees, eliminating flowering would prevent the possibility of uncontrolled cross-pollination from transgenic pollen into natural forest populations. Engineering sterility is feasible in forest trees because seed formation is not required for the desired product (wood) or for plant propagation (35). Engineering sterility by preventing flowering in conifers may have the added benefit of increasing yield, by diverting energy that would normally be spent on cone production into wood biomass (35). Most strategies for genetically engineering sterility propose to exploit the promoters of flower-specific genes, such as *LFY* or *AP1*, rather than the coding regions themselves. These promoters are active at high levels in floral meristems (23,24,26) and therefore, if fused to genes that produce cytotoxic substances, such as the dipththeria toxin A or ribonucleases (RNases), should cause cell death specifically in these tissues (for review of use of cytotoxic genes in plants, see Ref. 36). Specifically, transgenic plants containing a *LFY* promoter-RNase fusion have been produced (37). These plants developed normally until floral induction, at which time no flowers were produced, demonstrating the general feasibility of the proposed approach. Cone-specific cDNAs have been identified (11), and the promoters controlling the expression of these genes may useful for engineering sterility in conifers. It is noteworthy that neither of two *LFY*-related genes isolated from *P. radiata* is expressed specifically in reproductive cones (10,38). Consequently, the promoters of such genes are unlikely to be suitable for cone-specific cell ablation purposes.

## IV. REDESIGNING PLANT ARCHITECTURE

A mutation seemingly complementary to *lfy* has also been identified in *Arabidopsis*. *Terminal flower 1* (*tfl1*) mutants flower early and, as the name implies, produce a terminal flower after initiating a fixed number of lateral flowers (18). This phenotype is very similar to the one induced by overexpression of *LFY* (28). In contrast, overexpression of *TFL1* from the 35S promoter did not produce a phenotype resembling the *lfy* mutation (39). Instead of producing flowers having shootlike features (typical of *lfy* flowers), 35S:*TFL1* plants produced normal flowers. However, the switch from vegetative to reproductive phase was significantly delayed in these plants, resulting in a more highly branched growth pattern and the production of many more flowers than in wild-type controls (39). On the basis of these results, it was proposed that TFL1 regulates the rate at which the plant shoot progresses through all the different phases of the life cycle (39). This suggests a possible means of controlling plant shape by modulating TFL1 activity in transgenic plants. As already demonstrated in *Arabidopsis*, constitutive higher expression of *TFL1* leads to highly branched plants that eventually produce a large number of flowers. Such plants also take longer to complete their life cycle (39), a trait that may be undesirable in certain agricultural settings. This problem may be circumvented by targeting TFL1 expression to specific cell types or to specific times during development. For example, increased TFL1 activity in lateral meristems would promote branching, whereas increased TFL1 activity in the shoot apical meristem, coupled with reduced TFL1 activity in lateral meristems, may allow flower production along a single main axis. It may also be feasible to synchronize flower production by inducing a pulse of TFL1 activity at the appropriate time in development.

The recent identification of the *SELF-PRUNING* (*SP*) gene of tomato as a *TFL1* homologue (40) offers an example of how modifying the expression of this class of genes in a crop plant has had a beneficial agronomic impact. Loss-of-function *sp* mutants prematurely terminate the production of inflorescence units, thereby limiting shoot growth and resulting in bushier plants (40). This phenotype has been considered to be the "single most important genetic trait in the development of modern agrotechniques (for tomato) . . . because it facilitates mechanical harvest" (40). Expression of an antisense version of the 35S:*SP* transgene in plants with a functional *SP* gene produced a phenotype resembling the *sp* mutation, suggesting that it should be possible to generate similar transgenic effects by modifying *TFL1/SP* genes in other plant species.

## V. MAKING DESIGNER FLOWERS

### A. Floral Organ Identity Genes

A typical angiosperm flower consists of four distinct types of organs positioned in four concentric whorls (7). From the outermost whorl in, these are sepals (whorl 1), petals (whorl 2), stamens (whorl 3), and carpels (whorl 4). An important first step toward the eventual manipulation of floral organs is to gain a better understanding of the genes responsible for specifying their identities. To this end, several mutants that produce specific types of organs at inappropriate places (so-called homeotic mutants) have been studied. For example, the *Arabidopsis* mutant *apetala 3* (*ap3*) produces sepals in place of petals in whorl 2 and carpels in place of stamens in whorl 3; the identities of organs in whorls 1 and 4 are not altered (41). *ap3* is typical of mutations in a group of genes known as the *floral organ identity genes* (FOIs). These characteristically affect the identity of organs in two adjacent floral whorls. Three classes of FOIs exist: class A (e.g., *APETALA 2*) control the identity of organs in whorls 1 and 2; class B (e.g., *AP3*) control the identity of organs in whorls 2 and 3, class C (e.g., *AGAMOUS*) control the identity of organs in whorls 3 and 4. A simple model (the ABC model) has been formulated to describe how the identities of all four types of floral organs are specified by FOIs (7). According to the model, there are three homeotic functions or activities, termed A, B, and C, needed to specify the identity of the four different organ types. Each of these activities is restricted to two adjacent whorls of a normal (wild-type) flower, and the type of organ produced is determined by the sum of A, B, and C activities present in any particular whorl. Activity A alone specifies sepal identity, the combination of A + B specifies petal formation, B + C specifies stamens, and C alone specifies carpel identity. It follows that any type of organ is capable of being produced in any given whorl of a flower—the important factor is the sum of the A, B, C activities. An additional feature of the model is that the A and C activities are mutually exclusive: if one activity is missing, the other spreads to all four whorls.

Mutations in FOIs alter the distribution of A, B, or C activities, resulting in novel combinations of these activities in specific floral whorls and hence the production of novel organ types. For example, the *ap3* mutant described lacks the B activity and according to the model should have the following distribution of homeotic activities in whorls 1 to 4: A, A, C, C. If A alone specifies sepals and C alone specifies carpels, the resulting flower would be expected to produce flowers containing two outer whorls of sepals and two inner whorls of carpels, and this is what is observed (41).

The model appears to explain the phenotypes of single, double, and triple mutants analyzed to date accurately (7,42). Although the model was formulated independently of molecular information on the FOI genes, it appears to be consistent with available molecular data. For example, the *AP3* mRNA is specifically expressed in whorls 2 and 3 (41), as would be expected of a B class gene, and *AGAMOUS* (AG) mRNA is only expressed in whorls 3 and 4 (43), as would be expected for a C class gene. Furthermore, *AG* mRNA spreads to all four whorls of the flower in class A mutants (44), as would be expected from the mutual exclusion feature of the model.

### B. Creating New Arrangements of Floral Organs

Using the ABC model as a guide, it should be possible to produce transgenic flowers having novel distributions of homeotic gene activity, and thus to alter the identity of organs in any given floral whorl in a predictable manner. Results obtained to date suggest that this is in fact possible. For example, the class C gene *AG*, which is normally expressed only in whorls 3 and 4, has been expressed throughout the *Arabidopsis* sp. flower by placing its coding region under the control of the 35S promoter (45). According to the model, these transgenic flowers would be expected to possess C activity in all four whorls and to have no A activity in any of the whorls because of the A and C exclusion feature. The distribution of the B activity would not be expected to change from whorls 2 and 3. Therefore, the predicted ABC activities from whorls 1 to 4 should be C, BC, BC, C, specifying carpel, stamen, stamen, and carpel. This is the phenotype that was observed (45). Similar phenotypes were produced in transgenic tobacco (46) and tomato (47) plants expressing 35S fusions of the homologous *AG* genes, as well as in tobacco plants expressing 35S fusions of the *Brassica napus* (48) and rice (49) *AG* homologues, and *Arabidopsis* plants expressing 35S *AG* homologues from black spruce (11) and Norway spruce (50). Expression of *AG* from the *AP3* promoter, which is active in whorls 2 and 3, altered the identity of whorl 2 petals to stamens but did not affect the identity of sepals in the first whorl (51).

Expression of the B class activity in all four whorls also induced organ identity changes predicted by the ABC model (52). From the outermost whorl in, the predicted distribution of homeotic activities in these flowers would be AB, AB, BC, BC, specifying the expected (and observed) sepal, sepal, stamen, stamen. However, in this case, plants had to express 35S fusions of two genes, *AP3* and *PISTILLATA* (*PI*), before the desired phenotype was observed, suggesting that both genes are required and sufficient for full class B activity (52).

### C. Producing Apetalous Canola

What are the potential applications of modifying floral organ identity? One often-mentioned application is the production of new horticultural crops (53). Class C mutants lack sexual organs and produce additional petals, resulting in an attractive doubled flower phenotype. Similar phenotypes could readily be induced in horticultural crops using antisense or cosuppression technology (9). Another possibility is the production of canola lines that lack petals. Anyone who has seen a field of canola in bloom will have noticed its bright yellow color. This, of course, is caused by sunlight reflecting from the petals. It has been estimated that the canola flowers reflect or absorb around 50% of the total solar radiation and 20% of photosynthetically active radiation available to the plant (2). Cultivars that do not produce petals allowed much more light through the canopy (up to 70% more in one study) (54), increasing seed growth and overall yield (2). An additional benefit of the apetalous character may be the reduction of Sclerotinia stem rot. Petals appear to be an important infection site, and an apetalous mutant of *B. napus* was shown to be considerably more resistant to stem rot (55).

The inheritance of useful sources of the apetalous trait appears to be complex (56) and can be associated with undesirable agronomic traits (57), making it difficult to transfer into commercial cultivars. In contrast, transgenic approaches offer straightforward strategies for creating apetalous flowers. The simplest approaches involve expressing a cytotoxic gene in whorl 2, thus eliminating the production of petals. A number of cytotoxic genes (36) and a whorl 2–specific promoter (58) are available for this purpose. It is noteworthy that the whorl 2–specific promoter is a derivative of the *Arabidopsis ap3* promoter. Although the effect of linking this promoter to cytotoxic genes has not been reported, tobacco and *Arabidopsis* plants containing the diphtheria toxin A chain fused to a version of the *ap3* promoter that is expressed in whorls 2 and 3 failed to produce petals and stamens (59). Development in the second and third whorls was arrested early; growth and development of first and fourth whorl organs were not affected. Therefore, it should be possible to use the whorl 2–specific *ap3* promoter to eliminate petal production with no negative effect on the rest of the flower. An alternative strategy could involve the addition of C activity in whorl 2, to replace petals with stamens, an alteration that may be desirable for increasing pollen production.

## VI. NEW SYSTEMS FOR SEED AND FRUIT PRODUCTION

The production of seeds and fruits is arguably the most important agricultural function of flowers. In most cases, this requires sexual reproduction. Consequently, events leading to the formation of male and female gametes, fertilization, and embryo production represent prime targets for molecular genetic manipulation. Two examples of how transgenic technologies can be applied are considered. In both cases, success promises to have major impacts on yield and on the systems used for seed production and distribution.

### A. Hybrid Seed Production

Heterosis, or hybrid vigor, has been described in many plant species and has formed the basis for producing high-yielding cultivars in several crops, notably maize (60). However, the high cost of developing conventional hybrid systems, coupled with the fact that suitable systems do not exist for some important crops, has prompted several efforts to develop transgenic-based molecular hybridization strategies (61,62). Most of these strategies are based on the ability to inhibit pollen production or release reversibly. In one approach, this is accomplished by the specific expression of a cytotoxic RNase transgene called *Barnase* in the tapetal layer of cells that surround the developing pollen grain (62,63). Expression of Barnase results in the degradation of cellular RNA, thereby shutting off protein synthesis and causing cell death. Plants expressing the pollen-specific Barnase are male sterile. This ensures that the ovules will not be fertilized by self-pollen and permits fertilization with the pollen donor of choice. Once the hybrid seed is produced, it is desirable to restore fertility to ensure good seed production in the growers' fields. In the case of Barnase, this is accomplished by transforming the plants with a second transgene that produces a specific inhibitor of Barnase called *Barstar* (62,63). Plants expressing both Barnase and Barstar produce normal pollen and are male fertile.

Several private companies are actively developing molecular hybridization systems. Transgenic canola containing a hybridization system similar to the one described has been available commercially in Canada since 1996, under the name of Invigor. In coop trials (2 or 3 years), Invigor cultivars yielded 6.6–27.6% higher than check cultivars in the mid and long season zones of the Canadian prairies (64). It was estimated that more than 2 million acres of Invigor canola would be seeded in Western Canada in 1999 (Tom Schuler, AgrEvo Canada Inc., personal communications).

### B. Seeds Without Fertilization

Some plants have the ability to reproduce asexually through a process called *apomixis* (65). Recurrent apomixis, whereby seeds develop from sporophytic tissues of the ovules, generates progeny that are genetically identical to the maternal plant. The ability to render crop plants apomictic through transgenic approaches could lead to tremendous agricultural benefits. First, because the progeny are clones of the maternal plant, apomixis offers the possibility of fixing hybrid vigor (66). Because there is no need for pollination, male sterile lines can be grown, eliminating the possibility of outcrossing from genetically modified pollen to wild relatives. Apomixis would also simplify breeding schemes, since there would be no need to obtain homozygous material. Finally, because meiotic sterility would no longer be a problem, it has been proposed that apomixis would facilitate the production of new interspecific and intergeneric hybrids (66).

Although the molecular biology of apomixis is poorly understood at the present time, it is an area of intensive study (67,68). Several groups are using genetic approaches in attempts to create apomictic *Arabidopsis*. One strategy involves inducing mutations in plants defective in the FOI gene *PISTILLATA* (69). These plants produce small siliques devoid of seeds. By selecting for plants that produce larger siliques, mutants capable of limited seed development in the absence of fertilization (*FERTILIZATION INDEPENDENT SEED* [*FIS*]) were identified (69). Two of these genes have been cloned: *FIS2* encodes a putative transcription factor belonging to the zinc-finger family of proteins; *FIS3* encodes a Polycomb group protein (70). The *FIS3* gene was also cloned independently as the F644 (*MEDEA*) gene (71). A second Polycomb-related gene involved in fertilization-independent endosperm development, *FERTILIZATION-INDEPENDENT ENDOSPERM* (*FIE*), has also been isolated (72). The identification of genes such as *FIS* and *FIE* suggests that sexually reproducing plants have the genetic potential for apomixis. In *Drosophila* and mammals, Polycomb group proteins are involved in the long-term repression of homeotic genes (73). The *CURLY LEAF* gene of *Arabidopsis*, which represses expression of the homeotic gene *AG* in vegetative tissues, also encodes a Polycomb group protein (74). The identification of *FIS2* and *FIE* as Polycomb group proteins suggests that these may repress early stages of endosperm development until fertilization occurs.

By using a different approach, a gene capable of inducing embryo development in vegetative cells has also been cloned from *Arabidopsis* (75). Mutations in this gene, *LEAFY COTYLEDON 1* (*LEC1*), have pleiotropic effects, suggesting that the gene is involved in several aspects of late embryo development (76). The *LEC1* gene is normally expressed only during seed development in the embryo and endosperm (75). However, constitutive expression of *LEC1* from the 35S promoter resulted in the production of embryo-like structures on leaves of transgenic plants (75).

Manipulation of genes such as *FIS*, *FIE*, and *LEC1* may eventually allow the genetic engineering of apomixis. One possibility could be to transform plants with a *LEC1* transgene that is specifically expressed at a critical time during embryo sac development to trigger embryo formation without fertilization. A similar approach for inducing apomixis using a carrot receptor kinase gene whose expression is linked with embryogenic competence has been proposed (77). In this case, it is uncertain whether expression of the receptor kinase is sufficient to induce embryo formation. Additionally, it is not known whether expression of either LEC1 or the carrot receptor kinase by itself will induce other aspects of seed development (i.e., endosperm, seed coat).

## VII. PERSPECTIVES

The age of modifying plant growth and development by transgenic means is still very much in its infancy. Most transgenic material produced to date was generated primarily for research purposes. Results obtained clearly demonstrate that modulating the expression of individual flowering genes can have dramatic effects on development. However, many uncertainties and potential problems remain to be addressed. For example, in addition to inducing early flowering, overexpression of genes such as *CO* and *LFY* is often associated with undesirable phenotypes, such as reduced vegetative development, smaller leaves (15), and/or abnormal flower development including the inability to form fertile pollen (29). Clearly, these effects will have to be overcome before commercialization can be considered. In some cases, better control of gene expression or the use of genes encoding variant proteins may solve the problem. It is also uncertain how results obtained in the laboratory with *Arabidopsis* or other model plants will hold up in crops tested under field conditions. Nevertheless, our ability to alter plant development radically in agriculturally beneficial ways will continue to increase as we gain a better understanding of the genes currently cloned and continue to isolate and study new genes. These tasks will be greatly facilitated by the current thrust of genomics-based research (78) and the development of powerful new methods for evolving protein function (79).

## ACKNOWLEDGMENTS

I would like to thank Drs. Ravi Chibbar, Raju Datla, and John Mahon for critical comments on the manuscript. This publication is NRCC No. 43800.

## REFERENCES

1. DR Davies. The pea crop. In: R Casey, DR Davies, eds. Peas: Genetics, Molecular Biology and Biotechnology. Wallingford: CAB International, 1993, pp 1–12.
2. NJ Mendham, PA Salisbury. Physiology: Crop development, growth and yield. In: DS Kimber, DI McGregor, eds. Brassica Oilseeds: Production and Utilization. Wallingford: CAB International, 1995, pp 11–64.
3. C Law, C Dean, G Coupland. Genes controlling flowering and strategies for their isolation and characterization. In: BR Jordan, ed. The Molecular Biology of Flowering. Wallingford: CAB International, 1993, pp 47–68.
4. YY Levy, C Dean. The transition to flowering. Plant Cell 12:1973–1990, 1998.
5. ES Coen. Floral symmetry. EMBO J 15:6777–6788, 1996.
6. EM Meyerowitz. Genetics and molecular mechanisms of pattern formation in Arabidopsis flower development. J Plant Res 111:233–242, 1998.
7. ES Coen, EM Meyerowitz. The war of the whorls: Genetic interactions controlling flower development. Nature 353:31–37, 1991.
8. J Putterill, F Robson, K Lee, R Simon, G Coupland. The *CONSTANS* gene of Arabidopsis promotes flowering and encodes a protein showing similarities to zinc finger transcription factors. Cell 80:847–857, 1995.
9. RA Jorgensen, RG Atkinson, RLS Forster, WJ Lucas. An RNA-based information superhighway in plants. Science 279:1486–1487, 1998.
10. A Mouradov, T Glassick, B Hamdorf, L Murphy, B Fowler, S Marla, RD Teasdale. *NEEDLY*, a Pinus radiata ortholog of *FLORICAULA/LEAFY* genes, expressed in both reproductive and vegetative meristems. Proc Natl Acad Sci USA 95:6537–6542, 1998.
11. R Rutledge, S Regan, O Nicolas, P Fobert, C Cote, W Bosnich, C Kauffeldt, G Sunohara, A Seguin, D Stewart D. Characterization of an *AGAMOUS* homologue from the conifer black spruce (Picea mariana) that produces floral homeotic conversions when expressed in *Arabidopsis*. Plant J 15:625–634, 1998.
12. MA Haring, CMT Rommes, HJJ Nijkamp, J Hille. The use of transgenic plants to understand transposition mechanisms and to develop transposon tagging strategies. Plant Mol Biol 16:449–461, 1991.
13. V Sundaresan. Horizontal spread of transposon mutagenesis: New uses for old elements. Trends Plant Sci 1:184–190, 1996.
14. R Azpiroz-leehan, KA Feldmann. T-DNA insertion mutagenesis in *Arabidopsis*: Going back and forth. Trends Genet 13:152–156, 1997.
15. R Simon, MI Igeno, G Coupland. Activation of floral meristem identity genes in *Arabidopsis*. Nature 384:59–62, 1996.
16. R Macknight, I Bancroft, T Page, C Lister, R Schmidt, K Love, L Westphal, G Murphy, S Sherson, C Cobbett, C Dean. *FCA*, a gene controlling flowering time in Arabidopsis, encodes a protein containing RNA-binding domains. Cell 89:737–745, 1997.
17. ZY Wang, EM Tobin. Constitutive expression of the *CIRCADIAN CLOCK ASSOCIATED 1* (*CCA1*) gene disrupts circadian rhythms and suppresses its own expression. Cell 93:1207–1217, 1998.
18. S Shannon, DR Meeks-Wagner. A mutation in the Arabidopsis *TFL1* gene affects inflorescence meristem development. Plant Cell 3:877–892, 1991.
19. LS Robert, F Robson, A Sharpe, D Lydiate, G Coupland. Conserved structure and function of the *Ara-*

*bidopsis* flowering time gene *CONSTANS* in *Brassica napus*. Plant Mol Biol 37:763–772, 1998.
20. MS Pidkowich, JE Klenz, GW Haughn. The making of a flower: Control of floral meristem identity in *Arabidopsis*. Trends in Plant Science 4:64–70, 1999.
21. MF Yanofsky. Floral meristems to floral organs: Genes controlling early events in *Arabidopsis* flower development. Annu Rev Plant Physiol Plant Mol Biol 46:167–188, 1995.
22. E Huala, IM Sussex. *LEAFY* interacts with floral homeotic genes to regulate Arabidopsis floral development. Plant Cell 4:529–548, 1992.
23. MA Mandel, C Gustafson-Brown, B Savidge, MF Yanofsky. Molecular characterization of the *Arabidopsis* floral homeotic gene *APETALA1*. Nature 360:273–277, 1992.
24. D Weigel, J Alvarez, DR Smyth, MF Yanofsky, EM Meyerowitz. *LEAFY* controls floral meristem identity in Arabidopsis. Cell 69:843–859, 1992.
25. JT Odell, F Nagy, NH Chua. Identification of DNA sequences required for activity of the cauliflower mosaic virus 35S promoter. Nature 313:810–812, 1985.
26. MA Bláquez, L Soowal, I Lee, D Weigel. *LEAFY* expression and flower initiation in *Arabidopsis*. Development 124:3835–3844, 1997.
27. MA Mandel, MF Yanofsky. A gene triggering flower formation in *Arabidopsis*. Nature 377:522–524, 1995.
28. D Weigel, O Nilsson. A developmental switch sufficient for flower initiation in diverse plants. Nature 377:495–500, 1995.
29. Nilsson, D Weigel. Modulating the timing of flowering. Curr Opin Biotechol 8:195–200, 1997.
30. SG Southerton, SH Strauss, MR Olive, RL Harcourt, V Decroocq, X Zhu, DJ Llewellyn, WJ Peacock, ES Dennis. *Eucalyptus* has a functional equivalent of the *Arabidopsis* floral meristem identity gene *LEAFY*. Plant Mol Biol 37:897–910, 1998.
31. J Kyozuka, S Konishi, K Nemoto, T Izawa, K Shimamoto. Down-regulation of *RFL*, the *FLO/LFY* homolog of rice, accompanied with panicle branch initiation. Proc Natl Acad Sci USA 95:1979–1982, 1998.
32. J Kyozuka, R Harcourt, WJ Peacock, ES Dennis. *Eucalyptus* has functional equivalents of the *Arabidopsis AP1* gene. Plant Mol Biol 35:573–584, 1997.
33. H Albert, EC Dale, E Lee, DW Ow. Site-specific integration of DNA into wild-type and mutant lox sites placed in the plant genome. Plant J 7:649–659, 1995.
34. GFW Gocal. Flowers on demand. PBI Bulletin May, 1998, pp 2–4.
35. B Rutledge. Producing sterile trees. PBI Bulletin May, 1998, pp 16–18.
36. CD Day, VF Irish. Cell ablation and the analysis of plant development. Trends Plant Sci 2:106–111, 1997.
37. Nilsson, E Wu, DS Wolfe, D Weigel. Genetic ablation of flowers in transgenic *Arabidopsis*. Plant J 15:799–804, 1998.
38. EJ Mellerowicz, K Horgan, A Walden, A Coker, C Walter. PRFLL—a *Pinus radiata* homologue of *FLORICAULA* and *LEAFY* is expressed in buds containing vegetative shoot and undifferentiated male cone primordia. Planta 206:619–629, 1998.
39. OJ Ratcliffe, I Amaya, CA Vincent, S Rothstein, R Carpenter, ES Coen, DJ Bradley. A common mechanism controls the life cycle and architecture of plants. Development 125:1609–1615, 1998.
40. L Pnueli, L Carmel-Goren, D Hareven, T Gutfinger, J Alvarez, M Ganal, D Zamir, E Lifschitz. The *SELF-PRUNING* gene of tomato regulates vegetative to reproductive switching of sympodial meristems and is the ortholog of *CEN* and *TFL1*. Development 125:1979–1989, 1998.
41. T Jack, LL Brockman, E Meyerowitz. The homeotic gene *APETALA3* of Arabidopsis thaliana encodes a MADS box and is expressed in petals and stamens. Cell 68:683–697, 1992.
42. EM Meyerowitz, JL Bowman, LL Brockman, GN Drews, T Jack, LE Sieburth, D Weigel. A genetic and molecular model for flower development in *Arabidopsis thaliana*. Development Suppl 1:157–167, 1991.
43. MF Yanofsky, H Ma, JL Bowman, GN Drews, KA Feldmann, EM Meyerowitz. The protein encoded by the *Arabidopsis* homeotic gene *agamous* resembles transcription factors. Nature 346:35–39, 1990.
44. GN Drews, JL Bowman, EM Meyerowitz. Negative regulation of the Arabidopsis homeotic gene *AGAMOUS* by the *APETALA2* product. Cell 65:991–1002, 1991.

45. Y Mizukami, H Ma. Ectopic expression of the floral homeotic gene AGAMOUS in transgenic Arabidopsis plants alters floral organ identity. Cell 71:119–131, 1992.
46. SA Kempin, MA Mandel, MF Yanofsky. Conversion of perianth into reproductive organs by ectopic expression of the tobacco floral homeotic gene NAG1. Plant Physiol 103:1041–1046, 1993.
47. L Pnueli, D Hareven, SD Rounsley, MF Yanofsky, E Lifschitz. Isolation of the tomato AGAMOUS gene TAG1 and analysis of its homeotic role in transgenic plants. Plant Cell 6:163–173, 1994.
48. MA Mandel, JL Bowman, SA Kempin, H Ma, EM Meyerowitz, MF Yanofsky. Manipulation of flower structure in transgenic tobacco. Cell 71:133–143, 1992.
49. HG Kang, YS Noh, YY Chung, MA Costa, K An, G An. Phenotypic alterations of petal and sepal by ectopic expression of a rice MADS box gene in tobacco. Plant Mol Biol 29:1–10, 1995.
50. K Tandre, M Svenson, ME Svensson, P Engström. Conservation of gene structure and activity in the regulation of reproductive organ development of conifers and angiosperms. Plant J 15:615–624, 1998.
51. T Jack, L Sieburth, E Meyerowitz. Targeted misexpression of AGAMOUS in whorl 2 of Arabidopsis flowers. Plant J 11:825–839, 1997.
52. BA Krizek, EM Meyerowitz. The Arabidopsis homeotic genes APETALA3 and PISTILLATA are sufficient to provide the B class organ identity function. Development 122:11–22, 1996.
53. L Sieburth. Modifying floral organs. PBI Bulletin May, 1998, pp 6–8.
54. MJ Fray, EJ Evans, DJ Lydiate, AE Arthur. Physiological assessment of apetalous flowers and erectophile pods in oilseed rape (Brassica napus). J Agric Sci 127:193–199, 1996.
55. SR Rimmer, L Buchwaldt. Diseases. In: DS Kimber, DI McGregor, eds. Brassica Oilseeds: Production and Utilization. Wallingford: CAB International, 1995, pp 111–140.
56. GC Buzza. Plant Breeding. In: DS Kimber, DI McGregor, eds. Brassica Oilseeds: Production and Utilization. Wallingford: CAB International, 1995, pp 153–176.
57. MJ Fray, P Puangsomlee, J Goodrich, G Coupland, EJ Evans, AE Arthur, DJ Lydiate. The genetics of stamenoid petal production in oilseed rape (Brassica napus) and equivalent variation in Arabidopsis thaliana. Theor Appl Genet 94:731–738, 1997.
58. TA Hill, CD Day, SC Zondlo, AG Thackeray, VF Irish. Discrete spatial and temporal cis-acting elements regulate transcription of the Arabidopsis floral homeotic gene APETALA3. Development 125:1711–1721, 1998.
59. CD Day, BF Balgoci, VF Irish. Genetic ablation of petal and stamen primordia to elucidate cell interactions during floral development. Development 121:2887–2895, 1995.
60. PBE McVetty. Cytoplasmic male sterility. In: KR Shivanna, VK Sawhney, eds. Pollen Biotechnology for Crop Production and Improvement. Cambridge: Cambridge University Press, 1997, pp 155–182.
61. P Arnison. Hybrid seed—new systems for the next century. PBI Bulletin, January 1997, pp 106.
62. ME Williams, J Leemans, F Michiels. Male sterility through recombinant DNA technology. In: KR Shivanna, VK Sawhney, eds. Pollen Biotechnology for Crop Production and Improvement. Cambridge: Cambridge University Press, 1997, pp 237–258.
63. RB Goldberg, TP Beals, PM Sanders. Anther development: Basic principles and practical applications. Plant Cell 5:1217–1229, 1993.
64. P Thomas. Herbicide resistant canola varieties. Available at http://www.agric.gov.ab.ca/crops/canola/herb_ind.html
65. AM Koltunow. Apomixis: Embryo sacs and embryos formed without meiosis or fertilization in ovules. Plant Cell 5:1425–1437, 1993.
66. JP Vielle Calzada, CF Crane, DM Stelly. Apomixis: The asexual revolution. Science 274:1322–1323, 1996.
67. AM Koltunow, RS Bicknell, AM Chaudhury. Apomixis: Molecular strategies for the generation of genetically identical seeds without fertilization. Plant Physiol 108:1345–1352, 1995.
68. AM Chaudhury, S Craig, E Dennis, W Peacock. Ovule and embryo development, apomixis and fertilization. Curr Opin Plant Biol 1:26–31, 1998.
69. AM Chaudhury, L Ming, C Miller, S Craig, ES Dennis, WJ Peacock. Fertilization-independent seed development in Arabidopsis thaliana. Proc Natl Acad Sci USA 94:4223–4228, 1997.
70. M Luo, P Bilodeau, A Koltunow, ES Dennis, WJ Peacock, AM Chaudhury. Genes controlling fertilization-independent seed development in Arabidopsis thaliana. Proc Natl Acad Sci USA 96:296–301, 1999.

71. T Kiyosue, N Ohad, R Yadegari, M Hannon, J Dinneny, D Wells, A Katz, L Margossian, JJ Harada, RB Goldberg, RL Fischer. Control of fertilization-independent endosperm development by the *MEDEA* polycomb gene in *Arabidopsis*. Proc Natl Acad Sci USA 96:4186–4191, 1999.
72. N Ohad, R Yadegari, L Margossian, M Hannon, D Michaeli, JJ Harada, RB Goldberg, RL Fischer. Mutations in FIE, a WD Polycomb group gene, allow endosperm development without fertilization. Plant Cell 11:407–416, 1999.
73. V Pirrotta. Polycombing the genome: PcG, trxG, and chromatin silencing. Cell 93:333–336, 1998.
74. J Goodrich, P Puangsomlee, M Martin, D Long, EM Meyerowitz, G Coupland. A Polycomb-group gene regulates homeotic gene expression in *Arabidopsis*. Nature 386:44–51, 1997.
75. T Lotan, M Ohto, KM Yee, MA West, R Lo, RW Kwong, K Yamagishi, RL Fischer, RB Goldberg, JJ Harada. Arabidopsis *LEAFY COTYLEDON1* is sufficient to induce embryo development in vegetative cells. Cell 93:1195–1205, 1998.
76. MAL West, KM Yee, J Danao, JL Zimmerman, RL Fisher, RB Goldberg, JJ Harada. *LEAFY COTYLEDON1* is an essential regulator of late embryogenesis and cotyledon identity in Arabidopsis. Plant Cell 6:1731–1745, 1994.
77. SC DeVries, ED Schmidt, GJ Van Holst, V Hecth. Production of apomictic seed. PTC WO 97/43427, 1997.
78. M Bevan, I Bancroft, HW Mewes, R Martienssen, R McCombie. Clearing a path through the jungle: Progress in Arabidopsis genomics. Bioessays 21:110–120, 1999.
79. S Harayama. Artificial evolution by DNA shuffling. Trends Biotechnol 16:76–82, 1998.

# 13
# Genetic Technology in Peas for Improved Field Performance and Enhanced Grain Quality

**Roger Leslie Morton, Stephanie Gollasch, Hart E. Schroeder, Kaye S. Bateman, and Thomas J. Higgins**
*CSIRO Plant Industry, Canberra, Australian Capital Territory, Australia*

| | | |
|---|---|---|
| I. | INTRODUCTION | 197 |
| II. | PEA TRANSFORMATION PROTOCOL | 198 |
| | A. Strains and Plasmids | 198 |
| | B. Source Material | 198 |
| | C. Agrobacterium Infection | 198 |
| | D. Plant Regeneration | 201 |
| | E. Expected Results | 202 |
| | F. Other Pea Transformation Methods | 202 |
| III. | PROGRESS TOWARD IMPROVED AGRONOMIC AND QUALITY TRAITS IN PEAS BY TRANSGENESIS | 203 |
| IV. | FUTURE RESEARCH | 203 |
| V. | CONCLUDING REMARKS | 204 |
| | REFERENCES | 204 |

## I. INTRODUCTION

The pea (*Pisum sativum*) crop is an important source of protein for animal and human nutrition. The productivity of pea could be greatly increased by the introduction of pest and disease resistance, improved protein quality, and herbicide tolerance traits. Plant genetic engineering provides an opportunity to introduce such traits from previously unavailable sources. The first production of fertile transgenic pea plants was reported in 1992 (1). Since then several improved methods of pea transformation have been developed. In this chapter we provide a detailed protocol for efficient pea transformation and review the current status of projects aimed at pea improvement by the transgenic approach.

## II. PEA TRANSFORMATION PROTOCOL

This protocol is based on our original published method (2) with some recently introduced improvements.

### A. Strains and Plasmids

The binary vector pTAB10 (3) and a modified version of this plasmid (pKSB10.MCS.ori2) (Fig. 1a) have been used successfully for pea transformation. Both of these vectors contain the *bialaphos resistance* (bar) gene from *Streptomyces hygroscopicus*, which allows selection of transformed tissue on phosphinothricin (PPT; also called glufosinate ammonium). For selection of transformed tissues on kanamycin a new vector (pRM58) (Fig. 1b) was constructed. pRM58 contains the Subclover Stunt Virus segment 7 promoter (SV7) (4) linked to the nptII gene. This vector has three sites for cloning and replicates to high a copy number in *Escherichia coli* because of the presence of the ColE1 origin of replication. AGL1 (5), a RecA(−) supervirulent strain of *Agrobacterium tumefaciens*, was used for the plant transformations. We found that this strain gave approximately twofold higher transformation frequencies than strain LBA4404.

### B. Source Material

*Pisum sativum* plants were grown in the glasshouse and immature pods containing seeds at 2 to 5 days beyond maximal fresh weight were harvested. At this stage the pod has begun to change from bright green to yellow and the embryonic axis is uniformly beige in color. The pods were sterilized in 70% (v/v) ethanol (1 min) followed by 1% (w/v) sodium hypochlorite (20 min) and three washes with sterile distilled water.

We have also used mature pea seeds from hand-harvested plots as an alternative source of explant material. We found that the level of microbial contamination of explants derived from machine-harvested mature peas was unacceptably high. Presumably this was because the material contained seeds with cracks harboring microorganisms that were not killed by the surface sterilization procedure. Dry seed was placed in a 500-ml bottle such that it was half-full. The bottle was filled with 70% (v/v) ethanol for 1 min, followed by 7.3M orthophosphoric acid, and incubated at room temperature until the seed coat developed a wrinkled and loose appearance (1 to 2 h). The seeds were then washed with five changes of sterile water followed by three to four further washes with sterile water over 3 to 4 hours. The seeds were left to imbibe in water at room temperature for the next day. If the embryos did not show enlarged radicles through the seed coat at this stage the seeds were incubated overnight at room temperature; otherwise they were left overnight at 4°C. The total time from sterilization of dry seed to the start of the *Agrobacterium* sp. cocultivation phase is 48 h.

### C. Agrobacterium Infection

Under aseptic conditions, seeds were removed from the pods, and the testa were excised. Explants for transformation were cut from the embryonic axes of these seeds. To facilitate manipulation, the embryonic axis was left temporarily attached to one of the cotyledons (Fig. 2a). The root end was cut off and the remainder of the axis was sliced longitudinally into three to five segments (Fig. 2b) with a scalpel blade that was wet with a suspension of A. tumefaciens containing the plant transformation vector. Segments were then fully immersed in the bacterial suspension (approximately $3 \times 10^9$ cells ml$^{-1}$; OD$_{600}$ of 1:10 dilution = 0.1–0.2). After 30 to 40 min with shaking at

# Genetic Technology in Peas

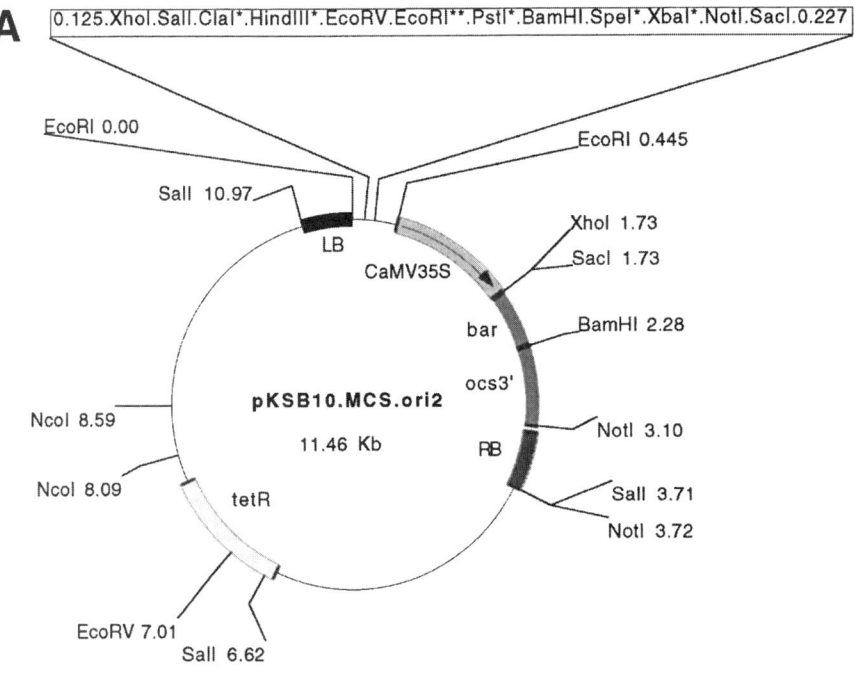

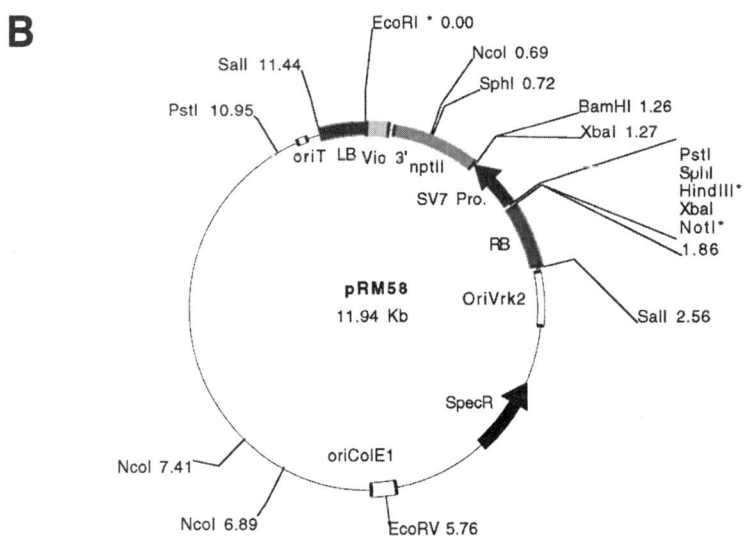

**Figure 1** Plasmids used in pea transformation. *Unique restriction sites; **, effectively unique sites; Scale in kilobase pairs. LB and RB, left and right borders, respectively, of the *Agrobacterium* T-DNA; CaMV35S, 35S promoter from CaMV; bar, bar gene from *Streptomyces hygroscopicus*; ocs 3′, 3′ untranslated region from the octopine synthase gene of *A. tumefaciens*; tetR, region conferring tetracycline resistance to bacteria; SV7 pro., promoter from segment 7 of the Subclover stunt virus; nptII, neomycin phosphotransferase II gene; vic 3′, 3′ untranslated region from the pea vicilin gene; SpecR, region conferring spectinomycin resistance to bacteria; oriColE1, origin of replication from the ColE1 plasmid; OriVrk2, origin of replication from the RK2 plasmid; oriT, origin of conjugal transfer of the RK2 plasmid.

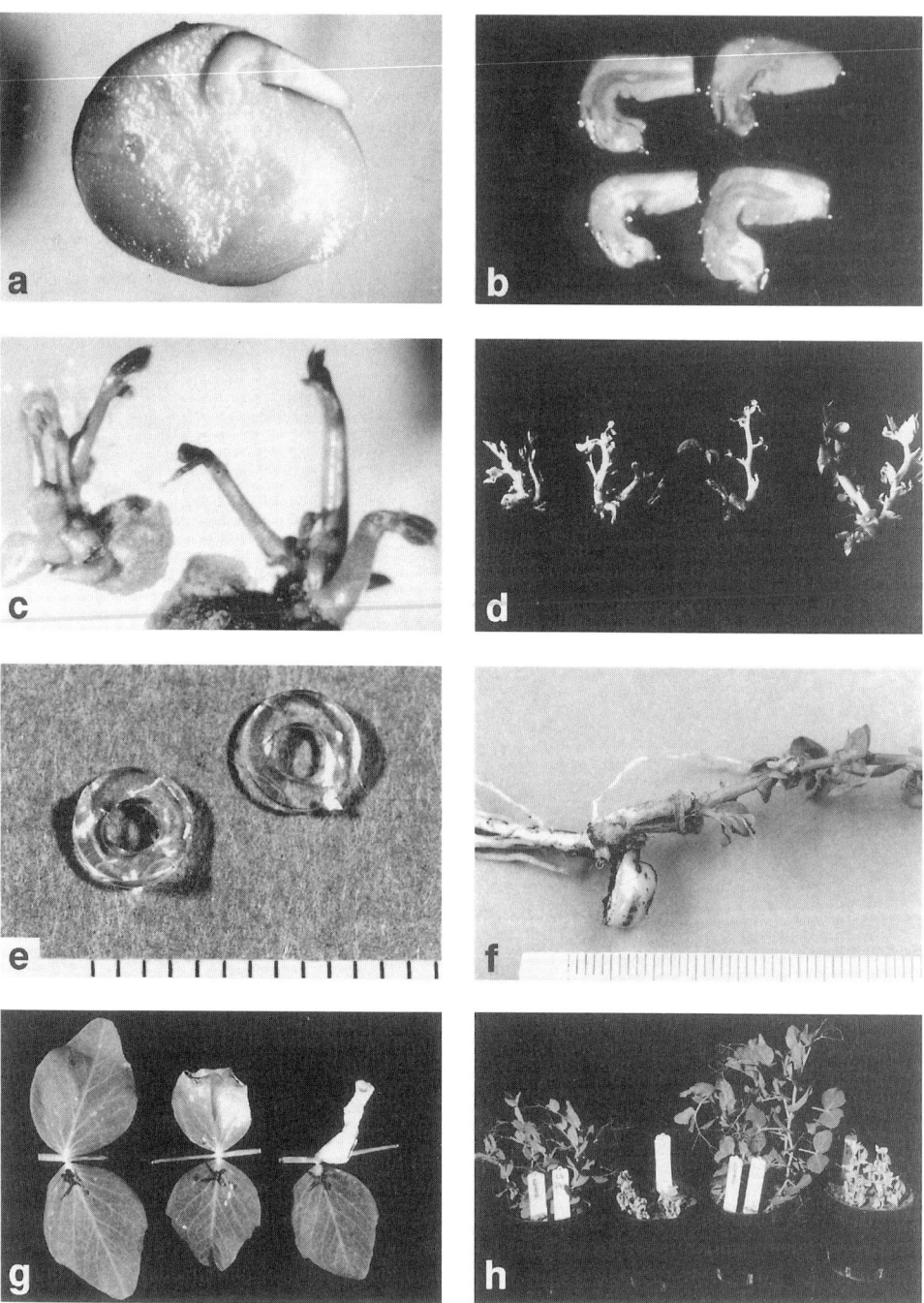

**Figure 2** Pea transformation procedure. (a) Embryonic axis attached to one cotyledon; (b) explant segments derived from embryonic axis; (c) multiple shoots developing on P245 medium; (d) distinguishing PPT-resistant (dark) and PPT-susceptible shoots (light) on P21 medium with 10 mgL$^{-1}$ PPT; (e) Silicone ring used for grafting (scale in mm); (f) graft junction 1 month after grafting procedure. The rootstock is toward the left and the scion toward the right. The 'V' shaped cut and the silicone ring are visible in the center of the image (scale in mm); (g) leaf painting test: pea leaves 5 days after application of Basta at 0.6 gL$^{-1}$ active ingredient to the upper leaflet; the leaves on the left and center are from transformed plants, and the leaf on the right is from a nontransformed line; (h) spray test for PPT resistance. Left to right: transgenic cv Rondo; nontransformed cv Rondo; transgenic cv Greenfeast; nontransformed cv Greenfeast 14 days after spraying plants with Basta at 1.4 g/L active ingredient.

room temperature the excess *Agrobacterium* sp. was removed by tilting the plate and aspirating with a Pasteur pipette. Wet segments were then plated on $B_5h$ medium (6) and cultured at 23°C under fluorescent light with a 16-hour photoperiod.

## D. Plant Regeneration

After 4 days the segments were removed and placed on P245 medium (table 1) and incubated at 23°C under fluorescent light with a 16-hour photoperiod for 15 days. At this time two thirds of the callus was removed from the base of the clumps of green shoots (Fig. 2c) and the shoot clump was incubated for a further 20 days on fresh P245 medium. At this stage green shoots were transferred as a clump of two or three to fresh P21 medium (Table 1). After 15–20 days multiple green shoots were separated from each other and transferred to fresh P21 medium. During passaging on P21 medium multiple shoots formed at the base of the plantlet. Some of these died as a result of selection and others remained green (Fig. 2d). Every 15–20 days for five to six passages the green shoots were cut away from the dead ones and transferred singly to fresh P21 medium. When the developing shoots were more than 20 mm long and capable of surviving selection as a single shoot they were grafted onto root stocks in the greenhouse.

The grafting procedure is modified from the method of Murfet (7). Pea seeds (cv. Greenfeast) were sown 25 mm deep to promote a suitably long shoot for grafting. After 6 days, when the shoot was just emerging from the soil surface, the top layer of soil was removed to expose the shoot. The shoot tip was cut horizontally under the lowest bract. A longitudinal 5 mm cut down the length of the stock was made and a ring of silicone tube (2-mm bore × 1-mm wall × 1 mm long) (Fig. 2e) placed over the cut stock and pushed to the base of the cut using hooked forceps. A long V shape cut was made in the end of the transformed shoot, which was then inserted into the cut stock. The silicone ring was slid up and around the graft and the pot covered with a plastic bag and shade cloth. Any shoots that emerged below the graft were immediately removed with a scalpel. After 10 days the plastic bag was loosened and at 14 days it was removed.

For transformation with nptII vectors, kanamycin sulfate was substituted for PPT in the

**Table 1** Media Used

| Medium | Ingredients |
|---|---|
| P245 | Murashige-Skoog macro- and micronutrients (17) |
|  | $B_5$ vitamins (18) |
|  | 4.5 mgL$^{-1}$ 6-benzylaminopurine |
|  | 0.02 mgL$^{-1}$ naphthalene acetic acid |
|  | 3% (w/v) sucrose |
|  | 100 mgL$^{-1}$ myoinositol |
|  | 0.7 gL$^{-1}$ 2-[N-morpholino]ethanesulfonic acid (MES) |
|  | 150 mgL$^{-1}$ timentin |
|  | 10 mgL$^{-1}$ phosphinothricin (PPT) |
|  | pH adjusted to 5.8 |
|  | 0.75% Difco agar |
| P21 | As above but with |
|  | sucrose 2% |
|  | 1 mgL$^{-1}$ 6-benzylaminopurine |
|  | 0.2 mgL$^{-1}$ naphthalene acetic acid |

media at levels between 75 and 150 mgL$^{-1}$. In other respects the transformation procedure is the same.

### E. Expected Results

The time from explants to grafting averaged 5 and 6 months for PPT and kanamycin selection, respectively. Grafted plants produced mature seed within 4 months. Replacing the root induction method we previously used (2) with the grafting technique decreases the time taken to produce mature transgenic plants by 4 to 6 weeks. The grafting technique also has the advantage that it does not suffer from the between-cultivar variation in efficiency observed for the root induction method.

Between 0.5% and 2.5% of the starting embryo slices gave rise to transformed plants. Putative transformed plants were tested by a simple leaf-painting test. The upper surfaces of leaflets were painted with Basta (a PPT herbicide with 200 g L$^{-1}$ PPT) diluted (333-fold) to 0.6 g L$^{-1}$ PPT in water. After 5 days leaflets on untransformed plants showed complete necrosis (Fig. 2g, right), whereas transformed plants had mild symptoms (Fig. 2g, middle) or were unaffected (Fig. 2g, left). The results of leaf painting with Basta correlate well with measurements of phosphinothricin acetyl-transferase activity on leaf protein extracts (2).

We believe that multiple passaging of the green shoots with selection reduces the chances that chimeric plants will be produced. Almost invariably, primary transgenic plants that were PPT-resistant using the leaf painting test transmitted this trait to their progeny. The trait was usually inherited in a mendelian fashion in the first generation. However, in some lines, the bar gene appears to have been subjected to gene silencing in subsequent generations.

Putative nptII transformed plants were screened by leaf painting with geneticin (G418). The upper surface of the pea leaf was painted with a 0.3% or 1% (w/v) geneticin in a 0.3% (v/v) solution of the surfactant Agrol$_{600}$. Wild-type leaves wrinkle and die within 5 days, whereas transformed tissue shows only minor damage. The lower geneticin level distinguishes nptII-expressing immature leaves from wild type. Higher levels of the antibiotic are required to genotype older leaves. The results from leaf painting with geneticin correlated well with nptII enzyme activity measured on leaf protein extracts (8).

The plant hormone regime described here differs from the one previously reported (2). We have found that, in contrast to the previous protocol, the reported modifications allow transformation and regeneration of most cultivars of pea.

### F. Other Pea Transformation Methods

Bean and Davies and coworkers have produced transformed peas, using dry pea seed as a starting material and selection with kanamycin (9) or PPT (10). Their method also differs from the one presented here in that they remove both the shoot apex and the root from the germinated pea seeds. In their method, cells in the lateral cotyledonary meristem are the most likely targets of transformation. In the transformation method described here both the apical shoot meristems and the cotyledonary meristems are present and both are potential targets for transformation. However, from microscopic examination of the regenerating plant material, we believe that the transformed tissue produced in our experiments is usually derived from the apical shoot meristem.

Grant and colleagues (11,12) have also developed a method able to produce fertile transgenic peas by using PPT selection or kanamycin selection. In common with our method, the explant source is immature pea seeds. However, they use the cotyledons of the seeds as the explant material.

# Genetic Technology in Peas

## III. PROGRESS TOWARD IMPROVED AGRONOMIC AND QUALITY TRAITS IN PEAS BY TRANSGENESIS

Transformation of pea is now routine, and researchers have begun to introduce genes for new traits into peas, by using this technology. New traits that have been successfully introduced into peas by trangenesis include agronomically useful levels of herbicide tolerance by the use of the bacterial bar gene, resistance to pea weevils by using the α-amylase inhibitor gene from kidney bean, resistance to pea seed-borne mosaic virus by using the replicase gene from the virus, and alteration of seed amino acid composition by using the 2S sunflower seed albumin gene (SSA). See Table 2 for details.

## IV. FUTURE RESEARCH

The most successful and reliable pea transformation systems have made use of the patented bar gene as the selectable marker. Since we need "freedom to operate" in order to commercialize any new pea varieties, we are optimizing our pea transformation system by using the nptII gene.

The weevil-resistant peas are undergoing field trials in preparation for possible commercial release.

In order to have a useful improvement in seed quality a greater than 10% increase in the level of sulfur amino acids will be necessary (Table 2). We are screening transgenic peas for lines

**Table 2** Improving Agronomic and Quality Traits in Peas by Transgenesis

| Trait | Gene | Progress |
|---|---|---|
| Herbicide resistance | Bar | Field trials indicate Basta at 3 L Ha$^{-1}$ kills untransformed plants, whereas transformed plants were unaffected by an application at 7 L Ha$^{-1}$ (2) |
| Insect resistance | α-Amylase inhibitor (αAI) | In αAI-containing peas none of 2300 weevil-infested seeds developed adults. The weevil larvae did not develop past the first instar stage. In contrast, on 1280 of 1620 weevil-infested wild-type peas the larvae completed development into adults (21). There are no adverse effects on rats eating αAI-containing peas (19). |
| Improved seed quality | Sunflower seed albumin (SSA) | Pea seeds expressing the sulfur-rich SSA at 1% of total seed protein have 10% more carbon-bonded-sulfur than wild-type peas |
| Virus resistance | Virus replicase | Three lines containing pea seed-borne mosaic virus (PSbMV) replicase gene were highly resistant to PSbMV (20) |
| Virus resistance | Virus coat protein | Alfalfa Mosaic Virus (AMV) and pea seed-borne mosaic virus (PSbMV) coat protein genes have been introduced into peas (12), and these lines are being screened for virus resistance |
| Ascochyta resistance | Various | We are engineering peas with various antimicrobial genes and testing them for resistance to Ascochyta blight. So far no lines have improved resistance. |

with higher levels of SSA expression. We have data from other grain legumes indicating that when SSA expression is high, the supply of sulfur amino acids to the seed may be a limiting factor controlling the level of sulfur amino acids in the seed protein. Consequently, future research will be directed to manipulating sulfur amino acid biosynthesis in grain legume seeds. Certain antimicrobial transgenes act synergistically to protect plants from pathogens (13–16). We are transforming peas with genes encoding three different antimicrobial proteins in order to test whether this synergistic effect occurs in peas.

## V. CONCLUDING REMARKS

Pea transformation is now routine in a number of laboratories around the world. The first commercially useful pea produced by transgenesis may be the weevil-resistant peas, which have already been tested in two successful field trials. These plants have a bean seed protein expressed in the pea seed using a seed specific promoter from beans. We have found that most transformants containing this gene produce significant amounts of the bean protein. Similarly, most pea plants transformed with the sunflower seed protein linked to a seed-specific promoter produce the sunflower protein at more than 0.5% of the seed protein. By contrast, attempts to express antimicrobial proteins in the leaves of peas produce many plants with no detectable expression, some with low levels, and none with high levels. This variation may be due to the potentially phytotoxic nature of the gene products, or it may be due to other factors such as the nature of the promoters or the fact that seeds are natural protein storage organs. It may be possible to generate peas producing transgene-encoded proteins at high levels in the leaves by screening a large number of independent transformants. However, considering the relatively low transformation frequency obtained in pea transformation, this is a difficult task. We have never observed "silencing" of the seed-specific genes in our pea lines, but we frequently encounter silencing of the strong constitutive CaMV35S promoter. The other constitutive viral promoter we have tested (SV7) has shown silencing in the third generation in the single transgenic line we have tested. These problems mean that altering whole plant traits will be more difficult than altering seed traits. More sophisticated approaches will be required, such as the use of inducible promoters or the subcellular targeting of expressed proteins.

## REFERENCES

1. Puonti-Kaerlas, T Eriksson, P Engstrom. Inheritance of a bacterial hygromycin phosphotransferase gene in the progeny of primary transgenic pea plants. Theor Appl Genet 84:443–450, 1992.
2. HE Schroeder, AH Schotz, T Wardley-Richardson, D Spencer, TJV Higgins. Transformation and regeneration of two cultivars of Pea (*Pisum sativum* L.). Plant Physiol 101:751–757, 1993.
3. LM Tabe, T Wardley-Richardson, A Ceriotti, A Aryan, W McNabb, A Moore, TJ Higgins. A biotechnological approach to improving the nutritive value of alfalfa. J Anim Sci 73:2752–2759, 1995.
4. P Boevink, PWG Chu, P Keese. Sequence of subterranean clover stunt virus DNA: Affinities with the geminiviruses. Virology 207:354–361, 1995.
5. GR Lazo, PA Stein, RA Ludwig. A DNA transformation-competent *Arabidopsis* genomic library in *Agrobacterium*. BioTechnol 9:963–967, 1991.
6. DCW Brown, A Atanassov. Role of genetic background in somatic embryogenesis in *Medicago*. Plant Cell Tissue Org Culture 4:111–122, 1985.
7. IC Murfet. Flowering in *Pisum:* Reciprocal grafts between known genotypes. Aust J Biol Sci 24:1089–1101, 1971.
8. RE McDonnell, RD Clark, WA Smith, MA Hinchee. A simplified method for the detection of

neomycin phosphotransferase II activity in transformed plant tissues. Plant Mol Biol Rep 5:380–386, 1987.
9. DR Davies, J Hamilton, P Mullineaux. Transformation of peas. Plant Cell Rep 12:180–183, 1993.
10. SJ Bean, PS Gooding, PM Mullineaux, DR Davies. A simple system for pea transformation. Plant Cell Rep 16:513–519, 1997.
11. JE Grant, PA Cooper, AE McAra, TJ Frew. Transformation of peas (Pisum sativum L.) using immature cotyledons. Plant Cell Rep 15:254–258, 1995.
12. JE Grant, PA Cooper, BJ Gilpin, SJ Hoglund, JK Reader, MD Pither-Joyce, GM Timmerman-Vaughan. Kanamycin is effective for selecting transformed peas. Plant Sci 139:159–164, 1999.
13. QZ Maher, S Masoud, RA Dixon, CJ Lamb. Enhanced protection against fungal attack by constitutive co-expression of chitinase and glucanase genes in transgenic tobacco. Bio technology 12:807–812, 1994.
14. PJM van den Elzen, E Jongedijk, LS Melchers, BJC Cornelissen. Virus and fungal resistance: From laboratory to field. Philos Trans R Soc Lond B Biol Sci 342:271–278, 1993.
15. G Jach, B Gornhardt, J Mundy, J Logemann, E Pinsdorf, R Leah, J Schell, C Maas. Enhanced quantitative resistance against fungal disease by combinatorial expression of different barley antifungal proteins in transgenic tobacco. Plant J 8:97–109, 1995.
16. E Jongedijk, H Tigelaar, SC Roekel, SA Bres-Vloemans, I Dekker, PJM van den Elzen, JC Cornelissen, LS Melchers. Synergistic activity of chitinases and beta-1,3-glucanases enhances fungal resistance in transgenic tomato plants. Euphytica 85:173–180, 1995.
17. T Murashige, F Skoog. A revised medium for rapid growth and bioassays with tobacco tissue cultures. Physiol Plant 15:473–497, 1965.
18. OL Gamborg, RA Miller, K Ojima. Nutrient requirements of suspension cultures of soybean root cells. Exp Cell Res 50:151–158, 1968.
19. A Pusztai, G Grant, S Bardócz, R Alonso, MJ Chrispeels, HE Schroeder, LM Tabe, TJV Higgins. Expression of insecticidal bean α-amylase inhibitor transgene has minimal detrimental effect on the nutritional value of peas in the rat at 30% of the diet. J Nutr 129:1597–1603, 1999.
20. AL Jones, IE Johansen, SJ Bean, I Bach, AJ Maule. Specificity of resistance to pea seed-borne mosaic potyvirus in transgenic peas expressing the viral replicase (Nlb) gene. J Gen Virol 79:3129–3137, 1998.
21. RL Morton, HE Schroeder, KS Bateman, MJ Chrispeels, E Armstrong, TJV Higgins. Bean α-amylase inhibitor 1 in transgenic peas (*Pisum sativum*) provides complete protection from pea weevil (*Bruchus pisorum*) under field conditions. Proc. Natl. Acad. Sci. USA 97:3820–3825, 2000.

# 14

# Genetic Engineering for Levels of Select Phytonutrients Affecting Human Health

**George G. Khachatourians**
*University of Saskatchewan, Saskatoon, Saskatchewan, Canada*

|      |                                                  |     |
|------|--------------------------------------------------|-----|
| I.   | INTRODUCTION                                     | 207 |
| II.  | FOOD CROPS AND HUMAN NUTRITION                   | 208 |
|      | A. Plant Micronutrients and Metabolites          | 209 |
|      | B. Plant Micronutrients and Human Health         | 210 |
|      | C. Engineering of Micronutrient Content          | 211 |
| III. | MINING OF PLANT GENOMICS FOR HUMAN HEALTH FACTORS | 212 |
| IV.  | CONCLUDING REMARKS                               | 212 |
|      | REFERENCES                                       | 213 |

## I. INTRODUCTION

Food is fundamental to human life and maintenance of optimal health. Improving nutritional quality of food and its ingredients is one of the high-priority areas of research worldwide. Significant strides in improving the quality and volume of food grade substrates, ingredients, or products have been made through microbial biotechnology (1). Connected to these advances are a large number of microbial products and processes contextual to applied microbiology and biotechnology that have had a positive impact on the economics of food and beverage industries.

One of the highest-priority areas of research in the United States is improvements in the population health through diet, nutrition, and foods (2,3). Strategically important issues in this regard are the value of plant nutrients or phytonutrients and micronutrients and the role that genetic engineering can play. This chapter focuses on the nature and activities of some of the phyto- and micronutrients of plants and the record of genetic engineering of food plants for human health.

The term phytonutrients refers to those secondary compounds in plant foods that are generated through complex biosynthetic pathways known to be controlled by genetic and environmental factors. Phytonutrients and classically defined nutrients can provide benefits beyond the prevention of dietary deficiencies (4). Plant leaves, fruits, seeds, tubers, and roots can be valuable sources of nutrients, foods, and medicinals. There are three groups of materials (a) macronutrients (carbohydrates, fats, and proteins); (b) some 17 minerals (Ca, P, Cl, K, S, Mg, Na, Se, Fe,

Cu, Zn, Co, Cr, Mn, Mo, I, and F) and fat-soluble and water-soluble vitamins (respectively; A, D, E, and K and vitamins $B_1$, thiamin; $B_2$, riboflavin; niacin; pantothenic acid; $B_6$, pyridoxine; biotin; folic acid; $B_{12}$, cobalamin; and C), which make up the micronutrients or organic and inorganic compounds; and (c) essential ingredients (fiber, carotenoids, bioflavenoids) that can profoundly affect our well-being or risk of disease throughout our lives from pregnancy to lactation, childhood, adolescence, and old age (5).

Carbohydrates and proteins or amino acids make up the bulk of foodstuff and are used primarily as an energy supply. Embryonic development of the nervous system is regulated in part by retinoids and cobalamin. Vitamin A controls cell differentiation. Vitamins C and E, selenium- and sulfur-containing amino acids, β-carotene, zinc, and copper help with the prevention of oxidative damage and free radical accumulation. Throughout life we need antioxidants, or substances that when present at low concentrations compared to those of an oxidizable substrate significantly delay or prevent oxidation of that substrate (6). Antioxidants are not strictly required in the diet, yet they are linked to the promotion of good health, longevity, and vitality. Some plant chemicals, such as phenolic compounds, are widely distributed in the plant kingdom. Plant tissues synthesize the phenolic compounds' resveratrol, flavonoids, and furanocoumarins. Some of these phenolic compounds are toxic to humans (7). Others have beneficial effects, for example, on low-density lipoproteins and aggregation of platelets, because they reduce the risk of coronary heart disease (4).

There is a wide range of pharmaceutically and medically valuable compounds that plant cells produce in culture or in whole plants and that are extracted for use from large volumes of material (8). In these situations quality assurance and reasonable price are not guaranteed. Research into the biochemical genetics of these micronutrients suggests that we can enhance or increase production through genetic manipulation of plant metabolism. The recent attention focused on genetic engineering of micronutrients therefore should produce no surprise (6,7,9). Plant genetics and genetic engineering can enhance nutritional quality and composition and nutritionally and medically important material within plants. By learning about the finer aspects of biosynthetic pathways and modifying gene expression levels or transmission of other genetic controls into such host plants we can enhance the production of these metabolites. This should have a direct impact for both overfed and underfed populations whether in the developing world or industrialized countries.

## II. FOOD CROPS AND HUMAN NUTRITION

Nutritional composition of food crops for human consumption and its modification are being viewed as urgent worldwide health issues (9). Therapeutic levels of many essential nutrients needed for health and those that could be a part of our diet could be obtained through additional food fortification or direct genetic modification of micronutrient levels in food crops. This issue arises because the basic nutritional needs of much of the world's population are unmet. The dietary composition of foods consumed by people in developing countries resides mostly in a few staple foods, such as cassava, wheat, rice, and corn, which are poor in both macro- and micronutrients. A large gap in the quality of diets for over 800 million people includes 250 million children with vitamin A deficiency that can lead to blindness, 2 billion people at risk for iron deficiency, and 1.5 billion people at risk for iodine deficiency. To improve this huge nutritional deficiency problem, vitamin A and iron deficiency has been addressed by fortification of rice seed by soybean ferritin gene and vitamin A precursor synthesis (9–13). Even in industrialized nations, where both food abundance and variety are present and daily caloric intake is often excessive, micronutrient deficiencies are surprisingly common as a result of poor eating habits.

Today not only in the United States but in many nations the concentrations of essential vitamins and minerals are defined in terms of the recommended diet allowance (RDA) (14). Over the past 50 years, some of the RDA values have changed and others have remained constant (15). However, RDAs may not reflect the optimal levels of micronutrients for health. Indeed the University of California Berkeley Wellness Letter and other authorities recommend much higher levels of micronutrients. It is a practice for some processed foods to be fortified with additional macro- or micronutrients. Indeed to alleviate specific nutritional disorders there is growing support for increased intake of some micronutrients, for example, vitamins B and C, carotenoids, and selenium. Greater intake of this group significantly reduces the risk of certain cancers, cardiovascular diseases, and chronic degenerative diseases associated with aging (16–20). Cruciferous vegetables such as cabbage, cauliflower, Brussel sprouts, and broccoli, and oilseed plants, e.g., mustard, are rich sources of micronutrients. Selenium after conversion to organic selenium compounds and at least 15 glucosinolates that modify the activity of enzymes affecting carcinogen clearance, estrogen metabolism, and estrogen-related concerns of aging women and men, increase the apoptosis in cancerous cell lines (21,22). A 1998 report by Orser and associates (23) indicates that *Brassica juncea*, or Indian mustard, grown under hydroponic cultivation can contain 2000 ppm selenium. Certainly all three, culturing technology, metabolic engineering, and transgenic food crop constructs, or combinations thereof, should present new means of production of food crops that will enhance health.

### A. Plant Micronutrients and Metabolites

Food crops' micronutrients and their levels should be important features of balancing micronutrient nutrition and enhancing human health. This is because several micronutrients are specifically involved with gene expression mechanisms (24). Both traditional plant breeding and molecular approaches to increasing the micronutrient density in edible portions of food crops are useful (25,26). However, a great deal more could be learned from manipulation by molecular genetic techniques. Applications of this knowledge could be used to alter micronutrient density, accumulation, and uptake in edible portions of food crops. Schachtman and coworkers (11) suggest two approaches: first, the use of deoxyribonucleic acid (DNA) markers as genetic tags for the introgression of desired traits, and second, the introduction of defined genetic material, which is the process of genetic engineering.

Plants, as much as fungi and certain bacteria, are unique in that after their exponential growth a high level of production of various primary metabolites and synthesis of a large number of secondary metabolites, such as hormones and drugs, including antibiotics and toxins, occur (27). These metabolites are made in response to development, various environmental stress responses, and defense against wounding and phytopathogenic microorganisms. Phytochemicals are often specific to species or genera. Notwithstanding the role of phytochemicals for plants, many have important consequences for animals and humans, including medicinal or health-promoting value (28,29).

Macronutrients and health-promoting compounds are associated with seeds, roots, bark, leaves, glandular tissue, and pollen (30). Qualitative improvements of many of these nutrients, e.g., storage proteins, amino acid composition, production of novel carbohydrates, changes in fatty acid composition, and reduction in antinutrients, although complex in terms of genetic manipulations, nonetheless are being actively pursued (31,32). Medicinal properties of garlic (*Allium sativum*) arise from at least 15 biologically active compounds, the best known of which are allicin, diallyl disulfide, and diallyl trisulfide (33). Although the chemical and botanical aspects of garlic and its nutritional and medicinal aspects are well documented (34), its genetics is less studied. With some prerequisite genetic knowledge of a large number of target-specific actions in relation to particular constituents—allicin, DADS, and DAS, among others—it should be possible to construct transgenic food crops with antimicrobial, antihypertensive, lipid-lowering, anticancer, and fibrinolytic activities. Hop plant (*Humulus lupulus* L.) cones' contribution to their characteristic bitterness is due to the α- and β-bitter acids (e.g., humulone, colupomulone) of beer-brewing process. These secondary metabolites and their novel polyketide synthase, phlorisovalerophenon synthase, have been characterized and should lead to its genetic characterization and manipulation (35).

Seeds are a rich source of macronutrients, carbohydrates, proteins, and oil, but also minerals and vitamins. Molecular dissection and improvements of the nutritional and functional properties are yielding specific knowledge of development and genetics, as well as solutions to the problems often associated with their underutilization (36). Plant seeds such as those from legumes and cereal make up over 50% of the per capita energy and protein intake worldwide and 63–65% in the developing countries. We now have a better understanding of specific forms of starch and oil body development and control of partitioning of carbon among the starch-, oil-, and protein-storage components (37–40). Because of our understanding and manipulative technologies of *Brassica* sp. oil production and oleosins, in the oil storage bodies of seeds, many useful products, including recombinant proteins and other pharmaceutically important materials such as hirudin and interlukins, can be engineered and extracted economically (38–41).

Leguminous plants, especially the pea, present problems related to incomplete protein and starches and low protein digestibility. These problems have been addressed with mutants for seed-specific starch-branching enzyme. Here the ratio of amylose to amylopectin and, hence, the properties of the isolated starch have been greatly influenced. Biochemical genetics and convenient and sensitive isozyme assays have revealed the role of specific isozymes in maize, potato, and pea starch biosynthesis (42). Microstructure and ingredients of starch seed have profound effects on the flour and dough produced and their handling and baking properties (43). Introduction into potato of a mutant *Escherichia coli* adenosine diphosphate (ADP)-glucose pyrophosphorylase enzyme, thought to be a rate-limiting enzyme for starch biosynthesis, enhanced the amount of starch accumulated. Changes in the starch biosynthesis enzymes should influence the physical structures and properties of the extracted starch of the starch grains. These changes are valuable to many industries (44). Likewise the roles of specific desaturases, thioesterases, and hydroxylases in *Arabidopsis* and *Brassica* spp. and soybean fatty acid biosynthesis could be modified by alterations in these genes (45). Insertion of genes encoding new fatty acid desaturases under the control of seed-specific promoters in canola produces new oil constituents. A whole series of cultivars containing low and high levels of oleic, palmitic, linolenic, and linoleic acids are emerging (46, 47). Griffiths and coworkers (48) found that sesamin, a lignan present in sesame (*Sesamum indicum*) oil, specifically inhibits D5 desaturation and formation of arachidonic acid from dihomo γ-linolenic acid in cell free extracts and reduced cell proliferation in cell cultures derived from canine prostatic tissues of epithelial origin from neoplastic and nonneoplastic tissues. Genetic engineering of oil plants to contain vitamins and other factors to prevent proliferation of cancerous cells or tissues including metastatic cells should be a significant step.

The ability to create transgenic food plants offers the opportunity to program plants to accumulate macronutrients and energy sources for humans. Native and foreign proteins in seeds, roots, or leaves enhance their quality and value and provide a new source of valuable medicinals (36). There seems to be little reason why plants should not be adopted as sources for human therapeutic metabolites, enzymes, vaccines, and other compounds (49–51).

### B. Plant Micronutrients and Human Health

Micronutrients account for up to 30% of a tissue's dry weight, whereas individual micronutrients are generally much less then 0.1% of a tissue's dry weight. Micronutrients including vitamins are involved in modulation of transcription, translation, and posttranslational modification of certain proteins (24). In addition micronutrients can be anti- and coantimutagenic or mutagenic and comutagenic (52). The question becomes whether or not significant increases in micronutrient levels are possible. The answer in part depends on (a) the particular compound(s) under consideration and (b) whether excessive dietary intake could have unintended negative health consequences. The positive effects of select micronutrients, e.g., iron, calcium, selenium, iodine, and the vitamins folic acid and vitamins E, B, and A, are in optimal human health (9). However, these micronutrients are present in limited quantities in foods or diets worldwide. In micronutrient-rich diets an upper safe level of intake for these minerals and vitamins is possible. Mechanisms of action, significance, multiple side effects, and levels of tolerance of the antioxidant vitamins A, E, and C are well known (53–58). Fortuitously, plants only synthesize provitamin A carotenoids, which are used as substrates for retinol synthesis by humans (9). The overall process is highly regulated, and as a consequence the upper safe intake level for β-carotene (the most active provitamin A carotenoid in plants) is 20 times that of retinol or 100 times the RDA for vitamin A. On this basis, ideally manipulating provitamin A carotenoid synthesis in plants, rather than attempting to introduce retinol synthesis, has become the target for engineering of plants for human health.

Certain plant metabolites are known for their anticarcinogenic properties and their other biochemical and nutritional effects (58). At least 100 distinct glucosinolates are present in plants (58). The glucosinolates of cruciferous vegetables increase carcinogen detoxification and lower carcinogen–deoxyribonucleic acid (DNA) interactions (59,60). The phytoestrogens, such as genistein and daidzein, are isoflavones that are particularly abundant in soybeans and have health-promoting attributes. Individuals with soy-rich diets have significantly lower occurrences of some cancers, osteoporosis, and coronary heart disease when compared to individuals with-low-soy diets (61). However, Lappe and associates (62) report that phytoestrogen levels in transgenic herbicide-tolerant soybeans are lower. Intake of carotenoids (both pro– and non–provitamin A) is shown to reduce the risk of a number of health problems, e.g., certain types of cancers, cardiovascular disease, and age-related macular degeneration (9,63). Although the benefits of glucosinolates, isoflavones, and carotenoids to human health are recognized, Western diets are generally poor in foods containing the highest levels of these compounds.

## C. Engineering of Micronutrient Content

Food crop micronutrient contents have been modified by traditional and modern genetics (11). Traditional crop breeding has increased productivity, yields, and occasionally and unexpectedly micronutrient composition, yet similar successes through molecular breeding have not occurred. Efforts in breeding food crops for ease of processing, malting, baking, and extruding, which determine industrial utility, must accompany efforts to enhance nutritional qualities for human consumption (61). However, the strategy for and selection of these traits have not had a strong biochemical genetic basis nor an explanation of phenotypic differences (63), with the result that limited progress has been made in engineering of micronutrient levels for human health by molecular techniques.

The last decade has shown rapid progress in the identification of developmental and biochemical genetics for the biosynthesis of precursors to vitamin A, E, and C of the model plant *Arabidopsis thaliana* (64–68). Vitamins C and E and the vitamin A precursor carotinoid have important functions as antioxidant vitamins, and successful results have been obtained in engineering vitamin biosynthesis genes in rice (69) and other plants (9). Oxidative reactions to lipids in stressed tissues and cells are defended by antioxidants through both enzymatic (e.g., superoxide dismutase, GSH-peroxidase, and -reductase) and nonenzymatic (micronutrient and vitamin) mechanisms (54,55). As a result the reduction of free radicals through multiple effects of antioxidant vitamins becomes significant in dietary intake and protection against oxidative stress (53,57). Determination of the biochemical genetics of the carotenoid biosynthetic pathway, iron uptake, and biotin, thiamin, and vitamin E synthesis has made significant strides, including gene cloning, sequencing, and heterologous expression systems (69,70).

## III. MINING OF PLANT GENOMICS FOR HUMAN HEALTH FACTORS

Recent developments in plant genomics, proteomics, and bioinformatics have made significant and rapid headway in the identification of metabolically and biosynthetic important sequences and comparisons with those of other organisms (36,71–73). Because of sequence homology and conservation and structure-function similarity, certain micronutrients (essential vitamins and minerals) and primary and secondary metabolic genes are identifiable. New research methodology, gene probes, high-throughput screening (microarray), and automated DNA sequencing are making the dissection of nutritional genomics a certainty (74). Genomic and proteomic databases and use of bioinformatics should make the establishment of metabolic pathways attainable (75). Further, because of plant gene resources, comparison of gene and DNA sequence differences can be developed in the short term. An alternate approach is the use of the genomic data from microorganisms and the universality of certain biosynthetic systems, e.g., from lower eukaryotes, filamentous fungi, and yeasts or bacteria. This strategy was used for the α-tocopherol (vitamin E) biosynthetic pathway in *Arabidopsis* sp. (76). Genetic data for the first step of the 10-gene pathway were isolated from *Arabidopsis* sp. and the photosynthetic bacterium *Synechocystis* sp. (PCC6803). Because the genomes of both *A. thaliana* and the cyanobacterium *Synechocystis* sp. (PCC6803) have been sequenced (77,78) it has been shown that the two have a 35% amino acid identity (76). Although α-tocopherol content within the plant oils is low, using the strategy of cloning of γ-tocopherol methyltransferase (γTMT) from either *Synechocystis* or *Arabidopsis* sp., Shintani and DellaPenna (76) used carrot seed–specific promoter to overexpress the γTMT in *A. thaliana* and increase the α-tocopherol 80-fold and that of vitamin E 9-fold when compared to those of wild type (76).

The preceding examples demonstrate the power of applying genomics to dissect vitamin biosynthesis and its content. It should be possible to apply the same strategy for other food crop constructions. Further use of technologies such as bioinformatics and microarray expression systems, expressed sequence tags, and other automated systems for isolation and elucidation of pathway, specific genes for many phytochemicals and their gene mapping and orthologues in a variety of food crops, e.g., corn and soybean, should become a reality (9,71,73,79).

## IV. CONCLUDING REMARKS

A UNICEF report on the state of the world's children (80) discusses in detail the prevalence and causes of worldwide child malnutrition and recommends steps that must be taken. According to this report malnutrition is implicated in more than 6 million or half of all deaths of children below 5 years of age worldwide. This tragedy in terms of magnitude is unmatched by any infectious disease since the Black Death (80). In many instances, the report cites examples of micronutrient deficiencies that can be remedied by increasing levels of phytonutrients. To remedy this situation coordinated efforts for research on five major staple food crops—rice, wheat, maize, beans, and cassava—are needed.

Research to improve the nutritional quality of plants has historically been limited by a lack of basic knowledge of plant metabolism and the often daunting task of selection of levels of nutrients for increased micronutrients. Therefore, improvement in the levels of phytonutrients will require an interdisciplinary approach and collaboration of professionals in the natural, agriculture, food, and health sciences (4,81). The use of both conventional and modern genetic techniques and biotechnology, associated with the arrival of genomics, proteomics, and informatics, should allow greater integrative approaches to plant-based foods and phytonutrients. In addition, new goals for the production (both pre- and post harvest), handling, and storage of the phytonutrient content of foods and food economics and policy programs (4,82,83) must be identified, researched, and implemented (80). This type of food production will cross barriers of species, family, and phylum (9,26).

As a result of the increase in our basic knowledge of microbial contributions and phytonutrients and their genetic basis, the opportunities in food production during the coming decade will be truly unparalleled (1). Furthermore, these developments will place plant and food science researchers in the position of being able to modify the nutritional content of major crops to improve aspects of human health. For essential minerals and vitamins that are limited in world diets, the need and way forward are clear, and improvement strategies should be pursued with attention to the upper safe limit of intake for each phytonutrient (56,82). However, for many other health-promoting phytochemicals, decisions will need to be made regarding which crops to modify to achieve the precise compound(s) and their metabolites for nutritional impact and health benefits (84).

Decisions regarding genetic engineering of plants for types and levels of phytonutrients will require strong interdisciplinary collaborations among scientists and communication professionals. We have learned that emerging information, requires communication, discussions, and dialogue with the broader audience of consumers, decision makers, and decision influencers. In short, biotechnology of food will be the new paradigm for ensuring a safe and healthful food supply to serve our needs in this century.

## ACKNOWLEDGMENTS

I am thankful for my reviewers, Lorraine M. Khachatourians, Dr. Robert Tyler, and Dr. Adrienne Woytowich.

## REFERENCES

1. Y-H Hui, GG Khachatourians. Food Biotechnology: Microorganisms. New York: VCH Press, 1995, pp 937.
2. CM Weaver, MK Schmidt, CE Woteki, WR Bidlack. Research needs in diet, nutrition, and health: America's food research needs into the 21st century: A report of the Research Committee of the Institute of Food Technologists. Food Technol 47:14S–17S, 25S, 1993.
3. ME Sanders, B Wasserman, EA Foegeding. Research needs in biotechnology: America's food research needs into the 21st century: A report of the Research Committee of the Institute of Food Technologists. Food Technol 47:18S–21S, 1993.
4. CR Fjeld, RH Lawson. Food, phytonutrients, and health. Proceedingss of the Forum and Workshops, College Park, MD. Nutr Rev 57:S1–S52, 1999.
5. D Bhatia. Vitamins. Part II. General considerations. In: Y-H Hui, ed. Encyclopaedia of Food Science and Technology. New York: Wiley-Interscience, 1991, pp 2687–2697.
6. B Halliwell. Antioxidant characterization and mechanism. Biochem Pharmacol 49:1341–1348, 1995.
7. O Daniel, MS Meier, J Schlatter, P Frischknecht. Selected phenolic compounds in cultivated plants: Ecologic functions, health implications, and modulation by pesticides. Environ Health Perspect 107 (suppl 1):109–114, 1999.

8. WGW Kurz, F Constable. Plant cell cultures, a potential source of pharmaceuticals. Adv Appl Microbiol 25:209–240, 1979.
9. D DellaPenna, Nutritional genomics: Manipulating plant micronutrients to improve human health. Science 285:375–379, 1999.
10. T apRees. Prospects of manipulating plant metabolism. Trends Biotechnol 13:375–387, 1995.
11. DP Schachtman, SJ Barker, RM Welch, RD Graham. Molecular approaches for increasing the micronutrient density in edible portions of food crops. In: RD Graham, ed. Sustainable Field Crop Systems for Enhancing Human Health: Agricultural Approaches to Balanced Micronutrient Nutrition. Field Crops Res 60:81–92, 1999.
12. DH Catoway. Human Nutrition: Food and Micronutrient Relationships Washington, DC: International Food Policy Research Institute, 1995.
13. F Goto, T Yoshihara, N Shigemoto, S Toki, F Takaiwa. Iron fortification of rice seed by soybean ferritin gene. Nat Biotechnol 17:282–285, 1999.
14. Food and Nutrition Board, National Research Council. Recommended Dietary Allowances. Washington, DC: National Academy Press, 1989.
15. PA Lachance. International perspective: Basis, need, and application of recommended dietary allowances. Nutr Rev 56:S2–S4, 1998.
16. LM DeLuca, K Kosa, F Andreola. The role of vitamin A in differentiation and skin carcinogenesis. Nutr Biochem 8:426–437, 1997.
17. M Hronek, Z Zadak, D Solichova, P Jandik, B Melichar. The association between specific nutritional antioxidants and manifestation of colorectal cancer. Nutrition 16:189–191, 2000.
18. DA Cooper, AL Eldridge, JC Peters. Dietary carotenoids and certain cancers, heart disease, and age-related macular degeneration: A review of recent research. Nutr Rev 57:201–214, 1999.
19. JP Kehrer, CV Smith. Free radicals in biology: Sources, reactivities, and roles in the etiology of human diseases. In: S Frei, ed. Natural Antioxidants in Human Health and Diseases. San Diego: Academic Press, 1994.
20. AIFCR. Food, Nutrition and the Prevention of Cancer: A Global Perspective: World Cancer Research Fund. Washington, DC: American Institute for Cancer Research, 1997.
21. GR Fenwick, RK Heaney, WJ Mullin. Glucosinolates and their breakdown products in food and food plants. CRC Crit Rev Food Sci Nutr 18:123–201, 1983.
22. R McDanell, AEM McLean, AB Hanley, RK Heaney, GR Fenwick. Differential induction of mixed-function oxidase activity in rat liver and intestine by diets containing processed cabbage: Correlation with cabbage levels of glucosinolates and glucosinolate hydrolysis products. Food Chem Toxicol 25:363–368, 1987.
23. CS Orser, DE Salt, IJ Pickering, RC Prince, A Epstein, BD Ensley. *Brassica* plants to provide enhanced human mineral nutrition: Selenium phytoenrichment and metabolic transformation. J Med Food 1: 253–261, 1998.
24. CD Berdanier. Nutrient-gene interactions. Nutr Today 35:8–17, 2000.
25. MW Farnham, PW Simon, JR Stommel. Improved phytonutrient content through plant genetic improvement. Nutr Rev 57:S19–S26, 1999.
26. C Somerville, S Somerville. Plant functional genomics. Science 285:380–383, 1999.
27. JB Harborne. Introduction to Ecological Biochemistry. San Diego: Academic Press, 1993.
28. E Conn. The world of phytochemicals. In: DL Gustine, HL Flores, eds. Phytochemicals and Health. Rockville, MD: American Society of Plant Physiologists 1995, pp 1–14.
29. BD Oomah G Maza. Health benefits of phytochemicals from selected Canadian crops. Trends Food Sci Technol 10:193–198, 1999.
30. D McCaskill, R Croteau. Strategies for bioengineering the development and metabolism of glandular tissues in plants. Nat Biotechnol 17:131–136, 1999.
31. IE Liener. Implications of antinutritional components in soybean foods. Crit Rev Food Sci Nutr 34:31–67, 1993.
32. DS Brar, R Ohtani, H Uchimya. Genetically engineered plants for quality improvements. Biotechnol Genet Eng Rev 13:167–179, 1995.
33. RA Nagourney. Garlic: Medicinal food or nutritious medicine? J Med Food 1:13–28, 1998.

34. JA Milner. Garlic: Its anticarcinogenic and antitumorogenic properties. Nutr Rev 54:S82–S86, 1996.
35. NB Paniego, KWM Zuurbier, S-Y. Fung, R van der Heijdenm, JJC Scheffer, R Verpoorte. Phlorisovalerophenon synthase, a novel polyketide synthase from hop (*Humulus lupulus* L.) cones. Eur J Biochem 262:612–6161, 1999.
36. BO deLumen. Molecular approaches to improving the nutritional and functional properties of plant seeds as food sources: Developments and comments. J Agric Food Chem 38:1779–1788, 1990.
37. L Tabe, TJV Higgins. Engineering plant protein composition for improved nutrition. Trends Plant Sci 3:282–286, 1998.
38. DJ Murphy. Engineering oil production in rapeseed and other oil crops. Trends Biotechnol 14:206–213, 1996.
39. GJH van Rooeijen, MM Maloney. Plant seed oil-bodies as carriers for foreign proteins. Biotechnology 13:72–77, 1995.
40. GJH van Rooeijen, MM Maloney. Structural requirements of oleosin domain for subcellular targeting to oilbody. Plant Physiol 109:1353–1361, 1995.
41. MM Maloney, LA Holbrook. Subcellular targeting and purification of recombinant proteins in plant production systems. In: MP Tomb, ed. Biotechnology and Genetic Engineering Reviews. Vol 14. Andover, England: Intercept, 1997, pp 321–336.
42. CT Larsson, P Hofvander, J Khoshnoodi, B Ek, L Rask, H Larsson. Three isoforms of starch synthase and two isoforms of branching enzymes are present in potato tuber starch. Plant Sci 117:9–16, 1996.
43. K Autio, T Laurikainen. Relationship between flour/dough microstructure and dough handling and baking properties. Trends Food Sci Technol 8:181–185, 1997.
44. MK Beatty, A Rahman, H Cao, W Woodman, M Lee, AM Myers, MG James. Purification and molecular characterization of ZPU1, a pullulans-type starch-debranching enzyme from maize. Plant Physiol 119:255–266, 1999.
45. MT Facciotti, PB Berain, L Yuan. Improved strarate phenotype in transgenic canola expressing a modified acyl-acyl carrier protein thioesterases. Nat Biotechnol 17:593–597, 1999.
46. CE Palmer, WA Keller. Transgenic oilseed Brassicas. In: GG Khachatourians, A McHughen, WK Nip, R Scorza, YH Hui, eds. Transgenic Plants and Crops. New York, Marcel Dekker 2001.
47. O Sayanova, MA Smith, P Lapinskas, AK Stobart, G Dobson, WM Christie, PR Shrewry, JA Napier. Expression of a borage desaturase cDNA containing an N-terminal cytochrome *b*5 domain results in the accumulation of high levels of delta-6 desaturated fatty acids in transgenic tobacco. Proc Natl Acad Sci USA 94:4211–4216, 1997.
48. G Griffiths, HE Jones, CL Eaton, AK Stobart. Effect of sesamin on growth and arachidonic acid content of neoplastic and non-neoplastic prostate epithelial cell cultures. Phytother Res 12:417–421, 1998.
49. CJ Arntzen. High-tech herbal medicine: Plant-based vaccines. Nat Biotechnol 15:221–222, 1997.
50. K Ma, A Hiatt, M Hein, ND Vine, F Wang, T Stabila, C vanDolleweerd, K Mostov, T Lehner. Generation and assembly of secretory antibodies in plants. Science 268:716–719, 1995.
51. HS Mason, CJ Arntzen. Transgenic plants as vaccine production system. Trends Biotechnol 13:388–392, 1995.
52. CK Chow. Mutagenesis and micronutrient relationship. Food Addit Contam 7(supp 1):S44–S47, 1980.
53. D Kitts. An evaluation of the multiple effects of the antioxidant vitamins. Trends Food Sci Technol 8: 198–203, 1997.
54. B Haliwell, S Chirico. Lipid peroxidation: Its mechanisms, measurement and significance. Am J Clin Nutr 57:715S–725S, 1993.
55. B Haliwell. Free radicals and antioxidants: A personal review. Nutr Rev 52:253–265, 1994.
56. PA Lachance. Overview of key nutrients: Micronutrient aspects. Nutr Rev 56:S34–S39, 1998.
57. AA Woodall, G Briton, MJ Jackson. Dietary supplementation with carotenoids: Effects on α-tocopherol levels and susceptibility of tissues to oxidative stress. Br J Nutr 76:307–317, 1996.
58. SF Vaughn. Glucosinolates as natural pesticides. In: HG Cutler, SJ Cutler, eds. Biologically Active Natural Products: Agrochemicals. Boca Raton, FL: CRC Press, 1999, pp 81–91.
59. GR Fenwick, RK Heaney, WJ Mullin. Glucosinolates and their breakdown products in food and food plants. CRC Crit Rev Food Sci Nutr 18:123–201, 1983.
60. R McDanell, AEM McLean, AB Hanley, RK Heaney, GR Fenwick. Differential induction of mixed-

function oxidase activity in rat liver and intestine by diets containing processed cabbage: Correlation with cabbage levels of glucosinolates and glucosinolate hydrolysis products. Food Chem Toxicol 25: 363–368, 1987.
61. MA Zeligs. Diet estrogen status: The cruciferous connection. J Med Food 1:67–82, 1998.
62. MA Lappe, EB Bailey, C Childres, KDR Setchell. Alterations in clinically important phytoestrogens in genetically modified, herbicide-tolerant soybeans. J Med Food 1:241–245, 1998.
63. JL Charieux, Dietary phytoestrogens. Nutr Rev 54:S109–S114, 1996.
64. R Graham, D Senadhira, S Beebe, C Iglesias, I Monasterio, RM Welch. Breeding for micronutrient density in edible portions of staple food crops: Conventional approaches. In: RD Graham, ed. Sustainable field crop systems for enhancing human health: Agricultural approaches to balanced micronutrient nutrition. Field Crops Res 60:57–80, 1999.
65. SR Norris, TR Barrette, D DellaPenna. Genetic dissection of carotenoid synthesis in Arabidopsis defines plastoquinones as an essential component of phytoene desaturation. Plant Cell 7:2139–2149, 1995.
66. B Pogson, K McDonald, M Truong, G Britton, D DellaPenna. Arabidopsis carotenoid mutants demonstrate that lutein is not essential for photosynthesis in higher plants. Plant Cell 8:1627–1639, 1996.
67. MA Grusak, D DellaPenna, RM Welch. Physiologic processes affecting the content and distribution of phytonutrients in plants. Nutr Rev 57:S27–S33, 1999.
68. PL Conklin, JE Pallanca, RL Last, N Smirnoff. Ascorbic acid metabolism in the ascorbate-deficient *Arabidopsis* mutant *vtc*1. Plant Physiol 115:1277–1285, 1997.
69. PK Burkhardt, P Berger, J Wunn, A Kloti, GA Armstrong, M Schledz, J vonLintig, I Potrykus. Transgenic rice (*Coryza sativa*) endosperm expressing daffodil (*Narcissus pseudonarcissus*) phytoene synthase accumulates phytoene, a key intermediate of provitamin A biosynthesis. Plant J 11:1071–1078, 1998.
70. T Guru. New genes boost rice nutrients. Science 285:994–995, 1999.
71. FX Cunningham, E Gantt. Genes and enzymes of carotenoid biosynthesis in plants. Annu Rev Plant Physiol Plant Mol Biol 49:557–583, 1998.
72. SR Norris, X Shen, D DellaPenna. Complementation of the *Arabidopsis pds*1 mutation with the gene encoding *p*-hydroxyphenylpyruvate dioxygenase. Plant Physiol 117:1317–1323, 1998.
73. V Walbot. Genes, genomes, genomics: What can plant biologists expect from the 1998 National Science Foundation Plant Genome Research Progr? Plant Physiol 119:1151–1155, 1999.
74. B Mazur, E Krebbers, S Tingey. Gene discovery and product development for grain quality traits. Science 285:372–375, 1999.
75. MH Saier Jr. Genome sequencing and informatics: New tools for biochemical discoveries. Plant Physiol 49:151–171, 1998.
76. D Shintani, D DellaPenna. Elevating the vitamin E content of plants through metabolic engineering. Science 282:2098–2100, 1998.
77. DW Meinke, JM Cherry, C Den, SD Rounley, M Koornneef. *Arabidopsis thaliana*: A model plant for genome analysis. Science 282:662–682, 1998.
78. H Kotani, S Tabata. Lessons from sequencing of the genome of a unicellular cyanobacterium, *Synechocystis* sp. PCC6803. Annu Rev Plant Physiol Plant Mol Biol 49:151–171, 1998.
79. UNICEF. The State of the world's children 1998. New York: UNICEF, 1999.
80. MR Pounce, P Robles, JL Micol. High-throughput genetic mapping in *Arabidopsis thaliana*: Mol Gen Genet 261:408–415, 1999.
81. LV Kochian, DF Garvin. Agricultural approaches to improving phytonutrient content in plants: An overview. Nutr Rev 57:S13–S18, 1999.
82. BA Underwood, S Smitasiri. Micronutrient malnutrition: Policies and programs for control and their implications. Annu Rev Nutr 19:303–3244, 1999.
83. IL Goldman, AA Kader, C Heintz. Influence of production, handling, and storage on phytonutrient content of foods. Nutr Rev 57:S46–S52, 1999.
84. F Khachik, JS Bertram, M-T Huang, JW Fahey, P Talalay. Dietary carptenoids and their metabolites as potentially useful chemoprotective agents against cancer. In: L Packer, M Hiramatsu, T Yoshikawa eds. Antioxidant Food Supplements in Human Health. Orlando, FL: Academic Press. 1999, pp 203–229.

# 15
# Genetic Engineering and Resistance to Viruses

**Marc Fuchs**
*Institut National de la Recherche Agronomique, Colmar, France*

**Dennis Gonsalves**
*Cornell University, Geneva, New York*

|      |                                                                              |     |
|------|------------------------------------------------------------------------------|-----|
| I.   | INTRODUCTION                                                                 | 217 |
|      | A. Background                                                                | 217 |
|      | B. Objectives and Scope of the Review                                        | 218 |
| II.  | PATHOGEN-DERIVED RESISTANCE                                                  | 218 |
|      | A. Concept                                                                   | 218 |
|      | B. Application to Virus Resistance in Plants                                 | 218 |
|      | C. Characterization of Engineered Resistance to Viruses                      | 218 |
|      | D. Probable Mechanism(s) Underlying Engineered Resistance Against Viruses    | 219 |
|      | E. Factors to Consider in Developing Virus-Resistant Transgenic Plants       | 221 |
| III. | APPLICATIONS: A FEW SUCCESS STORIES                                          | 222 |
|      | A  Field Evaluation                                                          | 222 |
|      | B. Commercialization                                                         | 222 |
|      | C. Benefits to Agriculture                                                   | 223 |
| IV.  | ENVIRONMENTAL RISK ISSUES                                                    | 223 |
|      | A. Potential Impact                                                          | 223 |
|      | B. Opposition to Transgenic Plants                                           | 224 |
|      | C. Risk Assessment Studies: Scientific Facts                                 | 224 |
| V.   | DISCUSSION AND FUTURE PROSPECTS                                              | 226 |
|      | REFERENCES                                                                   | 227 |

## I. INTRODUCTION

### A. Background

There has been tremendous progress in agricultural biotechnology in recent years. For example, transgenic plants resistant to insects, herbicides, and diseases have been produced, field-tested, and commercialized. In the case of viruses, significant breakthroughs opened new avenues to engineering resistance in crops. Resistance to viruses has been achieved by transforming susceptible plant varieties with genes or gene sequences derived from viral genomes. This approach is known as *pathogen-derived resistance* (PDR) (1). Advances continue to be made in understanding the cellular and molecular mechanisms of PDR, identifying effective virus-derived gene constructs, and developing virus-resistant transgenic crops.

### B. Objectives and Scope of the Review

Many reviews have been written recently on engineered resistance against viruses in plants (2–9). Here we describe how the concept of pathogen-derived resistance has been applied to engineer resistance to plant viruses in crops. Then we summarize our current knowledge on the mechanisms underlying engineered resistance. We will also highlight progress on the commercialization of virus-resistant transgenic crops and address issues related to their potential environmental impact. Finally, future prospects of engineered resistance against plant viruses are discussed.

## II. PATHOGEN-DERIVED RESISTANCE

### A. Concept

The majority of virus-resistant transgenic plants result from the application of the concept of PDR (1). This concept is based on the use of virus-derived genes and gene segments as the source of resistance. Various constructs, including full-length, untranslated, and truncated coding and noncoding complementary deoxyribonucleic acids (cDNAs), in sense or antisense orientation, have been employed to engineer resistance to viruses in plants. Virus genes that confer resistance include constructs encoding coat proteins (CPs), replicases, movement proteins, proteinases, defective interfering ribonucleic acid (RNA), and satellite RNAs (5–7).

### B. Application to Virus Resistance in Plants

The success of PDR in conferring resistance was first demonstrated with bacteriophage (1). Beachy and colleagues were the first to report on the application of PDR to engineer resistance against viruses in plants (10). These authors showed that transgenic tobacco that accumulates CP of *Tobacco mosaic virus* (TMV) is protected from infection by TMV, and by closely related tobamoviruses. Subsequently, PDR has been applied against a wide range of viruses in many plant species (5,6).

### C. Characterization of Engineered Resistance to Viruses

The degree of engineered resistance to viruses varies among plant lines. In some cases there is no detectable accumulation of the target virus(es) anywhere in the inoculated plants. This extremely high level of resistance, often defined as *immunity*, is of practical use.

In other cases the resistance in slightly weaker and virus(es) can accumulate on inoculated leaves but subsequent movement of the virus seems to be blocked. Also, plants can be initially susceptible to the virus; however, symptoms are attenuated or absent in the upper leaves and there is little or no virus accumulation (11). This resistance phenotype is referred to as *recovery*. This type of resistance is also of practical importance.

Resistance has also been described as delay in the onset of disease symptoms. Transgenic plants are symptomatic and accumulate the challenge virus(es) after some delay compared to nontransgenic controls. This type of resistance can also have, although limited, practical value.

## D. Probable Mechanism(s) Underlying Engineered Resistance Against Viruses

There are at least two distinct types of mechanisms underlying engineered resistance against viruses: one requiring expression of the transgene-derived protein (2–4,8,9) and the other dependent only on the presence of transgene-derived messenger ribonucleic acid (mRNA) (2,12,13). Protein-mediated resistance seems to confer resistance to a broader range of virus strains and species (14), whereas RNA-mediated resistance seems to provide higher levels of resistance to a specific virus strain (15).

Protein-mediated resistance was first reported for TMV in tobacco (10). The resistance is greater to TMV than to tobamoviruses that have CP genes more distantly related to the transgene (14). Accumulation of CP in transgenic plants is indispensable for protection against TMV. It appears that the CP of TMV expressed by transgenic plants interferes with disassembly of TMV particles (16,17). It may also interact with a host component or directly with the viral RNA to prevent replication, translation, or assembly into virions (2–4). A correlation is proposed between CP subunit-subunit interactions and CP-mediated resistance against TMV. Indeed, when using mutants of the CP of TMV that affect subunit-subunit interactions in transgenic *Nicotiana tabacum*, an increased resistance is achieved compared to that of wild-type CP as a result of strong interaction between CP subunits expressed by transgenic plants and challenge virus CP subunits (18). Sometimes, protein-mediated resistance confers protection only to the virus from which the transgene is derived, whereas in other cases it provides protection against related viruses. It is unclear why some CP provides broad and strong degrees of resistance whereas other CP provides only narrow or weak resistance.

On the other hand, evidence shows that RNA-mediated resistance occurs through a posttranscriptional gene silencing (PTGS) process. This phenomenon was first observed with transgenic plants not carrying viral genes (19,20). PTGS leads to a marked reduction in the accumulation of the transgene mRNAs and the degradation of the RNA of challenging virus. Dougherty and colleagues were the first to show a relationship between virus resistance in transgenic plants and PTGS involving a RNA turnover (11). Transcription run-on experiments with isolated nuclei from transgenic plants established that silencing of virus-derived transgenes occurs posttranscriptionally. In virus-resistant transgenic lines, relatively low and undetectable steady-state accumulation of transgene RNA, along with little or no protein product, is observed. The mechanisms of PTGS are not fully understood, although there is strong evidence of targeted degradation of RNA in the cytoplasm. Indeed, it appears that PTGS suppresses in trans the accumulation of viral RNA which have homology in the coding region of the transgene, thereby conferring homology-dependent virus resistance.

Silencing of the viral transgene can occur prior to virus infection but can also be activated only after virus infection. If silencing occurs prior to virus infection, plants do not accumulate the challenge virus. On the other hand, if silencing occurs after infection, it becomes active in upper leaves after infection of the lower leaves. In this case, the transgene is initially expressed at a high level and accumulation of transgene mRNA is suppressed in asymptomatic tissue, thereby providing resistance to secondary infection if the challenge virus is homologous to the transgene mRNA.

PTGS requires transgene expression (21), and its apparent strength increases with the transcriptional state and copy number of the transgene. Goodwin and colleagues (22) showed that transgenic *N. tabacum* expressing multiple copies of an untranslated CP gene of *Tobacco etch virus* (TEV) are highly resistant to TEV, whereas transgenic plants expressing single copies of the same transgene express the recovery phenotype. Correlation between copy number of the transgene and virus resistance is, however, not observed in all cases of PTGS.

Methylation of the transgene has been associated to some extent with PTGS (23–26). Methylation is usually concentrated at the 3' end of the transgene coding region (21) and may be responsible for the production of prematurely terminated transgene mRNAs (aberrant RNAs) or transgene mRNAs that are truncated or improperly processed. In 1999 Guo and coworkers (27) showed that methylation, which was found all along the transgene sequence, is associated with establishment and maintenance of PTGS, and therefore with virus resistance. These authors proposed that PTGS involves RNA signals, either from the silenced transgene and/or from the challenge virus, which activate a specific cytoplasmic RNA degradation pathway and induce changes, DNA methylation in particular, in homologous transgenes (27). These changes switch transgene from an active to a silenced status.

The role of RNA signals as target and initiator of PTGS is becoming more compelling (28). The target RNA may be transgene mRNAs or viral RNAs that have sequence homology to the sense RNA product of the transgene. Accumulation of transgene and viral RNA could reach a certain RNA threshold, thus suppressing RNA accumulation in a sequence-specific manner and mediating specific degradation (2). What triggers the RNA turnover is not known, but an overproduction of RNA such that a threshold concentration is overcome has been implicated. Also, the synthesis of aberrant transgene mRNAs could play an important role (29). PTGS has been shown to correlate with abundant RNA degradation intermediates (30). The RNA target of PTGS is particularly the 3' region (24,25,31); however, it can also be located in the central coding region (32) or in the 5' region of the transgene (31,33).

To account for the activation and specificity of PTGS, it seems that RNA products of the transgene are important. The sequence requirements for triggering gene silencing may differ from those involved in the degradation process in PTGS (34). Ruiz and associates (35) showed that initiation of virus-induced silencing is dependent on the challenge virus and maintenance of virus-induced silencing is virus-independent.

PTGS is influenced by the development stage of the plants (31,36,37). Even highly resistant—previously considered immune—transgenic lines may accumulate a high level of transgene RNA at early times in development and are susceptible to TEV infection (31). Thus, it appears that a common mechanism of gene silencing and virus resistance occurs in transgenic lines exhibiting high resistance and recovery resistance phenotypes.

In the majority of examples of PTGS sense RNA is a target of homology-dependent silencing (11,24,26,37–41); however, the negative-strand RNA was also shown to be the target (32). Is the mechanism of PTGS RNA strand-specific? It does not seem the general rule; however, there are examples of resistance that has been achieved although an abundance of transgene RNA is produced (42). In this case, it appears that homology-dependent resistance and PTGS are not related. Overall the sense RNA transcript of the transgene could act as mediator of the resistance mechanism eventually by annealing to complementary RNA in viral replication intermediates and promoting degradation of duplexed RNA (43). The silence state is characterized by reduced levels of full-length mRNA and the appearance of specific low-molecular-weight RNA fragments (22,31). These low-molecular-weight RNA fragments consist of both the 5' and 3' portions of the CP transgene that could emerge through endonucleolytic cleavages. These low-molecular-weight RNA fragments that result from degradation of mRNA and aberrant RNA could trigger silencing.

Looking at the silencing effects and variants that have been observed in virus-resistant transgenic plants, it seems that virus-derived transgenes become inactivated as a result of diverse defense systems that are designed to neutralize invasive viruses. Similarities exist between PTGS and natural virus defense mechanisms. PTGS has been described as a natural resistance to *Cauliflower mosaic virus* (CaMV) in nontransgenic brassicas (44). Similarly, *Tomato black ringspot virus* infection in nontransgenic plants induces a resistance similar to transgene-induced silencing (45). In these two studies, nontransgenic plants inoculated with viruses were able to overcome infection by initiating turnover of replicating viral genomes in the cytoplasm. It is hypothesized that virus-derived transgenes subject to PTGS may produce RNA resembling replicating viruses, such as double-stranded features (45). It seems that viral genomes can undergo alterations that give rise to RNA susceptible to degradation in a manner identical to what is occurring in transgenic plants. The extent to which plants have adapted these defense mechanisms to control expression of genes and viruses in not yet known. However, Ratcliff and coworkers (46) in 1999 demonstrated that PTGS of transgenes is functionally the same as PTGS-like defense response to virus infection and that the latter is a manifestation of a natural defense mechanism that is also induced by *Tobacco rattle virus*. Determining the molecular basis of triggering, maintaining, regulating, and turning off silencing is a major challenge with direct application in engineered virus resistance in crops.

In 1998 the P1-HC-Pro polyprotein of TEV was shown to suppress PTGS (47,48). Thus, plant viruses may enhance susceptibility within a host through inhibition of a potent defense response (49). Similarly, *Cucumber mosaic virus* (CMV) counteracts PTGS (50), indicating that CMV can inhibit cellular factors involved in the RNA degradation step of PTGS or inhibit the systemic spread of the silencing signal. Also, PTGS is suppressed by the HC-Pro of *Potato virus Y* (PVY) and the 2b of CMV but not by *Potato virus X* (PVX) (51). This effect is protein-rather than RNA-mediated. The 2b protein of *Tomato aspermy virus* (TAV) but not of CMV has been shown to induce hypersensitive cell death and virus resistance in *N. tabacum,* whereas in *N. benthamiana* the TAV 2b protein suppresses PTGS (52). The dual functionality of the TAV 2b protein indicates that plants may react to virus-encoded suppressor of PTGS by inducing an independent resistance mechanism.

Plant mutants impaired in the triggering of PTGS have been obtained (53). Such plants should help in identifying genetic loci governing gene silencing mechanisms in plants. Two loci, sgs1 and sgs2, have been shown to be affected in PTGS and silencing-related transgene methylation (53).

## E. Factors to Consider in Developing Virus-Resistant Transgenic Plants

Since the first report of engineered resistance to a plant virus (10), PDR has been applied to numerous viruses and crops. A number of transgenic crops have been tested in the field, and some of them have even been commercialized (5,54).

Several factors need to be considered when developing virus resistance in plants. Since the cellular and molecular mechanisms underlying engineered resistance are not well understood, it is almost impossible to design virus-derived genes that induce predictable phenotypes. In other words, it is difficult to predict how a given transgene will affect the level of resistance.

Among the driving forces for our work on engineered resistance against viruses are a strong interest in identifying resistant lines as early as possible in the plant development process and the desire to test for resistance under natural conditions of virus infection and plant growth. Therefore, once putative transformed plants are established in the greenhouse and their transgenic status assessed, resistance screening experiments are undertaken. If resistant lines are identified, they are further propagated and analyzed under field conditions. Field trials are often more effective than greenhouse evaluation for examining disease symptoms, and identifying and eliminating plants with undesirable phenotypes (55).

## III. APPLICATIONS: A FEW SUCCESS STORIES

### A. Field Evaluation

A number of transgenic plants with virus-derived gene constructs have been extensively evaluated under field conditions and shown to be valuable to control viral diseases. For example, transgenic tomato, potato, squash, melon, cucumber, and papaya provide practical resistance to viruses under natural exposure (5,54,56). More recently, resistance to *Potato leafroll virus* (PLRV) was observed in transgenic potato plants expressing the PLRV CP gene through 5 years of field trials (57). Also, Thomas and coworkers (55) showed that aphid-mediated spread of PLRV was restricted in fields of transgenic potato plants expressing the PLRV CP gene. The markedly reduced secondary transmissions of PLRV are likely explained by lower virus titer in transgenic plants, thus decreasing virus content of aphid vectors and reducing transmission efficacy. Field resistance of transgenic potato plants to mechanical inoculation by several strains of PVY has also been demonstrated (58). More field tests of transgenic crops engineered for virus resistance are under way.

A number of transgenic crops exhibit a high degree of resistance under greenhouse conditions (5). It was shown in 1999 that transgenic rice plants expressing the CP genes (CP1, CP2, and CP3) of *Rice tungro spherical virus* (RTSV) are protected from leafhopper-mediated inoculation (59). Resistance was expressed as a reduction in RTSV accumulation and a significant delay in RTSV infection in transgenic lines that accumulated transgene mRNA. The delay in infection by RTSV may be sufficient to allow rice to develop enough to escape disease by a late infection.

Several reports showed that transgenic tomato plants expressing the CP gene of CMV are resistant to natural spread of CMV by indigenous aphid populations (60–62). Similarly, transgenic tomato plants expressing an ameliorative satellite RNA of CMV exhibited resistance to CMV infection in the field (63). Transgenic plants had mild or no symptoms, low virus titer, and up to 83% higher marketable fruit yield than nontransgenic plants. In addition, risk assessment studies showed low levels of satellite RNA transmission within the test site and no evidence of damage caused by the satellite RNA on surrounding plants. Similarly, transgenic plants of hot pepper expressing the CMV satellite RNA showed symptom attenuation upon mechanical inoculation by CMV (64).

Other transgenic crops have been engineered with virus-derived gene sequences and shown to be protected from virus infection. For example, transgenic pea plants expressing the *Pea enation mosaic virus* (PEMV) CP gene display attenuated symptoms and delayed PEMV multiplication upon mechanical inoculation (65). Also, transgenic peas resistant to *Pea seed-borne mosaic virus* have been developed (34). Transgenic rice plants expressing the replicase gene of *Rice yellow mottle virus* (RYMV) are resistant to mechanical inoculation by RYMV (66). Transgenic plum trees expressing the CP gene of *Plum pox virus* (PPV) are resistant to aphid inoculation and chip budding challenge by PPV (67). Transgenic sugarcane resistance to *Sorghum mosaic virus* was developed in 1999 (68). It will be interesting to see whether the resistance of these plants holds in the field.

### B. Commercialization

Virus-resistant transgenic plants have been commercialized in the United States. These comprise a vegetable and a fruit crop. The first commercial virus-resistant transgenic crop was a summer squash line resistant to *Watermelon mosaic virus* (WMV) and *Zucchini yellow mosaic virus* (ZYMV). This line expresses the CP genes of WMV and ZYMV and is highly resistant to mixed aphid-vectored infection by these two viruses (69,70). It was commercially released as Freedom II in the spring of 1995 by Asgrow Seed. Transgenic papaya containing the CP gene of *Papaya ringspot virus* (PRSV) and resistant to PRSV (71) was the second virus-resistant transgenic crop commercially released in 1998 (54).

Outside the United States, virus-resistant transgenic plants, including tomato, pepper, tobacco, potato, and soybean, are also deployed on a large scale in the People's Republic of China.

Given the efficacy of the PDR strategy to control viral diseases and the number of virus-resistant transgenic crops showing excellent performance in small-scale field experiments, more of them will likely reach the market in the near future.

## C. Benefits to Agriculture

Virus-resistant transgenic plants offer many benefits to agriculture (5). Benefits are of agronomical importance in particular when other sources of resistance have not been identified and when host resistant genes cannot be easily transferred into elite cultivars by classical breeding. In this case, engineered resistance may be the only approach to develop virus-resistant varieties. Also, the development of varieties with multiple-virus resistance has been facilitated by the PDR strategy. Benefits are of economic importance when transgenic crops increase yield and improve crop quality. This can be critical for subsistence farmers who rely on a limited food supply. Benefits are of epidemiological importance since transgenic plants do not serve as a virus source for secondary spread, thereby reducing epidemics to neighbor fields. For example, reduced PLRV infection rate and lower virus titers have been achieved in transgenic potato plants (55). Lower levels of PLRV should reduce acquisition frequencies and transmission within and between potato fields. Also, lower levels of PLRV should reduce the use of insecticides to control aphid vectors. Benefits are also of environmental importance since the use of insecticides to control vectors is highly reduced. Thus, chemical residues in food and water supplies should be limited and the protection of pesticide applicators improved.

## IV. ENVIRONMENTAL RISK ISSUES

### A. Potential Impact

Considering the novelty of the PDR strategy to engineer resistance to viruses, the large-scale use of virus-resistant transgenic crops has raised legitimate concerns about their potential incidence in the environment. The potential environmental impacts of virus-resistant transgenic plants has been extensively reviewed (5,72–76).

One potential risk involves the encapsidation of the genome of challenge viruses within the CP subunits expressed by transgenic plants (77). This phenomenon is called *heterologous encapsidation*. Since the interaction between a virus and its vector is often dependent on the properties of the CP, newly encapsidated virions may be acquired by other vectors than those that naturally transmit the challenge virus. For example, heterologous encapsidation may assist the spread of a virus that is defective in transmission by its vectors. However, since heterologous encapsidation does not alter the genome of the challenge virus, changes in virus-vector specificity should be a single-generation event with temporary consequences. Therefore, heterologous encapsidation should not cause long-term environmental problems.

A second potential concern is that transgene mRNAs may be involved in RNA recombination with the RNA genome of challenge viruses (78,79). This process involves the exchange of RNAs during virus replication and the development of chimeric species that combine two originally distinct RNAs. In other words, all or a portion of the viral transgene may be incorporated into the genome of challenge viruses. Chimeric RNA molecules may arise with new biological properties. Since recombination alters the genome of challenge viruses, virions with new biological properties, including expansion of host range, increased pathogenicity, and changes in vector specificity, could emerge.

Complementation is another potential concern. Virus-resistant transgenic plants could serve as reservoirs of functional proteins for challenge viruses. Consequently, challenge viruses that are deficient in the synthesis of some proteins or are producing dysfunctional proteins may acquire new properties through complementation. For example, transgenic plants expressing a functional movement protein could complement the cell-to-cell diffusion of a virus with a defective movement protein. Complementation occurs at the plant level; therefore, it may cause more severe infections and eventually economic losses. However, since complementation does not alter the genome of challenge viruses, it is expected to have limited environmental impact.

Synergism between virus-derived transgene products and challenge viruses may increase virus titers and symptom severity. Thus, economic losses are potential consequences of synergism. This process does not affect the genome of challenge viruses. Therefore, it is not envisioned to cause environmental hazard.

Gene flow between virus-resistant transgenic plants and wild relatives can lead to the movement and establishment of virus-derived transgenes in populations of wild species (80). Wild plants that acquire the transgenes could have a competitive advantage and exhibit increased weediness potential. Thus, wild plants could become invasive in natural habitats.

Most of the potential risks associated with transgenic plants are similar to those of non-transgenic plants that are subjected to mixed virus infection (5,76,78). Thus, the baseline for risk assessment studies is the current situation in the absence of transgenic plants. An appropriate issue to be addressed is, Do the potential risks associated with transgenic plants occur beyond those of background events? In other words, do recombination, heterologous encapsidation, complementation, synergism, and gene flow in transgenic plants present additional risks compared to those in nontransgenic plants?

### B. Opposition to Transgenic Plants

Biotechnology and virus-resistant transgenic crops can impart environmental benefits through reduced pesticide use and increased productivity; however, they pose potential environmental risks. Along with uncertainties regarding long-term health and environmental impacts, increasingly moral and economic concerns have been advanced. Such concerns have reached a point where intense antibiotechnology crusades are frequent. Also, in an alarming number of events opponents of biotechnology advocate and practice illegal means of physical attack on property, especially in Europe.

The current climate of opposition highlights the need for a wide range of consultation, dialogue, and exchange of ideas. Such consultations should be guided by research results. Therefore, science-based risk assessment studies are timely for national and international policy-making agencies and organizations that are engaged in formulating issues, opinions, and recommendations. Where do we stand regarding risk assessment studies? Is there a clear line between facts generated by scientifically sound studies and emotional science fiction scenarios?

### C. Risk Assessment Studies: Scientific Facts

Risk assessment studies have been conducted to evaluate the potential impact on the environment of virus-resistant transgenic plants (5). These studies have been performed mainly in the laboratory. Laboratory studies are valuable to identify potential hazards, examine their frequency of occurrence, and understand their mechanisms. Subsequently, they can help design trangenes with maximized resistance and limited environmental impact. For example, Lecoq and colleagues (81) demonstrated that the CP gene of PPV is able to mediate the spread of an aphid-nontransmissible strain of ZYMV by heterologous encapsidation. Subsequently, Jacquet and associates (82) showed that a truncated CP gene of PPV, which was unable to form viruslike particles because of a deletion in the 5′ end, confers resistance to PPV but does not assist the spread of the aphid-nontransmissible strain of ZYMV. One has to remember that heterologous encapsidation has also been documented when nontransgenic plants are coinfected (5).

Greene and Allison (83) showed that the 3′ untranslated region of *Cowpea chlorotic mottle virus* (CCMV) is involved in recombination with deletion mutants of CCMV, thus restoring systemic spread. These authors also demonstrated that transgenes without the 3′ untranslated region are less likely to be involved in recombination events (79,84).

Also, recombinant viruses developed in transgenic *N. bigelovii* expressing CaMV gene VI upon CaMV infection (85). The challenge virus was gradually replaced with a recombinant virus that acquired the transgene. Double recombination between the CP transgene of *Tomato bushy stunt virus* (TBSV) in transgenic *N. benthamiana* and a TBSV mutant with a defective CP gene was shown to restore wild-type virus (86). Recombination has also been shown when transgenic *N. benthamiana* expressing the CP gene of *African cassava mosaic virus* (ACMV) are challenged with a CP deletion mutant of ACMV (87). It is noteworthy that recombination also occurs in the case of mixed infection of nontransgenic plants (5,88,89).

Complementation has been shown to occur in transgenic plants (5). More recently, transgenic *N. benthamiana* expressing the p51 gene of the triple gene block of *Peanut clump virus* (PCV) that is involved in cell-to-cell movement complements deletion mutant PCV RNA2 transcripts that are deficient in p51, thereby restoring systemic virus multiplication (90). This complementation acts exclusively in cis since a defective mutant of *Beet necrotic yellow vein virus*, another pecluvirus, did not cause systemic symptoms on transgenic plants with the p51 protein of PCV. Similarly, tobacco plants expressing RNA1 of CMV complement RNA2 and RNA3 in viral movement, thereby promoting long-distance movement, systemic infection, and virus multiplication (91,92). Complementation for movement is known also in the case of nontransgenic plants (5).

Synergism has also been reported for transgenic plants expressing the 5′ terminal region of potyviruses and for nontransgenic plants subject to mixed infection with a potyvirus (5).

So far, limited information on risk assessment is available from field experiments. Field experiments, however, are paramount to evaluate potential risks to the environment because they relate to agricultural practice and to natural dynamics of vector population. Also, field studies can be carried out under conditions of little to no selection pressure, which are critical when assessing environmental risks. In 1998 several transgenic lines of potato plants expressing either the CP gene or the replicase gene of PLRV were exposed to virus infection in the field over a 6-year period and tested for potential impact on transmission characteristics, serological properties, host range, and symptoms of challenge viruses (93). Results show that modified viruses with altered characteristics or novel viruses with new properties were not detected in field-exposed plants. Also, transgenic melon and squash expressing CP genes from aphid-borne viruses failed to mediate the transmission of an aphid-nontransmissible strain of CMV over a 2-year field study (94).

Virus-derived gene constructs can be transferred from virus-resistant transgenic crops into wild species as are any other conventional or engineered genes. We monitored the dispersal of CP genes from virus-resistant transgenic squash into a free-living relative that is commonly known as Texas gourd (95). Field experiments showed that the CP genes can provide a selective advantage to the wild squash if they are grown under intense disease pressure (Fuchs and Gonsalves, unpublished results).

These studies and others (96) suggest so far that transgenic crops expressing CP genes of aphid-transmissible viruses are likely to have little, if any, detectable environmental impact beyond those of natural background level (5).

## V. DISCUSSION AND FUTURE PROSPECTS

The past decade has witnessed an explosion in the development of virus-resistant crops. To a large extent these advances have been made possible through the application of the concept of PDR. The PDR strategy is a powerful approach to develop virus-resistant plants. A variety of PDR strategies (CP, replicase, movement protein, proteinase, satellite RNA, defective interfering RNA) have been used to develop virus resistance. The large majority of transgenic plants engineered for virus resistance express CP genes. A number of transgenic crops have been tested for virus resistance under field conditions. Some of them have even been commercialized and are deployed on a large scale in the United States and in the People's Republic of China.

Tremendous progress has been made toward understanding the underlying mechanisms of PDR in relation to viruses in plants. CP-mediated resistance is observed in some viral species, but recent evidence suggest that RNA-mediated resistance is the form of PDR. Further research is necessary to gain insights into the complexity of PTGS and engineered resistance to viruses. It will be interesting to identify features of RNA signals that influence the triggering of PTGS, its efficacy, and its maintenance. Such information will be valuable in designing virus-derived gene constructs that induce predictable resistance phenotypes and increased levels of sustainable resistance.

An approach for durable and effective pathogen-derived resistance against viruses is to combine different types of engineered resistance. To achieve broad resistance against different viruses or different virus strains, multiple-virus-derived gene constructs have been combined and used in a single plant line. For example, genes encoding nucleoproteins from three different strains of *Tomato spotted wilt virus* (TSWV) were combined in a single construct to engineer resistance against the three strains of TSWV (97). Similarly, the CP of PVY and PVX was used to engineer resistance to mixed infections by PVY and PVX (98). Also, the CP genes of WMV and ZYMV were combined to engineer resistance against these two viruses in squash (69). And the same approach was used to engineer resistance against CMV, WMV, and ZYMV in squash (69). To improve the degree and breadth of resistance, combining host-derived genes and virus-derived transgenes seems to be a strategy of choice (99).

A few virus-resistant transgenic crops have been commercially released. Given their efficacy at controlling viral disease, more virus-resistant transgenic crops are likely to reach the market in the near future. Virus-resistant transgenic crops offer numerous benefits to agriculture and the environment. Legitimate concerns have been expressed about the large-scale use of virus-resistant transgenic plants. Identifying potential risks and assessing their impact are necessary for the safe deployment of virus-resistant transgenic crops. Thus, science-based risk assessment studies are important to ascertain benefits versus risks and help regulatory agencies make decisions for the safe release of transgenic crops. So far, risk assessment studies performed in the field suggest that virus-resistant transgenic crops have limited detectable environmental impact. For example, there seems to be consensus that the benefits offered by virus-resistant transgenic plants outweigh negative consequences of evolution of novel hybrid viruses with destructive disease potential (5,100). The proper introduction and monitoring of virus-resistant transgenic crops, after due consideration and evaluation of potential environmental risks, are the safest approach to protect the environment. Should more risk assessment experiments be carried out with virus-resistant transgenic crops? Is there a point when broad conclusions can be drawn on their safety? Can consensus be achieved among countries on environmental safety issues? Focusing our efforts on understanding selection processes in nature should help extend our knowledge on the potential impact of virus-resistant transgenic crops.

## ACKNOWLEDGMENTS

We are grateful to Dr. L. M. Yepes for critical reading of the manuscript and to Dr. C. Schmitt for consideration.

## REFERENCES

1. JC Sanford, SA Johnston. The concept of parasite-derived resistance deriving resistance genes from the parasite's own genome. J Theor Biol 113:395–405, 1985.
2. DC Baulcombe. RNA as target and an initiator of post-transcriptional gene silencing in transgenic plants. Plant Mol Biol 32:79–88, 1996.
3. RN Beachy. Mechanisms and application of pathogen-derived resistance in transgenic plants. Curr Opin Plant Biotechnol 8:215–220, 1997.
4. RN Beachy. Coat-protein-mediated resistance to tobacco mosaic virus: Discovery mechanisms and exploitation. Philos Trans R Soc Lond B 354:659–664, 1999.
5. M Fuchs, D Gonsalves D. Genetic engineering. In: NA Rechcigl JE Rechcigl, eds. Environmentally safe approaches to crop disease control. Boca Raton, FL: CRC Lewis, 1997, pp 333–368.
6. GP Lomonossoff. Pathogen-derived resistance to plant viruses. Annu Rev Phytopathol 33:323–343, 1995.
7. CA Malpica, MT Cervera, C Simoens, M van Montagu. Engineering resistance against viral diseases in plants. Subcell Biochem 29:287–320, 1998.
8. ED Miller, C Hemenway. History of coat protein–mediated protection. Methods Mol Biol 81:25–38, 1998.
9. U Reimann-Philipp. Mechanisms of resistance: Expression of coat protein. Methods Mol Biol 81:521–532, 1998.
10. AP Powell Abel, RS Nelson, B De, N Hoffmann, SG Rogers, RT Fraley, RN Beachy. Delay of disease development in transgenic plants that express the tobacco mosaic virus coat protein gene. Science 232:738–743, 1986.
11. JA Lindbo, L Silva-Rosales, WM Proebsting, WG Dougherty. Induction of a highly specific antiviral state in transgenic plants: Implications for regulation of gene expression and virus resistance. Plant Cell 5:1749–1759, 1993.
12. P de Haan. Mechanisms of RNA-mediated resistance to plant viruses. Methods Mol Biol 81:533–546, 1998.
13. T van den Boogaart, GP Lomonossoff, JW Davies. Can we explain RNA-mediated resistance by homology-dependent gene silencing? Mol Plant Microbe Interact 11:717–723, 1998.
14. A Nejidat, RN Beachy. Transgenic tobacco plants expressing a coat protein gene of tobacco mosaic virus are resistant to some other tobamoviruses. Mol Plant Microl Interact 3:247–251, 1990.
15. M Zaitlin, JM Anderson, KJ Perry, L Zhang, P Palukaitis. Specificity of replicase-mediated resistance to cucumber mosaic virus. Virol 201:200–205, 1994.
16. JC III Register, RN Beachy RN. Resistance to TMV in transgenic plants results from interference with an early event in infection. Virology 166:524–532, 1988.
17. JK Osbourn, JW Watts, RN Beachy, TAM Wilson. Evidence that nucleocapsid disassembly and a later step in virus replication are inhibited in transgenic tobacco plants expressing the coat protein gene of tobacco mosaic virus. Virology 158:126–132, 1989.
18. M Bendahmane, JH Fitchen, G Zhang, RN Beachy. Studies of coat protein–mediated resistance to tobacco mosaic tobamovirus: Correlation between assembly of mutant coat proteins and resistance. J Virol 71:7942–7950, 1997.

19. DR Gallie. Controlling gene expressing in transgenics. Plant Biotech 1:166–172, 1998.
20. MA Matzke, AJM Matzke. Epigenetic silencing of plant transgenes as a consequence of diverse cellular defense responses. Cell Mol Life Sci 54:94–103, 1998.
21. JJ English, GF Davenport, T Elmayan, H Vaucheret, DC Baulcombe. Requirement of sene transcription for homology-dependent virus resistance and trans-inactivation. Plant J 12:597–603, 1997.
22. J Goodwin, K Chapman, S Swaney, TD Parks, EA Wernsman, WG Dougherty. Genetic and biochemical dissection of transgenic RNA-mediated virus resistance. Plant Cell 8:95–105, 1996.
23. DC Baulcombe, JJ English. Ectopic pairing of homologous DNA and post-transcriptional gene silencing in transgenic plants. Curr Opin Cell Biol 9:373–382, 1996.
24. JJ English, E Mueller, DC Baulcombe. Suppression of virus accumulation in transgenic plants exhibiting silencing of nuclear genes. Plant Cell 8:179–188, 1996.
25. T Sijen, J Wellink, JB Hiriart, A van Kammen. RNA-mediated virus resistance: Role of repeated transgenes and delineation of targeted regions. Plant Cell 8:2277–2294, 1996.
26. HA Smith, SL Swaney, TD Parks, EA Wernsman, WG Dougherty. Transgenic plant virus resistance mediated by untranslatable sense RNAs: Expression, regulation, and fate of nonessential RNAs. Plant Cell 6:1441–1453, 1994.
27. HS Guo, JJ Lopez-Moya, JA Garcia. Mitotic stability of infection-induced resistance to plum pox potyvirus associated with transgene silencing and DNA methylation. Mol Plant Microbe Interact 12: 103–111, 1999.
28. M Wassenegger, T Pelissier. A model for RNA-mediated gene silencing in higher plants. Plant Mol Biol 37:349–362, 1998.
29. MF Mette, J van der Winden, MA Matzke, AJM Matzke. Production of aberrant promoter transcripts contributes to methylation and silencing of unlinked homologous promoters in trans. EMBO J 18:241–248, 1999.
30. GJ van Eldik, K Litiere, JJMR Jacobs, M van Montagu, M Cornelissen. Silencing of β-1,3-glucanase genes in tobacco correlates with an increased abundance of RNA degradation intermediates. Nucleic Acids Res 26:5176–5181, 1998.
31. MM Tanzer, WF Thompson, MD Law, EA Wernsman, S Uknes. Characterization of post-transcriptionally suppressed expression that confers resistance to tobacco etch virus infection in tobacco. Plant Cell 9:1411–1423, 1997.
32. MR Marano, D Baulcombe. Pathogen-derived resistance targeted against the negative-strand RNA of tobacco msaic virus: RNA strand-specific gene silencing? Plant J 13:537–546, 1998.
33. S Sonoda, M Mori, M Nishiguchi. Homology-dependent virus resistance in transgenic plants with the coat protein gene of sweet feathery mottle potyvirus: Target specificity and transgene methylation. Phytopathology 89:385–391, 1999.
34. AI Jones, IE Johansen, SJ Bean, I Bach, AJ Maule. Specificity of resistance to pea seed–borne mosaic potyvirus in transgenic peas expressing the viral replicase (NIb) gene. J Gen Virol 79:3129–3137, 1998.
35. MT Ruiz, O Voinnet, DC Baulcombe. Initiation and maintenance of virus-induced gene silencing. Plant Cell 10:937–946, 1998.
36. M Moreno, JJ Bernal, I Jimenez, E Rodriguez-Cerezo. Resistance in plants transformed with the P1 and P3 gene of tobacco vein mottling potyvirus. J Gen Virol 79:2819–2927, 1998.
37. SZ Pang, FJ Jan, K Carney, J Stout, DM Tricoli, HD Quemada, D Gonsalves. Post-transcriptional transgene silencing and consequent tospovirus resistance in transgenic lettuce are affected by transgene dosage and plant development. Plant J 9:899–909, 1996.
38. P de Haan, JJL Gielen, M Prins, IG Wijkamp, A van Schepen, D Peters, MQJM van Grinsven, RW Goldbach. Characterization of RNA-mediated resistance to tomato spotted wilt virus in transgenic tobacco plants. Bio-technology 10:1133–1137, 1992.
39. HS Guo, JA Garcia. Delayed resistance to plum pox potyvirus mediated by a mutated RNA replicase gene: Involvement of a gene-silencing mechanism. Mol Plant Microbe Interact 10:160–170, 1997.
40. E Mueller, J Gilbert, G Davenport, G Brigneti, D Baulcombe. Homology-dependent resistance: Transgenic virus resistance in plants related to homology-dependent gene silencing. Plant J 7:1001–1013, 1995.

41. M Prins, RD Resende, C Anker, A Vanschepen, P de Haan, R Goldbach. Engineered RNA-mediated resistance to tomato spotted wilt virus is sequence-specific. Mol Plant Microbe Interact 9:416–418, 1996.
42. F Tenllado, I Garcia-Luque, MT Serra, JR Diaz-Ruiz. *Nicotiana behthamiana* plants transformed with the 54-kDa region of pepper mild mottle tobamovirus replicase gene exhibit two types of resistance reponses against viral infection. Virology 211:170–183, 1995.
43. PM Waterhouse, MW Graham, MB Wang. Virus resistance and gene silencing in plants can be induced by simultaneous expression of sense and antisense RNA. Proc Natl Acad Sci USA 95:13959–13964, 1998.
44. SN Covey, NS Al-Kaff, A Langara, DS Turner. Plants combat infection by gene silencing. Nature 385:781–782, 1997.
45. F Ratcliff, BD Harrison, DC Baulcombe. A similarity between viral defense and gene silencing in plants. Science 276:1558–1560, 1997.
46. FG Ratcliff, SA MacFarlane, DC Baulcombe. Gene silencing without DNA: RNA-mediated cross-protection between viruses. Plant Cell 11:1207–1215, 1999.
47. R Anandalakshmi, GJ Pruss, X Ge, R Marathe, A Mallory, TH Smith, VB Vance. A viral suppressor of gene silencing in plants. Proc Natl Acad Sci USA 95:13079–13084, 1998.
48. KD Kasschau, JC Carrington. A counterdefensive strategy of plant viruses: Suppression of posttranscriptional gene silencing. Cell 95:461–470, 1998.
49. DR Smyth. Gene silencing: Plants and viruses fight it out. Curr Biol 9:R100–R102, 1999.
50. C Beclin, R Berthome, JC Palauqui, M Tepfer, H Vaucheret. Infection of tobacco or *Arabidopsis* plants by CMV counteracts systemic post-transcriptional silencing of nonviral (trans)genes. Virology 252:313–317, 1998.
51. G Brigneti, O Voinnet, WX Li, LH Ji, SW Ding, DC Baulcombe. Viral pathogenicity determinants are suppressors of transgene silencing in *Nicotiana benthamiana*. EMBO J 17:6739–6746, 1998.
52. HW Li, AP Lucy, HS Guo, WX Li, LH Ji, SM Wong, SW Ding. Strong host resistance targeted against a viral suppressor of the plant gene silencing defense mechanism. EMBO J 18:2683–2691, 1999.
53. T Elmayan, S Balzergue, F Beon, V Bourdon, J Daubremet, Y Guenet, P Mourrain, JC Palauqui, S Vernhettes, T Vialle, K Wostrikoff, H Vaucheret. *Arabidopsis* mutants impaired in cosuppression. Plant Cell 10:1747–1757, 1998.
54. D Gonsalves. Control of papaya ringspot virus in papaya: A case study. Annu Rev Phytopath 36:415–437, 1998.
55. PE Thomas, WK Kaniewski, EC Lawson. Reduced field spread of potato leafroll virus in potatoes transformed with the potato leafroll virus coat protein gene. Plant Dis 81:1447–1453, 1997.
56. WK Kaniewski, PE Thomas. Field testing resistance of transgenic plants. Methods Mol Biol 81:497–507, 1998.
57. LM Kawchuk, DR Lynch, RR Martin, GC Kozub, B Farries. Field resistance to the potato leafroll luteovirus in transgenic and somaclone potato plants reduces tuber disease symptoms. Can J Plant Pathol 19:260–266, 1997.
58. KL Hefferon, H Khalilain, MG Abou-Haidar. Expression of the PVYO coat protein (CP) under the control of the PVX CP gene leader sequence: Protection under greenhouse and field conditions against PVYO and PVYN infection in three potato cultivars. Theor Appl Genet 94:287–292, 1997.
59. E Sivamani, H Huet, P Shen, CA Ong, A de Kochko, C Fauquet, RN Beachy. Rice plant (*Oryza sativa* L.) containing rice tungro spherical virus (RTSV) coat protein transgenes are resistant to virus infection. Mol Breed 5:177–185, 1999.
60. W Kaniewski, V Ilardi, L Tomassoli, T Mitsky, J Layton, M Barba. Extreme resistance to cucumber mosaic virus (CMV) in transgenic tomato expressing one or two viral coat proteins. Mol Breed 5:111–119, 1999.
61. JF Murphy, EJ Sikora, B Sammons, WK Kaniweski. Performance of transgenic tomatoes expressing cucumber mosaic virus CP gene under epidemic conditions. HortSci 33:1032–1035, 1998.
62. L Tomassoli, V Ilardi, M Barba, W Kaniewski. Resistance of transgenic tomato to cucumber mosaic cucumovirus under field conditions. Mol Breed 5:121–130, 1999.

63. JR Stommel, ME Tousignant, T Wai, R Pasini, JM Kaper. Viral satellite RNA expression in transgenic tomato confers field tolerance to cucumber mosaic virus. Plant Dis 82:391–396, 1998.
64. SJ Kim, SJ Lee, BD Kim, KH Paek. Satellite-RNA-mediated resistance to cucumber mosaic virus in transgenic plants of hot pepper (*Capsicum annuum* cv. Golden Tower). Plant Cell Rep 16:825–830, 1997.
65. GM Chowrira, TD Cavileer, SK Guptaa, PF Lurquin, PH Berger. Coat protein–mediated resistance to pea enation mosaic virus in transgenic *Pisum sativum* L. Transgenic Res 7:265–271, 1998.
66. YM Pinto, RA Kok, DC Baulcombe. Resistance to rice yellow mottle virus (RYMV) in cultivated African rice varieties containing RYMV transgenes. Nat Biotechnol 17:702–707, 1999.
67. M Ravelonandro, R Scorza, JC Bachelier, G Labonne, L Levy, V Damsteegt, AM Callahan, J Dunez. Resistance to transgenic *Prunus domestica* to plum pox virus infection. Plant Dis 81:1231–1235, 1997.
68. IL Ingelbrecht, JE Irvine, TE Mirkov. Posttranscriptional gene silencing in transgenic sugarcane: Dissection of homology-dependent virus resistance in a monocot that has a complex polyploid genome. Plant Physiol 119:1187–1197, 1999.
69. DM Tricoli, KJ Carney, PF Russell, JR McMaster, DW Groff, KC Hadden, PT Himmel, JP Hubbard, ML Boeshore, HD Quemada. Field evaluation of transgenic squash containing single or multiple virus coat protein gene constructs for resistance to cucumber mosaic virus, watermelon mosaic virus 2, and zucchini yellow mosaic virus. Biotechnology 13:1458–1465, 1995.
70. M Fuchs, D Gonsalves. Resistance of transgenic hybrid squash ZW-20 expressing the coat protein genes of zucchini yellow mosaic virus and watermelon mosaic virus 2 to mixed infections by both potyviruses. Biotechnology 13:1466–1473, 1995.
71. S Lius, RM Manshardt, MMM Fitch, JL Slightom, JC Sanford, D Gonsalves. Pathogen-derived resistance provides papaya with effective protection against papaya ringspot virus. Mol Breed 33:323–343, 1997.
72. GA de Zoeten. Risk assessment: Do we let history repeat itself? Phytopathology 81:885–886, 1991.
73. K Harding, PS Harris. Risk assessment of the release of genetically modified plants: A review. Agro Food Industry Hi-Tech 8:8–13, 1997.
74. R Hull. Virus resistant plants: Potential and risks. Chem Ind 17:543–546, 1990.
75. M Tepfer. Viral genes and transgenic plants: What are the potential environmental risks? Biotechnology 11:1125–1132, 1993.
76. M Tepfer, E Balazs. Virus-Resistant Transgenic Plants: Potential Ecological Impact. Berlin: Springer Verlag, 1997, p 126.
77. P Palukaitis. Virus-mediated genetic transfer in plants. In: M Levin, H Strauss, eds. Risk Assessment in Genetic Engineering. New York: McGraw-Hill, 1991, pp 140–162.
78. R Aaziz, M Tepfer. Recombination in RNA viruses and in virus-resistant transgenic plants. J Gen Virol 80:1339–1346, 1999.
79. RF Allison, WL Schneider, AE Greene. Recombination in plants expressing viral transgenes. Semin Virol 7:417–422, 1996.
80. JF Hancock, R Grumet, SC Hokanson. The opportunity for escape of engineered genes from transgenic crops. HortSci 31:1080–1085, 1996.
81. H Lecoq, M Ravelonandro, C Wipf-Scheibel, M Monsion, B Raccah, J Dunez. Aphid transmission of a non-aphid transmissible strain of zucchini yellow mosaic potyvirus from transgenic plants expressing the capsid protein of plum pox potyvirus. Mol Plant Microbe Interact 6:403–406, 1993.
82. C Jacquet, B Delecolle, B Raccah, H Lecoq, J Dunez, M Ravelonandro. Use of modifed plum pox virus coat protein genes developed to limit heteroencapsidation-associated risks in transgenic plants. J Gen Virol 79:1509–1517, 1998.
83. AE Greene, RF Allison. Recombination between viral RNA and transgenic plant transcripts. Science 263:1423–1425, 1994.
84. AE Greene, RF Allison. Deletions in the 3' untranslated region of cowpea chlorotic mottle virus transgene reduce recovery of recombinant viruses in transgenic plants. Virology 225:231–234, 1996.
85. L Kiraly, JE Bourque, JE Schoelz. Temporal and spatial appearance of recombinant viruses formed between cauliflower mosaic virus (CaMV) and CaMV sequences present in transgenic *Nicotiana bigelovii*. Mol Plant Microbe Interact 11:309–316, 1998.

86. M Borja, T Rubio, HB Scholthof, AO Jackson. Restoration of wild-type virus by double recombination of tombusvirus mutants with a host transgene. Mol Plant Microbe Interact 12:153–162, 1999.
87. T Frischmuth, J Stanley. Recombination between viral DNA and the transgenic coat protein gene of African cassava mosaic geminivirus. J Gen Virol 79:1265–1271, 1998.
88. A Gal-On, E Meiri, B Raccah, V Gaba. Recombination of engineered defective RNA species produces infective potyvirus in planta. J Virol 72:5268–5270, 1998.
89. C Masuta, S Ueda, M Suzuki, I Uyeda. Evolution of a quadripartite hybrid virus by interspecific exchange and recombination between replicase components of two related tripartite RNA viruses. Proc Natl Acad Sci USA 95:10487–10492, 1998.
90. M Erhardt, E Herzog, E Lauber, C Fritsch, H Guilley, G Jonard, K Richards, S Bouzoubaa. Transgenic plants expressing the TGB1 protein of peanut clump virus complement movement of TGB1-defective peanut clump virus but not of TGB1-defective beet necrotic yellow vein virus. Plant Cell Rep 18:614–619, 1998.
91. M Suzuki, C Masuta, Y Takanami, S Kuwata. Resistance against cucumber mosaic virus in plants expressing the viral replicon. FEBS Lett 379:26–30, 1996.
92. T Canto, P Palukaitis. Transgenically expressed cucumber mosaic virus RNA1 simultaneously complements replication of cucumber mosaic virus RNAs 2 and 3 and confers resistance to systemic infection. Virol 250:325–336, 1998.
93. PE Thomas, S Hassan, WK Kaniewski, EC Lawson, JC Zalewski. A search for evidence of virus/transgene interactions in potatoes transformed with the potato leafroll virus replicase and coat protein genes. Mol Breed 4:407–417, 1998.
94. M Fuchs, FE Klas, JR McFerson, D Gonsalves. Transgenic melon and squash expressing coat protein genes of aphid-borne viruses do not assist the spread of an aphid non-transmissible strain of cucumber mosaic virus in the field. Transgenic Res 7:449–462, 1998.
95. M Fuchs, D Gonsalves. Risk assessment of gene flow associated with the release of virus-resistant transgenic crop plants. In: M Tepfer, E Balazs, eds. Virus-resistant transgenic plants: Potential ecological impact. Berlin: Springer Verlag, 1997, p 114–120.
96. M Fuchs, A Gal-On, B Raccah, D Gonsalves. Epidemiology of an aphid nontransmissible potyvirus in fields of nontransgenic and coat protein transgenic squash. Transgenic Res 8:429–439, 1999.
97. M Prins, P de Haan, R Luyten, M van Veller, MQJM van Ginsven, R Goldbach. Broad resistance to tospoviruses in transgenic tobacco plants expressing the tospoviral nucleoprotein gene sequences. Mol Plant Microbe Interact 8:85–91, 1995.
98. C Lawson, W Kaniewski, L Haley, R Rozmann, C Newall, P Sanders, NE Tumer. Engineered resistance to mixed virus infection in a potato cultivar: Resistance to potato virus X and potato virus Y in transgenic Russet Burbank. Biotechnology 8:127–134, 1990.
99. A Palucha, W Zagorski, M Chrzanowska, D Hulanicka. An antisense coat protein gene confers immunity to potato leafroll virus in a genetically engineered potato. Eur J Plant Pathol 104:287–293, 1998.
100. T Rubio, M Borja, HB Scholthof, AO Jackson. Recombination with host transgenes and effects on virus evolution: An overview and opinion. Mol Plant Microbe Interact 12:87–92, 1999.

# 16
# Genetic Engineering for Resistance to Nematodes

**Daguang Cai, Urs Wyss, Christian Jung, and Michael Kleine**
*University of Kiel, Kiel, Germany*

| | | |
|---|---|---|
| I. | INTRODUCTION | 233 |
| II. | LIFE CYCLE AND INFECTIOUS PROCESS OF NEMATODES | 234 |
| | A. Cyst Nematodes | 234 |
| | B. Root Knot Nematodes | 235 |
| III. | NATURAL RESOURCES OF RESISTANCE GENES | 236 |
| | A. Natural Resistance | 236 |
| | B. Marker Assisted Selection and Gene Cloning | 236 |
| | C. Structure and Function of Nematode Resistance Genes | 242 |
| IV. | ARTIFICIAL NEMATODE RESISTANCE SYSTEMS | 242 |
| | A. Effector Genes | 242 |
| | B. Disrupture of Feeding Cells | 244 |
| V. | CONCLUSION | 244 |
| | REFERENCES | 244 |

## I. INTRODUCTION

Nematodes play an important role as parasites of humans, animals, and plants. In agriculture the economic losses caused by plant-parasitic nematodes worldwide are estimated to amount to ~U.S.$77 billion a year (1). Typically, plant-parasitic nematodes have a highly diversified range of plant parasitism (2,3). Some spend their whole life cycle outside the root, feeding on the surface (browsing ectoparasites) or deeper tissues (sedentary ectoparasites); others have evolved the capability to invade the root and to feed from cortical (migratory endoparasites) or stelar cells (sedentary endoparasites). Economically most relevant are sedentary endoparasites of the genera *Heterodera* and *Globodera* (cyst nematodes), and of the genus *Meloidogyne* (root-knot nematodes). They represent the most advanced level of root parasitism as they induce and maintain specific nurse cell structures as a continuous source of food for development and reproduction. Agronomically important species of the cyst nematodes, common mainly in temperate regions of the world, are *G. rostochiensis* and *G. pallida* on potato, *H. glycines* on soybean, *H. schachtii* on sugar beet, and *H. avenae* on cereals. In contrast, root-knot nematodes, with *M. incognita* as one of the most important representative, have a very broad host range and are adapted to warm and hot climates.

Nematodes can be controlled by crop rotation, by fumigation with nematicides, or by growing of resistant crops. Wide crop rotation is difficult to achieve and chemical control, because of its environmental and toxicological hazards, is opposed by increasing limitations. Therefore, breeding of resistant varieties offers the most promising alternative. In the past, breeders have successfully introduced nematode resistance into crop species, often by species hybridization with wild relatives. New genetic variability is needed because of the lack of resistance genes in crop species (e.g., sugar beet) and also because of pathotypes that break commonly used monogenic resistance. Molecular markers have gained increasing importance for the introduction of resistance genes into valuable breeding material and for positional cloning of resistance genes. Recent progress in molecular cloning of nematode resistance genes will promote our understanding of host-pathogen interaction and plant-specific defense. In addition, this will open new avenues to the genetic engineering of resistance to nematodes by using either natural or artificial resistance genes.

Here, we present a description of the life cycles of cyst and root-knot nematodes. Furthermore, a summary of molecular cloning procedures of naturally occurring nematode resistance genes is given and alternative approaches for the generation of artificial resistance are discussed.

## II. LIFE CYCLE AND INFECTION PROCESSES OF NEMATODES

### A. Cyst Nematodes

Cyst nematodes invade plant roots as infective second-stage juveniles (J2) that hatch from eggs retained in the protective cyst of the dead female. These juveniles are equipped with a robust stylet, by means of which they invade the root and migrate to the differentiating vascular cylinder, where they finally induce and maintain multinucleate syncytia that arise from expanding cambial cells whose protoplasts fuse after partial cell wall dissolution. Fully developed syncytia, maintained by egg-producing females, can contain more than 200 integrated cells representing a large nurse cell unit with metabolically highly active cytoplasm (4–7).

The majority of cyst nematode species reproduce by cross-fertilization. Sex determination is most likely controlled by the amount and quality of food available after syncytium induction (5,7). Much more food is required by a female than by a male juvenile; consequently the volume of syncytia maintained by males is considerably smaller than that by females (Fig. 1). Male nematodes feed only until the end of the third developmental stage (Fig. 1). They become vermiform again while they molt to the J4 stage, and after the last molt they emerge through the juvenile cuticles in search of females that attract them with sex pheromones for copulation (Fig. 1).

The molecular triggers involved in feeding-site induction are not yet known. It is, however, generally believed that secretory proteins released through the nematode's stylet orifice play a decisive role in the induction process. Obviously, nematodes use their stylets not only to cut slits into the cell walls, through which they pass into neighboring cells (8), but also to assist hatching and root penetration. Smant and coworkers (9) reported for the first time that the J2 of the potato cyst nematodes synthesize $\beta$-1,4-endoglucanases in the two subventral glands that are secreted through the stylet. These cellulases are thought to soften root cell walls and thus facilitate intracellular migration.

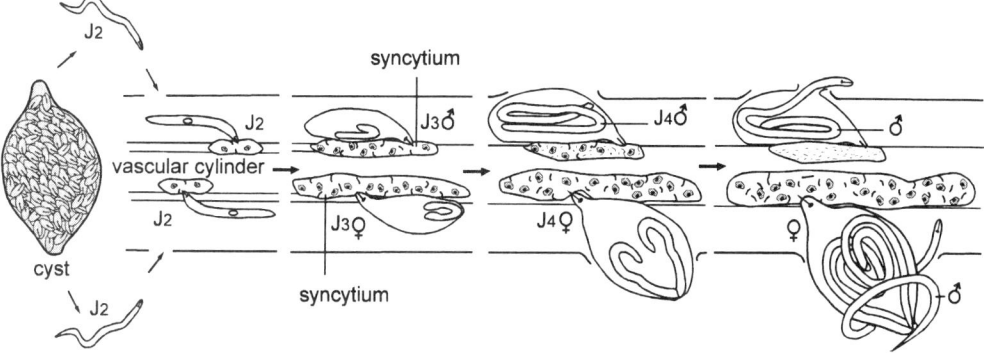

**Figure 1** Schematic representation of the life cycle of a cyst nematode.

## B. Root-Knot Nematodes

Most root-knot nematodes reproduce by mitotic parthenogenesis (10). In contrast to cyst nematodes with generally a restricted host range, root diffusates are not involved in the hatching process of the infective root-knot J2 juveniles, which are retained in a gelatinous egg sac produced by the females (11). However, as in cyst nematodes, these compounds attract and direct the J2 juveniles to the root tip of their host plant, where they generally enter the root. Inside the root the juveniles migrate intercellularly to the vascular cylinder, where they become sedentary after nurse cell induction. In *Meloidogyne* species these cells develop by the rapid expansion of about half a dozen cambial cells that develop a multinucleate state by repeated synchronous mitoses in the absence of cytokinesis. Mature giant cells are metabolically highly active, increased deoxyribonucleic acid (DNA) content of their nuclei is caused by alterations during the cell cycle (12,13).

Differentiation of giant cells is accompanied by a pronounced galling of the surrounding root tissue (Fig. 2) while pericyle and cortex cells enlarge and divide. The J2 feed for several days from the expanding giant cells, become saccate, and finally stop feeding at the end of the J2 stage. After that they enter an expanded molting cycle (Fig. 2), in which they molt three times in succession to adult females. The females resume feeding from the giant cells, which now function as xylem-related transfer cells and are also supplied with nutrients from the phloem (14). They are metabolically highly active to provide the females with sufficient food to produce many hundred

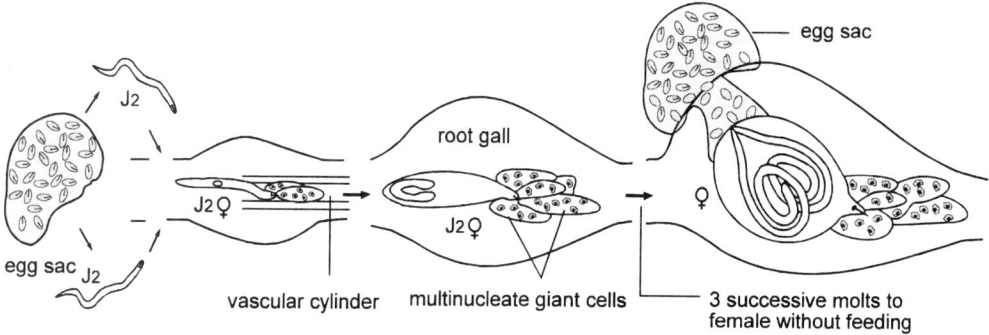

**Figure 2** Schematic representation of the life cycle of a root-knot nematode.

eggs. Under adverse nutritional conditions, feeding J2 of mitotically parthenogenetic species undergo complete or partial sex reversal and develop as males.

The behavior of *M. incognita* J2 from root invasion until giant cell induction has been documented inside roots of *Arabidopsis thaliana* with the aid of video-enhanced contrast light microscopy and time lapse studies (15). Hatched J2 usually invade in the region of elongation close to the meristematic zone. The walls of epidermal and subepidermal cells are weakened and finally destroyed by continuous head rubbings and stylet movements followed by metacorpal bulb pumpings of a few seconds in duration. This behavior indicates that wall degrading enzymes might be involved in root invasion. Support is provided by the detection of a novel cellulose binding protein (MI-CBP-1), immunolocalized in the subventral glands and secreted through the stylet of preparasitic and parasitic J2s of *M. incognita* (16).

## III. NATURAL RESOURCES OF RESISTANCE GENES

### A. Natural Resistance

Resistance to nematodes is described as the ability of host plants to restrict or prevent nematode reproduction. Host-parasite-specific defense reactions often follow the "gene-for-gene" relationship. However, in nematology it is difficult to locate and to characterize virulence genes because of the parasitic features of the nematode, which completes its life cycle within the host, making controlled crosses and analysis of progeny functionally impossible.

Different mechanisms of plant-specific defense responses conferring resistance to root-knot and cyst nematodes have been reported. One principle is based on a hypersensitive response of the root tissue during nematode invasion that leads to the death of the nematode (17). The second feature of resistance against nematodes that does not protect the plants from nematode invasion is that the induction of the feeding site inside the root is inhibited or that initially established feeding structures desintegrate in early stages of nematode development. Resistance to root-knot nematodes of the genus *Meloidogyne*, mediated by a single dominant gene *Mi* that was introduced into cultivated tomato (*Lycopersicon esculentum*) from its wild relative *L. peruvianum*, is characterized by an immediate, localized necrosis or hypersensitive response of the cells serving as feeding structures for the nematode (18). In contrast, *H1*-mediated resistance in potato against the cyst nematode *Globodera rostochiensis* is accomplished by necrosis of cells surrounding the induced feeding structure, which leads to the isolation and final breakdown of the feeding site (19,20). A third type of feeding cell disruption can be observed in sugar beet (*Beta vulgaris*) conferring resistance to the beet cyst nematode *Heterodera schachtii*. The resistance gene $Hs1^{pro-1}$ (21) had been introduced into sugar beet from the wild beet *B. procumbens* (22). The resistance response has been studied on the cellular and ultrastructural levels. J2 juveniles are able to invade the root of resistant plants and proceed to the vascular cylinder to induce formation of syncytia. However, syncytia do not develop regularly, suffering from the formation of specific membrane aggregations that condense to distinct bodies filling large parts of the syncytium, consequently causing the degradation of the syncytia and the death of the juveniles (23).

### B. Marker-Assisted Selection and Gene Cloning

Genes for nematode resistance have been introduced into elite breeding lines either from the gene pool of the cultivated species or from related wild species. Molecular markers with tight genetic linkage to the gene provide a means to accelerate the selection procedure (Table 1). In addition, molecular markers have been used as tools to identify and isolate resistance genes by the application of positional cloning strategies. The subsequent transfer of isolated resistance genes to

**Table 1** Mapping and Cloning of Genes for Nematode Resistance from Crop Species

| Plant species | Nematode | Resistance gene | Marker system[a] | Cloning strategy | Refs. |
|---|---|---|---|---|---|
| Potato | *Globodera rostochiensis* | *Gro1* | RFLP[a], AFLP[b], and RAPD[c] | PCR-based resistance gene tagging | 24 |
| | *G. rostochiensis* | *H1* | RFLP and SCAR[d] | | 25 |
| | *G. pallida* | *Gpa2* | RFLP and AFLP | Positional cloning (the gene has been cloned) | 26 61 |
| Soybean | *Heterodera glycines* | *rhg1* | RFLP | Positional cloning | 27 |
| | *H. glycines* | *Rhg4* | RFLP | Positional cloning | 28 |
| Barley | *H. avenae* | *Ha2, Ha3* | RFLP | | 29 |
| | *H. avenae* | *Ha4* | RFLP | | 30 |
| Wheat | *H. avenae* | *Cre1* | RFLP, RAPD, and STS[e] | | 31 32 |
| | *H. avenae* | *Cre3* | RFLP | Positional cloning | 33 |
| Rye | *H. avenae* | *CreR* | RFLP | | 34 |
| Peanut | *Meloidogyne arenaria* | *Mae, Mag* | RAPD, SCAR, and RFLP | | 35 |
| Sugar beet | *H. schachtii* | *Hs1$^{pro-1}$* | RFLP, RAPD, and AFLP | Positional cloning (the gene has been cloned) | 21 |
| | | *Hs2* | RFLP | | 36 |
| Tomato | *M. incognita* | *Mi-1* | RFLP, RAPD, and AFLP | Positional cloning (the gene has been cloned) | 37 |
| | | *Mi-3* | RAPD and RFLP | | 38 |
| | *G. rostochiensis* | *Hero* | RFLP | Positional cloning | 39 |
| Tobacco | *M. incognita* | *Rk* | RAPD | | 40 |

[a]RFLP, restriction fragment length polymorphism; AFLP, amplified fragment length polymorphism; RAPD, random amplified polymorphic DNA; SCAR, sequence characterized amplified region; STS, sequence tagged site; PCR, polymerase chain reaction.
*Source*: Modified from C. Jung et al., 1998 (80).

breeding lines will enlarge the genetic base for the breeding material and open new ways in the breeding procedure, e.g., the combination of different resistance genes to generate durable resistant crops. Until now, three genes for nematode resistance have been cloned from their chromosomal position and further characterized by genetic complementation (Table 1).

### 1. Sugar Beet

The beet cyst nematode is a major pest of sugar beet. Genes for nematode resistance are lacking in the gene pool of cultivated beet species; however, complete resistance has been reported from the wild beet species *B. procumbens, B. patellaris,* and *B. webbiana* (41). A single dominant gene $Hs1^{pro-1}$ was introduced into sugar beet from *B. procumbens* chromosome 1 via species hybridizations and backcrossing (42,43). Evidence was given from isozyme (27) and molecular marker (44) analysis that at least one more resistance gene is present on chromosome 7 of *B. procumbens*. Meanwhile, a virulent pathotype of *H. schachtii* was found that is able to overcome the resistance from chromosome 1 ($Hs1^{pro-1}$) but not from chromosome 7 of *B. procumbens* (45), indicating a gene-for-gene relationship. The $Hs1^{pro-1}$ gene has been mapped to a complex wild beet translocation at the end of sugar beet chromosome IX (36).

For cloning of the gene a novel approach has been applied by the use of genome-specific satellite markers and chromosomal breakpoint analysis. A YAC-contig spanning the $Hs1^{pro-1}$ locus has been generated (46) and three different complementary deoxyribonucleic acids (cDNAs) were isolated with the aid of the YAC clones. Finally, one cDNA (1832) corresponding to the $Hs1^{pro-1}$ gene was identified by genetic complementation in roots of susceptible beet under the control of the *CaMV35S* promotor. The same incompatible reaction as in resistant plants was observed (Fig. 3A,21), demonstrating for the first time the potential of natural resistance genes for breeding resistant crops. Southern analysis revealed that only one copy of the $Hs1^{pro-1}$ sequence was present in the wild beet genome and this sequence was completely absent from the genome of cultivated beet species (Fig. 3B). The gene was found to be active in root tissue only with a slightly enhanced expression upon nematode infection (Fig. 3C). Sequence analysis of the $Hs1^{pro-1}$ gene revealed an open reading frame of 846 base pairs encoding a gene product of 282 amino acids. The predicted polypeptide can be dissected into different subdomains (Fig. 4) with motifs common to resistance gene products recently cloned (Fig. 5).

### 2. Tomato

The second nematode resistance gene that has been cloned is *Mi,* which had been introduced from the wild species *L. peruvianum* into cultivated tomato by conventional breeding. It is responsible for the hypersensitive reaction of tomato root cells after infection with *Meloidogyne* spp. (*M. incognita, M. javanica,* and *M. arenaria*). This resistance proved to be durable and is present in all modern tomato cultivars. The gene has been tightly linked to restriction fragment length polymorphism (RFLP) (37), random amplified polymorphic DNA (RAPD), and amplified fragment length polymorphism (AFLP) marker loci and cloned from its position on tomato chromosome 6. Genetic complementation using either whole cosmids or the genomic DNA, including the entire *Mi*-coding sequence and its own regulatory region, resulted in resistant plants (47,48). Surprisingly, *Mi* also confers resistance to a totally unrelated parasite, the potato aphid *Macrosiphum euphorbiae* (49). The gene is located within a cluster of at least eight genes and has features typical of previously cloned disease resistance genes, such as a nucleotide binding site and a leucine-rich region. However, there is little similarity in the gene products of *Mi* and $Hs1^{pro-1}$. Another candidate for positional cloning is the *Hero* gene from *Lycopersicon pimpinellifolium*, a wild relative of tomato. It confers resistance to *G. rostochiensis* and was mapped in tomato with tightly linked RFLP markers. Meanwhile, YACs have been isolated from the gene-bearing region. This

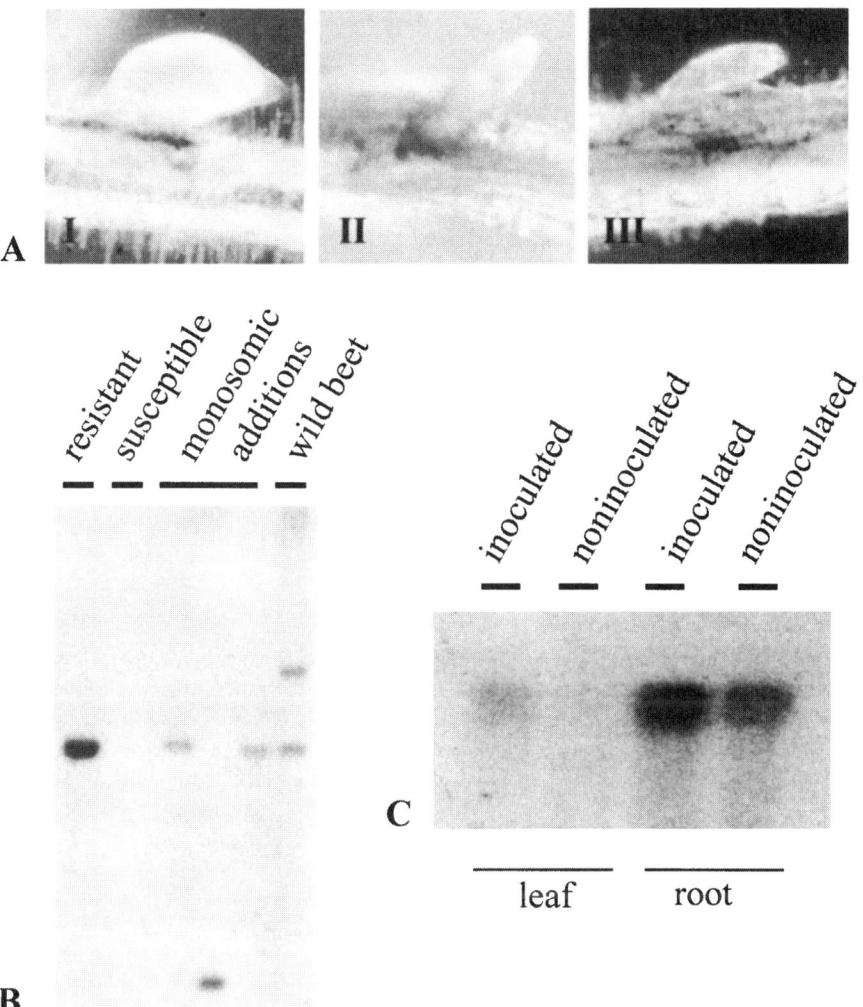

**Figure 3** Functional analysis of the $Hs1^{pro1}$ sequence. (A) To identify the resistance gene, genetic complementation was performed in susceptible sugar beet hairy roots by use of *A. rhizogenes*–mediated transformation: I, 3–4 Weeks after inoculation with *H. schachtii* a compatible reaction as indicated by a fully developed female on the susceptible control line 93161p; II, incompatible reaction 6 weeks after inoculation as indicated by stagnating female on the resistant control line A906001; III, the transgenic line containing cDNA 1832 exhibited the same resistance reaction as the resistant line. (B) Southern and (C) Northern analysis of cDNA 1832 (21).

gene is also of interest for potato breeding because its resistance mechanism is different from that of any of the previously described genes of potato (39).

## 3. Potato

Breeding for nematode resistance in potato has a long tradition. Major loci as well as quantitative trait loci (QTL) for polygenic resistance have been mapped with molecular markers (50,51). The

| A | 1   | MRRCGYSLGLGEPNLDG |                |
|---|-----|-------------------|----------------|
| B | 18  | KPNLDYDAVCRPS     |                |
| C | 31  | ELHALKKGALDYIQNSENQI |             |
|   | 51  | LFTIHQIFESWIFSSKK L |              |
|   | 69  | LDRISE RISKEE FTKAA | DDCW         |
|   | 90  | ILEKIWKL LEEIENLHL L | MDPDD       |
|   | 113 | FLH LKTQ LRMKT VADSE | TFCFRSK     |
|   | 137 | GLIEVTKLSKD LRHKVPKI |              |
|   | 156 | LG VE VDPMGGPVIQESA | MELY         |
|   |     | xLxxaxxaxLxxLxxaxxxL | (Konsensus) |
| D | 177 | REKRRYEK          |                |
| E | 185 | IHLLQAFQGVESAVK   |                |
| F | 200 | GFFFNYKQLLVIMMGSL |                |
| G | 217 | EAKANFAVIGGSTESSDLLAQLFLEPTYYP |   |
|   | 247 | SLDGAKTFIGDCWEHDQAVGSGLDC |        |
| H | 272 | RHHRK<u>NRT</u>AKQ |               |

**Figure 4** The predicted amino acid (aa) sequence of cDNA 1832. The aa sequence of the $Hs1^{pro-1}$ gene corresponds to a 33-kDa (282-aa) protein. Numbers indicate aa position. The polypeptide can be dissected into eight domains (A–H): (**A**) putative signal peptide, (**B**) a subdomain of no specified features, (**C**) a leucine-rich region consisting of 24% leucines/isoleucines arranged into seven imperfect repetitive units (LRRs) of 20 aa contributing to 63% of all leucines in the predicted protein, (**D**) a hydrophilic region, (**E**) a subdomain of no specified features, (**F**) a hydrophobic region (24) with a predicted α-helical secondary structure suggestive of a transmembrane span. The charged aa (Lys, Glu) flanking the hydrophobic domain is typical for transmembrane segments. (**G**) A subdomain with no specified features and (**H**) a basic C-terminal tail with a putative N-glycosylation signal (underlined).

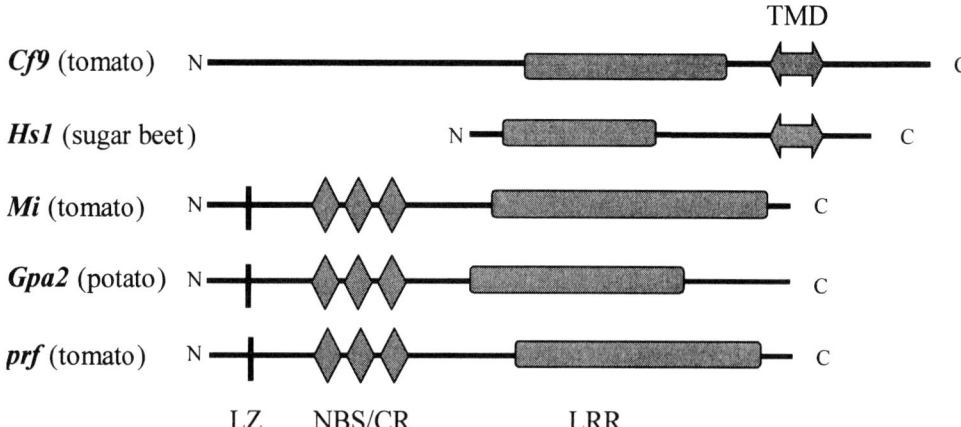

**Figure 5** Comparison of plant nematode resistance gene products with the plant resistance gene products of *Cf9* and *prf*. LZ, leucine zipper; NBS, nucleotide binding site; LRR, leucine-rich repeat; TMD, transmembrane domain; CM, conserved region.

gene *Gro1*, conferring resistance to all pathotypes of the root-cyst nematode *Globodera rostochiensis*, has been fine-mapped with RFLP, AFLP, and RAPD markers (24,52). There is evidence that *Gro1* is among those sequences that have been amplified with the help of primers derived from sequence motifs conserved between different resistance genes that have been previously cloned (25). At least five copies of these sequences carrying a nucleotide binding site (NBS) motif are located at the *Gro1* locus. However, the final proof will be given by genetic complementation experiments with transgenic potato, which are on the way (C. Gebhardt, personal communication). For use in breeding programs, polymerase chain reaction (PCR) assays diagnostic for RFLP marker alleles closely linked to alleles of *Gro1* and *H1*, a second resistance locus, have been developed (26). The function of *H1* occurs according to the gene-for-gene model of resistance, leading to necrosis around the feeding site.

The third nematode resistance gene (*Gpa2*) has recently been cloned (61). This gene confers resistance to certain pathotypes of the the potato cyst nematode *G. pallida* and was mapped to the same 6 cM genetic interval on chromosome 12 of potato as the virus resistance gene *Rx* (53,54). A PCR-based screening of four overlapping BAC clones spanning the *Rx/Gpa2* interval led to the identification of four candidate resistance gene homologues (RGH1–4), which were selected for complementation analysis in a susceptible potato genotype. Plants transformed with one of the four homologues showed the same incompatible interaction with *G. pallida* as the resistant control plants. This particular homologue was therefore designated as *Gpa2*. The *Gpa2* polypeptide consisting of 912 amino acids is a member of the leucine zipper, nucleotide binding, leucine-rich repeat family of plant genes. *Gpa2* and *Rx* share a high degree of homology (61).

## 4. Other Important Crop Species

The soybean cyst nematode (SCN), (*Heterodera glycines*) is economically the most important pathogen of soybean. Different types of resistance to *H. glycines* were mapped with molecular markers (28,55). A major partial resistance locus (*rhg1*) is located on chromosome G of soybean, explaining up to 50% of the genetic variation (27). A BAC contiguous segment of DNA (contig) was constructed around this locus with the help of tightly linked markers (56), making *rhg1* a candidate as the first nematode resistance gene to be cloned from this species. Closely linked markers are also available for the second resistance gene, *Rhg4*, on linkage group A of soybean, and these were used to identify clones from a BAC library (B. F. Matthews, unpublished). The Southern root-knot nematode (*Meloidogyne incognita*) also belongs to the major group of pathogens of soybean, but resistance is inherited in a quantitative manner. Two major QTLs for resistance to this nematode have been mapped with RFLP markers (57).

In some regions of the world, such as Australia, wheat and barley crops suffer heavily from infection with the cereal cyst nematode (CCN) *Heterodera avenae*. In barley, the nematode resistance loci *Ha1* and *Ha2* (allelic to *Ha3*) have been mapped to chromosome 2 (29), a new gene, *Ha4*, was mapped to chromosome 5 (30). In wheat, two loci were mapped with RFLP, RAPD, and sequence tagged site (STS) markers: the *Cre1* locus on chromosome 2B (31) and *Cre3* (*Ccn-D1*) from *Triticum tauschii* (32). In 1997 a gene originating from the *Cre3* locus that has similar domains to known resistance genes was isolated (33).

In peanut (*Arachis hypogaea*), the root-knot nematode *Meloidogyne arenaria* causes major problems. Two dominant genes, *Mae* and *Mag*, from *Arachis cardenasii*—reducing nematode egg number or restricting galling—have been tightly linked to RAPD, sequence characterized amplified region (SCAR), and RFLP loci (35). A marker-based selection program was started for introgression of these genes into elite lines. In *Arabidopsis* sp., which serves as a model system for studying the interactions between beet cyst nematode and its host (58), no resistance genes have been found in spite of intense screening studies.

## C. Structure and Function of Nematode Resistance Genes

The cloning of different nematode resistance genes demonstrated that host-parasite-specific defense reactions fit into the gene-for-gene model. But the molecular mechanisms responsible for the nematode resistance of the gene-for-gene type function more broadly than expected. The cloned nematode resistance genes *Mi* and *Gpa2* belong to the cytoplasmically located R genes sharing some common structural motifs, such as a nucleotide-binding site and a leucine-rich repeat region, that are characteristic of a family of plant disease-resistance genes against viruses, bacteria, and fungi (Fig. 5). *Mi* shares considerable sequence homology to the *prf* gene of tomato and to the *Rpm1* gene from *Arabidopsis* sp., both required for resistance to *Pseudomonas syringae*, and to two fungal resistance genes *I2C-1* and *I2C-2*. The *Mi* gene is able to trigger resistance to two unrelated parasites, a nematode (*Meloidogyne* spp.) and an aphid (*Macrosiphum euphorbiae*). Although *Gpa2* and *Rx* confer resistance to unrelated pathogens, *G. pallida* and potato virus X, they seem to have a common ancestral gene as they are members of the same gene cluster with a high degree of homology. It seems reasonable that these genes, although conferring resistance to different pathogens, might share components of downstream reactions in the resistance response, e.g., the signal transduction pathway. In contrast, the $Hs1^{pro-1}$ gene encodes a 282 protein with a putative N-terminal extracellular leucine-rich region (LRR), followed by a membrane-spanning domain and a C-terminal region, assuming that the protein may be anchored in the cell membrane and act as a receptor binding elicitors from the nematode (Fig. 5). Alternatively, it may be located within the cytoplasm, binding exudates of the nematode delivered into the cell via the stylet. However, the protein domains of the $Hs1^{pro-1}$ gene product described here share only weak homologies to other consensus features often found in plant R genes. In addition, $Hs1^{pro-1}$ is not a member of a gene family, with only one copy present in the haploid wild beet genome. Homologues were identified only in two resistant wild species, *B. patellaris* and *B. webbiana*, with 93% and 96% sequence homology, respectively (44). Related sequences are also identified in cultivated beet and *A. thaliana*. However, the function of these sequences is unknown. These findings imply that $Hs1^{pro-1}$ probably represents a new class of a disease-resistance gene.

## IV. ARTIFICIAL NEMATODE RESISTANCE SYSTEMS

Transgenic approaches offer the opportunity to pursue alternative strategies for establishing nematode resistance in plant species. The aim of artificial resistance (59,60) is to generate a durable and efficient system that controls a broad range of plant-parasitic nematodes. This can be achieved by introducing effector genes into the host plant that have a direct nematicidal impact or substances that cause the breakdown of specific feeding structures (Fig. 6).

### A. Effector Genes

The efficient application of antinematode genes to generate artificial resistance in the plant relies on the specificity for the nematode, proper and sufficient expression at the target site, and nonphytotoxicity to the host. Thus, the employment of compounds produced by the plant itself will be the most promising. In this respect, proteinase inhibitors (PIs), often found in plants in response to wounding or herbivory, have been investigated for their nematicidal activity. Cysteine proteinases are involved in protein metabolism and digestion of dietary protein and could be shown to be present in nematodes (62,63). They are known to be inhibited by cystatins, which are small PIs. Therefore, the oryzacystatin I gene from rice was modified by mutagenesis to improve

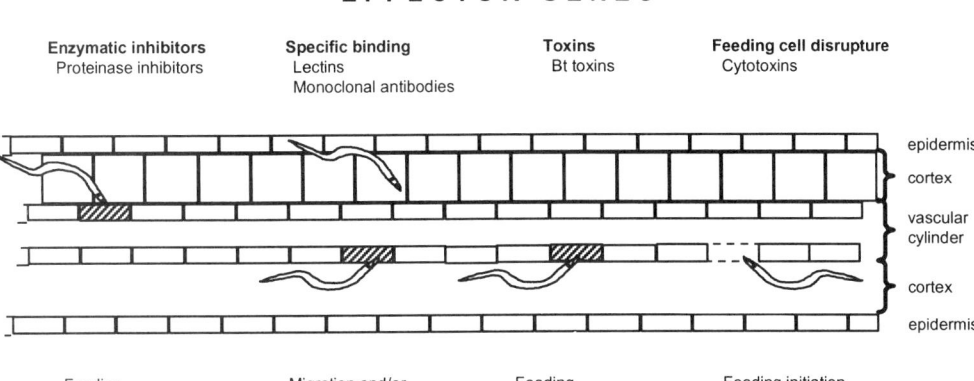

**Figure 6** Schematic presentation of the root tissue invaded by infective juveniles of stage J2 showing the impact of different effectors against root parasitic nematodes of the genera *Heterodera*, *Globodera*, and *Meloidogyne*. Hatched cells indicate feeding sites selected by the nematode (61).

the efficacy of the inhibitor and was subsequently introduced and expressed, under the control of the *CaMV35S* promotor, into tomato hairy roots and *Arabidopsis* sp. plants. The expression of cystatin in the transgenic plants prevented female nematodes from developing properly, causing reduced size and fecundity (64). In addition, dual proteinase inhibitor constructs have been designed to enhance resistance (65). The advantage of this type of approach is that because the PIs are nontoxic, they can be expressed in every cell—leading to the control of a broad range of nematode species. However, the evolution of PI-tolerant nematodes has to be taken into consideration.

Lectins are another class of putative antinematode proteins. These carbohydrate-binding proteins could be targeted to interact with the nematode at different sites: within the intestine, the surface coat (66), or with amphidial secretions (which would mean that the chemosensory perception of the nematode and therefore the ability to orientate within the root was disturbed) (67). The gene encoding the snowdrop (*Galanthus nivalis*) lectin GNA was introduced into potato plants and expressed at a maximum of approximately 0.5% of total root protein; a reduction of *G. pallida* females of up to 80% resulted. In contrast, a higher expression of GNA resulted in an increased number of females, indicating that only a critical level of GNA expression might have an effect on the nematode (68).

The nematotoxic properties of *Bt* toxins from *Bacillus thuringiensis* strains were thoroughly studied with the free-living nematode *Caenorhabditis elegans*, revealing a different mode of action when compared with insecticidal strains (69). A preliminary study with transgenic tomato plants expressing the *Bt* endotoxin CryIab after inoculation with *Meloidogyne* resulted in a reduction in egg mass per gram of root of about 50% (70). Further studies and subsequent field trials have to be performed to corroborate the impact of *Bt* toxin on plant-pathogenic nematodes.

Expression of monoclonal antibodies within the plant that are directed against nematode-specific proteins should lead to a specific inhibition of nematode parasitism. This concept has been tested by expressing a monoclonal antibody specific to stylet secretions of *M. incognita* in tobacco plants (71) and by transiently expressing single-chain antibodies in tobacco protoplasts (72). However, although the antibody was expressed in the appropriate plant tissue, it appeared to have no influence on the parasitism by the nematode, indicating that antibodies might not be a sufficient means of controlling nematodes (71).

## B. Disrupture of Feeding Cells

The breakdown of the specific feeding structures is the second target for engineering an artificial resistance system. Therefore, an approach aiming at the introduction of genes encoding phytotoxic compounds that disrupt and desintegrate these specialized cells has been developed. This can only be realized by selecting promoters exclusively induced in the cells of the feeding structure, thus limiting the expression of "suicide" genes to the target tissue. Therefore, the promoter trapping technique has been applied to detect up- and down-regulation of genes within, or in the vicinity of, feeding cells (73). In 1998 the first nematode-responsive DNA sequences related to differentiation of feeding structures in *A. thaliana* were isolated by using a T-DNA-tagging approach based on the expression of a randomly integrated promoterless β-glucuronidase gene (74). However, such regulating elements are difficult to find, because except for natural resistance genes a plant might not have evolved genes specific for a pathogen-induced tissue. To circumvent this problem, a two-component system consisting of a plant-toxic gene under the control of a nematode-responsive promoter and a second detoxifying gene under the control of a constitutive promoter has been developed (75). Thus, "leaky" expression of the cytotoxic gene can be neutralized by the antagonistic gene. This system makes use of the *barnase* and *barstar* genes, first applied to generate and subsequently to restore male sterility in plants (76). Meanwhile different strategies have been applied in the search for nematode-responsive promoters that are up-regulated in the feeding tissue. Differential cDNA screening led to the identification of *Lemmi9*, a tomato gene that is up-regulated in giant cells after *M. incognita* infection (77). The tobacco gene *TobRB7* is up-regulated, and deletion analysis of the promoter revealed that a 300-bp fragment conferred high GUS expression solely in giant cells, but not in syncytia (78). GUS fusion analysis of *hmg2*, which encodes a key enzyme of isoprenoid metabolism, indicated that the promoter of the gene might be up-regulated in tobacco giant cells (79). However, employment of this system with the specific promoters described here remains a challenge.

## V. CONCLUSION

The molecular identification of natural nematode resistance genes provides the opportunity to understand the basic biological characteristics of plant resistance to a parasitic animal and the relationship to other pathogen resistance genes (R genes). On the other hand, the R genes and the genes involved downstream of the resistance response could be directly transferred into crop species for which no genetic resources have been identified so far. Further understanding of natural plant resistance genes, their products, and the molecular interactions responsible for both early recognition and activation of resistance will have a major impact on both classical breeding and genetic engineering. The combination of resistance genes with artificial resistance mechanisms in one plant offers the opportunity to breed varieties with a broad resistance, a process that is still a challenge because nematode populations often display variation of virulent pathotypes. Nematode-resistant varieties that will be available in the near future open new alternatives for crop production and will help to increase yields on heavily infested soils. The aim is cost-effective, durable, and environmentally friendly disease control.

## REFERENCES

1. JN Sasser, DW Freckmann. A world perspective on nematology: The role of the society. In: JA Veech, DW Dickson, eds. Vistas on Nematology. Society of Nematologists, 1987, pp 7–14.
2. PC Sijmons, HI Atkinson, U Wyss. Parasitic strategies of root nematodes and associated host cell responses. Annu Rev Phytopathol 32:235–259, 1994.

3. U Wyss. Root parasitic nematodes: An overview. In: C Fenoll, FMW Grundler, S Ohl, eds. Cellular and Molecular Aspects of Plant-Nematode Interactions. Dordrecht: Kluwer Academic, 1997, pp 5–22.
4. A Böckenhoff, DAM Prior, FMW Grundler, Oparka KJ Induction of phloem unloading in *Arabidopsis thaliana* roots by the parasitic nematode *Heterodera schachtii*. Plant Physiol 112:1421–1427, 1996.
5. FMW Grundler, A Böckenhoff. Physiology of nematode feeding and feeding sites. In: C Fenoll, FMW Grundler, S Ohl, eds. Cellular and Molecular Aspects of Plant-Nematode Interactions. Dordrecht: Kluwer Academic, 1997, pp 107–119.
6. W Golinowski, M Sobczak, W Kurek, G Grymaszewska. The structure of syncytia. In: C Fenoll, FMW Grundler, S Ohl, eds. Cellular and Molecular Aspects of Plant-Nematode Interactions. Dordrecht: Kluwer Academic, 1997, pp 80–97.
7. RS Hussey, FMW Grundler. Nematode parasitism of plants. In: RN Perry, DJ Wright, eds. The Physiology and Biochemistry of Free-Living and Plant-Parasitic Nematodes. Wallingford: CAB International Press, 1998, pp 213–243.
8. U Wyss, U Zunke. Observations on the behaviour of second stage juveniles of *Heterodera schachtii* inside host roots. Rev Nematol 9:153–165, 1986.
9. G Smant, JPWG Stokkermans, Y Yitang, JM de Boer, TJ Baum, X Wang, RS Hussey, FJ Gommers, B Henrissat, EL Davis, J Helder, A Schoots, J Bakker. Endogenous cellulases in animals: Isolation of β-1, 4-endoglucanase genes from two species of plant-parasitic cyst nematodes. Proc Natl Acad Sci USA 95:4906–4911, 1998.
10. AAF Evans. Reproductive mechanisms. In: RN Perry, DJ Wright, eds. The Physiology and Biochemistry of Free-Living and Plant-Parasitic Nematodes. Wallingford: CAB International Press, 1998, pp 133–154.
11. PW Jones, GL Tylka, RN Perry. Hatching. In: RN Perry, DJ Wright, eds. The Physiology and Biochemistry of Free-Living and Plant-Parasitic Nematodes. Wallingford: CAB International Press, 1998, pp 181–212.
12. J de Almeida Engler, V De Vleesschauwer, S Burssens, JL Celenza, D Inzé, M Van Montagu, G Engler, G Gheysen. Molecular markers and cell cycle inhibitors show the importance of cell cycle progression in nematode-induced galls and syncytia. Plant Cell 11:793–808, 1999.
13. T Bleve Zacheo, MT Melillo. The biology of giant cells. In: C Fenoll, FMW Grundler, S Ohl, eds. Cellular and Molecular Aspects of Plant-Nematode Interactions. Dordrecht: Kluwer Academic, 1997, pp 65–79.
14. R Dorhout, FJ Gommers, C Kolloffel. Phloem transport of carboxyfluorescein through tomato roots infected with *Meloidogyne incognita*. Physiol Mol Plant Pathol 43:1–10, 1993.
15. U Wyss, FMW Grundler, A Münch. The parasitic behaviour of second-stage juveniles of *Meloidogyne incognita in roots of Arabidopsis thaliana*. Nematologica 38:98–111, 1992.
16. X Ding, J Shields, R Allen, RS Hussey. A secretory cellulose-binding protein cDNA cloned from the root-knot nematode (*Meloidogyne incognita*). Mol Plant Microbe Interact 11:952–959, 1998.
17. FMW Grundler, M Sobczak, S Lange. Defence responses of *Arabidopsis thaliana* during invasion and feeding site induction by the plant-parasitic nematode *Heterodera glycines*. Physiol Mol Plant Pathol 50:419–429, 1997.
18. RE Paulson, JM Webster. Ultrastructure of hypersensitive reaction in roots of tomato, *Lycopersicon esculentum* L., to infection by the root-knot nematode *Meloidogyne incognita*. Physiol Plant Pathol 2:227–234, 1972.
19. SL Rice, BSC Leadbeater, AR Stone. Changes in cell structure in roots of resistant potatoes parasitized by potato cyst-nematodes. 1. Potatoes with resistance gene *H1* derived from *Solanum tuberosum* ssp andigena. Physiol Plant Pathol 27:219–234, 1985.
20. SL Rice, BSC Leadbeater, AR Stone. Changes in cell structure in roots of resistant potatoes parasitzed by potato cyst nematodes. 2. Potatoes with resistance derived from Solanum vernei. Physiol Mol Plant Pathol 31:1–14, 1987.
21. D Cai, M Kleine, S Kifle, H Harloff, NN Sandal, KA Marcker, RM Klein-Lankhorst, EMJ Salentijn, W Lange, WJ Stiekema, U Wyss, FMW Grundler, C Jung. Positional cloning of a gene for nematode resistance in sugar beet. Science 275:832–834, 1997.
22. M Kleine, D Cai, R Klein-Lankhorst, N Sandal, EMJ Salentijn, H Harloff, S Kifle, KA Marcker, WJ Stiekema, C Jung. Breeding for nematode resistance in sugar beet: A molecular approach. In: C Fenoll,

FMW Grundler, S Ohl, eds. Cellular and Molecular Aspects of Plant-Nematode Interactions. Dordrecht: Kluwer Academic, 1997, pp 176–190.
23. B Holtmann, M Kleine, FMW Grundler. Ultrastructure and anatomy of nematode induced syncytia in roots of susceptible and resistant sugar beet. Protoplasma 211:39–50, 2000.
24. A Ballvora, J Hesselbach, J Niewöhner, D Leister, F Salamini, C Gebhardt. Marker enrichment and high resolution map of the segment of potato chromosome VII harbouring the nematode resistance gene *Gro1*. Mol Gen Genet 249:82–90, 1995.
25. D Leister, A Ballvora, F Salamini, C Gebhardt. A PCR based approach for isolating pathogen resistance genes from potato with potential for wide application in plants. Nat Genet 14:421–429, 1996.
26. J Niewöhner, F Salamini, C Gebhardt. Development of PCR assays diagnostic for RFLP marker alleles closely linked to alleles *Gro1* and *H1*, conferring resistance to the root cyst nematode *Globodera rostochiensis* in potato. Mol Breed 1:65–78, 1995.
27. VC Concibido, DA Lange, RL Denny, JH Orf, ND Young. Genome mapping of soybean cyst nematode resistance genes in "Peking," PI 90763, and PI 88788 using DNA markers. Crop Sci 37:258–264, 1997.
28. DM Webb, BM Baltazar, AP Rao-Arelli, J Schupp. Genetic mapping of soybean cyst nematode race-3 resistance loci in the soybean PI 437.654. Theor Appl Genet 91:574–581, 1995.
29. JM Kretschmer, KJ Chalmers, S Manning, A Karakousis, AR Barr, AKMR Islam, SJ Logue, YW Choe, SJ Barker, RCM Lance, P Langridge. RFLP mapping of the *Ha2* cereal cyst nematode resistance gene in barley. Theor Appl Genet 94:1060–1064, 1997.
30. AR Barr, KJ Chalmers, A Karakousis, JM Kretschmer, S Manning, RCM Lance, J Lewis, SP Jeffries, P Langridge. RFLP mapping of a new cereal cyst nematode resistance locus in barley. Plant Breed 117:185–187, 1998.
31. KJ Williams, JM Fisher, P Langridge. Development of a PCR-based allele-specific assay from an RFLP probe linked to resistance to cereal cyst nematode in wheat. Genome 39:798–801, 1996.
32. RF Eastwood, ES Lagudah, R Appels. A directed search for DNA sequences tightly linked to cereal cyst nematode resistance genes in *Triticum tauschii*. Genome 37:311–319, 1994.
33. ES Lagudah, O Moullet, R Appels. Map-based cloning of a gene sequence encoding a nucleotide-binding domain and a leucine-rich region at the *Cre3* nematode resistance locus of wheat. Genome 40:659–665, 1997.
34. C Taylor, KW Shepherd, P Langridge. A molecular genetic map of the long arm of chromosome 6R of rye incorporating the cereal cyst nematode resistance gene, *CreR*. Theor Appl Genet 97:1000–1012, 1998.
35. GM Garcia, et al. Identification of RAPD, SCAR, and RFLP markers tightly linked to nematode resistance genes introgressed from *Arachis cardenasii* into *Arachis hypogaea*. Genome 39:836–845, 1996.
36. R Heller, J Schondelmaier, G Steinrücken, C Jung. Genetic localization of four genes for nematode (*Heterodera schachtii* Sch.) resistance in sugar beet (*Beta vulgaris* L.). Theor Appl Genet 92:991–997, 1996.
37. MT Ganal, SD Tanksley. Recombination around the *Tm2a* and *Mi* resistance genes in different crosses of *Lycopersicon peruvianum*. Theor Appl Genet 92:101–108, 1996.
38. J Yaghoobi, I Kaloshian, Y Wen, VM Williamson. Mapping a new nematode resistance locus in *Lycopersicon esculentum*. Theor Appl Genet 91:457–464, 1995.
39. MW Ganal, R Simon, S Brommonschenkel, M Arndt, MS Phillips, SD Tanksley, A Kumar. Genetic mapping of a wide spectrum nematode resistance gene (*Hero*) against *Globodera rostochiensis* in tomato. Mol Plant Microbe Interact 8:886–891, 1995.
40. HY Yi, RC Rufty, EA Wernsman, MC Conkling. Mapping the root-knot nematode resistance gene (*Rk*) in tobacco with RAPD markers. Plant Dis 82:1319–1322, 1998.
41. C Price. Breeding sugar beets for resistance to the cyst nematode *Heterodera schachtii*. J Am Soc Sug Beet Technol 13:354–361, 1965.
42. H. Löptien. Breeding nematode-resistant beets. I. Development of resistant alien additions by crosses between *Beta vulgaris* L. and wild species of the section *Patellares*. Z Pflanzenzüchtung 92:208–220, 1984.

43. C Jung, G Wricke. Selection of diploid nematode-resistant sugar beet from monosomic addition lines. Plant Breed 98:205–214, 1987.
44. M Kleine, H Voss, D Cai, C Jung. Evaluation of nematode resistant sugar beet (*Beta vulgaris* L.) lines by molecular analysis. Theor Appl Genet 97:896–904, 1998.
45. W Lange, J Müller, TSM De Bock. Virulence in the beet cyst nematode (*Heterodera schachtii*) versus some alien genes for resistance in beet. Fundam Appl Nematol 16:447–454, 1993.
46. M Kleine, D Cai, C Eibl, RG Herrmann, C Jung. Physical mapping and cloning of a translocation in sugar beet (*Beta vulgaris* L.) carrying a gene for nematode (*Heterodera schachtii*) resistance from *B. procumbens*. Theor Appl Genet 90:399–406, 1995.
47. P Vos, G Simons, T Jesse, J Wijbrandi, L Heinen, R Hogers, A Frijters, J Groenendijk, P Diergaarde, M Reijans, J Fierens-Onstenk, M de Both, J Peleman, T Liharska, J Hontelez, M Zabeau. The tomato *Mi-1* gene confers resistance to both root-knot nematodes and potato aphids. Nat Biotechnol 16:1365–1369, 1998.
48. SB Milligan, J Bodeau, J Yaghoobi, I Kaloshian, Zabel, VM Williamson. The root knot nematode resistance gene *Mi* from tomato is a member of the leucine zipper, nucleotide binding, leucine-rich repeat family of plant genes. Plant Cell 10:1307–1320, 1998.
49. M Rossi, FL Goggin, SB Milligan, I Kaloshian, DE Ullman, VM Williamson. The nematode resistance gene *Mi* of tomato confers resistance against the potato aphid. Proc Natl Acad Sci USA 95:9750–9754, 1998.
50. CM Kreike, JRA De Koning, JH Vinke, JW Van Ooijen, C Gebhardt, WJ Stiekema. Mapping of loci involved in quantitatively inherited resistance to the potato cyst-nematode *Globodera rostochiensis* pathotype Ro1. Theor Appl Genet 87:464–470, 1993.
51. CM Kreike, JRA De Koning, JH Vinke, JW Van Ooijen, C Gebhardt, WJ Stiekema. Quantitatively-inherited resistance to *Globodera pallida* is dominated by one major locus in *Solanum spegazzinii*. Theor Appl Genet 88:764–769, 1994.
52. JME Jacobs, K Horsman, PFP Arens, Verkerk, B Bakker, E Jacobsen, A Pereira, WJ Stiekema, HI Van Eck. Mapping of resistance to the potato cyst nematode *Globodera rostochiensis* from the wild potato species *Solanum vernei*. Mol Breed 2:51–60, 199.
53. J Rouppe van der Voort, K Kanyuka, E van der Vossen, A Bendahmane, P Mooijman, R Klein-Lankhorst, W Stiekema, D Baulcombe, J Bakker. Tight physical linkage of the nematode resistance gene *Gpa2* and the virus resistance gene *Rx* on a single segment introgressed from the wild species *Solanum tuberosum* subsp. *andigena* CPC 1673 into cultivated potato. Mol Plant Microbe Interact 12:197–206, 1999.
54. A Bendahmane, K Kanyuka, DC Baulcombe. The *Rx* gene from potato controls separate virus resistance and cell death responses. Plant Cell 11:781–792, 1999.
55. RA Vierling, J Faghihi, VR Ferris, JM Ferris. Association of RFLP markers with loci conferring broad-based resistance to the soybean cyst nematode (*Heterodera glycines*). Theor Appl Genet 92:83–86, 1996.
56. D Danesh, S Penuela, J Mudge, RL Denny, H Nordstrom, JP Martinez, ND Young. A bacterial artificial chromosome library for soybean and identification of clones near a major cyst nematode resistance gene. Theor Appl Genet 96:196–202, 1998.
57. JP Tamulonis, BM Luzzi, RS Hussey, WA Parrott, HR Boerma. RFLP mapping of resistance to Southern root-knot nematode in soybean. Crop Sci 37:1903–1909, 1997.
58. PC Sijmons, FMW Grundler, N von Mende, PR Burrows, U Wyss. *Arabidopsis thaliana* as a new model host for plant-parasitic nematodes. Plant J 1:245–254, 1991.
59. G Gheysen, W van der Eyken, N Barthels, M Karimi. The exploitation of nematode-responsive plant genes in novel nematode control methods. Pestic Sci 47:95–101, 1996.
60. HJ Atkinson, CJ Lilley, PE Urwin, MJ McPherson. Engineering resistance to plant parasitic nematodes. In: RN Perry, DJ Wright, eds. The Physiology and Biochemistry of Free-Living and Plant-Parasitic Nematodes. Wallingford: CAB International Press, 1998, pp 381–413.
61. E van der Vossen, J Rouppe van der Voort, K Kanyuka, A Bendahmane, H Sandbrink, DC Baulcombe, J Bakker, WJ Stiekema, RM Klein-Lankhorst. Homologues of a single resistance-gene cluster in potato confer resistance to distinct pathogens: a virus and a nematode. Plant J 23:567–576, 2000.

62. VM Koritsas, HJ Atkinson. Proteinases of females of the phytoparasite *Globodera pallida* (potato cyst nematode). Parasitology 109:357–365, 1994.
63. CJ Lilley, PE Urwin, MJ McPherson, HJ Atkinson. Characterization of intestinally active proteinases of cyst nematodes. Parasitology 113:415–425, 1996.
64. PE Urwin, JC Lilley, MJ McPherson, HJ Atkinson. Resistance to both cyst and root-knot nematodes conferred by transgenic *Arabidopsis* expressing a modified plant cystatin. Plant J 12:455–461, 1997.
65. PE Urwin, MJ McPherson, HJ Atkinson. Enhanced transgenic plant resistance to nematodes by dual proteinase inhibitor constructs. Planta 204:472–479, 1998.
66. J Spiegel, MA McClure. The surface coat of plant parasitic nematodes: Chemical composition, origin, and biological role—a review. J Nematol 27:127–134, 1995.
67. MA McClure. Lectin binding sites on the amphidial exudates of *Meloidogyne*. J Nematol 20:1–326, 1995.
68. PR Burrows, ADP Barker, CA Newell, WDO Hamilton. Plant-derived enzyme inhibitors and lectins for resistance against plant-parasitic nematodes in transgenic crop. Pestic Sci 51:35–43, 1997.
69. G Borgonie, M Claeys, F Leyns, G Arnaut, D de Waele. Effect of a nematicidal *Bacillus thuringiensis* strain on free-living nematodes. 1. Light microscopic observations, species and biological stage specificity and identification of resistant mutants of *Caenorhabditis elegans*. Fundam Appl Nematol 19:391–398, 1996.
70. PR Burrows, D de Waele. Engineering resistance against plant parasitic nematodes using anti-nematode genes. In: C Fenoll, FMW Grundler, S Ohl, eds. Cellular and Molecular Aspects of Plant-Nematode Interactions. Dordrecht: Kluwer Academic, 1997, pp 217–236.
71. TJ Baum, A Hiatt, WA Parrot, LH Pratt, RS Hussey. Expression in tobacco of a functional monoclonal antibody specific to stylet secretions of the root-knot nematode. Mol Plant Microbe Interact 9:382–387, 1996.
72. MN Rosso, A Schouten, J Roosien, T Borst-Vrenssen. Expression and functional characterisation of a single chain FV antibody directed against secretions involved in plant nematode infection process. Biochem Biophys Res Commun 220:255–263, 1996.
73. N Barthels, FM van der Le, JC Klap, OJM Goddijn, M Karimi, P Puzio, FMW Grundler, SA Ohl, K Lindsey, L Robertson, WM Robertson, M van Montagu, G Gheysen, PC Sijmons. Regulatory sequences of *Arabidopsis* drive reporter gene expression in nematode feeding structures. Plant Cell 9:2119–2134, 1997.
74. PS Puzio, D Cai, S Ohl, U Wyss, FMW Grundler. Isolation of regulatory DNA regions related to differentiation of nematode feeding structures in *Arabidopsis thaliana*. Physiol Mol Plant Pathol 53:177–193, 1998.
75. SA Ohl, FM van der Lee, PC Sijmons. Anti-feeding structure approaches to nematode resistance. In: C Fenoll, FMW Grundler, S Ohl, eds. Cellular and Molecular Aspects of Plant-Nematode Interactions. Dordrecht: Kluwer Academic, 1997, pp 250–261.
76. C Mariani, V Gossele, M De Beuckeleer, M de Block. A chimaeric ribonuclease-inhibitor gene restores fertility to male sterile plants. Nature 357:384–387, 1992.
77. W Van der Eycken, J Almeida Engler, D Inze, M Van Montagu, G Gheysen. A molecular study of root-knot nematode induced feeding sites. Plant J 9:45–54, 1996.
78. CH Opperman, CG Taylor, MA Conkling. Root-knot nematode–directed expression of a plant root–specific gene. Science 263:221–223, 1994.
79. CL Cramer, D Weissenborn, CK Cottingham, CJ Denbow. Regulation of defense-related gene expression during plant-pathogen interactions. J Nematol 25:507–518, 1993.
80. C Jung, D Cai, M Kleine, Engineering nematode resistance in crop species. Trends in Plant Science 3:266–271, 1998.

# 17
# Genetic Engineering and Resistance to Insects

**Dwayne D. Hegedus, Margaret Y. Gruber, and Lorraine Braun**
*Agriculture and Agri-Food Canada, Saskatoon, Saskatchewan, Canada*

**George G. Khachatourians**
*University of Saskatchewan, Saskatoon, Saskatchewan, Canada*

| | | |
|---|---|---|
| I. | INTRODUCTION | 249 |
| II. | OVERVIEW OF PLANT DEFENSE MECHANISMS | 250 |
| III. | ANTIBIOSIS | 252 |
| | A. Microbial Protein–Based Antibiotic Strategies | 253 |
| | B. Plant Protein–Based Antibiotic Strategies | 256 |
| IV. | ANTIXENOSIS | 258 |
| | A. Plant Morphological Characteristics | 258 |
| | B. Phytochemical Profiles | 259 |
| V. | TOLERANCE | 262 |
| VI. | PEST MANAGEMENT | 262 |
| | A. Multitrophic Considerations | 262 |
| | B. Crossover Strategies: Transgenic Insect Resistance and Integrated Pest Management | 263 |
| | REFERENCES | 265 |

## I. INTRODUCTION

In determining the most appropriate "solution" to an insect problem, diverse but inextricably interrelated factors must be taken into consideration. These factors range from the relative importance of the affected crop to the economic viability and security of a region's agriculture industry to how specific details of insect or plant biochemical and physiological characteristics might be exploited to provide effective and durable insect resistance. In this chapter, we review transgenic pest control strategies currently under investigation or implementation and attempt to address their overall compatibility in relation to each other and to integrated management systems. Several comprehensive reviews of this subject matter are also available (1–6).

The interactions that occur between plant and pest are highly dynamic and evolved and can span multiple trophic levels. At a primary level, plants constitutively produce a plethora of chemicals that serve to attract, stimulate, or deter insects, be they beneficial or antagonistic. Once the insect begins feeding, the plant responds by synthesizing inhibitory and antinutritional compounds and proteins to make itself as unpalatable as possible. Predatory and parasitic insects may then be attracted to their respective prey or hosts by the "scent" emanating from a wounded plant or pest excreta. However, predators and parasites in turn may be affected by any of the inhibitory substances consumed by their prey. Other features such as color, spectral patterns, and morphological characteristics also greatly influence the degree to which these insect/host plant interactions occur.

From the perspective of developing host plant resistance, additional factors that are required for an insect to complete its life cycle may be considered. For example, carbohydrate, amino acid, sterol, and elemental needs vary greatly among different insect developmental stages. In some cases it may be better to reduce the pest's reproductive ability rather than target the stage that is causing the most damage; the latter usually corresponds to the period when the insects are most robust and difficult to injure. Such an approach would require a concerted effort by producers within a given geographic area to prevent infestation from adjacent unprotected fields. This example clearly underlines the need to develop socially acceptable, responsible strategies that will undoubtedly involve organic, biological, agrochemical, and transgenic approaches.

The contextual framework of this article relates to the complex interactions among plants, pests and predators and is divided into three broad categories that are being investigated to develop insect resistance in crop plants. They are based on the general mechanisms used by plants to withstand insect attack. Three mechanisms are cited: *Antibiosis* refers to the vast array of molecules and macromolecules (proteins) that plants produce to reduce the fitness of the pest. *Antixenosis* relates to mechanisms, both phytochemical and morphological, that serve to deter insect attraction or feeding. *Tolerance* is the ability of the plant to withstand and overcome a limited degree of damage. The choice of which type of strategy to employ is best made only after considering all of the final uses of the crop, including its use as animal or human food.

## II. OVERVIEW OF PLANT DEFENSE MECHANISMS

Plants are under constant threat of attack from pathogens as well as vertebrate and invertebrate herbivores. To cope with this onslaught, they have evolved elaborate mechanisms to perceive the attack and a responsive "signalling language" to tie the reception of information to defense strategies in a coordinated fashion. The mechanisms by which plants respond to threats, be they pathogen, pest, or mechanical wounding, overlap to a large degree but also exhibit significant differences (Fig. 1). Plants resist attack from pathogens if they possess specific receptors that recognize ligands, in the form of metabolites, proteins and specific carbohydrates, secreted or shed by the pathogen (7,8). This interaction initiates a signal transduction cascade and gives rise to both localized and systemic responses. The key primary response in pathogen-affected cells is a self-destructive, hypersensitive reaction involving the release of toxic oxygen species. This process eventually leads to cell apoptosis and lysis, serving to block the replication of intracellular pathogens, such as viruses, and preventing movement of pathogens to adjacent uninfected tissue (9,10).

Damage to plant tissues from herbivory, or other forms of mechanical wounding, causes the release of cell wall pectic fragments. These polygalacturonides are powerful activators of systemic wound responses when applied exogenously either alone (11) or in the form of insect regurgitate (12). Tissues within the localized area are signaled of the attack and generate a response consisting of physiological changes in cell wall architecture and the induction of defense proteins, hyperoxidized chemicals and secondary metabolites at the wound site (13).

Induction of a secondary systemic response is mediated by signaling molecules, such as jasmonic acid (14), abscisic acid (15), ethylene (16), systemin (17), and salicylic acid (18). The

# Resistance to Insects

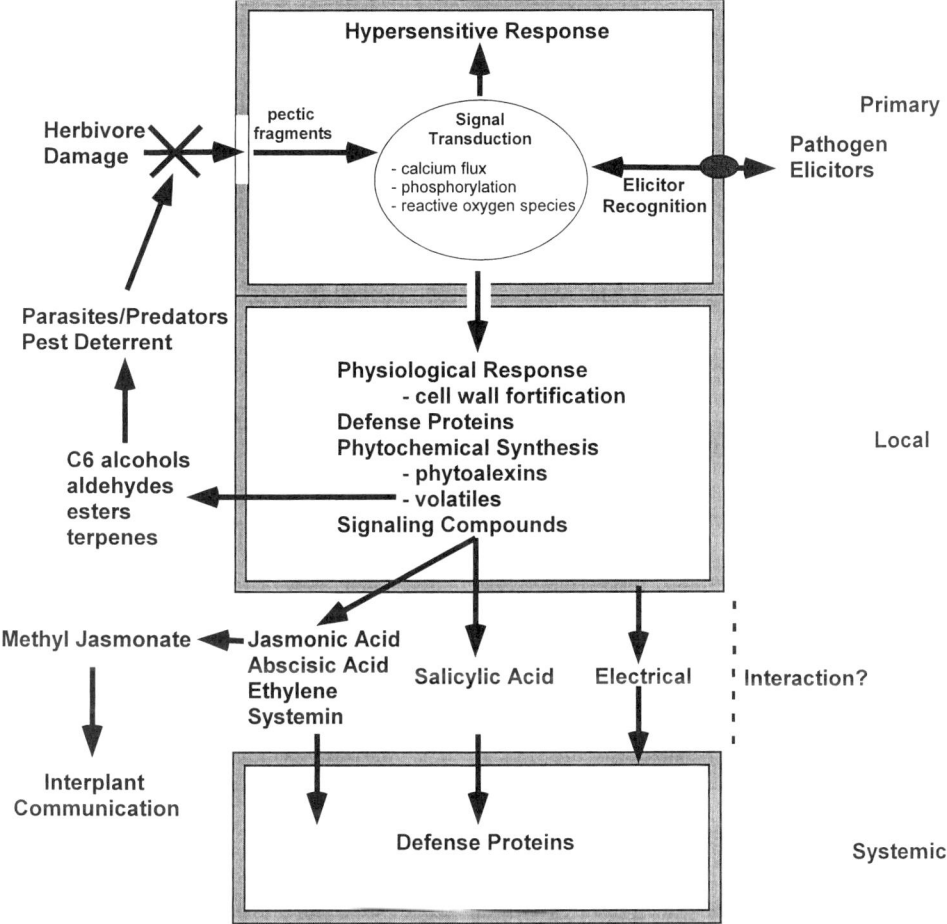

**Figure 1** Overview of plant defense responses to insect pests and pathogens. (Adapted from Ref. 13.)

concentration of these compounds increases greatly in locally affected tissues, and they are exported throughout the plant via the phloem (19). Electrical signals passed through the continuous network of plasmodesmata may also play a role in long-distance signaling (20). Volatile derivatives of the octadecanoid pathway, methyl jasmonate and ethylene, act synergistically to sensitize other plant tissues (21–23) and may induce defense responses in adjacent plants (24). There is some evidence suggesting that signaling of mechanical trauma tends to be associated with the jasmonic acid pathway and that salicylic acid serves as the intermediary for pathogen-associated interactions (25). Indeed, a significant degree of interaction (synergism and antagonism) and coordination exists between the various pathways (26,27).

There is also new evidence indicating that two pathways, either jasmonate- or salicylate-dependent, are responsible for pathogen resistance. Each pathway appears to be used for a distinct group of microbial pathogens (28). A subset of defense proteins, the pathogenesis-related (PR) proteins, exhibit activity toward both insects and pathogens and are synthesized primarily in response to salicylic acid and its analogues (29). A larger and more diverse group of proteins responding to jasmonic acid, as well as to other signaling intermediaries, include: antimicrobials (defensins/thionins, ribosome-inactivating proteins); enzymes that generate phenolic derivatives;

protease, amylase, and polygalacturonidase inhibitors; and enzymes that either generate or degrade toxic compounds. Some of these proteins that have been examined for their ability to confer insect resistance in plants are discussed later.

Another emerging area of study with potential to develop insect resistance is the relationship between insect behavior and alterations in the plant's phytochemical profile. Clearly, the plant itself has much to offer in terms of providing researchers with alternatives to develop pest control strategies. However, the inherent intricacy of plant-pest relationships, combined with the amazing ability of insects to adapt rapidly to changing environments, calls into question the practice of relying on single-gene strategies (e.g., *Bacillus thuringiensis* δ-endotoxin) for pest control.

## III. ANTIBIOSIS

Antibiosis results in an antagonistic situation in which one organism produces a compound or metabolite that functions to the detriment of another. A common theme when developing antibiotic pest control strategies using transgenic plants has been to disrupt insect midgut physiological features or digestive biochemical traits (Fig. 2). Here only a thin, porous membrane, termed the *peritrophic matrix*, protects the midgut epithelial cells while serving to compartmentalize digestion and allow nutrients and some proteins to pass through. Small molecules, viruses, and pathogens are able to penetrate this barrier and gain access to the underlying exposed cells (30). Others molecules affect insect health indirectly by limiting its ability to digest plant macromolecules. Although the outcome of ingestion of relatively large amounts of the various toxins is death of the pest, lower doses may simply impair the ability of the insect to sequester and assimilate nu-

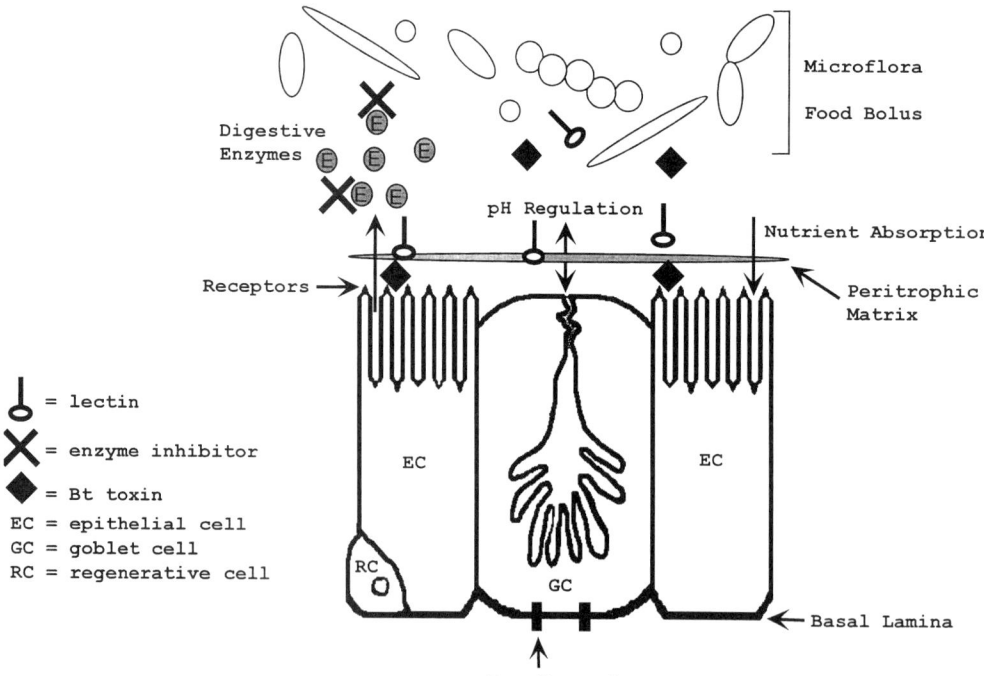

**Figure 2** Bisect of typical lepidopteran insect midgut showing the sites of action of and potential for synergism between various transgenic pest control strategies.

## Resistance to Insects

trients. Thus, the definition of resistance needs to include reducing insect fitness, growth rate, and development. The following subsections review microbial and plant-derived proteins that are being used to develop insect resistant transgenic plants.

### A. Microbial Protein–Based Antibiotic Strategies

#### 1. *Bacillus thuringiensis* Toxins

An increasing number of insect-pathogenic microorganisms have been registered for use as biological control agents; however, few have been successfully incorporated into integrated insect management systems for insect pests. The insect pathogenic bacterium *Bacillus thuringiensis* (*Bt*) is the most widely used and successful biological insecticide; products containing some form of *Bt* constitute 80–90% of the microbial pesticides purchased and used worldwide (31). *Bacillus thuringiensis* is a ubiquitous gram-positive soil bacterium found in habitats ranging from tropical jungles to the Arctic tundra (32). Strains of this bacterium were first registered in the United States in 1959, however, the isolation of the HD-1 strain in 1970 provided a commercial strain with a marked increase in activity against the larval stages of a wide variety of lepidopteran pest species under field conditions. The HD-1 type is the basis of numerous commercial products and is still the strain most often used in topically applied products. It produces a variety of insecticidal proteins, the most important of which form the insecticidal crystal protein (ICP). These crystals contain protoxins that are converted by proteolytic enzymes in the midgut of susceptible insects to form smaller toxic peptides. Activated toxins bind to specific receptors on midgut epithelial cells, creating pores in cell membranes leading to cell lysis and eventually insect death (33).

The most effective method for delivery of insecticidal proteins is that used by the plant itself since the toxin directly affects only those herbivorous insects causing damage. To date, the only commercial insect-resistant transgenic plants express genes encoding various forms of the *B. thuringiensis* δ-endotoxin, termed *cry* (Table 1) (34–77). At least 10 such genes have been introduced into various plants, including those encoding proteins specific for lepidopteran and coleopteran insects (5). Plants so far transformed include major crops (tobacco, cotton, canola, alfalfa, soybean, maize, and rice), specialty crops (tomato, sweet corn, potato, petunia, peanut, white clover, rutabaga, cabbage, and broccoli), and trees (poplar, white spruce, and walnut). Estimates from 1997 indicate that more than 6 million acres of *Bt*-corn, 3 million acres of *Bt*-cotton and 40,000 of *Bt*-potatoes were planted in the United States (78).

Initially while both full length and truncated *cry* genes could be introduced into tobacco and tomato by *Agrobacterium tumefaciens*-mediated transformation, reliable insect resistance was not achieved until expression levels were increased through the use of modified genes possessing optimized plant codon usage (67,42). The first generation of transgenic plants containing *cry* genes provided high levels of δ-endotoxin in all plant tissues. Through replacement of constitutive promoters, such as the CaMV 35S promoter, with wound-inducible (63), chemically-inducible (66) and tissue-specific promoters (42), the second generation *Bt*-crops will incorporate some aspects required to address resistance management (79). Development of resistance to *Bt* toxins is one of the main concerns related to use of *Bt*-expressing transgenic plants. Laboratory selection for resistance to *Bt* δ-endotoxin has been demonstrated for lepidoptera (80–81), coleoptera (82), and diptera (83). However, to date the diamondback moth, *Plutella xylostella*, a pest of cruciferous plants, is the only insect reported to have developed high levels of resistance in the field (84–85). Transgenic plants expressing active toxins directly remove requirements for specific gut conditions required to activate the protoxin; this could potentially expand the range of non-target hosts (86). In addition, transformed plants containing single, or even multiple, *cry* genes lack the advantages provided by the entire, intact bacterium, such as spores and other *Bt* toxins that contribute to insect mortality, including β-exotoxin, α-exotoxin, and phospholipase C (lecithinase).

**Table 1** Commercial Insect-Resistant Transgenic Plants

| Host plant | Gene | Target insect | Reference |
|---|---|---|---|
| Alfalfa | cry1Ca | Spodoptera sp. | 34 |
| Apple | cry1Ac | Cydia pomonella | 35 |
| Arabidopsis | cry1Ca | Lepidoptera | 34 |
| Broccoli | cry1Ac | Lepidoptera | 36 |
| Cabbage | cry1Ac | Lepidoptera | 36 |
| Canola | cry1Ac | Plutella xylostella, Heliothis zea, Trichoplusia ni, Spodoptera exigua | 37 |
| Cotton | cry1Ab | Lepidoptera | 38 |
|  | cry1Ab, cry1Ac | H. zea, S. exigua, T. ni | 39 |
| Maize | cry1A | Ostrinia nubilalis | 40 |
|  | cry1A | H. zea | 41 |
|  | cry1Ab | O. nubilabis | 42 |
|  | cry1H | Lepidoptera | 43 |
|  | cry9C | Lepidoptera | 44 |
| Peanut | cry1Ac | Lepidoptera | 45 |
| Petunia | cry1Ac | S. exigua, Manduca sexta, T. ni | 46 |
| Poplar | cry1Aa | Lymantria dispar | 47 |
|  | cry1Aa | L. dispar, Malacosoma disstria | 48 |
|  | cry3A | Chrysomela tremulae | 49 |
| Potato | cry1Ab | Phthorimaea operculella | 50 |
|  | cry1Ab | P. operculella | 51 |
|  | cry3A | Leptinotarsa decemlineata | 52 |
|  | cry3A | L. decemlineata | 53 |
| Rice | cry1Ab | Chilio suppressalis, Scirpophaga incertulas, C. suppressalis | 54 |
|  | cry1Ab | Marasmia patnalis, Cnaphalocrosis medinalis | 55 |
|  | cry1Ab, cry1Ac | Stem borers | 56 |
|  | cry1Ab | S. incertulas | 57 |
| Rutabaga | cry1Aa | Lepidoptera | 58 |
| Soybean | cry1Ab | H. zea, H. virescens, Pseudoplusia includens | 59 |
|  | cry1Ac | H. zea, H. virescens, P. includens | 60 |
| Sweet corn | cry1Ab | H. zea, Spodoptera frugiperda | 61 |
| Tobacco | cry1Aa | M. sexta | 62 |
|  | cry1Ab | M. sexta | 63 |
|  | cry1Ab | H. virescens, M. sexta | 64 |
|  | cry1Ab | M. sexta, H. virescens | 65 |
|  | cry1Ab | M. sexta | 66 |
|  | cry1Ab, cry1Ac | M. sexta | 67 |
|  | cry1Ac | M. sexta | 68 |
|  | cry1Ac | H. zea | 69 |
|  | cry1Ac (chloroplasts) | H. virescens, H. zea, S. exigua | 70 |
|  | cry1C | Spodoptera littoralis | 71 |
|  | cry1Ca | Spodoptera sp. | 34 |
|  | cry2Aa (chloroplasts) | H. virescens, H. zea, S. exigua | 72 |
| Tomato | cry1Ab | H virescens | 73 |
|  |  | Pinworm | 74 |
| Walnut | cry1Ac | C. pomonella | 35 |
| White clover | cry1Ba | Wiseana spp. | 75 |
|  | cry1Ab | Lepidoptera | 76 |
| White spruce | cry1Ab | Choristoneura fumiferana | 77 |

Concomitant with the introduction of *Bt* crops has been concern about the adverse effects of *Bt* plants on nontarget and beneficial insects (87–88). It is generally accepted that the benefits arising from the judicious use of *Bt* crops far surpasses the severe negative impact of chemical insecticides on non-target insects. However, reports of potential detrimental effects, such as those described very recently for monarch butterfly larvae feeding upon *Bt*-corn pollen (89), highlight the need for continued refinement of the technology (90). For example, use of a chloroplast protein expression system resulted in expression levels of 2 to 3% of total soluble protein (72). This would provide sufficient toxin to satisfy a high-dose strategy to delay resistance development (91), would avoid expression in pollen, and would reduce the potential for out-crossing of genes to other plants as a result of maternal inheritance of the chloroplast genome.

## 2. Vegetative Insecticidal Protein

Despite access to thousands of isolates and hundreds of genes the assumption that traditional *Bt* toxin genes would be a panacea for all pest-associated problems has simply not held true. Screening efforts continue in a fervent effort to identify additional genes and novel toxic activities. A screen of bacterial isolates for activity against two root-associated corn pests, the corn rootworm and the black cutworm, revealed that some strains of *Bacillus cereus* and *B. thuringiensis* produce insecticidal protein toxins during the vegetative growth phase, quite unlike *Bt* δ-endotoxin, which is synthesized during sporulation. Two classes of proteins and their corresponding genes have been characterized, the vegetative insecticidal protein 1 and 2 (Vip1-Vip2) binary complex, which encode proteins of 52 kDa and 100 kDa, respectively, and Vip3, an 88-kDa protein (92). The relative toxicity of Vip3a exceeds that of δ-endotoxin and exhibits broad-spectrum activity against lepidopteran insects (93). Similarly to that of δ-endotoxin, the primary mode of action of Vip3a is to disrupt and lyse midgut cells; thus specificity is determined by the ability of the toxin to recognize receptors and bind to these cells (94). To date, there are no published reports that these genes have been expressed in transgenic plants.

## 3. *Photorhabdus luminescens* toxins

The ability of certain nematodes to infect and kill insects has been known for many decades and numerous unsuccessful attempts have been made to develop these organisms as biological control agents. Recently, it was discovered that the insecticidal activity was attributable to toxins released from *Photorhabdus luminescens*, a bacterial symbiont that is released by the nematode upon penetration into the hemocoel. The toxic activities were found to be associated with at least four large protein complexes, *tca-d*, each approaching 1,000,000 kDa. Insects administered the toxin, either orally or via injection into the hemocoel, exhibited histopathological symptoms affecting the midgut that were surprisingly similar to that produced by *Bt* δ-endotoxin (95,96). Fractionation of the complexes revealed an assortment of 10–14 protein subunits ranging in size from 30 to 200 kDa. Smaller complexes consisting of only a few of these proteins retained insecticidal activity (97). Both *tca* and *tcb* are tetramers of a 280-kDa monomer that are processed by *P. luminescens* metalloproteases to provide broad-spectrum activity. The genes for all of the *tc* proteins have been cloned (98), and it is only a matter of time before their effects in transgenic plants are determined.

## 4. Lipid Disrupting Proteins

Other bacterial and plant-derived proteins also target the insect midgut epithelium, although generally in a less insect-specific manner. Cholesterol oxidase, first discovered in *Streptomyces*, sp., filtrates causes lysis of midgut epithelial cells, possibly through the disruption of integral membrane sterols (99). This enzyme has been expressed in plants (100) and is effective against a wide variety of lepidopteran insects as well as the boll weevil (101). In addition, plant-derived lipid acyl hydrolases disrupt midgut membrane phospholipid architecture, leading to epithelial cell disruption. They also possess broad-spectrum activity against most orders of insect pests, but to date there are no reports of introduction into transgenic plants (102).

## B. Plant Protein–Based Antibiotic Strategies

### 1. Protease Inhibitors

Plants accumulate reserves of carbohydrate and nitrogen in the form of protein and starch. To protect these reserves and prepare for attack, plants often compartmentalize massive amounts of inhibitors in their storage organs, seeds, and roots to reduce the activity of insect digestive enzymes required for the breakdown of plant macromolecules (103–105). In this regard, protease and amylase inhibitors have proved effective against diverse insect pests when expressed in transgenic plants. For example, protease inhibitors confer resistance when expressed in transgenic tobacco (106), rice (107), cotton (108), alfalfa (109), strawberry (110), poplar (111), canola (112), and other plants.

Protease inhibitors can also act to reinforce the action of *Bt* toxins (113,114); however, the specific nature of insect protease/protease inhibitor/*Bt* toxin interactions must first be considered. All Bt δ-endotoxins are "activated" in the insect midgut by a combination of high pH and serine protease-mediated cleavage of the propeptide (115). Thus, synergistic interactions can occur when the protease inhibitor and toxin work in concert and activation of *Bt* toxin is not affected (113), for example, when protease inhibitors prevent the degradation of *Bt* toxin by midgut proteases (116,117). The key to achieving synergism lies in the ability to design protease inhibitors that do not affect the efficacy of *Bt* toxin but are still capable of inhibiting digestive processes. This is clearly exemplified by contradictory reports that the action of *Bt* toxin is potentiated by protease inhibitors in the Colorado potato beetle (113) but not by either susceptible or resistant diamondback moth larvae (118). Attempts are now under way to increase protease inhibitor specificity by using site-directed mutagenesis of the protease binding domain (119,120) in combination with phage display selection of novel variants (121).

Unfortunately, the utility of protease inhibitors is complicated by the observation that insects compensate for the loss of proteolytic activity by increasing the synthesis of affected proteases and by expressing novel proteases insensitive to the inhibitor (122–125). This epigenetic regulation is thought to be responsible for the curious finding that insect growth and development are stimulated when they are exposed to low doses of seemingly toxic proteases inhibitors (126,127). In other instances, high levels of gut proteolytic activity result in protease inhibitor degradation and inherent resistance of some insects to these compounds (128,129). The regulation of insect digestive proteases may also have consequences for *Bt* δ-endotoxin efficacy, since altered proteolytic processing of the toxin has been implicated in the specificity of (130) and resistance to the toxin (131). For example, *Plodia interpunctella*, has developed resistance to *Bt* toxin through loss of a toxin-activating gut serine protease (132). It is possible that the regulation of protease activity by protease inhibitors could be exploited to resensitize *Bt* toxin–resistant insects by inducing the production of new proteases that would activate the toxin.

### 2. Amylase Inhibitors

Plant seeds contain high concentrations of starch that can be rapidly mobilized to provide the energy required for germination and growth prior to root and shoot development. Consequently, many insects feed almost exclusively on this rich source of nutrients. Bruchid beetles, common pests of stored grains, feed on the seeds of leguminous plants and are particularly susceptible to amylase inhibitors (133). Seeds of the common bean, *Phaseolus vulagaris*, contain a multisubunit 45-kDa glycoprotein, the α-amylase inhibitor (α-AI). When larvae of the *Callosobruchus* beetle, a pest of azuki bean, were fed seeds expressing α-AI, development was completely inhibited. Conversely, *Zabrotes subfasciatus*, a bruchid pest of the common bean, has developed resistance to this protective mechanism of its natural host and was not affected by ingestion of the transgenic seeds (134).

Several species of weevil also rely on starch as a primary source of carbohydrate and burrow into the seedpod to feed upon the soft tissue of immature seeds. Incorporation of α-AI into pea weevil diet increased larval development time (135), although there is some controversy as to whether the bean lectin, phytohemagglutinin, may also be involved in the inhibitory mechanism (136). When expressed at high levels in transgenic peas, α-AI completely prevented weevil development on seeds and neutralized the effects of weevil damage on seed yield (137).

## 3. Lectins

Over two decades ago lectins, which are proteins having affinity for distinct carbohydrate moieties, were shown to possess insecticidal properties (136). Lectins are thought to exert their toxic effect by interacting with glycoproteins embedded within the peritrophic matrix, in turn disrupting digestive processes and nutrient assimilation. The peritrophic matrix is a highly organized, semipermeable array of chitin, proteins, and aminoglycans lining the insect midgut. It serves several vital functions, including facilitating nutrient adsorption, protecting underlying epithelial cells from digestive enzymes, regulating water and ion movement, and protecting the insect from pathogenic bacteria (toxins), viruses, and fungi (30). The mannose-specific lectin GNA, derived from the snowdrop, *Galanthus nivalis*, is highly toxic to sap-feeding insects such as aphids when incorporated into diet (138) or expressed in transgenic plants (139). Insect toxicity of GNA has also been reported for the lepidopteran *Lacanobia oleracea* (tomato moth): larval growth was reduced by 50% when reared upon transgenic potato (140). However, it is difficult to draw broad conclusions about the efficacy of any specific lectin for a particular type of pest without prior testing. Lectin toxicity was shown to correlate with differences in midgut carbohydrate profiles (141) and with susceptibility to proteolytic degradation (142); these factors that can vary greatly even between closely related insect species.

## 4. Chitinases

Since the peritrophic matrix plays such an important role in insect digestion, plants have evolved mechanisms in addition to lectins that impair its function or actively destroy it. The scaffold underlying the peritrophic matrix consists mainly of chitin, making it susceptible to attack by the actions of chitinases. Several distinct classes of chitinases from all kingdoms have been characterized biochemically; Kramer and Muthukrishnan (143) in 1997 reviewed their utility as pest control agents. Brandt and coworkers (144) were the first to demonstrate that perforations in the peritrophic matrix were formed upon treatment with chitinase; this effect was later confirmed in vivo (145). Mosquitoes that had fed upon blood containing *Streptomyces griseus* chitinase were unable to complete formation of the peritrophic matrix upon feeding. Transgenic tobacco expressing an insect-derived chitinase gene from the tobacco hornworm, *Manduca sexta*, was shown to be effective in reducing larval weight gain approximately sixfold (146). Some reports show that chitinases of insect origin may be more effective than plant or bacterial enzymes (146, 147), although more evidence is required to support such generalizations.

Chitinases can act in synergy with other toxins to induce insect resistance. This is evident from the observation that bean endochitinase I expressed alone in transgenic potato was unable to reduce feeding by the tomato moth (148), but in concert with GNA was able to decrease aphid fecundity (149). There are also indications that chitinase may potentiate the action of *Bt* toxins. For example, a sixfold increase in toxicity toward *Spodoptera littoralis* occurred when *Cry*1C δ-endotoxin was combined with bacterial endochitinase (150). In addition, both growth rate and feeding of the tobacco budworm, *Heliothis virescens*, and the tobacco hornworm, *Manduca sexta*, were reduced when they were fed transgenic chitinase-expressing plants that were topically treated with sublethal doses of *Bt* toxin (146).

## IV. ANTIXENOSIS

Plants are held in specific partnerships with insects as a result of physical and chemical attributes that elicit insect behavior and plant responses. These interactions offer opportunities to alter plant cues in order to deter insect attraction or feeding behavior; this is termed *antixenotic resistance*. As more detailed information of this nature becomes available, strategies that involve inducing feeding and egg-laying deterrents and eliminating feeding stimulants will become more prominent in the "toolbox" of resistance traits available to the molecular breeder. Conversely, removal of the large spectrum of volatile compounds that attract an insect to a host plant may prove impractical (151). However, reducing the window of opportunity that a pest has to interact with a host plant reduces the potential for development of resistant insect populations. Strategies may be tailored to a specific insect pest, reducing the potential to harm innocuous or beneficial organisms, such as birds or insect pollinators. As with all resistance strategies, though, development of antixenosis may potentiate or alter interactions to create new pests and, thus, should be accompanied by an in-depth assessment of its potential to promote ecological change.

### A. Plant Morphological Characteristics

#### 1. Trichomes

Trichomes (plant hairs) prevent an insect from grasping a plant and detecting its surface shape and chemical features or from positioning its feeding and egg-laying apparatuses close enough to the plant to carry out their respective activities. The stiffness and frequency of trichomes as well as the excretion of glandular chemicals control how these organs will affect insects. The effect of trichomes on insects and the variation in trichome structure have been documented for many plant species, but only a few plant systems that are being dissected genetically are discussed here.

Potatoes have two types of glandular secretory trichomes, the short S type, containing oxidases that cause a "browning" reaction when in contact with phenolics (152), and the taller B type, which continuously produce droplets of sucrose ester (153). Glandular trichomes are responsible for resistance to Colorado potato beetle and leaf aphids in the wild potato, *Solanum berthaultii*, and in cultivated hybrids (154,155). In potato, quantitative trait loci analysis of trichome type and abundance revealed a large number of independent loci (156).

In contrast, crucifer trichomes are simple but do offer some resistance to flea beetles, leaf hoppers, and diamondback moth larvae (157,158). However, egg laying by some lepidopteran species is higher on trichome-bearing (pubescent) crucifer lines than on smooth (glabrous) lines. Several *Arabidopsis* sp. genes that control trichome development have been isolated or genetically characterized. These include three positive regulatory genes termed *glabrous (gl)* (159–162), the cell cycle regulatory gene *tryptycon (try)* (163–165), and *cotyledon trichome* (COT1), an initiator of trichome production in late leaf formation (164). *Transparent testa glabrous (ttg)* is an unusual gene that simultaneously stimulates the expression of trichomes, anthocyanins, and seed mucilage in *Arabidopsis* sp. (162).

Several trichome regulatory genes are now available; however, their behavior in heterologous crop plants is only now being examined. Nonetheless, several nontrichome regulatory genes have the potential to affect trichome production when inserted into plants. The *myb* gene, *mixta*, that controls conical cell formation in *Antirrhinum majus*, resulted in excess trichomes on leaves and floral organs when introduced into tobacco under control of a constitutive promoter (166). These experiments indicate that conical cells and trichomes are produced by a common developmental pathway that is sensitive to the time of expression of the regulatory gene. The *Arabidopsis ttg* mutant has been complemented with a maize anthocyanin *myc*-like regulatory gene, *Lc*, to produce both trichomes and leaf anthocyanins (167). A more extreme version of this phenotype was obtained when *Lc* alone was inserted into a wild-type *Arabidopsis* sp. line and into *Brassica napus* (167,168). In contrast, the *gl*1 gene requires downstream genes such as *try* to function (160,169). However, the concept of insect resistance based on producing a plant with a dense surface of trichomes still requires further evaluation.

## 2. Surface Characteristics

Waxes, surface secretions, and surface texture combine to form a physical barrier that can inhibit insects from coming into direct contact with a plant; however, not all insects are deterred by these types of surface modifications. For example, low wax (*glossy*) lines of cabbage, *Brassica oleracea*, or lines in which the wax bloom is physically disturbed, are more susceptible to flea beetle damage than cabbage with an undisturbed wax bloom (*glaucous*) (170,171). However, changes to surface chemical features can result in the production of feeding deterrents in low-wax lines. This is likely the case with low-wax cabbage phenotypes that are less susceptible to feeding damage by diamondback moth and imported cabbageworm larvae (170,172).

More than 25 complementation groups for altered wax layers, termed *eceriferum (cer)*, have been identified in *Arabidopsis* sp. (173–178). Unfortunately, introduction of several of these *cer* genes into *B. napus* has not generated any altered wax phenotypes and did not inhibit flea beetle feeding (S. Gleddie, personal communication). These include *cer 2*, a potential regulatory gene for stem wax, and *cer 3*, a gene regulating the release of fatty acids from elongase complexes. In contrast, large morphological changes in the wax layer were observed when the *Arabidopsis* sp. *cut*1 gene, which encodes an enzyme involved in the condensing step of very long-chain fatty acid biosynthesis, was expressed in *Arabidopsis* sp. (179).

## B. Phytochemical Profiles

Typically, but not exclusively, plant chemical deterrents and stimulants that affect insect behavior include secondary metabolites and hyperoxidized compounds. The complex biochemical pathways that produce this diverse spectrum of chemicals are developmentally regulated in concert with the responses to the many different kinds of stress the plant encounters. Specific elicitation mechanisms and the responsive chemical signaling mechanisms tie the reception of information from the surrounding environment to these pathways in a highly dynamic fashion. This inherent complicity causes the behavior of plant genes involved in these interactions to be somewhat unpredictable, particularly when a member of a multigene family or a regulatory gene is inserted into heterologous plants (180,181). In addition, only a few genes are available for secondary metabolite-generating enzymes that divert chemical precursors at points near the end of a branch pathway to form new structures where a more controlled outcome could be predicted (182). Enzymes functioning at earlier steps of a pathway, that feed metabolites into pathways or display broad substrate specificity, increase the potential for production of a diverse spectrum of chemicals (183,184).

## 1. Wounding and Oxidative Stress

Hyperoxidative responses underlie many types of plant stress and involve the production of free radicals and the induction of oxidizing enzymes, such as lipoxygenase, peroxidases, nicotinamide-adenine dinucleotide– (NAD-) and nicotinamide-adenine dinucleotide phosphate– (NADP-) dependent oxidases, and polyphenol oxidases. Insect-associated wounding of soybean, tomato, wheat, and barley has been shown to induce lipoxygenase, ascorbate oxidase, peroxidases, and polyphenol oxidases (185–188). In addition to its role in the production of reactive oxygen and reactive lipid radicals, lipoxygenase stimulates the biosynthesis of jasmonate. Plants respond to this hyperoxidative state in such a way that resistance is achieved with as little overall negative consequence to the host plant as possible.

Hyperoxidation and other wounding responses can be exploited as a means to develop insect resistance by presensitizing plants to induce these responses (189,190). As well, common features of hyperoxidative responses highlight the potential for achieving cross-protection to several types of stress. For example, prior treatment of plants with chemical and biological elicitors of systemic acquired resistance (SAR), such as paraquat and aciflurfen, substantially improved resistance to both copper and paraquat (191). Parallel experiments have shown that transformation of potato plants with the "oxidative burst" gene, glucose oxidase, resulted in protection from the potato soft rot fungus; this effect was neutralized by the addition of the antioxidant enzyme, catalase (192).

The effect of presensitization on plant biochemical characteristics is pattern-specific, and the manner in which it is achieved can have diverse effects on an insect community. For example, prior feeding by aphids on tomatoes induced peroxidase and lipoxygenase, but not polyphenol oxidase and proteinase inhibitors, causing them to become a better host for noctuids such as *Spodoptera exigua* (193). Conversely, prior feeding on tomato by the corn earworm, *Helicoverpa zea*, had the opposite effect, resulting in increased resistance to a wider community of insects including aphids, mites, noctuids, and the phytopathogen *Pseudomonas syringae*. Addition of soybean lipoxygenase to insect diets resulted in significant antimetabolic effects on rice brown planthopper, *Nilaparvata lugens*, nymphs and *H. zea* larvae (185,194). Under the same conditions, no effect was observed on the rice green leafhopper, *Nephotettix cinciteps* (194). Although these experiments tie insect behavior to plant toxicity after presensitization, the effect of presensitization on long-term plant fitness requires further investigation, particularly since growth and development converge with wound response signals in plants (195). This was highlighted by the observation that jasmonate induction of plant defenses in tobacco lowered seed production (196).

The few transgenic plants that have been generated by using a presensitization strategy have, for the most part, not exhibited spectacular results and underscore our lack of knowledge in this area. For example, introduction of the gene encoding allene oxide synthase led to increased jasmonic acid levels in transgenic potato, but jasmonate-responsive genes were not induced (197). Sweetgum, *Liquidambar styraciflua*, transformed with tobacco anionic peroxidase developed leaves resistant to caterpillars and beetles, but was more susceptible to *H. zea* (198). The resistance was more effective with smaller insects when either tobacco and tomato plants were transformed with the same peroxidase gene.

## 2. Secondary Metabolites

Secondary metabolite that have been described as insect deterrents include a wide variety of compounds broadly classified as phenolics, terpenes and their cyclic derivatives, alkaloids, glucosinolates, cyanogenic compounds, and nonprotein amino acids. A large body of literature exists to document secondary metabolite interactions with specific insect pests in both in vitro and in planta feeding studies. For example, sesquiterpene lactones have been shown to protect against stored product pests (199), and coumarins (phenolics) inhibit feeding of adult sweet potato weevil, *Cylas formicarius elegantulus*, in vitro (200). Flavonoid phenolics are known to confer resistance to *H. zea* (201) and glycoalkaloids provide protection to potato tubers from the wireworm, *Agriotes obscurus*, and from the Colorado potato beetle (202,203). Levels of hydroxamic acids in cereals (wheat, rye, and barley) are correlated with resistance to the aphid *Metopolophium dirhodum* (204).

The action of insect feeding can release deterrents that are present in derivatized noninhibitory forms in plants. For example, *Sorghum bicolor* accumulates a sufficient amount of hydrolases and phenolic acid esters in separate organelles in mature leaves such that when they are mixed by grasshopper feeding the esters are hydrolyzed and reduce feeding (205). Presensitization with wounding signals can also induce secondary metabolites in a manner described for hyperoxidized metabolites. For example, jasmonate induces nicotine production in wild tobacco, *Nicotiana attenuata*, improving resistance to the grasshopper *Trimerotropis pallidipennis*, as well as other insect herbivores (196).

Phenylpropanoid biosynthesis and flavonoid biosynthesis are the best characterized of all the secondary metabolite pathways, with biochemical and regulatory genes available from several plant species (181,206–208). Other secondary metabolite pathways, such as the isoprenoid pathway, have also been studied but to a lesser extent (183,209–214). In addition, a wide range of recombinant plant lines and deoxyribonucleic acid (DNA) insertion mutants are available in industrial and public laboratories to assist in gene isolation (215–217). Although transgenic plants developed by using genes involved in secondary metabolite synthesis are becoming more common, few accounts record their effects on insects. Still, expression of a tryptophan decarboxylase gene in tobacco was correlated with reduced whitefly fecundity (218,219). As well, cytokinin-responsive secondary metabolites may be the mechanism underlying resistance to the tobacco hornworm, *M. sexta*, larvae and the green peach aphid, *Myzus persicae*, when tobacco was transformed with the bacterial isopentenyl transferase gene (220).

Other types of plant metabolites have been shown to affect insect growth and development. For example, depletion of $\delta 5$-phytosterols, such as sistosterol, campesterol, and stigmasterol, in *B. napus* by using the fungicides genpropimorph and tridemorph followed by replacement by $4\alpha$-methyl and 4-desmethyl sterols, severely affected the growth, pupal development, morphological features, and survival of the Bertha armyworm, *Mamestra configurata* (221).

The successful exploitation of plant secondary metabolism as a strategy to alter insect behavior or plant toxicity requires a detailed understanding of insect-host plant interactions. Insects are often more responsive host plant chemical signals after prior exposure to the host during rearing (151,222). Also, young larvae are usually more susceptible to chemicals than older larvae (223). As well, a secondary metabolite can act as a feeding stimulant at low concentrations or as a specific derivative, but act as a deterrent in high concentration, or act as a closely related structure, or exhibit altered activity in the presence of other chemicals (224,225). Chemical transformations that may occur within an insect or soil must also be considered.

## 3. Insect Biochemical Characteristics and Plant Resistance

The impact of insect biochemical processes on the toxicity of plant chemicals is often neglected when assessing the effect of secondary metabolites on insects. It is important to recognize that resistance to insects is often a function of chemical oxidation that occurs when oxidizing enzymes are mixed with plant chemicals in the insect gut during digestion (226). A prime example of this is found in tomato, where 40% of the chlorogenic acid present in leaves is oxidized by leaf phenolic oxidases to form the potent antinutritive alkylator chlorogenoquinon, in the moderately alkaline gut of *H. zea* and *Spodoptera exigua* (227). When covalently bound to dietary protein, chlorogenoquinone can chemically degrade essential amino acids in the insect gut, subsequently starving the insect. However, phenol oxidases are unlikely to be a useful source of resistance to the Colorado potato beetle since alkylatable groups of proteins ($NH_2$ and $SH_2$) are in a nonreactive, protonated state in the acidic environment of the coleopteran gut (228). The effect of chlorogenic acid content on *Manduca sexta* was also reduced by the extreme alkalinity of this larval gut (229).

Baculovirus and *Bt* δ-endotoxin were both inactivated by chlorogenic acid–rich tomato leaf extracts, clearly indicating the consequences that plant food sources and insect biochemical process can have on other control agents (230,231). Insect adaptation over the longer term can also reduce the effectiveness of a particular metabolic strategy. For example, some insects have adapted to feeding on seemingly toxic plants by constitutively producing antioxidative enzymes in their gut and salivary fluid (232). Hence, the role of salivary enzymes in reducing toxic reactions is also important when evaluating the damage done by insects.

## V. TOLERANCE

The overall health and vigor of a seedling play a major role in the successful establishment of the crop and in the ability of a plant to produce enough biomass to tolerate wound damage and to undergo adequate regrowth after insect feeding. Plant hormones play a major role in the establishment of seedlings and in overall growth and development. However, the manipulation of plant growth hormones can have very broad effects (233,234), and the overall outcome may be too difficult to control to result in an agriculturally useful plant.

In 1999, the expression of the iron-binding protein ferritin in transgenic tobacco plants conferred tolerance to viral and fungal necrotic lesions (235). Sequestration of intracellular iron and protection against reactive hydroxyl radicals through a Fenton reaction were proposed to be the primary mechanisms by which tolerance was achieved in these plants. Although these results appear to contradict the presensitization strategy with hyperoxidation genes, they highlight the importance of developing plants with moderate, rather than dramatic changes in chemical characteristics to achieve durable insect resistance.

## VI. PEST MANAGEMENT

### A. Multitrophic Considerations

Insects are exceedingly sensitive to volatile compounds and exploit this sensory function to locate mates, host plants, and prey. However, the relationships between a given host plant and its associated insect complex are often multifaceted and have likely evolved to stabilize plant/pest/predator interactions. In some cases these interactions benefit only the plant; for example, certain volatile terpenes serve to attract predatory mites (236) and parasitic wasps (237) to their respective prey and can also repel aphid pests (238). In other situations the overall benefit is not so easily defined. Many cruciferous plant species release volatile thio- and iso-thiocyanates that attract as well as deter pests (239–241).

A discussion of any pest control strategy would not be complete unless problematic "side effects" that impact on registration safety or efficacy of engineered plants are considered. Most concerns center on potential deleterious effects on beneficial insects, such as pollinators, parasites, and predators, and on the consumers of the crop, whether humans or animals. Schuler and colleagues (242) reviewed many of the more recent studies that attempted to determine the ill effects of transgenic plants on natural insect enemies and concluded that the potential exists for certain strategies to influence this natural balance. Overall enemy fitness could be reduced as a direct result of ingesting toxins first consumed by the insect host or when a host is no longer able to provide sufficient nutritional sustenance for complete parasitic larval development. For example, ladybirds fed aphids reared upon GNA-expressing potato exhibited a reduction in fecundity, egg viability, and longevity, although no acute toxicity was observed (243). Conversely, parasitism of the aphid *Myzus pericae* by the parasitoid *Diaeretiella rapae* was not affected when reared upon transgenic oilseed rape (244). Similarly, neither olfactory learning behavior nor longevity was affected by long-term, low-dose ingestion of protease inhibitors (245), and detrimental effects were only observed at doses 100 times higher than that which could be generated in transgenic plants (246).

In addition to the insect and animal community directly associated with the plant, the relevance of external trophic interactions needs to be considered. In 1999 it was reported that monarch butterfly larvae were adversely affected when fed milkweed tainted with an unspecified amount of corn pollen expressing *Bt* δ-endotoxin (89). Although actual pollen levels on milkweed growing adjacent to transgenic corn crops, which are an important parameter in determining the outcome, were not established, this report contributed to immediate reactions by importers of these commodities (247). It was reported in 1999 that pink cotton bollworm larvae resistant to *Bt* δ-endotoxin exhibited delayed development. This effect could skew the expected ratio of susceptible versus resistant alleles in the population if resistant insects become sexually mature later in the season and mate predominantly with other resistant insects. As a result, the overall benefit of refugia in preventing resistant homozygous insects from arising could be reduced (248).

### B. Crossover Strategies: Transgenic Insect Resistance and Integrated Pest Management

In this chapter, we have cited several examples of certain proteins, enzyme inhibitors, and lectins, that have a protective role within storage tissues and seeds, that have insecticidal properties. However, in only a few cases have the genes for such traits been cloned and transferred to other plants to show insect resistance. In other cases transgenic plants based on insecticidal toxic proteins can also be a useful strategy in combating insect pests of plants. The basic research and discovery of toxins of entomopathogenic fungi should also contribute genetic material for strategies to generate insect-resistant transgenic plants (249–251). Other natural resistance mechanisms involving secondary metabolism, physical structures of plant tissues, and plants engineered to contain insect pest–specific feeding deterrents are among options yet to be developed. Also awaiting contextual exploitation is the perspective that engineering of the pest itself (252) has a place in the practice of plant-injurious-insect management. Overall, genetic engineering to increase indigenous resistance of plants to insect attack has matured from theory to commercial reality.

These examples of transgenic plants could complement other forms of crop protection. Paradoxically, however, these technologies, in spite of their potential, have not moved farmers or the practice of farming closer to ecologically sustainable practices, whether in developed or developing countries. In part the reason for the paradox is that single-gene transgenic plant strategies have had their own limited success when compared to prior pest management practice. A classic example is the *Bt* transgenic crops, which under monoculture and repeated cropping practices generate *Bt*-resistant insects.

An alternative strategy to the use of single genes has been the use of gene combinations, whose products are targeted to different biochemical and physiological processes within the insect. The so-called pyramiding genes for a strategy that allows a multiple mechanistic form of insect control to provide more durable resistance is a rather recent concept (253). This concept relies on the ability of plant biotechnologists/breeders to overlay agriculturally desirable traits in the form of packages of different genes into crops. It is conceivable that such a package introduced into crops will increase the protective efficacy, spectrum of activity, and durability of resistance to insects and will be designed to treat different crops and particular insect pests at any one place.

As indicated (in Sec. III.B.1) protease inhibitors should be particularly valuable because, in addition to having insecticidal effects, they could protect other introduced gene products from premature digestion in the insect gut. Gene pyramiding by the introduction of both cowpea trypsin (serine protease) inhibitor (CpTI) and pea lectin (P-Lec) into tobacco represents one of the first examples of this concept (252). Here, transgenic plants were obtained by cross-breeding plants derived from the two primary transformed lines (254). The insecticidal effects of the CpTI and the P-Lec in the transgenic tobacco plants were additive, not synergistic. Insects feeding on the double expressed transgenic tobacco plants showed a biomass reduction of approximately 90% compared with the figure of 50% from those feeding on plants expressing either CPTI or P-Lec alone. Also, leaf damage was least on the double expressing plants. Gatehouse and coworkers (149) generated potato plants expressing the snowdrop lectin, GNA, and bean chitinase (BCH) that exhibited enhanced aphid resistance, as evidenced by a 95% reduction in fecundity when compared with control plants. Fecundity on plants expressing GNA alone was reduced by about 70%, whereas reduction in fecundity on those expressing BCH alone was small and statistically insignificant. These findings clearly demonstrate the gains from gene pyramiding in which two compatible traits can display a synergistic insect control effect.

We can put the threat of insects to our food crops in a historical context. Farmers ever since recorded history of farming and cropping practices have found insects, diseases, and weeds to be their constant companions. Whether they attributed them to the anger of gods or occurrence of natural events, farmers have had to protect their agricultural practices and products against these elements of nature. Shared collective experiences helped people to understand some cause-and-effect relationships; that understanding along with articulation of the fundamentals of agrology led to the development of control and prevention of insect pest damage to crops. The introduction and intensive use of synthetic chemical pesticides were the next major achievements in agriculture. This euphoria was short-lived as negative consequences of the practice, most significantly the development and rise of insecticide-resistant insects, became evident. A different thought and approach to the pest management gave rise to the impetus for integrated pest management (IPM) and its rooting in global agriculture (255). IPM and its ecological approach to the pest problem have remained the unmatched and dominant paradigm in crop protection against pests. A central tenet of IPM relies on reducing or even eliminating the use of pesticides by management of pests based on the time when the pest populations have reached certain practical or economic thresholds. The practice of IPM is based on a thorough and extensive knowledge of the agroecosystem, that is, technical and socioeconomic research founded on the science of applied ecology (255). IPM is a philosophy created by people living within a particular social context.

Practice of IPM in its relationship to biotechnology of microbial pesticides is complex both in its nature and in its implementation (256,257). Few people understand the distinctions between pest "control" and pest "management." In traditional insect pest management, we practice control including the annihilation of the pest. In management, a reduction of pest numbers and hence their effect on cropping can be achieved through understanding of the role of pesticides, measurement of pest population through sampling, understanding of the economic threshold phenomenon, and integration of pest management in relation to effects on nontarget organisms, beneficial insects, and other means. It is clear to the practitioner of IPM that the ideal conditions under which its implementation is to be attempted can and do differ.

How IPM and transgenic insect-resistant plants can contextually relate to each other is a question that we are just beginning to answer. There are significant differences in the trends in industry needs for profit from transgenic seeds in the developed world, the practicality of IPM in heavily capitalized industrial agroecosystems, and that in resource-poor farming systems in developing countries. Nonetheless, we can offer some observations as a starting point for addressing the role of transgenic insect-resistant plants within IPM.

There are complex relationships between different stores of knowledge and how they are influenced by their social context (255). It is these relationships that are the key to the consideration of transgenic plants as a part of IPM practice. An IPM program that contains transgenic insect resistance(s) traits must be considered from the perspective of farmers' awareness. Historically, the process of knowledge transfer has been based on a goal of the "public good" and has supported agricultural scientists to develop the "technology" or knowledge of practice, which through extension services was provided to farmers. Transfer of technology (TOT), which is a linear approach to technology (knowledge) transfer, has become the dominant norm. Here, knowledge or research products (from technologically advanced nation or point) are disseminated through a number of intermediary steps (such as extension services to extension officers) to farmers.

Present day agriculture research and transfer of the information have shifted to generation of knowledge or technology for "private good," although and arguably the boundaries of public vs. private good are tightly bound with the notion of return on the investment. The rules have changed! Where once the traditional TOT had neither a contract nor a source of huge profit for the seed producer, today a seed with transgenic traits, such as insect control, has either or both a significantly higher price (and a high profit) and a contract that defines terms of reference for seed use. Researchers developing new transgenic insect-resistant plants nowadays do this under a grant, contract in academe, or government labs, often in collaboration of one kind or another with strong agriculture biotechnology companies. The idea, the technique, the process, or the product is the technology, which intellectual property law and patents protect. Extension plays a diminishing role, increasingly replaced by education of farmers through corporate marketing and sales, neither of which is free. Most growers who are older, nearing retirement, are used to traditional TOT via extension. Progressive or younger farmers who are more likely to be receptive to new ideas, on the other hand, accept both the traditional and the commercial TOT. As a result there is a bifurcation of source of information and its audience. The sources and manners of adoption of new technology or information will have their own consequences, such as helping or hindering of trust building relations with its providers. The same practice occurred in human medicine as more and more pharmaceutical companies entered the business of "drug education." In this instance, initially the physician's advisory role diminished, only to return after several incidents of questionable media releases.

As suggested by Altieri (258), there is a lack of training in "holistic thinking" so far as crop protection is concerned; those in the profession are trained too narrowly and with a focus within disciplines. Plant protection research tends to depend on the use of component technologies. There is much research on pesticides, biological control, and pest-resistant varieties but relatively little research on linking these components. It is this linking that is needed both in IPM research and in transfer of know-how. Most researchers and scientists working in crop protection are quite specialized and work on components. IPM programs require extensive crosscutting knowledge and research skills. Sociologists and economists who focus on agronomic practice will have to work with narrowly trained molecular geneticists in order to consider the more difficult issues of IPM in the context of transgenics, TOT in practice, and the site- and time-specific context that IPM requires. The powers and promises of transgenic insect control strategies are congruent with the powerful but unrealized IPM strategies. The two should cross-feed and cross-cut to create a remarkable synergy in the production of food crops worldwide.

## REFERENCES

1. N Carozzi, M Koziel. Advances in Insect Control: The Role of Transgenic Plants. London: Taylor and Francis, 1997.

2. JJ Estruch, NB Carozzi, N Desai, NB Duck, GW Warren, MG Koziel. Transgenic plants: An emerging approach to pest control. Nat Biotechnol 15:137–141, 1997.
3. AMR Gatehouse, JA Gatehouse. Identifying proteins with insecticidal activity: Use of encoding genes to produce insect-resistant transgenic crops. Pestic Sci 52:165–175, 1998.
4. L Jouanin, M Bonade-Bottino, C Girard, G Morrot, M Giband. Transgenic plants for insect resistance. Plant Sci 131:1–11, 1998.
5. TH Schuler, GM Poppy, BR Kerry, I Denholm. Insect-resistant transgenic plants. TIBTECH 16:168–175, 1998.
6. VA Hilder, D Boulter. Genetic engineering of crop plants for insect resistance—a critical review. Crop Protect 18:177–191, 1999.
7. KE Hammond-Kosack, JDG Jones. Resistance gene–dependent plant defense responses. Plant Cell 8:1773–1791, 1996.
8. B Baker, P Zambryski, B Staskawicz, SP Dinesh-Kumar. Signaling in plant-microbe interactions. Science 276:726–733, 1997.
9. JT Greenberg, A Guo, DF Klessing, FM Ausubel. Programmed cell death in plants: A pathogen-triggered response activated coordinately with multiple defense functions. Cell 77:551–563, 1994.
10. RN Goodman, AJ Novacky. The Hypersensitive Response in Plants to Pathogens: A Resistance Phenomenon. St. Paul, MN: American Phytopathological Society, 1996.
11. JF Thain, HM Doherty, DJ Bowles, DC Wildon. Oligosaccharides that induce proteinase inhibitor activity in tomato plants cause depolarization of tomato leaf cells. Plant Cell Environ 13:569–574, 1990.
12. KL Korth, RA Dixon. Evidence for chewing insect–specific molecular events distinct from a general wound response in leaves. Plant Physiol 115:1299–1305, 1997.
13. F Mourgues, MN Brisset, E Chevreau. Strategies to improve plant resistance to bacteria diseases through genetic engineering. TIBTECH 16:203–210, 1998.
14. RA Creelman, JE Mullet. Biosynthesis and action of jasmonates in plants. Annu Rev Plant Physiol Plant Mol Biol 48:355–381, 1997.
15. JAD Zeevaart. Metabolism and physiology of abscisic acid. Annu Rev Plant Physiol Plant Mol Biol 39:439–473, 1988.
16. T Boller. Ethylene in pathogenesis and disease resistance. In: AK Mattoo, JC Suttle, eds. The Plant Hormone Ethylene. Boca Raton, FL: CRC Press, 1991, pp 293–314.
17. G Pearce, D Strydom, S Johnson, CA Ryan. A polypeptide from tomato leaves induces wound-inducible proteinase inhibitor proteins. Science 253:895–898, 1991.
18. J Durner, J Shah, DF Kessig. Salicylic acid and disease resistance in plants. Trends Plant Sci 2:266–274, 1997.
19. CA Ryan. The search for the proteinase inhibitor–inducing factor, PIIF. Plant Mol Biol 19:123–133, 1992.
20. K Roberts. Potential awareness of plants. Nature 360:14–15, 1992.
21. Y Xu, PF-L Chang, D Liu, ML Narasimhan, KG Raghothama, PM Hasegawa, RA Bressan. Plant defense genes are synergistically induced by ethylene and methyl jasmonate. Plant Cell 6:1077–1085, 1994.
22. J Bruin, MW Sabelis, M Dicke. Do plants tap SOS signals from their infested neighbours? Trends Ecol Evol 10:167–170, 1995.
23. M McConn, RA Creelman, E Bell, JE Mullet, J Browse. Jasmonate is essential for insect defense in *Arabidopsis*. Proc Natl Acad Sci USA 93:5473–5477, 1997.
24. EE Farmer, CA Ryan. Interplant communication: Airborne methyl jasmonate induces synthesis of proteinase inhibitors in plant leaves. Proc Natl Acad Sci USA 87:7713–7716, 1990.
25. J Ryals, S Uknes, E Ward. Systemic acquired resistance. Plant Physiol 104:1109–1112, 1994.
26. ER Ward, SJ Uknes, SC Williams, SS Dincher, DL Wiederhold, DC Alexander, P Ahl-Goy, J-P Metraux, JA Ryals. Coordinate gene activity in response to agents that induce systemic acquired resistance. Plant Cell 3:1085–1094, 1991.
27. GI Jenkins. Signal transduction networks and the integration of responses to environmental stimuli. Adv Bot Res Incorporating Adv Plant Pathol 29:53–73, 1999.
28. B Thomma, K Eggermont, I Penninckx, B Mauch-Mani, R Vogelsang, B Cammue, W Broekaert.

Separate jasmonate-dependent and salicylate-dependent defense-response pathways in *Arabidopsis* are essential for resistance to distinct microbial pathogens. Proc Natl Acad Sci USA 95:15107–15111, 1998.
29. A Stintzi, T Heitz, V Prasad, S Wiedemann-Merdinoglu, S Kauffmann, P Geoffroy, M Legrand, B Fritig. Plant "pathogenesis-related" proteins and their role in defense against pathogens. Biochimie 75:687–706, 1993.
30. RL Tellam. The peritrophic matrix. In: MJ Lehane, PF Billingsley, eds. Biology of the Insect Midgut. London: Chapman and Hall, 1996, pp 86–114.
31. M Levin. Microbial pesticides: Safety considerations. In: GT Tzotzos ed. Genetically Modified Organisms: A Guide to Biosafety. Wallingford, UK: CAB International. United Nations Industrial Development Organization, 1995, pp 93–109.
32. PAW Martin, RS Travers. Worldwide abundance and distribution of *Bacillus thuringiensis* isolates. Appl Environ Microbiol 55:2437–2442, 1989.
33. H Höfte, HR Whiteley. Insecticidal crystal proteins of *Bacillus thuringiensis*. Microbiol Rev 53: 242–255, 1989.
34. N Strizhov, M Keller, J Mathur, Z Koncz-Kálmán, D Bosch, E Prudovsky, J Schell, B Sneh, C Koncz, A Zilberstein. A synthetic *cryIC* gene, encoding a *Bacillus thuringiensis*-endotoxin, confers *Spodoptera* resistance in alfalfa and tobacco. Proc Natl Acad Sci 93:15012–15017, 1996.
35. AM Dandekar, GH McGranahan, SL Uratsu, C Leslie, PV Vail, JS Tebbets, D Hoffmann, J Driver, P Viss, DJ James. Engineering for apple and walnut resistance to codling moth. Proceedings of the BCPC Pests and Diseases, 1992, pp 741–747.
36. TD Metz, R Dixit, ED Earle. *Agrobacterium tumefaciens*–mediated transformation of broccoli (*Brassica oleracea* var. *italica*) and cabbage (*B. oleracea* var. *capitata*). Plant Cell Rep 15:287–292, 1995.
37. CN Stewart Jr, MJ Adang, JN All, PL Raymer, S Ramachandran, WA Parrott. Insect control and dosage effects in transgenic canola containing a synthetic *Bacillus thuringiensis cryIAc* gene. Plant Physiol 112:115–120, 1996.
38. JH Benedict, ES Sachs, DW Altman, WR Deaton, RJ Kohel, DR Ring, SA Berberich. Field performance of cottons expressing transgenic Cry1A insecticidal proteins for resistance to *Heliothis virescens* and *Helicoverpa zea* (Lepidoptera: Noctuidae). J Econ Entomol 89(1):230–238, 1996.
39. FJ Perlak, RW Deaton, TA Armstrong, RL Fuchs, SR Sims, JT Greenplate, DA Fischoff. Insect resistant cotton plants. Bio/Technology 8:939–943, 1990.
40. CL Armstrong, GB Parker, JC Pershing, SM Brown, PR Sanders, DR Duncan, T Stone, DA Dean, DL DeBoer, J Hart, AR Howe, FM Morrish, ME Pajeau, WL Petersen, B Reich, R Rodriguez, CG Santino, SJ Sato, W Schuler, SR Sime, S Stehling, LJ Tarochione, ME Fromm. Field evaluation of European cornborer control in progeny of 173 transgenic corn events expressing an insecticidal protein from *Bacillus thuringiensis*. Crop Sci 35:550–557, 1995.
41. S Sims, JC Pershing, BJ Reich. Field evaluation of transgenic corn containing Bt Berliner insecticidal protein gene against *Helicoverpa zea* (Lepidoptera: Noctuidae). J Entomol Sci 31:340–346, 1996.
42. MG Koziel, GL Beland, C Bowman, NB Carozzi, R Crenshaw, L Crossland, J Dawson, N Desai, M Hill, S Kadwell, K Launis, K Lewis, D Maddox, K McPherson, MR Meghji, E Merlin, R Rhodes, GW Warren, M Wright, SV Evola. Field performance of elite transgenic maize plants expressing an insecticidal protein derived from *Bacillus thuringiensis*. Bio/Technology 11:194–200, 1993.
43. N Duck, S Evola. Use of transgenes to increase host plant resistance to insects: Opportunities and challenges. In: N Carozzi, M Koziel, ed. Advance in Insect Control: The Role of Transgenic Plants. London: Taylor and Francis, 1997, pp 1–20.
44. M Peferoen. Insect control with transgenic plants expressing *Bacillus thuringiensis* crystal proteins. In: N Carozzi, M Koziel ed. Advance in Insect Control: The Role of Transgenic Plants. London: Taylor and Francis, 1997, pp 21–48.
45. AF Krattiger. Insect resistance in crops: A case study of *Bacillus thuringiensis* and its transfer to developing countries. International Service for the Acquisition of Agri-Biotech Applications: Briefs 2:1–42, 1997.
46. AD Omer, J Granett, AM Dandekar, JA Driver, SL Uratsu, FA Tang. Effects of transgenic petunia

expressing *Bacillus thuringiensis* toxin on selected lepidopteran pests. Biocontrol Sci Technol 7: 437–448, 1997.
47. BH McCown, DE McCabe, DR Russell, DJ Robison, KA Barton, KF Raffa. Stable transformation of *Populus* and incorporation of pest resistance by electric discharge particle acceleration. Plant Cell Rep 9:590–594, 1991.
48. KW Kleiner, DD Ellis, BH McCown, KF Raffa. Field evaluation of transgenic poplar expressing a *Bacillus thuringiensis cry1A(a) d*-endotoxin gene against forest tent caterpillar (Lepidoptera: Lasiocampidae) and gypsy moth (Lepidoptera: Lymantriidae) following winter dormancy. Environ Entomol 24:1358–1364, 1995.
49. D Cornu, JC Leplé, M Bonadé-Bottino, A Ross, S Augustin, A Delplanque, L Jouanin, G Pilate. Expression of a proteinase inhibitor and a *Bacillus thuringiensis* delta-endotoxin in transgenic poplars. Proceedings of the IUFRO Meeting on Somatic Cell Genetics and Molecular Genetics of Trees. Dordrecht: Kluwer, 1996, pp 131–136.
50. M Peferoen. Engineering of insect-resistant plants with *Bacillus thuringiensis* crystal protein genes. In: ed. AMR Gatehouse, VA Hilder, and D Boulter. Biotechnology in Agriculture. Vol. 7. Plant Genetic Manipulation for Crop Protection. London: CAB International, 1992, pp 135–153.
51. S Jansens, M Cornelissen, R De Clercq, A Reynaerts, M Peferoen. *Phthorimaea operculella* (Lepidoptera: Gelichiidae) resistance in potato by expression of the *Bacillus thuringiensis* Cry1A(b) insecticidal crystal protein. J Econ Entomol 88:1469–1476, 1995.
52. MJ Adang, MS Brody, G Cardineau, N Eagan, RT Roush, CK Shewmaker, A Jones, JV Oakes, KE McBride. The reconstruction and expression of a *Bacillus thuringiensis cryIIIA* gene in protoplasts and potato plants. Plant Mol Biol 21:1131–1145, 1993.
53. FJ Perlak, TB Stone, YN Muskopf, LJ Petersen, GB Parker, SA McPherson, J Wyman, S Love, G Reed, D Biever, DA Fischhoff. Genetically improved potatoes: Protection from damage by Colorado potato beetles. Plant Mol Biol 22:313–321, 1993.
54. H Fujimoto, K Itoh, M Yamamoto, J Kyozuka, K Shimamoto. Insect resistant rice generated by introduction of a modified delta-endotoxin gene of *Bacillus thuringiensis*. Bio/Technology 11:1151–1155, 1993.
55. J Wünn, A Klöti, PK Burkhardt, GC Ghosh Biswas, K Launis, VA Iglesias, I Potrykus. Transgenic Indica rice breeding line IR58 expressing a synthetic *cry1A(b)* gene from *Bacillus thuringiensis* provides effective insect pest control. Bio/Technology 14:171–176, 1996.
56. X Cheng, R Sardana, H Kaplan, I Altosaar. *Agrobacterium*-transformed rice plants expressing synthetic *cryIA(b)* and *cryIA(c)* genes are highly toxic to striped stem borer and yellow stem borer. Proc Natl Acad Sci USA 95:2767–2772, 1998.
57. I Potrykus, GA Armstrong, P Beyer, S Bieri, PK Burkhardt, H Ding Chen, GC Ghosh Biswas, SK Datta, J Fütterer, A Klöti, G Spangenberg, R Terada, J Wünn, H Zhao. Transgenic indica rice for the benefit of less developed countries: Toward fungal, insect, and viral resistance and accumulation of beta-carotene in the endosperm: Proceedings of the Third International Rice Genetics Symposium Proc Third Int Rice Genetics Symp, 1996, pp 179–187.
58. X Li, H-Z Mao, Y-Y Bai. Transgenic plants of rutabaga (*Brassica napobrassica*) tolerant to pest insects. Plant Cell Rep 15:97–101, 1995.
59. WA Parrott, JN All, MJ Adang, MA Bailey, HR Boerma, CNJ Stewart Jr. Recovery and evaluation of soybean (*Glycine max* [L.] Merr.) plants transgenic for a *Bacillus thuringiensis* var. *kurstaki* insecticidal gene. In Vitro Cell Dev Biol 30P:144–149, 1994.
60. CN Stewart Jr, MJ Adang, JN All, HR Boerma, G Cardineau, D Tucker, WA Parrott. Genetic transformation, recovery, and characterization of fertile soybean transgenic for a synthetic *Bacillus thuringiensis cryIAc* gene. Plant Physiol 112:121–129, 1996.
61. RE Lynch, BR Wiseman, D Plaisted, D Warnick. Evaluation of transgenic sweet corn hybrids expressing Cry1A(b) toxin for resistance to corn earworm and fall armyworm (Lepidoptera: Noctuidae). J Econ Entomol 92:246–252, 1999.
62. KA Barton, HR Whiteley, N-S Yang. *Bacillus thuringiensis* delta-endotoxin expressed in transgenic *Nicotiana tabacum* provides resistance to lepidopteran insects. Plant Physiol 85:1103–1109, 1987.
63. M Vaeck, A Reynaerts, H Höfte, S Jansens, M De Beuckeleer, C Dean, M Zabeau, M Van Montagu, J Leemans. Transgenic plants protected from insect attack. Nature 328:33–37, 1987.

64. GW Warren, NB Carozzi, N Desai, MG Koziel. Field evaluation of transgenic tobacco containing a *Bacillus thuringiensis* insecticidal protein gene. J Econ Entomol 85:1651–1659, 1992.
65. NB Carozzi, GW Warren, N Desai, SM Jayne, R Lotstein, DA Rice, S Evola, MG Koziel. Expression of a chimeric CaMV 35S *Bacillus thuringiensis* insecticidal protein gene in transgenic tobacco. Plant Mol Biol 20:539–548, 1992.
66. S Williams, L Friedrich, S Dincher, N Carozzi, H Kessmann, E Ward, J Ryals. Chemical regulation of *Bacillus thuringiensis* δ-endotoxin expression in transgenic plants. Bio/Technology 10:540–543, 1992.
67. FJ Perlak, RL Fuchs, DA Dean, SL McPherson, DA Fischhoff. Modification of the coding sequence enhances plant expression of insect control protein genes. Proc Natl Acad Sci USA 88:3324–3328, 1991.
68. MJ Adang, E Firoozabady, J Klein, D DeBoer, V Sekar, JD Kemp, E Murray, TA Rocheleau, K Rashka, G Staffeld, C Stock, D Sutton, DJ Merlo. Expression of a *Bacillus thuringiensis* insecticidal crystal protein gene in tobacco plants. In: CJ Arntzen, C Ryan, ed. Molecular strategies for crop protection. New York: Alan R. Liss, 1987, pp 345–353.
69. MP Hoffmann, FG Zalom, LT Wilson, JM Smilanick, LD Malyj, J Kiser, VA Hilder, WM Barnes. Field evaluation of transgenic tobacco containing genes encoding *Bacillus thuringiensis* delta-endotoxin or cowpea trypsin inhibitor: Efficacy against *Helicoverpa zea* (Lepidoptera: Noctuidae). J Econ Entomol 85:2516–2522, 1992.
70. KE McBride, Z Svab, DJ Schaaf, PS Hogan, DM Stalker, P Maliga. Amplification of a chimeric *Bacillus* gene in chloroplasts leads to an extraordinary level of an insecticidal protein in tobacco. Bio/Technology 15:362–365, 1995.
71. M Mazier, J Chaufaux, N Sanchis, D Lereclus, M Giband, J Tourneur. The *cryic* gene from *Bacillus thuringiensis* provides protection against *Spodoptera littoralis* in young transgenic plants. Plant Sci 127:179–190, 1997.
72. M Kota, H Daniell, S Varma, SF Garczynski, F Gould, WJ Moar. Overexpression of the *Bacillus thuringiensis* (Bt) Cry2Aa2 protein in chloroplasts confers resistance to plants against susceptible and Bt-resistant insects. Proc Natl Acad Sci USA 96:1840–1845, 1999.
73. DA Fischhoff, KS Bowdish, FJ Perlak, PG Marrone, SM McCormick, JG Niedermeyer, DA Dean, K Kusano-Kretzmer, EJ Mayer, DE Rochester, SG Rogers, RT Fraley. Insect tolerant transgenic tomato plants. Bio/Technology 5:807–813, 1987.
74. X Delannay, BJ LaVallee, RK Proksch, RL Fuchs, SR Sims, JT Greenplate, PG Marrone, RB Dodson, JJ Augustine, JG Layton, DA Fischhoff. Field performance of transgenic tomato plants expressing *Bacillus thuringiensis* var. *kurstaki* insect control protein. Bio/Technology 7:1265–1269, 1989.
75. EPJ Burgess, AMR Gatehouse. Engineering for insect pest resistance. In: BD McKersie, DCW Brown, ed. Biotechnology and the Improvement of Forage Legumes. London: CAB International, 1997, pp 229–258.
76. CR Voisey, DWR White, PJ Wigley, CN Chilcott, PG McGregor, DR Woodfield. Release of transgenic white clover plants expressing *Bacillus thuringiensis* genes: An ecological perspective. Biocontrol Sci Technol 4:475–481, 1994.
77. DD Ellis, DE McCabe, S McInnis, R Ramachandran, DR Russell, KM Wallace, BJ Martinell, DR Roberts, KF Raffa, BH McCown. Stable transformation of *Picea glauca* by particle acceleration. Bio/Technology 11:84–89, 1993.
78. B Kantz. Bt crops: Year two. AgConsultant 53:21–22, 1997.
79. ME Whalon, WH McGaughey. Insect resistance to *Bacillus thuringiensis*. In: L Kim ed. Advanced Engineered Pesticides. New York: Marcel Dekker, 1993, pp 215–232.
80. WH McGaughey. Insect resistance to the biological insecticide *Bacillus thuringiensis*. Science 229:193–195, 1985.
81. BE Tabashnik, N Finson, MW Johnson. Managing resistance to *Bacillus thuringiensis:* Lessons from the diamondback moth (Lepidoptera: Plutellidae). J Econ Entomol 84:49–55, 1991.
82. ME Whalon, DL Miller, RM Hollingworth, EJ Grafius, JR Miller. Selection of a Colorado potato beetle (Coleoptera: Chrysomelidae) strain resistant to *Bacillus thuringiensis*. J Econ Entomol 86:226–233, 1993.

83. IF Goldman, J Arnold, BC Carlton. Selection for resistance to *Bacillus thuringiensis* subspecies *israelensis* in field and laboratory populations of the mosquito *Aedes aegypti*. J Invertebr Pathol 47:317–324, 1986.
84. BE Tabashnik, NL Cushing, N Finson, MW Johnson. Field development of resistance to *Bacillus thuringiensis* in diamondback moth (Lepidoptera: Plutellidae). J Econ Entomol 83:1671–1676, 1990.
85. J Ferré, MD Real, J Van Rie, S Jansens, M Peferoen. Resistance to the *Bacillus thuringiensis* bioinsecticide in a field population of *Plutella xylostella* is due to a change in a midgut membrane receptor. Proc Natl Acad Sci USA 88:5119–5123, 1991.
86. JA Addison. Persistence and nontarget effects of *Bacillus thuringiensis* in soil: A review. Can J For Res 23:2329–2342, 1993.
87. MT Johnson, F Gould. Interaction of genetically engineered host plant resistance and natural enemies of *Heliothis virescens* (Lepidoptera: Noctuidae) in tobacco. Environ Entomol 2:586–597, 1992.
88. VJ Mascarenhas, RG Luttrell. Combined effects of sublethal exposure to cotton expressing the endotoxin protein of *Bacillus thuringiensis* and natural enemies on survival of bollworm (Lepidoptera: Noctuidae) larvae. Environ Entomol 26:939–945, 1997.
89. JE Losey, LS Raynor, ME Carter. Transgenic pollen harms monarch larvae. Nature 399:214, 1999.
90. R Brousseau, L Masson, DD Hegedus. Insecticidal transgenic plants: Are they irresistible? AgBiotechNet 1:1–10, 1999.
91. R Roush. Managing resistance to transgenic crops. In: N Carozzi, M Koziel, eds. Advances in Insect Control: The Role of Transgenic Plants. London: Taylor and Francis, 1997, pp 271–294.
92. GW Warren. Vegetative insecticidal proteins: Novel proteins for control of corn pests. In: N Carozzi, M Koziel, eds. Advances in Insect Control: The Role of Transgenic Plants. London: Taylor and Francis, 1997, pp 109–121.
93. JJ Estruch, GW Warren, MA Mullins, GJ Nye, JA Craig, MG Koziel. Vip3A, a novel *Bacillus thuringiensis* vegetative insecticidal protein with a wide spectrum of activities against lepidopteran insects. Proc Natl Acad Sci USA 93:5389–5394, 1996.
94. CG Yu, MA Mullins, GW Warren, MG Koziel, JJ Estruch. The *Bacillus thuringiensis* vegetative insecticidal protein Vip3A lyses midgut epithelium cells of susceptible insects. Appl Environ Microbiol 63:532–536, 1997.
95. M Blackburn, E Golubeva, D Bowen, RH Ffrench-Constant. A novel insecticidal toxin from *Photorabdus luminescens*, toxin complex a (Tca) and its histopathological effects on the midgut of *Manduca sexta*. Appl Environ Microbiol 64:3036–3041, 1998.
96. DJ Bowen, TA Rocheleau, M Blackburn, O Andreev, E Golubeva, R Bhartia, RH Ffrench-Constant. Insectidial toxins from the bacterium *Photorhabdus luminescens*. Science 280:2129–2132, 1998.
97. DJ Bowen, JC Ensign. Purification and characterization of a high-molecular-weight insecticidal protein complex produced by the entomopathogenic bacterium *Photorhabdus luminescens*. Appl Environ Microbiol 64:3029–3025, 1998.
98. R Ffrench-Constant, D Bowen. *Photorhabdus* toxins: Novel biological insecticides. Curr Opin Micribiol 2:284–288, 1999.
99. JP Purcell, JT Greenplate, MG Jennings, JS Ryerse, JC Pershing, SR Sims, MJ Prinsen, DR Corbin, M Tran, RD Sammons. Cholesterol oxidase: A potent insecticidal protein active against boll weevil larvae. Biochem Biophys Res Comm 196:1406–1413, 1993.
100. DR Corbin, JT Greenplate, EY Wong, JP Purcell. Cloning of an insecticidal cholesterol oxidase gene and its expression in bacteria and in plant protoplasts. Appl Environ Microbiol 60:4239–4244, 1994.
101. DR Corbin, JT Greenplate, MG Jennings, JP Purcell, RD Sammons. Method of controlling insects. US Patent 5,554,369, 1996.
102. TA Walsh, RA Houtchens, JA Strickland, GL Orr, DJ Merlo. Insecticidal proteins and method for plant protection. US Patent 5743477, 1998.
103. GH Feng, M Richardson, MS Chen, KJ Kramer, T Morgan, GR Reeck. Alpha-amylase inhibitors from wheat: Amino acid sequences and patterns of inhibition of insect and human alpha-amylases. Insect Biochem Mol Biol 26:419–426, 1996.
104. AV Leberre, GC Bompard, F Payan, P Rouge. Characterization of functional properties of the alpha-

amylase inhibitor (alpha-AI) from kidney bean (*Phaseolus vulgaris*) seeds. Biochim Biophys Acta 1343:31–40, 1997.

105. AP Giri, MS Kachole. Amylase inhibitors of pigeonpea (*Cajanus cajan*) seeds. Phytochemistry 47:197–202, 1998.

106. R Johnson, J Narvaez, A Gynheung, C Ryan. Expression of proteinase inhibitors I and II in transgenic tobacco plants: Effects on natural defense against *Manduca sexta* larvae. Proc Natl Acad Sci USA 86:9871–9875, 1989.

107. XL Duan, XG Li, QZ Xue, M Aboelsaad, DP Xu, R Wu. Transgenic rice plants harboring an introduced potato proteinase inhibitor II gene are insect resistant. Nat Biotechnol 14:494–498, 1996.

108. JC Thomas, DG Adams, VD Keppenne, CC Wasman, JK Brown, MR Kanost, HJ Bohnert. Protease inhibitors of *Manduca sexta* expressed in transgenic cotton. Plant Cell Rep 14:758–762, 1995b.

109. JC Thomas, CC Wasman, C Echt, RL Dunn, HJ Bohnert, TJ McCoy. Introduction and expression of an insect proteinase inhibitor in alfalfa (*Medicago sativa* L.). Plant Cell Rep 14:31–36, 1994.

110. J Graham, RJ McNicol, K Greig. Towards genetic based insect resistance in strawberry using the cowpea trypsin inhibitor gene. Ann Appl Biol 127:163–173, 1995.

111. NB Klopfenstein, HS McNabb, ER Hart, RB Hall, RD Hanna, SA Heuchelin, KK Alen, NO Shi, RW Thornburg. Transformation of *Populus* hybrids to study and improve pest resistance. Silvae Genet 42:86–90, 1993.

112. C Girard, M Bonade-Bottino, MH Pham-Delegue, L Jouanin. Two strains of cabbage seed weevil (Coleoptera: Curculionidae) exhibit differential susceptibility to a transgenic oilseed rape expressing oryzacystatin I. J Insect Physiol 44:569–577, 1998a.

113. SC MacIntosh. Potentiation of *Bacillus thuringiensis* insecticidal activity by serine protease inhibitors. J Agric Food Chem 38:1145–1152, 1990.

114. C Pannetier, M Giband, P Couzi, V Letan, M Mazier, J Tourneur, B Hau. Introduction of new traits into cotton through genetic engineering: Insect resistance as example. Euphytica 9:163–166, 1997.

115. RE Andrews, MM Bibilos, LA Bulla. Protease activation of the entomocidal protoxin of *Bacillus thuringiensis* subsp. *kurstaki*. Appl Environ Microbiol 50:737–742, 1985.

116. M Keller, B Sneh, N Strizhov, E Prudovsky, A Regev, C Koncz, J Schell, A Zilberstein. Digestion of delta-endotoxin by gut proteases may explain reduced sensitivity of advanced instar larvae of *Spodoptera littoralis*. Insect Biochem Mol Biol 26:365–373, 1996.

117. Z Shao, Y Cui, X Liu, H Yi, J Ji, Z Yu. Processing of delta-endotoxin of *Bacillus thuringiensis* subsp. *Kurstaki* HD-1 in *Heliothis armigera* midgut juice and the effects of protease inhibitors. J Invertebr Pathol 72:73–81, 1998.

118. BE Tabashnik, N Finson, MW Johnson. Two protease inhibitors fail to synergize *Bacillus thuringiensis* in diamondback moth (Lepidopteran: Plutellidae). J Econ Entomol 85:2082–2087, 1992.

119. BL Roberts, W Markland, AC Ley, RB Kent, DW White, SK Guterman, RC Ladner. Directed evolution of a protein: Selection of potent neutrophil elastase inhibitors displayed on M13 fusion phage. Proc Natl Acad Sci USA 89:2429–2433, 1992.

120. PE Urwin, HJ Atkinson, DA Waller, MJ McPherson. Engineered oryzacystatin-I expressed in transgenic hairy roots confers resistance to *Globodera pallida*. Plant J 8:121–131, 1995.

121. H Koiwa, RE Shade, K Zhu-Salzman, L Subramanian, LL Murdock, SS Nielsen, RA Bressan, PM Hasegawa. Phage display selection can differentiate insecticidal activity of soybean cystatins. Plant J 14:371–379, 1998.

122. MA Jongsma, PL Bakker, J Peters, D Bosch, WJ Stiekema. Adaptation of *Spodoptera exigua* larvae to plant proteinase inhibitors by induction of gut proteinase activity insensitive to inhibition. Proc Natl Acad Sci USA 92:8041–8045, 1995.

123. RM Broadway. Dietary regulation of serine proteinases that are resistant to serine proteinase inhibitors. J Insect Physiol 43:855–874, 1997.

124. JA Gatehouse, AL Shannon, EPJ Burgess, JT Christeller. Characterization of major midgut proteinase cDNAs from *Helicoverpa armigera* larvae and changes in gene expression in response to four proteinase inhibitors in the diet. Insect Biochem Mol Biol 27:929–944, 1997b.

125. MA Bonade-Bottino, J Lerin, B Zaccomer, L Jouanin. Physiological adaptation explains the insen-

sivity of *Baris coerulescens* to transgenic oilseed rape expression oryzacystatin I. Insect Biochem Mol Biol 29:131–138, 1999.
126. F De-Leo, MA Bonade-Bottino, LR Ceci, R Gallerani, L Jouanin. Opposite effects *on Spodoptera littoralis* larvae of high expression level on a trypsin proteinase inhibitor in transgenic plants. Plant Physiol 118:997–1004, 1998.
127. C Girard, ML Metayer, B Zaccomer, E Bartlet, I Williams, M Bonade-Bottino, M-H Pham-Delegue, L Jouanin. Growth stimulation of beetle larvae reared on a transgenic oilseed rape expressing a cysteine proteinase inhibitor. J Insect Physiol 44:263–270, 1998.
128. C Girard, ML Metayer, M Bonade-Bottino, M-H Pham-Delegue, L Jouanin. High level of resistance to proteinase inhibitors may be conferred by proteolytic cleavage in beetle larvae. Insect Biochem Mol Biol 28:229–237, 1998.
129. AP Giri, AM Harsulkar, VV Deshpande, MN Sainani, VS Gupta, PK Ranjekar. Chickpea defensive proteinase inhibitors can be inactivated by podborer gut proteinases. Plant Physiol 116:393–401, 1998.
130. MZ Haider, BH Knowles, DJ Ellar. Specificity of *Bacillus thuringiensis* var. *colmeri* insecticidal delta-endotoxin is determined by differential proteolytic processing of the protoxin by larval gut proteases. Eur J Biochem 156:531–540, 1986.
131. B Oppert, KJ Kramer, DE Johnson, SC MacIntosh, WH McGaughey. Altered protoxin activation by midgut enzymes from a *Bacillus thuringiensis* resistant strain of *Plodia interpunctella*. Biochem Biophys Res Comm 198:940–947, 1994.
132. B Oppert, KJ Kramer, RW Beeman, D Johnson, WH McGaughey. Proteinase-mediated insect resistance to *Bacillus thuringiensis* toxins. J Biol Chem 272:23473–23476, 1997.
133. CM Reis, MM Calvet, MP Sales, KVS Fernandes, VM Gomes, J Zavier. Alpha-amylase inhibitors of legume seeds and their involvement in the resistance to Bruchid beetles. Arq Biol Tecnol 40:413–418, 1997.
134. M Ishimoto, T Sato, MJ Chrispeels, K Kitamura. Bruchid resistance of transgenic asuki bean expressing seed alpha-amylase inhibitor of common bean. Entomol Exp Appl 79:309–315, 1996.
135. JE Huesing, RE Shade, MJ Chrispeels, LL Murdock. Alpha-amylase inhibitor, not phytohemagglutinin, explains resistance of common bean seed to cowpea weevil. Plant Physiol 96:993–996, 1991.
136. DH Janzen, HB Juster, IE Liener. Insecticidal action of the phytohemagglutinin in black beans on a bruchid beetle. Science 192:795–796, 1976.
137. HE Schroeder, S Gollasch, A Moore, LM Tabe, S Craig, DC Hardie, MJ Chrispeels, D Spencer, TJV Higgins. Bean alpha-amylase inhibitor confers resistance to the pea weevil (*Bruchus pisorum*) in transgenic peas (*Pisum sativum* L.). Plant Physiol 107:1233–1239, 1995.
138. N Sauvion, Y Rahbe, WJ Peumans, EJM van Damme, JA Gatehouse, AMR Gatehouse. Effects of GNA and other mannose binding lectins on development and fecundity of the peach-potato aphid *Myzus persicae*. Entomol Exp Appl 79:285–293, 1996.
139. VA Hilder, KS Powell, AMR Gatehouse, JA Gatehouse, Y Shi, WDO Hamilton, A Merryweather, CA Newell, JC Timans, WJ Peumans, E van Damme, D Boulter. Expression of snowdrop lectin in transgenic tobacco plants results in added protection against aphids. Transgenic Res 4:18–25, 1995.
140. E Fitches, AMR Gatehouse, JA Gatehouse. Effects of snowdrop lectin (GNA) delivered via artificial diet and transgenic plants on the development of tomato moth (*Lacanobia oleracea*) larvae in laboratory and glasshouse trials. J Insect Physiol 43:727–739, 1997.
141. AMR Gatehouse, SJ Shackley, KA Fenton, J Bryden. Mechanism of seed lectin tolerance by a major insect storage pest of *Phaseolus vulgaris*, *Acanthoscelides obtectus*. J Sci Food Agric 47:269–280, 1989.
142. K Zhu-Salzman, RE Shade, H Koiwa, RA Salzman, M Narasimhan, RA Bressan, PM Hasegawa, LL Murdock. Carbohydrate binding and resistance to proteoloysis control insecticidal activity of *Griffonia simplicifolia* lectin II. Proc Natl Acad Sci USA 95:15123–15128, 1998.
143. KJ Kramer, S Muthukrishnan. Insect chitinases: Molecular biology and potential use as biopesticides. Insect Biochem Mol Biol 27:887–900, 1997.
144. CR Brandt, MJ Adang, KD Spence. The peritrophic membrane: Ultrastructural analysis and function

as a mechanical barrier to microbial infection in *Orgyia pseudotsugata*. J Invertebr Pathol 32:12–24, 1978.
145. M Shahabuddin, T Toyoshima, M Aikawa, DC Kaslow. Transmission-blocking activity of a chitinase inhibitor and activation of malarial parasite chitinase by mosquito protease. Proc Natl Acad Sci USA 90:4266–4270, 1993.
146. X Ding, B Gopalakrishnan, LB Johnson, FF White, X Wang, TD Morgan, KJ Karmer, S Muthukrishnan. Insect resistance of transgenic tobacco expressing an insect chitinase gene. Transgenic Res 7:77–84, 1997.
147. X Wang, X Ding, B Gopalakrishnan, TD Morgan, L Johnson, F White, S Muthukrishnan, KJ Kramer. Characterization of a 46 kDa insect chitinase from transgenic tobacco. Insect Biochem Mol Biol 26:1055–1064, 1996.
148. AMR Gatehouse, GM Davison, CA Newell, A Merryweather, WDO Hamilton, EPJ Burgess, RJC Gilbert, JA Gatehouse. Transgenic potato plants with enhanced resistance to the tomato moth, *Lacanobia oleracae*: Growth room trials. Mol Breed 3:49–63, 1997.
149. AMR Gatehouse, RE Down, KS Powell, N Sauvion, Y Rahbe, CA Newell, A Merryweather, WDO Hamilton, JA Gatehouse. Transgenic potato plants with enhanced resistance to the peach-potato aphid *Myzus persicae*. Entomol Exp Appl 79:295–307, 1996.
150. A Regev, M Keller, N Strizhov, B Sneh, E Prudovsky, I Chet, I Ginzberg, Z Koncz-Kalman, C Koncz, J Schell, A Zilberstein. Synergistic activity of *Bacillus thuringiensis* delta-endotoxin and a bacterial endochitinase against *Spodoptera littoralis* larvae. Appl Environ Microbiol 62:3581–3586, 1996.
151. JJA van Loon, WH Frentz, FA Eeuwijk. Electroantennogram responses to plant volatiles in two species of *Pieris* butterflies. Entomol Exp Appl 62:253–260, 1992.
152. P Gregory, WM Tingey, DA Ave, P Bouthyette. Potato glandular trichomes: A physicochemical defense mechanism against insects. In: MB Green, PA Hedin, eds. Natural Resistance of Plants to Pests. Washington, DC: The American Chemical Society Symposium Series No. 296, 1986, pp 160–167.
153. RB King, RP Singh, A Boucher. Variation in sucrose esters from the B glandular trichomes of certain wild potato species. Am Potato Res 64:524–534, 1987.
154. WM Tingey. Potato glandular trichomes defensive activity against insect attack. In: PA Hedin, ed. Naturally occurring pest bioregulators. ACS Symposium Series No. 449. Washington, DC: American Chemical Society, 1991, pp 126–858.
155. D McCaskill, R Croteau. Strategies for bioengineering the development and metabolism of glandular tissues in plants. Nat Biotechnol 17:31–36, 1999.
156. MW Bonierbale, RL Plaisted, O Pineda, SD Tanksley. QTL analysis of trichome-mediated insect resistance in potato. Theor Appl Genet 87:973–987, 1994.
157. P Palaniswamy, RP Bodnaryk. A wild *Brassica* from Sicily provides trichome-based resistance against flea beetles, *Phyllotreta cruciferae* (Goeze) (Coleoptera: Chrysomelidae). Canadian Entomologist 126:1119–1130, 1994.
158. RJ Lamb. Hairs protect pods of mustard (*Brassica hirta* Gisilba) from flea beetle feeding damage. Can J Plant Sci 60:1439–1440, 1980.
159. DG Oppenheimer, PL Herman, S Sivakumaran, J Esch, MD Marks. A *myb* gene required for leaf trichome differentiation in *Arabidopsis* is expressed in stipules. Cell 67:483–493, 1991.
160. JC Larkin, DG Oppenheimer, S Pollock, MD Marks. *Arabidopsis glabrous*1 gene requires downstream sequences for function. Plant Cell 5:1739–1748, 1993.
161. MD Marks, KA Feldman. Trichome development in *Arabidopsis thaliana*. I. T-DNA tagging of the GLABROUS1 gene. Plant Cell 1:1043–1050, 1989.
162. M Korneef. The complex syndrome of TTG mutants. Arabidopsis Inform Serv 18:45–51, 1981.
163. DB Szymanski, MM Marks. *GLABROUS*1 overexpression and *TRYTPYCHON* alter the cell cycle and trichome cell fate in *Arabidopsis*. Plant Cell 10:2047–2062, 1998.
164. DB Szymanski, DA Klis, JC Larkin, MM Marks. *COT*1: A regulator of *Arabidopsis trichome* initiation. Genetics 149:565–577, 1998.
165. A Schnittger, G Juergens, H Martin. Tissue layer and organ specificity of trichome formation are regulated by *BLABRA1* and *TRYPTYCON* in *Arabidopsis*. Development 125:2283–2289, 1998.

166. BJ Glover, M Perez-Rodriguez, C Martin. Development of several epidermal cell types can be specified by the same MYB-related plant transcription factor. Development 125:3497–3508, 1998.
167. AM Lloyd, V Walbot, RW Davis. Anthocyanin production in dicots activated by maize anthocyanin-specific regulators, R and C1. Science 258:1773–1775, 1992.
168. AV Babwah, GG Brown, CS Waddell. Development of selectable and screenable markers in *Brassica napus*. Abstracts: Brassica Genetic Technologies for the Future. 11th International Crucifer Genetics Workshop, Quebec, 1998, pp 31.
169. DB Szymanski, RA Jilk, SM Pollock, MM Mark. Control of *GL2* expression in *Arabidopsis* leaves and trichomes. Development 125:1161–1171, 1998.
170. KA Stoner. Glossy leaf wax and plant resistance to insects in *Brassica oleracea* under natural infestation. Environ Entomol 19:730–739, 1990.
171. RP Bodnaryk. Leaf epicuticular wax, an antixenotic factor in Brassicaceae that affects the rate and pattern of feeding of flea beetles, *Phyllotreta cruciferae* (Goeze). Can J Plant Sci 72:1295–1303, 1992.
172. SD Eigenbrode, KE Espelie, AM Shelton. Behavior of neonate diamondback moth larvae (*Plutella xylostella* L.) on leaves and on extracted leaf waxes of resistant and susceptible cabbages. J Chem Ecol 17:1691–1704, 1991.
173. JP McNevin, W Woodward, A Hannoufa, KA Feldmann, B Lemieux. Isolation and characterizaton of *eceriferum (cer)* mutants induced by *T-DNA* insertions in *Arabidopsis thaliana*. Genome 36:610–618, 1993.
174. A Hannoufa, J McNevin, B Lemieux. Epicuticular wax of *eceriferum* mutants of *Arabidopsis thaliana*. Phytochem 33:851–855, 1993.
175. MA Jenks, HA Tuttle, SD Eigenbrode, KA Feldman. Leaf epicuticular waxes of the Eceriform mutants in *Arabidopsis*. Plant Physiol 108:369–377, 1995.
176. MGM Aarts, CJ Keijzer, WJ Stiekma, A Pereira. Molecular characterization of the CER1 gene of *Arabidopsis* involved in epicuticular wax biosynthesis and pollen fertility. Plant Cell 7:2115–2127, 1995.
177. A Hannoufa, V Negruk, E Galina, B Lemieux. The CER3 gene of *Arabidopsis thaliani* is expressed in leaves, stems, roots, flowers and apical meristems. Plant J 10:459–467, 1996.
178. Y Xia, BJ Nikolau, PS Schnable. Developmental and hormonal regulation of the *Arabidopsis CER2* gene that codes for a nuclear-localized protein required for the normal localization of cuticular waxes. Plant Physiol 115:925–937, 1997.
179. L Kuntz, S Iyer, S Zachgo, M Giblin, D Taylor, T Millar. Molecular genetic manipulation of very long chain fatty acid biosynthetic pathways in *Arabidopsis*. Abstract. Proceedings of the Canadian Society for Plant Molecular Biology Annual Meeting. Edmonton, Alberta, June 16–19, 1998.
180. M Leech, K May, D Hallard, R Verpoorte, V de Luca, P Christon. Expression of two consecutive genes of a secondary metabolite pathway in transgenic tobacco: Molecular diversity influences levels of expression and product accumulation. Plant Mol Biol 38:765–774, 1998.
181. M Gruber, H Ray, P Auser, B Skadhauge, J Falk, KK Thomsen, J Stougaard, A Muir, G Lees, B McKersie, S Bowley, D von Wettstein. Genetic systems for condensed tannin biotechnology. In: GG Gross, R Hemingway, T Yoshida, eds. Plant Polyphenols. 2. Chemistry and Biology. Plenum Press, New York, pp 315–341, 1999.
182. CP Joshi, VL Chiang. Conserved sequence motifs in plant *S*-adenosyl-L-methionine-dependent methyl transferases. Plant Mol Biol 37:663–674, 1998.
183. CL Steele, J Crock, J Bohlmann, R Croteau. Sesquiterpene synthases from grand fir (*Abies grandis*)—comparison of constitutive and wound-induced activities, and cDNA isolation, characterization and bacterial expression of delta-selinene synthase and gamma-humulene synthase. J Biol Chem 273:2078–2089, 1998.
184. RR Sederoff, JJ MacKay, J Ralph, RD Hatfield. Unexpected variation in lignin. Curr Opin Plant Biol 2:145–152, 1999.
185. GW Felton, JL Bi, CB Summers, AJ Mueller, SS Duffey. Potential role of lipoxygenases in defense against insect herbivory. J Chem Ecol 20:651–666, 1994.
186. MJ Stout, J Workman, SS Duffey. Differential induction of tomato foliar proteins by arthropod herbivores. J Chem Ecol 20:2575–2594, 1994.

187. AJ van der Westhuizen, X-M Qian, A-M Botha. Differential induction of apoplastic peroxidase and chitinase activities in susceptible and resistant wheat cultivars by Russian wheat aphid infestation. Plant Cell Rep 18:132–137, 1998.
188. CP Constabel, CA Ryan, C.A. A survey of wound- and methyl jasmonate–induced leaf polyphenol oxidase in crop plants. Phytochemistry 47:507–511, 1998.
189. QJ Groom, MA Torres, AP Fordham-Skelton, KE Hammond-Kosack, NJ Robinson, JDG Jones. RbohA, a rice homologue of the mammalian gp91phox respiratory burst oxidase gene. Plant J 10: 515–522, 1996.
190. H Willekens, D Inze, M van Montagu, W van Camp. Catalases in plants. Mol Breed 1:207–228, 1995.
191. NE Strobel, JA Kuc. Chemical and biological inducers of system resistance to pathogens protect cucumber and tobacco plants from damage caused by paraquat and cupric chloride. Phytopathology 85:1306–1310, 1995.
192. G Wu, BJ Shortt, DM Shah. Disease resistance conferred by expression of a gene encoding $H_2O_2$-generating glucose oxidase in transgenic potato plants. Plant Cell 7:1357–1368, 1995.
193. MJ Stout, KV Workman, RM Bostock, SS Duffey. Specificity of induced resistance in the tomato, *Lycopersicon esculentum*. Oecologia 113:74–81, 1998.
194. KS Powell, AMR Gatehouse, VA Hilder, JA Gatehouse. Antimetabolic effects of plant lectins and plant and fungal enzymes on the nymphal stages of two important rice pests, *Nilaparvata lugens* and *Nephotettix cinciteps*. Entomol Exp Appl 66:119–126, 1993.
195. L Zhou, JC Jang, TL Jones, J Sheean. Glucose and ethylene signal transduction cross-talk revealed by an *Arabidopsis* glucose-insensitive mutant. Proc Natl Acad Sci USA 95:10294–10299, 1998.
196. IT Baldwin, CA Preston. The ecophysiological complexity of plant responses to insect herbivores. Planta 208:137–145, 1999.
197. K Harms, R Atzorn, A Brash, H Kuhn, C Wasternack, L Willmitzer, H Pena-Cortes. Expression of a flax allene oxide synthase cDNA leads to increased endogenous jasmonic acid (JA) levels in transgenic potato plants but not to a corresponding activation of JA-responding genes. Plant Cell 7:1645–1654, 1995.
198. PF Dowd, LM Lagrimini, DA Herms. Differential leaf resistance to insects of transgenic sweetgum (*Liquidambar styraciflua*) expressing tobacco anionic peroxidase. Cell Mol Life Sci 54:712–720, 1998.
199. J Nawrot, Z Smitalova, M Holub. Deterrent activity of sesquiterpene lactones from the Umbelliferae against storage pests. Biochem Syst Ecol 11:243–245, 1983.
200. LAD Williams, MJ Anderson, YA Jackson. Insecticidal activity of synthetic 2-carboxybenzofurans and their coumarin precursors. Pestic Sci 42:167–171, 1994.
201. CA Elliger, BC Chan, AC Waiss Jr. Flavonoids as larval growth inhibitors: Structural factors governing toxicity. Naturwissenschaften 67:358–360, 1980.
202. JJ Hlywka, GR Stephenson, MK Sears, RY Yada. Effects of insect damage on glycoalkaloid content in potatoes (*Solanum tuberosum*). J Agric Food Chem 42:2545–2550, 1994.
203. T Jonasson, K Olsson. The influence of glycoalkaloids, chlorogenic acid and sugars on the susceptibility of potato tubers to wireworm. Potato Res 37:205–216, 1994.
204. VH Argando, JG Luza, HM Niemeyer, LJ Corcuera. Role of hydroxamic acids in the resistance of cereals to aphids. Phytochem 19:1665–1668, 1980.
205. S Woodhead, G Cooper-Driver. Phenolic acids and resistance to insect attack in *Sorghum bicolor*. Biochem Syst Ecol 7:309–310, 1979.258.
206. PJ Lea, ed. Enzymes of secondary metabolism. In: PM Dey, JB Harborne, eds. Methods in Plant Biochemistry. Vol. 9. New York: Academic Press, 1994.
207. N Lewis. A 20th century roller coaster ride: A short account of lignification. Curr Opin Plant Biol 2:153–162, 1999.
208. RW Whetton, JJ MacKay, RR Sederoff. Recent advances in understanding lignin biosynthesis. Annu Rev Plant Physiol Plant Mol Biol 49:585–609, 1998.
209. JV Shanks, R Bhadra, J Morgan, S Fijhwani, S Vani. Quantification of metabolites in the indole alkaloid pathways of *Catharanthus roseus*: Implications for metabolic engineering. Biotech Bioeng 58:333–338, 1998.

210. J Bohlmann, GG Meyer, R Croteau. Plant terpenoid synthesis: Molecular biology and phylogenetic analysis. Proc Natl Acad Sci USA 95:4120–4133, 1998.
211. HK Lichtenthaler. The plants' 1-deoxy-D-xylulose-5-phosphate pathway for biosynthesis of isoprenoids. FETT-LIPID 100:128–138, 1998.
212. G Bringmann, M Ruecker, M Wenze., C Guenther, K Wolf, J Holenz, J Schlauer. Biological activities and biosynthetic origin of acetogenic isoquinoline alkoids. Pharm Pharmacol Lett 8:5–7, 1998.
213. K Hauschild, HH Pauli, TM Kutcher. Isolation and analysis of a gene *bbe*1 encoding the berberine bridge enzyme from the California poppy *Eschschohzia californica*. Plant Mol Biol 36:473–478, 1998.
214. WM Chou, TM Kutchem. Enzymatic oxidations in the biosynthesis of complex alkaloids. Plant J 15:289–300, 1998.
215. C Lister, C Dean. Recombinant inbred lines for mapping RFLP and phenotypic markers in *Arabidopsis thaliana*. Plant J 4:745–750, 1993.
216. J Keller, E Lim, DW James Jr, HK Dooner. Germinal and somatic activity of the maize element *Activator (Ac)* in *Arabidopsis*. Genetics 131:449–459, 1992.
217. K Feldman. T-DNA insertion mutagenesis in *Arabidopsis:* Mutational spectrum. Plant J 1:71–82, 1991.
218. JC Thomas, DG Adams, CL Nessler, JK Brown, HJ Bonhert. Tryptophan decarboxylase, tryptamine and reproduction of the whitefly. Plant Physiol 109:771–720, 1995.
219. DD Songstad, V De Luca, N Brisson, WGW Kurz, CL Nessler. High levels of tryptamine accumulation in transgenic tobacco expressing tryptophan decarboxylase. Plant Physiol 94:1410–1413, 1990.
220. A Smigocki, JW Neal Jr, I McCanna, L Douglass. Cytokinin-mediated insect resistance in *Nicotiana* plants transformed with the *ipt* gene. Plant Mol Biol 23:325–335, 1993.
221. RP Bodnaryk, L Ma, L Kudryk. Effects of modifying the phytosterol profile of canola, *Brassica napus* L., on growth, development, and survival of the Bertha armyworm, *Mamestra configurata* Walker (Lepidoptera: Noctuidae), the flea beetles, *Phyllotreta cruciferae* (Goeze) (Coleoptera: Chrysomelidae) and the aphids, *Lipaphis erysimi* (Kaltenback). Can J Plant Sci 77:677–683, 1997.
222. JJA van Loon. Chemoreception of phenolic acids and flavonoids in larvae of two species of *Pieris*. J Comp Physiol A 166:889–899, 1990.
223. MB Isman, SS Duffey. Toxicity of tomato phenolic compounds to the fruitworm, *Heliothis zea*. Ent Exp Appl 31:370–376, 1982.
224. RP Bodnaryk. Developmental profile of sinalbin (p-hydroxybenzyl glucosinolate) in mustard seedlings, *Sinapis alba* L., and its relationship to insect resistance. J Chem Ecol 17:1543–1556, 1991.
225. H Eichenseer, CA Mullin, S Chyb. Antifeedent discrimination thresholds for two populations of western cornworm. Physiol Entomol 23:220–226, 1998.
226. SS Duffy, MJ Stout. Antinutritive and toxic compounds of plant defense against insects. Arch Insect Biochem Physiol 32:3–37, 1996.
227. GW Felton, KK Donato, RM Broadway, SS Duffey. Impact of oxidized plant phenolics on the nutritional quality of dietary protein to a noctuid herbivore, *Spodoptera exigua*. J Insect Phyiol 38:277–285, 1992.
228. GW Felton, J Workman, SS Duffey. Avoidance of antinutritive plant defense: Role of midgut pH in Colorado potato beetle. J Chem Ecol 18:571–583, 1992.
229. GW Felton, SS Duffey. Reassessment of the role of gut alkalinity and detergency in insect herbivory. J Chem Ecol 17:1821–1836, 1991.
230. GW Felton, SS Duffey. Inactivation of baculovirus by quinones formed in insect-damaged plant tissues. J Chem Ecol 16:1221–1236, 1990.
231. CT Ludlum, GW Felton, SS Duffey. Plant defenses: Chlorogenic acid and polyphenol oxidase enhance toxicity of *Bacillus thuringiensis* subsp. *kurstaki* to *Heliothis zea*. J Chem Ecol 17:217–237, 1991.
232. GW Felton, SS Duffey. Protective action of midgut catalase in lepidopteran larvae against oxidative plant defenses. J Chem Ecol 17:1715–1732, 1991.

233. S Kusaba, Y Kano-Murakami, M Matsuoka, M Tamaoki, T Sakamoto, I Yamaguchi, M Fukumoto. Alteration of hormone levels in transgenic tobacco plants overexpressing the rice homeobox gene OSH1. Plant Physiol 116:471–476, 1998.
234. P Roeckel, T Oancia, JR Drevel. Phenotypic alterations and component analysis of seed yield in transgenic *Brassica napus* plants expressing the tzs gene. Physiol Planta 102:243–249, 1998.
235. M Deak, GV Horvath, S Davletova, K Torok, L Sass, I Vass, B Barna, Z Kiraly, D Dudits. Plants ectopically expressing the iron-binding protein, ferritin, are tolerant to oxidative damage and pathogens. Nat Biotechnol 17:192–196, 1999.
236. M Dicke, TA Van Beek, MA Posthumus, N Ben Dom, H Van Bokhoven, A De Groot. Isolation and identification of a volatile kariomone that affects acarine predator-prey interactions. J Chem Ecol 16:381–396, 1990.
237. TCJ Turlings, JH Loughrin, PJ McCall, USR Rose, WJ Lewis, JH Tumlinson. How caterpillar-damaged plants protect themselves by attracting parasitic wasps. Proc Natl Acad Sci USA 92:4169–4174, 1995.
238. RW Gibson, JA Pickett. Wild potato repels aphids by release of aphid alarm pheromone. Nature 302:608–609, 1983.
239. DC Griffiths, AJ Hick, BJ Pye, LE Smart. The effects on insect pests of applying isothiocyanate precursors to oilseed rape. In: MFB Dale, AM Dewar, RJ Froud-Williams, TJ Hocking, D Gareth Jones, BL Rea, eds. Production and Protection of Oilseed Rape and Other *Brassica* Crops. Wellesbourne: The Association of Applied Biologists, 1989, pp 359–364.
240. KA Evans, LJ Allen-Williams. Electroantennogram responses of the cabbage seed weevil, *Ceutorhynchus assimilis*, to oilseed rape, *Brassica napus*, ssp. *Oliefera*, volatiles. J Chem Ecol 18:1641–1659, 1992.
241. KA Pivnick, RJ Lamb, D Reed. Response of flea beetles, *Phylotreta* spp., to mustard oils and nitriles in field trapping experiments. J Chem Ecol 18:863–873, 1992.
242. TH Schuler, GM Poppy, BR Kerry, I Denholm. Potential side effects of insect-resistant transgenic plants on arthropod natural enemies. TIBTECH 17:210–216, 1999.
243. ANE Birch, IE Geoghegan, MEN Majerus, JW McNicol, CA Hackett, AMR Gatehouse, JA Gatehosue. Tri-trophic interactions involving pest aphids, predatory 2-spot ladybirds and transgenic potatoes expressing snowdrop lectin for aphid resistance. Mol Breed 5:75–83, 1999.
244. TH Schuler, GM Poppy, RPJ Potting, I Denholm, BR Kerry. Interactions between insect tolerant genetically modified plants and natural enemies. Gene Flow Agric Rel Transgenic Crops 72:197–202, 1999.
245. C Girard, LA Picard-Nizou, E Grallien, B Zaccomer, L Jouanin, MH Pham-Delegue. Effects of proteinase inhibitor ingestion on survival, learning abilities and digestive proteinases of the honeybee. Transgenic Res 7:239–246, 1998.
246. MA Bonade-Bottino, C Girard, L Jouanin, M Le Metayer, AL Picard-Nizou, G Sandoz, MH Pham-Delegue, J Lerin. Effects of transgenic oilseed rape expressing proteinase inhibitors on pest and beneficial insects. Acta Hortic 459:235–239, 1998.
247. A Saegusa. Japan tightens rules on GM crops to protect the environment. Nature 399:719, 1999.
248. Y-B Liu, B Tabashnik, TJ Dennehy, AL Patin, AC Bartlett. Development time and resistance to *Bt* crops. Nature 400:519, 1999.
249. GG Khachatourians. Production and use of biological pest control agents. Trends Biotechnol 4:120–124, 1986.
250. GG Khachatourians. Physiology and genetics of entomopathogenic fungi. In: DK Arora, L Ajello, KG Mukerji, eds. Handbook of Applied Mycology. Vol. 2. Humans, Animals, Insects. New York: Marcel Decker, 1991, pp 613–661.
251. GG Khachatourians. The relationship between biochemistry molecular biology of entomopathogenic fungal insect diseases. In: DH Howard, JD Miller, eds. Animal Human Relationships. The Mycota, K Esser, PA Lemke, Series ed. Vol. VI. Berlin: Springer Verlag, 1996, pp 331–362.
252. TA Pfeifer, TA Grigliatti. Future perspectives on insect pest management: Engineering of the pest. J Invertebr Pathol 67:109–119, 1996.
253. AMR Gatehouse, Biotechnological applications of plant genesin production of insect-resistant

crops. In: SL Clement, SR Quisenberry, ed. Global Resources for Insect-Resistant Crops. Boca Raton, FL: CRC Press, 1999, pp 263–280.
254. D Boulter, GA Edwards, AMR Gatehouse, JA Gatehouse, VA Hilder. Additive protective effects of incorporating two different higher plant derived insect resistance genes in transgenic tobacco plants. Crop Protect 9:351–354, 1990.
255. S Morse, W Buhler. Integrated pest management: Ideals and realities in developing countries. Boulder, CO: Lynne Reiner, 1997, pp 7–78.
256. E Valencia, GG Khachatourians. Integrated pest management and entomopathogenic fungal biotechnology in the Latin Americas. I. Opportunities in a global agriculture. Rev Acad Colomb Cienc 22:193–202, 1998.
257. GG Khachatourians, E Valencia. Integrated pest management and entomopathogenic fungal biotechnology in the Latin Americas. II. Key Research and Development Prerequisites. Rev Acad Colomb Cienc 23: pp 491–496, 1999.
258. MA Altieri. Agroecology: The scientific basis of alternative agriculture. London: IT Publications. 1987.

# 18
# Intellectual Property Protection for Transgenic Plants

**Brain G. Kingwell and Joy D. Morrow**
*Smart and Biggar/Fetherstonhaugh and Co., Vancouver, British Columbia, Canada*

|       |                                                              |     |
|-------|--------------------------------------------------------------|-----|
| I.    | INTRODUCTION                                                 | 279 |
| II.   | PATENTS                                                      | 280 |
|       | A. International Framework                                   | 280 |
|       | B. General Patent Principles                                 | 280 |
|       | C. The Claims Define the Invention                           | 281 |
|       | D. Patentability of Plants: United States Versus European Union | 283 |
|       | E. Inventorship                                              | 284 |
|       | F. Patent Infringement                                       | 285 |
|       | G. Plant Patents in the United States                        | 286 |
| III.  | PLANT VARIETY PROTECTION                                     | 287 |
| IV.   | CONCLUSION                                                   | 290 |
|       | BIBLIOGRAPHY                                                 | 290 |
|       | NOTES                                                        | 291 |

## I. INTRODUCTION

An important objective of intellectual property law is to provide a reliable framework of rules, so that important commercial decisions about investments in research may be made with a degree of certainty about how the products of the research will be protected. In keeping with this requirement for predictability, intellectual property laws are typically drafted in general terms and applied to new technologies on the basis of established principles.

This chapter sets out to describe the established principles governing various forms of intellectual property and tells the not-yet-complete story of how those principles are expected to be applied to transgenic plants. The historical context of ever-increasing protection for plant innovations, coupled with ongoing efforts to achieve international harmonization, provides an optimistic backdrop to some of the uncertainties that surround present efforts to obtain strong intellectual property protection for the full scope of innovations in the area of transgenic plants.

## II. PATENTS

### A. International Framework

The harmonization of patent protection across jurisdictional boundaries has been pursued through various initiatives in international law over a considerable period of time. The Paris Convention for the Protection of Industrial Property[1] has been the subject of a number of revisions since its initial text was agreed to in 1883. The impetus to harmonize international patent laws is often seen today in the efforts of the World Intellectual Property Organization (WIPO) (http://www.wipo.org), an agency of the United Nations.

Under the Paris Convention, member states (convention countries) are entitled to maintain independent criteria with respect to patentability, and the principle of *national treatment* operates to make patent rights available to foreign applicants and nationals equally. This national treatment is augmented by a system that allows applicants to claim *foreign priority* in convention countries for up to 1 year after the filing of their first application in a convention country. The intended effect of a claim to foreign priority is that any application filed in a convention country within the convention year will be treated for most purposes as if it had been filed on the date when the first application was filed in a convention country.

Although the Paris Convention marks the beginning of what is now a long-standing trend in the internationalization of intellectual property, its lack of uniform standards for patentability leaves open the possibility that different jurisdictions may adopt dissimilar standards for assessing the patentability of inventions. Further efforts toward the harmonization of patent systems have resulted in more recent regional and international agreements, such as the European Patent Convention (EPC), the North American Free Trade Agreement (NAFTA), the Trade-Related Aspects of Intellectual Property Rights (TRIPs) component of the Uruguay round of the General Agreement on Tariffs and Trade (GATT), and the Patent Cooperation Treaty, under which WIPO administers an international patent application system.

Under the TRIPs agreement, as an exception to a general rule that patents are to be made available without discrimination as to the field of the technology, Article 27 permits member countries to exclude from patentability, among other things, plants and essentially biological processes for the production of plants (other than microbiological processes). Article 27 does, however, also provide that members must ensure the protection of plant varieties either by patents, by an effective sui generis system, or by any combination of such systems. In many jurisdictions, the implementation of the International Union for the Protection of New Varieties of Plants (UPOV) Convention (discussed later) provides such a sui generis system for the protection of plant varieties.

The language of the TRIPs agreement leaves considerable room for its implementation either as a broad bar to patenting plants or as a narrow exclusion from patentability of plants that are protectable under a sui generis system. Mexico, for example, has changed its laws to allow patenting of plant varieties while excluding plants per se—an approach that is essentially the opposite of the approach suggested in the recent European Biotechnology Directive (discussed later).

### B. General Patent Principles

The generally recognized criteria for patentability, as for example set out in Article 33 of the Patent Cooperation Treaty, are that an invention must

1. Be novel
2. Involve an inventive step (i.e., it must not be obvious)
3. Be useful or industrially applicable

## C. The Claims Define the Invention

The requirements for patentability are generally judged with respect to the *claimed* invention[2]. A patentee is free to define the invention broadly in the claims and to cover various aspects of the invention, provided the requirements for patentability can be met with respect to each claimed aspect of the invention. For example, aspects of transgenic plant innovations may include the following:

1. The plant itself
2. A species, genus, or other range of plants, transformed to include the new trait
3. Progeny of the transformed plant
4. Plant parts (plant cells, protoplasts, cell tissue cultures, plant calli, embryos, pollen, flowers, kernels, tubers, etc.)
5. Plant products, such as oils or fibers
6. Methods of making (transforming) the plant (or plant cells)
7. Methods of making a product by using the plant
8. Methods of reproducing the plant.

Genomic innovations in the field of transgenic plants may include genes that are novel in the sense that they have not previously been identified; genes that are novel only in the sense that their function has not previously been identified; new combinations of genetic material, such as a recombinant gene made up of a tissue-specific promoter and a coding sequence with which it would not normally be associated; newly identified alleles of known genes; expressed sequence tags; and vectors for transforming plant cells.

In some circumstances, a transgenic plant may be broadly defined in claims using functional language, rather than specific sequence limitations. For example, U.S. Patent No. 5,639,947 includes a claim to

> a transgenic plant comprising: (a) plant cells containing nucleotide sequences encoding immunoglobulin heavy- and light-chain polypeptides that each contain an immunoglobulin leader sequence forming a secretion signal; and (b) immunologically active immunoglobulin molecules encoded by said nucleotide sequences.

Alternatively, a plant may be more narrowly defined by reciting a specific nucleotide sequence. For example, where the novelty resides in a new promoter, the promoter may be recited in combination with an undefined sequence of interest, such as "a transgenic plant containing a chimeric gene comprising: (a) a raspberry *dru1* promoter, and (b) a DNA sequence encoding a product of interest, where said DNA sequence is heterologous to said promoter and said DNA sequence is operably linked to said promoter to enable constitutive expression of said product[3]."

### 1. Novelty

To be new, an invention generally must not previously have been made available to the public, either through human activities or by virtue of its occurrence in nature. Some jurisdictions provide grace periods, during which an inventor's own disclosure of the invention will not act as a bar to patentability, such as the 1-year grace periods provided by the United States and Canada. In other jurisdictions, such as in European member states of the EPC, any disclosure of an invention before filing of a patent application may act as a bar to patentability.

Typically genetic inventions must be carefully characterized in claims in ways that distinguish them from a naturally occurring product. An agronomically important allele, for example, may be discovered in an existing variety and claimed as an "isolated" sequence, or as a component of a vector, or as a recombinant component of a transgenic plant, all of which may be embodiments of the allele that did not previously exist in nature. The allele itself may be narrowly defined in terms of its entire sequence, or more broadly characterized as requiring only certain functionally important portions of its sequence.

## 2. Obviousness/Inventive Step

Whether an innovation involves an inventive step, or, on the contrary, is obvious, is typically assessed by determining whether each of the components of the invention may be found in the prior art, either as identifiable pieces of publicly available information or as part of the common general knowledge of someone skilled in the art of the invention. Typically, to support a finding of obviousness there must be some reasonable basis for combining the prior art references to obtain the invention. Many inventions are combinations of preexisting knowledge in a way that yields an unexpected, and hence patentable, result. For example, the transformation of a known plant species with a known gene may give rise to unexpected phenotypic changes in the transgenic plant that provide a basis for patentability.

In biological systems, often the consequences of a particular genetic transformation are not obvious until the transformation is carried out. In the context of U.S. law, for example, this principle is reflected in decisions to the effect that an invention is not considered obvious merely because the prior art suggests that the claimed composition or device should be made without revealing a reasonable expectation of success in making the claimed invention.[4]

## 3. Utility/Industrial Applicability

The requirement that a claimed invention must be useful gives rise to the practice of reciting in a claim all of the component parts of an invention that are required for operation, even though the essence of the invention may lie in only some or one of the parts. As a result, even a claim that claims an invention broadly may recite a significant number of components, where those components are defined primarily in functional terms, for example[5]:

> 1. Transgenic *Brassica* species cells and progeny thereof comprising an expression cassette, wherein said cells are characterized as oncogene-free and capable of regeneration to morphologically normal whole plants, and wherein said expression cassette comprises, in the 5'-3' direction of transcription:
>    1. a transcription initiation region functional in *Brassica* species cells;
>    2. a [deoxyribonucleic acid] DNA sequence comprising an open reading frame having an initiation codon at its 5' terminus or a nucleic acid sequence complementary to an endogenous transcription product; and
>    3. a transcription termination region functional in *Brassica* species cells;
>
> wherein at least one of said transcription initiation region and transcription termination region is not naturally associated with said DNA sequence or said nucleic acid sequence; and wherein said expression cassette imparts a detectable trait when said *Brassica* species cells are grown under conditions whereby said DNA sequence or said nucleic acid sequence is expressed.

## 4. Disclosure

The requirement that a patentee must provide an enabling disclosure of the claimed invention creates a link between the breadth of allowable claims and the extent of disclosure. In general, a sufficient number of examples must be presented to provide a reasonable basis for concluding that an invention has been enabled across its entire scope. For example, the disclosure of a single working example, in which a gene has been introduced into a single plant species, may be inadequate to support a claim to the transformation and expression of that gene in any plant.[6] On the other hand, where an applicant is able to establish that a transformation method is generally effective in a particular species, it may be possible to obtain broad claims to transgenic plants of that species without limitation as to the nature of the transforming gene.[7]

In circumstances in which it may be difficult to provide a written description that enables others to make and use a transgenic plant, a biological deposit may be made in a recognized depository to provide others with materials that may be necessary to practice the invention. In many jurisdictions, such deposits are recognized through the implementation of the Budapest Treaty on the International Recognition of the Deposit of Micro-Organisms for the Purposes of Patent Procedure (the Budapest Treaty). In most jurisdictions, deposits must be made prior to filing an application. In the United States, it may be possible to rely upon a deposit made after the filing date of the application provided the deposit is made before the patent issues. Deposits of biological material for patent purposes are maintained by recognized depositories such as the American Type Culture Collection (ATCC) under terms designed to ensure that the deposit remains viable so that the public can practice the invention after the expiration of the patent term.

## D. Patentability of Plants: United States Versus European Union

### 1. United States

In the United States, a full range of patent protection is available for plants, from extremely broadly defined groups of transgenic plants (exemplified by U.S. Patent No. 5,159,135, reading on transgenic cotton) to narrow coverage somewhat comparable in scope to a breeder's right (exemplified by U.S. Patent No. 5,602,318). Narrow patent claims, like breeder's rights, typically refer to the variety itself to define the scope of the subject matter. For example, the subject matter of U.S. Patent No. 5,602,318 is defined in part as follows:

1. Seed of a maize inbred line, designated PHDG1, and having ATCC Accession No. 97663.
2. A maize plant and its parts produced by the seed of claim 1 and its plant parts.

For several years after a seminal 1980 U.S. Supreme Court decision in which patents on life forms were approved (*Diamond v. Chakrabarty*[8]) the U.S. Patent and Trademark Office restricted applicants to claiming hybrid plants, because hybrids could not be protected as varieties under the Plant Variety Protection Act (PVPA)[9]. This changed as a result of a 1985 administrative decision in the Patent Office, which held that the existence of an overlap in protection with the PVPA was not a bar to the patentability of plants per se[10]. Despite the now long-established practice of granting patents in the United States on both transgenic and nontransgenic plants, the patentability of plants was only confirmed by a court in 1998[11]. The patents in issue in that case relate to hybrid corn plants developed by Pioneer HI-Bred, and the defendant asserts that the patents are invalid because the plants are improper subject matter for utility patents, arguing in essence that plants are to be protected exclusively by the Plant Patent Act[12] or by the Plant Variety Protection Act[13]. At the trial level, the U.S. court applied the *Chakrabarty* decision and found that plants were among the very broad range of subject matter that was appropriate for utility patents. It remains to be seen what the results will be of the appeal from this trial level decision.

## 2. European Union

In Europe, there are differing national patent laws, as well as the umbrella of the regional patent system under the European Patent Convention (EPC). In member states of the EPC, the EPC takes precedence over national laws, and national laws must be harmonized with the EPC. Further European regional complexities arise through legislation enacted by the European Community or Union through directives or regulations. For example, the Directive on the Legal Protection of Biotechnological Inventions (the "Biotechnology Directive") was passed by the European Parliament on June 16, 1998, and went into force on July 30, 1998.

Article 15 of the Biotechnology Directive obliges member states to put it in force by no later than July 30, 2000. Although the Biotechnology Directive is applicable to member states of the European Union, not all contracting states of the EPC belong to the European Union. The uncertainty with respect to how the European Patent Office would implement the Biotechnology Directive has to a certain extent been alleviated with the adoption of new regulations under the EPC.[14]

The patentability of transgenic plants in Europe rests in principle upon the interpretation of Article 53(b) of the EPC, which excludes from patentability "plant or animal varieties or essentially biological processes for the production of plants or animals; this provision does not apply to microbiological processes or the products thereof." The purpose of the exception in the first clause of Article 53 was to prevent double protection of plant varieties with both patents and plant variety rights.[15] Interpretation of this provision was the subject of extensive analysis in a case involving Novartis' European Patent Application 91810144.5.[16] In simple terms, the result of this case may be taken as establishing that claims to transgenic plants are in principle patentable, provided specific plant varieties are not individually claimed (and irrespective of the way in which the varieties are produced). Claims to processes for producing transgenic plants are also considered patentable subject matter, and will be evaluated without regard to the fact that such claims are deemed to cover products (plants) obtained by such processes.

## E. Inventorship

In most jurisdictions, patents are granted to the first applicant to file a patent application. In the United States, however, the entitlement to a patent is based on priority of invention, i.e., a first-to-invent system. The practice in the United States of granting patents to the first inventor gives rise to a uniquely American procedure called an *interference*, within the United States Patent & Trademark Office, for determining priority of invention.

In general terms, invention in the United States has two components: (a) conception and (b) reduction to practice. Conception involves the formulation in the inventor's mind of a complete idea for an operative invention and a method of making the invention, together with disclosure of the idea. Reduction to practice involves either constructing a working embodiment of the invention or filing a patent application that fully describes a working embodiment. Typically, the first person to conceive of an invention will be the first inventor, provided that person is diligent in reducing the invention to practice.

This focus on inventorship in the United States makes it important for inventors to retain clear records of the inventive process, particularly dates of conception and reduction to practice. The evidentiary rules that are typically applied in the United States in assessing inventorship generally favor documentary evidence of inventorship corroborated by noninventors. These peculiarities of U.S. practice became more relevant outside the United States with the changes to U.S. law that accompanied the implementation of NAFTA and the Uruguay round of GATT. These changes have made it possible for Canadian and Mexican inventors to prove dates of invention with reference to activities that occurred outside the United States as early as December 8, 1993. For other World Trade Organization member countries, the right to prove a date of invention with reference to acts occurring outside the United States has been extended back to January 1, 1996. Inventors in WTO countries now have good reason to document the innovation process carefully, to provide evidence that may be necessary to establish priority in the United States.

### F. Patent Infringement

The exclusive right conferred by a patent in most jurisdictions includes the right to exclude others from making, using, and selling the invention defined by the claims for the term of the patent. Generally, patents have a term of 20 years from the filing date of the patent application[17]. The remedies available for a patentee against an infringer typically include an injunction (an order that a party discontinue acts of infringement), damages (the monetary measure of the patentee's loss as a result of the infringement), and/or the recovery of the infringer's profits. A punitive award may be available in circumstances in which the infringement is particularly egregious, as for example is available in the United States through an award of triple damages in cases in which infringement is willful.

In the context of transgenic plants, there are several interesting questions with respect to what will constitute infringement of certain claims. For example, claims to genetic material may be considered to be infringed by the reproduction, use, or sale of plants containing that genetic material. This raises difficult issues in jurisdictions where plants are not per se patentable. On the one hand, there is an argument to the effect that the exclusion from patentability of plants should not be subverted by the enforcement of claims to genetic material. On the other hand, the logical application of the traditional analysis of infringement suggests that the reproduction, use, or sale of a plant also constitutes the reproduction, use, and sale of the plant's genetic material, including patented genetic material. It is entirely possible that different jurisdictions will resolve these competing points of view in different ways. In Europe, Article 9 of the Biotechnology Directive addresses this point as follows:

> The protection conferred by a patent on a product containing or consisting of genetic information shall extend to all material, save as provided in Article 5(1) [i.e., excluding the human body and its parts], in which the product is incorporated and in which the genetic information is contained and performs its function.

Even in jurisdictions where plants per se are not patentable, claims may generally be obtained to processes for making a novel plant. It is well established in many jurisdictions that process claims may be infringed by making, using, or selling the product of a patented process, even where claims for the product itself are not granted. In the past, this principle has been applied in the chemical and pharmaceutical fields to prevent the importation and sale of a compound made abroad using a patented process. This principle is, for example, enshrined in the European Patent Convention as follows: "If the subject matter of the European patent is a process, the protection conferred by the patent shall extend to the products directly obtained by such a process"[18].

A further issue arises with respect to whether subsequent generations of a genetically modified plant will be considered to be products of the initial process of genetic modification. In Europe, the directive answers these questions in the affirmative in Article 8, which also provides similar protection to the propagated products of patented biological materials, as follows:

> 1. The protection conferred by a patent on a biological material possessing specific characteristics as a result of the invention shall extend to any biological material derived from that

biological material through propagation or a multiplication in an identical or divergent form and possessing those same characteristics.
2. The protection conferred by a patent on a process that enables a biological material to be produced possessing specific characteristics as a result of the invention shall extend to biological material directly obtained through that process and to any other biological material derived from the directly obtained biological material through propagation or multiplication in an identical or divergent form and possessing those same characteristics.

In most jurisdictions, relief is available to a patentee to prevent the importation of a product made abroad by a patented process. For example, in the United States, the International Trade Commission is empowered to issue an exclusion order of a product made, produced, or processed abroad if the product was made by a claimed process[19], and infringement includes the act of importation into the United States of a product that was made by a process patented in the United States[20]. The burden of proving that a product is foreign-produced by a patented process has been shifted in many jurisdictions, in some cases as a result of Article 34 of the TRIPs agreement, which in effect states a presumption that a product is made by a patented process, where a defendant is unable to prove otherwise.

The principle of patent infringement extending to the propagated products of a patented plant or a patented process for producing a plant may be contrasted with the doctrine of *exhaustion of rights*. In general, under the exhaustion doctrine, the rights of the patentee cease, or are exhausted, by the first sale of the product embodying the invention. In Europe, the Biotechnology Directive enshrines the principle of exhaustion throughout the European Community in Article 10 and further provides a form of farmer's exemption in Article 11, as follows:

*Article 10*

The protection referred to in Articles 8 and 9 shall not extend to biological material obtained from the propagation or multiplication of biological material placed on the market in the territory of a Member State by the holder of the patent or with his consent, where the multiplication or propagation necessarily results from the application for which the biological material was marketed, provided that the material obtained is not subsequently used for other propagation or multiplication.

*Article 11*

By way of derogation from Articles 8 and 9, the sale or other form of commercialization of propagating material to a farmer by the holder of the patent or with his consent for agricultural use implies authorization for the farmer to use the product of his harvest for propagation or multiplication by him on his own farm.

With respect to patenting of expressed sequence tags (ESTs) and other genetic information, the directive indicates that "a mere DNA sequence without indication of a function does not contain any technical information and is therefore not a patentable invention," and the "industrial application of a sequence or partial sequence must be disclosed in the patent application as filed"[21].

## G. Plant Patents in the United States

The 1930 Plant Patent Act made intellectual property protection available for a distinct subset of new and distinct plant varieties in the United States, namely, asexually reproducible plant varieties. The reason for the restriction of plant patent protection to this kind of subject matter was the prevailing notion that sexually reproduced plants could not be reproduced true to type through seedlings[22]. Through amendments in 1954, the subject matter of plant patent protection was clarified to include "a distinct and new variety of plant, including cultivated sports, mutants, hybrids,

# Intellectual Property Protection

and newly found seedlings, other than a tuber propagated plant or a plant found in an uncultivated state"[23]. Legislative reports at the time of the enactment of the Plant Patent Act are clear in indicating that it was not intended to cover "varieties of plants which exist in an uncultivated or wild state, but are newly found by plant explorers or others"[24]. The exclusive right granted by a plant patent corresponds to its restrictive scope of subject matter, inasmuch as a patentee may exclude others from asexually reproducing the plant or selling or using the plant that has been asexually reproduced. It is not an infringement to reproduce the plant sexually or to breed a similar variety independently.

The exclusive right granted to plant patents has recently been expanded to include "the right to exclude others from asexually reproducing the plant, and from using, offering for sale, or selling the plant so reproduced, or any of its parts, throughout the United States, or from importing the plant so reproduced, or any parts thereof, into the United States"[25]. Because of the different novelty requirements, it may be possible to obtain a plant patent on an asexually reproduced plant that is discovered growing in an area under cultivation, whereas a regular U.S. patent would not generally be available for such a "product of nature."

The innovation protectable under a plant patent can be seen as the combination of a three-step process: (a) cultivation or discovery of the plant, (b) identification of the new and distinct characteristics of the plant, and (c) asexual reproduction of the plant. However, only the second and third of these steps are required to establish the right to a plant patent. A person who carries out the first step, without participating in the other two, is not an inventor[26]. The proof of actual distinctness through asexual reproduction, prior to filing an application, is a requirement for plant patent protection[27]. Where asexual reproduction of a plant is routine, the person who merely carries out this third aspect of the innovation, without participating in the recognition of the distinct characteristics, may not be even a joint inventor of the new plant[28].

## III. PLANT VARIETY PROTECTION

In jurisdictions outside the United States, in the years following the enactment of the Plant Patent Act in 1930, a number of forms of intellectual property protection were introduced to provide protection for plant varieties as an incentive for systematic plant breeding. Some patents were also granted under the preexisting patent systems of some countries. The desire for a uniform approach to granting intellectual property rights to plant breeders, as well as an interest in seeing that rights would be respected across jurisdictional boundaries, led a number of European states to enter into discussions between 1957 and 1961 that led to the adoption of the International Convention for the Protection of New Varieties of Plants, signed in Paris on December 2, 1961. The parties to that convention make up the International Union for the Protection of New Varieties of Plants, generally known by the abbreviation UPOV (based on the initials of its name in French, Union pour la Protection des Obtentions Végétales; see *http://www.upov.int*).

The protection available for plant varieties has been harmonized to a certain extent under the UPOV Convention. There are, however, significant differences (discussed late) in the provisions in the amended 1978 UPOV Convention to which many member states still adhere[29], and the most recent 1991 amendments that have been adopted in a growing number of member states[30]. In UPOV member states, applicants are entitled to national treatment and a right of foreign priority for plant variety rights applications filed within a year of the filing date of the first application within a member country[31].

What constitutes a protectable plant variety under the UPOV Convention involves four preconditions:

> 1. **Novelty.** The variety must be new in a commercial sense, inasmuch as it must not have been offered for sale in the jurisdiction where protection is sought. Member states are permitted to

allow an exception for prior sales within the jurisdiction for up to one year prior to filing an application. The 1991 version of the UPOV Convention makes the one year grace period mandatory and further defines the requirements by indicating that "propagating or harvested material of the variety" must not have been "sold or otherwise disposed of to others" outside of the stipulated grace periods. The variety must also not have been on sale for more than four years in any other state (a period which is extended to six years in the case of grape vines and trees, including rootstocks).

2. **Distinctiveness.** The variety must be clearly distinguishable by one or more defined characteristics from any other variety whose existence is a matter of common knowledge. This standard of course leaves open the possibility of protection for plants which have previously existed in nature as "unrecognized" varieties.

3. **Uniformity.** The variation between individual plants within the variety must be limited, typically with respect in particular to the characteristics which make the variety distinct.

4. **Stability.** The variability in the relevant characteristics of the variety through repeated propagation must be limited. As with uniformity, this will typically be assessed with respect in particular to the characteristics which make the variety distinct.

What constitutes a plant that is protectable under varietal rights may vary. For example, in the United States, under the Plant Variety Protection Act, fungi and bacteria are excluded. As an example of how the requirements of uniformity and stability may be applied, in the United States first-generation hybrids are generally ineligible for protection because they are deemed to be inherently genetically unstable and thereby unable to contain sufficient uniformity and stability to qualify for plant variety protection.

The requirements for plant variety protection are typically assessed through examination establishing that the variety for which protection is sought is sufficiently distinct, homogeneous, and stable. Examination is typically based on growing tests carried out by public authorities or by the breeder seeking protection. UPOV has promulgated guidelines for examination, in an effort to harmonize international standards. Nevertheless, it generally remains the case that an applicant must carry out growing trials in each jurisdiction in order to satisfy the national authorities that their requirements for plant variety protection are met.

Protection is available to the breeder of a new variety, or an entity that has acquired rights from the breeder. In the 1991 convention *breeder* is defined to include a person who breeds or discovers and develops a variety or the employer of that person. Under this definition, the 1991 UPOV Convention differs from the 1978 convention by requiring the development of a variety, rather than the mere discovery of a variety.

According to the 1978 UPOV Convention, the minimum period of protection is 18 years for grape vines and trees, including rootstocks, and at least 15 years for all other plants. Under the 1991 convention, these periods are extended to a minimum of 25 years for grape vines and trees, including rootstocks, and at least 20 years for all other plants.

The genera and species of plants that may be protected in each member state under its particular implementation of plant variety rights may vary. However, member states may not make the protection granted to a given plant variety dependent upon the protection of the same variety in any other state. In the United States, for example, amendments in 1994 made plant variety protection available to any sexually reproduced or tuber propagated plant. The 1978 convention specified that plant variety protection may be made available to all genera and specie but did not require member states to implement such expansive protection. The 1991 convention requires member states to provide protection for all plant genera and species.

The 1991 changes to the UPOV Convention introduced substantial changes to the scope of protection afforded by plant variety rights. Under the 1978 UPOV Convention, rights are only granted to a breeder in respect of the reproductive or vegetative propagating material[32] *as such*, i.e., only when such material is to be used for reproductive reproduction or vegetative propagation. The 1991 UPOV Convention extends the plant variety rights to harvested material, including whole plants and parts of plants (provided that the harvested material has been obtained through the unauthorized use of propagating material and that the breeder has had no reasonable opportunity to exercise the right in relation to the propagating material). The 1991 UPOV Convention also specifies as an optional class of protectable material the products made directly from harvested material[33].

In addition to expanding the nature of the material protected by plant variety rights, the 1991 UPOV Convention broadens the range of acts in respect of such material that are the exclusive right of the breeder. Under the 1978 UPOV Convention, there are three exclusive rights granted to the breeder in respect of the reproductive or vegetative propagating material:

1. The production for the purposes of commercially marketing such material
2. The offering for sale of such material
3. The marketing of such material

The rights under the 1978 UPOV Convention do not prevent others from using a new variety as the initial source of variation for creating other new varieties, or marketing the new varieties. Also, the breeders' authorization is not required under the 1978 UPOV Convention for the production of propagating material that is not for commercial marketing, as such, thus in effect creating the "farmers' exemption" for the repeated use of farm-saved seed. The categories of activities reserved to the breeder in the 1991 UPOV Convention are considerably broader:

1. Production or reproduction (multiplication)
2. Conditioning for the purpose of propagation
3. Offering for sale
4. Selling or other marketing
5. Exporting
6. Importing
7. Stocking for any of the foregoing purposes

The 1991 UPOV Convention extends the breeders' rights well beyond the limited right granted in the 1978 UPOV Convention to prevent others from repeatedly using a variety for commercial production of another variety. In the 1991 UPOV Convention, right are extended beyond the protected variety itself, to cover

1. Varieties that are essentially derived from the protected variety
2. Varieties that are not clearly distinguishable from the protected variety
3. Varieties whose production requires the repeated use of the protected variety

Under the 1991 UPOV Convention, a variety is considered to be essentially derived from an initial variety when the new variety is predominantly derived from the initial variety and retains the expression of the essential characteristics of the initial variety, even where the derived variety is clearly distinguishable from the initial variety. As examples of methods of derivation, the 1991 UPOV Convention recited selection of natural or induced mutants, somaclonal variants, selection of variant individuals from the initial variety, backcrossing, or transformation by genetic engineering of the initial variety. Given the definition of an essentially derived variety, it would appear that in most circumstances a transgenic plant developed through the use of a gene obtained from an initial variety will not be considered to have been essentially derived from the initial variety.

The 1991 UPOV Convention specifies three compulsory exceptions to the breeders' rights:

1. Acts done privately and for noncommercial purposes

2. Acts done for experimental purposes
3. Acts done for the purposes of breeding and exploiting other varieties (provided such other varieties are not essentially derived varieties)

The 1991 UPOV Convention also includes as an optional exception a permission to farmers to use farm-saved seed for propagating purposes, on their own holdings. In implementing the 1991 UPOV Convention, the United States has, for example, continued to allow farmers to save seed for replanting on their own farm, while eliminating an exception that allowed farmers to sell or share saved seed in certain circumstances.

The 1991 UPOV Convention has removed the prohibition that existed in previous versions of the treaty against concurrent protection of a variety under a patent and a plant breeder's right. This change to the UPOV Convention is especially interesting in light of the interpretation of Article 53(b) of the European Patent Convention, where great difficulties have been encountered in defining a variety eligible for plant variety rights as distinct from plant innovations that are eligible for patent protection.[34]

Both the 1978 and 1991 UPOV Convention protect the denomination of a variety as the generic designation of the variety. The denomination must be sufficiently distinctive to allow a variety to be identified, while not misleading with respect to the characteristics of the variety. Because a denomination must be generic, it cannot also be the proprietary trademark of any party. A party is by definition distinctive of the commercial source of a product or service, and trademarks in general become invalid if they become the generic name applied to a particular product. The UPOV Convention permits a trademark to be used in association with varietal denominations. In making any such association, a trademark owner should be careful to ensure that its mark does not become, like the denomination, a generic term for the variety.

## IV. CONCLUSION

A fundamental distinction between patent rights and plant variety rights arises in part from the way in which the subject matter of these rights is defined. In a patent, the patentee uses the claims to capture the abstract idea of an invention as broadly as is possible in view of the prior art. Patents thereby afford innovators an opportunity to define for themselves the full scope of their monopoly, as discussed. In contrast, plant variety rights are generally based upon a narrative and/or pictorial description of a variety that is itself a physical embodiment of the innovation. As such, the variety in a sense defines itself with only limited leeway given to a breeder to characterize the distinctive features of the variety.

Patents clearly offer the most flexible form of intellectual property protection for transgenic plants. Even in jurisdictions where patent claims to plants per se are not available, patents offer an important form of protection for other aspects of inventions in this field. Nevertheless, the 1991 changes to the UPOV Convention lend strength to plant variety rights, and this form of intellectual property protection should not be overlooked as a mechanism for securing meaningful protection. Similarly, plant patent protection in the United States is a unique form of protection that should be considered for asexually reproduced plants.

## BIBLIOGRAPHY

HJ Aenishen, The European Patent Office's Case Law on the Patentability of Biotechnology Inventions. Carl Huimanns Verlag, 1997.
DS Chism, Chism on Patents. Release 71. Matthew Bender & Company, 1999.
Intellectual Property Rights in Agricultural Biotechnology. Erbisch, Maredia, eds. CAB International, 1998.

## NOTES

1. March 20, 1883, as rev. at Brussels, December 14, 1900; at Washington on June 2, 1911; at The Hague on November 6, 1925; at London on June 2, 1934; at Lisbon on October 31, 1958; and at Stockholm on July 14, 1967.
2. The claims are typically found as the numbered paragraphs at the end of the text of the patent.
3. U.S. Patent No. 5,783,394.
4. *Amgen Inc. v. Chugai Phamzaceutical Co.,* 18 USPQ 2(d) 1016.
5. U.S. Patent No. 5,463,174.
6. See, for example, *In Re Goodnian et aL,* 29 USPQ 2(d) 2010 (Fed. Cir. 1993).
7. See U.S. Patent Nos. 5,159,135 and 5,463,174, relating, respectively, to cotton and Brassica.
8. 206 U.S.P.Q. 193 (1980), establishing the patentability of life-forms in principle, and categorizing the scope of subject matter broadly as "everything under the sun made by man."
9. The PVPA, which is in effect the implementation of the UPOV Convention in the United States.
10. *Ex part Hibberd* (1985), 227 U.S.P.Q. 443, a decision of the Board of Patent Appeals and Interferences.
11. *Pioneer HI-Bred Intemational Inc. v. J.E.M. Ag. Supply Inc.,* 49 USPQ 2d 1813 (N. District Iowa, 1998).
12. 35 USC 161–164.
13. 7 USC 2321 at seq.
14. Decision of the Administrative Council of June 16, 1999, amending the Implementing Regulations to the European Patent Convention.
15. A fact that is reflected in the decision of *Ciba Geigy,* O.J. EPO 1984, 112/T49/83.
16. Trangenic Plant/Novartis II G0001/98.
17. Harmonized internationally under the TRIPs agreement.
18. EPC Article 64(2).
19. U.S.C. 1337 (1991).
20. 35 U.S.C. § 271 (g).
21. Preamble Clauses 22 and 23, as well as Article 5(3).
22. *Diamond v. Charabarty,* (1980) 206USPQ (United States Supreme Court): "[S]exually reproduced plants were not included under the 1930 Act because new varieties could not be reproduced true-to-type through seedling."
23. 35 U.S.C. Section 161.
24. S. Rep. No. 315, 71st Cong. 2d Sess. (1930).
25. 35 U.S.C. Section 163; *Plant Patent Amendments Act of 1998,* H.R. 1197.
26. *Ex parte Moore* (1957) 115 U.S.P.Q. 145 (Pat. Off. Bd. App. 1957).
27. *Dunn v. Ragin* (1941) 50 U.S.P.Q. 472 (Pat. Off. Bd. of Int'f. 1941).
28. *Ex parte Kluis* (1946 70 U.S.P.Q. 165) (Pat. Off. Bd. App. 1945).
29. Such as Argentina, Australia, Brazil, Canada, Chile, China, France, Italy, Mexico, New Zealand, Poland, Portugal, South Africa, and the Ukraine (as of June 29, 1999).
30. Such as Denmark, Germany, Israel, Japan, Netherlands, Russian Federation, Sweden, United Kingdom, and the United States (as of June 29, 1999).
31. The concepts of national treatment and foreign priority are discussed earlier.
32. Under the 1978 UPOV Convention, vegetative propagating material is deemed to include whole plants, and to extend to ornamental plants or parts thereof normally marketed for purposes other than propagation *when they are used commercially as propagating material in the production of ornamental plants or cutflowers.*
33. As with harvested material, this optional protection is available provided the products have been obtained through the unauthorized use of harvested material and the breeder has had no reasonable opportunity to exercise his or her right in relation to the products made directly from the harvested material.
34. The subject of the Plant Genetic Systems case before the Technical Board of Appeals in decision T35693 (O.J. EPO 1995, 545), and the corresponding decision of the Enlarged Board of Appeal reported as G3/95 (O.J. EPO 1996, 169).

# 19
# Public Perceptions of Transgenic Plants

**Thomas Jefferson Hoban**
*North Carolina State University, Raleigh, North Carolina*

| | | |
|---|---|---|
| I. | INTRODUCTION | 293 |
| II. | PUBLIC PERCEPTIONS OF TECHNOLOGY | 294 |
| III. | OVERVIEW OF RESEARCH PROJECTS | 295 |
| IV. | BIOTECHNOLOGY AWARENESS AND KNOWLEDGE | 296 |
| V. | PUBLIC ACCEPTANCE OF BIOTECHNOLOGY | 298 |
| VI. | CONCLUSIONS AND IMPLICATIONS | 301 |
| | REFERENCES | 303 |

## I. INTRODUCTION

Since the late 1990s a very rapid diffusion of transgenic plants has occurred throughout much of North America. Farmer acceptance of these crops has been phenomenal. Given today's agricultural economy, farmers are looking for any competitive edge they can get. The early products of biotechnology offer that edge to many farms. Transgenic plants will continue to be developed and grown in the future—provided that a number of social and political constraints can be overcome.

We have recognized for a number of years that consumer acceptance is the ultimate determinant of the success of transgenic plants (1). There have been a number of efforts in North America to ensure that acceptance. However, we also now realize that products must be acceptable in the international marketplace. The situation in Europe has proved to be much more difficult for transgenic plants. Some activist groups have been able to build widespread fear and opposition to transgenic plants. That makes it vital that we understand the level of public awareness and acceptance in the different world markets (2,3).

Education is very important, but it must be based on an accurate understanding of public knowledge and public attitudes. This article reviews trends of public awareness and consumer acceptance of biotechnology as we have been tracking them since the early 1990s (4,5). The chapter also presents a comparative assessment of the latest available information about consumer attitudes and awareness around the world (3,6,7). First, it will be useful to put the issue of public perceptions of transgenic plants into a broader conceptual framework.

## II. PUBLIC PERCEPTIONS OF TECHNOLOGY

To understand public perceptions of transgenic plants, it is helpful to understand public perception of technology more generally. Scientists, public policymakers, and educators have long been interested in how people perceive and understand risks associated with health and environmental issues (8,9). This interest has continued to grow along with the controversies over new technologies (10,11).

Risk perception must be considered in its social and cultural context (12,13). In fact, the widespread public concern over risks appears to be a relatively recent phenomenon, as are many of the technological risks themselves (14). In fact, even the idea that we deserve a risk-free life is a relatively localized and recent attitude that is most common in the Western industrialized world. Risk perceptions tend to be very personal (12,14). People have different and values. Factors such as early experiences, education, and personality affect individual perceptions of risk.

Scientists, government officials, and industry spokespersons often lament the lack of public understanding about risk (15). Disagreements between scientists and the public represent problems for each side. Scientists often become frustrated in their attempts to communicate with the public about technical risk assessment and management. Technical experts feel the public misunderstands the "real" nature of different risks. It is true that many people minimize some important risks (e.g., high levels of dietary fat or alcohol consumption) and exaggerate less important risks (e.g., flying in commercial airplanes or eating produce with trace levels of pesticide residue). However, experts often adopt the arrogant and counterproductive attitude that they are right and the public is wrong and use complex arguments and technical jargon that tend to alienate the lay public further.

Members of the general public are often angered by technical experts who appear cold and impersonal in their use of complex statistical probabilities and bureaucratic jargon (8,16). The majority of citizens rely on intuitive risk judgments, typically called *risk perceptions*. Their experience with hazards arises largely from indirect sources (i.e, the news media), rather than direct experience. Most people are convinced they face more risk today than in the past. They also express concerns that future risks will be even greater. These views contrast sharply with those of most professional risk assessors (17).

The lay public's definition includes a variety of qualitative and subjective factors (e.g., ethics, control, or the involuntary nature of exposure). In fact, the public generally doesn't understand technical risk assessments (16). The common public response to an unknown or new risk is "I'm not sure what it is about that stuff; but I know I don't like it!" Sandman concludes that the public has a broader and more relevant conception of what is meant by risk than the technical experts because technical assessments cannot incorporate the types of concerns the public expresses (16).

Many problems associated with risk management and communication result from the differences between scientists and the public (13,15). The technical concept of risk is too narrow and ambiguous to serve as the crucial yardstick for policy-making. Public perceptions, however, are the product of intuitive biases, economic interests, emotions, and cultural values. All of these are hard to measure and evaluate. Technical risk must, therefore, be viewed in combination with psychological, social, and cultural processes that can heighten or reduce public perceptions of risk.

It is important to understand the types of factors that are most important to the lay public when a particular risk is evaluated (13,16). For the most part, these factors are largely ignored by risk assessment professionals in their calculations of technical hazard. For example, some factors make the potential risks of transgenic appear more serious and therefore, less unacceptable to the public. One of the major factors that make a risk unacceptable is the extent to which people feel they face a risk involuntarily. Genetically modified ingredients in food are perceived as more risky because they are considered to pose an involuntary risk. Many people believe they have very little control over food production and processing. They may doubt that the government or industry is willing or able to control food safety or environmental risks.

Risks that are perceived to be unfair are also considered to be less acceptable. For example, people who think they will bear the potential risks of transgenic plants (e.g., European consumers) may feel they are not receiving the benefits from these technologies (which go to U.S. farmers). People find known risks (i.e., those that are understood) generally more acceptable than those that are unknown. The public will find little comfort in the fact that the government and scientific community cannot assure them about the level of risks with 100% certainty. Likewise, people will be worried if they are told that the "experts" do not agree about the long-term risks associated with transgenic plants. That clearly is one reason why all the opponents must do is raise doubt in order to promote opposition to biotechnology.

As can be seen from this discussion, many factors may cause the average citizen to become fairly alarmed about the perceived risks of transgenic plants. They base their perceptions on intuition, emotion, and selective perception of uncertain information. Furthermore, people are often concerned with secondary effects (e.g., impacts on quality of life, wildlife, and future generations) that the experts are unable to evaluate. Because political leaders have many of the same perceptions as other citizens, they are likely to base policy decisions on subjective factors, as well.

Public perception of risk is also influenced by public attitudes toward science and technology in general (12,15). Public confidence in science and technology has diminished in recent years, especially in Europe. More people distrust new and unfamiliar technologies than in the past. Many people have low levels of scientific literacy. These problems are particularly serious as related to agriculture, because many people are no longer personally familiar with farming. Most have little understanding of how food is produced. They have little appreciation for the historic role of science in assuring an abundant and relatively low cost food supply. All these various factors combined make for shaky ground on which to grow transgenic plants.

## III. OVERVIEW OF RESEARCH PROJECTS

The U.S. Department of Agriculture (USDA) was the sponsor of the first study we did in 1992 (1), a national telephone survey in the United States of over 1200 people. Eight focus groups were also conducted. Then I had a chance to follow up a couple of years later with another national telephone survey of 1000 U.S. consumers for the Grocery Manufacturers of America, focusing on what, at the time, was a hot topic, bovine somatotropin (18). The Food Marketing Institute included some questions on biotechnology on their U.S. surveys (19,20) and included the same biotechnology questions on their survey in Europe during 1995 (21).

Then in March 1997, the International Food Information Council did another survey of American consumers' attitudes (5). In fact, that survey was repeated in February 1999 (22). There is also some relatively recent information from a comparative international study. A team of European researchers conducted a survey in Europe of over 16,000 consumers (7). Another researcher conducted the same survey with 1000 Canadian consumers (6). Jon Miller and I recently conducted a survey in the United States of over 1000 consumers that included most of the same questions (3). This is some very interesting information, particularly given the fact that we have common questions in all these countries.

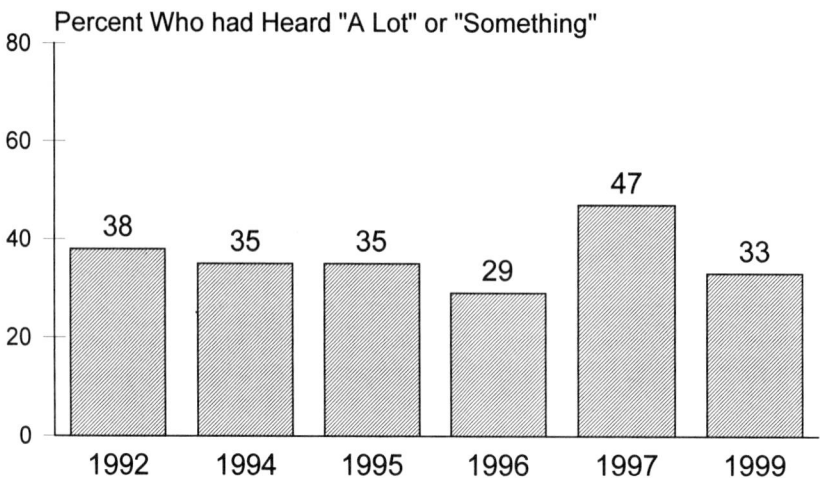

Figure 1  American consumers' awareness of biotechnology.

## IV. BIOTECHNOLOGY AWARENESS AND KNOWLEDGE

Surveys in the United States have been tracking public awareness over time (3,4,5,20,22). Respondents have been asked, "How much have you heard or read about biotechnology?" (Fig. 1). Two thirds of U.S. consumers had heard little or nothing about biotech between 1992 and 1996. In fact, awareness seemed even to have gone down a little in 1996. In March 1997, almost half of all respondents reported "a lot" of or "some" awareness. Awareness in the United States had increased with all the news on cloning and Dolly. However, awareness dropped again in early 1999. It is likely that awareness rose again later that year as a result of the expanded media coverage.

Respondents to the 1996–1998 surveys in the United States, Canada, and Europe were asked, "Have you heard or read anything about biotechnology in the past three months?" (3,6,7). In the United States and Canada just over half the people said they had (Table 1). Awareness was highest in Austria and Finland; most other countries were about the same as North America. This was likely due to relatively low levels of media coverage. This coverage does determine the extent to which biotechnology is an issue at any particular time.

If people are genuinely interested in a subject, such as biotechnology, they talk about it. That could be a family member, friend, physician, or even a scientist. Table 1 also shows how many claimed to have ever talked to someone about biotechnology in each of the countries. If they have not discussed biotechnology, chances are they are not all that concerned with or interested in it. There are some major differences. Denmark does tend to be highest, followed by Germany, Sweden, Austria, and Finland. When an issue becomes a social controversy, there is a tendency to talk about it more. Some of the countries (such as Ireland and Spain) had not had much controversy or discussion over biotechnology (at least at the time of the survey).

Several knowledge questions were asked on the recent European, Canadian, and U.S. surveys (Table 2). These also reflect the types of impressions people have of biotechnology. Consumers need a basic understanding of how food is produced. Respondents were asked whether it was true or false that "Yeast for brewing beer consists of living organisms." Remember this is a random sample of consumers, not scientists. In the United States and Canada, three out of four people got the "right" answer. There is more variation in Europe. For example, in Spain and Portugal, less than half of the people recognized this as true. Consumers in several Scandinavian countries and the United Kingdom tend to be among the highest in terms of their understanding

**Table 1** Extent of Consumers' Awareness of Modern Biotechnology

|  | Read or heard anything about biotechnology | Talked with someone about biotechnology |
|---|---|---|
| United States | 55% | 55% |
| Canada | 54% | 56% |
| Austria | 74% | 54% |
| Belgium | 45% | 40% |
| Denmark | 61% | 74% |
| Finland | 72% | 54% |
| France | 54% | 45% |
| Germany | 60% | 61% |
| Greece | 30% | 28% |
| Ireland | 37% | 34% |
| Italy | 51% | 41% |
| Luxembourg | 60% | 52% |
| Netherlands | 52% | 44% |
| Portugal | 39% | 31% |
| Spain | 40% | 33% |
| Sweden | 61% | 58% |
| United Kingdom | 55% | 48% |

**Table 2** Indicators of Respondents' Understanding of Biology and Biotechnology

|  | Yeast is living organism[a] | Only GMO plants have genes[b] | GMO fruit can change humans[c] |
|---|---|---|---|
| United States | 75% | 45% | 61% |
| Canada | 76% | 52% | 62% |
| Austria | 58% | 34% | 29% |
| Belgium | 71% | 31% | 51% |
| Denmark | 90% | 44% | 57% |
| Finland | 77% | 44% | 54% |
| France | 66% | 32% | 52% |
| Germany | 75% | 36% | 38% |
| Greece | 50% | 20% | 36% |
| Ireland | 70% | 20% | 34% |
| Italy | 65% | 35% | 58% |
| Luxembourg | 67% | 40% | 48% |
| Netherlands | 64% | 51% | 74% |
| Portugal | 42% | 27% | 32% |
| Spain | 46% | 28% | 40% |
| Sweden | 86% | 46% | 62% |
| United Kingdom | 85% | 40% | 55% |

[a]Percentage recognizing the following as *true*: "Yeast for brewing beer consists of living organisms."
[b]Percentage recognizing following as *false*: "Ordinary tomatoes do not contain genes while genetically modified ones do."
[c]Percentage recognizing following as *false*: "By eating a genetically modified fruit, a person's genes can be changed."

of basic biological principles. Knowledge levels in Austria on this question and the others were relatively low.

The questions got more difficult, as well as more specific to biotechnology. Consumers were asked whether the following statement was true or false: "Ordinary tomatoes do not contain genes, while genetically modified ones do." As shown in Table 2 there is a lot of misunderstanding on this particular question. Not many people gave the correct answer. In fact, many claimed they did not know, including almost half the Americans. This has important implications because if people think genetically modified tomatoes have something "different" in them, that idea is going to raise some concerns. It is necessary to address such misperceptions early in the educational process. Again major differences can be noted among European countries.

Another question is also quite interesting. Scientists would tend to agree that eating genetically modified food will not change a person's genes. Table 2 shows a great amount of variation in response to this statement. There tends to be better understanding in some countries (like the Netherlands, Sweden, Canada, and the United States). On the other hand, over 70% of the people in Austria believe it to be true or said they did not know. That impression would explain some of the perceived risks and fears people have about transgenic plants.

One of the key issues about education is to identify and use sources of information that consumers trust (1,3,18). U.S. consumers have been asked several times whom they would trust to give them information about biotechnology. There is a very encouraging pattern, which represents part of the reason why the United States has been so calm regarding food biotechnology. The American Medical Association, Food and Drug Administration, American Dietetics Association, and university scientists (which are third-party scientific groups) tend to be the most trusted. Groups like TV news reporters, biotech companies, food manufacturers, chefs, activist groups, and grocery stores tend to have lower credibility. In the European countries, this pattern is basically reversed (7). The environmental and consumer groups are at the top of the list; government and industry are both quite low in credibility. This is not surprising in light of the "mad cow" controversy and other problems that have occurred recently.

## V. PUBLIC ACCEPTANCE OF BIOTECHNOLOGY

Next, we turn our attention to public attitudes about biotechnology. First is a comparison of results from 1992, 1994, and 1998 in the United States (1,3,18). A very interesting trend throughout this and other data is the remarkable stability of people's opinions on biotechnology in the United States. These results are as close to identical as you can find on a series of surveys. In all 3 years, the interviews included the question "Tell me whether you support or oppose the use of biotechnology in agriculture and food production." It was asked toward the end of each interview, as a summary comment. In 1992, 70% said they supported it, a few did not know, and less than 20% were opposed. We repeated this question in 1994, during the height of the U.S. controversy over bovine somatotropin (BST). In that year, 72% said they supported it. Then in 1998 we again found 72% support.

It is important to determine whether there are differences among demographic groups (23). There are a couple of key influences that stand out. One involves a gender gap. Men are clearly more positive than women in their evaluation of biotechnology. We have found that tendency over the years on a variety of questions. This difference is important because when it comes to food, women continue to set the family food policy. They serve as food gatekeepers in our society as far as what is acceptable or not to feed the family. There are also significant differences in terms of formal educational level. Respondents with college degrees tend to be more likely to support biotechnology. College tends to provide an opportunity to be exposed to a variety of different ideas. However, not all people who graduate from college have a good understanding of science.

There are several other sets of findings that show results over time in the United States (5,19,20,22). These surveys clearly show that three of four people would be willing to buy potatoes or tomatoes developed through biotechnology to be protected from insect damage and requiring fewer pesticides (Fig. 2). This finding has also stayed remarkably consistent over time. On another application (produce that tasted fresher and better) we also found acceptance to be fairly high. Another question asked whether people thought they or their families would benefit from the use of biotechnology over the next 5 years (1,5,18,22). As shown in Fig. 3, between two thirds and three quarters of U.S. consumers are optimistic about future benefits.

The surveys in Europe, Canada, and the United States asked consumers to evaluate six different applications of biotechnology along four dimensions (3,6,7). Two applications related to

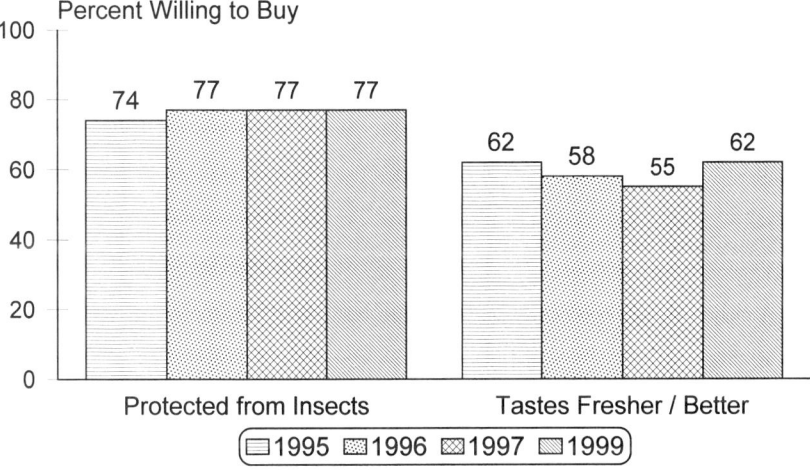

**Figure 2** American consumers' willingness to buy produce developed through biotechnology.

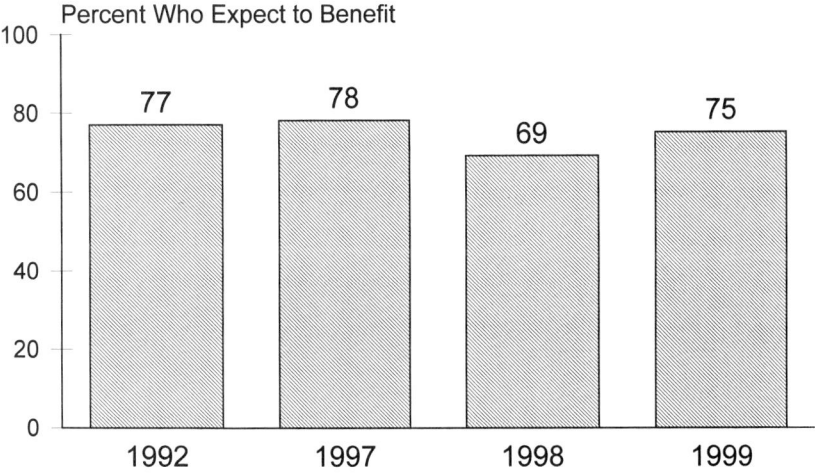

**Figure 3** American consumers' belief that biotechnology will provide benefits in next 5 years.

food, two involved animals, and two related to human health care. As expected, human health care is acceptable to and seen as valuable by about 85% of people. The insect-protected crop plants were seen as third most acceptable, right after human medicine. Animal biotechnology was much less acceptable.

Table 3 presents details on how consumers from the different countries rated the insect-protected plants along each of the four dimensions. The table also provides a bottom line summary by asking whether or not consumers agreed or disagreed that insect-protected crops developed through biotechnology should be encouraged. The results are actually quite encouraging. Canadian and U.S. consumers were very positive, as were respondents from the Netherlands, Italy, the United Kingdom, and Finland. Even half the German citizens agreed that these products should be encouraged. That is a very different story from the perception that all of Europe is negative on biotechnology. In fact, the data show that Austria was the only country predominantly very negative at the time of these surveys.

It is important to put attitudes about biotechnology into a comparative perspective. Fig. 4 shows how U.S. consumers rated the risks of biotechnology compared to five other concerns (19,20). The topic that gets the most attention in the media now (and probably the one most consumers should worry about) is microbial contamination. Three of four people said that it is a serious hazard, followed by pesticide residues, a proportion that actually has declined in recent years. Antibiotics and hormones, irradiated foods, additives, and preservatives are next in line. In the United States, foods developed through biotechnology have consistently been the lowest on the list of potential public concerns.

When we turn our attention back to the European consumers they were also putting biotechnology into a reasonable perspective (Fig. 5). In fact, "genetic engineering" was seen as slightly more risky than artificial coloring, nitrites, cholesterol, and fat (21). This is certainly well below some of the other issues that were of most concern to the European consumers. So again, it is important to keep these issues in perspective. If you only listen to Greenpeace, you would probably

**Table 3** Respondents' Agreement That "Using Biotechnology to Insert Genes from One Plant into a Crop Plant" Has Certain Characteristics

|  | Useful to society | Risky to society | Morally acceptable | Should be encouraged |
|---|---|---|---|---|
| United States | 74% | 40% | 68% | 66% |
| Canada | 82% | 38% | 79% | 77% |
| Austria | 36% | 49% | 28% | 23% |
| Belgium | 71% | 39% | 68% | 62% |
| Denmark | 70% | 62% | 54% | 48% |
| Finland | 80% | 31% | 70% | 72% |
| France | 71% | 52% | 65% | 67% |
| Germany | 62% | 44% | 54% | 50% |
| Greece | 68% | 38% | 59% | 60% |
| Ireland | 66% | 45% | 57% | 53% |
| Italy | 76% | 44% | 70% | 70% |
| Luxembourg | 57% | 49% | 52% | 44% |
| Netherlands | 80% | 64% | 73% | 66% |
| Portugal | 76% | 47% | 72% | 73% |
| Spain | 65% | 43% | 59% | 56% |
| Sweden | 62% | 55% | 59% | 54% |
| United Kingdom | 75% | 54% | 64% | 59% |

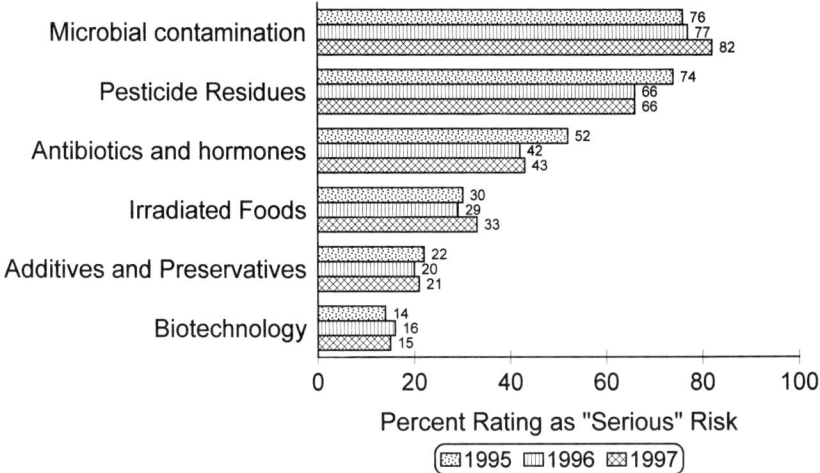

**Figure 4** American consumers' perceptions of food attributes as a "serious risk."

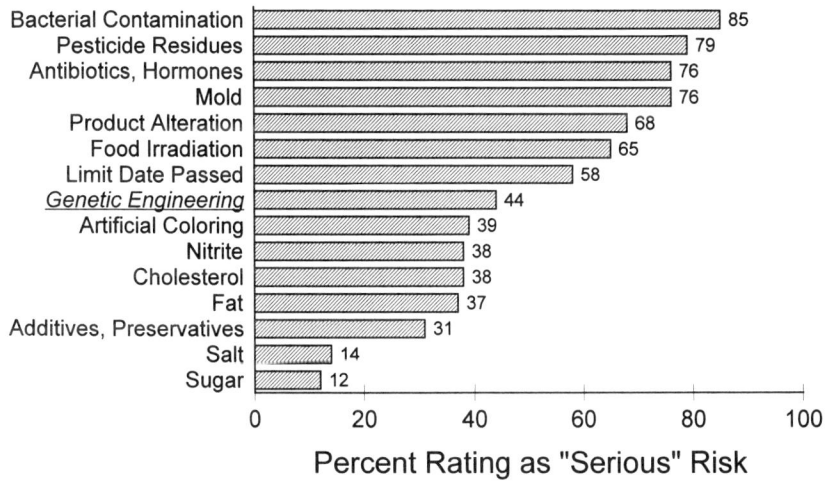

**Figure 5** European consumers' perceptions of food attributes as a "serious risk."

think that genetic engineering is the major concern of European consumers, but it is not. Again it is interesting to note the country-by-country differences (Fig. 6). There are some striking differences; Sweden, Austria, and Germany tend to have the higher level of concern. In most other countries, less than half the consumers saw biotechnology as a serious hazard.

## VI. CONCLUSIONS AND IMPLICATIONS

The majority of North American consumers have positive attitudes about biotechnology. They perceive benefits and will buy the products. Market tests have also reflected that trend. The Nature Mark potato did quite well in market tests when they were labeled and placed next to others. Consumers perceived a benefit of the reduced use of pesticides. Also, there had been a very clear

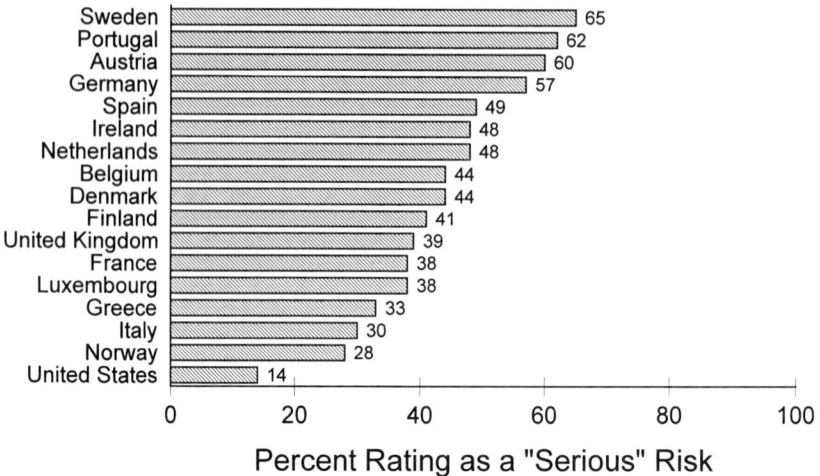

**Figure 6** Perception of genetic engineering as a serious food risk.

preference among British consumers for tomato paste from genetically modified tomatoes. However, when faced with activist pressure the UK supermarkets felt compelled to remove the consumers' choice to buy that successful product.

It is important to keep in mind that transgenic plants are not a top-of-mind issue for the vast majority of consumers. When studies ask, "What is the major problem facing our country?" no one says, "Biotechnology," maybe 2% will say, "The environment," and nobody even says "food." The major concerns are generally issues such as crime, the economy, and breakdown of moral values. Scientists, politicians, and others read books like this and attend conferences because they are interested in the subject. However, average consumers are really not so interested or concerned.

Consumers' attitudes about biotechnology are closely related to their general beliefs about science, technology, and food. In the United States, there has always been strong public support for and appreciation of science. People recognize that they have gained a lot of benefits from science and technology. Consumers may feel there is a potential downside, but overall they are very supportive of new developments. In fact, North American consumers are generally quite pragmatic about food. With any food product, consumers mainly want to know about taste, nutrition, convenience, safety, and price. Those are the main questions a consumer will want answered about food produced through biotechnology or any means. How the seed was produced will generally be of little relevance for most people (unless they are scared into thinking about it).

The future prospects for Europe are much less positive, at least in the short term. Farmers and food companies in the United States and Canada want to make sure something is done to stabilize the situation over there. The U.S. government is probably not going to mandate segregation of transgenic crops. That would be a logistical nightmare. The Europeans had a chance to buy elsewhere in 1996 when the crops came in. Now, South American farmers and others are starting to raise crops developed through biotechnology. In fact, more products are going to arrive on the market from around the world. European companies and consumers will soon have few options except to pay a premium price for a negligible, psychological benefit.

When we examine awareness in Europe, there is an apparent contradiction. Results reviewed earlier showed that the country that was least likely to find biotechnology acceptable was Austria. However, Austrian consumers also report relatively high awareness. This does not, however, mean that education does not work. It depends a lot on what people have heard or read. People who have done media analysis in Austria saw mostly negative reports, so consumers may have read more about biotechnology, but what they had heard or read was very negative. The opponents of biotechnology in Europe have had the chance to tell their side of the story for several years without much balance. There is some evidence that this has changed in the last year or so in terms of more positive media coverage (in some countries). However, the news coverage did become much more negative in the United Kingdom and elsewhere.

Educational efforts will continue to be very important. Such efforts are starting to take hold among European leaders and consumers. We had a meeting in Washington several years ago sponsored by the Georgetown Center for Food and Nutrition Policy. Leaders were invited from the European Union. They were hungry for information. These were some of the top European officials, but most of what they had heard about biotechnology had come from Greenpeace. European leaders expressed serious concerns about lost jobs, increasing food prices, and other economic costs that are going to result from rejection of biotechnology.

Statistical analysis has helped evaluate what influences people's acceptance of biotechnology (1,23). At the top of the list are awareness and knowledge. People should have at least some level of knowledge about biotechnology. They also need to recognize a societal benefit or feel there is something in it for them personally. They need to view it as ethically acceptable. Ultimately acceptance comes down to confidence in government and trust in information sources. There certainly are groups of consumers within each country (including the United States and Canada) who are negative. But these types of surveys represent a random sampling of citizens, not the opponents who get all the media attention.

The educational opportunities and challenges are very important. There still is a lot of work to do in Europe. In the United States, we have so far been able to reach consumers effectively by educating opinion leaders, including scientists and government officials. Through groups like the International Food Information Council (22) the media are provided with new, factual information on biotechnology. Finally, farmers and the food industry need more education. It is very important that food retailers and others who have direct contact with consumers have enough information.

Education needs to explain the benefits and the uses of biotechnology, to give people a better basis for evaluating products. We must address consumer concerns, including labeling, allergenicity, and other questions that are on people's minds. It is also important to tell consumers about third-party oversight and regulations. Consumers want to know that the government is regulating biotechnology. In the United States the Food and Drug Administration (FDA) and USDA have done a good job of keeping public confidence high. Europe has been a much different story.

Finally it is important to put biotechnology into a historical context. We need to tell people that we have been breeding plants for years. Some consumers seem surprised to learn that scientists and farmers have already changed plants. Overall, we need to increase consumer understanding of food production and processing. Most consumers simply think that food comes from the grocery store, or increasingly from restaurants. These are all part of the educational challenges and opportunities with biotechnology.

## REFERENCES

1. TJ Hoban, PA Kendall. Consumer Attitudes About Food Biotechnology. Raleigh, NC: North Carolina Cooperative Extension Service, 1993.
2. TJ Hoban. Consumer acceptance of biotechnology: An international perspective. Nat Biotechnol 15: 232–234, 1997.
3. TJ Hoban. International acceptance of agricultural biotechnology. Proceedings of the Annual Meeting of the National Agricultural Biotechnology Council, Ithaca, NY: National Agricultural Biotechnology Council, 1999, pp 59–73.

4. TJ Hoban. Trends in consumer attitudes about biotechnology. J Food Distrib Res 27:1–10, 1996.
5. TJ Hoban, L Katic. American consumers' views on biotechnology. Cereal Foods World 43:20–22, 1998.
6. E Einsiedel. Biotechnology and the Canadian Public. Calgary, Alberta: University of Calgary, 1997.
7. European Commission. European Opinions on Modern Biotechnology. Luxembourg: European Commission, 1997.
8. National Research Council (Committee on Risk Perception and Communication). Improving Risk Communication. Washington, DC: National Academy Press, 1989.
9. R Wilson, EA Crouch. Risk assessment and comparisons: An introduction. Science 236:267–270, 1987.
10. D Lupton. Risk. London: Routledge, 1999.
11. B Glassner. The Culture of Fear: Why Americans Are Afraid of the Wrong Things. New York: Basic Books, 1999.
12. CA Heimer. Social structure, psychology, and the estimation of risk. Annu Rev Sociol 14:491–519, 1988.
13. RE Kasperson, O Renn, P Slovic, HS. Brown, J Emel, R Goble, JX Kasperson, S Ratick. The social amplification of risk: A conceptual framework. Risk Anal 8:177–187, 1988.
14. HW Lewis. Technological Risk. New York: WW Norton, 1990.
15. WR Freudenburg. Perceived risk, real risk: Social science and the art of probabilistic risk assessment. Science 242:44–49, 1988.
16. PM Sandman. Explaining Environmental Risk: Some Notes on Environmental Risk Communication. Washington, DC: U.S. Environmental Protection Agency, 1986.
17. P Slovic. Perception of risk. Science 236:280–285. 1987.
18. TJ Hoban. Consumer Awareness and Acceptance of Bovine Somatotropin. Washington, DC: The Grocery Manufacturers of America, 1994.
19. Food Marketing Institute. Trends in the United States, Consumer Attitudes and the Supermarket. Washington, DC: Food Marketing Institute, 1996.
20. Food Marketing Institute. Trends in the United States, Consumer Attitudes and the Supermarket. Washington, DC: Food Marketing Institute, 1997.
21. Food Marketing Institute. Trends in Europe. Washington, DC: Food Marketing Institute, 1995.
22. International Food Information Council. Washington, DC: International Food Information Council. Available at http://ificinfo.health.org
23. TJ Hoban, E Woodrum, R Czaja. Public opposition to genetic engineering. Rural Sociol 57:476–493, 1992.

# 20
# Industry Perspectives

**Katherine A. Means**
*Produce Marketing Association, Newark, Delaware*

| | | |
|---|---|---|
| I. | INTRODUCTION | 305 |
| II. | PRODUCERS AND AGRONOMIC BENEFITS | 306 |
| III. | CONSUMERS AND CONSUMER BENEFITS | 306 |
| IV. | MARKETPLACE ISSUES | 307 |
| V. | CONCLUSION | 308 |

## I. INTRODUCTION

Marketers of any product begin seeking a newer, better, different version of their product as soon as it hits the marketplace. This is no less true of produce marketers. Outside the modern discussion of genetically modified fruits and vegetables, breeders have sought enhanced product qualities since the beginning of agriculture as an occupation. They seek what they need for the marketplace—better taste, longer shelf life, improved appearance, enhanced shipping qualities, traits that allow production in hostile environments, greater per-acre or per-tree production, and more. Biotechnology offers the allure of faster and more precise trait changes or enhancements that move this centuries-old practice of breeding forward. It also has the potential for consumer resistance.

The success of GMOs in the fresh produce marketplace in North America hinges on several factors:

- The existence of a benefit from any GMO that is evident or can be made evident to the buyer
- The education of the buyer (whether a trade buyer or a consumer) about the development (process) and benefits of GMOs (this includes the labeling debate)
- Successful interaction with opponents of GMOs (even though they were not vocal in North America through the 1990s)

At its simplest, the produce marketplace is an offer and an acceptance. For whatever reason, a buyer wants what a seller has, and a transaction takes place. The seller gets something of value from the buyer in exchange for this thing that the buyer wants. The buyer sees some benefit in making this exchange for the item. The greater the benefit, the greater the exchange may be.

But the marketplace is rarely that simple. Sellers surround their wares with marketing to make those wares more attractive to buyers. Buyers may become skeptical about the marketing efforts and want more information. What exactly is this product? Where did it come from? How was it produced? Is there anything objectionable in its history? What are the benefits of this product? Buyers want to know what they are buying, and they want to see some benefit from the purchase: i.e., it must satisfy a need. Take food as an example. It can satisfy the most basic need for survival. Or it may satisfy a more complex need, relating to quality of life, social interaction, or pleasure.

The ultimate buyer is the consumer, but there are many buyers along every distribution chain. In the food business, seed companies buy materials they need to produce their seeds. Growers buy seeds and plants from the seed companies. They also buy inputs they need to produce commodity crops. Manufacturers buy crops to produce value-added foods. Wholesalers and retailers buy those foods, or the commodity crops, to offer the consumer. The fresh produce industry is a very complex chain of distribution links and diverse channels (retail, food service, etc.).

Although many distinctions can be made about the buyers of genetically engineered foods, it is simplest to break them down into two groups: producers and consumers. Wholesalers, distributors, and retailers act as agents for consumers. They are likely to offer for sale the things their customers have indicated that they want or new things that they believe their customers will want.

Certainly, any benefit to one group may cross over or be viewed as a benefit to another group, but one can look at discrete benefits to producers and consumers.

## II. PRODUCERS AND AGRONOMIC BENEFITS

Genetic engineering offers many potential benefits to the producer—the ability to grow a crop in hostile conditions, time savings (reduced on-farm work of crop protection efforts or tillage), economic savings (reduced spending on crop/weather protection tools), economic gain (longer growing season, ability to grow high-revenue crops in previously hostile environments), and environmental benefit (fewer inputs into the environment). Where the grower sees that the benefit of a GMO is worth it, he will pay more for it. Take just four examples:

- Genetically engineered frost resistance would allow crops to be grown in areas previously inaccessible because of cold weather. It could extend the growing season in other areas.
- Genetically engineered drought resistance would allow production in hostile climates as well.
- Genetically engineered pest resistance can permit production in areas infested with a given pest (insect, disease, weed) and can reduce crop protection inputs, saving the grower money, time, and environmental impact.
- Chemical resistance genetically engineered into a given plant [to certain crop protection tools] allows its pests to be eradicated more easily without damage to the crop being grown.

## III. CONSUMERS AND CONSUMER BENEFITS

Like growers, consumers pay for benefits they find attractive—the more attractive, the more they are likely to pay. Consumers are not always the ones to call for any given benefit; they may see a benefit only after it is presented. Did consumers ask manufacturers to fortify foods with nutrients (e.g., calcium-fortified orange juice)? Or did manufacturers research consumer trends to determine likely benefits for which consumers would pay?

Consumers may see some of the on-farm benefits as beneficial to society at large, and a certain percentage of them might pay more for a product that can be grown with fewer crop protection chemicals. These same consumers would respond positively to consumer marketing of produce grown through integrated pest management or organic production methods. In the mainstream, however, consumers react to benefits that specifically affect them. These might include improved taste (e.g., a peach that always tastes like a ripe, just-picked peach), better nutrition (e.g., a carrot with enhanced beta carotene), or greater convenience (e.g., easier peeling citrus).

The produce industry, as any other industry, reacts to its customers' actual or potential desires. The desire for greater availability of produce led to breeding varieties that can be transported great distances, sometimes at the expense of flavor. It also led to harvesting practices that permit some crops to be harvested early, shipped more conveniently, then ripened at the end of the distribution chain.

Growers and researchers have worked to bring value-added items to the marketplace, such as seedless watermelon or seedless grapes. Those breeding pioneers who produce crops that have value-added benefits to consumers often reap the financial rewards. Patents protect the value, and brands or marketing information draws the value-added item to the attention of the consumer.

## IV. MARKETPLACE ISSUES

Genetic engineering allows breeding to take a huge step toward specificity in trait exchange. It permits quicker, effective development of new products with either agronomic or consumer benefits. It also raises concerns in the minds of some consumers and is a rallying point for some consumer advocacy groups.

For the fresh fruit and vegetable industry in the United States and Canada, genetic engineering or biotechnology was a hot topic in the late 1980s and early 1990s. The Flavr Savr™ tomato developed by Calgene caused a big stir for several reasons. It promised a value-added tomato that would allow it to remain firm (for shipping) while tasting great—like a backyard tomato. Consumers would like this trait because one of the greatest complaints about produce is that tomatoes shipped in from other places do not taste as good as local tomatoes grown in the summer. The Flavr Savr also held promise as a consumer testing ground. It would be one of the first genetically engineered products marketed under a brand name with consumer benefits. However, progress ran behind, and production never reached the level to provide answers to the consumer acceptance question.

After that, into the mid- and late 1990s, most talk about genetic engineering involved agronomic benefits. On the consumer benefit side, biotech became a secondary topic on industry programs, incidental to some other topic such as improved nutrition or better taste. Later in the decade, test marketing of genetically modified produce showed promise, including consumer acceptance of products offering agronomic benefits and consumer benefits.

When genetic engineering was the darling of the industry conference circuit, labeling was discussed as part of the overall issue. The industry understood the regulatory implications—that labeling would be required at the consumer level if a food was significantly changed or if a known allergen was introduced by genetic engineering. There was certainly some resistance to labeling as a burden to business, but there was a sense that no one would substantially change a food or introduce an allergen, so labeling from that perspective became a nonissue.

Labeling on the marketing side was something else entirely. Some in the industry saw educating the consumer as the way to go; others thought this was too complex an issue for industry to tackle, and perhaps it was government's role to educate. But because the Flavr Savr tomato was such an integral part of the biotech discussion at the time, the advantages of labeling were obvious. If a marketer wanted to sell a value-added product, such as the Flavr Savr, she or he certainly would want to label it in some way to convey the added value and induce the consumer to choose that product.

The crux of the labeling issue as it related to marketing was this: If you're going to spend lots of money on a better product, shouldn't you tell your customer about it? This certainly applied to the consumer benefits, such as taste, nutrition, and storage. There was less consensus about communicating agronomic benefits to consumers. Some in the industry believed that telling consumers that a special variety permitted the grower to use fewer chemical inputs would be a selling point. Others believed that communicating an agronomic benefit (e.g., drought resistance that cut the grower's irrigation costs) would not help the consumer make a decision to purchase that product.

Through the decade of the 1990s genetically engineered produce with consumer benefits was practically nonexistent, and the issue faded into the background. However, outside North America the labeling issue raged. In Europe, Australia, New Zealand, and parts of Asia and Africa, consumers and consumer advocacy groups were calling for mandatory labeling of genetically modified organisms. They perceived health threats and believed consumers needed to know what products had been genetically modified so they could have a choice. Whether the genetic enhancement benefited the industry or consumers, labeling was an issue. It became a major consumer issue (consumer issues are treated in Chapter 19), and any major consumer issue usually surfaces as a major trade issue.

Outside the fresh produce arena, many genetically modified crops are marketed. One U.S. grain grower said of his crops: "I don't think that at this point we are using anything that hasn't been genetically modified. We'd have to label everything." Yet on the other side of the Atlantic, consumers were calling for just that—mandatory labeling of all GMOs. They went further than just calling for labeling—some called for a ban on production, import, or marketing of GMOs. World trade groups began to be involved in trying to develop policies for international trade. Their issues range from perceived danger to the environment to perceived danger to their domestic producers. For many, the labeling goes beyond a sign in the retail store—it goes to declaring GMOs on menus in restaurants. Even in the United States, where acceptance of GMOs seems stronger than elsewhere, the government received many objections to allowing any GMO food to be labeled "organic" even if it meets all other requirements for the "organic" label.

## V. CONCLUSION

As the industry enters the new millennium, the following issues, among others, will continue to affect the marketplace as it relates to genetically modified foods.

1. Consumer resistance in the United States through the 1980s and 1990s was not noticeable. However, when the issue of GMOs and organics arose, comments showed significant opposition to considering GMOs "organic," regardless of production method. Also, consumer activism on the GMO issue was dormant in the United States at that time.

2. Labeling and consumer choice will continue to be debated. Whereas it might be obvious to label a product that offers an added value (especially if it is offered at an increased price) to convince consumers to purchase that item, labeling of all GMOs regardless of obvious consumer benefit will become an issue in North America, where it has not been before.

3. International trade will be affected by differing views among nations—views not just about labeling but also about acceptability of the GMOs at all. And those views will be translated into trading policies and perhaps trading barriers.

4. The role of educator will also be debated as groups look to biotech firms, growers, marketers, retailers, government, and researchers to fill it.

# 21
# Political and Economic Consequences

**Peter W. B. Phillips and George G. Khachatourians**
*University of Saskatchewan, Saskatoon, Saskatchewan, Canada*

|     |                                                              |     |
| --- | ------------------------------------------------------------ | --- |
| I.  | INTRODUCTION                                                 | 311 |
| II. | ECONOMIC IMPACT                                              | 313 |
|     | A. The Economics of Innovation                               | 313 |
|     | B. Scale and Scope Economies                                 | 314 |
|     | C. The Economics of Agrifood Industrial Structures           | 315 |
|     | D. Trade and Regional Growth                                 | 316 |
|     | E. Economic Winners and Losers                               | 316 |
| III.| POLITICAL CONSEQUENCES                                       | 317 |
|     | A. Politics of Intellectual Property Rights                  | 317 |
|     | B. Politics of Agrifood Policy                               | 319 |
|     | C. Politics of Regulation                                    | 320 |
|     | D. Politics of International Problems                        | 321 |
| IV. | A NEW FRAMEWORK FOR MANAGING COMPLEX POLITICAL AND ECONOMIC ISSUES | 323 |
| V.  | CONCLUSIONS                                                  | 324 |
|     | REFERENCES                                                   | 324 |

## I. INTRODUCTION

Some 40% of the world's market economy is based upon biological products and processes (1). Biotechnology is increasingly affecting the competitive base for much of that industry. As with most revolutionary technological changes, biotechnology has generated both economic and political responses.

    The impact of biotechnology in the agrifood world can be examined in two relatively distinct periods. Gestation spanned the period from 1973, when the Cohen-Boyer utility patent on deoxyribonucleic acid (DNA) cloning technology marked the beginning of modern biotechnology, to 1994, which marked the first year of widespread commercialization of food products modified with biotechnology. Most of the economic changes precipitated by biotechnology can be traced to this gestation period. The new technology has significantly altered the innovation process itself by converting it from a supply push to demand pull system, has changed the economies of scale and scope in the industry, and has generated significant new value, precipitating massive industrial restructuring. Thus, the comparative advantages of research and production have changed, resulting in a shift of the location of research and production. In short, the new technologies have created both winners and losers. Meanwhile, the political system adapted and responded to these changes. Governments moved to capitalize on the opportunities by extending private property rights into the sector and by beginning to redirect public research and development in support of private effort. Governments also either modified their existing regulatory systems or began to develop new special-purpose systems to manage the potential risks involved with using and commercializing biotechnology-based products.

In the mid-1990s the adoption phase began; it has not gone smoothly. In 1994 widespread commercial introduction of genetically modified (GM) agrifood products began, first with Monsanto's Posilac bovine somatotropin and Calgene's Flavr Savr tomato and now with GM corn, cotton, soybeans, canola, potatoes, and a wide variety of other products. By 2000, more than 40 transgenic modifications involving 13 products were approved and produced in 1 or more of 12 countries. James (2) estimates that total world production rose from less than 1 million acres in 1995 to approximately 100 million acres in 1999 (Table 1). Coinciding with the introduction of these products, regulatory systems came under increasing pressure and parts of the system failed. Although the European Union (EU) developed a special-purpose regulatory system for biotechnology, it has ceased to function because of uneven capacity at the national level. The United Kingdom suffered a regulatory meltdown in 1996 after the discovery that bovine spongiform encephalopathy (BSE, or "mad cow disease") was linked to a new human variant of Creutzfeldt-Jakob disease. As a result, the UK government has been unable to assuage consumer concerns about genetically modified foods. Similar regulatory failures in France and Belgium have gridlocked the regulatory system in Europe. In 1999 the European Commission was actually prosecuting France for not implementing a European Directive related to GM canola. Although regulators in other parts of the world have greater support from their citizens, their control remains tenuous. Given distrust in regulators everywhere, many consumers and opponents of biotechnology are demanding labels to identify GM foods in order to enable individuals to act on decisions they make by themselves with regard to their consumption of them. The resulting uncertainty of market access and consumer acceptance has raised doubts about the immediate future for biotechnology in the global agrifood system.

The next decade will be the critical period for the technology. Although many of the economic changes have begun, the transformation in the agrifood sector is far from complete. As new technologies are introduced and new products are commercialized, the impact of biotechnology will widen. To date the focus has been on introducing new varieties or technologies that improve the agronomic performance in mostly developed country agriculture (first-generation products). Already the research focus has expanded to include adding differentiable product attributes (second generation) and to developing new niche products (third generation). There is also significant potential for biotechnology to influence agriculture in developing countries; James (2) estimated that 18% of total transgenic crop production was in developing countries. More closely linking consumer and producer interests in both producing and consuming countries through second- and third-generation products, may actually diminish the current antagonism in the marketplace. If, on the other hand, consumer and citizen concerns are not addressed, adoption of the new technology could slow or reverse.

Over and above the future for any specific biotechnology products, this new technology has precipitated a new and much more complex framework for managing technological change, which will likely influence technological development in many other areas.

**Table 1** Worldwide Distribution of 98.6 Million Acres of Planted Transgenic Crops as a Percentage of the Total by Country, Crop, and Traits

| Distribution | Percentage of total |
|---|---|
| By country | |
| United States | 74 |
| Argentina | 15 |
| Canada | 10 |
| China | <1 |
| Australia | <1 |
| South Africa | <1 |
| Mexico | <1 |
| Spain | <1 |
| France | <1 |
| Portugal | <1 |
| Rumania | <1 |
| Ukraine | <1 |
| By crop | |
| Soybean | 52 |
| Corn | 30 |
| Cotton | 9 |
| Canola | 9 |
| Potato | <1 |
| Squash | <1 |
| Papaya | <1 |
| Flax | <1 |
| Rice | <1 |
| Tomato | <1 |
| Sugar beet | <1 |
| Melon | <1 |
| By trait | |
| Herbicide tolerance (HT) | 71 |
| Insect resistance (*Bt*) | 22 |
| Stacked HT/*Bt* | 7 |
| Virus resistance | <1 |
| Nutritional change | Trace |

*Source:* Adapted from Ref. 2

## II. ECONOMIC IMPACT

The introduction and use of biotechnology in the agrifood industry globally have caused a number of economic impacts. It is worthwhile to examine them separately to determine their long-run influence on the industry and society.

### A. The Economics of Innovation

The new technologies have fundamentally altered the innovation process itself. In the agrifood sector, the innovation process has historically been quite linear. It often started with curiosity-based research and moved through the development, production, and, ultimately, marketing of the new product. As a result, the research process was relatively narrowly defined and self-contained. The advent of biotechnology did two things: First, it created the potential to target the research more finely to specific market needs. Second, it made the research process far more complex, with no one individual or small group of individuals able to undertake the entire process. The innovation process has become more like a "chain-link model." The structure still involves a basically linear process moving from potential market to invention, design, adaptation, and adoption, but now it involves feedback loops from each stage to previous stages and opportunities for the innovator to seek out existing knowledge or to undertake or commission research to solve problems in the innovation process (3).

In this context, it is possible to identify four types of knowledge (4). "Know-why" knowledge (the understanding of scientific principles and laws of nature) is developed almost exclusively by universities and government research institutes. It is usually codified and published in academic or professional journals, which provide the base for all the rest of the research effort. "Know-what" knowledge involves facts and techniques, which usually can be codified, protected by intellectual property rights (e.g., patents), and exploited or transferred to others in the marketplace. "Know-how" refers to the skills or practical capacity to put the know-why and know-what knowledge together. This capacity is often learned by doing, making it more difficult to codify and transfer to others. Finally, "know-who," or knowing who knows what and who knows how to do what, is increasingly important as collaboration now is necessary to develop new products. Know-who knowledge is seldom codified but accumulates often within an organization or, at times, in communities where there is a cluster of public and private entities that are all engaged in the same type of research and development. Know-how and know-who provide the vital connective tissue that makes knowledge flow, which is a necessary ingredient to convert know-why knowledge into commercial applications. This new model of innovation has led to much greater specialization in the research business and a rise in the importance of networking and alliances.

## B. Scale and Scope Economies

The fragmentation of the innovation system into different structures of knowledge development and use has fundamentally affected the economies of scale and scope in the industry. Basic know-why research does not appear to exhibit any significant economies of scale, and, given that most research universities and labs are collaborating with others outside their home base, the economies of scope appear to be limited (5). Similarly, few biotechnology centers appear to exhibit significant economies of scale in the development of know-what knowledge. Most of the fundamental patentable technologies developed have come from disparate research programs around the world and have then been assembled and commercialized by private companies.

In contrast, there do appear to be significant economies of scale and scope in the know-how and know-who stages of development. At the most basic level, imperfect markets make smaller ventures invest relatively more of their resources in price discovery and contracting to undertake research; that tends to favor larger, more integrated firms. In this knowledge-based industry, firms also must invest heavily in "learning by doing" in order to enter the research business. This investment is a real barrier to entry as there is significant uncertainty that the resulting output from the research effort will have any commercial value. The entry problem is compounded by the investment required to satisfy regulators that the product is safe and conforms to industry standards. As a result of these relatively high fixed costs, firms face declining average costs over the feasible operating range. Thus, each unit of input yields successively higher volume of output. In the canola industry, for example, the global research budgets have risen more than 10-fold in real terms over the past 30 years, but the real average cost per variety has dropped sharply since the early 1980s and the marginal cost has remained relatively small.

Beyond the basic economies of scale in the industry, there is significant potential for externalities or economies of scope to influence the industry. On the positive side, the sharp drop in the average cost per variety is at least partly due to "mysteries being in the air" (6) as advancements are disseminated throughout the research community. A survey in early 1998 of all firms undertaking research into canola suggests that the flow of knowledge has a real impact on their operations. The survey asked companies about the proximity of competitors and/or collaborators as factors in locating their research efforts. Half of all the respondents, who represented the majority of larger private companies responding, acknowledged the importance of proximity to either collaborators or competitors. About 40% recognized the importance of being close to their collaborators, particularly the National Research Council and Agriculture and Agri-Food Canada research centers in Saskatoon.

There has been extensive research in other countries and in other product areas to determine the elements that influence the creation and capture of these economies of scope. Metcalfe (7) asserts that an innovative system comprises

> that set of distinct institutions which jointly and individually contribute to the development and diffusion of new technology and which provides the framework within which governments form and implement policies to influence the innovation process. As such it is a system of interconnected institutions to create, store and transfer the knowledge, skills and artifacts which define new technologies.

Zucker, Darby, and Brewer (8) showed that the location of research "stars" also positively influences the location and clustering of private commercial efforts.

## C. The Economics of Agrifood Industrial Structures

The introduction of biotechnology has precipitated a major industrial restructuring in the agrifood sector. The opportunity presented by biotechnology to manage the research process to deliver custom products has created investment potential for some private companies while creating a threat or risk to many existing businesses. The first-generation biotechnology innovations (e.g., herbicide tolerance and *Bacillus thuringiensis* [*Bt*] resistance) lowered the cost of production, and the second- and third-generation products (which change the value to the end consumer) create significant overall returns to the industry and society. Just and Hueth (9) point out that the chemical companies were quick to realize that the new ability to manipulate the genetic coding of crops changed both their markets and their concept of their business. The new technologies created the opportunity for them to develop new varieties tolerant to their patented herbicides and thereby extract greater rents from new markets for their product. At the same time, some companies decided that biotechnology-based activities had the potential to produce significantly higher profits than other activities. Some of the companies that were upstream or downstream of the farmer in the food system have restructured to capture this business, often divesting themselves of lower-return activities and replacing those activities with direct investments in, or acquisitions of, knowledge-based parts of the production system. Two trends are spurring further consolidations. The presence of significant economies of scale and scope in the product development end of the business has encouraged firms and research facilities to merge or ally themselves into larger "life science" enterprises. The difficulties in protecting and transferring technologies among partners, combined with the imperative to quality-assure the product through the production and marketing chain, have also driven those firms to integrate vertically. As a result, the production system from basic genetics right through to the consumer is increasingly tightly managed within a small group of industrial networks.

### D. Trade and Regional Growth

The share of produce that is traded is increasing because the new technologies are changing national and regional comparative advantages. Whereas agrifood production has been relatively capital-intensive for almost a generation, the sharp rise in private investment in the sector since 1990 has now made it relatively research-intensive. Economic theory suggests that production will tend to locate in regions of the world where the local factor endowments favorably match the factor requirements. Although in theory technology should flow as easily as products, in practice technologies do not disperse rapidly or completely. An increasing technology or productivity gap between low-technology countries and research-intensive countries is developing (10). In addition, the trade barriers that have impeded the redistribution of agrifood production to the most efficient sites are, with the World Trade Organization (WTO) agreement in 1995, being dismantled; that dismantling will allow agricultural production to be determined by factor endowments. As a result, production of research-intensive products is tending to concentrate in and around those research centers that provide economies of scope, which are mostly in developed countries. Canola production, for instance, is now significantly more concentrated in research-endowed countries than it was in the 1960s (5). Two other related trends are increasing the dependence on trade. Many of the new products being developed are superior goods, with rising per capita consumption strongly positively correlated with per capita income. At the same time, many second- and third-generation biotechnology innovations exhibit "product" attributes. These products in many cases will be manufactured only in one country (near the research center), then traded globally. Other countries will concentrate on other products, which also may be niche products. As a result, there is increasing potential that traditional net exporters may both export and import. All of these trends in economies of scale and scope, product attributes, and consumer demand increase the dependence of the industry on international trade.

### E. Economic Winners and Losers

The fundamental economic changes under way in the agrifood sector are creating both winners and losers. Economic studies done over the years show that research in agriculture has yielded relatively high private returns and even higher public returns (11), but that farmers tend to get a smaller share of the returns on innovations that improve yield rather than quality and their share is depressed when the related processing sector is imperfectly competitive. When the trend is juxtaposed with biotechnology-based products, four conclusions can be drawn. First, although a wide variety of economic studies of public research in the agrifood sector have estimated the internal rate of return ranges between 20% and 95% (11), it is likely that the returns are lower for biotechnology-based developments. Although some selected projects to breed high-value attributes into plants and animals may still have high returns, in aggregate the action of competitive private research "races," often subsidized by public programs, is likely to lead to excessive investment, accelerated creative destruction, and relatively low social returns. For example, the overall rate of return for canola research is estimated to have dropped by the end of the 1990s into the single-digit range (12). If the developments seen in the canola sector are repeated in other agrifood areas, similar results can be expected. Second, past studies have shown that the gains from yield-enhancing innovations are often bid away by competitive farmers, thereby translating into lower retail prices and higher consumer welfare. Hence first-generation biotechnology products are not expected to benefit farmers over the long term. Nevertheless, early adopters will often gain a "first mover" benefit and some individual farmers may gain as their specific agronomic and management circumstances allow them to profit from the new technologies (13). This will become even more relevant as new input and output traits are stacked in individual crops, making seeds more tailored to specific agronomic circumstances.

Meanwhile, many studies suggest that high concentration ratios (i.e., oligopolistic market structures) in the input and output sectors enable some firms to capture a higher share of the return (14). Given that many of the first-generation innovations link chemical products to plants, some individual chemical companies will certainly gain market share and profits. Although Moschini, Lapan, and Sobolevsky (15) conclude that intellectual property rights enable Monsanto to capture about half of the value created by Roundup Ready soybeans (leaving producers worse off and consumers with the rest of the benefit), there are compelling arguments that they may not be able to do as well as that. Incomplete property rights ensure that at least part of the benefit is not captured. Monsanto has estimated that approximately 25% of their benefit would be lost to the bin-run or brown-bagged seed markets without technology use agreements (TUAs). Even with TUA contracts, they estimate they still lose 10% of the benefit (16). In aggregate, it is also not clear that industry is gaining. Given the high usage of chemicals even before herbicide-tolerant and Bt crops were developed, gains by one company may simply be offset by the lost market share of others. Thus, the chemical industry may have gained little from these innovations when research expenses were deducted. Furthermore, Green (17) argues that the relatively short life cycles of the new varieties and products force the innovators to share some of the returns to ensure rapid market adoption. Finally, a variety of studies conclude that quality-enhancing innovations benefit the production system relatively more than yield improvements because they enable producers to segment the market and increase demand for their product, thereby offsetting the price-dampening effects of greater production. Second- and third-generation biotechnology products could therefore yield a higher return to the production system, some of which could accrue to farmers because they will need to be paid to produce and market in such a way that protects the quality of the product.

One could conclude with the adage "It was the best of times and it was the worst of times," Biotechnology has significant potential to continue to transform plants in many desirable ways. The economic base of the global agrifood industry has been fundamentally changed by the introduction of biotechnology. The ability to breed and grow crops with targeted traits selectively has opened the door for new innovation structures, a rapidly industrializing agriculture, and shifting production within and between countries. As a result, there are major winners and losers. As with any industry in which participants face either major gains or losses, there is increasing pressure for the state to respond and intervene in the market.

## III. POLITICAL CONSEQUENCES

Economic change often is precipitated by or compels political responses. This is especially true for changes in the agrifood sector, which is traditionally viewed as a politically important constituency that produces a strategic good of value as a geopolitical tool. Some have argued that agriculture is at least 50% politics. That relationship continues in the biotechnology-based agrifood industry.

### A. Politics of Intellectual Property Rights

The most significant government response to the introduction of biotechnology was the extension of intellectual property rights to processes and products of biotechnology. The main impetus for change came in the United States. As the single largest source of new innovations and the single largest market for those innovations, the United States plays a vital role in almost all innovation-based sectors. In 1970 the United States made the first substantive move to encourage increased private investment in agricultural research. Although the 1935 Plant Variety Protection Act (PVPA) had granted breeders 18 years exclusive rights to new asexually produced plant varieties (e.g., hybrid corn), it was only in 1970 that the U.S. government renewed the PVPA and granted the same property rights to sexually reproduced varieties such as open- or self-pollinated varieties (these rights were limited by both a farmers' exemption to save and reuse seed and a research exemption).

On the technology front, the U.S. Patent Office granted the Cohen-Boyer utility patent on DNA cloning technology in 1973, thereby starting the race to privatize agronomic technologies. Over the succeeding years, virtually all of the main technologies required to manipulate a plant or animal genetically have been patented in the United States and other prime markets. Those technologies that have not been patented globally still receive some effective protection from U.S. patents because any products incorporating unlicensed technologies can be blocked from entering the United States. This was followed in the 1980s with a number of landmark rulings related to patenting living organisms. In 1980 the U.S. Supreme Court ruled in *Diamond vs. Chakrabarty* that the U.S. patent law provides for patenting life-forms, in that case an oil-eating bacterium. In 1985 the first patent for a living plant was issued. Since then a wide variety of plants have been patented. Plant patents provide additional protection over plant variety protection rights in that plant patents do not provide exemptions for either researchers or farmer-saved seed.

These moves in the United States precipitated a response in most other developed countries. The International Convention for the Protection of New Varieties of Plants (UPOV), 1961, was revised in 1972, 1978, and 1991 to extend rights internationally. The convention requires member states to provide protection for new varieties of plants, including for plants bred in other member states. The agreement and the corresponding national laws grant breeders exclusive rights to market varieties for a set period (usually 15–20 years), often with exemptions to allow producers to save and reuse their seed and to permit breeders to use protected varieties as genetic resources in further breeding. As of 1999, 48 countries had some form of plant breeders' rights, representing 25% of the world. Of those, 38 had adopted UPOV standards; the other 10 had national policies.

As intellectual property rights (IPRs) were being developed both nationally and internationally, the agrifood sector increasingly came under the aegis of the World Intellectual Property Office (WIPO) and relevant treaties. As WIPO had 175 members as of September 1, 2000, the commitments to the fair and equal treatment of intellectual property had expanded. Even so, delivery was somewhat spotty because only a few countries have acceded to all of the agreements. Furthermore, although WIPO offers conciliation services, there are no formal dispute settlement systems within the various treaties. More recently, in the 1995 round of multilateral trade negotiations, the Trade-Related Intellectual Property (TRIPs) Agreement was developed to address the deficiencies of national, UPOV, and WIPO systems. The agreement, which came into effect on January 1, 1995, and has been ratified by 139 nations, covers a number of areas relevant to biotechnology, including trademarks; geographical indications, including appellations of origin; patents, including the protection of new varieties of plants; and undisclosed information, including trade secrets and test data. The TRIPs Agreement provides for certain basic principles, such as national and most-favored-nation treatment; sets out general rules to ensure that procedural difficulties in acquiring or maintaining IPRs do not nullify the substantive benefits that should flow from the agreement; and provides for appeal to the binding WTO disputes settlement process. The obligations under the Agreement will apply equally to all member countries but developing countries will have a longer period to phase them in.

Even with the apparent international expansion of property rights to intellectual property, the system is certainly not simple. As with the WTO trade rules, the absence of Russia and China from the agreement leaves a major hole in its coverage. In addition, many of the countries have yet to address the commitment to extend patent rights to whole organisms or to implement an effective sui generis (purpose-built) system. Although the United States and EU now patent living organisms, Canada and many other countries have so far failed to do so. Nevertheless, the dominant U.S. role in the trade system has tended to discourage private companies hoping to trade their products from using unlicensed intellectual property. The U.S. system of intellectual property rights is thereby effectively extended to nonconforming countries.

Both public and private breeders have expressed some concern that their "freedom to operate" is being narrowed by the increasingly proprietary nature of fundamental biotechnology and breeding technologies, genes, gene constructs, and germplasm. The fear is that corporate ownership of all of the technologies could shut down public or shut out private competing breeding programs, especially those targeted on agronomic improvements or smaller crops. Many public sector scientists suggest that the state should invest in the development of duplicate technologies or germplasm in order to provide an accessible base of technologies for smaller private and public breeders. Alternatively, some have advised that the state should use the powers vested in its intellectual property rights regimes or antitrust laws to encourage greater dissemination of nonrival, patented innovations in order to generate more access and, hence, greater positive spill-over effects (18).

## B. Politics of Agrifood Policy

Governments have been challenged to refashion their role in agrifood research and development. Agriculture has been one area in which the public sector historically has contributed a significant share of research resources and has directly undertaken a large share of the research effort. Except for those agrifood products with effective hybrids (e.g., maize corn), most of the effort has been undertaken by governments, publicly funded universities, or private companies funded by public grants. That relationship held until the 1980s in most product markets. Since then, however, new, proprietary technologies have been developed and most of the resulting crop innovations have been commercialized by private companies. As the germplasm, technologies, genes, and seeds industries have been privatized, the public sector's historical role as proprietor or lead innovator has been challenged. Now, the state acts more often as a partner and promoter, creating the basic economic structure for public and private investment in research and development (R&D) through direct investment and through a selection of fiscal measures targeted on the industry. This has involved a shift away from doing all of the varietal development in public or university labs toward doing more custom work and collaborations, often on precommercial or noncompetitive projects.

Although it is next to impossible to determine the exact impact of these policies on private research, industry competitiveness, and industrial location, there are a number of examples that demonstrate that, at least at the margin, these policies can and do influence private decisions about location of research effort. The benefits for governments are significant. If knowledge spillovers (e.g., know-how related to genetic transformations) are limited to a specific location (perhaps because the diffusion of the knowledge requires face-to-face interactions), then any scale or scope economies that result will be captured by the region that undertakes that activity. That creates the possibility that countries engaging in technological competition may endogenously generate comparative advantage. In short, "comparative advantage evolves over time" (19). If the final product of biotechnology is tradable but the innovation-based knowledge is a nontransferable intermediate factor of production, then the fact that innovation begins or is supported in one jurisdiction could indefinitely put that site on a higher trajectory of R&D and new product development (19). As a result, the high-technology share of gross domestic product (GDP) and of exports will be higher than otherwise, and society will be better off. Although the R&D benefits would thereby be captured in discrete jurisdictions, the spillovers and resulting accelerated innovation should ultimately benefit consumers wherever they are.

## C. The Politics of Regulation

Governments around the world have been urged to regulate biotechnology-based research and production. Some governments have regulated their domestic research, production, and marketing systems in an attempt to provide a set of rules and norms that ultimately determine the extent and scope of private initiative. States use a mix of research and development policies, tax incentives, intellectual property rights, competition policy, environmental laws, and various regulations to achieve their goal. In each case, the state attempts to balance the public interests of the general citizenry with the private interests of both domestic and foreign firms. Two basic approaches to regulating biotechnology have evolved (20). The United States and a wide variety of other countries have developed regulatory systems that focus on the potential risks of biotechnology in the resulting agrifood products. They have adopted the *essential equivalency* standard such that if a biotechnology product has the same molecular structure as a conventional product, then no incremental labeling or regulatory restrictions are necessary. Because the focus is on products, these countries have for the most part been able to use or extend existing legislation, regulation, and agencies to manage biotechnology. This has yielded relatively efficient and generally accepted regulatory decisions. In contrast, the European Union (EU) and some other nations focus their regulatory energies on the technological process and not on the end product attributes. In this context, the EU has adopted the *precautionary principle*, whereby products using biotechnology are assumed to be risky until proved otherwise. In addition, as of 1999 all products containing more than 1% genetically modified materials required labeling indicating their presence. Because the EU has chosen to regulate the technology and not the product, it has been forced to develop entirely new legislation and regulation, yet it leaves much of the review and compliance system in the hands of traditional national regulatory agencies. Initially, decisions were delayed because the system was incomplete. With the regulatory meltdown in the United Kingdom and problems in other EU member states, however, the review system has come to a complete standstill. Although 18 genetically modified products were provisionally approved for use as of April 1998, the four most recent applications were rejected. In addition, Denmark, Britain, and France have implemented temporary halts on GMO approvals in their countries, Austria, Luxembourg, France, and the United Kingdom have all imposed unilateral bans on certain new crops and the EU is not expected to approve any new GMOs before 2002 (21). Most of the other Organization for Economic Co-operation and Development (OECD) countries have adopted one of the two models for regulating their industries, whereas virtually no developing country has any effective system. This incomplete and at times conflicting regulatory system has created some pressures for change.

Almost all domestic regulators are faced with rising consumer concern about the health, safety, environmental, economic, and ethical implications of biotechnology in the agrifood system. From a public policy perspective, environmental and food safety assessments should help public perception. However, such assessments are not perceived to adequately deal with a broad range of secondary, unintended effects that could affect allergenicity, the environmental and plant nutrient and toxicant levels (22). Already there is evidence that some transgenic material does not maintain itself but instead must be controlled by specific plant selection techniques (23). This simply contributes to the large information gap that exists between producers and consumers, especially for biotechnology-based products, and creates a market failure that opens the way for government action. Whenever governments intervene, they have tried to address the information gap but have often ended up inserting noneconomic public objectives into the market. Although all of the currently operating domestic regulatory systems use objective, science-based criteria, there is significant pressure for subjective criteria to be formally incorporated in domestic regulatory systems. In the first instance, this has involved many nongovernmental organizations' calling for either outright bans or stringent regulations based on strict interpretation of the precautionary principle. Especially in the EU, where citizens have lost faith in their regulators, there has been significant pressure for compulsory labeling of genetically modified products. In addition to the national bans and embargoes, many European retailers and food processors have responded to consumer concerns and have developed "GM-free" own-label products. Seven large supermarket chains have joined forces to eliminate GM ingredients in their own-label products: J. Sainsbury and Marks and Spencer in the United Kingdom, Carrefour in France, Migros in Switzerland, Delhaize in Belgium, Superquinn in Ireland, and Effelunga in Italy (24). The UK response has been greater, as Tesco and Sainsbury, Britain's two largest chains, have been joined by Marks and Spencer, ASDA, and Iceland in adopting similar GM-free strategies. Meanwhile, Unilever and Nestle in the United Kingdom, both large food processors, announced in 1999 that they would remove genetically modified elements from their products (25). This has extended to other markets. In 1999 Japan introduced proposals for a new labeling system, Australia and New Zealand announced they would have compulsory labeling by 2001, and South Korea planned to have a system by the end of 1999. Meanwhile, the agrifood industry in the United States and Canada began in late 1999 to develop voluntary labeling standards for GM foods. This move to labeling poses a conundrum for companies wishing to place new products in the market because market acceptance requires market access and regulatory approval, which in turn appears increasingly to require market acceptance (20). Governments everywhere are pondering how to manage this challenge to the system.

### D. Politics of International Problems

Both opponents and supporters of agricultural biotechnology have increasingly looked to international organizations for solutions to concerns that either cannot be or are not being addressed within domestic systems (Table 2). Seven international organizations are currently involved internationally in regulating to some degree aspects of biotechnology (26). These bodies are in two categories: those that are science- and health-based, and those that have objectives of facilitating international trade, environmental considerations, or various other social and political goals. The three science-based institutions are the International Plant Protection Convention (IPPC), which regulates plant health; the International Office of Epizootics (IOE), which manages animal health issues; and the Codex Alimentarius Commission (Codex), which establishes common standards for food safety and labeling. The key broad-based institution is the World Trade Organization, which provides for national and most-favored-nation treatment for trade among its 139 members. The 1995 agreement also establishes common procedures for handling issues related to sanitary and phytosanitary standards, by drawing on the technical standards developed at the IPPC, the IOE, and Codex and technical barriers to trade, such as labeling standards. Meanwhile, the 29-member Organization for Economic Co-operation and Development has been working over the past decade to develop consensus documents for use in national regulatory systems, various regional processes are seeking technical resolutions (e.g., the U.S.-EU Trans-Atlantic Economic Partnership and Canada-U.S. bilateral negotiations on agricultural biotechnology), and more than 170 countries negotiated a BioSafety Protocol to protect the environment from risk of biotechnology by regulating transborder movements of living modified organisms.

**Table 2** The Main Distinguishing Features of the Regulatory Systems

| Country | IPR system | Production system | Biotechnology regulations | Marketing regulations |
|---|---|---|---|---|
| United States | Process, utility, and life patents | PVPA; conform to UPOV 1978; no seeds registration | Product-based; no special legislation | No labeling required |
| Canada | Process and utility patents; no life patents | PBR Act, 1990; conform to UPOV 1978; Seeds Act regulates registration | Product-based; no special legislation | No labeling required |
| European Union | Process, utility, and life patents | Conform to UPOV 1978 or 1991; have seeds registration systems | Process-based; special legislation | Mandatory GMO labeling rules |
| Japan | Process and utility patents; no life patents | Conform to UPOV 1978 | Process-based; special legislation | Mandatory GMO labeling rules |
| India | Weak process and utility patents; no life patents | Not currently a member of UPOV | None | None |
| Australia | Process, utility, and life patents | Conform to UPOV 1978 as of 1989; seeds registration | Product-based | Mandatory labeling rules |
| China | No effective patent protections | Entered negotiations to join UPOV; seeds registration | None | None |

The detail of each institution is beyond the scope of this chapter. Suffice it to conclude that there currently is no one particular institution created for the explicit purpose of regulating internationally the products of biotechnology. Similarly, there does not appear to be an obvious choice for fulfilling this role, at least among the seven existing institutions. In the absence of any overarching international agreement on regulating products of biotechnology, differing national standards provide the minimum standard. Nevertheless, the EU and many countries in the developing world are looking to various international negotiations to address the wide array of social concerns around biotechnology (e.g., ethics and political concerns). Some saw the attempted Seattle Round of the WTO as an ideal instrument to manage biotechnology trade issues because of its binding dispute settlement mechanism, which helps to equalize power between large and small member states. In addition to gaining access to the dispute settlement process, many of these countries seek amendments or new agreements that directly address their subjective ethical, environmental, social, and consumer concerns. Essentially, they are asking other member states to bargain away their existing rights either to retaliate or to seek compensation when another country impedes trade for subjective purposes. Failure to launch new WTO negotiations in Seattle in 1999 shifted most of the attention to the BioSafety Protocol, which was unable to resolve the issue so the search continues.

The future international structure for regulating biotechnology is unclear. The existing international regulatory system relating to biotechnology still has some problems, as many countries just, if at all, comply with their commitments. In addition, China, the first country to commercialize a biotechnology-based crop and potentially one of the largest users of biotechnology, and potential key markets like Russia are waiting to join these international agreements. Although negotiations for accession have begun, it will take years for them to conclude and an even longer time for adjustments to be completed.

Partly because all of the existing domestic and international regulatory systems were largely designed to nurture and support development of biotechnology-based commerce, citizens and consumers have challenged governments to consider and respond to subjective concerns. It is too early to say how governments and international institutions will respond.

## IV. A NEW FRAMEWORK FOR MANAGING COMPLEX POLITICAL AND ECONOMIC ISSUES

One might be tempted to argue that the traditional institutions and disciplines could handle the challenges of biotechnology, just as they handled the green revolution. Biotechnology has, however, irrevocably complicated an already complex system. In the 1940–90 period, the green revolution entered a market that was dominated by the U.S. economy and U.S. surplus food production. Farmers, executive governments, and a handful of international institutions largely managed the impact. That manageable world is gone forever: many commodity markets have been deregulated; decolonization has led to a rapid proliferation of sovereign states; international trade, labor, and capital markets have been liberalized by international agreement; many markets are now linked by multinational corporations; and public interest groups have become transnational.

Now, along come biotechnology and an array of transgenic plants, animals, and microbes, which raise a whole new set of issues. Stonehouse (27) has postulated that some issues require broad, comprehensive, and integrative thinking, which involves consideration of spatial, temporal, scale, quantitative, and qualitative factors. In our view, transgenic food plants present one such case. Biotechnology challenges consumers, producers, industry, markets, and governments to handle and resolve a large number of complex issues. In short, one can argue that since 1990 the rules and terms of reference for agricultural politics and economics have changed. These new issues cannot be addressed by the same approaches used in traditional cases (i.e., before the 1990s). Isolated academic disciplines' and single-purpose institutions' inability to handle the array of issues jeopardizes their reputation. Both the disciplines and the institutions fail because they construct artificial and partial models, reducing their relevance and usefulness to the agrifood industry and policymakers. Instead, they need to develop new approaches to handle complex societal problems and issues. The problems of biotechnology (as is true of many problems today) cannot be adequately addressed by single disciplines: they cross over multiple disciplines. Today's economic-political analyses require a much broader theoretical understanding of underlying and complex social problems. Therefore, there is need for broadening the terms of disciplinary reference. Most of those involved in the policy debate (e.g., academics, analysts, managers, scientists, and bureaucrats) must learn to position their work and commentaries in context with the multidimensional and complex mixture of science, society, the environment, and the market. This will be a real challenge because most social or natural scientists are not well grounded in the complementary and multidisciplinary sciences and as a result do not have any common concepts, vocabularies, or working knowledge to handle modern food science issues.

Certainly no one approach can solve the present issues related to GM and transgenic food plants. A toolbox is required. The important transgenic food crops issues reach beyond the bounds of one, two, and even three disciplines. Stonehouse (27) argues for linking specialists from many disciplines to develop meaningful policies and cautions that this is a complex process, requiring a great many balancing acts, e.g., (a) anthropocentricity with ecocentric approaches; (b) temporal, shorter-term imperatives against longer-term needs; and (c) short-term economic efficiency against long-term sustainability.

## V. CONCLUSIONS

The pendulum in economic or political debates swings at different speeds and for durations. Old constraints will have to be reconsidered and discarded if necessary. As with any evolving system, technologies and the rules of the game can and will change. This makes prediction difficult. Nevertheless, a number of trends likely have some time before they are exhausted.

Biotechnology will continue to alter the economic foundation of the agrifood industry. It has already accelerated the industrialization of the food system, relocated production among regions of the world, and created winners and losers. So far most of the existing regulatory and marketing systems have encouraged market forces to prevail. If trends continue, production of transgenic food plants will continue to rise in North America and Argentina as new crop varieties enter the market, with total transgenic production peaking at something less than 100% of the total acreage. The rest of the world is more difficult to forecast. Already we are seeing some adoption of the technology in the EU (three member states reported some GM crop production in 1999) and beachheads in Africa, Asia, Latin America, and Australia. There are a number of potential trends. If the regulators and industry fail to address consumer and citizen concerns, production could come to a halt in many of these markets. If industry and the regulators are able to muddle through (or even do better), then production in those areas could grow. Given the level of concern in many of those markets, one could expect the adoption rates to lag those observed in North America, with peak production 5–10 years from now.

The biggest uncertainty about the future is that consumer and citizen concerns have not been adequately accommodated. As the value of these products ultimately lies with consumers, industry and governments will need to find some way to bridge the gulf. So far consumers do not believe any benefit will flow to them from the new technologies and believe they bear all the risks (i.e., economic, environmental, health, and safety). Some in the industry have responded with information campaigns, which so far have failed. Unlike for many products, greater awareness does not appear to translate into greater acceptance and willingness to buy. In fact, greater awareness is strongly negatively correlated with acceptance (20). The industry will need to engage more constructively in the debate. For the first 5 years of commercial production, Monsanto was synonymous with the new technology. The recent restructuring of companies in the industry may help to broaden the base of companies engaged in the public debate, and should provide better perspective on the issues and opportunities. If the conflict between producers and consumers is not resolved, political processes and consumer choices may limit the use of the technology in the agrifood sector. Under somewhat similar circumstances, during the 1970s food irradiation technologies were effectively mothballed in most developed countries because of poor political choices and widespread consumer concern. Alternatively, if adequate market-based or political responses to citizen and consumer concerns can be found, then this unease may only be a temporary problem. Ultimately, the challenge is to rebuild trust in our agrifood systems. This requires both managing and being seen to manage the objective, science-based risks of the new technology and providing mechanisms for handling more subjective concerns. Only then will the agrifood sector be able to address the subjective uncertainties and concerns many consumers, citizens, and governments have raised. In the meantime, conflict will be the order of the day.

## REFERENCES

1. R Gadbow, T Richards, ed. Intellectual property rights and global consensus, global conflict? Boulder, CO: Westview Press, 1990.
2. C James. ISAAA Briefs: Global Review of Commercialized Transgenic Crops: 1998, 1999. Available at http://www.isaaa.org/frbrief8.htm.

3. OECD. The Knowledge Based Economy. Paris: OECD, 1996.
4. E Malecki. Technology and Economic Development: The Dynamics of Local, Regional and National Competitiveness. Toronto: Longman, 1997.
5. PWB Phillips, GG Khachatourians. The biotechnology revolution in global agriculture: Innovation, invention and investment in the global canola sector. London: CAB International, 2001.
6. A Marshall. Principles of economics. London: Macmillan, 1890.
7. J Metcalfe. Technology systems and technology policy in an evolutionary framework. Camb J Econ 19:25–46, 1995.
8. L Zucker, M Darby, M Brewer. Intellectual human capital and the birth of U.S. biotechnology enterprises. Am J Agric Econ 88:290–306, 1998.
9. RE Just, DL Hueth. Multimarket exploitation: The case of biotechnology and chemicals. Am J Agric Econ 75:936–945, 1993.
10. R Lucas. On the mechanics of economic development. J Mon Econ 22:30–42, 1988.
11. J Alston. Research Returns Redux: A Meta-Analysis of the Returns to Agricultural R&D. Discussion Paper No. 38. Washington, DC: International Food Policy Research Institute, 1998.
12. RS Gray, Malla, PWB Phillips. Gains to Yield Increasing Research in the Evolving Canadian Canola Research Industry. Proceedings of the ICABR Conference on the Shape of the Coming, Agricultural Biotechnology Transformation: Strategic Investment and Policy Approaches from an Economic Perspective, University of Rome Tor Vergata, 1999.
13. ME Fulton, L Keyowski. The producer benefits of herbicide-resistant canola. AgBioForum 2:2, 1999. Available at http://www.agbioforum.missouri.edu.
14. J Alston, G Norton, P Pardey. Science Under Scarcity: Principles and Practice for Agricultural Research Evaluation and Priority Setting. New York: CAB International, 1998.
15. G Moschini, H Lapan, A Sobolevsky. Trading technology as well as final products: Roundup Ready soybeans and welfare effects in the soybean complex. Proceedings of the ICABR Conference on the Shape of the Coming Agricultural Biotechnology Transformation: Strategic Investment and Policy Approaches from an Economic Perspective, University of Rome, Tor Vergat, 1999.
16. M Roth. Comments made on a panel discussion at the NE-165 Conference: Transitions in Agbiotech: Economics of Strategy and Policy in Washington, DC June 24–25, 1999.
17. C Green. The Industrial Economics of Biotechnology, 1997. Available at http://strategis.ic.gc.ca/SSG/ca00913e.html.
18. W Lesser. Global interdependence and the private sector. In: NBAC Report 6: Agricultural Biotechnology and the Public Good. New York: NBAC, 1994.
19. M Grossman, E Helpman. Innovation and Growth in the Global Economy. London: MIT Press, 1991.
20. G Isaac, PWB Phillips. Market access and market acceptance for agricultural biotechnology products, Proceedings of the ICABR conference on the Shape of the Coming, Agricultural Biotechnology Transformation: Strategic Investment and Policy Approaches from an Economic Perspective. University of Rome, Tor Vergata, June 17–19, 1999.
21. GMO familiarity may breed comfort. Western Producer, Saskatoon, April 22, 1999.
22. O Kappeli, L Auberson. How safe is safe enough in plant genetic engineering? Trends Plant Sci 3:276–28, 1998.
23. CE Palmer, WA Keller. Transgenic oilseed brassicas. In: GG Khachatourians, A McHughen, WK Nip, R Scorza, YH Hui, eds. Transgenic Plants and Crops. New York: Marcel Dekker, 2001.
24. Supermarkets in move on modified food. The Financial Times. March 18, 1999.
25. Bowditch Group. The AgBiotech Newsletter 174 April 30, 1999. Available at http://www.bowditch-group.com/index.html/index2.htm.
26. D Buckingham, R Gray, PWB Phillips, T Roberts, J Bryce, B Morris, D Stovin, G Isaac, B Anderson. The International Coordination of Regulatory Approaches to Products of Biotechnology. Research report submitted to Agriculture and Agri-Food Canada, June 1999.
27. PD Stonehouse. A new modus operandi for the agricultural economics profession. J Agric Environ Ethics 3:55–67, 1997.

# 22
# Introduction and Expression of Transgenes in Apples

**Abhaya M. Dandekar**
*University of California, Davis, California*

| | | |
|---|---|---|
| I. | INTRODUCTION | 327 |
| II. | TRANSFORMATION OF APPLE | 328 |
| III. | ALTERNATIVES TO CHEMICAL PESTICIDES: GENES ENCODING INSECTICIDAL PROTEINS | 329 |
| IV. | RESISTANCE DEVELOPMENT AND MANAGEMENT STRATEGIES | 331 |
| V. | ENGINEERING DISEASE-RESISTANT APPLES | 332 |
| VI. | CARBOHYDRATE METABOLISM IN APPLES | 332 |
| VII. | ETHYLENE BIOSYNTHESIS IN APPLES | 334 |
| VIII. | OTHER QUALITY TRAITS | 336 |
| | REFERENCES | 337 |

## I. INTRODUCTION

The development of a variety of horticultural practices and the application of plant breeding techniques over the last several millennia have resulted in the domestication and production of a wide range of valuable tree fruit and nut crop species. This has created opportunities for nurseries and growers and a demand for their products by consumers. Interest in fruit and nut crops has been heightened in recent times by a variety of findings that link various macro- and micronutrients with beneficial effects on human health. There now seems to be good evidence for the old adage, "An apple a day keeps the doctor away."

The 21st century will likely pose a new set of challenges for growers and nurseries, as an increased demand for tree fruit products will require novel production practices that sustain yield and quality while minimizing damage to the environment and the food chains vital to human health. Additionally, the rising cost of land and increasing cost of labor will provide an incentive to grow high-value crops like the fruit and nut tree crops. Sustaining both supply and demand in the future will depend more heavily on the development and deployment of a range of new technologies, including biotechnology. The technology to produce transgenic plants will have an important and powerful impact on some of the immediate problems of tree crops, such as disease and pests, and could reduce dependence on chemical pesticides and fungicides. However, commercial success in the long term will depend upon the impact that this technology will have on improving various quality and nutritional traits. This review focuses on apple, with an emphasis on the transgenic technology being developed and the potential areas of impact.

The apple belongs to the genus *Malus* of the Rosaceae family, which includes some of the most prominent fruit crop species (pear, plum, peach, nectarine, prune, cherry, apricot, strawberry, raspberry, and blackberry). The genus *Malus* consists of at least 20 to 30 different species; most of the domestic cultivars are derived from interspecific hybridization because of self-incompatibility. The generally accepted scientific name for apple is *Malus* × *domestica*. The cultivation of apple dates back to a few centuries B.C. and was practiced by the Greeks and Romans, who in turn spread the cultivation of apples through Asia and Europe. Many horticultural practices, such as budding, grafting, hedging, and selection, were developed in medieval times when apples were grown around religious houses. The first scientific breeding efforts with apple date back to Thomas Andrew Knight (1759–1835).

On the basis of most recent estimates apple is one of the four most widely grown fruit in the world, with citrus, grapes, and bananas (citrus, 102,822,056 metric tons; grapes, 57,397,245 metric tons; bananas, 58,618,083 metric tons; apple, 56,059,564 metric tons). The annual worldwide production of apple is about 56,059,564 metric tons (125,573,420,000 pounds). This production occurs in a total area of 7,088,233 hectares and is worth over $20 billion (grower return based on average U.S. 1997 figures). The total production in the United States is worth about $1.7 billion, of which the majority is accounted for by just a handful of cultivars, such as 'Red Delicious' and 'Golden Delicious,' 'Granny Smith,' 'McIntosh,' and 'Rome Beauty.' The production in the United States over the past 10 years has been steady, with reasonably good consumer acceptance of individual apple cultivars including a variety of new cultivars. This trend would support the transfer of particular traits, e.g., pest resistance; disease resistance; and improved quality, color, and nutritional quality of fruit, into leading apple cultivars. The success of apple varieties like 'Fuji' indicates that new varieties of apples with moderate appearance but superior quality can compete with apple varieties that have superior appearance/color but variable quality.

## II. TRANSFORMATION OF APPLE

Apple transformation was first reported by James and coworkers (James et al., 1989), who used the *Agrobacterium*-mediated transformation of leaf disks from the apple cultivar 'Greensleeves' with the binary vector pBin6. A detailed procedure for the transformation of apple was subsequently published for the transformation of 'Greensleeves,' which is an excellent experimental system for apple (James and Dandekar, 1991). Genes incorporated via *Agrobacterium*–mediated transformation were found to be stably incorporated and inherited in a simple mendelian fashion (James et al., 1994, 1996). Transformation was possible because key tissue culture procedures such as micropropagation and adventitious bud formation had been previously developed in apple. Adventitious bud formation on leaf disks is a vital component of any apple transformation system because it maintains clonal identity of the cultivar being transformed. Unfortunately, in the case of apple, regeneration is highly variable among different cultivars. Factors that appear to affect regeneration frequencies include medium (inorganic content), presence of phytohormone (both type and amount), physiological traits of the explant source, and cultivar of apple used. Regeneration has been reported in several different apple cultivars and rootstocks, including 'McIntosh,' 'Triple Red Delicious' (Fasolo et al., 1989), 'M26,' 'M25,' 'M9,' and 'Greensleeves' (James, 1987; James et al., 1988). Transgenic apple plants have now been regenerated in several cultivars and rootstocks, including 'Delicious' (Maximova et al., 1998; Puite and Schaart, 1996; Sriskandarajah and Goodwin, 1998; Sriskandarajah et al., 1994), 'Elstar' (Puite and Schaart, 1996), 'Gala' (Maximova et al., 1998; Puite and Schaart, 1996; Yao et al., 1995), 'Greensleeves' (James et al., 1989; 1996; Maximova et al., 1998), 'M26' (Holefors et al., 1998; Lambert and Tepfer, 1992; Maheswaran et al., 1992; Norelli et al., 1994), 'McIntosh' (Bolar et al., 1997), and 'Pink Lady' (Sriskandarajah and Goodwin, 1998).

A significant problem is the low efficiency of transformation observed in apple (James et al., 1989; James and Dandekar, 1991). Studies from the late 1980s and the 1990s indicate that several factors can influence the transformation efficiency of apple, including the type and physiological characteristics of explant tissue used, the strain of *Agrobacterium* sp., and the design of plasmid vectors (James et al., 1988; James and Dandekar, 1991; Dandekar et al., 1990; De Bondt et al., 1994; Sriskandrajah et al., 1994; and Sriskandrajah and Goodwin, 1998). *Agrobacterium* sp. strains containing pTiBo542 or pTiC58 and their derivatives were found to be most virulent on apple (Dandekar et al., 1990; De Bondt et al., 1994; Martin et al., 1990). Virulence of the *Agrobacterium* strain could be enhanced either genetically by introducing additional copies of the virulence genes (Dandekar et al., 1990) or physiologically by growing the bacterium under conditions that induce the virulence genes (James et al., 1993). Addition of sugars like glucose to the cocultivation medium were stimulatory (De Bondt et al., 1994). Transformation was found to be highly variable among different apple culivars, possibly because of differences in infectability or integration of deoxyribonucleic acid (T-DNA) (De Bondt et al., 1994; Puite and Schaart, 1996). Physiological status of the explant and its source is important for transformation, i.e., leaves from rooted shoots (James et al., 1990), age of explant (De Bondt et al., 1994), type of leaf wounding (Norelli et al., 1996), explants from etiolated shoots (Liu et al., 1998), and conditioning of explants in liquid medium (Sriskandarah and Goodwin, 1998). Genetic selection of transformed tissue is another critical stage, with the amount of antibiotic varying with the apple cultivar and type of gelling agent being used during the selection phase are critical factors. The antibiotic kanamycin has been the major selective agent used in most if not all studies, and the gelling agents were agar, gelrite, or a combination of the two (James et al., 1989; Maheswaran et al., 1992; Norelli and Aldwinckle, 1993; Yepes and Aldwinckle, 1994). Different antibiotic combinations need to be tested to optimize selection of transformants as well as to remove persistent *Agrobacterium* sp. (Hammerschlag et al., 1997; Norelli and Aldwinckle, 1993; Yepes and Aldwinckle, 1994)

The first field trial of transgenic apple plants was conducted in California in spring 1992 (Dandekar et al. permit no. 91-218-03 of the U.S. Department of Agriculture's Animal Plant Health Inspection Service). Subsequently, many more have been conducted in the United States and a few in the United Kingdom and New Zealand. These trials indicated no adverse effects of the transformation process and normal fertile plants were produced. Yield and quality trials have as yet to be conducted. It will be important to demonstrate via field testing that the introduction of a single new trait by transformation does not alter the quality or productivity of the enhanced apple cultivar.

## III. ALTERNATIVES TO CHEMICAL PESTICIDES: GENES ENCODING INSECTICIDAL PROTEINS

Insects are a significant problem in leading apple production areas like the Pacific Northwest in the United States. Colding moth (CM) (*Cydia pomonella*) is the major pest and could be regarded as the number one insect pest worldwide in apple-producing areas. It attacks fruit at the 1- to 2-cm stage hence the name *codling* ("young one/baby"). Codling moth is a lepidopteran insect that lays eggs on fruit or leaf clusters near fruit. Larvae that emerge feed on fruit and cause considerable economic damage. CM larvae do not feed on the surface but burrow into the apple fruit, thus making them difficult to control with chemical pesticides. The key pesticides azinphos methyl (Guthion) and methyl parathion (penncapN) used to control CM have now been banned in the United States under a 1996 law inspired to protect children who have been shown to be particularly vulnerable to the poisonous pesticide residues (Doyle, 1999). Clearly, alternatives are needed that will kill the pests and avoid chemical pesticides. One such approach in tree crops is to use genes that encode resistance to insect pests from a variety of sources (Escobar and Dandekar, 2000)

A predominant strategy to engineer resistance to CM in apple has been the expression of genes from *Bacillus thuringiensis* (*Bt*) that encode insecticidal proteins. The insecticidal activity of *Bacillus thuringiensis* resides in the bacterium as a parasporal crystalline inclusion body containing one or more insecticidal crystal proteins (ICPs) (previously referred to as *delta-endotoxins*) (reviewed in Whiteley and Schnepf, 1986; Schnepf et al., 1998; Höfte and Whiteley, 1989). The insecticidal properties of crystal/spore suspensions of *Bacillus thuringiensis* have been exploited commercially for more than 40 years in products such as Dipel (Abbot Laboratories), Javelin (Sandoz), and Thuricide (Sandoz). When the ICP crystal is ingested by the target insect it is solubilized in the alkaline pH of the insect midgut and acted upon by midgut protease(s), releasing the active N-terminal fragment of the insecticidal crystal protein (ICPF) and killing the insect (Höfte and Whiteley, 1989). In vitro studies have shown that the ICPF binds to specific cell receptors on the brush border membrane of midgut epithelial cells with high specificity and high affinity (Hofmann et al., 1988a, 1988b; Van Rie et al., 1990; Schnepf et al., 1998). In vivo, this binding correlates with the formation of pores and membrane lesions that lead to swelling, leakage, and lysis of the epithelium, ultimately causing in the death of the insect through starvation and septicemia (Knowles and Ellar, 1987; Schnepf et al., 1998). Nontarget organisms presumably do not possess the specific cell receptors and as a result are unaffected by this protein.

The genes encoding ICPs from different *Bacillus* species can be categorized into different groups on the basis of the host range of their activity and DNA sequence homology (Crickmore et al., 1998; Höfte and Whiteley, 1989). The *cry*I family of genes are specific for lepidopteran species, and *cry*IAc and *cry*IAb have been shown to be toxic to CM larvae (Vail et al., 1991). Initial studies on the expression of *cry*IAc in apple were performed by using unmodified gene sequences obtained from *Bacillus thuringiensis*. This approach was based upon the success achieved in tobacco (Vaeck et al., 1987) and tomato (Fischoff et al. 1987) against the tobacco hornworm and the tobacco budworm. However, transgenic apple plants containing wild-type *cry*IAc sequences from *Bacillus thuringiensis* expressed levels of ICP insufficient to cause a significant mortality rate of CM larvae (Dandekar et al., 1992). Similar results were obtained in walnut against CM larvae (Dandekar et al., 1994). The *cry*IAc sequences from *Bacillus thuringiensis* were found to have several problems that interfered with their expression in plants. The most significant problem was a strong codon bias needed for expression in *Bacillus thuringiensis*. This bias involves a preference for A/T in the third position rather than the G/C preferred in plants (Murray et al., 1989; Campbell and Gowri, 1990). Many of the codons within the sequence also contain CG or AT in the second and third positions, and in most eukaryotes there is an absence of XCG and XAT codons (Beutler *et al.*, 1989). In addition to the incompatibility observed at the codon usage level, the gene sequences from *Bacillus thuringiensis* contained many sequences that cause messenger ribonucleic acid (mRNA) instability, including killer sequence, ATTTA (Shaw and Kamen, 1986); polyA-like signals, AATAA (Dean et al., 1986); T-rich regions, e.g., TTTPTR or >TTTT; and A-rich regions, e.g., >AAAA (Goodall and Filipowicz, 1989). A strategy to circumvent these problems was reported by Perlak ad coworkers (1991); it involved making alterations of this gene through chemical synthesis. The gene product has the identical amino acid sequence; the major differences are

## Transgenes in Apples

at the nucleic acid level. This altered gene functioned very well in plants, having levels of expression >500-fold higher than the wild-type coding region from the bacteria (Perlak et al., 1991). Chemically synthesized versions of *cry*IAc have been successfully introduced into apple, where they confer high levels of mortality to CM larvae (Dandekar et al., unpublished). These trees are currently being field-tested (USDA/APHIS permit no. 97-028-02r). Similar experiments have also been done in walnut (Dandekar et al., 1998) and persimmon (Tao et al., 1997), where excellent efficacy has been observed against target insect larvae by using chemically synthesized versions of *cry*IAc. Transgenic tissues expressing the cryIAc protein at levels as low as 0.02% of total cellular protein produced a 100% mortality rate for CM larvae (Dandekar et al., 1998).

The genes encoding ICPs of *Bacillus thuringiensis* can be used against some of the other commercially significant lepidopteran pests that plague apple orchards everywhere. These include orange tortrix (*Argyrotaenia citrana*), apple pandemis (*Pandemis prysuana*), oblique banded leafroller (*Chroistoneura rosaceana*), fruittree leafroller (*Archips argyrosphila*), omnivorous leafroller (*Platynota sultana*), western tussock moth, fruit worms (*Orthosia hibisi, Amphipyra pyramidoies*), and leafminers (*Phyllomorycter* spp.). Typically these lepidopteran insects are taken care of by chemical pesticides and are more of a problem with organically grown apples. However, the ICPs of *Bacillus thuringiensis* display a strong differential toxicity and therefore it would be important to check, in addition to CM, that these other caterpillars are also susceptible. All apple orchard locations are different and may have a range of pests in addition to CM. Aphid mite and scale problems are usually associated with the use of chemical pesticides because they kill the predators for these insects. One of the great benefits of a transgenic orchard expressing ICPs would be the greatly enhanced opportunity to integrate biocontrol strategies. The predators for mites, aphids, and scale would flourish in a transgenic apple orchard because no chemical pesticides would be sprayed, and this would mean less aphid, mite, and scale problems and less chemical used to control them.

## IV. RESISTANCE DEVELOPMENT AND MANAGEMENT STRATEGIES

Given their persistence and stability in the ecosystem, trees expressing a single resistance gene that may have a strong potential to induce the selection of resistance in an insect populations (Raffa, 1989). The biopesticides containing the *Bacillus thuringiensis*–encoded ICPs have enjoyed a spotless 30+ year commercial record and safe history of usefulness in the field with no evidence of resistance development. However, this situation has changed since the early 1990s with several findings of resistance development (Ferré et al., 1991; Sims and Stone, 1991; Tabashnik et al., 1990, 1991). Reports have appeared on the development of resistance under postharvest conditions of extreme selection pressure in stored grain products (McGaughey, 1985; McGaughey and Johnson, 1987 and McGaughey and Beeman, 1988), and under deliberate selection conditions in the laboratory (Sims and Stone, 1991; Stone et al., 1989). Studies on the mechanism of resistance indicate that it may be due to a decrease in the binding affinity between the insect's midgut brush border membrane and the insecticidal protein (Van Rie et al., 1990; MacIntosh et al., 1991; Ferré et al., 1991). In most cases resistance was recessive, suggesting that strategies involving the presence of refugia would be useful to prevent the development of resistance. In these strategies, untransformed plants serve as refugia and are planted within the transgenic population to help maintain a viable susceptible population that would breed out resistance (Gould, 1998; Roush, 1998). Another resistance management strategy is to have plants expressing high levels of ICP that will kill not only the susceptible insects but also the heterozygous resistant insects. The Environmental Protection Agency (EPA) in 1998 defined the high dose as 25 times the level required to kill 99% of susceptible insects (Gould, 1998). In 1999 an incompletely dominant resistance allele was reported in *Ostrinia nubilalis* (European corn borer) (Huang et al., 1999). This is the first report of a dominant ICP resistance allele, and this type of resistance will be difficult to manage with the two strategies described. Dominant resistance to ICPs could be controlled by expressing additional insecticidal genes that do not share the same mode of action, gene pyramiding. Clearly, research is needed to clarify the mechanism of resistance and then develop and deploy a resistance management strategy for codling moth in apple.

## V. ENGINEERING DISEASE-RESISTANT APPLES

Among the most significant diseases in apple are scab, caused by the fungus *Venturia inaequalis*, and fire blight, caused by the bacteria *Erwinia amylovora*. Both diseases are spread in cool, moist conditions and cause significant losses worldwide. The two general strategies to obtain resistance to pathogens are (a) to clone disease resistance genes from resistant varieties and (b) to test individual genes that directly affect growth and multiplication of the pathogen. Resistance genes trigger a plant defense response through a gene-for-gene interaction (reviewed in Maleck and Lawton, 1998; Staskawicz et al., 1995). Resistance to apple scab, referred to as the V*f* gene, is present in *Malus floribunda* and has been introgressed into several apple varieties, such as 'Prima.' In 1998 the map position of the V*f* gene was determined in a mapping population that included progeny from a cross between the apple cultivars 'Prima' and 'Fiesta' (Maliepaard et al., 1998). Direct genetic approaches using simple sequence repeats to identify markers that will link the V*f* gene to unique DNA fragments are also being investigated (Gianfranceschi et al., 1998). About 16 markers have been identified from a molecular genetic analysis of 19 *Malus × domestica* (Borkh.) cultivars or selections each that have germplasm from *Malus floribunda* (Gianfranceschi et al., 1996, 1998). Tightly linked random amplified polymorph DNA (RAPD) markers have also been developed for a V*f* gene that gives resistance to five races of *Venturia inaequalis* (Hemmat et al., 1998). Once the V*f* gene is cloned, transformation experiments can be used to introduce this into apple cultivars of choice and to verify resistance to *Venturia inaequalis*. Alternative approaches to engineer resistance have involved the expression of exochitinase (Bolar et al., 1998) and endochitinase (Bolar et al., 1997). Transgenic apple plants expressing high levels of endochitinase show resistance to *Venturia inaequalis;* however, the growth and development of the transgenic plants are stunted (Bolar et al., 1997). Approaches to obtain resistance to the fire blight–causing organism *Erwinia amylovora* in apple have involved expression of genes encoding lytic peptides. Success was obtained with the expression of the lytic peptide attacin E in the apple rootstock Malling 26, providing good resistance to the pathogenic bacteria (Norelli et al., 1994). In addition to scion cultivars, rootstocks are particularly sensitive to *Erwinia amylovora*. Unfortunately, the transgenic Malling 26 turned out to be Malling 7, which are naturally resistant to fire blight. Additional genes are currently being tested.

## VI. CARBOHYDRATE METABOLISM IN APPLES

In addition to sucrose and starch, apple plants accumulate sorbitol. Sorbitol, the sugar alcohol of glucose, is a predominant sugar found in apple. It is synthesized in mature leaves and translocated to fruit, where it is converted to fructose. The sugar alcohol sorbitol appears to be widely distributed in nature, where it can be found in species of bacteria, insects, animals, yeasts, algae, fungi, and higher plants (Touster and Shaw 1962; Bieleski 1982). However, certain woody members of the Rosaceae family, including *Malus*, *Pyrus*, *Prunus*, and *Sorbus* spp., appear to be unique in the entire plant kingdom with respect to their ability to synthesize, accumulate, and degrade sorbitol

(Bieleski 1982; Loescher 1987). In apple trees, sorbitol is believed to be the major photosynthetic product, translocated from the mature leaves to growing tissues such as fruits and young leaves (Bieleski 1969; Webb and Burley 1962; Zimmermann and Ziegler, 1975).

At the biochemical level, as shown in Fig. 1, sorbitol metabolism occurs as a result of two significant reactions that represent the rate-limiting steps in this pathway. One results in the formation of sorbitol 6-phosphate from glucose 6-phosphate by aldose 6-phosphate reductase (EC 1.1.1.200) in photosynthetic tissues (mature leaves), referred to here as *sorbitol 6-phosphate dehydrogenase* (S6PDH). The second results in the conversion of sorbitol to fructose by sorbitol dehydrogenase (SDH) in sink tissues (fruit and young developing leaves). These two enzymes appear to be the critical steps in sorbitol metabolism, and this view is supported by the following pieces of evidence: (a) The labeling pattern of photosynthates in mature apricot leaves suggests that sorbitol synthesis occurs through the intermediate sorbitol 6-phosphate (Bieleski and Redgwell, 1977, Negm and Loescher 1981; Ridgwell and Bieleski, 1978). (b) The purification of S6PDH from leaves of apple (Kanayama and Yamaki, 1993) and loquat (Hirai, 1981) and the detection of the enzyme in leaves of pear, peach, apricot (Negm and Loescher, 1981) and eight other species of plants from the three subfamilies of Rosaceae (Hirai, 1981) show this enzyme is widespread in plants synthesizing sorbitol. (c) Unlike mature leaves, which mainly produce [$^{14}$C] sorbitol from $^{14}CO_2$, very young apricot leaves mainly produce [$^{14}$C] sucrose and no [$^{14}$C] sorbitol. The young apricot leaves, however, apparently import sorbitol by translocation from older leaves (Bieleski and Redgwell, 1985). (d) Seasonal changes in sorbitol level coincide with the amount of S6PDH (Hirai, 1983; Sakanishi et al., 1998; Yamaki and Ishikawa 1986). (e) The S6PDH from apple was purified, and the complementary DNA (cDNA) coding for this enzyme was cloned (Kanayama et al., 1992, Kanayama and Yamaki, 1993). Expression of this cDNA in tobacco was sufficient for the synthesis of sorbitol (Tao, et al., 1995). Expression of sense and antisense cDNA encoding S6PDH in apple results in the suppression of sorbitol accumulation in transgenic apple plants (Dandekar et al., USDA-APHIS permit no. 97-037-02r), indicating that S6PDH is the key enzyme involved in sorbitol biosynthesis. (f) SDH was detected in apple fruit (Negm and Loescher, 1979; Yamaki, 1980) and was purified and cDNA-identified (Kanayama and Yamaki,

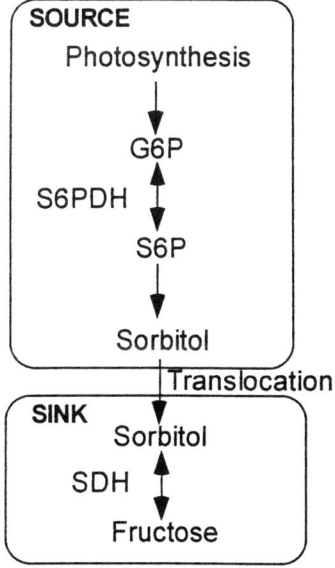

**Figure 1** Sorbitol metabolism in apple (Rosaceae).

1994; Yamada et al., 1998). The levels of SDH are highly regulated during the development of apple fruit (Yamaguchi et al., 1996). (g) Very young apple leaves contain SDH but no S6PDH activity, whereas the mature leaves mainly have S6PDH (Negm and Loescher, 1981); this was related to leaf carbohydrate levels and net photosynthesis, indicating that sorbitol metabolism in apple is tightly controlled and may be related to mechanisms regulating partitioning of source and sink activity (Loescher et al., 1982). (h) A phosphatase enzyme (sorbitol 6-phosphatase) removes the phosphate moiety from sorbitol 6-phosphate to form the translocated sorbitol (Grant and Ap Rees, 1981). (i) In 1998 the genomic sequences of S6PDH characterized from *Malus domestica* cv. Greensleeves' (apple) revealed that they are highly conserved among various members of the family Rosaceae (Bains et al., 1998). Taken together, these data provide evidence that in many of the Rosaceae the sorbitol produced as a result of photosynthesis occurs exclusively through S6PDH, whereas the utilization is primarily through SDH. These observations support the idea that the anabolic and catabolic pathways for sorbitol accumulation are different and under separate genetic and metabolic control.

The partitioning of assimilated carbon limits both rate and pattern of vegetative and reproductive growth in plants. One of the genetic approaches currently being used to study carbon partitioning in apple involves using transgenic plants expressing antisense/sense S6PDH or SDH (Dandekar et al., USDA-APHIS permit no. 97-037-02r). The results of these experiments could help link genetically the key carbon metabolic pathways with the regulatory enzymatic steps. In turn this information will help provide a better understanding of the relationship in apple between source organs involved in partitioning assimilated carbons like sucrose and sorbitol and various sink organs utilizing these translocated sugars. Additionally, learning more about these pathways and developing the means to manipulate them will have an impact not only on the yield of biomass but also on the yield of important quantitative traits of the apple crop.

Sorbitol may play a role in the ability of apple plants to respond to environmental stress. Several physiological roles have been proposed for sugar alcohols in general in higher plants; they may act as compatible solutes, protectants to stabilize membranes, and carbon storage compounds (Lewis, 1984; Schobert, 1977). The most important role reported is that sugar alcohols serve as osmolytes in response to water, salt, and other abiotic stress (Tarczynski et al., 1993; Wang et al., 1995; Wang and Stutte, 1992). The accumulation of sugar alcohols helps the cell gain more water through osmotic adjustment (Turner and Jones, 1992; Morgan, 1984). Much attention has been given to mannitol and its involvement in salt tolerance, but relatively little information exists on the importance of the sugar alcohol sorbitol in higher plants. So far it has also been implicated in the process of osmotic adjustment to overcome decreasing water potential created by environmental stress in cherry (Ranney et al., 1991), apple leaves in response to drought stress (Wang et al., 1995; Wang et al., 1996; Wang and Stutte, 1992), cold hardiness (Raese et al., 1978; Whetter and Taper, 1966), increased photosynthesis via elevated $CO_2$ level (Pan and Quebedeaux, 1995), and protein stabilization during dessication (Wimmer et al., 1997). In 1997 Moing and coworkers suggested that sorbitol variability is related to the geographical origin of *Prunus* species. The study of transgenic apple plants expressing antisense transgenes that inhibit the synthesis or degradation of sorbitol will provide genetic approaches to study some of these issues.

## VII. ETHYLENE BIOSYNTHESIS IN APPLES

Ethylene is an important plant growth regulator that affects diverse plant processes, including fruit ripening, senescence, and response to biotic and abiotic stress (Ecker, 1995; Ecker and Davis, 1987; Kieber and Ecker, 1993; McKeon and Yang, 1987; Reid, 1987). At a molecular level ethylene has been shown to mediate the expression of specific genes involved in ripening (Fluhr and Mattoo, 1996; Theologis, 1994) and stress (Ecker, 1995; Kieber and Ecker, 1993, Jackson 1997).

The physiological and biochemical characteristics of apple ripening have been reviewed comprehensively by Hulme (1958, 1971). Much of the classical work on fruit ripening was done with apple. Kidd and West (1922) observed the characteristic rise in respiration when apples are detached from the tree and stored at normal ripening temperatures. They called the phenomenon *climacteric* (Kidd and West, 1925). In 1932, Kidd and West (1933) showed that vapors produced from ripe apples caused unripe apples to go through the characteristic climacteric rise in respiration and to ripen. Gane (1934, 1935) demonstrated that the active component of the vapors was ethylene. The gaseous hormone ethylene plays an important role in the ripening of apples, with an excellent correlation between the level of ethylene in the apple fruit and the extent of its shelf life and eating quality. Selections with inherently low ethylene production can be generated in progeny of a cross that involves a low ethylene producer as one of the parents (Stow et al., 1993). 1-Amino cyclopropane-1-carboxylate (ACC) is a key intermediate in the biosynthesis of ethylene and the rate-limiting step (Adams and Yang, 1979; McKeon and Yang, 1987). The biosynthetic pathway from methionine involves the enzyme ACC synthase, which converts SAM to ACC. ACC is then converted to ethylene by the ethylene-forming enzyme (EFE) or ACC oxidase, as shown in Fig. 2.

Complementary DNA corresponding to messenger ribonucleic acid (mRNA) of both ACC synthase (Sato and Theologis, 1989; Van Der Straeten et al., 1990; Huang et al., 1991) and EFE/ACC oxidase (Hamilton et al., 1991; Spanu et al., 1991) has been cloned. The cDNA from these genes expressed in the antisense orientation causes a decrease in the biosynthesis of ethylene and a delay in ripening (Hamilton et al., 1990; Oeller et al., 1991; Theologis, 1994). The ACC synthase enzyme has been purified from apple (Yip et al., 1991), and the cDNA encoding its mRNA has also been isolated (Dong et al., 1991). The EFE or ACC oxidase has been characterized from apple (Fernandez-Maculet and Yang, 1992), and the cDNA encoding the mRNA has also been isolated (Dong et al., 1992). Alternative approaches to control the level of ethylene are through

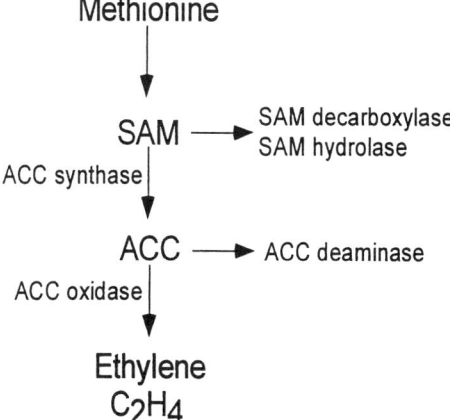

**Figure 2** Metabolic conversion of methionine to ethylene: key steps for the control of ethylene biosynthesis.

the expression of an enzyme that metabolizes ACC via expression of an enzyme from bacteria called *ACC deaminase* (Klee et al., 1991) or through the tissue-specific expression of the enzyme SAM hydrolase (Good et al., 1994) or SAM decarboxylase (Metha et al., 1997, 1999), which metabolizes SAM (see Fig. 2). The latter approaches render the precursor, SAM or ACC, unavailable for ethylene biosynthesis (Fluhr and Mattoo, 1996; Theologis, 1994).

Many physiological, genetic, and developmental factors influence the storability and eating quality of apples. Two types of indices are used to determine eating quality. One is visual appeal, the other is physiological maturity, which includes acidity, ethylene or $CO_2$ concentration, soluble solids, flesh firmness, and starch level. These have been standardized for each cultivar and are also used to predict harvest dates and to determine storage time. The concentration of ethylene is key to all and has a profound influence on all the maturity indices. Therefore, the control of ethylene biosynthesis in apple could be the key to control not only shelf life but also eating quality. This is being investigated through the expression of antisense mRNA or through the expression of other genes that interfere in the synthesis of ethylene as explained (Fig. 2). Currently, transgenic apple plants with alteration in ethylene biosynthesis are being field-tested using the following approaches: (a) expression of antisense/sense ACC synthase (Dandekar et al., USDA-APHIS permit no. 97-037-02r) and (b) expression of antisense/sense EFE (Dandekar et al., USDA-APHIS permit no. 97-037-02r). It will be some time before the impact of these approaches on apple shelf life and quality can be fully determined.

## VIII. OTHER QUALITY TRAITS

Skin color in apples is an important trait, as a deep red color is preferred commercially and aesthetically. The skin color of apple is a blend of a variety of different components that include anthocyanins/flavanols, carotenoids, and chlorophyll. The yellow/green color is due to the synthesis of chlorophyll and carotenoids in the plastids. However, the red color is due to the synthesis of anthocyanins and flavonols that accumulate in the vacuole. The anthocyanins are highly regulated by the environment (Saure, 1990). There is good information available on the exact biochemical nature of the pigments present in apple (Lancaster and Dougall, 1992). Engineering skin color may be possible, as the basic pathways for pigment biosynthesis have been worked out in flowers (Meyer et al., 1987; Mol et al., 1989). However, more specific aspects of the pathways need to be worked out in apple, for example, the possible pathways for the synthesis of cyanidin glycosides, proanthocyanidins (like catechin), and flavonols (like quercetin glycoside). Some of this information has been accumulated (Lister et al., 1997). Additionally, skin color is highly regulated during fruit development and ripening (Lister et al., 1996). Therefore, modifications in apple skin color will need to be engineered in a tissue-specific and developmental stage–specific pattern, and this process would require the identification of specific regulatory sequences.

Changes in texture are an important feature in the eating quality of apple; a firm breaking texture is desirable. Understanding the softening process could lead to better control against "mealiness." In many climacteric fruits dramatic changes in texture and cell wall structure take place during the ripening process, dominated by the autocatalytic synthesis of ethylene. Therefore, some of the changes in texture and quality can be attributed to the direct action of ethylene. The regulation of endopolygalacturonase in apple has been studied and may be involved in softening (Wu et al., 1993).

Controlling metabolic activity during storage is especially important to control a variety of abiotic problems in apple such as watercore, bitter pit, and storage scald. It will be interesting to study antisense S6PDH apple plants that have very low levels of sorbitol in their fruit to see whether their potential to develop watercore has been altered. Watercore areas are high in sorbitol, and it is thought that the accumulation of sorbitol leads to cellular damage by osmotic lysis of cells accumulating sorbitol. Metabolic activity caused by stress in storage fruit can lead to the formation of reactive oxygen species (ROIs), leading to problems like storage scald. Oxidative stress has been implicated in causing cellular damage when tissues are exposed to a variety of stressful conditions, including high light intensity, extremes of temperature, salinity, heavy metals, herbicides, and a variety of toxins. One of the important ROI scavenging systems in plants is the enzyme superoxide desmutase (SOD). Transgenic plants expressing elevated levels of SOD were protected from the damaging effects of ROIs (Allen et al., 1997). SODs are a class of metalloproteins that catalyze the dismutation of the superoxide radical to hydrogen peroxide and oxygen and are present as various isoenzyme forms in different fruit, including apple (Manganaris and Alston, 1997).

Studies in 1997 showed that polyphenol constituents derived from fruits like apple are more effective antioxidants in vitro than vitamins C and E and thus may be more valuable for protection in vivo. Therefore, the active ingredient in "an apple a day keeps the doctor away" may well be the phytochemical component of the apple fruit. The phytochemical components include flavonoids, phenylpropanoids, and phenolic acids and have been highlighted as important contributing factors in the antioxidant activity in our diet (Rice-Evans et al., 1997). In apple, flavonols like quercetin and flavones like rutin may be important (Rice-Evans et al., 1997). As more is learned about the pathways for the synthesis of these compounds their amounts can then be modulated. Some of these compounds are responsible for color, so modification of their levels may also influence fruit appearance.

A significant problem with "fresh cut" or "juice" processed apples is enzymatic browning, a widespread problem in the food industry. This is a readily observable problem in bruised and wounded fruit. The browning reaction is due to enzymatic action and not a Maillard-type nonenzymatic browning. The enzyme involved is polyphenol oxidase (PPO; EC 1.10.3.1), also referred to as *catechol oxidase*, *phenolase*, and *o-diphenol oxygen oxidoreductase*, it is widely distributed among terrestrial and aquatic plant species (Sherman et al., 1991). Another enzyme, tyrosinase (monophenol, dihydroxyphenalanine: oxygen oxidoreductase, EC 1.14.18.1), also catalyzes both the orthohydroxylation of monophenols and the oxidation of *o*-diphenols to *o*-quinones (Lerch, 1981). This copper-containing enzyme is also widely distributed in nature, where it is responsible for the formation of melanin pigments (Bell and Wheeler, 1986; Walker and Ferrar, 1998). Whereas several important functions and properties can be ascribed to melanin pigments (Bell and Wheeler, 1986), biological and physiological studies have provided few answers toward the understanding of PPO function and expression in plants. Evidence from the 1990s suggests that these enzymes may have a role in disease (Walker and Ferrar, 1998) and pest resistance (Steffens et al., 1994). What is known is that PPO mRNA is highly regulated both developmentally and temporally in plants (Hunt et al., 1993), and different forms are present in the skin and flesh of apple fruit (Robinson SP, personal communication). Antisense expression of the PPO in apple is a potential approach to reduce the enzymatic browning of apple flesh. Antisense expression will need to be targeted to fruit tissue to prevent interference with PPO expression elsewhere in the plant that may be important for disease/pest resistance.

## REFERENCES

Adams DO, Yang SF. 1979. Ethylene biosynthesis: Identification of 1-aminocyclopropane-1-carboxylic acid as an intermediate in the conversion of methionine to ethylene. Proc Natl Acad Sci USA 76:170–174.

Allen RD, Webb RP, Schake SA. 1997. Use of transgenic plants to study antioxidant defences. Free Rad Biol Med 23:473–479.

Bains HS, Tao R, Uratsu SL, Dandekar AM, 1998. Genomic Nucleotide Sequence of a NADP sorbitol-6-phosphate dehydrogenase gene from apple (Accession No. AF057134) (PGR 98–193). Plant Physiol 118:1533.

Bell AA, Wheeler MH. 1986. Biosynthesis and functions of fungal melanins. Annu Rev Phytopathol 24: 411–451.

Beutler E, Gelbart T, Han J, Koziol JA, Beutler B, 1989. Evolution of the genome and the genetic code: Selection at the dinucleotide level by methylation and polyribonucleotide cleavage. Proc Natl Acad Sci USA 86:192–196.

Bieleski RL. 1969. Accumulation and translocation of sobitol in apple phloem. Aust J Biol Sci 22:611–620.

Bieleski RL. 1982. Sugar alcohols. In Loewus F, Tanner W, eds. Encyclopedia of plant physiology. New Series. Vol. 13A. Berlin: Springer-Verlag, pp 158–192.

Bieleski RL, Redgwell RJ. 1977. Synthesis of sorbitol in apricot leaves. Aust J Plant Physiol 4:1–10.

Bieleski RL, Redgwell RJ. 1985. Sorbitol versus sucrose as photosynthesis and translocation products in developing apricot leaves. Aust J Plant Physiol 12:657–668.

Bolar JP, Aldwinckle HS, Harman GE, Norelli JL, Brown SK. 1997. Endochitinase-transgenic McIntosh apple lines have increased resistance to scab. Phytopathology 87(suppl):S10.

Bolar JP, Norelli JL, Aldwinckle HS, Harman GE, Brown SK. 1998. Expression of an exochitinase gene from *Trichoderma harzianum* in transgenic apple lines. Phytopathology 88(suppl):S8–S9.

Campbell WH, Gowri G. 1990. Codon usage in higher plants, green algae, and cyanobacteria. Plant Physiol 92:1–11.

Crickmore N, Zeigler DR, Feitelson J, Schnepf E, Van Rie J, Lereclus D, Baum J, Dean. 1998. Revision of the nomenclature for the *Bacillus thuringiensis* pesticidal crystal proteins. Microbiol Mol Biol Rev 62:807–813.

Dandekar AM, Uratsu SL, Matsuta N. 1990. *Agrobacterium*-mediated transformation of apple: Factors influencing virulence. Acta Hortic 280:483–494.

Dandekar AM, McGranahan GH, Uratsu SL, Leslie C, Vail PV, Tebbets SJ, Hoffman D, Driver J, Viss P, James DJ. 1992. Engineering for apple and walnut resistance to codling moth. Proceedings of the British Council for Crop Protection Conferences—Pests and Diseases, Nov. 23–26. Vol. 2 Cambridge: pp 741–747.

Dandekar AM, McGranahan GH, Vail PV, Uratsu SL, Leslie CA, Tebbets JS, Hoffman DJ. 1994. Low levels of expression of *cry*IA(c) sequences of *Bacillus thuringiensis* in transgenic walnut. Plant Sci 96:151–116.

Dandekar AM, McGranahan GH, Vail PV, Uratsu SL, Leslie CA, Tebbets JS. 1998. High level of expression of full length *cry*IA(c) gene from *Bacillus thuringiensis* in transgenic somatic walnut embryos. Plant Sci 131:181–193.

Dean C, Tamaki S, Dunsmuir P, Favreau M, Katayama C, Dooner H, Bedbrook J. 1986. mRNA transcripts of several plant genes are polyadenylated at multiple sites in vivo. Nucleic Acids Res 14:2229–2240.

De Bondt A, Eggermont K, Druart P, De Vil M, Goderis I, Vanderleyden J, Broekaert WF. 1994. *Agrobacterium*-mediated transformation of apple (*Malus x domestica* Borkh.): An assessment of factors affecting gene transfer efficiency during early transformation steps. Plant Cell Rep 13:587–593.

De Bondt A, Eggermont K, Penninckx I, Goderis I, Broekaert WF. 1996. *Agrobacterium*-mediated transformation of apple (*Malus x domestica* Borkh.): An assessment of factors affecting regeneration of transgenic plants. Plant Cell Rep 15:549–554.

Dong J-G, Kim WT, Yip WK, Thompson GA, Li L, Bennett AB, Yang S-F. 1991. Cloning of a cDNA encoding 1-aminocyclopropane-1-carboxylate synthase and expression of its mRNA in ripening apple fruit. Planta 185:38–45.

Dong J-G, Olsen DB, Silverstone A, Yang S-F. 1992. Sequence of a cDNA encoding for a 1-aminocyclopropane-1-carboxylate oxidase homolog from apple fruit. Plant Physiol 98:1530–1531.

Doyle M. 1999. EPA bans fruit-crop pesticide to protect children. The Sacramento Bee, August 3, pp A1, A7.

Ecker JR. The ethylene signal transduction pathway in plants. Science 268:667–675.

Ecker J, Davis RW. 1987. Plant defense genes are regulated by ethylene. Proc Natl Acad Sci USA 84:5202–5206.

Escobar M, Dandekar AM. 2000. Development of insect resistance in fruit and nut tree crops. In: Jain SM, Minocha SC, eds. Molecular Biology of Woody Plants. Vol. 2. Dordrecht, Netherlands: Kluwer.

Fasolo F, Zimmerman RH, Fordham I. 1989. Adventitious shoot formation on excised leaves of in vitro grown shoots of apple cultivars. Plant Cell Tissue Organ Cult 16:75–87.

Fernandez-Maculet JC, Yang SF. 1992. Extraction and partial characterization of the ethylene-forming enzyme from apple fruit. Plant Physiol 99:751–754.

Ferré J, Real MD, van Rie J, Jansens S, Peferon M. 1991. Resistance to the *Bacillus thuringiensis* bioinsecticide in a field population of *Plutella xylostella* is due to a change in a midgut membrane receptor. Proc Natl Acad Sci USA 88:5119–5123.

Fischhoff DA, Bowdish KS, Perlak FJ, Marrone PG, McCormick SM, Niedermeyer JG, Dean DA, Kusano-Kretzmer K, Mayer EJ, Rochester DE, Rogers SG, Fraley RT. 1987. Insect tolerant transgenic tomato plants, Biotechnology 5:807–813.

Fluhr R, Mattoo AK. 1996. Ethylene: Biosynthesis and perception. Crit Rev Plant Sci 15:479–523.

Gane R. 1934. Production of ethylene by some ripening fruits. Nature 134:1008.

Gane R. 1935. The formation of ethylene by plant tissue and its significance in the ripening of fruit. J Pomol Hortic Sci 13:351–358.

Gianfranceschi L, Koller L, Seglias N, Kellerhals N, Gessler C. 1996. Molecular selection in apple for resistance to scab caused by *Venturia inaequalis*. Theor Appl Genet 93:199–204.

Gianfranceschi L, Seglias N, Tarchini R, Komjanc M, Gessler C. 1998. Simple sequence repeats for the genetic analysis of apple. Theor Appl Genet 96:1069–1076.

Good X, Kellogg JA, Wagoner W, Langhoff D, Matsumura W, Bestwick RK. 1994. Reduced ethylene synthesis by transgenic tomatoes expressing *S*-adenosylmethionine hydrolase. Plant Mol Biol 26:781–790.

Goodall GJ, Filipowicz W. 1989. The AU-rich sequences present in the introns of plant nuclear pre-mRNAs are required for splicing. Cell 58:473–483.

Gould F. 1998. Sustaining the efficacy of Bt toxins. In Hardy RWF, Segelken JB, eds. National Agricultural Biotechnology Report 10, pp 77–86.

Grant CR, Ap Rees T. 1981. Sorbitol metabolism by apple seedlings. Phytochemistry 20:1505–1511.

Hamilton AJ, Lycett GW, Grierson D. 1990. Antisense gene that inhibits synthesis of the hormone ethylene in transgenic plants. Nature 346:284–287.

Hamilton AJ, Bouzayen M, Grierson D. 1991. Identification of a tomato gene for the ethylene-forming enzyme by expression in yeast. Proc Natl Acad Sci USA 88:7434–7437.

Hammerschlag FA, Zimmerman RH, Yadava UL, Hunsucker S, Gercheva P. 1997. Effect of antibiotics and exposure to an acidified medium on the elimination of *Agrobacterium tumefaciens* from apple leaf explants and on shoot regeneration. J Am Soc Hortic Sci 122:758–763.

Hemmat M, Weeden NF, Aldwinckle HS, Brown SK. 1998. Molecular markers for the scab resistance (Vf) region in apple. J Am Soc Hort Sci 123:992–996.

Hirai M. 1981. Purification and characteristics of sorbitol-6-phosphate dehydrogenase from loquat leaves. Plant Physiol 67:221–224.

Hirai M. 1983. Seasonal changes in sorbitol-6-phosphate dehydrogenase in loquat leaf. Plant Cell Physiol 24:925–931.

Hofmann C, Luthy P, Hutter R, Pliska V. 1988a. Binding of the delta-endotoxin from *Bacillus thuringiensis* to brush border membrane vesicles of the cabbage butterfly (*Pieris brassicae*). Eur J Biochem 173:85–91.

Hofmann C, Vanderbruggen H, Hofte H, Van Rie J, Jansens S, Van Mellaert H. 1988b. Specificity of *Bacillus thuringiensis* δ-endotoxin is correlated with the presence of high-affinity binding sites in the brush border membrane of target insect midguts. Proc Natl Acad Sci USA 85:7844–7848.

Höfte H, Whiteley HR. 1989. Insecticidal crystal proteins of *Bacillus thuringiensis*. Microbiol Rev 53:242–255.

Holefors A, Zhongtian X, Welander M. 1998. Transformation of the apple rootstock M26 with the *rol*A gene and its influence on growth. Plant Sci 136:69–78.

Huang F, Buschman LL, Higgins RA, McGaughey WH. 1999. Inheritance of resistance to *Bacillus thuringiensis* toxin (Dispel ES) in the European corn borer. Science. 284:965–967.
Huang P-L, Parks JE, Rottmann WH, Theologies A. 1991. Two genes encoding 1-aminocyclopropane-1-carboxylate synthesis in zucchini (*Cucurbita pepo*) are clustered and similar but differentially regulated. Proc Natl Acad Sci USA 88:7021–7025.
Hulme AC. 1958. Adv Fd Res 8, 297.
Hulme AC. 1971. The Biochemistry of Fruits and Their Products. Vol. 2. London and New York. Academic Press, pp 333–373.
Hunt MD, Eannetta NT, Yu H, Newman SM, Steffens JC. 1993. cDNA Cloning and expression of potato polyphenol oxidase. Plant Mol Biol 21:59–68.
Jackson M. 1997. Hormones from roots as signals for the shoots of stressed plants. Trends Plant Sci 2:22–28.
James DJ. 1987. Cell and tissue culture technology for the genetic manipulation of temperate fruit trees. In: Russell GE, ed. Biotechnology and Genetic Engineering Reviews. Vol. 5, pp 33–79.
James DJ, Passey AJ, Rugini E. 1988. Factors affecting high frequency plant regeneration from apple leaf tissues cultured in vitro. J Plant Physiol 132:148–154.
James DJ, Passey AJ, Barbara DJ, Bevan MW. 1989. Genetic transformation of apple (*Malus pumila* Mill.) using a disarmed Ti-binary vector. Plant Cell Rep 7:658–661.
James DJ, Passey AJ, Barbara DJ. 1990. Regeneration and transformation of apple and strawberry using disarmed Ti-binary vectors. Acta Hortic 280:495–502.
James DJ, Dandekar AM. 1991. Regeneration and transformation of apple (*Malus pumila* Mill.) In: Lindsey K, eds. Dordrect Plant tissue culture mannual: Fundamentals and applications. Kluwer Academic, 1–18.
James DJ, Uratsu SL, Cheng J, Negri P, Viss P, Dandekar AM. 1993. Conditions that induce *Agrobacterium* Vir genes also enhance apple cell transformation. Plant Cell Rep 12:559–563.
James DJ, Passey AJ, Barker SA. 1994. Stable gene expression in transgenic apple tree tissues and segregation of transgenes in the progeny—preliminary evidence. Euphytica 77:119–121.
James DJ, Passey AJ, Baker SA, Wilson FM. 1996. Transgenes display stable patterns of expression in apple fruit and Mendelian segregation in the progeny. Biotechnology 14:56–60.
Jones OP. 1976. Effect of phloridzin and phloroglucinol on apple shoots. Nature 262:392–393.
Kanayama Y, Mori H, Imaseki H, Yamaki S. 1992. Nucleotide sequence of a cDNA encoding NADP-sorbitol-6-phosphate dehydrogenase from apple. Plant Physiol 100:1607–1608.
Kanayama Y, Yamaki S. 1993. Purification and properties of NADP-dependent sorbitol-6-phosphate dehydrogenase from apple seedlings. Plant Cell Physiol 34:819–823.
Kanayama Y, Yamaki S. 1994. Purification and properties of NAD-dependent sorbitol dehydrogenase from apple fruit. Plant Cell Physiol 35:887–892.
Kanayama Y, Sakanishi K, Mori H, and Yamaki S. 1996. Expression of the gene for NADP-dependent sorbitol-6-phosphate dehydrogenase in apple seedlings. Plant Cell Physiol 36:1139–1141.
Kidd F, West C. 1922. Report of the Food Investigation Board, London, 1921. Gt Brit Dept Sci Ind Res Food Invest Bd Rept. pp 14–16.
Kidd F, West C. 1925. The course of respiratory activity throughout the life of an apple. Gt Brit Dept Sci Ind Res, Food Invest Bd Rept for 1924, pp 27–33.
Kidd F, West C. 1933. The effects of ethylene and apple vapours on the ripening of fruits. Gt Brit Dept Sci Ind Res Food Invest Bd Rept for 1932, pp 55–58.
Kieber JJ, Ecker, JR. 1993. Ethylene gas: It's not just for ripening any more. Trends Genet 9:356–362.
Klee HJ, Hayford MB, Kretzmer KA, Barry GF, Kishore GM. 1991. Control of ethylene synthesis by expression of a bacterial enzyme in trangenic tomato plants. Plant Cell 3:1187–1193.
Knowles BH, Ellar, DJ. 1987. Colloid-osmotic lysis is a general feature of the mechanism of action of *Bacillus thuringiensis* δ-endotoxins with different insect specificity. Biochem Biophys Acta 924:509–518.
Lambert C, Tepfer D. 1992. Use of *Agrobacterium rhizogenes* to create transgenic apple trees having an altered organogenic response to hormones. Theor Appl Genet 85:105–109.
Lancaster JE, Dougall DK. 1992. Regulation of skin color in apples. Crit Rev Plant Sci 10:487–502.
Lerch K. 1981. Copper monooxygenases: Tyrosinase and dopamine β-monooxygenase, In: Siegel H, ed. Metal Ions in Biological Systems. Vol 13. New York: Marcel Dekker, pp 143–186.

Lewis DH. 1984. Occurrence and distribution of storage carbohydrates in vascular plants. In: Lewis DH, ed. Storage Carbohydrates in Vascular Plants. Cambridge: Cambridge University Press, pp 1–52.
Lister CE, Lancaster JE, Walker JRL. 1996. Phenylalanine ammonia-lyase (PAL) activity and its relationship to anthocyanin and flavonoid levels in New Zealand–grown apple cultivars. J Am Soc Hortic Sci 121:281–285.
Lister CE, Lancaster JE, Sutton KH, Walker JRL. 1997. Aglycone and glycoside specificity of apple skin flavonoid glycosyltransferase. J Sci Food Agric 75:378–382.
Liu Q, Salih S, Hammerschlag F. 1998. Etiolation of 'Royal Gala' apple (*Malus x domestica* Borkh.) shoots promotes high-frequency shoot organogenesis and enhanced β-glucuronidase expression from stem internodes. Plant Cell Rep. 17 (in press)
Loescher WH. 1987. Physiology and metabolism of sugar alcohols in higher plants. Physiol Plant 70:533–557.
Loescher WH, Marlow GC, Kennedy RA. 1982. Sorbitol metabolism and sink-source interconversions in developing apple leaves. Plant Physiol 70:335–339.
MacIntosh SC, Stone TB, Jokerst RS, Fuchs RL. 1991. Binding of *Bacillus thuringiensis* proteins to a laboratory-selected line of *Heliothis virescens*. Proc Natl Acad Sci USA 88:8930–8933.
Maheswaran G, Welander M, Hutchinson JF, Graham MW, Richards D. 1992. Transformation of apple rootstock M26 with *Agrobacterium tumefaciens*. J Plant Physiol 39:560–568.
Maleck K, Lawton K. 1998. Plant strategies for resistance to pathogens. Curr Opin Biotechnol 9:208–213.
Maliepaard C, Alston FH, van Arkel G, Brown LM, Chevreau E, Dunemann F, Evans KM, Gardiner S, Guilford P, van Heusden AW, Janse J, Laurens F, Lynn JR, Manganaris AG, den Nijs APM, Periam N, Rikkerink E, Roche P, Ryder C, Sansavini S, Schmidt H, Tartarini S, Verhaegh JJ, Veielink-van Ginkel M, King GJ. 1998. Aligning male and female linkage maps of apple (*Malus pumilla* Mill.) using multi-allelic markers. Theor Appl Genet 97:60–73.
Manganaris AG, Alston FH. 1997. Genetics of superoxide dismutase in apple. Theor Appl Genet 95:484–489.
Martin GC, Miller AN, Castle LA, Morris JW, Morris RO, Dandekar AM. 1990. Feasibility studies using β-glucuronidase as a gene fusion marker in apple, peach, and radish. J Amer Soc Hortic Sci 115:689–691.
Maximova SN, Dandekar AM, Guiltinan MJ. 1998. Investigation of *Agrobacterium*-mediated transformation of apple using green fluorescent protein: High transient expression and low stable transformation suggest that factors other than T-DNA transfer are rate limiting. Plant Mol Biol 37:549–559.
McGaughey WH. 1985. Insect resistance to the biological insecticide *Bacillus thuringiensis*. Science 229:193–195.
McGaughey WH, Johnson DE. 1987. Toxicity of different serotypes and toxins of *Bacillus thuringiensis* to resistant and susceptible Indianmeal moths (Lepidoptera: Pyralidae). J Econ Entomol 80:1122–1126.
McGaughey WH, Beeman RW. 1988. Resistance to *Bacillus thuringiensis* in colonies of Indianmeal moth and almond moth (Lepidoptera: Pyralidae). J Econ Entomol 81:28–33.
McKeon T, Yang SF. 1987. Biosynthesis and metabolism of ethylene. In: Davies PJ, ed. Plant Hormones and Their Role in Plant Growth and Development. Martinus Nijhoff, pp 94–112.
Mehta R, Handa A, Mattoo A. 1997. Interactions of ethylene and polyamines in regulating fruit ripening. In: Kanellis AK, Chang C, Kende H, Grierson D, eds. Biology and Biotechnology of the Plant Hormone Ethylene. Boston: Kluwer Academic, pp 321–326.
Mehta R, Handa A, Mattoo AK. 1999. Ethylene, polyamines and fruit ripening. In: Altman A, Ziv M, Izhar S, eds. Plant Biotechnology and In Vitro Biology in the 21st Century. Boston: Kluwver Academic, pp 591–595.
Meyer P, Heidmann I, Forkmann G, Saedler H. 1987. A new petunia flower color generated by transformation of a mutant with a maize gene. Nature 330:677–678.
Moing A, Langlois N, Svanella L, Zanetto A, Gaudillere JP. 1997. Variability in sorbitol: Sucrose ratio in mature leaves of different *Prunus* species. J Am Soc Hortic Sci 122:83–90.
Mol JNM, Stuitje AR, vander Krol A. 1989. Genetic manipulation of floral pigmentation genes. Plant Mol Biol 13:287–294.
Morgan JM. 1984. Osmoregulation and water stress in higher plants. Annu Rev Plant Physiol 35:299–319.

Murray EE, Lotzer J, Eberle M. 1989. Codon usage in plant genes. Nucleic Acids Res 17:477–498.
Negm FB, Loescher WH. 1979. Detection and characterization of sorbitol dehydrogenase from apple callus tissue. Plant Physiol 64:69–73.
Negm FB, Loescher WH. 1981. Characterization and partial purification of aldose-6-phosphate reductase (alditol-6-phosphate: NADP-1-oxidoreductase) from apple leaves. Plant Physiol 67:139–142.
Norelli JL, Aldwinckle HS. 1993. The role of aminoglycoside antibiotics in the regeneration and selection of neomycin phosphotransferase-transgenic apple tissue. J Am Soc Hortic Sci 118:311–316.
Norelli JL, Aldwinckle HS, Destefano-Beltran L, Jaynes JM. 1994. Transgenic 'Malling 26' apple expressing the attacin E gene has increased resistance to *Erwinia amylovora*. Euphytica 77:123–128.
Norelli JL, Mills J, Aldwinckle HS. 1996. Leaf wounding increases efficiency of *Agrobacterium*-mediated transformation of apple. Hortscience 3:1026–1027.
Oeller PW, Min-Wong L, Taylor LP, Pike DA, Theologis A. 1991. Reversible inhibition of tomato fruit senescence by antisense RNA. Science 254:437–439.
Pan QY, Quebedeaux B. 1995. Effects of elevated $CO_2$ on sorbitol partitioning in sink and source apple leaves. Hortscience 30:770.
Perlak FJ, Fuchs RL, Dean DA, McPherson SL, Fischhoff DA. 1991. Modification of the coding sequence enhances plant expression of insect control protein genes. Proc Natl Acad Sci USA 88:3324–3328.
Puite KJ, Schaart JG. 1996. Genetic modification of the commercial apple cultivars Gala, Golden Delicious and Elstar via an *Agrobacterium tumefaciens*–mediated transformation method. Plant Sci 119:125–133.
Raese JT, Williams MW, Billingsley HD. 1978. Cold hardiness, sorbitol, and sugar levels of apple shoots as influenced by controlled temperature and season. J Am Soc Hortic Sci 103:796–801.
Raffa K. 1989. Genetic engineering of trees to enhance resistance to insects. Bioscience 39:524–534.
Ranney TG, Bassuk NL, Whitlow TH. 1991. Osmotic adjustment and solute constituents in leaves and roots of water-stressed cherry (*Prunus*) trees. J Am Soc Hortic Sci 116:684–688.
Reid M. 1987. Ethylene in plant growth and development and senescence. In: Davies PJ, ed. Plant Hormones and Their Role in Plant Growth and Development. Martinus Nijhoff, pp 94–112.
Rice-Evans CA, Miller NJ, Paganga G. 1997. Antioxidant properties of phenolic compounds. Trends Plant Sci 2:152–159.
Ridgwell RJ, Bieleski RL. 1978. Sorbitol-1-phosphate and sorbitol-6-phosphate in apricot leaves. Phytochemistry 17:407–409.
Roush RT. 1998. Two-toxin strategies for management of insecticidal transgenic crops: Can pyramiding succeed where pesticide mixtures have not? Phil Trans R Soc Lond B 353:1777–1786.
Sakanishi K, Kanayama Y, Mori M, Yamada K, Yamaki S. 1998. Expression of the gene for NADP-dependent sorbitol-6-phosphate dehydrogenase in peach leaves of various developmental stages. Plant Cell Physiol 39:1372–1374.
Sato T, Theologis A. 1989. Cloning the mRNA encoding 1-aminocyclopropane-1-carboxylate synthase, the key enzyme for ethylene synthesis in plants. Proc Natl Acad Sci USA 86:6621–6625.
Saure MC. 1990. External control of anthocyanin formation in apple. Sci Hortic 42:181–218.
Schobert B. 1977. Is there an osmotic regulatory mechanism in algae and higher plants? J Theor Biol 68:17–26.
Schnepf E, Crickmore N, Van Rie J, Lerecus D, Baum J, Feitelson J, Zeigler DR, Dean DH. 1998. *Bacillus thuringiensis* and its pesticide crystal proteins. Microbiol Mol Biol Rev 62:775–806.
Shaw G, Kamen R. 1986. A conserved AU sequence from the 3' untranslated region of GM-CSF mRNA mediates selective mRNA degradation. Cell 46:659–667.
Sherman TD, Vaughn KC, Duke SO. 1991. A limited survey of the phylogenetic distribution of polyphenol oxidase. Phytochemistry, 30:2499–2506.
Sims SR, Stone TB. 1991. Genetic basis of tobacco budworm resistance to an engineered *Pseudomonas fluorescens* expressing the delta-endotoxin of *Bacillus thuringiensis kurstaki*. J Invertebr Pathol 57:206–210.
Spanu P, Reinhardt D, Boller T. 1991. Analysis and cloning of the ethylene-forming enzyme from tomato by functional expression of its mRNA in *Xenopus laevis* oocytes. EMBO J 10:2007–2013.
Sriskandarjah S, Goodwin PB, Speirs J. 1994. Genetic transformation of apple scion cultivar 'Delicious' via *Agrobacterium tumefaciens*. Plant Cell Tissue Organ Cult 36:317–329.

Sriskandarjah S, Goodwin PB. 1998. Conditioning promotes regeneration and transformation in apple leaf explants. Plant Cell Tissue Organ Cult 53:1–11.
Staskawicz BJ, Ausubel FM, Baker BJ, Ellis JG, Jones JDG. 1995. Molecular genetics of plant disease resistance. Science 268:661–667.
Steffens JC, Harel E, Hunt MD. 1999. Polyphenol oxidase. In: Ellis BE, ed. Genetic Engineering of Plant Secondary Metabolism. New York: Plenum Press, pp 275–312.
Stone TB, Sims SR, Marrone PG. 1989. Selection of tobacco budworm for resistance to a genetically engineered *Pseudomonas fluorescens* containing the delta-endotoxin of *Bacillus thuringiensis* subsp. *kurstaki*. J Invertebr Pathol 53:228–234.
Stow J, Alston F, Hatfield S, Genge P. 1993. New selections with inherently low ethylene production. Acta Hortic 326:85–92.
Tabashnik BE, Cushing NL, Finson N, Johnson MW. 1990. Field development of resistance to *Bacillus thuringiensis* in diamondback moth (Lepidoptera: Plutellidae). J Econ Entomol 83:1671–1676.
Tabashnik BE, Finson N, Johnson MW. 1991. Managing resistance to *Bacillus thuringiensis*: Lessons from the diamondback moth (Lepidoptera: Plutellidae). J Econ Entomol 83:1671–1676.
Tao R, Uratsu SL, Dandekar AM. 1995. Sorbitol synthesis in transgenic tobacco with apple cDNA encoding NADP-dependent sorbitol-6-phosphate dehydrogenase. Plant Cell Physiol 36:525–532.
Tao R, Dandekar AM, Uratsu SL, Vail PV, Tebbets JS. 1997. Engineering genetic resistance against insects in Japanese persimmon using the *cry*IA(c) gene of *Bacillus thuringiensis*. J Am Soc Hortic Sci 122(6):764–771.
Tarczynski MC, Jensen RG, Bohnert HJ. 1993. Stress protection of transgenic tobacco by production of the osmolyte mannitol. Science 259:508–510.
Theologis T. 1994. Control of ripening. Curr Opin Biotechnol 5:152–157.
Touster O, Shaw DRD. 1962. Biochemistry of acyclic polyols. Physiol Rev 42:181–225.
Turner MC, Jones MM. 1992. Turgor maintenance by osmotic adjustment: A review and evaluation. In: Turner NC, Kramer PJ, eds. Adaptation of Plants to Water and High Temperature Stress. pp 87–103.
Vaeck M, Reynaerts A, Hofte H, Jansens S, DeBeuckeleer M, Dean C, Zabeau M, Van Montagu M, Leemans J. 1987. Transgenic plants protected from insect attack. Nature 328:33–37.
Vail PV, Tebbets JS, Hoffmann DF, Dandekar AM. 1991. Response of production and postharvest walnut pests to *Bacillus thuringiensis* insecticidal crystal protein fragments. Biol Control 1:329–333.
Van Der Straeten D, Van Wiemeersch L, Goodman H, Van Montagu M. 1990. Cloning and sequence of two different cDNAs encoding 1-aminocyclopropane-1-carboxylate synthase in tomato. Proc Natl Acad Sci USA 87:4859–4863.
Van Rie J, McGaughey WH, Johnson DE, Barnett BD, Van Mellaert H. 1990. Mechanism of insect resistance to the microbial insecticide *Bacillus thuringiensis*. Science 247:72–74.
Walker JRL, Ferrar PH. 1998. Diphenol oxidases, enzyme-catalysed browning and plant disease resistance. In: Tombs MP, ed. Biotechnology and Genetic Engineering Reviews. Vol 15. Andover: Intercept, pp 457–498.
Wang Z, Quebedeaux B, Stutte GW. 1995. Osmotic adjustment—effect of water stress on carbohydrates in leaves, stems and roots of apple. Aust J Plant Physiol 22:747–754.
Wang Z, Quebedeaux B, Stutte GW. 1996. Partitioning of (14C)glucose into sorbitol and other carbohydrates in apple under water stress. Aust J Plant Physiol 23:245–251.
Wang Z, Stutte GW. 1992. The role of carbohydrates in active osmotic adjustment in apple under water stress. J Am Hortic Sci 117:816–823.
Webb KL, Burley JWA. 1962. Sorbitol translocation in apple. Science 137:766.
Whetter JM, Taper CD. 1966. Seasonal occurrence of sorbitol (D-glucitol) in buds and leaves of Malus. Can J Bot 41:175–175.
Whiteley HR, Schnepf HE. 1986. The molecular biology of parasporal crystal body formation in *Bacillus thuringiensis*. Annu Rev Microbiol 40:549–576.
Wimmer R, Olsson M, Petersen MTN, Hatti-Kaul R, Petersen SB, Muller N. 1997. Towards molecular understanding of protein stabilization: Interaction between lysozyme and sorbitol. J Biotechnol 55:85–100.
Wu Q, Szakacs-Dobozi M, Hemmat M, Hrazdina G. 1993. Endopolygalacturonase in apples (*Malus domestica*) and its expression during fruit ripening. Plant Physiol 102:219–225.

Yamada K, Oura Y, Mori H, Yamaki S. 1998. Cloning of NAD-dependent sorbitol dehydrogenase from apple fruit and gene expression. Plant Cell Physiol 39:1375–1379.

Yamaguchi, H, Kanayama Y, Yamaki S. 1994. Purification and properties of NAD-dependent sorbitol dehydrogenase from apple fruit. Plant Cell Physiol 35:887–892.

Yamaguchi, H, Kanayama Y, Soejima J, Yamaki S. 1996. Changes in the amounts of the NAD-dependent sorbitol dehydrogenase and its involvement in the development of apple fruit. J Am Soc Hortic Sci 121:848–852.

Yamaki S. 1980. Property of sorbitol-6-phosphate dehydrogenase and its connection with sorbitol accumulation in apple. Hortscience 15:268–270.

Yamaki S, Ishikawa K. 1986. Roles of four sorbitol-related enzymes and invertase in the seasonal alterations of sugar metabolism in apple tissues. J Am Soc Hortic Sci 111:134–137.

Yao J, Cohen D, Atkinson R, Richardson K, Morris B. 1995. Regeneration of transgenic plants from the commercial apple cultivar Royal Gala. Plant Cell Rep 14:407–412.

Yepes L, Aldwinckle HS. 1994. Factors that affect leaf regeneration efficiency in apple, and effect of antibiotics in morphogenesis. Plant Cell Tissue Organ Cult 37:257–269.

Yip W-K, Dong J-G, Yang SF. 1991. Purification and characterization of 1-aminocyclopropane-1-carboxylate synthase from apple fruits. Plant Physiol 95:251–257.

Zimmermann MH, Ziegler H. 1975. List of sugars and sugar alchols in sieve-tube exudates. In: Zimmermann MH, Miburn JA, eds. Transport of Plants. I. Phloem transport. Heidelburg: Springer-Verlag, pp 480–502.

Zimmerman RH. 1983. In: Moore JN, Janick J, eds. Methods in Fruit Breeding. West Lafayette, IN: Purdue University Press, pp 124–135.

# 23
# Genetic Transformation of Avocado

**Richard E. Litz and Witjaksono**
*University of Florida, Homestead, Florida*

| | | |
|---|---|---|
| I. | INTRODUCTION | 345 |
| | A. Taxonomy | 345 |
| | B. History of the Crop | 346 |
| II. | GENETICS AND BREEDING | 347 |
| | A. Scion Cultivars | 348 |
| | B. Rootstock Cultivars | 350 |
| III. | SOMATIC CELL GENETIC MANIPULATION | 351 |
| | A. Somatic Embryogenesis | 351 |
| | B. Genetic Transformation | 352 |
| IV. | CONCLUSION | 353 |
| | REFERENCES | 353 |

## I. INTRODUCTION

The avocado *Persea americana* Mill. (family Lauraceae) is a major fruit crop of the tropics and subtropics. World production is nearly 2,000,000 metric tons, and the avocado ranks 10th in total production among fruit crops after *Musa* spp. (banana and plantain), *Citrus* spp. (all types), grape, apple, mango, pear, plum, peach, and papaya (1). The major production areas are in the Western Hemisphere, with the exception of Australia, South Africa, Israel, and Indonesia. The most important producing countries are Mexico, the United States, Brazil, Dominican Republic, and Indonesia. In the United States, annual production of avocado ranks after that of *Citrus* spp. (all types), grape, apple, peach, and pear (2) with an annual value of ca. $210 million. There are two major production areas of avocados in the United States: Southern California and Miami-Dade County, Florida. Mexico, the United States, Israel, South Africa, and Australia are the leading avocado exporting countries. Although most avocados are consumed as fresh fruit, there is also a limited demand for fruit for processing.

### A. Taxonomy

The genus *Persea* consists of species in two subgenera, *Eriodaphne* and *Persea* (3). The avocado and most of the large-seeded *Persea* spp. are included in the subgenus *Persea*. Central America, most probably the area that includes Guatemala, Honduras, and the southernmost part of Mexico, e.g., the state of Chiapas, is considered to be the probable center of origin of the avocado (3). There are three horticultural varieties of avocado, Mexican, Guatemalan, and West Indian, and they appear to have evolved under different ecogeographical conditions (4). The Guatemalan type is a native of the highlands of Guatemala; the Mexican type possibly originated in the highlands of south central Mexico; the West Indian avocados, despite the name, may have originated in the Pacific lowlands of Central America (4). Williams (5,6) considered that the West Indian avocado might have evolved from the Mexican type, becoming adapted to a warmer, tropical climate.

## B. History of the Crop

The avocado was first domesticated in Mexico. Ancestors of cultivated avocados were utilized as a food by hunter-gatherers as early as 7000 B.C. Avocado seeds that have been dated from this period have been collected from caves in the Tehuacan region (7). These ancient seeds are substantially smaller than those of modern fruit, indicating that there has been continuous selection over thousands of years for fruit size.

Currently, avocados are produced in tropical and subtropical North and South America and Australia, Israel and Indonesia (Asia), and subtropical parts of Africa. Mexico is by far the largest producer of avocados, followed by the United States, Dominican Republic, Indonesia, Brazil, Israel, and Colombia (Fig. 1). Because different varieties of avocados are suited to different climate conditions, there is no single cultivar that is grown throughout its range of cultivation. 'Hass,' a complex hybrid with Mexican and Guatemalan lineage, and 'Fuerte,' a Mexican × Guatemalan hybrid (8), are the most important cultivars; they are grown in the tropical highlands, in the subtropics, and in areas with a Mediterranean climate, e.g., California (United States), Mexico, Australia, South Africa, and Israel. Both cultivars originated as chance seedling trees (9). Guatemalan avocados are usually grown at medium elevations in the tropics. The West Indian (e.g., 'Simmonds' and 'Waldin') and West Indian × Guatemalan hybrids (e.g., 'Choquette' and 'Monroe') predominate in the lowland tropics throughout the Americas, including Florida (United States).

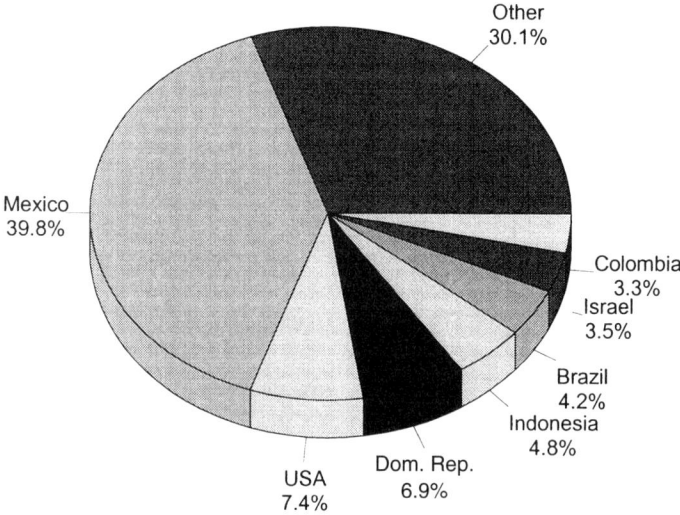

**Figure 1** Leading avocado-producing countries. (From Ref. 1.)

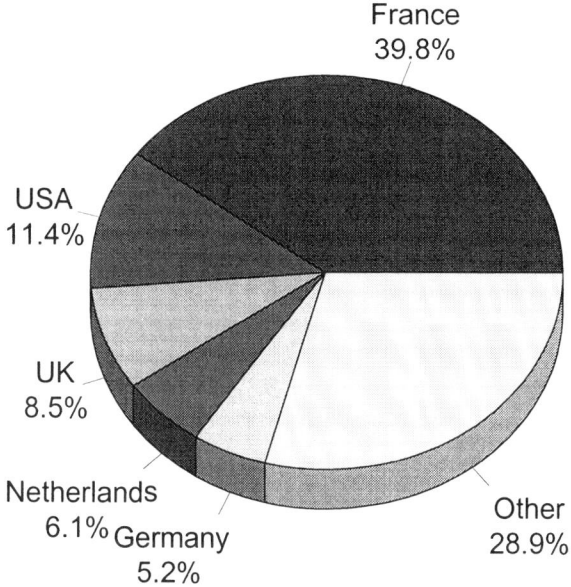

**Figure 2** World trade of avocados—major importing countries. (From Ref. 1.)

The export and U.S. market for avocados is dominated by 'Hass.' France is the leading importer of fresh avocados, followed by the United States, United Kingdom, Netherlands, and Germany (Fig. 2). West Indian and West Indian × Guatemalan cultivars have failed to gain a large market share in the United States largely because of postharvest problems (discussed later). The value of avocado exports has been estimated to be approximately U.S.$233,627,000 (1).

## II. GENETICS AND BREEDING

Modern avocado cultivars are selections that have been made from openly pollinated trees. There has been very little conventional breeding of avocados because of their heterogeneity and long juvenile period. A few scion variety selections are currently under evaluation, including 'Gwen,' 'Lamb Hass,' and 'Sir Prize.' The mode of inheritance of only one genetic trait of avocado has been identified: the dwarfing character of *P. schiedeana*, which has been attributed to a single gene (10). According to Lavi and coworkers (11,12), morphological traits in avocado are probably coded by several loci with several alleles in each locus.

Molecular tools for marker identification and mapping are increasingly being exploited. Using restriction fragment length polymorphism (RFLP) and microsatellite markers, Davis and colleagues (13) are attempting to determine the paternal origins of important avocado cultivars and the frequency of outcrossing in populations of avocados in southern California. The relationship among avocado cultivars and among different *Persea* species has been explored by Mhameed and associates (14), using microsatellite markers. Mhameed and coworkers (15) have reported a deoxyribonucleic acid (DNA) marker that is associated with fruit color. Linkage of a DNA marker with flesh fiber has also been reported (16). Such correlations together with a linkage map should eventually allow marker assisted selection for avocado improvement.

Avocado improvement programs have focused on both scion and rootstock cultivar development. Currently, the most active breeding programs are located at the University of California (Riverside), in the United States, and the Volcani Center, Bet Dagan, in Israel.

## A. Scion Cultivars

The world export market for avocados is based almost exclusively upon 'Hass' and 'Fuerte.' A number of cultivars that closely resemble 'Hass' have been released in recent years in order to replace this selection, partially, including 'Gwen,' 'Jim,' 'Reed,' and 'Lamb Hass'; however, none of these selections has been able to compete effectively with 'Hass.' The primary breeding objectives include both fruit characters and tree size and form. A dwarf habit and precocity of bearing are considered to be very important, e.g., the 'Colin V-33' selection (17). Resistance to major fruit and foliar diseases, e.g., anthracnose caused by *Colletotrichum gloeosporioides* Penz., alternaria rot caused by *Alternaria alternata* (FR.: FR.) Keissel, and avocado sun blotch caused by the avocado sun blotch viroid, is imperative.

### 1. Control of Fruit Ripening

Mexican-type avocado fruit have been demonstrated to be strongly climacteric (18), having a biphasic ripening phase. A fixed climacteric phase has been shown to be preceded by a variable lag phase (19). Low levels of endogenous ethylene accumulate during the lag phase and possibly trigger the climacteric as sensitivity to ethylene increases (20,21). Once initiated, however, ripening cannot be arrested. Mature fruit remain on the tree during the lag phase, and ripening begins to occur only after the fruit are picked (22). Mexican and Guatemalan types of avocados can be maintained on the tree for 2–6 months, as the fruit continues to accumulate oil (23). Since ripening of these avocado types is initiated only after picking, avocado fruit can be "stored" on the tree until the price of fresh fruit is high enough. A consequence of this strategy of prolonged "on-tree-storage," however, is loss of production during the subsequent year, i.e., alternate bearing.

Fruit of West Indian and West Indian × Guatemalan hybrids cannot be stored on the trees for more than 2–3 months, and this trait is strongly cultivar-dependent (22). Consequently, in tropical zones, several avocado cultivars must be grown to ensure availability of fruit through much of the year. In Florida, for example, approximately 30 avocado cultivars are commercially grown in order to ensure fruit availability from the end of May through the beginning of March (24). Establishing uniform market standards for West Indian and Guatemalan × West Indian avocados has therefore been difficult. Extending the on-the-tree storage life of West Indian and Guatemalan × West Indian avocados could revolutionize the industry in tropical countries, resulting in fewer cultivars, higher fruit quality, and easier marketing.

According to Burg and Burg (25), ripening of avocado fruit can be initiated by exposure of mature fruit to 0.1 ppm ethylene, although Biale (22) has indicated that 10 ppm ethylene was required to initiate ripening of 'Fuerte' fruit. Studies have indicated that ripening can be slowed, although not stopped, by removal of ethylene from the storage atmosphere (26). The shelf life of fresh fruit can be manipulated to some degree by blocking the action of ethylene through the use of controlled atmospheres or ethylene absorbents, e.g., $KMnO_4$ or activated charcoal or vanadium oxide; however the effectiveness of ethylene removal is variable. Like most fruit, the avocado ripens naturally from the inner to the outer mesocarp, whereas ethylene-induced ripening triggers ripening from the outside to the inside of the fruit (27). Thus, ethylene treatments to induce fruit ripening often result in unequally distributed ripening. Low temperature (6–13°C, depending on the cultivar) storage can also slow ripening; however, fruit quality and appearance can be negatively affected, particularly for Guatemalan × West Indian cultivars (28).

Ripening of avocado fruits is associated with softening of the flesh. This is probably caused by the concerted action of different cell wall hydrolases, whose activity changes during ripening, thereby altering the properties of many cell wall constituents. Knee and Bartley (29), for example, showed that fruit cells walls contain relatively more pectic and less hemicellulosic material than other plant cell walls, with a larger proportion of middle lamella to primary cell wall than other organs. Ripening of avocado and other fruits has been associated with the increased mobilization of polygalacturonases and cellulase (30–33). The presence of cell wall–degrading enzymes in fruit tissue may not necessarily imply that the enzymes are implicated in the ripening process but may represent a general feature that involves a metabolically active cell wall (31).

Fruit cells retain the capacity for protein synthesis during ripening (33). During ripening, avocado fruit synthesize ribosomal ribonucleic acid (rRNA), transfer RNA (tRNA), and poly A+ RNA (34,35), and these can be translated in vitro. Christoffersen and coworkers (36) demonstrated that different messenger RNAs (mRNAs) increase during the climacteric rise in respiration and ethylene production of ripening avocado fruit. The changes in mRNA and protein populations that occur during ripening reflect changes in gene expression. Among the mRNAs that have been isolated from ripening avocado fruits, Christoffersen and associates (37) showed that the message for cellulase was encoded, Bozak and coworkers (38) identified the message for cytochrome P-450 oxidase and McGarvey and colleagues (39) isolated a protein, pAVOe3, which was homologous with pTOM13 that was isolated from ripening tomato fruit (40). The application of exogenous ethylene can cause increased expression of many of these genes (30); Buse and Laties (41) demonstrated that ethylene mediates posttranscriptional regulation of cellulase, cytochrome P-450 oxidase, polygalacturonase, and 1-amino cyclopropane-1-carboxylic acid (ACC) oxidase during avocado fruit ripening. Dopico and coworkers (42) and Kutsunai coworkers (43) have isolated polygalacturonase cDNA and have demonstrated that ripening-related genes are either absent or present at a low level in immature fruit. Ethylene is therefore pivotal for controlling ripening of avocado fruit.

The immediate precursor of ethylene is ACC, which is derived from $S$-adenosylmethionine (SAM). The two key enzymes that catalyze the conversion of SAM to ACC and of ACC to ethylene are ACC synthase and ACC oxidase, respectively (44). $S$-adenosylmethionine is also pivotal for several other biosynthetic pathways, which include polyamine production, phospholipid synthesis, and DNA methylation.

The expression of certain genes in transgenic plants can be accomplished by introducing the gene constructed to generate antisense RNA (45), causing the expression of specific genes to be diminished and permitting their identification and assessment of their function during ripening. Genetic transformation has been utilized successfully to interfere with ethylene action. Suppression of ethylene biosynthesis in fruit during their development should also suppress ripening. In fact, Hamilton and associates (46) and Oeller and coworkers (47) transformed tomato with ACC synthase and ACC oxidase in the antisense, respectively, and were able to suppress fruit ripening. In the presence of ethylene, the inhibition of ethylene biosynthesis could be overcome, resulting in normally ripened fruit. This could have a significant impact on handling of avocado fruit post harvest, particularly of the West Indian type, which have poor shelf life and cannot be stored on the tree.

## 2. Avocado Sun Blotch Disease

Sun blotch disease is caused by a viroid and is pervasive in many germplasm collections of avocado and in commercial plantings. Not until in 1997 was a reliable screening procedure for avocado sunblotch viroid developed (48). Conventional strategies for producing disease-indexed plants from bacteria- or virus-infected plants have never been confirmed for a disease caused by a viroid; however, Sano and colleagues (49) have demonstrated a strategy for conferring resistance of potato to potato spindle tuber viroid based upon transformation with yeast-derived double-stranded RNA-specific ribonuclease pac 1. Such a strategy could be effective for controlling avocado sunblotch disease.

## B. Rootstock Cultivars

### 1. Phytophthora Root Rot Resistance

The major constraint for avocado production throughout much of its range is a serious root rot disease caused by the soilborne pathogen *Phytophthora cinnamomi* Rands. It has been estimated that phytophthora root rot (PRR) causes annual losses in the United States alone of approximately $30 million (50). There is no true resistance to the disease within the species, although a few rootstock selections that have good tolerance of PRR have been made (51,52). Tolerance of PRR by 'Duke 7' is associated with rapid growth of the feeder root system, which outpaces destruction of the root system by the pathogen. 'Thomas' is a recent rootstock selection that outperforms 'Duke 7' in many avocado producing areas of southern California that are heavily affected by PRR (53–55). 'Martin Grande,' a hybrid of avocado and *P. schiedeana*, also appears to have good tolerance of PRR in certain locations; however, it is currently under field trial. The rootstock selection 'PP4,' a seedling of 'Barr Duke' tentatively named 'Zentmyer,' appears to be more tolerant of PRR than 'Thomas' in many locations in southern California (54,55). Phytophthora root rot–tolerant rootstocks have generally been selected from mature trees that have survived in soils that are heavily infested with the pathogen (escapes). Their performance under controlled conditions has been determined after vegetative propagation by the specialized etiolation technique (56). This technique involves graftage of rootstock (as scion) onto a young seedling, followed by etiolation. Rooting of the etiolated shoot is induced by wounding and the rooted shoot is utilized as clonal rootstock.

According to Zentmyer and Schroeder (57), *Persea* species within the subgenus *Eriodaphne* are highly resistant to PRR. The *Eriodaphne* species, e.g., *P. borbonia*, *P. cinerascens*, and *P. pachypoda*, are generally small-seeded. The PRR-resistant species are sexually and graft-incompatible with *Persea* species in the subgenus *Persea* (58). Therefore, this source of resistance to the pathogen is inaccessible for use as rootstock using conventional genetic approaches.

The potential for using somatic hybridization to make wide interspecific crosses between avocado and the PRR-resistant species in the subgenus *Eriodaphne* was proposed by Pliego-Alfaro and Bergh (59) as a way of developing a new generation of PRR-resistant rootstocks. Witjaksono and coworkers (60) demonstrated that avocado could be regenerated from protoplasts isolated from embryogenic cultures and has carried out preliminary fusion studies (61).

There is considerable likelihood that PRR of avocado can be controlled by using one or more transgenic approaches. Transformation of existing rootstock or scion cultivars with *rol* genes from *Agrobacterium rhizogenes* could result in more vigorous growth of the root system (62,63). Faster root regeneration and growth have been associated with PRR resistance (64).

Another strategy could involve the transformation of avocado with genes that are associated with the constitutive expression of pathogensis-related (PR) proteins, e.g., β-1,3-glucanase, plant defensin, and ribosome inactivating proteins (65,66). Plants that have been transformed with these genes generally show enhanced tolerance to infection by pathogens (67). The hyphal walls of *P. cinnamomi*, an oomycete, are composed largely of glucan with β-1,3 and β-1,6 linkages (68). Overexpression of soybean β-1,3-glucanase in tobacco has been demonstrated to confer resistance to fungal disease (69); however, Yi and Huang (70) demonstrated that both β-1,3-glucanase and chitinase are induced after infection of soybean by *Phytophthora megasperma* f.sp. *glycinea* and suggested that both PR proteins may be necessary for control of growth of the pathogen. In fact, a combination of two antifungal proteins in transgenic plants has been shown to have a synergistic effect (71,72). Genetic transformation of the best avocado rootstock cultivars, e.g., 'Duke 7' or 'Thomas,' with a combination of antifungal proteins should augment existing tolerance of PRR in these selections. Moreover, genetic transformation of scion selections with the same genes under the control of a root-specific promoter should permit these cultivars to be grown on their own roots.

## 2. Salinity Tolerance

According to Embleton and associates (73), Ben-Ya'acov (74), and Gustafson and coworkers (75), the resistance to salinity is greater in West Indian avocado roots than in Mexican or Guatemalan avocado roots. The productivity of plants that are grown under high soil salinity include reduced yields and tree size, poor root growth, lower photosynthesis, and chloride toxicity. The major avocado production areas of the United States (California) and Israel increasingly are dependent upon irrigation water that contains high levels of sodium chloride. Therefore, identification of rootstock selections that are highly resistant to saline conditions is essential for the sustainability of these industries.

## III. SOMATIC CELL GENETIC MANIPULATION

### A. Somatic Embryogenesis

Avocado has been regenerated from embryogenic cultures that have been induced from zygotic embryo, i.e., nonclonal, explants (76–79) and from the nucellus of a few cultivars, including the important rootstock 'Thomas' and the most important scion cultivar 'Hass' (61).

### 1. Induction of Embryogenic Cultures

*a. Zygotic Embryo Explants.* The procedure for induction of embryogenic cultures was originally described by Mooney and van Staden (71) and Pliego Alfaro and Murashige (72) and later adapted by Witjaksono and Litz (73). Avocado fruitlets, approximately 0.3–0.5 cm long without the calyx, are at the appropriate stage of development for explanting the zygotic embryo. After removal of sepals and peduncles, the fruitlets are surface-disinfested for 10–20 minutes with 10–20% commercial bleach containing 1–2 drops of Tween 20 per liter. After rinsing of the fruits with two changes of sterile, deionized water, they are bisected along the longitudinal axis under axenic conditions. The zygotic embryo from each immature seed is removed and plated on the induction plant growth medium.

*b. Nucellar Explants.* Witjaksono (61) demonstrated that embryogenic cultures could be induced from the nucellus removed from 0.3- to 0.5-cm-length avocado fruit (see previous discussion). The immature fruit are surface-sterilized and bisected under axenic conditions. The zygotic embryo is removed and discarded, and the nucellus with attached ovule parts is explanted onto induction plant growth medium.

*c. Induction Medium and Growing Conditions.* The induction medium consists of B5 major salts (80), supplemented with MS minor salts (81), 4.14 µM picloram, and (in milligrams per liter) thiamine HCl (4), *myo*-inositol (100), sucrose (30,000), and 8 g l$^{-1}$ TC agar. The pH is adjusted to 5.7–5.8 prior to addition of agar and autoclaving at 121°C at 1.1 kg cm$^{-2}$ for 15 minutes. The plant growth medium is generally dispensed in 10-ml aliquots into sterile disposable petri dishes (60 × 15 mm), and cultures are maintained in darkness at 25°C.

### 2. Maintenance of Embryogenic Cultures

Embryogenic cultures, which consist of proembryonic masses (PEMs), are usually visible 3–4 weeks after explanting and are transferred onto fresh semisolid maintenance medium (complete

MS basal medium) for one to two subcultures. Thereafter, they are subcultured at 2- to 4-week intervals, using the smallest PEMs (0.2–0.5 mm in diameter) as the inoculum. Embryogenic suspension cultures can be established by inoculating 100–300 mg of 8- to 10-day-old PEMs from semisolid maintenance medium into filter-sterilized 40-ml liquid maintenance medium in 125-ml Erlenmeyer flasks; however, the maintenance of PEMs in suspension is highly genotype-dependent (79,82). Cotyledonary somatic embryos develop under inductive, i.e., maintenance, conditions for most genotypes. Embryogenic suspension cultures are subcultured at biweekly intervals into filter-sterilized medium and are maintained on a rotary shaker at 120 rpm at 25°C in darkness.

### 3. Maturation of Somatic Embryos and Plant Recovery

In order to initiate somatic embryo development from embryogenic PEM cultures, it is necessary to subculture them onto semisolid MS medium without picloram (65,66,68); however, most of the early cotyledonary stage somatic embryos are physiologically abnormal, i.e., hyperhydric, and cannot develop to maturity. The frequency of recovery of opaque, white somatic embryos (>0.8-cm diameter) can be increased if relatively high-gellan gum concentrations are used (6–7 g l$^{-1}$) (83). The cultures are maintained in darkness at 25°C and subcultured onto fresh medium at 2 to 3-month intervals. Mature somatic embryos are approximately 1–2 cm in diameter and usually germinate after embryo enlargement has ceased, usually 4–5 months after transfer onto maturation medium. Germinating somatic embryos are subcultured onto medium in 150- by 25-mm glass test tubes that have been closed with Suncaps. When root or shoot growth is apparent after 4–5 months, the cultures are transferred to 16-h light provided by cool white fluorescent tubes (80–100 µmol m$^{-2}$ s$^{-1}$) with 20,000 ppm $CO_2$ in a nitrogen carrier in order to optimize plant recovery and photoautotropy (84). Even so, the recovery of plants from somatic embryos is low (71–74).

Because of the low germination frequency of avocado somatic embryos, shoots that emerge from each somatic embryo are usually decapitated 1–1.5 cm from the tip, cultured on avocado shoot proliferation medium, and subcultured at 8-week intervals for several passages (83). Shoot proliferation medium consists of semisolid MS medium, supplemented with 4.44 µM BA and 1 $NH_4NO_3$:2 $KNO_3$ (total $N = 40$ mM). Rooting of individual shoots is accomplished by pulsing shoot cuttings for 3 days in a root induction medium containing 122.6 µM IBA, followed by subculture on semisolid MS medium with 1 g l$^{-1}$ activated charcoal.

## B. Genetic Transformation

The recovery of genetic transformants of avocado has depended on the establishment of highly embryogenic suspension cultures that consist of PEMs that proliferate by the process of secondary somatic embryogenesis or budding (see earlier discussion). Cruz-Hernandez and colleagues (85) demonstrated that growth of embryogenic suspension cultures of the PEM type can be suppressed by approximately 50% in the presence of 50 mg l$^{-1}$ kanamycin sulfate, whereas, 50% growth suppression occurred on semisolid medium containing 100 mg l$^{-1}$ kanamycin sulfate. Complete suppression of embryogenic cultures occurs on semisolid medium with 200 mg l$^{-1}$ kanamycin sulfate. Cruz-Hernandez and coworkers (85) utilized a two-step selection process to recover genetically transformed embryogenic cultures that were resistant to kanamycin and that expressed the β-glucuronidase (GUS) gene.

### 1. Coculture of Embryogenic Cultures with Genetically Engineered *Agrobacterium tumefaciens*

Embryogenic cultures on semisolid medium are gently abraded with a soft camel hair brush. The abraded cultures together with acetosyringone-activated *Agrobacterium tumefaciens* (strain 9749

ASE2 containing a cointegrate vector pMON9749 with a selectable kanamycin-resistant marker [*npt*II] and a reportable marker [GUS]) are inoculated into liquid maintenance medium (discussed previously) and cocultured for 3 days at 100 rpm. Superficial *A. tumefaciens* is eliminated by incubating the cultures in maintenance medium supplemented with 50 mg l$^{-1}$ kanamycin sulfate and 200 mg l$^{-1}$ cefotaxime. Rapidly growing, putatively kanamycin-resistant proembryonic masses are thereafter subcultured in liquid medium with filter-sterilized 100 mg l$^{-1}$ kanamycin sulfate and 200 mg l$^{-1}$ cefotaxime.

Cultures are initially selected in maintenance medium containing 50 mg l$^{-1}$ kanamycin sulfate for 2–4 months. This is followed by more intensive selection in the presence of 100 mg l$^{-1}$ kanamycin sulfate followed by 200 mg l$^{-1}$ filter-sterilized kanamycin sulfate with weekly transfers into fresh medium of the same formulation(s) in order to eliminate chimaeras. Somatic embryo development is initiated by subculture onto maturation medium (see earlier discussion) without selection, followed by subculture onto maturation medium containing kanamycin sulfate. Transformed somatic embryos can be recovered that stain positively for GUS in the X-GLUC reaction (86), and the integration of *npt*II and GUS into the avocado genome has been confirmed by polymerase chain reaction (PCR) and Southern hybridization (87,88). Transgenic plants have not been regenerated. Using the same approach, Cruz-Hernandez and coworkers (personal communication) have genetically transformed embryogenic cultures of avocado with the gene for the PR-related protein chitinase.

## IV. CONCLUSION

Improvement of perennial crop species that have long juvenile periods, such as the avocado, is difficult and time-consuming if conventional breeding is followed (59). Genetic transformation of elite avocado selections is feasible, and this should radically alter the way that this species, including scion and rootstock selections, is manipulated in the future.

## ACKNOWLDGEMENTS

The authors are grateful for support provided by the California Avocado Society, the California Avocado Commission, the Research and Development Center for Biology (LIPI) (Indonesia), and the Agency for Assessment and Application of Technology (BPPT) (Indonesia). Florida Agricultural Experiment Station Journal Series No. R-06911

## REFERENCES

1. FAOSTAT DATABASE RESULTS. FAO of the United Nations, Rome, 1997. Available at http://www.fao.org.
2. Noncitrus fruits and nuts summary, July 1992. Agricultural Statistics Board, NAA, USDA, 1992.
3. L Kopp. A taxonomic revision of the genus *Persea* in the western hemisphere (*Persea-Lauracea*). Mem NY Bot Gdn 14:1–120, 1966.
4. RW Scora, BO Bergh. The origin and taxonomy of avocado (*Persea americana* Mill.) Lauraceae. Acta Hortic 275:387–394, 1990.
5. LO Williams. The botany of avocado and its relatives. In: JW Sauls, RL Phillips, LK Jackson, eds. Proceedings of the First International Tropical Fruit Short Course: The Avocado. Gainesville: Florida Cooperative Extension Service, Institute of Food and Agricultural Sciences, University of Florida, 1976, pp 9–15.

6. LO Williams. The avocado, a synopsis of the genus *Persea*, subg. *Persea.* Econ Bot 31:315–320, 1977.
7. CE Smith, Jr. Archeological evidence for selection in avocado. Econ Bot 20:169–175, 1966.
8. BO Bergh. In: J Janick, JN Moore eds. Advances in Fruit Breeding. Lafayette, In: Purdue University Press, 1975, pp 541–567.
9. BO Bergh. Avocado breeding and selection. In: JW Sauls, RL Phillips, LK Jackson, eds. Proceedings of the First International Tropical Fruit Short Course: The Avocado. Gainesville: Florida Cooperative Extension Services, Institute of Food and Agricultural Sciences, University of Florida, 1976, pp 24–33.
10. BO Bergh, E Lahav. Avocados. In: J Janick, JN Moore, eds. Fruit Breeding. Vol I. Subtropical and Tropical Fruits. New York: John Wiley & Sons, 1996, pp 113–166.
11. U Lavi, E Lahav, C Degani, S Gazit, J Hillel. Genetic variance components and heretabilities of several avocado traits. J Am Soc Hortic Sci 118:40–404, 1993.
12. U Lavi, E Lahav, C Degani, S Gazit. Genetics of flower color, flowering group and anise scent in avocado. J Hered 84:82–84, 1993.
13. J Davis, D Henderson, M Kobayashi, MT Clegg, MT Clegg. Genealogical relationships among cultivated avocado as revealed through RFLP analyses. J Hered 89:319–323, 1998.
14. S Mhameed, D Sharon, D Kaufman, E Lahav, J Hillel, C Degani, U Lavi. Genetic relationship within avocado (*Persea americana* Mill.) cultivars and between *Persea* species. Theor Appl Gen 94:279–284, 1997.
15. S Mhameed, J Hillel, E Lahav, D Sharon, U Lavi. Genetic association between DNA fingerprint fragment and loci controlling agriculturally important traits in avocado (*Persea americana* Mill.). Euphytica 81:81–87, 1995.
16. D Sharon, J Hillel, S Mhameed, PB Cregan, E Lahav, U Lavi. Association between DNA markers and loci controlling avocado traits. J Am Soc Hortic Sci 123:1016–1022, 1998.
17. S Sanchez Colin. Colin V-33 una nueva variedad de aguacate en Mexico. Metepec: Servicios Agricolas Integrados en el Estado de Mexico, Unidad Communicacional Conjunto Codagem, 1980.
18. I Adato, S Gazit. Postharvest response of avocado fruits of different maturity to delayed ethylene treatments. Plant Physiol 53:899–902, 1977.
19. IL Eaks. Respiratory rate, ethylene production and ripening response of avocado fruit to ethylene or propylene following harvest maturities. J Am Soc Hortic Sci 105:744–747, 1980.
20. EJ McMurchie, WB McGlasson, IL Eaks. Treatment of fruit with propylene gives information about the biogenesis of ethylene. Nature 237:235–236, 1972.
21. BC Peacock. Role of ethylene in the initiation of fruit ripening. Q J Agric Anim Sci 29:687–690, 1972.
22. JB Biale. The postharvest biochemistry of tropical and subtropical fruits. Adv Food Res 10:293–354, 1960.
23. AW Whiley. Persea americana Mill. In: EWM Verheij, RE Coronel, eds. Plant Resources of Southeast Asia. Wageningen: Pudoc, 1992, pp 249–254.
24. JH Crane, CF Balerdi, CW Campbell. The Avocado. Gainesville: Circular 1034, Florida Cooperative Extension Service, Institute of Food and Agricultural Sciences, University of Florida, 1996.
25. SP Burg, EA Burg. Role of ethylene in fruit ripening. Plant Physiol 37:179–189, 1962.
26. SP Burg, EA Burg. Fruit storage at subatmospheric pressures. Science 153:314–315, 1966.
27. MA Gomez-Lim. Postharvest physiology. In: RE Litz, ed. The Mango Botany Production and Uses. Wallingford: CAB International, 1998, pp 425–445.
28. TT Hatton Jr, PL Harding, WF Reeder, CW Campbell. Ripening and storage of Florida avocados. Agric Res Serv US Dept Agric Mrktg Res Rep No. 697, 1965.
29. M Knee, IM Bartley. Composition and metabolism of cell wall polysaccharides in ripening fruits. In: J Friend, MJC Rhodes, eds. Recent Advances in the Biochemistry of Fruits and Vegetables. New York: Academic Press, 1981, pp 133–148.
30. RL Fischer, AB Bennett. Role of cell wall hydrolases in fruit ripening. Annu Rev Plant Physiol Plant Mol Biol 42:1146–1153, 1991.
31. Huber, DJ. The role of cell wall hydrolases in fruit softening. Hortic Rev 5:169–219, 1983.
32. E Pesis, Y Fuchs, G Zauberman. Cellulase activity and fruit ripening in avocado. Plant Physiol 61:416–419, 1978.

33. GA Tucker, D Grierson. Fruit ripening. In: D Davies, ed. The Biochemistry of Plants. Vol 12. New York: Academic Press, 1987, pp 265–319.
34. DA Starrett, GG Laties. The effect of ethylene and propylene pulses on respiration, ripening advancement, ethylene-forming enzyme and 1-aminocyclopropane-1-carboxylic acid synthase activity in avocado fruit. Plant Physiol 95:921–927, 1991.
35. ML Tucker, GG Laties. Interrelationship of gene expression, polysome prevalence, and respiration during ripening of ethylene and/or cyanide-treated avocado fruit. Plant Physiol 74:307–315, 1984.
36. RE Christoffersen, E Warm, GG Laties. Gene expression during fruit ripening in avocado. Planta 155: 52–57, 1982.
37. RE Christoffersen, ML Tucker, GG Laties. Cellulase gene expression in ripening avocado fruit: The accumulation of cellulase mRNA and protein as demonstrated by cDNA hybridization and immunodetection. Plant Mol Biol 3:385–391, 1984.
38. KR Bozak, H Yu, R Sirevag, RE Christoffersen. Cloning and sequence analysis of ripening-related cytochrome P-450 cDNAs from avocado fruit. Proc Natl Acad Sci USA 87:3904–3908, 1990.
39. DG McGarvey, R Sirevag, RE Christoffersen. Ripening related gene from avocado fruit. Plant Physiol 98:554–559, 1990.
40. MJ Holdsworth, CR Bird, J Ray, W Schuch, D Grierson. Structure and expression of anethylene-related mRNA from tomato. Nucleic Acids Res 15:731–739, 1987.
41. EL Buse, GG Laties. Ethylene-mediated posttranscriptional regulation in ripening avocado (*Persea americana*) mesocarp discs. Plant Physiol 102:417–423, 1993.
42. B Dopico, AL Lowe, ID Wilson, C Merodio, D Grierson. Cloning and characterization of avocado fruit mRNAs and their expression during ripening and low temperature storage. Plant Mol Biol 21:437–449, 1993.
43. S Katsunai, AC Lin, FW Percival, GG Laties, RE Christoffersen. Ripening related polygalacturonase cDNA from avocado. Plant Physiol 103:289–290, 1993.
44. H Kende. Ethylene biosynthesis. Annu Rev Plant Physiol Plant Mol Biol 44:283–307, 1993.
45. Y Eguchi, T Itoh, J Tomizawa. Antisense RNA. Annu Rev Biochem 60:631–652, 1991.
46. AJ Hamilton, GW Lycett, D Grierson. Antisense gene that inhibits synthesis of the hormone ethylene in transgenic plants. Nature 346:284–287, 1990.
47. PW Oeller, LM Wong, LP Taylor, DA Pike, A Theologis. Reversible inhibition of tomato fruit senescence by antisense RNA. Science 254:37–439, 1991.
48. RJ Schnell, DN Kuhn, CM Ronning, D Harkins. Application of RT-PCR for indexing avocado sun blotch viroid. Plant Dis 81:1023–1026, 1997.
49. T Sano, A Nagayama, T Ogawa, I Ishida, Y Okada. Transgenic potato expressing a double-stranded RNA-specific ribonuclease is resistant to potato spindle tuber viroid. Biotechnology 15:1290–1294, 1997.
50. MD Coffey. Phytophthora root rot of avocado. Plant Dis 71:1046–1052, 1987.
51. GA Zentmyer, WA Thorn. Resistance of the Duke variety of avocado to Phytophthrora root rot. Calif Avo Soc Yrb 40:169–173, 1956.
52. GA Zentmyer, WA Thorn, RM Burns. The Duke avocado. Calif Avo Soc Yrb 47:28–36, 1963.
53. J Menge. Screening and evaluation of new rootstocks with resistance to *Phytophthora cinnamomi*. Proceedings of the California Avocado Research Symposium, Riverside, University of California, 1997, pp 35–47.
54. J Menge. Screening and evaluation of new rootstocks with resistance to *Phytophthora cinnamomi*. Proceedings of the California Avocado Research Symposium, Riverside, University of California, 1998, pp 41–43.
55. J Menge. Screening and evaluation of new rootstocks with resistance to *Phytophthora cinnamomi*. Proceedings of the California Avocado Research Symposium, Riverside, University of California, 1999, pp 69–72.
56. RG Platt. Current techniques of avocado propagation. In: JW Sauls, RL Phillips, LK Jackson, eds. Proceedings of the First International Tropical Fruit Short Course: The Avocado. Gainesville: Florida Cooperative Extension Service, Institute of Food and Agricultural Sciences, University of Florida, 1976, pp 92–95.

57. GA Zentmyer, CA Schroeder. Test of *Persea* species for resistance to *Phytophthora cinnamomi*. Calif Avo Soc Yrb 38:163–164, 1953/1954.
58. EF Frolich, CA Schroeder, GA Zentmyer. Graft compatibility in the genus *Persea*. Calif Avo Soc Yrb 58:102–105, 1958.
59. F Pliego-Alfaro, BO Bergh. Avocado. In: FA Hammerschlag, RE Litz, eds. Biotechnology of Perennial Fruit Crops. Wallingford, CAB International, 1992, pp 323–333.
60. Witjaksono, RE Litz, JW Grosser. Isolation, culture and regeneration of avocado (*Persea americana* Mill.) protoplasts. Plant Cell Rep 18:235–242, 1998.
61. Witjaksono. Development of protocols for avocado tissue culture: Somatic embryogenesis, protoplast culture, shoot regeneration and protoplast fusion. Ph.D. dissertation, University of Florida, Gainesville, 1997.
62. E Rugini, A Pellegrineschi, M Mencuccini, D Mariotti. Increase of rooting ability in the woody species kiwi (*Actinidia deliciosa* A. Chev.) by transformation with *Agrobacterium rhizogenes rol* genes. Plant Cell Rep 10:291–295, 1991.
63. TPM van der Salm, R Bauwer, AJ van Dijk, LCP Keizer, CH Hanischtencate, LHW van der Plas, JJM Dons. Stimulation of scion bud release by *rol* gene transformed rootstock of *Rosa hybrida* L. J Exp Bot 49:847–852, 1998.
64. JH Graham. Root regeneration and tolerance of citrus rootstocks to root rot caused by *Phytophthora nicotianae*. Plant Dis 85:111–117, 1995.
65. CJ Lamb, JA Ryals, ER Ward, RA Dixon. Emerging strategies for enhancing crop resistance to microbial pathogens. Biotechnology 10:1436–1445, 1992.
66. BJC Cornelissen, LS Melchers. Strategies for control of fungal diseases with transgenic plants. Plant Physiol 101:709–712, 1993.
67. DJ Yun, RA Bressan, PM Hasgawa. Plant antifungal proteins. Plant Breed Rev 14:39–84, 1997.
68. S Bartnicki-Garcia, E Lippman. Enzymatic digestion and glucan structure of hyphal walls of *Phytophthora cinnamomi*. Biochim Biophys Acta 136:533–543, 1967.
69. M Yoshikawa, M Tsuda, Y Takeuchi. Resistance to fungal diseases in transgenic tobacco plants expressing the phytoalexin elicitor-releasing factor β-1,3-endoglucanase from soybean. Naturwissenschaften 80:417–420, 1993.
70. SY Yi, BK Hwang. Differential induction and accumulation of β-1,3-glucanase and chitinase isoforms in soybean hypocotyls and leaves after compatible and incompatible infection with *Phytophthora megasperma* f.sp. *glycinea*. Physiol Mol Plant Pathol 48:179–192.
71. F Mauch, B Mauch-Mani, T Boller. Antifungal hydrolases in pea tissue. II. Inhibition of fungal growth by combinations of chitinase and β-1,3-glucanases. Plant Physiol 88:936–942, 1988.
72. LS Melchers, AS Ponstein, MB Sela-Buurlage, SA Vloemans, BJC Cornelissen. In vitro anti-microbial activities of defense proteins and biotechnology. In: B Fritag, M Legrand, eds. Mechanisms of Plant Defense Response. Dordrecht: Kluwer Academic, 1993, pp 401–410.
73. TW Embleton, M Matsumura, WB Storey, MJ Garber. Chlorine and other elements in avocado leaves influenced by rootstock. J Am Soc Hortic Sci 80:230–236, 1955.
74. AD Ben-Ya'acov. Characteristics associated with salt tolerance in avocados grafted on Mexican and West Indian rootstocks. Proc 18th Intern Hort Cong Vol 1, 1970, p 135.
75. CD Gustafson, A Kadman, AD Ben Ya'acov. Sodium-22 distribution in inarched grafted avocado plants. Proc 18th Intern Hort Cong Vol 1, 1970, p 135.
76. PA Mooney, J van Staden. Induction of embryogenesis in callus from immature embryos of *Persea americana*. Can J Bot 65:622–626, 1987.
77. F Pliego-Alfaro, T Murashige. Somatic embryogenesis in avocado (*Persea americana* Mill.) *in vitro*. Plant Cell Tissue Org Cult 12:61–66, 1988.
78. A Raviv, RA Avenido, LF Tisalona, OP Damasco, EMT Medoza, Y Pinkas, S Zilkah. Callus and somatic embryogenesis of *Persea* species. Plant Tissue Cult Biotechol 4:196–206, 1998.
79. Witjaksono, RE Litz. Induction and growth characteristics of embryogenic avocado (*Persea americana* Mill.) cultures. Plant Cell Tissue Org Cult 58:19–29,1999.
80. OL Gamborg, RA Miller, K Ojima. Plant cell cultures. I. Nutrient requirements of suspension cultures of soybean root cells. Exp Cell Res 50:151–158, 1968.

81. T Murashige, F Skoog. A revised medium for rapid growth and bioassays with tobacco tissue cultures. Physiol Plant 15:473–497, 1962.
82. Witjaksono, RE Litz, F Pliego-Alfaro. Somatic embryogenesis in avocado (*Persea americana* Mill.). In: SM Jain, PK Gupta, RJ Newton, eds. Somatic Embryogenesis in Woody Plants. Vol. 5. Dordrecht: Kluwer Academic, 1999, pp 197–214.
83. Witjaksono, RE Litz. Maturation of avocado (*Persea americana* Mill.) somatic embryos and plant recovery. Plant Cell Tissue Org Cult 58:141–148.
84. Witjaksono, B Schaffer, A Colls, RE Litz, PA Moon. Avocado shoot culture, plantlet development and net $CO_2$ assimilation in an ambient and enhanced $CO_2$ environment. In Vitro Cell Dev Biol 35:238–244.
85. Cruz-Hernandez, A Witjaksono, RE Litz, MA Gomez-Lim. *Agrobacterium tumefaciens*–mediated transformation of embryogenic avocado cultures and regeneration of somatic embryos. Plant Cell Rep 17:497–503, 1998.
86. RA Jefferson. Assaying chimeric genes in plants: The gus fusion system. Plant Mol Biol Rep 5:387–405, 1987.
87. JJ Doyle, JL Doyle. Isolation of plant DNA from fresh tissue. Focus 12:12–14, 1990.
88. JH Miller. Experiments in molecular genetics. Cold Spring Harbor, NY: Cold Spring Harbor Laboratory Press, 1972.

# 24
# Production of Transgenic Banana (*Musa* species)

**László Sági, Serge Remy, Juan Bernardo Pérez Hernández, and Rony Swennen**
*Catholic University of Leuven, Leuven, Belgium*

| | | |
|---|---|---|
| I. | INTRODUCTION | 359 |
| | A. Production and Significance | 359 |
| | B. Main Constraints of Banana Production | 360 |
| II. | GENETIC TRANSFORMATION OF BANANA | 361 |
| | A. In Vitro Cell and Protoplast Culture | 361 |
| | B. Genetic Transformation | 362 |
| III. | PROSPECTS AND CONCLUSION | 365 |
| | A. Resistance to Pathogens | 365 |
| | B. Fruit Quality Control | 366 |
| | C. Potential for Pharmaceutical Applications | 366 |
| | REFERENCES | 366 |

## I. INTRODUCTION

### A. Production and Significance

The term *banana* covers here dessert, starchy (cooking banana and plantain), and beer bananas. Banana is the most important fruit crop on Earth with an annual production of close to 90 million tons, which accounts for approximately 20% of total primary fruit production worldwide (1).

In more than 120 banana-producing countries, which are mainly located in the Third World, banana is consumed either fresh, prepared by cooking or boiling, or processed to beer, juice, chips, puree, and other food products. The nutrient composition and nutritional value of banana are very similar to those of potato. Because of its high starch content, banana fruit is a major source of dietary carbohydrate uptake, but it also contains significant amounts of potassium, magnesium, and phosphorus, as well as the vitamins $B_6$ and C (2). In particular, plantain is also rich in vitamin A, which is often deficient in human diet in the tropics. In addition, banana fruit has been extensively used in folk medicine and is reported to exert a cholesterol-lowering effect (3) as well as to decrease the risk of colorectal cancer (4). Banana is a staple food for at least 400 million people and it is the fourth major food commodity (after rice, milk, and wheat) for the developing world.

Some 15% of banana production is exported (1) and serves as dessert for many more millions of people. The majority of export banana has been produced on large commercial estates and by smallholders in Latin America and the Caribbean. Export banana trade has a gross value of more than U.S.$5 billion (1), and banana cultivation is of great socioeconomic importance for many less developed countries that rely on banana export to provide employment opportunities and to generate a significant part of their foreign exchange.

The distribution of production is relatively even among Latin America and the Caribbean (36%), Africa (34%), and Asia with the Pacific (29%). However, a considerable difference exists in the banana types cultivated in the different continents. More than 80% of dessert banana is produced in Latin America and the Caribbean, whereas cooking banana is cultivated mainly in the Asia-Pacific region. Plantains are produced in West and Central Africa and in Latin America, whereas beer banana production is limited to East Africa.

## B. Main Constraints of Banana Production

Most cultivated landraces are parthenocarpic and sterile triploids, characteristics that constitute a major bottleneck to running successful hybridization programs. These obstacles, coupled with the relatively long generation time and poor agronomic practice, make it very difficult for banana breeders to release new cultivars with resistance to the numerous viral, bacterial, and fungal pathogens as well as nematode and insect pests.

The most devastating banana disease is the Sigatoka complex, which is caused by the ascomycetes fungal pathogens *Mycosphaerella fijiensis* (black Sigatoka) and *M. musicola* (yellow Sigatoka). Sigatoka leaf disease presents a serious economic problem for the banana industry, and yield losses may amount to 30–50% worldwide (5,6). The annual cost of fungicide control in commercial estates may range between U.S.$600 and U.S.$1800 per hectare, accounting for up to 30% of total production costs. In addition to their pressure on the environment, application of fungicides is economically questionable in backyards and small farms, where banana cultivation mainly takes place. The recent success in breeding of plantain (7) and other types of banana (8) for Sigatoka resistance is still relatively limited.

Another major fungal disease is Panama disease or banana wilt, which is caused by *Fusarium oxysporum* f.sp. *cubense*. This soil-inhabiting fungus caused one of the most destructive epidemics in history when in the 1950s and 1960s its race 1 destroyed approximately 40,000 hectares of commercial plantation and nearly knocked out the banana industry (9). Race 4 (10), discovered in 1978, presently threatens export banana production in the subtropics (11) as well as in the tropics, as hitherto no source resistant to it has been identified. The fungus infects lateral roots and blocks the plant vascular system, resulting in typical wilt symptoms. Since no effective fungicide control is available for this disease, production is maintained by new plantings in clean soil.

At present, perhaps the most serious viral disease that affects banana is bunchy top disease. The causal agent, banana bunchy top virus (BBTV), is a multicomponent single-stranded deoxyribonucleic acid (DNA) virus, which represents a new group of plant viruses, the nanaviruses. BBTV is transmitted by the aphid *Pentalonia nigronervosa* in a persistent manner or by vegetative propagation (12) and is widely distributed in the Asia-Pacific region. In Africa, BBTV is present in a few countries only, and in Latin America it is still absent. Growth of heavily infected plants is significantly reduced and yields can be entirely lost. In addition, no BBTV-resistant source has yet been identified.

Banana streak badnavirus, a pararetrovirus first described from Morocco in 1986 (13), is becoming another major threat to banana production all over the world. The virus was found in 1999 to be able to integrate into the banana genome (14), and the integrated sequence has been shown to reactivate to an episomal virus after tissue culture (15). This may have serious consequences as banana is one of the most intensively micropropagated plant species. Among the other pathogenic viruses, banana bract mosaic potyvirus is a quarantine threat in the Philippines and in India, and cucumber mosaic virus, which causes infectious chlorosis in all banana-producing regions, is difficult to control because of its broad host spectrum (16).

Among the other diseases and pests, the bacterial wilt or Moko disease should be mentioned; caused by the bacterium *Ralstonia* (earlier *Pseudomonas*) *solanacearum*, it is destructive in the humid areas of Central America and the Philippines. In some restricted regions of commercial production, the bacterial head rot disease caused by *Erwinia carotovora* ssp. *carotovora* can present significant problems.

Of the migratory endoparasitic nematodes, which are widespread pests of banana (e.g., the root-lesion nematodes *Pratylenchus* spp. and the spiral nematode, *Helicotylenchus multicinctus*), the burrowing nematode, *Radopholus similis*, is the most dangerous (17). This nematode causes uprooting of fruit-bearing plants, constituting a steadily increasing problem in commercial estates and small fields worldwide. Among the sedentary species, the root-knot nematodes *Meloidogyne* spp. are a major problem in Taiwan and cause root deformation and stunting.

Finally, the banana borer weevil, *Cosmopolites sordidus* (18), by laying its eggs close to the base of the pseudostem, after which the larvae burrow into the rhizome, causes serious damage in the tropics, especially in sub-Saharan Africa and in the Pacific.

## II. GENETIC TRANSFORMATION OF BANANA

### A. In Vitro Cell and Protoplast Culture

The beginning of banana in vitro culture occurred in the early 1970s (19); it has grown to an industry, with 20 million micropropagated plants annually, as estimated in 1991 (20) and more than 100 million plants by 1999 (E. Khayat, personal communication, 1999). However, the more advanced in vitro culture techniques such as somatic embryogenesis, embryogenic cell suspension (ECS), and protoplast cultures are less frequently used. At present, three main procedures have been established for somatic embryogenesis in banana, based on vegetative tissue explants such as rhizome fragments or leaf bases (21,22), in vitro proliferating meristems (23), and male flower bud primordia (24,25). These procedures resulted in plant regeneration from diverse cultivars through somatic embryogenesis, though at a different rate. Long-term regenerable ECS cultures were first established from proliferating meristems (26) and more recently from immature male flowers (27,28).

A common bottleneck of these procedures is that they are very time-consuming and labor-intensive. Somatic embryogenesis takes places at 3–6 months at a frequency of less than 5% of the inoculated explants, and ECS cultures are established in another 4–6 months. The presence of endophytic contaminations is another difficulty encountered during the maintenance of ECS cultures. Finally, the prolonged in vitro culture period required to obtain ECS cultures may have adverse effects on phenotypic uniformity and stability of the regenerated plants.

Similarly to that in other monocotyledonous species, the development of ECS cultures opened the way to the isolation of banana protoplasts with high embryogenic potential. Megia and coworkers (29) described protoplast isolation from banana ECS cultures, but only sustained division and callus formation was achieved. High-frequency plant regeneration through direct somatic embryogenesis was first reported for a cooking banana landrace (30). In this study, high plating efficiencies (between 20% and 40%) were obtained when protoplasts were cultured on a feeder layer of embryogenic banana cells or plated alone at a high ($10^6$ ml$^{-1}$) density. Cell suspensions initiated from a related landrace also proved to be a good source for regenerable protoplasts (31).

## B. Genetic Transformation

The development of regenerable ECS and protoplast cultures has made it possible to transfer foreign genes into banana. Successful genetic transformation was first demonstrated in electroporated protoplasts by the transient expression of the *gusA* reporter gene under control of derivatives of the 35S RNA promoter of the cauliflower mosaic virus (CaMV 35S) (32). Later, the first transgenic banana plants were produced by microparticle bombardment of ECS cultures with an in-house-developed particle inflow gun (33,34). Also, *Agrobacterium*-mediated transformation of in vitro meristems has been reported to result in regeneration of transgenic banana plants (35).

These initial studies have optimized the major factors that influence the frequency of transformation. In addition, the transient and stable expression pattern of standard and novel promoters in banana has been investigated. According to these studies, the promoter and first exon and intron of the maize ubiquitin (*ubi1*) gene proved to be the strongest, followed by the rice actin (*act1D*) gene promoter and the recombinant Emu promoter.

More recently, a 1369-bp DNA fragment isolated from a full-length clone of the sugarcane bacilliform badnavirus has been shown to have promoter activity in transient gene expression assays using banana and other monocot (maize, millet, sorghum) and dicot (canola, sunflower, and tobacco) plant species. Experiments with transgenic banana and tobacco plants have shown that this promoter could drive high expression of the *gusA* reporter gene in most cell types of vegetative tissues both in vitro and in the greenhouse, indicating a near-constitutive manner of expression. The expression levels were variable among different transgenic lines but were generally comparable with the activities of both the maize *ubi1* promoter and an enhanced CaMV 35S promoter. Thus, the promoter from sugarcane bacilliform badnavirus represents a useful tool for the high-level expression of foreign genes in both monocot and dicot transgenic plants and could be used similarly to the maize *ubi1* and the CaMV 35S promoter (36).

### 1. Microparticle Bombardment of Embryogenic Cell Suspension Cultures

Microparticle bombardment of cultures technology combines microparticle bombardment transformation with a regeneration protocol from suspension cultures containing fast-dividing and embryogenic somatic cells (33). The standard conditions for DNA delivery are the following: 25–50-mg fresh weight cells (4–6 days after subculture), approximately 800-ng plasmid DNA, and 500 μg tungsten (1.1- or 1.7-μm diameter) microparticles per plate, 8-bar helium pressure, 12-cm target distance, and 30-millisecond opening time. After a recovery period of 7 days, the bombarded cells are selected in the presence of 50 mg $l^{-1}$ hygromycin or geneticin for about 2 months until transformed embryos are formed (Fig. 1). According to our experience, the technology is only limited by the capacity for plant regeneration of the available cell cultures. A cell suspension culture with good morphogenic potential regenerates on average two to three transgenic plants per plate under the conditions described. In total, 100 plates can be routinely bombarded in 1 day with our particle inflow gun. Six months later the plants can be established in the greenhouse and grown to maturity. Fig. 2 demonstrates the normal bunch formation and fruit production of a transgenic dessert banana plant.

Up to now, close to 900 independent transgenic lines have been produced in the authors' laboratory from the Horn-type plantain 'Three Hand Planty' and the Cavendish-type dessert banana 'Williams.' These plants were transformed with either of two selectable marker genes conferring resistance to the antibiotics hygromycin (*hpt*) and geneticin (*neo*) and a number of chimeric genes including three different antifungal genes and several genes isolated from banana viruses such as the coat protein gene of banana bract mosaic virus (37). Large-scale molecular characterization demonstrated that the majority of these plants contained the foreign genes, thus indicating the efficiency of the selection scheme. For example, polymerase chain reaction (PCR)

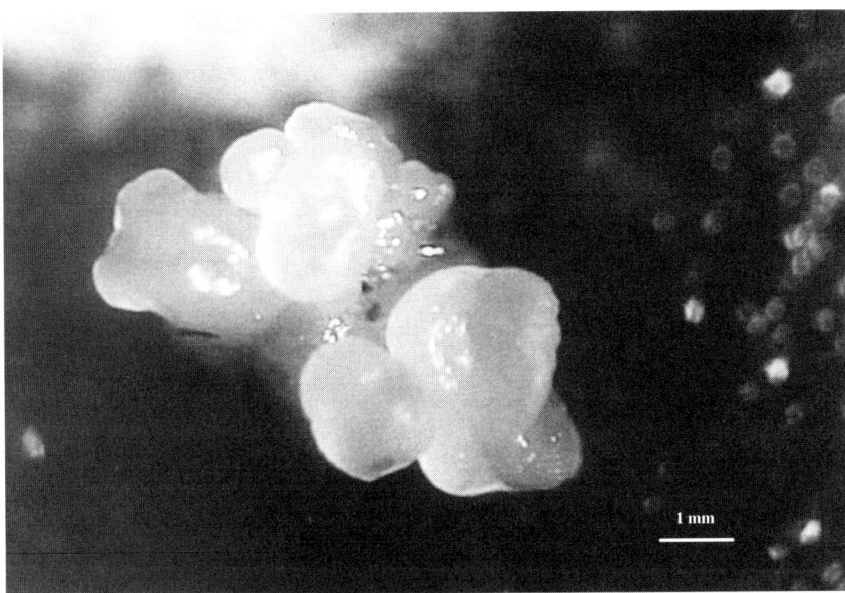

**Figure 1** Transformed embryo of the dessert banana landrace 'Williams' at 2-month selection on geneticin after microparticle bombardment of an embryogenic cell suspension culture. Bar, 1 mm. (Photograph courtesy of L. Sági.)

screening of 776 plants confirmed that more than 90% of them contained the selectable marker gene. There was no major difference in this respect between the two landraces (93% in 'Three Hand Planty' and 98% in 'Williams') and the two genes used for selection (90% and 94% for the *hpt* and the *neo* gene, respectively). Similar results were obtained by genomic Southern analysis of 56 plants, which showed that 89.3% had the selectable marker gene integrated (Remy in preparation).

Since the banana landraces transformed are sterile and propagate vegetatively, analysis of the transgenes in the sexual progeny is excluded, and the integration pattern and expression level in five subsequent asexual generations have been found to be stable (38). However, this containment of the transgenes makes banana very suitable for field tests without the risk of foreign gene escape in the environment.

Durable resistance to multiple pathogens by molecular improvement is likely to be achieved via simultaneous integration of several genes with different targets or modes of action into the plant genome. Such simultaneous gene transfer can be performed by coprecipitating a mixture of chimeric gene constructs onto microparticles prior to bombardment.

PCR screening and Southern hybridization analysis of more than 40 transgenic plants revealed that, as could be predicted, linked genes cointegrated at a high frequency that ranged between 90% and 100% in three independent experiments. Also, as expected, cotransformation with another, unlinked gene resulted in a lower cotransformation frequency than that with linked genes. However, this frequency was still remarkably high, between 70% and 80%, probably as a result of efficient coprecipitation of the DNA molecules on the surface of the microparticles (39). As a result of more complex cotransformation experiments we now have a population of transgenic banana plants containing up to six different genes (S. Remy, I. Deconinck, and L. Sági, unpublished). These observations thus indicate that simultaneous bombardment of different plasmid molecules may be a convenient way to introduce multiple genes into banana.

**Figure 2** Bunch formation and fruit production of a transgenic dessert banana plant in the greenhouse. (Photograph courtesy of W. Dillemans.)

## 2. Agrobacterium-Species–Mediated Transformation

The published method is based on the cocultivation of in vitro meristematic tissues with *Agrobacterium tumefaciens* (35). A major limitation of this technique is that chimeric plants can potentially be regenerated at a high frequency because of the highly differentiated nature of the explants. In addition, we have found that wild-type agrobacteria are not able to induce tumors or swellings on intact plants or cultured tissues (unpublished data).

In order to analyze the possible recalcitrance of banana to *Agrobacterium* sp. systematically we investigated two early steps in the plant-bacterium interaction: chemotaxis to exudates from excised and/or wounded tissues and attachment to cells or tissues. Chemotaxis was studied by a swarm agar plate assay, and the results demonstrated that leaf, rhizome, root, and in vitro proliferating tissues from different banana landraces were all able to induce a positive chemotactic reaction of the *A. tumefaciens* strains EHA101, LBA4404, and Chry5 (40). In addition, massive physical binding of agrobacteria to banana cells isolated from ECS cultures or mesophyll tissue as well as to root hairs or isolated in vitro meristems was demonstrated by light and fluorescent microscopy as well as by scanning electron microscopy, respectively (40). It is concluded that the interaction of *A. tumefaciens* with banana is not impaired in these early steps, and this finding can be a basis for the development of efficient *Agrobacterium* sp.–mediated transformation protocols in this crop. Indeed, more recently we have been able to generate a high number of transgenic plants via *Agrobacterium* sp. cocultivation of plantain ECS cultures followed by selection on geneticin or herbicide (Basta). Screening of 360 transgenic plants has revealed that 98% of them expressed the introduced *gusA* intron construct or the *gfp* reporter gene. The transgenic nature of independent plants has been confirmed by extensive PCR and Southern analysis (Pérez Hernández et al., submitted).

## III. PROSPECTS AND CONCLUSION

### A. Resistance to Pathogens

Pathogenic fungi cause the most devastating diseases in banana and thus present the greatest economic as well as environmental concern for the banana industry. Until genes conferring resistance to fungal pathogens is available in banana (41), heterologous genes encoding proteins with antifungal activity are the primary targets for transgenic expression. Perhaps the most promising candidates are antimicrobial proteins (AMPs) of plant origin (42) as they have exerted high in vitro activity to *Mycosphaerella fijiensis* and *Fusarium oxysporum* f. sp. *cubense* without being toxic to human or banana cells (43).

With our microparticle bombardment technology (33), several hundreds of independent transgenic lines, which express different genes encoding defensin-type AMPs and a nonspecific lipid transfer protein alone or in combinations, have been produced of a plantain and a Cavendish dessert banana landrace. Molecular analysis by PCR, reverse-trancriptase PCR, and Southern and Northern hybridization of these lines confirmed that a vast majority of them contained and expressed the introduced genes (see earlier discussion).

By using specific antibodies, the concentration of two antifungal proteins in total leaf proteins was determined in transgenic plant extracts; it ranged from 0.05% to more than 1%, depending on the promoter used. Addition of such extracts to germinating spores of a field isolate of *M. fijiensis,* the causal agent of black Sigatoka disease, resulted in a significant inhibition of fungal growth, which makes the respective plants interesting candidates for field testing.

Approximately 100 transgenic lines have been micropropagated for testing the stability and effect of the transgenes in different field environments. In addition, close to 200 individual plants have been transferred to the greenhouse, where 50 of them were evaluated by a simple and reproducible leaf disk bioassay for fungus tolerance. A differential disease response was observed among independent transgenic plants, which ranged from no resistance to a high degree of tolerance compared to that of the control plants. Computer image capturing and software based area calculation allowed for the precise measurement of the infected leaf area and for the classification of independent transformants according to their tolerance (44).

To date, no transgenic bananas have yet entered the phase of extensive field trials, mainly as a result of the lack of biosafety guidelines and/or functional regulatory bodies in many tropical countries where banana is grown. This situation is a major bottleneck to progress in transgenic field testing of tropical crops, especially in sub-Saharan Africa.

Banana transformation also provides a tool to engineer resistance against parasitic nematodes. In principle, three strategies can be applied: (a) expression of toxic proteins for direct effects on nematodes, (b) expression of nematode resistance genes, and (c) expression of proteins that compromise the nematode-plant interaction such as plantibodies against essential proteins secreted by the nematodes. At present, the first strategy appears to be the most perspective against the migratory endoparasitic nematodes of banana because resistance genes to this group of nematodes have not even been studied and the migratory nematode-plant interaction is not well understood. Currently, we are exploring an approach based on the regulated expression of barnase, a cytotoxic ribonuclease (RNase), under the control of a potato *gst1* gene promoter that was shown to be inducible by infection with sedentary nematodes (45).

### B. Fruit Quality Control

Since bananas are climacteric fruits and their ripening can be induced by ethylene, genetic modification of ethylene biosynthesis is a logical strategy to produce fruit with an extended green life both for the customer as well as for shipment. Ethylene reduction may also decrease the detrimental biochemical reactions caused by fungal pathogens or mechanical wounding.

In the late 1990s, significant advances were made in understanding the changes in gene expression that are associated with banana fruit ripening (46,47). Both papers have simultaneously reported the isolation and identification of transcripts that are differentially expressed during fruit ripening. Since then, efforts have been directed to the isolation and characterization of the regulatory elements of these genes, which may be useful for fruit-specific expression of foreign proteins in transgenic bananas.

In addition, an abundant 31-kDa protein from the pulp (the edible part) of banana fruit has also been characterized and turned out to be highly homologous with previously described chitinases (48). The protein was shown to be highly pulp-specific and preferentially degraded during fruit ripening. It has been hypothesized that the physiological role of this protein may be that of a storage protein rather than that of plant protection.

### C. Potential for Pharmaceutical Applications

A potential application of banana plants engineered for fruit-targeted transgene expression can be molecular farming for the production of recombinant proteins of pharmaceutical importance. Of these proteins, the most interesting candidates could be vaccines, provided that they remain active after oral administration.

The first of the candidate vaccines was the hepatitis B surface antigen (HBsAg), which was produced in transgenic tobacco plants (49) and retained immunogenic properties after injection in mice (50). Further investigations with other candidate vaccines indicated that they may invoke an oral immune response in mice after feeding on transgenic potato tubers (51). Such oral vaccines may be used against diarrheal diseases that occur very frequently in the tropics (52). Transgenic bananas expressing such vaccines may be useful complements to traditional vaccination programs.

### ACKNOWLEDGMENTS

Financial support from the INIBAP ProMusa program and BADC (Belgian Administration for Development Cooperation) is acknowledged. The authors thank Ilse Deconinck and Greet Schoofs for technical assistance.

### REFERENCES

1. FAO Statistical Databases, 1999. Available at http://www.fao.org
2. USDA Nutrient Database, 1999. Available at http://www.nal.usda.gov/fnic/cgi-bin/nut_search.pl 1999.

3. T Horigome, E Sakaguchi, C Kishimoto. Hypocholesterolaemic effect of banana (*Musa sapientum* L. var. Cavendishii) pulp in the rat fed on a cholesterol-containing diet. Br J Nutr 68:231–244, 1992.
4. H Deneo-Pellegrini, E de Stefani, A Ronco. Vegetables, fruits, and risk of colorectal cancer: A case-control study from Uruguay. Nutr Cancer 25:297–303, 1996.
5. RH Stover. Effet du *Cercospora* noir sur les plantains en Amérique Centrale. Fruits 38:326–329, 1983.
6. KN Mobambo, F Gauhl, D Vuylsteke, R Ortiz, C Pasberg-Gauhl, R Swennen. Yield loss in plantain from black sigatoka leaf spot and field performance of resistant hybrids. Field Crops Res 35:35–42, 1993.
7. D Vuylsteke, R Swennen, R Ortiz. Registration of 14 improved tropical *Musa* plantain hybrids with black sigatoka resistance. Hortscience 28:957–959, 1993.
8. PR Rowe. Banana plant 'FHIA-01.' U.S. Patent No. PP8,983, 1994.
9. NW Simmonds. Bananas. 2nd ed. London: Longmans, 1966, p 512.
10. E-J Sun, H-J Su, WH Ko. Identification of *Fusarium oxysporum* f.sp. *cubense* race 4 from soil or host tissue by cultural characters. Phytopathology 68:1672–1673, 1978.
11. RC Ploetz. Fusarium wilt of banana. St. Paul, MI: APS Press, 1990, p 140.
12. JE Thomas, MK Smith, AF Kessling, SD Hamill. Inconsistent transmission of banana bunchy top virus in micropropagated bananas and its implication for germplasm screening. Aust J Agric Res 46:663–671, 1995.
13. BEL Lockhart. Purification and serology of a bacilliform virus associated with banana streak disease. Phytopathology 76:995–999, 1986.
14. G Harper, JO Osuji, JS Heslop-Harrison, R Hull. Integration of banana streak badnavirus into the *Musa* genome: Molecular and cytogenetic evidence. Virology 255:207–213, 1999.
15. T Ndowora, G Dahal, D LaFleur, G Harper, R Hull, NE Olszewski, B Lockhart. Evidence that badnavirus infection in *Musa* can originate from integrated pararetroviral sequences. Virology 255:214–220, 1999.
16. Z Singh, RAC Jones, MGK Jones. Identification of cucumber mosaic virus subgroup I isolates from banana plants affected by infectious chlorosis disease using RT-PCR. Plant Dis 79:713–716, 1995.
17. S Gowen, P Quénéhervé. Nematode parasites of bananas, plantains and abaca. In: M Luc, RA Sikora, J Bridge, eds. Plant Parasitic Nematodes in Subtropical and Tropical Agriculture. Wallingford: CAB International, 1990, pp 431–460.
18. HE Ostmark. Economic insect pests of bananas. Annu Rev Entomol 19:161–176, 1974.
19. SS Ma, CT Shii. In vitro formation of adventitious buds in banana shoot apex following decapitation. J Chin Soc Hortic Sci 18:135–142, 1972.
20. Y Israeli, E Lahav, O Reuveni. In vitro culture of banana. In: S Gowen, ed. Bananas and Plantains. London: Chapman & Hall, 1995, pp 147–178.
21. FJ Novak, R Afza, M Van Duren, M Perea-Dallos, BV Conger, X Tang. Somatic embryogenesis and plant regeneration in suspension cultures of dessert (AA and AAA) and cooking (ABB) bananas (*Musa* spp.). Biotechnology 7:154–159, 1989.
22. KS Lee, FJ Zapata-Arias, H Brunner, R Afza. Histology of somatic embryo initiation and organogenesis from rhizome explants of *Musa* spp. Plant Cell Tissue Organ Cult 51:1–8, 1997.
23. D Dhed'a, F Dumortier, B Panis, D Vuylsteke, E De Langhe. Plant regeneration in cell suspension cultures of the cooking banana cv. 'Bluggoe' (*Musa* spp. ABB group). Fruits 46:125–135, 1991.
24. JV Escalant, C Teisson, F Cote. Amplified somatic embryogenesis from male flowers of triploid banana and plantain cultivars (*Musa* spp.). In Vitro Cell Dev Biol 30P:181–186, 1994.
25. C Navarro, RM Escobedo, A. Mayo. In vitro plant regeneration from embryogenic cultures of a diploid and a triploid, Cavendish banana. Plant Cell Tissue Organ Cult 51:17–25, 1997.
26. D Dhed'a. Culture de suspensions cellulaires embryogéniques et régénération en plantules par embryogénèse somatique chez le bananier et le bananier plantain (*Musa* spp.). PhD dissertation, Katholieke Universiteit Leuven, 1992.
27. FX Cote, R Domergue, S Monmarson, J Schwendiman, C Teisson, JV Escalant. Embryogenic cell suspensions from the male flower of *Musa* AAA cv. Grand nain. Physiol Planta 97:285–290, 1996.
28. A Grapin, J Schwendiman, C Teisson. Somatic embryogenesis in plantain banana. In Vitro Cell Dev Biol 32P:66–71, 1996.

29. R Megia, R Haicour, L Rossignol, D Sihachakr. Callus formation from cultured protoplasts of banana (*Musa* sp.). Plant Sci 85:91–98, 1992.
30. B Panis, A Van Wauwe, R Swennen. Plant regeneration through somatic embryogenesis from protoplasts of banana (*Musa* spp.). Plant Cell Rep 12:403–407, 1993.
31. R Megia, R Haicour, S Tizroutine, V Bui Trang, L Rossignol, D Sihachakr, J Schwendiman. Plant regeneration from cultured protoplasts of the cooking banana cv. Bluggoe (*Musa* spp., ABB group). Plant Cell Rep 13:41–44, 1993.
32. L Sági, S Remy, B Panis, R Swennen, G Volckaert. Transient gene expression in electroporated banana (*Musa* spp., cv. 'Bluggoe', ABB group) protoplasts isolated from regenerable embryogenic cell suspensions. Plant Cell Rep 13:262–266, 1994.
33. L Sági, B Panis, S Remy, H Schoofs, K De Smet, R Swennen, BPA Cammue. Genetic transformation of banana and plantain (*Musa* spp.) via particle bombardment. Biotechnology 13:481–485, 1995.
34. L Sági, S Remy, B Verelst, R Swennen, B Panis. Stable and transient genetic transformation of banana (*Musa* spp.) protoplasts and cells. In: YPS Bajaj, ed. Biotechnology in Agriculture and Forestry. Vol. 34. Plant Protoplasts and Genetic Engineering VI. Berlin, Heidelberg: Springer-Verlag, 1995, pp 214–227.
35. GD May, R Afza, HS Mason, A Wiecko, FJ Novak, CJ Arntzen. Generation of transgenic banana (*Musa acuminata*) plants via *Agrobacterium*-mediated transformation. Biotechnology 13:486–492, 1995.
36. PM Schenk, L Sági, T Remans, RG Dietzgen, MJ Bernard, MW Graham, JM Manners. A promoter from sugarcane bacilliform badnavirus drives transgene expression in banana and other monocot and dicot plants. Plant Mol Biol 36:1221–1230, 1999.
37. MF Bateson, JL Dale. Banana bract mosaic virus: Characterisation using potyvirus specific degenerate PCR primers. Arch Virol 140:515–527, 1995.
38. R Swennen, I Van den houwe, S Remy, L Sági, H Schoofs. Biotechnological approaches for the improvement of Cavendish bananas. Acta Hortic 490:415–423, 1998.
39. S Remy, I François, BPA Cammue, R Swennen, L Sági. Co-transformation as a potential tool to create multiple and durable resistance in banana (*Musa* spp.). Acta Hortic 461:361–365, 1998.
40. JB Pérez Hernández, S Remy, V Galán Saúco, R Swennen, L Sági. Chemotactic movement and attachment of *Agrobacterium tumefaciens* to banana cells and tissues. J Plant Physiol 155:245–250, 1999.
41. I Wiame, R Swennen, L Sági. Towards PCR-based cloning of candidate disease resistance genes from banana (*Musa acuminata*). Acta Hortic 521:51–58, 2000.
42. WF Broekaert, BPA Cammue, M De Bolle, K Thevissen, G De Samblanx, RW Osborn. Antimicrobial peptides from plants. Crit Rev Plant Sci 16:297–323, 1997.
43. BPA Cammue, MFC De Bolle, FRG Terras, WF Broekaert. Fungal disease control in *Musa:* Application of new antifungal proteins. Proceedings of International Symposium on Genetic Improvement of Bananas for Resistance to Diseases and Pests, Montpellier, 1992, pp 221–225.
44. S Remy, I Deconinck, R Swennen, L Sági. Assessment of fungus tolerance in transgenic banana using a leaf disc assay. Proceedings of International Symposium on the Molecular and Cellular Biology of Banana, Ithaca, NY, 1999, p 38.
45. G Strittmatter, G Gheysen, V Gianinazzi-Person, K Hahn, A Niebel, W Rohde, E Tacke. Infections with various types of organisms stimulate transcription from a short promoter fragment of the potato *gst1* gene. Mol Plant Microbe Interact 9:1–4, 1996.
46. SK Clendennen, GD May. Differential gene expression in ripening banana fruit. Plant Physiology 115:463–469, 1997.
47. R Medina-Suárez, K Manning, J Fletcher, J Aked, CR Bird, GB Seymour. Gene expression in the pulp of ripening bananas. Plant Physiol 115:453–461, 1997.
48. SK Clendennen, R López-Gómez, M Gómez-Lim, CJ Arntzen, GD May. The abundant 31-kilodalton banana pulp protein is homologous to class-III acidic chitinases. Phytochemistry 47:613–619, 1998.
49. HS Mason, DMK Lam, CJ Arntzen. Expression of hepatitis B surface antigen in transgenic plants. Proc Natl Acad Sci USA 89:11745–11749, 1992.

50. Y Thanavala, YF Yang, P Lyons, HS Mason, C Arntzen. Immunogenicity of transgenic plant–derived hepatitis B surface antigen. Proc Natl Acad Sci USA 92:3358–3361, 1995.
51. TA Taq, HS Mason, JD Clements, CJ Arntzen. Oral immunization with a recombinant bacterial antigen produced in transgenic plants. Science 268:714–716, 1995.
52. L Richter, HS Mason, CJ Arntzen. Transgenic plants created for oral immunization against diarrheal diseases. J Travel Med 3:52–56, 1996.

# 25
# Production of Transgenic Melon

**Ekaterini Papadopoulou and Rebecca Grumet**
*Michigan State University, East Lansing, Michigan*

| | | |
|---|---|---|
| I. | INTRODUCTION | 371 |
| II. | OBJECTIVES FOR MELON IMPROVEMENT | 372 |
| III. | REGENERATION AND TRANSFORMATION TECHNIQUES FOR MELON | 373 |
| | A. Regeneration of Melon | 373 |
| | B. Melon Transformation | 376 |
| IV. | CONCLUSION | 378 |
| | REFERENCES | 378 |

## I. INTRODUCTION

The melon species *Cucumis melo* L. includes a group of high-value crops such as muskmelon, cantaloupe, honeydew, casaba, and winter melon that are cultivated throughout the world (1). The mature melon fruits are appreciated as summer dessert fruit, and the immature fruits may be consumed as vegetables. The *C. melo* species has been subdivided into cultivar groups or botanical varieties by several authors, including Whitaker and Davis (2) and Robinson and Decker-Walters (3). Robinson and Decker-Walters (3) use the following groupings: *cantalupensis* (cantaloupe and muskmelon), *inodorus* (winter melon), *flexuosus* (snake melon), *conomon* (pickling melon), *dudaim* (pomegranate melon), and *momordica* (phoot, snap melon). The dessert melons, which are the most commercially important types, generally belong to the *cantalupensis* and *inodorus* groups. Members of the *cantalupensis* group include cantaloupe and Charentais-type melons; honeydew is the most popular member of the *inodorus* group, which also includes casaba, canary, and crenshaw melons. In other classifications (e.g., see Ref. 2), the muskmelon types are separated from the cantaloupes. The muskmelons, which are classified as var. *reticulatus*, are the most commercially important type in North America, whereas the cantaloupes (var. *cantaloupensis*) are grown mostly in Europe.

Since the 1990s, melon has been the target of transformation research. After initial development of transformation systems using marker genes in the early 1990s (e.g., 4–6), melon has been transformed with a variety of potentially economically useful traits such as virus resistance, ripening characteristics, salt tolerance, and altered protein composition (e.g., 7–11). Work with transgenic melons has moved beyond the laboratory and in several cases is nearing commercialization. The seed industry has been very active in this area; in the 1997–1999 seasons, there were more than 50 applications from at least five companies for field trials performed with transgenic melons in the United States (Table 1). In this chapter we describe the development and uses of melon transformation technology. Melon biotechnology also has been reviewed by Guis and coworkers (12).

## II. OBJECTIVES FOR MELON IMPROVEMENT

Integration of selected genetic traits into plants can be a powerful tool for crop improvement. In the past two decades plant transformation studies have focused on the development of practical, reliable, and efficient systems for the genetic transformation of a wide variety of plant species by introducing and expressing genes conferring desired characteristics. Extensive effort has been focused on major agronomic crops, e.g., soybean, corn, and cotton, which are currently being produced commercially on large acreages in the United States, but several horticultural crops, including cucurbit crops, also have received attention (13).

Objectives for melon improvement include increased yield per plant, more and heavier fruits, early harvest, high flesh proportion, improved fruit quality (primarily sucrose content), and increased postharvest life (14–16). Resistance to environmental stresses, mainly drought and salinity, is of special interest in the arid areas of cultivation, and as is true for most crops, reduction of production losses resulting from susceptibility to various insects and diseases is of high priority throughout the world. Important fungal diseases include powdery and downy mildew

**Table 1** Published Examples and Recent Field Tests of Transgenic Melons in the United States

| Published examples | | |
| --- | --- | --- |
| Traits | Genotype | Reference |
| Marker genes | Hale's Best Jumbo | 4 |
| | Orient Sweet | 5 |
| | Prince, EG360, Ajinomoto, Sunday Aki | 6 |
| | Amarillo Oro | 56 |
| | Sunday Aki | 19 |
| ZYMV CP[a] | Hale's Best Jumbo | 7 |
| CMV CP[a] | Burpee Hybrid, Hale's Best Jumbo, Topmark | 6,8 |
| CMV, ZYMV, WMV CP[a] | CZW-30 | 11 |
| Antisense ACO[b] | Vedrantais | 9 |
| Yeast salt tolerance | Pharo | 10 |
| U.S. field tests 1997–1999[c] | | |
| Trait | No. notifications | |
| Virus resistance | 39 | |
| Altered ripening | 14 | |
| Altered protein | 1 | |

[a]ZYMV, CMV, WMV CP, coat proteins of zucchini yellow mosaic virus, cucumber mosaic virus, and watermelon mosaic virus.
[b]ACO, 1-amino-cyclopropane carboxylic acid oxidase (ethylene biosynthetic enzyme).
[c]*Source*: USDA-APHIS records.

(*Sphaerotheca fuliginea* and *Pseudoperonospora cubensis*, respectively) and fusarium wilt (*Fusarium oxysporum* f.sp. *melonis* and *F. solani*). Severe yield losses also can be incurred by viral infections, such as the whitefly-transmitted geminivirus, squash leaf curl virus, and aphid-transmitted viruses, including cucumber mosaic virus and several potyviruses, e.g., zucchini yellow mosaic virus, watermelon mosaic virus, and the watermelon strain of papaya ringspot virus (17). Although the wild *Cucumis* species offer genes for resistance to a variety of pests and pathogens, interspecific transfer by conventional methods is inhibited due to sexual incompatibility (16).

Other goals for melon improvement include facilitation of breeding processes via development of stable monoecious or gynoecious genotypes (18). Availability of such breeding lines would reduce the cost of emasculation of the female parents currently required for production of hybrid seed. It also has been suggested by Shetty and colleagues (19) that melon fruits may be used for production of proteins such as thaumatin, lysozyme, proteases for meat tenderization, or chymosin for cheese making.

## III. REGENERATION AND TRANSFORMATION TECHNIQUES FOR MELON

Fundamental requirements for production of transgenic plants include plant tissue competent for transformation, a method for stable introduction of the novel genetic material, and, in the majority of the currently used transformation systems, a crop-specific procedure to regenerate whole plants from individually transformed cells. Since it is the regeneration procedure that is often the limiting factor in establishing efficient transformation systems (20), we will first discuss melon regeneration. Ideally, the regeneration procedure should be highly efficient, include minimal time in culture both to increase efficiency and to minimize somaclonal variation, and be applicable to a wide range of genotypes. Similarly, the goals for transformation include high efficiency, ease of selection, and stable, adequate gene expression.

### A. Regeneration of Melon

#### 1. Explant Source

Melon plants have been regenerated from tissue culture by using a variety of methods including organogenesis, embryogenesis, and regeneration from protoplasts (Table 2). Regeneration via organogenesis has been most efficient and has been achieved for explants from cotyledons, hypocotyls, leaves, and roots. During regeneration from cotyledons, the area close to the basal cut edge of the explant appears to be the most active site for initiation and development of adventitious shoot buds (8). A study of buds regenerated from cv. Galia revealed that the regeneration process begins with cell division in the epidermal layer (21). Subsequently, structures from multiple subepidermal cell layers are formed; the first shoot buds can be observed after 15 days in culture, and the first shoot 1 week later.

All of the published melon transformation systems have utilized organogenesis rather than embryogenesis-based methods, and the majority have used cotyledon-derived systems (Table 1). Although cotyledons are the most routinely responsive and are the preferred explant source for transformation in our lab, we also have been successful using leaf explants (22); the greater amount of leaf tissue per plant provides an advantage in cases in which seed supply is limited. Factors that can influence the responsiveness of cotyledon or leaf explants to regeneration include genotype, age of the donor plant, culture conditions (e.g., temperature, light), and presence of growth regulators, which are discussed in the following sections.

**Table 2** Melon Regeneration Systems

| Explant | Genotype (cv.) | Growth regulators | Reference |
|---|---|---|---|
| **Organogenesis** | | | |
| Cotyledon | Amarillo Oro | IAA, kinetin | 70 |
| | Cantaloup Charentais T, DM17187, Ogon no. 9, Doublon, Piboule, | IAA, kinetin | 47 |
| | Hale's Best Jumbo | IAA, BA, ABA | 29 |
| | Accent, Galia, Preco, Viva, 4215 (Diamex) | BA, IAA | 40 |
| | Topmark | BA, IAA | 71 |
| | Earl's Favorite Harukei No. 3 | BA, IAA, 2,4-D, NAA | 24 |
| | Pusa Madhuras | BAP | 72 |
| | Five inbred lines | $AgNO_3$ | 35 |
| | Sunday Aki | BA, proline, SA, aspirin | 73 |
| | Galia | IAA, BA, TDZ, PC | 31 |
| | Galia | BA, ancymidol | 28 |
| Hypocotyl | Pusa Sharbati | BAP, $ABA/GA_3$ | 74 |
| | Earl's Favorite Harukei No. 3 | BA, IAA, 2,4-D, NAA | 24 |
| Epicotyl | Pusa Madhuras | BAP, IAA, kinetin, $GA_3$ | 72 |
| Petiole | PMR (cantaloupe) | NAA, BA, zeatin dihydroside | 75 |
| | Earl's Favorite Harukei No. 3 | BA, IAA, 2,4-D, NAA | 24 |
| Leaf | Pusa Sharbati | BAP, 2iP | 46 |
| | Accent, Galia, Preco, Viva, 4215 | BA | 40 |
| | PMR 3680 (cantaloupe) | NAA, BA | 75 |
| | Hale's Best Jumbo, Ananas El Dokki | IAA, BA, $AgNO_3$, sulfonylurea | 36 |
| Root | Pusa Sharbati | BAP | 48 |
| **Somatic embryogenesis** | | | |
| Cotyledons | Sunday Akigata | 2,4-D, BA | 76 |
| | Preco, Charentais T | 2,4-D, BAP | 41 |
| (quiescent) | 51 commercial cv., Male Sterile A147 (breeding line) | 2,4-D, BA, kinetin, TDZ | 42 |
| | 14 commercial cv. | 2,4-D, kinetin | 44 |
| | Vedrantais | 2,3-D, BAP | 77 |
| Hypocotyl | Amarillo Oro | IAA | 70 |
| | Earl's Favorite Harukei No. 3 | IAA | 24 |
| Petioles | Earl's Favorite Harukei No. 3 | IAA | 24 |
| Leaves | Preco, Charentais T | 2,4-D, BAP | 41 |
| | Earl's Favorite Harukei No. 3 | IAA | 24 |
| **Regeneration from protoplasts** | | | |
| Cotyledon | Cantaloup Charentais | IAA, kinetin | 78 |
| (seed cotyledon) | Earl's Favorite Harukei No. 3 | 2,4-D, IAA, NAA | 24 |
| | Preco, Charentais T | 2,4-D, BAP | 41 |
| Hypocotyl | Amarillo Oro | IAA, kinetin | 70 |
| | Earl's Favorite Harukei No. 3 | 2,4-D, IAA, NAA | 24 |
| Leaves, petioles | Earl's Favorite Harukei No. 3 | 2,4-D, IAA, NAA | 24 |

ABA, abscisic acid; BA, benzyladenine; BAP, 6-benzylaminopurine; 2,3-D, 2,3-dichlorophenoxyacetic acid; 2,4-D, 2,4-dichlorophenoxyacetic acid; 2iP 6-(γ, γ-dimethylallylamino)-purine; NAA, α-naphthalene acetic acid; TDZ, thidiazuron; IAA, indole-3-acetic acid; $GA_3$ gibberellic acid; PC, paclobutrazol.

## 2. Media Components

As in many regeneration systems, auxin and cytokinin types and concentrations are critical factors (23). Low auxin concentrations (0–1 mg/l indoleacetic acid [IAA]) favored shoot induction from cotyledons, hypocotyls, leaves, and petioles, whereas higher concentrations (20–100 mg/l) resulted in somatic embryo formation (24). The balance between auxin(s) and cytokinin(s) also appears to be important. In general, for a given concentration of auxin, increasing cytokinin concentration increased organogenesis (25). The presence of cytokinins (benzylamino purine or benzyladenine) during in vitro culture of cotyledons caused mobilization and metabolism of storage proteins and stimulated shoot regeneration of cv. Galia explants, but only in the light; cytokinin alone was not sufficient to induce regeneration or to support growth responses in the dark (26–28).

Other phytohormones also have been found to influence organogenesis from melon explants. Abscisic acid (ABA) has been shown to enhance shoot regeneration (29), and gibberellin (GA), which can be antagonistic to ABA action (30), appeared to influence regeneration efficiency negatively (28,29). Addition of the antigibberellin agent ancymidol increased organogenesis of cotyledon explants; exogenous GA counteracted the effect of both ancymidol and BA and reduced the regeneration rate (28). Likewise, the growth regulators thidiazuron and paclobutrazol (anti-GA activity) appeared to mimic the effect of cytokinin and auxin, respectively, on organogenesis (31).

Ethylene also can negatively affect regeneration in many tissue culture systems, and silver ion, which is an inhibitor of ethylene action (32), can enhance regeneration in various systems, including melon (e.g., 33–36). Interestingly, the role of ethylene in regeneration competence of melon explants was examined in plants expressing an antisense *1-aminocyclopropane-1-carboxylate-oxidase (ACO)* gene (37). Normal ethylene biosynthesis is impaired in these plants as a result of reduction in the ACO enzyme that catalyzes the last step of ethylene biosynthesis, the conversion of 1-aminocyclopropane-1-carboxylate (ACC) to ethylene (38). Consistent with observations of the effect of including silver nitrate in the medium, explants derived from transgenic antisense *ACO* plants exhibited higher rates of shoot regeneration (37).

## 3. Explant Factors

The morphogenic response of explants during tissue culture is also greatly dependent on the genotype and physiological characteristics of the tissue used. A wide variation in regeneration efficiency has been observed, depending on the melon cultivar (29,39–44). In many cases, even though the explants generate a large number of buds, only a small portion develop into shoots; the shoot production percentage varied with different cultivars (8,29,39). In other cases, Gaba and coworkers (21) observed formation of leaves, or leaflike structures, that are not always accompanied by shoot apices, possibly because of genetic effects. Ficcadenti and Rotino (45) found that genotypes of *C. melo* var. *inodorous* were more consistently regenerative than those of var. *reticulatus*.

The physiological characteristics of the explant source also are critical. Cotyledon and leaf age and size have been found to influence regeneration (36,46–48). Although in general, younger tissue is more responsive, for leaf regeneration, it was possible to be too young (less than 3-cm diameter) for optimal regeneration rates (36). Interestingly, the donor plant environment also was found to be an important factor; leaves harvested from plants grown in pots in the greenhouse or growth chamber were much more responsive than leaves derived from plants grown in vitro in Magenta boxes (36). Both the external environment (growth chamber vs. culture room) and the internal environment (pot vs. Magenta box) had a highly significant effect on regeneration rates, suggesting that light, temperature, and gas exchange conditions all contribute to the responsiveness of the explants.

## 4. Somaclonal Variation

Cytological analyses have indicated that regenerated melon plants exhibit a range of ploidy levels. Tetraploidy was most common and has been observed with regeneration via somatic embryogenesis and organogenesis, and from both cotyledon and leaf explants (8,36,49–53). Tetraploidy can be a problem for future production of transgenic lines as tetraploids can exhibit reduced fruit quality and fertility (52,54). The explant source may contribute to the frequency of tetraploidy as different tissues vary in their natural proportion of tetraploid cells; for example, depending on the stage of development, melon cotyledons can have up to 60% tetraploid cells (12). Organogenesis from 2-day-old cotyledons resulted in 81% tetraploids, whereas regeneration from young leaves only gave 15% tetraploids. In other cases, however, possibly because of differences in the genotypes tested, high percentages of tetraploids also have been reported for plants regenerated from leaf tissue. Yadav and associates (36) reported that 47% of the tested regenerants were tetraploid, and Kathal and coworkers (51) observed that the percentage of polyploid regenerants from leaf tissue increased as the time of in vitro culture increased. Polyploidy ranged from 50% after 30 days in culture to 90% after 90 days in culture. Thus, as has been observed in numerous systems (55), minimal time in culture is important, both for efficiency of the process and for reduction of somaclonal variation.

## B. Melon Transformation

### 1. Transformation Technology

Transgenic melons primarily have been produced via *Agrobacterium tumefaciens*–mediated transformation (Table 1), although particle bombardment also has been used successfully in at least one case (8). Factors influencing success with *Agrobacterium*-mediated transformation include *Agrobacterium* concentration, inoculation duration and temperature, and cocultivation period and addition of virulence activating factors (4,5,19,56). Transformation has been successfully reported for a variety of North American, European, Middle Eastern, and Asian genotypes.

Selection strategies have generally relied on incorporation of the *neomycin phosphotransferase (NPTII)* gene conferring kanamycin resistance, although selection for methyltrexate resistance also has been used successfully (4,5). Escapes from selection are observed frequently, however, so secondary selection of regenerated shoots on liquid medium or on rooting medium has been valuable to reduce the number of individuals that must be subsequently screened by molecular analyses (4,5,7). In our hands, further screening of putatively transgenic plantlets by enzyme-linked immunosorbent assay (NPT-ELISA) has proved effective to eliminate additional escapes after secondary selection. More than 90% of the NPT-positive individuals subsequently proved positive by polymerase chain reaction (PCR) and Southern analyses.

Segregation analyses of $R_1$ seedlings showed that the introduced genes are inherited in a Mendelian fashion (e.g., (4,5)). Interestingly, however, Dong and coworkers (5) observed that the expression pattern of the introduced reporter genes (*dihydrofolate reductase, DHFR*, and *β-glucuronidase, GUS*) regulated by the cauliflower mosaic virus 35S promoter was not uniform and was modified during plant development. As the organs matured, activity of the reporter genes became limited to vascular tissue. In some cases, transgene silencing has been observed in transgenic melon (57). In screening a population of 4600 hybrid progeny (nontransgenic × homozygous transgenic $R_2$) produced in the field, several individuals were recovered that possessed the *NPTII* transgene but did not produce NPT protein as detected by NPT-ELISA. Thus, gene silencing, which has been observed in numerous systems (58,59), is another factor to consider in developing high-quality transgenic melon lines expressing the introduced trait.

## 2. Introduced Traits

The genetic characters introduced into *C. melo* plants, to date, include genes to confer disease and stress resistance and modify storage life of the fruits. Viral coat protein genes of zucchini yellow mosaic virus and cucumber mosaic virus have been shown to confer resistance to their respective viruses successfully (7,8). Especially promising is the ability to engineer multiple-virus resistance. Transgenic cantaloupe expressing coat protein genes of cucumber mosaic virus, watermelon mosaic virus, and zucchini yellow mosaic virus exhibited resistance to all three viruses in field trials (11). Since similar genes have been used to produce commercial multiple-virus-resistant transgenic squash cultivars (60,61), commercial transgenic melons are likely to follow soon. Since the late 1990s there has been extensive field testing of virus-resistant melons by industry (Table 1).

Another trait that has reached the point of field testing by industry is altered ripening, leading to prolonged postharvest life. Transgenic melons have been produced with the gene encoding *S*-adenosyl methione hydroxylase (SAM hydroxylase) (USDA-APHIS records). SAM hydroxylase can reduce ethylene accumulation by reducing levels of SAM, the immediate precursor of ACC in the ethylene biosynthetic pathway (38). Transgenic melons also have been produced that express an antisense *ACC oxidase (ACO)* gene (9). These plants exhibited reduced ethylene biosynthesis and had inhibited fruit ripening on the vine. The abscission zone did not develop, so the fruits remained attached to the plants and accumulated higher amounts of sugar. Quality characteristics of the fruit, such as coloration of the rind and flesh and acidity, were not affected by the lower ethylene production (15); however, there was a significant reduction in total volatile composition (62). Exogenous ethylene restored the effect of the transgene and fruits matured normally, suggesting increased potential to regulate ripening as needed during the marketing process (9,15). The transgenic melons showed greatly enhanced shelf life; wild-type melons rotted within 2 weeks of harvest while the transgenics retained high fruit quality (12).

A third area of interest for melon transformation has been stress tolerance. In 1997 cotyledon and leaf explants of melon cv. Pharo were transformed with the *HAL1* gene, which confers salt tolerance in yeast (10). In vitro cultured transgenic plants expressing the *HAL1* gene had higher tolerance to NaCl but exhibited slower growth.

## 3. Risk Assessment

Environmental release and commercial production of transgenic crops have raised public concerns regarding safety. One frequently mentioned concern is potential escape of transgenes into populations of wild relatives with resultant potential perturbation of natural habitats (20,63–65).

Field trials with transgenic melons have addressed some of these questions (57). Experiments were designed to compare pollen-mediated movement of native and engineered genes and to assess whether gene movement could be contained through the use of trap border plantings. When movement of the engineered kanamycin resistance gene (*NPTII*) and that of a native marker gene (green cotyledon) were compared, there was no case of movement of the transgene in the absence of movement of the associated morphological marker gene (57). To our knowledge, this was the first direct test comparing movement of native and engineered genes. Although these results would be predicted for genes that have been stably integrated into the genome, public concern about the nature of transgenes makes direct verification of this assumption desirable.

A second question was whether gene movement could be contained through the use of trap border plantings. Hokanson and coworkers (57,66) indicated that although presence of borders could significantly reduce the amount of gene movement out of an enclosed field, it would not be possible to prevent movement by this method. Even excessive ratios of trap plants to transgenic donors did not completely prevent escape in all cases. Thus, even in small plots, let alone agricultural fields, transgenes are likely to escape. Potential gene escape from agricultural fields, however, is not an event unique to transgenes (67–69); the recurring question is, Is it the method of gene introduction per se that should be of concern, or the specific crop and gene in question? Currently USDA-APHIS make evaluations on a case by case basis by asking, Are there wild, interfertile populations in the area of release, and if so, is the particular gene likely to be of ecological significance, i.e., persist in the wild population and confer a selective advantage? Under those conditions further analysis is warranted.

## IV. CONCLUSION

In summary, melons have been the subject of numerous biotechnology efforts during the past decade. Although increased transformation efficiency would be desirable, several public and private labs throughout the world have successfully produced transgenic melons with a variety of introduced genes. The initial focus has been primarily in the area of virus resistance and altered ripening, but as more potentially useful genes become available, we are likely to see the range of traits increase over the next several years. In addition to success in developing methodology and introducing economically valuable traits, transgenic melons have played a role in examining broader ecological issues associated with the deployment of genetically engineered crops.

## ACKNOWLEDGMENTS

We thank Drs. Ken Sink and Dave Douches for helpful reviews of the manuscript. This work was in part supported by USDA-NRI grant no. 69775 and by the Michigan Agriculture Experiment Station.

## REFERENCES

1. N Nayar, R Singh. Taxonomy, distribution and ethnobotanical uses. In: NM Nayar, TA More, eds. Cucurbits. Enfield, NH: Science Publishers, 1998, pp 1–18.
2. TW Whitaker, GN Davis. Cucurbits: Botany, Cultivation, and Utilization. 1st ed. London: Leonard Hill, 1962, pp 37–61.
3. RW Robinson, DS Decker-Walters. Cucurbits. New York: CAB International, 1997, pp 58–112.
4. G Fang, R Grumet. *Agrobacterium tumefaciens* mediated transformation and regeneration of muskmelon plants. Plant Cell Rep 9:160–164, 1990.
5. J Dong, M Yang, S Jia, N Chua. Transformation of melon (*Cucumis melo* L.) and expression from the cauliflower mosaic virus 35S promoter in transgenic melon plants. Biotechnology 9:858–863, 1991.
6. K Yoshioka, K Hanada, Y Nakazaki, Y Minobe, T Yakuwa, K Oosawa. Successful transfer of the cucumber mosaic virus coat protein gene to *Cucumis melo*. Jpn J Breed 42:277–285, 1992.
7. G Fang, R Grumet. Genetic engineering of potyvirus resistance using constructs derived from the zucchini yellow mosaic virus coat protein gene. Mol Plant Microbe Interact 6:358–367, 1993.
8. C Gonsalves, B Xue, M Yepes, M Fuchs, K Ling, S Namba, P Chee, JL Slightom, D Gonsalves. Transferring cucumber mosaic virus–white leaf strain coat protein gene into *Cucumis melo* L. J Am Soc Hortic Sci 119:345–355, 1994.

9. R Ayub, M Guis, B Amor, L Gillot, JP Roustan, A Latche, M Bouzayen, JC Pech. Expression of ACC oxidase antisense gene inhibits ripening of cantaloupe melon fruits. Nature Biotechnol 14:862–866, 1996.
10. M Bordas, C Montesinos, M Dabauza, A Salvador, L Roig, R Serrano, V Moreno. Transfer of the yeast salt tolerance gene *HAL1* to *Cucumis melo* L cultivars and in vitro evaluation of salt tolerance. Transgenic Res 6:41–50, 1997.
11. M Fuchs, JR McFerson, DM Tricoli, JR McMaster, RZ Deng, ML Boeshore, FJ Reynolds, PF Russell, HD Quemada, D Gonsalves. Cantaloupe line CZW-30 containing coat protein genes of cucumber mosaic virus, zucchini yellow mosaic virus, and watermelon mosaic virus 2 is resistant to these three viruses in the field. Mol Breed 3:279–290, 1997.
12. M Guis, JP Roustan, C Dogimont, M Pitrat, JC Pech. Melon biotechnology. Biotechnol Genet Engineer Rev 15:289–310, 1998.
13. R Grumet, G Akula, T Lanina-Zlatkina, M Whitaker. Transgenic cucurbits: Production, uses, and risk assessment. Proceedings of the XVII Mexican Congress Plant Genetics, Aculpoco, 1998.
14. JD McCreight, H Nerson, R Grumet. Muskmelon. In: G Kallo, BO Berg, eds. Genetic Improvement of Vegetable Crops. New York: Pergamon Press, 1992.
15. M Guis, R Botondi, M Ben-Amor, R Ayub, M Bouzayen, JC Pech, A Latche. Ripening-associated biochemical traits of cantaloupe Charentais melons expressing an antisense ACC oxidase transgene. J Am Soc Hortic Sci 122:748–751, 1997.
16. T More, V Seshadri. Improvement and cultivation: muskmelon, cucumber and watermelon. In: NM Nayar and TA More, eds. Cucurbits. Enfield, NH: Science Publishers, 1998, pp 169–186.
17. R Provvidenti. Resistance to viral diseases of cucurbits. In: MM Kyle, ed. Resistance to Viral Diseases of Vegetables. Portland: Timber Press, 1993, pp 8–43.
18. T More, V Seshadri. Genetic studies. In: NM Nayar, TA More, eds. Cucurbits. Enfield, NH: Science Publishers, 1998, pp 129–153.
19. K Shetty, M Ohshima, T Murakami, K Oosawa, Y Ohashi. Transgenic melon (*Cucumis melo* L.) and potential for expression of novel proteins important to food industry. Food Biotechnol 11:111–128, 1997.
20. R Birch. Plant transformation: Problems and strategies for practical application. Annu Rev Plant Physiol Plant Mol Biol 48:297–326, 1997.
21. V Gaba, E Schlarman, C Elman, O Sagee, AA Watad, DJ Gray. In vitro studies on the anatomy and morphology of bud regeneration in melon cotyledons. In Vitro Cell Dev Biol Plant 35:1–7, 1999.
22. R Grumet. Genetic engineering for crop virus resistance. HortScience 30:449–456, 1995.
23. C Coenen, TL Lomax. Auxin-cytokinin interactions in higher plants: Old problems and new tools. TIPS 2:351–356, 1997.
24. Y Tabei, T Kanno, T Nishio. Regulation of organogenesis and somatic embryogenesis by auxin in melon, *Cucumis melo* L. Plant Cell Rep 10:225–229, 1991.
25. V Moreno, A Roig. Somaclonal variation in cucurbits. In: YPS Bajaj, ed. Somaclonal Variation in Crop Improvement. I. Biotechnology in Agriculture and Forestry. Vol II. Berlin, Heidelberg: Springer-Verlag, 1990, pp 435–464.
26. B Leshem, R Ronen, E Soudry, S Lurie, S Gepstein. Cytokinin at a large range of concentrations determines rates of polypeptide metabolism and regeneration in cultured melon cotyledons. J Plant Physiol 143:330–336, 1994.
27. B Leshem, R Ronen, E Soudry, S Lurie, S Gepstein. Cytokinin and white light coact to enhance polypeptide metabolism and shoot regeneration in cultured melon cotyledons. J Plant Physiol 145:291–295, 1995.
28. V Gaba, C Elman, AA Watad, DJ Gray. Ancymidol hastens in vitro bud development in melon. HortScience 31:1223–1224, 1996.
29. R Niedz, S Smith, K Dunbar, C Stephens, H Murakishi. Factors influencing shoot regeneration from cotyledonary explants of *Cucumis melo*. Plant Cell Tissue Org Cult 18:313–319, 1989.
30. JAD Zeevart, RA Creelman. Metabolism and physiology of abscisic acid. Annu Rev Plant Physiol Plant Mol Biol 39:439–473, 1988.
31. B Leshem, E Ronen, S Lurie. Thidiazuron and paclobutrazol appear to mimic cytokinin and auxin in-

fluences on organ regeneration and protein profiles in cultured melon cotyledons. J Plant Physiol 143: 344–348, 1994.
32. EM Beyer. Silver ion: A potent anti-ethylene agent in cucumber and tomato. HortScience 11:195–196, 1976.
33. S Hammar, R Grumet. Regeneration and *Agrobacterium tumefaciens* mediated transformation of cucumber (*Cucumis sativus* L.). HortScience 25:1070, 1990.
34. KMB Chraibi, JC Castelle, A Latche, JP Roustan, J Fallot. A genotype independent system of regeneration from cotyledons of sunflower (*Helianthus annuus* L.): The role of ethylene. Plant Sci 86:215–221, 1992.
35. JP Roustan, A Latche, J Fallot. Enhancement of shoot regeneration from cotyledons of *Cucumis melo* by $AgNO_3$, an inhibitor of ethylene action. J Plant Physiol 140:485–488, 1992.
36. RC Yadav, MT Saleh, R Grumet. High frequency shoot regeneration from leaf explants of muskmelon. Plant Cell Tissue Org Cult 45:207–214, 1996.
37. M Amor, M Guis, A Latche, M Bouzayen, JC Pech, JP Roustan. Expression of an antisense 1-aminocyclopropane-1-carboxylate oxidase gene stimulates shoot regeneration in *Cucumis melo*. Plant Cell Rep 17:586–589, 1998.
38. PR Johnson, JR Ecker. The ethylene gas signal transduction pathway: A molecular perspective. Annu Rev Genet 32:227–254, 1998.
39. M Orts, B Garcia-Sogo, M Roche, L Roig, V Moreno. Morphogenetic response of calli derived from primary explants of diverse cultivars of melon. HortScience 22:666, 1987.
40. R Dirks, M van Buggenum. In vitro plant regeneration from leaf and cotyledon explants of *Cucumis melo* L. Plant Cell 12:37–40, 1989.
41. I Debeaujon, M Branchard. Induction of somatic embryogenesis and caulogenesis from cotyledon and leaf protoplast-derived colonies of melon (*Cucumis melo* L.). Plant Cell Rep 12:37–40, 1992.
42. D Gray, D McColley, M Compton. High-frequency somatic embryogenesis from quiescent seed cotyledons of *Cucumis melo* cultivars. J Am Soc Hortic Sci 118:425–432, 1993.
43. RV Molina, F Nuez. Correlated response of in vitro regeneration capacity from different source of explants in *Cucumis melo*. Plant Cell Rep 15:129–132, 1995.
44. SE Kintzios, N Taravira. Effect of genotype and light intensity on somatic embryogenesis and plant regeneration in melon (*Cucumis melo* L.). Plant Breed 116:359–362, 1997.
45. N Ficcadenti, GL Rotino. Genotype and medium affect shoot regeneration of melon. Plant Cell Tissue Org Cult 40:293–295, 1995.
46. R Kathal, S Bhatnagar, SS Bhojwani. Regeneration of plants from leaf explants of *Cucumis melo* cv. Pusa sharbati. Plant Cell Rep 7:449–451, 1988.
47. L Bouabdallah, M Branchard. Regeneration of plants from callus cultures of *Cucumis melo* L. Z Pflanzen 96:82–85, 1986.
48. R Kathal, S Bhatnagar, SS Bhojwani. Plant regeneration from the callus derived from root explants of *Cucumis melo* L. cv. Pusa sharbati. Plant Sci 96:137–142, 1994.
49. H Ezura, H Amagi, K Yoshioka, K Oosawa. Highly frequent appearance of tetraploidy in regenerated plants, a universal phenomenon in tissue cultured melon. Plant Sci 85:209–213, 1992.
50. G Fassuliotis, DV Nelson. Regeneration of tetraploid muskmelons from cotyledons and their morphological differences from two diploid muskmelon genotypes. J Am Soc Hortic Sci 117:863–866, 1992.
51. R Kathal, S Bhatnagar, SS Bhojwani. Chromosome variations in the plants regenerated from leaf explants of *Cucumis melo* L. cv. Pusa sharbati. Caryologia 45:51–56, 1992.
52. J Adelberg, B Rhodes, H Skorupska. Generating tetraploid melons from tissue culture. Acta Hortic 336:373–380, 1993.
53. RC Yadav, R Grumet. Tendrils as an alternate tissue source for chromosome visualization. J Am Soc Hortic Sci 119:850–852, 1994.
54. I Susin, JM Alvarez. Fertility and pollen tube growth in polyploid melons (*Cucumis melo* L.). Euphytica 93:369–373, 1997.
55. M Lee, RL Phillips. The chromosomal basis of somaclonal variation. Annu Rev Plant Physiol Plant Mol Biol 39:413–437, 1988.

56. MP Valles, JM Lasa. *Agrobacterium*-mediated transformation of commercial melon (*Cucumis melo* L., cv. Amarillo Oro). Plant Cell Rep 13:145–148, 1994.
57. SC Hokanson, JF Hancock, R Grumet. Direct comparison of pollen-mediated movement of native and engineered genes. Euphytica 96:397–403, 1997.
58. J Finnegan, D McElroy. Transgene inactivation: Plants fight back. Biotechnology 12:883–888, 1994.
59. H Vaucheret, C Beclin, T Elmayan, F Feuerbach, C Godon, JB Morel, P Mourrain, JC Palauqui, S Vernhettes. Transgene-induced gene silencing in plants. Plant J 16:651–659, 1998.
60. M Fuchs, D Gonsalves. Resistance of transgenic hybrid squash ZW-20 expressing the coat protein genes of zucchini yellow mosaic virus and watermelon mosaic virus 2 to mixed infections of both potyviruses. Biotechnology 13:1466–1473, 1995.
61. DM Tricoli, KJ Carney, PF Russell, JR McMaster, DW Groff, KC Hadden, PT Himmel, JP Hubbard, ML Boeshore, HD Quemada. Field evaluation of transgenic squash containing single or multiple virus coat protein gene constructs for resistance to cucumber mosaic virus, watermelon mosaic virus 2, and zucchini yellow mosaic virus. Biotechnology 13:1458–1465, 1995.
62. AD Bauchot, DS Mottram, AT Dodson, P John. Effect of aminocyclopropane-1-carboxylic acid oxidase antisense gene on the formation of volatile esters in cantaloupe Charentais melon (cv. Vendradais). J Agric Food Chem 46:4787–4792, 1998.
63. J Rissler, M Mellon. Perils amidst the promise: Ecological risks of transgenic crops in a global market. Union of Concerned Scientists. Cambridge, MA, 1993.
64. JF Hancock, R Grumet, SC Hokanson. The opportunity for escape of engineered genes from transgenic crops. Hortscience 31:1080–1085, 1996.
65. J Kling. Could transgenic supercrops one day breed superweeds? Science 274:180–181, 1996.
66. SC Hokanson, JF Hancock, R Grumet. Effect of border rows and trap/donor ratios on pollen-mediated gene movement. Ecol Appl 7:1075–1081, 1997.
67. RE Arriola, NC Ellstrand. Crop to weed gene flow in the genus *Sorghum* (Poaceae): Spontaneous interspecific hybridization between johnsongrass, *Sorghum halepense*, and crop sorghum, *S. bicolor*. Am J Bot 83:1153–1159, 1996.
68. JR Renno, T Winkel, F Bonnefous, G Bezancon. Experimental study of gene flow between wild and cultivated *Pennisetum glaucum*. Can J Bot 75:925–931, 1997.
69. CR Linder, I Taha, GJ Seiler, AA Snow, LH Rieseberg. Long-term introgression of crop genes into wild sunflower populations. Theor Appl Genet 96:339–347, 1998.
70. V Moreno, M Garcia Sogo, I Granell, B Garcia-Sogo, A Roig. Plant regeneration from calli of melon (*Cucumis melo* L. cv. Amarillo Oro). Plant Cell Tissue Org Cult 5:139–146, 1985.
71. PP Chee. Plant regeneration from cotyledons of *Cucumis melo* 'Topmark'. Hortscience 26:908–910, 1991.
72. J Jain, TA More. In vitro regeneration in *Cucumis melo* cv. Pusa Madhuras. Cucurbit Genet Coop 15:62–64, 1992.
73. K Shetty, G Shetty, Y Nakazaki, K Yoshioka, Y Asano, K Oosawa. Stimulation of benzyladenine-induced in vitro shoot organogenesis in *Cucumis melo* L. by proline, salycilic acid, and aspirin. Plant Sci 84:193–199, 1992.
74. R Kathal, S Bhatnagar, SS Bhojwani. Regeneration of shoots from hypocotyl callus of *Cucumis melo* cv. Pusa sharbati. J Plant Physiol 126:59–62, 1986.
75. Z Punja, N Abbas, G Sarmento, F Tang. Regeneration of *Cucumis sativus* var. *sativus* and *C. sativus* var. *hardwickii*, *C. melo* and *C. metuliferus* from explants through somatic embryogenesis and organogenesis. Plant Cell Tissue Org Cult 21:93–102, 1990.
76. T Oridate, K Oosawa. Somatic embryogenesis and plant regeneration from suspension callus culture in melon (*Cucumis melo* L.). Jpn J Breed 36:424–428, 1986.
77. M Guis, A Latche, JC Pech, JP Roustan. An efficient method for production of diploid cantaloupe charentais melon (*Cucumis melo* L. var. *cantalupensis*) by somatic embryogenesis. Sci Hortic 69:199–206, 1997.
78. LA Roig, L Zubeldia, MC Orts, MV Roche, V Moreno. Plant regeneration from cotyledon protoplasts of *Cucumis melo* L. cv. Cantaloup Charentais. Cucurbit Genet Coop 9:74–76, 1986.

# 26
# Cranberry Transformation and Regeneration

**James J. Polashock and Nicholi Vorsa**
*Rutgers University, Chatsworth, New Jersey*

| | | |
|---|---|---|
| I. | INTRODUCTION | 383 |
| II. | IN VITRO CULTURE AND PLANT REGENERATION FROM SOMATIC TISSUE | 384 |
| | A. Growth Regulator Effects | 386 |
| | B. Effect of Explant Age and Orientation on the Medium | 386 |
| III. | TRANSFORMATION OF CRANBERRY TISSUE | 387 |
| IV. | SELECTION AND SCREENING OF PUTATIVE TRANSFORMANTS | 390 |
| | A. Transgenic Plant Selection Systems | 390 |
| | B. Transgenic Plant Screening Systems | 391 |
| V. | EXPRESSION OF GENES WITH POTENTIAL COMMERCIAL APPLICATION IN CRANBERRY | 391 |
| VI. | SUMMARY AND FUTURE PROSPECTS | 394 |
| | REFERENCES | 395 |

## I. INTRODUCTION

The American cranberry, *Vaccinium macrocarpon Ait.*, is a recently domesticated native of North America. *V. macrocarpon* is a diploid ($2n = 2x = 24$) long-lived woody perennial adapted to acid soils and a temperate climate, requiring approximately 1200 hours of chilling to break winter dormancy. Indigenous peoples were reported to collect the berries for various food and medicinal products (for brief review, see Ref. 1). Settlers from Europe began utilizing the fruit to help prevent scurvy. The first reported attempt to cultivate the cranberry was made in 1810 in Cape Cod, Massachusetts (1). Since that time, the cranberry industry in the United States has grown to a current production of approximately 550 million pounds annually (2). Today, cranberry fruit is utilized mostly for juices. Cranberry juice has been reported to help prevent urinary tract infections. Recent research in this area has suggested that compounds in the juice prevent or inhibit the adhesion of *Escherichia coli* in the urinary tract (3–5).

Although cranberries have been cultivated on a large scale since the mid-1800s, many of the varieties currently grown are wild selections from over 100 years ago. Varieties from controlled crosses were released between the mid-1950s and 1970 (1), one of which, 'Stevens,' has become the most popular cultivar. The release and widespread cultivation of new varieties are relatively slow in cranberries (6). This is probably due to two major factors. The first is that the American cranberry requires 3–5 years to achieve full production from planting, imposing a lengthy cycle for breeding and selection. The second is the cost associated with renovating an existing cranberry bed to plant a new variety. The return on investment has been reported to be a minimum of 8 years (7). Thus an introduced variety must be far superior to existing cultivars to allow renovation to be a viable option.

Not only do breeding and selection for superior traits in cranberry constitute a lengthy process, but the available genes are limited to the primary gene pool within *V. macrocarpon*. The diversity and genetic differentiation within wild *V. macrocarpon* populations appear to be low compared to those of other species (8). Although interspecific crosses within the same *Vaccinium* subsection offer opportunities, intersectional hybrids within the *Vaccinium* genus are generally sterile (9). The members of other species within the cranberry section *Oxycoccus* are largely polyploid (e.g., *V. oxycoccus* L.), imposing limitations on accessing this gene pool. Interploid hybridization, particularly when the desired introgression is from polyploid to diploid levels, is difficult.

Biotechnology as a method of crop improvement has become a valuable tool to complement existing breeding programs. Techniques, particularly transformation, offer several opportunities that are unattainable through conventional breeding. Direct transformation offers the opportunity for wide interspecies gene transfer.

In cranberry, the most serious threats to crop production are insect pests and fungal diseases. This has become particularly important in recent years with the loss of registered fungicides and pesticides. Fruit rot is a serious problem in some growing regions (particularly the Northeast) and is caused by a complex of organisms (reviewed in Ref. 10). Insertion of human-safe antifungal genes such as chitinases could provide sufficient resistance to allow for reduced reliance on traditional fungicides. Genes are also being identified that may be effective against cranberry insect pests. The insertion and expression of these genes in elite genotypes would offer new opportunities in the development of insect-resistant cultivars.

Although interspecific gene transfer is attractive, manipulation of the existing cranberry genome offers potential benefits as well. Most plants have natural insect and defense genes that could be manipulated to be more effective against common pests. Other genes may be manipulated to improve agronomic traits or to increase desirable compounds in the fruit.

Despite all of the potential advantages of cranberry biotechnology, there are few published reports regarding transgenic cranberry. One reason is that cranberry is a minor crop, with limited resources available to pursue the problems associated with cranberry biotechnology. Much of the progress that has occurred in cranberry is due in part to the work published in other species of *Vaccinium*, particularly blueberry (see, for example, Refs. 11–16).

Serres and McCown reviewed the state of the art in cranberry biotechnology up to 1995 (17). This paper addresses what we perceive to be the existing hurdles in cranberry biotechnology as well as advances since 1995.

Three basic steps are required for a viable biotechnology system in cranberry: (a) a reliable method for adventitious plant regeneration from somatic tissues of elite genotypes; (b) a scheme for gene transfer, integration, and expression in the cranberry genome; and (c) a system to select transformed cells and plants. Each of these basic steps can be further divided and optimized.

## II. IN VITRO CULTURE AND PLANT REGENERATION FROM SOMATIC TISSUE

In vitro culture of cranberry was first reported in 1984 (18); however, this work as well as that of Marcotrigiano and McGlew (19) focused on micropropagation from stem explants and indicated that many of the shoots formed in their system are probably adventitious. Scorza and Welker (20) reported adventitious shoot proliferation from cranberry stems and leaves. They listed cytokinin as a requirement and noted that shoot development and shoot proliferation were inversely related, indicating that the conditions (primarily growth regulator combinations) that induce shoot proliferation do not allow for shoot elongation.

Shoot regeneration from adventitious buds on tissue cultured stem sections was achieved by using a basic woody plant medium supplemented with 1 µM thidiazuron (TDZ) and 10 µM $N^6$-[2-isopentenyl] adenine (2ip) (21). This report focused on transformation of the regenerating stem segments and did not discuss efficiency or optimization of the regeneration process. Shoot regeneration from leaves was specifically examined by Marcotrigiano and coworkers (22). These researchers also used a woody plant medium (although different from that used by Serres et al. [21]) supplemented with 10 µM TDZ and 0 or 1.0 µM α-naphthaleneacetic acid (NAA). Elongation of the regenerated shoots was reported to be a key problem with this system, similar to that discussed in Scorza and Welker (20).

Our laboratory has taken a systematic approach to improving cranberry adventitious bud production and elongation since an efficient system is required if comprehensive transformation studies are to be explored. We have decided to concentrate on leaves as an explant since leaves are routinely chosen for *Agrobacterium*–mediated transformation studies. In addition, the only report of cranberry transformation used particle bombardment of stem sections (21). If bombardment continues to be the transformation method of choice, leaves should offer a larger target than stems.

To establish an efficient tissue culture and reliable regeneration system, we examined four basic variables, (a) genotype, (b) growth regulator combination(s), (c) explant age, and (d) explant orientation. All of these variables have been shown to be important in other tissue culture systems.

To address potential genotype differences, five cultivars were chosen for the study, 'Ben Lear,' 'Early Black,' 'Pilgrim,' 'Stevens,' and 'No. 35.' 'Ben Lear' and 'Early Black' are wild selections released in 1901 and 1852, respectively. Both are still widely cultivated. 'Pilgrim' and 'Stevens' are the result of crosses of wild selections from a USDA breeding program begun in 1929. These cultivars were released in 1961 and 1950, respectively. Both of these cultivars are widely grown, and 'Stevens' is the most common choice for new plantings. The last variety, 'No. 35,' is not widely cultivated; it is a cross between wild selections and is still being evaluated for desired horticultural traits. Cranberry cultivars have been shown in some cases to be heterogeneous genotypes (23,24). All of the clones used in our studies were DNA-fingerprinted to confirm genetic identity and homogeneity.

The selection of growth regulators to be tested was based on the successful regeneration of cranberry reported by Serres and coworkers (21) and Marcotrigiano and associates (22), namely, TDZ, NAA, and 2ip. Although extensive experiments were not conducted on optimization, the basal medium consisted of Anderson's major and micro salts (25), MS organics (26), and 2% sucrose (hereafter referred to as *basal medium*) for all experiments. The medium was adjusted to pH 5.3 with a solution of potassium hydroxide (0.5 N) before autoclaving (121°C, 15 minutes) and solidified with 5 g/l Agargel (Sigma Chemical Company, St Louis, MO).

Explant age was examined by parallel experiments utilizing leaves nearest the shoot apical meristem (the "youngest"—excluding the apical leaves less than $1.3 \times 2.5$ mm) down to 10 leaves from the apical meristem (the "oldest"). Leaf age was an important parameter to examine since it is commonly observed that aging of explants has a significant negative effect on regeneration (27).

Finally, effects of explant orientation on the medium were examined. This was expected to be an important consideration since leaf regeneration in blueberry was reported to arise only from the abaxial side (11). However, Marcotrigiano and colleagues (22) reported shoot regeneration from cranberry leaves occurred on both sides of the explants.

Cranberry stems are easily disinfected for maintenance in sterile culture. Deleafed stems are washed for 30 seconds in 70% ethanol followed by 10 minutes in 20% commercial bleach and rinsed in sterile water. Maintenance is on the basal medium described, in Magenta GA7 containers. Culture chambers are maintained at 25 ± 1°C and a 16-hour daily photoperiod provided by cool-white fluorescent tubes at 40 µmol · m$^2$ · s$^{-1}$. Plants grow normally although the leaves tend to be considerably smaller than on greenhouse or field grown plants.

All genotypes tested responded similarly in our experiments. Considering that we used both wild selections and first-generation hybrids, it is probably safe to assume that genotype-specific regeneration protocols will not have to be devised for transformation procedures. Varieties, however, varied considerably (up to four times) in their regeneration capacity (i.e., the number of shoot meristems formed).

## A. Growth Regulator Effects

Our research suggests that initially high levels (10 µM) of TDZ in the medium in combination with 5.0 µM 2ip allowed efficient regeneration in all cultivars tested (Table 1). In some cases, the explants were completely covered with adventitious shoots within 30 days in culture (Fig. 1). In our studies, NAA even at low concentration (0.5 µM) severely retards regeneration and induces proliferation of callus (28). This is in contrast to the results of Marcotrigiano and coworkers (22), which showed a general increase in adventitious shoot production when 1.0 µM NAA was added to the medium. This difference could be due to a number of factors such as explant genotype and basal medium composition. Our results are consistent with those of Billings and associates (11) in blueberry, another *Vaccinium* species.

The problem of adventitious shoot elongation that was encountered by other researchers was also an issue in our laboratory. No elongation of shoot meristems was observed when the explants were maintained on the regeneration medium (basal medium containing 10.0 µM TDZ and 5.0 µM 2ip). Transfer to lower TDZ (1.0 µM) and higher (10.0 µM) 2ip did not alleviate the problem. If, however, the explants are transferred to basal medium (without growth regulators) soon after bud induction (after about 40 days in regeneration medium), shoot elongation occurs in about 2 weeks (Fig. 2). Rooting of the elongated shoots has proved to be quite easy. All shoots transplanted to solid basal medium or moist sphagnum moss root within 15 days.

## B. Effects of Explant Age and Orientation on the Medium

The age of source leaves for explants (i.e., the oldest leaves from a shoot versus the youngest) has no apparent significant effect on regeneration. Thus, large-scale studies could utilize nearly all of the leaves on in vitro grown shoots.

**Table 1** Adventitious Shoot Regeneration and Elongation of Cranberry Leaf Explants

| Genotype | Explant number | Mean no. shoot meristems/leaf[a] | Mean no. elongated shoot meristems/leaf[a] |
|---|---|---|---|
| Early Black | 50 | 54.8a | 18.2a |
| Pilgrim | 50 | 51.2a | 13.6b |
| Ben Lear | 41 | 21.1b | 7.8c |
| Stevens | 50 | 14.2c | 6.9c |
| No. 35 | 49 | 12.2c | 6.5c |

[a]Mean values followed by the same letter are not significantly different (Duncan's test, $p = 0.05$).

**Figure 1** Pilgrim explant after ~20 days on elongation medium (transferred after ~40 days on regeneration medium).

Explant orientation on the medium has a profound effect on observed regeneration. When the adaxial side is in contact with the medium, about half the number of adventitious shoots are formed, as compared to when the adaxial side was up (away from the medium). This observation is not surprising since the shoot regeneration in our system occurs almost exclusively on the adaxial side for all genotypes on all media tested. This is in contrast to blueberry, in which Billings and coworkers (11) reported that shoots originated only from the abaxial side, and to cranberry (22), in which shoot regeneration was reported to occur on both sides of the leaves.

## III. TRANSFORMATION OF CRANBERRY TISSUE

With an efficient regeneration system, the various potential transformation methods can be explored. A reliable transformation system can be divided into the following basic steps: (a) a method of DNA delivery into the cells, (b) selective regeneration of transformed cells, and (c) a method of screening the putative transformed plants.

Several methods of DNA transfer into plant cells have been developed. The only published report of transformation in cranberry used electric discharge particle bombardment (21). Although successful, this system appears to generate a relatively low frequency (0.15%) of recovered transclones. Biolistic particle delivery systems tend to have low efficiency, can lead to

**Figure 2** Elongation of shoot tips from Pilgrim explants after transfer to basal medium.

chimeric plants, and are sometimes associated with insertion of multiple copies of the desired genes that can lead to expression problems such as gene silencing.

Transformation using *Agrobacterium tumefaciens* harboring disarmed Ti plasmids is well established. Wild-type *Agrobacterium* sp. strains are known to infect blueberries naturally. Rowland (29) tested five common tumorigenic strains on blueberry and found that at least three of those tested (T37, C58, and A281) could form galls. Serres and McCown (17), also using tumorigenic strains, found that three of the strains they tested (A348, A6, and A282/pTVK281) could from galls on cranberry. Interestingly, *A. tumefaciens* strains T37 and C58 were included in their research and were not reported to cause galls on cranberry within 3 weeks of inoculation. More recently, blueberry transformation has been achieved by using at least two different disarmed *Agrobacterium* sp. strains (15,16). The latter report compared efficiency of two common strains (LBA4404 and EHA105) of *A. tumefaciens* for transformation. Strain EHA105 (a supervirulent disarmed derivative of A281) (30) was found to be much more effective and was used in all of our studies. In order to allow better access of the plant cells to *Agrobacterium*, various methods of wounding can be used, including sonication (31) and digestive enzyme treatment.

One other method that has been successful in some plants is called *whisker transformation* (32,33). Silicon carbide "whiskers" are shaken with the explant in a solution containing the DNA to be inserted. The whiskers penetrate the cells and allow the DNA to enter. Some of the DNA becomes integrated into the plant genome, giving rise to transformants.

Since particle-mediated transformation of cranberry has been successful, we are testing a comparable method of bombardment, in which particles are accelerated by a helium burst (e.g., Bio-Rad, PDS1000). Similarly to the experiments detailed in Serres and coworkers (21), we have examined the effects of preculture period (0–20 days), particle size (0.75 and 1.0 µm), and distance between the stopping screen and the target. All of the experiments were performed with leaf explants on regeneration medium. A few blue sectors have been detected in some treatments using GUS transient assays, but confirmation of stable transformation requires further verification.

Since wild-type *A. tumefaciens* infection has also been reported, our lab has concentrated on using *Agrobacterium*–mediated transformation as an alternative system. Using Strain EHA105, we have confirmed expression of a GUS reporter gene. Transient assays show that about 50% of the treated explants have GUS activity. This activity is indicated by the presence of 1–12 blue spots/explant. The frequency of stable transformants to date is quite low, however, with only two confirmed transclones in over 100 leaf explants. We propose that one problem is simply exposure of the totipotent cells in the leaf to *Agrobacterium* infection. Since wounding is known to be a requirement for infection, efforts have focused on this aspect. Various techniques such as shaking with sand, sonication, whisker wounding, and lytic enzyme treatment need to be explored. However, any method that wounds the tissue too much causes death of the explant or a reduction of regeneration. Sonication as a method of microwounding appears to be a possibility (31,34). At least one report suggests that sonication alone may be sufficient for transformation (35). Preliminary evaluation using sonication is under way.

Experiments in our laboratory using sand and lytic enzymes to weaken the cell wall (various concentrations and incubation periods were tested) showed a negative effect on regeneration (probably due to severe wounding or degradation) and no increases in transient expression as judged by histochemical GUS assays (data not shown). These methods do not appear to hold promise for cranberry.

Silicon carbide "whiskers" have been used as a method of direct gene transfer (as detailed). This method was explored, but transient GUS expression was not observed, nor were stable transformants recovered. However, since the fibers could puncture cells, it was thought that "whisker" treatment might be a suitable wounding technique (i.e., for *Agrobacterium* sp. transformation). Results thus far have been negative.

Experiments to enhance *Agrobacterium*–mediated transformation efficiency include using sonication for wounding, exploring other *Agrobacterium* sp. strains, and using variations of cultivation techniques such as length of cocultivation, acetosyringone addition to the medium, preincubation of the explants, and bacterium density.

One problem often encountered with *Agrobacterium*–mediated transformation is clearing of the explants of bacteria after cocultivation, particularly when supervirulent strains are used (see, e.g., Refs. 36,37). After 4–6 days of cocultivation, rinsing with a solution containing a high level of carbenicillin (300 mg/L) for 30 minutes, followed by transfer to regeneration medium containing 300 mg/L carbenicillin (with or without kanamycin for selection) effectively eliminated bacteria from 95% of the explants within 14 days. This was determined by culturing the explants in LB (a rich bacterial growth medium) with no antibiotics for at least 10 days and observing for evidence of bacterial growth.

## IV. SELECTION AND SCREENING OF PUTATIVE TRANSFORMANTS

Selection of the transformed cells and subsequent regenerated plants is the next requirement. The most commonly used antibiotic for selection of transformed higher plants is kanamycin. Resistance to this antibiotic is conferred by insertion and expression of a gene encoding neomycin phosphotransferase II (NPT II). Kanamycin selection in cranberry has been successful (21). Other antibiotic resistance genes and in some cases herbicide resistance genes have been useful for selection. We have tested one other gene for selection that encodes resistance to the antibiotic hygromycin B.

The activity of many of the genes of interest that are inserted by transformation, especially those that are not constitutively expressed, is difficult to assay. For this reason, screenable markers or reporter genes are widely used, particularly when transformation procedures are being optimized. The most commonly used in higher plants is gusA, a gene encoding β-glucuronidase (GUS). Expression and activity of this enzyme are easily detectable by using both histochemical and fluorigenic substrates (38). More recent versions of this gene contain a plant intron to ensure expression only in plant cells and not in *Agrobacterium* sp. (39). This reporter gene was used successfully in cranberry transformation (21), and expression was shown to be influenced by various factors (40).

### A. Transgenic Plant Selection Systems

A key step in the transformation process entails the recovery and selective growth of the transformants. Kanamycin selection conferred by expression of the NPT II gene (under the control of a nopaline synthase promoter) allows selection in cranberry, but this system to date is not adequate. Serres and colleagues (21) used a flooding technique and 300 mg/l of kanamycin but routinely autoclaved the medium containing the antibiotic. Using filter-sterilized kanamycin (added to partially cooled media), we find that varieties we are testing are sensitive to even lower concentrations than that reported (21). Although this could be due to genotype differences used in our experiments, it is possible some kanamycin activity is lost as a result of autoclaving. Regardless, transformants still require extended periods to grow, and selection across the explant is heterogeneous, possibly because of variable absorption and transport of the antibiotic from the medium. These variable effects are thought to be partially alleviated by the flooding technique described.

Another antibiotic, hygromycin B, was tried in our experiments. Unlike kanamycin, which causes bleaching of the tissue, hygromycin B caused death of the explants at the concentration tested (100 mg/l). Since the adventitious buds are forming on the top of the explants, death of the surrounding explant tissue was detrimental to selection and growth of putative transformants.

The laboratory of Brent McCown (University of Wisconsin-Madison) has examined the potential for using herbicide resistance as a selectable marker. The *bar* gene, encoding phosphinothricine acetyltransferase (PAT), was transformed into cranberry plants and the transgenic plants were evaluated for herbicide selection in vitro. Bar gene expression confers resistance to glufosinate-ammonium (commercial formulation Liberty, Agrevo Corp.). This herbicide is a potent and irreversible inhibitor of glutamine synthetase.

Glufosinate-ammonium for selection did not work well. Even nontransformed plants had resistance to increasing levels of this herbicide in the medium (plants survived at a level of 20 mg/l). This was presumed to be due to an induction of glutamine synthetase gene expression and/or a natural tolerance for ammonia buildup in cranberry (ammonia buildup is a result of glutamine synthetase inhibition). Other herbicides and resistance genes have not yet been tested for in vitro selection utility.

## B. Transgenic Plant Screening Systems

Our work and other published work on cranberry transformation have been limited to expression of the GUS gene for screening putative transformants. This simple detection system has been invaluable in quickly evaluating transformation techniques through transient assays as well as expression in stable transformants. Unfortunately, as with kanamycin selection, this system is still not adequate. One problem is that the assay is destructive. Any explants used in transient assays are sacrificed. This is not a problem with putative transformants since only a small amount of tissue, such as a leaf fragment, is required for the assay. Gus detection in transformants by histochemical staining was found to be random and unpredictable (40). In addition, these researchers also found that detectable levels of Gus expression varied among transclones as well as within a transclone, depending on the developmental and physiological state of the tissue. The physical growth environment was also found to play a role. These phenomena have been noted in other plants such as blueberry (16).

Polyphenolic compounds, which are present in high concentration in cranberry tissue, were noted to decrease fluorigenic detection of GUS activity. The addition of polvinylpolypyrrolidone (PVPP) to the cranberry extract helped alleviate this problem (40). The polyphenolics in the tissue are presumed to be a problem in the histochemical assay as well. Substrate penetration may also reduce detectable activity in intact tissue. Leaves from our transformants expressing GUS typically turn blue near the cut surface and in zones of accidental wounding. Over time, the whole leaf stains blue. Vacuum infiltration of the substrate into the leaves does not seem to help.

Although GUS detection has been useful, refinement of the procedure is necessary to realize the full potential. Perhaps other selectable markers such as green fluorescent protein (GFP) will prove to be useful in cranberry. One advantage of GFP detection systems is the nondestructive attribute.

Aside from the problems associated with detecting enzyme activity, expression of any gene inserted into the plant is first regulated at the transcriptional level. Although a host of promoters have been used in transgenic plants, cranberry research has been limited to the CaMV 35s promoter and the *Agrobacterium* sp. nopaline synthase promoter. Both of these promoters are considered to be constitutively expressed in transgenic plants. Other promoters will be required in order for cranberry biotechnology to progress more efficiently. When several genes are driven by the same promoter, as is often the case (e.g., the CaMV 35s promoter is used for GUS expression as well as the gene of interest), cosuppression can occur. In addition, temporal and spatial regulation is a requirement for the expression of some useful genes in cranberry. Work in our lab is progressing toward isolating and testing other promoters for transgenic cranberry.

## V. EXPRESSION OF GENES WITH POTENTIAL COMMERCIAL APPLICATION IN CRANBERRY

Herbicide resistance is one of the most commonly explored applications of plant biotechnology. A crop resistant to a broad-spectrum herbicide could allow for weed control without crop damage. The success of this approach is evident in the commercial release of transgenic plants that are resistant to many available herbicides (e.g., Monsanto's Roundup Ready soybeans, corn, etc., and AgrEvo's Liberty Link soybeans, corn, etc.).

Researchers at the University of Wisconsin-Madison have tested transgenic 'Pilgrim' cranberry plants for field resistance to the herbicides glufosinate-ammonium and bialophos (Serres and McCown, unpublished data; also see Ref. 41). Cuttings from the transgenic plant with the highest expression of the *bar* gene were rooted and acclimatized to outdoor growth in a cold frame. Herbicide tolerance tests were performed in early July, late July, and September. The test in early July showed that the transformed plants were significantly more resistant than the nontransformed plants to the herbicide glufosinate-ammonium. Nontransformed plants were killed at 200–300 ppm, whereas the transformed plants survived 500 ppm. The transformed plants did, however, show some delay in growth and some shoot tip death at 500 ppm. In late July, a similar test, but with higher levels of glufosinate-ammonium (up to 1000 ppm), was conducted. Some carrier damage, noted at the highest level, may limit the amount of the herbicide that can be used, but even at 1000 ppm the transformed plants were not killed (Fig. 3), although the injury seen at 500 ppm in the earlier study was more severe. In September, a different, more translocatable form of the herbicide, bialophos, was tested. Levels tested (800 and 1600 ppm) were roughly equivalent to 500 and 1000 ppm glufosinate-ammonium. The transformed cranberry showed excellent resistance to this form of the herbicide with only minor damage at the highest concentration. The nontransformed plants also showed some resistance, although most shoot tips were killed and no flower buds were set. Self-pollinated progeny to generate plants homozygous for the resistance gene are being grown for evaluation.

Insect predation on plants is another problem cranberry growers face. Many insecticides kill beneficial insects as well as the pests. In addition, insecticides, like fungicides, are being banned from use as stricter regulations are put into place. Probably the most widely explored gene for transgenic resistance to insects is *Bt*. *Bt* is a gene from the bacterium *Bacillus thuringiensis* that encodes a δ-endotoxin. When ingested by certain insects, the δ-endotoxin is cleaved into an insecticidal compound. Various forms of the gene that impart different insect toxin specificity are now available.

Transgenic cranberry expressing the *Bt* gene were generated and tested for insecticidal activity (Serres and McCown, unpublished data; also see Ref. 41). None of the 64 transclones ex-

**Figure 3** Herbicide tolerance in transgenic cranberry conferred by *bar* gene expression. Plants were treated with 1000 ppm glufosinate-ammonium. The flat on the left is CaMV/bar transformed Pilgrim; the flat on the right is untransformed. Photograph was taken 1 month after spraying. (Photograph courtesy of R. Serres and B. McCown.)

pressing the gene produced significant blackheaded fireworm (BHFW, *Rhopobota naevana*) larval mortality in in vitro feeding assays. BHFW is a lepidopteran pest of cranberry. Field tests, however, using four transclones, over 2 years showed an increase in BHFW mortality rate ranging from a 12% to a 37% reduction in the number of insects reaching maturity (Fig. 4). Significant differences were also noted in the two cultivars tested, 'Pilgrim' and 'Stevens.' This was hypothesized to be due to differences in endogenous phenolic levels. Inhibition of *Bt* δ-endotoxin insecticidal activity due to tannins has been reported in other plant systems (42,43). More detailed studies were conducted to determine whether in fact cranberry phenolic compounds (such as flavonoids) could inhibit the δ-endotoxin activity to explain in part the low effectiveness and variability of the toxin on insect mortality rate in laboratory and field assays.

Crude cranberry extracts mixed with a modified commercial preparation of *Bt* were shown to inhibit insecticidal activity in droplet feeding assays using BHFW (Fig. 5) and soybean looper (*Pseudoplusia includens*). Extracts from cranberry to eliminate proteins and enhance flavonolic constituents had similar effects to the crude extract, suggesting endogenous cranberry flavonoids as the inhibitory compounds. These data are similar to those observed for β-glucuronidase (discussed previously) produced in transgenic plants (40).

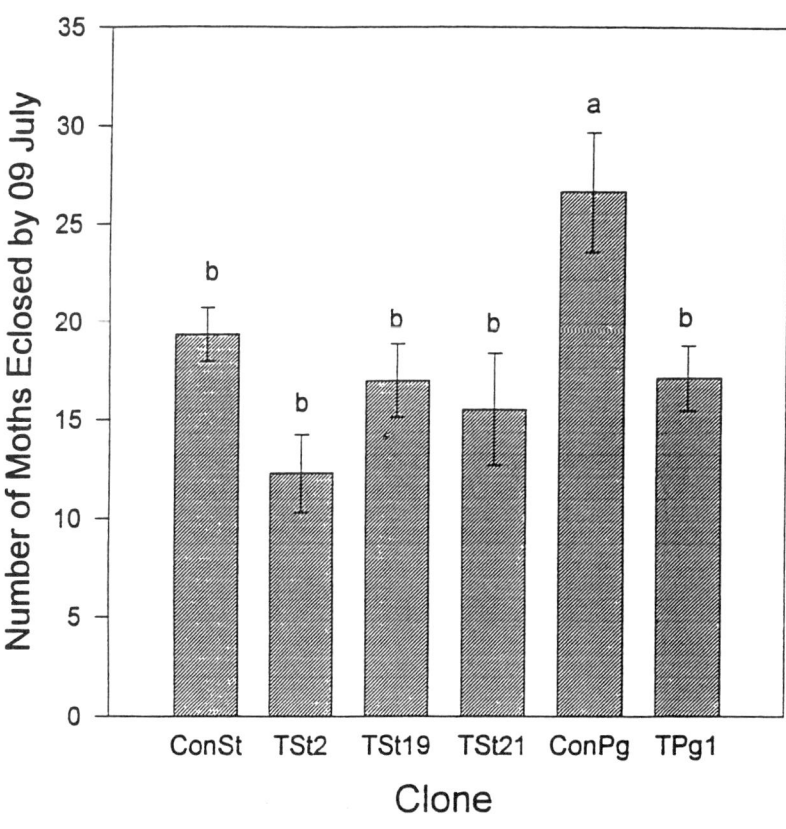

Figure 4 Emergence of adult BHFW moths on transclones of cranberry. ConSt, Stevens control; TSt(#), independent Stevens transclones; ConPg, Pilgrim control; TPg1, Pilgrim transclone. Bars with different letters are significantly different. (Courtesy of R. Serres and B. McCown.)

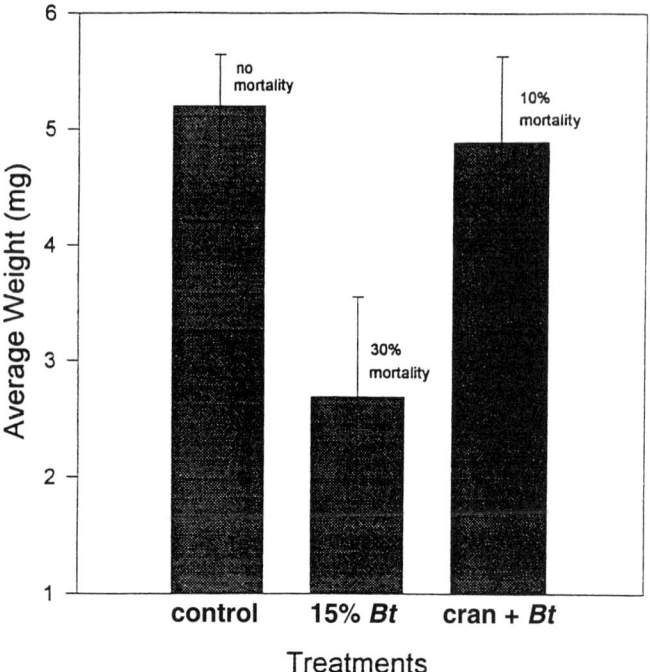

**Figure 5** Effect of the *Bt* δ-endotoxin alone (15% *Bt*) and mixed with cranberry leaf extracts (cran + *Bt*) on the growth of BHFH. (Courtesy of R. Serres and B. McCown.)

## VI. SUMMARY AND FUTURE PROSPECTS

In our opinion, the regeneration procedure developed in our lab from cranberry leaves is efficient, reliable, and applicable to all genotypes tested. Thus, regeneration from somatic tissue is no longer a hurdle or bottleneck in biotechnological approaches to cranberry improvement. The transformation procedure, however needs to be further improved. Our preliminary data using *A. tumefaciens* are encouraging. If, however, this method proves to be unreliable, particle bombardment techniques will be further examined and optimized. The success of Serres and coworkers (21), using a bombardment technique, indicates that this is a viable, even if not the ideal method of transformation.

Expression of all genes in transgenic cranberry needs to be further examined on all levels. Promoters for better temporal and spatial expression need to be isolated and tested. The most serious hurdle, using current technology, is developing selection methods to recover shoots from transgenic cells. Some inadequacies in this area, as well as screening procedures (such as GUS detection), are due in part to inhibition of activity by high levels of phenolics in cranberry, as described. Herbicide resistance may yet prove to be a viable alternative to antibiotic selection. Many other herbicide resistance genes and herbicides with different modes of action are available and warrant testing. Selection alternatives such as D-xylose utilization, conferred by expression of the xylose isomerase gene from *Thermoanaerobacterium thermosulfurogenes*, also merit exploration (44).

Public acceptance of transgenic food crops such as cranberry and the legal hurdles associated with licensing various genes (promoters, etc.) for use in a minor crop will continue to be important considerations. Efficacy and safety of all transgenic plants and products need to be diligently pursued to allow for the continued use of this powerful and indispensable technology.

## ACKNOWLEDGMENTS

We would like to thank Brent McCown and Rodney Serres for providing unpublished data. We would also like to thank Luping Qu, a postdoctoral research associate in our lab, for contributing recent data.

## REFERENCES

1. P Eck. The American Cranberry. New Brunswick, NJ, London: Rutgers University Press, 1990.
2. SS Rod DeSmet. 1997 Cranberry Highlights. Trenton, NJ: New Jersey Agricultural Statistics Service, 1997.
3. I Ofek, J Goldhar, D Zafriri, H Lis, R Adar, N Sharon. Anti–*Escherichia coli* adhesion activity of cranberry and blueberry juices. N Engl J Med 324:1599, 1991.
4. S Ahuja, B Kaak, J Roberts. Loss of fimbrial adhesion with the addition of *Vaccinium macrocarpon* to the growth medium of P-fimbriated *Escherichia coli*. J Urol 159:559–562, 1998.
5. AB Howell, N Vorsa, AD Marderosian, LY Foo. Inhibition of the adherence of P-fimbriated *Escherichia coli* to uroepithelial-cell surfaces by proanthocyanadin extracts from cranberries. N Engl J Med 339:1085–1086, 1998.
6. G Galletta. Blueberries and cranberries. In: Janick J, Moore J, eds. Advances in Fruit Breeding. In: Purdue University Press, West Lafayette, 1975, pp 154–196.
7. W Frantz, J Tullie. To renovate or not? Harvest, Spring 1996, pp 10–11, 15.
8. L Bruederle, M Hugan, J Dignan, N Vorsa. Genetic variation in natural populations of the large cranberry, *Vaccinium macrocarpon* Ait. (Ericaceae). Bull Torrey Bot Club 123:41–47, 1996.
9. N Vorsa. On a wing: The genetics and taxonomy of *Vaccinium* species from a pollination perspective. In: Yarborough D, Smagula J, eds. Proceedings of the Sixth International Symposium on *Vaccinium* Culture, ISHS, Orono, ME, 1997, pp 59–66.
10. PV Oudemans, FL Caruso, AW Stretch. Cranberry fruit rot in the Northeast: A complex disease. Plant Dis 82:1176–1184, 1998.
11. SG Billings, CK Chin, G Jelenkovic. Regeneration of blueberry plantlets from leaf segments. HortScience 23:763–766, 1988.
12. IM Dweikat, PM Lyrene. Adventitious shoot production from leaves of blueberry cultivated in vitro. HortScience 23:629, 1988.
13. P Callow, K Haghighi, M Giroux, J Hancock. In vitro shoot regeneration on leaf tissue from micropropagated highbush blueberry. Hortscience 24:373–375, 1989.
14. LJ Rowland, EL Ogden. Use of a cytokinin conjugate for efficient shoot regeneration from leaf sections of highbush blueberry. HortScience 27:1127–1129, 1992.
15. J Graham, K Greig, R McNicol. Transformation of blueberry without antibiotic selection. Ann Appl Biol 128:557–564, 1996.
16. X Cao, Q Liu, L Rowland, F Hammerschlag. GUS expression in blueberry (*Vaccinium* spp.): Factors influencing *Agrobacterium*-mediated gene transfer efficiency. Plant Cell Rep 18:266–270, 1998.
17. RA Serres, BH McCown. Genetic transformation in *Vaccinium macrocarpon* Ait. (cranberry). In: Bajaj YPS, ed. Biotechnology in Agriculture and Forestry. Berlin, Heidelberg: Springer-Verlag, 1995, pp 299–308.
18. R Scorza, W Welker, L Dunn. The effects of glyphosate, auxin, and cytokinin combinations on in vitro development of cranberry node explants. HortScience 19:66–68, 1984.
19. M Marcotrigiano, SP McGlew. A two-stage micropropagation system for cranberries. J Am Soc Hortic Sci 116:911–916, 1991.
20. R Scorza, WV Welker Jr. Cranberries (*Vaccinium macrocarpon* Ait.). In: Bajaj YPS, ed. Biotechnology in Agriculture and Forestry II. Crops. Berlin, Heidelberg: Springer-Verlag, 1988.
21. R Serres, E Stang, D McCabe, D Russell, D Mahr, B McCown. Gene transfer using electric discharge particle bombardment and recovery of transformed cranberry plants. J Am Soc Hortic Sci 117:174–180, 1992.

22. M Marcotrigiano, SP McGlew, G Hackett, B Chawla. Shoot regeneration from tissue-cultured leaves of the American cranberry (*Vaccinium macrocarpon*). Plant Cell Tissue Organ Cult 44:195–199, 1996.
23. RG Novy, C Kobak, J Goffreda, N Vorsa. RAPDs identify varietal misclassification and regional divergence in cranberry [*Vaccinium macrocarpon* (Ait.) Pursh]. Theor Appl Genet 88:1004–1010, 1994.
24. RG Novy, N Vorsa. Identification of intracultivar genetic heterogeneity in cranberry using silver-stained RAPDs. HortScience 30:600–604, 1995.
25. WC Anderson. Propagation of rhododendrons by tissue culture. Part 1. Development of a culture medium for multiplication of shoots. Proc Int Plant Prop Soc 25:129–134, 1975.
26. T Murashige, F Skoog. A revised medium for rapid growth and bioassays with tobacco tissue cultures. Physiol Plant 15:473–479, 1962.
27. RLM Pierik. In vitro culture of higher plants. Dordrecht, The Netherlands: Martinus Nijhoff, 1987.
28. L Qu, J Polashock, N Vorsa. A highly efficient in vitro cranberry regeneration system using leaf explants. HortScience 35:948–952, 2000.
29. LJ Rowland. Susceptibility of blueberry to infection by *Agrobacterium tumefaciens*. HortScience 25:1659, 1990.
30. E Hood, S Gelvin, R Fraley, M Chilton. The hypovirulence of *Agrobacterium tumefaciens* A281 in encoded in the region of pTiBo542 outside the T-DNA. J Bacteriol 168:1291–1301, 1986.
31. H Trick, J Finer. SAAT: Sonication assisted *Agrobacterium* transformation. Transgenic Res 6:329–334, 1997.
32. B Frame, P Drayton, S Bagnall, C Lewnau, W Bullock, H Wilson, J Dunwell, J Thompson, K Wang. Production of fertile transgenic maize plants by silicon carbide whisker-mediated transformation. Plant J 6:641–948, 1994.
33. J Thompson, K Wang, P Drayton. Whisker-mediated plant transformation: An alternative technology. In Vitro Cell Dev Biol 31:101, 1995.
34. C Meurer, R Dinkins, G Collins. Factors affecting soybean cotyledonary node transformation. Plant Cell Rep 18:180–186, 1998.
35. L-J Zhang, L-M Cheng, N Xu, N-M Zhao, C-g Li, J Yuan, S-R Jia. Efficient transformation of tobacco by ultrasonication. Biotechnology 9:996–997, 1991.
36. G Maheswaran, M Welander, J Hutchinson, M Graham, D Richards. Transformation of apple rootstock M26 with *Agrobacterium tumefaciens*. J Plant Physiol 139:560–568, 1992.
37. FA Hammerschlag, RH Zimmerman, UL Yadava, S Hunsucker, P Gercheva. Effect of antibiotics and exposure to an acidified medium on the elimination of *Agrobacterium tumefaciens* from apple leaf explants and on shoot regeneration. J Am Soc Hortic Sci 122:758–763, 1997.
38. R Jefferson, T Kavanagh, M Bevan. GUS fusions: β-Glucuronidase as a sensitive and versatile gene fusion marker in higher plants. EMBO J 6:3901–3907, 1987.
39. G Vancanneyt, R Schmidt, A O'Connor-Sanchez, L Willmitzer, M Rocha-Sosa. Construction of an intron-containing marker gene: Splicing of the intron in transgenic plants and its use in monitoring early events in *Agrobacterium*-mediated plant transformation. Mol Gen Genet 220:245–250, 1990.
40. R Serres, B McCown, E Zeldin. Detectable β-glucuronidase activity in transgenic cranberry is affected by endogenous inhibitors and plant development. Plant Cell Rep 16:641–646, 1997.
41. RA Serres. Genetic transformation of cranberry. PhD thesis, University of Wisconsin, Madison, 1993, p 165.
42. P Luthy, C Hofmann, F Jaquet. Inactivation of delta-endotoxin of *Bacillus thuringiensis* by tannin. FEMS Microbiol Lett 28:31–33, 1985.
43. A Navon, J Hare, B Federici. Interactions among *Heliothis virescens* larvae, cotton condensed tannin and the CrylA(c) δ-endotoxin of *Bacillus thuringiensis*. J Chem Ecol 19:2485–2499, 1993.
44. A Haldrup, S Petersen, F Okkels. The xylose isomerase gene from *Thermoanaerobacterium thermosulfurogenes* allows effective selection of transgenic plant cells using D-xylose as the selection agent. Plant Mol Biol 37:287, 1998.

# 27
# Transgenic Grapevines

**Dennis J. Gray, Subramanian Jayasankar, and Zhijian Li**
*University of Florida, Apopka, Florida*

**John Cordts**
*Profigen, Inc., Paterson, Washington*

**Ralph Scorza and C. Srinivasan**
*U.S. Department of Agriculture, Kearneysville, West Virginia*

|      |                                                        |     |
|------|--------------------------------------------------------|-----|
| I.   | INTRODUCTION                                           | 397 |
| II.  | REGENERATION SYSTEMS                                   | 398 |
| III. | SELECTABLE MARKERS                                     | 399 |
| IV.  | GENE DELIVERY SYSTEMS                                  | 400 |
|      | A. Particle Bombardment                                | 400 |
|      | B. *Agrobacterium* Species                             | 400 |
| V.   | PROGRESS IN CREATION OF TRANSGENIC GRAPEVINE CLONES    | 402 |
|      | REFERENCES                                             | 403 |

## I. INTRODUCTION

Grape (*Vitis* sp.) is the world's most valuable fruit crop as a result of its unique use for wine and as a fresh fruit and raisin. For wine, phenotype is particularly important and varietal fidelity is of paramount concern. Development of new varieties is hampered because grape is a woody perennial species with a life cycle of 2 to 5 years. It exhibits inbreeding depression so that homozygous lines cannot be developed and, in a breeding program, vigor must be scrupulously maintained by outcrossing. All of these factors conspire so that grape varieties must be propagated vegetatively to maintain clonal fidelity, and use of breeding is not a particularly viable method for improvement (although it is used successfully to breed new seedless table grapes). In all other instances, varietal improvement is achieved incrementally by selection of variant phenotypes from stands of established varieties. Such "clonal selection" (1) has been effectively used to produce recognizable subclones of many varieties. However, clonal selection is a random, "hit-or-miss" approach that does not allow intentional introduction of desirable traits, such as disease resistance. Like most crops, grape is subject to continual disease pressures; it would be advantageous to utilize genetic improvement to maintain or improve established production standards in existing viticultural regions and allow wider environmental adaptation so that grape production can be expanded into previously inhospitable regions. Genetic transformation is an attractive alternative for genetic improvement because it potentially alleviates the shortcomings of conventional approaches by allowing addition of specific traits into otherwise desirable varieties.

Despite development of resources and procedures that allow transformation in a wide range of crops species, grape has remained difficult, suggesting that other factors were at play. For example, only recently have stably transformed grapevines of important varieties been obtained This chapter reviews the literature on grape transformation from the perspective of technology development. Thus, we discuss regeneration systems, gene delivery systems, selectable markers, culture/selection systems, functional genes, and, finally, the current state of progress in the field.

## II. REGENERATION SYSTEMS

In studies in which the objective was mainly to insert and express a gene of interest, rather than obtaining transgenic plants, a regeneration system was not required, such that nonmorphogenic leaves and young stem segments could be used as target tissue (2). Both in vitro and greenhouse grown tissues have been used in such studies; however, in vitro grown tissues are preferred because of their aseptic nature. Recovery of transgenic plants requires a regeneration system, from which plants are ultimately derived from single cells. Without such a system, transformants would invariably be chimeric (i.e., only segments of the tissues and/or organs would be composed of transgenic cells). Potential choices for grapevine include meristem, organogenic, and embryogenic regeneration systems.

Transformation of shoot apical meristems has been successful for a few crop species. To develop transgenic grapes in this system, a stringent selection system is required, since the newly transformed cells usually are already differentiated and part of a developing organ. Without selection, chimeras may occur. Transformation of grape via use of meristem tissues is an attractive prospect because of the ready availability of in vitro adventitious shoot cultures for a wide range of varieties, however, to date, there has been no progress in this area.

The first report of stably transformed grape plants utilized an organogenic system of *Vitis rupestris* 'St. George' (3). In this, adventitious buds induced from the hypocotyls of somatic embryos served as the targets for gene delivery. Several transgenic vines were recovered from these cultures. Concurrent study of adventitious cultures from *Vitis vinifera* cultivars 'Cabernet Sauvignon' and 'Chardonnay,' which were initiated from vegetative nodes (not somatic embryos), resulted in transgenic buds, but plants were not recovered. Although shoot organogenesis was reported from leaves of a number of species and cultivars (4–6), this regeneration system, other than in one report, was never again shown to function in transgenic plant recovery, and the approach seems to have been abandoned.

Most progress in grape transformation has involved embryogenic cultures (7–9). Since the embryos appear to arise from single cells, somatic embryogenesis represents an ideal system for transformation. In grape, embryogenic cultures are most commonly induced from floral organs, such as stamens, and, less commonly, from vegetative organs, such as leaves. In some instances, cultures have been maintained for several years (10), allowing a potentially inexhaustible supply of tissue for transformation. Proliferation of somatic embryos occurs by two related developmental pathways—indirect and/or direct embryogenesis. With indirect embryogenesis, cells within unorganized masses of embryogenic tissue, termed *proembryonal complexes* (PEMs), give rise to somatic embryos (7). With direct embryogenesis, well-developed somatic embryos give rise to new embryos, primarily from epidermal or subepidermal cells located at the radicle (or root)/hypocotyl boundary. Thus, direct embryogenesis results in clusters of concurrently developing embryos that are attached at their bases (the site where the radicle is barely exposed from encasing hypocotyl tissue) or, in germinating embryos, where root meets hypocotyl. Both indirect and direct embryogenesis can occur in the same culture, and apparently both regeneration pathways have been utilized for transformation, although it is sometimes difficult to determine from the descriptive narratives provided in published protocols. Scorza and coworkers (11,12) clearly utilized isolated small embryos for transformation and recovered transformants by direct embryogenesis. Embryogenic callus was used in several other successful procedures (13–16). However, in these studies, it is difficult to ascertain the embryogenic regeneration pathway. Kikkert and colleagues (17) utilized an embryogenic suspension culture, suggesting more clearly that indirect secondary embryogenesis was the mode of regeneration. The primary handicap of embryogenic culture for grape transformation is that embryogenic cultures are difficult to induce and the response remains very genotype-dependent. Currently it is not possible to induce embryogenic cultures from many desirable varieties, and, hence, it is not possible to transform them.

## III. SELECTABLE MARKERS

Selectable markers permit resistant transformed cells to outgrow their nontransformed counterparts while under selection pressure and facilitate the successful recovery of transformants. Reporter markers, on the other hand, provide transformed cells with a visible phenotype that facilitates early identification of transformants at various stages of cell growth and development. Reporter markers can also be used as indicators of transgene integration that can be quantitatively scored for monitoring transformation conditions and for fine-tuning the parameters that affect transformation efficiency.

The most commonly used selectable marker for grape is the neomycin phosphotransferase II (NPT II) gene, which, when expressed, detoxifies and thus provides resistance to a number of antibiotics. The most commonly used of these antibiotics, kanamycin, is applied at 25–50 mg/L to select transgenic grapes successfully (e.g., 11–13,17). Interestingly, controversial and contradictory reports state that grape may be either too sensitive to kanamycin or not sensitive enough (9). Colby and Meredith (18) reported that 7 mg/L of kanamycin inhibited adventitious shoot formation in nontransgenic grape, causing them to speculate that even NPT II–positive cultures may not be resistant enough to select. Conversely, Perl and coworkers (15) reported that with their culture/transformation system kanamycin did not inhibit embryogenesis in control cultures at levels as high as 500 mg/L, so that the NPT II gene could not be used as a selectable marker. To circumvent this problem, genes for resistance to the antibiotic hygromycin and the herbicide Basta were used to select transgenic plants successfully. Differences in conclusions reached by these two studies may have been due to the testing of adventitious buds (not embryogenic cells) in the former study and the use of specific antioxidants in the latter study. Regardless of the contradictory reports, it clearly is possible to utilize NPT II gene expression, along with kanamycin, to select transgenic grape plants (11,12).

Mullins and associates (3) used the β-glucuronidase (GUS) gene as a reporter marker to demonstrate that in the absence of the selectable agent kanamycin, no transformants were recovered from various explants of both *V. rupestris* and *V. vinifera* genotypes previously inoculated with *Agrobacterium* sp. In contrast, GUS-positive transformants were produced exclusively in the presence of 15–20 mg/L kanamycin. However, nonchimeric transgenic plantlets of *V. rupestris* 'St. George' were obtained on a selection medium containing 10 mg/L kanamycin (3). Analysis of GUS activity in a large number of putative transformants of *V. rupestris* and a hybrid indicated that only about 30% of recovered plantlets selected on 10 mg/L kanamycin were nonchimeric transformants, and the remainder were either escaped nontransformed plantlets or chimeras (13). The GUS gene was also extensively utilized in studies of transformation of *V. vinifera* for identification and analysis of transgenic plants (11,12) and in studies to develop a biolistic transformation system in grapevine (17,19).

In a similar approach, Berres and associates (2) used a GUS histochemical assay to show that the T-6b gene derived from pTiTm4 was capable of stimulating the growth of transformed grape cells after *Agrobacterium* sp.–mediated transformation. The transformation eventually led to the formation of chimeric shoot buds from leaf explants in part as a result of the multicellular origin of adventitious buds and the lack of selection pressure to suppress growth of nontransformed cells.

More recently, the green fluorescent protein gene (GFP), isolated from the Pacific Northwest jellyfish (*Aequorea victoria*), has become a widely used reporter marker to monitor transgene expression in plants. GFP fluoresces in vivo when it receives light energy rather than as a result of a chemical reaction. GFP is unique in that its use does not require phytotoxic substrates as do most other reporter systems, including GUS. Therefore, GFP allows continuous monitoring of transgene expression without the need to sacrifice valuable tissue for analysis. Li and coworkers (20) used GFP to characterize factors that affect *Agrobacterium* sp.–mediated transformation of embryogenic cultures of various grapevine genotypes. In their study, susceptibility of 18 genotypes to *Agrobacterium* sp. infection and their ability to form stably transformed calli were determined. Equally importantly, the use of GFP facilitated early identification of nonchimeric transformants and permitted the recovery of transgenic grape somatic embryos and plantlets within a relatively short time. It is likely that GFP will find additional use as a reporter marker, not only to optimize transformation efficiency, but also to study gene expression and regulation in transgenic grape systems.

## IV. GENE DELIVERY SYSTEMS

A number of gene delivery systems have been tried to produce transgenic plants, including electroporation, electrophoresis, and piercing of cells with fibers. However, only two systems, particle bombardment– and *Agrobacterium* sp.–mediated transformation, have been successfully used for grape.

### A. Particle Bombardment

Particle bombardment, wherein microscopic gold or tungsten particles are coated with deoxyribonucleic acid (DNA), are explosively discharged into target tissues, has been extensively used to insert desirable genes into a range of crop species (21). Gray and associates (22) reported extensive transient expression of the GUS gene in embryogenic cultures of *V. vinifera* 'Thompson Seedless,' which rendered the bombarded side of somatic embryos dark blue in color. However, it was not until 1996 that transgenic plants of the *Vitis* hybrid 'Chancellor' were obtained by particle bombardment of embryogenic suspension cultures (17); this represents the only successful report of plant recovery by this approach. Difficulty in obtaining transgenic plants via particle bombardment may be due to the tendency of delivering multiple or partial gene copies into cells or to the trauma of mechanical damage to cells by the biolistic bombardment. Recovery of transgenic plants may become difficult because damaged cells are prone to abnormalities, including reduction of vigor, or, in the case of multicopy inserts, gene silencing may occur and transgenic cells may not express desired traits.

### B. *Agrobacterium* Species

*Agrobacterium* sp.–mediated transformation is the most commonly used method of inserting genes into a wide range of plants. Grape is a natural host for *Agrobacterium tumefaciens*, and crown gall is a commonly encountered, often serious, disease (23), initially suggesting that transformation would be a simple task. In practice, however, grape has been difficult to transform as evidenced by numerous failed efforts since the 1990s. Early failures led researchers to examine possible obstacles. The possible need to optimize *Agrobacterium* sp. strains was examined, since biovar 3 is the strain that infects grape, whereas biovar 1 is used commonly for transformation. However, both biovars produced galls on in vitro–grown shoots (24), suggesting that broad differences in strains were not problematic. Transgenic root cultures that produced opines (25) and an introduced marker gene (26) were recovered after *Agrobacterium rhizogenes* inoculation.

1. Culture of *Agrobacterium* Species for Transformation

Transformation has been accomplished by using *Agrobacterium* sp. strains LBA 4404, EHA 101, and EHA 105. However, the bacterial culture conditions preceding cocultivation vary widely. Generally, the bacterium is grown overnight in LB medium containing appropriate antibiotics at 28°C to a density of 0.2–0.6 OD at 600 nm. The culture is then pelleted and resuspended in a liquid plant growth medium containing 2–3% sucrose. This diluted bacterial culture in plant growth medium is used for transformation (15,27). Nakano and colleagues (14) added 20 µM acetosyringone as an induction agent in the *A. rhizogenes* cultures. Similarly, Scorza and associates (11,12) recommended resuspending overnight grown bacterial cultures in Murashige and Skoog medium containing 100 µM acetosyringone, 200 µM betaine phosphate, and 2% sucrose and incubation at 20°C for 6 hours to induce *Agrobacterium* sp. virulence genes prior to inoculation. Whether or not there was any beneficial effect of culturing the bacteria in medium with this chemical inducer was not explored.

2. Culture of Target Tissue Before Transformation

In several transformation systems, preculture of target tissue on a specified medium has been reported to increase transformation efficiency. Perl and coworkers (15) reported that culturing embryogenic calli in MS medium containing NOA (2 mg/L) for 7 days resulted in proembryogenic calli that are highly suitable for transformation. They indicated that substituting NOA for 2, 4-D resulted in the formation of better proembryogenic tissue for transformation. Perhaps the nascent proembryogenic mass that formed after 7 days of subculture consisted of more actively dividing cells and, thus, was a better target tissue for transformation. Scorza and coworkers (11,12) cultured target tissue in ER proliferation medium (28) for several weeks prior to transformation. In addition to having high levels of cysteine, a sulfur-containing amino acid with strong antioxidant properties, this medium also has a high gel strength and lesser duration of time specified for autoclaving. Whether or not these factors contributed to increased transformation efficiency is not known. Nakano and associates (14) used leaf-derived embryogenic callus in a modified Nitsch and Nitsch medium containing 2,4-D and 4PU. They also recommended using freshly subcultured tissue for transformation. Hebert and associates (19), who utilized fine suspension–derived embryogenic tissue of 'Chancellor' grapes and obtained consistently high transformation rates, maintained suspension cultures in MS medium containing 2,4-D.

3. Cocultivation and Selection of Transformed Cultures

Cocultivation is typically started in a suspension of bacterial cells and embryogenic tissue, after which the cultures are plated on solid medium. The duration of cocultivation varies from just 1 hour (2) to 3 days (14). For *V. vinifera* cultivars, cocultivation for 2 days appears to be ideal (11,15). Most of these studies suggest cocultivating on filter paper over a solid medium containing acetosyringone. Selection of transformed cultures typically began within 2 days (17). However, Scorza and colleagues (11) cultured the putative transformants for up to 6 weeks without selection, then transferred actively growing cultures to selection medium. In addition to kanamycin, the growth medium was supplemented with antibiotics such as carbenicillin and cefotaxime to suppress *Agrobacterium* sp. for several weeks after cocultivation. Whenever a marker gene, such as GUS, was used in the construct, periodic assays were performed to monitor and optimize the selection process.

The first stably transformed grape plants were obtained by cocultivation of the adventitious buds of rootstock *V. rupestris* 'St. George' with *A. tumefaciens* (3). *A. rhizogenes* was used to produce the first transgenic plants from somatic embryos of *V. vinifera* 'Koshusanjaku', which were resistant to the kanamycin and expressed the GUS (14). *A. tumefaciens* was used to transform embryogenic callus of *V. rupestris* and *V. rupestris* × *V. berlandieri* '110 Richter' (13). Resulting plants expressed both NPT II and GUS and, significantly, the grapevine fanleaf nepovirus coat protein gene. Scorza and associates (11) combined particle bombardment wounding with *A. tumefaciens* to obtain transgenic plants from zygotic embryo–derived somatic embryos of *V. vinifera*. Presumably, the particle wounding treatment increased access of target cells to bacteria. In addition, the cocultivation and plant regeneration media contained a high concentration of cysteine. Resulting plants contained the NPT II and GUS genes. Subsequently, the particle wounding/*Agrobacterium* transformation procedure was used to obtain plants of *V. vinifera* 'Thompson Seedless,' which, in addition to the NPT II and GUS genes, expressed the Shiva-1 synthetic lytic peptide gene (12). A different approach utilized treatment 0.25% PVPP, an antioxidant, for 1 week after cocultivation to obtain transgenic plants from embryogenic cultures of *V. vinifera* 'Superior Seedless' (15). As in the work of Scorza and coworkers (11,12), antioxidants were shown to prevent culture browning during cocultivation and selection, allowing better transgenic plant recovery. Perl and coworkers (16) used this protocol to obtain transgenic plants of *V. vinifera* 'Red Globe.'

## V. PROGRESS IN THE CREATION OF TRANSGENIC GRAPEVINE CLONES

Reports of functional genes (excluding antibiotic resistance genes) that are expressed in grapevine cultures have been infrequent. An early example (12) documented the insertion of a gene encoding a lytic peptide with antimicrobial properties into 'Thompson Seedless.' However, progress in the practical use of molecular genetics for grapevine improvement is accelerating, as evidenced by reports presented at the Seventh International Symposium on Grapevine Genetics and Breeding, held in Montpellier, France, in 1998. Numerous oral and written presentations demonstrated a variety of global interests as well as successes in grapevine transformation.

Major needs expressed by grape growers concerned issues of pathogen resistance (fungus, virus, bacteria), insect resistance, and qualities/physiological characteristics, such as seedlessness, antioxidant levels, and browning control. Several groups reported having transgenic plants in lab or field evaluation programs to test for resistance to grapevine fanleaf virus (GFLV), a damaging and widespread nepovirus found in most grape growing regions of the world (29–35). Two groups developed transgenic plants containing other coat protein genes for resistance to grapevine viruses A and B (GVA and GVB), believed to be involved in Kober stem grooving and corky bark (29,36). Other groups are interested in developing resistance to arabis mosaic virus (29,34) and grapevine leafroll virus (31).

Transgenic clones of rootstock 'Couderc 3309' (*V. riperia* × *V. rupestris*), 'Richter 110' (*V. rupestris* × *V. berlandieri*), and 'Teleki 5C' (*V. berlandieri* × *V. riparia*) containing the mutant *virE2* del B gene from *A. tumefaciens* are being tested for resistance to crown gall (*A. vitis* and *A. tumefaciens*) (32). Harst and coworkers (37) and Kikkert and coworkers (38) developed transgenic plants with chitinase/endochitinase genes, which are being evaluated for resistance to grey mold and powdery mildew. Reustle and Matt (39) have also used an endochitinase gene, but did not recover transgenic plants.

Levenko and Rubtsova (40) used Bar gene to develop resistance of grapevines to the herbicide phosphinothricin (Basta). Similarly, Gollop and colleagues (41) are studying proanthocyanidin biosynthesis by transforming with the genes for dihydroflavonal reductase (DFR) and leucoanthocyanidin dioxygenase (LDOX). Transgenic vines containing the Bar gene were resistant to 20 ml/L Basta, a concentration useful for weed control. Franks and coworkers (42) are working to silence polyphenol oxidase (PPO) genes in plants to control browning in raisins. Plants in the field are waiting to be evaluated for PPO activity and dried fruit color. Finally, vines containing antifreezing genes have been developed (35). The nature of the genes was not clear, but the intention is presumably to expand the useful growing regions for grape and/or increase their hardiness to localized freezing conditions.

The skills and techniques necessary to transform, select, and regenerate transgenic grapevines are in place. A number of groups are currently placing potentially useful genes into existing varieties. In the near future, plants with useful attributes will be identified and the persistence of the traits over an adequate period in perennial grape will be tested. Careful testing of clonal fidelity will be necessary, in particular, to determine the ultimate usefulness of transgenic wine grape cultivars.

## REFERENCES

1. JM Rantz. International Symposium on Clonal Selection. Portland, OR: American Society for Enology and Viticulture, 1995, pp 1–163.
2. R Berres, L Otten, B Tinland, E Malgarini-Clog, B Walter. Transformation of *Vitis* tissue by different strains of *Agrobacterium tumefaciens* containing the T-6b gene. Plant Cell Rep 11:192–195 1992.
3. MG Mullins, FCA Tang, D Facciotti. *Agrobacterium*-mediated genetic transformation of grapevines: Transgenic plants of *Vitis rupestris* Scheele and buds of *Vitis vinifera* L. Biotechnology 8:1041–1045, 1990.
4. Z-M Cheng, BI Reisch. Shoot regeneration from petioles and leaves of *Vitis* × *labruscana* 'Catawba.' Plant Cell Rep 8:403–406, 1989.
5. JA Stamp, SM Colby, CP Meredith. Direct shoot organogenesis and plant regeneration from leaves of grape (*Vitis* spp.). Plant Cell Tissue Org Cult 22:127–133, 1990.
6. JA Stamp, SM Colby, CP Meredith. Improved shoot organogenesis from leaves of grape. J Am Soc Hortic Sci 115:1038–1042, 1999b.
7. DJ Gray. Non-zygotic embryogenesis. In: RN Trigiano, DJ Gray, eds. Plant Tissue Culture Concepts and Laboratory Exercises. CRC Press, 1996, pp 133–147.
8. DJ Gray, CP Meredith. Grape. In: FA Hammerschlag, RE Litz, eds. Biotechnology of Perennial Fruit Crops. CAB International, 1992, pp 229–262.
9. C Srinivasan, R Scorza. Transformation of somatic embryos of trees and grapevine. In: SM Jain, PK Gupta, NJ Newton, eds. Somatic Embryogenesis in Woody Plants. Vol. 5. London: Kluwer Academic, 1999, pp 313–330.
10. DJ Gray. Effects of dehydration and exogenous growth regulators on dormancy, quiescence and germination of grape somatic embryos. In Vitro Cell Dev Biol 25:1173–1178, 1989.
11. Scorza, JM Cordts, DW Ramming, RL Emershad. Transformation of grape (*Vitis vinifera* L.) zygotic-derived somatic embryos and regeneration of transgenic plants. Plant Cell Rep 14:589–592, 1995.
12. Scorza, JM Cordts, DJ Gray, D Gonsalves, RL Emershad, DW Ramming. Producing transgenic 'Thompson Seedless' grape (*Vitis vinifera* L.) plants. J Am Soc Hortic Sci 12:616–619, 1996.
13. S Krastanova, M Perrin, P Barbier, G Demangeat, P Cornuet, N Bardonnet, L Otten, L Pinck, B Walter. Transformation of grapevine rootstocks with the coat protein gene of grapevine fanleaf nepovirus. Plant Cell Rep 14:550–554, 1995.
14. M Nakano, Y Hoshino, M Mii. Regeneration of transgenic plants of grapevine (*Vitis vinifera* L.) via *Agrobacterium rhizogenes*–mediated transformation of embryogenic calli. J Exp Bot 45:649–656, 1994.

15. A Perl, O Lotan, M Abu-Abied, D. Holland. Establishment of an *Agrobacterium*-mediated transformation system for grape (*Vitis vinifera* L.): The role of antioxidants during grape-*Agrobacterium* interactions. Nature Biotechnol 14:624–628, 1996.
16. A Perl, R Gollop, A Lipsky, D Holland, N Sahar, E Or, R Elyasi. Regeneration and transformation of grape (*Vitis vinifera* L). Plant Tissue Cult Biotechnol 2:187–193, 1996.
17. R Kikkert, D Hebert-Soule, PG Wallace, MJ Striem, BI Reisch. Transgenic plantlets of 'Chancellor' grapevine (*Vitis* sp.): from biolistic transformation of embryogenic cell suspensions. Plant Cell Rep 15:311–316, 1996.
18. SM Colby, CP Meredith. Kanamycin sensitivity in cultured tissues of *Vitis*. Plant Cell Rep 9:237–240, 1990.
19. D Hebert, JR Kikkert, FD Smith, BI Reisch. Optimization of biolistic transformation of embryogenic grape cell suspensions. Plant Cell Rep 12:585–589, 1993.
20. ZJ Li, S Jayasankar, NJ Barnett, DJ Gray. Characterization of *Agrobacterium*-mediated transformation of grapevine (*Vitis* spp.) using the green fluorescent protein (GFP) gene. In Vitro Cell Dev Biol Plant 35:42A, 1999.
21. DJ Gray, JJ Finer. Development and operation of five particle guns for introduction of DNA into plant cells. Plant Cell Tissue Org Cult 33:219, 1993.
22. DJ Gray, DD Songstad, ME Compton. Expression of GUS in bombarded grape somatic embryos and cells, 1993 World Congress on Cell and Tissue Culture. In Vitro Cell Dev Biol 29A:65A, 1993.
23. TJ Burr. Crown gall. In: RC Pearson, AC Goheen, eds. Compendium of Grape Diseases, St. Paul: APS Press, 1988, pp 41–42.
24. PR Hemstead, BI Reisch. In vitro production of galls induced by *Agrobacterium tumefaciens* and *Agrobacterium rhizogenes* on *Vitis* and *Rubus*. J Plant Physiol 120:9–17, 1985.
25. I Gribaudo, A Schubert. Grapevine root transformation with *Agrobacterium rhizogenes*. Proceedings Fifth International Symposium on Grape Breeding, Vitis Special Issue, 1990, pp 412–418.
26. D Guellec, C David, M Brabchard, J Tempe. *Agrobacterium rhizogenes*–mediated transformation of grapevine (*Vitis vinifera* L.). Plant Cell Tissue Org Cult 20:211–215, 1990.
27. EW Stover, TJ Burr, HJ Swartz. Transformation of crown gall resistant and susceptible *Vitis* genotypes with *Agrobacterium vitis*. Vitis 35:29–33, 1996.
28. RL Emershad, DW Ramming. Somatic embryogenesis and plant development from immature zygotic embryos of seedless grapes (*Vitis vinifera* L.). Plant Cell Rep 14:6–12, 1994.
29. P Barbier, M Perrin, P Cobanov, B Walter. Probing pathogen-derived resistance against the fanleaf virus in grapevine: An analysis of some more transformants (poster). VIIth International Symposium on Grapevine Genetics and Breeding, Montpellier, France, July 6–10, 1998.
30. R Golles, A da Camara Machado, A Minafra, V Savino, G Saldarelli, GP Martelli, H Puhringer, H Katinger, M Laimer da Camara Machado. Transgenic grapevines expressing coat protein gene sequences of grapevine fanleaf virus, arabis mosaic virus, grapevine virus A and grapevine virus B. VIIth International Symposium on Grapevine Genetics and Breeding, Montpellier, France, July 6–10, 1998.
31. S Krastanova, KS Ling, HY Zhu, B Xue, TJ Burr, D Gonsalves. Development of transgenic grapevine rootstocks with genes from grapevine fanleaf virus and grapevine leafroll associated closteroviruses 2 and 3 (poster). VIIth International Symposium on Grapevine Genetics and Breeding, Montpellier, France, July 6–10, 1998.
32. B Xue, KS Ling, CL Reid, S Krastanova, M Sekiya, EA Momol, S Sule, J Mozsar, D Gonsalves, TJ Burr. Transformation of five grape rootstocks with plant virus genes and a virE2 gene from *Agrobacterium tumefaciens*. In Vitro Cell Dev Biol Plant 35:226–231, 1999.
33. MC Mauro, B Walter, L Pinck, L Valat, P Barbier, M Boulay, P Coutos-Thevenot. Analysis of 41B grapevine rootstocks for grapevine fanleaf virus resistance. VIIth International Symposium on Grapevine Genetics and Breeding, Montpellier, France, July 6–10, 1998.
34. A Spielmann, S Krastanova, V Douet-Orhant, S Marc-Martin, MH Prince Sigrist, P Gugerli. Resistance to nepoviruses in grapevine: Expression of several putative resistance genes in transgenic plants (poster). VIIth International Symposium on Grapevine Genetics and Breeding, Montpellier, France, July 6–10, 1998.
35. I Tsvetkov, V Tsolova, A Atanossov. Gene transfer for stress resistance in grape (poster). VIIth International Symposium on Grapevine Genetics and Breeding, Montpellier, France, July 6–10, 1998.

36. L Martinelli, N Buzkan, A Minafra, P Saldarelli, D Costa, V Poletti, S Festi, A Perl, GP Martelli. Genetic transformation of grape for resistance to viruses related to the rugose wood disease complex. VIIth International Symposium on Grapevine Genetics and Breeding, Montpellier, France, July 6–10, 1998.
37. M Harst, B-A Bornoff, E Zyprian, R Topfer. Regeneration and transformation of different explants of *Vitis vinifera*. VIIth International Symposium on Grapevine Genetics and Breeding, Montpellier, France, July 6–10, 1998.
38. JR Kikkert, GM Reustle, GS Ali, PW Wallace, BI Reisch. Expression of a fungal chitinase in *Vitis vinifera* L.: Merlot and Chardonnay plants produced by biolistic transformation. VIIth International Symposium on Grapevine Genetics and Breeding, Montpellier, France, July 6–10, 1998.
39. GM Reustle, A Matt. First steps to use grapevine protoplasts for breeding purposes. VIIth International Symposium on Grapevine Genetics and Breeding, Montpellier, France, July 6–10, 1998.
40. BA Levenko, MA Rubtsova. Herbicide resistant transgenic plants of grapevine. VIIth International Symposium on Grapevine Genetics and Breeding, Montpellier, France, July 6–10, 1998.
41. R Gollop, Y Eshat, A Perl. Proanthocyanidins production in tissue culture: Regulation of dihydroflavonal reductase and leucoanthocyanidin dioxygenase promoter-GUS-intron fusions in grape (poster). VIIth International Symposium on Grapevine Genetics and Breeding, Montpellier, France, July 6–10, 1998.
42. T Franks, P Iocco, M Thomas. Antisense and sense suppression of polyphenol oxidase in transgenic *Vitis vinifera* cv. Sultana (poster). VIIth International Symposium on Grapevine Genetics and Breeding, Montpellier, France, July 6–10, 1998.

# 28
# Genetic Transformation of Kiwifruit (*Actinidia* species)

## M. Margarida Oliveira
*Faculdade de Ciências de Lisboa, Lisbon, and Instituto de Biologia Experimental e Tecnológica, Oeiras, Portugal*

## M. Helena Raquel
*Instituto de Biologia Experimental e Tecnológica, Oeiras, Portugal*

|  |  |  |
|---|---|---|
| I. | INTRODUCTION | 407 |
| II. | REGENERATION SYSTEMS IN *ACTINIDIA* SPECIES | 408 |
| III. | DEVELOPMENT OF DEOXYRIBONUCLEIC ACID TRANSFER PROTOCOLS | 409 |
|  | A. *Agrobacterium* Species–Mediated Transformation | 409 |
|  | B. Direct Deoxyribonucleic Acid Transfer | 415 |
| IV. | TRANSGENE INTERACTION AND STABILITY IN PROGENY | 417 |
| V. | GENETIC IMPROVEMENT | 417 |
| VI. | CONCLUSIONS | 418 |
|  | REFERENCES | 419 |

## I. INTRODUCTION

Kiwifruit (*Actinidia deliciosa* [A. Chev.] C.F. Liang et A.R. Ferguson var. *deliciosa*) (1) is an amazing example of a fruit that in about one century was domesticated and conquered the world market.

*Actinidia*, a genus containing about 60 species of dioecious vines (2), has its center of evolution probably located in southwestern China. In the 19th century, *Actinidia chinensis* was introduced into Europe, America, and New Zealand from China (3–5).

*Actinidia chinensis* includes diploid and tetraploid varieties and is possibly the progenitor of *Actinidia deliciosa*, the cultivated hexaploid species (6,7).

Compared with other fruits, those of *Actinidia* have high levels of vitamins, mineral salts, and fibers. These qualities, together with the unusual strongly pigmented flesh (usually green, but also yellow or red, depending on the variety), made kiwifruit very attractive to the consumer.

In the beginning of the 20th century, growers in New Zealand raised seedlings from which superior fruiting types were selected and propagated (4). It was after the first exports from New

Zealand to America that the name *kiwifruit* was chosen, in substitution for *Chinese gooseberry,* as it was known in Western countries. Kiwifruit invokes New Zealand's most famous flightless bird, *Apterix* sp. (kiwi), which is an emblem of the country (8).

The female cultivar 'Hayward,' obtained from a selection program in New Zealand in the 1930s, has been the source of the vast majority of the world's cultivated kiwifruit (5). However, to handle the potential problems of kiwifruit monoculture, the development of alternative superior genotypes is recommended (3,9).

Since the 1970s China has become particularly interested in kiwifruit for its economic value, and extensive investigations were carried out on native germplasm, leading to the selection of good *Actinidia* sp. lines (10).

In New Zealand, one of the most important world producers of kiwifruit, the improvement of *Actinidia* species, as an alternative to *A. deliciosa* but with equivalent postharvest characteristics and identical organoleptic qualities, is also one of the present research interests. To meet this goal, in the eighties, a controlled breeding program was initiated in New Zealand. However, for woody species, the production of new varieties can be extremely hard and expensive because of the difficulty in screening seedlings for the important characteristics. Application of a screening strategy at an early stage requires a good knowledge of the genetic basis and the development of molecular markers for important characteristics. By using random amplified polymorphic DNA (RAPD)-based bulk segregant analysis it was already possible to identify two sex-linked markers with inheritance patterns in diploid species that show a X and Y linkage, the male corresponding to the heterogametic sex (2). These studies are of particular importance for the early identification of male and female vines, which was not possible previously because of the absence of differentiating morphological characters before flower production, and because of the small size and large number of kiwifruit chromosomes ($2n = 174$ in hexaploid species) and the fact that they are cytologically indistinguishable (2).

To support the breeding program, New Zealand is presently developing a kiwifruit mapping project based on microsatellites (L. Fraser, personal communication, 1999).

Other improvement goals are, for instance, the development of increased resistance or tolerance to biotic or abiotic stresses. Pests and diseases still require wide application of chemicals (11), and low temperatures in winter and strong wind account for severe damage (3).

In parallel with the conventional breeding strategies, other strategies based on genetic engineering have emerged since the early 1990s, using tissue culture work already available, and are now starting to give the first promising results.

## II. REGENERATION SYSTEMS IN *ACTINIDIA* SPECIES

Being a dioecious plant, *Actinidia* sp. is highly heterozygous, complicating the propagation of selected plants through seeds. However, the rapid propagation of nursery material could be easily achieved by node culture, and this potential was explored as a substitute for the traditional cuttings and graftings.

In a short time, tissue culture work in *Actinidia* sp. expanded greatly, and a variety of explant types and regeneration strategies (meristem multiplication, adventitious shoot or root induction and regeneration, and somatic embryogenesis) are presently available. Plant regeneration was accomplished from shoot tips or axillary meristems, stem segments, leaves, filaments, anthers, endosperm, cotyledons, rescued hybrid embryos, calluses, cell suspensions, and protoplasts. The plants used in these studies included several *Actinidia* species, namely *A. chinensis*, *A. deliciosa*, *A. arguta*, *A. eriantha*, and *A. polygama*, as well as different genotypes among male and female clones (for reviews see Refs. 10–17).

## III. DEVELOPMENT OF DEOXYRIBONUCLEIC ACID TRANSFER PROTOCOLS

Deoxyribonucleic acid (DNA) transfer protocols in kiwifruit are primarily based on *Agrobacterium* sp. as transformation vector (Tables 1 and 2, Fig. 1). Direct DNA transfer to protoplasts, however, has also been applied, both for transient expression analyses and for stable transformation (11) (Table 1, Figs. 2 and 3). In spite of the increasing importance of particle bombardment for genetic transformation, this system has not yet been applied to *Actinidia* sp.

Most work was dedicated to *A. deliciosa* var. *deliciosa* cv. 'Hayward,' but male clones of this species were also used, as well as various genotypes of *A. chinensis* (Tables 1 and 2). As the regeneration system and the availability of explant material are among the most important factors for establishment of a transformation protocol, most authors have used leaves of micropropagated plantlets as explants for *Agrobacterium* sp. infection. Leaves provide a large amount of tissue with a reasonably good regeneration potential, and those from in vitro cultures usually respond better and avoid the contamination problems often found on explants taken from field material. To a lesser extent, stem segments and hypocotyl slices have also been used.

### A. *Agrobacterium* Species–Mediated Transformation

Several *Agrobacterium* sp. strains have been used for kiwifruit transformation; the most important are LBA4404 (11,18–21), C58 (7,18,19), and A281 (7,19,22,23). *Agrobacterium* sp. strains C58 and A281 have the disadvantage of being virulent and therefore possibly also transferring the opine and hormone synthesizing genes located on the transfer DNA (T-DNA), but their use has already provided interesting results and allowed the definition of efficient strategies for *Agrobacterium* sp.–mediated kiwifruit transformation.

Gall induction tests in kiwifruit have shown little difference among several strains, while yielding only small and irregular galls (18). However, callus independent growth, with nopaline production, could be achieved by using C58 (24).

The first report on regeneration of transgenic kiwifruit plants came from Japan (25), where strain EHA101 (corresponding to the disarmed A281) was used both for germinants and for explants taken from 'Hayward' field material. Infection was performed by tissue immersion in the *Agrobacterium* sp. suspension with vigorous shaking for 20 minutes, followed by cleaning on sterilized paper and cocultivation for 1 day. Selection was applied 2 weeks after cocultivation, on media supplemented with 25 mg/l kanamycin. These authors tested two culture media for shoot induction and had the best results when using B5 (26) with 3 mg/l zeatin, solidified with 2 g/l gelrite.

Important features in *Agrobacterium* sp.–mediated transformation of *A. deliciosa* arose from the work of Janssen (27) and Janssen and Gardner (19), who identified some critical conditions that were later used with success for *A. chinensis* transformation (7). According to Janssen and Gardner (19) the primary obstacle in kiwifruit transformation was the low frequency T-DNA transfer to the explant cells. In these infection experiments, leaf strips were taken from micropropagated plantlets, immersed in the bacterial culture (LBA4404, A281, EHA101, or C58) for 1 minute, gently blotted dry, rinsed in sterile water, placed on regeneration medium, and cocultivated for 2 days. Selection on 100 mg/l kanamycin was initiated immediately after the cocultivation period on medium containing 0.275 g/l phytagel, 5 mg/l zeatin, and 0.1 mg/l NAA. To improve T-DNA transfer Janssen and Gardner (19) designed a number of transformation experiments, from which they could draw a number of conclusions: (a) the source plants should be maintained on Phytagel and frequently transferred to fresh medium; (b) the infected explants should be placed on moist filter paper during the cocultivation period; (c) leaf preculture should

Table 1
Development of Gene Transfer Conditions and Stable Transformation Protocols

| Species/cultivar | Plant material | DNA transfer system | Plasmid/transgenes | Results | To retain | Ref. |
|---|---|---|---|---|---|---|
| Actinidia chinensis, A. deliciosa var. deliciosa cv. 'Hayward' | Stems of greenhouse and micropropagated plantlets | Agrobacterium sp. A722, C58, ICMP8302, ICMP8326, ID1576, LBA4404, A4T | pKIWI110 (uidA, nptII) | Hairy root induction by A4T; crown gall induction by all the other strains except LBA4404 GUS expression analyzed. | First published report on transgene expression in Actinidia sp. | 18 |
|  | Hypocotyl slices, from: 'Hayward' seedlings and stem cuttings of cv. 'Hayward' | Agrobacterium EHA101, 1 day cocultivation | pLAN411, pLAN421 (uidA, nptII) | Shoots were regenerated under a selection pressure of 25 µg/ml kan, GUS and NPT II expression (dot blot) were obtained for 85% of the green shoots regenerated from hypocotyl, molecular confirmation of transgene integration was not provided | First report of transgenic kiwifruit plants, variable GUS activity in transgenic plants, most green shoots of hypocotyl transformation expressed NPT II | 25 |
|  | Protoplasts isolated from petiole-derived callus | Direct DNA transfer using PEG 4000 (20% or 30%) | pDW2 (cat) | CAT transient expression analyzed by TLC and GC-MS; 30% PEG and heat shock increased transient expression but drastically reduced viability; carrier DNA reduced CAT transient expression | All the tested direct DNA transfer conditions yielded transient expression, meaning that those allowing higher cell viability should be chosen for stable transformation experiments | 34 |
|  | Leaves of micropropagated plantlets | Agrobacterium sp. LBA4404, C58, A281, EHA101 | pKIWI105, pKIWI110, pKIWI109, pLAN421 (uidA, nptII) | Evaluation of factors rulling transformation frequency (transient GUS as indicator): gene transfer rates slightly higher with A281 than with the other strains; pLAN421 yielded much higher percentage of GUS-positive plants than pKIWI110; transgenic shoots were resistant to 50 µg/ml kan; transformation was confirmed by GUS expression, PCR, Southern blotting | Optimized conditions: source plants maintained on Phytagel and frequently transferred to fresh medium; use of a filter paper layer during cocultivation; preculture detrimental; 20 µM acetosyringone increased transient GUS expression; transgenic shoots were efficiently recovered | 19,27 |
|  | Protoplasts isolated from petiole-derived callus | Direct DNA transfer (electroporation: 4 pulses 200–1300 V/cm; PEG treatment: PEG4000 20%) | p35SGUS (uidA) pDW2 (cat), pGP6 (nptII), pB1121 and pTi35SGUS (uidA, nptII) | Higher CAT and GUS transient expression (up to 100-fold) with PEG-treated protoplasts; microcolonies selected on 25 µg/ml kan; 70% of the shoots regenerated from electroporated protoplasts (200 V/cm) and tested for NPT II expression (dot blot) were positive; molecular confirmation of transgene integration was not provided; PCR detection of NPT II in Fig. 3 | PEG treatment yields much higher transient expression levels than electroporation, but electroporation at low voltage (200 V/cm) allowed higher cell viability (80%) and regeneration of kanamycin resistant NPT II–positive shoots; from PEG treatment no NPT II–positive regenerated shoots were detected | 11,28 |

# Genetic Transformation of Kiwifruit

| Plant material | Explant | Transformation method | Vector (genes) | Results | Comments | Ref. |
|---|---|---|---|---|---|---|
| *Actinidia deliciosa* var. *deliciosa* cv. 'Hayward' | Petiole, stem, and root segments of micropropagated plantlets | *Agrobacterium* sp. LBA4404 | p35SGUSINT (*uidA* intron, *nptII*) | Selection on 25–100 μg/ml kan-resistant shoots expressed NPT II (dot blot); GUS expression was usually absent or very weak; molecular confirmation of transgene integration was not provided | Explant preculture increased GUS expression 20 days after cocultivation; kanamycin-resistant NPT II–positive shoots were regenerated | 11,28 |
| *Actinidia chinensis* (four genotypes) *A. deliciosa* (four genotypes) | Young half-expanded leaves, petiole, or stem segments of micropropagated plantlets | *Agrobacterium* sp. C58, A281 | PKIWI105 (with *uidA* and *nptII*) | Transgenic plants regenerated from all genotypes; transformation confirmed by GUS expression, PCR amplification, and Southern blotting (Phytagel was used in all cultures and a filter paper layer was placed on the surface of each cocultivation plate) | No marked differences between C58 and A281 in efficiency of transformation; in the tested conditions *A. chinensis* regenerated better (faster and in higher numbers) than *A. deliciosa* | 7 |
| *Actinidia deliciosa* var. *deliciosa* cv. 'Hayward' | Protoplasts isolated from leaves of micropropagated plantlets | Direct DNA transfer using PEG 4000 (20%) | p35SGUS (*uidA*) | Transient GUS expression analyzed by HPLC; leaf preculture on Z-containing medium (as opposed to NAA), increased 2- to 2.5-fold the yield of viable protoplasts; higher GUS expression on protoplasts from epidermis and leaf veins as compared to those from mesophyll with a 10-fold increase in expression after heat shock | Leaf preculture on zeatin-containing medium significantly improved the yield of viable protoplasts, especially those from epidermis and leaf veins (fraction 1), may be efficiently used to study the expression of new promoters | 15 |
| *A. deliciosa* var. *deliciosa* male and female clones | Seeds obtained from field crosses between plants transformed by Janssen and Gardner (19) | (*Agrobacterium* sp. A281) | (pLAN421, a single border binary vector with *uidA* and *nptII*) | Seeds were germinated on 250 μg/ml kan and later transferred to the containment greenhouse; GUS expression varied in some progeny plants, suggesting gene silencing; on two female primary transgenic plants, analyzed in detail, no actively expressing copies of *nptII* and *uidA* were linked | Inheritance and expression of transgenes were demonstrated in kiwifruit progeny; large numbers of rearranged T-DNA copies and transgenes silencing was found (in primary transgenic plants and in some progeny), probably as a result of single-border binary vector | 37 |

**Table 2** Genetic Transformation in *Actinidia* for Growth Improvement (Increased Rooting) or Quality Improvement (Stress Resistance/Tolerance or Ethylene Control in Fruits)

| Species/ cultivar | Plant material | DNA transfer system | Plasmid/ transgenes | Results | Conclusions | Ref. |
|---|---|---|---|---|---|---|
| *Actinidia deliciosa* cv. 'Hayward' and a male clone | Leaf disks | *Agrobacterium* sp. LBA4404 | pBIN19 with *npt*II, and *rol* A, B, C | Shoots were recovered on 50–200 μg/ml kan; NPTII expression (dot blot) found for all tested shoots; Southern blot confirmed gene integration in the clone tested | Regenerated shoots showed increased rooting ability; after transfer to soil, plants tended to loose the aerial "hairy root" characteristics | 20 |
| *Actinidia deliciosa* cv. 'Hayward' | Leaf explants of micropropagated plantlets | *Agrobacterium* sp. A281 | pKIWI700 (antisense ACC oxidase), pKIWI710 (sense ACC oxidase), pKIWI720 (deleted sense ACC oxidase) (*nos* and *npt*II), pKIWI105 (*uid*A, *npt*II) | Studies of ethylene production after leaf wounding revealed that intact antisense ACC oxidase led to stronger reduction in ethylene production than deleted sense gene; PCR analyses revealed frequent loss of *nos* gene (on the right border) | First results on reduction of ethylene production, further studies required on the effect of the sense construct | 22 |
| *Actinidia chinensis* cv. 'Earligold' | Leaf explants of micropropagated plantlets | *Agrobacterium* sp. A281 | Two constructs with tandem ACC synthase and oxidase genes (one in sense and another in antisense) for cosuppression or antisense suppression of ethylene production | Most transformants contained intact T-DNAs; no reduction in wound ethylene biosynthesis was observed on transformants with sense or antisense constructs, but at least one of the two transgenes was transcribed | Possible insufficient level of endogenous ACC oxidase silencing, as a consequence of the low abundance of transgene transcripts; further transformants required; more analyses to perform on ethylene production on leaf wounds and fruits. | 23 |
| *Actinidia deliciosa* staminate cv. 'GTH,' pistillate cv. 'Hayward' | Leaf disks | *Agrobacterium* sp. LBA4404 | pBIN19 with *npt*II, and *rol* ABC, *rol* B, or a 35S-driven *osmotin* gene | Greenhouse and field grown plants with *rol*ABC showed stable expression, dwarf phenotype, and reduced number of flowers/plant; when used as rootstocks, these increased branching on the scion; plants carrying *rol*B alone were similar to untransformed ones; progeny of untransformed 'Hayward,' fertilized by a transgenic staminate, showed transgene inheritance and expression | *rol*ABC-transformed plants showed higher susceptibility to winter frost and higher resistance to drought; some transgenic 'Hayward' for the *osmotin* gene showed less susceptibility to *Botrytis cinerea*; the high resistance level of staminate GTH to *Pseudomonas* was reduced by *rol*ABC | 21 |

# Genetic Transformation of Kiwifruit

**Figure 1** *Agrobacterium* species–mediated genetic transformation of *Actinidia deliciosa*. (a) 10 µm; (b)–(d) 1 cm. (a) Gus staining of a leaf disk, 10 days after *Agrobacterium* sp. infection. (b) Transgenic shoot induction from leaf disks on selection media. (c) Transgenic *A. deliciosa* plantlets growing on selection media. (d) View of a transgenic field trial at the HortResearch Orchard, Kumeu, New Zealand. (Images kindly provided by the Horticulture and Food Research Institute of New Zealand, Ltd.)

be avoided; (d) 20 µM acetosyringone should be added to the *Agrobacterium* sp. growth medium and cocultivation plates for increased transient GUS expression.

The strategy adopted by Rugini and associates (20), also with success, was however significantly different. Leaf disks (1 cm in diameter) were excised from in vitro grown plantlets, immersed in the bacterial suspension (LBA 4404), and gently scratched on both surfaces by pressing them on carborundum granules with a copper wire brush. After 1-hour incubation, the disks were blotted and transferred to callus induction medium with the abaxial surface in contact with the medium for a 3-day cocultivation period. In these experiments the authors induced callus dedifferentiation (on MS medium with 5 mg/l NAA and 0.1 mg/l BAP) in the absence of selection pressure and proliferated callus tissue on increasing kanamycin concentrations (from 50 to 200 mg/l). After two transfers on 200 mg/l kanamycin, calluses were transferred to shoot induction medium (MS with 1 mg/l BAP) and transgenic shoots were regenerated.

Oliveira and coworkers (11) selected for transformation petiole segments, excised from micropropagated plantlets and plated in vertical position in appropriate orientation, i.e., basal side up, for differentiation of shoots on the top surface (28). A diluted bacterial suspension was applied on the top sectioned surface of each explant and cocultivation proceeded for 48 hours. Selection started at days 0, 8, 15, 21 and 45 after cocultivation with several concentrations of kanamycin (25, 50, and 75 mg/l) on medium containing 2 mg/l zeatin and 0.025 mg/l IAA. Green shoots were recovered in experiments in which kanamycin was applied at a late stage (after 45 days). These shoots were subcultivated on the same fresh selection medium for further growth, then transferred to H or H2 medium (11) with 15 or 100 mg/l of kanamycin. From these experiments kanamycin-

**Figure 2** Electroporation-mediated genetic transformation of *Actinidia deliciosa* cv. 'Hayward.' (a–c) 100 μm; (d, f), 500 μm; (e) 1 cm; (g, h) 2 cm. (a) Protoplasts, isolated from friable hyaline callus, after electroporation (200 V/cm, four pulses). (b, c) Aspects of first divisions in electroporated protoplast cultures. (d) Protoplast-derived microcolonies growing without selection pressure. (e) Kanamycin-resistant calluses differentiating shoots on 20 μg/ml kanamycin. (f) Scanning electron microscopic aspect of an emerging shoot bud. (g, h) NPT II–positive rooted plants growing in soil.

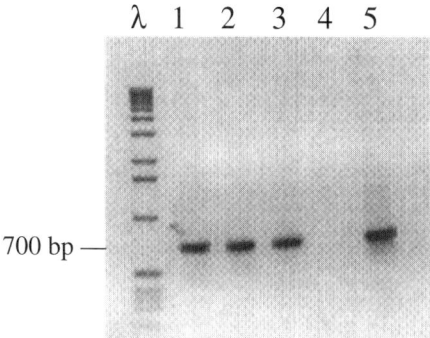

**Figure 3** PCR amplification of the *npt*II gene transferred through electroporation (200 V/cm, four pulses) to *A. deliciosa* cv. 'Hayward' protoplasts. The analyses were performed on the regenerated kanamycin-resistant NPT II–positive shoots. Samples: 1, 2, 3, tested shoots, 4, control shoot (nontransgenic), 5, positive control (plasmid DNA, p35SGUSINT).

resistant shoots, able to express NPT II in dot blot analyses, were recovered, although sustained multiplication on 100 mg/l kanamycin was not achieved. The *Agrobacterium* sp. infection was found to delay the organogenic response with the shoot initials arising after 4 weeks, instead of the usual 3 weeks. In the preliminary experiments using LBA 4404/p35SGUSINT, the histochemical detection of GUS activity was performed 20 days after transformation and culture under no selective conditions. The explants precultivated for 1 day before infection showed much more blue product, after the histochemical GUS analyses, than those inoculated immediately after sectioning. Moreover, in the former case, GUS analysis performed 3 days after cocultivation revealed GUS activity in meristematic cells at the top cutting surface. In the subsequent experiments the explants were cultivated for a period of 1 or 2 days prior to *Agrobacterium* sp. infection. However, the recovery of kanamycin-resistant shoots by using this method was more difficult than by using those described by Janssen and Gardner (19) or Uematsu and coworkers (25). These differences may be due to a number of factors, such as the regeneration conditions used and the antibiotics applied for *Agrobacterium* sp. elimination.

There are some interesting features described in the reports, such as the correlation of higher zeatin concentration with higher kanamycin selection pressure (25 mg/l kanamycin/3 mg/l zeatin used by Uematsu et al. [25] and 100 mg/l kanamycin/5 mg/l zeatin used by Janssen and Gardner [19].) Rugini and associates (20) used 50–200 mg/l kanamycin with 5 mg/l NAA for callus proliferation, and later 1 mg/l BAP for shoot regeneration. Fraser and associates (7), working with several *Actinidia* sp. genotypes, have tested various growth regulators in a wide range of concentrations (0.1–10 mg/l), and from these studies they excluded TDZ and kinetin and selected zeatin, alone or preferably in combination with BA, combined with 0.1 mg/l NAA. The authors didn't mention the zeatin concentration selected but indicated that the higher concentration tended to produce hyperhydric shoots. Selection pressure of 100 mg/l kanamycin was applied at the end of cocultivation period.

Regarding the antibiotics applied for *Agrobacterium* sp. elimination, it is interesting to verify that they were usually not the most appropriate ones. According to more recent studies (29), carbenicillin is somewhat more effective in controlling the growth of EHA101 (and therefore also the strain from which it derives, A281), a strain quite resistant to antibiotics, whereas for LBA4404, cefotaxime is the most effective antibiotic. Nevertheless, Uematsu and coworkers (25) and Janssen and Gardner (19) both used cefotaxime at 500 mg/l for EHA101 and A281 (19,25), Fraser and coworkers (7) used 300 mg/l cefotaxime for A281, and Rugini and associates (20) and Oliveira and coworkers (11) have used carbenicillin, respectively, at 200 and 500 mg/l for LBA4404. Although Rugini (20) states that no visible effects were evoked by carbenicillin, the negative effect of this antibiotic on shoot regeneration in fruiting species has been widely documented (30–32). Carbenicillin has been indicated as an antibiotic with auxin-like effects (33), inducing callus formation and inhibiting regeneration (32). The use of a high carbenicillin concentration together with insufficient concentration of growth regulators, mainly of cytokinin, could have played an important role in reducing shoot regeneration efficiency/survival in the experiments of Oliveira and associates (11).

The *Agrobacterium* sp.–mediated transformation strategy has also led to transgenic plants of *Actinidia arguta* (Harvey, unpublished results). Compared to *A. chinensis*, this species exhibits a higher resistance to kanamycin and requires different regeneration protocols (K Richardson, personal communication, 1999).

## B. Direct Deoxyribonucleic Acid Transfer

Direct permeabilization of DNA to kiwifruit protoplasts was attempted by using polyethylene glycol (PEG) and electroporation with rectangular pulses (11,15,34). Most experiments used protoplasts isolated from friable hyaline calluses (11,34) but also from leaves of micropropagated plantlets (15).

Analyses of transient expression of reporter genes (*cat*, chloramphenicol acetyl transferase gene, and *uid*A, β-glucuronidase gene) were carried out to evaluate the appropriate conditions for DNA transfer and revealed that in the conditions tested PEG treatment was much more effective than electroporation (Table 1). A number of experiments, performed by using different plasmid constructs, consistently yielded much higher transient expression when using PEG treatment, although it was associated with a drastic reduction in cell viability (always below 50%) (11,34). The use of heat shock (45°C, 5 minutes) prior to membrane permeabilization or an increase in PEG 4000 concentration (from 20% to 30%) both accounted for an increase in transient expression while strongly reducing cell viability (11). However, electroporation (four pulses of 40 μs and 200 V/cm of field strength), although providing no significant or reproducible levels of transgene transient expression, resulted in the regeneration of transgenic shoots (Fig. 2). Above 600 V/cm the protoplast viability fell to levels identical to or lower than those obtained with PEG treatment, an effect that is probably related to the large protoplast size (90 μm in diameter). The efficiency of electroporation may be related to its absence of toxicity and eventually also to an effect in increasing protoplast regeneration ability, as described for several species (for a review see Ref. 35), including woody fruiting species (36). In contrast, PEG toxicity, in spite of the careful washing of protoplasts, accounted for cell death, increased browning, and low plating efficiency, even in the absence of selection pressure (11).

Electroporated protoplasts always showed strong variations in shape after cell wall regeneration, as well as asymmetrical divisions at culture initiation (Fig. 2b, c), probably as a result of deep changes in the cytoskeleton caused by the electric pulses. However, no other differences were observed during culture, except the high number of shoot-regenerating colonies in control experiments (28).

Selection for kanamycin resistance was only applied in protoplast-derived cultures when colonies had regenerated, in order to allow rescue of resistant ones for shoot induction. However, the continuous presence of kanamycin in the culture media extended the time required for shoot induction and regeneration and progressively decreased the number of survivals.

In these experiments Oliveira and coworkers (11) applied long periods of selection, 5–8 months, followed by omission of the selective agent to facilitate shoot elongation.

Nearly 80% of the shoots regenerated from electroporated protoplasts were positive for NPT II expression analyzed by dot blot (11), as opposed to what was found for PEG-regenerated shoots, in which no strong or consistent NPT II expression was observed. In the electroporation-derived and NPT II–positive shoots, the presence of the *npt*II gene could be confirmed by polymerase chain reaction (PCR) amplification (Fig. 3).

Chimeric shoots were occasionally found (11). This situation could be related to the culture system used, in which the protoplasts were not immobilized (immobilization here proved to be detrimental) but instead laid free on a thin layer of liquid culture medium on top of an agarose-solidified bead. Under these conditions transformed and untransformed protoplasts could aggregate and give rise to colonies, and, under mild selection conditions, each could contribute to the regeneration of one shoot.

The long period still required to regenerate plants from cv. 'Hayward' protoplasts is a big disadvantage when compared to the period necessary for plant regeneration from organs like leaves. However, the use of leaves as a protoplast source (15) may constitute an alternative source of cells for transformation that, combined with electroporation, may provide large numbers of regenerants. Moreover, this protoplast source may be easily and efficiently used to monitor new promoters driving reporter genes.

## IV. TRANSGENE INTERACTION AND STABILITY IN PROGENY

The expression and inheritance of transgenes in *A. deliciosa* were carefully studied by Fung and colleagues (37). This work is of particular importance, considering that the introduction of genes of interest in kiwifruit is directed not only to one or two vegetatively propagated cultivars, but is also to wider application in the breeding programs under way in several countries. The work was conducted by using male and female transgenic plants of *A. deliciosa,* previously obtained by Janssen and Gardner (19) and growing in a field trial. The seeds obtained from these crosses were germinated under selective conditions and raised in the containment green-house for evaluation. It was found that seven of nine of the GUS-positive plants transferred to soil remained positive after 3 years under field conditions. However, the other two plants, which in culture showed irregular and pale blue staining, lost GUS expression. Moreover, it was found that at least some of the copies of *npt*II and *uid*A genes transferred to kiwifruit were integrated in different genetic loci, with independent segregation demonstrated in the progeny. The two female primary transgenic plants in which transgene inheritance was monitored in detail both contained large numbers of T-DNA insertions, as demonstrated through Southern hybridization, which could have facilitated rearrangements and transgene silencing. Fung and coworkers (37) suggested that the use of the single-border binary vector could have been the cause of these unexpected segregation patterns. The loss of *npt*II and *uid*A genes was observed in some progeny plants, although transgene silencing was only detected for *uid*A.

The behavior of transgenic kiwifruit plants in the field was also monitored by Rugini and associates (21), who already have about 100 transgenic plants with several genes introduced (E. Rugini, personal communication, 1999) (Table 2). In these plants, stable expression of *rol*ABC genes was shown for the 8-year-old staminate plants and 5-year-old pistillate 'Hayward' plants analyzed (21). The progeny derived from 'Hayward' plants pollinated with transgenic pollen (carrying *rol*ABC) have shown transgene inheritance and expression (21).

## V. GENETIC IMPROVEMENT

Genetic improvement of kiwifruit through genetic engineering is being attempted to modify the growth behavior of the plant and increase abiotic stress tolerance (*rol* genes) (20,21), to increase pathogen resistance (*osmotin* gene) (21), and to reduce ethylene production in the fruit (ACC synthase and oxidase genes, in sense or antisense orientation) (22,23).

The *rol*ABC genes, driven by their own promoters, in transgenic kiwifruit plants caused the characteristic dwarf phenotype, with reduced flower production due to the reduced number of ramified peduncles. Moreover, they exhibited slightly higher susceptibility to winter frost and higher resistance to drought, even when grafted on wild rootstocks (21). However, in the case of a staminate cultivar, GTH, it was verified that *rol*ABC introgression accounted for a reduction in the natural resistance to *Pseudomonas syringae* pv. *syringae* and *P. viridiflavae*. This effect was also confirmed in the offspring carrying the transgenes. These transgenic plants, when used as rootstocks, induced increased branching on the scion.

Some 'Hayward' genotypes carrying the 35S-driven *osmotin* gene showed less susceptibility to inoculation with the necrotrophic fungus *Botrytis cinerea* (21).

The studies being conducted to control ethylene production in the fruit have been directed to *A. deliciosa* (22) and *A. chinensis* (23) (Table 2). In *A. deliciosa*, the development of an ethylene biosynthetic mutant was aimed at defining the effects of ethylene and the role of climacteric in ripening fruit (22). For *A. chinensis* the control of endogenous ethylene production may increase its storage properties to levels similar to those of kiwifruit (23). As has been done for other plants, control of ethylene production was attempted here through introduction of sense or antisense copies of endogenous ACC synthase or oxidase genes, single or in tandem, full length or deleted sequences, for antisense suppression or sense cosuppression of their expression (Table 2). The regenerated transgenic plants were assayed for the presence and expression of the transgenes and tested for wound-induced ethylene biosynthesis. In *A. deliciosa* it was verified that the full-length antisense ACC oxidase caused a stronger reduction in ethylene production in wounded leave than the deleted sense gene (22). In *A. chinensis*, in spite of the successful introduction of T-DNA in most transformants and the transcription of at least one of the transgenes, no reduction in wound ethylene biosynthesis was observed (23). As suggested by the author, this may be due to insufficient silencing of ACC oxidase because of low abundance of transcripts, although further analyses are necessary and further transformants required.

## VI. CONCLUSIONS

Since the first report on transgene expression in *Actinidia* sp., several important achievements were attained and transgenic kiwifruit plants were regenerated by using both *Agrobacterium* sp.–mediated transformation and direct gene transfer.

As described in this review, important parameters for *Agrobacterium* sp.–mediated genetic transformation have been defined for some genotypes, although these still require further optimization particularly in what concerns the bacterial strain to use, the antibiotics for its elimination, and the most appropriate plasmid constructs.

Direct gene transfer also needs to be further explored, in order to achieve high numbers of fully transgenic plants in a shorter period. Electroporation seems to be much more promising for stable transformation than PEG treatment. However, the reduced number of experiments and variations performed with PEG, and its high efficiency in transient expression analyses, do not allow, at this stage, rejection of this strategy for transgenic shoot regeneration.

Transgenic plants derived from *Agrobacterium* sp. infection were grown to maturity, cross-pollinated, and their progeny analyzed. Inheritance and expression of the transgenes were shown in the progeny.

The manipulation of traits with agronomic importance is under study in an attempt to improve *Actinidia* sp. genotypes through genetic engineering. Genes leading to fruit quality improvement, through controlled ethylene production, to increased rooting or to stress resistance/tolerance (diseases, drought, cold, etc.) represent some of the main targets for kiwifruit genetic improvement. The introduction of such genes in some *Actinidia* sp. varieties still with limited commercialization may provide them competitive advantage over the 'Hayward' cultivar and contribute to increase the number of high-quality genotypes available for production and commercialization.

## NOTE ADDED IN PRESS

The introduction of disease tolerance in kiwifruit was reported by Nakamura and associates (38). The authors used *Agrobacterium* sp. LBA4404 with *npt*II as the selectable marker to introduce in kiwifruit cv. 'Hayward,' the soybean β-1,3-endoglucanase complementary DNA (cDNA) under the control of the 35S caMV promoter. The presence of the transgene was demonstrated in all the regenerated PCR-positive plants, with one to five integration sites. Transgene expression was demonstrated by immunoblot and enzyme assays with laminarin as substrate. In the plants showing higher enzyme activity (up to six fold that of controls) the inoculation of *Botrytis cinerea* led to necrotic lesions with an area significantly smaller than that of control plants.

The authors refer to previous work on kiwifruit transformation published at the J Jpn Soc Hort Sci, Jpn J Breed, and the Fruit Tree Research Station, Tsukuba, Ibaraki, Japan.

## ACKNOWLEDGMENTS

We gratefully acknowledge the Horticulture and Food Research Institute of New Zealand, Ltd., for kindly providing the images presented in Fig. 1. Drs. Lena Fraser, Catherine Harvey, and Kim Richardson are also gratefully acknowledged for their precious contribution to this paper and for their kindness and availability.

## REFERENCES

1. C-F Liang, AR Ferguson. Emendation of the Latin name of *Actinidia chinensis* Pl. var. *hispida* C.F. Liang. Guihaia 4:181–182, 1984.
2. CF Harvey, GP Gill, LG Fraser, MA McNeilage. Sex determination in *Actinidia*. 1. Sex-linked markers and progeny sex ratio in diploid *A. chinensis*. Sex Plant Reprod 10:149–154, 1997.
3. GS Lawes. The need for plant selection in the development of the kiwifruit (*Actinidia chinensis*) industry. Gartenbauwissenschaft 44:182–184, 1979.
4. CA Schroeder, WA Fletcher. The Chinese gooseberry (*Actinidia chinensis*) in New Zealand. Econ Bot 21:81–89, 1967.
5. S Young. The fruit that launched a thousand ships. New Sci 15:36–39, 1985.
6. RN Crowhurst, R Lints, RG Atkinson, RC Gardner. Restriction fragment length polymorphism in the genus *Actinidia* (Actinidiaceae). Plant Systematics Evol 172:193–203, 1990.
7. LG Fraser, J Kent, CF Harvey. Transformation studies of *Actinidia chinensis* Planch. NZ J Crop Hortic Sci 23:407–413, 1995.
8. AR Ferguson. Kiwifruit: A botanical review. Hortic Rev 6:1–64, 1984.
9. P Blanchet. L'amélioration génétique du kiwi et des Actinidias au lycée agricole de Montaubaun. L'Arboric Fruitière 456:46–51, 1992.
10. Z-G Huang, CY Tan. Kiwi fruit. In: Z Chen et al., eds. Handbook of Plant Cell Culture. Vol. 6. Perennial Crops. New York: McGraw-Hill, 1990, pp 407–417.
11. MM Oliveira, J Barroso, M Martins, MS Pais. Genetic transformation in *Actinidia deliciosa* (kiwifruit). In: YPS Bajaj ed. Biotechnology in Agriculture and Forestry. Vol. 29. Plant Protoplasts and Genetic Engineering V. Berlin, Heidelberg: Springer-Verlag, 1994, pp 189–210.
12. Z-G Huang, CY Tan. Chinese gooseberry, kiwifruit (*Actinidia* spp.). In: YPS Bajaj, ed. Biotechnology in Agriculture and Forestry. Vol. 6. Crops II. Berlin: Springer-Verlag, 1988, pp 166–180.
13. Mu SK, LG Fraser, CF Harvey. Rescue of hybrid embryos of *Actinidia* species. Sci Hortic 44:97–106, 1990.
14. Mu SK, LG Fraser, CF Harvey. Initiation of callus and regeneration of plantlets from endosperm of *Actinidia* interspecific hybrids. Sci Hortic 44:107–117, 1990.
15. MH Raquel, MM Oliveira. Kiwifruit leaf protoplasts competent for plant regeneration and direct DNA transfer. Plant Sci 121:107–114, 1996.
16. MM Oliveira. Somatic embryogenesis in kiwifruit (*Actinidia deliciosa*). In: SM Jain, PK Gupta, RJ Newton, eds. Somatic embryogenesis in woody plants. Vol 5. The Netherlands: Dordrech, Kluwer Academic, 1999, pp 181–195.
17. M Rey, T Fernández, V González, R Rodríguez. Kiwifruit micropropagation through callus shoot-bud induction. In Vitro Cell Dev Biol 28P:148–152, 1992.
18. RG Atkinson, CJ Candy, RC Gardner. *Agrobacterium* infection of five New Zealand fruit crops. N Z J Crop Hortic Sci 18:153–156, 1990.

19. B-J Janssen, RC Gardner. The use of transient GUS expression to develop an *Agrobacterium*-mediated gene transfer system for kiwifruit. Plant Cell Rep 13:28–31, 1993.
20. E Rugini, A Pellegrineschi, M Mencuccini, D Mariotti. Increase of rooting ability in the woody species kiwi (*Actinidia deliciosa* A. Chev.) by transformation with *Agrobacterium rhizogenes rol* genes. Plant Cell Rep 10:291–295, 1991.
21. E Rugini, M Muganu, M Pilotti, GM Balestra, L Varvaro, P Magro, R Bressan, C Taratufolo. Genetic stability, transgene hereditability and agronomic evaluation of transgenic kiwi (*Actinidia deliciosa* A. Chev.) plants for *rol*ABC, *rol*B and *osmotin* gene (abstract). Fourth International Symposium on Kiwifruit, Santiago, Chile, February 11–14, 1999, p 26
22. CWB MacDiarmid. Kiwifruit ACC oxidase genes. MSc dissertation, Auckland University, Auckland, New Zealand, 1993.
23. DJ Whittaker. Etylene biosynthetic genes in *Actinidia chinensis*. PhD dissertation, Auckland University, Auckland, New Zealand, 1997.
24. CJ Candy. Transformation of kiwifruit using *Agrobacterium*. MSc dissertation, Auckland University, Auckland, New Zealand, 1987.
25. C Uematsu, M Murase, H Ichikawa, J Imamura *Agrobacterium*-mediated transformation and regeneration of kiwi fruit. Plant Cell Rep 10:286–290, 1991.
26. OL Gamborg, RA Miller, K Ojima, Plant cell cultures. 1. Nutrient requirements of suspension cultures of soybean root cells. Exp Cell Res 50:151–158, 1968.
27. B-J Janssen. *Agrobacterium*-mediated gene transfer into kiwifruit. PhD dissertation, Auckland University, Auckland, New Zealand, 1991.
28. MM Oliveira. *Actinidia deliciosa* var. *deliciosa* (kiwi): Condições de cultura in vitro e bases para a transformação genética. PhD dissertation, Lisboa University, Lisboa, Portugal, 1992.
29. NJ Schackelford, CA Chlan. Identification of antibiotics that are effective in eliminating *Agrobacterium tumefaciens*. Plant Mol Biol 14:50–57, 1996.
30. JL Norelli, HS Aldwinckle. The role of aminoglycoside antibiotics in the regeneration and selection of neomycin phosphotransferase-transgenic apple tissue. J Am Soc Hortic Sci 118:311–316, 1993.
31. SL Sain, KK Oduro, DB Furtek. Genetic transformation of cocoa leaf cells using *Agrobacterium tumefaciens*. Plant Cell Tissue Org Cult 37:243–251, 1994.
32. LM Yepes, HS Aldwinckle. Factors that affect leaf regeneration efficiency in apple, and effect of antibiotics in morphogenesis. Plant Cell Tissue Org Cult 37:257–269, 1994.
33. JJ Lin, N Assad-Garcia, J Kuo. Plant hormone effect of antibiotics on the transformation efficiency of plant tissues by *Agrobacterium tumefaciens* cells. Plant Sci 109:171–177, 1995.
34. MM Oliveira, J Barroso, MS Pais. Direct gene transfer into kiwifruit protoplasts: Analysis of transient expression of the CAT gene using TLC autoradiography and a GC-MS based method. Plant Mol Biol 17:235–242, 1991.
35. SJ Ochatt, EM Patat-Ochatt. Protoplast technology for the breeding of top-fruit tress (*Prunus, Pyrus, Malus, Rubus*) and woody ornamentals. Euphytica 85:287–294, 1995.
36. SJ Ochatt, PK Chand, EL Rech, MR Davey, JB Power. Electroporation-mediated improvement of plant regeneration from colt cherry (*Prunus avium* × *Pseudocerasus*) protoplasts. Plant Sci 54:165–169, 1988.
37. R Fung, BJ Janssen, BA Morris, RC Gardner. Expression and inheritance of transgenes in kiwifruit. N Z J Crop Hortic Sci 26:169–179, 1998.
38. Nakamura, Y, H Sawada, S Kobayashi, I Nakajima, M Yoshikawa. Expression of soybean β-1,3-endoglucanase cDNA and effect on disease tolerance in kiwifruit plants. Plant Cell Rep 18:527–532, 1999.

# 29
# Genetic Transformation of Mango (*Mangifera indica* L.)

**Richard E. Litz**
*University of Florida, Homestead, Florida*

**Miguel A. Gomez-Lim**
*CINVESTAV Unidad Irapuato, Irapuato, Mexico*

|      |                                                          |     |
|------|----------------------------------------------------------|-----|
| I.   | INTRODUCTION                                             | 421 |
|      | A.  Taxonomy                                             | 422 |
|      | B.  History of the Crop                                  | 423 |
| II.  | GENETICS AND BREEDING                                    | 423 |
|      | A.  Limitations of Conventional Breeding                 | 423 |
|      | B.  Breeding Objectives                                  | 424 |
|      | C.  Breeding Accomplishments                             | 428 |
| III. | SOMATIC CELL GENETIC MANIPULATION                        | 429 |
|      | A.  Somatic Embryogenesis                                | 429 |
|      | B.  In Vitro Manipulation of Embryogenic Cultures of Mango | 430 |
| IV.  | CONCLUSIONS                                              | 432 |
|      | REFERENCES                                               | 432 |

## I. INTRODUCTION

The mango *Mangifera indica* L. (family Anacardiaceae) is one of the leading fruit crops of the world. It is estimated to be fifth in total world production after *Musa* (bananas and plantain), *Citrus* sp. (all types), grape, and apple (1,2). The mango is the most important fruit crop of Asia and is grown throughout the tropics and subtropics. Currently, India has the highest production, with approximately 50% of the world's total (Fig. 1). China, Thailand, Mexico, and Indonesia are also important producing countries, each with annual production exceeding 1 million MT. The export market for fresh mangoes and mango products has expanded rapidly since 1990, and Mexico is the leading exporter of fresh mangoes, currently worth approximately U.S. $160 million (1). Mangoes are also important exports of Brazil, South Africa, Pakistan, Philippines, and India, with an estimated annual value of U.S. $346 million. The leading export destinations include North America, the European Union, Hong Kong, and the United Arab Emirates (Fig. 2). The value of

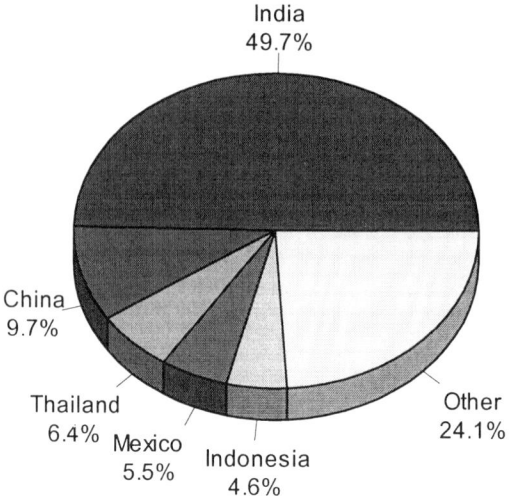

**Figure 1** World mango production in 1998. (From Ref. 1.)

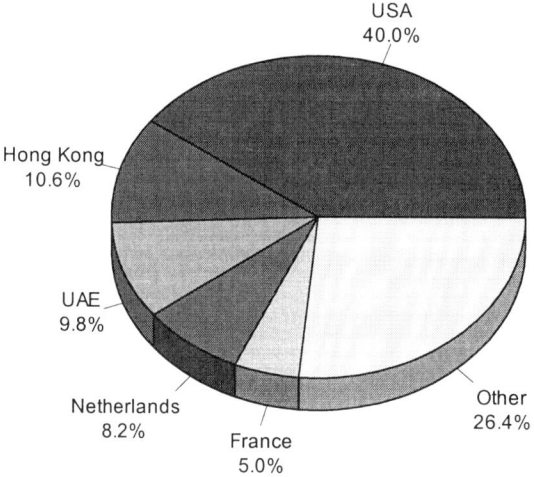

**Figure 2** Major mango importing countries. (From Ref. 1.)

mango exports from Mexico to the United States and Canada alone has been estimated to be approximately U.S.$138 million (1).

### A. Taxonomy

The center of origin of the genus *Mangifera* is considered to be in Southeast Asia (3), and the greatest *Mangifera* species diversity occurs in this region, particularly on the island of Borneo, the Indonesian archipelago, and the Malay peninsula. Approximately 69 species have been identified in the genus (4). *Mangifera indica* is considered to have originated along the northeastern boundary of the genus, possibly in present-day northern Thailand, Myanmar, Bangladesh, and eastern India. Within *Mangifera indica*, there are two distinct ecogeographical races: a polyembryonic type that is tropical and of probable Southeast Asian origin and a monoembryonic type that is subtropical and probably evolved in eastern India, Bangladesh, and Myanmar (3).

## B. History of the Crop

Domestication of mango probably occurred at different times and in different places on the Indian subcontinent and elsewhere in Southeast Asia. According to De Candolle (5), mango cultivation in India can be dated to ca. 4000 B.C. Today the mango tree and its fruit remain important Indian religious and cultural symbols (2). It has been suggested that the Portuguese, who established trading outposts along the west coast of India in the 15th century, introduced vegetative propagation methods into India, and these were used for the first time to clone superior monoembryonic trees (2). The most important mango cultivars of India, e.g., 'Alphonso,' 'Dashehari,' and 'Langra,' are selections that were made at the time of Ahkbar the Great and have therefore been propagated vegetatively for several hundred years (2,6). It is uncertain when cultivation of polyembryonic mangoes originated in Southeast Asia. According to a linguistic analysis of the various common names for mango in southeast Asia, Bompard and Schnell (3) concluded that polyembryonic mango was probably domesticated independently by several indigenous cultures in that region. The traditional cultivars of Southeast Asia, e.g., 'Arumanis' in Indonesia, and 'Carabao' in the Philippines, continue to be propagated as nucellar (true-to-type) seedlings in the region.

The spread of mango from South and Southeast Asia to tropical and subtropical areas of the world came about as a result of the Portuguese and Spanish voyages of exploration from the end of the 15th century onward. Generally speaking, the Southeast Asian polyembryonic types were carried across the Pacific Ocean along the Spanish trade routes to Peru, Central America, and Mexico. The most important mango cultivar of Mexico today is the polyembryonic 'Manila.' The Portuguese carried Indian mangoes westward to their outposts in Africa, and later to Brazil.

## II. GENETICS AND BREEDING

## A. Limitations of Conventional Breeding

Traditional breeding has had relatively little impact on mango cultivar development, and most cultivars are selections made among seedlings of openly pollinated trees. Controlled crosses between mango trees are difficult to achieve because of the low frequency of fruit set (<0.01%) (6) and the occurrence of polyembryony in Southeast Asian types. Mango trees are extremely heterogeneous as a result of outcrossing and polyploidy ($2n = 4x = 40$) (7). Their long juvenile period (at least 7 years after germination) and the time required for evaluation (12–15 years after germination) have also been significant barriers to progress in breeding. Cultivars that have been released from modern breeding programs have not found acceptance in the marketplace of traditional South and Southeast Asian mango growing countries, as their fruit quality has generally been inferior to that of traditional selections.

The international market standards for mango exports have been based upon the "Florida mangoes," e.g., 'Haden,' 'Tommy Atkins,' 'Keitt,' 'Van Dyke,' and 'Kent,' that have been derived from a secondary center of mango diversity that was established in Florida, in the United States, in the early 20th century. This center of diversity included introductions of monoembryonic and polyembryonic mango germplasm from throughout Asia (3,8). Several generations of Florida mango cultivars have resulted from open pollinations within this population. Many of the Florida mangoes are notable for bearing annually, having moderate levels of resistance to anthracnose, and producing fruit with an attractive blush favored by consumers in nontraditional mango consuming countries (9).

## B. Breeding Objectives

### 1. Scion Cultivars

Several horticultural traits that involve mango fruit quality and tree form have been targeted for improvement (6). These include regularity of bearing, particularly in biennial bearing Indian-type mangoes; precocity; and dwarf habit. The foliage and fruit should demonstrate resistance to a number of pests and diseases. The most important diseases affecting mango production and postharvest include anthracnose (caused by *Colletotrichum gloeosporioides* [Penz] Penz & Sacc. In Penz) and alternaria rot (caused by *Alternaria alternata* [FR.:FR.] Keissel) (10,11). Fruit should have an attractive blush and pleasant taste and aroma, and the pulp should be free of fiber. Since mango has become an important export commodity, an extended shelf life and freedom from physiological disorders that cause precocious fruit ripening, e.g., "jelly seed" and "soft nose," in certain important export cultivars like 'Tommy Atkins,' are also important.

*a. Mango Fruit Ripening.* Ripe mango fruit are highly perishable as a result of over-ripening and increased susceptibility to development of anthacnose. They show pronounced changes in texture during ripening, and the rate at which these changes take place has a significant effect on fruit storage life. Mangoes are frequently harvested when immature, so that either they fail to ripen or the quality and flavor of the ripened fruit is inferior (12). Fruit very often suffer from extensive handling damage during all stages of the postharvest chain. Estimates of total postharvest losses are controversial and range widely generally from about 10% to 80%.

The physiological characteristics of tropical fruit in general are similar to those of temperate and subtropical fruit. Differences can involve major substrates involved in ripening, the rate of ripening and senescence, and the order in which various components of ripening occur (13). The unique aspects of tropical fruit physiological features include their chilling sensitivity, rapid ripening (of climacteric fruit) compared to that of temperate fruit, and exposure of tropical fruit to high temperatures or other stresses during quarantine disinfestation in postharvest handling.

Softening of fruit flesh is thought to be due to the hydrolysis of various cell wall components and is accompanied by solubilization of pectin (14,15) and a net loss of arabinose, galactose, and galacturonic acid (16). A number of different cell wall hydrolytic enzymes increase in activity during softening (17). Mango fruit at the mature green stage contain some accumulated starch (18), which is mobilized during ripening (19). The primary consequence of starch hydrolysis is an increase in total sugars, with glucose and fructose as the main monosaccharides (20) and sucrose as the predominant sugar (21); however, total sugar content varies among cultivars: 'Carabao' has one of the highest values (22) and 'Golek' one of the lowest (cited in Ref. 23). Amylase and sucrose phosphate synthase increase at least 4-fold and 10-fold in activity, respectively, during ripening (21,24).

Mango fruit are climacteric and have a moderate respiration rate (25), with $CO_2$ production of 70–150 mg/kg/h. As there seems to be a relationship between respiration rate and storage life, there is a clear need to reduce respiration and thereby increase postharvest life. Temperature management is the main method for controlling the respiration rate, though this is limited in most tropical fruit by their chilling sensitivity. All climacteric fruits show a significant increase in ethylene production during ripening, which may be concurrent with the climacteric. The difference in temporal patterns between respiration and ethylene production in various climacteric fruit suggests a different underlying physiological attribute, which may involve differential sensitivities of fruit species to ethylene. In mango, the climacteric and the ethylene peak occur simultaneously (26).

The two key enzymes in the ethylene biosynthesis pathway are those catalyzing the conversion of *S*-adenosyl methionine (SAM) to 1-aminocyclopropane-1-carboxylic acid (ACC) and ACC to ethylene, called ACC synthase and ACC oxidase or ethylene forming enzyme (EFE), respectively (27,28). Burg and Burg (29) showed that a small but notable peak of ethylene occurs during ripening in monoembryonic 'Haden' and 'Kent,' and this has also been shown in polyembryonic 'Carabao' (30) and 'Manila' (31). Ethylene production is low in preclimacteric mango fruit but increases considerably during the climacteric. Cua and Lizada (30) showed that 'Carabao' mangoes produce ethylene prior to full maturity and that the highest rate (125 nl $g^{-1}$ $h^{-1}$) occurred in the outer mesocarp, although the levels of ethylene were comparable in the entire mesocarp. The levels of the EFE substrate, 1-aminocyclopropane-1-carboxylic acid (ACC), were also very similar throughout the mesocarp.

One of the major constraints for enhancing the yield of mango has been premature, ethylene-induced ripening, which occurs before they are shipped or consumed. Several methods are currently used to prevent or reduce this spoilage, e.g., harvesting of mature green fruit, use of ethylene adsorbents (potassium permanganate or activated charcoal/vanadium oxide) (32), and holding of fruit at an optimal storage temperature. The optimal storage temperature is dependent upon cultivar, preharvest environment (33), maturity at harvest (25), and storage conditions (34). Storage at temperatures ≥8–12°C causes ripening and senescence, and lower temperatures cause chilling injury.

Most of the postharvest technologies for mangoes are designed for disease and insect control and protection against injury during packaging and transport. Mangoes have poor storage quality and technologies for longer-term storage, i.e., controlled or modified atmospheres, are associated with physiological disorders (35). Prolonged cool storage can have deleterious effects on skin, flesh, and ripening attributes; the range of storage temperatures is limited (36). Symptoms of chilling injury of mango fruit include pitting, darkening, failure to ripen completely, and increased susceptibility to decay. Storage life and sensitivity to chilling injury seem to be dependent on fruit maturity and storage temperature. Fruit harvested at an earlier stage of maturation ripen more slowly at a given temperature and are more prone to chilling injury and other storage-related disorders (34). Fruit harvested when nearly ripe are less prone to chilling injury (37,38) and can be stored for up to 21 days at ≤8°C without deterioration in quality; however, the fruit may deteriorate rapidly after removal from storage (36).

Control or modification of storage atmosphere has not been commercially successful for mango fruit, which often show poor color and eating quality and presence of undesirable off-flavors. In addition, there is no clear evidence that mango ripening can be delayed satisfactorily (35). Kader (34) has suggested some regimens for modified atmosphere storage of mango, including $O_2$ 3–5%, $CO_2$ 5–10%, and 10–15°C, which can delay ripening with increased firmness retention.

The genetic manipulation of fruit ripening requires the availability of relevant genes, and genes coding for various ripening-related enzymes have been isolated from different fruits. The messenger ribonucleic acids (mRNAs) for virtually all the ripening-related genes isolated so far have been shown to be absent or present at a low level in immature fruit, to increase during ripening, and to decline as ripening progresses. Changes in mRNA and protein content occur during ripening (31). Some of the mRNAs that decline in quantity probably encode photosynthetic enzymes, as it is known that the photosynthetic apparatus is dismantled during ripening (39).

A complementary deoxyribonucleic acid (cDNA) library has been prepared from climacteric mango fruits and screened by using differential hybridization (17). A series of cDNA clones has been isolated whose common feature is the ripening-specific pattern of expression. Two cDNA clones identified coded for enzymes involved in ethylene biosynthesis, ACC synthase, and ACC oxidase (40). Their expression during ripening was studied in pulp and peel in northern-type experiments. ACC synthase message is undetectable in unripe fruit and starts to appear in turning fruit, reaching a maximum in ripe fruit (40). This pattern of expression is similar in the peel and in the pulp; however, the message appears in the pulp before the peel. ACC oxidase message shows similar kinetics in both types of tissue, but the message is clearly detectable before any ACC synthase message becomes detectable (40). These results suggest that ACC oxidase is expressed before ACC synthase and that ripening starts on the inside of fruit and proceeds outward. Ethylene-treated fruits show a different pattern of expression, with ACC oxidase and ACC synthase appearing first in the peel.

A cDNA for thiolase was also identified as having a ripening-specific pattern (41). Several fatty acids, particularly linoleic and oleic acids, decrease in concentration during ripening. Fatty acids in plants are metabolized by the β-oxidation pathway and the products are mainly used in the synthesis of carotenoids and terpenoid volatiles and are released by ripe fruits. These enzymes might be involved in metabolism of fatty acids to produce volatile compounds (23,41). Recently a cDNA clone for acyl coenzyme A (CoA) oxidase, the key enzyme in the β-oxidation pathway, was isolated from mango fruit and shown to behave identically to thiolase (A Nila-Mendez and MA Gómez-Lim, unpublished data).

Alternate oxidase is involved in the cyanide-resistant respiratory pathway. It has been studied mainly in thermogenic species, and its activity is correlated with heat production, necessary to volatilize foul-smelling compounds to attract insect pollinators. There is a significant participation of this pathway in the climacteric of many fruit. A cDNA coding for mango alternate oxidase has been isolated, and by Northern blot analysis the message was detected in unripe fruit and shown to increase substantially in ripe fruit (42). These results were correlated with similar increases in enzyme activity and protein accumulation. The temperature in ripe monoembryonic 'Alfonso' fruit is up to 10°C higher than in unripe fruit and this characteristic has been attributed to the activity of alternate oxidase (43). This extra heat might also serve to volatilize aroma-giving compounds.

Two cDNAs code for enzymes related to the cell wall, cellulase and β-galactosidase. Mango cellulase increases in activity during ripening (M. A. Gómez-Lim, unpublished data). β-Galactosidase is expressed in ripe fruit at high levels (S. Parra-Arenas and M. A. Gómez-Lim, unpublished data). A receptor to ethylene has been isolated from tomato and has been used to probe for a mango homologue (43a). The message seems to be present at low levels in unripe fruit and to increase as the fruit ripens. Genomic clones to the ethylene receptor, β-galactosidase, and alternate oxidase have also been identified and are currently being characterized (S. Parra-Arenas, P. Gutierrez-Martinez, and M. A. Gómez-Lim, unpublished data). These studies aim to identify and isolate fruit-specific and cell wall–related promoters to be utilized in transformation.

It is possible to turn off the expression of specific genes in transgenic plants by introducing a gene engineered to generate antisense RNA (44). This allows expression of specific genes to be diminished, permitting identification and assessment of their function. One approach to control fruit ripening is to depress ethylene content, using different strategies: inhibition of genes encoding ethylene biosynthetic enzymes by transformation with antisense genes and lowering of the level of ACC, the ethylene precursor, by deamination or hydrolysis.

ACC synthase and ACC oxidase have been modified by antisense technology in tomato with expression of the transgenes driven by the constitutive 35S promoter (45,46). Antisense fruit left on the vine or harvested and left in air had a 10- to 20-day delay in the onset of normal chlorophyll degradation, attained only an orange color, and failed to soften or develop a "ripe" aroma. Exogenous application of ethylene reverted the phenotype and the fruits were able to ripen normally. Similar results have been obtained in cantaloupe melon fruits, in which transformation with antisense ACC oxidase significantly delayed ripening (47), although color change was not inhibited (45,47).

Another strategy to reduce ethylene production has been the introduction of genes coding for enzymes that metabolize ACC. Klee and coworkers (48) cloned a *Pseudomonas* sp. ACC deaminase gene and inserted it into tomato. Ethylene biosynthesis was reduced to 3% to that of wild type, and the resultant transgenic plants did not show any apparent phenotypic abnormalities. Fruits from these plants exhibited significant delays in ripening, and the mature fruits remained firm for at least 6 weeks longer than the nontransgenic control fruit. Similarly, a gene from the bacteriophage T3 encoding the enzyme SAM hydrolase (SAMase), driven by a tomato fruit–specific promoter, was used to generate transgenic tomato plants that produce fruit with a reduced capacity to synthesize ethylene and a longer storage life (49). SAMase catalyzes the conversion of SAM to methylthioadenosine and homoserine.

*b. Other Fruit Traits.* Genetic engineering of the lipid composition of mango fruit could improve its nutritional value and enable the tree and fruit to withstand low-temperature stress. The identification of desaturases, enzymes that introduce a cis-double bond in saturated fatty acids, has led to the production of plants with an increased level of polyunsaturated fatty acids (50) and increased chilling tolerance (51). Although mango fruits do not store lipids, these compounds have been linked with color and flavor development of mango during ripening. The vast majority of lipids found in mango are esters of long-chain fatty acids. There is increased fatty acid oxidizing activity during mango fruit ripening (52). Several fatty acids (particularly linoleic and oleic acids) decrease in concentration during mango ripening. Products of β-oxidation are utilized in the synthesis of carotenoids and terpenoid volatiles, which are important aroma components of mango fruit. The peroxisomal thiolase identified in 1995 (41) and other enzymes of the β-oxidation pathway may represent possible targets for altering fatty acid metabolism and volatile compounds involved in fruit flavor and aroma.

No major advances have been made to identify flavor components and their enzymes in tropical fruits. This might be a reflection of the complexity of this trait. Since mangoes contain high quantities of carbohydrates, particularly sucrose, genetic manipulation of the sucrose-metabolizing enzymes may provide a way to alter sugar content and fruit flavor. Manipulation of different enzymes involved with carbohydrate metabolism (53) has been reported, and resultant plants showed an altered carbohydrate content. Because of the effect this process may have on plant metabolism, it will be important to use fruit-specific promoters to target the transgenes to fruit cells.

Fruit flavor could also be manipulated through the use of sweet-tasting proteins, i.e., monellin or thaumatin. They elicit a sweet flavor by binding specifically with taste receptors and are approximately 100,000 times sweeter than sugar on a molar basis (54). Transgenic tomato fruits containing monellin have been generated (55). Thaumatin has been shown to be strongly induced in ripening fruits (56,57).

The external color of fruit is an important factor in consumer preference. Ripe mango peel shows a wide range of color from green to greenish yellow, red, violet, and yellow, which probably originate from chlorophyll degradation and accumulation of anthocyanins and total carotenoids (58). The principal pigments in mango fruit are chlorophylls, carotenes, xanthophylls, and anthocyanins, which are synthesized via the terpenoid or phenylpropanoid pathways. There is only one example of genetic manipulation of fruit color. Phytoene synthase catalyzes the dimerization of two molecules of geranylgeranyl pyrophosphate to form phytoene, the first C40 carotene in the carotenoid synthesis pathway. Expression of phytoene synthase in antisense in tomatoes produced pale yellow flowers and fruits that ripen to a yellow color (59). Lycopene could not be detected in those fruits, although other aspects of ripening, i.e., polygalacturonase accumulation, were unaffected. Pigment synthesis manipulation represents an interesting choice to modify the color of a fruit to make it more attractive.

c. *Plant Architecture.* Robson and coworkers (60) produced transgenic plants that overexpress phytochrome A. The phytochromes are a family of photoreceptors that function as photoreversible pigments in plants. The transgenic plants were tested in the field and were reduced in stature; their leaves contained up to 30% more chlorophyll (which may result in increased photosynthesis) and significantly more biomass accumulated per plant (61). Mango plants with reduced stature would facilitate harvesting, reduce postharvest losses, and increase production per hectare.

d. *Flowering.* Flowering of many traditional mango cultivars is irregular and/or biennenial. At least three classes of homeotic genes control the determination of floral meristems and organ identity in higher plants (62). Transgenic plants that overexpress the *LEAFY* gene (63) or the *CONSTANS* gene (64) have been produced; in these plants these genes determine floral fate in lateral shoot meristems with the consequence that flower development is induced precociously. This phenotype was also obtained with plants overexpressing phytochrome genes (60). Cloning and engineering homologous genes in perennial crops with erratic flowering, i.e., mango, could lead to better control of flowering.

e. *Disease Resistance.* Genetic transformation is an environmentally sustainable alternative way to produce mango plants with resistance to fungal pathogens, particularly *Colletotrichum gloeosporioides*, the cause of anthracnose. Antifungal proteins have been utilized to confer resistance against different plant pathogens (65,66). Transgenic plants that constitutively express different pathogenesis-related (PR) proteins, i.e., chitinase and β-1,3-glucanase, show resistance to fungal infection (67); however, higher protection is achieved when more than one PR protein is employed (68).

Other antifungal proteins include ribosome inactivating proteins (RIPs), which inhibit protein synthesis by specific RNA *N*-glycosidase modification of 28S ribosomal RNA (rRNA). RIPs inactivate ribosomes of distantly related species, including fungi. Transgenic plants containing RIPs are resistant against fungal pathogens (69). Transformation with chitinase, β-1,3-glucanase, and RIPs leads to increased fungal protection (70) compared with the protection levels obtained with corresponding transgenic lines expressing a single barley transgene. The data indicate synergistic protective interaction of the coexpressed antifungal proteins in vivo.

Antifungal defensins are cysteine-rich low-molecular-weight (5-kD) proteins that have been shown to be potent elements for plant defense against fungal attack (71). Work in progress in several laboratories around the world is aimed at producing transgenic plants containing the defensin genes.

### 2. Rootstock Cultivars

The requirements for mango rootstocks are somewhat different from those for scion cultivars (6). Ideal rootstocks should be graft-compatible with scion cultivars and also confer dwarfing to the scion. The rootstocks should also be able to tolerate calcareous soils and saline conditions, which limit production in parts of north Africa, western Asia, and Pakistan. In order that they can be propagated with ease, rootstock cultivars should also be polyembryonic. Currently, there are no mango rootstocks that have been developed exclusively as rootstocks. Polyembryonic mango rootstocks '13-1' and 'Turpentine' are used extensively in Israel and in the Western Hemisphere, respectively.

### C. Breeding Accomplishments

Conventional breeding programs exist in India at several locations, in South Africa, and in Israel. Several cultivars have been developed and released in recent years, although none of them is grown on a large scale. These include 'Amrapalli,' 'Arka Anmol,' 'Arka Aruna,' 'Arka Puneet,' 'Au

Rumani,' 'Mallika,' 'Manjiri,' 'Neeleshan,' 'Ratna,' and 'Swarnajehangir' (India) and 'Ceriese,' 'Heidi,' 'Neldawn,' and 'Neldica' (South Africa) (6). There is strong consumer preference in traditional mango growing countries for traditional mango cultivars.

## III. SOMATIC CELL GENETIC MANIPULATION

Because of problems that are associated with application of conventional breeding to mango cultivar improvement and the conservatism of consumers in traditional mango growing countries, genetic engineering strategies hold considerable promise for improving existing, but in some way flawed, mango cultivars. Specific horticultural traits could be altered in traditional cultivars by using genetic transformation or mutation induction approaches.

### A. Somatic Embryogenesis

The prerequisites for genetically engineering existing mango cultivars include (a) a de novo regeneration procedure for mature phase mango selections in vitro and (b) a method for transforming the in vitro cultures. The procedure for induction of embryogenic cultures of mature phase mango has been well documented since the first report by Litz and associates (72). It is possible to induce embryogenic cultures from the nucellus of both polyembryonic and monoembryonic cultivars (72–74). The currently utilized procedure involves the extraction of the nucellus from the immature seeds of young mango fruit, under sterile conditions, 30–40 days after pollination (75–77). Embryogenic cultures are induced from the excised nucellus on sterile semisolid plant growth medium in petri dishes. The induction medium (hereafter referred to as MEM) consists of modified B5 (78) major salts, MS (79), minor salts and organic components, 400 mg l$^{-1}$ glutamine, 2.0 g l$^{-1}$ gellan gum, 60 g l$^{-1}$ and 4.5–9.0 µM 2,4-dichlorophenoxyacetic acid (2,4-D) (80,81). The nucellar cultures are incubated in darkness at 25°C and recultured onto freshly prepared induction medium at daily intervals until the accumulation of oxidative products in the medium has ceased (usually 3–4 days). An embryogenic response is normally evident 30–50 days after explanting, depending on the cultivar. Embryogenic cultures consist of proembryonic cells and masses (PEMs) (76). The induction response is genotype-dependent and is unrelated to the seed type (i.e., monembryony vs. polyembryony). There is a temporal requirement for exposure of mango cultures to 2,4-D in order to induce maximal efficiency of embryogenic competence (82). The minimal requirement for induction of embryogenic competence of 'Carabao' nucellar cultures on MEM is 7–14 days; however, the optimal response is obtained after a 56-day pulse with 2,4-D (81). The embryogenic competence of the explanted mango nucellus of some difficult-to-induce mango cultivars can be enhanced by using highly embryogenic mango cultures as a nurse (83), e.g., 'Nam doc Mai' on MEM with embryogenic 'Parris' as a nurse layer.

Maintenance of embryogenic cultures is optimal in suspensions; however, the ability to form embryogenic suspension cultures is also genotype-dependent (76,77), and embryogenic cultures of some important cultivars cannot be maintained in suspension. The medium for maintaining suspension cultures is induction formulation without gellan gum. Normally, 300–400 mg of an embryogenic culture is inoculated into 40-ml maintenance medium in 125-ml Erlenmeyer flasks and kept on a rotary shaker at 100 rpm in darkness at 25°C. Subculture into fresh liquid medium is essential at 5- to 7-day intervals, in order to prevent accumulation of oxidation products in the medium. Typical embryogenic suspensions consist of PEMs of different sizes. The smallest PEMs are in fact single globular somatic embryos, whereas the large PEMs are globular embryos whose normal pattern of development has been arrested by their continued exposure to inductive conditions (84). The large PEMs have lost the ability to organize as single embryos and secondary somatic embryos develop from the protoderm. Therefore, in order to recover single somatic embryos from maintenance conditions, it is necessary to subculture the smallest fraction of PEMs (520–860 µm), facilitated by passage through filtration fabric, into maturation medium. The largest PEM fraction is utilized as the stock and is repeatedly subcultured in maintenance medium.

The development of somatic embryos from PEMs in embryogenic cultures is stimulated by their subculture to medium without 2,4-D. This response is mediated in different ways by different cultivars. Embryogenic suspension cultures of some cultivars can form heart stage somatic embryos after transfer into liquid medium without 2,4-D (hereafter referred to as LEM) (e.g., 'Hindi,' 'Tommy Atkins'). More typically, it is necessary (a) to subculture from LEM onto semisolid medium without 2,4-D (hereafter referred to as SEM) in order for somatic embryo development to occur (e.g., 'Carabao') or (b) to subculture from semisolid induction medium onto SEM. Addition of 4.7–13.9 µM kinetin or 6-benzylaminopurine to SEM or LEM has been demonstrated to stimulate the early formation of cotyledonary stage somatic embryos (81). Hyperhydricity of late heart stage somatic embryos derived from suspension cultures is commonly observed, and these developmentally abnormal somatic embryos cannot mature normally, usually becoming necrotic at the late heart or torpedo stage of development. Hyperhydricity can be reversed either by partial desiccation of the somatic embryos under low relative humidity or by increase in the concentration of gellan gum in the plant growth medium from 2.0 to 6.0 g l$^{-1}$ (85).

The development of mango somatic embryos to maturity occurs on basal medium (discussed previously) enriched with 20% (w/v) filter-sterilized coconut water and containing 20 g l$^{-1}$ sucrose and 2.0 g l$^{-1}$ gellan gum (hereafter referred to as MMI) (81,86). Developing somatic embryos are incubated on MMI in darkness at 25°C until germination, at which time they are transferred to light conditions, consisting of a 16-h photoperiod supplied by cool white fluorescent lights (60–80 µmol m$^{-2}$ s$^{-1}$). The length of mango somatic embryos at the time of germination is approximately 4–6 cm. Exposure to light during somatic embryo maturation results in precocious germination of the somatic embryos and poor survival. Survival is optimal after transfer of plantlets into conditions that encourage photoautotropic development (87). Optimal conditions for plant recovery consist of basal medium, containing 0–5 g l$^{-1}$ sucrose and 0.5 g l$^{-1}$ activated charcoal (hereafter referred to as MMII), and incubation at 25°C with a 16-h photoperiod (160–180 µmol m$^{-2}$ s$^{-1}$) provided by cool white fluorescent lights in an environment containing 20,000 ppm $CO_2$ in a nitrogen carrier (76,81).

### B. In Vitro Manipulation of Embryogenic Cultures of Mango

The regeneration of mango cultivars from nucellar cultures is of critical importance for the application of somatic cell genetic approaches to the improvement of elite selections. Two approaches have been developed for manipulation of mango germplasm: in vitro selection after mutation induction and genetic transformation.

#### 1. In Vitro Selection

The application of in vitro mutation induction and selection for mango cultivar improvement is limited by the absence of effective selection and screening agents that can be used to target a specific trait within a large embryogenic cell population. Litz and coworkers (88) demonstrated that the culture filtrate produced by *Colletotrichum gloeosporioides* could potentially be used to select phytotoxin-resistant cells in embryogenic mango cultures. Embryogenic cultures of two important cultivars, 'Carabao' from the Philippines and 'Hindi' from Egypt, have been selected for resistance to the purified phytotoxin and culture filtrate of *C. gloeosporioides*, and phytotoxin-resistant cultures have been demonstrated to be strongly antifungal, as demonstrated in dual culture experiments with the fungal pathogen (89). Newly expressed extracellular pathogenesis-related proteins, including isozymes of chitinase and glucanase, were produced by the phytotoxin-resistant, antifungal embryogenic cultures, somatic embryos, and regenerants (89). Selection for resistance was also accompanied by significant alterations in the DNA banding patterns, as observed in RAPD gels (90). Unfortunately, other currently available in vitro selective agents have little utility for mango improvement, thereby effectively limiting this approach.

## 2. Genetic Transformation

In order to transform mango genetically successfully, it is essential that the in vitro culture of a cultivar can be maintained as a suspension culture, and this growth response has been demonstrated to be genotype-dependent with mango (see earlier discussion). When mango is maintained on semisolid induction medium, growth of embryogenic cultures is relatively slow, and the cultures rapidly become oxidized if they are not transferred to fresh medium regularly. The normal blackening of cultures makes it very difficult to distinguish living and dead PEMs. The first reports of the recovery of mango genetic transformants were by Mathews and associates (91,92), with 'Hindi' and an openly pollinated line derived from 'Keitt,' respectively. All genetic transformation studies involving mango have been mediated by disarmed strains of *Agrobacterium tumefaciens*. Preliminary transformation studies involved the use of strain C58CI, carrying the binary plasmid vector pGV3850::1103 with the NPT II gene, which was used for transforming embryogenic cultures derived from zygotic embyo explants of 'Keitt' (92). Embryogenic cultures of 'Hindi' that were derived from the nucleus were transformed with disarmed *A. tumefaciens* strain A208 containing the cointegrate plasmid vector pTiT37-SE::pMON9749 (9749 ASE) with NPT II and GUS (31). The most recent report described the use of disarmed *A. tumefaciens* strain LBA4404, containing the binary plasmid vector pBI121 with NPT II, GUS, and different genes in the antisense that mediate ethylene biosynthesis (93). The genes in the antisense included ACC oxidase from an *Arabidopsis thaliana* library, alternative oxidase from a ripe mango mesocarp library and ACC synthase from mango leaves. All of the transgenes were regulated by the 35S promoter.

The procedure that has been utilized for mango transformation involves the gentle abrasion of approximately 3 g of the largest PEM fraction (≥820 μm) from an embryogenic suspension culture with a sterile camel hair brush under axenic conditions. The abraded PEMs are then suspended in 50-ml liquid maintenance medium containing 0.05 ml of a log phase culture of acetosyringone-activated *A. tumefaciens* in 125-ml Erlenmeyer flasks (94). Suspensions are maintained at 120 rpm at 25°C in semidarkness and subcultured daily for 3 days.

The strategy for the recovery of transformed embryogenic cultures involves sequential transfers in order to stimulate the growth of completely transformed PEMs from partially transformed or chimeral cell clusters. Matthews and Litz (95) demonstrated that the growth of embryogenic mango suspension cultures is inhibited by 12.5 μg ml$^{-1}$ kanamycin, whereas the later stages of somatic embryo development are inhibited by 200 μg ml$^{-1}$ kanamycin sulfate on semisolid medium. According to the stepwise strategy that was developed for mango, partially transformed PEMs were recovered from suspension cultures in liquid maintenance medium containing 200 μg ml$^{-1}$ kanamycin and 500 μg cefotaxime. These cultures were plated on semisolid maintenance medium supplemented with 400 μg ml$^{-1}$ kanamycin sulfate and 500 μg cefotaxime. Transformed proembryos develop from genetically transformed cells on the surface of PEMs (32). Subculture of putative transformants from selection medium containing kanamycin sulfate and cefotaxime into liquid maintenance medium results in more rapid growth of the cultures and formation of more transformed proembryos. On the basis of the X-GLUC reaction for the presence of GUS (96), completely transformed PEMs and globular somatic embryos can be recognized after this procedure. Somatic embryo development can occur on maturation medium (discussed previously).

## IV. CONCLUSIONS

In vitro manipulation of mango as a means for improving existing cultivars is feasible; however, the embryogenic response of mango is not easily manipulated because of the strong genotype-dependent response of cultivars in suspension cultures and the rapid oxidation of cultures in suspension and on semisolid medium. Genetic transformation of mango has focused on control of fruit ripening, which has significant implications for postharvest handling and shipping of fruit from the field to packing houses and to international markets and for the prevention of postharvest spoilage due to anthracnose and alternaria rot. With the tools available, it is currently possible to alter practically any gene, as long as a suitable molecular probe is available, and to analyze the function during fruit ripening. There are several possible strategies for genetic manipulation of fruit ripening. Genes in the antisense or sense orientation can be utilized to inhibit specific genes. Gene repression resulting from homologous sense gene expression is termed *cosuppression*. There is no clear superiority of antisense versus cosuppression techniques. The current lack of knowledge regarding mechanisms for either antisense or cosuppression makes it difficult to ascertain which technique will be more effective in reducing expression of a particular target gene.

Currently, the control of anthracnose in humid production regions involves the weekly application of fungicide from the time of flowering until harvesting (10). This practice is increasingly viewed as being unsustainable. It should be possible to target control of anthracnose directly by transforming scion cultivars with genes for pathogenesis-related proteins. This could revolutionize mango production in the humid tropics.

## ACKNOWLEDGMENTS

The authors are grateful for support provided by Florida Mango Forum (REL) and the International Plant Genetic Resources Institute (IPGRI, formerly IBPGR) (REL). Florida Agricultural Experiment Station Journals Series no. R-06912.

## REFERENCES

1. FAOSTAT DATABASE RESULTS. http://www.fao.org/waicent/agricul.htm, 1998
2. SK Mukherjee. Introduction: Botany and importance. In: RE Litz, ed. The Mango Botany, Production and Uses. Wallingford, Oxon: CAB International, 1998, pp 1–19.
3. JM Bompard, RJ Schnell. Taxonomy and systematics. In: RE Litz, ed. The Mango Botany, Production and Uses. Wallingford, Oxon: CAB International, 1998, pp 21–47.
4. AFGH Kostermans, JM Bompard. The Mangoes: Botany, Nomenclature, Horticulture and Utilization. London: Academic Press, 1993.
5. AP DeCandolle. Origin of Cultivated Plants. London: Kegan Paul, 1884.
6. CPA Iyer, C Degani. Classical breeding and genetics. In: RE Litz, ed. The Mango Botany, Production and Uses. Wallingford Oxon: CAB International, 1998, pp 49–68.
7. SK Mukherjee. Mango: Its allopolyploid nature. Nature 150:196–197, 1950.
8. RJ Knight Jr, RJ Schnell. Mango introduction in Florida and the Haden cultivar's significance to the modern industry. Econ Bot 48:139–145, 1994.
9. RJ Knight Jr. Important mango cultivars and their descriptors. In: RE Litz, ed. The Mango Botany, Production and Uses. Wallingford Oxon: CAB International, 1998, pp 545–565.
10. JC Dodd, D Prusky, P Jeffries. Fruit Diseases. In: RE Litz, ed. The Mango Botany, Production and Uses. Wallingford Oxon: CAB International, 1998, pp 257–280.
11. RC Ploetz & O Prakash, In: RE Litz, ed. The Mango Botany, Production and Uses. Wallingford Oxon: CAB International, 1998, pp 281–325.

12. AP Medlicott, M Bhogol, SB Reynolds, SW New, AK Thompson. Harvest maturity effects on mango fruit ripening. Trop Agric (Trinidad) 65:153–157, 1988.
13. RE Paull. Tropical fruit physiology and storage potential. In: BR Champ, E Highley, GI Johnson, eds. Postharvest Handling of Tropical Fruits. Canberra ACT 2601: ACIAR Proceedings No. 50, 1994, pp 198–204.
14. B Roe, JH Bruemmer. Changes in pectic substances and enzymes during ripening and storage of Keitt mangoes. J Food Sci 46:186–189, 1981.
15. H Lazan, Z Mohd, LK Wah, J Voon, GR Chaplin. The potential role of polygalacturonase in pectin degradation and softening of mango fruit. ASEAN Food J 2:93–141, 1986.
16. K Brinson, PM Dey, MA John, JB Pridham. Postharvest changes in *Mangifera indica* mesocarp cell walls and cytoplasmic polisaccharides. Phytochemistry 27:719–723, 1988.
17. MA Gómez Lim. Postharvest physiology. In: RE Litz, ed. The Mango Botany, Production and Uses. Wallingford Oxon: CAB International, 1997, pp 423–443.
18. H Subramanyam, S Gouri, S Krishnamrthy. Ripening behaviour of mango fruits graded on specific gravity basis. J Food Sci Technol 13:84–86, 1976.
19. NS Morga, AD Lustre, MM Tuuac, AH Balagot, MR Soriaou, Physico-chemical changes in Philipine Carabao mangoes during ripening. Food Chem 4:225–254, 1979.
20. Y Selvaraj, R Kumar, DK Pal. Changes in sugars, organic acids, amino acids, lipid constituents and aroma characteristics of ripening mango (*Mangifera indica* L.) fruit. J Food Sci Technol 26:308–313, 1989.
21. M Castrillo, NJ Kruger, FR Whatley. Sucrose metabolism in mango fruit during ripening. Plant Sci 84:45–51, 1992.
22. BC Peacock, BI Brown. Quality comparison of several mango varieties. Proceedings of the First Australian Mango Research Workshop, 1984, pp 334–339.
23. MCC Lizada. Mango. In: G Seymour, J Taylor, G Tucker eds. Biochemistry of Fruit Ripening. London: Chapman & Hall, 1993, pp 255–271.
24. AK Mattoo, T Murata, EB Pantastico, K Chachin, K Ogata, CT Phan. Chemical changes during ripening and senescence. In: EB Pantastico ed. Postharvest Physiology, Handling and Utilization of Tropical and Subtropical Fruits and Vegetables. Westport, CT: AVI Publishing, 1975, pp 130–156.
25. AA Kader, RF Kasmire, FG Mitchell, MS Reid, NF Sommer, JF Thompson. Postharvest technology of horticultural crops. University of California Cooperative Extension Special Publication # 3311, 1985.
26. GA Tucker, D Grierson. Fruit ripening. In: PK Stumpf, EE Conn eds. The Biochemistry of Plants. Vol. 12. New York: Academic Press, 1987, pp 265–313.
27. DO Adams, SF Yang. Ethylene biosynthesis: Identification of 1-aminocyclopropane-1-carboxylic acid as an intermediate in the conversion of methionine to ethylene. Proc Natl Acad Sci USA 76:170–174, 1979.
28. H Kende. Ethylene biosynthesis. Annu Rev Plant Physiol Mol Biol 44:283–307, 1993.
29. SP Burg, EA Burg. Role of ethylene in fruit ripening. Plant Physiol 37:179–189, 1962.
30. AU Cua, MCC Lizada, Ethylene production in the 'Carabao' mango (*Mangifera indica L.*) fruit during maturation and ripening. Acta Hortic 269:169–179, 1990.
31. R López-Gómez, MA Gómez-Lim. Changes in mRNA and protein synthesis during ripening in mango fruit. J Plant Physiol 141:82–87, 1992.
32. T Maekawa. On the mango CA storage and transportation from subtropical to temperate regions in Japan. Acta Hortic 269:367–371, 1990.
33. ML Arpaia. Preharvest factors influence the postharvest fruit quality of tropical and subtropical fruit. Hortscience 29:982–985, 1993.
34. AA Kader. Modified and controlled atmosphere storage of tropical fruits. In: BR Champ, E Highley, GI Johnson, eds. Postharvest Handling of Tropical Fruits. Canberra ACT 2601, ACIAR Proceedings No. 50, 1994, pp 239–249.
35. GR Chaplin. Advances in postharvest physiology of mango. Acta Hortic 231:639–648, 1989.
36. GI Johnson, JL Sharp, DL Milne, SA Oosthuyse. Postharvest technology and quarantine treatments. RE Litz, ed. The Mango Botany, Production and Uses. Wallingford Oxon: CAB International, 1997, pp 447–507.

37. AP Medlicott, M N'Diaye, JMM Sigrist. Harvest maturity and concentration and exposure time to acetylene influence initiation of ripening in mangoes. J Am Soc Hortic Sci 115:426–430, 1990.
38. AP Medlicott, JMM Sigrist, O Sy. Ripening of mangoes following low temperature storage. J Am Soc Hortic Sci 115:430–434, 1990.
39. B Piechulla, KR Chonoles-Imlay, W Gruissem. Plastid gene expression during fruit ripening in tomato. Plant Mol Biol 5:373–378, 1985.
40. MA Gómez Lim. Mango fruit ripening: Physiology and molecular biology. Acta Hortic 341:484–499, 1993.
41. G Bojorquez, MA Gómez Lim. Peroxisomal thiolase mRNA is induced during mango fruit ripening. Plant Mol Biol 28:811–820, 1995.
42. A Cruz Hernandez, MA Gómez Lim. Alternative oxidase from mango (*Mangifera indica* L.) is differentially regulated during fruit ripening. Planta 197:569–576, 1995.
43. S Kumar, BC Patil, SK Sinha. Cyanide resistant respiration is involved in temperature rise in ripening mangoes. Biochem Biophys Res Commun 168:818–822, 1990.
43a. P Gutierrez-Martinez, R Lopez-Gómez, MA Gómez-Lim. Identification of an etr1-homologue from mango expressing during fruit ripening and wounding. J Plant Physiol 158:101–108, 2001.
44. Y Eguchi, T Itoh, J Tomizawa. Antisense RNA. Annu Rev Biochem 60:631–652, 1991.
45. AJ Hamilton, GW Lycett, D Grierson. Antisense gene that inhibits synthesis of the hormone ethylene in transgenic plants. Nature 346:284–287, 1990.
46. PW Oeller, LM Wong, LP Taylor, DA Pike, A Theologis. Reversible inhibition of tomato fruit senescence by antisense RNA. Science 254:437–440, 1991.
47. R Ayub, M Guis, M Ben-Amor, L Gillot, JP Roustan, A Latché, M Bouzayen, JC Pech. Expression of ACC oxidase antisense gene inhibits ripening of cantaloupe melon fruits. Biotechnology 14:862–866, 1996.
48. HJ Klee, MB Hayford, KA Kretzmer, GF Barry, GM Kishore. Control of ethylene synthesis by expression of a bacterial enzyme in transgenic tomato plants. Plant Cell 3:1187–1192, 1991.
49. X Good, JA Kellogg, W Wagoner, D Langhoff, W Matsumura, RK Bestwick. Reduced ethylene synthesis by transgenic tomatoes expressing S-adenosylmethionine hydrolase. Plant Mol Biol 26:781–786, 1994.
50. V Arondel V, Lemieux, I Hwang, S Gibson, HM Goodman, CR Somerville. Map-based cloning of a gene controlling omega-3 fatty acid desaturation in *Arabidopsis*. Science 258:1353–1355, 1992.
51. O Ishizaki-Nishikawa, T Fujii, M Azuma, K Sekiguchi, N Murata, T Ohtani, T Toguri. Low-temperature resistance of higher plants is significantly enhanced by a nonspecific cyanobacterial desaturase. Biotechnology 14:1003–1006, 1996.
52. SM Baqui, AK Mattoo, VV Modi. Glyoxylate metabolism and fatty acid oxidation in mango fruit during development and ripening. Phytochemistry 13:2049–2055, 1977.
53. KE Koch. Carbohydrate-mediated gene expression in plants. Annu Rev Plant Physiol Mol Biol 47:509–540, 1996.
54. H Van der Wel, K Arvidson Qualitative phycophysical studies on the gustatory effect of the sweet tasting proteins thaumatin and monellin. Chem Senses Flavor 3:291–299, 1978.
55. L Peñarrubia, R Kim, J Giovannoni, SH Kim, RL Fischer. Production of the sweet protein monellin transgenic plants. Biotechnology 10:561–564, 1992.
56. BR Fils-Lycaon, PA Wiersma, KC Eastwell, P Sautiere. A cherry protein and its gene, abundantly expressed in ripening fruit, have been identified as thaumatin-like. Plant Physiol 111:269–273, 1996.
57. DB Tattersall, R van Heeswijck, PB Hoj. Identification and characterization of a fruit-specific, thaumatin-like protein that accumulates at very high levels in conjunction with the onset of sugar accumulation and berry softening in grapes. Plant Physiol 114:759–769, 1997.
58. AP Medlicott, M Bhogol, SB Reynolds. Changes in peel pigmentation during ripening of mango fruit (*Mangifera indica*, var. Tommy Atkin). Ann Appl Biol 109:651–656, 1986.
59. CR Bird, JA Ray, JD Fletcher, JM Boniwell, AS Bird, et al. Using antisense RNA to study gene function: Inhibition of carotenoid biosynthesis in transgenic tomatoes. Biotechnology 9:635–638, 1991.
60. PHR Robson, AC McCormac, AS Irvine, H Smith. Genetic engineering of harvest index in tobacco through overexpression of a phytochrome gene. Biotechnology 14:995–998, 1996.

61. PHR Robson, H Smith. Fundamental and biotechnological applications of phytochrome transgenes. Plant Cell Environ 20:831–839, 1997.
62. D Weigel, EM Meyerowitz. The ABCs of floral homeotic genes. Cell 78:203–209, 1994.
63. D Weigel, O Nilsson. A developmental switch sufficient for flower initiation in diverse plants. Nature 377:495–500, 1995.
64. R Simon, MI Igeno, G Coupland. Activation of floral meristem identity genes in Arabidopsis. Nature 384:59–62, 1996.
65. CJ Lamb, JA Ryals, ER Ward, RA Dixon. Emerging strategies for enhancing crop resistance to microbial pathogens. Biotechnology 10:1436–1445, 1992.
66. BJC Cornelissen, LS Melchers. Strategies for control of fungal diseases with transgenic plants. Plant Physiol 101:709–712, 1993.
67. K Broglie, J Chet, M Holliday, R Cressman, P Biddle, S Knowlton, CJ Mauvais, R Broglie. Transgenic plants with enhanced resistance to the fungal pathogen *Rhizoctonia solani*. Science 254:1194–1197, 1991.
68. Q Zhu, EA Mahler, S Masoud, RA Dixon, CJ Lamb. Enhanced protection against fungal attack by constitutive co-expression of chitinase and glucanase genes in transgenic tobacco. Biotechnology 12:807–811, 1994.
69. J Logemann, G Jach, H Tommerup, J Mundy, J Schell. Expression of a barley ribosome-inactivating protein leads to increased fungal protection in transgenic tobacco plants. Biotechnology 10:305–308, 1992.
70. G Jach, B Gornhardt, J Mundy, J Logemann, E Pinsdorf, R Leah, J Schell, C Maas. Enhanced quantitative resistance against fungal disease by combioatorial expression of different barley antifungal proteins in transgenic tobacco. Plant J. 8:97–109, 1995.
71. WF Broekaert, FR Terras, BP Cammue, RW Osborn. Plant defensins: Novel antimicrobial peptides as components of the host defense system. Plant Physiol 108:1353–1358, 1995.
72. RE Litz, RK Knight, S Gazit. Somatic embryos from cultured ovules of polyembryonic *Mangifera indica* L. Plant Cell Rep 1:264–266, 1982.
73. RE Litz, RJ Knight, S Gazit. In vitro somatic embryogenesis from *Mangifera indica* L. callus. Sci Hortic 22:233–240, 1983.
74. RE Litz. In vitro somatic embryogenesis from nucellar callus of monoembryonic *Mangifera indica* L. Hortscience 19:715–717, 1984.
75. RE Litz, U Lavi. Biotechnology. In: RE Litz, ed. The Mango Botany, Production and Uses. Wallingford Oxon: CAB International, 1998, pp 401–423.
76. RE Litz, PA Moon, MJ Monsalud, S Jayasankar, H Mathews. Somatic embryogenesis in *Mangifera indica* L. (mango). In: SM Jain, PK Gupta, RJ Newton, eds. Somatic Embryogenesis in Woody Plants. Dordrecht: Kluwer Academic, 1995, pp 341–356.
77. RE Litz, MA Gómez Lim, S Jayasankar, L de Moura Barros. Biotechnology for mango improvement and propagation. In: RP Srivastava, ed. Recent Advances in Mango Research. New Delhi: Society for the Advancement of Horticulture, 1999.
78. OL Gamborg, RA Miller, K Ojima. Plant cell cultures. I. Nutrient requirements of suspension cultures of soybean root cells. Exp Cell Res 50:151–158, 1968.
79. T Murashige, F Skoog. A revised medium for rapid growth and bioassays with tobacco tissue cultures. Physiol Plant 15:473–497, 1962.
80. SG DeWald, RE Litz, GA Moore. Optimizing somatic embryo production in mango. J Am Soc Hortic Sci 114:712–716, 1989.
81. RE Litz, VH Mathews, PA Moon, F Pliego-Alfaro, C Yurgalevitch, SG DeWald. Somatic embryos of mango (*Mangifera indica* L.). In: K Redenbaugh, ed. Synseeds: Applications of Synthetic Seeds to Crop Improvement. Boca Raton, FL: CRC Press, 1993, pp 409–425.
82. BL Lad, S Jayasankar, F Pliego-Alfaro, PA Moon, RE Litz. Temporal effect of 2,4-D on induction of embryogenic nucellar cultures and somatic embryo development of 'Carabao' mango. In Vitro Cell Dev Biol 33:253–257, 1997.
83. RE Litz, RC Hendrix, PA Moon, VM Chavez. Induction of embryogenic mango cultures as affected by genotype, explanting, 2,4-D and embryogenic nurse culture. Plant Cell Tissue Org Cult 53:13–18, 1998.

84. EG Williams, G Maheshwaran. Somatic embryogenesis: Factors influencing coordinated behaviour of cells as an embryogenic group. Ann Bot 57:443–462, 1986.
85. MJ Monsalud, H Mathews, RE Litz, DJ Gray. Control of hyperhydricity of mango somatic embryos. Plant Cell Tissue Org Cult 42:195–206, 1995.
86. SG DeWald, RE Litz, GA Moore. Maturation and germination of mango somatic embryos. J Am Soc Hortic Sci 114:837–841, 1989b.
87. T Kozai. Micropropagation under photoautotrophic conditions. In: PC Debergh, RH Zimmerman, eds. Micropropagation Technology and Application. Dordrecht: Kluwer Academic, 1991, pp 447–469.
88. RE Litz, VH Mathews, RC Hendrix, C Yurgalevitch. Mango somatic cell genetics. Acta Hort 291:133–140, 1991.
89. S Jayasankar, RE Litz. Characterization of embryogenic mango cultures selected for resistance to *Colletotrichum gloeosporioides* culture filtrate and phytotoxin. Theor Appl Genet 96:823–831, 1998.
90. S Jayasankar, RE Litz, RJ Schnell, A Cruz-Hernandez. Embryogenic mango cultures selected for resistance to *Colletotrichum gloeosporioides* culture filtrate show variation in random amplified polymorphic DNA (RAPD) markers. In Vitro Cell Dev Biol 34:112–116, 1998.
91. H Mathews, RE Litz, DH Wilde, S Merkel, HY Wetzstein. Stable integration and expression of β-glucuronidase and NPT II genes in mango somatic embryos. In Vitro Cell Dev Biol 28P:172–178, 1992.
92. H Mathews, RE Litz, HD Wilde, HY Wetzstein. Genetic transformation of mango. Acta Hortic 341:93–97, 1993.
93. A Cruz Hernandez, MA Gómez Lim, RE Litz. Transformation of mango somatic embryos. Acta Hortic 455:292–298, 1997.
94. H Mathews, N Bharathan, RE Litz, KR Narayanan, PS Rao, CR Bhatia. The promotion of *Agrobacterium*-mediated transformation in *Atropa belladonna* L. by acetosyringone. J Plant Physiol 136:404–409, 1989.
95. H Mathews, RE Litz. Kanamycin sensitivity of mango somatic embryos. Hortscience 25:965–966, 1990.
96. RA Jefferson. Assaying chimeric genes in plants: The GUS gene fusion systems. Plant Mol Biol 5:387–405, 1987.

# 30
# Transgenic Papayas in Hawaii: A Useful Tool for New Cultivar Development and Clonal Propagation

**Maureen M. M. Fitch**

*U.S. Department of Agriculture, Aiea, Hawaii*

| | | |
|---|---|---|
| I. | INTRODUCTION | 437 |
| II. | TRANSFORMATION | 439 |
| | A. Methods | 439 |
| | B. Traits | 439 |
| | C. Plant Materials and Papaya Culture Initiation | 440 |
| | D. Plasmids | 440 |
| | E. Inoculation of 'Kamiya' $R_0$ Plants with a PRSV Kahuku Isolate | 441 |
| | F. Field Evaluation of 'Kamiya' $R_0$ Plants | 442 |
| III. | RESULTS AND DISCUSSION | 443 |
| | A. Selection on Kanamycin | 443 |
| | B. Symptom Development on Test Plants Grown Under Fluorescent Lights | 444 |
| | C. Molecular Analysis and Field Evaluation of Transgenic 'Kamiya' Papayas | 444 |
| | D. Field Evaluation of Transgenic Hybrids Derived from Deregulated, Commercialized 'Sun Up' Papayas | 445 |
| | E. Field Evaluation of Clonally Propagated Transgenic Hybrid Lines | 445 |
| IV. | SUMMARY AND CONCLUSIONS | 445 |
| | REFERENCES | 446 |

## I. INTRODUCTION

Papaya (*Carica papaya* L.) is an important fruit crop in the tropics and subtropics because of its pleasant flavor, year-round bearing habit, and short growth cycle of 10–12 months from seed to ripe fruit. Ripe papayas are consumed at breakfast or as dessert, or the immature fruit is cooked with meat as squash is. Green fruit serve as a daily staple in papaya salad in Southeast Asia and in Africa are tapped for the latex, which contains papain, the main ingredient in meat tenderizer. Ripe fruit are a rich source of vitamins A and C (Arriola et al., 1980) and provide enzymes that assist in healthy digestion. The crop is grown in a wide range of cropping systems from small home gardens to large-scale plantation agriculture in the Americas, Asia, Australia, and Africa. In

Hawaii, papayas constitute the sixth most important agricultural crop after pineapples, sugarcane, macadamia nuts, coffee, and foliage crops. The farm-gate value is about $18 million annually. It is raised on about 500 small farms for a total of about 3000 acres, on the islands of Hawaii (Big Island), Oahu, Kauai, Molokai, and Maui. Prior to near destruction of the industry by an epidemic of papaya ringspot virus (PRSV) in 1992, 95% of the industry was located in Puna on the Big Island.

Market expansion has been limited by fruit availability and postharvest quality, the former condition most heavily impacted by diseases such as PRSV, a worldwide problem; fungal diseases; and postharvest damage from insect larvae. Solutions to the virus disease problems have included isolation, surveillance, sanitation by removing infected individuals, planting of tolerant cultivars, and, most recently, genetically engineered virus resistance (Fitch et al., 1992; Lius et al., 1997; Gonsalves, 1998). PRSV-resistant papayas do not exist, although some wild species from among the other 21 different caricas are resistant. Attempts to introgress resistance genes from *C. pubescens*, *C. quercifolia*, *C. stipulata*, and *C. cauliflora* resulted in $F_1$s with field resistance from *C. quercifolia* and *C. cauliflora* but the plants were sterile (Manshardt and Wenslaff, 1989). More recently, R. Drew identified 11 interspecific hybrids between papaya and *C. quercifolia* that were PRSV-resistant and showed some fertility (unpublished results); thus the potential exists that resistance genes from wild species can be introgressed into papaya over time. Tolerance exists in a Colombian papaya accession that was used to create the cultivar 'Cariflora' in Florida (Conover et al., 1985), which in turn was introgressed into large-fruited Thai papayas of commercial quality with attenuated virus symptoms (V. Prasartsee, unpublished results). Prasartsee determined that rigorous selection for tolerance in each seed generation was extremely important to prevent loss of tolerance. In Hawaii, F. Zee and S. Ferreira produced introgressed lines from the Colombian accession that were of acceptable quality (unpublished results).

The newly commercial Hawaiian transgenic papayas 'Sun Up' and 'Rainbow,' immune in the field to Hawaiian PRSV, were produced by collaboration of the University of Hawaii and Cornell University. The homozygous 'Sun Up,' a red-fleshed cultivar that was derived from the original $R_0$ transgenic line, 55-1, is immune to the Hawaiian PRSV and to other PRSV strains from around the world except for the Thailand and Taiwan strains, which show the greatest base pair divergence from the Hawaiian isolate (Tennant, 1996; Souza, 1999). 'Rainbow,' on the other hand, is resistant once the seedlings become established in the field after 2 months (S. Ferreira, unpublished results). Other groups attempted transformation of local cultivars with resistant genes from local PRSV strains (Tennant, 1996; Souza, 1999; Gonsalves, 1998). The Hawaiian virus-resistant papaya represents the first fruit tree crop deregulated and licensed for commercialization in the United States and the world. In the United States, it has the distinction of being the first transgenic commercial crop produced by public institutions.

After PRSV, the most important pests on papaya in Hawaii are fungal diseases, which include phytophthora fruit and root rot (*Phytophthora palmivora*), anthracnose (*Colletotricum gloeosporioides*), and powdery mildew (*Oidium caricae*); fruit rots caused by *Rhizopus stolonifera*, *Stemnophyllum lycopersici*, and *Phomopsis* sp. are also important postharvest concerns minimized by fungicide treatment in the field prior to harvest and by postharvest disinfestation. Disinfestation consists of forced hot air incubation (Armstrong et al., 1995) for about 4 hours. Transformation of papaya for fungal resistance is ongoing in our laboratory (M. Fitch and J. Zhu). We have produced transgenic papayas expressing a rice chitinase gene (Lin et al., 1995), which inhibited growth of *Phytophthora palmivora* mycelia in an in vitro assay and may show suppressed fungal symptoms. Papayas containing a grapevine stilbene synthase gene and dahlia, radish, and onion antimicrobial peptide genes, and the pathogenesis-related protein control gene NPR1 from *Arabidopsis* sp. are being regenerated (J. Zhu, unpublished results).

The fruit fly (Oriental fruit fly, *Bactrocera dorsalis* [Hendel]; Mediterranean fruit fly, *Cer-*

*atitis capitata* [Wiedemann]; and melon fly, *Bactrocera cucurbitae* [Coquillett]) are pests that limit export of papayas to the U.S. mainland and Japan. The decontamination procedure is a forced hot air disinfestation that was developed to kill eggs and young larvae. We have obtained constructs that may impart insect resistance to papaya as well. Transformation experiments have been initiated.

## II. TRANSFORMATION

### A. Methods

The embryogenic callus transformation method based on walnut transformation by McGranahan et al. (1988) that we initially developed is being utilized by most groups, using either *Agrobacterium* (Neupane, 1996) or the particle gun (Cabrera-Ponce et al., 1995; Tennant, 1996; Souza, 1999; Cai et al., 1999; Mahon et al., 1996; de la Fuente et al., 1997). Several groups reported on improvements to our original published method (Fitch et al., 1992), which included using rapidly growing somatic embryos (Cabrera-Ponce et al., 1995; Cai et al., 1999). Cabrera-Ponce reported a 50-fold improvement over our results, or 60 transgenic lines per gram fresh weight vs. our 1.14 lines. The reliability of papaya transformation will improve after these findings.

### B. Traits

#### 1. Virus Resistance

After publication of our results, researchers in other tropical areas reported on successes with their PRSV strains and cultivars. Gonsalves's laboratory became an international center for engineered virus resistance, coordinating students and researchers, their respective papaya cultivars, and PRSV strains. Since homologous PRSV coat protein (CP) genes appeared to give the strongest resistance, lines suitable for each locality were created. The group in Taiwan field-tested resistant lines for 4 years with fair protection, although some plants showed symptoms in the field. Subsequent lines were as immune as the Hawaiian 'Sun Up' (S.-D. Yeh, unpublished results). Yeh developed infectious ribonucleic acid (RNA) strains that give mild strain cross-protection, possibly eliminating the necessity to transform the plants (unpublished results). In Australia, transgenic papayas (Mahon et al., 1996) remained virus-free in the field (R. Lines, D. Persley, R. Drew, unpublished results). Two of the three most promising lines developed leaf abnormalities that can probably be eliminated by crossing with nontransgenic lines. In Jamaica, 'Sunrise' type plants transformed with the Jamaican PRSV translatable and untranslatable CP genes are being evaluated for deregulation and licensing. In Brazil and Thailand, transgenic 'Sunrise' and Thai papaya cultivars, respectively containing translatable and nontranslatable CP genes of their respective PRSV strains, await clearances for field testing (D. Gonsalves, M. Souza, personal communication).

#### 2. Delayed Ripening

Neupane (1997) transformed papaya by using both the particle gun and *Agrobacterium* sp. for delayed ripening with antisense 1-amino cyclopropane-1-carboxylic acid (ACC) synthase and ACC oxidase genes from papaya. Two of five lines expressed the antisense message in Northern blots, but neither showed delayed ripening (K. Neupane, unpublished results). The $R_1$ generation did not show delayed ripening, the $R_2$ generation, some of which were homozygous for antiethylene genes, are being evaluated along with about 25 additional $R_0$ lines. J. Botella is attempting the same type of work in Australia.

## 3. Herbicide Resistance

Cabrera-Ponce and coworkers (1995) reported on herbicide-tolerant papaya from transformation with the particle gun. The papayas were selected on 100 mg/l kanamycin + 4 mg/l phosphinothricin (PPT) and were resistant to the PPT herbicide. From bombardment to rooted plants, the process took about 12 months. Control plants suffered total necrosis 2 weeks after application of 0.1% (w/v) PPT, whereas transgenic plants withstood up to 3% PPT. A high degree of resistance was attained since the prescribed application rate is 0.5% to 1.0%. It is not known whether PPT is registered for use on papaya in Mexico. More than 100 GUS/NPTII (fusion protein)–positive clones were isolated in this experiment. These results differed from ours, in which only one third of the lines contained β-glucuronidase (GUS) or the PRSV CP gene, whereas all lines contained neomycin phosphotransferase (NPTII) (Fitch et al., 1992). We assume that all lines contained the PPT gene since both PPT and kanamycin were used in selection. Two selectable markers for cobombardment resulted in excellent cotransformation.

## 4. Aluminum Tolerance

De la Fuente et al. (1997) reported on increasing the aluminum tolerance of papaya, a sensitive crop. The particle gun was used to deliver a chimeric citrate synthase gene from *Pseudomonas aeruginosa* driven by a 35S promoter and terminated by nos 3' sequences. Overproduction of citrate two- to three-fold higher than in controls resulted in resistant plants that rooted and grew normally in 300 μM aluminum (Al), whereas nontransgenic plants failed to root in 50 μM and higher concentrations of Al. Bombardment protocols are described in Cabrera-Ponce and colleagues (1995).

## C. Plant Materials and Papaya Culture Initiation

Somatic embryos derived from either immature embryos (Fitch and Manshardt, 1990) or hypocotyls (Fitch, 1993) were grown on embryo proliferation medium containing 10 mg/l of 2,4-dichlorophenoxy-acetic acid (2,4-D) in 1/2 MSO medium (callus induction medium, Fitch et al., 1992). Immature zygotic embryos about 90 days old were excised and plated on callus induction medium 1–2 weeks prior to bombardment. Somatic embryos developed on the apical domes and were spread on filter paper on induction medium. Since mature seeds are more readily available than immature ones, the hypocotyl protocol was adopted for generating embryogenic callus. It was quicker and yielded embryos in 1.5 to 3 months. Two- to four-month-old cultures were bombarded. Although cultures older than 6 months were discarded, we still obtained a high frequency of abnormal and dwarf plants, about 10/100, among the transgenic 'Kamiya' regenerants (M. Fitch, unpublished results). Either the cultivar is not stable or exposure to the high 2,4-D concentration is causing the ploidy abnormality. In field tests of $F_1$ and backcrossed hybrids of the 'Kamiya' line, similar abnormal plants were observed among seedlings at frequencies of 1/100–300 plants. Among the 'Sunset' and 'Kapoho' transformants, only 1/100 abnormal plants were observed.

## D. Plasmids

Plasmids ranged in size from about 10 kb for the PRSV replicase (NIB), antimicrobial, and rice chitinase constructs to nearly 20 kb for the translatable (PRV4) and untranslatable (UT) PRSV constructs. The target genes were flanked by the 35S promoter and terminator sequences from cauliflower mosaic virus, and the selection gene NPTII was driven and terminated by nos sequences. The rice chitinase gene was driven by 35S with a nos terminator, as were the stilbene synthase, antimicrobial, and NPRI gene constructs. Hygromycin phosphotransferase was the selection gene on the rice chitinase construct and NPTII on the other three constructs.

## 1. Plasmid Preparation and Use with the Particle Gun

Plasmids in *Escherichia coli* were isolated by Qiagen column purification (Qiagen, Inc, Valencia, CA). Ten micrograms of plasmid DNA were precipitated onto 3 mg of gold (1.6 µm in diameter). Two micrograms of deoxyribonucleic acid (DNA)/0.6 mg of gold were bombarded at 1550 psi to a 2.5-cm target of embryogenic calli 6.5 cm from the macrocarrier disk launch point by using a Biolistics PDS-1000/He device (Bio-Rad, Hercules, CA). Gap distance between the rupture disk and macrocarrier was 6.5 cm. Each plate was bombarded three times for a total of 6 µg of plasmid DNA per 2.5 cm target. The somatic embryos that made up the target weighed approximately 500 mg.

Clusters of bombarded somatic embryos were transferred immediately after bombardment to recovery plates of induction medium for about 1 week, mainly to observe them for fungal and bacterial contamination. After recovery they were placed onto a suboptimal selection concentration, 75 mg/l kanamycin for 1 month, then to 150 mg/l with monthly subculturing. By the end of the third month, selectively growing embryos were observed. They were plated separately onto fresh plates of 150 mg/l kanamycin, and if they showed rapid growth, a quarter of the isolate was transferred to 300 mg/l kanamycin, a quarter retained on 150 mg/l, and the rest regenerated on growth regulator-free medium.

## 2. DNA Extraction and Polymerase Chain Reaction

DNAs were extracted by the CTAB method (Doyle and Doyle, 1987). Polymerase chain reaction (PCR) primers for NPTII were reported in Chee and associates (1989) and PRV CP in Slightom (1991). The NIB forward primer was 5′-ATC ATT CCA TGG GTA GTC GTT GGC TC-3′ and NIB reverse was 5′-AGC TAA CCA TGG TTA CTG ATG AAA CC-3′. Standard conditions for PCR were 1-minute denaturation at 95°C, 1-minute annealing at 58°C, and 2-minutes extension at 72°C. Lines were screened for the NPTII or hygromycin selection genes and target genes, PRV4, UT, and NIB.

## 3. Southern Hybridization

Southern hybridizations were performed as reported in Fitch and coworkers (1992) on putative transgenic lines containing pGA482GG/PRV4 (Ling et al., 1991), pGA482GG/UT, or pBin19/PRVNib-2. The pGA482GG/UT construct contains the PRV4 gene sequence plus three additional stop condons at the 5′ end of the sequence to give a nontranslatable RNA. The pBin19/PRVNib-2 construct contains a 1.13-kb HindIII fragment of the PRSV replicase (NIB) gene. A 500-bp EcoRI fragment described in Fitch and coworkers (1992) was used to probe for the former two constructs, the 1.13-kb HindIII fragment was used to probe for NIB-transformed lines. The digoxygenin nonradioactive method (Boehringer-Mannheim, Indianapolis, IN) was used for hybridizations and detection. DNAs from the rice chitinase, stilbene synthase, antimicrobial, and NPR1 gene transformants are currently being extracted for Southern hybridization analysis.

## E. Inoculation of 'Kamiya' $R_0$ Plants with a PRSV Kahuku Isolate

Shoots from Kamiya lines containing the virus resistance genes were propagated in vitro and acclimated to laboratory conditions under illumination with fluorescent lights (35 mmol/m$^2$/s). Fifteen- to twenty-five-cm-tall plants were pretreated in the dark overnight to increase leaf succulence. A PRSV-infected papaya leaf section (Kahuku strain) about 6 cm in diameter was macerated in pH 7 phosphate buffer and applied to two of the youngest fully expanded test plant leaves which had been lightly dusted with powdered carborundum. The plants were grown under lights at about 28°C and observed for symptom development. Micropropagated hemizygous 'Rainbow' plants, containing only 1 copy of the resistance gene, were used as positive controls, although resistance could be broken under laboratory conditions. Untransformed seedlings served as negative controls.

## F. Field Evaluation of 'Kamiya' $R_0$ Plants

'Kamiya' transgenic isolates were planted in two different APHIS-approved fields starting in 1997. Four lines, one containing PRV4, two containing NIB, and one line transgenic for NPTII only, were planted under moderately high virus pressure. Two lines containing the untranslatable CP RNA sequence, UT9-8A (resistant) and UT9-9A (susceptible), were planted in an APHIS-approved field in Kunia in April 1999. Since virus pressure at the Kunia site is very low, it is possible to maintain the susceptible lines for making selfed crosses and outcrosses between transgenic and nontransgenic plants. We are evaluating one apparently normal, resistant 'Kamiya' line, UT9-8A, for horticultural qualities along with another abnormal, tetraploidlike female line, UT9-9A, making selfed pollinations or outcrossing the female with untransformed 'Kamiya' pollen to try to produce homozygous lines that could have even greater resistance. As other lines containing the replicase, UT, or PRV4 genes become available, they are planted in the field for seed production and evaluation for horticultural traits.

### 1. Field Evaluation of Transgenic Hybrids Derived from the Deregulated, Commercialized 'Sun Up'

We embarked upon a backcrossing project utilizing the commercialized 'Sun Up' virus resistance gene after discovery of abnormal 'Kamiya' transformants because we wished to produce a useful virus-resistant 'Kamiya' in a timely manner. The objectives of the backcrossing project were (a) to produce PRSV resistant plants with nearly the same qualities as the virus susceptible lines that are well adapted to certain locations and (b) to determine whether the inbred line, $F_1$ progeny, or one of the backcross generations is superior to the inbred line. Hybrids between 'Rainbow' $F_2$ progeny and the inbred 'Kamiya' were planted in two commercial locations on the island of Oahu in Hawaii in 1998 to evaluate horticultural characteristics and virus resistance. The 'Kamiya' $F_1$ hybrids were planted in Laie (about 300 trees) and Kahaluu (about 150 trees). Tree height and diameter, height of first flower, yield, and fruit quality data were collected monthly. Major selection criteria were firm flesh, pleasing flavor, good fruit set, good fruit shape, minimal carpellody, and minimal sterility. Superior selections are being clonally propagated and will be evaluated with growers in several locations. In addition, the 'Kamiya' lines will be advanced to backcross generations of at least 94% parent type (at least five generations), then selfed to select for homozygous plants that can be propagated by seed if growers desire that option. In each planting, nontransgenic parents, $F_1$, and backcross plants will be compared. Selections of superior individuals will be clonally propagated and evaluated in replicated plots.

'Kapoho' was the industry standard because of its excellent shipping and marketing qualities—firm fruit and high sucrose content. At the same time the 'Kamiya' backcross project was initiated, we also undertook 'Kapoho' backcrossing because the cultivar had been the most important in the industry. Since 'Rainbow' is 50% 'Kapoho,' we obtained backcross$_1$ ($BC_1$) hybrids by crossing 'Rainbow' $F_2$ progeny with 'Kapoho.' Seedlings were planted in on the island of Oahu in Poamoho (about 100 trees) and Helemano (about 50 trees) and on the island of Hawaii in Keaau (about 100 trees) and Kapoho (about 100 trees). Monthly growth, yield, and quality data were collected on 'Kapoho,' 'Rainbow,' and the $BC_1$ (~75% 'Kapoho') at the Poamoho planting. Crosses were made between 'Rainbow' $F_2$ that were homozygous for the CP gene and 'Line 8,' a cultivar with excellent flavor but poor shelf life; two large-fruited Thai-style papayas; and a medium-sized cultivar called 'Maunawili Sweet' selected for good growth in a high-rainfall area. The objective of these additional crossing projects was to determine whether our strategy to cross a hybrid with an inbred of interest would lead to the inbred type sooner than the more standard method of crossing two inbreds followed by backcrossing to one parental type.

## 2. Field Evaluation of Clonally Propagated Transgenic Hybrid Lines

Papaya seedlings segregate for sex expression, either one female to two hermaphrodites from selfed hermaphrodites, or one female to one hermaphrodite or "male" in outcrosses (Storey, 1953). Typically seeds or seedlings are multiple planted then thinned to a single hermaphrodite or female, depending on the location. The waste associated with multiple planting, in addition to occasional scarcity of seeds, led to experiments to determine the feasibility of clonal propagation of papaya. Clonal propagation of hermaphrodite 'Rainbow,' 'Kamiya' $F_1$, and 'Kapoho' $BC_1$ plants was established by two methods, micropropagation in tissue culture and rooted cuttings. Micropropagation was initiated by using in vitro germinated seedlings and lateral shoots of mature hermaphrodite trees selected for quality characteristics. In vitro germinated seedlings (Fitch, 1993) were screened for the hermaphrodite type using PCR primers designed from randomly amplified polymorphic DNAs (RAPDs, Sondur et al., 1996). Shoot proliferation occurred on a modified Murashige and Skoog medium (MSO, Fitch et al., 1990, GIBCO/BRL, Grand Island, NY) containing 0.2 mg/l benzylaminopurine and 0.2 mg/l naphthalene acetic acid (MBN) and 2.5 g/l Phytagel (Sigma, St. Louis, MO). Shoots 1 cm or taller were removed from clusters and incubated in the dark for 5 days on MSO containing 1 mg/l indole butyric acid and solidified with 2.5 g/l Phytagel. The shoots were rooted under fluorescent lights (35 µmol/m$^2$/s PAR) at 27°C in jars containing 25 ml vermiculite moistened with 30 ml 1/2 MSO and 3% sucrose. Rooted plants were planted in Sunshine Mix#4 and acclimated under clear plastic domes. Acclimated plants were repotted into 3.5-inch plastic pots, grown in the greenhouse, acclimated to sunlight, and field planted or were retained as stock plants from which cuttings were removed every 3–6 weeks.

Cuttings from greenhouse-grown stock plants, 8–22 cm in length and 6–10 mm in diameter, were rooted in medium-grade 50% vermiculite/50% perlite medium under intermittent mist (5 seconds every 10 minutes continuously) with 29°C bottom heating (M. Fitch, unpublished results). Cuttings rooted whether or not cut ends were dipped in 0.1% IBA (Hormex 11, Teufel Nursery, Inc, Portland, OR). Approximately 90% of the cuttings rooted in 2–6 weeks. Plants were removed from the mist bench, fertilized with time release granules, and acclimated in the greenhouse for 2–4 weeks, in sunlight about 2–4 weeks, and transplanted in the field.

Micropropagated and cutting-derived clonal 'Rainbow' hermaphrodites, hermaphrodite seedlings screened by PCR for sex expression, and multiple-planted seedlings (five/hole) were installed in a randomized complete block experiment with four replications. The experiment was installed on April 30, 1998, on Hawaii in Keaau, south of Hilo, where annual precipitation averages 130 inches a year. Data on height, diameter, height of first bud, and fruit yield were collected and analyzed by an analysis of variance. The second experiment, comparing cuttings, micropropagated plants, and multiple-planted seedlings, was installed by the grower on October 23, 1998, in Helemano, Oahu. A third planting with the same treatments as the first was installed in Mokuleia, Oahu, on March 10, 1999.

## III. RESULTS AND DISCUSSION

### A. Selection on Kanamycin

By using NPTII as the selection gene, satisfactory results were obtained by using kanamycin at levels ranging from 75 mg/l to 300 mg/l. We found that escapes were eliminated by moving isolates recovered after the third month of selection to 300 mg/l kanamycin and retaining only those that grew vigorously on the higher antibiotic concentration.

## B. Symptom Development on Test Plants Grown Under Fluorescent Lights

Symptoms on untransformed control and transgenic susceptible plants developed 10–14 days post inoculation when the plants were grown under fluorescent lights in the laboratory. Small control seedlings died within 1 month and leaves on larger susceptible plants turned yellow and abscised, and the plants eventually died after 2 months. 'Rainbow' plants with one copy of the transgene developed symptoms after 16 days, but the older leaves remained green. Early symptoms included downward or upward curving of new leaf margins, mottling of new leaves, and distortion. The resistant transgenic line UT9-8A did not show symptoms 6 weeks after the first set of inoculations and was inoculated a second time along with a second UT9-8A plant and untransformed control. Neither plant showed PRSV symptoms 2 weeks after the second series of inoculations; thus we concluded that UT9-8A was immune to the Kahuku PRSV strain.

## C. Molecular Analysis and Field Evaluation of Transgenic 'Kamiya' Papayas

PCR and Southern hybridizations confirmed the presence of the transgenes. PCR for the NPTII selection, translatable and untranslatable CP, and NIB genes was used to identify putative resistant lines. Southern hybridizations revealed specific bands for PRV4, UT, and NIB at the predicted molecular weights, 1.7 kb for the CP genes as shown in our earlier studies (Fitch et al., 1992, 1994) and 1.0 kb for the replicase gene. For practical, rapid screening, molecular characterization using PCR was sufficient to identify lines that were then mechanically inoculated with a Kahuku PRSV strain. We identified three immune lines, one with the translatable CP gene (PRV4-1), one with the untranslatable CP gene (UT9-8A), and one with the NIB gene (NIB 1-1). Other lines showed field resistance or delayed symptom expression (NIB 3-1); other lines were not resistant. The untranslatable gene isolate UT9-8A is the most promising because the plants appear to be normal, diploid, and hermaphrodite. We are propagating the line for small-scale field testing and will complete southern and northern hybridizations to confirm the presence and functioning of the transgenes. The other two lines, PRV4-1 and NIB 1-1, are apparently tetraploid, highly sterile, and not immediately useful. Barring fertility and agronomic problems, we believe that the UT9-8A line is a good candidate for deregulation and licensing, as was line 55-1, which was the progenitor of 'Sun Up' and 'Rainbow.'

The plant line containing only NPTII showed virus symptoms within 4 months of field planting, whereas the three lines carrying resistance genes were immune for about 2 years. One of the three lines exhibited a 2-month delay in symptom development in the laboratory but was immune under field conditions. The other two lines may have multiple copies of the resistance gene since they never exhibited virus symptoms in the laboratory or field. However, they all appeared to be tetraploid and set small numbers of miniature fruit that contained only one or two seeds. Attempts to embryo rescue and germinate the rare embryos were not successful.

Other than a high degree of carpellody in one individual, normal for this cultivar at certain times of the year, the plants appeared to be normal. Other UT9-8A plants and the corresponding nontransgenic 'Kamiya' plants in the same field did not exhibit the degree of carpellody as high as that of the one UT9-8A individual; thus, the problem may be overfertilization or some other environmental factor. The susceptible UT9-9A, a female, is abnormal and appears to be tetraploid with larger than normal petioles and leaves but small petals and ovaries. Some buds pollinated with another transgenic line are enlarging; thus, the ovaries may not be sterile. If seeds develop, it will be interesting to determine the ploidy and fertility of the seedlings.

## D. Field Evaluation of Transgenic Hybrids Derived from Deregulated, Commercialized 'Sun Up' Papaya

In Hawaii, hemizygous 'Sun Up'–based hybrids have been field planted at several sites and are showing excellent production and no breakdown except in rare instances in which small plants became infected and did not recover, for example, 1/800 in Kapoho, 1/400 in Keaau, and 1/300 in Kahuku. We kept the plants for 3–5 months and removed them when they failed to recover.

The backcrossing experiments were greatly simplified with presence of the GUS gene in the donor plants. Production data from the first rounds of backcrossing the Kapoho $BC_1$ and 'Kamiya' $F_1$ plants showed that the deregulated, licensed transgene in the new hybrid backgrounds was acceptable. We determined that the Kapoho $BC_1$ plants had fruit size and firmness intermediate between those of the inbred, untransformed Kapoho and 'Rainbow.' Sweetness was approximately the same. We anticipate that we will be able to select for increased firmness in the next backcross generation, $BC_2$, at the same time selecting for the larger size and pleasant aroma that we detected in the $BC_1$ lines.

We believe that our protocol to backcross an inbred line like 'Kamiya' with a hybrid like 'Rainbow' instead of more directly, with another inbred, resulted in a hybrid with a greater proportion of the target inbred's characteristics. Although the $F_1$ fruit were only 50% 'Kamiya,' they resembled the 'Kamiya' parent, retaining the deep orange fruit color, coconutlike flavor, low bearing habit, root rot tolerance, and large fruit size characteristics. The presence of a higher percentage of the 'Kamiya' genome may have resulted in $F_1$ progeny that exhibited the backcross parent phenotype in the shortest period possible. We have early indications from comparisons between the 'Kamiya' $F_1$ and 'Kamiya' $BC_1$ plants that hybrid vigor in the 'Kamiya' $F_1$s may make those progenies more desirable than advanced backcrosses, which could exhibit some 'Kamiya' problems like carpellody, sterility, and seasonal fruit bitterness and unpleasant odor. Carrying on the 'Kamiya' hybrid evaluation, one of our new projects is to compare $F_1$ progeny from 'Kamiya' × 'Rainbow' $F_2$ with 'Kamiya' × 'Sun Up,' (inbred transgenic) to determine whether the 'Rainbow' $F_2$ hybrids are more similar to 'Kamiya' than are the 'Sun Up' hybrids.

Our data crossing 'Line 8' with the 'Rainbow' $F_2$ support our hypothesis. We found that the poor shelf life trait of 'Line 8' was still pronounced in the 'Line 8' $F_1$ hybrids. We are further testing our hypothesis by crossing three other inbred lines that have large fruit size, 'Kuala Lumpur Yellow,' a Thai-style red, and 'Maunawili Sweet.'

## E. Field Evaluation of Clonally Propagated Transgenic Hybrid Lines

The asexual propagation experiments comparing clonal hermaphrodites with seedlings yielded ripe fruits earlier and lower on the trunk, for potential harvest savings compared to single- or multiple-planted seedlings. The multiple-planted trees produced comparable yields about 3 months later, but the 2 months of additional yield adds income for the grower and reduces costs by eliminating labor for thinning. An additional benefit not predicted in this experiment was shorter trees over the harvest cycle, delaying the use of harvest poles, ladders, and fork or scissor lifts for harvest. Harvest is the most costly operation in papaya production.

## IV. SUMMARY AND CONCLUSIONS

The deregulated, commercial transgenic virus-resistant 'Sun Up' and 'Rainbow' cultivars have been useful additions for papaya improvement in the state of Hawaii. They provided researchers with virus-resistant cultivars for field experimentation, a practice that had been limited in the past by losses from PRSV infection. We were able to determine the feasibility of producing and planting asexually propagated trees and to evaluate new hybrids in several important papaya-growing regions in the state. We determined that an asexual propagation protocol that utilizes micropropagation and rooted cuttings could allow growers about 2 months of earlier production compared to use of standard practice multiple-planted seedlings. An additional benefit is the lower bearing height of the trees, which translates into savings in harvest time.

We have continued transformation experiments of papaya, producing another transgenic virus-resistant cultivar, 'Kamiya,' containing one of three PRSV virus resistance genes. Our data show that the translatable and untranslatable PRSV coat protein genes and the replicase gene impart resistance to 'Kamiya' papaya. D. Gonsalves, S.-P. Yeh, and R. Drew (unpublished results) also report on successful virus-resistant papayas transformed with PRSV CP genes. Papaya transformation also resulted in herbicide and aluminum resistance. The impact of the first virus-resistant papayas on Hawaiian agriculture is significant, the industry was saved from destruction due to the virus. We expect that transgenic papayas in other regions of the world will impact those local economies similarly. Herbicide- and aluminum-tolerant papayas may also prove to be valuable commercially. Moreover, new genes for improvement of the crop are now available to us, for example, genes for fungal and insect resistance, for broader solutions to agricultural problems. In working directly with our growers while testing our new cropping system and hybrids, we found that technology transfer has led to better understanding and teamwork among scientists, growers, and other agricultural industry members. The growers are more comfortable with us, the scientists, when they see the products of our experiments translating directly into useful products for their livelihoods.

## REFERENCES

Armstrong JW, BKF Hu, SA Brown (1995) Single-temperature forced hot-air quarantine treatment to control fruit fly (Diptera: Tephritidae) in papaya. J Econ Entomol 88:678–682.

Arriola M, J Calzada, J Menchu, C Rolz, R Garcia, S Cabrera (1980) Papaya. In: S Nagy, P Shaw, eds. Tropical and Subtropical Fruits. AVI Publishers, Westport, Connecticut. pp 326–340.

Cabrera-Ponce JL, A Vegas-Garcia, L Herrera-Estrella (1995) Herbicide resistant transgenic papaya plants produced by an efficient particle bombardment transformation method. Plant Cell Rep 15:1–7.

Cai W, C Gonsalves, P Tennant, G Fermin, M Souza, N. Sarindu, F Jan, H Zhu, D Gonsalves (1999) A protocol for efficient transformation and regeneration of *Carica papaya* L. In Vitro Cell Biol Plant 35: 61–69.

Chee PP, KA Fober, JL Slightom (1989) Transformation of *soybean (Glycine max)* by infecting germinating seeds with *Agrobacterium tumefaciens*. Plant Physiol 91:1212–1218.

Cheng, YH, JS Yang, SD Yeh (1996) Efficient transformation of papaya by coat protein gene of papaya ringspot virus mediated by *Agrobacterium* following liquid-phase wounding of embryogenic tissues with carborundum. Plant Cell Rep 16:127–132.

Conover RA, RE Litz, SE Malo (1986) 'Cariflora'—A papaya ringspot virus tolerant papaya for South Florida and the Caribbean. Hort Science 21:1072.

de la Fuente JM, V Ramirez-Rodriguez, JL Cabrera-Ponce, L Herrera-Estrella (1997) Aluminum tolerance in transgenic plants by alteration of citrate synthesis. Science 276:1566–1568.

Doyle JJ, JL Doyle (1987) A rapid DNA isolation procedure for small quantities of fresh leaf tissue. Phytochem Bull 19:11–15.

Fitch MMM (1993) High frequency somatic embryogenesis and plant regeneration from papaya hypocotyl callus. Plant Cell Tissue Org Cult 32:205–212.

Fitch MMM (1995) Somatic embryogenesis in papaya (*Carica papaya* L.). In: YBS Bajaj, ed. Biotechnology in Agriculture and Forestry. Vol. 30. Somatic Embryogenesis and Synthetic Seed I. Berlin: Springer-Verlag, pp 260–279.

Fitch MMM, RM Manshardt (1990) Somatic embryogenesis and plant regeneration from immature zygotic embryos of papaya (*Carica papaya* L.). Plant Cell Rep 9:320–324.

Fitch MMM, RM Manshardt, D Gonsalves, JL Slightom, JC Sanford (1990) Stable transformation of papaya via microprojective bombardment. Plant Cell Rep 9:189–194.

Fitch MMM, RM Manshardt, D Gonsalves, JL Slightom, JC Sanford (1992) Virus resistant papaya plants derived from tissues bombarded with the coat protein gene of papaya ringspot virus. Biotechnology 10:1466–1472.

Fitch MMM, RM Manshardt, D Gonsalves, JL Slightom (1993) Transgenic papaya plants from *Agrobacterium*-mediated transformation of somatic embryos. Plant Cell Rep 12:245–249.

Fitch MMM, S-Z Pang, JL Slightom, S Lius, P Tennant, RM Manshardt, D Gonsalves (1994) Genetic transformation in *Carica papaya* L. (papaya). In: YPS Bajaj ed. Biotechnology in Agriculture and Forestry. Vol. 29. Plant Protoplasts and Genetic Engineering V Berlin: Springer-Verlag, pp 236–256.

Gonsalves D (1998) Control of papaya ringspot virus in papaya: A case study. Annu Rev Phytopathol 36: 415–437.

Lin W, CS Anuratha, K Datta, I Potrykus, S Muthukrishan, S Datta (1995) Genetic engineering of rice for resistance to sheath blight. BioTechnology 13:686–691.

Ling K, S Namba, C Gonsalves, JL Slightom, D Gonsalves (1991) Protection against detrimental effects of potyvirus infection in transgenic tobacco plants expressing the papaya ringspot virus coat protein gene. BioTechnology 9:752–758.

Lius S, RM Manshardt, MMM Fitch, JL Slightom, JC Sanford, D Gonsalves (1997) Pathogen-derived resistance provides papaya with effective protection against papaya ringspot virus. Mol Breed 3:161–168.

Mahon RE, MF Bateson, DA Chamberlain, CM Higgins, RA Drew, JL Dale (1996) Transformation of an Australian variety of *Carica papaya* using microprojectile bombardment. Aust J Plant Physiol 23: 679–685.

Manshardt RM (1998) 'UH Rainbow' papaya. Univ Hawaii Coll Trop Agric Hum Resour: Germplasm G-1. http://www/2ctahr.hawaii.edu.

Manshardt RM (1992) Papaya. In: FA Hammerschlag, RE Litz, eds. Biotechnology of Perennial Fruit Crops. Wallingford, England: CAB International, pp 489–511.

Manshardt RM, TF Wenslaff (1989) Interspecific hybridization of papaya with other *Carica* species. J Amer Soc Hort Sci 114:689–694.

McGranahan GH, CA Leslie, SL Uratsu, LA Martin, AM Dandekar (1988) *Agrobacterium*-mediated transformation of walnut somatic embryos and regeneration of transgenic plants. BioTechnology 6:800–804.

Neupane KR (1997) Genetic engineering of papaya (*Carica papaya* L.) for modified ethylene biosynthesis. PhD. Dissertation, University of Hawaii, Honolulu, Hawaii. pp. 173.

Pang SZ, JC Sanford (1988) *Agrobacterium*-mediated gene transfer in papaya. J Am Soc Hortic Sci 113: 287–291.

Slightom JL (1991) Custom polymerase chain reaction engineering of a plant expression vector. Gene 100:251–255.

Sondur S, RM Manshardt, J Stiles (1996) A genetic linkage map of papaya based on randomly amplified polymorphic DNA markers. Theor. Appl. Gen. 93:547–553.

Souza, MT (1999) Analysis of the resistance in genetically engineered papaya against papaya ringspot potyvirus, partial characterization of the PRSV. Brazil.Bahia isolate, and development of transgenic papaya for Brazil. PhD dissertation, Cornell University, Ithaca, NY.

Storey WB (1953) Genetics of papaya. J Hered 44:70–78.

Tennant PF (1996) Evaluation of the resistance of coat protein transgenic papaya against papaya ringspot potyvirus isolates and development of transgenic papaya for Jamaica. PhD dissertation, Cornell University, Ithaca, NY.

Tennant P, M Fitch, R Manshardt, J Slightom, D Gonsalves (1997) Resistance against papaya ringspot virus isolates in coat protein transgenic papaya is affected by transgene dosage and plant development (abstract). Phytopathology S96.

Tennant PF, C Gonsalves, KS Ling, M Fitch, R Manshardt, J Slightom, D Gonsalves (1994) Differential protection against papaya ringspot virus isolaes in coat protein gene transgenic papaya and classically cross-protected papaya. Phytopathology 84:1359–1366.

# 31
# Genetic Engineering of Strawberries and Raspberries

**Robert R. Martin**
*U.S. Department of Agriculture, Corvallis, Oregon*

|      |                                                                              |     |
|------|------------------------------------------------------------------------------|-----|
| I.   | INTRODUCTION                                                                 | 449 |
| II.  | REGENERATION METHODS FOR *FRAGARIA* AND *RUBUS* SPECIES                      | 450 |
| III. | DEVELOPMENT OF TRANSFORMATION PROTOCOLS FOR *FRAGARIA* AND *RUBUS* SPECIES   | 451 |
| IV.  | INCORPORATION OF SPECIFIC TRAITS—INSECT, NEMATODE, AND MITE RESISTANCE       | 452 |
| V.   | VIRUS RESISTANCE                                                             | 454 |
| VI.  | FUNGAL RESISTANCE                                                            | 456 |
| VII. | FRUIT QUALITY                                                                | 457 |
| VIII.| GENOMICS                                                                     | 458 |
|      | A. Gene Discovery                                                            | 458 |
|      | B. Promoters                                                                 | 458 |
|      | C. Gene Flow                                                                 | 459 |
|      | REFERENCES                                                                   | 459 |

## I. INTRODUCTION

Breeding of *Fragaria* × *ananassa* Duch. (strawberry) began in the early 1800s (Scott and Lawrence, 1975) and *Rubus* breeding (raspberry, blackberry, and hybrid berries) has been going on since the late part of the 19th century (Dale et al., 1989). Despite the relatively long history of breeding with these crops and the many successes in breeding for disease and pest resistance, there remain disease problems for which there are no known sources of resistance in the available germplasm. In other cases, pests and pathogens have evolved to overcome host resistance that breeders have incorporated in new cultivars. Phytophthora root rot (*Phytophthora fragariae* f.sp. *fragariae*) (Nickerson and Jamieson, 1995) in strawberry and raspberry bushy dwarf virus (RB strain) in red raspberry (Barbara et al., 1984) are two examples of pathogens overcoming employed resistance genes. Raspberry aphids, *Amphorophora idaei* in Europe and *A. agathonica* in North America, have evolved to overcome resistance genes used widely in raspberry breeding programs to control aphid-borne viruses (Daubeny, 1996). Overuse and/or misuse of fungicides and insecticides has resulted in the development of pathogens and pests resistant to chemicals. Some chemicals are withdrawn for use on these minor crops because the costs associated with keeping them on the label are prohibitive or the work to get a chemical registered on these minor crops is not done, again as a result of cost considerations. The Food Quality Protection Act in the United States will result in a review of all pesticides and very likely will result in the loss of many of those used currently for control of insects and pathogens of these crops.

With the advent of biotechnology, specifically genetic engineering, the potential to introduce novel types of disease and pest resistance from sources unrelated to these crops became available. Many of the disease- and pest-resistant traits being tested in *Fragaria* and *Rubus* spp. have been shown to be effective in other crops or in model systems that are easier to transform and regenerate than members of the Rosaceae (Estruch et al., 1997). Strategies being used for virus resistance have been shown to provide some level of resistance with other viruses in other crops prior to being used in the small fruit crops (Martin, 1995; Wilson, 1993). Strawberry and rubus are generally much more difficult to regenerate and transform than tobacco or tomato and other plants used in model systems to develop these technologies. In addition to pest and disease resistance genes that are involved in fruit quality characteristics are being isolated (APHIS website). The first commercial transgenic crop, the 'Flavr Savr' tomato, was designed to alter fruit softening associated with fruit ripening. Although technologically a success, 'Flavr Savr' was not accepted by the consumer for several reasons, including a lack of acceptance of genetically engineered products and the lack of a perceived improvement in fruit quality. Another lesson learned from this early tomato work was that to be successful, this new technology needs to be applied to the best cultivars of a given crop. The state of the technology today is limited in that only a few genes can be inserted into a plant at a time. Thus, most current work deals with single traits, which are ideally suited for pest and disease resistance. Genetic engineered resistance has been shown to be quite effective in controlling selected insects in commercial plantings of cotton, potatoes, and maize carrying a *Bacillus thuringiensis* (*Bt*) toxin gene. Herbicide resistance has been used widely in soybeans in the United States and in canola in Canada. Virus resistance based on coat protein genes has been shown to be quite effective in commercial squash, potatoes, and papaya. Virus resistance strategies have been tested in a wide range of crops and many of these will likely be commercialized over the next few years.

## II. REGENERATION METHODS FOR *FRAGARIA* AND *RUBUS* SPECIES

Plant tissue culture was used for virus eradication in strawberry as early as 1962 (Belkengren and Miller, 1962) and for micropropagation as early as 1974 (Boxus, 1974). Plant regeneration of strawberry from callus was reported in 1977 (De Assis and Hildebrandt, 1977; Lee and DeFossard, 1977). Most tissue culture and micropropagation of strawberry have been directed to horticultural purposes, primarily virus elimination and rapid propagation. Recently, regeneration has been studied much more intensively as a first step in transformation to introduce desired genes.

Most groups found that media containing macro- and microsalts after Murashige and Skoog (1962) and $B_5$ vitamins according to Gamborg coworker (1968) was useful as a base medium for strawberry regeneration. The addition of casein hydrolysate or tryptone at 400 mg/l enhanced regeneration (Liu and Sanford, 1988; Finstad and Martin, 1995). Early regeneration from strawberry callus tissue (Jones et al., 1988) showed that callus was produced on MS medium containing inositol, thiamine HCl, and benzylamine purine (BA) at 0.9. Regeneration of shoots from leaf or petiole pieces was dependent on the addition of plant hormones to the tissue culture medium. Parameters for optimal regeneration can vary, depending on whether the leaf tissue was derived from plantlets grown in vitro or in a greenhouse (Liu and Sanford, 1988; Finstad and Martin, 1995). The hormones and their concentrations have varied considerably among cultivars that have been regenerated successfully. Conditions that were optimal for 'Allstar' (in vitro cultures, 2.5 mg benzyladenine ([BA])/l and 0.5 mg indolebutyric acid [IBA]/l; greenhouse cultures [3.0 mg/l BA and 0.1 mg/l IBA], or 'Redcoat' (10 mg/l BA and 1 mg/l naphthaleneacetic acid [NAA] (Nehra et al., 1990a) failed to produce any shoots with 'Hood' and 'Totem' (Finstad and Martin, 1995). Thidiazuron was the most effective cytokinin for regeneration of 'Totem' strawberry (Finstad and Martin, 1995). Conditions for regeneration of 'Marmolada' and 'Onebar' were similar to those for 'Allstar' (Martinelli et al., 1997). In studies with 'Chandler' strawberry (Barcelo et al., 1998), both IBA and BA were required for regeneration. In this case, incubation of leaf disks in darkness for 1 to 4 weeks increased the rate at which regeneration occurred but did not affect the final percentage of disks with shoots. The level of irradiance after the dark treatment did not affect the regeneration rate of 'Chandler' leaf pieces (20–80 µmol m$^{-2}$ s$^{-1}$ was tested). Regeneration efficiency (percentage of explants that developed shoots) approached 100% under some conditions. One can see from the work that has been done with strawberry regeneration that at this time there is not a procedure that is useful for the regeneration of a broad range of strawberry cultivars. In contrast, there is a regeneration and transformation protocol that was successful with a range of potato cultivars (Kawchuk et al., 1991). Carbenicillin was inhibitory to regeneration of strawberry (Finstad and Martin, 1995; Graham et al., 1995), whereas Timentin had no effect on the percentage of explants that developed shoots in 'Totem' when used at 100–200 µg/ml. Timentin was also more effective at eliminating *Agrobacterium* spp. from the cultures than carbenicillin and/or cefotaxime and it was also less costly.

Fewer people have been successful with *Rubus* sp. regeneration and in most cases the percentage regeneration is much lower for *Rubus* sp. than for *Fragaria* sp. when conditions are optimized. Owens y de Novoa (1992) found that stem and root explants of *R. idaeus* grew more callus but only leaf explants produced shoots in a medium that consisted of macronutrients of MS micronutrients with 2× iron plus MS vitamins and 30 g/l sucrose. Optimal regeneration was with 2 µM BA plus 0.5 µM NAA or both hormones at 5 µM. McNicol and Graham (1990) were able to get regeneration from both leaf and internodal segments when using a modified MS medium containing the phytohormone combination 0.2 mg/l BA and 0.2 mg/l 2,4-D. They regenerated two hybrid berries, 'Sunberry' (46% regeneration) and 'Tayberry' (27%); 'Loch Ness' blackberry (27%); and 'Autumn Bliss' red raspberry (7.5%). Peeled internodal stem segments produced more regenerants than unpeeled segments, and conditions for regeneration of stem segments were much less critical than for leaf pieces. Thidiazuron was reported to be more effective than BA for induction of organogenesis in *Rubus* sp. (Fiola et al., 1990; Cousineau and Donnelly, 1991; Owens y de Novoa and Conner, 1992a). Mathews and associates (1995) were able to get regeneration from petiole, leaf, and internode explants of several different red raspberry cultivars by using 0.1 mg/l IBA, 0.1–1.0 mg/l Thidiazuron, and 10 mg/l silver nitrate. For 'Meeker' red raspberry (50% regeneration), they used 3% D-glucose as the carbon source, whereas with 'Canby' (1%) and 'Chilliwack' (8%) the carbon source was 3% sucrose. Transformation of arctic bramble (*Rubus arcticus* L.) was successful on media containing 2 mg/l BA, and 0.2 mg/l NAA and was much more successful when using internodal segments (15–27%) compared with leaf pieces (1%). As with strawberry the development of a regeneration protocol for *Rubus* sp. is quite empirical, with optimal hormone and medium conditions dependent on cultivar and source and type of explants.

## III. DEVELOPMENT OF TRANSFORMATION PROTOCOLS FOR *FRAGARIA* AND *RUBUS* SPECIES

Development of transformation protocols requires several decisions, beginning with whether to use biolistics or *Agrobacterium* sp.–mediated technologies and choice of *Agrobacterium* sp.

strains if this is the method of choice. The *Agrobacterium* sp. strain can greatly influence the transformation success. A wide range of strains of *A. tumefaciens* and *A. rhizogenes* were capable of causing tumors in strawberry, though the efficiency of tumor formation varied widely (Uratsu et al., 1991). In a test of 35 strains of *A. tumefaciens* on 10 genotypes of red raspberry Owens y de Novoa and Conner (1992b) found that strains A208, A281, ACH5, C58, and TR105 were the most effective at inducing tumors on leaf disks in vitro. In a separate study, LAB4404, a disarmed strain of C58, was the most efficient of seven *A. tumefaciens* strains at forming tumors on leaf disks in vitro (de Faria et al., 1997). *A. tumefaciens* strains LBA4404, EHA101, MP90, and EHA105 have been the most widely used for transformation in these crops. *Agrobacterium tumefaciens*–mediated transformation of *F. × ananassa* (Nehra et al., 1990; James et al., 1990; Finstad and Martin, 1995; Graham et al., 1995; Mathews et al., 1995; Barcelo et al., 1998), *F. vesca* (El-Mansouri et al., 1996; Haymes and Davis, 1997), and *Rubus* (Graham et al., 1990; Hassen et al., 1993; Mathews et al., 1995; Kokko and Karenlampi, 1998) has been developed by a number of groups.

Choice of a selectable marker is critical to the success of a transformation protocol. Transgenic plants have been grown from strawberry protoplasts electroporated with a plasmid carrying β-glucuronidase (GUS) and selectable marker genes (Nyman and Wallin, 1992). It is also likely that particle bombardment could be used to transform strawberry and raspberry since transient expression of the green fluorescent protein (GFP) has been observed in several cultivars of *F. × ananassa* and *R. idaeus* (P. Kohnen, personal communication). However, most transformation in strawberry and *Rubus* sp. has been done with leaf disks or petioles by using antibiotic resistance genes as selectable markers. In the development of transformation protocols, β-glucuronidase (GUS) (Jefferson, 1987) has been used most widely as a marker since expression can be monitored by a simple staining procedure to confirm that tissue has been transformed. More recently, the green fluorescent protein (GFP) (Oparka et al., 1995) has been used to monitor gene expression since this marker can be visualized in a nondestructive assay. The *nptII* gene, which confers resistance to kanamycin, has been the most widely used antibiotic marker for *Fragaria* and *Rubus* sp. transformation (Nehra et al., 1990; James et al., 1990; Finstad and Martin, 1995; El-Mansouri et al., 1996; Barcelo, 1998; Graham et al., 1990; Graham et al., 1995; Haymes and Davis, 1997; Mathews et al., 1995;) The *hpt* gene, which confers resistance to the antibiotic hygromycin, has also been used successfully for selection in these crops (Mathews et al., 1995; Nyman and Wallin, 1992) Kanamycin selection is usually done at between 25 and 50 µg/ml and hygromycin selection at 10 µg/ml. In our experience, there are fewer escapes with hygromycin selection than with kanamycin.

Since *Rubus* and *Fragaria* sp. are vegetatively propagated crops, the primary transgenics can be used directly. With many other crops, the R1 and R2 generations are tested for gene function and the process of going through true seed eliminates most if not all chimerism from the transgenic plants. Using GUS as a marker gene, Mathews and coworkers (1998) have shown that many primary transgenics are chimeric. By performing iterative cultures with increasing selection pressure they were able to eliminate chimerism from transformed plants as demonstrated by GUS assays on multiple leaves from rooted plants and by Southern hybrization.

## IV. INCORPORATION OF SPECIFIC TRAITS—INSECT, NEMATODE, AND MITE RESISTANCE

Many different protease inhibitors have been inserted into plants in efforts to develop resistance to insects, nematodes, and mites. The effectiveness of an inhibitor depends on the major protease that the target pest contains in its digestive system (Ryan, 1990). The most widely used protease inhibitor to date has been the serine inhibitor from cowpea (cowpea trypsin inhibitor [CpTi]), which has been inserted into a wide range of crop plants, including strawberry (Hilder et al., 1987; James et al., 1992; Watt et al., 1995). Cysteine proteinase inhibitor gene oryzacystatin (OCI) from rice has also been used to make transgenic strawberry and *Rubus* sp. (Vrain, 1997). When the potato multicystatin (PMC), a cysteine protease inhibitor, was inserted into corn (Orr et al., 1994), it was shown to be very effective against western corn rootworm but much less effective against southern corn rootworm. The cysteine proteases make up 92% of the total protease activity in western corn rootworm but only 75% of the protease activity of the southern corn rootworm. With the employment of protease inhibitors, it will be interesting to see how quickly insects can evolve to change the relative importance of different proteases in their digestive system as this rate will greatly influence the long-term effectiveness of protease inhibitors as a means of insect control.

When CpTi was inserted into 'Rapella' strawberry and 10 kanamycin-resistant transformants, they were shown to have positive results for the inserted genes by Southern hybridization and exhibited phosphotransferase activity (James et al., 1992). Eight plants of six of the clones were tested in insect feeding assays using the black vine weevil (*Otioryhnchus sulcatus*). Twenty larvae, less than 3 days old, were added to each potted plant. After 2 months the survival rate of larvae was 80–90% on the transformed clones and the controls. The CpTi has also been incorporated into 'Melody' strawberry (Graham et al., 1995; Graham et al., 1996; Watt et al., 1999). In feeding studies with black vine weevil in which eight eggs were applied to the base of each test plant and allowed to develop for 20 weeks, the best transgenic line had between 240% and 360% more root weight than the nontransgenic control or GUS transgenic plants. The transgenic plants exhibited a range of gene expression levels based on percentage reduction in trypsin activity. In pots where the plants were not challenged with vine weevils, the root weight of control and transgenic plants was the same. Some of the transgenic plants showed a gradual decline in tissue culture, which was possibly due to the presence of the transgene. This is not surprising since plants have proteases to "recycle" their proteins and a protease inhibitor could interfere with normal plant metabolism, and it is possible some essential gene was interrupted in these lines. Surprisingly, the transgenic lines that were the highest expressors of the CpTi were not the ones that declined in tissue culture, suggesting site of insertion as the probable cause of the decline. Very few of the eggs developed to the adult stage on the control or the transgenic plants in the feeding experiments. A similar reaction to that observed with the pot feeding studies was observed after 2 years of trials in the field with strawberries carrying the CpTi transgene (Graham et al., 1999), in which two cultivars with the CpTi gene demonstrated the same high level of resistance to the black vine weevil under field conditions that was observed in greenhouse trials. The sharp contrast in effectiveness between these two trials may be due to expression levels of CpTi in the transgenic plants. In the first study (James et al., 1992), the level of activity of the CpTi was not determined. In the trials by Graham and colleagues (1997), one transgenic line (4.25) did not have any effect on black vine weevil as compared to the controls and this line had the lowest activity of the lines tested in their second feeding study. However, line 4.25 did have higher levels of activity than line 4.1 from their first feeding study, which had the least amount of feeding damage of all the plants they tested. These studies suggest that there is not a direct correlation between expression levels of CpTi and protection from black vine weevil damage.

The rice cysteine proteinase inhibitor oryzacystatin (OCI) has been incorporated into strawberry and red raspberry plants (Vrain et al., 1997) Prior to using the OCI gene from rice to transform plants they researchers tested its activity against the digestive proteases of root weevil and found it to inhibit most protease activity; it also inhibited papain activity in gel assay tests. When the OCI from rice was incorporated into strawberry and raspberry (*R. idaeus*), the presence of the gene was demonstrated by selection on kanamycin and Southern hybridization. When the plants were challenged with *Pratylenchus penetrans* (nematodes), *Tetranychus urticae* (two-spotted spider mite), or aphids there was no significant reduction in the number of any of these pests on the transgenic plants compared to that on the control plants. The plants have not yet been tested for resistance to weevils.

Lectins are being evaluated as potential control agents for several insect pests and nematodes (Vrain et al., 1997). Most lectins are toxic to many insects and mammals. The lectin from snowdrop (*Galanthus nivalis*) has little or no toxicity against mammals but has been shown to be toxic against several insects and against root knot and root lesion nematodes (Hilder et al., 1995). The protease inhibitors have activity against various chewing insects but have not been shown to have any effect on aphids, whereas the snowdrop lectin offers some protection against aphids. Lectins have a different mode of action than protease inhibitors so the potential exists to combine two control strategies into a single plant that would reduce the chance of the pest's developing resistance to the transgenic control measure.

*Bacillus thuringiensis* (*Bt*) is a bacterium that produces toxins that are specific for certain insect larvae, primarily in the lepidopteran and coleopteran species. There are different forms of *Bt* toxin, including *Btk* toxin (from *B. thuringiensis* var. *kurstaki*), which is effective primarily against lepidopteran insects, and *Btt* toxin (from *B. thuringiensis* var. *tenebrionis*), which is effective against a range of beetles. In addition, other *Bt* toxins have been isolated with different specificities including dipterans. Transgenic crops (corn, cotton, potatoes) containing the *Btk* and *Btt* toxins have been grown commercially in the United States. At this point, the use of these crops has resulted in considerable reduction in insecticide application compared to that is control plantings. The problem of development of resistance to *Bt* toxin is much more of a concern with transgenic crops than when *Bt* is used as an applied insecticide. In the transgenic crops, the insects are exposed to the toxin all the time. Another concern is that organic growers can use *Bt* as an insecticide; if resistance develops the organic growers will not have an alternative control measure whereas conventional growers can go back to using other insecticides. Lepidopteran larvae are primarily problems as fruit contaminants during harvest in raspberry. The *Bt* toxins have not been used in *Fragaria* or *Rubus* sp. at this time, however. *Bt* toxin is an approved topical insecticide to reduce larvae numbers prior to harvest, and there may be potential for its employment as a transgene for control of the leafrollers in red raspberry.

## V. VIRUS RESISTANCE

The first demonstration of genetically engineered resistance to a plant pathogen was resistance to tobacco mosaic virus (TMV) conferred by the incorporation of the TMV coat protein in tobacco (Powell-Abel et al., 1986). Since that time that coat protein (CP) of many different plant viruses has been used with varying degrees of success to engineer virus resistance (Martin, 1995). As more examples of CP-mediated resistance were studied, it became clear that in some examples there was a positive correlation between CP expression level and resistance to TMV (Powell-Able et al., 1986) and cucumber mosaic virus (Cuozzo et al., 1988), whereas in other cases this correlation did not exist (Ling et al., 1991; Stark and Beachy, 1989). Several strategies have been developed for engineering virus resistance since the early work with CP genes. These include the use of mutated polymerase (Golemboski et al., 1990), mutated movement proteins (Beck et al., 1994), nontranslatable ribonucleic acid (RNA) (Lindbo et al., 1993), and plantibodies (Taviadorakl, et al., 1993). The nontranslatable RNA strategy is RNA-based rather than protein-based, and it is likely that many of the CP-mediated resistant plants that have been developed are actually resistant as a result of the RNA and not the CP. Since the work by Lindbo and coworkers (1993), many people have reported sense constructs giving high levels of resistance even when the CP and transcript are not detectable. For more detailed information on engineered virus resistance, readers are referred to several reviews that have been written on engineering virus resistance in plants (Martin, 1995; Wilson, 1993) and a 1998 book (Hadidi et al., eds., 1998) that has six chapters covering this area. Pathogen-derived viral resistance can reduce spread of virus in a field situation even though the transgenic plants are not immune to the target virus (Thomas et al., 1997). In the study by Thomas and coworkers potato lines expressing the CP of potato leafroll virus that were only moderately resistant to the virus had significantly less secondary spread due to aphid transmission.

In the small fruit crops, there has been limited work published on engineering virus resistance. However, there are several reports of engineering genes of viruses that infect small fruit crops into various *Nicotiana* spp., which are easier to work with and thus good choices to determine whether a specific strategy will be useful. Strawberries have been engineered to contain the coat protein gene of strawberry mild yellow edge potexvirus (SMYEPV) (Finstad and Martin, 1995). Three of the transgenic lines showed some delay in symptom development in greenhouse trials, but in field trials the delay was not significant when considering strawberry is multiyear crop in the Pacific Northwest region of North America. The lack of resistance may be due to the nature of the virus as it now appears that the transmission of this virus is facilitated by a helper virus (Lamprecht and Martin, 1999) since SMYEPV is not aphid-transmissible from plants infected with a full-length infectious clone of the virus. The virus from the infectious clone is transmissible when coinoculated with wild-type virus. If the SMYEPV is being transmitted to plants in the coat protein of a second virus, then the uncoating process would not be restricted by the transgene product. In other systems with CP-mediated resistance, the plants were not resistant to infection by viral RNA. Further work on engineering resistance to SMYEPV is being carried out, using mutated polymerase, mutated movement protein, and nontranslatable RNA strategies, in my laboratory.

Coat proteins of nepoviruses, arabis mosaic (ArMV), strawberry latent ringspot (SLRV), and tomato ringspot (TomRSV) have been transformed into tobacco plants (Cooper et al., 1994; Kreiah et al., 1996; Yepes et al., 1996). These viruses are important in raspberry (Ellis et al., 1991) and strawberry (Maas, 1998), and the information obtained in the tobacco system should be applicable to these crops. It is easier to work with tobacco or some other herbaceous hosts of a virus for testing the effectiveness of a resistance strategy rather than strawberry or *Rubus* sp. SLRV produces two coat proteins, and both proteins, as well as the precursor of the two proteins, were expressed in *Nicotiana clevelandii*. The inoculated leaves of all the transgenic plants became infected; however, systemic invasion of the virus in these plants was only 39%, 32%, and 24% for the small, large, and precursor proteins, respectively. When plants were challenged with SLRV by using viruliferous nematodes, only plants with the large coat protein subunit became infected. The plants expressing the small subunit or the precursor protein were not infected in the nematode transmission tests.

The CP of ArMV was expressed in *N. tabacum* 'Xanthi' (Bertioli et al., 1991) and in *N. tabacum* 'White Burley' (Steinkellner et al., 1992). Level of expression varied widely from 0.03% to 0.8% of total plant soluble protein. The plants expressing the CP developed local lesions when inoculated with ArMV but no systemic symptoms when inoculated with either virus particles or viral RNA (Bertioli et al., 1992). The uninoculated leaves did contain virus as determined by bioassay but between one and two orders of magnitude less than that of control plants. When the transgenic plants were exposed to viruliferous field populations of *Xiphinema diversicaudatum* none became infected, whereas many of the control plants did become infected systemically (Cooper et al., 1994). The coat protein of ArMV has been inserted into strawberry (McNicol et al., 1997); however, the status of these plants with respect to virus resistance has not been published.

The coat protein gene of TomRSV, in both the sense and the antisense orientation, has been incorporated into *N. benthamiana* and *N. tabacum*, a systemic and local lesion host, respectively.

When the transgenic *N. benthamiana* lines were challenged with TomRSV, the reaction in the transgenic plants ranged from complete resistance (16/166) to delayed symptom development or a moderation of symptom severity. Of the completely resistant lines, seven had sense and nine had antisense constructs. The CP was undetectable by enzyme-linked immunosorbent assay (ELISA) in the lines carrying the CP sense constructs, and RNA levels were low or undetectable by Northern blot on resistant sense and antisense lines, suggesting an RNA-mediated resistance mechanism. In the transgenic *N. tabacum* 2/16 lines were completely resistant and the remaining 14 developed local lesions ranging in number from similar to that of wild type to significantly fewer lesions. TomRSV CP gene has also been transformed into grape (Scorza et al., 1996).

Raspberry bushy dwarf virus (RBDV) is the most serious virus disease of raspberry worldwide. The virus is pollen-borne; thus vector control is not an option in managing this virus. There is a single dominant resistance gene (*Bu*) to the common strain of RBDV that has been used widely in breeding programs, especially in the United Kingdom. In 1984 a new strain of the virus that overcomes the *Bu* resistance was identified in Europe (Barbara et al., 1984). In field situations, RBDV can spread through large plantings of susceptible cultivars in less than 6-years (Martin, 1998). Cultivars containing the *Bu* gene are not widely planted in the Pacific Northwest of North America because of horticultural and fruit quality characteristics, and many growers are opting for a 6-year replanting schedule rather than planting of resistant cultivars.

Groups in Scotland (Jones et al., 1998) and in the United States (Martin and Mathews, unpublished data) are developing engineered resistance to RBDV. The group in Scotland is working with a resistance-breaking strain of RBDV. The CP gene in the positive and antisense orientation and as a nontranslatable RNA as well as the putative polymerase gene in the positive sense orientation were inserted into *N. tabacum* 'Samsun' and *N. benthamiana*. Most transgenic lines with the three different constructs showed some resistance to RBDV. Some lines showed only 10–35% the number of local lesions, but the virus titers based on ELISA readings were 65–94% of those of the control plants. This finding suggests that the transformants may have some resistance to infection but once an infection is established there is little or no resistance to virus multiplication. This is similar to the work done with TMV coat protein (Powell-Abel et al., 1986). In other lines, with CP gene in either positive or negative sense and with nontranslatable CP gene, ELISA values were reduced (0–74% of controls). Also, in some cases the inoculated leaves became infected but there was an absence of symptoms in uninoculated leaves or a decrease in symptom severity. Many of the transgenic lines also showed a decrease in symptom severity.

The group in Oregon is working with an S isolate (non-resistance-breaking) that they have sequenced (Taylor and Martin, 1999). The CP gene, movement protein, three different mutations of the movement protein, and nontranslatable RNA were transformed into 'Meeker' red raspberry (Martin and Mathews, unpublished). Fifty transformants were produced for each construct. Plants were tested by polymerase chain reaction (PCR) for the presence of the inserted genes and five plants with each construct tested for copy number by Southern hybridization tests. In early grafting experiments, 47 of the 153 plants tested to date remained free of the virus after being grafted twice with a RBDV source. These plants are now being evaluated in field experiments for fruit quality, yield, and field resistance to RBDV.

## VI. FUNGAL RESISTANCE

One of the most important fungal diseases of *Rubus* and *Fragaria* spp. is gray mold caused by *Botrytis cineria*. Gray mold is difficult to control because it has a very broad host range and there is inoculum present during the entire growing season. Also, there is not a good source of resistance available to breeders to work with in their programs. Under favorable conditions for infection by *Botrytis* sp. growers must spray their crops frequently to control this disease. In addition, the fungus has developed resistance to a number of chemicals commonly used in control programs. Cultural practices to thin the canopy and increase air movement can reduce disease incidence, but when weather conditions are ideal, chemical control is the only successful means to control this disease. Any form of resistance that can be developed through genetic engineering to help control *Botrytis* sp. would be of great utility. It is unlikely that a genetic engineering solution will be successful on its own, but it could be a very important part of an integrated approach to control *Botrytis* sp.

Polygalacturonase inhibiting protein (PGIP) has been used in strawberry and respberry in attempts to develop some level of resistance to *Botrytis* sp. (Ramanathan et al., 1995; Mathews et al., 1995). Several companies are working with antifungal strategies in strawberry, but the details are not public knowledge at this time. They are trying to engineer resistance to several different fungi, including *Verticillium*, *Colletrotrichum*, and *Fusarium* sp. (APHIS website). Some of the strategies are making use of chitinases, glucanases, PGIP, and possibly other methods. Agritope Inc. (Portland, OR) is working with *S*-adensylmethionine hydrolase gene to affect fruit softening (Mathews et al., 1995). If this does impact fruit softening it should decrease susceptibility to *Botrytis* sp.

Chitinases, glucanases, and ribosome inhibiting proteins (RIPs) have been quite successful in other crops (Punja and Zhang, 1993; Jach et al., 1995) and therefore should increase fungal resistance in strawberry and *Rubus* sp. Each of these genes conferred some level of resistance to infection with *Rhizoctonia solani* in barley (Jach et al., 1995). Combining chitinases and glucanases or chitinase with RIP increased the levels of resistance obtained (Jach et al., 1995).

The chitinases should have an effect against a wide range of fungi in the ascomycetes and basidiomycetes since their cell walls contain large amounts of chitin. It is unlikely, however, that this strategy would be effective against *Phytophthora* sp. root rot, which is a serious problem in *Rubus* sp. and strawberries.

## VII. FRUIT QUALITY

After disease and pest resistance most efforts at genetic engineering of plants are designed to improve quality aspects of the crop (APHIS website). In strawberry and *Rubus* sp., several different transgenic strategies have been employed to improve fruit quality. When *S*-adensylmethionine hydrolase (SAMase) was incorporated into tomato the results were reduced ethylene production and extended shelf life (Good et al., 1994). The same strategy has been applied to raspberry and strawberry for increasing the postharvest shelf life of these fruits. SAMase has been incorporated into 'Totem' strawberry and 'Meeker,' 'Chilliwack,' and 'Canby' red raspberry (Mathews et al., 1995). Though strawberry and raspberry are not considered typical climacteric fruit (Kader, 1991), there is evidence indicating that the removal of ethylene could reduce decay in fresh berries. Treatment of 'Chandler' strawberry with ethylene absorbant resulted in increased firmness and reduced colonization by fungi from 26.3% to 10% (De la Plaza and Merodia, 1989). The application of ethylene to green fruit of raspberry enhances the rate of fruit ripening (Iannetta et al., 1998).

The incorporation of PGIP into strawberry and raspberry (Mathews et al., 1995; Ramanathan et al., 1995) to increase fruit firmness is supported by the increase in polygalacturonase activity during ripening of raspberry fruit (Iannetta et al., 1998; Jones et al., 1998). The effect on fruit quality as a result of incorporating the PGIP into these fruits has not been reported. This is similar to the strategy that was used on the first genetically modified fruit, the 'FlavrSavr' tomato produced by Calgene Inc. In the case of the tomato, an insert to transcribe an antisense RNA to the polygalacturonase was used to reduce the level of the polygalacturonase activity in the fruit and delay fruit softening during the ripening process.

Invertase converts sucrose to glucose and fructose. The invertase genes from potato, both cell wall and vacuolar forms, have been incorporated into 'Symphony' and 'Senga Sengana' strawberries (Bachelier et al., 1997). The cell wall form is thought to be involved in unloading of sucrose from the phloem and control assimilation at the various sinks including fruit. The vacuolar form may be involved in converting sucrose to hexoses within a cell and modifying the osmotic potential within a cell and the turgor pressure within the fruit. The goal is to be able to modify the balance of sugars within fruit to improve flavor and processing quality. The results of the use of the invertase gene on the strawberry fruit have not yet been reported.

## VIII. GENOMICS

### A. Gene Discovery

Many of the genes used for disease and pest resistance, with the exception of the viral genes, have been adopted from work done on other crops. The use of "off the shelf" genes makes the development of transgenic plants a much easier task and, in the case of disease and pest resistance, is a very useful approach. When trying to modify fruit quality parameters it is necessary to know which genes are expressed in the fruit of interest and at which stage of fruit maturation the various genes are expressed. In some cases, the genes of interest will have homologues that have been characterized from other crops, and these genes may be quite useful (e.g., PGIP and SAMase). In other cases the best genes available for modifying fruit maturation, flavor, texture, processing qualities, and other characteristics, will be derived from the crop whose fruit is being modified.

Jones and associates (1998) have identified 20 genes that are upregulated during fruit ripening in 'Glen Clova' red raspberry. These genes have a range of functions ranging from cell wall hydrolysis (polygalacturonase and pectin methyl esterase) to ethylene formation (1-aminocyclopropane-1-carboxylic acid oxidase), a protease inhibitor, latexlike protein, metallothioneinlike protein, and genes with unknown functions. A low-molecular-weight heat shock protein appears to be expressed at elevated levels in the achenes during seed maturation and in the receptacle at the W2 stage during which most changes occur that lead to fruit ripening (Medina-Escobar et al., 1998). The function of any of this heat shock protein during fruit ripening is unknown, but its importance as well as that of other isolated genes can be elucidated through transgenic technologies. In the very near future, the entire genome of *Arabidopsis* sp. will be sequenced; however, it may take much longer for function to be assigned to many of the sequenced genes. As gene functions are elucidated, it may be possible to isolate homologous genes from strawberry and raspberry. Randomly amplified polymorphic DNA (RAPD) markers closely linked to genes controlling traits of interest are being used to assist in selection for difficult to screen traits (Melenbacher, 1995; Antonius-Klemola, 1999).

### B. Promoters

Most transgenes in strawberry and *Rubus* sp. that have been expressed to date were placed behind the 35S promoter of cauliflower mosaic virus. Tissue-specific promoters will be required when trying to modify fruit characteristics and would be desirable for many pest and disease resistance traits as well. When engineering resistance to *Phytophthora fragariae* var. *rubi*, nematodes, or nematode-borne viruses in raspberry it is only necessary to have the traits expressed in the root tissue. Resistance to aphid-borne viruses need only be expressed in the tops of plants, and resistance to RBDV, a pollen-borne virus, need be expressed only in flowers. The fruit-specific promoter E4 was used to drive the SAMase gene incorporated into raspberry and strawberry (Mathews et al., 1995). Wound-specific promoters would be ideal for genes that control feeding by chewing insects provided the response time is sufficient to effect the insect.

## C. Gene Flow

For many of the agronomic crops grown in the United States and Canada there are no sexually compatible relatives in the native flora. This greatly reduces the problems with gene flow into native vegetation that may increase the weediness or invasiveness of native relatives. With strawberry and *Rubus* spp. there are sexually compatible native relatives in most growing regions. Thus, we can expect that transgenes introduced into these crops will make their way into native species over time (Luby and McNicol, 1995; Hokanson et al., 1997). Gene flow to native relatives may not happen in a few years, but once transgenic plants are released into the environment the process begins. We may not detect any movement of genes to native populations, but unless there is a readily visible marker the likelihood of finding a rare outcrossing is small (Luby and McNicol, 1995). The real task for risk assessment is to do our best to evaluate the impact of a gene in the native vegetation, since in these crops it is very likely the genes will make their way to the native relatives. The first points to consider are, Does the trait exist in the environment already, and, if so, what do we know about its impact? 'Willamette' was the major red raspberry cultivar grown in the Pacific Northwest from the mid-1940s to about 1990. This cultivar has the *Bu* gene for RBDV resistance, which gives us a considerable history with this trait. Aphid resistance in red raspberry has been used extensively in the breeding program in British Columbia, again giving some experience with this trait. Also, aphid resistance has been identified in *R. strigosus*, the native red raspberry in North America, which means this trait has been in the native vegetation for many, many years. Often we will not have any direct experience from native species or from conventionally bred cultivars with a trait that we are dealing with in transgenic plants. Many *Rubus* species are very invasive and difficult to control, and, therefore, herbicide resistance may not be a suitable trait to incorporate into raspberry or blackberry (Traynor and Westwood, 1999).

At a 1999 conference on the ecological effects of pest resistance genes in managed ecosystems it was concluded that the genetically engineered pest resistance traits currently being field tested or commercially released present no fundamental differences, with regard to ecological risks, from similar traits bred into crops through conventional breeding (Traynor and Westwood, 1999). There was also agreement that crops engineered with multiple pest resistances present more complex ecological questions. Continued discussions among ecologists, environmentalists, weed scientists, horticulturists, entomologists, pathologists, and concerned citizens will be needed to evaluate the ecological risks associated with each generation of transgenic plants that is developed. Food safety issues are dealt with by the Food and Drug Administration and the U.S. Dept. of Agriculture in the United States and equivalent agencies in other countries. Public perception and acceptance of transgenic crops are difficult to predict. At the present time public perception in Canada and the United States is much more favorable to transgenic foods than in any other part of the world. An unbiased explanation of relative risks needs to be presented to the public. Who should do this is unclear, but it should not be left to the companies that are developing the transgenic foods as their credibility will be questioned since they have a biased interest in the products. Universities and government agencies are the likely candidates to develop and present the information to the public, but at the present time this work is sorely lacking.

## REFERENCES

APHIS Web site: www.aphis.usda.gov

Antonius-Klemola, K. 1999. Molecular markers in *Rubus* (Rosaceae) research and breeding. J. Hortic. Sci. Biotech. 74:149–160.

Bachelier, C., Graham, J., Machray, G., Du Manoir, J., Roucou, J.F., McNicol, R.J. and Davies, H. 1997. Integration of an invertase gene to control sucrose metabolism in strawberry cultivars. Acta Hortic. 439:161–164.

Barbara, D.J., Jones, A.T., Henderson, S.J., Wilson, S.C. and Knight, V.H. 1984. Isolates of raspberry bushy dwarf virus differing in *Rubus* host range. Ann. Appl. Biol. 105:49–54.

Barcelo, M., El-Mansouri, I., Mercado, J.A., Quesada, M.A. and Alfaro, F.P. 1998. Regeneration and transformation via *Agrobacterium tumefaciens* of the strawberry cultivar Chandler. Plant Cell Tissue Org. Cult. 54:29–36.

Beck, D.L., Van Dolleweerd, C.J., Lough, T.J., Balmori, E., Voot, D.M. Anderson, M.T., O'Brien, I.E.W. and Forster, R.L.S. 1994. Disruption of virus movement confers broad-spectrum resistance against systemic infection by plant viruses with a triple gene block. Proc. Natl. Acad. Sci. USA 91:10310–10314.

Belkengren, R.O. and Miller, P.W. 1962. Culture of apical meristems of *Fragaria vesca* strawberry plants as a method of excluding latent A virus. Plant Dis. Rep. 46:119–121.

Bertioli, D.J., Harris, R.D., Edwards, M.L., Cooper, J.I. and Hawes, W.S. 1991. Transgenic plants and insect cells expressing the coat protein of arabis mosaic virus produce empty viruslike particles. J. Gen. Virol. 72:1801–1809.

Bertioli, D.J., Cooper, J.I., Edwards, M.L. and Hawes, W.S. 1992. Arabis mosaic nepovirus coat protein in transgenic tobacco lessens disease severity and virus replication. Ann. Appl. Biol. 120:47–54.

Boxus, P. 1974. The production of strawberry plants by in vitro micro-propagation. J. Hortic. Sci. 49:209–210.

Cooper, J.I., Edwards, M.L., Rosenwasser, O. and Scott. N.W. 1994. Transgenic resistance genes from nepoviruses: Efficacy and other properties. N. Z. J. Crop Hortic. Sci. 22:129–137.

Cousineau, J.C. and Donnelly, D.J. 1991. Adventitious shoot regeneration from leaf explants of tissue cultured and greenhouse-grown raspberry. Plant Cell Tiss. Org. Cult. 27:249–255.

Cuozzo, M., O'Connell, K., Kaniewski, W.K., Fang, R.X., Chua, N.H., and Tumer, N. 1988. Viral protection in transgenic tobacco plants expressing the cucumber mosaic virus coat protein or its antisense RNA. Bio/Technology 6:549–557.

Dale, A., McNicol, R.J., Moore, P.P. and Sjulin, T.S. 1989. Pedigree analysis of red raspberry. Acta Hortic. 262:35–39.

Daubeny, H.A. 1996. Brambles. In: Fruit Breeding. Volume II. Vine and Small Fruits (Janick, J. and Moore, J.N., Eds.). John Wiley & Sons, New York, pp 109–190.

De Assis, M. and Hildebrandt, A.G. 1977. In-vitro growth of apical meristems of strawberry. In Vitro 13:145–149.

de Faria, M.J.S.S., Donnelly, D.J. and Cousineau, J.C. 1997. Adventitious shoot regeneration and *Agrobacterium*-mediated transformation of red raspberry. Arq Biol Technol 40:518–529.

De la Plaza, J.L. and Merodio, C. 1989. Effect of ethylene chemisorption on refrigerated strawberry fruit. Acta Hortic. 265:427–433.

Ellis, M.A., Converse, R.H., Williams, R.N. and Williamson, B. (Eds.) 1991. Compendium of Raspberry and Blackberry Diseases and Insects. APS Press, St. Paul, MN.

El-Mansouri, I., Mercado, J.A., Valpuesta, V., Lopez-Aranda, J.M., Pliego-Alfaro, F. and Quesada, M.A. 1996. Shoot regeneration and *Agrobacterium*-mediated transformation of *Fragaria vesca* L. Plant Cell Rep. 15:642–646.

Estruch, J.J., Carozzi, N.B., Desai, N., Duck, N.B., Warren, G.W. and Koziel, M.G. 1997. Transgenic plants: An emerging approach to pest control. Nature Biotechnol. 15:137–141.

Finstad, K. and Martin, R.R. 1995, Transformation of strawberry for virus resistance. Acta Hortic. 385: 86–90.

Fiola, J.A., Hassan, M.A., Swartz, H.J., Bors, R.H. and McNicol, R. 1990. Effect of thidiazurson, light fluence rates and kanamycin on in vitro shoot organogenesis from excised *Rubus* cotyledons and leaves. Plant Cell Tissue Org. Cult. 20:223–228.

Gamborg, O.L., Miller, R.A. and Ojima, K. 1968. Nutrient requirement of suspension cultures of soybean root cultures. Exp. Cell. Res. 50:151–158.

Goemboski, D.F., Lomonossoff G.P., and Zaitlin, M. 1990. Plants transformed with a tobacco mosaic virus nontructural gene sequence are resistant to the virus. Proc. Natl. Acad. Sci. USA 87:6311–6315.

Good, X., Kellogg, J.A., Wagoner, W., Langhoff, D., Matsumura, W. and Bestwick, R.K. 1994. Reduced ethylene synthesis by transgenic tomatoes expressing *S*-adenosylmethionine hydrolase. Plant Mol. Biol. 26:781–790.

Graham, J., McNicol, R.J. and Kumar, A. 1990. Use of the GUS gene as a selectable marker for *Agrobacterium*-mediated transformation of *Rubus*. Plant Cell Tissue Org. Cult. 20:35–39.

Graham, J., McNicol, R.J. and Greig, K. 1995. Towards gentic based insect resistance in strawberry using the cowpea trypsin inhibitor gene. Ann. Appl. Biol. 127:163–173.

Graham, J., Gordon, S.C. and Williamson, B. 1996. Progress towards the use of transgenic plants as an aid to control soft fruit pests and diseases. Brighton Crop Protection Conference—Pests and Diseases, pp 777–782.

Graham, J., Gordon, S.C. and McNicol, R.J. 1997. The effect of the CpTi gene in strawberry against attack by vine weevil (*Otiorhynchus sulcatus* F. Coleoptera: Curculionidae). Ann. Appl. Biol. 131:133–139.

Hadidi, A., Khetarpal, R.K. and Koganezawa, H. 1998. Plant Virus Disease Control. APS Press, St. Paul, MN.

Hassen, M.A., Swartz, H.J., Inamine, G. Mullineaux, P. 1993. *Agrobacterium tumefaciens*–mediated transformation of several *Rubus* genotypes and recovery of transformed plants. Plant Cell Tissue Org. Cult. 33:9–17.

Haymes, K.M. and Davis, T.M. 1997. *Agrobacterium*-mediated transformation of 'Alpine' *Fragaria vesca*, and transmission of transgenes to R1 progeny. In: Molecular genetic studies in *Fragaria* species: *Agrobacterium*-mediated transformation and fine mapping of the *Phytophthora fragariae* resistance gene *Rpf1* (K.M. Haymes, Ph.D. Thesis, Wageningen Agricultural University), pp 83–96.

Hilder, V.A., Gatehouse, A.M.R., Sheerman, S.E., Barker, R.F. and Boulter, D. 1987. A novel mechanism of insect resistance engineered into tobacco. Nature 12:160–163.

Hilder, V.A., Powell, K.S., Gatehouse, A.M.R. Gatehouse, J.A., Hamilton, W.D.O., Merryweather, A., Newell, C.A., Timans, J.C., Pneumans, W.J., Vandamme, E. and Boulter, D. 1995. Expression of snowdrop lectin in transgenic tobacco plants results in added protection against aphids. Transgenic Res. 4:18–25.

Hokanson, S.C., Grumet, R. and Hancock, J.F. 1997. Effect of border rows and trap/donor ratios on pollen-mediated gene movement. Ecol. Appl. 7:1075–1081.

Iannetta, P., Van den Berg, J., Wheatley, R., McMillan, G., McNicol, R. and Davies, H. 1998. A causal role for ethene in raspberry fruit ripening. VIIth International Rubus-Ribes Symposium, Queen's College, University of Melbourne, Melbourne, Australia, Jan 9–16, 1998 (Abstract).

Jach, G., Gornhardt, B., Mundy, J., Logemann, J. Pinsdorf, E. Leah, R., Schell, J. and Maas, C. 1995. Enhanced quantitative resistance against fungal disease by combinatorial expression of different barley antifungal proteins in transgenic tobacco. Plant J. 8:97–109.

James, D.J., Passey, A.J. and Barbara, D.J. 1990. *Agrobacterium*-mediated transformation of apple and strawberry using disarmed Ti-binary vectors. Acta Hortic. 280:495–502.

James, D.J., Passey, A.J., Easterbrook, M.A., Solomon, M.G. and Barbara, D.J. 1992. Progress in the introduction of transgenes for pest and disease resistance into strawberries. Phytoparasitica 20: 83–87.

James, D.J., Passey, Webster, A.D., Barbara, D.J., A.J., Dandekar, A.M., Uratsu, S.L. and Viss, P. 1993. Transgenic apples and strawberries: Advances in transformation, introduction of genes for insect resistance and field studies of tissue cultured plants. Acta Hortic. 336:179–184.

Jefferson, R.A. 1987. Assaying chimeric genes in plants: The GUS gene fusion system. Plant Mol. Biol. Rep. 8:79–94.

Jones, A.T., Angel-Diaz, J.E., Mayo, M.A., Brennan, R.M., Ziegler, A., McGavin, W.J., deNova, C., Graham, J. and Lemmety. 1998. Recent progress towards control of two important viruses and their variants in small fruit crops in Europe. Acta Hortic. 471:87–92.

Jones, C.S., Davies, H.V., McNicol, R.J. and Taylor, M.A. 1998. Analysis of genes differenially espressed during fruit ripening in raspberry (*Rubus idaeus* cv. Glen Clova). VIIth International Rubus-Ribes Symposium, Queen's College, University of Melbourne, Melbourne, Australia, Jan 9–16, 1998 (Abstract).

Jones, O.P., Waller, B.J. and Beech, M.G. 1988. The production of strawberry plants from callus cultures. Plant Cell Tissue Org. Cult. 12:235–241.

Kader, A.A. 1991. Quality and its maintenance in relation to the postharvest physiology of strawberry. In: The Strawberry into the 21st Century (Dale, A. and Luby, J.J. Eds.). Timber Press, Portland, OR, pp 145–152.

Kawchuk, L.M., Martin, R.R. and McPherson, J. 1991. Sense and antisense RNA-mediated resistance to potato leafroll virus in Russet Burbank potato plants. Molec. Plant Microbe Interact. 4:247–253.

Kokko, H.I. and Karenlampi, S.O. 1998. Transformation of arctic bramble (*Rubus arcticus* L.) by *Agrobacterium tumefaciens*. Plant Cell Rep. 17:822–826.

Kreiah, S., Edwards, M.L., Hawes, W.S., Jones, A.T., Brown, D.J.F., McGavin, W.J. and Cooper, J.I. 1996. Some coat protein constituents from strawberry latent ringspot virus expressed in transgenic tobacco protect plants against systematic invasion following root inoculation by nematode vectors. Eur. J. Plant Pathol. 102:297–303.

Lamprecht, S. and Martin, R.R. 1999. Aphid transmission of a SMYE-potexvirus full-length clone. APS Meeting Montreal Canada, Aug. 8–12 (Abstract).

Lee, E.C.M. and DeFossard, R.A. 1977. Some factors affecting multiple bud formation of strawberry (*Fragaria × ananassa* Duchesne) in vitro. Acta Hortic. 78:187–195.

Lindbo, J.A., Silva-Rosales, L., Proebsting, W.M. and Dougherty, W.G. 1993. Induction of a highly specific antiviral state in transgenic plants: Implications for regulation of gene expression and virus resistance. Plant Cell 5:1749–1759.

Ling, K., Namba, S., Gonsalves, C., Slightom, J.L. and Gonsalves, D. 1991. Protection against detrimental effects of potyvirus infection in transgenic tobacco plants expressing the papaya ringspot virus coat protein gene. Bio/Technology 9:752–758.

Liu, Z.R. and Sanford, J.C. 1988. Plant regeneration by organogenesis from strawberry leaf and runner tissue. HortScience 23:107–159.

Luby, J.J. and McNicol, R.J. 1995. Gene flow from cultivated to wild raspberries in Scotland: Developing a basis for risk assessment for testing and deployment of transgenic cultivars. Theor. Appl. Genet. 90:1133–1137.

Maas, J.L. (Ed.) 1998. Compendium of Strawberry Diseases, 2nd edition. APS Press, St. Paul, MN.

Manning, K. 1997. Ripening enhanced genes of strawberry: Their expression, regulation and function. Acta Hortic. 439:165–168.

Martin, R.R. 1995. Alternatives to the use of coat protein for engineering virus resistance in plants. Acta Hortic. 385:18–28.

Martin, R.R. 1998. Raspberry viruses in Oregon, Washington and British Columbia. Acta Hortic. 471:71–74.

Martinelli, A., Gaiani, A. and Cella, R. 1997. *Agrobacterium*-mediated transformation of strawberry cultivar Marmolada onebar. Acta Hortic. 439:169–174.

Mathews, H., Wagoner, W., Kellog, J. and Bestwick, R. 1995. Genetic transformation of strawberry: Stable integration of a gene to control biosynthesis of ethylene. In Vitro Cell. Dev. Biol. 31:36–43.

Mathews, H., Dewey, V., Wagoner, W. and Bestwick, R.K. 1998. Molecular and cellular evidence of chimaeric tissues in primary transgenics and elimination of chimaerism through improved selection protocols. Transgenic Res. 7:123–129.

McNicol, R.J. and Graham, J. 1989. Genetic manipulation in *Rubus* and Ribes. Acta Hortic. 262:41–46.

McNicol, R.J. and Graham, J. 1990. In vitro regeneration of *Rubus* from leaf and stem segments. Plant Cell Tissue Org. Cult. 21:45–50.

McNicol, R.J., Graham, J. and Kerby, N.W. 1997. Recent advances in strawberry breeding and product development at SCRI. Acta Hortic. 439:129–138.

Medina-Escobar, N., Cardenas, J., Munoz-Blanco, J. and Caballero, J.L. 1998. Cloning and molecular characterization of a strawberry fruit ripening-related cDNA corresponding a mRNA for a low-molecular-weight heat-shock protein. Plant Mol. Biol. 36:33–42.

Mehlenbacher, S.A. 1995. Classical and molecular approaches to breeding fruit and nut crops for disease resistance. HortScience 30:466–477.

Millan-MNehra, N.S., Chibbar, R.N., Kartha, K.K., Datla, R.S.S., Crosby, W.L. Stushnoff, C., Endoza, B., and Graham, J. 1999. Organogenesis and micropropagation in red raspberry using forchlorfenuron (CPPU). J. Hortic. Sci. Biotech. 74:219–223.

Murashige, T. and Skoog, F. 1962. A revised medium for rapid growth and bioassay with tobacco tissue cultures. Physiol. Planta. 15:473–497.

Nehra, N.S., Chibbar, R.N., Kartha, K.K., Datla, R.S.S., Crosby, W.L. and Stushnoff, C. 1990. *Agrobac-*

*terium*-mediated transformation of strawberry calli and recovery of transgenic plants. Plant Cell Rep. 9:10–13.
Nehra, N.S., Chibbar, R.N., Kartha, K.K., Datla, R.S.S., Crosby, W.L. and Stushnoff, C. 1990. Genetic transformation of strawberry by *Agrobacterium tumefaciens* using leaf disk regeneration system. Plant Cell Rep. 9:293–298.
Nickerson, N.L. and Jamieson, A.R. 1995. Canadian races of the red stele root rot fungus, *Phytophthora fragariae* var. *fragariae*. Adv. Strawberry Res. 14:31–35.
Nyman, M. and Wallin, A. 1992. Transient gene expression in strawberry (*Fragaria* × *ananassa* Duch.) protoplasts and the recovery of transgenic plants. Plant Cell Rep. 11:105–108.
Nyman, M. and Wallin, A. 1993. Regeneration of plants from protoplasts of cultivated strawberry (*Fragaria* × *ananassa*) and wild strawberry (*Fragaria vesca*). Biotechnol. Agric. For. 23:32–42.
Oparka, K.J., Roberts, A.G., Prior, D.A.M., Chapman, S., Baulcombe, D. and Santa Cruz, S. 1995. Imaging the green fluorescent protein in plants—viruses carry the torch. Protoplasma 189:133–141.
Orr, R.L., Strickland, J.A. and Walsh, T.A. 1994. Inhibition of *Diabrotica* larval growth by multicystatin from potato tubers. J. Insect Physiol. 40:893–900.
Owen y de Novoa, C. 1992. Empirical evaluations of in vitro media components for cell growth and shoot regeneration from *Rubus* explants. N. Z. Nat. Sci. 19:79–86.
Owens y de Novoa, C., and Conner, A.J. 1992a. Comparison of in vitro shoot regeneration protocols from *Rubus* leaf explants. N. Z. J. Crop Hortic. Sci. 20:471–476.
Owens y de Novoa, C., and Conner, A.J. 1992b. Responses of *Rubus* genotypes to strains of *Agrobacterium*. J. Genet. & Breed. 45:359–368.
Pasquale, R. 1993. Recent trends in strawberry production and research: An overview. Acta Hortic. 348:23–44.
Powell-Abel, P., Nelson, R.S., De, B., Hofman, N., Rogers, S.G., Fraley, R.T. and Beachy, R.N. 1986. Delay of disease development in transgenic plants that express the tobacco mosaic virus coat protein gene. Science 232:738–743.
Punja, Z.K. and Zhang, Y.Y. 1993. Plant chitinases and their roles in resistance to fungal diseases. J. Nematol. 25:526–540.
Ramanthan, V., Simpson, C.G., Johnston, D.J., Iannetta, P.M.M., Thow, G., Graham, J., McNicol, R.J. and Williamson, B. 1995. A strategy for control of raspberry grey mould (*Botrytis cinerea*) involving a poygalacturonase-inhibiting protein. Proc. 10th Biennial Australasian Plant Pathol. Soc. Conf., Lincoln Univ. New Zealand.
Reddy, A.S.N., Jena, P.K., Mukherjee, S.K. and Poovaiah, B.W. 1990. Molecular cloning of cDNAs for auxin-induced mRNAs and developmental expression of the auxin-inducible genes. Plant Mol. Biol. 14:643–653.
Ryan, C.A. 1990. Protease inhibitors in plants: Genes for improving defenses against insects and pathogens. Annu. Rev. Phytopathol. 28:425–449.
Scorza, R., Cordts, J.M., Gray, D.J., Gonsalves, D., Emershad, R.I. and Ramming, D.W. 1996. Producing transgenic 'Thompson Seedless' grape (*Vitis vinifera* L.) plants. J. Am. Soc. Hortic. Sci. 121:616–619.
Scott, D.H. and Lawrence, F.J. 1975. Strawberries. In: Advances in Fruit Breeding (Janick, J., and Moore, J.N., Eds.). Purdue University Press, West Lafayette, IN, pp 71–97.
Stark, D.M. and Beachy, R.N. 1989. Protection against potyvirus infection in transgenic plants: Evidence for broad spectrum resistance. Bio/Technology 7:1257–1262.
Steinkellner, H., Weinhausl, A., Laimer, M., Machado, A. de C. and Hermann, K. 1992. Identification of the coat protein of arabis mosaic nepovirus and its expression in transgenic plants. Acta Hortic. 308:37–41.
Swartz, H.J. and Stover, E.W. 1996. Genetic transformation in raspberries and blackberries (*Rubus* species). Biotechnol. Agric. For. 38:297–307.
Taviadorkl, P., Benvenuto, E., Trinca, S., De Martinis, D., Cattaneo, A., and Galeffi, P. 1993. Transgenic plants expressing a functional single-chain Fv antibody are specifically protected from virus attack. Nature 366:469–472.
Taylor S. and Martin, R.R. 1999. Sequence comparison between common and resistance breaking strains of raspberry bushy dwarf virus. APS Meeting, Montreal, Canada, Aug 8–12 (Abstract).

Thomas, P.E., Kaniewski, W.K. and Lawson, E.C. 1997. Reduced field spread of potato leafroll virus in potatoes transformed with the potato leafroll virus coat protein gene. Plant Dis 81:1447–1453.

Traynor, P.L. and Westwood, J.H. (eds.) 1999. Ecological effects of pest resistance genes in managed ecosystems. Proc. of a Workshop, Jan 31–Feb 3, 1999, Bethesda, MD.

Uratsu, S.L., Ahmadi, H. Bringhurst, R.S. and Dandekar, A.M. 1991. Relative virulence of *Agrobacterium* strains on strawberry. HortScience 26:196–199.

Vrain, T.C., Bhagwat B., Wilson, S., Whalgren, S. and Raworth, D. 1997. Engineering genetic resistance to root weevils, two-spotted spider mites, and root-lesion nematodes in red raspberry. Annual Report to Washington Red Raspberry Commission, pp 52–56.

Wallin, A., Skjoldebrand, H. and Nyman, M. 1993. Protoplasts as tools in *Fragaria* breeding. Acta Hortic 348:414–421.

Watt, K., Graham, J., Gordon, S.C., Woodhead, M. and McNicol, R.J. 1999. Current and future transgenic control strategies to vine weevil and other insect resistance in strawberry. J Hortic Sci Biotechnol. 74:409–421.

Wettstein, D. von. 1989. Perspectives for the genetic engineering of plants for agriculture, horticulture and industry. Plant Mol. Biol 13:313–317.

Wilson, T.M.A. 1993. Strategies to protect crop plants against viruses: Pathogen-derived resistance blossoms. Proc Natl Acad Sci USA 90:3134–3141.

Yepes, L.M., Fuchs, M., Slightom, J.L. and Gonsalves, D. 1996. Sense and antisense coat protein gene constructs confer high levels of resistance to tomato ringspot nepovirus in transgenic *Nicotiana* species. Phytopathology 86:417–424.

# 32
# Progress in Asparagus Biotechnology

**Amnon Levi**
*U.S. Department of Agriculture, Charleston, South Carolina*

**Kenneth C. Sink**
*Michigan State University, East Lansing, Michigan*

|  |  |  |
|---|---|---|
| I. | INTRODUCTION | 465 |
| II. | SOMATIC EMBRYOGENESIS IN ASPARAGUS | 466 |
| III. | HAPLOID EMBRYOS THROUGH ANTHER CULTURE | 469 |
| IV. | GENETIC TRANSFORMATION OF ASPARAGUS | 470 |
| V. | ASPARAGUS PROTOPLASTS | 470 |
|  | REFERENCES | 471 |

## I. INTRODUCTION

*Asparagus officinalis* L. is considered one of the oldest garden vegetable plants and is a crop with high value now grown worldwide. In 1990, there were over 140,000 ha of asparagus worldwide (49). In recent years there has been an increased demand for green fresh asparagus year round, particularly in affluent countries in the Northern Hemisphere (49). *Asparagus officinalis* L. is a native European-Siberian continental plant. The northeast Mediterranean is its center of origin, where it is well adapted to sandy soils (43). The genus *Asparagus* belongs to the Liliaceae family. It comprises over 150 herbaceous and woody perennial species that thrive in the temperate and tropical regions of the world. Among these species *Asparagus officinalis* L. is the only known edible one (35). *Asparagus officinalis* L. is a herbaceous perennial monocot, and a cool season plant for which 24°–29°C day and 13°–19°C night temperatures are optimal for productivity and longevity (51). Young asparagus crowns establish a rhizome structure with a large root system and stems. Each rhizome has a few lateral buds that develop into succulent fleshy shoots (spears). The spears are triangular and comprise short internodes and lateral buds. Expansion of individual spears occurs foremost at the base internode and then in successive internodes, resulting in dynamic growth of the spear into a 4- to 6-feet tall fern (51). *Asparagus officinalis* L. is a dioecious species with male and female plants (sex ratio 1:1). The male flowers have perfect anthers and an aborted ovary. The female flowers are smaller than those of male plants and have nonfunctional anthers, a perfect ovary and style, and three stigmas. The asparagus fruit is a red berry at maturity and contains up to six small globular seeds. The seed is mostly endosperm tissue surrounding a small embryo (55).

Well-maintained asparagus fields can be productive for 20–25 years, producing annual yields of 3000 pounds per acre. In recent years, asparagus yields have been declining significantly, and asparagus fields are being removed after 8–12 years. In addition to the decreased productivity and longevity of established plantings, reestablishment of asparagus in fields where it was previously grown is difficult. This difficulty is described as the "asparagus decline and replant syndrome" (25,34). The common soil-borne fungi *Fusarium oxysporum* and *Fusarium moniliforme* are considered the principal causes for the root and crown rot diseases that result in the decline of asparagus fields (7,25). Therefore, the development of varieties with resistance to *Fusarium* spp. is a primary objective in asparagus breeding. Grogan and Kimble (23) found one asparagus line with limited tolerance to fusarium. Takatori and Souther (59) in California and Ellison (16) in New Jersey also screened asparagus accessions but could not find any accession with fusarium resistance. Stephens and coworkers (57) screened 90 cultivars and breeding lines for resistance to fusarium and found one cultivar, 'Lucullus 234,' that has resistance to *F. oxysporum* and *F. moniliforme*. At present only two ornamental species, *Asparagus densiflorus* (Kanto) jessop (cvs.) 'Sprengerii,' and *A. plumosus,* have been found to be highly resistant to *F. oxysporum* and *F. moniliforme* (40,57). However, both species are sexually incompatible with *A. officinalis* (16).

Asparagus has breeding impediments caused by its dioecious reproductive mode and its nature as a perennial grass with a long life cycle. Asparagus plants begin to yield after 4 to 5 years in the field, whereas the average harvesting period is about 10–15 years. Therefore, the development of new cultivars by traditional breeding procedures requires many years (16).

A primary breeding goal is to increase genetic uniformity of asparagus crowns combined with high productivity and longevity. The production of double haploid plants through polyembryonic seeds or anther culture is currently the only avenue to obtain pure line plants that can be used for further breeding and production of uniform hybrid lines (14). Thus, in recent years, many attempts have been made to establish an anther culture system for asparagus.

Further confounding breeding efforts is the situation that there are major differences in the growth pattern of asparagus female versus male plants. Male plants produce a larger number of spears and have greater longevity than females. Furthermore, male plants do not manifest seedlings that may harbor diseases and compete with the crowns that are already established in the field. Thus, there is a great interest in the development of all-male varieties. The sexual mode in asparagus is controlled by a single gene. The female plants are homozygous *mm*, whereas the male plants are *Mm* (42). Therefore, double haploid plants originated through polyembryonic seeds or anther culture should be homozygous female (mm) or supermale (MM) plants. Such plants can be used for further breeding and production of genetically uniform male hybrid (Mm) plants (14).

## II. SOMATIC EMBRYOGENESIS IN ASPARAGUS

Asparagus has a low propagation rate using traditional methods. Experiments in dividing mature crowns and in establishing rooted cuttings have had only limited success (1). Yang and Clore (64–66) developed a procedure for propagating aerial crowns of asparagus stems in potted plants kept under high moisture. After enlargement, shoots and roots are formed and are separated manually to form new plants. However, this technique is slow and not practical for mass propagation. Conversely, tissue culture techniques developed for the multiplication and rooting of asparagus plants in vitro resulted in greater efficiencies (6,26,54,62,64–66). Furthermore, the production of clonal plants through somatic embryogenesis is estimated to be at least 5- to 10-fold higher in comparison to in vitro propagation of adventitious shoots.

Somatic embryogenesis in asparagus can be used as a tool for rapid production of elite asparagus crowns, selected hybrid varieties, and buildup of the parents of such hybrids. It can also be useful for clonal propagation of double haploid plants that may be used as parents of $F_1$ hybrids (14). As in other plant species, and in monocots in particular, the development of a reproducible somatic embryogenesis system is an essential step for the genetic transformation of asparagus with valuable genes.

Asparagus was the first monocotyledonous species to be reported for successful formation of somatic embryos in vitro (63). In this primary study, hypocotyl-derived callus formed a large number of globular embryos when placed on Linsmaier and Skoog (41) (LS) basal medium + 2,4-dichlorophenoxyacetic acid (2,4-D) (1 mg/l) and kinetin (0.3 mg/l). The embryos differentiated in liquid suspension of the same composition, and after the removal of the growth regulators the embryos germinated shoots and roots. Additional studies were devoted to the development and efficiency of somatic embryogenesis systems in asparagus. Steward and Mapes (58) reported the formation of somatic embryos from asparagus stem segments. The segments were induced to form callus in White's liquid medium supplemented with coconut milk (10%) and naphthalene acetic acid (NAA) followed by replacement of NAA with 2,4-D. The callus was induced to form embryos by being placed in high-salt medium supplemented with NAA. The embryos were induced to form shoot and root primordia upon their transfer to medium supplemented with coconut milk and indole-3-acetic acid (IAA). Reuther (53) reported that callus derived from shoot tips and shoot segment–derived callus, formed on LS + NAA (1 mg/l) and kinetin (1 mg/l), could give rise to somatic embryos when transferred to LS + IAA (1 mg/l) + benzyladenine (BA) (0.1 mg/l). Subsequently, the embryos developed into plants when transferred to LS, or to LS + IAA (0.5 mg/l).

In our study (37) we observed that somatic embryogenesis in asparagus can take place when using various explants and media. Spear cross sections, in vitro crowns, and lateral buds were all suitable explants for the initiation of embryogenic callus. When cultured on MS medium + NAA (0.1 mg/l) + kinetin (0.01 mg/l), the crowns established in vitro and the lateral buds occasionally formed new shoots and roots. These were accompanied with the formation of a yellowish, friable embryogenic callus. After two subcultures, at 4-week intervals, the calli were transferred to embryo induction medium of MS + 2,4-D or NAA + kinetin (0–10 mg/l). After 4 weeks calli were transferred to embryo development medium MS + NAA (0.1 mg/l) and kinetin (0.01 mg/l). The results in this study implied that NAA (0.3–10 mg/l) was sufficient for the induction of asparagus somatic embryos in callus derived from lateral buds or from crowns established in vitro. The use of 2,4-D resulted in a vigorous induction of embryonic callus as compared with that produced with NAA. On the other hand, induction with NAA resulted in a higher number of normal embryos that could germinate into plants. Embryos developed and converted to plants within 4 weeks on MS + NAA (0.01 mg/l) and kinetin (0.1 mg/l) (37,38). Saito and associates (56) also examined the use of NAA versus 2,4-D for the induction of embryogenic callus from asparagus epicotyls. In their study, 2,4-D was preferred since it induced callus with a higher number of embryos when compared with NAA. Saito and coworkers (56) also reported that the use of medium containing Gelrite and flasks with ventilative filters (as compared with aluminum foil) was most favorable with respect to embryo desiccation and conversion to plants. In our study (36) we found that carbohydrate source and concentration had a marked effect on the formation and development of asparagus somatic embryos. Sucrose, fructose, or glucose levels of 5–6% were optimal for the development of asparagus somatic embryos; transfer of embryos to medium with a lower sugar level of 2% promoted embryo conversion to plants.

In another study (39), we examined the production of asparagus somatic embryos through suspension cultures (Fig. 1). Also, we examined the effect of carbohydrate source and concentra-

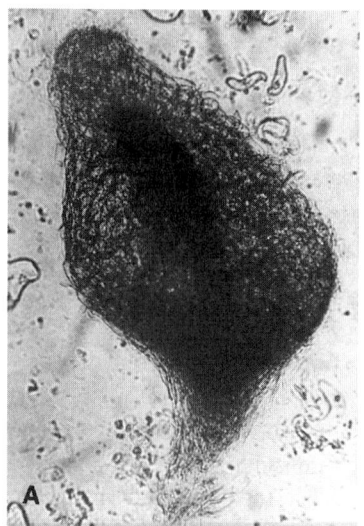
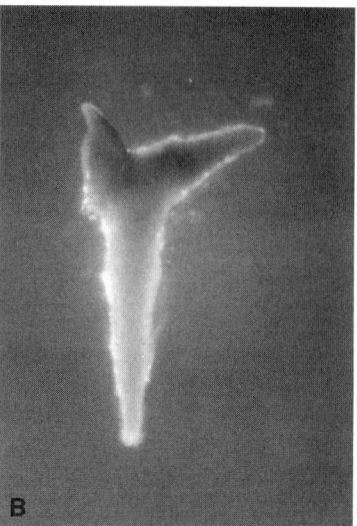

**Figure 1** (A) Bipolar asparagus somatic embryo in suspension culture (MS + 10 mg/l NAA). (B) Asparagus somatic embryo germinating on MS medium containing 2% sucrose, following transfer from MS medium that contained 8% sucrose. (C) Somatic embryo–derived plants in a field trial.

tion on conversion of the embryos to plants. Callus derived from stems of an elite asparagus crown was used for the establishment of embryo suspension cultures in liquid MS + NAA (10–20 mg/l). Subsequently, the development and conversion of embryos to plants were enhanced two- to four fold, after transfer of embryos from suspension to solid MS with high carbohydrate levels (4–10%), for 2 weeks, followed by transfer of the embryos to solidified MS with a low carbohydrate level (2%). These results implied that asparagus somatic embryos require high carbohydrate levels during their formation and maturation and a lower carbohydrate level (2%) for rapid development of the embryo to plant. Using this protocol a few hundred plants of asparagus male line (M.S.U. 88-10) were produced through embryo suspension culture and were successfully transplanted to the field.

Odake and coworkers (50) also examined the production of asparagus embryos through suspension culture. In their study the combination of NAA (3 mg/l) + kinetin (1 mg/l) was solely used for the induction and maintenance of embryonic cultures. Their study implied that full-strength MS medium was most effective for embryo induction, whereas half-strength MS was more suitable for embryo conversion and development to plants. Cytological examination of 33 embryo-derived plants revealed that most of them (over 90%) were tetraploids. In this case the donor plant was a diploid. The authors considered the possibility that the continuous use of growth regulators in the medium caused chromosome doubling.

Kohmura and associates (31) induced the formation of embryogenic callus from bud clusters. This callus was derived through sequential subcultures on MS medium + 10 mg/l ancymidol, followed by placement on LS + 0.2 mg/l 2,4-D. Upon transfer to liquid medium free of growth regulators the embryos developed an elongated shape. After transfer to solid medium the mature embryos developed into plants. The authors successfully transplanted to the field over 2000 plants derived from such somatic embryos. Cytological examination of these plants indicated a normal chromosome number ($2n = 2x = 20$).

Somatic embryogenic systems in asparagus provide a useful tool for rapid propagation of asparagus elite crowns and for the tissue culture component for genetic transformation with genes of valuable agronomic traits. The literature indicates somatic embryogenic systems in asparagus can be established with various explant types and medium conditions. However, lateral buds and in vitro established crowns were most suitable explants for initial embryogenic callus.

Conversion rate of embryos to plants is a limiting step. The use of 1) MS (48) or LS medium (41) containing Gelrite and 5–6% sucrose in flasks with aeration filters (as compared with aluminum foil) should be most favorable with respect to embryo desiccation. Half-strength MS or LS medium with a low carbohydrate level (2%) should be favorable for conversion of embryos to plants (31).

## III. HAPLOID EMBRYOS THROUGH ANTHER CULTURE

The development of haploid asparagus plants through anther culture can significantly enhance the development of all-male hybrid varieties and supermales (*MM*) that are highly valuable in asparagus breeding programs, as well as the development of double haploids that can be used for the production of genetically and phenotypically uniform hybrid lines. Hsu and colleagues (29,30) reported the development of haploid plants from anther culture. In their study less than 2% of anthers produced embryogenic calli. Feng and Wolyn (19) developed an improved method for obtaining a high frequency of haploid asparagus embryos through anther culture. Anthers were placed on MS + casein hydrolysate (500 mg/l), glutamine (800 mg/l), NAA (2 mg/l), BA (1 mg/l), and sucrose (5%) at 32°C in the dark for 3–4 weeks. Calli and embryos were transferred to MS + sucrose (6%), NAA (0.1 mg/l) and kinetin (0.1 mg/l), and ancymidol (0.65 mg/l) for maturation at 25°C. More than 50% of mature embryos germinated on MS + $GA_3$. Their procedure yielded about 50 haploid plants per 100 anthers. They observed that high temperature was critical for the induction of embryogenic callus and that anthers with microspores at the late-uninucleate stage produced the highest frequency of embryogenic calli. In a related study, Feng and Wolyn (20) developed a procedure for production of haploid asparagus embryos from liquid cultures of anther-derived calli by using the growth retardant ancymidol. The same authors also reported the recovery of haploid plants from asparagus microspore culture, which increases the efficiency of the haploid production procedure and decreases chimeric plants that may derive from anther culture (20). The production and use of all-male hybrids derived from anther culture are becoming essential and integral parts of asparagus breeding programs. Such hybrids are highly productive and more uniform than those derived from heterozygous clones (18).

## IV. GENETIC TRANSFORMATION OF ASPARAGUS

*Agrobacterium tumefaciens* has the ability to attach to asparagus cells (15,60), and asparagus exhibits a wound response after inoculation with *Agrobacterium* sp. (24). *Asparagus officinalis* was the first monocotyledonous plant to be reported for the successful formation of hormone-independent and opine-producing crown gall tissue (27). In their experiment the authors inoculated sections of young asparagus spears with the *A. tumefaciens* oncogenic strain C58. The infected explants were cultured on MS + NAA and BA (1 mg/l) and cefotaxime (500 mg/l). After 4 weeks, calluses that proliferated from inoculation sites were placed on basal medium. About 10% of the callus colonies (mainly those that originated from the upper portion of a young spear) continued to proliferate on the basal medium, giving rise to a hormone-free callus. A few of these hormone-free calli were tested and found to contain opines, in contrast with the control colonies. This experiment confirmed the first successful genetic transformation of a monocotyledonous plant using *A. tumefaciens*. Using Southern analysis Bytebier and coworkers (4) showed complete integration of transfer deoxyribonucleic acid (T-DNA) in the *Asparagus officinalis* genome. This provided the first molecular evidence for the integration of T-DNA in a monocotyledonous plant. Conner and coworkers (8) tested 4 *A. tumefaciens* oncogenic strains (A4T, A281, A722, and C58) against 17 genotypes of *Asparagus officinalis*. Only strain C58 induced tumor formation, and only 3 of the 17 asparagus genotypes ('Violet of Albenga,' 'Limbras 18,' 'Roem Van Brunswijk') had tumors containing nopaline. This result occurred exclusively in response to strain C58. After 7 months on selection medium containing kanamycin, calli were formed on 3 of 125 explants. These calli retained kanamycin resistance and produced nopline after subculture on kanamycin selection medium. One callus retained kanamycin resistance and nopline synthesis for 18 months. Shoots that were regenerated from this callus also produced nopline and retained kanamycin resistance. Southern analysis confirmed that the kanamycin resistance gene had been integrated into the asparagus genome. In further studies, using the *A. tumefaciens* strain C58, Prinsen and associates (52) investigated the levels of endogenous auxin (IAA) and cytokinins in asparagus crown gall cells grown on medium free of growth regulators. They detected no increase in IAA but a significant increase in cytokinins in tumor cell lines as compared with untransformed callus.

Delbreil and colleagues (10) examined long-term embryogenic callus as a target for agrobacterium-mediated genetic transformation. The results showed a relatively low transformation frequency. However, use of a large number of embryos can compensate for the low transformation efficiency. The results indicated that elongated embryos were more receptive to transformation than globular embryos.

At present there has been only one report of direct gene-mediated transformation in asparagus. This was through electroporation of plasmid into callus-derived protoplasts (45–47). Heat shock treatment and the presence of polyethylene glycol in the electroporation solution enhanced transient GUS expression in the protoplasts. The electroporated protoplasts formed kanamycin-resistant calluses with positive GUS expression.

## V. ASPARAGUS PROTOPLASTS

Protoplasts of asparagus were among the earliest of plant species to be studied. Bui-Dang-Ha and Mackenzie (2) in the first report on the genus isolated protoplasts from cladodes of cv. 'Marche de Malines' and cultured them in MS + NAA and zeatin. Only about 4% of the alive protoplasts at day 7 divided once; sometimes division was not noted until day 17. Glutamine at 1000 mg/l was found critical to sustained division. Small colonies formed that were transferred to solid MS and regenerated roots on NAA + zeatin, or shoots when zeatin was replaced with BA together with 40 mg/l adenine. However, when Bui-Dang-Ha and coworkers (3) used the undifferentiated protocalli in further experiments, whole plants were obtained via a somatic embryogenic route. Such calli cultured on media with a high cytokinin to auxin ratio (zeatin : NAA) or on one of equal molarity (BA : NAA) formed new friable tissues that contained embryos at various developmental stages. Only a few plants were obtained after transfer of the embryoids to solid or semiliquid media containing IAA and zeatin.

Subsequent studies concentrated on increasing the plating efficiency (PE), media regime, and genotypes to allow asparagus to be an efficient system. Kong and Chin (32) used seedling-derived callus as a source of protoplasts and by droplet culture in liquid media or agarose droplets on porous polypropylene membranes increased PE to about 10%. Plants were regenerated via organogenesis by a two-step scheme using both a cytokinin and an auxin at each step with ancymidol in the last step to promote rooting. In an attempt to use mature asparagus plants in a protoplast scheme, Elmer and colleagues (17) initiated callus of four genotypes on MS containing auxins and a cytokinin, subcultured the calli, and cultured the isolated protoplasts in a modified liquid Kao and Michayluk (KM) medium. PE was 6–7%, and only one genotype sustained division past 8 weeks to form microcalli and eventually regenerate via organogenesis. Kunitake and Mii (33) used embryogenic calli as the donor cells for isolating protoplasts and after liquid culture for 30 days; after growth on gellan gum solidified MS + 2,4-D callus was produced. The callus was subsequently converted to somatic embryos that germinated into plantlets at a 30–40% rate. Dan and Stephens (9) used cv.'Lucullus 234' and cultured callus-derived protoplasts in KM medium solidified with agarose; a PE of 19% was obtained and plants were regenerated via organogensis. Chen and associate (5) combined isolated protoplasts from in vitro grown etiolated shoots, agarose culture with an asparagus cell feeder layer, and regeneration by embryogenesis. A PE of 20% was obtained and about 25% of the colonies produced somatic embryos, of which 20% yielded plants, which were found to be diploids.

In summary, the key biotechniques are established for asparagus, but modification of protocols for all of them still seems necessary for success with particular genotypes (44). Breeding of asparagus may be reaching a plateau in terms of increasing yield, and incorporating single-gene-based traits is difficult if not impossible through traditional breeding schemes. Thus, introduction of transgenes will play a key role in the future development of new cultivars. Indeed, genes having potential for beetle, virus, and herbicide resistances are on hand as well as some that would increase product quality. The remaining question is whether these new traits will be introduced more effectively to growers through cloned or seed-produced hybrids.

## REFERENCES

1. DC Anderson, JH Ellison. Root initiation of stem tip cuttings from mature asparagus plants. Proc Am Soc Hort-ic Sci 90:158–162, 1967.
2. D Bui-Dang-Ha, IA Mackenzie. The division of protoplasts from *Asparagus officinalis* L. and their growth and differentiation. Protoplasma 78:215–221, 1973.
3. D Bui-Dang-Ha, B Norreel, A Masset. Regeneration of *Asparagus officinalis* L. through callus cultures derived from protoplasts. J Exp Bot 26:263–270, 1975.
4. B Bytebier, F Dedoeck, H De Greve, M Van Montagu, J Hernalsteens. T-DNA organization in tumor cultures and transgenic plants of the monocotyledon *Aspragus officinalis*. Proc Nat-l Acad Sci USA 84:5345–5349, 1987.
5. GY Chen, AJ Conner, MC Christy, AG Fautrier, RJ Field. Culture and regeneration of protoplasts from shoots of asparagus cultures. Int J Plant Sci 158:543–551, 1997.
6. CK Chin. Promotion of shoot and root formation in asparagus in vitro by ancymidol. Hortscience 17:590–591, 1982.

7. SI Cohen, FD Heald. A wilt and root rot of asparagus caused by *Fusarium oxysporum* (Schlecht). Plant Dis Rep 25:503–509, 1941.
8. AJ Conner, MK Williams, SC Deroles, RC Gardner. *Agrobacterium*-mediated transformation of asparagus. In: KS McWhirter, RW Downes, BJ Read, eds. Proceeding of the 9th Australian Plant Breeding Conference, Wagga Wagga, Australia, pp 1988, 131–132.
9. Y Dan, CT Stephens. Studies of protoplast culture types and plant regeneration from callus-derived protoplasts of *Asparagus officinalis* L. cv. Lucullus 234. Plant Cell Tissue Org Cult 27:321–331, 1991.
10. B Delbreil, P Guerche, M Jullian. *Agrobacterium*-mediated transformation of *Asparagus officinalis* L. long-term embryogenic callus and regeneration of transgenic plants. Plant Cell Rep 12:129–132, 1993.
11. B Delbreil, I Goebel-Tourand, C Lefrancois, M Jullien. Isolation and characterization of long-term embryonic lines in *Asparagus officinalis* L. J Plant Physiol 144:194–200, 1994.
12. B Delbreil, M Jullien. Evidence for in vitro induced mutation which improves somatic embryogenesis in *Asparagus officinalis* L. Plant Cell Rep 13:372–376, 1994.
13. Y Desjardins. Micropropagation of asparagus (*Asparagus officinalis* L.). In: YPS Bajaj ed. Biotechnology in Agriculture and Forestry. Vol. 19. High-Tech and Micropropagation III. Springer-Verlag Berlin, 1992, pp 26–41.
14. C Dore. Asparagus anther culture and field trials of dihaploids and F1 Hybrids. In: YPS Bajaj ed. Biotechnology in Agriculture and Forestry Vol. 12. Heidelberg: Springer-Verlag Berlin, 1990, 322–345.
15. J Draper, IA Mackenzie, MR Davey, JP Freeman. Attachment of *Agrobacterium tumefaciens* to mechanically isolated asparagus cells. Plant Sci Lett 29:227–236, 1983.
16. JH Ellison. Asparagus Breeding. In: MJ Bassett, ed. Breeding Vegetable Crops. Westport, CT: Avi Publishing, 1986, pp 523–569.
17. WH Elmer, T Ball, M Volokita, CT Stephens, KC Sink. Plant regeneration from callus-derived protoplasts of asparagus. J Am Soc Hortic Sci 114:1019–1024, 1989.
18. A Falavigna, PE Casali, MG Tacconi. Advances in asparagus breeding following in vitro anther culture. Acta Hortic 415:137–142, 1996.
19. XR Feng, DJ Wolyn. High frequency production of haploid embryos in asparagus anther culture. Plant Cell Rep 10:574–578, 1991.
20. XR Feng, DJ Wolyn. Development of haploid asparagus embryos from liquid cultures of anther-derived calli is enhanced by ancymidol. Plant Cell Rep 12:281–285, 1993.
21. XR Feng, DJ Wolyn. Recovery of haploid plants from *Asparagus* microspore culture. Can J Bot 72:296–300, 1994.
22. AA Franken. Sex characteristics and inheritance of sex in asparagus. Euphytica 19:277–287, 1970.
23. RG Grogan, KA Kimble. The association of *Fusarium* wilt with the asparagus decline and replant problem in California. Phytopathology 49:122–125, 1959.
24. K Harikrishna, E Paul, R Darby, J Draper. Wound response in mechanically isolated asparagus mesophyll cells: A model monocotyledon system. J Exp Bot 42:791–799, 1991.
25. AC Hartung. Allelopatic potential of asparagus (*Asparagus officinalis* L.). PhD thesis, Michigan State University, East Lansing, 1987.
26. PM Hasegawa, T Murashige, FH Takatori. Propagation of asparagus through shoot apex culture. II. Light and temperature requirements, transplantability of plants, and cyto-histological characteristics. J Am Soc Hortic Sci 8:143–148, 1973.
27. JP Hernalsteens, L Thia-Toong, J Schell, M Van Montagu. An *Agrobacterium*-transformed cell culture from the monocot *Asparagus officinalis*. EMBO J 3:3039–3041, 1984.
28. RC Herner, G Vest. Asparagus Workshop Proceedings, Department of Horticulture, Michigan State University, E. Lansing, MI, 1974, p 79.
29. JY Hsu, CC Yeh, HS Tsay. Proc 4th symposium on Asparagus Research in Taiwan. 1988, pp 153–163.
30. JY Hsu, CC Yeh, TP Yang, WC Lin, HS Tsay. Initiation of cell suspension cultures and plant regeneration from protoplasts of asparagus. Acta Hortic 271:135–143, 1990.
31. H Kohmura, S Chokyu, T Harada. An effective micropropagation system using embryogenic calli induced from bud clusters in *Asparagus officinalis* L. J Jpn Soc Hortic Sci 63:51–59, 1994.
32. Y Kong CK Chin. Culture of asparagus protoplasts on porous polypropylene membrane. Plant Cell Rep 7:67–69, 1988.

33. H Kunitake, M Mii. Somatic embryogenesis and plant regeneration from protoplasts of asparagus (*Asparagus officinalis* L.) Plant Cell Rep 8:706–710, 1990.
34. RJ Lake, PG Falloon, DWM Cook. Replant problem and chemical components of asparagus roots. NZ J Crop Hortic Sci 21:53–58, 1993.
35. GHM Lawrence. Taxonomy of Vascular Plants. New York: Macmillan 1982, pp 823.
36. A Levi, KC Sink. Differential effects of sucrose, glucose and fructose during somatic embryogenesis in asparagus. J Plant Physiol 137:184–189, 1990.
37. A Levi, KC Sink. Somatic embryogenesis in asparagus: The role of explants and growth regulators. 10:71–75, 1991.
38. A Levi, KC Sink. Histology and morphology of asparagus somatic embryos. Hortscience 10:1322–1324, 1991.
39. A Levi, KC Sink. Asparagus somatic embryos: Production in suspension culture and conversion to plants on solidified medium as influenced by carbohydrate regime. Plant Cell Tissue Org Cult 31:115–122, 1992.
40. GD Lewis, PB Shoemaker. Resistance of asparagus species to *Fusarium oxysporum* f. *asparagi*. Plant Dis Rep 46:364–365, 1964.
41. E Linsmair, F Skoog. Organic growth factor requirements of tobacco tissue cultures. Physiol Planta 18:100–127, 1965.
42. H Loptien. Identification of the sex chromosome pair in asparagus (*Asparagus officinalis* L.) Z Pflanzenzuchtg 82:162–173, 1979.
43. J Luzny. The history of asparagus as a vegetable, the tradition of its growing in Czechoslovakia (CSSR) and prospect of its further propagation and breeding. In: G Reuther, ed. Proceedings of the 5th International Asparagus Symposium. Eucarpia Section Vegetables. Geisenheim Forschungsanstalt, Germany, 1979.
44. RA May, KC Sink. Genotype and auxin influence direct somatic embryogenesis from protoplasts derived from embryogenic cell suspensions of Asparagus officinalis L. Plant Sci 108:71–84, 1995.
45. S Mukhopadhyay, Y Desjardins. A comparative study on mode of culture and plant regeneration from protoplast-derived somatic embryos of two genotypes of *Asparagus officinalis* L. Plant Sci 100:97–104, 1994.
46. S Mukhopadhyay, Y Desjardins. Plant regeneration from protoplast-derived somatic embryos of *Asparagus officinalis* L. J Plant Physiol 144:94–99, 1994.
47. S Mukhopadhyay, Y Desjardins. Direct gene transfer to protoplasts of two genotypes of *Asparagus officinalis* L. by electroporation. Plant Cell Rep 13:421–424, 1994.
48. T Murashige, F Skoog. A revised medium for rapid growth and bioassays with tobacco tissue cultures. Physiol Planta 15:473–497, 1962.
49. MA Nichols. Asparagus: The world scene. Acta Hortic 271:25–31, 1990.
50. Y Odake, A Udagawa, H Saga, M Mii. Somatic embryogenesis of tetraploid plants from internodal segments of a diploid cultivar of *Asparagus officinalis* L. grown in liquid culture. Plant Sci 94:173–177, 1993.
51. LC Peirce. Vegetables, Characteristics Production and marketing. New York: John Wiley & Sons, 1987.
52. E Prinsen, B Bytebier, JP Hernalsteens, J De Greef, H Van Onckelen. Functional expression of *Agrobacterium tumefaciens* T-DNA *onc*-genes in asparagus crown gall tissues. Plant Cell Physiol 31:69–75, 1990.
53. G Reuther. Adventitious organ formation and somatic embryogenesis in callus of asparagus and iris and its possible application. Acta Hortic 78:217–224, 1977.
54. G Reuther. Asparagus. In: RS Sharp, DA Evans, PV Ammirato, Y Yamada, eds. Handbook of Plant Cell Culture. Vol. 2. New York: Macmillan, 1984, pp 211–242.
55. SM Riviere. Les activites meristemitiques durant lontogenese d'une plantule de monocotyledon a germination hypogee. 1. *Asparagus officinalis* L. (liliacees). C R Acad Sci Paris D 277:293–296, 1973.
56. T Saito, S Nishizawa, S Nishimura. Improved culture conditions for somatic embryogenesis from *Asparagus officinalis* L. using an aseptic ventilative filter. Plant Cell Rep 10:230–234, 1991.
57. CT Stephens, RM De Vries, KC Sink. Evaluation of *Asparagus* species for resistance to *Fusarium oxysporum* f. sp. *asparagi* and *F. moniliforme*. Hortscience 24:365–368, 1989.

58. FC Steward, MO Mapes. Morphogenesis and plant propagation in aseptic cultures of asparagus. Bot Gaz 132:70–79, 1971.
59. F Takatori, FD Souther. Asparagus Workshop Proceedings. Department of Plant Science, University of California, Riverside, 1978, p 100.
60. N Terouchi, S Hasegawa, H Matsushima, Y Kaneko, K Syono. Observation by SEM of the attachment of *Agrobacterium tumefaciens* to the surface of vinca, asparagus and rice cells. Bot Mag Tokyo 103: 11–23, 1990.
61. JMM Vanbakel, JJA Kerstens. Foot rot in asparagus caused by *Fusarium oxysporium* f. sp. *asparagi*. Neth J Plant Pathol 76:320–325, 1970.
62. M Volokita, A Levi, KC Sink, A revised protocol for micropropagation of asparagus. Asparagus Res Newslett 4:8–17, 1987.
63. C Wilmar, M Hellendoorn. Growth and morphogenesis of asparagus cells cultured in vitro. Nature 217:369–370, 1968.
64. HJ Yang. Tissue culture technique developed for asparagus propagation. Hortscience 12:140–141, 1977.
65. HJ Yang, WJ Clore. Aerial crowns in *Asparagus officinalis* L. Hortscience 8:33, 1973.
66. HJ Yang, WJ Clore. In vitro reproductiveness of asparagus stem segments with branch-shoots at the node. Hortscience 10:411–412, 1975.

# 33
# Generation of Transgenic Bean (*Phaseolus* species) Plants for Improvement of Nutritional Quality

**Jae-Whune Kim**
*Microplants Co., Ltd., Chollabukdo, Korea*

**Chee H. Harn**
*Biotechnology Center, Nong Woo Bio Co., Ltd., Kyonggi-do, Korea*

| | | |
|---|---|---|
| I. | INTRODUCTION | 475 |
| II. | TISSUE CULTURE OF BEANS | 476 |
| III. | METHODS FOR TRANSFORMATION OF BEANS | 477 |
| | A. *Agrobacterium* Species–Mediated Transformation | 477 |
| | B. Direct Gene Transfer | 478 |
| | C. Particle Bombardment | 478 |
| IV. | IMPROVEMENT OF THE NUTRITIONAL QUALITY OF BEAN SEEDS BY GENE TRANSFER | 479 |
| V. | CONCLUSIONS | 482 |
| | REFERENCES | 482 |

## I. INTRODUCTION

Legume seeds contain a large amount of storage protein and are one of the essential food sources for humans and animals. Soybean provides a key nutritional food supply for Asian countries and peas are an important food for African and Latin American countries (1). Like soybeans and peas, beans such as French, common, garden, kidney, and snap are also important food sources on a worldwide basis and their genetic stock, favored for dry seeds, constitutes about 30% of the total world legume production.

The major cultivated representative of the bean is *Phaseolus vulgaris* L. Over a period of at least 7000 to 8000 years beans have evolved from wild growing vines distributed in the highlands of Middle America and the Andes into a major leguminous food crop, grown worldwide in a broad range of environments and cropping systems (2,3). Beans are rich in proteins, complex carbohydrates, and dietary fibers, especially water-soluble fibers. They are low in fat and sodium and contain no cholesterol. Recently, the bean plant has become a target plant for genetic engineering through possible modification of the seed composition. Both qualitative and quantitative improvement of the seed components is possible by engineering biochemical pathways of protein, fat, and starch biosynthesis and/or by increasing resistance to insects, various pathogens, and broad-spectrum herbicides.

Tissue culture systems for bean plants have been extensively studied. Bean plants have been recovered from tissue cultures of both apical and axillary meristems (4,5). However, success in inducing organogenesis and somatic embryogenesis has been limited. Therefore, regeneration after gene transfer into beans has not been feasible. We review regeneration techniques and transformation protocols for beans and also discuss current strategies to improve the nutritional quality of seeds.

## II. TISSUE CULTURE OF BEANS

It is important to have a proper regeneration system by which transformed cells containing one or more foreign genes grow into explants using proper media. Several groups have been working on bean regeneration by using different tissues and methods. Martins and Sondahl (6) used the shoot apex with three to four leaf primordia of 1- to 2-mm length from 7-day-old seedlings of 33 bean cultivars to establish primary cultures on B5 medium containing 0.05 µM benzyladenine (BA). These cultivars formed multiple buds during the primary culture. Allavena and Rossetti (4) achieved micropropagation of apical and axillary explants of four different genotypes. Genotypic differences were found with respect to multiple shoot formation on Murashige and Skoog (MS) medium supplemented with combinations of BA and naphaleneacetic acid (NAA) after 28 days of culture. McClean and Grafton (5) regenerated shoots from cotyledon and epicotyl nodal tissue of 12- to 14-day-old seedlings on MS medium supplemented with 5 µM BA. Shoots were rooted on B5 medium. Three-day-old seedling explants that contained a cotyledon and a small portion of the embryonic axis split in half were regenerated. A compact, shiny light-green-colored meristematic ring appeared at the base of the axillary bud. New buds were formed from this ring and continued to produce shoots in the media containing vitamins, 1% sucrose, 2 µM BA, and 15 µM $GA_3$. For high-frequency shoot formation, intact bean seedling cultures were used to form adventitious shoots from tissues in the axillary bud region at the cotyledonary node and areas surrounding the shoot apex of the seedlings (7). However, most of these methods used for shoot formation are applicable only to tissues containing the meristem.

Organogenesis has been attempted for bean regeneration in our laboratory (8). Organogenesis was induced from the basal part of the excised embryonic axis (3 mm in length) with the apical meristem of seeds staged at 3 to 5 days after germination. Multiple shoot formation was observed from the excised embryonic axis on MS medium containing 2 µM BA after 4 weeks of culture (Fig. 1). Regeneration was successful, with an efficiency of approximately 40%. This method may also apply to other bean cultivars because it is easy to obtain multiple shoots from the embryonic axis of seedlings.

Other approaches for regeneration of bean plants from callus and protoplast cultures have also been attempted. Westhuizen and Groenewald (9) obtained callus growth and subsequent root formation from bean radicles by using MS medium supplemented with combinations of auxins 2,4-dichlorophenoxyacetic acid ([2,4-D], NAA), cytokinins (kinetin, 2ip, BA), and $GA_3$. A method for isolation of protoplasts from the cotyledonary leaves of bean cultivars has also been developed (10). They obtained viable green calli from protoplasts, but these calli could not regenerate shoots. Despite many attempts, regeneration of bean plants from callus and protoplast cultures has been unsuccessful.

**Figure 1** Multiple shoots (arrowheads) induced from the basal part of the excised embryonic axis with apical meristem (arrow) of seedling at 3 days after germination.

Somatic embryogenesis is also an alternative for bean regeneration. In vitro culture of globular and heart-shaped embryos has been attempted (11). Somatic embryos were replicated in the manner of embryoids, but neither the shoot nor the root apex developed. So far, the many attempts to regenerate plants from somatic embryos of beans have failed.

## III. METHODS FOR TRANSFORMATION OF BEANS

Several protocols for bean transformation are now available, and these may be powerful enough to complement traditional breeding programs for crop improvement (12,13). Transformation of beans has been achieved by three methods: *Agrobacterium* species–mediated transformation, electroporation, and particle bombardment.

### A. *Agrobacterium* Species–Mediated Transformation

Embryonic axes, hypocotyl, epicotyl, cotyledonary nodes, primary leaf nodes, and primary leaves from bean seedlings were used as explants in transformation experiments in order to establish an optimal routine protocol for *Agrobacterium* species–mediated transformation of beans. Mariotti and coworkers (14) attempted to transform beans with *Agrobacterium* sp. LBA 4404 harboring the vector pBI121 with the β-glucuronidase (GUS) report gene. Gene transfer was achieved by cocultivation of the primary leaf node. They were able to produce some shoots that expressed kanamycin resistance and GUS gene activity. McClean and associates (15) developed a germinating seed assay to determine the susceptibility of various bean genotypes to infection by disarmed *Agrobacterium* sp. that contained the NPTII gene between T-DNA borders of the binary plasmid. From the infected tissues they obtained callus and root tissues that were capable of growth in the presence of kanamycin, but transgenic plants did not regenerate. Lewis and Bliss (16) stab-inoculated intact shoot tips of germinating bean seeds with *Agrobacterium* sp. They obtained tumor formation and GUS gene expression in the meristem region. Stab inoculation is an attractive method of transformation because it does not require tissue culture regeneration. It creates a wound site and exposes the interior cell layers of the meristem to *Agrobacterium* sp. The addition of acetosyringone (100 µM) to the medium, which has been suggested to improve transformation (17,18), yielded higher tumor proliferation in a number of cultivars. Becker and colleagues (19) transformed cotyledonary nodes and leaf nodes of bean cultivars with *Agrobacterium* sp. and compared different culture conditions in order to improve the efficacy of transformation and regeneration of transformed plants. It was found that the selection medium strongly influences the further growth of explants. Fifty milligrams of kanamycin inhibited growth within only 3 weeks, and at higher kanamycin concentrations (100–200 mg/l) the material survived for just 1 to 2 weeks. A general problem of these *Agrobacterium* species–mediated transformation methods is the inability to regenerate explants after infection.

## B. Direct Gene Transfer

Polyethylene glycol (PEG)-mediated gene transfer coupling with electroporation has been applied to gene transformation directly from protoplast cultures of legumes. Leon and associates (20) demonstrated transient gene expression of GUS and luciferase (LUC) by direct gene transfer to protoplasts of bean cultivars by using electroporation treatment with PEG. They established optimal conditions for the method. As an example, for electroporation the input voltage was varied while the pulse length was kept constant at 12 ms and PEG 4000 was used in a buffer of 0.1 M $Ca(NO_3)_2$ and 0.4 M mannitol. Similar levels of expression were also obtained by electroporation of protoplasts without PEG, suggesting that PEG is not an absolute requirement (21). Dillen and associates (22) used electroporation-mediated transformation to deliver DNA into intact seedling tissues of several cultivars. Transient gene expression was achieved in hypocotyl and epicotyl tissue and occurred in sectors of variable size. None of these studies was successful in producing transgenic bean plants. It seems that the foreign gene must enter the bean meristem tissue cells in order to produce transgenic plants.

## C. Particle Bombardment

Gene transfer by particle bombardment with DNA-coated particles is a rapid and simple means of transforming intact cells. This method has been widely used to introduce foreign DNA into plant cells (23,24). Successful transformation of bean plants by particle bombardment has been limited because systems for the regeneration of whole plants from transformed cells have not been developed. The efficiency of gene delivery by bombardment has been tested by transient GUS gene expression in cotyledons and embryonic axes of immature bean seeds (25). Approximately 60% of apical meristems showed at least one GUS-expressing unit after three treatments of particle bombardment. However, stability of transformation and the capacity for regeneration were not studied in this case. Aragao and coworkers (26) assessed the pressure of a helium shock wave and the distance between the particle plate and the sample as they influence transient gene expression by using a calcium spermidine procedure for coating gold particles with DNA. Russell and colleagues (27) reported the production of stable transgenic bean plants using electric discharge particle acceleration applied to embryonic axes and multiple shoot regeneration. Transgenic plants expressing GUS, herbicide, and virus resistance were recovered and maintained in subsequent generations. The introduced bar gene was shown to confer strong resistance to transgenic beans against herbicide application in a greenhouse. The yield of transgenic plants was, however, not more than 0.03% and the copy number of the transferred gene varied. Also, position effects could not be excluded. Aragao and coworkers (28) bombarded mature bean embryos with a methionine-rich 2S albumin gene from the Brazil nut (*Bertholletia excelsa*). Transient expression of the protein was detected by western blotting and chimeric tissues were obtained. However, stable transformants were not recovered. The bean was cotransformed by Aragao's group by using particle bombardment with the methionine-rich 2S gene isolated from Brazil nut and antisense sequences of genes of the bean golden mosaic gemini virus. The linked genes were inherited in a mendelian fashion, showing a high efficiency of transformation (up to 100%) (29).

A new approach that has been tested extensively is to use the embryonic axes of mature seeds as target tissue for direct gene transfer using particle bombardment. Kim and Minamikawa (30) attained stable bean transformation by this particle bombardment process. Various bombardment conditions were tested by bombarding the shoot apical meristem of embryonic axes with the plasmid carrying concanavalin A, the jackbean storage protein, with GUS (ConA::GUS) gene coated particles. The most important factor for higher transient expression was the time of bombardment. Longer bombardment time was the major cause of cell injury. A time of 0.015-seconds in this device reduced injury from bombardment. After particle bombardment under optimal conditions, the shoot apex of embryonic axes was cut in a cross-series section at 10-µm thickness and the particle distribution was examined by light microscope (Fig. 2). Many particles were delivered into the shoot apex cells by the following optimization conditions: (a) pressure of helium shock wave, 3 kgf/cm$^2$; (b) distance between the particle plate and sample, 4 cm; (c) amount of particles per bombardment, 0.05 mg; (d) amount of DNA per bombardment, 0.2 µg; and (e) chamber vacuum, 600 mm Hg. The typical blue color indicated a successful delivery of the plasmid carrying concanavalin A and the GUS gene into bombarded cells of the shoot apex (Fig. 3A). After 3 weeks of culture, embryonic axes developed into plantlets with two to three leaves (Fig. 3B). The regenerated plants flowered after 6 weeks in pots and subsequently produced seeds (Fig. 3C). GUS activity was observed in the cotyledons and seed coats but not in the pods of the transgenic plants (Fig. 3D). Also, stable expression of the canavalin and GUS fusion was examined in transgenic bean seeds at maturation stages (31). GUS expression was positive in 40 of 54 seeds collected from five transgenic plants analyzed. This developing technology was designed to circumvent problems of conventional tissue culture and plant regeneration.

## IV. IMPROVEMENT OF THE NUTRITIONAL QUALITY OF BEAN SEEDS BY GENE TRANSFER

The development of transgenic plants with increased levels of nutrients may offer increased health benefits. Sulfur deficiency caused a marked change in the relative content of seed storage proteins in legumes (32). The sulfur deficiency resulted in a reduced level of a sulfur-rich protein (33). These results indicate that the levels of methionine and cysteine in seeds influence the synthesis of sulfur-rich proteins. Plants take up sulfur in the form of sulfate and convert it to sulfide through adenosine-5'-phosphosulfate and sulfite (34). Cysteine is formed from sulfide and *O*-acetylhomoserine and reacts with *O*-phosphohomoserine to generate cystathionine (35). Methionine is formed from cystathionine via homocystathionine. An increase in the level of sulfur-containing amino acids in seeds possibly induces the synthesis of a larger amount of sulfur-rich

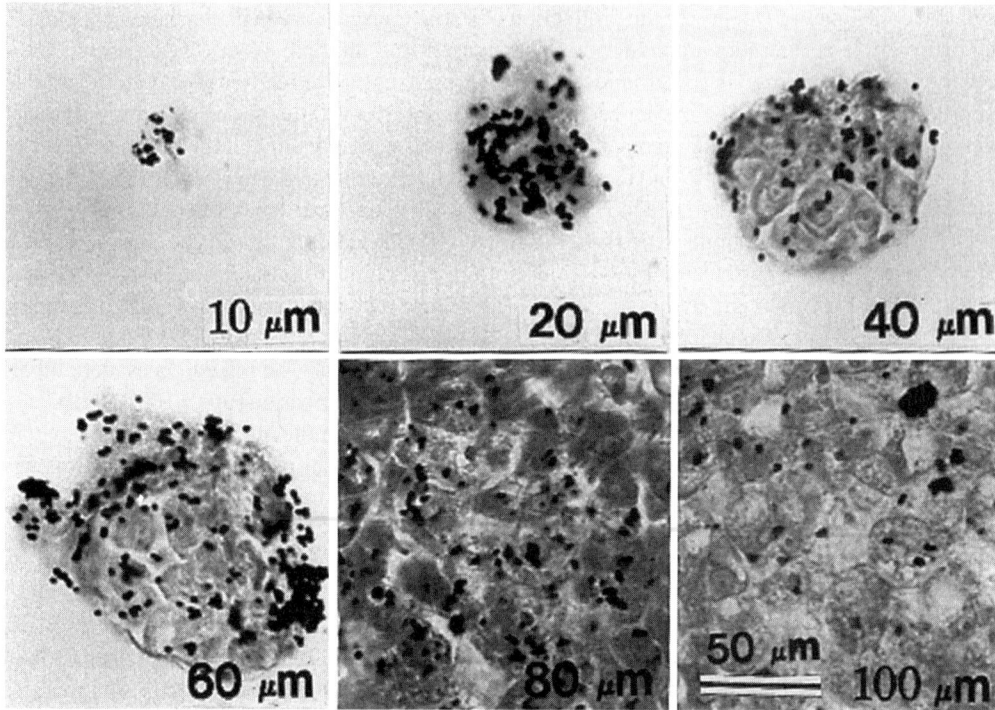

**Figure 2** Shoot apex of embryonic axis was cut with cross-series section at 10-μm thickness and the distribution of particles was examined by light microscope. 10- to ~100-μm depth from the top of shoot apex.

proteins. However, the methionine content in phaseolin, a storage protein in bean, is much lower than in other cereals (36). Production of a larger amount of methionine-rich proteins in the bean seed would be expected to improve the nutritional quality of the seed protein. Control of the regulation of methionine biosynthesis by manipulating the proteins in bean seeds increases the methionine content in legume seed proteins. Phaseolin constitutes approximately 40% of the total protein in the bean seeds but contains only three methionine residues (37). To improve the nutritional quality of phaseolin, a modified phaseolin was synthesized by insertion of a 45-bp oligonucleotide containing six methionine codons into the third exon of the phaseolin gene (38). However, only a small amount of the modified phaseolin accumulated in the transgenic tobacco seeds.

Analysis of the amino acid sequence of a 2S albumin in seeds of the Brazil nut revealed the presence of 20% methionine residues (39). To increase the methionine content of the tobacco seed protein, complementary DNA (cDNA) of the 2S albumin was ligated to the promoter region of the phaseolin gene, and the chimeric gene was introduced into tobacco plants (40). The accumulation of the 2S albumin resulted in a 30% increase in the methionine content of the transgenic tobacco seeds. However, the Brazil nut 2S albumin was found to be an allergen (41). Therefore, it may not be possible for humans and animals to utilize seeds in which the Brazil nut 2S albumin gene is expressed. Instead of 2S albumin, a 10-kDa albumin in sunflower seed can be used because it is not an allergen (42) and contains 16% methionine residues (43).

Manipulation of the carbohydrate content in bean seed is also a possibility to improve nutritional quality. About 61% of the total dry weight of bean seed is carbohydrate. Starch is the major storage carbohydrate and is composed of the two glucose polymers linear amylose and

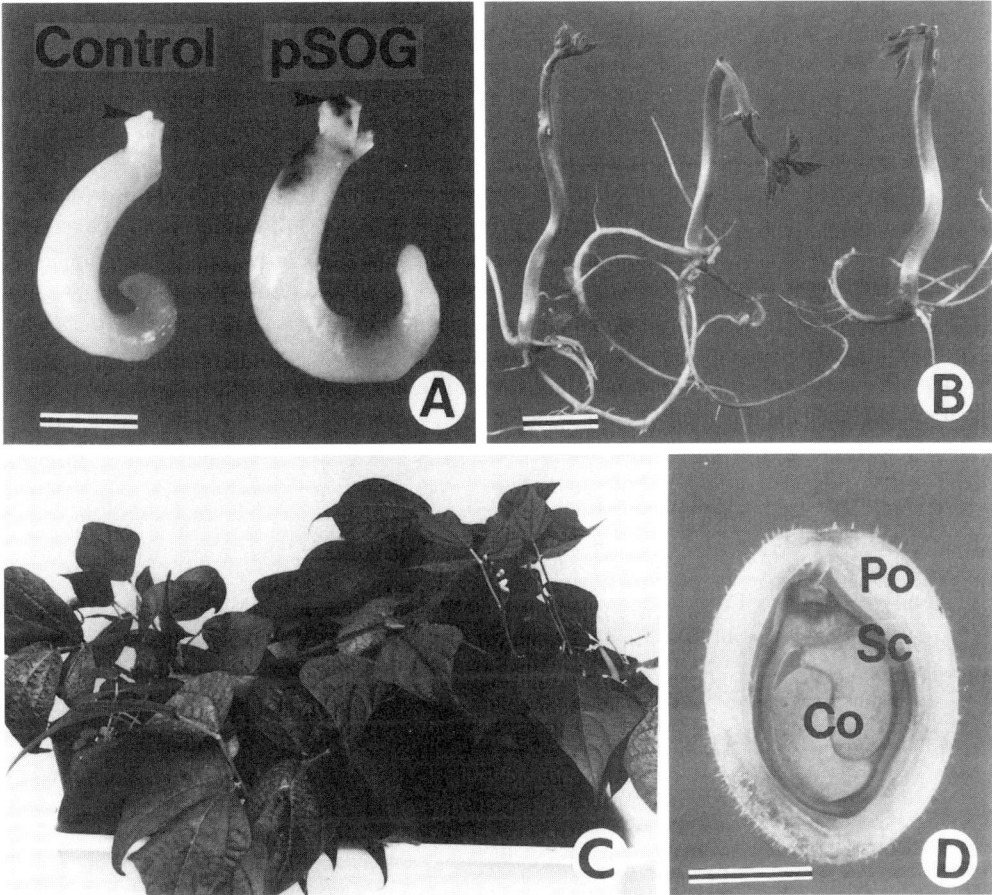

**Figure 3** GUS staining and plant regeneration after particle bombardment. (A) Transient GUS expression in the embryonic axes after particle bombardment. Control, nonbombarded embryonic axis; pSOG, embryonic axis bombarded with concanavalin A plasmid-coated particles. Arrowheads point to shoot apexes. (B) Plantlets recovered after 3 weeks of culture. (C) Plants regenerated from bombarded embryonic axes. (D) Expression of the concanavalin A and GUS gene in a seed of transgenic bean plant. Co, cotyledon; Sc, seed coat; Po, pod. Bars = 0.5 cm.

branched amylopectin, which are arranged into a highly ordered crystalline structure of starch granules. Several enzymes, such as adenosine diphosphate (ADP) glucose pyrophosphorylase (AGPase), starch synthase, starch branching enzyme, and starch debranching enzyme catalyze chain elongation, branching, and trimming during the starch biosynthesis process. Providing that sufficient metabolic precursors are present, significant increases in starch levels should be possible by controlling the expression of these enzymes. Much of this work has been done with AGPase. Elevated levels of AGPase increased starch yields in potato tuber up to 30% (44). Conversely, transgenic potato plants transformed with AGPase cDNA in the reverse orientation express little AGPase with 70% reduction of starch levels and increased sugars (45). This approach in the long term will result in manipulations of starch yield and quality. Among legumes only pea has been studied for characterization of isoforms of the specified enzymes. In view of the importance of bean corps as a major food source, modification of the carbohydrate composition of the bean is foreseen.

## V. CONCLUSIONS

Various methods for plant regeneration and genetic transformation of the bean (*Phaseolus vulagaris*) have been discussed in this chapter. Regeneration of bean plants is not an easy task and is dependent mainly on the tissues selected for use. Induction of shoot formation of bean was achieved by various explants of seedlings, such as axillary bud and shoot apex. One of the few breakthroughs for regeneration was due to organogenesis of the embryonic axis. However, regeneration attempts with callus and protoplast culture have not been successful. Since the use of protoplast and callus is necessary for maintaining cultures and transformation, regeneration methods using these cultures should be improved. Although success in transformation has been limited, a major accomplishment for stable transformation of bean was obtained by particle bombardment of the shoot apical meristem of embryonic axes. Since legumes are important crops for food, manipulation of genes encoding value-added materials has the potential to change the compositions of seeds quantitatively and qualitatively. Genetically modified transgenic bean plants would be of great benefit to humankind.

## ACKNOWLEDGMENT

We would like to thank Drs. T Minanikawa and D. Yamauchi for suggestions and technical help to J. W. Kim. We also want to express our thanks to Microplants Co., Ltd., for financial support.

## REFERENCES

1. NS Yang. Transgenic Plants. 2: Transgenic Plants from Legumes. New York: Academic Press 1993, pp 79–102.
2. P Gepts, D Debouck. Domestication and evolution of the common bean (*Phaseolus vulgaris* L.). UK Indian Acad Sci 7–53, 1991.
3. D Debouck, DGO Schmidt, SCD Vries. Early events in higher-plant embryogenesis. Plant Mol Biol 22:367–377, 1993.
4. A Allavena, L Rossetti. Micropropagation of bean (*Phaseolus vulgaris* L.): Effect of genetic, epigenetic and environmental factors. Sci Hortic 30:37–46, 1986.
5. P McMclean, KF Grafton. Regeneration of dry bean (*Phaseolus vulgaris* L.) via organogenesis. Plant Sci 60:117–122, 1989.
6. IS Martins, MR Sondahl. Multiple shoot formation from shoot apex cultures of *Phaseolus vulgaris* L. J Plant Physiol 115:205–208, 1984.
7. KA Malik, PK Saxena. Regeneration in *Phaseolus vulgaris:* Promotive role of N6-benzylaminopurine in cultures from juvenile leaves. Planta 184:148–150, 1992.
8. JW Kim, SK Han, SY Kwon, HS Lee, SS Kwak. High frequency shoot induction and plant regeneration from the cut end of hypocotyls of seed plant seedlings. Annual Meeting of Korean Plant Tissue Culture Society. Kwangju, 1998, p 15.
9. AJ Westhuizen, EG Groenewald. Root formation and attempts to establish morphogenesis in callus tissues of beans (*Phaseolus vulgaris* L.). S Afr J Bot 56:271–273, 1990.
10. L Crepy, LMG Barros, VRN Valentine. Callus production from leaf protoplasts of various cultivars of bean (*Phaseolus vulgaris* L.). Plant Cell Rep 5:124–126, 1986.
11. JP Baudoin, B Lecomte, G Mergeai. Contribution for improved *in vitro* culture of globular and early heart-shaped embryos in *Phaseolus*. Annu Rep Bean Improv Coop (BIC) 38:103–104, 1995.
12. NV Raikhel, RL Last. The wide world of plant molecular genetics. Plant Cell 5:823–829, 1993.
13. MD Block. The cell biology of plant transformation: Current state, problems, prospects and the implications for the plant breeding. Euphytica 71:1–14, 1993.

14. D Mariotti, GS Fontana, L Santini. Genetic transformation of grain legumes: *Phaseolus vulgaris* L. and *P. coccineus* L. J Genet Breed 43:77–82, 1989.
15. P McClean, P Chee, B Held, J Simental, RF Drong, J Slighton. Susceptibility of dry bean (*Phaseolus vulgaris* L.) to *Agrobacterium* infection: Transformation of cotyledonary hypocotyl tissues. Plant Cell Tissue Org Cult 24:131–138, 1991.
16. ME Lewis, FA Bliss. Tumor formation and β-glucuronidase expression in *Phaseolus vulgaris* inoculated with *Agrobacterium tumefaciens*. J Am Soc Hortic Sci 119:361–366, 1994.
17. KH Joao, TA Brown. Enhanced transformation of tomato co-cultivated with *Agrobacterium tumefaciens* C58C1Rifr: pGSFR 1161 in the presence of acetosyringone. Plant Cell Rep 12:422–425, 1993.
18. S Nishibayashi, H Kaneko, Hayakawa T. Transformation of cucumber (*Cucumis sativus* L.) plants using *Agrobacterium tumefaciens* and regeneration from hypocotyl explants. Plant Cell Rep 15:809–814, 1996.
19. J Becker, T Vogel, J Iobal, W Nagl. *Agrobacterium* mediated transformation of *Phaseolus vulgaris*: Adaptation of some conditions. Annu Rep Bean Improv Coop USA (BIC) 37:127–128, 1994.
20. P Leon, F Planckaert, V Walbot. Transient gene expression in protoplasts of *Phaseolus vulgaris* isolated from a cell suspension culture. Plant Physiol 95:968–972, 1991.
21. M Bustos. Transgenic gene expression in *Phaseolus vulgaris* by direct gene transfer to protoplasts. Plant Mol Biol Rep 9:322–332, 1991.
22. W Dillen, G Engler, M van Montagu, G Angenon. Electroporation-mediated DNA delivery to seedling tissues of *Phaseolus vulgaris* L. (common bean). Plant Cell Rep 15:119–124, 1995.
23. JJ Finer, MD McMullen. Transformation of cotton (*Gossypium hirsutum* L.) via particle bombardment. Plant Cell Rep 8:586–589, 1990.
24. VR Bommineni, RSS Datla, EWT Tsang. Expression of GUS in somatic embryo cultures of black spruce after microprojectile bombardment. J Exp Bot 45:491–495, 1994.
25. A Genga, A Allavena. Factors affecting morphogenesis from immature cotyledons of *Phaseolus coccineus* L. Plant Cell Tissue Organ Cult 27:186–196, 1991.
26. FJL Aragao, MF Grossi de Sa, MR Davey, ACM Brasiliro, JC Faria, EL Rech. Factors influencing transient gene expression in bean (*Phaseolus vulgaris* L.) using an electrical particle acceleration device. Plant Cell Rep 12:483–490, 1993.
27. DR Russell, K Wallace, J Bathe, B Martinell, D McCabe. Stable transformation of *Phaseolus vulgaris* via electric discharge mediated particle acceleration. Plant Cell Rep 12:165–169, 1993.
28. FJL Aragao, MF Grossi de Sa, ER Almeida, ES Gander, EL Rech. Particle bombardment–mediated transient expression of a brazil nut methionine-rich albumin in bean (*Phaseolus vulgaris*). Plant Mol Biol 20:357–359, 1992.
29. FJL Aragao, LMG Barros, ACM Brasiliero, SG Ribeiro, FD Smith, JC Sanford, JC Faria, EL Rech. Inheritance of foreign genes in transgenic bean (*Phaseolus vulgaris* L.) co-transformed via particle bombardment. Theor Appl Genet 93:142–150, 1996.
30. JW Kim, T Minamikawa. Transformation and regeneration of French bean plants by the particle bombardment process. Plant Sci 117:131–138, 1996.
31. JW Kim, T Minamikawa. Stable delivery of a concanavalin promoter-β-glucuronidase gene fusion into French bean by particle bombardment. Plant Cell Physiol 38:71–75, 1997.
32. TJV Higgins. Synthesis and regulation of major proteins in seeds. Annu Rev Plant Physiol 35:191–221, 1984.
33. PM Chandler, TJV Higgins, PJ Randall, D Spencer. Regulation of legumin levels in developing pea seeds under conditions of sulfur deficiency: Rates of legumin synthesis and levels of legumin mRNA. Plant Physiol 71:47–54, 1983.
34. RJ Schmidt, K Jager. Open questions about sulfur metabolism in plants. Annu Rev Plant Physiol Plant Mol Biol 43:325–349, 1992.
35. BJ Miflin, PJ Lea. Amino acid metabolism. Annu Rev Plant Physiol 28:299–329, 1977.
36. D Yamauchi, T Minamikawa. Improvement of the nutritional quality of legume seed storage proteins by molecular breeding. J Plant Res 111:1–6, 1998.
37. JL Slightom, SM Sun, TC Hall. Complete nucleotide sequence of a French bean storage protein gene: Phaseolin. Proc Natl Acad Sci USA 80:1897–1901, 1983.

38. LM Hoffman, DD Donaldson, Herman EM. A modified storage protein is synthesized, processed, and degraded in the seeds of transgenic plants. Plant Mol Biol 11:717–729, 1988.
39. C Ampe, J van Damme, LAB de Castro, MJAM Sampaio, M van Montagu, J Vandekerckhove. The amino-acid sequence of the 2S sulphur rich proteins from seeds of Brazil nut (*Bertholletia excelsa* H.B.K.). Eur J Biochem 159:597–604, 1986.
40. SB Altenbach, KW Pearson, G Meeker, LC Staraci, SSM Sun. Enhancement of the methionine content of seed proteins by the expression of a chimeric gene encoding a methionine-rich protein in transgenic plants. Plant Mol Biol 13:513–522, 1989.
41. JA Nordlee, SL Taylor, JA Townsend, LA Thomas, RK Bush. Identification of a Braizil-nut allergen in transgenic soybeans. N Engl J Med 334:688–692, 1996.
42. L Molvig, LM Tabe, BO Eggum, AE Moore, S Craig, D Spencer, TJV Higgins. Enhanced methionine levels and increased nutritive value of seeds of transgenic lupins (*Lupinus angustifolius* L.) expressing a sunflower seed albumin gene. Proc Natl Acad Sci USA 94:8393–8398, 1997.
43. AA Kortt, JB Caldwell, GG Lilley, TJV Higgins. Amino acid and cDNA sequences of a methionine-rich 2S protein from sunflower seed (*Helianthus annuus* L.). Eur J Biochem 195:329–334, 1991.
44. DM Starck, KP Timmerman, GF Barry, J Preiss, GM Kishore. Regulation of the amount of starch in plant tissues by ADP glucose pyrophosphorylase. Science 258:287–292, 1992.
45. B Muller-Rober, U Sonnerwald, L Willmitzer. Inhibition of the ADP glucose pyrophosphorylase in transgenic potatoes leads to sugar-storing tubers and influences tuber formation and expression of tuber storage protein genes. EMBO J 11:1229–1238, 1992.

# 34
# Genetic Engineering of Beet and the Concept of the Plant as a Factory

**Robert Sévenier, Andries J. Koops, and Robert D. Hall**
*Plant Research International, Wageningen, The Netherlands*

|     |                                                              |     |
| --- | ------------------------------------------------------------ | --- |
| I.  | INTRODUCTION                                                 | 485 |
|     | A. The Sugar Beet Crop                                       | 485 |
|     | B. Engineering Agronomic Traits                              | 486 |
|     | C. Biosafety Aspects                                         | 489 |
| II. | METHODOLOGIES TO ENGINEER SUGAR BEET                         | 489 |
|     | A. *Agrobacterium* Species                                   | 489 |
|     | B. Particle Bombardment/Electroporation/Sonication           | 490 |
|     | C. Guard Cell Protoplasts                                    | 490 |
| III.| SUGAR BEET AS A FACTORY: FRUCTAN PRODUCTION IN SUGAR BEET    | 491 |
|     | A. Sugar Beet as a "Plant as a Factory"                      | 491 |
|     | B. The Fructans                                              | 492 |
|     | C. Oligofructan Production in Sugar Beet                     | 493 |
| IV. | CONCLUSIONS                                                  | 498 |
|     | REFERENCES                                                   | 498 |

## I. INTRODUCTION

### A. The Sugar Beet Crop

Sugar beet is one of the 13 species of *Beta* (Chenopodiaceae). Like all other cultivated beets, sugar beet belongs to the species *Beta vulgaris* ssp. *vulgaris*. Cultivated beets are derived most probably from an ancestral maritime population (*Beta vulgaris* spp. *maritima*)(1).

Before the 19th century, the beet was cultivated for its leaves, which were used as a vegetable or as cattle feed; the roots were devoted to medicinal purposes (2). As a result of the European political events of the end of the 18th century, the trade of cane sugar from overseas became limited. As a consequence of these restrictions some previous work showing the presence of sugar in the taproot of sugar beet was highlighted (3). In 1575, Olivier de Serres, a French agronomist, noted the presence of sugar in the root of *Beta vulgaris* (3). In 1745, the German chemist Marggraf extracted and solidified sucrose from sugar beet extracts. Forty years later, these findings were resumed by Achard, who also developed sugar beet farming. His work resulted in 1802 in the establishment of the first experimental sugar factory (3).

Since then the culture of sugar beet for the production of sucrose has expanded rapidly. In 1900, 9.6 million metric tons of sugar was produced, 54% extracted from sugar beet. In 1998, 125 million metric tons of sugar was extracted, one third from sugar beet. Sugar beet is grown in 55 countries throughout Europe, North America, and extensive areas of Asia on a total of 6.96 million ha.

The sugar beet has a biennial growth habit. During the first year, the plant produces a rosette of leaves sustained by a large fleshy taproot. Nowadays, sugar beet varieties can develop taproots weighing 1 kg and composed of 18% sucrose (fresh weight) and yielding up to 15 tons sucrose per hectare (4). In the second year, after a low-temperature/long-day exposure, the plants start flowering. Flowers develop on the terminal ends of the main axis as well as on lateral branches. These flowers, pollinated by wind, give rise to mono- or multigerm seeds.

## B. Engineering Agronomic Traits

### 1. Genetic Engineering in the Framework of Classical Breeding

As sugar beet became a crop, breeding strategies led to a dramatic improvement in the sugar content of the taproot, from 5% to 6% sucrose at the beginning of the 19th century, to ±20% sucrose today (5,6). One of the objectives of sugar beet breeders is to improve these figures even further. To allow the beet to express its potential fully, optimal conditions should be present during the farming and harvesting of the taproot (5). Because this ideal situation is elusive, genetic improvement will allow sugar beet to compensate for unfavorable conditions. The different characteristics being targeted concern the seed production steps as well as taproot development.

Almost all conventional beet production is based on triploid F1 hybrid varieties (1). Seed is obtained by crossing tetraploid fertile plants with pollen-sterile cytoplasmic male sterility (CMS), diploid plants from which the hybrid seed is later harvested.

The universal use of only one source of CMS leads to a very narrow cytoplasmic genetic base (1,6). Two main methods have been studied for the purpose of broadening the source of cytoplasm useful for sugar beet breeding. Somaclonal variation should allow the creation of new cytoplasm as well as new genetic characteristics (6), and protoplast fusion could help save time during the introduction of new cytoplasm in breeding schemes (1). As for other species clonal micropropagation has been addressed with the aim of developing efficient methods for rapid micropropagation of valuable breeding lines (7). To facilitate the production of $F_1$ hybrids, haploid plant production was studied and was revealed to be useful but nevertheless laborious and not very efficient (7).

Optimal conditions for taproot production would include resistance to important diseases such as cercosporiosis, downy mildew, black rot, viruses, and rhizomania. Tolerance to herbicides and environmental factors (cold, drought, and soil salinity) is also desirable. Classical breeding techniques have been and are still being applied to improve sugar beet for these characteristics. However, the process remains slow (7). Therefore, since the mid-1980s attempts have been made to exploit biotechnological techniques. Although these new techniques have proved to be powerful tools for other species, their use appeared to be difficult for sugar beet (7).

Many efforts have been made to develop an efficient method of genetic transformation for sugar beet, initially with little or no success (8). However, two main protocols are now routinely used to obtain transgenic sugar beet (see Methodologies for technical details).

The main goals of the genetic engineering of sugar beet concern the development of herbicide-tolerant sugar beet, development of sugar beet resistant to pests and diseases, and improvement of sugar (beet) quality.

## 2. Herbicide-Tolerant Sugar Beet

Weeds in commercial sugar beet fields constitute a major problem, when not controlled, they compete with sugar beet, thereby reducing yield (9). No single herbicide can effectively control all weed species without inflicting some harm to the sugar beet crop (10,11). Weed control is nowadays realized without major difficulty, but the programs are quite complex and comprise different pre- and postemergence applications of mixes of one to three different herbicide specialties (12). To achieve proper control of the weed population, farmers must spend time surveying the development of the weeds and applying the herbicides. This necessity creates extra costs.

The culture of sugar beet varieties engineered for herbicide tolerance should overcome these problems. It should also be possible to introduce genes conferring resistance to wide-spectrum herbicides such as glyphosate, with low impact on the environment.

In the early 1990s several reports detailed field trials of herbicide-resistant sugar beet (13–15). At least three different classes of herbicide were concerned: phosphinotricin (glufosinate, bialaphos), glyphosate, and sulfonylurea. The genes introduced into sugar beet in order to confer resistance to these molecules encoded phosphinotricin acetyl transferase (PAT), enolpyruvylshikimate 3-phophate synthase (ESPS), and acetolactate synthase (ALS), respectively (13). ESPS was also associated with glyphosate oxidase reductase (GOX) (11). The transformed plants obtained were able to resist agronomic doses of herbicide during field trials. Tolerance to the herbicides appeared to be inversely correlated to the number of copies of the transgenes (11,14,16). Transformed lines resistant to glyphosate ammonium permitted a greater flexibility in the application of the herbicide (14). Glyphosate- and glufosinate-resistant sugar beet have been thoroughly tested under field conditions (10,17). The first transformed herbicide-resistant sugar beet varieties were expected to reach commercialization in the year 1999 (10) in the United States.

## 3. Engineering Sugar Beet for Pest and Disease Resistance

The most advanced and spectacular results concern sugar beet engineered to resist either the beet necrotic yellow vein virus (BNYVV) or nematodes.

*a. BNYVV.* The BNYVV is a soil-borne furovirus transmitted by the fungus *Polymixae betae* (18). The virus is responsible for rhizomania, a disorder characterized by a massive proliferation of rootlets that causes a severe stunting of the root. This results in a reduction in sugar yield (19) that can be as dramatic as 50% of the expected harvest (20). In 1992, it was estimated that 10% of the sugar beet acreage planted in Western Europe was affected by rhizomania (19).

Breeders have been improving the sugar beet tolerance to BNYVV since the 1980s and varieties displaying varying levels of resistance have resulted (20).

Biotechnology, and especially gene transfer, is nowadays being used to fight this disease. It is possible to combat viral infection by producing the coat protein of the virus in the plant, by expressing the viral RNA, or by interfering with the movement of the virus within the plant (21). The expression of the BNYVV coat protein in sugar beet protoplasts has been shown to confer a high level of protection to these cells (22). This protective effect was confirmed by using the hairy root test system, for which *Agrobacterium rhizogenes* is used to induce proliferation of sugar beet roots in vitro (23). Transformed sugar beet plants expressing the BNYVV coat protein have also been produced. Greenhouse and field trials confirmed the protective effect of the transgene against the development of rhizomania. Multiplication of the virus in the transformed plants was significantly reduced (19). Field trials with sugar beet hybrids combining both the engineered resistance and the classical tolerance showed higher resistance than that of varieties with classical tolerance alone (24).

*b. Nematodes.* *Heterodera schachtii* is the most important nematode pest of beet (25). This cyst nematode develops on the root by inducing highly specialized feeding structures within the vascular system. To control the development of nematodes, sugar beet growers must rotate crops and use nematicides and resistant varieties. However, nematode eggs can survive unfavorable conditions for many years, and nematicides are now forbidden in many countries because of their high toxicity (26). There is no useful resistance against nematodes in the species *B. vulgaris*, but *B. procumbens* and *B. webbiana* do possess resistance genes against this pest (26–28). These genes have been located by creating alien monosomic additions in *B. vulgaris*. Genetic localization resulted in the identification of three resistance genes: $Hs1^{web-1}$, $Hs1^{web-7}$ from *B. webbiana*, and $Hs1^{pro-1}$ from *B. procumbens* (29). In 1997 the $Hs1^{pro-1}$ gene was successfully cloned, a "first" for a gene conferring resistance against nematodes (27). The isolated gene was able to confer nematode resistance to susceptible *B. vulgaris* lines, as tested by the hairy roots system. Transgenic plants expressing $Hs1^{pro-1}$ have been obtained and will be tested in the field (30).

## 4. Quality Improvement via Genetic Engineering

Besides the introduction of genes conferring pest, disease, or herbicide resistances, genetic engineering could also be exploited to improve the quality of the sugar beet crop. This improvement can be achieved by reducing the amount of undesirable compounds that interfere with sucrose processing or by increasing the total yield of sucrose.

Among the undesirable compounds, the most prominent are betaine, raffinose, and invert sugar, which interfere with sucrose extraction and crystallization (31,32). Betaine is a major impurity; reduction of the amount of betaine can be achieved via an antisense strategy directed against the choline monoxygenase or the betaine dehydrogenase enzymes both involved in betaine biosynthesis (31). Raffinose, a trisaccharide, accounts for 1% of sucrose. It has been proposed to use bacterial α-galactosidase in order to degrade raffinose in sucrose and galactose (32). An antisense strategy directed against vacuolar invertase present in the storing cells of the taproot should reduce the hydrolysis of sucrose in glucose and fructose and limit both the amounts of this sugar and the loss of sucrose (32). The sucrose content of the taproot is affected by numerous factors and thus could be very difficult to target by using gene transfer technology. However, single factors are likely to be rate-limiting and could be easier targets for a transgenic approach (31).

The transport of sucrose through the membranes is mediated by carrier proteins that could limit the uptake. Engineering the transport capacity of these transporters could provide a way to increase the sink strength of the taproot (31).

Another approach to increase the sucrose content of the taproot would be to manipulate the root ultrastructure (33,34). High levels of sucrose are associated with the presence of numerous parenchymatous zones in the root. If all the secondary cambial rings could contribute to the production of parenchymatous zones, this could result in an increase in the sink capacity of the root and in the development of small storage cells in close proximity to the active phloem delivering the sucrose. The distribution of sucrose would then also be more homogeneous within the root. Several sugar beet genes involved in the cell cycle have been isolated, but in order to obtain the desired new root morphologic characteristics, the expression of these genes will have to be precisely monitored. The promotors able to fine-tune this expression are, however, not yet available.

Although sugar beet requires vernalization to flower, some plants are able to bolt during the first year. The root loses its sugar and becomes lignified. This decreases both harvesting efficiency and sugar yield (4). In northern latitudes sugar beet cannot be sown before the last frost to eliminate any risk of bolting in the following summer (34). Bolting in a sugar beet crop has been proposed to be due to contamination by pollen from wild annual beets in sugar beet seed production areas (35). A better understanding of the molecular basis of the bolting phenomenon could allow the development of genetic solutions.

## C. Biosafety Aspects

Although sugar beet is used as a source of food, few criticisms have been raised against transformed sugar beet in this respect. This can be explained because of the process required to extract sucrose out of sugar beet taproot. The sucrose extraction includes a diffusion step. The taproots are sliced and the sucrose is allowed to diffuse out of the cells into warm water. This is followed by a precipitation/filtration step (CaO and $CO_2$ are added to the juices in order to provoke precipitation of the impurities). The sucrose is then allowed to crystallize after concentration of the juices. Each step of the process proved to be very efficient in terms of DNA degradation and protein elimination (36). White sugar obtained from transformed sugar beet and that from conventional beet are indistinguishable with respect to purity (36).

However, other biosafety aspects, which concern the introduction of transformed sugar beets into the environment, have been studied. *Beta vulgaris* spp. *maritima*, the wild beet, occurs naturally in coastal areas of European countries (37). The weed beet, supposedly an offspring of *B. vulgaris* and *B. vulgaris* spp. *maritima*, is a noxious weed in Western Europe, notably in Britain and France (37). Pollen exchange between sugar beet and wild beet is a well-known phenomenon that sugar beet breeders have to prevent during seed production (35). A distance of 1000 m is advised between beet devoted to seed production and any other cultivated beet or wild beet plants (35,37). However, pollen has been shown to be transported up to 3200 m (38). The exchange of pollen is possible in both directions (39). The bolting of sugar beet (flowering during the first year of culture) is likely to be caused by accidental pollination of beets by weed beets in seed production areas (35,40,41).

Hand crossing of wild beet and spinach beet with pollen produced by herbicide-resistant beets showed transfer of the transgene to the offspring. The level of expression of the transgene was sufficient to confer herbicide resistance to the hybrid (42). Transfer of herbicide resistance to weed beets could seriously impair the control of this weed (41). BNYVV-resistant transgenic beets did not show any significant increase of competitiveness when compared to classically bred tolerant sugar beets (43). However, competitiveness of beets expressing PR proteins such as chitinase could be different (44). Beets and other members of the Chenopodiaceae are nonmycorrhizal, but transformed beets expressing PR proteins could induce a decrease in mycorrhizal populations resulting in a depletion of the following crop in the rotation. The transfer of such a gene to the wild beet could modify the invasiveness of this plant by conferring both disease resistance and advantage to the mycorrhizal plants of the ecosystem. Each new characteristic introduced in sugar beet should therefore be studied separately, and risk assessment should focus on the effect of the introduced gene on the whole ecosystem (37).

## II. METHODOLOGIES TO ENGINEER SUGAR BEET

### A. *Agrobacterium* Species

Genetic transformation of sugar beet is reported to be difficult. The essential requirement of cells competent for both regeneration and transformation is only partially fulfilled (45). Several types of explant have been tested. Cotyledons (45–47), hypocotyl (46), shoots bases (48), petioles (45), leaf disks (49,50), and seedling-derived calli (14) have all proved to be useful with specific protocols. Although callus formation can be induced for practically all genotypes of sugar beet, few genotypes can produce regenerable callus (14). The susceptibility of transformation by *Agrobacterium tumefaciens* is also genotype-dependent (46,48). This susceptibility could be increased by preculturing the explants before cocultivation with the bacteria or by using acetosyringone (45).

Transformation frequencies were increased in the presence of 2,4-D, but regeneration was then severely reduced (45). Several *A. tumefaciens* strains have been used: AGLO (45), LBA (48–50), C58C1 (14), and EHA101 (47,50). Trials with "shooter" LBA strains did not prove to be superior (49). Selection of transformants can be achieved in the presence of antibiotics (kanamycin, G418) (45,46,48,50), herbicides (phosphinotricin, chlorsulfuron) (14,50), or mannose (47) although the proportion of escapes varies among agents. The nature of the promotor driving the selectable marker sequence determines the level of selection to apply (14). Using cotyledonary nodes a high regeneration frequency could be obtained (70%), but transgenic sugar beet plants were only observed at a frequency $\leq 1\%$ (45). The same frequency was obtained after selection on mannose, whereas selection on kanamycin allowed a frequency of only 0.1% with a high proportion of escapes (47). In 1998 Zhang and collaborators reported an average transformation frequency of 3% and a maximum of 9% in certain experiments (50).

### B. Particle Bombardment/Electroporation/Sonication

DNA uptake by sugar beet protoplasts has been achieved by electroporation (51–53). Alternating current pulses prove to be the most efficient for transient expression of chloramphenicol acetyltransferase (CAT) when compared to rectangular or exponentially decaying pulses (51). Joersbo developed a formula to estimate the efficiency of electroporation as a function of the electroporation parameter based on data collected with sugar beet cell suspension protoplasts (52). Although the regeneration of electroporation-mediated transgenic sugar beet has not been reported, stably transformed calli have been obtained at a frequency of $2–6 \times 10^{-5}$ (53). Although only reported once, mild sonication, which is based on transferring DNA by exposing cells to ultrasonic waves, has mediated transient expression of CAT (54). The maximal level of expression was 7 to 15 times that obtained with electroporation. The surviving protoplasts divided, and their plating efficiency was not affected. Transient expression has equally been observed after particle bombardment of cell suspensions (55,56) and apices (57) of sugar beet. The rate of transient expression in apices was very low, but precise targeting of the different cell layers of the meristem was possible by choosing the correct particle size and acceleration pressure (57). Particle bombardment has also been reported for the transformation of chloroplasts (56). Such a possibility could be a valuable tool to prevent sexual transfer of transgenes to wild beets.

### C. Guard Cell Protoplasts

Although protoplast-based regeneration/transformation systems have been quite successful for a wide range of species, such systems for sugar beet were for many years extremely recalcitrant (1). However, the discovery that sugar beet guard cell protoplasts apparently have a much higher propensity to divide and regenerate in culture than all other cell types (58,59) paved the way for the development of a unique guard cell–based transformation system for this crop (16). Heterogeneous sugar beet leaf protoplast populations have division frequencies of 0.01–0.5%, a value that could be significantly improved by using purified epidermis as the starting material instead of whole leaves. In this case the protoplast population was enriched specifically for guard cells (Fig. 1) from 0.5% to a level approaching 95% and the division frequencies increased to $\leq 50\%$ (16). Converting this system into one for transformation was relatively simple, and the use of PEG-mediated DNA transfer, combined with herbicide (bialaphos) selection conferred by the *pat* gene, resulted in suitable transformation frequencies in the $10^{-4}$ range (Fig. 1). This entails that from a single protoplast isolation using 1.5 g leaf material >30 transformed plants can be produced after 5–8 weeks. As a result of the watertight nature of the selection protocol, combined with a somatic embryogenesis–based regeneration protocol, all regenerated plants in these ex-

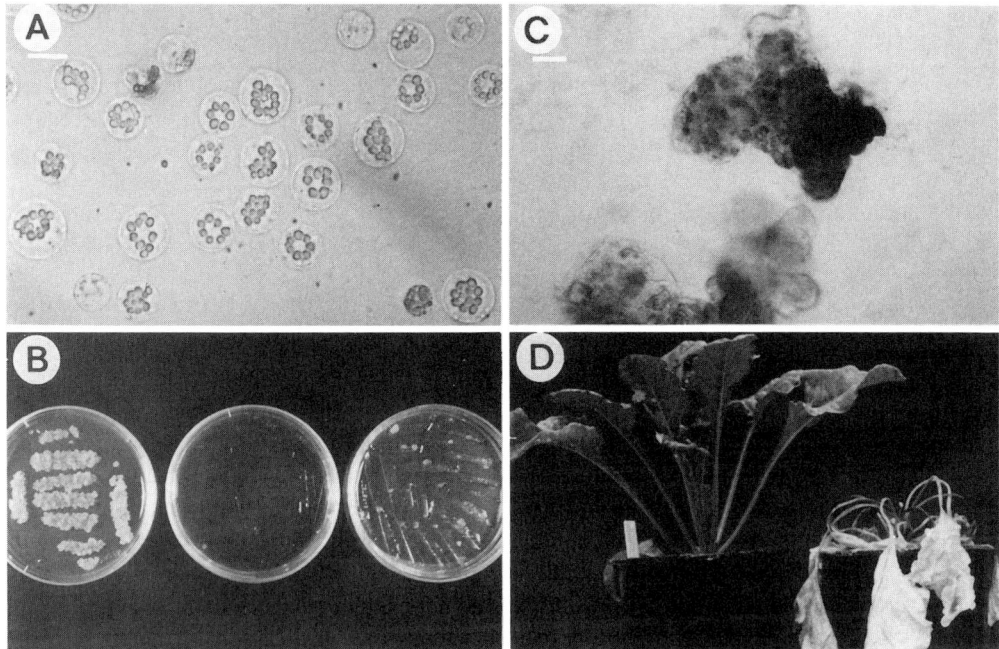

**Figure 1** (A) Enriched population of guard cell protoplasts directly after isolation (bar = 20 μm). (B) Effect of 0.25 mg/l bialaphos on guard cell callus growth: left, transformed cells without selection pressure; middle, nontransformed cells with selection pressure; right, pPG5 (16)–transformed cells with selection pressure. (C) Callus tissue derived from a single guard cell protoplast after transformation with pPG5 DNA stained for GUS activity (bar = 100 μm). (D) Nontransformed control plant (right) and pPG5 transgenic line (left), 14 days after spraying with twice the recommended dose of glufosinate. (From Ref. 16.)

periments are nonchimeric transgenics with no escapes. Somaclonal variation is also almost absent, although approximately 25% of regenerants become tetraploid. It has been estimated that for this protocol, after eliminating all plants unsuitable for breeding purposes (i.e., tetraploids, multiple copy plants, somaclonal variants), at least half remain for subsequent use (60). This system is now in commercial use and the first varieties produced are planned for commercialization in the very near future.

## III. SUGAR BEET AS A FACTORY: FRUCTAN PRODUCTION IN SUGAR BEET

### A. Sugar Beet as a "Plant as a Factory"

Sugar beet, which produces 10 tons of sucrose per hectare, is one of the most productive crops in temperate climatic zones. Because of its high carbohydrate yield, established cultivation, and advanced industrial processing, sugar beet is considered as an industrial crop. Therefore, sugar beet is a good candidate for the development of a plant able to produce, on a large scale, economically important products. However, the choice of the product should be made wisely. The specific characteristics of the sugar beet should be taken into account in order to assure the desired result. These characteristics include different aspects such as the agronomic characteristics of the culture, the physiological characteristics of the plant, the biochemical aspects of sucrose metabolism, and the industrial processing.

Among the products that can be considered are carotenoids, simple sugar derivatives, or trehalose (32,61), and fructans (62). It is also possible to modify pectins and cellulose in the taproot in order to increase the value of the molasses (32). Another way to use sugar beet as a plant factory would be to introduce genes under the control of chemically induced promotors, allowing the production of alcohol or other microbial fermentation products after harvest (32). The first example of the use of sugar beet as a factory for novel compounds is detailed in the discussion that follows.

### B. The Fructans

Fructans are polymers of fructose that are used as storage carbohydrates by 15% of all flowering plant species. Fructans are synthesized in unrelated species widespread throughout the plant kingdom (63). Among the angiosperms that store fructans, the Asteraceae are most noted, as well as the Cyperaceae, Poaceae, and Liliaceae. Three types of fructans can be distinguished on the basis of the binding of fructose units to one of the three primary hydroxyl groups of sucrose. The inulin-type fructans of the Asteraceae are extensions of sucrose through β-2,1-linked fructosyl units (64). The inulin neoseries of the Liliaceae is produced by transfer of a fructosyl unit from a fructan to the 6-position of the glucosyl moiety of sucrose, forming neokestose (65). The graminan- and phlein-type fructans of the Poaceae result from extension of sucrose through β-2,6-linked fructosyl units, with or without branched β-2,1-linked fructosyl units (66). The degree of polymerization (DP) (i.e., the number of hexose moieties building up the fructan molecule) of plant fructans can range from 10 to 250. In addition to their role as a plant carbohydrate store, fructans may also be involved in osmotic adjustment and in drought and cold tolerance (67). Unlike starch that is stored as an insoluble carbohydrate in the plastids, fructans are soluble and accumulate in the vacuole. Besides plants, many microorganisms are also capable of producing fructans. Bacterial fructans are usually of a very high molecular weight, consisting of over 100,000 fructose units bound through β-2,6 linkage type with occasional β-2,1 branching.

#### 1. Fructan Biosynthesis Among the Asteraceae

In the Asteraceae, inulin biosynthesis occurs via the combined action of two enzymes located in the vacuole (68). Sucrose sucrose fructosyltransferase (1-SST) catalyzes the transfer of a fructosyl unit from a sucrose to another sucrose molecule through a β-2,1 linkage (68). This results in the synthesis of 1-kestose, the shorter inulin-type fructan molecule, and in the release of one molecule of glucose. SST can also catalyze the synthesis of longer fructans, but with a lower affinity (68).

$$GF + GF \rightarrow GFF + G$$
$$GFF + GFF \rightarrow GFFF + GF$$

where G represents glucose, GF represents sucrose, GFF represents kestose, and GFFF represents nystose.

The synthesis of longer molecules, such as the fructans encountered in Jerusalem artichoke, which have a degree of polymerization of up to 50, is catalyzed by the action of the fructan fructan fructosyltransferase (1-FFT) (69). FFT catalyzes the transfer of fructosyl units from one fructan molecule to another fructan molecule.

$$GF(F)n + GF(F)m \rightarrow GF(F)n+1 + GF(F)m-1 \quad n \text{ and } m > 0$$

## 2. Industrial Application for Fructans

Fructans can be used either in the food industry or in the nonfood industry. In the former, two main applications can be described. First, after hydrolysis fructans can be used as a fructose source to produce high-fructose syrups. Second, fructans can be used as food additives. Fructans are currently used as low-calorie sweetener (70) and as bulking agent (71). These two applications present different requirements concerning the fructan molecular structure. A low-calorie sweetener is achieved by using short-chain fructans (DP3 to DP5), whereas a bulking effect is associated with a higher DP (>10). Low-molecular-weight fructans presently on the market (Meioligo, Japan; Nutraflora, in the United States; Actilight in Europe) are produced enzymatically from sucrose by using a fructosyl transferase isolated from a selected *Aspergillus* sp. strain (70). Longer fructan molecules are derived from inulin extracted from chicory or Jerusalem artichoke (71).

Because of their beneficial effect on human health they are also qualified as functional food ingredients (72,73). Roberfroid and Delzenne in 1998 reviewed the evidence that sustains this claim (73). Fructans are not digested in the upper part of the gastrointestinal tract and thus reach the colon, where they can be used by the endogeneous bacteria. Furthermore, it has been established that in the colon, fructan hydrolysis favored the development of bifidobacteria populations.

For nonfood applications, fructans need to be chemically modified. Modifications such as methacrylation (74), acetylation methylation (75), cyanoethylation (76), oxidation, and carboxymethylation (77) alter the structure of the fructose polymer and thus its properties. These new properties make these molecules useful for the pharmaceutical industry (hydrogel allowing intestinal targeting of drugs) and the detergent industry ($Ca^{2+}$ chelating properties).

Despite great potential application, fructan usage is limited, mainly because of the lack of an abundant and cheap production source (78). Reactor-based synthesis of short-chain fructans incurs high production costs, whereas long-chain fructans are extracted from crop plants with low agronomic value. If we could take advantage of a productive crop such as sugar beet for the production of fructans, an efficient system would be established, allowing widespread fructan utilization (62).

## C. Oligofructan Production in Sugar Beet

With respect to producing fructans in an established crop, sugar beet is obviously a serious candidate. From a biochemical point of view, sugar beet is especially well equipped for this purpose as the vacuolar concentration of sucrose is approximately 0.5 M. Another advantage of producing fructans in a crop unable to synthesize them naturally is the freedom to manipulate the structure of the newly accumulated compounds. This can be achieved by the judicious choice of genes and regulatory elements introduced into the transformed plant.

### 1. The Strategy

The goal was to transform sugar beet into a fructan production facility. Three prerequisites had to be met:

A reliable transformation protocol had to be developed.
The genes involved in fructan biosynthesis had to be cloned.
Prototype sugar beet transgenics expressing the fructan genes had to be produced.

First, a novel and efficient protocol for sugar beet transformation was developed (16). This protocol based on guard cell protoplast transformation, has been described previously.

## 2. Cloning of Plant Fructosyl Transferase Genes

While other groups were working with genes derived from microorganisms, we chose to focus on genes derived from plants. In plants, fructan biosynthesis occurs via the action of at least two enzymes. SST is responsible for short-chain fructan biosynthesis, long-chain fructan results from the catalytic activity of FFT. In microorganisms only one enzyme is responsible for the synthesis of very-high-DP fructans. Thus, the use of plant enzymes should allow fine control of the fructan molecular structure. Both SST and FFT enzymes from Jerusalem artichoke were purified (68) and the genes cloned (79). In control tests, SST and FFT enzymes encoding sequences were transformed into petunia in which the expression of these two genes resulted in fructan accumulation (79).

## 3. Transformation of Sugar Beet with the *1-sst* from Jerusalem Artichoke

Once the functions of the cloned genes were verified it was possible to envisage the transformation of sugar beet. In a first step, to assess the feasibility of the concept, we chose to transform sugar beet with *1-sst* alone (62).

   *a. Acquisition of Transformed Sugar Beet.* A construct harboring the *1-sst* sequence under the control of an enhanced 35S CaMV promotor and the *pat* gene, which confers resistance to the herbicide bialaphos, was prepared. This construct was used to transform sugar beet guard cell protoplasts (16), resulting in the acquisition of a number of independent transformed lines. The transgenic plants harboring the *1-sst* gene showed no visual differences in leaf shape or color under greenhouse conditions as compared with nontransformed sugar beet (Fig. 2).

**Figure 2** Young, nontransformed sugar beet plant (left) and 1-*sst*-transformed sugar beet plant (right) grown in pots under greenhouse conditions.

b. *Molecular Analysis of the* 1-sst *Harboring Sugar Beet.* Southern blot analysis revealed that all the lines have incorporated one or more copies of the *1-sst* transgene. As expected with a transformation protocol making use of direct transfer of naked DNA into plant cells, the copy number varied between the different lines. Of the different lines analyzed, four contained a single locus. In three lines we detected less than five loci, and in one line more than 10 sites of integration were found. Partial integration or recombination was also detected for certain sites of integration for three of the lines. Transcripts of the transgene were detected in six lines.

c. *Biochemical Analysis of the* 1-sst *Sugar Beet.* The content of soluble carbohydrates in the transformed taproot was analyzed by using thin layer chromatography (TLC) and high-pressure anion exchange chromatography (HPAEC) coupled to a pulse amperometric detector (HPAEC-PAD) (62). TLC analysis of taproot extracts indicated the presence of fructose-containing compounds migrating at the same rates as the standard compounds $GF_2$, $GF_3$, and $GF_4$ in four of the lines analyzed. One of these transformed lines was further analyzed by HPAEC-PAD (Fig. 3). This analysis confirmed the presence of low-molecular-weight fructans in the taproot of the transformed sugar beet. The chromatogram resulting from the analysis of the control taproot showed a major peak corresponding to sucrose and a relatively small peak for raffinose. However, the chromatographic trace obtained with the transgenic taproot extract showed dramatic changes. The peak corresponding to sucrose was radically reduced and three new peaks, identified as $GF_2$, $GF_3$, and $GF_4$, appeared. In leaves, the changes in soluble carbohydrate content were less spectacular, but small peaks of $GF_2$, $GF_3$, and $GF_4$ were detectable.

In order to quantify the amount of oligofructans accumulated in the *1-sst*-harboring sugar beet, HPAEC analysis was performed on extracts prepared from five plants from the transformed line previously analyzed and five control plants, all grown under greenhouse conditions (62). The total soluble carbohydrate content as expressed in hexose equivalents of the transformed taproot (460 ± 101 µmol/g fresh weight [FW]) was not significantly different from that of the control taproot (499 ± 155 µmol/g FW) (Table 1). The sucrose content in the control roots was 246.1 µmol/g FW and only 23.1 µmol/g FW in the transformed roots. However, in the *1-sst* sugar beet taproot, $GF_2$, $GF_3$, and $GF_4$ accumulated up to 73.8, 33.7, and 5.7 µmol/g FW, respectively.

d. *Superiority of Sugar Beet for the Production of Fructan.* The introduction of *1-sst* cDNA from Jerusalem artichoke into sugar beet resulted in a dramatic change of the main type of storage carbohydrate, and sucrose was nearly totally converted into low-molecular-weight fructans. In the plants analyzed 90% of the sucrose imported by the taproot was converted into low-molecular-weight fructans.

This yield can be compared with that in other published studies dealing with fructan production in plants transformed with fructosyl transferases of plant or bacterial origin (Table 2). Genes originating from microorganisms and from plants both proved to be useful for fructan production in transformed plants. However, the vacuole of the plant cell appeared to be the most appropriate compartment to achieve fructan accumulation. Targeting of the introduced gene to the cytoplasm (80–82) or to the apoplasm (81) could result in significant accumulation of fructan but was always associated with reduced organ growth, reduced organ dry weight, and necrosis. The reasons for these phenotypic impairments were not elucidated. The presence of the introduced enzyme itself (81) and the diversion of sucrose away from the starch biosynthesis pathway (80) have been proposed as putative explanations. As expected, potato (81,83,84), maize (82), and sugar beet (62), three crops recognized for their carbohydrate accumulating capacity, displayed higher fructan accumulation than tobacco (66,67,85). Fructan synthesis is accompanied by glucose release. It has been shown previously that high hexose levels could induce dramatic changes with respect to plant development and phenotype (86,87). Expression of the *1-sst* of *Cynara* sp. in potato is associated with a significant increase of the hexose (84). Likewise, starch-deficient potatoes expressing a bacterial levan sucrase contained a large amount of glucose stoichiometrically

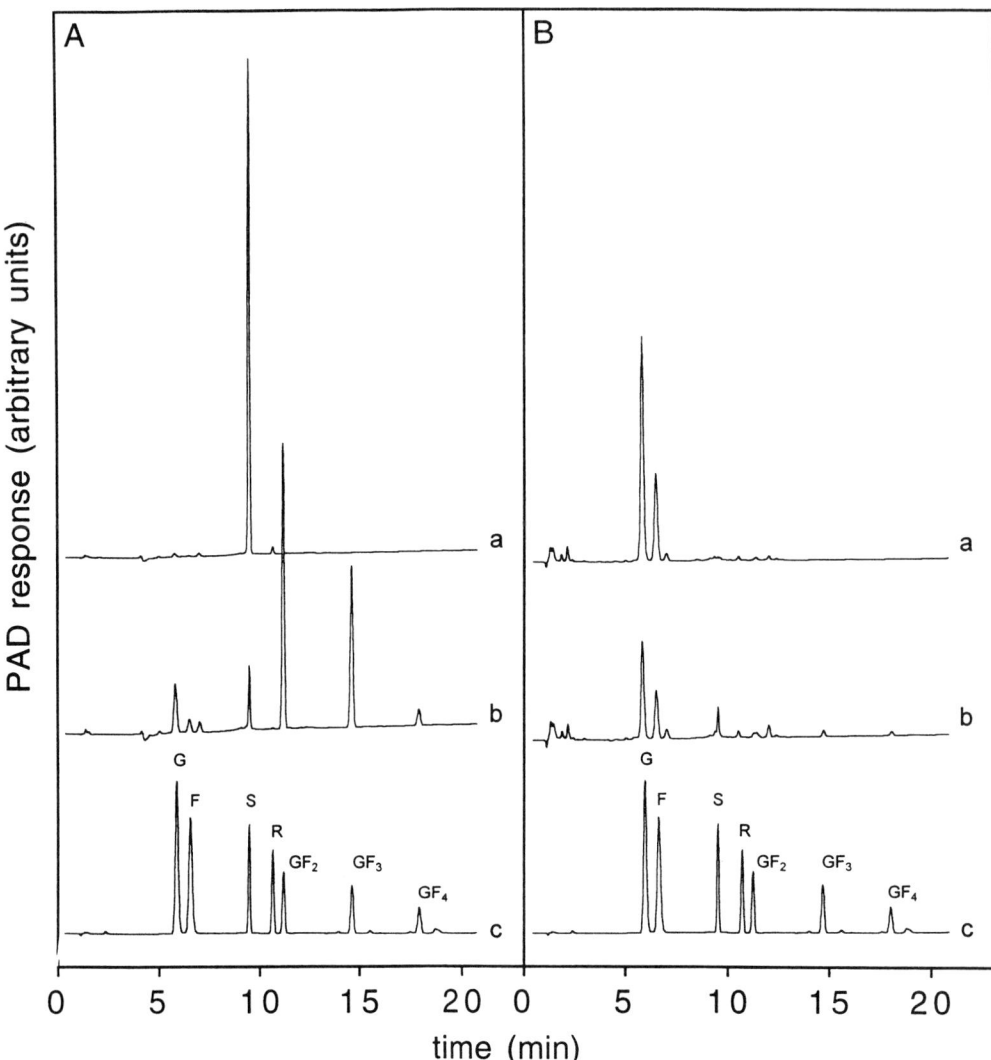

**Figure 3** High-pressure anion exchange chromatography/pulsed amperometric detection analysis of the soluble carbohydrates of (A) taproot and (B) leaf extract: a, nontransformed sugar beet; b, sugar beet transformed with the 1-*sst* gene; c, mixture of 20 mg/l glucose (G), fructose (F), sucrose (S), raffinose (R), 1-kestose ($GF_2$), nystose ($GF_3$), $1^f$-fructofuranosyl nystose ($GF_4$). (From Ref. 62.)

**Table 1** Soluble Carbohydrate Content of *1-sst* Sugar Beet Line and Control Sugar Beet[a]

| $\mu mol\ g^{-1}$ FW | | Glucose | Fructose | Sucrose | $GF_2$ | $GF_3$ | $GF_4$ |
|---|---|---|---|---|---|---|---|
| Line 1 | Leaf | 4.8 ± 2.2 | 2.7 ± 1.7 | 1.8 ± 0.5 | 0.1 ± 0.1 | 0.3 ± 0.1 | 0.5 ± 0.2 |
|  | Root | 25.1 ± 12.7 | 3.1 ± 1.2 | 23.1 ± 13.0 | 73.8 ± 16.8 | 33.7 ± 2.1 | 5.7 ± 0.5 |
| Control | Leaf | 10.6 ± 7.3 | 5.1 ± 3.7 | 0.6 ± 0.2 | n.d. | n.d. | n.d. |
|  | Root | 3.5 ± 1.3 | 0.6 ± 0.5 | 246.1 ± 76.3 | n.d. | n.d. | n.d. |

[a]Mean values ± SD from 5 plants. n.d., not detectable.
*Source*: Ref. 62.

**Table 2** Fructan Production in Genetically Engineered Crops[a]

| Crop transformed | Gene introduced | Gene origin | Gene targeting | Tissue analyzed | Sucrose | Fructan | Glucose | Remarks | Ref. |
|---|---|---|---|---|---|---|---|---|---|
| Tobacco | sacB | Bacillus subtilis | Vacuole | Leaves | 0.14 mg/g FW | 2.8 mg/g FW | 1.5 mg/g FW | — | (85) |
| Tobacco | sacB | B. subtilis | Vacuole | Leaves | 0.3–0.9 mg/g FW | 0.05–0.3 mg/g FW | 0.5–0.7 mg/g FW | — | (67) |
| Tobacco | 6-sft | Hordeum vulgare | Vacuole | Leaves | 5–10 mg/g DW | 0.05–0.3 mg/g DW | N.R. | — | (66) |
| Tobacco | 6-sft | H. vulgare | Vacuole | Roots | 50 mg/g DW | 0.5–3 mg/g DW | N.R. | — | (66) |
| Tobacco | sacB | B. amyloliquefaciens | Cytoplasm (inducible promotor) | Leaves | 7 mg/g FW | 4 mg/g FW | 6 mg/g FW | Necrosis appeared after induction | (80) |
| Potato | sacB | B. subtilis | Vacuole | Leaves (old) | | 350 mg/g DW | | reduced starch | (83) |
| Potato | sacB | B. subtilis | Vacuole | Microtubers | | 50 mg/g DW | | reduced starch | (80) |
| Potato | sacB | B. amyloliquefaciens | Cytoplasm | Microtubers | N.R. | 5–50 mg/g DW | N.R. | reduced starch and tuber DW | (80) |
| Potato (starch-deficient) | 1-sst | Cynara scolymus | Vacuole | Microtubers | 14 μmol/g FW | 19 μmol/g FW | N.R. | — | (84) |
| Potato (starch-deficient) | lsc | Erwinia amylovora | Vacuole | Tubers | N.R. | 70–120 mg/g DW | 82–201 mg/g DW | — | (81) |
| Potato (starch-deficient) | lsc | E. amylovora | Apoplasm | Tubers | N.R. | 190 mg/g DW | 50 mg/g DW | reduced tuber FW | |
| Potato | lsc | E. amylovora | Cytoplasm | Tubers | N.R. | 0 mg/g DW | N.R. | | |
| Maize | sacB | B. amyloliquefaciens | Vacuole | Seeds | N.R. | 10–80 mg/g DW | N.R. | — | (82) |
| Maize | sacB | B. amyloliquefaciens | Cytoplasm | Seeds | N.R. | 16–18 mg/g DW | N.R. | Severe reduction of seed DW | |
| Sugar beet | 1-sst | Helianthus tuberosus | Vacuole | Leaves | 1.8 μmol/g FW | 0.9 μmol/g FW | 4.8 μmol/g FW | — | (62) |
| Sugar beet | 1-sst | H. tuberosus | Vacuole | Roots | 23.1 μmol/g FW | 110 μmol/g FW | 25 μmol/g FW | — | |

[a] Tobacco, Potato, Maize and Sugar Beet do not natively accumulate fructan.
N.R., not reported.

related to levan production (81). The glucose released during fructan synthesis in these plants might have been metabolized only to a minor extent. In contrast, when sugar beet expressed *1-sst* of Jerusalem artichoke, the amount of glucose accumulated is equivalent to only 15% of the expected amount released during fructan synthesis (62).

In all the published data concerning transgenic plants expressing fructosyl transferase, sugar beet expressing the *1-sst* of Jerusalem artichoke has shown the highest efficiency of fructan accumulation and the highest rate of glucose remetabolization in the absence of any phenotypic abnormalities. The results obtained therefore suggest that sugar beet is one of the most promising candidates for industrial production of fructans in crop plants (62,78,88).

## IV. CONCLUSIONS

It is now possible to engineer sugar beet efficiently through the introduction of foreign DNA via guard cell protoplasts. This has allowed the concept of the plant as factory to be developed for this highly productive crop.

The results obtained with the introduction of the *1-sst* gene from Jerusalem artichoke into the genome of sugar beet clearly demonstrate the feasibility of fructan production in this normally sucrose storing plant. The high conversion rate of sucrose into fructans as observed in greenhouse-grown beets is a promising indication of future agricultural success. The second step will be to test the feasibility of synthesizing alternative fructans with a higher DP in sugar beet. The introduction of both *1-sst* and *1-fft* from *H. tuberosus* will be carried out in order to test this. A third step will be to modify the structure of the fructans synthesized in transformed sugar beet. The introduction of genes encoding for different combinations of fructosyl transferase activities should permit us to manipulate the structure of the fructan molecule produced. Using transformed sugar beet expressing the desired gene combinations at the appropriate phase of plant development will allow us to produce fructan molecules fulfilling the particular requirements of specific industrial applications.

## ACKNOWLEDGMENT

In 1997, R. Sévenier was supported by a Lavoisier grant from the Ministère des Affaires Etrangères, Paris, France.

## REFERENCES

1. RD Hall, C Pedersen, FA Krens. Regeneration of plants from protoplasts of *Beta vulgaris* (sugar beet). In: YPS Bajaj, ed. Biotechnology in Agriculture and Forestry. Vol. 29. Plant protoplasts and genetic engineering V. Berlin: Springer-Verlag, 1994, pp 16–37.
2. TSM de Bock. The genus Beta: Domestication, taxonomy and interspecific hybridization for plant breeding. Acta Hortic 182:335–343, 1986.
3. C Winner. History of the crop. In: DA Cooke, RK Scott, eds. The Sugar Beet Crop: Science into Practice. London: Chapman and Hall, 1993, pp 1–36.
4. MC Elliott, GD Weston. Biology and physiology of the sugar-beet plant. In: DA Cooke, RK Scott, eds. The Sugar Beet Crop: Science into Practice. London: Chapman and Hall, 1993, pp 37–65.
5. NO Bosemark. Genetics and breeding. In: DA Cooke, RK Scott, eds. The Sugar Beet Crop: Science into Practice. London: Chapman and Hall, 1993, pp 66–119.
6. JW Saunders, WP Doley, JC Theurer, MH Yu. Somaclonal variation in sugarbeet. In: YPS Bajaj, ed. Biotechnology in Agriculture and Forestry. Vol. 11. Somaclonal Variation in Crop Improvement. Berlin: Springer-Verlag, 1990, pp 465–490.

7. AI Atanassov. Sugar beet. In: DA Evans, WR Sharp, PV Ammirato, eds. Handbook of Plant Cell Culture, Vol. 4. Techniques et Applications. New York: Macmillan, 1986, pp 652–680.
8. S Kaffka, PG Lemaux. Sweeter times ahead for sugarbeet growers. Nature Biotechnol 14:1088, 1996.
9. EE Schweizer, MJ May. Weeds and weed control. In: DA Cooke, RK Scott, eds. The Sugar Beet Crop: Science into Practice. London: Chapman and Hall, 1993, pp 487–518.
10. I Brants, H Harms. Herbicide Tolerant Sugar Beet. Brussels: International Institute for Sugar Beet Research, 1998, pp 195–204.
11. M Mannerlof, S Tuvesson, P Steen, P Tenning. Transgenic sugar beet tolerant to glyphosate. Euphytica 94:83–91, 1997.
12. B Cailliez. The art of designing an adequate programme [L'art de composer le programme adequat]. Cultivar 417:36–38, 1997.
13. CH Bornman. Molecular Biology in Sugar Beet. Brussels: International Institute for Sugar Beet Research, 1990, pp 31–38.
14. K D'Halluin, M Bossut, E Bonne, B Mazur, J Leemans, J Botterman. Transformation of sugarbeet (*Beta vulgaris* L.) and evaluation of herbicide resistance in transgenic plants. Biotechnology 10:309–314, 1992.
15. P Tenning. Genetic engineering in sugarbeet [Genteknik i sockerbeta]. Sveriges Utsadesforenings Tidskrift 105:164–166, 1995.
16. RD Hall, T Riksen Bruinsma, GJ Weyens, IJ Rosquin, PN Denys, IJ Evans, JE Lathouwers, MP Lefebvre, JM Dunwell, A van Tunen, FA Krens. A high efficiency technique for the generation of transgenic sugar beets from stomatal guard cells. Nature Biotechnol 14:1133–1138, 1996.
17. A Loock, JR Stander, J Kraus, R Jansen. Performance of Transgenic Glufosinate Ammonium (Liberty) Tolerant Sugarbeet Hybrids. Brussels: International Institute for Sugar Beet Research, 1998, pp 339–344.
18. MJC Asher. Rhizomania. In: DA Cooke, RK Scott, eds. The Sugar Beet Crop: Science into Practice. London: Chapman and Hall, 1993, pp 311–346.
19. M Mannerlof, BL Lennerfors, P Tenning. Reduced titer of BNYVV in transgenic sugar beets expressing the BNYVV coat protein. Euphytica 90:293–299, 1996.
20. AJ Buchting. Experiences from release experiments with rhizomania resistant sugarbeet. Zuckerindustrie 120:138–142, 1995.
21. E Lauber, C Bleykasten Grosshans, H Guilley, S Bouzoubaa, K Richards, G Jonard. Strategies for Producing Pathogen-Derived Resistance to Rhizomania in Transgenic Sugarbeets. Brussels: International Institute for Sugar Beet Research, 1998, pp 205–220.
22. J Kallerhoff, P Perez, S Bouzoubaa, S Ben Tahar, J Perret. Beet necrotic yellow vein virus coat protein-mediated protection in sugarbeet (*Beta vulgaris* L.) protoplasts. Plant Cell Rep 9:224–228, 1990.
23. U Ehlers, U Commandeur, R Frank, J Landsmann, R Koenig, W Burgermeister. Cloning of the coat protein gene from beet necrotic yellow vein virus and its expression in sugar beet hairy roots. Theor Appl Genet 81:777–782, 1991.
24. W Mechelke, J Kraus. Field Results of Sugar Beets with Transgenic Rhizomania Resistance. Brussels: International Institute for Sugar Beet Research, 1998, pp 351–356.
25. DA Cooke. Pests. In: DA Cooke, RK Scott, eds. The Sugar Beet Crop: Science into Practice. London: Chapman and Hall, 1993.
26. C Jung, D Cai, M Kleine, DG Cai. Engineering nematode resistance in crop species. Trends Plant Sci 3:266–271, 1998.
27. D Cai, M Kleine, S Kifle, HJ Harloff, NN Sandal, KA Marcker, RM Klein Lankhorst, EMJ Salentijn, W Lange, WJ Stiekema, U Wyss, FMW Grundler, C Jung. Positional cloning of a gene for nematode resistance in sugar beet. Science 275:832–834, 1997.
28. H Paul, JEM van Deelen, B Henken, TSM de Bock, W Lange, FA Krens. Expression in vitro of resistance to *Heterodera schachtii* in hairy roots of an alien monotelosomic addition plant of *Beta vulgaris*, transformed by *Agrobacterium rhizogenes*. Euphytica 48:153–157, 1990.
29. R Heller, J Schondelmaier, G Steinrucken, C Jung. Genetic localization of four genes for nematode (*Heterodera schachtii* Schm.) resistance in sugar beet (*Beta vulgaris* L.). Theor Appl Genet 92:991–997, 1996.

30. C Jung. Cloning and Breeding Utility of the Gene Hs1 for Nematode Resistance from *Beta procumbens*. Brussels: International Institute for Sugar Beet Research, 1998, pp 221–227.
31. JB Nichols, CC Dalton, GA Todd, NW Broughton. Genetic engineering of sugar beet. Zuckerindustrie 117:797–800, 1992.
32. E Kuhn. Transgenic sugar beets as industrial plants [Transgene Zuckerruben als industriepflanzen]. Zuckerindustrie 123:28–34, 1998.
33. MC Elliott, DF Chen, MR Fowler, MJ Kirby, M Kubalakova, NW Scott, A Slater. Transgenesis–a scheme for improving sugar beet productivity. Russian J Plant Physiol 43:544–551, 1996.
34. TH Thomas, KMA Gartland, A Slater, MC Elliott. Opportunities for manipulation of growth and development. In: DA Cooke, RK Scott, eds. The Sugar Beet Crop: Science into Practice. London: Chapman and Hall, 1993, pp 521–550.
35. S Santoni, A Berville. Evidence for gene exchanges between sugar beet (*Beta vulgaris* L.) and wild beets: Consequences for transgenic sugar beets. Plant Mol Biol 20:578–580, 1992.
36. J Klein, J Altenbuchner, R Mattes. Nucleic acid and protein elimination during the sugar manufacturing process of conventional and transgenic sugar beets. J Biotechnol 60:145–153, 1998.
37. A Kapteijns. Risk assessment of genetically modified crops: Potential of four arable crops to hybridize with the wild flora. Euphytica 66:145–149, 1993.
38. BV Ford Lloyd. Transgenic risk is not too low to be tested. Nature 394:715, 1998.
39. H van Dijk, B Desplanque. Gene Exchange Between Wild and Cultivated Beet: Risks Associated with the Use of Transgenic Beets. Brussels: International Institute for Sugar Beet Research, 1998, pp 255–268.
40. P Boudry, M Morchen, P Saumitou Laprade, P Vernet, H Van Dijk. The origin and evolution of weed beets: Consequences for the breeding and release of herbicide-resistant transgenic sugar beets. Theor Appl Genet 87:471–478, 1993.
41. K Harding, PS Harris. Risk assessment of the release of genetically modified plants: A review. Agro Food Industry Hi Tech 8:8–13, 1997.
42. D Bartsch, M Pohl Orf. Ecological aspects of transgenic sugar beet: Transfer and expression of herbicide resistance in hybrids with wild beets. Euphytica 91:55–58, 1996.
43. D Bartsch, M Schmidt, M Pohl Orf, C Haag, I Schuphan. Competitiveness of transgenic sugar beet resistant to beet necrotic yellow vein virus and potential impact on wild beet populations. Mol Ecol 5:199–205, 1996.
44. RM Miller. Nontarget and ecological effects of transgenically altered disease resistance in crops—possible effects on the mycorrhizal symbiosis. Mol Ecol 2:327–335, 1993.
45. FA Krens, A Trifonova, LCP Keizer, RD Hall. The effect of exogenously-applied phytohormones on gene transfer efficiency in sugarbeet (*Beta vulgaris* L.). Plant Sci 116:97–106, 1996.
46. B Jacq, O Lesbore, RS Sangwan, BS Sangwan Norreel. Factors influencing T-DNA transfer in *Agrobacterium*-mediated transformation of sugarbeet. Plant Cell Rep 12:621–624, 1993.
47. M Joersbo, I Donaldson, J Kreiberg, SG Petersen, J Brunstedt, FT Okkels. Analysis of mannose selection used for transformation of sugar beet. Mol Breed 4:111–117, 1998.
48. K Lindsey, P Gallois. Transformation of sugarbeet (*Beta vulgaris*) by *Agrobacterium tumefaciens*. J Exp Bot 41:529–536, 1990.
49. FA Krens, C Zijlstra, W van der Molen, D Jamar, HJ Huizing. Transformation and regeneration in sugar beet (*Beta vulgaris* L.) induced by 'shooter' mutants of *Agrobacterium tumefaciens*. Euphytica (suppl):185–194, 1988.
50. CL Zhang, DF Chen, S Kubis, A Mc Cormac, M Kubalakova, J Zhang, MZ Bao, NW Scott, A Slatter, JS Heslop-Harrison, MC Elliot. Improved Procedures for Transformation of Sugar Beet. Brussels: International Institute for Sugar Beet Research, 1998, pp 381–392.
51. M Joersbo, J Brunstedt. Direct gene transfer to plant protoplasts by electroporation by alternating, rectangular and exponentially decaying pulses. Plant Cell Rep 8:701–705, 1990.
52. M Joersbo, J Brunstedt, F Floto. Quantitative relationship between parameters of electroporation. J Plant Physiol 137:169–174, 1990.
53. K Lindsey, MGK Jones. Stable transformation of sugarbeet protoplasts by electroporation. Plant Cell Rep 8:71–74, 1989.

54. M Joersbo, J Brunstedt. Direct gene transfer to plant protoplasts by mild sonication. Plant Cell Rep 9:207–210, 1990.
55. JC Ingersoll, TM Heutte, LD Owens. Effect of promoter-leader sequences on transient expression of reporter gene chimeras biolistically transferred into sugarbeet (*Beta vulgaris*) suspension cells. Plant Cell Rep 15:836–840, 1996.
56. H Daniell, R Wu. Foreign Gene Expression in Chloroplasts of Higher Plants Mediated by Tungsten Particle Bombardment: Methods in Enzymology: Recombinant DNA. Part H. San Diego: Academic Press, 1993, pp 536–556.
57. A Mahn, A Matzk, C Sautter, J Schiemann. Transient gene expression in shoot apical meristems of sugarbeet seedlings after particle bombardment. J Exp Bot 46:1625–1628, 1995.
58. RD Hall, HA Verhoeven, FA Krens. Computer-assisted identification of protoplasts responsible for rare division events reveals guard-cell totipotency. Plant Physiol 107:1379–1386, 1995.
59. RD Hall, T Riksen-Bruinsma, GJ Weyens, MP Lefèbvre, JM Dunwell, FA Krens. Stomatal guard cells are totipotent. Plant Physiol 112:889–892, 1996.
60. RD Hall. Biotechnological applications for stomatal guard cells. J Exp Bot 49:369–375, 1998.
61. G Kidd, J Devorak. Trehalose is a sweet target for agbiotech. Biotechnology 12:1328–1329, 1994.
62. R Sévenier, RD Hall, IM van der Meer, HJC Hakkert, AJ van Tunen, AJ Koops. High level fructan accumulation in a transgenic sugar beet. Nature Biotechnol 16:843–846, 1998.
63. G Hendry. Evolutionary origins and natural functions of fructans—a climatological, biogeographic and mechanistic appraisal. New Phytol 123:3–14, 1993.
64. H Meier, J Reid. Reserve polysaccharides other than starch in higher plants. In: F Loewus, W Tanner, eds. Encyclopedia of Plant Physiology. New Series. Berlin: Springer-Verlag, 1982, pp 418–471.
65. I Vijn, A van Dijken, M Luscher, A Bos, E Smeets, P Weisbeek, A Weimken, S Smeekens. Cloning of sucrose:sucrose 1-fructosyltransferase from onion and synthesis of structurally defined fructan molecules from sucrose. Plant Physiol 117:1507–1513, 1998.
66. N Sprenger, L Schellenbaum, K van Dun, T Boller, A Wiemken. Fructan synthesis in transgenic tobacco and chicory plants expressing barley sucrose:fructan 6-fructosyltransferase. FEBS Lett 400: 355–358, 1997.
67. EAH Pilon Smits, MJM Ebskamp, MJ Paul, MJW Jeuken, PJ Weisbeek, SCM Smeekens. Improved performance of transgenic fructan-accumulating tobacco under drought stress. Plant Physiol 107:125–130, 1995.
68. AJ Koops, HH Jonker. Purification and characterization of the enzymes of fructan biosynthesis in tubers of *Helianthus tuberosus* Colombia. II. Purification of sucrose:sucrose 1-fructosyltransferase and reconstitution of fructan synthesis in vitro with purified sucrose:sucrose 1-fructosyltransferase and fructan:fructan 1-fructosyltransferase. Plant Physiol 110:1167–1175, 1996.
69. AJ Koops, HH Jonker. Purification and characterization of the enzymes of fructan biosynthesis in tubers of *Helianthus tuberosus* 'Colombia'. I. Fructan:fructan fructosyl transferase. J Exp Bot 45:1623–1631, 1994.
70. JW Yun. Fructooligosaccharides: Occurence, preparation, and application. Enzyme Microb Technol 19:107–117, 1996.
71. P Coussement, A Franck. New food applications for inulin. Agro Food Industry Hi Tech 9:26–28, 1998.
72. GR Gibson, MB Roberfroid. Dietary modulation of the human colonic microbiota: Introducing the concept of prebiotics. J Nutr 125:1401–1412, 1995.
73. MB Roberfroid, NM Delzenne. Dietary fructans. Annu Rev Nutr 18:117–143, 1998.
74. L Vervoort, G van den Mooter, P Augustijns, I Vinckier, P Moldenaers, R Kinget. Development of inulin hydrogels as carriers for colonic drug targeting. Leuven: 7th Seminar on Inulin, 1998, pp 138–146.
75. F Damian, G van den Mooter, C Samyn, P Augustijns, R Kinget. In vitro degradation of acetyl and methyl inulins: implication for colon drug targeting. Leuven: 7th Seminar on Inulin, 1998, pp 147–151.
76. DL Verraest, JA Peters, HC Kuzee, HWC Raaijmakers, H van Bekkum. Selective complexation of metal ions by inulin derivatives: Modification of inulin with amidoxime groups and coordination with copper(II) ions. Leuven: 7th Seminar on Inulin, 1998, pp 52–56.

77. HWC Raaijmakers, HC Kuzee, MEB Bolkenbaas. New inulin-based polycarboxylates. Leuven: 7th Seminar on Inulin, 1998, pp 64–67.
78. S Smeekens. A convert to fructans in sugar beet. Nature Biotechnol 16:822–823, 1998.
79. IM van der Meer, AJ Koops, JC Hakkert, AJ van Tunen. Cloning of the fructan biosynthesis pathway of Jerusalem artichoke. Plant J 15:489–500, 1998.
80. PG Caimi, LM McCole, TM Klein, HP Hershey, CJ Pollock, JF Farrar. Cytosolic expression of the *Bacillus amyloliquefaciens* SacB protein inhibits tissue development in transgenic tobacco and potato. Third International Fructan Symposium, Logan, UT, July 21–24, 1996, 136:19–28, 1997.
81. M Rober, K Geider, B Muller Rober, L Willmitzer. Synthesis of fructans in tubers of transgenic starch-deficient potato plants does not result in an increased allocation of carbohydrates. Planta 199:528–536, 1996.
82. PG Caimi, LM McCole, TM Klein, PS Kerr. Fructan accumulation and sucrose metabolism in transgenic maize endosperm expressing a *Bacillus amyloliquefaciens* SacB gene. Plant Physiol 110:355–363, 1996.
83. IM van der Meer, MJM Ebskamp, RGF Visser, PJ Weisbeek, SCM Smeekens. Fructan as a new carbohydrate sink in transgenic potato plants. Plant Cell 6:561–570, 1994.
84. EM Hellwege, D Gritscher, L Willmitzer, AG Heyer. Transgenic potato tubers accumulate high levels of 1-kestose and nystose: Functional identification of a sucrose sucrose 1-fructosyltransferase of artichoke (*Cynara scolymus*) blossom discs. Plant J 12:1057–1065, 1997.
85. MJM Ebskamp, IM van der Meer, BA Spronk, PJ Weisbeek, SCM Smeekens. Accumulation of fructose polymers in transgenic tobacco. Biotechnology 12:272–275, 1994.
86. A Von Schaewen, M Stitt, R Schmidt, U Sonnewald, L Willmitzer. Expression of a yeast-derived invertase in the cell wall of tobacco and *Arabidopsis* plants leads to accumulation of carbohydrate and inhibition of photosynthesis and strongly influences growth and phenotype of transgenic tobacco plants. EMBO J 9:3033–3044, 1990.
87. U Sonnewald, M Brauer, A Von Schaewen, M Stitt, L Willmitzer. Transgenic tobacco plants expressing yeast-derived invertase in either the cytosol, vacuole or apoplast: A powerful tool for studying sucrose metabolism and sink/source interactions. Plant J 1:95–106, 1991.
88. AG Heyer, JR Lloyd, J Kossmann. Production of modified polymeric carbohydrates. Curr Opin Biotechnol 10:169–174, 1999.

# 35
# Transgenic Carrots with Enhanced Tolerance to Fungal Pathogens

**Zamir K. Punja**
Simon Fraser University, Burnaby, British Columbia, Canada

|      |                                                                                 |     |
|------|---------------------------------------------------------------------------------|-----|
| I.   | INTRODUCTION                                                                    | 503 |
| II.  | GENETIC IMPROVEMENTS THROUGH BREEDING                                           | 504 |
| III. | TISSUE CULTURE OF CARROT                                                        | 504 |
| IV.  | GENETIC ENGINEERING OF CARROT                                                   | 504 |
|      | A. Materials and Methods                                                        | 504 |
|      | B. Results                                                                      | 507 |
| V.   | OTHER REPORTS OF *AGROBACTERIUM* SPECIES–MEDIATED TRANSFORMATION                | 511 |
| VI.  | STRATEGIES TO ENHANCE TOLERANCE OF CARROTS TO FUNGAL DISEASES                   | 515 |
| VII. | EVALUATING CARROT LINES EXPRESSING CHITINASES FOR ENHANCED TOLERANCE TO FUNGAL PATHOGENS | 515 |
| VIII.| OTHER REPORTS OF CARROT TRANSFORMATION FOR DISEASE TOLERANCE                    | 517 |
| IX.  | CONCLUSIONS AND FUTURE PROSPECTS                                                | 517 |
|      | REFERENCES                                                                      | 519 |

## I. INTRODUCTION

Carrot (*Daucus carota* L. subsp. *sativa*), a member of the family Umbelliferae, is grown worldwide for its edible taproot. Carrot roots are marketed as fresh whole or baby carrots, for use in processing (in frozen or canned foods, soups, and juice), and as frozen products. Nutritionally, carrots are known to be high in β-carotene (provitamin A) as well as in vitamin $B_1$ and C, and are a good source of dietary fiber. The average per capita consumption of fresh carrots in the United States in 1997 was 5.7 kg (1). The carrot industry worldwide is located throughout Europe (42% of total acreage), Asia (37.5%), North America (14%), and South America (6.5%) (2), and the total worldwide acreage is around 1.4 million acres. Sales of carrot seed provide additional revenues, estimated at $92 million worldwide (2).

Carrot is a biennial plant, which flowers after exposure to a period of cold temperatures, generally during the second year of growth. Premature flowering (bolting) in the first year of growth can occur in some cultivars grown in certain regions. High temperatures (30–35°C) during growth can result in lighter-colored roots with reduced levels of β-carotene, whereas low temperatures (5–10°C) early in the season can result in poor seedling establishment. The optimal temperature range for growth and root development is 15–20°C (3). In addition to environmental stresses, biotic agents including fungi, bacteria, viruses, phytoplasmas, and nematodes can cause significant losses in yield and quality of carrots worldwide (3–5).

## II. GENETIC IMPROVEMENTS THROUGH BREEDING

Development of carrot cultivars through breeding has been a major effort since the 1980s and has resulted in significant improvements to yield and quality (3,5,6). Root shape, length, and color, smooth skin, flavor, early maturity, and resistance to various diseases are high priorities. Most of these genetic improvements in carrot require long-term effort, and many are multigenic traits and thus are not amenable to genetic engineering. In contrast, disease resistance and a range of other traits are monogenic. Much of the carrot acreage currently comprises of $F_1$ hybrids, produced by using a system of cytoplasmic male sterility that makes crossing of inbred lines achievable and economical (5,7,8). Inbred lines with proprietary genes introduced by genetic engineering techniques can therefore be marketed under licensing agreements.

## III. TISSUE CULTURE OF CARROT

Carrot is a model system for use in tissue culture, and extensive research has been conducted in a number of areas—somatic embryogenesis, bioreactor scale-up of suspension cultures, protoplast culture and fusion, and somaclonal variation (3,9,10). For somatic embryogenesis, the most commonly used media are Murashige and Skoog (MS) (11), Uchimya and Murashige (12), and B5 media (13), containing 2,4-dichlorophenoxyacetic acid (2,4-D) and kinetin or benzyladenine (3). Transfer of proembryonic masses to liquid medium can result in highly embryogenic synchronous cultures (3). In the absence of auxin, somatic embryos continue to develop into plantlets. Protoplasts can be isolated from these suspension cultures or somatic embryos, or from mature plant tissues. A number of variants have been observed in long-term tissue cultures of carrot and also after selection, including altered morphological characteristics, resistance to specific amino acid analogues and antibiotics (3,14), and resistance to the herbicide glyphosate (15). Protoplast fusion studies using carrot have resulted in somatic hybrids with altered characteristics (3,14). Neither artificial selection in tissue culture nor somatic hybridization studies have yielded significant improvements to the presently grown cultivars of carrot.

## IV. GENETIC ENGINEERING OF CARROT

Summarized in the following sections are the procedures used in our laboratory to derive transgenic carrot lines. The variables that can influence transformation efficiency are described here and elsewhere (16,17).

### A. Materials and Methods

#### 1. Explant Source and Tissue Culture Conditions

Carrot seeds are frequently heavily contaminated with fungal organisms and may require vigorous disinfection prior to use. The use of aseptically grown plants as the explant source is advantageous. Seeds of cultivars 'Golden State,' 'Danvers Half Long,' 'Nantes Long,' 'Scarlet Nantes,' and 'Nanco' were disinfected by washing in detergent (Liqui-Nax, (Alconox, Inc. NY), followed by dipping in 70% EtOH for 30 s, then in a 5% solution of commercial bleach (Javex, 6.25% sodium hypochlorite), followed by a 1% solution of benzalkonium chloride (in 10% EtOH) (Calbiochem Co., La Jolla, CA), 2 min in each, and finally rinsing three times in sterile distilled water (1 min each). The seeds were transferred to water agar plates containing 10 g l$^{-1}$ glucose, 100 mg l$^{-1}$ benomyl (as Benlate, 50% WP), 100 mg l$^{-1}$ dichloran (as Botran, 75% WP), and 100 mg l$^{-1}$ ampicillin and incubated at 23–28°C. After germination (about 10–15 days), seedlings were transferred to Magenta boxes (Magenta Corp., Chicago, IL) containing approximately 30 ml of full-strength growth regulator–free MS medium (11) with myo-inositol (100 mg l$^{-1}$), thiamine HCl (0.8 mg l$^{-1}$), 30 g l$^{-1}$ sucrose, and 10 g l$^{-1}$ tissue culture agar (Sigma). The pH of the medium was adjusted to 5.8 prior to autoclaving at 15 psi for 15 min. Sterile ampicillin was added at a concentration of 100 mg l$^{-1}$ to prevent bacterial contamination. The seedlings were grown at 23–28°C under cool-white fluorescent lamps (intensity of 450 µmol m$^{-2}$ s$^{-1}$) with a 16-h light and 8-h dark cycle.

The medium used for callus induction and somatic embryo formation was full-strength MS, containing thiamine, sucrose, agar, and ampicillin at the concentrations specified. The influence of 2,4-D concentrations was determined at 0.5, 2.0, or 4.5 µM. Epicotyl segments, approximately 1 cm long, from 2-week-old aseptic seedlings (cv. 'Golden State' and 'Nanco') were placed in petri dishes (100 × 15 mm), 8–10 explants per dish, with four to six replicate dishes per treatment. All dishes were sealed with Parafilm and incubated under the same conditions as those for seedling growth. The effect of cultivar on frequency of somatic embryogenesis was evaluated by using epicotyls from seedlings of cultivars 'Nanco,' 'Golden State,' and 'Danvers Half Long.' Approximately 8–10 explants were placed on MS medium containing 4.5 µM 2,4-D, with five to six replicate dishes per cultivar.

## 2. Bacterial Strains and Plasmids

Disarmed *Agrobacterium tumefaciens* strains MOG 101 (octopine type) and EHA 105 (leucinopine, supervirulent type) (courtesy of MOGEN Int. nv, The Netherlands) were used. Each strain harbored one of three binary plasmids. Two plasmids, pMOG196 (14.0 kb) and pMOG198 (14.2 kb) (MOGEN Int. nv), contained between the T-DNA borders a petunia acidic chitinase gene (*Pach 1*) or tobacco basic chitinase gene (*Tbch 1*), respectively, driven by the CaMV 35S promoter (18). A third plasmid, pGA-492-CHN (14.8 kb), was constructed by excising the 2.5-kb *EcoR l-Cla l* fragment of pk35CHN641 (19), consisting of the bean endochitinase (CH5B) coding region fused to the CaMV 35S promoter, ligating into a *EcoR l-Cla l* linearized plasmid pGA492 (20), and cloning into *Escherichia coli* DH5α. Plasmid pGA492-CHN was mobilized from *E. coli* into *A. tumefaciens* strains MOG101 and EHA105 via triparental matings. All three plasmids also contained a transferable selectable marker for kanamycin resistance, the neomycin phosphotransferase gene (*npt II*) driven by the nopaline synthase (*nos*) promoter. A single colony of each of the *Agrobacterium* sp. strains was inoculated into 50 ml of Luria broth (21) medium (with kanamycin at 100 mg l$^{-1}$) and grown overnight at 29°C on a rotary shaker at 200 rpm. The cultures were then diluted 1:50 in minimal medium (MM) (22) (with kanamycin at 100 mg l$^{-1}$) and grown overnight under the same conditions. The bacterial suspensions were diluted in MM (without kanamycin) to a density of $A_{600}$ = 0.15, placed at 29°C at 200 rpm, and allowed to grow to a density of $A_{600}$ = 0.25–0.30. Aliquots (15 ml) of each of the cultures were centrifuged at 3700 × g for 5 min at ambient room temperature. The supernatant was discarded and the pellets were resuspended in MS medium (pH 5.3 when using EHA105 and pH 5.8 when using MOG101) containing 100 µM acetosyringone and diluted to a final density of $A_{600} = 0.05$ (approximately 1 × $10^8$ cells ml$^{-1}$) (22).

## 3. Transformation Procedure

Epicotyl explants (1 cm long) from 2- to 5-week-old aseptic carrot seedlings were precultured for 2 days on MS medium supplemented with 4.5 µM 2,4-D prior to the transformation experiments. After the preculture period, the explants were immersed in 10–25 ml of the bacterial suspension for 4 min, rinsed in liquid MS medium, dried briefly on sterile filter paper, and placed back on preculture medium. Explants were cocultivated in the dark for 2–4 days at 26°C. The infected explants were subsequently rinsed in sterile water, blotted dry, and placed on selective medium (MS medium containing 4.5 µM 2,4-D, 400 mg l$^{-1}$ carbenicillin, and 25 mg l$^{-1}$ kanamycin). The tissue culture plates (60 × 15 mm), each containing 8–10 explants, were incubated in the dark at ambient temperatures of 23–28°C. Both positive and negative controls were included in the experiments. The positive control consisted of noncocultivated explants cultured on antibiotic-free regeneration medium (MS containing 4.5 µM 2,4-D), and the negative controls consisted of noncocultivated explants cultured on selective medium. After 4 weeks, all explants were transferred to fresh selective medium with a higher concentration of kanamycin (100 mg l$^{-1}$) and placed under cool-white fluorescent lamps with a 16-h light and 8-h dark cycle at ambient temperatures. Calli were subcultured to fresh selective medium every 4–6 weeks, over a 4 to 6-month period. Upon appearance of somatic embryos, these calli were either placed into 50 ml of liquid MS medium containing 0.5 µM 2,4-D and 50 mg l$^{-1}$ of kanamycin or subcultured directly onto MS medium lacking plant growth regulators or kanamycin (MSO). Suspension cultures (in 50 ml of medium, incubated at 150 rpm) were subcultured to fresh medium every 2 weeks. Once the suspension cultures were established, approximately 0.5 ml of each suspension was spotted onto MSO medium and allowed to grow for 4–6 weeks. When embryos were produced, they were transferred to MS medium containing 50 mg l$^{-1}$ kanamycin and no 2,4-D. After 4 weeks, rooted plantlets were transferred into sterile potting medium and placed in a growth chamber maintained at 26°C, 85% relative humidity, and a 16-h photoperiod.

## 4. Data Analysis

The data on percentage of somatic embryo production after different treatments from the replications and repetitions were pooled and analyzed for significant differences by using chi-square analysis ($P \leq 0.05$). In addition, the standard deviation was calculated and used to differentiate means from different treatments. In all experiments that evaluated each of the parameters influencing *Agrobacterium* sp. transformation, the minimal number of explants included ranged from 33 to 404. All experiments were repeated at least once; some were repeated four to five times over a 2-year period.

## 5. Confirmation of Transformation

*a. Polymerase Chain Reaction Analysis.* Total nucleic acids were isolated from leaf tissue (1–2 g fresh weight) of transgenic and nontransgenic plants according to the protocol described by Kanazawa and Tsutsumi (23) and were subsequently treated with 50 µg mlms1 of ribonuclease (RNAse). The DNA was then extracted with phenol:chloroform:isoamyl alcohol (25:24:1) and ethanol-precipitated. Two specific sequences of the *npt* II coding region were used for polymerase chain reaction (PCR) amplification of this gene in the genomic DNA. Oligomer A, a 17mer with 5'-3' sequence GATGGATTGCACGCAGG, was located 15 bp upstream from the start codon, and oligomer B, a 17mer with 5'-3' (bottom strand) sequence GAAGGCGATA-

GAAGGCG, shared identity with the 3' region of the NPT II gene, 17 bp 5' of the stop codon. These primers amplified a region about 800 bp in size. Each PCR reaction (25 µl overlaid by 50 µl of mineral oil) consisted of 1× Taq buffer (MgCl$_2$-free), 2.5 mM MgCl$_2$, 200 µM dNTPs, 0.5 µM of each oligonucleotide primer, and approximately 20 ng of template DNA. Each PCR reaction was incubated at 92°C for 5–10 min and then 1.25 units of Taq polymerase (Promega) was added and the mixture quickly placed on ice. The temperature cycling for the PCR was as follows: 29 cycles at 94°C for 1 min, 54°C for 2 min, and 72°C for 3 min. The 30th cycle was the same except that DNA synthesis at 72°C was carried out for 10 min. The PCR products were analyzed electrophoretically on 2.0% agarose gels. The oligomers were also tested by amplifying the characteristic 800-bp region of the *npt II* gene by using 10 ng of total *Agrobacterium* sp. DNA containing pMOG 196 as positive control template.

 *b. Southern Hybridization Analysis.* Approximately 5–12 µg of total genomic DNA was digested with *Hind III* (BRL) according to manufacturer's specifications and electrophoresed on a 0.7% agarose gel. DNA fragments were blotted to positively charged nylon membranes (Boehringer Mannheim) by capillary action with 10× SSC and fixed by ultraviolet (UV) crosslinking. Filters were hybridized against a Digoxigenin-UTP PCR-labeled probe, which was made by amplifying the 800-bp fragment containing the coding region of the *npt II* gene by using the PCR temperature cyclings and conditions specified. However, for this reaction, the dTTP concentration was changed to 17 µM and Dig-UTP (Boehringer Mannheim) was added to a final concentration of 8.5 µM. Hybridizations were conducted by using about 15 ng of the labeled probe per milliliter of hybridization solution containing 2% blocking buffer (Boehringer Mannheim) and 50% formamide (redistilled nucleic acid grade) and incubated at 40–42°C for at least 15 h. Washes and chemiluminescence detection of hybridized filters were done according to Boehringer Mannheim's instructions. The blots were exposed to radiographic films (Kodak OMAT-K) for 1–24 h.

 *c. ImmunoBlot Analysis.* Samples of transgenic carrot tissues (callus or leaves) were frozen in liquid nitrogen immediately after collection and finely ground into a powder by using a mortar and pestle. All extractions were performed at 4°C. The powder was extracted with 0.1 M sodium citrate buffer, pH 5.0, and filtered through four layers of cloth (Miracloth) and centrifuged at 20,000 g for 30 min. The supernatant was put onto an ultrafiltration unit (Amicon YM10 filter) for protein concentration adjustment. Protein samples applied to pretreated nitrocellulose membranes and immunostained by using the Bio-Rad Immun-Blot GAR-AP assay kit (Cat. No. 170–6509). The antibody raised against tobacco basic chitinase (provided by Dr. B. Fritig, Institut de Biologie Moleculaire des Plantes, France) and the bean chitinase (provided by R. Broglie, DuPont, USA) were used at 1:2000 dilutions.

## B. Results

### 1. Tissue Culture

When epicotyl explants were placed on MS medium containing 0.50, 2.0, or 4.5 µM 2,4-D, calli developed at different rates, but the overall frequency of somatic embryo formation was only affected in cv. 'Nanco' and was highest at 4.5 µM 2,4-D. In cv. 'Golden State,' 2,4-D concentrations had no effect (Table 1). In all treatments, after 4 weeks of incubation, calli were homogeneous and friable and frequently had somatic embryos and small shoots. The effect of cultivar on frequency of somatic embryogenesis was also determined by comparing the final percentage of embryogenic calli after 2–4 months of incubation on 4.5 µM 2,4-D. The cultivar 'Nanco' had the highest frequency (81.4%), followed by cv. 'Golden State' (77.2%) and 'Danvers Half Long' (55.3%). Therefore, carrot cultivar may be a variable in transformation experiments as a result of differing responses in tissue culture.

**Table 1** Effect of 2,4-D Concentration on Somatic Embryo Formation in Carrot Cultivars 'Golden State' and 'Nanco'

| Carrot cultivar[a] | Embryogenic calli (%) following exposure to 2,4-D[b] | | |
|---|---|---|---|
| | 0.5 µM | 2.0 µM | 4.5 µM |
| Golden State | 13.2 (38)[c] | 10.3 (39) | 14.7 (34) |
| Nanco | 7.0 (57) | 19.2 (52) | 25.4 (59) |

[a]Epicotyl segments (1-cm-long) from 2-week-old aseptic seedlings were used.
[b]Percentage of calli out of the total plated that developed somatic embryos was rated after 2–4 months in culture.
[c]Numbers in parentheses refer to total number of explants plated.
*Source*: Ref. 16.

## 2. Results from Transformation Experiments

After cocultivation, epicotyl explants were placed on selective medium with 25 mg $l^{-1}$ of kanamycin for 4 weeks. Subsequently, healthy calli were subcultured to fresh selective medium containing 100 mg $l^{-1}$ kanamycin. This two-step procedure was selected to minimize potential inhibition of regeneration by kanamycin (24,25). Initially, cocultivated and control explants expanded to about the same size during selection on 25 mg $l^{-1}$ kanamycin. Once on 100 mg $l^{-1}$, only the cocultivated explants and positive controls developed additional callus, whereas the negative control did not increase in size and was bleached. Transformation efficiency was calculated as the total number of cocultivated explants that produced embryogenic calli on selective medium relative to the total number of cocultivated explants originally plated on medium containing 100 mg $l^{-1}$ of kanamycin.

*a. Effect of Explant Age.* Epicotyl explants from 2-, 3-, 4- and 5-week-old aseptic seedlings of the carrot cultivar 'Golden State' were cocultivated for 2–3 days with the supervirulent *Agrobacterium* sp. strain EHA 105 (containing either pMOG196 or pMOG198). The transformation frequency ranged from 5.4% to 8.6% for 2- to 5-week-old seedlings, respectively (Fig. 1). These different explant ages were not statistically different ($P = 0.91$, chi-square test); therefore, explant age was not considered to be a variable that could affect transformation frequency.

*b. Effect of* Agrobacterium *Species Plasmid.* To determine whether transformation efficiency was influenced by the *Agrobacterium* sp. plasmid, explants of cultivar 'Golden State' were cocultivated for 2–3 days with *Agrobacterium* sp. strain EHA 105 with either pMOG196 or pMOG198. In addition, explants of the cultivar 'Danvers Half Long' were cocultivated for 3 days with *Agrobacterium* sp. strain MOG 101 harboring one of plasmids pMOG196, pMOG198, or pGA492-CHN. With cv. 'Golden State,' the transformation efficiency was 5.6% when pMOG196 was used, and 6.4% with pMOG198. With cv. 'Danvers Half Long,' transformation efficiency was 7.3% (pMOG196), 2.4% (pMOG198), and 1.0% (pGA492-CHN). Chi-square analysis ($P \leq 0.05$) indicated that these transformation efficiencies were not significantly different. Therefore, the plasmid used was not considered to be a variable in transformation experiments.

*c. Effect of Cocultivation Period.* Cocultivation periods of 2, 3, and 4 days were evaluated. Epicotyl explants of cultivars 'Golden State' and 'Danvers Half Long' were cocultivated for these periods after infection by *Agrobacterium* sp. strain EHA 105 containing either pMOG196 or pMOG198. The results showed that 2 or 3 days of cocultivation yielded similar transformation frequencies, but these were significantly higher than for 4-day cocultivation (Table 2). Therefore, shorter periods of cocultivation are recommended for transformation experiments.

*d. Effect of Carrot Cultivar.* Epicotyl explants of four carrot cultivars were cocultivated for 2–3 days with *Agrobacterium* sp. strain MOG101 (containing pMOG196, pMOG198, or

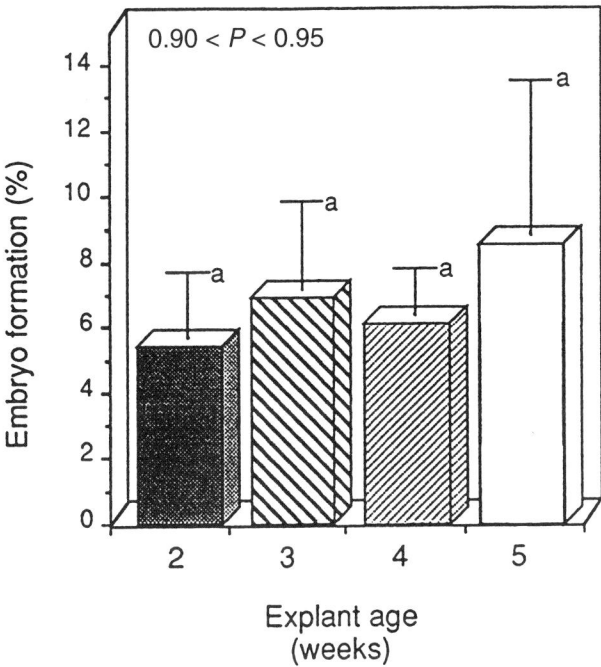

**Figure 1** Influence of explant age on transformation efficiency of carrot cv. 'Golden State' cocultivated with *Agrobacterium* sp. strain EHA 105 for 2–3 days. (From Ref. 16.)

**Table 2** Effect of Cocultivation Period of Carrot Epicotyl Explants with *Agrobacterium tumefaciens* Strain EHA 105 on Transformation Frequency

| Carrot cultivar[a] | Embryogenic calli (%) following a cocultivation period of[b] | | |
|---|---|---|---|
|  | 2 days | 3 days | 4 days |
| Golden State | 6.5 | 6.0 | 0.0 |
| Danvers Half Long | —[c] | 1.8 | 0.0 |

[a]Epicotyl segments (1-cm-long) were taken from aseptic seedlings.
[b]Percentage of calli out of the total plated following cocultivation that developed somatic embryos was rated after 2–4 months in culture.
[c]Not determined.
*Source*: Ref. 16.

pGA492-CHN). 'Scarlet Nantes' had the highest transformation rate (6.2%), followed by 'Danvers Half Long' (5.1%), 'Nanco' (0.5%), and 'Nantes Long' (0.0%). All cultivars were found to be significantly different ($P = 0.05$) (Fig. 2A). In a second experiment, cultivars 'Golden State,' 'Nanco,' and 'Danvers Half Long' were cocultivated with *Agrobacterium* sp. strain EHA 105 (containing pMOG196, pMOG198, or pGA492-CHN). 'Nanco' had the highest transformation efficiency (12.1%), followed by 'Golden State' (6.1%) and 'Danvers Half Long' (1.8%) (Fig. 2B). These results indicate that significant cultivar × strain interactions were present.

 e. *Effect of* Agrobacterium *Species Strain.* Explants of the cultivars 'Danvers Half Long' and 'Nanco' were cocultivated for 3 days with *Agrobacterium* sp. strain MOG 101 or EHA 105. The transformation efficiency was significantly affected by the *Agrobacterium* sp. strain (Table 3), and a marked strain-cultivar interaction was observed.

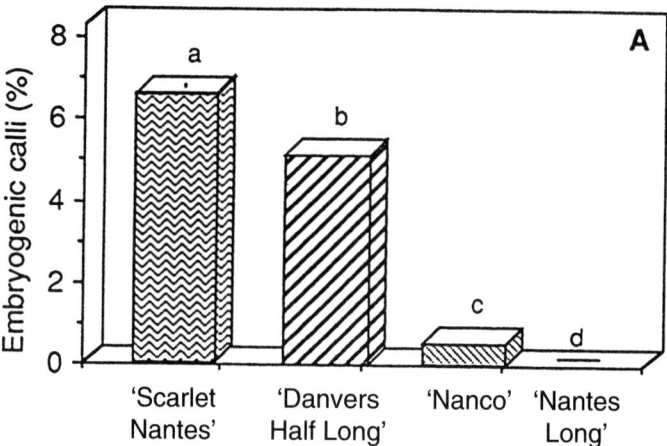

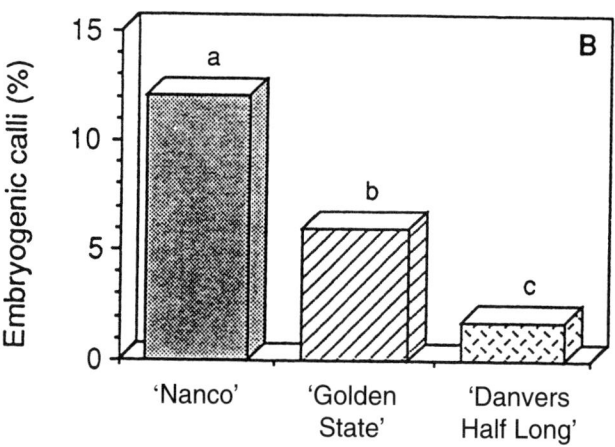

**Figure 2** Influence of carrot cultivar on transformation efficiency after cocultivation with *Agrobacterium* sp. strain MOG 101 (A) or strain EHA 105 (B). Means followed by different letters are significantly different ($P = 0.05$, chi-square test). (From Ref. 16.)

### 3. Regeneration of Transformed Plants

After production of somatic embryos (Fig. 3A), calli were either placed in liquid MS medium (0.5 µM 2,4-D, 50 mg l$^{-1}$ kanamycin) or subcultured onto MS medium without growth regulators or kanamycin (MSO). Once suspension cultures became established, approximately 0.5 ml of each suspension was plated onto MSO. Within 4–6 weeks, these somatic embryos developed into shoots (Fig. 3B), which were transferred to MS medium containing 50 mg l$^{-1}$ kanamycin. Plantlets that rooted and developed leaves (Fig. 3C) were transferred to potting mix and placed in a growth chamber, where they grew into mature plants (Fig. 3D).

### 4. Confirmation of Transformation

*a. Polymerase Chain Reaction Analysis.* Polymerase chain reaction (PCR) amplification of the *npt II* coding region was conducted with cultivars 'Nanco' and 'Golden State' transformed

**Table 3** Effect of *Agrobacterium* Strain on Transformation Frequency of Carrot

| Carrot cultivar[a] | Embryogenic calli (%) following exposure to *Agrobacterium* strain[b] | |
|---|---|---|
| | MOG 101 | EHA 105 |
| Danvers Half Long | 5.1 (39)[c] | 1.8 (386) |
| Nanco | 0.5 (396) | 12.1 (146) |

[a]Epicotyl segments (1-cm-long) from 2–4-week-old aseptic seedlings were used.
[b]Percentage of calli out of the total plated following co-cultivation that developed somatic embryos was rated after 2–4 months in culture.
[c]Numbers in parentheses refer to total number of explants plated. Data from the replications and repetitions were pooled.
*Source*: Ref. 16.

with strain EHA 105 containing either pMOG196 or pMOG198, and cultivar 'Danvers Half-Long' transformed with strain EHA 105 containing pGA492-CHN. An 800-pb band that was obtained was visible by exposing the ethidium bromide stained gel to UV light, control plants did not yield this band (Fig. 4A). Among all of the plantlets recovered through the tissue culture process, none that did not contain the *npt II* insert was found (data not shown).

    *b. Southern Hybridization Analysis.* Plants of cv. 'Golden State' transformed with *Agrobacterium* sp. strain EHA 105 (pMOG198) and 'Danvers Half Long' transformed with strain EHA 105 containing pGA492-CHN were analyzed. Since the DNA isolated was readily digested by restriction endonucleases, it did not require further purification. *Hind III* digestion of DNA from three different transformed plants, originating from different calli, yielded from one to four fragments that hybridized with the *npt II* probe, whereas no hybridization was detected in the control plant DNA (Fig. 4B). DNA from pMOG198 was used as positive control.

    *c. Immunoblot Analysis.* Dot blot analysis was performed on total soluble protein extracted from tissues of confirmed transgenic plants, as well as from control (untransformed) plants. By using the antibody against tobacco chitinase, the protein was detected in leaf tissue and callus cultures of cultivar 'Nanco' (Fig. 4C). By using the antibody against bean chitinase, the protein was detected in leaves of cultivar 'Danvers Half Long.' No reaction was observed with protein extracts from untransformed carrot plants.

## V. OTHER REPORTS OF *AGROBACTERIUM* SPECIES–MEDIATED TRANSFORMATION

Carrot roots have been inoculated with wild-type *A. rhizogenes* to produce transformed root tissue, from which plants were successfully regenerated via somatic embryogenesis (26). Although these plants had altered root and leaf morphological features, they were used to quantify rates of transfer of introduced genes through successive generations. *A. rhizogenes* was also used to introduce the maize transposable element *Ac* into carrot (27). Upon regeneration of plants, the characteristic hairy root phenotype developed, but it did not impede study of *Ac* activity. Hairy root carrot cultures have also been utilized to study host-mycorrhizal interactions in vitro (28).

    Scott and Draper (12) utilized proembryogenic carrot suspension cells and inoculated them with a nononcogenic *A. tumefaciens* strain C58C1 carrying the plasmid pGV3850::1103 with a chimaeric kanamycin resistance gene (*nos-npt II*). Plants were regenerated from transformed cells via somatic embryogenesis in the presence of 100 mg lms1 kanamycin; transformation efficiency was estimated to be 62–74%. Southern blot analysis showed that copy number ranged

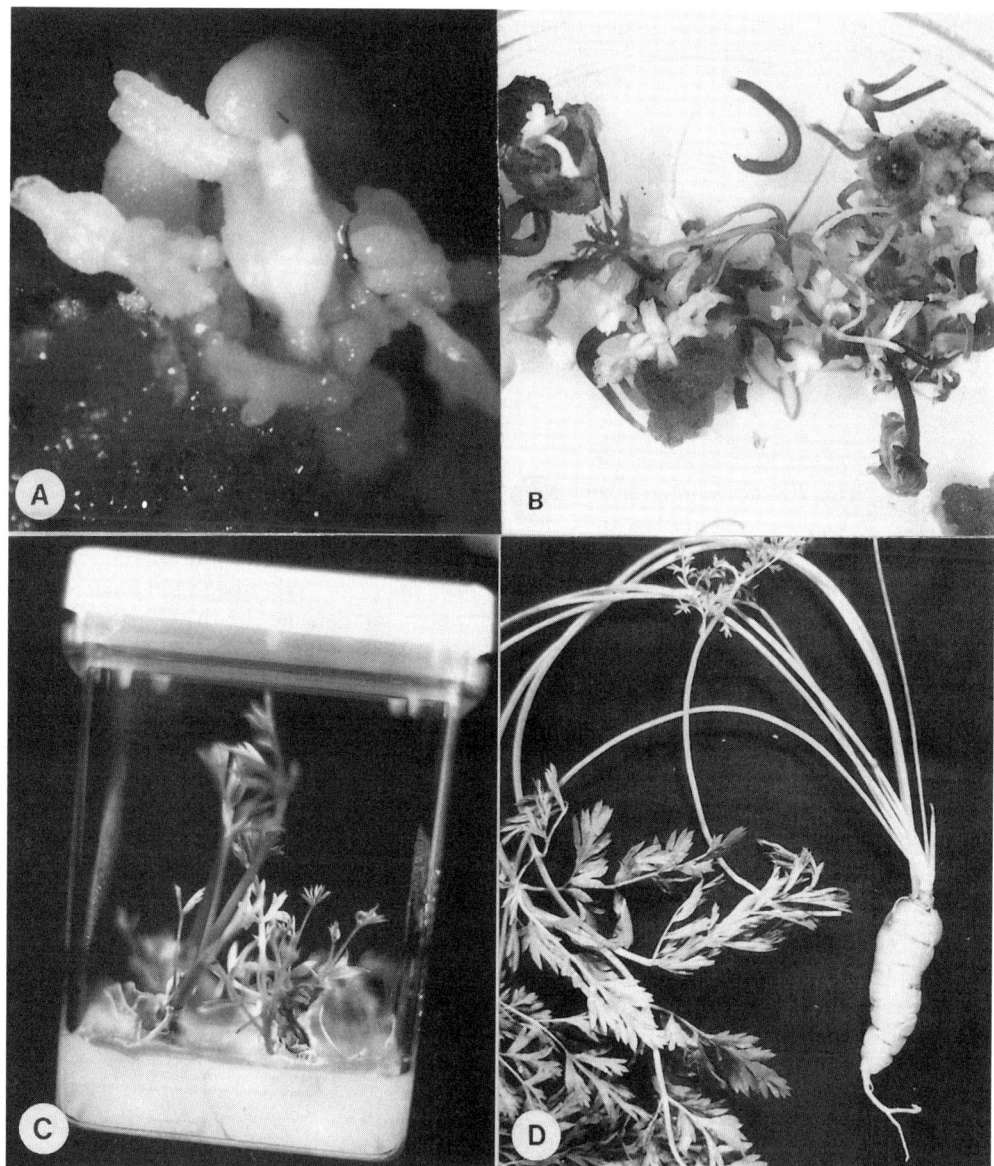

**Figure 3** Regeneration of plantlets after *Agrobacterium* species–mediated transformation of epicotyl explants of carrot. (A) Somatic embryo formation on medium containing 100 mg l$^{-1}$ of kanamycin; (B) development of plantlets on growth-regulator-free medium with kanamycin; (C) growth of transformed plantlets on kanamycin-containing medium; (D) taproot of transgenic plant after 2 months of growth in potting medium. (From Ref. 17.)

from one to eight. Western blot analysis confirmed the expression of the *npt II* gene and indicated that there was a single, full-length polypeptide (12).

In a study by Wurtele and Bulka (29), carrot callus cells were transformed with several strains of *A. tumefaciens* that possessed the same C58 chromosomal background. Each strain contained binary vector pGA472, which comprised the *nos-npt II* construct. Callus pieces (1 month old) were divided into 0.5 g (fresh-weight) aliquots, mixed with a suspension of *A. tumefaciens*,

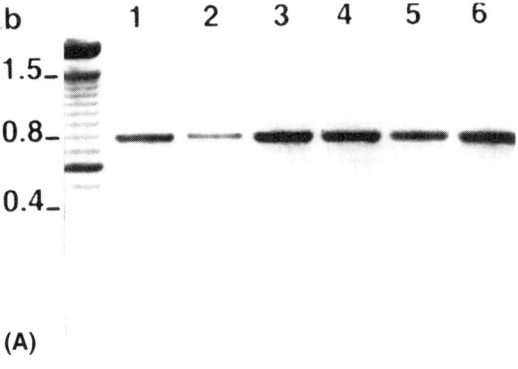

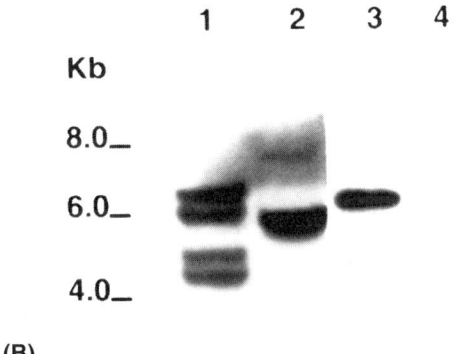

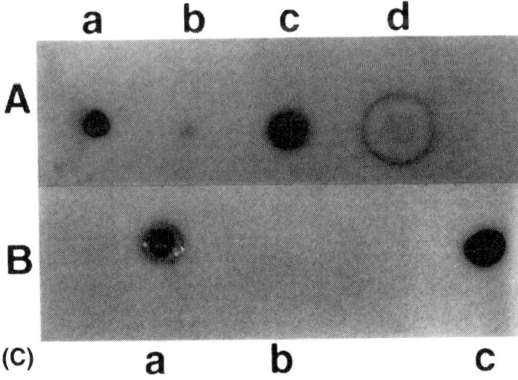

**Figure 4** Analysis of transgenic carrot plants. (A) Polymerase chain reaction amplification of the *npt II* coding region from plasmid pMOG196 (lane 1) and five transgenic plants (lanes 2–6). (B) Southern analysis using an *npt II* probe of three transgenic plants (lanes 1–3) and a nontransformed control (lane 4). (C) Protein dot blot of tissue extracts reacted against chitinase antibodies. Panel A, tobacco chitinase antibody with positive control (a), negative control (b), transgenic leaves (c), transgenic suspension culture (d). Panel B, positive control (a), negative control (b), transgenic leaves (c). (From Ref. 17.)

cocultivated for 3 days, then transferred to regeneration medium. Plants were regenerated via somatic embryogenesis on medium containing 300 mg l$^{-1}$ kanamycin. The effects of p-hydroxybenzoic acid, acetosyringone, and mechanical wounding of the cells at the time of cocultivation were studied. The results showed that these treatments had little or no effect on the timing or frequency of kanamycin-resistant clump formation (29). Southern blot analysis, using the binary vector as a probe, showed that the transgenic plants contained 1 to 15 copies of the introduced gene (29).

Carrot hypocotyl segments from sterile 1-week-old seedlings were used as the explant source by Thomas and coworkers (13). The explants were preincubated on tissue culture medium (B5 with 9 µM 2,4-D) for 2 days prior to infection with *A. tumefaciens* LBA4404 containing CaMV 35S-GUS construct on a binary vector (pRGUSII). The plasmid also contained the *npt II* gene for kanamycin resistance. The results showed that preincubation of explants was essential, since no transformation occurred in the absence of the preculture treatment (13). Different carrot varieties were surveyed, and it was found that production of kanamycin-resistant calli was variety-dependent and ranged from 0.9% to 5.8%. The transformed calli (up to 1 cm in diameter) were used to establish suspension cultures, and transformed plants were regenerated via somatic embryogenesis. Southern blot analyses confirmed that selected lines were transformed, with one to three copies of the GUS gene (13).

Pawlicki and associates (30) studied the factors influencing *A. tumefaciens*–mediated transformation of carrot. The parameters evaluated were carrot variety, age and type of explants, cocultivation and precultivation times, and effect of acetosyringone. Explants from five varieties were cocultivated with a nononcogenic *A. tumefaciens* strain C58C1 carrying the plasmid pGSTRN943 or pGSGluc1. Both plasmids contained a chimeric *npt II* gene, and pGSGluc1 also contained the β-glucuronidase gene. Petioles, cotyledons, hypocotyls, and roots were used as explant sources and were compared for their transformation efficiency. Cocultivation periods were varied from 1 to 7 days, and precultivation periods (1, 2, 3, or 7 days or 1 month) prior to infection, and the presence (100 µM) or absence of acetosyringone, were examined. Plants were regenerated from embryogenic calli (with 100 mg l$^{-1}$ kanamycin) and rooted plantlets were potted and transferred to the greenhouse. The results from this study showed that transformation efficiency was dependent on carrot variety and ranged from 0% to 46.7%. Explant type was also found to be important: petiole explants were more conducive to transformation than cotyledon, hypocotyl, or root explants. The age of explants was also found to be a variable: a 3- to 4-week-old seedling was optimal. Cocultivation periods of 2 or 3 days gave higher transformation frequencies than periods of 1 or 7 days. Last, it was found that a preculture period or the presence of acetosyringone did not influence transformation efficiency. Transformation was confirmed by histochemical detection of β-glucuronidase activity in transformed cells, and by Southern hybridization analysis. Guivarch and colleagues (31) reported that the presence of acetosyringone significantly increased the transformation efficiency of cells in carrot root disks as measured by GUS expression.

Carrot root disks were utilized by Gogarten and coworkers (32) to study the effect of an antisense construct containing a vacuolar H$^+$ adenosine triphosphatase (ATPase) subunit cDNA. The *npt II* gene was used as the selectable marker. Kanamycin-resistant calli were grown in liquid medium containing 2,4-D and benzylaminopurine. Shoots and roots were induced on solid media containing appropriate growth regulators, and plantlets were recovered.

In a 1998 study by Hardegger and Sturm (33), β-glucuronidase expression driven by the 35S promoter was shown to be high in transgenic carrot root tissues. The explants used for cocultivation were hypocotyls, and calli were produced on B5 medium with naphthaleneacetic acid and 6-benylaminopurine. *Agrobacterium* sp. strain GV3101 was found to be more efficient than LBA 4404 in the frequency of transformation, giving an average frequency of 20%.

Direct DNA uptake by carrot protoplasts by chemical and electrical means has been utilized to study gene expression, for example, the effects of promoters, enhancer elements, and introns on transcriptional efficiencies. These studies were made possible by coupling sequences of interest with the coding regions of reporter genes, such as the firefly luciferase (*lux*), neomycin phosphotransferase (*npt II*), chloramphenicol acetyl transferase (*cat*), and β-glucuronidase (GUS) (34,35). A gene for herbicide resistance was introduced into carrot via direct gene transfer (36). The resultant transgenic plants contained the phosphinothricin-*N*-acetyltransferase gene and were shown to be resistant to the herbicide L-phosphinothricin.

## VI. STRATEGIES TO ENHANCE TOLERANCE OF CARROTS TO FUNGAL DISEASES

The availability of cloned genes encoding for pathogenesis-related (PR) proteins, antifungal proteins (AFPs), and a number of antimicrobial compounds, derived from a range of sources (37–40), provides opportunities to enhance the tolerance of carrot to a broad spectrum of fungal pathogens after *Agrobacterium* species–mediated transformation. Enhanced disease resistance has been reported in a range of plants after gene introduction and expression of AFPs (41–43), PR proteins (44–49), phytoalexins (50), hydrogen peroxide (51), defensins (52), and thionins (53).

Chitinases are well-characterized PR proteins. Their expression in plants is increased by pathogen attack and certain abiotic stresses (54,55), and they accumulate intracellularly in the central vacuole (basic chitinases) or extracellularly in the intercellular space (acidic chitinases) (18,19). Some evidence for the role of chitinases in the defense response of plants is available (19,44,46,56). Chitinases catalyze the hydrolysis of chitin, a substrate that is not found in plants but is a component of the cell walls of many fungi (57). Thus, exposure of fungal cells to chitinases has been shown to cause them to lyse (54,58). To determine whether enhanced expression of chitinases in plants could also have a similar effect on fungi *in planta*, different chitinase encoding genes were introduced into carrot. These included chitinases cloned from petunia (18), tobacco (18), and bean (19). Since carrot is naturally susceptible to a number of important fungal pathogens (3,5), the overexpression of this protein was evaluated for enhanced protection against several fungal pathogens.

## VII. EVALUATING CARROT LINES EXPRESSING CHITINASES FOR ENHANCED TOLERANCE TO FUNGAL PATHOGENS

Transgenic lines of carrot cvs. 'Golden State' and 'Nanco,' containing either an acidic chitinase gene from petunia or a basic chitinase gene from tobacco, were evaluated. The transgenic plants were regenerated from transformed calli after *A. tumefaciens*–mediated transformation and multiplied by using a cell suspension culture system as described earlier. Nontransgenic controls, which consisted of either seedlings grown in tissue culture or plantlets derived from embryogenic calli, were obtained by using the same tissue culture procedures as for transformation. All seedlings and plantlets were transferred to sterile potting mix after 4–6 weeks of growth in vitro and maintained at 22–26°C under a 16-h photoperiod provided by lamps (Gro-Lux) at an intensity of 140 $\mu EM^{-2} s^{-1}$. Plants were evaluated at various times during their growth, ranging from 4 to 12 months after transplanting.

The pathogens used were *Alternaria radicini*, *Botrytis cinerea*, *Rhizoctonia solani*, *Sclerotium rolfsii*, and *Thielaviopsis basicola*. All cultures were maintained on 15% V8 juice agar and incubated at 22–28°C on the laboratory bench. Colonies were subcultured prior to use in the inoculation experiments. Petioles from transgenic and control plants of each cultivar were selected from similar-aged plants. The petioles were excised at the crown, the leaves were trimmed off, and 10- to 16-cm-long sections were washed under running tap water for 3–5 min. Petioles of similar diameter were grouped and cut into 8-cm-long sections; six segments were combined and assayed together as one sample. Petioles from transgenic and control plants were inserted upright into actively growing colonies (7–14-days-old, in 100 × 15 mm petri dishes) of the fungal pathogens; the distal ends of the petioles were grouped and held together with tape (Fig. 5). Where possible, samples of six petiole segments from transgenic and control plants were placed on the same fungal colony, at opposite ends. The petri dishes with the petioles were sealed in plastic bags lined with moistened paper towels. The extent of lesion development from the base of the petiole upward was measured after 2, 4, and 7 days of incubation at 22–24°C. Lesion length was measured to the nearest millimeter on each petiole, and the mean lesion length for each colony was determined. In each experiment, two to three cultures of each inoculum age were used to assess each cultivar/transgenic line. The experiment was conducted five times, each with two to three replicate colonies of each pathogen, using petioles from the same transgenic plants collected at different times during growth. Data from all of the experiments were subjected to analysis of variance (ANOVA). Untransformed data were analyzed by the general linear modeling program (GLM) in the SAS statistical software package (SAS Institute, Cary, NC). Pairwise comparisons of treatment means (lesion size) were made for transgenic versus control tissues for both cultivars at each of the three assessment dates and for the two inoculum types. Significant differences are indicated at $P = 0.01$ using the LSD test.

The detached petiole assay was a sensitive and reproducible method for rating the extent of

**Figure 5** Detached petiole assay used to evaluate response of transgenic and nontransgenic plants to fungal inoculation. Petioles were placed upright into a colony of *Botrytis cinerea*; left group of petioles are from control plants, right group are from transgenic plants. Photograph was taken after 3 days, note lesion development and collapse of tissues at the base of petioles of the control.

fungal development on transgenic and control carrot tissues. In general, significant differences were not apparent until 4 days after inoculation for most of the pathogens tested. After extensive fungal colonization, there was collapse of the petioles within 5–7 days due to extensive cell maceration. To prevent this, each group of petioles was supported by a central wooden stake, which held them upright until the experiment was completed (Fig. 5). For the transgenic line expressing the acidic petunia chitinase, statistical analyses of the data did not show a significant difference between transgenic and nontransgenic plants in several experiments (data not shown). The transgenic line expressing the basic tobacco chitinase was also challenged with the pathogens and comparisons were made with control plants. With *B. cinerea*, lesion size was significantly ($P = 0.01$) smaller in transgenic plants of both 'Golden State' and 'Nanco' at 4 and 7 days after inoculation (Fig. 6A). With *R. solani*, lesion development was significantly reduced in the transgenic cultivar 'Nanco' at all evaluation times (2, 4, and 7 days), whereas in 'Golden State,' a significant reduction was seen only after 7 days (Fig. 6B). With *S. rolfsii.*, overall lesion sizes were smaller in transgenic cultivar 'Nanco' at all evaluation times, whereas 'Golden State' had lower disease only at day 7 (Fig. 6C). The greater reduction in lesion size at later evaluation times is indicative of a cumulative effect of the antifungal activity due to the expressed chitinases. Thus, chitinase expression in transgenic carrot plants has the potential to delay appearance of disease symptoms and result in lower overall disease incidence.

## VIII. OTHER REPORTS OF CARROT TRANSFORMATION FOR DISEASE TOLERANCE

Several antifungal proteins from tobacco were constitutively expressed in transgenic carrot plants when using constructs from tobacco containing cDNAs and genomic fragments after *Agrobacterium* sp. transformation (59,60). The genes encoded AP 24 (PR-5, osmotin), chitinase, and glucanase. Transformed lines were regenerated and self-pollinated, and seed was used for field trials in 2 years. Selected lines showed increased resistance against four pathogens: *Alternaria dauci*, *A. radicini*, *Cercospora carotae*, and *Erysiphe heraclei*. The most promising lines were those expressing a combination of a chitinase and a glucanase gene, and they exhibited broad-spectrum tolerance (59,60). A human lysozyme gene was constitutively expressed in two carrot cultivars and some of the transgenic lines showed enhanced resistance to *Erysiphe heraclei* and *Alternaria dauci* (61).

## IX. CONCLUSIONS AND FUTURE PROSPECTS

It is evident from all of the published work on carrot transformation to date that several different protocols are available. The variables that appear to affect the frequency of transformation include carrot cultivar, explant source, bacterial strain, tissue culture medium, and growth regulators. In spite of this, recovery of transgenic carrot plants has been reported from several laboratories. Surprisingly, however, there are few reports on the introduction of potentially useful agronomic traits in carrot, with the exception of fungal disease resistance. When compared to other vegetable crops, such as tomato and cucumber, carrot appears to have lagged behind in the level of interest for genetic transformation for the introduction of novel traits. This is surprising for a plant species that responds well in tissue culture and is a horticulturally valuable crop.

Although breeding efforts have provided and will continue to provide superior cultivars for commercial use, there are many unexploited opportunities for future targets in genetic engineering of carrot. The feasibility of introducing virus resistance, resistance to insect pests, tolerance of environmental stress, and modulation of the carotenoid biosynthetic pathway and the ratio of

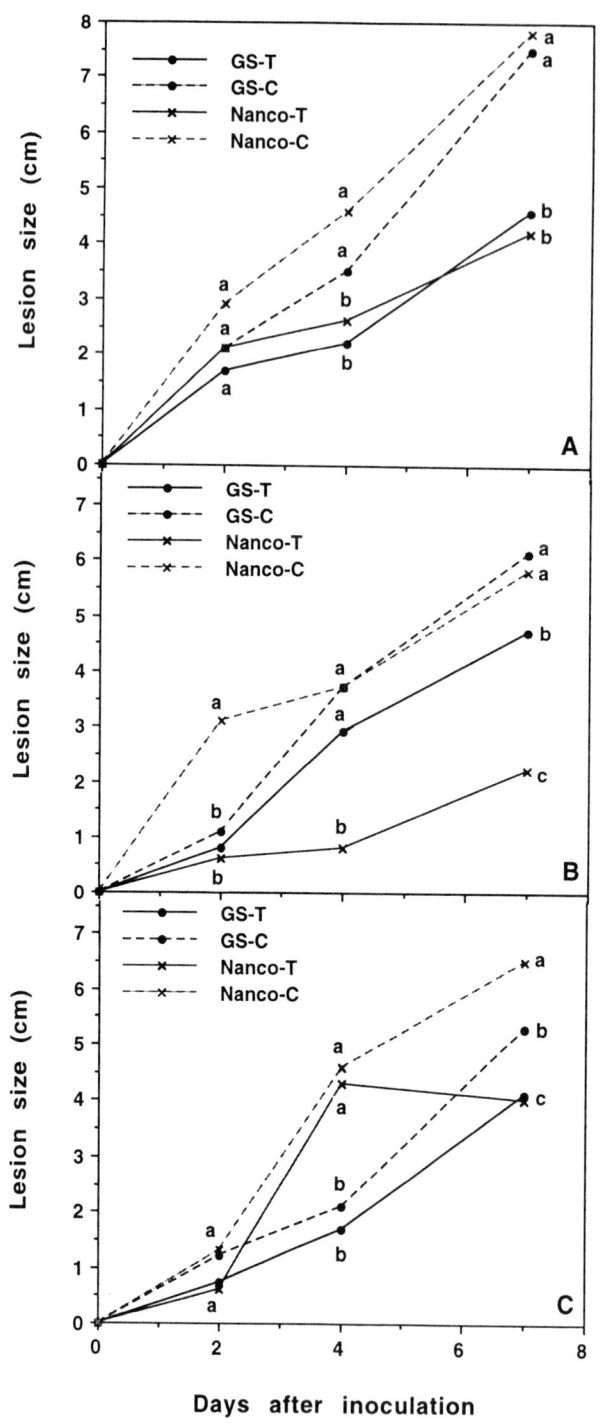

**Figure 6** Lesion development on transgenic and control petioles of two carrot cultivars at 2, 4, and 7 days after inoculation with different pathogens. (A) *Botrytis cinerea*. (B) *Rhizoctonia solani*. (C) *Sclerotium rolfsii*. Means followed by different letters are significantly different ($P = 0.01$, LSD test) within each time of inoculation.

various sugars for quality enhancement has yet to be explored. Confirmation of expression of the desired trait under field conditions and transmission of the introduced genes to the progeny will be required.

## REFERENCES

1. Vegetables and Specialties Situation and Outlook Report. Economic Research Service. Washington, DC: United States Department of Agriculture. July, 1998.
2. L Baker. Carrot seed market could see dramatic growth. In: J Hager, ed. Carrot Country. Vol 6. Yakima, WA: Columbia, 1998, pp 16–17.
3. PV Ammirato. Carrot. In: DA Evans, WR Sharp, PV Ammirato, eds. Handbook of Plant Cell Culture. Vol 4. New York: Macmillan, 1986, pp 457–499.
4. G Agrios. Plant Pathology, 4th ed. New York: Academic Press, 1997.
5. CE Peterson, PW Simon. Carrot breeding. In: MJ Basset, ed. Breeding Vegetable Crops. Westport, CT: AVI, 1986, pp 321–356.
6. PW Simon. Carrot genetics. Plant Mol Biol Rep 2:54–63, 1984.
7. MD St. Pierre, RJ Bayer. The impact of domestication on the genetic variability in the orange carrot, cultivated *Daucus carota* spp. *sativus* and the genetic homogeneity of various cultivars. Theor Appl Genet 82:249–253, 1991.
8. RW Allard. Principles of Plant Breeding. New York: John Wiley & Sons, 1960.
9. PV Ammirato. Organizational events during somatic embryogenesis. In: CE Green, DA Somers, WP Hackett, DD Biesboer, eds. Plant Tissue and Cell Culture. New York: Alan R Liss, 1987, pp 57–81.
10. JL Zimmerman. Somatic embryogenesis: A model for early developments in higher plants. Plant Cell 5:1411–1423, 1993.
11. T Murashige, F Skoog. A revised medium for rapid growth and bioassays with tobacco tissue cultures. Physiol Plant 15:473–497, 1962.
12. RJ Scott, J Draper. Transformation of carrot tissues derived from proembryogenic suspension cells: A useful model system for gene expression studies in plants. Plant Mol Biol 8:265–274, 1987.
13. JC Thomas, MJ Guiltinan, S Bustos, T Thomas, C Nessler. Carrot (*Daucus carota*) hypocotyl transformation using *Agrobacterium tumefaciens*. Plant Cell Rep 8:354–357, 1989.
14. ZR Sung, D Dudits. Carrot somatic cell genetics. In: NK Panopoulos, ed. Genetic Engineering in the Plant Sciences. New York: Praeger, 1981, pp 11–37.
15. M Murata, J-H Ryu, S Caretto, D Rao, H-S Song, JM Widholm. Stability and culture medium limitations of gene amplification in glyphosate resistant carrot cell lines. J Plant Physiol 152:112–117, 1998.
16. MO Hernandez. Tissue culture and *Agrobacterium*-mediated transformation of carrot (*Daucus carota* L.). MSc thesis, Simon Fraser University, Burnaby, British Columbia, 1994.
17. MO Gilbert, YY Zhang, ZK Punja. Introduction and expression of chitinase encoding genes in carrot following *Agrobacterium*-mediated transformation. In Vitro Cell Dev Biol Plant 32:171–178, 1996.
18. HJM Linthorst, LC van Loon, CMA van Rossum, A Mayer, JF Bol, SC van Roekel, EJS Melenhoff, BJC Cornelissen. Analysis of acidic and basic chitinases from tobacco and petunia and their constitutive expression in transgenic tobacco. Mol Plant Microbe Interact 3:252–258, 1990.
19. K Broglie, I Chet, M Holliday, R Cressman, P Biddle, S Knowlton, CJ Mauvais, R Broglie. Transgenic plants with enhanced resistance to the fungal pathogen *Rhizoctonia solani*. Science 254:1194–1197, 1991.
20. G An. Binary Ti vectors for plant transformation and promoter analysis. Methods Enzymol 153: 292–305, 1987.
21. J Sambrook, EF Fritsch, T Maniatis. Molecular cloning. A Laboratory Manual. 2nd ed. Cold Spring Harbor, New York: Cold Spring Harbor Laboratory Press, 1989.
22. SCHJ Turk, LS Melchers, H den Dulk-Ras, AJG Regensburg-Tuink, PJJ Hooykaas. Environmental conditions differentially affect *vir* gene induction in different *Agrobacterium* strains. Role of the Vir A sensor protein. Plant Mol Biol 16:1051–1059, 1991.

23. A Kanazawa, N Tsutsumi. Extraction of restrictable DNA from plants of the genus *Nelumbo*. Plant Mol Biol Rep 10:316–318, 1992.
24. SR Jia, M-Z Yang, R Ott, N-H Chua. High frequency transformation of *Kalanchoe laciniata*. Plant Cell Rep 8:336–340, 1989.
25. MF Van Wordragen, HJM Dons. *Agrobacterium tumefaciens*-mediated transformation of recalcitrant crops. Plant Mol Biol Rep 10:12–36, 1992.
26. D Tepfer. Transformation of several species of higher plants by *Agrobacterium rhizogenes:* Sexual transmission of the transformed genotype and phenotype. Cell 37:359–367, 1985.
27. MA Van Sluys, J Tempe. Behaviour of the maize transposable element *Activator* in *Daucus carota* L. Mol Gen Genet 219:313–319, 1989.
28. G Bécard, JA Fortin. Early events of vesicular-arbuscular mycorrhiza formation on Ri T-DNA transformed roots. New Phytol 108:211–218, 1988.
29. ES Wurtele, K Bulka. A simple, efficient method for the *Agrobacterium*-mediated transformation of carrot callus cells. Plant Sci 61:253–262, 1989.
30. N Pawlicki, RS Sangwan, BS Sangwan-Norreel. Factors influencing the *Agrobacterium tumefaciens*-mediated transformation of carrot (*Daucus carota* L.). Plant Cell Tissue Org Cult 31:129–139, 1992.
31. A Guivarch, J-C Caissard, S Brown, D Marie, W Dewitte, H Van Onckelen, D Chriqui. Localization of target cells and improvement of *Agrobacterium*-mediated transformation efficiency by direct acetosyringone pretreatment of carrot root discs. Protoplasma 174:10–18, 1993.
32. JP Gogarten, J Fichmann, Y Braun, L Morgan, P Styles, SL Taiz, K DeLapp, L Taiz. The use of antisense mRNA to inhibit the tonoplast $H^+$ ATPase in carrot. Plant Cell 4:851–864, 1992.
33. M Hardegger, A Sturm. Transformation and regeneration of carrot (*Daucus carota* L.). Mol Breed 4:119–127, 1998.
34. R Bower, RG Birch. Transformation in *Daucus carota* L. (Carrot). In: YPS S Bajaj, ed. Biotechnology in Agriculture and Forestry. Plant Protoplasts and Genetic Engineering III. Vol. 22. Berlin, Heidelberg: Springer-Verlag, 1993, pp 172–181.
35. JO Rasmussen, OS Rasmussen. PEG mediated DNA uptake and transient GUS expression in carrot, rapeseed and soybean protoplasts. Plant Sci 89:199–207, 1993.
36. W Dröge, I Broer, A Pühler. Transgenic plants containing the phosphinothricin-*N*- acetyltransferase gene metabolize the herbicide L-phosphinothricin (glufosinate) differently from untransformed plants. Planta 187:142–151, 1992.
37. BJC Cornelissen, LS Melchers. Strategies for control of fungal diseases with transgenic plants. Plant Physiol 101:709–712, 1993.
38. S Kamoun, CI Kado. Genetic engineering for plant disease resistance. In: L Kim, ed. Advanced Engineered Pesticides. New York: Marcel Dekker, 1993, pp 165–198.
39. CJ Lamb, JA Ryals, ER Ward, RA Dixon. Emerging strategies for enhancing crop resistance to microbial pathogens. Bio technology 10:1436–1445, 1992.
40. PJM Van den Elzen, E Jongedijk, LS Melchers, BJC Cornelissen. Virus and fungal resistance: From laboratory to field. Phil Trans R Soc Lond B 342:271–278, 1993.
41. R Broglie, K Broglie, D Roby, I Chet. Production of transgenic plants with enhanced resistance to microbial pathogens. In: S-D Kung and R Wu, eds. Transgenic Plants. Vol. 1. Engineering and utilization. New York: Academic Press, 1993, pp 265–276.
42. G Jach, B Görnhardt, J Mundy, J Logemann, E Pinsdorf, R Leah, J Schell, C Maas. Enhanced quantitative resistance against fungal diseases by combinatorial expression of different barley antifungal proteins in transgenic tobacco. Plant 8:97–109, 1995.
43. J Logemann, G Jach, H Tommerup, J Mundy, J Schell. Expression of a barley ribosome-inactivating protein leads to increased fungal protection in transgenic tobacco plants. Bio technology 10:305–308, 1992.
44. R Broglie, K Broglie. Chitinase gene expression in transgenic plants: A molecular approach to understanding plant defence responses. Philos Trans R Soc Lond B 342:265–270, 1993.
45. W Howie, L Joe, E Newbigin, T Suslow, P Dunsmuir. Transgenic tobacco plants which express the *chi*A gene from *Serratia marcescens* have enhanced tolerance to *Rhizoctonia solani*. Transgenic Res 3:90–98, 1994.

46. ZK Punja, SHT Raharjo. Response of transgenic cucumber and carrot plants expressing different chitinase enzymes to inoculation with fungal pathogens. Plant Dis 80:999–1005, 1996.
47. Y Tabei, S Kitade, Y Nishizawa, N Kikuchi, T Kayano, T Hibi, K Akutsu. Transgenic cucumber plants harboring a rice chitinase gene exhibit enhanced resistance to gray mould (*Botrytis cinerea*). Plant Cell Rep 17:159–164, 1998.
48. T Terakawa, N Takaya, H Horiuchi, M Koike, M Takagi. A fungal chitinase gene from *Rhizopus oligosporus* confers antifungal activity to transgenic tobacco. Plant Cell Rep 16:439–443, 1997.
49. Q Zhu, EA Maher, S Masoud, RA Dixon, CJ Lamb. Enhanced protection against fungal attack by constitutive co-expression of chitinase and glucanase genes in transgenic tobacco. Biotechnology 12:807–812, 1994.
50. R Hain, H Reif, E Krause, R Langebartels, H Kindl, B Vornam, W Wiese, E Schmelzer, P Schreier, R Stocker, K Stenzel. Disease resistance results from foreign phytoalexin expression in a novel plant. Nature 361:153–156, 1993.
51. G Wu, B Shortt, EB Lawrence, EB Levine, KC Fitzsimmons, DM Shah. Disease resistance conferred by expression of a gene encoding $H_2O_2$-generating glucose oxidase in transgenic potato plants. Plant Cell 7:1357–1368, 1995.
52. FRG Terras, K Eggermont, V Kovaleva, NV Raikhel, RW Osborn, A Kester, SB Rees, S Torrekens, F Van Leuven, J Vanderleyden, BPA Cammue, WF Broekaert. Small cysteine-rich antifungal proteins from radish: Their role in host defense. Plant Cell 7:573–588, 1995.
53. P Epple, K Apel, H Bohlmann. Overexpression of an endogenous thionin enhances resistance of *Arabidopsis* against *Fusarium oxysporum*. Plant Cell 9:509–520, 1997.
54. ZK Punja, Y-Y Zhang. Plant chitinases and their roles in resistance to fungal diseases. J Nematol 25:526–540, 1993.
55. DA Samac, DM Shah. Developmental and pathogen-induced activation of the *Arabidopsis* acidic chitinase promoter. Plant Cell 3:1063–1072, 1991.
56. N Benhamou, K Broglie, R Broglie, I Chet. Antifungal effect of bean endochitinase on *Rhizoctonia solani*: ultrastructural changes and cytochemical aspects of chitin breakdown. Can J Microbiol 39:318–328, 1993.
57. LC Van Loon, YAM Gerritsen, CE Ritter. Identification, purification and characterization of pathogenesis-related proteins from virus-infected Samsum NN tobacco leaves. Plant Mol Biol 9:593–609, 1987.
58. F Mauch, B Mauch-Mani, T Boller. Antifungal hydrolases in pea tissue. Plant Physiol 88:936–942, 1988.
59. H Tigelaar, MH Stuiver, L Molendijk, E Troost-van Deventer, MB Sela-Buurlage, J Storms, L Plooster, F Sijbolts, J Custers, M Apotheker-de Groot, LS Melchers. Broad spectrum fungal resistance in transgenic carrot plants (abstract). Phytopathology 86:S57, 1996.
60. LS Melchers, MH Stuiver, Novel genes for disease-resistance breeding. Curr Opin Plant Biol 3:147–152, 2000.
61. M Takaichi, K Oeda. Transgenic carrots with enhanced resistance against two major pathogens, *Erysiphe heraclei* and *Alternaria dauci*. Plant Sci 153:135–144, 2000.

# 36
# Transgenic Cassava for Food Security and Economic Development

**Nigel J. Taylor, M. V. Masona, and Claude M. Fauquet**
*ILTAB/Donald Danforth Plant Science Center, St. Louis, Missouri*

**Christian Schöpke**
*ValiGen, San Diego, California*

| | | |
|---|---|---|
| I. | INTRODUCTION | 523 |
| II. | AGRONOMIC CHARACTERISTICS AND SOCIOECONOMIC ROLE OF CASSAVA | 524 |
| III. | THE ROLE OF BIOTECHNOLOGY IN CASSAVA IMPROVEMENT | 526 |
| IV. | DEVELOPMENT AND PRESENT STATE OF TRANSGENIC TECHNOLOGIES IN CASSAVA | 527 |
| V. | APPLICATIONS OF GENETIC ENGINEERING IN CASSAVA | 531 |
| | A. Virus Resistance | 531 |
| | B. Bacterial Blight Resistance | 533 |
| | C. Resistance to Insect Pests | 534 |
| | D. Herbicide Resistance | 535 |
| | E. Starch and Polymers | 536 |
| | F. Cyanogenesis | 537 |
| | G. Postharvest Deterioration | 537 |
| | H. Other Applications | 538 |
| VI. | PROMOTERS AND TRANSGENE EXPRESSION | 539 |
| VII. | INTEGRATED GENOMIC AND GENETIC TRANSFORMATIONS | 540 |
| VIII. | SUSTAINING PROGRESS AND APPLICATION OF CASSAVA TECHNOLOGY | 541 |
| | A. Collaboration and the Role of the Cassava Biotechnology Network | 541 |
| | B. Technology Transfer | 541 |
| IX. | CONCLUSION | 542 |
| | REFERENCES | 543 |

## I. INTRODUCTION

The tropical root crop cassava (*Manihot esculenta* Crantz) is, after rice and maize, the most important staple food crop cultivated in the tropics and the primary source of dietary calories in many of the world's low-income countries. Cassava is cultivated in 80 tropical countries over a total of 16 million hectares and eaten by an estimated 500 million people daily (1). Addressing constraints to cassava production through the application of transgenic biotechnologies could improve the quality of life for hundreds of millions of people and contribute to future food security for billions more. Despite this, reports of the first confirmed transgenic cassava plants in 1996 were met with mixed enthusiasm. Although they credited the technical achievements, they raised concerns as to the relevance and value of the breakthroughs in a crop with undefined commercial value (2).

Others view the development of transgenic technologies for cassava differently and believe that along with that of other tropical food crops, its improvement should command the highest priority. The projected total number of people who must be supported by the world's agricultural systems by the year 2050 varies according to source, but is estimated at between 8 and 11 billion (3). What is not disputed is that 90% of the population increase from present levels will occur in the less developed countries (LDCs), requiring a doubling in food production in these regions (4). With 820 million people already classed as chronically undernourished (3,4), critical questions are being asked about the ability of tropical agriculture to fulfill future nutritional and economic requirements in the LDCs. Cassava's high productivity under suboptimal and low-input agricultural conditions has identified it as an important factor in attaining sustainable food security and economic development within many tropical countries (1,5). In order to meet these challenges all available tools, including biotechnology, should be brought to bear to address its major yield constraints and to enhance product quality.

This chapter outlines the progress made in developing and applying transgenic technologies for the agronomic improvement of cassava. First, the crop and its socioeconomic importance are described given in order to highlight the perceived requirement for biotechnology in cassava improvement programs. The development of genetic transformation technologies and their application to address the major limiting factors in cassava cultivation are then described. It will become apparent that transgenic capabilities in cassava are lagging behind those of most other major food and cash crops. Although genetic transformation protocols are now in place, at the time of writing no transgenic cassava plants have undergone field testing. The technical reasons for this are discussed, but cassava's position as primarily a subsistence crop raises numerous issues that must be addressed before transgenic approaches for its improvement can be implemented and sustained.

Nevertheless, it is hoped the reader will appreciate that significant breakthroughs have occurred in cassava biotechnology since the mid-1990s. Genetically engineered cassava plants expressing marker genes (6–8) and transgenes for virus resistance (9), modified starch synthesis (10), herbicide resistance (11), and reduced cyanogenic qualities (12) have been recovered. Mapping projects are under way (1) and numerous BAC and complementary deoxyribonucleic acid (cDNA) libraries are now available. In relation to the resources committed, progress has been good and an initial period of frustration has given way to one of optimism and opportunity.

## II. AGRONOMIC CHARACTERISTICS AND SOCIOECONOMIC ROLE OF CASSAVA

Cassava, otherwise known as *tapioca*, *yuca*, *manioc*, or *mandioca*, is an outcrossing, monecious member of the family Euphorbiaceae. Considered an allotetraploid ($2n = 36$), it is a highly heterozygous, semiwoody perennial shrub varying from 1 to 4 m in height, depending on the cultivar (Fig. 1A). Propagation takes place vegetatively, via lignified stem cuttings. After planting, new roots are formed and axillary buds sprout to form the shoot system. Three to four months later, photosynthates produced by the established leaf canopy are diverted to the root system,

# Transgenic Cassava

**Figure 1** Cultivation and processing of cassava. (A) Cassava field in Indonesia; (B) harvested cassava tubers, 30–40 cm, awaiting processing; (C) Cameroonian villagers processing cassava into gari, a type of cassava flour.

where the excess energy is converted to starch and stored in the parenchyma of greatly thickened storage roots, generally referred to as tubers (13,14) (Fig. 1B). Although some peoples consume the leaves as a green vegetable, the crop is cultivated primarily for the high calorific value of its storage roots, which are processed to produce a large range of food products (14) (Fig. 1C).

*M. flabelifollia* has now been confirmed as the most likely progenitor of cultivated cassava, with a center of origin along the southern border of Amazonia (15). The Portuguese transported the plant to western Africa during the 16th century; from there it spread eastward and eventually reached Madagascar. Introduction to Asia took place in the 17th century, and by the end of the 1800s it was established in all the world's tropical regions (5). Cultivation continued to increase throughout the 20th century, most noticeably in Africa, where the colonial powers often encouraged its cultivation as a famine reserve and where it gradually replaced traditional but less robust root crops such as yam (*Diosiorea* sp). At the beginning of the 21st century, Africa is the largest cassava producing region, harvesting 85.9 million tons fresh weight tubers from 10.3 million hectares, followed in production by Asia and South America (3) (Table 1). On a country scale, Nigeria, Thailand, Indonesia, and Brazil are the major cassava producers (16); however, 11 of the

**Table 1** World Cassava Production by Region

| Region | Area harvested (ha × $10^6$) | Average yield (tons/ha)[a] | Production (tons × $10^6$) |
|---|---|---|---|
| Africa | 10.3 | 8.3 | 86.0 |
| Asia | 3.5 | 13.3 | 46.5 |
| South America | 2.2 | 12.4 | 27.4 |
| Others | 0.2 | — | 1.1 |
| World | 16.2 | 9.9 | 161.0 |

[a]Weights expressed as wet weight of harvested root tissues.
*Source*: Ref. 3.

12 countries that consume the most cassava per capita are in sub-Saharan Africa, where 40% of the population rely on it as their major staple food (17).

Cassava's success is based on its unusually adaptable and productive nature. Although greatest yields are obtained on loamy, sandy soil under humid tropical conditions, it is highly tolerant of marginal or eroded soils and adapts easily to the acidic oxisols prevalent throughout much of the tropics. An efficient photosynthetic system enables it to accumulate as much as $250 \times 10^3$ cal/ha/day, a rate 25% and 40% greater than that of rice and maize, respectively (18). Coupled with a high harvest index of up to 0.8 dry weight tubers/total plant dry weight (5,18) this makes cassava one of the most productive crop plants and the cheapest known source of starch. When grown on fertile soils under high radiation, cassava yields are comparable on a dry weight basis to rice, maize, and sugarcane. However, under conditions of biotic and abiotic stress cassava excels, outyielding all other staple food crops (1,5). Once established, cassava plants are able to withstand considerable periods of drought or defoliation by insect pests by mobilizing reserves laid down in the root system to reestablish the leaf canopy and resume growth.

Cassava's reliability, flexible harvest time, and amenability to intercropping make it highly attractive to smallholder farmers. These characteristics, combined with the need to process cassava tubers within 48 hours of harvesting, have kept the majority of cassava cultivation and commerce small-scale and local. Most cassava is grown for human consumption, with an average of 88 kg consumed per person per year in Africa (1). Cassava's economic value lies primarily at the village and town levels, where farmers trade fresh and processed cassava products in local markets. Its perception as purely a subsistence crop is somewhat misleading, as an estimated 40% of the cassava planted in African fields in 1996 was intended for sale (19). This proportion is even higher in West Africa and South America, where there are established and expanding markets for animal feed, alcohol production, and industrial starch from cassava processing. Thailand and Indonesia are the major exporters of cassava products, accounting for more than 80% of the $435 m worth of cassava dried pellets and starch traded on world markets in 1997 (3). World fresh production for 1998 was valued at over $16 billion (20), but the monetary value of the crop on a global scale is less important than its role within the microeconomies of the LDCs.

There is no doubt that cassava is of central importance to a significant proportion of the world's population. However, because of its perception in the North as a subsistence crop with little or no economic value, it has only recently become the subject of concerted research and improvement programs. This means that cassava still retains an unknown capacity for yield and quality improvement, one that is likely to be greater than that of other major food crops. With increasing population pressure requiring greater productivity from marginal and degraded soils under suboptimal farming conditions, cassava is poised to become one of the most important crops of the 21st century.

## III. THE ROLE OF BIOTECHNOLOGY IN CASSAVA IMPROVEMENT

Despite its robust nature, cassava is susceptible to a number of biotic and abiotic constraints that must be addressed if it is to fulfill its yield potential reliably. These vary in importance from one location to another but include preharvest losses to viral and bacterial diseases, pests, and weeds. Production in Africa is most adversely affected, resulting in average yields of 8.3 tons fresh weight tubers per hectare, 60% below those achieved in Asia and almost 10 times lower than that recorded under optimal field trial conditions (5). Rapid postharvest deterioration is a ubiquitous problem that limits marketability of the crop and hinders the development of medium- to large-scale commercial exploitation. Lack of high-quality planting material, low-protein content of cassava products, and presence of cyanogenic compounds in the storage roots are also considered limiting factors. Addressing these restrictions would have a significant impact on food production in the tropical regions.

The diverse nature of the constraints, different socioeconomic requirements for cassava among cultural groups, problems encountered in disseminating information to small farmers, and lack of resources for agrochemical applications, all mean that the production of improved germplasm remains the most effective manner in which to address limitations to cassava's cultivation and exploitation. Conventional breeding for cassava improvement was initiated in the 1920s (21), and concerted programs have been ongoing since the 1960s at the CGIAR centers with a mandate for cassava: the Centro Internacional de Agricultura Tropical (CIAT, Colombia) and the International Institute for Tropical Agriculture (IITA, Nigeria). Traditional breeding systems are hindered, however, by cassava's highly heterozygous nature, its shy and asynchronous flowering, and inbreeding depression. Cassava breeding programs are bulky and lengthy, requiring screening of between 50,000 and 100,000 seedlings after the first sexual crossing (14) and at least 10 years for the improved product to reach the farmer. In addition, as a result of wide segregation of desirable traits after sexual crossing, any new variety will differ in numerous traits from the genotype originally selected by the farmer as best suited to his or her agronomic conditions and socioeconomic needs.

The development and application of transgenic technologies in cassava could circumvent many of the problems inherent in traditional improvement programs. Transfer of beneficial traits from cultivated varieties and wild relatives could be rapidly integrated into a given cultivar without compromising its existing desirable agronomic characteristics (22,23). Equally important, such capabilities would facilitate the introduction of heterologous genes, such as those imparting pathogen-derived resistance, insect toxins, antisense traits, and apomixis, something not possible by conventional breeding, that could potentionally revolutionize the cultivation and utilisation of cassava. Although priorities were set regarding the application of genetic engineering to address the major constraints outlined here (23), it was obvious that protocols for the production of transgenic cassava plants must first be developed and established.

## IV. DEVELOPMENT AND PRESENT STATE OF TRANSGENIC TECHNOLOGIES IN CASSAVA

Concerted efforts to develop a genetic transformation system for cassava were initiated in the early 1990s, but as a result of restricted funding and the crop's recalcitrance to in vitro manipulation, effective protocols remained elusive until 1996. Somatic embryogenesis was first reported in cassava in 1982 (24), after which the original embryogenic protocol was improved and extended into several dozen cultivars (25–27). In all cases, embryogenic structures possessing a high degree of organization between and within their epidermal and subepidermal cell layers (28) are induced to develop from immature leaf explants after 3 weeks culture on medium supplemented with auxin. As the only plant regeneration system available, these structures, and the secondary embryos that can be produced from them, were employed as target tissues in early attempts to develop a genetic transformation capability in cassava. However, the multicellular nature of the morphogenic events and the highly organized nature of the embryogenic structures made efficient selection for the transgenic cells difficult and resulted, at best, in the recovery of chimeric tissues (29).

Achievement of a genetic transformation capability for cassava resulted from breakthroughs in manipulating the embryogenic culture systems. Three different methods have now been reported for recovering transgenic cassava plants, with integration of the transgenes confirmed by Southern blot analysis in each case (6–8,11,12). Fig. 2 provides a schematic represen-

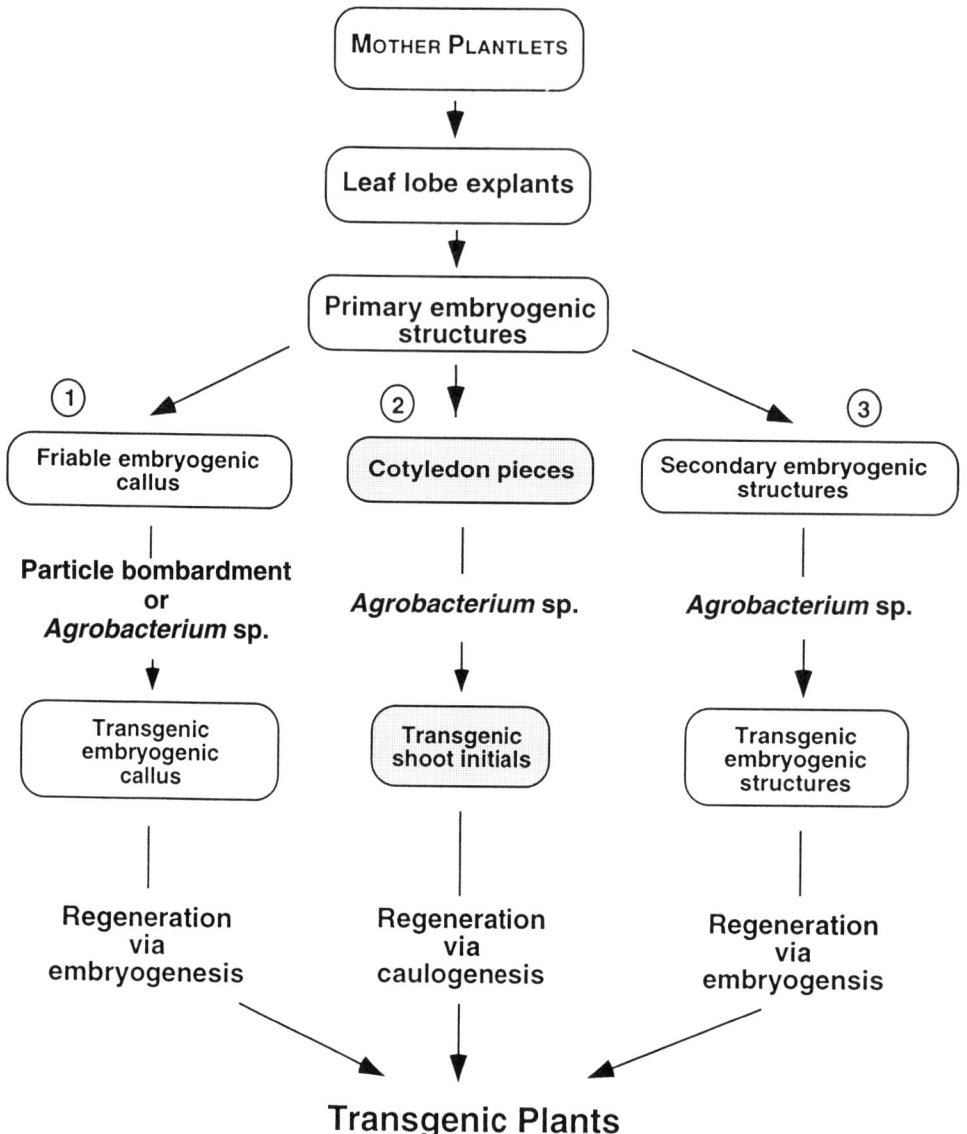

**Figure 2** Schematic representation of three methods for recovering transgenic cassava plants.

tation of the genetic transformation and tissue culture systems now used for the crop. In the first method, procedures were developed for the induction of friable embryogenic callus and suspension cultures (30). Such tissues are recognized as excellent target tissues for transgene insertion and are employed as the basis for genetic transformation programs in numerous crop species including the cereals (31,32). In most cases zygotic embryos are used as the initial explant. However, in cassava the heterozygous nature of this tissue makes it an unsuitable starting material, and so somatic embryos are produced first from leaf tissues with friable embryogenic callus (FEC) subsequently induced from these totipotent structures.

Two laboratories have confirmed regeneration of transgenic plants from the African cultivar TMS 60444 after particle bombardment of embryogenic suspension tissues. At the International Laboratory for Tropical Agricultural Biotechnology (ILTAB), the marker genes *npt*II and *uid*A were inserted; selection on media containing the antibiotic paromomycin followed (6). Workers at the Agricultural University, Wageningen employed similar procedures but utilized the marker gene *luc* to allow visual identification and selection of transgenic tissues after application of the substrate luciferine (7). Transgenic tissues were isolated and proliferated; afterwhich regeneration of transgenic plants took place through embryo development, maturation, and germination (Fig. 3). The potential of this system to produce transgenic plants has been further realized by recovery of transgenic plants after genetic transformation of embryogenic suspension tissues with *Agrobacterium* (33) and use of the callus as the target tissue in place of suspension-derived tissues.

Simultaneously with the initial report by ILTAB, the Swiss Federal Institute of Technology (ETH) confirmed recovery of transgenic cassava plantlets from the South American cultivar 'M.Col 22' (8). In this system organized embryogenic structures initiated from leaf explants were allowed to mature and develop green cotyledons on medium lacking auxin. The foliose organs were excised from the mature embryos, cut into small pieces, and infected with *Agrobacterium* carrying the *npt*II (or *hph*) and *uid*A marker genes. Shoot regeneration was induced by culturing the explants on medium containing the cytokinin benzylaminopurine and the selection agent geneticin or hygromycin. The third documented transformation system also exploits the morphogenic potential of somatic embryo–derived cotyledons. In this case a new cycle of somatic embryos was induced from fragments of these organs infected by *Agrobacterium*, followed by regeneration and germination of the embryos on medium containing the relevant antibiotics. Researchers at both CIAT (11) and, more recently, Ohio State University (12) have reported the recovery of transgenic cassava in this way from cvs. 'M.Peru 183' and 'M.Col 2215,' respectively. For further details on the tissue culture and genetic transformation transfer systems in cassava the reader is referred to a number of recent reviews (1,16,34,35).

In most crop species genetic transformation systems are initially developed and optimized for genotypes that are most amenable to manipulation by the available tissue culture protocols. These then act as valuable models for the study of transgene integration, expression, and imparted phenotype, with further efforts required to expand the technology into economically important cultivars. Cassava is no exception, but with more than 1700 cassava cultivars grown throughout the tropics, the challenge in this crop is significant. It will not be possible, or even desirable, to

**Figure 3** Regeneration of transgenic cassava plants. (A) Cotyledon-stage somatic embryos, regenerating from friable embryogenic callus; (B) transgenic cassava plantslets, developing from a germinated somatic embryo; (C) greenhouse-grown transgenic cassava plant cv. TMS 60444.

develop genetic transformation capabilities for every cultivar, but the technology must be expanded into the regionally most important varieties if the benefits of genetic engineering are to reach the small- to medium-scale farmer. Efforts to develop cultivar-independent transformation protocols by direct gene transfer to proliferating meristems have proved unsuccessful (36), leaving the production of embryogenic tissues a prerequisite for transformation in any given cassava cultivar. Fig. 1 illustrates how all three established transformation systems are reliant on the ability to induce and then proliferate embryogenic tissues from immature leaf tissues. Consequently, considerable effort has been directed at improving induction procedures from this organ (25–27, 37). Encouragingly, competence for somatic embryogenesis appears widespread in *M. esculenta* and it has been possible to induce embryogenic tissues in almost all 65 cultivars tested to date (1). Nevertheless, embryogenic efficiencies are strongly genotype-dependent, varying from 100% to less than 5% (26,27), making further research necessary to improve our understanding of the factors controlling embryogenic competence in cassava and to allow incorporation of more cultivars into the transgenic protocols.

Despite genotypic variation, results indicate that genetic transformation in a significant range of cassava cultivars will be possible in the near future. Friable embryogenic callus has been established in 16 cultivars and plant regeneration achieved in 9 (9,37). Transgenic plants have been recovered in the three African cultivars 'TMS 60444,' 'Bonoua Rouge,' and 'L2,' and the South American cultivar 'M.Col 1505' after particle bombardment of embryogenic suspension tissues. The production and regeneration of transgenic plants of cv. 'TMS 60444' are now routine, taking 4–6 months from the time of gene insertion to plantlet establishment. At the time of writing approximately 125 transgenic plant lines containing transgene constructs coding for marker genes and for putative resistance to virus and bacterial diseases have been recovered at ILTAB (9). A similar number of transgenic plant lines containing visual marker genes and an antisense starch gene by researchers in the Netherlands have been recovered (10,37).

Plant recovery via organogenesis from cotyledon fragments has also been improved and extended into six cultivars (38). Recovery of transgenic plantlets from transgenic cotyledon tissues can be achieved in a slightly shorter time than from the disorganized embryogenic tissues, whether by somatic embryogenesis or caulogenesis. Both these systems are simpler in concept and require less tissue culture expertise than regeneration from FEC but necessitate greater labor inputs if large numbers of transgenic plants are to be recovered. The lack of a disorganized growth phase is desirable as it minimizes the chance of culture-induced variation but also raises as yet unresolved questions regarding the likelihood of inadvertently recovering chimeric plants.

Considerable debate has centered on which of the genetic transformation systems are superior and/or preferable for use in cassava transformation programs. As for other species, criteria assessed are reproducibility, number of independent transgenic plant lines recoverable, time required to recover plants, quality of the regenerants, and possible chimerism (39). For cassava, however, attention must also be paid to transferability of the protocols to other laboratories, most especially those in the LDCs. Future requirements in transgenic programs should also be considered. The ability to pyramid genes for disease resistance and manipulation of biosynthetic pathways will require integration and expression of multiple transgenes in a manner similar to that demonstrated in 1998 in rice (40). Full and rapid exploitation of genomic tools will also necessitate insertion of large regions (up to several hundred thousand base pairs) of deoxyribonucleic acid (DNA), such as those carried in bacterial artificial chromosomes (BACs). Each of the transformation methods reported has its strengths and possible drawbacks, such that at this time, it is not possible to say which will become the preferred method. Indeed, the resources available in any given location, the target cultivars, the amount of DNA being inserted, the number of plant lines required, and the resources and expertise of the respective laboratories may mean that all have a role to play in the application of transgenic technologies in cassava.

## V. APPLICATIONS OF GENETIC ENGINEERING IN CASSAVA

With reproducible genetic transformation systems now established, cassava biotechnology has entered its second phase, in which transgenes of potential agronomic interest are being integrated, expressed, and studied in cassava. Table 2 lists the agronomic traits identified as targets for transgenic cassava and gives an estimated time scale for their achievement. Transgenic programs under way reflect a combination of the major constraints to cassava production, the genes available to tackle them, and the goals and expertise of the respective laboratories driving the research.

### A. Virus Resistance

The application of genetic transformation for increased resistance to the major cassava viruses is a major priority. Pathogen-derived resistance strategies have proved effective in other crops (41) and are being developed against both major viral diseases of cassava: cassava common mosaic virus (CsCMV) and African cassava mosaic disease (ACMD).

#### 1. Cassava Common Mosaic Virus

CsCMV is a potexvirus capable of causing yield losses in excess of 30% in Brazil, Peru, Colombia, and Paraguay (42) and in 1995 a new strain in Venezuela was reported (43). Fauquet and coworkers found that expression of CsCMV coat protein (CP) gene in *N. benthamiana* plants was correlated with increased resistance to the homologous Brazilian strain of the virus, such that transgenic plants required 10–100 times greater concentration of virus innoculum to elicit the same symptoms as control plants (44). The CP transgene also provided some resistance against

**Table 2** Goals and Projected Time Spans for Genetically Engineered Cassava

| Goal | Estimated time to field trial (Years) |
|---|---|
| *Virus resistance* | |
| Common cassava mosaic virus | 1–2[a] |
| African cassava mosaic virus | 1–2[a] |
| *Bacterial resistance* | |
| Cassava bacterial blight | 1–2[a] |
| *Insect resistance* | |
| Stem borer | 1–2[a] |
| Hornworm | 3–5 |
| Mealy bug | 3–5 |
| Nematode resistance | 3–5 |
| Herbicide resistance | 1–2[a] |
| Modified starch | 1–2[a] |
| Reduced cyanogenic properties | 1–2[a] |
| Improved postharvest traits | 3–5 |
| Elevated root proteins | 3–5[a] |
| Extented leaf life | 1–2[a] |
| Bioplastic production | >10 |
| Apomixis (clonal true seed) | >10 |
| Modified plant architecture | 5–10 |
| Elevated root dry matter | 3–5 |

[a]Projects under way.

CsCMV strains from Colombia and Paraguay. With such clear demonstration of transgenic resistance, insertion of the CsCMV CP was undertaken as soon as reliable protocols became available for producing genetically transformed cassava plants. Particle bombardment of embryogenic suspension cultures at ILTAB has resulted in the production of several hundred callus lines resistant to the antibiotic paromomycin, from which 30 independent transgenic plant lines containing the CP gene were regenerated. Western blot analysis confirmed varying levels of CP expression in leaf tissues from 12 of the 30 plant lines (45). The most robust of these plants are now established in the greenhouse and are being assessed to determine whether expression of the transgene in cassava tissues imparts effective resistance to the virus.

## 2. African Cassava Mosaic Disease

African cassava mosaic disease is the most serious disease of cassava and considered the most important vector-borne disease of any African crop. Production is adversely affected in every sub-Saharan country. In West Africa, which has been subjected to the most detailed study, 80–100% of all plants are infected (20,46) by the virus. The disease was initially thought to be caused by African cassava mosaic virus (ACMV), a geminivirus transmitted by the whitefly *Bemisia tabaci* and spread through the use of infected cuttings. However, recent investigations have shown the situation to be more diverse and complicated than previously considered, as at least two other species, East African mosaic virus (EACMV) (46) and South African mosaic virus (SACMV) (47), are present on the continent, along with many strains and variants. Mixed infections can exist within the same field and indeed the same plant, leading to very severe symptoms. The virus is not usually lethal but suppresses plant vigor, causing yield reductions of between 15% and 30% continent wide, estimated to be equivalent to 25 million tons of food or $2.3 billion per year (20). Most worryingly, recombination between these viruses appears to be driving the evolution of the disease, leading to the appearance of new and highly virulent strains that have devastated cassava production in recent years and caused severe hardship in areas of Uganda and neighboring countries (48). In addition, *Bemisia tabaci* has been reported in Cuba and Brazil, raising concern that the disease could become established in Latin America (W. Roca, personal communication, 1999). Because there is little known resistance to ACMD in the South American germplasm, introduction of the virus into this continent would have potentially devastating effects on food and commercial cassava production, germplasm collections, and native *Manihot* species.

Breeding programs at IITA have produced cultivars with elevated resistance to ACMD, mostly through incorporation of traits from the wild species *M. glaziovii* (49), but the improved cultivars do not suit the cultural needs of all African farmers. There is a desire, therefore, to utilize genetic transformation to introduce new sources of resistance directly into specific cassava cultivars. Three such strategies have proved effective in suppressing ACMV replication and movement in transgenic *N. benthamiana* test plants. Employing a pathogen-derived replicase strategy, Hong and colleagues found that integration and expression of the ACMV AC1 (Rep) gene driven by the enhanced 35S promoter imparted elevated resistance to infection by this virus (50). Likewise, Sangare and colleagues showed that expression of a mutated AC1 gene delayed symptom appearance and severity and reduced virus movement (51). In an alternative approach, advantage has been taken of naturally occurring low levels of defective interfering (DI) particles associated with infections of ACMV. The DIs consist of a mixed population of subgenomic DNA B deletion mutants, approximately half the size of the full-length component, in which both genes have been disrupted. When DI dimers were integrated into the *N. benthamiana* genome, they became mobilized in the presence of ACMV, replicating to high numbers as extrachromosomal copies, interfering with viral replication, and imparting elevated resistance to the virus through symptom attenuation and delay (52). In a third approach, ACMV infection was reduced by employing a gene coding for the toxic ribosome inactivating protein (RIP) isolated from *Dianthus caryophyllus* and cloned downstream from an ACMV DNA A viron sense (AV1) promoter. After transformation into *N. benthamina*, expression of the transgene only took place when presence of the virus transactivated the AV1 promoter. Expression of the toxic protein caused death of the infected cells in a hypersensitive-like manner, significantly reducing virus accumulation and systemic movement (53).

Transgenic cassava plants of cv. 'TMS 60444' containing AC1 and DI sequences have been regenerated at ILTAB. Insertion of the DI was straightforward, but when the AC1 gene driven by the cassava vein mosaic virus (CsVMV) promoter was bombarded into embryogenic suspension tissues, very few transgenic lines could be recovered compared to those of controls transformed with a promoterless version of the gene (ILTAB, unpublished results). Analysis of the recovered transgenic AC1 callus lines indicated that expression of the wild-type ACMV AC1 gene is toxic to cassava cells. However, spontaneously produced, nonlethal versions of the AC1 transgene recovered from surviving callus lines have been identified, cloned, and used to produce further generations of transgenic plants. Genetically transformed plants containing a point mutation in the NTP binding domain of the AC1 gene have been produced in this way, with expression confirmed at the ribonucleic acid (RNA) level. Challenge of transgenic plants containing DI dimers and the AC1 gene is ongoing. Cultivar 'TMS 60444' is known to be highly susceptible to the virus (54) and should provide a good model for determining the effectiveness of these pathogen-derived strategies.

These studies represent only the first efforts to develop transgenic resistance to ACMD. Further questions must be addressed, including durability of any imparted resistance, especially under field conditions and over numerous vegetative generations. Integration of the DI from ACMV imparted elevated resistance to both Nigerian and Kenyan isolates of this virus in *N. benthamiana* (50), but it is not known whether it, or the AC1 gene, will maintain resistance to other ACMD-causing virus species, or against mixed infections. In such cases it may be necessary to insert pathogen-derived transgenes from more than one virus, depending on the region of intended cultivation, or to develop broad-based resistance strategies against all the ACMD-causing geminiviruses. It is also hoped that mapping and genomic studies will allow identification and cloning of natural virus resistance genes within the next 5 years. Once isolated, these genes can be transferred directly to farmer-preferred cultivars or used as the basis for conventional breeding programs.

## B. Bacterial Blight Resistance

Cassava bacterial blight (CBB), causal agent *Xanthomonas axonopodis* pv. manihotis (*Xam*), is endemic to Africa and Latin America and constitutes a threat to the crop in both continents. A systemic and epiphytic pathogen, *Xam* produces infection that causes a variety of symptoms, including angular leaf spots, vascular necrosis, wilting, and dieback. Occurrence and severity of the disease are unpredictable and vary with the cultivar and climatic conditions. CBB is spread mostly through the use of infected cuttings with losses of up to 100% occurring in severe cases, and control is limited to good sanitation and the use of resistant varieties. Resistance was initially bred into cultivated varieties from the wild relative *M. glaziovii* and is thought to be polygenic (55).

Efforts to develop a transgenic approach to CBB resistance are still in their early stages but hold great promise. Two approaches are being taken. In the first, the bacterial resistance gene *Xa21*, isolated from a wild relative of rice, which imparts resistance to bacterial leaf blight caused by *Xanthomonas oryzae* pv. oryzae (*Xoo*), is being inserted into the cassava genome. *Xa21* is an R gene encoding a putative receptor kinase that provides resistance to a significant geographical range of *Xoo* isolates when expressed as a transgene in cultivated rice varieties (56). Both wild-type and complementary DNA (cDNA) versions of the gene, driven by the natural and 35S promoters, respectively, have been inserted into cassava cv. 'TMS 60444.' It is yet to be determined whether *Xa21* is expressed in cassava tissues or whether any cross-protection will be imparted against *Xam*. Promisingly, however, PCR primers designed for *Xa21* have been used to amplify homologous sequences from CBB-resistant, but not CBB-susceptible cassava cultivars. An amplified sequence has been used to screen the BAC library at CIAT; it has allowed five clones that contain the *Xa21* homologue to be identified (M. Fregene, personal communication, 1999). Fingerprinting revealed three different restriction fragment length polymorphism (RFLP) patterns from the five clones, and information from the mapping population has allowed the trait to be correlated to known resistance against at least one CBB strain with agronomic importance in northern South America (V. Verdier, personal communication, 1998). The three BACs have been transferred to ILTAB, where they are being cobombarded with a selectable marker gene into embryogenic TMS 60444 tissues. Recovered transgenic tissues will be screened with the respective primers to identify which have incorporated the desired sequence, and plants will be regenerated from these lines. Methods for challenging young plantlets by stem inoculation have been developed at CIAT (V. Verdier, personal communication, 1998), demonstrating good correlation between seedling and adult resistance in CBB (49).

This approach marks an important move forward for cassava biotechnology as it represents for the first time the combination of functional genomics, mapping, and transgenic technologies in an integrated manner to combat an important agronomic constraint. If resistance to CBB can be transferred to 'TMS 60444,' other relevant cultivars will be transformed in the same way or with the cloned gene(s) responsible for imparting resistance. As for ACMD it is likely that future efforts will be directed at pyramiding and expressing multiple resistance transgenes to ensure durable and widespread resistance to the different CBB strains.

### C. Resistance to Insect Pests

Because of its long planting to harvest time, cassava is susceptible to severe infestation by insect pests. The use of infected cuttings in subsequent generations can then result in accumulating pest populations, significant yield reductions, and even crop failure. In Africa, mealybug (*Phenacoccus manihoti*) and cassava green mite (*Mononychellus tanajoa*), both introduced from South America and therefore without natural predators, are considered the most important pests affecting the crop. In Latin America the range of insect pests is greater; it includes mites, cutworms, scale insects, mealybugs, and lace bugs (57). However, the hornworm (*Erinnyis ello*) and stem borers (various species of Coleoptera, Lepidoptera, and Hymenoptera) pose the most serious threat. The use of chemical control systems is not generally recommended (57) because of their toxicity and damage to natural enemies of the pest populations, while the purchase of chemicals is not economically feasible for many smallholders. As for other constraints, the use of resistant planting material is seen as the most effective way to combat many of cassava's insect pests.

Transgenic plants with elevated resistance to insect damage have proved effective in a number of crops (58,59) and were planted over 3.7 million hectares worldwide in 1998 (60). In most cases, one or more *cry* genes from the bacterium *Bacillus thuringiensis* (*Bt*) coding for insecticidal (*Bt*) toxins are expressed by the plant genome. Ingestion of plant tissues containing these toxins has proved effective for controlling a variety of insect families. In response to farmer-identified concerns regarding the stem borer *Chilomima clarkei*, a program for the production of transgenic cassava expression *cry* genes has been initiated at CIAT. Losses to this insect are increasing in northern Colombia, causing as much as 60% reduction in tuber yield and 100% loss of planting material. Infestations identified in one location in 1995 had spread to three states by

1998, threatening commercial and subsistence farmers growing the susceptible cultivar 'M.Col 2215' (Venezolana) in this region and in neighbouring Venezuela and Ecuador (P. Chavarriaga, personal communication, 1999). The *cry1A (b)* gene has been obtained and is being transferred into 'M.Col 2215' by using *Agrobacterium* and particle bombardment transformation systems. Further work will then assess levels of expression and accumulation of the toxin in cassava tissues, especially under field conditions, and determine what degree of toxicity will be required to control the insects. Depending on the success of this study and further investigations on the pest-plant interaction, pyramiding *Bt* genes and management of transgenic plantings will be designed in order to counteract accumulation of resistance within the pest population (61).

A similar *Bt* approach may also be effective to combat the lepidopterous cassava hornworm, as spraying with *Bacillus thuringiensis* has been shown to be effective against this pest (22). The use of lectins is being considered to control the Coleoptera (16). Application of these approaches in cassava will obviously require significant and sustained effort before a fully effective product is available to the farmer, but with initial steps now under way the potential is significant. In addition to the use of heterologous genes, mapping and genomic studies may allow the identification of naturally occurring resistance genes within the cassava genome, or that of its wild relatives. Recent success in identifying host plant resistance to the whitefly *Aleurotachelus socialis* holds promise in this area. Whiteflies pose a production threat in Colombia and act as the vector for ACMD (62). Cloning insect resistance genes in cassava would open new avenues for their control, helping to control ACMD epidemics as well as having possible implications for combating geminivirus infections in other crops.

## D. Herbicide Resistance

Weed control is essential in newly planted cassava fields. Between two and six weedings are required over a 3- to 4-month period until developing cassava plants establish an effective canopy between the rows. Failure to achieve this reduces root numbers and size, depressing overall yields by as much as 40% in early and 70% in later or nonbranching cultivars. Weeds compete with the young cassava plants for water, nutrients, and light, and in Africa *Imperata cylindrica* also causes direct effects on the harvestable product by damaging tubers and facilitating the onset of root rot (14).

Weeding takes place by mechanical or chemical means or a combination of the two (13). The use of herbicides is increasing, especially in Asia and South America, and carries many benefits for the farmer if he or she can afford the applications. Development of transgenic herbicide resistance cassava cultivars is considered by some to be an undesirable application for cassava biotechnology because purchase of the chemicals would not be possible by the poorest farmers. However, with the proven success of herbicide-resistant transgenic soybean, maize, and rapeseed in the United States and in developing countries such as Mexico and Argentina (60), the value of herbicide-resistant cassava is being reassessed. Some persuasive arguments can be made in its favor. Medium- to large-scale cassava producers would benefit from significant cost savings and improved yields, and the use of herbicide applications would allow low- or zero-tillage systems to be adopted in marginal soils and hillsides, reducing labor requirements and soil erosion. From a technical standpoint, unlike strategies for resistance to disease and insects or for improved product value, herbicide-resistant transgenics can be viewed as a short-term, achievable goal. The technology is proven and simple, requiring insertion and suitable expression of only one transgene. In addition, the two most commonly used herbicide-resistant genes, imparting protection to Basta and glyphosate (Roundup), respectively, are known to be suitable for use with cassava. Both herbicides are already applied to cassava in the field (13) and both have been shown to be suitable chemical selection agents for in vitro selection of transgenic tissues.

Transgenic cassava plants expressing resistance to Basta have been produced at CIAT (11) and a program to develop herbicide resistance in the relevant agronomic cassava cultivars has been initiated at that institute. At this time, however, gaining access to suitable transgenes for large-scale production remains a problem. Although it is hoped that this can be resolved, it raises the first of many possible future issues concerning cassava, property rights, commercial interests, and their influence on the development of transgenic food crops in the LDCs.

### E. Starch and Polymers

Cassava's storage roots consist of between 73% and 85% starch on a dry weight basis (63). This is utilized for numerous applications within the LDCs and for export to the industrialized countries. Cassava starch products are used in the food and beverage industries as thickeners and as sources of glucose, dextrin, high-fructose syrup, sorbitol, and monosodium glutamate, and in nonfood applications for the manufacture of paper, adhesives, textiles, and packaging materials; as fillers in paint; and as raw material for the chemical industry (64). Cassava starch is of good quality, possessing desirable characteristics such as high clarity, neutral flavor, excellent thickening, and easy digestion (65). Demand from food processing industries, coupled with requirements for renewable sources of biodegradable polymers, ensures that world starch markets will expand over the coming decades. Genetic transformation of the major starch crops will be applied to meet these requirements by facilitating increased yields, but also by adding value to the agricultural product. At present, most starch modification takes place after harvesting and extraction, adding costs and producing environmentally damaging effluent (64). The ability to adapt and manipulate starch quality and quantity *in planta*, by up- or down-regulating genes within the biosynthetic pathways, offers significant opportunities for LDCs to increase cassava's role in world starch markets and to create new products through the development of novel high-value storage compounds within its roots. The use of antisense genes to reduce starch accumulation and increase sugar content of the storage roots also holds promise for producing a sweet cassava and improving the crop as a source of industrial alcohol.

In progress toward these aims, five cDNAs from the cassava starch biosynthetic pathway have been cloned at the Agricultural University Wageningen, three corresponding to ADP glucose pyrophosphorylase genes (ADPase), one branching enzyme (63), and one granule-bound starch synthase (GBSS) (66). Potato plants transformed with cassava ADPase B and GBSS cDNAs in antisense orientation have been shown to down-regulate starch accumulation significantly; in the case of GBSS almost amylose-free potato tubers are the result (66). More recently, 21 transgenic cassava plant lines were recovered from embryogenic tissues bombarded with the AGPase B antisense sequence fused to the CaMV 35S promoter (10). The transgenic plants contained between one and seven copies of the insert, and although no storage roots were formed, an established assay for starch accumulation in the stems of in vitro plantlets revealed that more than 70% of the plant lines accumulated reduced levels of starch compared to those of the controls. Interestingly, the frequency of transgenic plants displaying the desired phenotype was significantly higher than that observed in potato (5%) transformed with the same construct, possibly because of the homologous genetic background in cassava. This work demonstrates that applying transgenic technologies to adapt cassava's major economic product is possible. With improved procedures for the recovery of transgenic cassava plants and more genes from biosynthetic pathways becoming available, interest is growing in both the private and the public sector in the opportunities offered by this technology.

*M. esculenta's* high productivity and ability to store large amounts of material within membrane-bound organelles in its storage roots may make molecular farming technologies economically attractive in cassava, once ongoing research has overcome technical problems in other model species. For example, if starch accumulation can be eliminated by the use of antisense sequences, it may be possible to engineer cassava genetically to synthesize and accumulate high-value bioplastic polymers, such as polyhydroxybutyrate (PHB), within its storage roots (67).

## F. Cyanogenesis

Cassava is the most important of the cyanogenic crop plants, accumulating cyanogenic glycosides in all its tissues. This capability has led to its reputation in the North as an unsafe and undesirable food. However, the problems associated with cyanide poisoning from cassava consumption are insignificant considering the huge amounts eaten daily and the calorific value thereby obtained (17). Cassava cultivars can be split into sweet and bitter varieties, containing low and high amounts of cyanogenic glycosides, respectively. Sweet cultivars are grown to be eaten as fresh snacks or vegetables, but the great majority of cultivated cassava consists of bitter varieties that must be processed prior to consumption in order to prevent exposure to the cyanide-generating compounds. Processing can take the form of soaking-pounding, grating-fermenting-pressing, or chipping-drying. In each case the aim is to cause detoxification by breakdown of cyanogenic compounds and release of gaseous cyanide prior to ingestion. Biochemical analysis of the processes involved has revealed that the cyanogenic glycosides linamerin and lotaustralin are synthesized and accumulated in the vacuole, whereas enzymes required for their degradation are located in the cell wall. Maceration of the tissues therefore facilities contact, such that linamarin (which constitutes 95% of the glycosides) is hydrolysed to acetone cyanohydrin and glucose by linamerase, followed by conversion of the acetone cyanohydrin to cyanide and acetone by hydroxynitrile lyase (HNL) (68,69). The evolutionary/ecological importance of the cyanogenic compounds in the plant is not known, but they may play a role in nitrogen storage or in defense against pest and pathogen attack. Certainly, bitter varieties are preferred by most subsistence farmers for reasons of food security, as they are more resistant to pests and diseases and are less attractive to large herbivores and vermin (70).

For these reasons some controversy has centered on the necessity or desirability of producing acyanogenic cassava (71). Nevertheless, developing transgenic cassava with reduced cyanogenic content would have advantages in medium- to large-scale commercial operations by minimizing processing time and costs and by reducing or eliminating problems of toxic effluent (72). Possible strategies involve the use of cloned DNA of P450 cytochrome in the antisense orientation in order to down-regulate the first step in linamarin synthesis, or to manipulate linamarase or HNL expression to increase the activity of these degrading enzymes (73). In 1998 findings at Ohio State University showed that linamarase and HNL are both present in leaf tissues, but that HNL is absent, or present at undetectable levels, in cassava tubers, leading to inefficient detoxification of acetone cyanohydrin in processed roots (69). These workers utilized this knowledge to produce genetically transformed plants of cv. 'M.Col 2215' expressing cDNA of the HNL gene. The transgenic plants were found to have two to three times higher activity of this enzyme in their leaves as compared to the control (12). Further studies are ongoing to determine the presence and action of this enzyme in transgenic root material and to assess whether elevated HNL activities will promote more rapid and complete degradation of linamerase during tuber processing.

As for all applications of genetic transformation in cassava to date, these findings require further work before a useful product can be released to the farmer. However, the results are promising and indicate that cassava cultivars with reduced cyanogenic content, or even acyanogenic tubers, could be available in the near future.

## G. Postharvest Deterioration

Cassava storage roots are unavoidably damaged at harvesting and deteriorate rapidly, becoming inedible as soon as 48 hours after removal from the soil (74). This characteristic restricts marketability of the crop and necessitates immediate processing of the harvested tubers, either on farm or close to the site of production. Urban centers within the LDCs are growing rapidly, presenting large markets for cassava products. Satisfying this demand completes a cycle of food supply and income generation for rural development but is frustrated by the rapid postharvest perishability of cassava storage roots. Large-scale commercialization of the crop is hindered by the necessity to invest capital in processing plants within the growing areas, rather than transporting the harvested product to the factories. Improving the postharvest characteristics of cassava is seen as a high priority for the development of the crop and its contribution to economic development within the tropics (1,22,75).

Initial deterioration of cassava storage roots is characterized by discoloration of the vascular and storage parenchyma tissues and by biochemical changes typical of the wound response in other plant tissues. In such systems, wound sealing and development of a wound periderm act to suppress signals liberated from damaged cells at the wound surface. This reaction prevents the resulting cascade of damaging biochemical reactions involving lytic enzymes, protease inhibitors, hydroxyproline-rich glycoproteins, and numerous phenolic compounds from extending throughout the plant (75). In cassava roots, however, the suberization/lignification healing process fails or is inadequate, causing physiological deterioration to spread throughout the root and leaving the wound open to infection by microorganisms. Sealing the roots in plastic bags to ensure high humidity facilitates wound healing, allowing the tubers to be stored for considerable periods (75). Spraying the tubers with wax is also effective. Although neither process is economically feasible for improving shelf life, they illustrate that the postharvest physiological characteristics of cassava tubers may be amenable to genetic engineering if important aspects of the deterioration process can be manipulated and/or the healing process up-regulated. Variation in susceptibility to postharvest deterioration between cultivars also means that mapping for such traits may lead to the identification of the genes or gene complexes involved.

Much work remains, but whereas complexity of the deterioration process frustrated initial work, recent application of molecular techniques has allowed numerous genes involved in the wounding process to be cloned and identified. At the University of Bath, United Kingdom, a cDNA library has been constructed from messenger RNA (mRNA) isolated from tuber tissues 48 hours post harvesting. After screening with heterologous cDNA and genomic probes, several clones for genes with putative action in the wound response have been isolated and sequenced, including a hydroxyproline-rich glycoprotein, a β-1,3-glucanase, and an ACC oxidase (76,77). In addition, three phenylalanine (PAL) genes have been identified in cassava. Two have been fully sequenced, including a 900-bp region of the promoter from the cPAL2 gene (76). In an alternative approach, researchers at Agricultural University Wageningen are using cDNA-AFLP differential display to study gene expression after tuber harvesting. Around 90 transcripts have been isolated and are being examined to determine which are linked to the deterioration phenomenon (78).

These studies represent the first steps to application of genetic engineering to address postharvest deterioration in cassava. Any transgenic approach must, however, ensure specificity to the harvested product and not affect wound healing or defense mechanisms in other organs or the intact root. For this reason future efforts are likely to focus on specific ways to direct and enhance localized healing response in wounded tuber tissues, possibly by manipulating hydroxyproline-rich proteins and related genes under the control of root-specific and wound-inducible promoters.

## H. Other Applications

Transgenic approaches have been suggested to address other limiting factors in cassava agronomy, including resistance to fungal root rots and nematodes (22), and to prolong leaf life for enhanced photosynthetic area and as a green vegetable (79). However, two proposed targets, elevating protein content in the tuber tissues and apomixis, deserve further attention, as, although technically demanding, both could potentially revolutionize the cultivation and commercial exploitation of cassava.

Cassava tubers have a very low protein content (1–2% dry weight), meaning that a diet consisting of cassava must be accompanied with protein from fish, meat, or legumes if deficiencies are to be prevented. Although a varied diet is obviously best, increasing overall protein content within the storage roots would improve the nutritional quality of cassava and benefit the consumer's health. Cassava leaf tissues have a relatively high protein content, and so an approach aimed at manipulating source-sink relationships may be required. Greater impact, however, could be achieved by genetically transforming cassava to produce high-molecular-weight glutenins (80). At this time, substituting wheat flour with more than 30% cassava flour significantly reduces the quality of the resulting bread. Engineering cassava for improved bread-making qualities could give cassava access to existing and expanding markets within the LDCs, impacting the macroecomomies of these countries by reducing their expenditure on imported cereals.

Developing a true seed propagation system for cassava was identified as a high-priority objective in 1989 (22,23). Cassava planting material is bulky, is difficult to store, and as for all vegetatively propagated crops requires careful management to prevent accumulating problems from pests and diseases. Propagation via zygotic seed is not a viable option because of high heterozygosity in *Manihot* sp. Engineering cassava with genes imparting apomitic traits would revolutionize the crop, allowing a small, disease-free clonal propagule to be produced and stored. Farmers could be provided with certified, high-quality planting material and transportation of elite cassava propagation material to counter serious disease and pest outbreaks made possible over considerable distances. Although such goals seemed far-fetched a decade ago, recent confirmation of natural apomictic tendencies in cassava (81), genomic studies, and the availability of genetic transformation systems have brought this goal closer to reality. Although no project is yet under way to introduce such traits into cassava, it remains an attractive, if long-term application goal for cassava biotechnology.

## VI. PROMOTERS AND TRANSGENE EXPRESSION

All the transgenic strategies described require effective promoters to drive transgene expression in the desired manner. Cassava has particular and demanding needs in this area. Transgene expression must be maintained over considerable periods and numerous vegetative cycles under what in many cases will be conditions of water stress, poor nutrition, and pest or disease pressure. In addition, producing plants homozygous for the transgene(s) is not feasible; nor is a sexual cycle available to reverse the effects of gene silencing in the adult plants. As a result of the relatively recent development of a transgenic capability, no information is available concerning long-term transgene expression in cassava under field conditions. However, with the required biosafety regulations for transgenic plants established in Colombia, the first trails under tropical conditions were scheduled at CIAT in 2000–2001.

The development of tissue-specific and inducible promoters is of central importance to future transgenic crop programs and should command significant research input. To date all the transgenic cassava plants recovered have employed forms of the CaMV 35S or cassava vein mosaic virus (CsVMV) constitutive promoters to drive the transgenes and selectable markers. The CsVMV promoter and its derivatives were isolated at ILTAB (82) and tested in tobacco and rice, in which constitutive and root- and phloem-specific versions were identified (83). Transgenic cassava plants containing these promoters fused to the *uid*A gene have been regenerated and are presently being investigated for expression of the marker gene under greenhouse conditions. Root-specific promoters have obvious application in modifying the harvestable product; a phloem-specific promoter should direct pathogen-derived resistance to that tissue, thereby combating the systemic spread of viral infections.

Successful gene cloning from the cassava genome is also leading to the isolation of new endogenous promoters with potential use in transgenic programs. The *cbg3* promoter has been isolated from a genomic library after screening for homology to a linamarase cDNA clone. Subsequent studies of transient GUS expression driven by this promoter have revealed its strong root-specific action (84). Likewise a 840-bp PAL2 promoter has been shown to drive stable expression in cassava tissues (ILTAB, unpublished results) and is being investigated to determine whether it possesses wound- or pathogen-inducible properties.

## VII. INTEGRATED GENOMIC AND GENETIC TRANSFORMATIONS

If the benefits of transgenic programs are to be realized fully, genetic transformation technologies should operate hand in hand with the information and products generated by genomics (85). Indeed the preceding sections have contained numerous references to the use of genomic studies to identify and provide cloned genes and promoters of agronomic interest. As for genetic transformation technologies, restricted funding for genomic studies in cassava slowed initial progress in this field. However, significant advances that have been made over the last few years now offer exciting opportunities for cassava biotechnology. A number of tools have been developed (reviewed in Ref. 1), including a genetic linkage map derived from parents chosen to segregate for ACMV and CBB resistance, cooking quality, photosynthetic rate, and resistance to mealybug. The maps (male and female) presently consist of 230 restriction fragment length polymorphisms (RFLPs) 100 randomly amplified polymorphic DNAs (RAPDs) (86), and microsatellite markers (87). A number of cDNA and genomic libraries have been generated from leaf and tuber tissues, and most recently a library of 70 000 BAC clones has been generated. A further two BAC libraries are planned and more ambitious plans have been tabled to raise funding required to commence sequencing of ESTs or the entire cassava genome.

For the first time a fully integrated, multidimensional approach can be envisaged for cassava biotechnology, which will bring together cassava breeders, agronomists, phytopathologists, physiologists, and entomologists with the biotechnologists. Development of high-resolution genetic maps will become an increasingly important tool for the identification, cloning, and study of genes conferring desirable agronomic traits in *M. esculenta* cultivars. Screening for homology within collections of related wild species can allow the full diversity of particular genes or gene families to be identified and exploited. BAC clones will provide rapid access to the desirable genetic sequences so that transformation can be employed to transfer these into farmer-preferred cultivars. Results from the CIAT CBB gene-tagging project represent an early example of the potential benefits of such an approach, in which a region of the genome responsible for 80% of known variation for resistance to a South American strain of the disease has been identified (V. Verdier, personal communication, 1999).

## VIII. SUSTAINING PROGRESS AND APPLICATION OF CASSAVA TECHNOLOGY

### A. Collaboration and the Role of the Cassava Biotechnology Network

Developing and sustaining transgenic biotechnologies in cassava create challenges other than the purely technical. Agricultural biotechnology is being driven largely by the private sector of the

North, with the majority of research and development programs directed at the temperate crops or at tropical species such as pineapple, banana, and rubber, from which a financial reward is expected. In addition, many of the tools and protocols being developed are protected by patents and confidentiality agreements. Developing world subsistence crops, such as cassava, sweet potato, yam, and millet, have little or no place in such market-oriented activities, meaning that the resources required to maintain cassava biotechnologies must be found elsewhere (88). To date, this has come largely from nongovernmental development agencies and from charitable organizations such as the Rockefeller Foundation. Such funding is always limited, however, making building research teams and maintaining a critical mass of expertise in this field ongoing problems.

The progress in cassava gene biotechnologies described has been achieved in less than 10 laboratories, located almost exclusively in Europe, in the United States, and at CIAT, Colombia. Collaboration and communication among the major groups have been essential to the breakthroughs achieved and were greatly facilitated by the coordinating efforts of the Cassava Biotechnology Network (CBN). The CBN was established in 1988 after an initiative by several individual scientists who believed in the potential of biotechnology for cassava improvement and sought support from the CGIAR centers at CIAT and IITA. Financed in large part by the Dutch government's Special Programme on Biotechnology and Development Cooperation (DGIS/BIOTECH), the CBN had grown to more than 400 members by 1997 (1,89). The CBN acted as the most important coordinating factor in cassava research throughout the 1990s, maximizing the limited funding available by initiating communication and collaboration among the advanced laboratories and the LDCs who wished to acquire transgenic capabilities for cassava. Through its priority setting surveys it has also helped ensure that farmer-identified needs maintained a high profile and that the development of cassava biotechnology remained as focused as possible on the needs of the intended end users. At the time of writing, the future of the CBN is in doubt as a result of lack of a suitable sponsor. However, among its other achievements it has been an instrumental force in coordinating the development and success of the molecular mapping, genomic tools, and genetic transformation projects described in this review. These are the major technical advances that provide cassava biotechnology with a springboard for the future.

With proven transgenic capabilities now established, commercial backing is increasing for cassava biotechnology research in Europe, the United States, and South Africa. Although it is to be welcomed as an important new source of funding and for improving cassava's profile, it raises dilemmas regarding the free exchange of information and possible protection of cassava-related biotech products. Commercially funded activities will also be directed at improving cassava as a marketable product and therefore aimed at medium- to large-scale production. The challenge will be to ensure that the smallholder cassava growers can also gain from such activities, have access to the products of the new technologies and opportunity to use them to improve their standard of living.

## B. Technology Transfer

An important factor in delivering the benefits of transgenic technologies to the tropical farmer is the transfer of the relevant knowledge, tools, and capability to the LDCs. Although cassava is a tropical crop, development of its transgenic capabilities has taken place largely in laboratories within the industrialized countries, and thus is remote from the intended end users. Different countries face different problems regarding cassava cultivation and utilization and require an indigenous capacity with which to address their specific constraints. Technology transfer to the LDCs shortens the distance between the biotechnologist and the farmer and empowers these countries to seek funding from international agencies, negotiate with multinational companies, and generate self-sustaining indigenous biotech industries (90).

Technology transfer has been a principal aim of the major funding bodies, the CBN, and the laboratories involved in developing cassava biotechnology. From a technical point of view, an important issue is "transferability" of the transformation protocols to operating conditions prevalent in laboratories within the tropical countries. For example, the use of luciferase as a visual selection system requires sophisticated optical equipment not affordable in the LDCs. Likewise, the desire for simple protocols has caused some to favor the use of *Agrobacterium* sp.–based transformation systems. However, many countries, such as Brazil, Mexico, Malaysia, Thailand, India, Zimbabwe, and South Africa, have well-equipped labs and considerable experience in handling embryogenic cultures and particle bombardment protocols in other crops such as rice. Adapting these to cassava should not be difficult. The limiting factor in transferring transgenic technologies to developing countries is in most cases the lack of suitably trained scientists and technicians and funding required to sustain research programs until a useful product can be realized (91). Unlike in rice biotechnology, for which the Rockefeller Foundation has funded the training of scores of scientists, no coordinated effort has been initiated for cassava. A number of developing country scientists have made important contributions to the development of cassava biotechnology while working in collaboration with the advanced laboratories. In addition, through the CBN sponsorship a handful of scientists from Asia, Africa, and South America have received training and returned to their home institutes to initiate cassava transformation programs (89,90). Nevertheless, this input is not sufficient to enable a self-sustaining base of suitably trained scientists to be established in the LCDs. Transfer of the required personnel and tools remains a significant challenge and will be central factor in the successful application of cassava biotechnology. Compared to those for other major crops such as rice, maize, or cotton, the technologies required to produce transgenic cassava are recent and relatively underdeveloped. Much therefore remains to be done before the full benefits can be transferred to the farmer. A central factor in achieving this goal will be the role of the International Centres at CIAT and IITA, who have historically carried the responsibility for ensuring effective technology transfer to smallholders in the LDCs.

## IX. CONCLUSION

There is little doubt that transgenic biotechnologies hold enormous potential for improved crop yields and enhanced agricultural products. Questions remain, however, as to whether the opportunities offered by these technologies will be applied to benefit the human population as a whole or whether the major beneficiaries will be producers and consumers within the industrialized countries. The majority of the human population live within the LDCs and rely on tropical and subtropical crops for their dietary and economic needs. Population growth means that improvement in the staple crops relevant to these regions must be achieved and sustained throughout the coming decades if the LDCs are to maintain their political and economic development. It is within these countries that agricultural biotechnology can have its greatest impact.

To date, the resources applied to developing biotechnologies for tropical food crops such as cassava are enormously out of proportion to their worldwide socioeconomic importance (4). Nevertheless, as described in this review, significant progress has been made in developing biotechnologies in cassava. An important stage has now been reached in which both the genetic transformation protocols and the genomic tools to provide genes and promoters with potential agronomic interest are in place. Many challenges remain, but there is great optimism that if the required levels of support can be maintained, both the technical and practical challenges can be overcome to ensure that the smallholder cassava farmer will also be a beneficiary of the biotechnology revolution.

## REFERENCES

1. AM Thro, M Fregene, NJ Taylor, CJJM Raemakers, J Pounti-Kaerlas, C Schöpke, RGF Visser, I Potrykus, CM Fauquet, W Roca, C Hershey. Genetic biotechnologies and cassava-based development. In: T Hohn, K Leisinger, eds. Biotechnology of Food Crops in Developing Countries. Berlin: Springer Verlag, 1999, pp 142–185.
2. Transforming the root of the problem (editorial). Nat Biotechnol 14:677, 1996.
3. FAO. FAOSTAT Statistical Database, Agriculture Data. Available at hppt//apps.fao.org
4. HW Kendall, RN Beachy, T Eisner, F Gould, R Herdt, PH Raven, JS Schell, MS Swaminathan. Bio-engineering of crops: Report of the World Bank Panel on transgenic crops. Washington, DC: The World Bank, 1997.
5. JH Cock. Cassava: A basic energy source for the tropics. Science 218:755–762, 1982.
6. C Schöpke, NJ Taylor, R Carcamo, NK Konan, P Marmey, GG Henshaw, RN Beachy, CM Fauquet. Regeneration of transgenic cassava plants (*Manihot esculenta* Crantz) from microbombarded embryogenic suspension cultures. Nat Biotechnol 14:731–735, 1996.
7. CJJM Raemakers, E Sofiari, NJ Taylor, GG Henshaw, E Jacobsen, RGF Visser. Production of transgenic cassava plants by particle bombardment using luciferase activity as the selection marker. Mol Breed 2:339–349, 1996.
8. HQ Li, C Sautter, I Potrykus, J Pounti-Kaerlas. Genetic transformation of cassava (*Manihot esculenta* Crantz). Nat Biotechnol 14:736–740, 1996.
9. NJ Taylor, C Schöpke, MV Masona, R Cárcamo, Thao Ho, RN Beachy, CM Fauquet. Production of transgenic cassava plants with marker genes and genes of interest by microparticle bombardment and *Agrobacterium*-mediated gene transfer. Brazilian Cassava Journal, Supplement, Fourth International Meeting Cassava Biotechnology Network, Salvador, Brazil, 1998, p 34.
10. TRI Munyikwa, CJJM Raemakers, M Schreuder, R Kok, M Schippers, E Jacobsen, RGF Visser. Pinpointing towards improved transformation and regeneration of cassava (*Manihot esculenta* Crantz). Plant Sci 135:87–101, 1998.
11. R Sarria, E Torres, F Angel, P Chavarriaga, WM Roca. Transgenic plants of cassava (*Manihot esculenta* with resistance to Basta obtained by *Agrobacterium*-mediated transformation. Plant Cell Rep 19:339–344, 2000.
12. DI Arias-Garzon, RT Sayre. Genetic engineering approaches to reducing cyanide toxicity in cassava (*Manihot esculenta* Crantz). Brazillian Cassava Journal, Supplement, Fourth International Meeting Cassava Biotechnology Network, Salvador, Brazil, 1998, p 29.
13. P Silvestre. Cassava: The Tropical Agriculturist. Hong Kong: Macmillan, 1989, pp 1–16.
14. IITA. Cassava in Tropical Africa: A Reference Manual. Wisbech, England: Bolding Mansel International, 1990.
15. K Olsen, BA Schaal. The origins of cassava: A phytogeographical analysis of *Manihot esculenta*. Proc Natl Acad Sci 1196:5586–5591, 1999.
16. J Pounti-Kaerlas. Cassava biotechnology. Biotech Genet Engin Rev 15:329–361, 1998.
17. H Rosling. Cassava toxicity and food security: A report for UNICEF. 2nd ed. Uppsala, Sweden: Tryck kontakt, 1988, pp 8–34.
18. WE Splittstoesser. Crop physiology of cassava. Hortic Rev 13:105–129, 1992.
19. FJ Nweke, JK Lynam. Cassava in Africa. Afr J Root Tuber Crops 2:10–13, 1997.
20. JM Thresh, GW Otin-Nape, M Thankappan, V Muniyappa. The mosaic diseases of cassava in Africa and India caused by whitefly-borne geminiviruses. Rev Plant Pathol 77:935–945, 1998.
21. DL Jennings. Breeding for resistance to African cassava mosaic geminivirus in East Africa. Trop Sci 34:110–122, 1994.
22. RB Bertram. Cassava. In: GJ Persley, Agricultural Biotechnology: Opportunities for International Development. Oxon: CAB International, 1990, pp 241–261.
23. CIAT. Report on the founding workshop for the advanced cassava research network. CIAT Working Document No. 52, 1989.
24. JA Stamp, GG Henshaw. Somatic embryogenesis in cassava. Z Pflanzenphysiol 105:183–187, 1982.

25. L Szabados, R Hoyos, WM Roca. In vitro somatic embryogenesis and plant regeneration of cassava. Plant Cell Rep 6:248–251, 1987.
26. CJJM Raemakers, M Amati, G Staritsky, E Jacobsen, RGF Visser. Cyclic somatic embryogenesis and plant regeneration in cassava. Ann Bot 71:289–294, 1993.
27. NJ Taylor, RJ Kiernan, C Davey, GG Henshaw, D Blakesley. Improved procedures for the production of embryogenic tissues across a range of African cassava cultivars: Implications for genetic transformation. Afr J Root Tuber Crops 2:200–204, 1997.
28. J Stamp. Somatic embryogenesis in cassava: The anatomy and morphology of the regeneration process. Ann Bot 59:451–459, 1987.
29. C Schöpke, C Franche, D Bogusz, P Chavarriaga, C Fauquet, RN Beachy. Transformation in cassava (*Manihot esculenta* Crantz). In: YPS Bajaj, ed. Biotechnology in Agriculture and Forestry. Vol 23. Plant Protoplasts and Genetic Engineering IV. New York: Springer, 1993, pp 273–289.
30. NJ Taylor, M Edwards, RJ Kiernan, C Davey, D Blakesley, GG Henshaw. Development of friable embryogenic callus and embryogenic suspension cultures in cassava (*Manihot esculenta* Crantz). Nat Biotechnol 14:726–730, 1996.
31. P Christou. Transformation technology. Trends Plant Sci 1:423–431, 1996.
32. IK Vasil. Molecular improvement of cereals. Plant Mol Biol 25:925–937, 1994.
33. AE González, C Schöpke, NJ Taylor, RN Beachy, CM Fauquet. Regeneration of transgenic plants (*Manihot esculenta* Crantz) through *Agrobacterium*-mediated transformation of embryogenic suspension cultures. Plant Cell Rep 17:827–831, 1998.
34. CJJM Raemakers, E Sofiari, E Jacobsen, RGF Visser. Regeneration and transformation in cassava. Euphytica 96:153–161, 1997.
35. C Schöpke, NJ Taylor, R Cárcamo, AE González, VM Masona, CM Fauquet. Transgenic cassava (*Manihot esculenta* Crantz). In: YPS Bajaj, ed. Biotechnology in Agriculture and Forestry. Vol. 47. Transgenic Crops II. Berlin: Springer Verlag, 2001, pp 234–254.
36. J Pounti-Kaerlas, P Frey, I Potrykus. Development of meristem gene transfer techniques for cassava. Afr J Root Tuber Crops 2:175–178, 1997.
37. CJJM Raemakers, M Schreuder, TRI Munyikwa, E Jacobsen, RGF Visser. Towards a routine transformation procedure for cassava. Proceedings Fourth International Meeting Cassava Biotechnology Network, Salvador, Brazil, 1998.
38. HQ Li, YW Huang, CY Liang, JY Guo, HX Liu, I Potrykus, J Pounti-Kaerlas. Regeneration of cassava plants via shoot organogenesis. Plant Cell Rep 17:410–414, 1998.
39. RG Birch. Plant transformation: Problems and strategies for practical application. Annu Rev Plant Physiol 48:297–326, 1997.
40. L Chen, P Marmey, NJ Taylor, J-P Brizard, C Espinoza, P D'Cruz, H Huet, S Zhang, A de Kochko, RN Beachy, CM Fauquet. Expression and inheritance of multiple transgenes in rice. Nat Biotechnol 16:1060–1064, 1998.
41. RN Beachy. Mechanisms and applications of pathogen-derived resistance in transgenic plants. Curr Opin Biotechnol 8:215–220, 1997.
42. LA Calvert, MI Cuervo, MD Ospina, CM Fauquet, B-C Ramirez. Characterisation of common mosaic virus and a defective RNA species. J Gen Virol 77:525–530, 1996.
43. E Marys, ML Izaguirre-Mayoral. Isolation and characteristaion of a new Venezuelan strain of cassava common mosaic virus. Ann Appl Biol 127:105–112, 1995.
44. CM Fauquet, C Schöpke, A Sangaré, P Chavarriaga, RN Beachy. Genetic engineering technologies to control viruses and their application to cassava viruses. Proceedings First International Scientific Meeting Cassava Biotechnology Network, Cartagena, Colombia, 1993, pp 190–207.
45. C Schöpke, NJ Taylor, MV Masona, R Cárcamo, Thao Ho, RN Beachy, CM Fauquet. Characterisation of transgenic cassava plants containing the coat protein gene of cassava common mosaic virus. Brazilian Cassava Journal, Supplement, Fourth International Meeting Cassava Biotechnology Network, Salvador, Brazil, 1998, p 33.
46. MM Swanson, BD Harrison. Properties, relationships and distribution of cassava mosaic geminiviruses. Trop Sci 34:15–25, 1997.
47. LC Berrie, K Palmer, EP Rybicki, SH Hiyadat, DP Maxwell, MEC Rey. A new isolate of African cassava mosaic virus in South Africa. Afr J Root Tuber Crops 2:49–52, 1997.

48. BD Harrison, X Zhou, GW Otim-Nape, Y Liu, DJ Robinson. Role of novel type of double infection in the geminivirus-induced epidemic of severe cassava mosaic in Uganda. Ann Appl Biol 131:437–448, 1997.
49. SK Hahn, ER Terry, IO Leuschner, C Akobundu, C Okali, R Lal. Cassava improvement in Africa. Field Crops Res 2:193–226, 1979.
50. Y Hong, J Stanley. Virus resistance in *Nicotiana benthamiana* conferred by African cassava mosaic virus replication-associated protein (AC1) transgene. Mol Plant Microbe Interact 9:219–225, 1996.
51. A Sangaré, D Deng, CM Fauquet, RN Beachy. Resistance to African cassava mosaic virus conferred by mutant of the putative NTP-binding domain of the Rep gene (AC1) in *Nicotiana bethamiana*. Mol Biol Rep 5:95–102, 1999.
52. T Frischmuth, J Stanley. African cassava mosaic virus DI DNA interferes with the replication of both genomic components. Virology 183:539–544, 1991.
53. Y Hong, K Saunders, MR Hartley, J Stanley. Resistance to geminivirus infection by virus-induced expression of dianthin in transgenic plants. Virology 220:119–127, 1996.
54. D Fargette, LT Colon, R Bouveau, CM Fauquet. Components of resistance to African cassava mosaic virus. Eur J Plant Pathol 102:645–654, 1996.
55. V Verdier, B Boher, H Mariate, JP Geiger. Pathological and molecular characterisation of *Xanthomonas campestris* strains causing diseases of cassava (*Manihot esculenta*). Appl Environ Microbiol 60:4478–4486, 1994.
56. PR Ronald. The molecular basis of disease resistance in rice. Plant Mol Biol 35:179–186, 1997.
57. JC Lozano, A Bellotti, JA Reyes, R Howeler, D Leihner, J Doll. Field problems in cassava Cali, Colombia, CIAT, 1981, pp 72–119.
58. TH Schuler, GM Poppy, BR Kerry, I Denholm. Insect-resistant transgenic plants. Trends Biotechnol 16:168–175, 1998.
59. J Schell. Cotton carrying the recombinant insect poison Bt toxin: No case to doubt the benefits of plant biotechnology. Curr Opin Biotechnol 8:235–236, 1997.
60. C James. Global review of commercialized transgenic crops: 1998. ISAAA Briefs No. 8. Ithaca, NY: ISAAA, 1998.
61. D Pink, I Puddephat. Deployment of disease resistance genes by plant transformation—a "mix and match" approach. Trends Plant Sci 4:71–75, 1999.
62. AC Bellotti, BV Arias, C Iglesias, E Barrera. Host plant resistance to whiteflies in cassava. Joint Annual Meeting, American Phytopathological Society of America, Las Vegas, 1998.
63. TRI Munyikwa, S Langeveld, SNIM Salehuzzaman, E Jacobsen, RGF Visser. Cassava starch biosynthesis: New avenues for modifying starch quantity and quality. Euphytica 96:65–75, 1997.
64. RGF Visser, E Jacobsen. Towards modifying plants for altered starch content and composition. Trends Biotechnol 11:63–68, 1993.
65. JMV Blanshard. Cassava starch, structure, properties and implications for contemporary processing. Proceedings Second International Scientific Meeting Cassava Biotechnology Network, Bogor, Indonesia, 1994, pp 625–638.
66. SNIM Salehuzzaman, E Jacobsen, RGF Visser. Isolation and characterization of a cDNA encoding granule-bound starch synthase in cassava (*Manihot esculenta* Crantz) and its antisense expression in potato. Plant Mol Biol 23:947–962, 1993.
67. OJM Goddijin, J Pen. Plants as bioreactors. Trends Biotechnol 13:379–387, 1995.
68. JM McMahon, WLB White, RT Sayre. Cyanogenesis in cassava (*Manihot esculenta* Crantz). J Exp Bot 46:731–741, 1995.
69. WLB White, DI Arias-Garzon, JM Mahon, RT Sayre. Cyanogenesis in cassava: The role of hydroxynitrile lyase in root cyanide production. Plant Physiol 116:1229–1225, 1998.
70. R Kapinga, N Mlingi, H Rosling. Reasons for bitter cassava in southern Tanzania. Afr J Root Tuber Crops 2:81–84, 1997.
71. H Rosling. Molecular anthropology of cassava cyanogenesis. In: BWS Sobral, ed. The Impact of Plant Molecular Genetics. New York: Birkhaeuser Boston 1996, pp 315–327.
72. M Bokanga, AJA Essers, N Poulter, H Rosling, O Tewe. Cassava safety: Lessons from an interdisciplinary workshop. Proceedings Second International Scientific Meeting Cassava Biotechnology Network, Bogor, Indonesia, 1994, pp 564–578.

73. MA Hughes, J Hughes, K Brown, S Liddle. Recent advances in molecular and biochemical studies of cyanogenesis in cassava: complexity of the cassava genome. Afr J Root Tuber Crops 2:77–81, 1997.
74. JE Ricard. Physiological deterioration of cassava roots. J Sci Food Agric 36:167–176, 1985.
75. JR Beeching, AD Dodge, KG Moore, HM Phillips, JE Wenham. Physiological deterioration in cassava: possibilities for control. Trop Sci 34:335–343, 1994.
76. H Li, Y Han, JR Beeching. Phenylalanine ammonium-lyase gene organisation, structure and activity in cassava. Brazilian Cassava Journal, Supplement, Fourth International Meeting Cassava Biotechnology Network, Salvador, Brazil, 1998, p 87.
77. LF Pereira, A Agyare-Tabbi, L Erikson. PCR amplification, cloning and expression of a phenylalanine ammonia-lyase gene in cassava. Afr J Root Tuber Crops 2:123–127, 1997.
78. J Huang, C Bachem, E Jacobsen, RGF Visser. Molecular analysis of postharvest deterioration in cassava. Brazillian Cassava Journal, Supplement, Fourth International Meeting Cassava Biotechnology Network, Salvador, Brazil, 1998, p 86.
79. HQ Li, I Potrykus, J Puonti-Kaerlas. Engineering leaf life in cassava. Brazilian Cassava Journal, Supplement, Fourth International Meeting Cassava Biotechnology Network, Salvador, Brazil, 1998, p 31.
80. PR Shewry, A Clowes, AS Tatham, JR Beeching. Opportunities for manipulating the protein and starch composition of cassava tuberous roots. Proceedings First International Scientific Meeting Cassava Biotechnology Network, Cartagena, Colombia, 1993, pp 252–254.
81. NMA Nassar, MA Viera, C Viera, D Grattapaglia. A molecular and embryonic study of apomixis in cassava (*Manihot esculenta* Crantz). Euphytica 102:9–13, 1998.
82. B Verdaguer, A de Kochko, RN Beachy, CM Fauquet. Isolation and expression in transgenic tobacco and rice plants of the cassava vein mosaic virus (CVMV) promoter. Plant Mol Biol 31:1129–1139, 1996.
83. B Verdaguer, A de Kochko, CI Fux, RN Beachy, CM Fauquet. Functional organisation of the cassava vein mosaic virus (CsVMV) promoter. Plant Mol Biol 37:1055–1067, 1998.
84. S Liddle, J Hughes, MA Hughes. Analysis of a cassava root specific β-glycodase promoter. Afr J Root Tuber Crops 2:158–162.
85. BJ Mazur, SV Tingey. Genetic mapping and introgression of genes of agronomic importance. Curr Opin Biotechnol 6:175–182, 1995.
86. MA Fregene, F Angel, R Gomez, F Rodriguez, W Roca, J Tohme, M Bonierbale. A molecular genetic map of cassava (*Manihot esculenta* Crantz). Theor Appl Genet 95:431–441, 1997.
87. P Chavarriaga-Aguirre, MM Maya, MW Bonierbale, S Kresovich, MA Fregene, J Tohme, G Kochert. Microsatellites in cassava (*Manihot esculenta* Crantz): Discovery, inheritance and variability. Trends Genet 97:493–501, 1998.
88. NJ Taylor, C. Schöpke, MV Masona, CM Fauquet. Development and potential impact of genetic engineering technologies in cassava. Biotechnol Int II:268–275, 1999.
89. AM Thro, NJ Taylor, CJJM Raemakers, J Pounti-Kaerlas, C Schöpke, RGF Visser, C Inglesias, MJ Sampaio, CM Fauquet, W Roca, I Potrykus. Maintaining the cassava biotechnology network. Nat Biotechnol 16:428–430, 1998.
90. NJ Taylor, CM Fauquet. Transfer of rice and cassava gene technologies to developing countries. Biotechnol Int 1:239–246, 1997.
91. GH Toenniessen. Plant biotechnology and developing countries. Trends Biotechnol 13:404–409, 1995.

# 37
# Transgenic Cauliflower with Insect Resistance

**K. Chengalrayan, Yih-Ming Chen, and Kai-Wun Yeh**
*National Taiwan University, Taipei, Taiwan*

**Po-Jen Wang**
*National Chung Hsing University, Taichung, Taiwan*

| | | |
|---|---|---|
| I. | INTRODUCTION | 547 |
| | A. Advantages of Genetically Engineered Insect-Resistant Plants | 548 |
| | B. *Bacillus thuringiensis* Toxins | 548 |
| | C. Proteinase Inhibitors | 548 |
| II. | SPORAMIN | 550 |
| III. | EXPRESSION OF TI GENE IN TOBACCO | 551 |
| IV. | TRANSFORMATION OF CAULIFLOWER WITH TI GENES | 553 |
| | A. Literature Survey | 553 |
| | B. Regeneration | 554 |
| | C. Bacterial Culture | 555 |
| | D. Transformation | 555 |
| V. | GENE EXPRESSION AND TI ACTIVITY | 555 |
| | A. Protein Extraction and TI Activity | 555 |
| | B. Immunoblotting | 556 |
| | C. Southern Hybridization | 556 |
| VI. | INSECT BIOASSAY | 556 |
| VII. | FUTURE DEVELOPMENTS | 559 |
| | REFERENCES | 559 |

## I. INTRODUCTION

Crop plants are susceptible to a wide range of herbivorous insects. Losses due to pests and diseases have been estimated at 37% of the agricultural production worldwide, with 13% due to insect pests each year (1). Present methods of crop protection rely mainly on the use of agrochemicals, and the cost associated with management practices and chemical control of insects approaches $10 billion annually. Moreover, insects have demonstrated a high capacity to develop resistance to a wide array of chemical insecticides (2). Therefore, it is necessary to develop a more environmentally friendly agriculture that will have decreased inputs in energy and chemicals and will not generate harmful outputs such as pesticide residues.

Some successes have been achieved by using conventional plant breeding techniques. However, the problems with traditional plant breeding for pest controls are that it is time-consuming and laborious. It is difficult to modify single traits and must rely on existing genetic variability. Plant genetic engineering offers the possibility of introducing resistance genes from foreign species into crop plants. Genetic engineering of the crop plants can make a major contribution to such inherently insect-tolerant varieties. Thus, genetic engineering opens up a virtually limitless source of germplasm for selection of insect control genes to introduce into elite crop varieties.

### A. Advantages of Genetically Engineered Insect-Resistant Plants

There are several advantages of genetically engineered insect-resistant plants:

1. Protection is provided continuously throughout the life cycle of a particular crop irrespective of weather conditions.
2. Protection is provided even to plant tissues such as roots, undersides of leaves, and insides of pods, which are difficult to protect by chemical sprays.
3. It allows transfer of insect resistance genes across the species of conventional breeding, such as different species, genera, and even kingdoms.
4. Insects are always treated at the most sensitive stage.
5. Only crop-eating insects are exposed.
6. The effective material is confined to the plant tissues expressing the transgene and therefore does not leach into the environment.

The advent of genetic engineering in the 1970s led to the possibility of using this technique to enhance crop resistance to pests. As a result, in 1994 the U.S. company Calgene obtained approval to commercialize the genetically modified Flavr Savr delayed ripening of tomato (3). From then on, the development and use of transgenic crops have gained momentum, with insect resistance traits ranking second only to herbicide resistance (3).

### B. *Bacillus thuringiensis* Toxins

*Bacillus thuringiensis* (*Bt*) is a commercially successful biological insecticidal agent. Different strains of *Bt* produce a variety of crystal toxins with distinct host ranges. At least 10 genes encoding different *Bt* toxins have been engineered into plants. These genes produce parasporal crystals that consist of about 130-kDa proteins known as δ-endotoxins (5), which are solubilized and processed in the insect midgut to active toxins of about 65–75 kDa comprising the N terminus of the proteins. They exert their toxicity by binding to midgut epithelial cells and ultimately by causing osmotic lysis through pore formation in the cell membrane that leads to the death of the insect (5). *Bt* toxins have been transferred and expressed in more than 26 different plant species (6). The levels of resistance they confer depend on whether native-bacterial or truncated, codon-optimized genes are used (7). Until now, codon-optimized genes have been transferred to a limited number of crops: cotton, maize, potato, broccoli, cabbage, and alfalfa. By far the greatest expenditure of research on plant genetic engineering has gone into insect-resistant transgenic crops expressing *Bt* toxin. However, there are obvious risks in reliance on this single resistance factor (8). Several insect species have evolved resistance to *Bt* gene–transformed crops (9–11). Therefore, many research projects are aimed at discovering non-*Bt* proteins to control insect pests.

### C. Proteinase Inhibitors

The alternative to the *Bt* toxin gene strategy is to use plant-derived insecticidal proteins that interfere with the nutritional needs of the insect. For example, polyphenol oxidases generate toxic compounds from dietary components (12), and proteinase inhibitors and α-amylase inhibitors (13) deprive the insect of nutrients by interfering with its digestive enzymes. Proteinase inhibitors have received ample attention because of their small size, abundance, and stability, which make them easy to work with (14). The introduction of specific proteinase inhibitors into plants is an alternative approach for obtaining crops that are resistant to insect attack. Serine, thiol, and aspartic proteinases and metalloproteinases are found in the plant kingdom. These proteinases are responsible for dietary protein in insects. Serine proteinases are often present as the main digestive enzymes when the pH of the insect midgut lumen content is neutral or alkaline, and cysteine and aspartic proteinases are present with more acidic gut contents (8).

It is advantageous to use these inhibitors as insect control agents because they are active against a wide range of insects. Moreover, they are inactivated during cooking and are not consumed through food. However, high levels of proteinase inhibitors are required for insect killing and there is a need to regulate expression of such inhibitors to specific tissues. Recently proteinase inhibitor proteins and their genes have received greater attention (Table 1).

Table 1  Proteinase Inhibitors That Have Been Engineered into Plants

| Proteinase inhibitors | Transformed plants | Reference |
|---|---|---|
| **Plant origin** | | |
| C-II (soybean serine proteinase inhibitor) | Potato | (15) |
| CMTI (squash trypsin inhibitor) | Tobacco | (16) |
| CpTI (cowpea trypsin inhibitor) | Lettuce | (17) |
| | Sweet potato | (18) |
| | Tobacco | (19) |
| | Tomato | (20) |
| 14K-CI (bifunctional cereal inhibitor of serine-proteinases and amylases) | Tobacco | (21) |
| OC-1 (rice cysteine proteinase inhibitor) | Poplar | (22) |
| | Tobacco | (23) |
| PI-IV (soybean serine proteinase inhibitor) | Potato | (15) |
| | Tobacco | (15) |
| Pot PI-II (potato proteinase inhibitor II) | Birch | (17) |
| | Rice | (23) |
| | Tobacco | (24), (25) |
| SKTI, Kti$_3$ (soybean Kunitz trypsin inhibitor) | Potato | (15) |
| Sweet potato sporamin (trypsin inhibitor) | Tobacco | (26) |
| | Cauliflower | (27) |
| Tomato proteinase inhibitor I | Alfalfa | (28) |
| | Nightshade | (28) |
| | Tobacco | (28), (24) |
| | Tomato | (29) |
| Tomato proteinase inhibitor II | Tobacco | (24) |
| | Tomato | (29) |
| **Animal origin** | | |
| Antichymotrypsin from *Menduca sexta* | Cotton | (30) |
| | Tobacco | (31) |
| Anti-elastase from *Manduca sexta* | Alfalfa | (32) |
| | Cotton | (30) |
| | Tobacco | (31) |

Search for plant species that are resistant to insects

↓

Identify compound (protein, secondary metabolites or others)

responsible for resistance

↓

Isolate and clone the gene(s) encoding the protein that

confers the resistance

↓

Introduce the gene into a crop and perform field trials

**Figure 1** Strategy for creating more resistant plants by using genetic engineering.

Since proteinase inhibitors are playing a key role, we searched for plants that were resistant to local insect pests, to isolate the gene responsible and to transfer it into commercially important crop(s). Following this strategy (Fig. 1), we cloned a trypsin inhibitor (called *sporamin*) gene from sweet potato tubers and transferred it to cauliflower.

## II. SPORAMIN

Sporamin is the major storage protein in sweet potato tuber (33). It accounts for 60–80% of the total soluble protein. Several reports suggested that a large amount of trypsin inhibitor (TI) is present in the tuber of sweet potato and total soluble protein concentrations in tubers are shown to be positively correlated with TI activity (34–38). It is known that the sporamin gene belongs to a gene family with more than 10 identified genes. Up to now, at least 8 complementary deoxyribonucleic acid (cDNA) clones and 4 genomic clones have been characterized by various groups (39,40). On the basis of nucleotide sequence homology, the sporamin gene can be divided into two subfamilies, called *sporamin A* and *sporamin B*. Moreover, the homology is approximately 90–98% within the subfamilies and 78–82% between subfamilies. Recent studies have demonstrated that the amino acid sequences predicted from sporamin cDNAs are homologous to the soybean TI and should be classified as plant Kunitz inhibitors (Fig. 2) (41). On the basis of the molecular mass (~21 kDa), two disulfide bonds, and further amino acid homology, it is classified as a serine type TI of the Kunitz family. Expression of the sporamin gene can be systematically induced in the leaf tissues of sweet potato by wounding. This suggests that sporamin may play some roles in the defense mechanism as a proteinase inhibitor. In order to identify the functional role of sporamin, we expressed cDNAs that encode preprosporamin, prosporamin, and sporamin constructed in *Escherichia coli* cells as fusion proteins (39). All three forms of sporamin were shown to have strong inhibitory activity against trypsin (39).

```
SPO  A    MKALTLALFLALSLYLLPNPAHSRFNPIRLPTTHEPASSE
SPO  B    MKALALF-FL-LSLYLLPNPAHSKFNPIRLRPAHETASSE
STI
ETI  a
ACTI


SPO  A    TPVLDINGDEVRAGGNYYMVSAIWGAGGGGLRLAHLDMMSK-CATDVIVS
SPO  B    TPVLDINGAEVRAGENYYIVSAIWGAGGGGLRLVRLDSSSNECASDVIVS
STI       DFVLDNEGEVVQNGGTYYILSDITAF-GG-IRAAPTGNER--CPLTVVQS
ETI  a    VLLDGNGEVVQNGGTYYLLPQVWAQGGG-VQLAKTGEET--CPLTVVQS
ACTI      KELLDATGTILRNGGAYYILPALRGKGGG-LTLAKTGDES--CPLTVVQA


SPO  A    PNDLDNGDPITITPATADPESTVVMA-STYQTFRFNIATNKLCVNNV-NW
SPO  B    RSDFNNGDPITITP--ADPESTVVMP-STFQTFRFNIATNKLCVNNV-NW
STI       RNELDKGIGTIISPSYRIRFIAEGHPLS--LKFDS-FAVIMLCVGIPTEW
ETI  a    PNELSNGKPIRIESRLRSAFIPDDD-----KVRIGFAYAPKCAPSP-WW
ACTI      QSTTKRGLPAVIWTPPKIAILTPGFYL----NFEFQPRDLPACLQKY-ST


SPO  A    GIQHDSASGQYFLKAGEFVSDN-SNQ-FKIELVDANLNS-YKLTYCQFGS
SPO  B    GIQHDSESGQYFVKAGEFVSDN-SNQ-FKIEVVNDNLNA-YKISYCQFGT
STI       SVVEDLPEGPAV-KIGENKDAM-DGW-FRLERVSDDEFNNYKLVFCPQQA
ETI  a    TVLEDEQEGLSV-KLSEDESTQFDYP-FKFEQVSDKLHS-YKLLYCEGKH
ACTI      LPWKVEGESQEV-KIAPKEKEQFLVGSFKIKPYRDD----YKLVYCEGNS


SPO  A    D--KCYNVGRFHDHM--LRTTRLALS-NSPF-VFVI-KPTDV
SPO  B    E--KCFNVGRYYDPL--TRATRLALS-NTPF-VFVI-KPTDM
STI       EDDKCGDIGISIDDDGHTR--RLVVSKNKPL-VVQFQK-LDKESL
ETI  a    E--KCASIGINRDQKGY-R--RLVVTEDNPLTVV-LKK--DESS
ACTI      DDESCKDLGISIDDENN-R--RLVVKDGHPLAVRFE-KAHRSG
```

**Figure 2** Alignment of the derived amino acid sequences of sporamin SPOA, sporamin SPOB, soybean Kunitz-type trypsin inhibitor (STI), *Acacia confusa* trypsin inhibitor (ACTI), and *Erythrina variegata* trypsin inhibitor (ETI). The sequences showing homology are boxed. SPOA is deduced from spTI-1 cDNA sequence, and SPOB deduced from pIMO 336 cDNA, which belongs to sporamin B gene family. (From Ref. 38.)

## III. EXPRESSION OF TI GENE IN TOBACCO

A full-length 0.93-kb cDNA encoding the TI was subcloned into the *Xba* I/*Bam*H I sites of the plant transformation vector PBI 121 under the control of the 35S cauliflower mosaic virus (CaMV) promoter and with 3′ Nos terminator (Fig. 3). The CaMV 35S promoter has been used in the majority of insect-resistant transgenic plants (6). Selectable marker genes are introduced

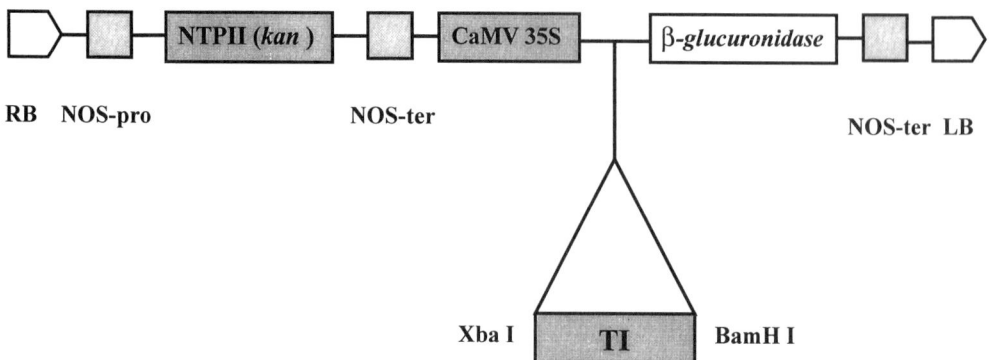

**Figure 3** Construct of pBI121/TI vector used in cauliflower transformation (TI 0.93 kb, pBI121 13 kb, pBI121/TI 13.93 kb).

alongside the insect resistance gene to allow selective growth of plant cells that have incorporated the new genes. We employed the most commonly used marker, the bacterial neomycin phosphotransferase II (npt II) gene, for the selection of transgenic plants.

Using standard molecular procedures, the sequence of the construct was determined to identify the precise orientation (42). Finally, the vector with the gene in correct orientation was transferred to *Agrobacterium tumefaciens* LBA 4404 and integrity of the plasmid was verified by Southern analysis.

Initially we conducted experiments on tobacco plants and transferred the gene to tobacco by using the leaf disk transformation method described by Horsch et al. (43). From 30 independent transformants, 6 transgenic lines that grew faster and developed vigorous roots in soil were chosen for further experiments such as TI activity and Western analysis. Northern blot analysis with $^{32}$P-labeled sporamin TI cDNA probe clearly showed that 0.9-kB transcripts were present in all transgenic tobacco lines (26). The TI gene under the control of the 35S promoter led to the production of trypsin inhibitor proteins up to 0.2% of the total protein.

To test the insect activity of the introduced TI gene, transgenic and control tobacco plants were produced from stem cuttings and grown in 20-cm pots with loam-based compost in a temperature-controlled growth chamber. When the plants were about 30 cm tall, they were individually infected with 50 early second instar larvae of tobacco cutworm. Larvae were distributed on leaves and the plants were surrounded with 1-mm nylon mesh. After an incubation period of 4 days, the larval population was monitored by counting and weighing all larvae present on each plant. The experiment was repeated to confirm the results. After an incubation period of 4 days, the leaves of the control tobacco plants were completely eaten. However, insecticidal bioassays of transgenic tobacco plants showed that the larval growth of *Spodoptera litura* (F.), the tobacco cutworm, was severely retarded, as compared to their growth on control plants (Table 2). The high level of TI expression in Tb-7 transformant tobacco leaves had deleterious effects on larvae, causing a 93.3% mortality rate, and decreasing mean larval weight was correlated with a decrease in the leaf surface eaten. However, larvae fed leaves from Tb-6 plants expressing TI at a low level did not die. These results showed that TI gene can be used for insect resistance in other crops. With this background, experiments to transfer TI gene into cauliflower were carried out.

**Table 2** Insect Bioassay on Different Lines of Transgenic Tobacco (Tb) Plants

| Tobacco plants | Insect survival (%) | Average body weight (mg) | Leaf area eaten per larva (cm$^2$) |
|---|---|---|---|
| Wild type | 100  | 33.4 ± 4.4 | 5.5 |
| Tb-1 | 43.5 | 15.7 ± 4.2 | 4.3 |
| Tb-2 | 77.8 | 17.8 ± 5.9 | 2.6 |
| Tb-3 | 96.4 | 12.0 ± 3.1 | 1.6 |
| Tb-6 | 96.4 | 12.0 ± 3.1 | 1.6 |
| Tb-7 | 6.7  | 6.1 ± 0.8  | 0.8 |
| Tb-8 | 91.3 | 16.3 ± 3.3 | 2.2 |

## IV. TRANSFORMATION OF CAULIFLOWER WITH TI GENE

The world annual production of cauliflower is over 12.7 million metric tons, of which more than 30% is produced in China (FAO, 1997). It is one of the common *Brassica* sp. vegetables grown in Taiwan, and the production of this crop is second only to production of cabbage. At least six varieties of this crop are grown in different seasons. However, one of the major problems in this crop is insect damage. Some of the insect pests are listed in Table 3 (44).

### A. Literature Survey

*Brassica* spp. and related species are fairly easy to handle in tissue culture systems, making this genus a suitable subject for exploiting techniques of genetic transformation for breeding purposes (45).

**Table 3** Important Insect Pests of Cauliflower

| | |
|---|---|
| Blossom beetles | *Meligethes* spp. |
| Brassica pod midge | *Dasineura brassica* |
| Cabbage aphid | *Brevicoryne brassicae* |
| Cabbage butterflies | *Pieris brassicae, P. rapae, P. napi* |
| Cabbage gall weevil | *Ceutorhynchus assimilis* |
| Cabbage leaf miner | *Phytomyza rufipes* |
| Cabbage moth | *Mamestra brassicae* |
| Cabbage root fly | *Delia radicum* |
| Cabbage stem flea beetle | *Psylliodes chrysocephala* |
| Cabbage stem weevil | *Ceutorhynchus quadridens* |
| Cabbage white fly | *Aleurodes proletella* |
| Cut worms | *Agrotis* spp. |
| Diamondback moth | *Plutella xylostella* |
| Flea beetles | *Phyllotreta* spp. |
| Garden pebble moth | *Evergestis forficalis* |
| Slugs | *Arion hortensis* |
|  | *Agriolimax agrestis* |
| Swede midge | *Contarina nasturtii* |
| Turnip gall weevil | *Ceutorhynchus pleurostigma* |
| Turnip root fly | *Delia floralis* |

**Table 4** Literature Survey on Genetic Transformation of Cauliflower

| Explant | Response | Mode of transformation | Source |
| --- | --- | --- | --- |
| Hypocotyl | Trangenic plants | *Agrobacterium rhizogenes*, (A4, 15834, 8196, TR101, TR7) | (46) |
| Leaf disk and petiole | Kanamycin-resistant plants | *A. tumefaciens* (C58, C1RiF) | (47) |
| Leaf disk | Kanamycin-resistant plants | *A. tumefaciens* (6042, 6044, 6046, 6048) | (48) |
| Mesophyll protoplasts | Kanamycin-resistant plants | PEG uptake Electroporation | (48) |
| In vitro plants | Cauliflower mosaic virus–resistant plants | *A. tumefaciens* (C58p, MP90) | (49) |
| Hypocotyl | Insect-resistant plants | *A. tumefaciens* (LBA 4404) | (27) |
| Hypocotyl | Basta-resistant plants | *A. tumefaciens* (C58 C1 RiF) | (50) |
|  | Transgenic plants with altered pollen stigma interaction and fully compatible when self-pollinated | *A. tumefaciens* (CIB542) | (51) |
| Leaf and petioles of intact cotyledons | Kanamycin-resistant plants | *A. rhizogenes* (A4T) | (52) |

Genetic engineering protocols for transferring marker, herbicide-resistant, and self-compatible genes to cauliflower have been developed by various groups (Table 4). However, there were no reports on generation of insect-resistant cauliflower; therefore, experiments to transfer the sporamin gene into Taiwan cauliflower cultivars to combat insect pests were initiated.

## B. Regeneration

Three key Taiwanese cauliflower cultivars, 'Known You Early No 2,' 'Snow Lady,' and 'Beauty Lady,' were used for regeneration. Seeds of all three cultivars were disinfected in 70% ethanol for 10 seconds, followed by 0.5% sodium hypochlorite for 50 minutes (with sonication during the initial 20 minutes). The seeds were germinated for 4 days on $B_5$ basal medium (53) with 3% sucrose and 0.8% Bacto gar. The incubation conditions for germination and other in vitro experiments were at 25°C and with a photoperiod of approximately 28 $\mu E\ m^{-2}\ s^{-1}$ for 16 hours. Three different explants (hypocotyls, cotyledonary petioles, and cotyledons) were excised and tested for their regeneration capacities in $B_5$ basal medium with 0.2 mg/l indole-3-acetic acid (IAA) and 0.5, 1.0, 2.0, or 4.0 mg/l benzyl adenine (BA).

Among three explants tested, the hypocotyl explants of all the cultivars displayed the highest regeneration rate (data not shown) after 3 weeks incubation; adventitious buds formed primarily at the cut ends adjacent to the apical meristems. The optimal BA concentration for shoot bud regeneration for 'Known You Early No 2' and 'Snow Lady' was 5.0 mg/l and for 'Beauty Lady' 1.0 mg/l. At these BA concentrations, the response was 74%, 68%, and 37% for 'Beauty Lady,' 'Snow Lady,' and 'Known You Early No. 2,' respectively. The further experiments were carried out by using hypocotyl explants.

A significant breakthrough was achieved when the explants were pretreated for 3 days in callus-inducing (CI) medium ($B_5$ basal medium supplemented with 1 mg/l 2,4-dichlorophenoxyacetic acid [2,4-D] and 0.5 mg/l Kinetin [Kn]) followed by 29.4 µM silver ion (prepared by mixing 1:3 molar concentration solutions of silver nitrate and sodium thiosulfate and adding them to culture medium via filter sterilization). The percentage of regeneration frequency in all three cultivars exceeded 95%. This work is the first successful case of applying an ethylene inhibitor to in vitro cauliflower regeneration (27). The stimulating effects of the silver ion and other ethylene inhibitors on in vitro regeneration of *Brassica* spp. have been documented (54–56). Added proof of ethylene's regulating in vitro shoot regeneration was provided in generating transgenic *Brassica* sp. plants with an antisense ACC gene (57).

## C. Bacterial Culture

*Agrobacterium tumefaciens* LBA 4404 with PBI 121 was cultured in YEP medium (beef extract 5 g/l, yeast extract 1 g/l, bactopeptone 5 g/l, sucrose 5 g/l, and $MgSO_4$ 2 mM) containing 100 mg/l kanamycin and incubated in a shaker at 240 rpm and 28°C for 2 days. The suspensions were then mixed with sterile glycerol (1:1 vol/vol) and stored at −70°C. The *Agrobacterium* sp. inoculum from glycerol stock was added to YEP medium containing 100 mg/l kanamycin (1:25 vol/vol) and incubated under the specified conditions. After 10 hours, 50 µM acetosyringone was added and incubated further for 8 hours. The bacterial culture was then centrifuged at 3500 rpm and the pellet was resuspended in an equal volume of liquid CI medium with 10 mM D-glucose. The culture was 100 times diluted for transformation.

## D. Transformation

The hypocotyl explants (precultured in callus-inducing medium for 3 days) were incubated with 10 ml bacterial supension for 1 hour in darkness. The explants were blot dried in sterile Whatman filter paper and incubated in solid CI medium at 25°C in dark for 72 hours. The cocultivated explants were washed with 1.5% D-mannitol with 1500 mg/l carbencillin, blotted dry, and placed in shoot induction medium supplemented with 150 mg/l carbenicillin and 50 mg/l kanamycin. By this method more than 100 kanamycin-resistant plants were obtained.

## V. GENE EXPRESSION AND TI ACTIVITY

Some of the transgenic plants growing in 50 mg/l kanamycin containing medium were used for in vitro TI activity, Southern blot, and Western blot assays.

### A. Protein Extraction and TI Activity

Soluble proteins were extracted from 0.3 g of fresh leaf of each putative transgenic and control plant, by grinding in liquid nitrogen with 3× volume of extraction buffer containing 30 mM Tris-HCl (pH 7), 1% PVP, and 1% vitamin C. After centrifugation at 12,500 rpm for 15 minutes, the supernatant containing soluble proteins was collected and stored at −70°C for in vitro TI activity and immunoblot assays.

The TI activity assay was based on the Geiger and Fritz (58) cell free trypsin assay method with N-benzoyl DL-arginine-4-nitroanilide hydrochloride (DL-BAPA) substrate. For TI assay, approximately 50 µg of protein sample (from crude extract) was added to 500 µl of reaction buffer (10 mM $CaCl_2$, 100 mM Tris-HCl, pH 8.0) containing 3 µg trypsin (step 1) and incubated at 37°C for 10 minutes. After adding different concentrations of BAPA (stock concentration, 50 mg BAPA dissolved in 1 ml DMSO) in 500 µl of reaction buffer (step 2), it was further incubated for 20–30 minutes. The reaction was terminated with 500 µl of 10% acetic acid (step 3) and measured for $OD_{405}$.

The trypsin hydrolyzes the substrate and forms 4-nitroaniline, which absorbs light at $OD_{405}$. TI activities were measured by the reduction in the readings due to presence of TI over the TI-free control (Table 5). All the tested putative transgenic plants showed significantly higher TI protein content than control plants.

**Table 5** In Vitro TI Activity Assay of Transgenic Cauliflower Plants (Sample 1296 'Snow Lady,' 1298 'Beauty Lady,' 1-1, 1-2, and others) and Nonresistant Control (CK) Plants[a]

| Sample | | $OD_{405}$ |
|---|---|---|
| 1296 | 1-1 | 1.256 ± 0.061 |
| | 1-2 | 1.120 ± 0.22 |
| | 1-3 | 1.265 ± 0.041 |
| | 2-1 | 0.934 ± 0.010 |
| 1290 | 2-2 | 1.101 ± 0.037 |
| | 3-1 | 1.175 ± 0.007 |
| | 3-2 | 1.127 ± 0.051 |
| | 3-3 | 1.184 ± 0.009 |
| | 5-1 | 1.074 ± 0.014 |
| | 5-3 | 0.831 ± 0.031 |
| CK | 1 | 1.387 ± 0.016 |
| | 2 | 1.358 ± 0.003 |

[a]Values are averages of three readings ± SD. The lower the OD value, the higher the TI level.

## B. Immunoblotting

Appropriate quantities of protein were fractionated by SDS-PAGE following the Laemmli (59) procedure and transferred to nitrocellulose by using Mini Trans Blot Cell (Bio Rad, Hercules, CA). The TI polypeptide was detected by using a rabbit anti-TI serum and goat antirabbit immunoglobulin G (IgG) coupled with horseradish peroxidase as the secondary antibody. The TI polypeptide was detected as a 21-kDa band that was present in all the six transgenic plants tested, whereas the control plant did not show such a band (Fig. 4).

## C. Southern Hybridization

DNA was isolated from fresh leaf tissues as described by Dellaporta and coworkers (60). Primers used in polymerase chain reaction (PCR) were the flanking sequences of the TI cDNA with 5'CATGAAGCCCTCACACTG3' at the 5' end and 5'CATTACACATCGGTAGGTTTG3' at the 3' end. PCR products and biotinylated probe of cDNA (prepared according to the Feinberg and Vogelstein (61) protocol) were used in Southern hybridization with the NEBlot Phototype kit and its procedures from New England Biolabs (Beverly, MA)

The genomic DNAs from all tested plants produced 0.66-kb DNA segments after PCR, whereas that of the control plant did not. Southern hybridization with TI cDNA probe confirmed that these 0.66-kb DNA segments contained TI gene sequences (Fig. 5).

## VI. INSECT BIOASSAY

Small-scale insect feeding trials were carried out. Leaves of two transgenic 'Snow Lady' plants were used to feed the first instar larvae of the common *Brassica* sp. lepidopteran pests (mandible), *Spodoptera litura*, and *Plutella xylostella* in closed containers. The remaining confirmed transgenic plants were cloned in vitro and used in an open infestation test carried out in a greenhouse with opened windows.

A high degree of insect protection was evident in our bioassays (Table 6), (Fig. 6A, B).

# Transgenic Cauliflower

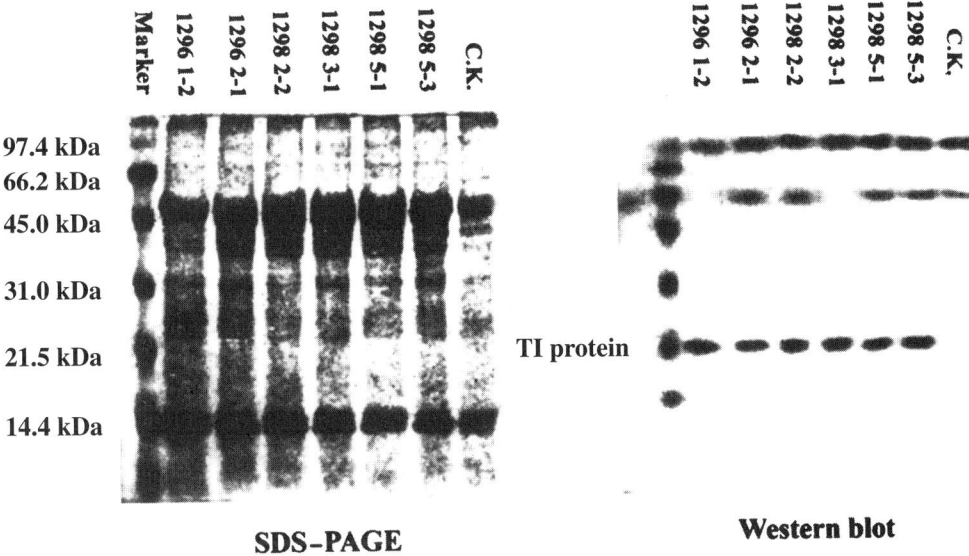

**Figure 4** PCR and Southern blot analysis of the selected kanamycin-resistant regenerant ($R_0$) cauliflower plants (1296 'Snow Lady,' 1298 'Beauty Lady') and control (CK) plants. (From Ref. 26.)

**Figure 5** Insect resistance of transgenic and control cauliflowers: (left) cauliflower transformed with sporamin gene and (right) control.

**Table 6** Insect Bioassays Using Two Clonal Plantlets of a TI Trangenic Cauliflower 'Snow Lady' Plant 1296 1-1 and Nontransgenic Control (CK) Plants

| Insect species | Host plant | Larvae survived (%) | Larvae reaching pupa stage (%) | Larvae reaching adult stage (%) |
|---|---|---|---|---|
| *Spodoptera litura* | 'Snow Lady' 1296 1-1 | 56.7 | 22.7 | 13.2 |
| | CK | 86.7 | 63.6 | 49.1 |
| *Plutella xylostella* | 'Snow Lady' 1296 1-1 | 60.0 | 36.0 | 21.6 |
| | CK | 100.0 | 100.0 | 100.0 |

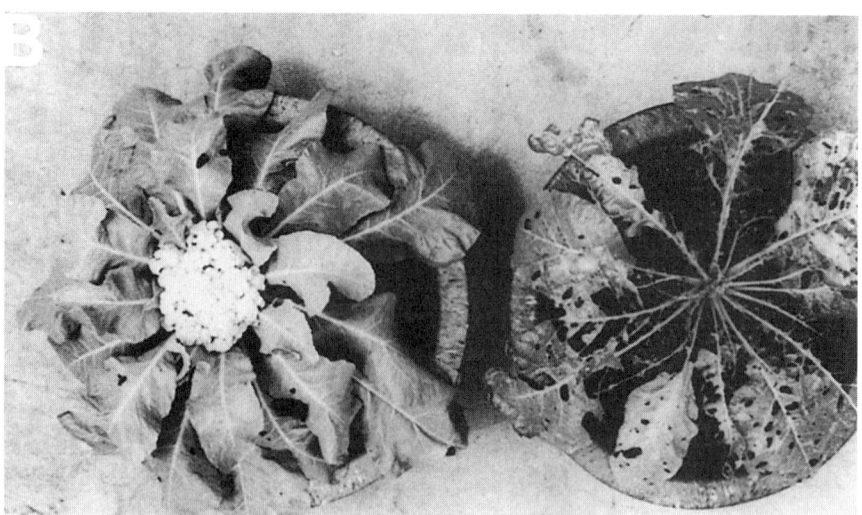

**Figure 6** Plant open infestation test carried out in a greenhouse with open windows (left, TI transgenic cauliflower plants in vitro cloned from $R_0$ plants). The insects were identified as *Pieris conidia* (A) and *Plutella xylostella* (B). (From Ref. 26.)

These results demonstrated that the TI proteins produced in the transgenic cauliflower plants were functionally active in planta. Our insect bioassays, although at small scales, also provided evidence that the introduction of a protease inhibitor gene into transgenic crop plants can be efficacious. Analysis of the resultant $R_1$ and $R_2$ generation plants as well as a large-scale field test of the clonal $R_0$ plants are currently in progress.

## VII. FUTURE DEVELOPMENTS

The search for the novel resistance genes will be continued. For example, we recently discovered another trypsin inhibitor from sweet potato leaves. Different genes will be combined in plants to increase the range of pests affected and to delay the development of insect resistance to the gene products. More specific and powerful promoters are needed to replace, at least partly, the CaMV 35S promoter. We expect to expand our technology to commercialize vegetable crops such as cauliflower, cabbage, and Chinese kale in the near future.

## REFERENCES

1. AMR Gatehouse, D Boulter, VA Hilder. Potential of plant derived genes in the genetic manipulations of crops for insect resistance. In: AMR Gatehouse, VA Hilder, D Boulter, eds. Biotechnology in Agriculture. No. 7. Plant Genetic Manipulations for Crop Protection. CAB International, 1992, pp 155–181.
2. WK Moberg. Understanding and combating agrochemical resistance. In: MB Green, MND Lemaranh, WK Mobverg, eds. Managing Resistance to Agrochemicals. ACS symposium series. Washington DC: 1990, pp 3–16.
3. C James. Global status and distribution of commercial transgenic crops in 1997. Biotechnol Dev Monit 35:9–12, 1998.
4. MG Koziel, NB Carozzi, TC Currier, GW Warren, SV Evola. The insecticidal crystal proteins of *Bacillus thuringiensis:* Past, present and future uses. Biotechnol Genet Eng Rev 11:171–228, 1993.
5. SS Gill, EA Cowles, PV Piclrantoniop. The mode of action *Bacillus thuringiensis* endotoxins. Annu Rev Entomol 37:615–636, 1992.
6. TH Schuler, GM Poppy, BR Kerry, I Denholm. Insect resistant transgenic plants. Trends Biotechnol 16:168–175, 1998.
7. M Peferoen. Progress and prospects for field use of Bt genes in crops. Trends Biotechnol 15:173–177, 1997.
8. D Boulter. Insect pest control by copying nature using genetically engineered crops. Phytochemistry 34:1453–1466, 1993.
9. Mcgaughey WH. Insect resistance to the biological insecticide *Bacillus thuringiensis*. Science 229: 193–195, 1985.
10. YB Liu, BE Tabashnik, E Bruce. PC Marianne. Field-evolved resistance to *Bacillus thuringiensis* toxin CrylC in diamond back moth (*Lepidoptera: Plutellidae*). J Econ Entomol 89:798–804, 1996.
11. YB Liu, BE Tabashnik. Inheritance of resistance to the *Bacillus thuringiensis* toxin Cry1C in the diamondback moth. Appl Environ Microbiol 63:2218–2223, 1997.
12. G Felton, K Donato, R Broadway, S Duffey. Impact of oxidized plant phenolics on the nutritional quality of dietary protein to noctuid herbivore, *Spodoptera exigua*. J Insect Physiol 38:277–285, 1992.
13. JE Husing, RE Shade, MJ Christpeels, LL Murdock. α-amylase inhibitor, not phytohemagglutinin, explains the resistance of common bean seeds to cowpea weevil. Plant Physiol 96:993–996, 1991.
14. C Ryan. Protease inhibitors in plants: Genes for improving defences against insects and pathogens. Annu Rev Phytopathol 28:425–446, 1990.
15. S Marchetti, A Giordana, AM Olivieri, C Fogher, M Delledonne. Genetic engineering for insect resistance in potato: Introduction of foreign proteinase inhibitor genes. Potato Res 37:450–451, 1994.

16. SC MacIntosh, GM Kishore, FJ Perlak, PG Marrone, TB Stone, SR Sims, RL Fuchs. Potentiation of *Bacillus thuringiensis* insecticidal activity by serine protease inhibitors. J Agric Food Chem 38:1145–1152, 1990.
17. AMR Gatehouse. Biotechnological application of plant genes in the production of insect resistant crops. In: SL Clement, SS Quinsenberry, eds. Global Plant Genetic Resources for Insect Resistant Crops. Boca Raton, FL: CRC Press, 1998.
18. CA Newell, JM Lowe, A Merryweather, LM Rooke, WDO Hamilton. Transformation of sweet potato (*Ipomoea batatas* (L.) Lam) with *Agrobacterium tumefaciens* and regeneration of plants expressing cowpea trypsin inhibitor and snowdrop lectin. Plant Sci 107:215–22, 1995.
19. VA Hilder, AMR Gatehouse, SE Sheerman, RF Barker, D Boulter. A novel mechanism of insect resistance engineered into tobacco. Nature 300:160–163, 1997.
20. SA Masoud, X Ding, LB Johnson, FF White, GR Reeck. Expression of a corn bifuntional inhibitor of serine proteinases and α-amylases in transgenic tobacco plants. Plant Sci 115:59–69, 1996.
21. JC Leplé, M Bonadé-Bottino, SM Augustin, GM Pilate, V Dumanios Lê Tân, A Delplanque, D Cornu L Jounin. Toxicity to Crysomela tremule (Coleoptera: Chrysomelidae of transgenic polars expressing cystein proteinase inhibitor. Mol Breed 1:319–328, 1995.
22. SA Masoud, LB Johnson, FF White, GR Reeck. Expression of a cysteine proteinase inhibitor (Oryzacystein-I) in transgenic tobacco plants. Plant Mol Biol 21:655–663, 1993.
23. X Duan, X Li, Q Xue, M Abo-El-Saad, D Xu, R Wu. Transgenic rice plants harboring an introduced potato proteinase inhibitor II gene are insect resistant. Nat Biotechnol 14:494–498, 1996.
24. R Johnson, JJ Narvaez, G An, GC Ryan. Expression of proteinase inhibitors I and II in transgenic tobacco plants: Effects of natural defence against *Manduca sexta* larvae. Proc Natl Acad Sci USA 86:9871–9875, 1989.
25. MA Jongsma, PL Bakker, J Peters, D Bosch, WJ Stiekema. Adaptation of *Spodoptera exigua* larvae to plant proteinase inhibitors by induction of proteinase activity insensitive of inhibition. Proc Natl Acad Sci USA 92:8041–8045, 1995.
26. KW Yeh, MI Lin, SJ Tsuan, YM Chen, CY Lin, SS Kao. Sweet potato (*Ipomoea batatas* L.) trypsin inhibitors expressed in transgenic tobacco plants confer resistance against *Spodoptera litura*. Plant Cell Rep 16:696–699, 1997.
27. LC Ding, CY Hu, KW Yeh, PJ Wang. Development of insect resistant transgenic cauliflower plants expressing the trypsin inhibitor gene isolated from local sweet potato. Plant Cell Rep 17:854–860, 1998.
28. J Narvaez-Vasquez, ML Orozco-Cardenas, CA Ryan. Differential expression of a chimeric CaMV-tomato proteinase Inhibitor I gene in leaves of transformed nightshade, tobacco and alfalfa plants. Plant Mol Biol 20:1149–1157, 1992.
29. B McGurl, M Orozco-Caradenas, G Pearce, CA Ryan. Overexpression of the prosystemin gene in transgenic tomato plants generates a systemic signal that constitutively induces proteinase inhibitor synthesis. Proc Natl Acad Sci USA 91:9799–9802, 1994.
30. JC Thomas, DG Adam, VD Keppenne, CC Wasmann, JK Brown, MR Kanost, HJ Bohnert. Proteinase inhibitors of *Menduca sexta* expressed in transgenic cotton. Plant Cell Rep 14:758–762, 1995.
31. JC Thomas, DG Adam, VD Keppenne, CC Wasmann, JK Brown, MR Kanost, HJ Bohnert. *Menduca sexta* encoded proteinase inhibitors expressed in *Nicotiana tabaccum* provide protection against insects. Plant Physiol Biochem 33:611–614, 1995.
32. JC Thomas, CC Wasmann, CM Echt, RL Dunn, HJ Bohnert, TJ McCoy. Introduction and expression of an insect protease inhibitor in alfalfa (*Medicago sativa* L.). Plant Cell Rep 14:31–36, 1994.
33. M Maeshima, T Sasaki, T Asahi. Characterization of major proteins in sweet potato tuberous roots. Phytochemistry 24:1899–1902, 1985.
34. JC Boukamp, SCS Tsou, SSM Lin. Genotype and environment effects on the relationship between protein concentration and trypsin inhibitor levels in sweet potatoes. Hortscience 20:886–889, 1989.
35. LF Dickey, WW Collin. Cultivar differences in trypsin inhibitors of sweet potato roots. J Am Soc Hortic Sci 109:750–754, 1984.
36. HS Li, Obak. Major soluble proteins of sweet potato roots and change in proteins after cutting, infection or storage. Agric Biol Chem 49:733–744, 1985.
37. YH Lin, JF Cheng, HY Fu. Partial purification and some properties of trypsin inhibitors of sweet potato (*Ipomoea batatas* L.) roots. Bot Bull Acad Sin 24:103–113, 1983.

38. YH Lin, BS Tsu. Some factors affecting levels of trypsin inhibitor activity of sweet potato (*Ipomoea batatas* L.) root. Bot Bull Acad Sin 28:139–149, 1987.
39. KW Yeh, JC Chen, MI Lin, YM Chen, CY Lin. Functional activity of sporamin from sweet potato (*Ipomoea batatas* L.): A tuber storage protein with trypsin inhibitor activity. Plant Mol Biol 33:565–570, 1997.
40. A Ishikawa, S Ohta, K Matsuoka, T Hattori, K Nakamura. A family of potato genes that encode Kunitz-type proteinase inhibitors: structural comparisons and differential expression. Plant Cell Physiol 35:303–312, 1994.
41. HD Bradshaw Jr, JB Hollick, TJ Parson, HRG Clark, MP Gordan. Systematically wound-responsive genes in poplar tree encode proteins similar to sweet potato sporamins and legume Kunitz trypsin inhibitors. Plant Mol Biol 14:51–59, 1989.
42. J Sambrook, EF Fritsch, T Maniatis. Molecular Cloning: A Laboratory Manual. Cold Spring Harbor, NY: Cold Spring Harbor Laboratory Press, 1989.
43. RB Horsch, JE Fry, NL Hoffman, DM Eicholtz, SG Rogers, RT Fraley. A simple and general method for transferring genes into plants. Science 227:1229–1231, 1985.
44. RB Maude, AR Thompson. Pests and diseases of vegetables. In: N Scopes, L Stables, eds. Pests and Disease Control Hand Book. BCPC, 1989, pp 261–321.
45. GB Poulsen. Genetic transformation of *Brassica*. Plant Breed 115:209–225, 1996.
46. C David, J Timpe. Genetic transformation of cauliflower (*Brassica oleracea* L. var. *Botrytis*) by *Agrobacterium rhizogenes*. Plant Cell Rep 7:88–91, 1988.
47. V Srivastava, AS Reddy, SG Mukherjee. Transformation and regenration of *Brassica oleracea* mediated by an oncogenic *Agrobacterium tumefaciens*. Plant Cell Rep 7:504–507, 1988.
48. K Eimert, F Siegemund. Transformation of cauliflower (*Brassica oleracea* L. var. *Botrytis*)—an experimental survey. Plant Mol Biol 19:485–490, 1992.
49. E Passelegue, C Kerlin. Transformation of cauliflower (*Brassica oleracea* L. var. *Botrytis*) by transfer of cauliflower mosaic virus genes through combined co-cultivation with virulent and avirulent strains of *Agrobacterium*. Plant Sci 113: 79–89, 1996.
50. M DeBlock, D Brouwer, P Tenning. Transformation of *Brassica napus* and *Brassica oleracea* using *Agrobacterium tumefaciens* and the expression of the bar and neo genes in the transgenic plants. Plant Physiol 91:694–701, 1989.
51. K Toriama, JC Stein, ME Nasrallah, JB Nasrallah. Transformation of *Brassica oleracea* with an S-locus gene from *B. campestris* changes the self-incompatibility phenotype. Theor Appl Genet 81:769–776, 1991.
52. MC Christey, BK Sinclair, RH Bruan, L Wyke. Regeneration of transgenic vegetable brassicas (*Brassica oleracea* and *Brassica campestris*) via Ri mediated transformation. Plant Cell Rep 16:587–593, 1997.
53. OL Gamborg, RA Miller, K Ojima. Nutrient requirements of suspension cultures of soybean root cells. Exp Cell Res 50:151–158, 1968.
54. GL Chi, DG Barfield, GE Sim, EC Pua. Effect of $AgNO_3$ and *aminoethoxyvinylglycine* on *in vitro* shoot organogenesis from seedling explants of recalcitrant *Brassica* genotypes. Plant Cell Rep 9:195–198, 1990.
55. GE Palmer. Enhanced shoot regeneration from *Brassica campestris* by silver nitrate. Plant Cell Rep 11:541–545, 1992.
56. L Burnett, M Arnold, B Huang. Enhancement of shoot regeneration from cotyledon explants of *Brassica rapa* ssp. *Olcifera* through pretreatment with auxin and cytokinin and use of ethylene inhibitors. Plant Cell Tissue Organ Cult 37:253–256.
57. EC Pua, JEE Lee. Enhanced *de novo* shoot morphogenesis *in vitro* by expression of antisense 1-aminocyclopropane-1-carboxylate oxidase gene in transgenic mustard plants. Planta 196:69–76, 1995.
58. R Geiger, H Fritz. Two proteinases and inhibitors-trypsin. In: J Bergmeyer, M Grabl, eds. Methods of enzymatic analysis. Vol. V. Enzymes 3: Peptidases, Proteinases and Their Inhibitors. Weinheim: Chemiew, 1984, pp 119–129.
59. UK Laemmli. Cleavage of structural proteins during the assembly of the head of bacteriophage T4. Nature 227:680–685, 1970.

60. SL Dellaporta, J Wood, JB Hicks. Maize DNA minipreperation. In: R Malmberg, J Messing, I Sussex, eds. Molecular Biology of Plants. Cold Spring Harbor, NY: Cold Spring Harbor Laboratory Press, 1984, pp 36–37.
61. AP Feinberg, B Vogelstein. A technique for radiolabelling DNA restriction endonuclease fragments to high specific activity. Anal Biochem 132:6–13, 1983.

# 38
# Virus-Resistant Chili Pepper Produced by *Agrobacterium* Species–Mediated Transformation

**Wen-qi Cai, Rong-Xiang Fang, Feng-li Zhang, Jiu-chun Zhang, Xiaoying Chen, Gui-ling Wang, and Ke-qiang Mang**
*Chinese Academy of Sciences, Beijing, China*

**Hong-sheng Shang, Xu Wang, and Yue-ren Li**
*Northwest Sci-Tech University of Agriculture and Forestry, Yangling, Shaanxi, China*

| | | |
|---|---|---|
| I. | INTRODUCTION | 563 |
| II. | MATERIALS AND METHODS | 564 |
| | A. The Binary Vector for Transformation | 564 |
| | B. Transformation and Regeneration of Chili Pepper | 564 |
| | C. Characterization of Transgenic Plants | 565 |
| | D. Protection Against Cucumber Mosaic Virus and Tobacco Mosaic Virus Infection | 566 |
| | E. Quality Analysis of Dry Fruit | 567 |
| III. | RESULTS AND DISCUSSION | 567 |
| | A. Regeneration of Transformed Chili Pepper Plants | 567 |
| | B. Analysis of Transgene Integration and Expression | 567 |
| | C. Resistance to Cucumber Mosaic Virus and Tobacco Mosaic Virus in Kanamycin-Resistant T1 Transgenic Plants | 572 |
| | D. Resistance to Different CMV Isolates in the T3 Progenies of the Homozygous Line 16-13 | 573 |
| | E. Field Test of the T2 Progeny of Line 16-13 | 574 |
| | F. Field Tests of the T3 and T4 Progeny of Line 16-13 | 575 |
| | G. Quality Analysis of Transgenic Dry Fruit | 576 |
| IV. | SUMMARY | 577 |
| | REFERENCES | 577 |

## I. INTRODUCTION

Chili pepper is an important vegetable that has been used as a spice and a source of red pigment. In many regions of the world, diseases limit the production of chili pepper. The most common diseases are caused by viruses, such as cucumber mosaic virus (CMV), tobacco mosaic virus (TMV), potato virus Y (PVY), and tomato spotted wilt virus (TSWV). In China, CMV and TMV are major viral pathogens in chili pepper production. Coat protein–mediated protection (CPMP) has been established in transgenic plants against many plant viruses, including TMV and CMV (1–4). We are using the same strategy to develop virus-resistant chili pepper transformed with the CMV CP and TMV CP genes. Pepper regeneration in vitro has been reported by several laboratories via seedling cotyledons or hypocotyl explants (5–14). In addition, there have been reports of *Agrobacterium* species–mediated transformation of pepper from cotyledons. Liu and coworkers (15) described transformed shoot buds and leaflike structures showing β-glucuronidase activity; however, attempts to regenerate whole transgenic plants were unsuccessful. Zhu and associates (16) reported transgenic sweet pepper transformed by *Agrobacterium tumefaciens* with the CMV CP gene. Two of the T1 plant lines expressed the CMV CP gene.

In this report we describe the regeneration of virus-resistant chili pepper plants by *Agrobacterium* species–mediated transformation of hypocotyl explants with TMV CP and CMV CP genes. The expression of transgenes in plants has also been determined. Field trials of the T3 and T4 progeny of the transgenic homozygous lines indicated that fruit yield was 47% to 110% higher than that of the control plants. Quality analysis showed that transgenic dry fruit is comparable to that of control plants.

## II. MATERIALS AND METHODS

### A. The Binary Vector for Transformation

The disarmed *Agrobacterium tumefaciens* LBA 4404 carrying the binary vector pBTC (3) was used to transform chili pepper. pBTC contains expression cassettes of CMV CP and TMV CP genes, each of which is flanked by the cauliflower mosaic virus 35S promoter and the rbcS transcription termination signals (Fig. 1).

### B. Transformation and Regeneration of Chili Pepper

Seeds of chili pepper (*Capsicum annuum* var. Longunt) cultivar '8212,' which has been used in chili pepper production in Northwest China, were obtained from a commercial source.

The seeds were surface-sterilized by soaking in a solution of 0.1% $HgCl_2$ with 0.1% Tween-20 for 1 minute, 30 seconds. The seeds were then rinsed with sterile distilled water and germinated on 1/2 Murashige and Skoog (MS) (17) medium containing 1/2 MS salts and 1/2 vitamins, 30% sucrose, and 0.65% agar in glass tubes (15 × 2.5 cm). The pH of the medium was adjusted to pH 5.8 and autoclaved for 15 minutes at 121°C. The tubes with seeds were maintained under dark conditions at 28°C for 9 days, then incubated at 25°C with a daily 16-hours light (2000 lux). The *Agrobacterium* sp. cells were cultured in 5 ml of liquid Luvia-Bertani (LB) medium plus

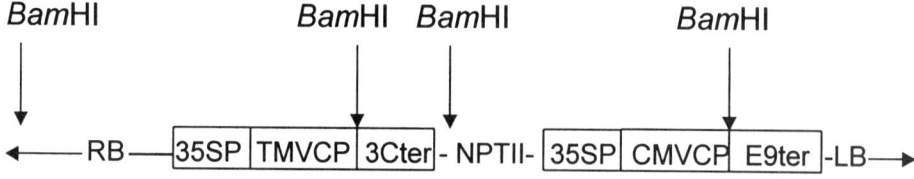

**Figure 1** A schematic diagram of the T-DNA region of pBTC plasmid.

30 mg/l of acetosyringone at 28°C and shaken at 200 rpm for 16 to 17 hours. The cells were diluted to $OD_{560}$ of 0.1 to 0.3 with liquid MS, pH 7.0, containing 30% sucrose and 30 mg/l acetosyringone. Apical hypocotyl sections (ca. 0.5 cm long) were cut from 12- to 14-day-old seedlings by using a surgical blade and placed into petri dishes on filter paper that had been wetted with sterile liquid MS. Excess medium was removed before inoculation. The hypocotyl sections were inoculated with a diluted culture of *Agrobacterium* sp. cells by pipetting 3 ml of the bacterial cell suspension onto the explants and gently mixed for 1–2 minutes. Inoculated hypocotyl sections were blotted dry on filter paper and placed horizontally onto the induction medium containing MS salts and vitamins plus 3% sucrose, 0.6% agar, 2 mg/l benzyladenine (BA) (Sigma), and 0.5 mg/l indole-3-acetic acid (IAA) (Sigma). The plates were sealed with parafilm and placed upside down in the dark at 25°C for 2 days of cocultivation. After cocultivation, explants were transferred to the selection and induction medium containing MS salts and vitamins, 3% sucrose, 0.6% agar, 2 mg/l BA, 0.5 mg/l IAA, plus 75 mg/l kanamycin (Sigma) and 300 mg/l carbenicillin and placed at 25°C under continuous cool-white fluorescent light. Three days later, all explants were moved onto the selection regeneration medium containing MS salts and vitamins, 3% sucrose, 0.6% agar, 0.5 mg/l BA, 0.05 mg/l IAA, plus 75 mg/l kanamycin and 300 mg/l carbenicillin in 50-ml glass flasks (10 explants per flask). The flasks were placed at 25°C under continuous cool-white fluorescent light. After 3 weeks, the explants were transferred onto selection regeneration medium supplemented with 50 mg/l of kanamycin and subcultured in fresh selection regeneration medium with 50 mg/l of kanamycin at 2- to 3-week intervals. Excised regenerated shoots were cultured in the selection elongation medium containing MS salts and vitamins, 3% sucrose, 0.6% agar, 0.1 mg/l IAA, plus 50 mg/l kanamycin and 300 mg/l carbenicillin. Elongated shoots were transferred to the selection rooting medium containing MS salts and vitamins, 3% sucrose, 0.6% agar, 0.1 mg/l indolebutyricacid (IBA) (Sigma), 50 mg/l kanamycin, and 300 mg/l carbenicillin for rooting. The shoots had rooted after 2 weeks. Finally, after another 2 weeks plantlets were removed from the medium and planted into vermiculite and fed with liquid MS medium that contained 0.1 mg/l IBA. Plants were potted in sterile soil and grown in a greenhouse after 4 or 5 days.

## C. Characterization of Transgenic Plants

### 1. Polymerase Chain Reaction Analysis

Genomic deoxyribonucleic acid (DNA) was extracted from young leaf tissues of transgenic and untransformed pepper plants by using the cetyltrimethylammonium bromide (CTAB) method (18). Oligonucleotides corresponding to the CaMV 35S promoter −78 to −59, sequence CGTAAGGATGACGCACAAT, and the complementary sequence of the termination signals of rbcS (E9 and 3C) TCGAAACCGGATGATACGAAC were used as primers for PCR amplification. DNA was amplified by 30 cycles of 1 minute at 93°C, 1 minute at 53°C, and 1 minute at 72°C, and finally 2 minutes at 72°C. PCR products were analyzed by electrophoresis on 1% agarose gel.

### 2. Southern Blot

The detection of integrated transgenes was performed by Southern blot analysis. Plant genomic DNA was extracted by the CTAB method from young leaves. About 20 μg of total DNA predigested with *Bam*HI was electrophoresed in 0.9% agarose gel and transferred by capillarity to Zeta Probe Blotting Memberane (BIO-RAD products) as described (19). The DNA fragments excised by *Bam*HI from pBTC were used as the positive control. The CP genes of the CMV and TMV amplified by PCR from pBTC and labeled with α-32p dCTP (Du Pont) were used as probes. Prehybridization, hybridization, and washes were done at 65°C and the filter was analyzed by autoradiography.

## 3. Reverse Transcription (RT)-Polymerase Chain Reaction

Detection of TMV CP and CMV CP gene transcripts was carried out by RT-PCR. Total ribonucleic acid (RNA) was purified as described by Chang and coworkers (20). First-strand complementary DNA (cDNA) synthesis was carried out on 5 µg of total RNA in 25 µl reaction mixtures with 5 units of avian/myeloblastosis virus reverse transcriptase (AMV RTase) (Amersham LIFE SCIENCE) according to the manufacturer's instruction. The PCRs were carried out by using 0.2 µl of RT reaction mixture. Oligonucleotides corresponding to the TMV (common strain) CP gene sequence 5′ CAGTTCGTGTTCTTGTCATC 3′ (from the 27th to 47th nucleotide) and the complementary sequence to the 406th to 426th nucleotide of the TMV CP gene 5′ CCGATTATAAGATCCGGTTC 3′ were used as primers for PCR amplification to detect the TMV CP gene transcript. DNA was amplified by 30 cycles of 30 seconds at 94°C, 30 seconds at 53°C, and 30 seconds at 72°C, and finally 10 minutes at 72°C. PCR primers corresponding to nucleotide 101 to 120 of the CMV (Shandong isolate) CP gene 5 minutes CAACTTTAGAGTCCTGTCG 3′ and the complementary sequence to nucleotide 563 to 582 of the CMV CP gene 5′ GAATACACGAGAACGGCG 3′ were used for identification of the CMV CP gene transcript. Amplifications were done by 30 cycles of 30 seconds at 94°C, 30 seconds at 45°C, and 30 seconds at 72°C, and finally 5 minutes at 72°C. The PCR products were analyzed by electrophoresis on a 2% agarose gel.

## 4. Indirect DAS-ELISA

Indirect double antibody sandwich enzyme-linked immunosorbent assay (DAS-ELISA) (21) was used to detect accumulation of CMV in field tests. Mouse antiserum against CMV was coated on 96 well plates at 1:500 dilution. Rabbit antiserum to CMV, preabsorbed by healthy tobacco sap, was used at 1:5000 dilution. Horseradish peroxidase conjugated goat antirabbit antibody (Beijing Institute of Biological Products) was used at 1:100 dilution. Leaves were ground in phosphate-buffered saline containing tween (PBST) buffer at 1:2 dilution.

## 5. Western Blot

For western blot analysis, 0.1 g of young leaf tissues was ground in 200 µl of protein extraction buffer (0.05 M $Na_2HPO_4$, pH 7.2, 0.5 M NaCl, 0.001 M ethylenediaminetetraacetic acid [EDTA], 0.001 M phenylmethylsulfonyl fluoride (PMSF), and 0.01 M 2-mercaptoethanol) and boiled for 5 minutes. The supernatant was added to an equal volume of the Laemmli buffer (22) and boiled for 5 minutes. One hundred micrograms of soluble protein (each lane) was loaded on a 12% sodium dodecyl sulfate-polyacrylamide gel (22). Proteins were transferred from the gel onto a polyvinylidene fluoride (PVDF) membrane (Millipore). Immunodetection was performed by using rabbit antiserum to CMV (1:3000 dilution) or rabbit antiserum to TMV (1:500 dilution), preabsorbed by healthy tobacco sap, and alkaline phosphatase conjugated goat antirabbit antibody (Promega), and visualized by nitro blue tetrazolium (NBT)-5-bromo-4-chloro-3-indolyl-phosphate (BCIP) staining.

## 6. Selection of Homozygous Lines

Primary polymerase chain reaction (PCR)-positive transformants (T0 generation) were self-pollinated. The resultant virus-resistant T1 plants were self-pollinated and their T2 progeny were germinated on MS medium with 300 mg/l kanamycin to select homozygous lines.

## D. Protection Against Cucumber Mosaic Virus and Tobacco Mosaic Virus Infection

1. The PCR-positive transformants were self-pollinated. The T1 and T2 progeny of homozygous lines, and the control pepper plants at the six- or seven-leaf stage were dusted with car-

borundum and rubbed with 20 µg/ml of homologous CMV (Shandong isolate) in the greenhouse. The plants were monitored for symptom development at 5- or 7-day intervals after inoculation. When tested for resistance to both CMV and TMV, the plants at the six- or seven-leaf stage were first inoculated with 20 µg/ml of homologous CMV, and 7 days later, two other leaves of the same plants were inoculated with 0.5 µg/ml of homologous TMV. Tobacco was used as the systemic host to check the infectivity of CMV or TMV inoculum. *Chenopodium quinoa* was used as the local lesion host for CMV and TMV to quantitate the level of virus replication. Two kinds of lesion were produced in this assay, which were caused by CMV and TMV.

2. For field testing of the T2, T3, and T4 progeny of the homozygous lines, the transgenic plants and control plants were planted in a CMV- and TMV-infested field in Yangling County, Shaanxi Province. The plants were subjected to natural pepper-growing conditions. Resistance was evaluated by monitoring symptom development. Disease severity was rated by the symptom scores: 0, no symptom; 1, slight mosaic; 2, deformed leaf; 3, mosaic leaf and dwarfing; 4, deformed leaves and dwarfing.

### E. Quality Analysis of Dry Fruit

The content of rough fiber, rough fat, total ash, and insoluble ash in HCl was determined according to Chinese National Standard GB10469, GB10065-89, GB8857, GB10473.

## III. RESULTS AND DISCUSSION

### A. Regeneration of Transformed Chili Pepper Plants

The sensitivity of untransformed 8212 hypocotyl tissue to kanamycin was examined by using shoot induction medium containing 0, 25, 50, or 75 mg/l of kanamycin. Normal calli and adventitious buds developed on the medium without kanamycin. However, no adventitious buds were produced when 50 mg/l or a higher level of kanamycin was used. A concentration of 75 mg/l of kanamycin was chosen for selection of transformed tissues during the first month of culture. Subsequently, 50 mg/l of kanamycin was used for selection of transformed plantlets. Generally, the buds arose from around the periphery of the cut surface at 3 to 4 weeks after explants were placed onto the selection medium (Fig. 2A). About 12% to 15% of explants developed buds in the selection medium. In most cases several buds emerged from one explant (Fig. 2B). However, only one or two buds developed to form normal shoots (Fig. 2C, D). To ensure reliable and consistent plantlet development, the small shoots should be kept in selection regeneration medium until they are at least 1.0 cm tall. Plantlets could be potted in soil 4 to 5 months after initiation of transformation (Fig. 2E). These plants appeared morphologically normal (Fig. 2F) and set fruit with viable seeds (Fig. 2G). In total we have obtained 49 normal kanamycin-resistant chili pepper plants from several transformation experiments.

### B. Analysis of Transgene Integration and Expression

#### 1. Selection of Transgenic Plants and Homozygous Lines

Seedling development in MS medium containing kanamycin was used to select for homozygous transgenic lines. Seeds of self-pollinated highly virus-resistant T1 transgenic pepper plants were germinated on MS medium with 300 mg/l of kanamycin. Kanamycin-resistant seedlings developed a white and long root with root hairs. However, kanamycin-sensitive or control seedlings produced only brown and short roots without root hairs. According to the results of rooting assays, T2 seedlings of six T1 lines (4-2, 12-2, 12-8, 16-2, 16-12, and 16-13) were 100% kanamycin-resistant (Table 1). The T3 progenies of line 16-13 were further tested for kanamycin resistance. These seedlings were also 100% kanamycin-resistant (data not shown).

**Figure 2** (A) The buds coming up from around the periphery of cut surface 3 to 4 weeks after explants were placed onto the selection medium; (B) several buds emerged from one explant; (C) only one or two buds developed to normal shoots; (D) elongated shoots on selection elongation medium; (E) the rooted plantlet on selection rooting medium; (F) the plantlet was potted to the vermiculite and soil 4 to 5 months after initiation of transformation; (G) transgenic chili pepper plant had set fruit; (H) PCR-positive T1 plant (left) and untransformed plants (right) were challenged with 20 µg/ml CMV and 0.5 µg/ml TMV.

The kanamycin-resistant T0 pepper plants were tested by PCR to determine whether each plant had been transformed. The PCR amplification results showed that at least 22 of the 49 kanamycin-resistant chili pepper plants had both the CMV CP and TMV CP genes integrated into the plant genome, as exemplified in Fig. 3. The PCR assay was also used to evaluate the homozygosity of the selected lines. For example, the two CP genes were detected in all the tested T3 progenies of line 16-13 as shown in Fig. 4.

**Table 1** Selection of Homozygotes on MS Medium with 300 mg/l Kanamycin

| Line | Number of resistant seedlings | Number of sensitive seedlings |
|---|---|---|
| 12-2 | 27 | 0 |
| 12-8 | 33 | 0 |
| 16-2 | 38 | 0 |
| 16-12 | 36 | 0 |
| 16-13 | 53 | 0 |
| 4-2 | 81 | 0 |

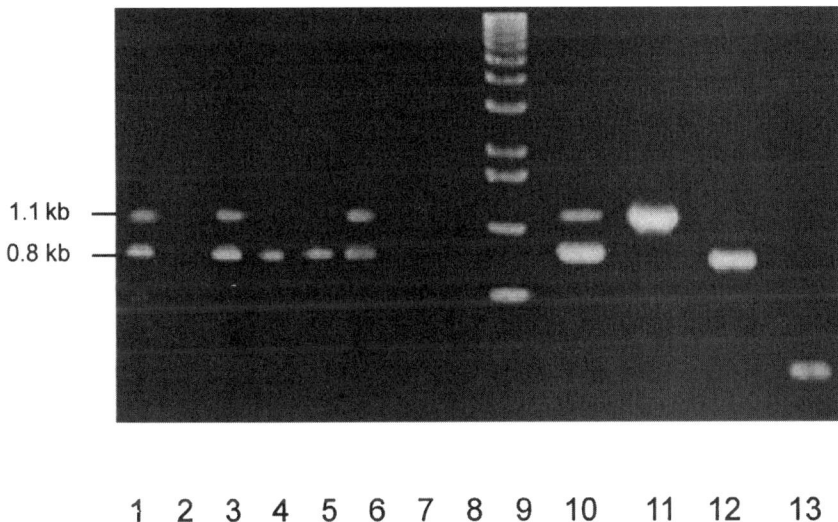

**Figure 3** PCR analysis of kanamycin-resistant T0 chili pepper plants. Lanes 1–7, seven independent kanamycin-resistant T0 plants; lane 8, untransformed plant; lane 9, 1-kb marker (BRL); lane 10, pBTC with CMV CP and TMV CP genes; lane 11, pBC with CMV CP gene; lane 12, pBT with TMV CP gene; lane 13, pB113 without CMV CP and TMV CP genes.

Southern blot analysis was also performed on a single T3 progeny from the homozygous line 16-13. Fig. 5 showed that *Bam*HI-digested 16-13 plant genomic DNA produced the same band pattern as did the *Bam*HI-digested plasmid pBTC (as a control) did, releasing a 3.8-kb (encompassing the TMV CP gene), a 2.6-kb (containing the CMV CP gene), and a 0.7-kb (with the 3′ end of the TMV CP gene) fragment (see Fig. 1). The Southern blot analysis demonstrated the presence of the T-DNA copy in the genome of the T3 progeny of line 16-13.

## 2. Transcription and Translation of TMV CP and CMV CP Genes in Line 16-13

The RT-PCR analysis was carried out on T3 progeny of line 16-13 to examine the transcription of the viral CP genes. Fig. 6A shows the presence of a fragment of ca. 0.4 kb, the expected size from the TMV CP gene transcript, and a fragment of ca. 0.5 kb (Fig. 6B), the expected size for the CMV CP gene transcript using primers, as described in the section, Materials and Methods.

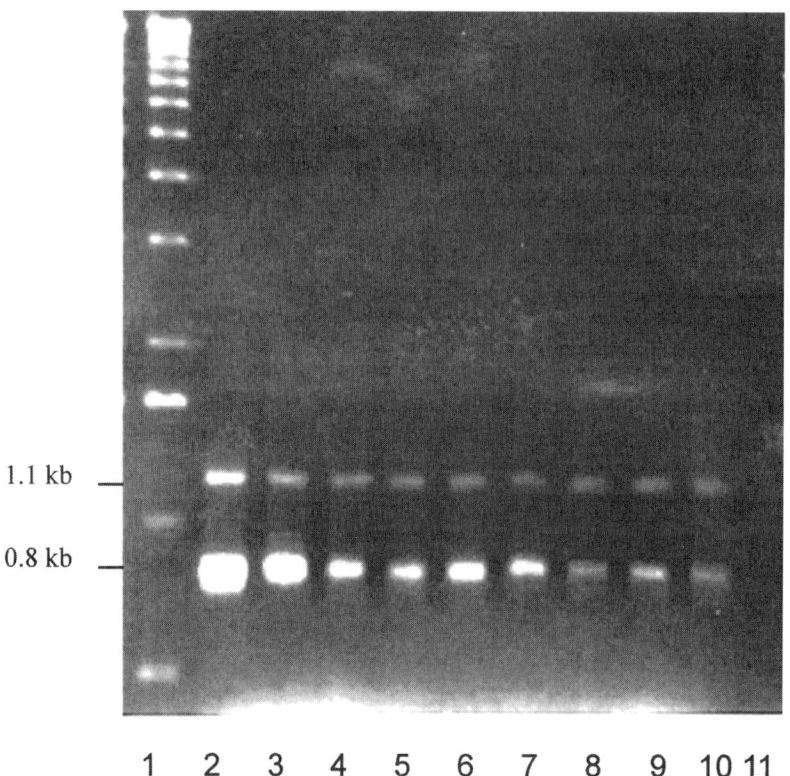

**Figure 4** PCR assay on T3 progeny of the homozygous Line 16-13. DNA fragments with expected size (1.1 kb for the CMV CP gene and 0.8 kb for the TMV CP gene) were observed in all eight T3 seedlings (lanes 3–10); no PCR product appeared in an untransformed '8212' (lane 11); lane 2 contains PCR product from plasmid DNA (pBTC); lane 1, 1-kb marker (BRL).

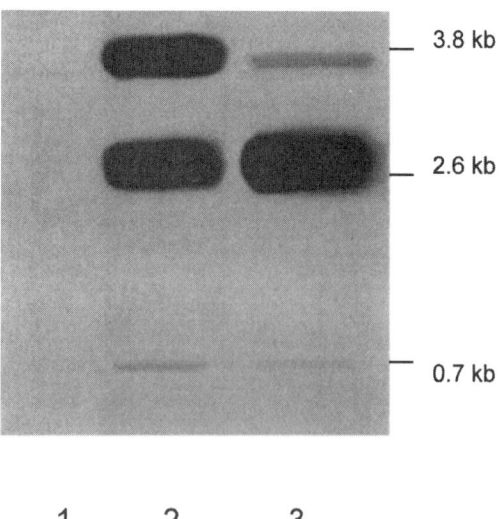

**Figure 5** Southern blot analysis of T3 progeny from homozygous Line 16-13. Lane 1, untransformed pepper plant; lane 2, plant in the T3 progeny from line 16-13; lane 3, plasmid pBTC.

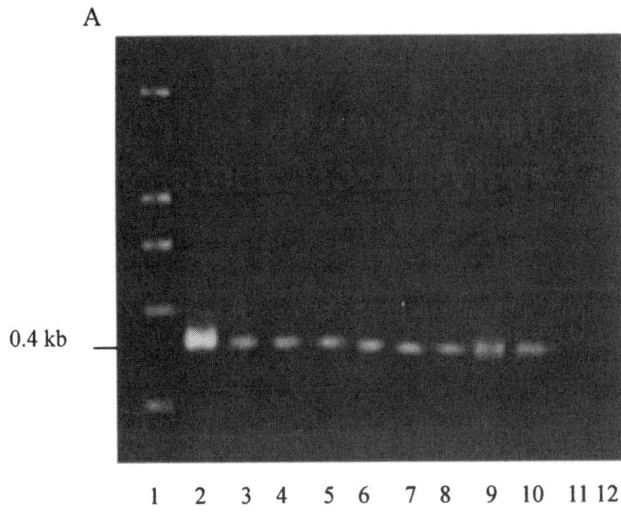

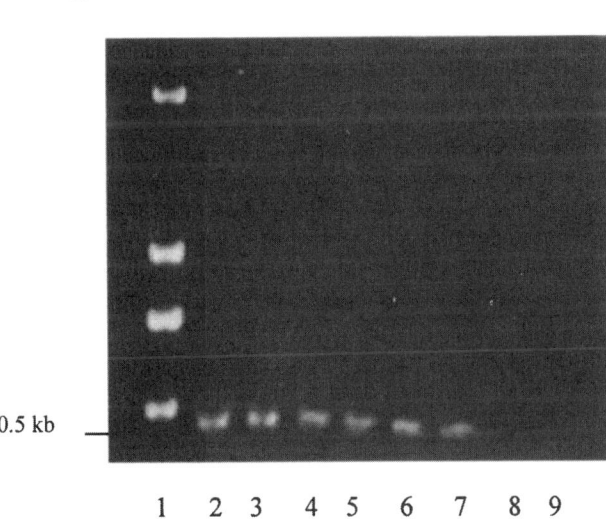

**Figure 6** Expression of TMV CP gene (A) and CMV CP gene (B) in T3 progeny of homozygous line 16-13 by RT-PCR. The RT-PCR was carried out on total RNA extracted from transformed and untransformed seedlings. Lane 1, DNA marker DL2,000 (Takara Biotech); lane 2, pBTC for PCR-positive control. (A) Lanes 3–10, plants from T3 progeny of line 16-13; lane 11, T3 progeny of line 16-13 without AMV RTase in reaction mixture; lane 12, untransformed plant; (B) Lanes 3–7, plants from T3 progeny of line 16-13; lane 8, plant from T3 progeny of line 16-13 without AMV RTase in reaction mixture; lane 9, untransformed plant.

Synthesis of TMV and CMV CPs in T3 progeny of line 16-13 was demonstrated by Western blot analysis. A 24-kDa protein specifically recognized by the CMV polyclonal antisera confirmed the expression of the CMV CP in seven T3 progeny of line 16-13. The amount of CMV CP accounted for 0.004–0.005% of the soluble protein in transgenic plants (Fig. 7A). A 17-kDa protein specifically reacted with the TMV polyclonal antisera indicated the expression of the TMV CP in all seven T3 progeny from line 16-13 (Fig. 7B). The expression level of the TMV CP was estimated at 0.002–0.003% of the soluble protein in transgenic plants.

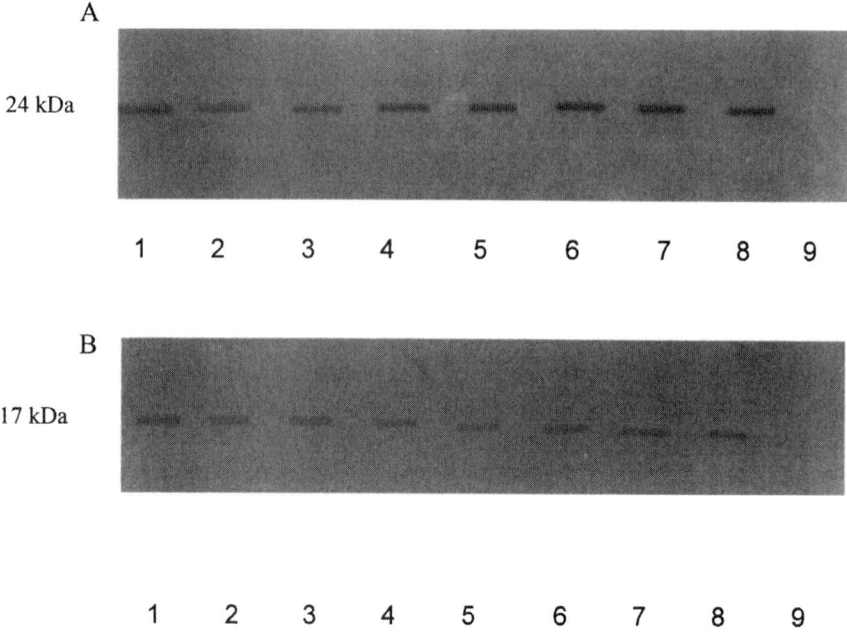

**Figure 7** Expression of CMV CP and TMV CP in T3 progenies of homozygous 16-13 by Western blot analysis. (A) Lane 1, 4.9 ng of CMV CP (the CMV CP is ca. 24-kDa); lanes 2–8, a 24-kDa protein was observed in T3 progenies of line 16-13; lane 9, no 24-kDa protein appeared in nontransformed control plant. (B) Lane 1, 3.8 ng of TMV CP (the TMV CP is ca. 17-kDa); lanes 2–8, a 17-kDa protein was observed in T3 progenies of line 16-13; lane 9, no 17-kDa protein appeared in the nontransformed control plant.

### C. Resistance to Cucumber Mosaic Virus and Tobacco Mosaic Virus in Kanamycin-Resistant T1 Transgenic Plants

Fifty T1 progeny of 4 self-pollinated PCR-positive T0 plants were inoculated with 20 µg/ml of purified CMV. Fifty days after inoculation, 50% of the progeny plants showed no or delayed symptoms; among them, 5 plants did not develop any symptoms until the fruiting period. However, all untransformed plants had developed systemic symptoms.

Sixty-three progeny of another 6 self-pollinated PCR-positive T0 plants were challenged successively with 20 µg/ml of CMV and 0.5 µg/ml of TMV. Eighty days after inoculation, 15 T1 plants showed resistance to virus infection (Fig. 2H, left), whereas untransformed pepper plants showed systemic symptoms (Fig. 2H, right). Some of the results of protection of T1 transgenic pepper plants against CMV and TMV infection are shown in Table 2. The T1 transgenic plants showed different levels of protection, as judged by symptoms and the infectivity assays on *Chenopodium quinoa*. The T1 plant lines 34-12, 10-11, 9-9, 34-2, 16-13, and 10-12 did not show any signs of infection on leaves, indicating high levels of resistance to CMV and TMV. Consistently, levels of virus multiplication in plants 34-12, 10-11, 9-9 were low, as the number of lesions produced were less than 10% of the number of lesions produced by extracts of nontransgenic plants. Virus multiplication in plants 34-2 and 10-12 were at the intermediate level, and more lesions appeared from the T1 plants 34-13, 5-6, 45-4, consistently with the more severe symptoms exhibited in the challenged plants.

**Table 2** Protection of T1 Transgenic Pepper Plants Against Cucumber Mosaic Virus and Tobacco Mosaic Virus Infection[a]

| Line | Lesions in *Chenopodium quinoa*, percentage[b] | Symptoms |
|---|---|---|
| 9-9 | 9.0 | − |
| 34-2 | 18.3 | − |
| 34-12 | 4.0 | − |
| 10-11 | 7.9 | − |
| 16-13 | NT[c] | − |
| 10-12 | 26.2 | − |
| 34-13 | 48.5 | + |
| 5-6 | 81.5 | ++ |
| 45-4 | 96.8 | +++ |
| Untransformed | 100.0 | +++ |

[a]The pepper plants were first inoculated with 20 µg/ml of homologous CMV, and 7 days later, two other leaves of the same plants were inoculated with 0.5 µg/ml of homologous TMV.
[b]The left half of the *C. quinoa* leaves was inoculated by the purified viruses as control, and the right half of leaves was inoculated by sap extracted from leaves of challenged transformed or untransformed pepper plants. For each plant, lesions from four halves of *C. quinoa* leaves were counted and averaged. The virus multiplication level is expressed as the ratio percentage of average lesion number for the transgenic pepper plants to the average lesion number for the untransformed pepper plants.
[c]NT, not tested.

## D. Resistance to Different CMV Isolates in the T3 Progenies of the Homozygous Line 16-13

To test the broadness of resistance, the T3 progenies of the line 16-13 and untransformed control plants were mechanically inoculated with three CMV isolates that induce different symptoms in cultivar 8212. Resistance was determined by observing symptom development until the fruiting stage. Table 3 showed the resistance of T3 plants of line 16-13 against three CMV isolates. The

**Table 3** Protection Against Cucumber Mosaic Virus Isolates of the T3 Progenies from the Homozygous Line 16-13

| Virus isolates[b] | NS, percentage[a] T3 plants | NS, percentage[a] Control plants |
|---|---|---|
| Shandong | 100 | 0 |
| Tomato | 67 | 0 |
| Anhui | 64 | 0 |

[a]NS, percentage of plants with no symptoms.
[b]Plants were inoculated with three CMV isolates in a greenhouse. Shandong isolate was the homozygous CMV that was isolated from tobacco in Shandong Province; tomato isolate was isolated from tomato; Anhui isolate was isolated from tobacco in Anhui Province. Symptoms on chili pepper cultivar '8212' caused by the isolates were as follows: Shandong isolate, vein clearing and leaf mosaic; tomato isolate, vein clearing and yellowing; Anhui isolate, yellowing and stunting.

plants were highly resistant to the Shandong isolate, the homozygous virus from which the transgene (CMV CP gene) was derived, and showed moderate resistance to the tomato and Anhui isolates.

### E. Field Test of the T2 Progeny of Line 16-13

The purpose of the field test was to evaluate coat protein–mediated protection during a complete plant growth cycle and against aphid-mediated infection. The seedlings of the T2 progenies of homozygous line 16-13 were planted in natural pepper-growing conditions before transplanting into the field in early July 1996. The results are shown by Fig. 8. Five days after transplantation the T2 plants showed 30% of disease incidence (DI) whereas untransformed plants were 62% of DI.

(A)      (B)

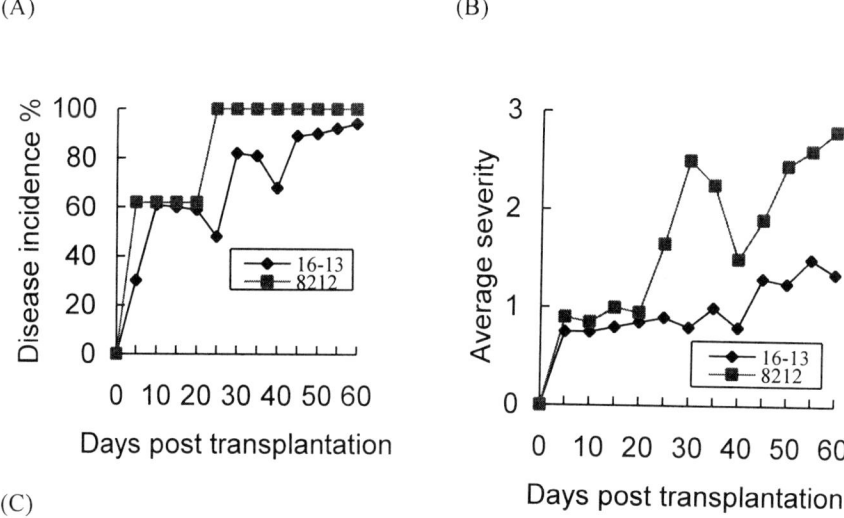

(C)

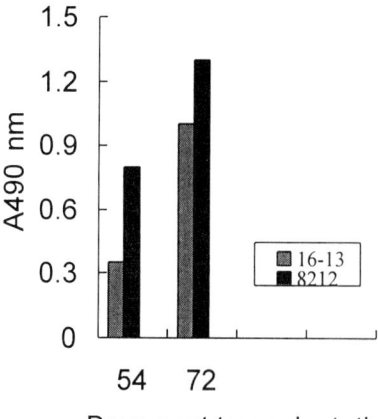

**Figure 8** The percentage disease incidence (A), average severity (B), and CMV accumulation (data from 10 random plants) (C) of T2 progeny of homozygous line 16-13 and control plants (8212) in a field test in 1996.

# Virus-Resistant Chili Pepper

Twenty days post transplantation the control plants became 100% infected, whereas the T2 plants showed 48% to 94% of DI (Fig. 8A). The T2 plants exhibited 1.35 of average severity (AS) whereas control plants exhibited 2.75 AS at 60 days post transplantation (Fig. 8B). Usually, when the AS exceeds 1.50, the plant becomes stunted and does not produce fruit. We did not observe stunted plants in T2 progenies of line 16-13. The CMV accumulation was also determined by indirect DAS-ELISA. Fifty-four days post transplantation the T2 plants of 16-13 showed a lower level of CMV accumulation when compared with untransformed cultivar '8212' (Fig. 8C).

## F. Field Tests of the T3 and T4 Progeny of Line 16-13

Three hundred T3 progenies of line 16-13 and untransformed control plants were tested in the field in 1997. The data from the 1997 field trial are summarized in Fig. 9 and Table 4. The DI and AS observed in 1997 were comparable to those in 1996. Seventeen of the tested 300 transgenic plants did not show any symptoms at harvest. A number of growth parameters are listed in Table

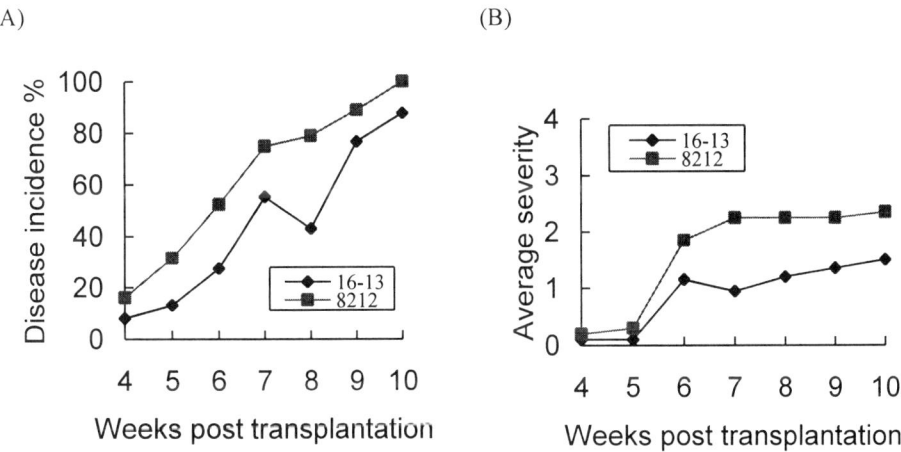

**Figure 9** The percentage disease incidence (A) and average severity (B) of T3 progenies of homozygous line 16-13 and control plants (8212) in the field test conducted in 1997.

**Table 4** Comparison of the Horticultural Characteristics and Fruit Yield of the T3 and T4 Progenies of Line 16-13 and Untransformed Chili Pepper in Field Trials

| Year | Cultivar | HT, cm[a] | SP, percentage[a] | NFP[c] | FWP, g[d] | FL, cm[e] | Yield, g[f] |
|---|---|---|---|---|---|---|---|
| 1997 | 16-13 | 57.3 | 5.0 | 25.9 | 86.9 | 10.4 | 7819.6 |
|  | 8212 | 49.3 | 33.5 | 19.1 | 59.1 | 8.1 | 5319.3 |
| 1998 | 16-13 | NT | NT | 42.0 | 189.0 | NT | 17010.0 |
|  | 8212 | NT | NT | 34.1 | 88.4 | NT | 7956.0 |

[a]HT, plant height at 8 weeks post transplantation.
[b]SP, stunted plant, estimated by symptom (scores 3 and 4) from 300 plants.
[c]NFP, number of fruit per plant.
[d]FWP, weight of fruit per plant.
[e]FL, length per fruit was measured from 300 fruits.
[f]Fruit yield was obtained by the total weight of 90 plants. HT, NFP, and FWP were averaged from 90 plants.

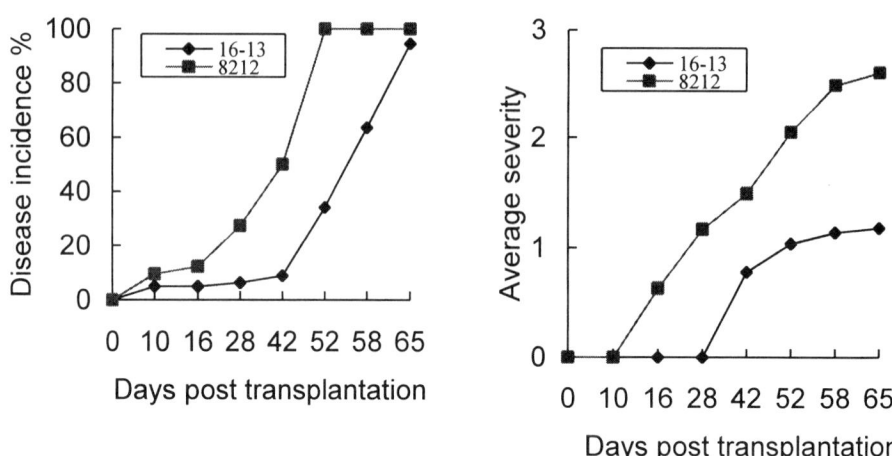

**Figure 10** The percentage disease incidence (A) and average severity (B) of T4 progenies of homozygous line 16-13 and control plants (8212) in a field test in 1998.

4. The T4 progenies (10,000 plants) of line 16-13 were further tested in the field in 1998, as shown in Fig. 10 and Table 4. The DI and AS observed in 1998 were similar to those of 1997. The transgenic chili pepper with the CMV CP and TMV CP genes exhibited high levels of viral resistance compared with untransformed plants under natural pepper-growing conditions. Fruit yields and vegetative growth of the progeny of homozygous plants averaged better than those of untransformed plants. The average fruit yield was 47% (1997) and 110% (1998) higher than that of the control plant (Table 4).

### G. Quality Analysis of Transgenic Dry Fruit

The quality analysis of transgenic fruits is shown in Table 5. The percentage of $H_2O$, rough fiber, rough fat, total ash, and insoluble ash (in HCl) of transgenic fruit is equivalent to that of the control plants. Changes in other major characteristics were not observed in transgenic fruits.

**Table 5** Dry Fruit Quality[a] of T3 Progenies of Line 16-13 Compared with Untransformed Chili Pepper '8212'

| Cultivar | Percentage | | | | |
|---|---|---|---|---|---|
| | $H_2O$ | Rough fiber | Rough fat | Total ash | Insoluble ash in HCl |
| 16-13 | 7.50 | 20.62 | 13.10 | 6.62 | 0.46 |
| 8212[b] | 6.26 ~ 10.64 | 20.17 ~ 26.08 | 7.11 ~ 12.94 | 5.80 ~ 7.50 | 0.11 ~ 0.49 |

[a]Quality analysis of dry fruit was determined according to the Chinese National Standard.
[b]Data from random sample of 10.

## IV. SUMMARY

Both cucumber mosaic virus (CMV) and tobacco mosaic virus (TMV) coat protein genes have been transferred into chili pepper (*Capsicum annuum* var. Longunt) cultivar 8212 mediated by *Agrobacterium* sp. Kanamycin resistant plantlets were regenerated from hypocotyls. Forty-nine independent transformants were obtained, 22 of which showed the presence of both of CMV and TMV coat protein genes by PCR analysis. Fifty progeny from 4 self-pollinated PCR-positive T0 plants were tested for resistance to infection by CMV, and 5 of them showed high levels of resistance. Sixty-three progeny of another 6 self-pollinated PCR-positive T0 plants were challenged with the inocula of CMV and TMV. Fifteen T1 plants were selected with high-level resistance to both CMV and TMV infection. Ten T1 lines were shown to be homozygotes on the basis of the results of kanamycin-resistance assay and PCR analysis of their T2 progenies. The T2, T3, and T4 progenies, respectively, of virus-resistant homozygotes were tested in natural CMV- and TMV-infested fields and by challenge with CMV and TMV in greenhouse tests. The progeny from a homozygous line 16-13 showed high levels of resistance and good-quality agronomic characteristics. Extensive molecular analyses such as PCR, Southern blot, RT-PCR, and Western blot were carried out on the T3 progenies of line 16-13, and the results indicated that the CMV CP and TMV CP genes were stably expressed.

## ACKNOWLEDGMENTS

The authors gratefully acknowledge Drs. Roger N. Beachy and Dennis Gonsalves and Mr. Augustine Gubba for their thoughtful review of the manuscript.

## REFERENCES

1. P Powell-Abel, RS Nelson, B De, N Hoffmann, SG Rogers, RT Fraley, RN Beachy. Delay of disease development in transgenic plant that express the tobacco mosaic virus coat protein gene. Science 232:738–743, 1986.
2. RS Nelson, SM McCormick, X Delannay, P Dube, J Layton, EJ Andoson, M Kaniewska, RK Proksch, RB Horsch, SG Rogers, RT Fraley, RN Beachy. Virus tolerance, plant growth, and field performance of transgenic tomato plants express coat protein from tobacco mosaic virus. Biotechnology 6:403–409, 1988.
3. RX Fang, YC Tian, GL Wang, XF Qin, MZ Yang, TY Li, BY Xu, KQ Mang, ZC Zhang, Q, Wu, RH Zhou, FL Wang, G Zhao, XD Han. Transgene tobacco plants resistant to infection of both TMV and CMV. Chin Sci Bull 36:524–525, 1990.
4. D Gonsalves, P Chee, R Provvidenti, R Seem, JL Slightom. Comparison of coat protein-mediated and genetically-derived resistance in cucumbers to infection by cucumber mosaic virus under field conditions with natural challenge inoculations by vectors. Biotechnology 10:1562–1570, 1992.
5. AL Gunay AL, PS Rao. In vitro plant regeneration from hypocotyl and cotyledon explants of red pepper (*Capsicum*). Plant Sci Lett 11:365–372, 1978.
6. M Fari, M Czako. Relationships between position and morphogenetic response of pepper hypocotyl explant culture in vitro. Sci Hortic 15:207–213, 1981.
7. GC Phillips, JF Hubstenberger. Organogenesis in pepper tissue cultures. Plant Cell Tissue Org Cult 4:261–269, 1985.
8. P Sripichitt, E Nawata, S Shigenaga. In vitro shoot-forming capacity of cotyledon explants in red pepper (*Capsicum annuum* L.cv. Yatsufusa). Jpn J Breed 37:133–142, 1987.
9. S Agrawal, N Chandra, SL Kothari. Plant regeneration in tissue cultures of pepper (*Capsicum annuum* L.cv. Mathania). Plant Cell Tissue Org Cult 16:47–55, 1989.

10. N Ochoa-Alejo, L Ireta-Moreno. Cultivar differences in shoot-forming capacity of hypocotyl tissue of chili pepper (*Capsicum annuum* L.) cultured in vitro. Sci Hortic 42:21–28, 1990.
11. LL Valera-Montero, N Ochoa-Alejo. A novel approach for chili pepper (*Capsicum annuum* L.) plant regeneration: Shoot induction in rooted hypocotyls. Plant Sci 84:215–219, 1992.
12. W Shao, JD Caponetti. Callus production, organogenesis, and pantlet formation in three cultivars of sweet pepper. J Tenn Acad Sci 68:113–118, 1993.
13. H Ezura, S Nishimiya, M Kasumi. Efficient regeneration of plants independent of exogenous growth regulators in bell pepper (*Capsicum annuum* L.). Plant Cell Rep 12:676–680, 1993.
14. AIA Ebida, CY Hu. In vitro morphogenetic responses and plant regeneration from pepper (*Capsicum annuum* L. cv. Early California Wonder) seedling explants. Plant Cell Rep 13:107–110, 1993.
15. W Liu, WA Parrott, DF Hildebrand, GB Collins, EG Williams. *Agrobacterium* induced gall formation in bell pepper (*Capsicum annuum* L.) and formation of shoot-like structures expressing introduced genes. Plant Cell Rep 9:360–364, 1990.
16. YX Zhu, Y Qu, J Wen, Y Zhang, ZL Chen. Transgenic sweet pepper plants from *Agrobacterium*-mediated transformation. Plant Cell Rep 16(1/2):71–75, 1996.
17. T Murashige, F Skoog. A revised medium for rapid growth and bioassays with tobacco tissue cultures. Physiol Plant 15:473–497, 1962.
18. SL Dellaparta, J Wood, JB Hicks. Maize DNAminiprep. In: R Malmber, J Messing, I Sussex, eds. Molecular Biology of Plant. Cold Spring Harbor, NY: Cold Spring Harbor Laboratory, 1985, pp 3–37.
19. J Sambrook, EF Fritsch, T Maniatis. Molecular cloning. 2nd ed. Cold Spring Harbor, NY: Cold Spring Harbor Laboratory Press, 1989, 19–22.
20. SJ Chang, J Puryear, J Cairney. A simple and efficient method for isolating RNA from pine tree. Plant Mol Biol Rep 11:113–116, 1993.
21. M Clark, A Adams. Characteristics of the microplate method of enzyme-linked immunosorbent assay (ELISA) for the detection of plant viruses. J Gen Virol 34:475–483, 1977.
22. UK Laemmli. Cleavage of structural proteins during the assembly of the head of bacteriophage T4. Nature 227:680, 1970.

# 39
# Transgenic Cucumber with Resistance to Cucumber Mosaic Virus

**Soryu Nishibayashi**
*Mitsubishi Space Software Co., Ltd., Amagasaki City, Japan*

|       |                                         |     |
|-------|-----------------------------------------|-----|
| I.    | INTRODUCTION                            | 579 |
| II.   | HISTORY OF CUCUMBER TRANSFORMATION      | 580 |
| III.  | PROTOCOL FOR TRANSFORMATION             | 580 |
|       | A. Plant Material                       | 580 |
|       | B. Culture of *Agrobacterium tumefaciens* | 580 |
|       | C. Inoculation and Coculture            | 581 |
|       | D. Cell Selection and Plant Regeneration | 581 |
| IV.   | EFFECT OF ACETOSYRINGONE ON TRANSFORMATION | 581 |
| V.    | CELL SELECTION                          | 581 |
| VI.   | EXPRESSION OF CaMV 35*s* PROMOTOR       | 582 |
| VII.  | CUCUMBER MOSAIC VIRUS RESISTANCE        | 583 |
| VIII. | SAFETY OF TRANSGENIC PLANTS             | 584 |
|       | REFERENCES                              | 584 |

## I. INTRODUCTION

Cucumber is one of the popular vegetable crops in America, Asia, and European countries. It suffers from viral diseases such as cucumber mosaic virus (CMV), zucchini yellow mosaic virus (ZYMV), and watermelon mosaic virus (WMV), which cause serious damage to both the plant body and the fruit yield. However, in many cases, new cucumber cultivars with the characteristics of viral resistance cannot be easily produced, as a result of the absence of useful plants with the characteristics of viral resistance for breeding purposes. If there is a wild plant with the characteristics of viral resistance, such a wild plant cannot be directly used for cucumber breeding, because of genetic barriers such as sexual incompatibility between donor and acceptor plants or the loss of desirable characteristics by breeding.

Using gene-engineering technologies, single dominant agronomic characteristics, including viral resistance, can be directly introduced into crops, without changing the desirable characteristics. Powell-Abell and coworkers (1) first proposed the possibility of plant breeding for viral resistance in transgenic tobacco plants, into which a TMV coat protein (*cp*) gene was introduced. The strategy of using coat protein genes has since been applied for plant protection against various viruses (2,3).

I describe here the method of *Agrobacterium tumefaciens*–mediated transformation in the hypocotyl explants of cucumber plants (4) and CMV resistance in transgenic cucumber plants with an introduced CMV-Y-*cp* gene (5).

## II. HISTORY OF CUCUMBER TRANSFORMATION

The transformation of cucumber was first achieved by Trulson and associates (6), who regenerated transgenic cucumber plants from roots that were induced by the inoculation of inverted hypocotyl explants with *Agrobacterium rhizogenes*, which harbored a binary vector plasmid with the NOS-*nptII* gene. Chee (7) cocultured cucumber cotyledonary explants with *Agrobacterium tumefaciens*, which harbored a binary vector plasmid with an NOS-*nptII* gene and obtained many transgenic plants by regeneration via somatic embryogenesis from the cotyledonary explants. Sarmento and coworkers (8) also used *Agrobacterium tumefaciens*, which harbored various binary vector plasmids with a CaMV 35*s*-*nptII* gene, and achieved transformation of cucumbers by using petiole and leaf explants. In these reports, an *nptII* gene was used as a selectable marker gene, and transgenic plants were regenerated on media containing kanamycin.

We established an *Agrobacterium tumefaciens*–mediated transformation system based upon hypocotyl explants of cucumber line 1021 (4). The transformation system is described the following section.

## III. PROTOCOL FOR TRANSFORMATION

### A. Plant Material

1. Remove the seed coats and sterilize the peeled seeds in 1.2% (v/v) sodium hypochlorite solution for 10–15 minutes.
2. Rinse with sterile distilled water several times.
3. Place on hormone-free Murashige and Skoog (MS) medium (pH 5.8) for 7–10 days.

All media that are used in the transformation experiments are supplemented with 3% sucrose and solidified with 0.2% gellite. All media are autoclaved at 120°C for 15 minutes. The culture conditions in all media are constant at 23°C under a 16-hour (light)–8-hour (dark) photoperiod in a culture room.

### B. Culture of *Agrobacterium tumefaciens*

1. Grow the EHA 101 strain of *Agrobacterium tumefaciens* (9) in a liquid YEB medium containing 50 mg/l kanamycin, 20 mg/l hygromycin B, and 25 mg/l chloramphenicol at 23°C for 3–7 days.
2. Dilute the suspension of *Agrobacterium tumefaciens* to 1/100 with liquid MS medium.

The EHA101 strain harbors binary plasmid pIG121-Hm (4) or pIG121-HmCP (5), which are derivatives of PBI121 (10). The pIG121-Hm has the neomycin phosphotransferase II gene (*nptII*), hygromycin phosphotransferase (*hph*) gene, and β-glucuronidase (*gus*) gene (11). The first intron sequence of the catalase gene of castor bean is inserted between the CaMV 35*s* promoter and β-glucuronidase gene. The plG121-HmCP is a derivative of the pIG121-Hm into which the CaMV 35*s*-CMV-O-*cp* gene was introduced.

## C. Inoculation and Coculture

1. Cut the hypocotyls of the seedlings into 1-mm segments with a razor blade.
2. Inoculate the hypocotyl segments with the 1/100 diluted *Agrobacterium tumefaciens* suspension.
3. Blot on sterilized filter papers.
4. Transfer to solidified MS coculture medium (pH 5.2) supplemented with 4 mg/l indole-3-acetic acid (IAA), 1 mg/l *N*6-(2-isopentyl)adenine (2ip), and 50 or 100 µM 3′, 5′-dimethoxy-4′-hydroxyacetophenone (Aldrich) and culture for 5 days.

## D. Cell Selection and Plant Regeneration

1. Transfer the cocultured hypocotyl explants to the solidified MS medium (pH 5.8) supplemented with 1.1 mg/l 2, 4-dichlorophenoxyacetic acid (2, 4-D), 0.23 mg/l 6-benzylaminopurine (BAP), 1% casein hydrolysate, 500 mg/l carbenicillin, 20–30 mg/l hygromycin B and culture for 4–6 weeks.
2. Transfer the tissues to solidified hormone-free MS medium supplemented with 500 mg/l carbenicillin and culture for 2–3 months. Change the medium at intervals of 4–6 weeks.
3. Transfer the regenerated plantlets to hormone-free MS medium supplemented with 500 mg/l carbenicillin in Plantaboxes.
4. Gradually acclimate plantlets for 1–2 months.
5. Transplant the acclimated plants into pots and grow in a closed greenhouse.

## IV. EFFECT OF ACETOSYRINGONE ON TRANSFORMATION

Sarmento and coworkers (8) show that exposing petiole segments of cucumbers to an *Agrobacterium tumefaciens* suspension with 20 µM acetosyringone has no significant effect in enhancing the frequency of kanamycin-resistant calli. In our study (4), the acetosyringone treatment during coculture enhances transformation efficiency on the cut surfaces of hypocotyl explants of cucumber. We suggest that the long exposure of the hypocotyl explants to *Agrobacterium tumefaciens* with acetosyringone is more effective than short times of exposure.

The cocultured explants are transferred to MS medium containing 20 mg/l hygromycin B for selection and plantlet regeneration. The appearance of hygromycin B–resistant calli indicating successful transformation was observed at high frequencies both with and without acetosyringone during coculture. Hygromycin B–resistant calli appeared on the cut surface in 100% of 100 explants cocultured with acetosyringone treatment and in 88% of 100 explants without the acetosyringone treatment 2 weeks after coculture (4).

In contrast, frequency of hygromycin B–resistant calli was 4% when explants were cocultured for 2 days without acetosyringone. Coculture for 2 days with acetosyringone resulted in callus on 60% of the explants.

These results show that acetosyringone treatment during coculture increases the number of transformed cells on the cut surface of cucumber hypocotyl explants.

## V. CELL SELECTION

The choice of antibiotic to select the transformed cells is very important in transformation studies. Kanamycin was usually used as the antibiotic in previous reports of cucumber transformation experiments (6–8). We compared the suitability of kanamycin and that of hygromycin B for selecting transformed cells (4).

In both control and *Agrobacterium tumefaciens*–infected explants treated with kanamycin, the hypocotyls grew to three to four times larger than the initial explants during a 2- to 3-week period. On the contrary, hypocotyl explants stopped growing in both control and *Agrobacterium tumefaciens*–infected samples treated with hygromycin B. Kanamycin killed the nontransformed cells slowly, and many nontransformed plantlets are regenerated from the calli. On the contrary, hygromycin B kills nontransformed cells quickly, and the frequency of regeneration of nontransformed plantlets is reduced to one third in comparison with kanamycin.

## VI. EXPRESSION OF CaMV 35s PROMOTER

CaMV 35s promoter is generally used for the expression of transgenes in transgenic crops. It is important to know the tissue cells in which the promoter is active. In transgenic cucumber plants (4) carring an introduced *gus* gene, the CaMV 35s promoter-*gus* gene was expressed in very young leaves that were regenerated from calli, but not in the mature leaves of both 20-cm- or 1.5-m-tall plants (Table 1). The promoter was active in the root meristem, young cotyledons of seedlings, and ovules of immature fruits and was silent in differentiated tissues cells of roots, hypocotyls, old cotyledons, stems, leaves, petioles, tendrils, male flowers, and fruits (female flowers) (Table 1) (4).

**Table 1** Expression of CaMV 35s-*gus* Gene in the Various Organs and Tissues of Transgenic Cucumber Plants

| Organ and tissue | Expression of CaMV 35s-*gus* gene[a] |
|---|---|
| Root | |
|   Meristem | +++ |
|   Differentiated | – |
| Stem | – |
| Leaf | |
|   Regenerated (from calli) | +++ |
|   Mature | – |
| Petiole | – |
| Tendril | – |
| Male flower | – |
| Fruit (female flower) | |
|   Ovule | |
|     Immature | +++ |
|     Mature | – |
|   Others | – |
| Hypocotyl of seedling | – |
| Cotyledon of seedling | |
|   Young | ++ |
|   Intermediate | + |
|   Old | – |
| Seed | +++ |

[a] +++, very strong; ++, strong; +, weak; –, no detection. Control cucumber plants did not display the expression of CaMV 35s-*gus* gene in the various organs and tissues.
*Source*: Refs. 3 and 4.

It is interesting that the CaMV 35s promoter is inactive in most of the differentiated tissues and organs of the transgenic cucumber plants, because it is generally considered that the CaMV 35s promoter functions in the meristematic cells of the roots, stems, leaves, and vascular bundles of various organs.

## VII. CUCUMBER MOSAIC VIRUS RESISTANCE

Transgenic cucumber plants (cv. 'Poinsett 76') with an introduced CMV-C-*cp* gene were first produced by Chee and Slightom (12). They and coresearchers investigated CMV disease development in the transgenic lines after inoculation of the CMV-CAT strain in the greenhouse and under field conditions (13). The inoculated plants responded with three general types of reaction: (a) susceptible (14%), (b) symptomless (36%), and (c) delayed symptoms (50%). Fruit yields and vegetative growth of transgenic lines averaged better than those of nontransgenic plants of the same cultivar under field conditions.

We introduced a gene encoding the CP of CMV-O strain into a line 1021, which has a low level of resistance to the CMV-Y strain (5). Two transgenic cucumber plants were cross-hybridized with cucumber cultivar 'Sharp 1,' which displays sensitivity to the CMV-Y strain. Plants (40/plant line) were inoculated with CMV-Y as described previously. The nontransgenic progeny of the parent plants displayed sensitivity to the CMV-Y strain, as did the cv. 'Sharp 1' (Fig. 1). The number of CMV-induced disease lesions on the first leaf in the nontransgenic progeny plants was half of that in the cultivar 'Sharp 1,' because the progeny plants were hemizygous with regard to the resistance gene in line 1021. On the other hand, the number of CMV-induced lesions on the nontransgenic progeny plants was nine times greater than that in the line 1021. On the contrary, the transgenic progeny plants that contain a *cp* gene displayed no symptoms during the 3 weeks after inoculation with 100 µg/ml CMV-Y strain (Fig. 1), whereas control plants did display symptom appearance (5).

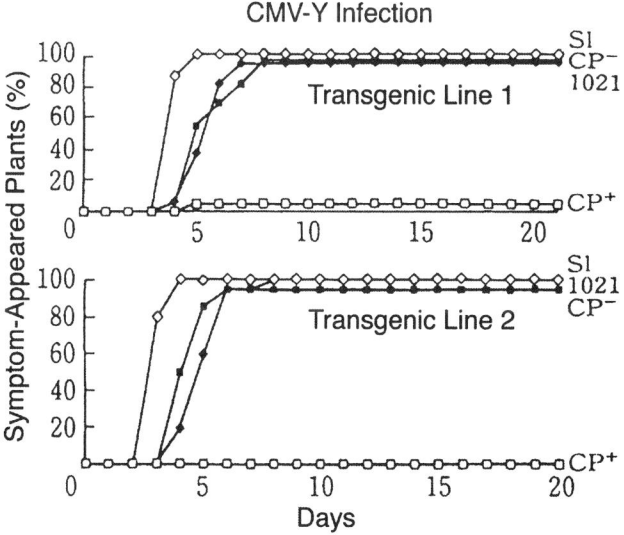

**Figure 1** Frequencies of symptom-appeared plants in two transgenic cucumber lines after inoculation with 100 µg/ml CMV-Y strain. S1 (cv. 'Sharp 1') and line 1021 as control plants. (From Ref. 5.)

The cucumber cv. 'Poinsett 76' used by Gonsalves and associates (13) was more susceptible than the cultivar 'Sharp 1' and the line 1021 (5).

Our results indicate that transgenic cucumber plants with the characteristics of very strong CMV resistance can be produced by combining natural genetic resistance and CMV-*cp* gene (5).

Coat protein was not detected in both leaves and cotyledons of the transgenic plants by Western blot analysis, whereas messenger ribonucleic acid (mRNA) was detected in them by Northern blot analysis (5). In recent studies, virus resistance in transgenic plants expressing a variety of viral genes is caused by the posttranscriptional mechanisms leading to degradation of viral RNA similar to the posttranscriptional gene silencing (14–16). Resistance in the transgenic plants may be caused by RNA-mediated resistance.

## VIII. SAFETY OF TRANSGENIC PLANTS

The transgenic cucumber plants are now being investigated for morphological characteristics, fruit yield, pollen, and seed fertility, in comparison with the control plants. There were no apparent differences in morphological characteristics, fruit yield, pollen, and seed fertility between the transgenic and control plants in greenhouse studies conducted over 2 years. Furthermore, there was no evidence of cross-hybridization between transgenic and control plants in the closed greenhouse. However, cucumber plants are usually hybridized by insect-mediated pollination. There is, therefore, a possibility of hybridization between nontransgenic and transgenic cucumber plants by insects in natural fields. The transgenic cucumber plants should be further investigated for their safety as food for human beings.

In the future, the transgenic cucumber plants may be demonstrated to be suitable for the market after sufficient investigation of their safety for human beings.

## ACKNOWLEDGMENTS

I thank Dr. Roger N. Beachy for critical reading of the manuscript. I also thank Dr. Ko Shimamoto for helpful discussion throughout the course of this study.

## REFERENCES

1. P Powell-Abell, RS Nelson, B De, N Hoffman, SG Rogers, TR Fraley, RN Beachy. Delay of disease development in transgenic plants that express the tobacco mosaic virus coat protein gene. Science 23: 738–743, 1986.
2. RN Beachy, S Loesch-Fries, NE Tumer. Coat protein-mediated resistance against virus infection. Annu Rev Phytopathol 28:451–474, 1990.
3. ED Miller, C Hemenway. History of coat protein-mediated protection. Methods Mol Biol 81:25–38, 1998.
4. S Nishibayashi, H Kaneko, T Hayakawa. Transformation of cucumber (Cucumis sativus L.) plants using Agrobacterium tumefaciens and regeneration from hypocotyl explants. Plant Cell Rep 15:809–814, 1996.
5. S Nishibayashi, T Hayakawa, T Nakajima, M Suzuki, H Kaneko. CMV protection in transgenic cucumber plants with an introduced CMV-O cp gene. Theor Appl Genet 93:672–678, 1996.
6. AJ Trulson, RB Simpson, EA Shahin. Transformation of cucumber (Cucumis sativus L.) plants with Agrobacterium rhizogenes. Theor Appl Genet 73:11–15, 1986.
7. PP Chee. Transformation of Cucumis sativus tissue by Agrobacterium tumefaciens and the regeneration of transformed plants. Plant Cell Rep 9:245–248, 1990.

8. GG Sarmento, K Alpert, FA Tang, ZK Punja. Factors influencing Agrobacterium tumefaciens mediated transformation and expression of kanamycin resistance in pickling cucumber. Plant Cell Tissue Org Cult 31:185–193, 1992.
9. EE Hood, GL Helmer, RT Fraley, MD Chilton. The hypervirulence of Agrobacterium tumefaciens A281 is encoded in a region of pTiBo542 outside of T-DNA. J Bacteriol 168:1291–1301, 1986.
10. RA Jefferson, TA Kavanagh, MW Bevan. GUS fusions: β-glucuronidase as a sensitive and versatile gene fusion marker in higher plants. EMBO J 6:3901–3907, 1987.
11. K Akama, H Shiraishi, S Ohta, K Nakamura, K Okada, Y Shimura. Efficient transformation of Arabidopsis thaliana: Comparison of the efficiencies with various organs, plant ecotypes and Agrobacterium strains. Plant Cell Rep 12:7–11, 1992.
12. PP Chee, JL Slightom. Transfer and expression of cucumber mosaic virus coat protein gene in the geneome of Cucumis sativus. J Am Soc Hortic Sci 116:1098–1102, 1991.
13. D Gonsalves, PP Chee, R Provvidenti, R Seem, JL Slightom. Comparison of coat protein-mediated and genetically-derived resistance in cucumbers to infection by cucumber mosaic virus under field conditions with natural challenge inoculations by vectors. Biotechnology 10:1562–1570, 1992.
14. HA Smith, SL Swaney, TD Parks, EA Wernsman, WG Dougherty. Transgenic plant virus resistance mediated by untranslatable sense RNAs: Expression, regulation, and fate of nonessential RNAs. Plant Cell 6:1441–1453, 1994.
15. M Prins, M Kikkert, C Ismayadi, W de Graauw, P de Haan, R Goldbach. Characterization of RNA-mediated resistance to tomato spotted wilt virus in transgenic tobacco plants expressing NS(M) gene sequences. Plant Mol Biol 33:235–243, 1997.
16. IL Ingelbrecht, JE Irvine, TE Mirkov. Posttranscriptional gene silencing in transgenic sugarcane: Dissection of homology-dependent virus resistance in a monocot that has a complex polyploid genome. Plant Physiol 119:1187–1198, 1999.

# 40
# Transgenic Parthenocarpic and Insect-Resistant Eggplant

**Giuseppe Leonardo Rotino**
*Research Institute for Vegetable Crops, Montanaso Lombardo, Italy*

**Francesco Sunseri**
*University of Basilicata, Potenza, Italy*

**Nazareno Acciarri**
*Research Institute for Vegetable Crops, Monsampolo del Tronto, Italy*

**Salvatore Arpaia**
*Metapontum Agrobios, Metaponto, Italy*

**Giuseppe Mennella**
*Research Institute for Vegetable Crops, Pontecagnano, Italy*

**Angelo Spena and Michela Zottini**
*University of Verona, Verona, Italy*

| | | |
|---|---|---|
| I. | INTRODUCTION | 588 |
| II. | BREEDING AND BIOTECHNOLOGICAL APPROACHES | 589 |
| | A. Main Breeding Goals | 589 |
| | B. Doubled Haploid in Breeding Program | 589 |
| | C. Somatic Hybridization | 589 |
| | D. Genetic Transformation | 589 |
| | E. Molecular Markers | 589 |
| III. | GENETIC ENGINEERING | 590 |
| | A. Insect Resistance | 590 |
| | B. Field Performance of Transgenic Lines | 591 |
| | C. Cry1 Gene and Shoot and Fruit Borer Resistance | 593 |
| IV. | PARTHENOCARPY | 593 |
| | A. Introduction | 593 |
| | B. Chimeric Gene DefH9-iaaM as Biotechnological Tool to Induce Parthenocarpy | 594 |
| | C. Characterization of the Primary Transformed Plant | 596 |
| | D. Greenhouse Evaluation of Engineered Parthenocarpic Hybrids | 597 |
| V. | PERSPECTIVES | 599 |
| | REFERENCES | 599 |

## I. INTRODUCTION

*Solanum melongena* L. ($2n = 24$), also known as eggplant, *aubergine,* or *brinjal,* is an important solanaceous vegetable crop mainly cultivated in the tropical Asian and Mediterranean countries.

Eggplant was cultivated for over 1500 years in Asia before being introduced in Europe by the Arabs (1), and the common name probably derived from the white egglike fruits. At present, the vegetable crop is cultivated on 1,130,000 hectares worldwide with an annual production of 17 million metric tons (2), an increase of 2.5 million metric tons in the last 4 years has been estimated (Table 1).

The most important eggplant-growing countries are China (the world's largest producer with 48.6% and 57.2% of harvested area and production, respectively), India, Japan, Indonesia, and Mediterranean countries such as Turkey, Egypt, Morocco, Italy, Greece, Spain, and France; and the United States (see Table 1). Eggplant, a good source of minerals and vitamins with a total nutritional value comparable to that of tomato, is an important and popular vegetable in the diet of these countries. It is usually freshly marketed, although the use of frozen eggplant is spreading in the developed countries.

The utilization of eggplant fruit and leaves for medicinal purposes such as treatment for diabetes, asthma, bronchitis, and dysuria has been reported (1,3); its tissue extracts are also reported to promote lowering of blood cholesterol levels (1).

Eggplant is a slow-growing perennial solanaceous crop in tropical climates, whereas in the temperate zones it behaves as an annual and its growing season can be lengthened under intensive greenhouse cultivation. At the end of the life cycle, the plant reaches a height ranging from 50 to 150 cm, bearing fruits of very different size, shape, and skin color. Reported as a day-neutral plant, it starts flowering at the 6th- to 10th-leaf stage, enduring for a long period. Considered an autogamous species, it has a frequency of natural cross-pollination reported to vary from 0.2% to 48% (4,5).

Since eggplant is highly responsive to in vitro manipulation (6), such biotechnological approaches as genetic transformation (7–9), doubled-haploid production (10), and somatic hybridization (11,12) may well help to solve several of its agronomic problems.

**Table 1** Worldwide Production of Eggplant

|  | Harvested area, ha | Yield, kg/ha | Production, metric tons |
|---|---|---|---|
| Egypt | 28,000 | 20,000 | 560,000 |
| Sudan | 4,500 | 24,444 | 110,000 |
| **Africa** | **42,320** | **18,487** | **782,385** |
| United States | 1,052 | 31,825 | 33,480 |
| **America** | **1,442** | **44,804** | **38,542** |
| China | 551,218 | 18,182 | 10,022,356 |
| India | 320,000 | 10,625 | 3,400,000 |
| Turkey | 32,600 | 26,074 | 849,999 |
| **Asia** | **1,060,814** | **15,106** | **16,024,802** |
| Italy | 11,000 | 28,609 | 314,700 |
| Spain | 3,500 | 31,429 | 110,000 |
| **Europe** | **25,650** | **23,692** | **607,700** |
| **World** | **1,133,940** | **15,442** | **17,510,483** |

*Source*: Based on FAO Production Yearbook. Rome: 1998.

## II. BREEDING AND BIOTECHNOLOGICAL APPROACHES

### A. Main Breeding Goals

Eggplant breeding is mainly focused on release of $F_1$ hybrids, which already have supplanted the open-pollinated varieties, particularly in the intensive growing areas. The main goal of breeding is to develop high-quality and pest-resistant varieties (13). In the countries where intensive and successive cropping is practiced, the development of genotypes resistant to soilborne diseases such as fungal wilt (*Verticillium* and *Fusarium* spp.), bacterial wilt (*Ralstonia solanacearum*), and nematode infestations (*Meloydogine* spp.) becomes necessary. Eggplant cultivations are severely damaged also by insects [e.g., Colorado potato beetle (*Leptinotarsa decemlineata* Say), fruit and shoot borer (*Leucinodes orbonalis* Guenee), greenhouse whitefly (*Trialeurodes vaporariorum* Westw), thrip (*Frankiniella occidentalis*), mites (e.g., *Tetranychus urticae*, *T. cinnaborinus*), and fruit rots (e.g., *Botrytis* and *Phomopsis* spp.)].

### B. Doubled Haploid in Breeding Program

Anther culture is an in vitro technique currently utilized in commercial breeding programs. The main advantage of doubled-haploid technique is its speeding up of the procedure to obtain pure lines for $F_1$ hybrid release (10). Another useful application is the generation of androgenetic dihaploid plants from tetraploid somatic interspecific hybrids; dihaploids can be more easily incorporated in an eggplant breeding program.

### C. Somatic Hybridization

Somatic hybrids have been obtained via protoplast fusion starting from several *Solanum* spp. as donor parent (12). Somatic hybridization could allow the recombination of both nuclear and cytoplasmic deoxyribonucleic acid (DNA). Thus, it is a powerful tool to expand the genetic variability and introgress useful genes from wild relatives into the eggplant gene pool.

### D. Genetic Transformation

Efficient *Agrobacterium tumefaciens*–mediated transformation protocols were developed for selected genotypes (8,9,14); however, other genotypes may require a refinement of the technique in order to regenerate transgenic plants.

### E. Molecular Markers

Allozyme and random amplified polymorphic DNA (RAPD) analyses performed on 52 *S. melongena* genotypes (27 and 25 lines of cultivated and weedy form var. *insanum*, respectively) were reported (15,16). The results showed a high degree of similarity between the accessions of each form tested ($I = 0.947$ by RAPD analysis); overall, the *S. melongena* var. *insanum* genotypes differed more from the cultivated eggplants (16).

In 1998 a RAPD-based linkage map of eggplant was presented (17), confirming the low degree of polymorphism detected in this species. The map was constructed on 168 $F_2$ segregating progenies obtained from a cross between an elite line and a bacterial wilt–resistant line. A total of 1232 primers were screened, resulting in only 95 primers (7.7%) detecting polymorphisms. The map consisted of 94 loci spanning 716 cM; the average linkage distance between markers of about 9 cM and the number of linkage groups observed (13) suggested that additional markers should be mapped (17).

These first reports indicated that allozymes and RAPD markers revealed a low degree of polymorphism in eggplant. This suggests that the use of more informative segregating populations and/or molecular marker classes, such as RFLP, amplified fragment length polymorphism (AFLP), and microsatellites, could be desirable in order to construct a genetic linkage map useful for assisted breeding and to study genome structure, organization, and introgressions.

## III. GENETIC ENGINEERING

### A. Insect Resistance

#### 1. Introduction

Modified *Bacillus thuringiensis* (*Bt*) Cry genes have been widely used to obtain transgenic plants resistant to insects (18). The level of *Bt* toxin expression in plants may differ, depending on plant species, age, tissues, and organs (19). The Colorado potato beetle (CPB, *Leptinotarsa decemlineata* Say) is the most important insect pest of eggplant in Europe and America, causing heavy economic losses (20–22). This insect has reached a prominent pest status because of its ability to develop resistance against insecticides (23). On the other hand, the shoot and fruit borer (SFB, *Leucinodes orbonalis* Guenee) is a specific insect that heavily infests eggplant in Asia (24).

Since wild-type Cry genes were unable to provide sufficient insect control because of their low expression (14,25), transgenic eggplants resistant to SFB and CPB by expression of codon-optimized Cry1 (26) and Cry3 genes (27–30) have been obtained, representing a very effective means of pest control.

#### 2. Cry3 Genes and Colorado Potato Beetle Resistance

A modified *Bt* gene (29,31) of *Bacillus thuringiensis* var. *Tolworthi* encoding a coleopteran insect-specific Cry3B toxin was mobilized to the female parent of the commercial $F_1$ hybrid 'Rimina' (27). The LBA4404 *Agrobacterium tumefaciens* strain supporting the plasmid pBinCry3B was employed in the transformation experiments, which were based essentially on the procedure described by Rotino and Gleddie (9), Rotino and coworkers (25), and Arpaia and associates (27). The plantlets were acclimated in a growth chamber, then transferred to the greenhouse. At flowering, the plants were hand-pollinated. Successful transformation was evident (158 transgenic plants were regenerated), on the basis of polymerase chain reaction (PCR) analysis for presence of the *Bt* and/or NPTII gene.

The expression of the *Bt* gene in 93 eggplants representing 22 different transformation events was tested by double antibody sandwich–enzyme-linked immunoassay (DAS-ELISA), which showed that about 60% of *S. melongena* plants were *Bt*-positive. It was estimated that approximately 320 ng/ml of toxin was present in extracts of high-expressing transgenic plants. In addition, 10 plants were also analyzed by Western blot analysis. In protein extracts of high-*Bt*-expressing plants a specific immunoreactive polypeptide of approximately 74 kDa comigrates with the standard Cry3 toxin generated in *Escherichia coli* (Fig. 1).

The toxicity of transformed plants was tested in several in vitro bioassays with CPB larvae through leaf disk experiments (Fig. 2) according to the protocol described by Rotino and colleagues (31). About 50% of the transgenic plants were toxic to neonate CPB larvae.

The inheritance of the transgene NPTII was verified by genetic analysis in several transformed $T_2$ progenies by spraying kanamycin solution and scoring for resistance/sensitivity by looking at absence/presence of bleaching sectors in the sprayed leaves (32).

Transgenic eggplants were also produced and tested for resistance to the CPB by using a synthetic *Bt* cry3A gene (30). Toxicity tests in planta and in vitro demonstrated that transformed

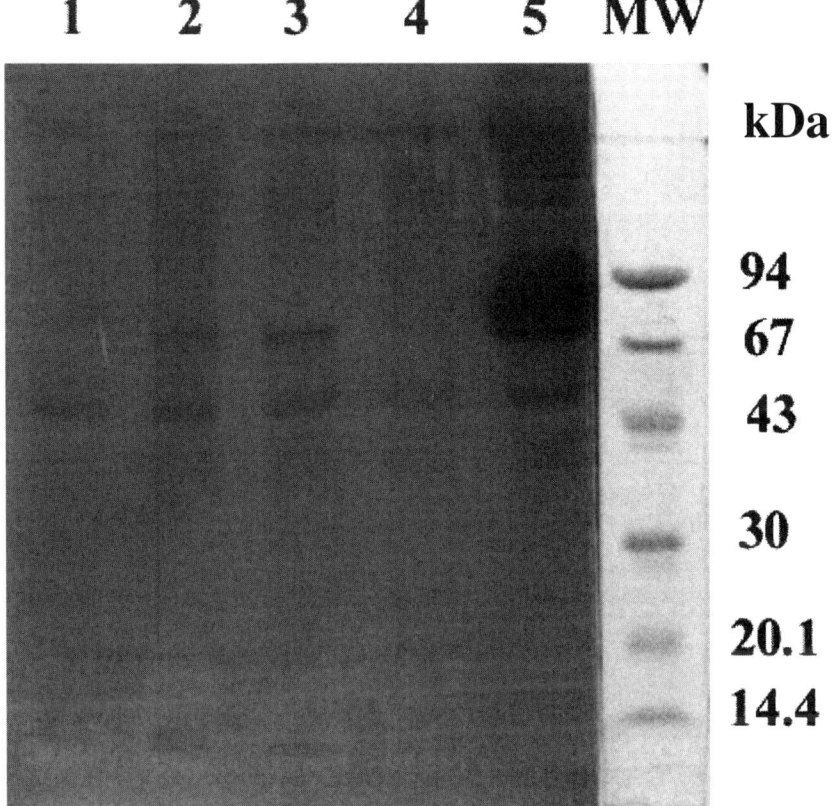

**Figure 1** Western blot analysis of *S. melongena* and *E. coli* extracts. Lane 1, low-expressing transgenic eggplant; lane 2–3, high-expressing trangenic eggplant; lane 4, untransformed control plant; lane 5, *Bt* toxin standard generated in *E. coli*; MW, molecular weight marker (Coomassie staining).

plants were resistant to neonate larvae and adult CPB. GUS expression and Southern and Northern analyses were performed to confirm the presence and the expression of transgenes. The primary transformants were field planted to assess their resistance to CPB. Plants having a single insert of the construct, upon selfing, produced progenies cosegregating for the cry3A gene at the expected 3:1 ratios. The resistant progeny showed a similar level of resistance to that of the parental genotypes. The $S_1$ progenies of the best genotypes were tested for agronomic performance in a replicated field trial. The experimental data showed that field populations of CPB and number of fruits per plant were significantly affected by the transgenic genotypes tested (30). Commercial eggplant varieties, developed from the homozygous resistant offspring, could represent a possible effective means for CPB control.

### B. Field Performance of Transgenic Lines

Field trials set at two different locations, Monsampolo del Tronto and Metaponto, were reported (33,34). Transgenic eggplant lines (#3-2, #6-1, and #9-8) expressing the modified Cry3B gene and the untransformed control line DR2 were evaluated, under natural CPB infestation, for their agronomic value with particular attention to the insect resistance character. Fields were prepared according to a randomized block design with four replications. Each plot measured 4.5 m by 9.6

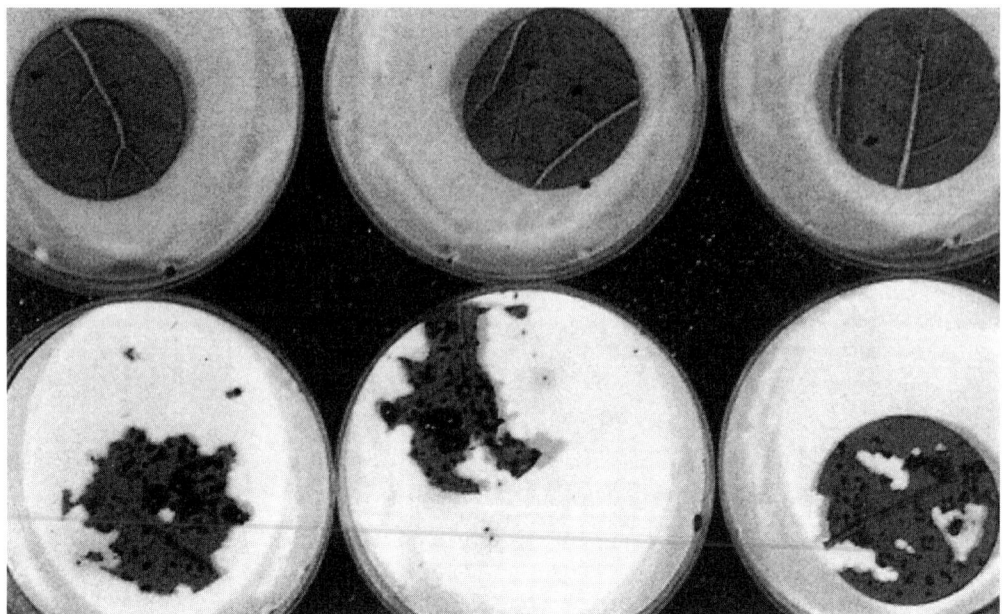

**Figure 2** Insect bioassay carried out on selfed progenies of plant #9-8. The upper row shows petri plates containing the larvae eating leaf disks cut from transgenic plants; the lower row shows larvae on leaf disks from untransformed plants.

m and contained 96 plants. During the growing season, the insect sampling program was carried out twice a week on 10% of randomly chosen plants, inspected for the number of CPB adults, egg masses, and small and large larvae; surveys were carried out for 8 weeks. For each harvest, fruit number and weight were recorded for data collection. Then, during the growing season, Cry3B protein was analyzed by DAS-ELISA, using different tissues of transgenic eggplants. At Monsampolo, the same experimental design was prepared in order to test a mix plot with 80% of transgenic and 20% control plants that should provide a refuge zone capable of reducing the selection pressure against the target insect.

In the field trial of Monsampolo the beetle's population was sizable. Whereas incoming beetles spread quite evenly on all plots, small and large larvae were mainly found on DR2 (control) and line #6-1, expressing a low amount of toxin (33). The analysis of variance confirmed that beetle population was significantly higher on control plants than on transgenic ones, both in the 100% field experiment (33,34) and in the 80% mixed field (Table 2) (Fig. 3).

Enzyme-linked immunosorbent assay (ELISA) results showed that the *Bt* toxin was evenly expressed in all the tissues. Significant reduced levels of *Bt* toxin were observed in line #6-1, when compared to corresponding tissues of lines #3-2 and #9-8.

In both field trials, all transgenic lines also had significantly higher fruit production (expressed as kilograms/plant) than the untransformed control (33,34); similar results were also obtained in the mixed field plot experiment (Table 2) (Fig. 4). Differences were more pronounced at Monsampolo, where the CPB infestation level was considerably higher. Line #6-1 confirmed its intermediate results in all cases.

The experimental data obtained in 1997 in Italy indicated that the resistance to CPB in transgenic eggplant, already evidenced in laboratory bioassays, was confirmed in field conditions. Transgenic plots, in fact, constantly showed lower presence of all CPB stages during the season and higher production in the presence of a high infestation level. The observations in the mixed

**Table 2** Statistical Analyses Related to the Data Collected at Monsampolo in the Mixed Plot (80% + 20%) Field Trial[a]

| Genotype | Egg masses, no.[b] | Small larvae, no.[b] | Large larvae, no.[b] | Adults, no.[b] | Production[c] |
| --- | --- | --- | --- | --- | --- |
| #3-2 | 0.148 B | 0.529 B | 0.369 B | 0.434 A | 0.178 A |
| #6-1 | 0.184 B | 0.818 B | 0.582 B | 0.550 A | 0.082 B |
| #9-8 | 0.096 B | 0.274 B | 0.138 B | 0.659 A | 0.189 A |
| DR2 | 0.299 A | 2.285 A | 2.303 A | 1.036 A | 0.062 B |

[a]Different letters indicate statistically significant differences [analysis of variance–(ANOVA-LSD, $\alpha = 0.05$)].
[b]Per plant.
[c]Kilograms per plant relative to only six data harvestings.

plot field, made by 80% transgenic plants and 20% refuge plants, showed that even in this case the crop can withstand natural infestation by *L. decemlineata* and yet provide higher production. Transgenic eggplant hybrids have been obtained; during 1999 they were under second-year field evaluation at three different Italian locations.

### C. Cry1 Gene and Shoot and Fruit Borer Resistance

Several lepidopteran-specific insecticidal *Bt* proteins were tested in order to evaluate their toxicity on the larvae of SFB (26). Three different Cry genes were found to be toxic at low concentration, but only a synthetic Cry1Ab gene, modified according to the plant codon usage, was utilized to perform genetic transformation of eggplant. A transgenic line that had fruits totally protected from larval injury was selected (26). A field trial using the resistant offspring of the genotype is planned. The use of fruit-specific promoters to express the *Bt* gene in eggplant was also considered as a possible next step for eggplant transformation.

## IV. PARTHENOCARPY

### A. Introduction

Parthenocarpy, the ability to set and develop fruits in absence of fertilization, is a desirable trait in eggplant. Parthenocarpic marketable fruits may be obtained under environmental conditions that normally are limiting for fruit setting and growth. Moreover, parthenocarpy may reduce the energy and labor necessary for off-season eggplant cultivation and enhance fruit quality.

Since the commercial ripeness of eggplant fruit precedes its physiological maturity, the presence of seeds considerably decreases the value of fruits for both the fresh and the processed market. The detrimental effects associated with the presence of seeds are: (a) a faster and more intense browning of the flesh after cutting, (b) an increment in the saponin and solasonin compounds also responsible for the bitter taste (35), and (c) a harder flesh. These factors make seed-containing berries unappetizing to consumers.

Outseason production of eggplant is usually achieved by cultivation in greenhouses. Hormonal treatment of flowers and/or heating is employed to counteract the negative effect of adverse environmental conditions on fruit set and growth (36). For such reasons, intense breeding activities are currently in progress to obtain parthenocarpic hybrids (37). Nevertheless, the parthenocarpic eggplant varieties available still require phytohormone treatments to produce fruits of marketable size (38).

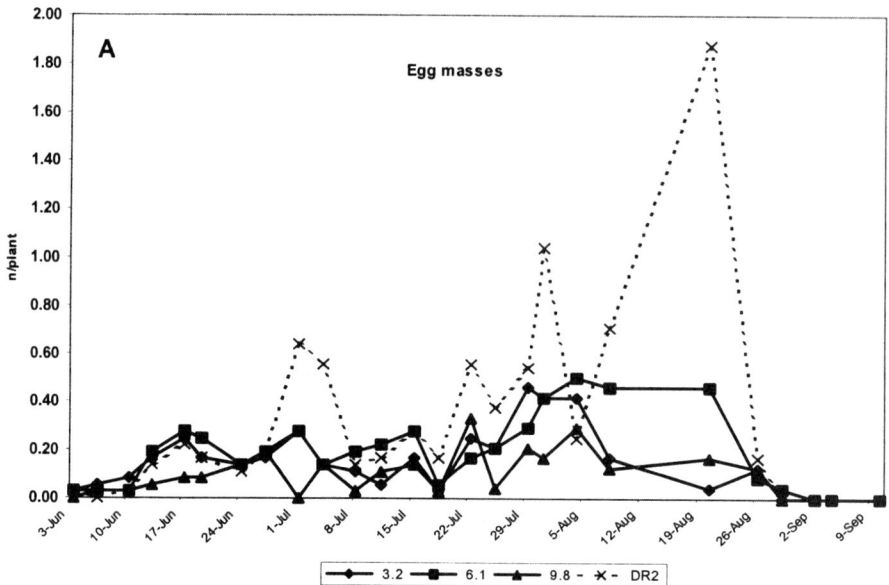

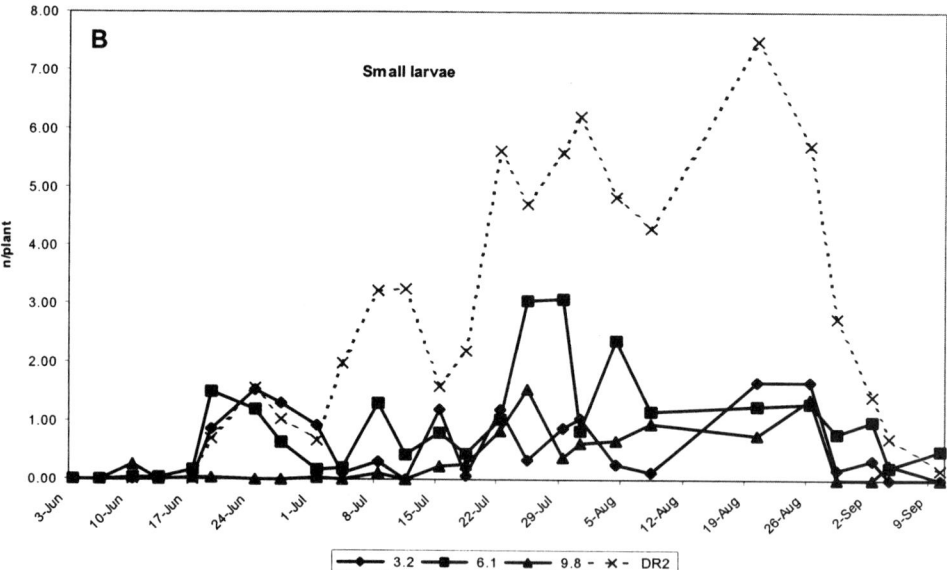

**Figure 3** Colorado potato beetle infestation course at Monsampolo field trial in a mixed plot experimental design (80% transgenic plant + 20% untransformed refuge plant): egg masses (A), small larvae (B), large larvae (C), and adults (D) per plant.

## B. Chimeric Gene DefH9-iaaM as a Biotechnological Tool to Induce Parthenocarpy

Fruit set and growth correlates with an increased level of auxin phytohormones in the developing ovules and embryos (39), and external application of phytohormones can replace the developing ovules to sustain fruit growth (40).

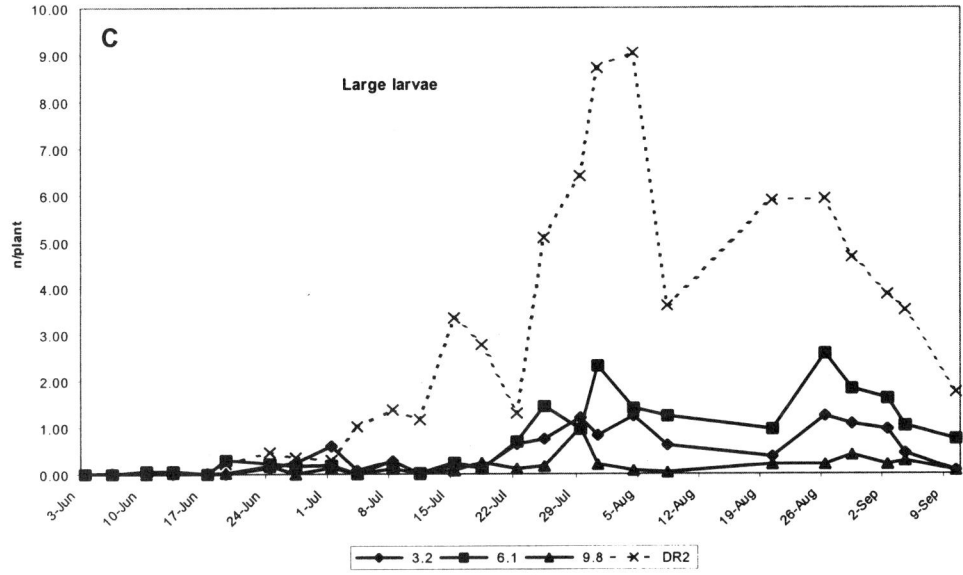

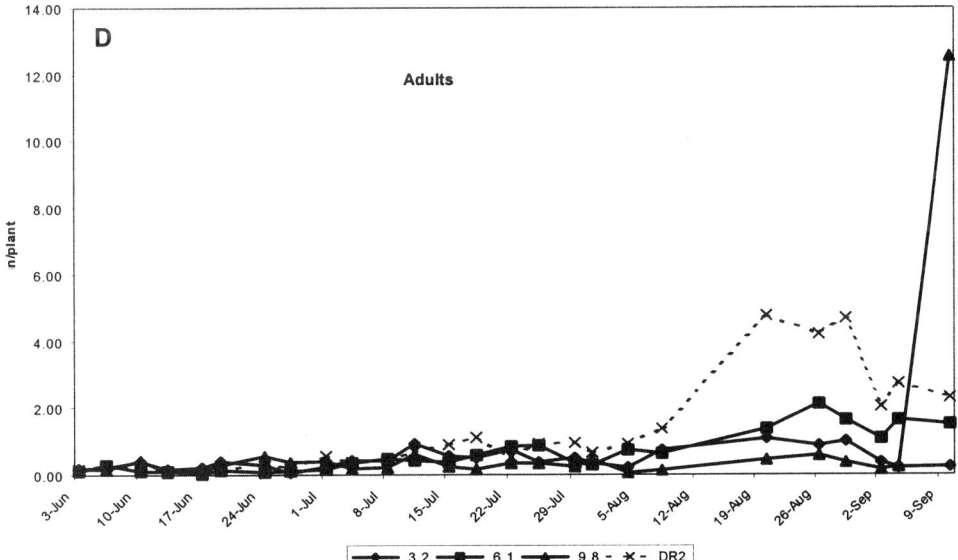

**Figure 3** (*Continued*).

Thus, to engineer parthenocarpic fruit development expression of a foreign gene responsible for phytohormone production has to take place during early floral and fruit development. The *DefH9,* deficient homologue 9, gene is a MADS-box containing gene from *Antirrhinum majus.* It is specifically expressed in the developing placenta and ovules. Its expression level in vegetative tissues is below the detection limit of Northern blot analysis (41). The promoter and the regulatory sequences of the *DefH9* gene were linked to the coding region of the *iaaM* gene from *Pseudomonas syringae* pv. *savastanoi* in order to obtain an increase in auxin content and/or ac-

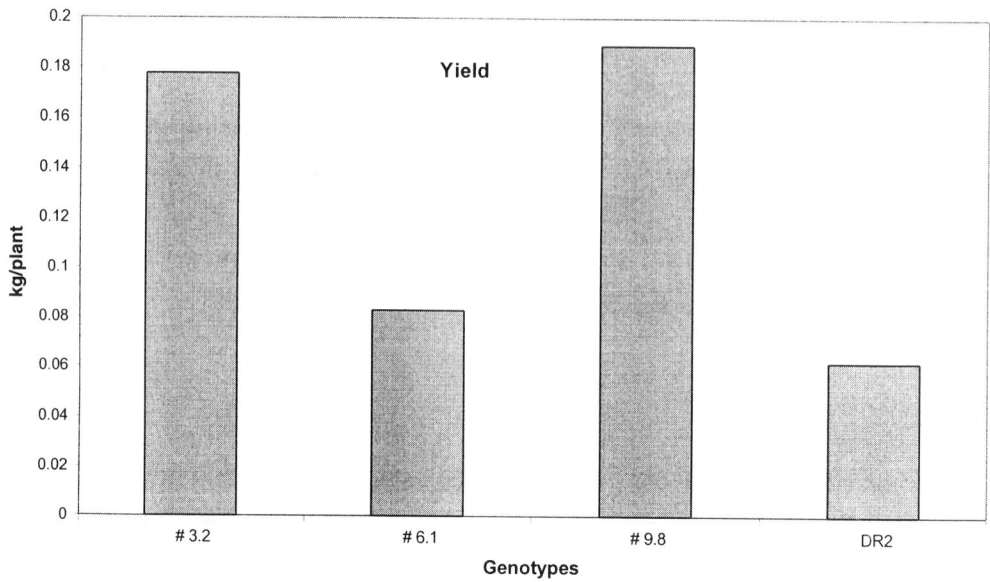

**Figure 4** Monsampolo field trial, mixed plot experimental design: eggplant production after the first six harvests.

tivity in cells of the female reproductive organ. The *iaaM* gene codes for the enzyme tryptophan monoxygenase, which converts tryptophan to indolacetamide, a percursor of indole-3-acetic acid (IAA), the major form of auxin in plants (42). Indolacetamide is slowly converted to IAA in plant cells, either chemically or enzymatically by hydrolases (43).

The chimeric gene *DefH9-iaaM* was inserted into either the plasmid pPCV002 (44) or pBin 19 (45), which also contain the selective marker NPTII conferring resistance to the antibiotic kanamycin. The recombinant plasmid was introduced into the *Agrobacterium tumefaciens* strain GV3101 for eggplant genetic transformation.

## C. Characterization of the Primary Transformed Plant

Parthenocarpic transgenic eggplants showed normal vegetative growth and a phenotype indistinguishable from that of seed-derived plants. This confirmed that the chimeric gene *DefH9-iaaM* does not interfere with the in vitro regeneration process and has no adverse effect on the phenotype of transgenic plants.

In all the transgenic parthenocarpic eggplants analyzed, fruit setting and normal fruit development were achieved from both emasculated and selfed flowers under very limiting environmental conditions (low temperature, low light intensity and duration). In the same conditions, the untransformed control plants did not set fruit even from hand-pollinated flowers (41).

Experiments performed under normal environmental conditions (spring and summer) allowed us to compare fruit setting and fruit characteristics in transgenic and untransformed plants. The weight and the size of parthenocarpic eggplant fruits obtained from unpollinated flowers were about four times higher than those of fruits derived from emasculated flowers of the corresponding untransformed controls. These results showed that engineered parthenocarpic eggplants were also superior to eggplants with a genetic tendency to parthenocarpy (Rotino, unpublished).

Expression of the *DefH9-iaaM* gene, analyzed by competitive reverse transcriptase polymerase chain reaction (RT-PCR) on polyA$^+$RNA in transgenic young flower buds (Fig. 5), opened

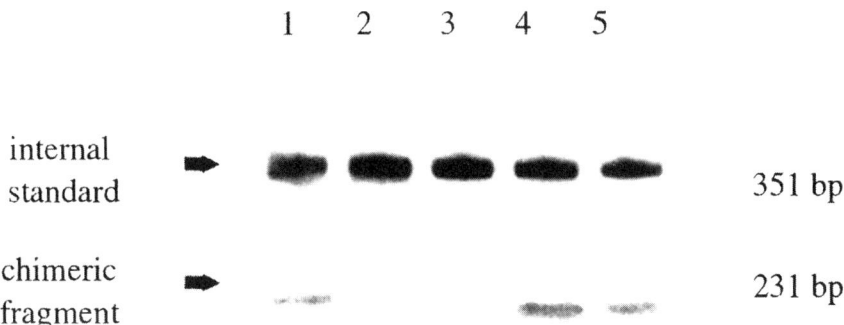

**Figure 5** Expression analysis of *DefH9-iaaM* gene by competitive RT-PCR. Analysis was performed with single-strand complementary (cDNA) synthesized from messenger RNA (mRNA) extracted from young transgenic eggplant flower buds of line Tal 1/1 #1-1 (lane 1) and line DR2 #33 and #35 (lanes 4 and 5, respectively). Lane 3, untransformed plant; lane 2, internal standard. An internal standard of 351 bp is present in all lines.

flowers, and fruit at various growth stages confirmed the presence of an amplicon corresponding to the 5′ end of the spliced *DefH9-iaaM* messenger ribonucleic acid (mRNA) in all the analyzed organs of the transgenic plants, whereas no amplicon was detected in the untransformed control plants (46). The steady-state level of expression in flower buds from independent transgenic plants has been estimated to be in the range of $1 \times 10^{-8}$ to $2 \times 10^{-9}$ of the total mRNA population. Thus, expression of the *DefH9-iaaM* gene starts during early flower development and is protracted to later stages of fruit development.

### D. Greenhouse Evaluation of Engineered Parthenocarpic Hybrids

Winter and early cultivation of transgenic parthenocarpic eggplant hybrids in greenhouse were performed to assess their agronomic performance in comparison to that of untransformed controls and/or commercial cultivars.

The winter trial was carried out in Sicily at Vittoria, which is a typical Mediterranean zone for this kind of production and every year about 1000 hectares of unheated plastic greenhouses is cultivated with eggplants. In this area, winter production is obtained by spraying eggplant flowers with phytohormones. Early production of transgenic hybrids was evaluated in a greenhouse trial carried out at Monsampolo del Tronto, Italy.

At Vittoria the cultural cycle lasted from October 22, 1997 (transplanting), to June 1, 1998 (the last harvest). The data regarding the performance of the transgenic hybrid P2, obtained by crossing the Tal 1/1 with DR2 iaaM #28-1 (41), are presented. The parthenocarpic transgenic hybrid was not treated with phytohormones, and the cultivar 'Black Bell' (as control) was either treated or not treated with phytohormones. This experiment was arranged according to a randomized block design with three replications, number of fruits and yield per plant were recorded.

Both in the early production (the first four harvests) and in subsequent harvests, the untransformed cultivar 'Black Bell' was positively affected by hormonal sprays; in fact, the number of fruit per plant and, consequently, the yield were significantly higher in the plots subjected to treatment than in the untreated ones (Table 3). In the early winter production the transgenic hybrid P2 furnished a number of fruit and a yield per plant not significantly different from those of the sprayed plant of the cultivar 'Black Bell.' This is in agreement with the finding that the treatment of floral buds with exogenous plant growth regulators did not affect set and growth of fruit and production of other transgenic parthenocarpic hybrids in the same trial (46). The hybrid P2

**Table 3**  Mean Values of Number of Fruits and Yield per Plant in the Transgenic Hybrid (P2) and in the Cultivar 'Black Bell' Subjected (A) or Not Subjected (B) to Phytohormone Treatments for the Winter Harvesting Periods Considered

| Genotypes | Treatments | Early winter production | | Production after March 3, 1998 | | Yield in the whole harvesting period, g/plant |
|---|---|---|---|---|---|---|
| | | Fruits/plant, no. | Yield/plant, g | Fruits/plant, no. | Yield/plant, g | |
| 'Black Bell' | A[a] | 3.4 ab | 838.0 a | 14.7 b | 3303 a | 4141 a |
| 'Black Bell' | B[b] | 1.4 b | 331.3 b | 7.4 b | 808 b | 1139 b |
| P2 | B[b] | 4.6 a | 751.0 a | 30.7 a | 5178 a | 5229 a |

For each trait, different letters indicate statistically significant differences (ANOVA-LSD, $\alpha = 0.05$).
[a]A, opened flowers treated with phytohormones.
[b]B, opened flowers not treated with phytohormones.

has long-shaped fruit with a lower average fruit weight than that of the pear-shaped fruit of the cultivar 'Black Bell.' The data of the subsequent harvests confirmed that marketable winter eggplant production was obtained by cultivating transgenic hybrid without hormonal treatment and under normal cultural practices.

At Monsampolo the cultural cycle lasted from March (transplanting) to July (the last harvest). The performance of four transgenic parthenocarpic hybrids (P3, P4, P7, and P8) and that of the untransformed commercial hybrid DRA 1205 (De Ruiter) were compared. The trial was arranged in a randomized block design with four replications; data were collected on flowering date, early (the first six harvests) and total number of fruits, and yield per plant. In each harvest a sample of fruits per plot was scored for the presence of seeds.

Among the tested hybrids, a maximal difference of 5 days in the flowering time was recorded; the untransformed hybrid showed a medium flowering date. In such a period the transgenic hybrids gave a similar yield that was significantly higher than that of the untransformed hybrids. On average, this yield difference was 170% over that of the hybrid DRA 1205 (Table 4). The significantly higher number of marketable fruits per plant accounted for the yield differences. Considering the yield obtained in the whole cultural cycle, all the transgenic hybrids had a higher production than the untransformed one, but only for the hybrid P8 this production was significantly different from that of the control.

**Table 4**  Mean Values of Number of Fruits, Yield per Plant, and Presence of Seeds in Fruit in Four Transgenic Hybrids (P3, P4, P7, and P8) and the Cultivar 'DRA 1205'

| Genotypes | Flowering date | Early production/plant | | Total production/plant | | Presence of seeds index[a] |
|---|---|---|---|---|---|---|
| | | Fruits, no. | Yield, g | Fruits, no. | Yield, g | |
| P3 | 5–18 | 1.4 a | 405.7 a | 15.0 ab | 3719 b | 0 |
| P4 | 5–15 | 1.3 a | 427.0 a | 16.2 a | 4163 b | 0 |
| P7 | 5–19 | 0.7 bc | 442.0 a | 10.2 b | 4146 b | 0 |
| P8 | 5–20 | 1.0 ab | 476.6 a | 13.0 ab | 5584 a | 0 |
| DRA 1205 | 5–18 | 0.3 c | 117.5 b | 13.7 ab | 3105 b | 2 |

[a]Presence of mature seeds index: 0, no seeds; 1, a few seeds; 2, medium presence of seeds; 3, high presence of seeds. For the values within each trait, at least a common letter indicates no significant difference according to Duncan's test ($\alpha = 0.05$).

The parthenocarpic hybrids produced seedless fruits through the entire harvest at both Vittoria and Monsampolo (Table 4); moreover, the flesh was evenly filled.

In greenhouse cultivation, engineered parthenocarpic hybrids offer the possibility of increasing winter productivity with a concomitant decrease of cultivation costs, allowing higher early production and better quality of fruit.

## V. PERSPECTIVES

Genetic engineering could be usefully employed both to enhance eggplant productivity, by introducing resistance to insects and diseases, and to improve the nutritional quality of fruits, by using genes encoding for proteins rich in essential amino acids, vitamins, and other elements. On the other hand, since the gene technology expands the range of genes available for breeding, it would be desirable to perform specific risk assessment studies before releasing genetically modified eggplant. A very important point to consider is that "eggplant is the major source of vegetables accessible to the poorest of poor in South and South East Asia" (24). Biotechnology applied to eggplant improvement will help to alleviate malnutrition, provided that this technology is more freely accessible to developing countries.

## ACKNOWLEDGMENTS

The original experimental data presented are from research partially financed by the Resistenze genetiche a stress biotici e abiotici, Biotechnologie Vegetali and Orticoltura programs of the Ministero Politiche Agricole (MiPA) and by the Biotecnologie program of the C.N.R. The authors thanks Genetic Initiative European Strategical Research in Agriculture (G.IN.E.ST.R.A) for encouragement and support.

## REFERENCES

1. R Khan. *Solanum melongena* and its ancestral forms. In: JG Hawkes, RN Lester, AD Skelding, eds. The Biology and Taxonomy of Solanaceae. London: Academic Press, 1979, pp 629–636.
2. FAO Production Yearbook. Rome: 1998.
3. S Porcelli. Problemi e prospettive della melanzana: Cenni introduttivi. Agric Ricerca 60:1–2, 1986.
4. L Quagliotti. Floral biology of *Capsicum* and *Solanum melongena*. In: JC Hawkes, RN Lester, AD Skelding, eds. The Biology and Taxonomy of Solanaceae. London: Academic Press, 1979, pp 399–419.
5. U Franceschetti, G Lepori. Natural cross pollination in eggplant. Sementi Elette 6:25–28, 1985.
6. K Hinata. Eggplant (*Solanum melongena* L.). In: YPS Bajaj, ed. Biotechnology in Agriculture and Forestry. Vol. 2. Berlin: Springer-Verlag, 1986, pp 363–370.
7. E Filippone, PF Larquin. Stable transformation of eggplant (*Solanum melongena* L.) by cocultivation of tissue with *Agrobacterium tumefaciens* carrying a binary plasmid vector. Plant Cell Rep 8:370–373, 1989.
8. A Guri, KC Sink. *Agrobacterium* transformation of eggplant. J Plant Physiol 133:52–55, 1988.
9. GL Rotino, S Gleddie. Transformation of eggplant (*Solanum melongena* L.) using a binary *Agrobacterium tumefaciens* vector. Plant Cell Rep 9:26–29, 1990.
10. GL Rotino. Haploidy in eggplant. In: SM Jains, SK Sopory, RE Veilleux, eds. In vitro haploid production in higher plants. Vol. 3. Dordrecht, The Netherlands: Kluwer Academic, 1996, pp 115–141.

11. A Guri, KC Sink. Interspecific somatic hybrid plants between eggplant *(Solanum melongena)* and *Solanum torvum.* Theor Appl Gen 76:490–496, 1988.
12. D Sihachakr, MC Daunay, I Serraf, MH Chaput, I Mussio, R Haicour, L Rossignol, G Ducreux. Somatic hybridization of eggplant *(Solanum melongena* L.) with its close relatives. Vol. 27. In: YPS Bajaj, ed. Biotechnology in Agriculture and Forestry. Berlin, Heidelberg: Springer-Verlag, 1994, pp 255–278.
13. GL Rotino, E Perri, N Acciarri, F Sunseri, S Arpaia. Development of eggplant varietal resistance to insects and diseases via plant breeding. Adv Hortic Sci 11:193–201, 1997.
14. Q Chen, G Jelenkovic, CK Chin, S Billings, J Eberhardt, JC Goffreda, P Day. Transfer and transcriptional expression of coleopteran cryIIIB endotoxin gene of *Bacillus thuringiensis* in eggplant. J Am Soc Hortic Sci 120:921–927, 1995.
15. JL Karihaloo, LB Gottlieb. Allozyme variation in the eggplant, *Solanum melongena* L. (Solanaceae). Theor Appl Genet 90:578–583, 1995.
16. JL Karihaloo, S Brauner, LB Gottlieb. Random amplified polymorphism DNA variation in the eggplant, *Solanum melongena* L. (Solanaceae). Theor Appl Genet 90:767–770, 1995.
17. T Nunome, T Yoshida, M Hirai. Genetic linkage map of eggplant *(Solanum melongena* L.). Proceedings of Xth Eucarpia Meeting on Genetics and Breeding on Capsicum and Eggplant. Capsicum News, special issue, Avignon, France, 1998, pp 239–242.
18. DA Fischoff. Insect resistant crop plants. In: GJ Perseley, ed. Biotechnology and Integrated Pest Management. Biotechnology in agriculture no. 15. Wallingford, England: CAB International, 1996, pp 214–227.
19. MG Koziel, GL Beland, C Bowman, NB Carozzi, R Crenshaw, L Crossland, J Dawson, N Desai, M Hill, S Kadwell, K Launis, K Lewis, D Maddox, K McPherson, MR Meghji, E Merlin, R Rhodes, GW Warren, M Wright, SV Evola. Field performance of elite transgenic maize plants expressing an insecticidal protein derived from *Bacillus thuringiensis.* Biotechnology 11:194–200, 1993.
20. S Cotty, JM Lashomb. Vegetative growth and yield response of eggplant to varying first-generation Colorado potato beetle densities. J NY Entomol Soc 90:220–228, 1982.
21. S Maini, G Nicoli, G Manzaroli. Evaluation of the egg parasitoid *Edovum puttleri* Grissel (Hym. Eulophydae) for biological control of *Leptinotarsa decemlineata* (Say) (Col. Chrysomelidae) on eggplant. Boll Ist Ent G Grandi, Univ Bologna 44:161–168, 1990.
22. S Arpaia, JM Lashomb, K Vail. Valutazione dell'attività trofica di *Leptinotarsa decemlineata* (Say) su melanzana. Inf Fitopatol 2:55–57, 1995.
23. JM Clarke, JA Argentine. Biochemical mechanism of insecticide resistance in the Colorado potato beetle. In: G Zenhder, RK Jansson, ML Powelson, KV Raman, eds. Potato pest management: a global perspective. St Paul, MN: APS, 1994, pp 294–308.
24. PA Kumar, AD Mandaokar, RP Sharma. Genetic engineering for the improvement of eggplant *(Solanum melongena* L.). Agbiotechnews Information 10:329N–332N, 1998.
25. GL Rotino, S Arpaia, R Iannacone, V Iannamico, G Mennella, V Onofaro, D Perrone, F Sunseri, Q Xike, F Sponga. *Agrobacterium*-mediated transformation of *Solanum* spp. using a Bt gene effective against coleopterans. Proceedings of the VIIIth Eucarpia Meeting on Genetics and Breeding of Capsicum and Eggplant, Rome, 1992, Capsicum Newsletter, special issue, pp 295–300.
26. PA Kumar, AD Mandaokar, K Sreenivasu, SK Chakrabarti, S Bisaria, RS Sharma, S Kaur, RP Sharma. Insect resistant transgenic brinjal plants. Mol Breed 4:33–37, 1998.
27. S Arpaia, G Mennella, V Onofaro, E Perri, F Sunseri, GL Rotino. Production of transgenic eggplant *(Solanum melongena* L.). resistant to Colorado potato beetle *(Leptinotarsa decemlineata* Say). Theor Appl Genet 95(3):329–334, 1997.
28. GC Hamilton, GL Jelenkovic, JH Lashomb, G Ghidiu, S Billings, JM Patt. Effectiveness of transgenic eggplant *(Solanum melongena* L.) against the Colorado potato beetle. Adv Hortic Sci 11:189–192, 1997.
29. R Iannacone, PD Grieco, F Cellini. Specific sequence modifications of a cry3B endotoxin gene result in high levels of expression and insect resistance. Plant Mol Biol 34(3):485–496, 1997.
30. G Jelenkovic, S Billings, Q Chen, J Lashomb, G Hamilton, G Ghidiu. Transformation of eggplant with synthetic cryIIIA gene produces a high level of resistance to the Colorado potato beetle. J Am Soc Hortic Sci 123(1):19–25, 1998.

31. GL Rotino, R Iannacone, MC Fiore, A Macchi, PD Grieco, S Arpaia, D Perrone, G Mennella, F Sunseri, F Cellini. Genetic engineering of eggplant (*Solanum melongena* L.). Acta Hortic 392:227–233, 1995.
32. F Sunseri, MC Fiore, F Mastrovito, E Tramontano, GL Rotino. In vivo selection and genetic analysis for kanamycin resistance in transgenic eggplant (*Solanum melongena* L.). J Genet Breed 47:299–306, 1993.
33. S Arpaia, N Acciarri, GM Di Leo, G Mennella, G Sabino, F Sunseri, GL Rotino. Field performance of Bt-expressing transgenic eggplant lines resistant to Colorado potato beetle. Proceedings of the Xth Meeting on Genetics and Breeding of Capsicum and Eggplant, Avignon, France, 1998, pp 191–194.
34. N Acciarri, S Arpaia, G Mennella, G Vitelli, F Sunseri, GL Rotino. Field evaluation of transgenic eggplant lines protected against the Colorado potato beetle by expression of a synthetic *Bacillus thuringiensis* gene. HortSci 35:722–725, 2000.
35. S Aubert, MC Daunay, E Pochard. Saponosides stéroidiques de l'aubergine (*Solanum melongena* L.). II. Variations des teneurs liées aux conditions de récolte, aux génotypes et à la quantité de graines des fruits. Agronomie 9:751–758, 1989.
36. J Nothman, I Rylski, M Spigelman. Flowering-pattern, fruit growth and colour development of eggplant during the cool season in a subtropical climate. Sci Hortic 11:217–222, 1979.
37. JW Hennart. Sélection de l'aubergine. PHM Rev Hortic 374:37–40, 1996.
38. C Leonardi, D Romano. Controllo della fruttificazione della melanzana in serra. Colture Protette 7/8:67–71, 1997.
39. DD Archbold, FG Dennis. Strawberry receptacle growth and endogenous IAA content as affected by growth regulator application and achene removal. J Am Soc Hortic Sci 110:816–820, 1985.
40. J Nitsch. Growth and morphogenesis of the strawberry as related to auxin. Am J Bot 37:211–215, 1950.
41. GL Rotino, E Perri, M Zottini, H Sommer, A Spena. Genetic engineering of parthenocarpic plants. Nat Biotechnol 15:1398–1401, 1997.
42. T Yamada, CJ Palm, B Brooks, T Kosuge. Nucleotide sequence of the *Pseudomonas savastanoi* indolacetic acid genes shows homology with *Agrobacterium tumefaciens* T-DNA. Proc Natl Acad Sci USA 82:6522–6526, 1985.
43. M Kawaguchi, M Kobayashi, A Sakurai, K Syono. The presence of an enzyme that converts indole-3-acetamide into IAA in wild and cultivated rice. Plant Cell Physiol 32(2):143–149, 1991.
44. C Koncz, J Schell. The promoter of TL-DNA gene 5 controls the tissue-specific expression of chimaeric genes carried by a novel type of *Agrobacterium* binary vector. Mol Gen Genet 204:383–396, 1996.
45. M Bevan. Binary *Agrobacterium* vectors for plant transformation. Nucleic Acids Res 12:8711–8721, 1984.
46. G Donzella, A Spena, GL Rotino. Transgenic parthenocarpic eggplants: superior germplasm for increased winter production. Mol Breed 6:79–86, 2000.

# 41

# Transgenic Cowpea, Lentil, and Chickpea with Reporter and Agronomically Relevant Genes

**Paul F. Lurquin**
*Washington State University, Pullman, Washington*

**Edgardo Filippone and Gabriella Colucci**
*University of Naples, Naples, Italy*

|      |                                              |     |
|------|----------------------------------------------|-----|
| I.   | INTRODUCTION                                 | 603 |
| II.  | COWPEA                                       | 604 |
|      | A. *Agrobacterium* Species–Mediated Gene Transfer | 604 |
|      | B. Direct Gene Transfer                      | 605 |
|      | C. Summary                                   | 606 |
| III. | LENTIL                                       | 607 |
|      | A. *Agrobacterium* Species–Mediated Gene Transfer | 607 |
|      | B. Direct Gene Transfer                      | 607 |
|      | C. Summary                                   | 607 |
| IV.  | CHICKPEA                                     | 608 |
|      | A. *Agrobacterium* Species–Mediated Gene Transfer | 608 |
|      | B. Direct Gene Transfer                      | 610 |
|      | C. Summary                                   | 610 |
| V.   | CONCLUSIONS                                  | 610 |
|      | REFERENCES                                   | 611 |

## I. INTRODUCTION

Chickpea (*Cicer arietinum* L.), cowpea (*Vigna unguiculata* Walp), and lentil (*Lens culinaris* Medik.) are considered "small crops" in the Western world and hence have received little or no attention from major biotechnology companies. Yet, these crops all have international importance and are a significant component of the livelihood of large numbers of farmers and consumers around the world, including many in developing countries. The world total production of pulses for the years 1995–1997 was, on average, 56 Mton and the harvested area was 72 MHa (1). During the same period, chickpeas were cultivated on 11.3 MHa, giving a total yield of 8.6 Mton, whereas lentils were cultivated on 3.3 MHa, giving a total yield of 2.9 Mton. Data on cowpeas are not available because this species is often confused with beans, and, hence, international statistics cannot be used for comparison. However, the total production of cowpea can be inferred from its prevalent areas of cultivation, Africa and India: figures of 7 MHa and a total production of 5 Mton represent good estimates. Finally, chickpea, lentil, and cowpea represent about 30% of total pulse production and harvested area.

The nutritional value of these three legumes is high, especially when combined with cereals. Given that most chickpea, lentil, and cowpea production as well as consumption occur outside the industrialized world, it is not surprising that advances in the genetic engineering of these plants have been lagging behind those of major crops such as maize, soybean, and rice. Nevertheless, these three legumes suffer from a host of bacterial, fungal, and viral diseases and insect damage. Conventional breeding programs have not been successful in adequately protecting these crops. In addition, lentils are poor competitors with weeds. Therefore, a genetic engineering approach to solve the problems specified is fully warranted. For example, the availability of insect resistance genes against bruchids could substitute for the absence of such traits in cross-compatible wild cowpea species. Also, fungal resistance genes against *Ascochyta rabiei* and *Fusarium oxysporum* f.sp. *ciceri* could have beneficial effects if transferred into the chickpea genome.

Unfortunately, compounding the apparent lack of interest in the genetic transformation of these plants is the paucity (and in several cases, the irreproducibility) of versatile tissue culture and regeneration techniques that plagues most legumes. Therefore, there are few reports dealing with the genetic transformation of these three genera. However, progress in this area is not absent; indeed, some international organizations such as the International Institute of Tropical Agriculture (IITA, Nigeria), the International Center for Agricultural Research in the Dry Areas (ICARDA, Syria), and the International Crops Research Institute for the Semi-Arid Tropics (ICRISAT, India) have been granted by some private companies or individual scientists permission to use patented gene coding for important resistance traits. Thanks to the active collaboration of several laboratories in Europe and the United States, experiments are now in progress to overcome problems related to the in vitro regeneration and genetic manipulation of these legumes.

As in most other cases, the transformation of these genera has been tackled via *Agrobacterium* species–mediated gene transfer as well as direct gene transfer. The aim of this chapter is to review the extent to which these approaches have been successful. We have used the terms *agroinfection* and *Agrobacterium species–mediated gene transfer* interchangeably in order to prevent monotony.

## II. COWPEA

### A. *Agrobacterium* Species–Mediated Gene Transfer

Among the three legumes considered here, cowpea was used first in genetic engineering attempts. For this, pieces of young leaves were cocultivated with disabled *Agrobacterium tumefaciens* strains harboring a plant-expressable *neo* gene and subsequently plated on kanamycin-containing medium. Kanamycin-resistant callus was obtained, but efforts to regenerate plants proved unsuccessful (2,3). Little was learned from these experiments, except that cowpea tissues were amenable to *Agrobacterium* species–mediated gene transfer, a characteristic that was not obvious at the time.

Given the problems associated with cowpea callus regeneration (which have not yet been solved), a tissue-culture-independent method was designed and tried (4). In this technique, mature embryonic axes were separated from cotyledons, sliced longitudinally, and cocultivated with fully armed *A. tumefaciens* (pTiBo 542) carrying the binary vector pGA472 able to confer kanamycin (geneticin) resistance to transformed plant tissues. The reasoning was that wounded embryos would be susceptible to agroinfection and hence T-DNA transfer. Up to 10% of the cocultivated embryos produced antibiotic-resistant calli. As expected, these calli could not be regenerated, given the concomitant transfer of fully armed T-DNA. Nevertheless, Southern blot analysis demonstrated presence of the *neo* gene in the resistant cells, indicating successful gene transfer into embryonic cells. Next, sliced embryonic axes were cocultivated with disarmed *A. tumefaciens* C58 carrying the p35SGUSINT binary vector containing the *uid*A reporter gene. In these experiments, up to 50% of the embryos developed chimeric GUS-positive shoots (4). However, small-scale micropropagation experiments did not lead to the isolation of homogeneously transformed plants (Penza and Filippone, unpublished), and selfing experiments were not conducted.

One group has claimed successful production of transgenic cowpea plants after cocultivation of mature, deembryonated cotyledon explants with *A. tumefaciens* (5). Adventitious shoots were regenerated by organogenesis after tissue incubation with *A. tumefaciens* LBA 4301 harboring the binary pUCD2340/pUCD2614 plasmid system. Plasmid pUCD2340 contained the hygromycin-resistant (*hpt*) gene driven by the *nos* promoter. Selection of transgenic shoots was done in shoot regeneration medium [B5 supplemented with $8 \times 10^{-6}$ 6-benzylaminopurine (BAP)] containing 25 mg/l hygromycin B. Up to 15–19% of the explants produced shoots under these conditions. Six surviving plants were obtained after rooting in the presence of hygromycin, of which four set seeds. None of the seeds showed germination. Presence of the transgene was demonstrated by Southern blot hybridization in an unspecified number of plant(s). It is unclear whether the transgenic plants were chimeric or uniformly transformed and there was no follow-up to this 1996 study. Thus, as far as we know, there is still no published evidence for heritable, stable transformation of cowpea by infection with *Agrobacterium* spp.

## B. Direct Gene Transfer

Direct gene transfer methodologies were developed in attempts to alleviate problems associated with plant regeneration from in vitro tissue cultures and (wrongly) assumed insensitivity of monocots to agroinfection. In fact, the first genetically transformed legume, soybean (*Glycine max*), was obtained by particle bombardment of embryonic axes (6).

Electroporation is the most versatile direct gene transfer technique as it applies equally well to prokaryotes, microscopic eukaryotes, animal cells, and plant protoplasts and cells (7). Early results have shown that the plant cell wall does not present an insurmountable barrier to DNA uptake by electroporation (8), an observation confirmed by later studies (9–11). On this basis, several groups have demonstrated transient expression of transgenes after electroporation of embryos from a variety of species (12–15). The same technique was applied to cowpea embryos (16). In this study, mature embryonic axes were electroporated at 666 V/cm (five pulses of 99-ms duration) with *uid*A constructs complexed with lipofectin or spermine and allowed to germinate and grow for 3 days on hormone-free medium. Results showed expression of GUS in a variety of cell types, including the meristematic cells in the apical dome, depending on the promoter driving the *uid*A gene. However, no efforts were made to determine whether GUS expression was only transient or continued as the plantlets grew, perhaps as the result of DNA integrative events.

Next, instead of embryos, cells present in the terminal nodal buds of cowpea plants were used as targets in electroporation experiments (17). For this approach, plants were allowed to grow for 3 to 4 weeks from seeds and the apical portion of the plants was decapitated close to the node of a fully expanded leaf. The stipule and petiole were removed to expose the nodal bud. All

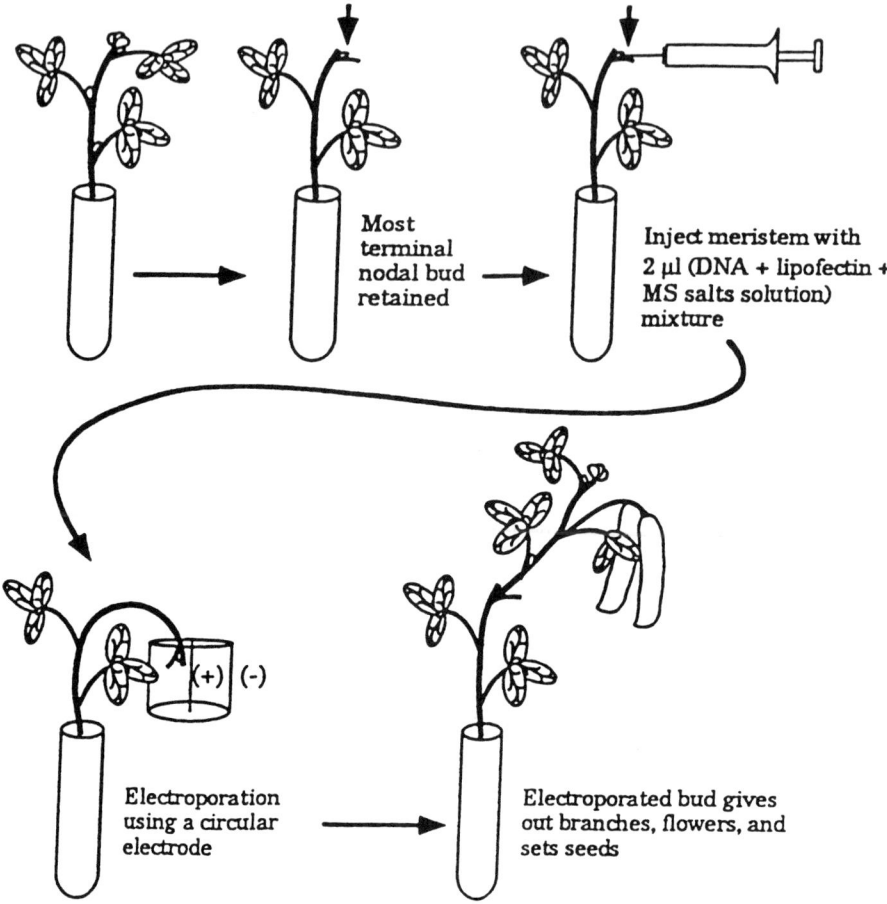

**Figure 1** Outline of in planta electroporation. (From Ref. 18.)

other meristematic buds below that nodal bud were also removed. The terminal buds were then injected with 2 μl lipofectin/pGPT 1.0 DNA (containing the *uid*A gene interrupted by an intron) and subsequently electroporated twice at 333 V/cm for 99 ms each, as shown in Fig. 1. A complete description of this protocol is given in Ref. 18. In one experiment, 40 electroporated meristems gave rise to 6 chimeric R0 branches. Three of these plants were allowed to self, and their R1 progeny analyzed for GUS expression. Of a total of 52 R1 plants, 9 were found to express GUS in all leaf cells examined. Further, two R1 plants obtained from independent GUS$^+$ R0 individuals were analyzed by Southern blot hybridization and shown to contain an integrated single intact copy of the *uid*A gene with different border fragments (17). No experiments were carried out beyond the R1 generation.

## C. Summary

In conclusion, there is presently no evidence that stable transgenic cowpea can be obtained by agroinfection. There is, however, no reason to believe that *Agrobacterium* spp.–mediated gene transfer in cowpea will never be a reality since there is good evidence that cowpea tissues are susceptible to *Agrobacterium* spp. and that reporter genes, selectable or screenable, can be expressed after cocultivation, including in regenerated plants. Rather, as has been the case with other legumes, success will depend on a judicious choice of target tissue and in vitro culture conditions.

In planta electroporation has led to the production of seemingly stable transgenic cowpea. However, genetic data are lacking and no transgenes of agronomical importance have been transferred with this methodology. Clearly, much more work is needed before convincing success can be claimed. It should be noted, however, that in planta electroporation has led to the production of pea enation mosaic virus–resistant pea lines (19).

## III. LENTIL

### A. *Agrobacterium* Species–Mediated Gene Transfer

As in the case of cowpea, there is no independently repeated, published evidence that transgenic lentil plants can be obtained via agroinfection. Nevertheless, conditions under which *Agrobacterium* spp. can transfer T-DNA to lentil tissues have been described (20–22). Contrary to cowpea, lentil can be regenerated both via organogenesis (23) and somatic embryogenesis (24). Thus, there is no a priori reason that would interfere with the development of transgenics in this organism. What, then, are the problems? First, it has been shown that different lentil cultivars (as well as pea cultivars) vary widely in their response to *A. tumefaciens* (22). This was demonstrated by cocultivating half-embryos of four different cultivars with *A. tumefaciens* C58 (pGV2260/p35SGUSINT) and scoring the proportion of $GUS^+$ plants as well as the number of $GUS^+$ foci present in positive plants. Both parameters were shown to vary widely by cultivar. On the other hand, use of hypervirulent *A. tumefaciens* EHA105 did not increase the proportion of $GUS^+$ plants or the number of $GUS^+$ foci per plant (22). Next, it was shown that $GUS^+$ foci are generally small and express the *uid*A gene at low levels. Finally, not all tissues used for cocultivation (shoot apex, epicotyl, root, cotyledon, and cotyledonary node explants) responded to agroinfection (21). Here again, it seems that what is needed is a massive and systematic exploration of cocultivation conditions, nature of the target tissues, regeneration methodology, and possibly *A. tumefaciens* strains and transformation vectors.

### B. Direct Gene Transfer

The in planta electroporation technique as described for cowpea was also used in the case of lentil, using again the *uid*A gene as a reporter (17). Eighteen R0 chimeric branches were obtained from 100 electroporated individuals. Twenty of 88 R1 individuals were found to be $GUS^+$ and 14 of 91 R2 plants (obtained by selfing the R1 generation) were also $GUS^+$. Two R1 individuals were shown by Southern blot hybridization to harbor a single integrated copy of the transgene.

The ratio of $GUS^+/GUS^-$ individuals in the R2 generation indicated a strong bias against the presence or expression of the transgene in this population. The reasons for this observation are not known; it may be due to loss of the transgene, its inactivation (silencing), or both.

### C. Summary

Progress toward the production of transgenic lentil is obviously in its infancy. It is encouraging to note that in vitro regeneration techniques have been developed in this genus and that transgene activity has been observed in tissues after *Agrobacterium* spp.–mediated transfer. As for in planta electroporation, even though putative transgenic R2 individuals have been detected, nonmendelian segregation ratios are disturbing and need to be explained.

## IV. CHICKPEA

### A. *Agrobacterium* Species–Mediated Gene Transfer

Cultivated chickpeas are represented by two cultigroups, Desi and Kabuli, each consisting of several genotypes. The former is characterized by wrinkled seeds with dark testa and is widely cultivated in India and Australia. The latter, yielding larger smooth seeds with brown yellowish testa, is cultivated in Europe, North Africa, the Middle East, and North America. These two cultigroups respond differently to in vitro culture; the Desi type is more amenable to manipulation in tissue culture.

Early results have shown that sliced chickpea mature embryos cocultivated with *Agrobacterium* spp. under conditions used for cowpea (see previous discussion and Ref. 4) gave rise to stained chimeric (GUS$^+$) germinating embryos (Fig. 2) and chimeric transgenic shoots (Fig. 3). Another group later demonstrated that not all chickpea cultivars are equally susceptible to various *A. tumefaciens* strains (25). For this, four chickpea genotypes, grown in either a soil-sand mixture or agar-solidified medium, were challenged with four different, fully virulent *Agrobacterium tumefaciens* strains (A281, A6, C58, and A348). Crown gall development on these plants was used to determine strength of interaction. All cultivars responded to all *A. tumefaciens* strains, but not equally; in general, in vitro cultured plants responded better than soil-grown plants and, in both cases, some cultivars responded better than others. "Hypervirulent" *Agrobacterium* sp. strain A281 (wrongly attributed a C58 chromosomal background in Ref. 25) proved superior to all other strains. *A. tumefaciens* A281, harboring the p35SGUSINT plasmid, was subsequently used by this group to show GUS expression in chickpea tissues derived from cocultivated zygotic embryos (26).

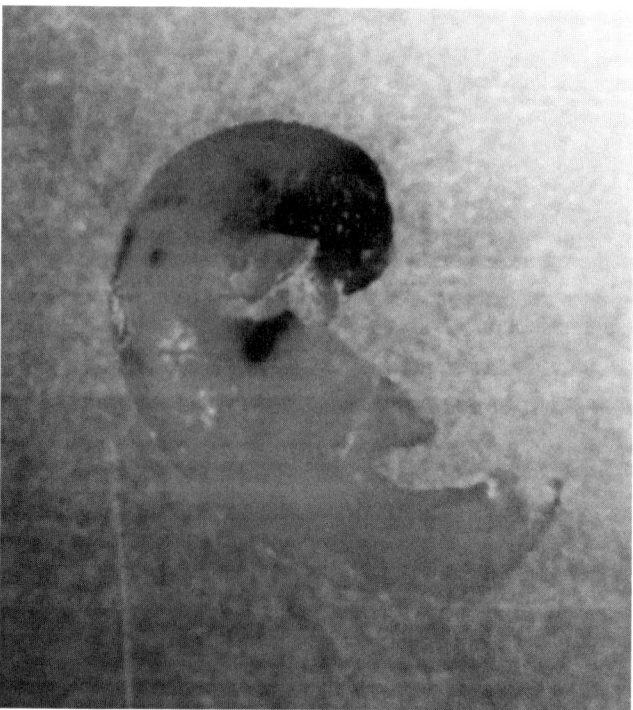

**Figure 2** One-week-old Kabuly cv. 'Sultano' chickpea embryo stained with X-gluc after cocultivation with *A. tumefaciens* C58 C1::p35SGUS/INT as in Ref. 4. Several transgenic sectors can be distinguished.

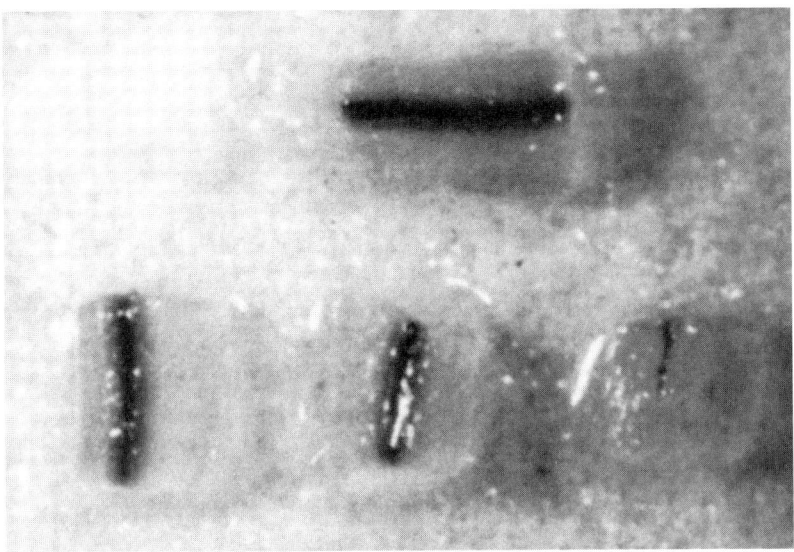

**Figure 3** GUS⁺ area in a chimeric transgenic chickpea stem. All sections correspond to one single cut-up stem obtained after *Agrobacterium* sp.–mediated gene transfer into a mature embryo as in Ref. 4.

Two reports have claimed success in the transformation of chickpea via agroinfection of mature embryos, followed by regeneration of plants (27,28). Both studies basically followed the same protocol: embryonic axes were isolated from imbibed seeds and the apex (27) or both apex and root meristem (28) severed. These explants were then cocultivated with *A. tumefaciens* LBA4404 containing the binary vector pBI121, which harbors both a screenable (*uid*A) and a selectable marker (*neo*). Cocultivated embryonic axes were then cultivated on medium favoring multiple shoot formation, that is, Murashige and Skoog (MS) medium supplemented with 1 mg/l kinetin (27) or 3 mg/l 6-benzylaminopurine (BAP) and 0.004 mg/l naphthalene acetic acid (NAA) in MS medium also containing 4 × MS micro salts (28). Emerged shoots were then transferred to rooting/selection medium in the presence of kanamycin. Transformation frequencies were 4% (number of transformed plants/initial embryos) (27) and 1–2% (number of regenerated shoots/number of transformed shoots), depending on the cultivar (28).

Both reports demonstrated presence of a transgene (*uid*A in Ref. 27 and *neo* in Ref. 28) in transformants and expression of both markers by enzyme assay. No genetic data regarding transmission of the marker genes to offspring were provided.

Recently, an *Agrobacterium* sp.–mediated transformation protocol involving the *bar* gene was used successfully with varieties of the Desi type (L. Molvig, CSIRO-Canberra, personal communication). As in the two previous reports, this method was also based on the cocultivation of mature embryos followed by induction of multiple shoots from preexisting meristems. Selection of the transformants was performed in the presence of phosphinothricin (PPT).

By using this procedure, transgenic chickpea plants were obtained (29); the plants expressed the *ech*-42 antifungal gene from the antagonistic fungus *Trichoderma harzianum* (30). This gene has already been transferred into tobacco and potato plants and shown to confer improved resistance to pathogens (31). The pBIN-based binary vector used in chickpea transformation experiments harbored the *ech*-42 gene under the control of the CaMV 35S promoter with a double enhancer and the *bar* gene under the same promoter with a single enhancer. Selection of putative transformed shoots was done in the presence of 5 mg/l PPT. The transformation frequency was low, near 0.1% (number of transformed plant/initial explant). Genetic data regarding transmission of the transgene to the progeny and results of pathogen inoculation of these plants are not yet available.

### B. Direct Gene Transfer

Particle bombardment has been used to detect transient foreign gene expression in chickpea mature, germinating embryos (26). Various chimeric constructs based on the *uid*A gene driven by an actin, CaMV 35S, or Win6 promoter were introduced into embryonic material and checked for expression in the growing plantlets. GUS activity was observed in epicotyls, hypocotyls, as well as in embryonic axes. The 35S promoter from cauliflower mosaic virus proved to be superior to the other two. No uniformly transformed plants were obtained in this study.

Transformation of chickpea via direct gene transfer has also been reported by Kar and coworkers (32). Transformation was achieved by using a He particle gun (Bio-Rad Biolistic 1000) with mature embryonic axes as targets. Two separate plasmids were cointroduced into the tissues, one with the selectable marker *nptII* and the second one harboring the *cryIA(c)* gene under control of the CaMV 35S promoter containing the tobacco mosaic virus $\Omega$ translational enhancer. Multiple shoot proliferation was induced in two Desi genotypes, in the presence of both BAP and NAA. An average of 20 shoots were produced per embryonic axis. Transformants were selected on kanamycin and, of a total of 24 resistant plants, all were positive by Southern analysis. The cotransformation rate of the *cryIA(c)* gene was estimated to be 46%. Transformation efficiency and genetic data regarding the transmission of the transgenes to the progeny were not reported. However, only two transgenic plants generated high levels of the full-length transcript. One of those was tested for resistance to pod-borer insects, and larval growth was shown to be strongly reduced relative to that of controls.

### C. Summary

There is convincing evidence that *Agrobacterium* sp.–mediated gene transfer into chickpea has been achieved. Undoubtedly, the keys to success were the choice of adequate plant material and formation of multiple shoots from it. Unfortunately, since the results of crosses with transgenics have not been published, it is not known whether transgenes can be stably transmitted to progeny plants. Also, it is not known whether the presence of multiple copies of the transgenes (28) resulted in different levels of gene expression. These questions must be addressed before engineering with agronomically important genes becomes an applied reality with chickpea.

## V. CONCLUSIONS

Among the three crops considered here, solid and independently reproduced evidence for genetic transformation has been obtained only in the case of chickpea. However, data regarding the inheritance of transgenes in progeny plants are totally absent in all cases. Projects to transfer fungal resistance genes and protein quality genes into valuable chickpea genotypes are now in progress in Europe and Australia. However, much more work is necessary to clarify the complex relationships among genotypes, culture media, and transformation procedures in order to obtain transformation frequencies comparable to those of other plant species. Transmission patterns of transgenes to progeny also need to be determined.

## ACKNOWLEDGMENT

PFL acknowledges the support of a USDA/CSFL grant.

## REFERENCES

1. FAO Production Yearbook. Vol. 51. 1997.
2. JA Garcia, J Hille, P Vos, R Goldbach. Transformation of cowpea (*Vigna unguiculata*) cells with an antibiotic resistance drug using a Ti plasmid–derived vector. Plant Sci 44:37–46, 1986.
3. EJ Perkins, CM Stiff, PF Lurquin. Use of *Alcaligenes eutrophus* as a source of genes for 2,4-D resistance in plants. Weed Sci 35:12–18, 1987.
4. R Penza, PF Lurquin, E. Filippone. Gene transfer by cocultivation of mature embryos with *Agrobacterium tumefaciens:* application to cowpea (*Vigna unguiculata* Walp). J Plant Physiol 138:39–43, 1991.
5. B Murthukumar, M Mariamma, K Veluthambi, A Gnanam. Genetic transformation of cowpea (*Vigna unguiculata* L. Walp) using *Agrobacterium tumefaciens*. Plant Cell Rep 15:980–985, 1996.
6. DE McCabe, WF Swain, BJ Martinell, P. Christou. Stable transformation of soybean (*Glycine max*) by particle acceleration. Biotechnology 6:923–926, 1988.
7. PF Lurquin. Gene transfer by electroporation. Mol Biotechnol 7:5–35, 1997.
8. RA Dekeyser, B Claes, RMU De Rycke, ME Habets, M Van Montagu, AB Caplan. Transient gene expression in intact and organized rice tissues. Plant Cell 2:591–602, 1990.
9. KD' Halluin, E Bonne, M Bossut, M De Beuckeleer, J Leemans. Transgenic maize plants by tissue electroporation. Plant Cell 4:1495–1505, 1992.
10. A Arencibia, PR Molina, G de la Riva, G Selman-Housein. Production of transgenic sugarcane (*Saccharum officinarum* L.) plants by intact cell electroporation. Plant Cell Rep 14:305–309, 1995.
11. N Sabri, B Pelissier, J Teissie. Transient and stable electrotransformations of intact black Mexican sweet maize cells are obtained after preplasmolysis. Plant Cell Rep 15:924–928, 1996.
12. DD Songstad, FG Halaka, DL DeBoer, CL Armstrong, MAW Hinchee, CG Ford-Santino, SM Brown, ME Fromm, RB Horsch. Transient expression of GUS and anthocyanin constructs in intact maize immature embryos following electroporation. Plant Cell Tissue Org Cult 33:195 201, 1993.
13. A Klöti, VA Iglesias, J Wünn, PK Burkhardt, SK Datta, I Potrykus. Gene transfer by electroporation into intact scutellum cells of wheat embryos. Plant Cell Rep 12:671–675, 1993.
14. A Chaudury, SC Maheshwari, AK Tyagi. Transient expression of *gus* gene in intact seed embryos of *Indica* rice after electroporation-mediated gene delivery. Plant Cell Rep 14:215–220, 1995.
15. W Dillen, G Engler, M Van Montagu, G Angenon. Electroporation-mediated DNA delivery to seedling tissues of *Phaseolus vulgaris* L. (common bean). Plant Cell Rep 15:119–124, 1995.
16. V Akella, PF Lurquin. Expression in cowpea seedlings of chimeric transgenes after electroporation into seed-derived embryos. Plant Cell Rep 13:110–117, 1993.
17. GM Chowrira, V Akella, PE Fuerst, PF Lurquin. Transgenic grain legumes obtained by *in planta* electroporation-mediated gene transfer. Mol Biotechnol 5:85–96, 1996.
18. GM Chowrira, V Akella, PF Lurquin. Electroporation-mediated gene transfer into intact nodal meristems in planta: generating transgenic plants without *in vitro* tissue culture. Mol Biotechnol 3:17–23, 1995.
19. GM Chowrira, TD Cavileer, SK Gupta, PF Lurquin, PH Berger. Coat protein–mediated resistance to pea enation mosaic virus in transgenic *Pisum sativum* L. Transgenic Res 7:265–271, 1998.
20. TD Warkentin, A McHughen. Crown gall transformation of lentil (*Lens culinaris* Medik.) tissues. Plant Cell Rep 10:489–493, 1991.
21. TD Warkentin, A. McHughen. *Agrobacterium tumefaciens*–mediated beta-glucuronidase (GUS) gene expression in lentil (*Lens culinaris* Medik.) tissues. Plant Cell Rep 11:274–278, 1992.
22. PF Lurquin, Z Cai, CM Stiff, EP Fuerst. Half-embryo cocultivation technique for estimating the susceptibility of pea (*Pisum sativum* L.) and lentil (*Lens culinaris* Medik.) cultivars to *Agrobacterium tumefaciens*. Mol Biotechnol 9:175–179, 1998.

23. DJ Williams, A McHughen. Plant regeneration of the legume *Lens culinaris* Medik. (lentil) in vitro. Plant Cell Tissue Org Cult 7:149–153, 1986.
24. PK Saxena, J King. Morphogenesis in lentil: plant regeneration from callus culture of *Lens culinaris* Medik. via somatic embryogenesis. Plant Sci 52:223–227, 1987.
25. R Islam, T Malik, T Husnain, S Riazuddin. Strain and cultivar specificity in the *Agrobacterium*-chickpea interaction. Plant Cell Rep 13:561–563, 1994.
26. T Husnain, T Malik, S Riazuddin, MP Gordon. Studies on the expression of marker genes in chickpea. Plant Cell Tissue Org Cult 49:7–16, 1997.
27. GS Fontana, L Santini, S Caretto, G Frugis, D Mariotti. Genetic transformation in the grain legume *Cicer arietinum* L. (chickpea). Plant Cell Rep 12:194–198, 1993.
28. S Kar, TM Johnson, P Nayak, SK Sen. Efficient transgenic plant regeneration through *Agrobacterium*-mediated transformation of chickpea (*Cicer arietinum* L.). Plant Cell Rep 16:32–37, 1996.
29. T De Martino, P Chiaiese, L Molvig, LM Tabe, G Colucci, E Filippone, TJ Higgins. Transgenic chickpea expressing a gene with antifungal potential. Proceeding of the XVth EUCARPIA Congress, 22, 1998.
30. CK Hayes, S Klemsdal, M Lorito, A Di Pietro, C Peterbauer, JP Nakas, A Tronsmo, GE Harman. Isolation and characterization of an endochitinase-encoding gene from a cDNA library of *Trichoderma harzianum*. Gene 138:143–148, 1994.
31. M Lorito, I Garcia, SL Woo, G Colucci, GE Harman, JA Pintor-Toro, E Filippone, S Muccifora, CB Lawrence, F Scala. Genes from mycoparasitic fungi as a novel source for improving plant resistance to fungal pathogens. Proc Natl Acad Sci USA 95:7860–7865, 1998.
32. S Kar, D Basu, S Das, NA Ramkrishnan, P Mukherjee, P Nayak, SK Sen. Expression of *cryI*A(c) gene of *Bacillus thuringiensis* in transgenic chickpea plants inhibits development of pod borer (*Heliothis armigera*) larvae. Transgenic Res 6:177–185, 1997.

# 42
# Genetic Manipulation of Lettuce

Michael Raymond Davey, Matthew Sean McCabe, Umaballava Mohapatra, and
J. Brian Power
*University of Nottingham, Nottingham, England*

| | | |
|---|---|---|
| I. | CLASSIFICATION AND ORIGIN OF CULTIVATED LETTUCE | 613 |
| II. | CONVENTIONAL BREEDING OF LETTUCE | 614 |
| | A. Sexual Hybridization | 615 |
| | B. Exploitation of the Wild Gene Pool | 615 |
| | C. Disease Resistance | 616 |
| | D. Insect Resistance | 616 |
| | E. Herbicide Resistance | 617 |
| | F. Product Quality | 617 |
| III. | APPLICATION OF BIOTECHNOLOGY TO LETTUCE BREEDING | 618 |
| | A. Protoplast Culture and Somatic Hybridization in Lettuce | 618 |
| | B. Somaclonal Variation | 619 |
| | C. Direct Deoxyribonucleic Acid Uptake into Isolated Protoplasts | 619 |
| | D. *Agrobacterium* Species–Mediated Gene Transfer | 619 |
| IV. | A ROUTINE PROCEDURE FOR LETTUCE TRANSFORMATION | 624 |
| V. | INACTIVATION OF TRANSGENE EXPRESSION IN LETTUCE | 629 |
| | A. Gene Promoters | 629 |
| | B. Transgene Copy Number | 630 |
| | C. Deoxyribonucleic Acid Methylation | 631 |
| VI. | CONCLUDING REMARKS | 631 |
| | REFERENCES | 632 |

## I. CLASSIFICATION AND ORIGIN OF CULTIVATED LETTUCE

Lettuce (*Lactuca sativa* L.) is an annual, self-fertile species of the family Asteraceae (formerly Compositae). The genus *Lactuca* contains approximately 100 species with 17 European, approximately 10 American, 33 tropical East African, and about 40 Asian species. *Lactuca sativa* L. is classified in section *Lactuca,* subsection *Lactuca. L. serriola* L. (1). The chromosome number for this diploid species is $2n = 2x = 18$, the haploid genome size can be calculated from $4C$ deoxyribonucleic acid (DNA) values (10.6–11.2 pg) as $2.56–2.7 \times 10^9$ bp (2).

Lettuce is classified on the morphological characteristics of its fleshy leaves; the tendency to form "heads" or "hearts" varies in different types. Seven morphological types, namely, (a) Crisphead, Iceberg or Cabbage; (b) Butterhead; (c) Cos; (d) Leaf or Cutting; (e) Latin; (f) Stem or Asparagus; and (g) the Oilseed group, are recognized by the International Code of Nomenclature for Cultivated Plants (1). Crisphead varieties with compact, large brittle-textured heads of tightly folded leaves have good shipping qualities. Butterhead types have crumpled, soft-textured leaves, giving small and loose heads, which are grown mostly in Europe. Cos types are characterized by long, oval, dark, upright green leaves, forming oblong hearts. Such plants are grown extensively in Southern Europe and Mediterranean regions. Leaf types produce a rosette of loose leaves instead of a compact head, whereas Latin lettuce cultivars are intermediate between the Butterhead and Cos types in forming loose heads with oval leaves. Such plants are cultivated mainly in Mediterranean and Latin American countries. Stem-type cultivars are grown in China. Their leaves are coarse and unpalatable, but the young fleshy stems are cooked prior to consumption.

Cultivated lettuce is mainly grown as a leaf vegetable, the principal centers of production and consumption are the United States and Europe. Economically, lettuce is the most important vegetable crop produced for the fresh market in the United States, where over 2 billion heads are harvested each year (3). Over 70% of the U.S. production is in California (4), where lettuce is grown as an outdoor crop. Italy, Spain, France, the Netherlands, and the United Kingdom are the main producers in Europe (5). As lettuce cannot tolerate extremes of temperature, its outdoor production in Northern Europe is confined to the summer months, and most crops are cultivated under glass. Other areas of cultivation include Canada, South America, northern Mexico, South Africa, the Middle East, Japan, China, and southeast Australia.

Nutritionally, lettuce contains only moderate amounts of phosphorus, iron, sodium, potassium, copper, ascorbic acid, and vitamin A and is ranked about 26th among common fruits and vegetables in terms of dietary contribution (3). However, in terms of bulk consumption, it ranks 4th behind tomato, orange, and potato in the United States (6). Lettuce also has some minor uses. Stems of an Asparagus type (e.g., 'Burpee's Celtuce') are cooked and eaten in China, nicotine-free cigarettes are made from lettuce leaves, and in Egypt, edible oil is extracted from seeds of some forms of lettuce. The wild species, *L. virosa*, is a source of lactucin and jacquinelin, which have sedative and analgesic properties (7).

Several possibilities have been proposed for the center of origin of cultivated lettuce. These include Egypt, the Mediterranean region, the Middle East, and southwest Asia. The most recent opinion is that cultivated lettuce probably originated in southwest Asia, from the area around the Euphrates and the Tigris Rivers (1). This is because the highest number of related species can be found between these rivers and evidence suggests that agriculture was practiced in this area long before its introduction to Egypt and the Mediterranean region. Cultivated lettuce probably spread from southwest Asia to Egypt, as suggested by evidence on wall paintings of about 2500 B.C. and, subsequently, to Greece and Rome, where it was described in literature dating from 550 B.C. These early lettuces were probably Cos and Leaf types. Evidence of lettuce cultivation in northwest Europe dates from around 1485, and the first descriptions of Cabbage lettuce appeared in 1543. Lettuce was introduced to America around 1500; that introduction ultimately led to development of the Crisphead lettuces (1).

## II. CONVENTIONAL BREEDING OF LETTUCE

An excellent account of lettuce breeding has been published (3); a summary is presented here. Human selection, leading to the domestication of wild *Lactuca* species, has resulted in a decrease in both latex content and bitter taste, loss of stem and leaf prickles, an extended vegetative growth period with head formation accompanied by tenderization of leaves, and increased seed size with nonshattering properties (1). Modern breeding objectives for cultivated lettuce include altering leaf shape and color and manipulating head formation from a loose to a compact, closed structure with delayed bolting. Other targets involve the introduction of male sterility; resistance to herbicides, diseases (particularly downy mildew and lettuce mosaic virus [LMV]), and insects; a shorter growing period, accompanied by improved quality under suboptimal growth conditions (such as reduced nitrate content during winter cultivation under glass); and extended postharvest shelf life (1).

## A. Sexual Hybridization

The practicalities of conventional lettuce breeding, i.e., effecting sexual hybridization, are hindered by autogamy. The structure of the lettuce flower and the sequence of events during anthesis result in a highly efficient self-fertilization system. The stamens are fused to form a tube (anther sheath), onto the inside of which pollen collects as the flower opens. As the style elongates and emerges from the anther sheath, pollen is shed onto the stigma. Lettuce flowers open once, in a process that takes 1–2 hours. Despite the efficiency of this self-fertilization system, techniques that can achieve 100% cross-hybridization have been developed. For example, anther sheaths can be removed prior to style emergence, but this process is tedious and time-consuming. Therefore, the most practical method is to wash the pollen grains from the stigma with a fine stream of water immediately after style emergence from the anther sheath. Foreign pollen can then be applied by touching another flower to the one that has been washed (3). It is desirable to have a seedling marker gene from the male parent, which codes for a dominant trait such as leaf color and/or leaf shape, to facilitate early identification of successful crosses.

## B. Exploitation of the Wild Gene Pool

The gene pool from which desirable traits can be acquired by conventional breeding is limited by the fact that cultivated lettuce is inbreeding and, to date, has been found to be sexually compatible with only a few of the 100 or so species within the genus *Lactuca*. The three wild relatives, *L. serriola*, *L. saligna*, and *L. virosa,* have been studied most extensively. *L. serriola* crosses freely with *L. sativa*. If *L. saligna* is used as a female parent, it can be crossed with *L. sativa* or *L. serriola* to produce fertile hybrids. *L. virosa* can be employed as a male or female parent in crosses with *L. sativa* and *L. serriola*, but the resulting hybrids are sterile. However, a hybrid resulting from a cross between *L. virosa* and *L. sativa* was made fertile by doubling its chromosome complement (3). Certain Asian and South African species have potential as parents for lettuce breeding, but studies on these wild relatives are incomplete (1,8).

One of the major successes in lettuce breeding utilizing a wild *Lactuca* species resulted in the release of the cv. 'Vanguard' in 1958 and, later, 'Vanguard 75' (9) and 'Salinas,' which are grown extensively in the United States. This was achieved by R.C. Thompson, who crossed PI 125130 (*L. virosa*) with a line derived from a complex *L. sativa*–*L. serriola* cross and treated the resulting $F_1$ hybrid with colchicine to produce a fertile amphidiploid. The latter was used in subsequent backcrosses to cultivated lettuce to achieve the diploid state and, eventually, led to the Crisphead cv. 'Vanguard.' Important traits derived from *L. virosa* included LMV resistance, reduced bitterness, an extensive root system, and tipburn resistance. Other useful traits have been identified in *L. serriola* (downy mildew and LMV resistance), *L. virosa* (leaf aphid resistance), and *L. saligna* (cabbage looper resistance) (10,11).

## C. Disease Resistance

Major emphases of lettuce breeding programs are the identification and introduction of novel genes for resistance to numerous viral and fungal pathogens (3,12). This is motivated by the continual appearance of new strains of pathogen that overcome previously developed resistance traits. In fact, constant selection of resistance to a certain pathogen inadvertently results in increased selection pressure for a pathogen that can overcome the resistance. For example, I. C. Jagger was the first to report the use of breeding techniques to generate a disease-resistant lettuce cultivar by selection of healthy survivors from crops of cv. 'New York' that had been extensively infected by the fungal disease brown blight. Crosses made from these highly resistant lines resulted in the release of the first brown blight–resistant 'Imperial' lines in 1926. However, a new strain of the fungus, capable of infecting all previously resistant cultivars, soon appeared in 1932. This would have probably destroyed the Lower Rio Grande lettuce industry, had it not been for the development of the new resistant cv. 'Valverde' in 1959 (3).

### 1. Downy Mildew Resistance

In the lettuce industry in Western Europe, the continuous cycle of new cultivar–new strain–new cultivar has been ongoing since 1925 with the fungal pathogen *Bremia lactucae*. This organism thrives in the cool wet climate of Northern Europe and causes downy mildew, which is the most destructive fungal disease of lettuce. Nearly all lettuce breeding programs incorporate mildew resistance, and, over the years, this fungus has led to the development of more than 120 cultivars as each new plant resistance gene is overcome by the pathogen (3,4). Interaction between *B. lactucae* and lettuce has been defined as a gene-for-gene system (13) in which dominant resistance genes (*Dm* genes) in lettuce correspond with avirulence genes in *B. lactucae*. Resistance based on *Dm* genes remains the most important strategy in breeding lettuce with resistance to *B. lactucae* (14). High levels of *Dm* gene–based resistance are found in wild *Lactuca* species, such as *L. virosa*; several crosses have been achieved between *L. virosa* and *L. sativa* (11,15). However, the resistance provided by this strategy is race-specific, and a strategy for a more durable broad, race-nonspecific resistance is still sought. A potential source of complete resistance, which is thought to be based on a different mechanism to *Dm* genes, also occurs in *L. saligna*. However, the mechanism and genetics of this type of resistance have yet to be studied in detail (14,16).

### 2. Lettuce Mosaic Virus Resistance

LMV is transmitted by aphids and is the most destructive viral disease of lettuce (17). Resistance of lettuce to LMV is determined by two closely linked or allelic recessive genes, called *mo* and *g*. The *g* gene was identified in the resistant cv. 'Gallega de Invierno' and was found to take the form of virus multiplication without symptoms (18). The resistant trait was found to be a single recessive gene (19). Similar resistance was found in three Egyptian lettuce lines, and it was assigned to the *mo* gene (20). The 'Gallega' source of resistance has been used in Europe to incorporate the *g* gene into many varieties of Butterhead, Cos, Crisphead, or Latin types. In the United States, the Egyptian source was used to introduce the *mo* gene into a few Crisphead cultivars; the 'Gallega' source was used for two new Butterhead- and Cos-type cultivars (21).

## D. Insect Resistance

Aphids and whitefly cause significant damage to lettuce and act as viral vectors. Hence, breeding for insect resistance can also confer viral resistance in lettuce hybrids. Resistance to pests and diseases has been identified in some cultivars of *L. sativa* and in wild species, such as *L. serriola* (22) and *L. saligna* (23). Resistance to *Nasonovia ribis nigri* (leaf aphid) has been identified in *L. virosa* and transferred into cultivated lettuce (24).

### E. Herbicide Resistance

Despite the use of herbicides, weeds caused an estimated loss of $29 million in field-grown lettuce in the United States during 1991. Under the same conditions, but without herbicides, the estimated loss would have been $135 million. During 1992, in eastern Canada, weeds accounted for an estimated loss of $2.4 million in lettuce (25). In other countries, where lettuce is grown primarily as an outdoor crop, weeds cause a significant reduction in crop production and quality. On a global basis, about 250 species are sufficiently troublesome (for crops in general) to be termed weeds. Of these, 40% are the members of the families Poaceae and Asteraceae. Undoubtedly, the use of herbicides has increased the efficiency of modern crop production, and herbicides are now available for the control of weeds in most crops. However, environmental and health considerations, together with the appearance of herbicide-resistant biotypes in a large number of weed species related to the sustained use of herbicides (26), have contributed to development of alternatives to chemical herbicides, or at least to reduction in their use. Herbicide-resistant biotypes have been reported from every continent (27). It has been estimated that in 1989, approximately 1 million hectares of land in the United States and 2 million hectares in Europe was infested with triazine-resistant weeds (28). To date, resistance to at least 15 classes of herbicides has been reported in more than 100 weed species (29).

The choice of herbicide and its active ingredient(s) depends upon the competing weeds; a number of active compounds and their premixes are recommended for outdoor lettuce (25). For example, the active ingredient pronamide is commonly used to kill annual bluegrass. Although most lettuce varieties are highly tolerant to this compound, it is a restricted-use herbicide because of its oncogenicity. Bensulfide is used for the preemergence control of annual grass and broadleaf weed species in lettuce crops. Although bensulfide inhibits root growth, its exact mode of action is not known. Benefin inhibits development of roots and shoots of grass and broadleaved weeds by interfering with cell division. Diclofop-methyl prevents biosynthesis of lipids in grasses. Other herbicides (active ingredients) commonly used for lettuce include chlorpropham, propyzamide, propachlor, and trifluralin. Sometimes a mixture of several active ingredients is used. For example, propham, diuron, and chlorpropham are mixed in the herbicide, Atlas Pink C (Atlas Crop Protection Ltd., Dewsbury, West Yorkshire, UK). The generation of herbicide-resistant lettuce could reduce the dependence on chemical control of weeds.

Conventional plant breeding has been used to transfer sulfonylurea resistance to cultivated lettuce from the wild species, *L. serriola* (30). However, the use of sexual hybridization to transfer herbicide resistance is restricted, primarily because of the unavailability of the resistance gene(s) in the gene pool and the incompatibility of herbicide-resistance source species within cultivated lettuce. Somatic hybridization by protoplast fusion offers an alternative approach; the limiting factor is the availability of the gene(s) in question in a *Lactuca* species that is amenable to somatic hybridization with *L. sativa*.

### F. Product Quality

#### 1. Reduced Accumulation of Nitrate

Accumulation of nitrate in leafy vegetables is undesirable, since its consumption produces nitrite in the human body. Nitrite combines with hemoglobin to form methemoglobin; the latter is incapable of binding oxygen (31). Blue baby syndrome in humans is associated with high nitrate concentrations in drinking water (32). In addition, nitrite may react with secondary amines in the stomach to produce carcinogenic nitrosamines, which, in the long term, may induce gastric cancers (33). Leafy vegetables form a major source of human nitrate intake, as the consumption of uncooked lettuce is high, as compared with that of other leafy vegetables. Thus, efforts are being made to limit the nitrate level in this crop (32,34). Lettuce grown without artificial light in winter under glass can contain 5–6 g of nitrate per kilogram fresh weight, which exceeds the maximal permissible levels (35) set by the European Union (EU) (Regulation No. 194/97), in winter-grown (2.5 g nitrate/kg f. wt.) and summer-grown (4.5 g nitrate/kg f. wt.) lettuce (36). If these limits were to be accepted, the lettuce industry could be in serious difficulty, unless nitrate concentration is reduced. However, there are contradictory unpublished claims that high nitrate content in leafy vegetables may be beneficial to human metabolism.

## 2. Uniformity of Size and Yield and Increased Postharvest Shelf Life

Commercially, uniformity of size at maturity and year-round cropping are important requirements, as uniformly sized heads facilitate handling and govern the unit price. Single cutting of the crop is a basic requirement for efficient mechanical harvesting. Since lettuce is consumed as a raw vegetable, its quality and appearance govern saleability. For example, immature heads lack firmness, and overmature heads often have cracked, discolored ribs and a bitter flavor. Because the crop is highly perishable and is often overproduced, an important trait for improvement is prolonged postharvest shelf life and, consequently, reliable shipping qualities. In this respect, Crisphead varieties were developed for shipping from the West Coast to the East Coast in the United States. In order to reduce importation costs, emphasis is being given to developing cultivars for nontraditional cropping areas (37). For example, in the United States, salt tolerance is an important character for introduction into lettuce, as salt accumulation in the soil may limit production. The increasing demand for winter lettuce in Northern Europe is now met by glasshouse-grown crops. Consequently, the introduction and breeding of high-yielding varieties of different morphological types, well adapted to glasshouse conditions, have become important objectives for breeders in Europe.

## III. APPLICATION OF BIOTECHNOLOGY TO LETTUCE BREEDING

Biotechnology has considerable potential for the genetic manipulation of lettuce and has been investigated for a number of years. *L. sativa* and its wild relatives are amenable to culture, and efficient shoot regeneration systems exist for cultured organs, tissues, cells, and isolated protoplasts of several species. One of the initial targets for biotechnology was to overcome the sexual incompatibility of cultivated lettuce with other *Lactuca* species by somatic hybridization, involving the fusion of isolated protoplasts. Tissue culture was also exploited to generate somaclonal variation in lettuce. The development of a reliable transformation system for lettuce has allowed rapid, specific introduction of an unlimited variety of potentially useful cloned genes into this crop (38). Techniques such as isozyme analysis, restriction fragment length polymorphism (RFLP) (1), and amplified fragment length polymorphism (AFLP) have also been used to investigate the lettuce genome and to identify potentially useful genes, such as those for disease resistance. Construction of a lettuce genomic library has allowed cloning and characterization of potentially useful genes from *L. sativa*. The following sections describe genetic manipulation of lettuce with particular reference to genetic engineering using *Agrobacterium tumefaciens*.

### A. Protoplast Culture and Somatic Hybridization in Lettuce

Protoplast fusion offers a method of circumventing both pre- and postzygotic barriers to conventional sexual hybridization, permitting the introgression of complex genetic information into the cultivated crop. Plant regeneration from isolated protoplasts is essential for such genetic manipulation. Although reports on protoplast culture and somatic hybridization are limited in lettuce and, indeed, in other members of the family Asteraceae, plants have been regenerated from protoplasts of *L. sativa* and some wild species, such as *L. serriola*, *L. saligna*, and *L. virosa* (39–44). The generation, through protoplast fusion, of somatic hybrids such as *L. sativa* [+] *L. serriola* (45), *L. sativa* [+] *L. deblis* or *L. indica* (46), *L. sativa* [+] *L. virosa* (47), and *L. sativa* [+] *L. tartarica* or *L. perennis* (48) has also been reported. Although somatic hybridization is a means of exploiting novel germplasm, the application of this technology in the development of marketable lettuce cultivars has not been fully realized. Somatic hybrids of *L. sativa* [+] *L. virosa* have been generated to transfer resistance against *Nasonovia ribis nigri* (leaf aphid), *Erysiphe cichoracearum* (powdery mildew), and *Pseudomonas cichorii* (bacterial rot) from wild species to cultivated species (47). Although these hybrids were more vigorous than either parent, they were sterile. Somatic hybrids of *L. sativa* [+] *L. tartarica* were sensitive to climatic conditions (48). The hybrids grew vigorously to form a rosette, but leaf development then ceased, roots became stunted, and leaves became necrotic. However, modification of the environmental conditions permitted growth in some of these plants with limited seed production after pollination of a large number of capitula with different cultivars of lettuce. Preliminary tests revealed transfer of downy mildew resistance from *L. tartarica* into cultivated species, but fertility was low. Further work is essential to generate a sufficient number of fertile lettuce somatic hybrid plants that exhibit characteristics from the more distantly related *Lactuca* species.

### B.  Somaclonal Variation

Somaclonal variant plants have been regenerated from explant-derived callus of lettuce (49,50), and protoclonal variation arising from leaf mesophyll–derived protoplasts of *L. sativa* cv. 'Climax' has been observed (51). However, to date, most of the somaclonal variants reported in this crop had reduced vigor, chlorophyll content, and fertility, together with other morphological deformities (4,5).

### C.  Direct Deoxyribonucleic Acid Uptake into Isolated Protoplasts

There is one early report of the generation of transgenic plants after DNA uptake into protoplasts. Thus, protoplasts isolated from expanded leaves of lettuce were electroporated in the presence of either pCAMV CAT, carrying the chloramphenicol acetyltransferase (*cat*) gene driven by the CaMV 35S promoter, or pABD1 with the neomycin phosphotransferase (*npt*II) gene (52). The *cat* gene was employed to demonstrate transient gene expression in this experimental system. Selection of kanamycin-resistant plants and analysis of their seed progeny showed that resistance to this antibiotic was inherited as a dominant Mendelian trait; integration of foreign DNA into the genomic DNA of antibiotic-resistant plants was confirmed by DNA-DNA hybridization. Overall, the use of direct DNA uptake into lettuce protoplasts has been superceded by simple and more reliable *Agrobacterium* sp.–based transformation systems.

### D.  *Agrobacterium* Species–Mediated Gene Transfer

Exploitation of the natural ability of the Gram-negative soil bacteria *Agrobacterium tumefaciens* and *A. rhizogenes* to engineer plant cells genetically has provided, undoubtedly, the major approach for gene transfer into a wide variety of crop plants of both dicotyledons and, more recently, monocotyledons, including the major cereals. Background information relating to the molecular basis of *Agrobacterium* sp.–mediated gene transfer into plants, the construction of chimeric genes, and the development of vectors for gene introduction into target species are summarized in several detailed reviews (53,54).

## 1. Introduction of Reporter and Selectable Marker Genes into Lettuce

Although *Agrobacterium* sp. are not usually considered to be natural pathogens of lettuce, the first successful *Agrobacterium* sp.–mediated gene transfer into this crop reported using wild-type octopine (ACH5) and nopaline (C58) strains of *A. tumefaciens*, as well as an engineered strain, GV3111, containing the cointegrate vector pTiB6S3, pMON120, or pMON200, or the binary vector pMON505 (38). Plasmid MON200 carried the chimeric *npt*II gene with a *nos* promoter and terminator (*nos.npt*II.*nos*); this selectable marker gene was absent in pMON120. Galls were induced on cotyledon explants from 4-day-old seedlings of the Butterhead lettuce cv. 'Cobham Green' grown aseptically in vermiculite and inoculated with suspensions of the wild-type bacterial strains ACH5 or C58 or the engineered strain GV3111. As expected, the unusual amino acids octopine and nopaline were produced by tumors incited by strains ACH5 and C58, respectively. Explants inoculated with the wild-type strain C58, or with *A. tumefaciens* strain GV3111, carrying pMON120, produced callus only in the absence of kanamycin. In contrast, explants transformed by GV1111, harboring either pMON200 or pMON505, produced callus on medium supplemented with 50 mg l$^{-1}$ of kanamycin sulfate. Shoots were regenerated from transformation experiments by using the disarmed pTiB6S3. Such experiments provided the first evidence that the lettuce genome could be manipulated by using *A. tumefaciens*–mediated gene delivery. However, a reliable transformation system was not achieved for the economically important Crisphead varieties (38).

Other workers were able to transform the lettuce cv. 'Kayser' by using *A. tumefaciens* strain LBA4404 with the binary vector pTRA415 (55). The vector carried the *npt*II gene, driven by the constitutive cauliflower mosaic virus 35S ribonucleic acid (RNA) (CaMV 35S) promoter and the β-glucuronidase (*gus*) reporter gene. The latter was driven either by the CaMV 35S promoter or by the stress- or salicylic acid–inducible pathogenesis-related (PR) 1a protein-encoding gene promoter isolated from tobacco. GUS activity was found to increase 3- to 50-fold in PR-*gus*–transformed plants, indicating the transgene was expressed normally under the regulated control of the PR 1a promoter. Cotyledon explants from 5-day-old seedlings were used as the source material for inoculation by *Agrobacterium* sp.

Workers in addition to Michelmore and associates (38) also exploited vectors developed by the agrochemical company Monsanto in early studies of lettuce transformation. For example, the lettuce cv. 'Lake Nyah' was transformed by inoculating cotyledons excised from axenically grown seedlings with *A. tumefaciens* carrying the cointegrate vector pMON200 (56). Transgenic plants of lettuce cv. 'South Bay' were generated by using *A. tumefaciens* strain A208, containing an engineered Ti plasmid, pTiT37SE, or the binary vector pMON9749 or pMON9793 (57). Transformed plants were selected on medium supplemented with 50–100 mg l$^{-1}$ kanamycin sulfate. As in earlier investigations, expression of the *gus* gene in transgenic plants was confirmed histochemically; DNA-DNA (Southern) hybridization confirmed the integration of the *gus* gene into the genome of transgenic plants. Integration of the *npt*II gene into the lettuce genome was confirmed by polymerase chain reaction (PCR).

Early transformation studies were limited to lettuce cultivars such as 'Cobham Green' and 'South Bay.' Importantly, Crisphead varieties showed very low transformation frequencies. However, a reliable, genotype-independent protocol for *A. tumefaciens*–mediated transformation of 13 lettuce cultivars, including Crispheads, was developed (58). Cotyledons excised from 7-day-old seedlings were inoculated with *A. tumefaciens* strain LBA4404 containing the binary vector pMOG23 (59) derived from pBIN19. The binary vector carried the chimeric *nos.npt*II.*nos* gene and a CaMV 35S *gus*-intron reporter gene (60) inserted into a multiple cloning site between the transfer DNA (T-DNA) borders, giving strain 0065. The introduction of pTOK47 carrying additional copies of the Ti plasmid virulence genes *vir B, vir C,* and *vir G* (61) resulted in strain 1065 with supervirulent properties. In addition to studying the role of the supervirulent plasmid in plant transformation, investigators (58) studied the influence of actively dividing cells harvested from suspensions of albino *Petunia hybrida* cv. 'Comanche' as nurse cultures and the effect of dilution of the bacterial suspension on the transformation frequency.

2. **Introduction of Agronomically Useful Genes into Lettuce by Using *Agrobacterium tumefaciens***

*a. Male Sterility.* In experiments to induce male sterility in lettuce, a pathogenesis-related β-1,3-glucanase gene linked to a tapetum-specific promoter, A9, was cloned into the binary vector pBIN19. The construct was introduced into *A. tumefaciens* carrying pGV2260 prior to transformation of the cultivar 'Lake Nyah' (62). Expression of the chimaeric gene caused dissolution of the callose wall of developing microspores in transgenic plants, resulting in male sterility in all the transformants. In contrast, plants regenerated from uninoculated cotyledons excised from seedlings of the same cultivar exhibited normal microspore development, including surface ornamentation as revealed by scanning electron microscopy. Such plants were fertile. These results indicated that callose degradation can be used to generate male sterile plants, which may be applicable to the production of male sterile $F_1$ hybrid seed in lettuce.

*b. Alteration of Plant Morphological Characteristics.* The *rolAB* genes from *Agrobacterium rhizogenes* were also introduced into the lettuce cultivar 'Lake Nyah' (63). Transgenic kanamycin-resistant plants exhibited more extensive root development than their nontransformed counterparts. Leaf explants from the transgenic plants were more responsive in terms of root and callus production and in their increased response to auxin supplied exogenously in the culture medium. Such transgenic plants could be more drought-tolerant; expression of the *rolAB* genes may increase the endogenous concentration of auxin, potentially reducing the incidence of leaf russett spotting in these plants.

*c. Coat Protein–Mediated Virus Resistance.* Studies have been conducted to introduce virus resistance into lettuce by using the coat protein–mediated approach. For example, a published protocol (38) was used to transfer the nucleocapsid (N) protein gene of the lettuce isolate of the tospovirus tomato spotted wilt virus (TSWV) into two lettuce breeding lines developed by the Asgrow Seed Company (64). Transgenic plants, which expressed the nucleocapsid (N) protein gene, were protected against TSWV isolates by transgenic N protein when the latter accumulated at high concentrations or by a N transgene silencing mechanism activated by overexpression of the gene. Posttranscriptional gene silencing was activated at a relatively early developmental stage in homozygous compared with hemizygous seed progeny. Consequently, the homozygous progenies showed a uniform suppression of N protein accumulation with high levels of virus resistance in all leaves of the silenced plants. In hemizygous plants, N protein accumulated at higher levels in the lower leaves and at reduced levels in successive leaves as measured from the base of the stems, as a result of transgene silencing, giving moderate levels of virus resistance. By using a similar experimental approach, the coat protein gene from LMV strain O (LMV-O) was introduced into the three virus-susceptible lettuce cultivars 'Girelle,' 'Jessy,' and 'Cocarde' (65). Several transformed plants accumulated detectable levels of LMV coat protein. The first-generation seed progeny of 12 transgenic plants in which the T-DNA was integrated at a single locus were studied for protection against LMV. The progeny from 5 of the transgenic plants exhibited resistance to LMV-O; the effectiveness of resistance to the virus was related to the developmental stage of the plants at the time of inoculation. The authors analyzed the first- (R1) and second-generation (R2) seed progeny from a cultivar 'Cocarde' transformant in detail and found that the homozygous, but not the hemizygous R1 plants were resistant to LMV-O, whereas the R2 progeny from one homozygous R1 plant were resistant to infection by LMV-O and to other strains of LMV. As in other examples of potyvirus sequence-mediated protection, some plants exhibited complete resistance, whereas in others this resistance was not always sustained, leading to a later development of viral symptoms. Other examples of the genetic modification of lettuce with respect to virus resistance are known (66,67); they are discussed later.

*d. Transposon Tagging of Downy Mildew Resistance.* Studies reporting the cloning of genes for resistance to downy mildew (*Dm*), initially found on *Dm3*, have been performed in order to clarify the molecular basis for resistance to *Bremia lactucae* in lettuce (68). These investigations have shown that several *Dm* genes are located in four clusters in the lettuce genome; *Dm3* is in the largest cluster, which contains at least nine *Dm* genes. One hundred and ninety-two primary transformed plants of the lettuce cultivar 'Diana' were generated by using *A. tumefaciens* carrying constructs with the maize *Ac* transposase and *Ds*. Several seed generation (R2) plants were screened for mutations at four *Dm* genes for their resistance to downy mildew. One family of plants (dm3t524) lost resistance to an isolate of *B. lactucae* expressing the avirulence gene *Av3*, with loss of resistance segregating as a single recessive allele of *Dm3*. The mutation was found not to be due to a large deletion, as all molecular markers flanking *Dm3* were present. Loss of *Dm3* activity cosegregated with a T-DNA from which *Ds* had been excised. Genomic DNA flanking the right border of this T-DNA was isolated by reverse polymerase chain reaction. This genomic sequence was found in four to five copies in the cultivar 'Diana,' but one copy was absent in all eight deletion mutants of *Dm3* and altered in dm3t524, indicating tight physical linkage to *Dm3*.

*e. Modification of Physiological Characteristics to Delay Senescence.* The T-*cyt* gene from the T-DNA of *A. tumefaciens*, carried on the binary vector pMOG23, was introduced into the lettuce cultivar 'Saladin' (69). The T-*cyt* gene, also known as the *ipt, tmr,* or gene 4, codes for dimethylallyl pyrophosphate:adenosine monophosphate (AMP) dimethylallyltransferase, an enzyme involved in the initial stages of cytokinin biosynthesis in plant cells transformed to tumorous growth after induction of crown gall disease. The *Agrobacterium* sp. strain used for transformation also carried pTOK47, giving a supervirulent phenotype. Interestingly, kanamycin-resistant shoots were initiated from inoculated cotyledon explants only when sites were deleted within the promoter of the T-*cyt* gene, probably because T-*cyt* overexpression with the intact promoter was phytotoxic. In culture, the kanamycin-resistant shoots exhibited several phenotypes, such as gall production, dwarfism, and hyperhydricity (vitrification). Rooted kanamycin-resistant plants recovered from their abnormal phenotype after transfer to the glasshouse and self-pollinated to produce viable seed. Transgenic plants exhibited increased cytokinin and chlorophyll contents in their leaves compared to nontransformed plants. Such results suggested that the introduction of the *ipt* gene into lettuce could provide a means to delay leaf senescence in this leafy vegetable, reducing the necessity for elaborate and costly postharvest controlled environmental conditions to prolong shelf life in the harvested crop. Thus, an approach similar to that already reported in tobacco (70) in which the *ipt* gene is linked to a senescence-specific promoter such as SAG12 from *Arabidopsis thaliana* could have application, in the longer term, in modifying the physiological features and, in turn, delaying senescence in lettuce (Fig. 1).

*f. Herbicide Resistance.* The introduction of herbicide resistance into lettuce has also been investigated, not only to generate herbicide-resistant plants per se, but also to use such reliable and readily screenable markers to study gene expression during plant development. In vitro seed germination and recallusing of tissue explants on herbicide-supplemented medium have been used to test the herbicide resistance of putative transgenic plants; the resistance of large numbers of glasshouse- or field-grown plants can be assessed by either spraying with the herbicide or its lo-

**Figure 1** Delayed senescence of the lower leaves of a lettuce plant (cv. 'Evola') transformed with the *ipt* gene linked to the *Arabidopsis* sp. senescence-specific promoter SAG12 (left). In contrast, senescence is well advanced in the lower leaves of an azygous segregant (center) and a nontransformed plant (right). All plants photographed 60 days after seed germination. Diameter of heads, 20 cm.

calized and controlled application to leaves. The advantage of localized application is that it is nonlethal. Since field-grown lettuce competes with a number of weeds (25), as already discussed, the introduction of resistance against a commonly used, efficient, broad-spectrum herbicide would provide a simple and relatively inexpensive procedure for weed management in this crop. Bialaphos, a natural tripeptide antibiotic synthesized by the Gram-positive soil bacterium *Streptomyces hygroscopicus,* is used as a nonselective herbicide, with the active moiety phosphinothricin (PPT), an analogue of L-glutamic acid and two L-alanine residues. PPT is also synthesized chemically as glufosinate ammonium. It is an irreversible inhibitor of glutamine synthetase, the only enzyme that detoxifies ammonia produced during nitrate reduction, photorespiration, and amino acid degradation in plant cells. After application of PPT to plants, ammonia accumulates to a phytotoxic concentration. *Streptomyces* sp., which synthesizes bialaphos, has evolved mechanisms to neutralize the toxicity of its own products. The well-studied *bar* gene, encoding bialaphos resistance, was introduced into the lettuce cv. 'Evola' by *Agrobacterium* sp.–mediated transformation of seedling cotyledons by using *A. tumefaciens* strains 0310 and 1310 (71). The latter strain carried the hypervirulent pTOK47 in addition to the binary vector with the *npt*II and *bar* genes. Primary transformants were selected on shoot regeneration medium supplemented with kanamycin sulfate. Interestingly, the hypervirulent pTOK47 in strain 1310 resulted in multiple insertions of T-DNA in some regenerated plants. In contrast, strain 0310 gave single-gene inserts in all plants analyzed by Southern blotting. Axenic seedlings grew on medium supplemented with glufosinate ammonium at 5 mg l$^{-1}$; glasshouse-grown plants were resistant to the herbicide at 300 mg l$^{-1}$. Stable expression of the *bar* gene was observed in the second-generation (R2) seed progeny. Both the *npt*II and *bar* genes segregated in a Mendelian fashion in some plant lines in the R1 generation; herbicide resistance segregated in the expected ratio in the R2 generation in most transgenic lines. This study confirmed that herbicide resistance (Fig. 2) can be introduced and stably expressed over several seed generations in lettuce.

*g. Secondary Metabolite Production.* Whereas most studies have focused on the transformation of *L. sativa*, there is a report of the use of the wild-type strain LBA9402 of *A. rhizogenes* to transform leaf explants from aseptically grown seedlings of *L. virosa* (72). The latter species was targeted, since it is a traditional medicinal plant with analgesic and sedative properties. Such properties are attributed to the presence of sesquiterpene lactones, which accumulate mainly in the latex of both roots and aerial parts of the plant. Transformed roots were induced and

**Figure 2** Nontransformed and transgenic plants of the lettuce cv. 'Evola' 5 weeks after seed germination and 14 days after a second application of Challenge (Finale) herbicide (AgrEvo Crop Protection) containing glufosinate ammonium at a final concentration of 300 mg $l^{-1}$. Left, nontransformed plants; center and right, second seed generation (T2) plants derived from a homozygous line (center) and a hemizygous line (right). Width of trays, 11.5 cm.

cultured in order to generate biomass and to facilitate analysis of secondary products. The authors succeeded in isolating eight sesquiterpene lactones from tissues of transformed roots.

## IV. A ROUTINE PROCEDURE FOR LETTUCE TRANSFORMATION

Several laboratories have generated transgenic lettuce plants by using *Agrobacterium* sp.–mediated gene delivery. The precise details of the procedure may differ among laboratories, but several features are common, including the use, in most cases, of seedling cotyledons as source materials for plant transformation. The advantage of such material is that it can be regenerated to plants readily and within a short time, seedling cotyledons are of uniform size, and the procedure is reliable with shoot regeneration from cotyledons established for a number of lettuce cultivars. A simple, reproducible procedure, developed originally to transform 13 cultivars of lettuce (58), has been used routinely in a number of subsequent transformation studies. It is illustrated in Figs. 3–6.

1. Seeds are surface-sterilized by immersion in a 10% (v/v) solution of bleach (0.5% sodium hypochlorite) for 30 minutes, followed by thorough rinsing with at least three changes of sterile distilled water.

2. The seeds are sown onto 20-ml aliquots of germination medium contained in 9-cm-diameter Petri dishes with approximately 40 seeds per dish. The germination medium consists of half-strength Murashige and Skoog (MS) (73) salts and vitamins with 10 g $l^{-1}$ sucrose, lacks growth regulators, and is semisolidified with 8 g $l^{-1}$ of agar (Sigma), pH 5.8. The dishes are sealed with Nescofilm (Bando Chemical Co., Kobe, Japan) and incubated for 7 days at 23° ± 2°C with

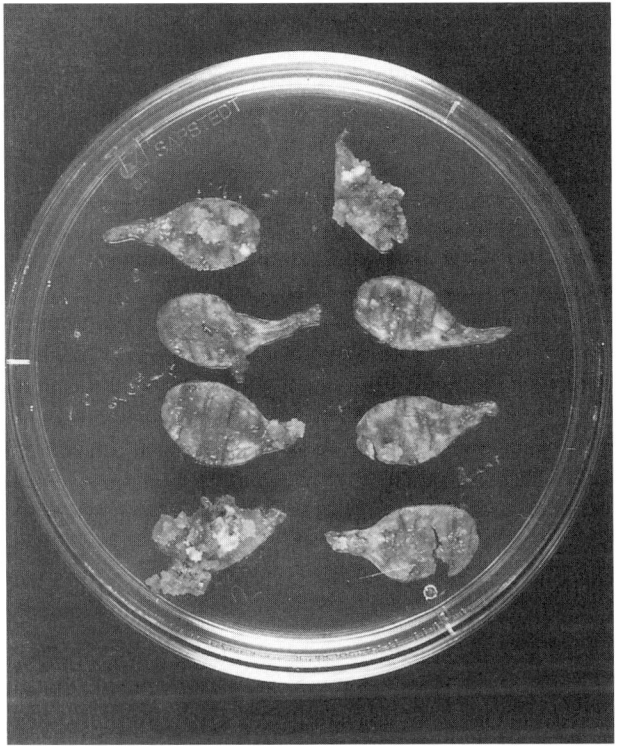

**Figure 3** Inoculated lettuce cotyledons producing callus and shoots after culture on MS-based shoot regeneration medium containing 30 g $l^{-1}$ sucrose, 0.04 mg $l^{-1}$ NAA, 0.5 mg $l^{-1}$ BAP, 500 mg $l^{-1}$ carbenicillin, 100 mg $l^{-1}$ cefotaxime, and 100 mg $l^{-1}$ kanamycin sulfate for 28 days. Diameter of Petri dish, 9 cm.

a 16-hour photoperiod provided by daylight fluorescent tubes, giving an irradiance of 200 μmol $m^{-2} s^{-1}$.

3. After 7 days, cotyledons are excised from the seedlings and wounded on their abaxial surfaces with a scalpel blade, giving shallow cuts about 1 mm apart at right angles to the midrib. Wounded cotyledons are ready for inoculation with *Agrobacterium* sp.

4. Bacterial cultures are initiated from −70°C glycerol stocks by plating onto a suitable medium, such as Luria broth (74), semisolidified with 1.5% (w/v) agar and supplemented with the appropriate combinations and concentrations of antibiotics to prevent loss of plasmid(s) from the bacterial cells (e.g., strain 1065 is grown in the presence of 100 mg $l^{-1}$ rifampicin, 50 mg $l^{-1}$ kanamycin sulfate, and 5 mg $l^{-1}$ tetracycline hydrochloride). Liquid cultures are established by transferring bacteria from agar plates to 20-ml aliquots of Luria broth in 100-ml conical flasks. The liquid medium also requires supplementation with the appropriate antibiotics (e.g., 40 mg $l^{-1}$ rifampicin, 50 mg $l^{-1}$ kanamycin sulfate, and 2 mg $l^{-1}$ tetracycline hydrochloride for strain 1065). Liquid cultures are incubated for 16 hours in the dark (28°C; the temperature must not exceed 30°C or plasmids will be lost from the bacterial cells) on a horizontal rotary shaker (150 rpm) and grown to an $OD_{600 \, nm}$ of 1.1–1.6, before being used to inoculate excised cotyledons.

5. Bacterial cultures are diluted 1:1 or 1:10 (v:v) with liquid Uchimiya and Murashige (UM) (75) medium consisting of MS salts and vitamins at full strength supplemented with 30 g $l^{-1}$ sucrose, 2 g $l^{-1}$ casein hydrolysate, 2 mg $l^{-1}$ 2,4-dichlorophenoxyacetic acid (2,4-D), 0.25 mg

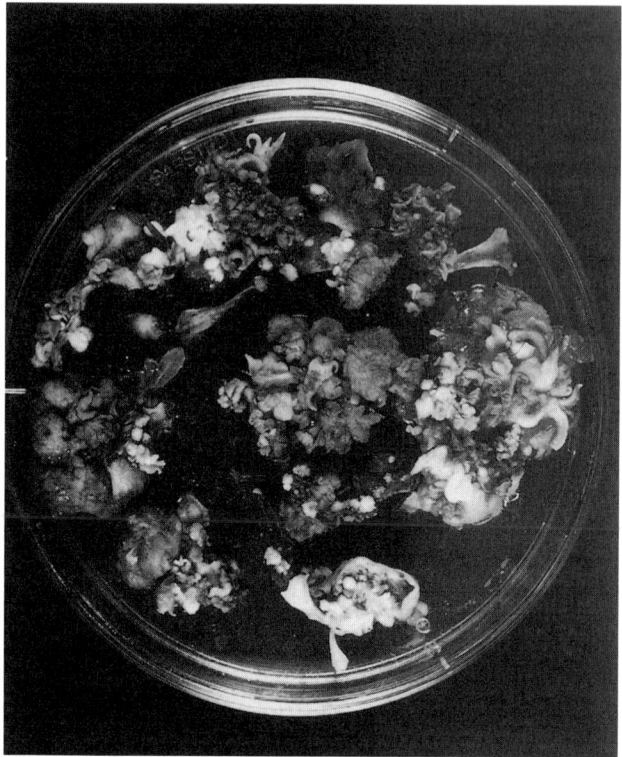

**Figure 4** Inoculated lettuce cotyledons producing callus and shoots after culture on MS-based shoot regeneration medium containing 30 g l$^{-1}$ sucrose, 0.04 mg l$^{-1}$ NAA, 0.5 mg l$^{-1}$ BAP, 500 mg l$^{-1}$ carbenicillin, 100 mg l$^{-1}$ cefotaxime, and 100 mg l$^{-1}$ kanamycin sulfate for 56 days. Diameter of Petri dish, 9 cm.

l$^{-1}$ kinetin, 9.9 mg l$^{-1}$ thiamine HCl, 9.5 mg l$^{-1}$ pyridoxine HCl, and 4.5 mg l$^{-1}$ nicotinic acid, pH 5.8.

6. Excised cotyledons are floated with their scored surfaces in contact with the 1:10 dilution of *Agrobacterium* sp. for 10 minutes or dipped (2–3 seconds) in the 1:1 dilution of the bacterial suspension. Controls (uninoculated cotyledons) are treated with liquid UM medium, but without bacteria.

7. Twenty-milliliter aliquots of agar-solidified (0.8% w/v) UM medium are dispensed into 9-cm Petri dishes and the surface of the medium covered with a sterile filter paper.

8. Cotyledons are blotted dry on sterile filter paper and placed with their wounded surfaces in contact with the filter paper overlying the UM medium in each Petri dish (eight cotyledons/dish). A sterile filter paper, dipped in UM liquid medium or sterile water, is placed over the cotyledons in each dish to keep the explants flat. The dishes are sealed with film (Nescofilm).

9. The explants are incubated for 2 days under the same conditions used for germinating seeds.

10. Cotyledons are transferred to 20-ml aliquots of MS-based shoot regeneration medium containing 30 g l$^{-1}$ sucrose, 0.04 mg l$^{-1}$ naphthalene acetic acid (NAA), 0.5 mg l$^{-1}$ 6-benzylaminopurine (BAP), and semisolidified with 0.8% (w/v) agar, pH 5.8. The shoot initiation medium used for *Agrobacterium* sp.–inoculated cotyledons also contains 500 mg l$^{-1}$ carbenicillin, 100 mg l$^{-1}$ cefotaxime, and either 50 or 100 mg l$^{-1}$ kanamycin sulfate (Fig. 3). Explants are

**Figure 5** An individual explant producing shoots after transfer to MS-based shoot regeneration medium (as in Figs. 3 and 4) but lacking carbenicillin. Diameter of jar, 5.5 cm.

subcultured three times on this medium every 14 days. Kanamycin sulfate at 100 mg $l^{-1}$ is more efficient for the selection of transgenic plants than is 50 mg $l^{-1}$ of the antibiotic.

11. Explants producing callus (Fig. 4) are transferred individually to 175-ml-capacity screw-capped Powder Round glass jars (Beatson Clarke, Rotherham, UK), each containing 40 ml of shoot initiation medium (Fig. 5). Carbenicillin is omitted at this stage from the medium used to culture cotyledons inoculated with *Agrobacterium* sp. Shoots normally appear from cultured cotyledons about 35 days after inoculation with *Agrobacterium* sp., but this interval varies with cultivar.

12. Regenerated shoots are excised when approximately 1 cm in height and transferred individually to 175-ml jars containing 40-ml aliquots of MS agar medium lacking growth regulators but with 30 g $l^{-1}$ sucrose, pH 5.8 (designated MSO). Shoots from *Agrobacterium* sp.–inoculated cotyledons are maintained on MSO agar medium with 100 mg $l^{-1}$ kanamycin sulfate in order to maintain a selection pressure on the transgenic shoots.

13. Rooted green plants, lacking any signs of bleaching (Fig. 6), are removed from their containers; their are roots washed free of culture medium, and individual plants are transferred to 9-cm-diameter plastic pots, the latter containing a mixture of Levington M3 compost (Fisons, Ipswich, UK), John Innes No. 3 compost (J. Bentley Ltd., Barrow-on-Humber, UK), and Perlite (Silvaperl, Gainsborough, UK) (3:3:2 by volume). Plants are enclosed in clear polythene bags and incubated at a light intensity of 360 µmol $m^{-2}$ $s^{-1}$ from daylight fluorescent tubes with a 16-hour photoperiod at $23° \pm 2°C$ for 7 days. After this time, one corner of each bag is removed, followed

**Figure 6** A cotyledon-derived rooted, regenerated shoot on MSO medium containing 100 mg l$^{-1}$ kanamycin sulfate. Diameter of jar, 5.5 cm.

by a second corner on day 14. Twenty-one days later, the plants are transferred to the glasshouse and grown to maturity.

14. The transgenic nature of kanamycin-resistant plants can be confirmed by routine molecular procedures [e.g., GUS histochemical and fluorometric assays; NPTII enzyme-linked immunosorbent assay (ELISA); and Southern, Western and Northern blotting], or by techniques specific to the gene(s) integrated into target plants.

15. Regenerated transgenic (T0) plants can be selfed and progeny from the first seed generation (T1 plants) and subsequent generations selected by germinating seeds on agar-solidified MS medium with 10 g l$^{-1}$ sucrose, lacking growth regulators, but containing 200 mg l$^{-1}$ kanamycin sulfate. Such experiments allow segregation analysis to be performed in order to confirm Mendelian inheritance.

Marked differences may occur in callus initiation and shoot regeneration from different lettuce cultivars (76). In some cultivars (e.g., cv. 'Flora'), hyperhydricity may impair shoot regeneration. The use of suspensions of albino *Petunia hybrida* cv. 'Comanche' as nurse cells during shoot regeneration from lettuce cotyledons has been assessed, but there was no beneficial effect on shoot regeneration from 13 lettuce cultivars evaluated (58). Other workers also reported that any beneficial effects of the use of suspensions of *Nicotiana plumbaginifolia* during shoot regeneration from cultured cotyledons of the cv. 'Cobham Green' were inconsistent (38). Some cultivars, such as 'Reflex,' were found to be transformed only when the *Agrobacterium* sp. strain carried the supervirulent pTOK47 (58). In general, a 1:10 (v:v) dilution of the bacterial suspension yields more kanamycin-resistant shoots than a 1:1 dilution, especially when cotyledons are in-

## V. INACTIVATION OF TRANSGENE EXPRESSION IN LETTUCE

Although the introduction of transgenes into a range of lettuce cultivars is now routine, the commercial exploitation of genetically engineered lettuce has been delayed by inconsistent transgene expression in several cultivars. Indeed, there is a substantial body of evidence suggesting that transgenes, which are expressed in model plants such as tobacco, may be poorly expressed in lettuce. Thus, in studies with lettuce infectious yellows virus (LIYV), plants of both tobacco and lettuce were transformed by the same three genes for resistance to this virus (67). LIYV resistance was readily detected in transgenic tobacco after analysis of only a relatively limited number of plants. In contrast, LIYV resistance was not observed in transgenic lettuce plants, in spite of numerous attempts to detect gene expression. Other workers have performed similar studies to confer virus resistance on lettuce. Thus, tobacco was transformed with a LMV coat protein (LMV-CP) gene, and heterologous resistance to potato virus Y was detected in transgenic plants (77). In later experiments, the same LMV-CP gene gave only poor resistance to LMV in transgenic lettuce plants of three European cultivars susceptible to LMV, namely, the cultivars 'Girelle,' 'Jessy,' and 'Cocarde.' These workers also reported that loss of resistance was more pronounced during subsequent seed generations, an observation made by others. In this respect, evidence was provided (66) that 39 T1 lettuce plants transformed by *Agrobacterium* sp. carrying either a LMV coat protein gene or an untranslatable LMV-CP gene between the T-DNA borders were resistant to LMV when inoculated mechanically with the virus. Subsequently, only 8 of the 39 LMV-resistant T1 plants produced LMV-resistant T2 seed progeny after selfing. Eighty percent of the T1 plants failed to transmit engineered LMV resistance to the T2 seed generation. Later generations of transgenic plants were not evaluated, and details of the promoter(s) used for the transgenes were not discussed. Several factors have been associated with transgene silencing, including the nature of the promoter driving transgene expression, DNA methylation, and transgene copy number (78). Surprisingly, little work has been carried out on transgene inactivation in lettuce. Some of the possible factors of transgene silencing in this vegetable are discussed later.

### A. Gene Promoters

In an extensive assessment of herbicide resistance conferred by the *bar* gene on the lettuce cv. 'Evola,' it was found that only 2.5% of T0 plants carrying the CaMV 35S-*bar* gene selected by their kanamycin resistance with the *npt*II gene on the same T-DNA as the *bar* gene transmitted herbicide resistance at high frequency to the T3 seed progeny, compared with 97% for plants transformed with the *bar* gene driven by the −784-bp plastocyanin promoter from pea (*petE*) (79). In CaMV 35S-*bar* transformants, only 16% of T1 plants, 22% of T2 plants, and 11% of T3 plants were resistant to glufosinate ammonium at 300 mg l$^{-1}$. In contrast, 63% of T1 plants, 83% of T2 plants, and 99% of T3 *petE-bar* transgenic plants were resistant to the same concentration of the herbicide. A CaMV 35S-*npt*II gene was present on the same T-DNA as the CaMV 35S-*bar* and *petE-bar* genes. NPTII protein, as determined by ELISA, was absent in 29% of the herbicide-resistant *petE-bar* T2 lines, indicating inactivation of the CaMV 35S promoter on the same T-DNA as an active *petE* promoter. In all cases, Southern hybridization confirmed the presence of the transgenes in kanamycin-resistant plants.

Other studies have also exploited the constitutive CaMV 35S promoter to express transgenes in lettuce (55,58,64,80), and some studies have utilized the *nos* (38,58,62,63,80,81), *Mac*

(82), *petE* (79,82), and ACT1 (McCabe et al., unpublished) promoters. Use of the tobacco pathogenesis-related protein gene promoter PR1a resulted in *gus* gene expression after induction by salicylic acid (55); in other investigations the *Mas* promoter gave root-specific expression (82). Additionally, the tapetum-specific promoter A9 was used, as already discussed, in the induction of male sterility in transgenic lettuce (62). More recently, McCabe and associates (unpublished) observed a 48-fold increase in zeatin riboside equivalents in primary leaves of 60-day-old lettuce plants of the cultivar 'Evola' after transformation with the *ipt* gene from *A. tumefaciens* linked to the senescence-specific promoter $P_{SAG12}$ from *Arabidopsis thaliana* (70).

### B. Transgene Copy Number

The effect of copy number on transgene expression in lettuce and, indeed, in other plants is generally ambiguous; many reports, even in the same target plants, are contradictory (83). T-DNA integration into lettuce is often complex (38,80) (Fig. 7). In a detailed investigation of the integration, expression, and inheritance of a CaMV 35S-*gus*-intron gene and a *nos.npt*II.*nos* gene on the same T-DNA, there was no clear correlation between gene copy number and transgene expression in the lettuce cv. 'Raisa' (80). T-DNA integration into the lettuce genome was complex after gene delivery by the supervirulent *A. tumefaciens* strain 1065. Interestingly and unexpectedly, truncation of the right side of the T-DNA, including the right border, was observed in the first seed generation plants from one line. However, complex T-DNA integration patterns did not always correlate with low transgene expression: about 30% of lines with a high T-DNA copy number showed high gene expression in the first seed generation, and high gene expression stably maintained at least to the fourth seed generation in selected plant lines. In a low-expressing line, trans-

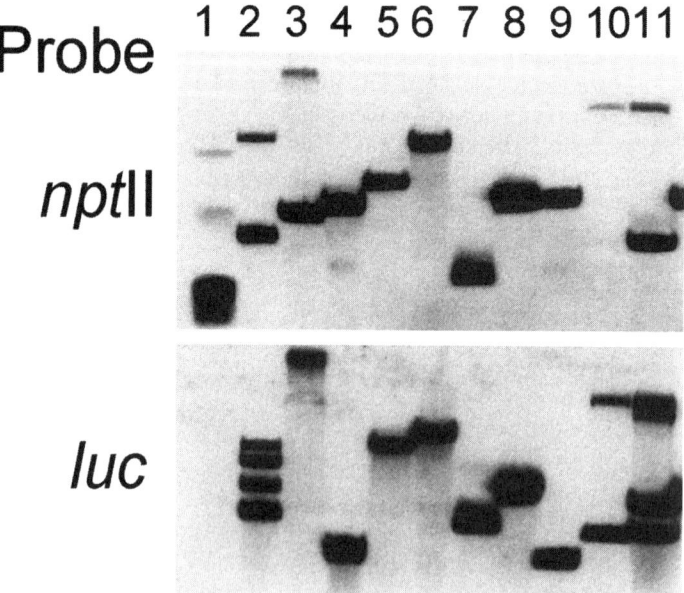

**Figure 7** T-DNA/plant DNA junction analysis, by nonradioactive Southern hybridization to left and right border sequences, in genomic DNA from transgenic lettuce plants (cv. 'Evola'). Plants were transformed with the *npt*II gene (adjacent to the left T-DNA border) and the *luc* (firefly luciferase) gene adjacent to the right T-DNA border. Lane 1, integration of the left T-DNA border and absence of the right border; lanes 2, 3, 10, and 11, multiple transgene insertions.

gene expression was lost in the second seed generation; another line exhibited complete gene silencing to the same generation. GUS activity showed a 50-fold difference, and 16-fold variation in NPTII protein content was detected in the first seed generation plants from different selected kanamycin-resistant parental plants.

In other studies, increased transgene dosage was reported to accelerate posttranscriptional transgene silencing in lettuce carrying a CaMV 35S-driven tomato spotted wilt virus N (TSWV N) coat protein gene; TSWV N protein accumulation decreased two to four times faster in homozygous than in hemizygous plants (64). Although several reports of a 3:1 segregation of kanamycin resistance to sensitivity among T1 plants of lettuce indicated single active T-DNA loci (38,55,65), Southern analysis has revealed that highly rearranged T-DNA insertions are common in transgenic lettuce. In this respect, 14 of 16 lettuce plants transformed by *A. tumefaciens* carrying binary or cointegrate vectors with the *nos.nptII.nos* gene contained T-DNA insertions that were rearranged (38). All 16 of these plants showed a 3:1 segregation of kanamycin resistance to sensitivity.

## C. Deoxyribonucleic Acid Methylation

DNA methylation has been correlated with transgene silencing in plants such as *Petunia* sp. (84,85), but little information is available for transgenic lettuce. Treatment of transgenic lettuce seedlings with the demethylating agent 5-azacytidine appeared to alleviate partial silencing of the CaMV 35S-*gus* gene in transgenic plants of the cultivar 'Raisa' (80), although in the same experiments reactivation of transgene activity did not occur in seedlings from a transgenic line in which the same *gus* gene was silenced. Susceptibility of the promoter to methylation may also account for transgene silencing. For example, the CaMV 35S promoter has approximately three times more methylation sites (59 per Kilobase) than the −784-bp *petE* promoter (18/kb), which could account for the susceptibility of the CaMV35S promoter to inactivation (79).

Overall, several factors relating to the precise nature of the DNA constructs employed probably contribute to transgene silencing in lettuce. Additionally, lettuce is reported to have a complex genome (86), with a high proportion of heterochromatic or repetitive DNA, which is likely to increase the inactivation of transgenes (87). An interesting observation is that lettuce genomic DNA cannot be digested fully with restriction endonucleases, such as *Sst*II, *Pst*I, and *Pvu*II, which contain CNG in their recognition sequence. This indicates that the lettuce genome is highly methylated. It is well known that the insertion of transgenes into highly methylated regions of a plant genome is associated with transgene methylation and inactivation, as in *Petunia* sp. (88). The same phenomenon may also occur in lettuce. One hypothesis as to why supervirulent plasmids, such as pTOK47, enhance transformation frequencies in lettuce is that the insertion of multiple T-DNA copies at different plant genomic loci may increase the probability of T-DNA integration at a site in the plant genome that is not susceptible to methylation and, consequently, gene silencing (80).

## VI. CONCLUDING REMARKS

Reliable *Agrobacterium* sp.–based transformation systems have been developed for several lettuce cultivars, yielding transgenic plants carrying reporter, selectable marker, and agronomically useful genes. Although most transformation protocols have focused on the use of seedling cotyledons for inoculation by *Agrobacterium* sp., there is also evidence that explants from the first true leaves may be preferable as source material for transformation (65). To date, genetically manipulated lettuce is not, apparently, in commercial production, although scientists advising the European Commission have given approval for production of herbicide-resistant radicchio produced by the Dutch

Company Bejo Zaden (89). If their advice is accepted, this will be the first genetically manipulated plant approved in the European Union for human consumption as a fresh, unprocessed product (90). However, Britain's Advisory Committee on Novel Foods and Processes (ACNFP) has expressed concern at the marketing of this product. Even if this approval is granted, there is a requirement to understand the unpredictable gene expression often observed in lettuce. Despite these difficulties, there remains considerable commercial investment by several international seed companies in the genetic engineering of leafy vegetables, particularly with respect to increased shelf life and disease resistance. It remains only a matter of time before such genetically manipulated produce is available to the consumer.

## ACKNOWLEDGMENTS

Some of the original work described in this review was supported by Van der Have Research (Ref. 58, 62, 63, 69) and EU A.I.R. Project Contract No. 92-0250 (Ref. 71, 79, 80) and performed under MAFF Plant Health Licences.

## REFERENCES

1. IM de Vries. Origin and domestication of *Lactuca sativa* L. Genet Res Crop Evolut 171:233–248, 1997.
2. MD Bennett, AV Cox, IJ Leitch. Angiosperm C-value database. Royal Botanic Gardens, Kew, 1997. Available at *http://www.rbgkew.org.uk./cval/database1.html*
3. EJ Ryder. Lettuce breeding. In: MJ Bassett, ed. Breeding Vegetable Crops. Westport, CT: AVI, 1986, pp 433–474.
4. DAC Pink, EM Keane. Lettuce *Latuca sativa* L. In: G Kalloo, BO Bergh, eds. Genetic Improvement of Vegetable Crops. Oxford: Pergamon Press, 1993, pp 543–571.
5. R Alconero. Lettuce (*Lactuca sativa*). In: YPS Bajaj, ed. Biotechnology in Agriculture and Forestry. Vol 6. Berlin: Springer-Verlag, 1988, pp 351–369.
6. CM Rick. The tomato. Sci Am 239:6–77, 1978.
7. D Gromek, W Kisiel, A Klodzinska, E Chojnackawojcik. Biologically-active preparations from *Lactuca virosa* L. Phytother Res 6:285–287, 1992.
8. D Zohary. The wild genetic resources of cultivated lettuce (*Lactuca sativa* L.). Euphytica 53:6–77, 1991.
9. EJ Ryder. "Vanguard 75" lettuce. Hortscience 14:284–285, 1997.
10. EJ Ryder, TW Whitaker. Lettuce *Lactuca sativa* (Compositae). In: NW Simmonds, ed. Evolution of Crop Plants. London: Longman Group, 1976, pp 39–41.
11. IM de Vries. Crossing experiments of lettuce cultivars and species (*Lactuca* sect. *Lactuca*. Compositae). Plant Syst Evolut 171:233–248, 1990.
12. Witsenboer, RV Kesseli, MG Fortin, Stanghellini, RW Michelmore. Sources and genetic structure of a cluster of genes for resistance to three pathogens in lettuce. Theor Appl Genet 91:178–188, 1995.
13. IR Crute, AG Johnson. The genetic relationship between races of *Bremia lactucae* and cultivars of *Lactuca sativa*. Ann Appl Biol 83:125–137, 1976.
14. FJM Bonnier, K Reinink, D Groenwold. New sources of major gene resistance in *Lactuca* to *Bremia lactucae*. Euphytica 61:203–211, 1992.
15. JM Smith, A Langton. A new source of resistance to downy mildew. Grower, September: 54–55, 1989.
16. A Lebeda, K Reinink. Histological characterisation of resistance *in Lactuca saligna* to lettuce downy mildew (*Bremia lactucae*). Physiol Mol Plant Pathol 44:125–139, 1994.
17. F Revers, H Lot, S Souche, O LeGall, T Candresse, J Dunez. Biological and molecular variability of Lettuce Mosaic Virus isolates. Phytopathol 87:397–403, 1997.
18. J Marrou. Confirmation de la laitue par les graines chez la variété Gallega de Invernio. Ann Phytopathol 1:213–218, 1969.

19. H Bannerat, L Boulidard, J Marrou, M Duteil. Etude de l'hérédité de la tolérance au virus de la mosaique de la laitue chez la variété Gallega de Invernio. Etudes de Virol. Ann Phytopathol 1:213–218, 1969.
20. EJ Ryder. The nature of resistance to lettuce mosaic. Proceedings of the Eucarpia Meeting on Leafy Vegetables, Wageningen, The Netherlands, 1976, pp 110–118.
21. S Dinant, H Lot. Lettuce mosaic virus. Plant Pathol 41:528–542, 1992.
22. AN Kishaba, TW Whitaker, PV Vail, HH Toba. Differential oviposition of cabbage looper in lettuce. J Am Soc Hortic Sci 98:367–368, 1973.
23. TW Whitaker, AN Kishaba, HH Toba. Host-parasite interactions of *Lactuca saligna* L. and the cabbage looper, *Trichopulsia ni* (Hubner). J Am Soc Hortic Sci 99:74–75, 1974.
24. AH Eenink, R Groenwold, FL Dielman. Resistance of lettuce (*Lactuca*) to the leaf aphid *Nasonovia ribis-nigris*: Transfer of resistance from *L. virosa to L. sativa* by interspecific crosses and selection of resistant breeding lines. Euphytica 3:291–300, 1982.
25. WP Anderson. Weed Science, Principles and Applications. Minneapolis: West, 1996, pp 1–388.
26. DJ Tonks, P Westra. Control of sulfonylurea-resistant kochia (*Kochia scorpia*). Weed Technol 11:270–276, 1997.
27. SR Moss, B Rubin. Herbicide-resistant weeds: A worldwide perspective. J Agric Sci 120:141–148, 1993.
28. HM LeBaron. Distribution and seriousness of herbicide-resistant weed infestations worldwide. In: JC Caseley, GW Cussans, RK Atkin, eds. Herbicide Resistance in Weeds and Crops. Oxford: Butterworth Heinemann, 1991, pp 27–55.
29. DL Shaner. Herbicide resistance: Where are we? How did we get here? Where are we going? Weed Technol 9:850–856, 1995.
30. CA Mallory-Smith, DC Thill, MJ Dial, RS Zemetra. Inheritance of sulfonylurea resistance in *Lactuca* spp. Weed Technol 4:787–790, 1990.
31. EG Steingrover, JW Steenhuizen, J Van der Boon. Effects of low light intensities at night on nitrate accumulation in lettuce grown on recirculating nutrient solution. Neth J Agric Sci 41:13–21, 1993.
32. A Gunes, WKN Post, EA Kirby, A Mehmet. Influence of partial replacement of nitrate by amino acid nitrogen or urea in the nutrient medium on nitrate accumulation in NFT grown winter lettuce. J Plant Nutr 17:1929–1938, 1994.
33. A Gunes, WHK Post, M Aktas. Effect of partial replacement of nitrate by NH4-N, urea-N and amino acid-N in nutrient solution on nitrate accumulatin in lettuce (*Lactuca sativa*). Agrochimica 39:326–333, 1995.
34. L Gaudreau, J Charbonneau, LP Vezina, A Gosselin. Effects of photoperiod and photosynthetic photon flux on nitrate content and nitrate reductase activity in greenhouse-grown lettuce. J Plant Nutr 18:437–453, 1995.
35. JPNL Roorda van Eysinga. Nitrate and glasshouse vegetables. Fertilizer Res 5:149–156, 1984.
36. P Santamaria. Occurrence of nitrate and nitrite in vegetables and total dietary intakes. Industrie Alimentari 36:1329–1334, 1997.
37. RW Michelmore, JA Eash. Tissue culture of lettuce. In: DA Evans, WR Sharp, P Amirato, eds. Handbook of Plant Cell Culture. Vol 4. London: Collier MacMillan, 1986, pp 512–551.
38. RW Michelmore, E Marsh, S Seely, B Landry. Transformation of lettuce (*Lactuca sativa*) mediated by *Agrobacterium tumefaciens*. Plant Cell Rep 6:439–442, 1987.
39. SF Berry, DY Lu, D Pental, EC Cocking. Regeneration of plants from protoplasts of *Lactuca sativa*. Z Pflanzenphysiol 108:31–38, 1982.
40. DE Engler, RG Grogan. Isolation, culture and regeneration of lettuce leaf mesophyll protoplasts. Plant Sci Lett 28:223–227, 1983.
41. C Brown, JA Lucas, JB Power. Plant regeneration from protoplasts of a wild lettuce species (*Lactuca saligna* L.). Plant Cell Rep 6:180–182, 1988.
42. S Enomoto, K Ohyama. Rgeneration of plants from protoplasts of lettuce and its wild species. In: YPS Bajaj, ed. Biotechnology in Agriculture and Forestry. Vol 8. Plant Protoplasts and Genetic Engineering I. Berlin, Heidelberg: Springer-Verlag, 1989, pp 217–226.
43. T Tanaka, T Matsumura, Y Morinaga. Studies on the protoplast culture I. Procedure of protoplast culture of lettuce (*Lactuca sativa* L.). Proc Faculty Agric Kyusyu Takai Univ 10:29–35, 1991.

44. CL Webb, MR Davey, JA Lucas, JB Power. Plant regeneration from mesophyll protoplasts of *Lactuca perennis*. Plant Cell Tissue Org Cult 38:77–79, 1994.
45. E Matsumoto. Production of somatic hybrids between *Lactuca sativa and L. serriola* by cell fusion. Jpn J Breed 35:134–135, 1987.
46. T Mizutani, XJ Liu, Y Tashiro, S Miyazaki, K Shimasaki. Plant regeneration and cell fusion of protoplasts from lettuce cultivars and related wild species in Japan. Bull Faculty Agric Saga University 67:109–118, 1989.
47. E Matsumoto. Interspecific somatic hybridisation between lettuce (*Lactuca sativa*) and wild species (*L. virosa*). Plant Cell Rep 9:531–534, 1991.
48. B Maisonneuve, MC Chupeau, Y Bellec, Y Chupeau. Sexual and somatic hybridisation in the genus *Lactuca*. Euphytica 85:281–285, 1995.
49. M Sibs. Expression of cryptic genetic factors in vivo and in vitro. Proceedings of the Conference for Broadening Genetic Base of Crops, Wageningen, 1979, pp 339–340.
50. C Brown, JA Lucas, IR Crute, DGA Walkey, JB Power. An assessment of genetic-variability in somacloned lettuce plants (*Lactuca sativa*) and their offspring. Ann Appl Biol 109:391–407, 1986.
51. DE Engler and RG Grogan. Variation in lettuce plants regenerated from protoplasts. J Hered 75:426–430, 1984.
52. MC Chupeau, C Bellini, P Guerche, B Maisonneuve, G Vastra, Y Chupeau. Transgenic plants of lettuce (*Lactuca sativa*) obtained through electroporation of protoplasts. Bio/Technology 7:503–508, 1989.
53. JR Zupan, P Zambryski. Update on plant transformation: Transfer of T-DNA from *Agrobacterium* to the plant cell. Plant Physiol 107:1041–1047, 1995.
54. B Tinland. The integration of T-DNA into plant genomes. Trends in Plant Sci 1:178–184, 1996.
55. S Enomoto, H Itohn, M Ohsima, Y Ohashi. Induced expression of a chimaeric gene construct in transgenic lettuce plants using tobacco pathogenesis-related protein gene promoter region. Plant Cell Rep 9:6–9, 1990.
56. CL Webb. Transformation and somatic hybridisation for lettuce (*Lactuca sativa*) improvement. PhD thesis, University of Nottingham, 1992.
57. AC Torres, DJ Cantliffe, B Laughner, M Bieniek, R Nagata, M Ashraf, RJ Ferl. Stable transformation of lettuce cultivar South Bay from cotyledon explants. Plant Cell Tissue Org Cult 34:279–285, 1993.
58. IS Curtis, JB Power, NW Blackhall, AMM de Laat, Davey MR. Genotype-independent transformation of lettuce using *Agrobacterium tumefaciens*. J Exp Bot 45:1441–1449, 1994.
59. PC Sijmons, BMM Dekker, B Schrammeijer, TC Verwoerd, PJM Van den Elsen, A Hoekema. Production of correctly processed human serum albumin in transgenic plants. Bio/Technology 8:217–221, 1990.
60. G Vancanneyt, R Schmidt, A O'Connor-Sanchez, L Willmitzer, M Rocha-Sosa. Construction of an intron-containing marker gene: splicing of the intron in transgenic plants and its use in monitoring early events in *Agrobacterium*-mediated plant transformation. Mol Gen Genet 220:245–250, 1990.
61. S Jin, T Komari, MP Gordon, E Nester. Genes responsible for the supervirulence phenotype of *Agrobacterium tumefaciens* A281. J Bacteriol 169:417–4425, 1987.
62. IS Curtis, C He, R Scott, JB Power, MR Davey. Genomic male sterility in lettuce, a baseline for the production of $F_1$ hybrids. Plant Sci 113:113–119, 1996.
63. IS Curtis, C He, JB Power, D Mariotti, AMM de Laat, MR Davey. The effects of *Agrobacterium rhizogenes rolAB* genes in lettuce. Plant Sci 115:123–135, 1996.
64. S-Z Pang, F-J Jan, K Carney, J Stout, DM Tricoli, HD Quemada, D Gonsalves. Post-transcriptional transgene silencing and consequent tospovirus resistance in transgenic lettuce are affected by transgene dosage and plant development. Plant J 9:899–890, 1996.
65. S Dinant, B Maisonneuve, J Albouy, Y Hupeau, M-C Chupeau, Y Bellec, F Gaudefroy, C Kusiak, S Souche, C Robaglia, H Lot. Coat protein gene-mediated protection in *Lactuca sativa* against lettuce potyvirus strains. Mol Breed 3:75–86, 1997.
66. RL Gilberton. Management and detection of LMV: Production of LMV resistant lettuce and LMV coat protein antibodies. Iceberg Lettuce Advisory Board Annual Report, 1996, pp 78–81.
67. BW Falk. Basic approaches to lettuce virus control. Iceberg Lettuce Advisory Board Annual Report, 1996, pp 70–74.

68. PA Okubara, R Arroyo-Garcia, KA Shen, M Mazier, BC Meyers, OE Ochoa, S Kim, C-H Yang, RW Michelmore. A transgenic mutant of *Lactuca sativa* (lettuce) with a T-DNA tightly linked to loss of downy mildew resistance. Mol Plant Microbe Interact 10:970–977, 1997.
69. IS Curtis, C He, WJR Jordi, E Davelaar, JB Power, AMM de Laat, MR Davey. Promoter deletions are essential for transformation of lettuce by the T-*cyt* gene: the phenotypes of transgenic plants. Ann Bot 83:559–567, 1999.
70. S Gan, RM Amasino. Inhibition of leaf senescence by auto-regulated production of cytokinin. Science 270:1986–1988, 1995.
71. U Mohapatra, MS McCabe, JB Power, F Schepers, A van der Arend, MR Davey. Expression of the *bar* gene confers herbicide resistance in transgenic lettuce. Trans Res 7:1–12, 1998.
72. W Kisiel, A Stojakowska, J Marlaz, S Kohlmünzer. Sequiterpene lactones in *Agrobacterium rhizogenes*-transformed hairy root culture of *Lactuca sativa*. Phytochemistry 40:1139–1140, 1995.
73. T Murashige, F Skoog. A revised medium for rapid growth and bioassays with tobacco tissue cultures. Physiol Plant 15:473–497, 1962.
74. J Sambrook, EF Fritsch, T Maniatis. Molecular Cloning. 2nd ed. Cold Spring Harbor, NY: Cold Spring Harbor Laboratory Press, 1989.
75. H Uchimiya, T Murashige. Evaluation of parameters in the isolation of viable protoplasts from cultured tobacco cells. Plant Physiol 54:936–944, 1974.
76. Z Xinrun, AJ Conner. Genotypic effects on tissue culture responses of lettuce cotyledons. J Genet Breed 46:287–290, 1992.
77. S Dinant, F Blaise, C Kusiak, S Astier-Manifacier, J Albouy. Heterologous resistance to Potato Virus Y in transgenic tobacco plants expressing the coat protein gene of lettuce mosaic potyvirus. Phytopathology 83:818–824, 1993.
78. J Finegan, D McElroy. Transgene inactivation: Plants fight back! Bio/Technology 12:883–888, 1994.
79. MS McCabe, F Schepers, A van der Arend, U Mohapatra, AMM de Laat, JB Power, MR Davey. Increased stable inheritance of herbicide resistance in transgenic lettuce carrying a *petE* promoter-bar gene compared with a CaMV 35S-*bar* gene. Theor Appl Genet 99:587–592, 1999.
80. MS McCabe, U Mohapatra, SC Debnath, JB Power, MR Davey. Integration, expression and inheritance of two linked T-DNA marker genes in transgenic lettuce. Mol Breed 5:329–344, 1999.
81. C-H Yang, B Carroll, S Schofield, J Jones, R Michelmore. Transinactivation of *Ds* elements in plants of lettuce (*Lactuca sativa*). Mol Gen Genet 241:389–398, 1993.
82. IS Curtis, JB Power, MS McCabe, AMM de Laat, MR Davey. Promoter-GUS fusions in lettuce. Abstr 4th Intern Congr Plant Molec Biol, Amsterdam. Abstract No. 1682, 1994.
83. SLA Hobbs, TD Warkentin, CMO Delong. Transgene copy number can be positively or negatively associated with transgene expression. Plant Mol Biol 21:17–26, 1993.
84. P Meyer, F Linn, I Heidmann, ZAH Meyer, I Niedenhof, H Saedler. Endogenous and environmental factors influence 35S promoter methylation of a maize A1 gene construct in transgenic petunia and its colour phenotype. Mol Gen Genet 231:345–352, 1992.
85. EC Ulian, JM Magill, CW Magill, RH Smith. DNA methylation and expression of NPTII in transgenic petunias and their progeny. Theor Appl Genet 92:976–981, 1996.
86. RW Michelmore. Genetic variation in lettuce. Iceberg Lettuce Advisory Board Annual Report, 1996, pp 62–65.
87. M Stam, JNM Mol, JM Kooter. The silence of genes in transgenic plants. Ann Bot 79:3–12, 1997.
88. F Pröls, P Meyer. The methylation patterns of chromosomal integration regions influence gene activity of transferred DNA in *Petunia hybrida*. Plant J 2:465–475, 1992.
89. Newswire. First for GM food. New Scientist, February 6, 1999, p 5.
90. T Dalyell. Thistle diary. More comment from Westminster. New Scientist, May 8, 1999, p 52.

# 43
# Maize Food and Feed: A Current Perspective and Consideration of Future Possibilities

**Bruce R. Hamaker**
*Purdue University, West Lafayette, Indiana*

**Brian A. Larkins**
*The University of Arizona, Tucson, Arizona*

|      |                                                              |     |
|------|--------------------------------------------------------------|-----|
| I.   | INTRODUCTION                                                 | 637 |
| II.  | MAIZE FROM ANCIENT TO CURRENT TIMES                          | 638 |
| III. | FOOD USES                                                    | 639 |
|      | A. Maize-Based Foods                                         | 639 |
|      | B. Maize Components Used as Food Ingredients                 | 640 |
| IV.  | GRAIN TYPES OPTIMAL FOR PROCESSING                           | 640 |
| V.   | FACTORS RELATED TO KERNEL TEXTURE                            | 641 |
| VI.  | BIOCHEMICAL COMPONENTS: THEIR STRUCTURE AND POTENTIAL FOR CHANGE | 641 |
|      | A. Starch                                                    | 641 |
|      | B. Protein                                                   | 644 |
|      | C. Lipid                                                     | 645 |
|      | D. Minor Components                                          | 646 |
| VII. | CREATING TRANSGENIC MAIZE                                    | 647 |
| VIII.| MODIFIED TRAITS IN TRANSGENIC MAIZE PLANTS                   | 649 |
| IX.  | FUTURE PROSPECTS                                             | 649 |
|      | REFERENCES                                                   | 650 |

## I. INTRODUCTION

Maize (*Zea mays*) is rated either second or third in terms of world cereal production, depending on crop year statistics. With maize production at 577 million metric tons (MT) in 1996, it is very close to that of wheat (585 million MT) and rice (562 million MT) (1). Nearly one half of the world production of maize is from North America (the United States, Mexico, Canada). China is the next largest producer at about 20% of the world's crop. Maize is the largest animal feed grain in the world, considering that wheat and rice are primarily food grains. However, maize is also a major food grain in many areas of Latin America and Africa and has a large food and industrial market in the United States. Currently, approximately 65% of the U.S. maize crop is used for animal feed; about 25% is exported, and the remaining 10% is further processed into food ingredients, nonfood coatings and adhesives, and ethanol (2). Among the many products derived from maize, high-fructose corn syrup (HFCS) and fuel alcohol predominated in 1998; each of these consumed approximately 14 million MT of corn. Starch (native and modified) and non-HFCS starch hydrolysates (glucose, oligoglucans, maltodextrins) consumed about 5.8 to 6.3 million MT each; beverage alcohol used about 3.2 million MT, and cereals and other dry milled products used about 3.6 million MT (2). In addition to these products, there are significant markets for by-products of the dry- and wet-milling processes, which include maize oil and gluten feed and meal.

In addition to the aforementioned "traditional" refined maize products, there are a number of nontraditional products used in food, pharmaceuticals, and industrial processes. Most of these substances are derived from maize starch. Polyols, or sugar alcohols, such as sorbitol, maltitol, mannitol, erythritol, and hydrogenated starch hydrolysates, function principally as nonnutritive and noncariogenic sweeteners. Cyclodextrins, or circular dextrins, are produced through an enzymatic process and function as carriers, or encapsulators, of food and pharmaceutical ingredients. A variety of organic acids, such as citric acid, lactic acid, itaconic acid, gluconic acid, and glucono-$d$-lactone, are derived from maize starch. In the United States and elsewhere, maize starch is processed in relatively small, but increasing, quantities to make resistant starch, which operates as dietary fiber.

Advances in biotechnology are changing maize from simply a source of nutrients and biopolymers for animal feed, food, and industrial applications to a biofactory that supplies novel or altered constituents. The following sections provide a detailed review of maize uses, potential areas for which transgenic work offers promise, traditional breeding and traits that are preserved, maize genetic engineering methodologies, engineered traits, and current and future objectives in transgenic maize research.

## II. MAIZE FROM ANCIENT TO CURRENT TIMES

Maize probably originated in Mexico and Central America, where it was first cultivated. From there, maize spread throughout South and North America; later it was taken to Europe. Maize is thought to have reached Africa and Asia via Europe. All maize is classified as *Zea mays*. Teosinte, *Euchlaena mexicana* Schrod, which is native to Mexico and Central America, is a close relative. Controversy still exists as to whether maize actually developed from teosinte (3,4). Selection of agronomic and, undoubtedly, quality traits by native populations produced the maize that forms the basis for modern varieties.

Native populations that subsisted on maize used the grain in a variety of food products. The most well-known are the alkaline-cooked foods, which are still popular today. These foods were prepared by mixing an alkali source—leached wood ash, lye, or lime—with water and whole kernel maize, then cooking and steeping the grain. Today, commercial processes use lime (calcium hydroxide) as the alkali source. Cooked and steeped kernels were washed to remove the pericarp or bran portion of the grain, resulting in what is known as *nixtamal*. The *nixtamal* was then ground in a stone mill to reduce particle size and produce the cohesive doughlike material termed *masa*. This intermediate was used to produce a variety of products, including tortillas, tamales, and atoles. In other parts of what is now Latin America, indigenous peoples used maize for their staple food in other types of alkaline-cooked products and in nonalkaline-cooked products, including flat breads, porridges, and beverages.

## III. FOOD USES

### A. Maize-Based Foods

Today, maize is used to make a multitude of food products, some of which date back to pre-Columbian civilizations. In the United States, masa-derived products (tortilla chips, corn chips, extruded snacks) are now a huge business with annual sales in the $4 billion range. Moreover, these snack products are gaining markets worldwide.

Other maize-derived food products in the United States are produced primarily from dry-milled maize fractions. Dry milling is a process in which tempered, or moistened, grain (to about 20% moisture content) is ruptured in a degerminator to produce germ, large vitreous endosperm pieces, and smaller particle size endosperm fractions. The dry miller sells the large vitreous fraction, free of bran and germ, as "flaking grits." Flaking grits are used to make corn flakes and have the highest market value of the dry-milled fractions. Other fractions, in order of reduced particle size, include brewer's grits, grits, meal, cones, and flour. A typical dry miller may sell upward of 10 dry-milled fractions of different particle size for further processing. Extruded, ready-to-eat breakfast cereals and snacks use a meal or flour as either the main or a minor ingredient.

Elsewhere in the world, maize is made into myriad food products, many of which provide the main source of energy, and sometimes protein, in the diets of the poor. A number of comprehensive reviews have been published on the common maize products eaten around the world (5,6). Serna-Saldivar and coworkers (6) have classified these foods according to type, as follows:

- Cooked, whole grain products—either alkaline-cooked, such as *hominy* (United States) and *pozole* (Mexico), or degerminated and debraned, such as *munuca* (Brazil)
- Thin unfermented porridges—such as *atole* (Mexico, Central America), which is a masa brought to a slurry with water, cooked, then blended with milk, sugar, and other flavorings; and *mingau* (Brazil), which is cooked maize grits or mashed immature kernels cooked in water
- Thin fermented porridges—such as *ogi* (Nigeria), produced from naturally fermented whole kernels that are ground, slurried, further fermented, separated from the clear liquid fraction, and boiled to a thin porridge; *ugi* (East Africa), produced by natural fermentation of slurried flour and water that is then boiled to a porridge and flavored; and *mahewu* (South Africa), a boiled maize meal with wheat flour added that is then fermented
- Thick porridges (unfermented and fermented)—such as *ugali* (East Africa), made from maize meal gelatinized to a thick paste; *to/tuwo/asida* (West Africa), also made in acid (with tamarand), alkali (with wood ash), or fermented versions; and *polenta* (South America/Europe), produced by cooking grits
- Snack foods—consumed worldwide, such as popcorn and corn and tortilla chips
- Steamed foods—such as *couscous* (West Africa and Brazil), produced by agglomeration of flour followed by steaming and drying; and *tamales* (Latin America), made from masa and other ingredients that are wrapped, usually in a leaf, and steamed
- Unfermented breads—such as *tortillas* (Mexico, Central America), produced by the masa process described; *arepas* (Venezuela), produced by grinding cooked grits to a dough, flattened, and baked; *roti/chapati* (India), a flat bread mostly made from wheat, sorghum, or millet, but also from maize; and *cornbread* (worldwide), made from either cornmeal or a blend of wheat flour and cornmeal with chemical leavening agents and flavorings

- Fermented breads—such as *injera* (Ethiopia), which is maize flour mixed with water and starter culture, fermented, and baked on one side to produce a flat bread with holes on the unbaked side
- Nonalcoholic and alcoholic beverages—such as *chicha morada* (South America), made from blue corn that is cooked in sugared water, filtered, and blended sometimes with a fruit juice; and a variety of alcoholic maize-based beverages, ranging from the opaque beer of Southern Africa to *chicha* of South America and *tesguino* of Mexico

### B. Maize Components Used as Food Ingredients

Maize is also wet milled into its component fractions [starch, protein, oil (germ), and fiber] and used "as is" or as starting material to produce a wide variety of ingredients that are used in food and nonfood products. The wet-milling process involves steeping the kernels in water injected with $SO_2$, followed by grinding and a series of separation steps primarily using hydrocyclones and sieves (7). Maize products derived from wet-milling include the following:

- Starch and starch-derived products—including maize starch, modified starches, dextrins (partially hydrolyzed starch), and starch refinery products, including glucose syrup, high-fructose syrup, dextrose, maize syrup solids, and maltodextrins. Ethanol is produced by using starch as the fermentative substrate.
- Protein and protein-derived products—including maize gluten feed/meal and maize oil meal, which are used as animal feed supplements, and a small amount of commercial purified zein.
- Oil—refined.
- Fiber—from steepwater and maize gluten feed/meal.

## IV. GRAIN TYPES OPTIMAL FOR PROCESSING

Although specialty maize crops, such as sweet corn and popcorn, are commonly consumed in some countries, the bulk of food corn is of the dent, floury, or flint endosperm types. Dent maize kernels have an outer layer of vitreous, hard endosperm and a core and crown of floury, soft endosperm. The characteristic "dent" is found at the crown. This normal maize type is used in the U.S. dry- and wet-milling industries. Flint maize, on the other hand, contains vitreous endosperm at the kernel crown and is generally harder than dent maize. Floury, or soft, kernels have no vitreous endosperm and are used for only a few kinds of prepared food.

In general, dry millers require a fairly hard maize kernel with a high percentage of vitreous endosperm to obtain the maximal amount of the desired flaking grit product. Accordingly, "food-type" hybrids have been developed in the United States to have harder, more vitreous kernels than the typical dent maize grown in the central Corn Belt region. Kernels, however, should not be so hard as to shatter during milling. Tests used to predict dry-milling performance include (a) the kernel density, determined by gas pycnometer or alcohol displacement; (b) the milling evaluation factor, which uses a scaled-down degerminator to measure flaking grit and total milled product yields; and (c) kernel hardness, assessed by the tangential abrasive dehulling method or by the Stenvert hardness test.

U.S. dent maize is typically somewhat soft for exporting, as it tends to break more easily than harder phenotypes during shipping and handling (8). Because of export considerations, as well as dry-milling performance, a number of public and private breeding programs have been actively developing dent germplasm with harder kernel texture (9). However, dent maize for general export has not seen large improvements in kernel hardness. Currently, farmers either grow food maize hybrids on contract or sell them with premiums to "short lists" of dry millers, large cereal processors, or commodity brokers. Through these avenues, preservation of grain identity is achieved. However, in the general marketplace, handling facilities for identity-preserved grain are not yet established.

In contrast to the dry-milling industry, the maize wet-milling industry has no interest in increased kernel hardness. In wet milling, the main objective is to obtain a maximal yield of starch. For this purpose, the desired kernel characteristics are a high starch content and efficient release of starch during the milling process. Wet millers currently purchase maize on the open market and thus do not discriminate on the basis of kernel type. However, a number of seed companies and university research programs are working on developing improved seed types for wet milling (10), through both classical and molecular approaches. It is generally thought that softer kernels with dense starch packing would result in high-starch wet-milling yields.

Maize kernel types that are optimal for alkaline-cooking processes are of medium, but consistent, hardness, with few stress cracks (microfissures that can appear during grain drying) (11). Ease of pericarp removal during the washing of the nixtamal is another important factor and is commonly selected for in breeding programs. Both yellow and white maize hybrids are used by masa processors.

## V. FACTORS RELATED TO KERNEL TEXTURE

Although the development of improved maize kernel types seems quite straightforward through selection for improved phenotypes, the molecular basis of kernel hardness is not yet well understood. A number of investigators (12–14) have studied the endosperm protein fraction as it relates to kernel hardness, under the assumption that the "matrix protein" that surrounds starch granules is responsible for formation of the vitreous endosperm portion of the kernel. This approach seems reasonable, because as the physiologically mature seed begins to dessicate, protein around starch granules should compress them, leaving virtually no air spaces and, consequently, yielding a compact, potentially translucent endosperm. On the other hand, in the center of the endosperm, either matrix protein is not as extensive as toward the periphery, or it is broken apart by internal forces within the seed during drydown. This leaves a floury, or opaque, center in a dent or flint kernel. As mentioned, dent maize also has a floury crown region. Mutant maize cultivars exist that are entirely floury, and starch packing in these phenotypes is not dense.

The best evidence that a particular maize endosperm protein influences kernel hardness is from studies on the changes that occur in the conversion of floury, *opaque2* maize to the hard kernel type known as *quality protein maize* (QPM). A correlation was shown between γ-zein content and amount of endosperm modification in QPM (15). In QPM, γ-zein, the storage protein found at the periphery of protein bodies, was found to be elevated to two to three times that of normal maize (16). The fact that γ-zein contains a large amount of cysteine (about 7 mol %) led to the speculation that increased levels of this protein create more disulfide bonds among the matrix proteins, thereby resulting in formation of vitreous endosperm. In general, however, the factor or factors controlling the formation of vitreous, hard endosperm in normal maize cultivars are not clear. The possibility of incorporating the modifying genes of QPM into normal maize genetic backgrounds has been explored (17).

## VI. BIOCHEMICAL COMPONENTS: THEIR STRUCTURE AND POTENTIAL FOR CHANGE

### A. Starch

Maize starch, like other cereal starches, is composed of two types of molecules, amylose and amylopectin. Amylose is the smaller molecule and has an average molecular weight in the range of $10^6$ Da (18). It is primarily a linear polymer of α-1,4 linked glucopyranosyl units with some α-1,6 linked branches of short linear chains. Despite the branches, amylose behaves in food systems much as a linear molecule, such that in its dispersed form after starch gelatinization it aligns and reassociates on cooling to form a gel. Amylose exists in an α-helical conformation that, because of its relatively hydrophobic interior, can form inclusion complexes with a number of lipid compounds, as well as some other amphipathic molecules. In normal maize, amylose comprises around 20–25%, by weight, of the total starch. Because it is much smaller in size than amylopectin, actual molar amounts of amylose are higher than those of amylopectin.

Amylopectin is a very large molecule on the order of $10^7$ or even $10^8$ Da and is highly branched, with α-1,6 linkages creating a "bushy," though fairly two-dimensional, molecule (18). The commonly accepted model of amylopectin (19) consists of three groups of linear α-1,4 linked glucopyranosyl chains (A, B, and C linear chains) of different length bound together by α-1,6 linked branch points. The A chains, with a degree of polymerization ($DP_n$) of about 15, are the shortest of the linear chains and are most abundant. They are found at the terminal nonreducing ends of the branched structure and, therefore, are linked by only one α-1,6 bond to a B chain. B chains are internal chains of larger and more variable length ($DP_n$ of about 45 to more than 100) (20). Short B chains exist within a "cluster" of branched chains in the molecule, whereas longer B chains span clusters. B chains contain two or more α-1,6 branch-point linkages. The single C chain contains the one reducing end of the molecule.

Starch molecules are assembled into granular structures in the amyloplast during seed development. Amylopectin forms crystallinelike growths within the starch granule, whereas amylose is thought to exist primarily in amorphous regions of the granule. Assembly and structure of the starch granule are reviewed by Gallant and coworkers (21).

Functionality of maize starch for food and industrial uses depends on the ratio of amylose to amylopectin, the fine structural details of the two components, interaction and complexation with other components, and the behavior of the starch granule itself. It has long been known that starches with varying amylose contents behaved differently. The high amylose, or *ae,* maize mutant contains upward of 60% amylose and is used in applications in which strong gels or films are desired. This is due to the high retrogradation tendency of amylose. It is now known that amylopectin fine structure of *ae* maize also differs from normal, in that it has a higher proportion of long B chains, which makes the molecule more "amyloselike" (22). On the other hand, *waxy (wx)* starch mutants are nearly devoid of amylose and form highly viscous pastes during gelatinization that tend to break down in food systems but do not gel on cooling. Waxy maize starches are commonly crosslinked by using phosphoryl choride or sodium trimetaphosphate to increase the integrity of the swollen, gelatinized granule, so it will not break down when shear force is present. In between these two wide ranges of amylose content are normal maize starches, whose functionality in food systems is influenced by less dramatic changes in amylose content. For instance, higher-amylose rice starches tend to produce harder, less sticky cooked rice than starches with somewhat low amylose content (23).

Recent research has focused on the question of how fine structural differences of amylopectin affect starch functionality and how changes in structure can be made to create starches that will behave optimally for specific applications. Starch wet-milling companies and starch processors have expressed interest in the creation of natural starches that behave as chemically modified starches. In a study conducted at Iowa State University by Johnson and Baumel (24) on estimates of economic gain due to genetic modifications to maize, a number of potentially useful starch modifications were noted. The authors did not take into account the feasibility of making such changes but instead concentrated on what the economic consequences would be. In terms of starch fine structure, they reported that lengthening the A and short B chains of amylopectin could improve gel formation, a modification that would be useful in photographic film and other applications; the potential market impact is about $1 billion per year. Introduction of aldehyde groups into amylose or amylopectin would provide potential crosslinking sites for improved resistance to paste breakdown by shear or acid, with application in the food and paper industries and a potential market impact of about $300 million per year. Introduction of acetylation into amylopectin to improve low-temperature paste stability and resistance to retrogradation and to improve adhesive properties has an estimated market impact of $240 million per year. Introduction of cationic groups into either amylose or amylopectin would increase paste viscosity and water-holding capacity and could be useful in paper sizing, wastewater treatment, and protein interactions in food and pharmaceuticals. This has a potential impact estimated at $60 million per year. Additional examples with lower estimates of economic impact include the following: the reduction of number of intermediate and long B chains in amylopectin (estimate of $44 million per year); introduction of 0-methyl, 0-acetyl, 0-succinyl, 0-glysocyl, or 0-galactosyl substituents in either amylose or amylopectin ($35 million per year); increase in molecular weight of amylose ($26 million per year); alteration of 1,4 linkages from α to β in amylopectin ($20 million per year); introduction of anionic groups into amylose or amylopectin ($20 million per year); increase in amylose to greater than 90% ($16 million per year); decrease in amylose to 5–10% ($16 million per year); increase in phytoglycogen (the highly branched precursor to amylopectin) to greater than 90% ($16 million per year); decrease in molecular weight of amylose ($14 million per year); shortening of A and short B chains of amylopectin ($5 million per year); replacement of starch with cyclodextrin ($2 million per year); and increase in phytoglycogen to greater than 25% ($2 million per year). In the animal feed area, the authors estimated that increasing starch digestibility to 100%, for example, by decreasing starch crystallinity, would have a potential market impact of almost $1.5 billion per year. Many of these alterations to starch would be impossible in terms of our current knowledge of starch biosynthesis (25). Yet, some specific changes to molecular fine structure and ratio of starch fractions have been achieved either through identification of naturally occurring maize starch mutants or through molecular techniques.

Naturally occurring maize starch mutants are many, and it is beyond the scope of this review to cover them in detail. A list of notable starch mutants follows:

- *ae*—amylose extender
- *wx*—waxy
- *bt*—brittle
- *sh*—shrunken
- *fl*—floury
- *du*—dull
- *su*—sugary

Most of the mutant phenotypes, as well as the resulting functionality changes, are caused by aberrations in the starch synthesizing enzymes that result in amylose/amylopectin ratio changes and/or fine structural changes in the starch molecules (26). *Ae* maize has a high amylose content and a higher proportion of long B chains in its amylopectin than normal maize starch, because of its lack of branching enzyme IIb (27). The *wx* mutant lacks granule-bound starch synthase (or waxy protein) responsible for the synthesis of amylose. Both the *bt* and *sh* maize mutations have low activity of adenosine diphosphate– (ADP)-glucose pyrophosphorylase, with the two genes encoding the two subunits of the enzyme (28,29). The *su*1 rice starch mutant produces highly branched structures, similar to phytoglycogen, that are caused by a reduction in the activity of the starch debranching enzyme (referred to as the *R-enzyme*) (30,31).

Double mutants have also been developed, including *ae, du; ae, su$_2$; ae, wx; du, h* (dull horny); *du,su$_2$*; and *du,wx*. Double mutants of waxy include *wx,fl$_1$* and *wx,sh$_1$*. A triple mutant, *ae,du,sh$_1$*, has purported special functional properties. Altering the starch synthesizing enzymes through such variables as activity, heat stability, and location will affect the functionality of the starch and ultimately can lead to novel starches as well as natural starches that are optimally suited for specific applications.

### B. Protein

In common maize, protein makes up, on average, about 9% of total kernel weight and about 8% of endosperm weight. Protein content, however, can vary considerably among maize cultivars. Because of the high amount of glutamine in the storage protein zein, the nitrogen-to-protein conversion factor of maize is a fairly low 5.7 (32). There are currently few commercial outlets for maize protein (gluten feed/meal), other than animal feed supplementation. A small market exists for purified zein, and it is produced and sold by one U.S. company. Zein has application in moisture-resistant coatings for foods and pharmaceuticals, as a result of its hydrophobic nature and ability to form films.

Maize endosperm protein, in the simplest and perhaps most meaningful classification scheme, consists of zeins and nonzeins (16,33). After synthesis, zeins are translocated into the lumen of the endoplasmic reticulum (ER), where they accumulate and are packaged into rather insoluble entities called *protein bodies*. These accretions of proteins retain their structural integrity during maturation and seed desiccation. Zeins comprise four basic subclasses of polypeptides, α, β, γ, and δ. Overall, zeins comprise about 70% of endosperm proteins in normal maize cultivars (33). The major storage protein is α-zein, which accounts for about 80% of the total zein fraction. The remaining zeins are mostly β and γ-zeins. The relatively more hydrophilic γ-zein is highly disulfide-linked (7 mole percent cysteine) and is located at the periphery of the protein body. It serves as a structural protein, promoting protein body formation and retention of α-zeins in the ER. The nonzein fraction is a heterogeneous group of proteins that comprises cytoplasmic and cell wall proteins that have various metabolic, transport, and structural roles in endosperm cells.

A number of maize mutants that affect zein proteins have been identified (reviewed in Ref. 34). The best known is the *opaque2* mutant, which causes elevated lysine and tryptophan content (35). The *opaque2* gene encodes a transcription factor that regulates synthesis of genes encoding the 22-kDa α-zein protein and a ribosomal inactivating protein (36,37), among others. The mutation causes reduced synthesis of zeins, resulting in a lower proportion of the major lysine-deficient protein in the seed. Additionally, a pleiotropic increase in lysine-containing nonzein proteins can occur (38). A higher nonzein content, together with lower zein content, produces seeds with appreciably higher levels of lysine and tryptophan. The enhanced synthesis of lysine-containing proteins was found to correlate with the levels of a cytoplasmic protein, elongation factor 1-α (eEF1A) (39). An enzyme-linked immunosorbent assay (ELISA) measuring eEF1A that permits rapid screening of breeding lines for lysine content has been developed.

The floury kernel phenotype of *opaque2* is unacceptable for preparation of most types of maize-based foods, and it increases susceptibility to insect and fungal infestation. Modifier genes were introduced by workers at CIMMYT in Mexico (40) and in South Africa (41) that effectively convert the soft *opaque2* kernel to the vitreous, hard phenotype. The modified *opaque2* mutants were designated *QPM*. As discussed, studies by Larkins and colleagues (15,16) showed that higher synthesis of γ-zein was related to the formation of the hard endosperm in these genotypes.

Few other protein mutants of maize have been as well characterized as *opaque2*. The *floury2* mutant produces abnormally shaped protein bodies that result from the accumulation of an unusual 22-kDa α-zein protein. Coleman and associates (42) showed this to be due to a mutation in the signal peptide sequence of a 22-kDa zein gene, preventing cleavage of the protein and its translocation into the lumen of the ER. *Floury2* causes a reduction in the synthesis of all classes of zein proteins, overexpression of BiP, an ER-resident chaperonin (43), and elevated lysine content. The presence of α-zein protein at the surface of the protein bodies results in somewhat higher in vitro protein digestibility. This was also observed in an *opaque* mutation in sorghum that causes highly irregularly shaped protein bodies (44).

Potential U.S. economic benefit of improving protein quality, quantity, and availability was estimated by Johnson and Baumel (24). The primary motivation for U.S. research into improving maize proteins is increasing the value of the grain for animal feed, although in developing countries with high maize consumption, improvements in nutritional value would benefit people as well. The economic predictive study found that increasing protein content would potentially lead to the greatest increase in value. If protein were increased from 8.7% to 16.7%, a potential market impact of $3.4 billion per year could be realized. Increasing protein digestibility of maize an additional 10 percentage points (84% to 94% in poultry, 82% to 92% in swine, and 73% to 83% in cattle) could add $840 million per year. If lysine content were doubled from 0.3% to 0.6% (seed basis) with a concomitant increase in protein by 0.3%, the authors predict a benefit of $490 million per year. Increasing lysine without increasing protein would reduce this to $370 million per year. An increase in the high-methionine-containing zeins could enhance value by $360 million per year. Increasing albumin protein and sulfur-containing amino acids could increase maize grain value by $310 and $260 million per year, respectively. As in the potential economic impact assessment of maize starch alterations, the authors qualified their predictions by noting that it is not clear that these changes can be made. In fact, in the area of proteins, many of these changes do appear possible, either through use of existing mutants or through genetic engineering.

## C. Lipid

Maize has between 3.5% and 5% oil in the kernel, but high-oil genotypes are commercially available with greater than 6% oil. Six percent is the content at which animals have shown a significant positive feed response compared to response to normal maize (45). Oil contents have been reported up to 24% in an inbred, but high-oil maize routinely ranges up to 9% or 10%. Although there is a sizable commercial market for refined maize oil, most oil is used in livestock feed. The advantage of high-oil maize is economic, based on replacement of the oil added to feed rations. The increased cost of high-oil hybrid maize seed must therefore weigh positively against the market cost of supplemental feed fats and oil. As a high-energy human food, particularly for children and other nutritionally vulnerable groups, high-oil maize would be valuable in developing countries.

High-oil maize has been shown in numerous feeding trials to be advantageous to poultry and livestock. For chickens, the weight gain and feed conversion ratio were superior in diets using 6% to 13% oil maize compared to 4.5% oil maize. In growing-finishing pigs, high-oil maize (7.5%) performed better in terms of average gain (kilograms) per kilogram feed compared to normal maize (3.5% oil) (46).

In terms of oil composition, maize oil has a favorable fatty acid profile, considering its high level of about 62% linoleic acid (18:2) and fairly high level of about 24% oleic acid (18:1) (46). Additionally, this is a stable oil, because of the low level of less than 1% of linolenic acid (18:3) and relatively high levels of the antioxidant tocopherols. Triacylglycerols make up nearly 80% of unrefined corn oil, the remainder consists of phospolipids/glycolipids (about 9%), sterols (4.5%), di- and monoacylglycerols (4%), hydrocarbons/sterylesters (3%), and about 1% free fatty acids. As oil content increases, the proportion of triacylglyerols increases, and the fatty acid composition tends toward a somewhat higher level of oleic acid. Although the bulk of lipids in the maize kernel are found in the germ, the endosperm contains about 1% lipid. Although seemingly insignificant, endosperm lipid is subdivided about equally into nonstarch and starch fractions. Endosperm lipids are different in composition from the predominating germ lipids and can affect starch gelatinization behavior. Nonstarch lipids originate either in the aleurone layer that contains mainly triacylglycerols or in other endosperm cells in the form of free fatty acids. Starch lipids are composed of monoacyl lipids, mainly free fatty acids and lysophospholipid. There is evidence that these lipids complex with helical sections of the amylose chains (47).

There exists a wide variation in fatty acid profiles of maize lipids that is ascribed to genotypic variability. In an extensive study, oleic acid levels were found to range from 14% to 64%, and linoleic acid levels ranged from 19% to 71% (48). The general trend was that increasing one of these two fatty acids resulted in lowering of the other.

Opportunities exist to improve maize lipid composition through molecular approaches (49,50). Recent research in the nutrition field has shown that fatty acid composition of oils in the diet ultimately affects fat composition in meat. Because maize is the largest feed crop in the world, it is conceivable that future development of maize with a more desirable fatty acid profile would improve meat fatty acid profiles.

## D. Minor Components

### 1. Phytate

Phytic acid is synthesized in maize and other seed grains as a phosphorus storage and resupply sink, as well as, in the phytate form, a macrostore for the cations K, Mg, and Ca and a microstore for Mn, Fe, and Zn (51). Phytic acid is *myo*-inositol 1,2,3,4,5,6-hex*akis*phosphate. It has been of interest principally to nonruminant animal and human nutritionists, because it can bind essential cationic minerals in the diet and make them unavailable for use. Its presence in the diet, therefore, can lead to deficiencies of phosphorus and other minerals. For example, iron absorption is strongly inversely correlated to phytate content (52). More recently, a concern of environmentalists to reduce phosphorus runoff from animal feedlots has heightened interest in reducing phytate content in maize. Johnson and Baumel (24) estimate that an increase in available phosphorus content from 20% available to 60% (achieved through reduction in phytate levels) would translate to increased value for the U.S. maize crop of about $200 million per year.

In the maize seed, the majority of phytate (about 90%) is found in the germ scutellum and the remainder in the aleurone. Up to 90% of stored phosphate is in the form of phytate, and it is deposited in protein bodies. Raboy (51) identified two low–phytic acid maize mutants with reductions in kernel phytic acid phosphorus of 33% and 66%. The latter mutant, *lpa1*, showed a molar equivalent increase in inorganic phosphorus that would be available to the animal and good agronomic and yield traits (53). Preliminary chick feeding results from the latter study suggested that phosphorus was more available and, correspondingly, there was less phosphorus in excreta. Enhanced knowledge of the synthetic pathway leading to the phytic acid end product will permit the manipulation of the amount of phytate in the maize seed through molecular biology techniques.

### 2. Minerals

There is currently a great deal of interest in finding ways to increase the amount of minerals and vitamins in grains. This new focus has arisen out of studies conducted in developing countries that showed that deficiencies in key micronutrients may profoundly affect the growth and development of children (CRSP studies). The minerals most needed in developing countries are iron, iodine, and zinc.

Maize germ contains the majority of minerals in the seed. Kernel minerals present in highest quantity are phosphorus (largely in the phytic acid form) and potassium. Iodine, iron, and zinc are present on average at about 400, 30, and 15 mg/kg grain, respectively. Broad ranges of mineral content have been shown for these three essential minerals, particularly for iodine and iron, and, accordingly, there is now research directed at increasing content of iron in maize that is used in high-maize-consuming developing countries (G. Edmeades, personal communication, 1998).

### 3. Vitamins

Key vitamins deficient in diets of the world's poor are likewise targets for improvement in maize, as well as in other grains. Maize is relatively high in the lipid-soluble tocopherols, or vitamin E, and the B family of vitamins, in particular niacin, thiamine, pantothenic acid, pyridoxine, and riboflavin. In maize, niacin is present in a bound form, making it unavailable, though it can be released with alkaline treatment, such as is used in the masa process. Maize also has a significant amount of β-carotene, the provitamin A. There is a sufficient range in β-carotene content among inbreds and hybrids to be of interest in breeding programs focused on improving maize nutritional quality.

## VII. CREATING TRANSGENIC MAIZE

Although *Zea mays* possesses significant genetic diversity, there are many traits that cannot be altered through conventional breeding. However, there is immense potential for improving the food and feed value of the seed through genetic engineering. In the past few years, the potential of this technology has been validated through the creation of insect-resistant and herbicide-tolerant maize hybrids. The enhancement or creation of traits influencing nutritional value, milling properties, and product development will soon follow.

An essential step in creating transgenic maize was the discovery of methods to introduce genes into single cells and regenerate them into fertile plants. There are several review articles describing maize transformation (54,55), and the 1999 synopsis by C. L. Armstrong (56) presents a nice overview of the developments that led to the successful transformation and regeneration of maize plants (57,58). Unfortunately, transformation and regeneration of monocots did not prove to be as facile as in dicots, in which cocultivation of leaf disks or young embryos with *Agrobacterium tumefaciens* containing a genetically engineered Ti plasmid allowed the efficient production of a number of different transgenic crop species (59). Indeed, it required patient and diligent research for nearly 15 years to identify the right set of conditions to regenerate maize plants from single cells, develop a procedure for the delivery of genes into these cells, and find selectable marker genes to identify transformation events among them. Today, the creation of genetically engineered maize plants is a routine, though not trivial procedure, and it is practiced in many laboratories. Nevertheless, there remain many opportunities to improve maize transformation by extending the range of transformation-competent genotypes, increasing the number (efficiency) of events per transformation, delivering a smaller number (one or two) of transgene copies or better yet obtaining gene replacement via homologous recombination, and identifying effective selectable markers that do not rely on antibiotic or herbicide resistance.

Although transformation of maize protoplasts was shown to be possible (60,61), the most widely used procedure for creating transgenic plants utilizes callus obtained from young (9 to 10 days after pollination) embryos. Thus far, the best genotype for obtaining this regenerable tissue is a hybrid derived from a cross between A188 and B73. After embryo culture on induction media, a few totipotent cells in the scutellum become competent for regeneration and form a friable callus that produces somatic embryos. This so-called type II embryogenic callus provides the starting material for creating transgenic plants.

As a monocot, maize is not easily infected by *Agrobacterium tumefaciens,* and early attempts to use the Ti plasmid for gene delivery into maize cells were not terribly efficient (62,63). Consequently, the first experiments to engineer maize genetically relied on the ballistic delivery of deoxyribonucleic acid– (DNA)-coated metal particles into embryogenic cells via the "gene gun" or "helium gun" (64,65). This procedure proved to be highly successful, and most of the transgenic maize lines in production today were created by this technique (66–69). Alternative procedures for forcibly delivering DNA into maize cells involve vortexing with silicon carbide fibers or electroporation. Fundamentally, silicon carbide "whiskers," as they are called, work in much the some way as bombarded metal particles in that they penetrate the cell wall and carry DNA into the protoplast. Several laboratories have reported routine success with this technique, e.g., Frame and coworkers (70), but the procedure has not been widely exploited. Electroporation of cells creates holes in the plasma membrane, allowing DNA to enter the cell. This procedure has been used to transform intact and wounded maize tissues (71–73), but the efficiency and reproducibility of electroporation are not high, and consequently it has not seen widespread application.

The forcible introduction of DNA into plant cells by biolistics or silicon carbide whiskers can lead to stable gene transformation. In fact, it is possible to introduce multiple genes on different plasmids simultaneously. Consequently, the transgene of interest can be contained on one vector, while the selectable marker gene is carried on another (66). Cells competent for DNA transformation do not discriminate! However, the gene in the vector DNA can undergo several fates once it enters the cell, and this leads to a variety of arrangements and rearrangements, such that it is not uncommon for there to be multiple copies of the transgenes in a variety of organizations. Unfortunately, the presence of too many copies of a transgene can lead to a phenomenon called *gene silencing,* which can affect expression of the transgene as well as related endogenous genes (cosuppression) (74). The basis of gene silencing and cosuppression is poorly understood, but they appear to be related to the number and expression level of the transgenes. In order to obtain effective, stable expression of transgenes, it is important to identify transformed plants that contain a relatively small number of transgenes, i.e., one to three.

The problem of gene silencing as a consequence of multiple transgene insertions is much less common with transformation events mediated by the Ti plasmid of *Agrobacterium tumefaciens.* This may be because transfer DNA (T-DNA) transformation is a more highly regulated process than what occurs with naked DNA (75). This fact, along with the observation that large DNA fragments can be transferred via the T-DNA, provided incentive to develop cereal transformation mediated by the Ti plasmid. Initial success with this technique was reported for studies with rice (76,77), and the approach was quickly adapted for barley (78) and maize (79–81). A number of laboratories are currently using this technique, and in the future it could be that T-DNA–mediated transformation of maize with *Agrobacterium tumefaciens* will become the standard procedure for making transgenic plants.

Several selectable marker genes have been shown to work effectively for identifying maize transformants. One of the most commonly used is the *bar* gene, which confers resistance to phosphinothricin and bialaphos (57,58). Other selection systems are based on resistance to hygromycin via the *hpt* gene (82), resistance to kanamycin and G418 via neomycin phosphotransferase II (71), and resistance to glyphosate via EPSP synthase genes (83,84). Concern has been raised about the safety of incorporating antibiotic resistance genes into crop plants, as it might lead to the spread of resistance genes to other organisms. However, vertical transmission of antibiotic resistance genes has not been observed. Although early versions of some transgenic crops contain antibiotic resistance markers, these genes are becoming less commonly utilized. The development of alternative selection strategies, such as the green fluorescent protein (85), continues to be an active area of research for developing genetically engineered crops.

Methods for obtaining site-specific gene insertions are being developed for plants, including maize. One approach makes use of prokaryotic systems of site-specific genetic recombination; the other relies on homologous recombination. The bacteriophage P1 contains Lox sequences that are subject to recombination through the activity of the Cre protein (86); the Flp recombinase system targets FRT sequences (87). With both of these examples, the target sequences for recombination, Lox or FRT, are placed in the genome of one plant to provide the insertion site for a genetically engineered trait. The gene encoding this trait, which has been transformed into a second plant, is introduced by mating the two. Experimental results with a nonregenerable maize cell culture using the Flp recombinase system have provided evidence of the utility of this approach (88). Gene targeting by homologous recombination has been demonstrated for several eukaryotes, but to date there has only been one successful report for plants (89). The development of an efficient system for gene targeting by homologous recombination in plants would clearly be a milestone, as it would eliminate the problems associated with gene silencing and cosuppression. It appears that one of the main factors limiting this procedure is transformation efficiency, as it was only possible to obtain homologous recombination when a very large number of transformation events was created (89).

## VIII. MODIFIED TRAITS IN TRANSGENIC MAIZE PLANTS

In 1990 the first transgenic maize plants were reported (57), but there have been only a few publications describing genetically engineered traits in maize. One of the first success stories was the creation of plants resistant to insects. This was accomplished by expressing a gene encoding the *Bt* toxin of *Bacillus thuringiensis* (84). The systemic production of *Bt* in corn was found to provide a significant level of protection against the European corn borer, with a commensurate enhancement in yield. Another valuable trait introduced into crop plants, including maize, is resistance to herbicides, such as glyphosate (59). Although insect and herbicide resistance is mainly of value to the farmer because it reduces production costs, these traits clearly benefit the consumer by reducing prices (better yield!) and by causing less harm to the environment than conventionally applied pesticides and herbicides. It has also been found that, as a consequence of reduced insect damage, *Bt* corn has a lower incidence of fungal infection, resulting in lower levels of fumonisin, a potent toxin that can cause fatal diseases in livestock and esophageal cancer in humans (90). Consequently, *Bt* corn could be healthier to eat than normal maize varieties.

Thus far, only a few genes have been engineered for expression in developing maize seeds. Russel and Fromm (91) used promoters from several endosperm-specific genes to target expression of a β-glucuronidase (GUS) reporter protein successfully. In so doing, they demonstrated the utility of this approach. However, there are only two reports of maize seeds used to produce a genetically engineered protein. Hood and associates (68) produced avidin, a glycoprotein found in eggs, and Zhong and colleagues (69) produced aprotinin, a serine protease inhibitor that can be used as a therapeutic agent, in transgenic maize seeds. Neither the avidin nor the aprotinin gene constructs used in these studies utilized a seed-specific (embryo or endosperm) promoter, and only a small amount (0.1%) of extractable protein was obtained. However, it appears these experiments were primarily conducted to demonstrate the potential for using maize seed as a bioreactor to obtain commercially valuable proteins.

## IX. FUTURE PROSPECTS

A number of laboratories are currently engaged in experiments to modify the biochemical pathways of seeds in order to alter the nature of the protein, starch, and oil they contain (92). We earlier noted the potential economic value of this for maize (24). In a preliminary report (93), scientists at DuPont described results of experiments to increase the lysine content of maize seed by expressing genes encoding feedback-insensitive aspartokinase and di-hydropicolinic acid synthase by using endosperm- and embryo-specific promoters. Although the endosperm promoter was not effective, the embryo promoter resulted in seeds with 50% to 100% higher levels of free lysine. Seeds of these plants would be valuable for feeding livestock, but the lysine would likely be lost from processed foods prepared from such genotypes.

Although there are few reports describing experiments to modify the starch, protein, and oil content of maize seeds, there is clearly a great deal of work being done to accomplish these goals. Perhaps the most advanced research is occurring in industrial laboratories; that could explain the sparse record of published results. However, the rate of progress could also be explained on the basis of the time required to develop materials for analysis. It takes several generations to move the transgene into a genetic background that is sufficiently uniform to assess the gene's impact. On the basis of the technical successes that have been demonstrated with reporter and model gene experiments, it is clear that genetic engineering of polymer synthesis and biosynthetic pathways in the maize endosperm and embryo is possible.

As previously noted, there are many opportunities to improve the techniques for creating transgenic plants and regulating the expression of the transgenes within them, and these topics will clearly be important areas for future research. In addition, there remains much to learn about the biochemical characteristics of starch (25), oil (49), and storage protein (Coleman, 1997) synthesis, and the ways these polymers interact to create the key phenotypic characteristics (hardness, vitreousness) of the mature kernel and influence its processing during wet and dry milling. Likewise, the biochemical aspect of the functional characteristics of maize flour is poorly understood. Understanding the biochemical and genetic basis of these traits will require the systematic analysis of normal and mutant genotypes that influence the key phenotypic traits, and it will no doubt require a multidisciplinary research approach involving geneticists, biochemists, and food scientists. Nevertheless, with the rapid development of genomics tools (94), which include data bases of gene sequences and tagged mutants, we can anticipate that our understanding of the genetic and biochemical basis of key agronomic and food processing traits will increase rapidly. The substantial potential for genetic engineering to improve the nutritional value and utility of maize for making food products is tremendous, and there are many opportunities, providing the consumer embraces the products afforded by this technology (95).

## REFERENCES

1. Food and Agriculture Organization. Production Yearbook: FAO Statistics. Vol. 49. Rome: Food and Agriculture Organization, 1996.
2. Corn Annual 1999. Washington, DC: Corn Refiners Association, Inc., pp 1–17.
3. Food and Agriculture Organization. Maize in Human Nutrition. Rome: Food and Agriculture Organization, 1992, pp 1–13.
4. B. Fussell. The Story of Corn. New York: Alfred A. Knopf, 1992, pp 77–85.
5. LW Rooney, SO Serna-Saldivar. Food uses of whole corn and dry-milled fractions. In: SA Watson, PE Ramstad, eds. Corn: Chemistry and Technology. St. Paul, MN: American Association of Cereal Chemists, 1987, pp 399–429.
6. SO Serna-Saldivar, MH Gomez, LW Rooney. Food uses of regular and specialty corns and their dry-milled fractions. In: AR Hallauer, ed. Specialty Corns. Boca Raton, FL: CRC Press, 1994, pp 263–298.
7. JB May. Wet milling: Process and products. In: SA Watson, PE Ramstad, eds. Corn: Chemistry and Technology. St. Paul, MN: American Association of Cereal Chemists, 1987, pp 377–397.
8. MR Paulsen, LD Hill, GC Shove, TJ Kuhn. Corn breakage in overseas shipments to Japan. Am Soc Agric Eng (Microfiche collection, fiche no. 87-6044), 1987, p 26.
9. KJ Cavanaugh, BE Zehr, WE Nyquist, BR Hamaker, PL Crane. Responses to selection for endosperm hardness and associated changes in agronomic traits after four cycles of recurrent selection in maize. Crop Sci 35:745–748, 1995.
10. BE Zehr, SR Eckhoff, WE Nyquist, PL Keeling. Heritability of product fractions from wet milling and related properties of maize grain. Crop Sci 36:1159–1165, 1996.
11. SO Serna-Saldivar, MH Gomez, LW Rooney. Technology, chemistry, and nutritional value of alkaline-cooked corn products. In: Y Pomeranz, ed. Advances in Cereal Science and Technology. St. Paul, MN: American Association of Cereal Chemists, 1990, pp 243–306.
12. JW Paulis, AJ Peplinski, JA Bietz, TC Nelsen, RR Bergquist. Relation of kernel hardness and lysine

to alcohol-soluble protein composition in quality maize hybrids. J Agric Food Chem 41:2249–2253, 1993.
13. RC Pratt, JW Paulis, K Miller, T Nelsen, JA Bietz. Association of zein classes with maize kernel hardness. Cereal Chem 72:162–167, 1995.
14. GH Eyherabide, JL Robutti, RS Borras. Effect of near-infrared transmission-based selection on maize hardness and composition of zeins. 73:775–778, 1996.
15. MA Lopes, BA Larkins. Gamma-zein is related to endosperm modification in Quality Protein Maize. Crop Sci 31:1655–1662, 1991.
16. JC Wallace, MA Lopes, E Paiva, BA Larkins. New methods for extraction and quantitation of zeins reveal a high content of γ-zein in modified opaque-2 maize. Plant Physiol 92:191–196, 1990.
17. GL Moro, MA Lopes, JE Habben, BR Hamaker, BA Larkins. Phenotypic effects of *opaque2* modifier genes in normal maize endosperm. Cereal Chem 72:94–99, 1995.
18. RL Whistler, JN BeMiller. Carbohydrate Chemistry for Food Scientists. St. Paul, Eagan Press, 1997, pp 117–151.
19. S Hizukuri. Polymodal distribution of the chain lengths of amylopectins, and its significance. Carbohydr Res 189:227–235, 1986.
20. Y Takeda, T Shitaozono, S Hizukuri. Molecular structure of corn starch. Starch/Staerke 40:51–54, 1988.
21. DJ Gallant, B Bouchet, PM Baldwin. Microscopy of starch: Evidence of a new level of granule organization. Carbohydr Polymers 32:177–191, 1997.
22. JD Klucinec, DB Thompson. Fractionation of high-amylose maize starches by differential alcohol precipitation and chromatography of the fractions. Cereal Chem 75:887–896, 1998.
23. BO Juliano. Varietal impact on rice quality. Cereal Foods World 43:207–222, 1998.
24. LA Johnson, CF Curtis. Economic values of genetic modifications for corn and soybeans. Available at www.exnet.iastate.edu/Pages/gram/98cornst/98cornst.html
25. LC Hannah. Starch synthesis in the maize seed. In: BA Larkins, IK Vasil, eds. Advances in Cellular and Molecular Biology of Plants. Vol. 4 Cellular and Molecular Biology of Plant Seed Development. Boston: Kluwer, 1997, pp 375–406.
26. Nelson, D Pan. Starch synthesis in maize endosperms. Annu Rev Plant Physiol Plant Mol Biol 46:475–496, 1995.
27. CD Boyer, J Preiss. Evidence for independent genetic control of the multiple forms of maize endosperm branching enzymes and starch synthases. Plant Physiol 67:1141–1145, 1981.
28. DB Dickinson, J Preiss. Presence of ADP-glucose pyrophosphorylase activity in *shrunken2* and *brittle2* mutants of maize. Plant Physiol 49:1058–1062, 1969.
29. LC Hannah, OE Nelson. Characterization of ADP-glucose pyrophosphorylase from *shrunken-2* and *brittle-2* mutants of maize. Biochem Genet 14:547–560, 1976.
30. Y Nakamura, T Umemoto, N Ogata, Y Kuboki, M Yano, T Sasaki. Starch debranching enzyme (R-enzyme or pullulanase) from developing rice endosperm: purification, cDNA chromosomal localization of the gene. Planta 199:209–218, 1996.
31. MK Beatty, A Rahman, H Cao, W Wookman, M Lee, AM Myers, MG James. Purification and molecular genetic characterization of APU1, a pullulanase-type starch-debranching enzyme from maize. Plant Physiol 119:255–266, 1999.
32. J Mosse. Nitrogen to protein conversion factor for ten cereals and six legumes or oilseeds: A reappraisal of its definition and determination: Variation according to species and to seed protein content. J Agric Food Chem 38:18–24, 1990.
33. BR Hamaker, AA Mohamed, JE Habben, CP Huang, BA Larkins. Efficient procedure for extracting maize and sorghum kernel proteins reveals higher prolamin contents than the conventional method. Cereal Chem 72:583–588, 1995.
34. CE Coleman, JM Dannenhoffer, BA Larkins. The prolamin proteins of maize, sorghum and *Coix*. In: BA Larkins, IK Vasil, eds., Cellular and Molecular Biology of Plant Seed Development. Boston: Kluwer Academic, 1997, pp 257–288.
35. ET Mertz, LS Bates, OE Nelson. Mutant gene that changes protein composition and increases lysine content of maize endosperm. Science 145:279–280, 1964.

36. RJ Schmidt, FA Burr, MJ Aukerman, B Burr. Maize regulatory gene opaque-2 encodes a protein with a 'leucine-zipper' motif that binds to zein DNA. Proc Natl Acad Sci USA 87:46–50, 1990.
37. S Lohmer, M Maddaloni, M Motto, N DiFonzo, H Hartings, F. Salamini, RD Thompson. The maize regulatory locus *Opaque-2* encodes a DNA-binding protein which activates the transcription of the b-32 gene. EMBO J 10:617–624, 1991.
38. JE Habben, AW Kirleis, BA Larkins. The origin of lysine-containing proteins in *opaque-2* endosperm. Plant Mol Biol 23:825–838, 1993.
39. JE Habben, GL Moro, BG Hunter, BR Hamaker, BA Larkins. Elongation factor-1α is highly correlated with the lysine content of maize endosperm. Proc Natl Acad Sci USA 92:8640–8644, 1995.
40. E Villegas, SK Vasal, M Bjarnason. Quality protein maize—what is it and how was it developed? In: ET Mertz, ed. Quality Protein Maize. St. Paul: American Association of Cereal Chemists, 1992, pp 27–48.
41. HO Gevers, JK Lake. Development of modified opaque-2 maize in South Africa. In: ET Mertz, ed. Quality Protein Maize. St. Paul: American Association of Cereal Chemists, 1992, pp 49–78.
42. CE Coleman, MA Lopes, JW Gillikin, RS Boston, BA Larkins. A defective signal peptide in the maize high-lysine mutant floury 2. Proc Natl Acad Sci USA 92:6828–6831, 1995.
43. RS Boston, EBP Fontes, BB Shank, RL Wrobel. Increased expression of the maize immunoglobulin binding protein homolog b-70 in three zein regulatory mutants. Plant Cell 3:497–505, 1991.
44. BR Hamaker, MP Oria, CA Weaver, JD Axtell. Improving sorghum nutritional quality. In: BA Larkins, ET Mertz, eds. Quality Protein Maize. Purdue University, 1997, pp 277–291.
45. RJ Lambert. High-oil corn hybrids. In: AR Hallauer, ed. Specialty Corns. Boca Raton, FL: CRC Press, 1994, pp 123–145.
46. EJ Weber. Lipids of the kernel. In: SA Watson, PE Ramstad, eds. Corn: Chemistry and Technology. St. Paul: American Association of Cereal Chemists, 1987, pp 311–349.
47. WR Morrison, TP Milligan, MN Azudin. A relationship between the amylose and lipid contents of starches from diploid cereals. J Cereal Sci 2:257–271, 1984.
48. MD Jellum. Fatty acid composition of corn (*Zea mays* L.) as influenced by kernel position on ear. Crop Sci 7:593–595, 1970.
49. J Browse. Synthesis and storage of fatty acids. In: BA Larkins, IK Vasil, eds. Cellular and Molecular Biology of Plant Seed Development. Boston: Kluwer Academic, 1997, pp 407–440.
50. E Krebbers, R Broglie, B Hitz, T Jones, N Hubbard. Biotechnological approaches to altering seed composition. In: BA Larkins, IK Vasil, eds. Cellular and Molecular Biology of Plant Seed Development. Boston: Kluwer Academic, 1997, pp 595–633.
51. V Raboy. Accumulation and storage of phosphate and minerals. In: BA Larkins, IK Vasil, eds. Cellular and Molecular Biology of Plant Seed Development. Boston: Kluwer Academic, 1997, pp 441–477.
52. JD Cook, MB Reddy, J Burri, MA Juillerat, RF Hurrell. The influence of different cereal grains on iron absorption from infant cereal foods. Am J Clin Nutr 65:964–969, 1997.
53. DS Ertl, KA Young, V Raboy. Plant genetic approaches to phosphorus management in agricultural production. J Environ Qual 27:299–304, 1988.
54. CJ Mackey, TM Spencer, TR Adams, AP Kausch, WJ Gordon-Kamm, PJ Lemaux, RJ Krueger. Transgenic Maize. In: S Kung, R Wu, eds. Transgenic Plants. Vol 2. Academic Press, 1993, pp 21–33.
55. P Christou. Genetic transformation of crop plants using microprojectile bombardment. Plant J 2:275–281, 1992.
56. CL Armstrong. The first decade of maize transformation: A review and future perspective. Maydica 44:101–109, 1999.
57. WJ Gordon-Kamm, TM Spencer, ML Mangano, TR Adams, RJ Danes, WG Start, JV O'Brien, SA Chambers, WR Adams Jr, NG Willetts, TB Rice, CJ Mackey, RW Krueger, AP Kausch, PG Lemaux. Transformation of maize cells and regeneration of fertile transgenic plants. Plant Cell 2:603–618, 1990.
58. ME Fromm, F Morrish, C Armstrong, R Williams, J Thomas, TM Klein. Inheritance and expression of chimeric genes in the progeny of transgenic maize plants. Biotechnology 8:833–839, 1990.
59. CS Gasser, RT Fraley. Genetic engineering plants for crop improvement. Science 244:1293–1299, 1989.
60. S Omirulleh, M Abraham, M Golovkin, I Stefanov, MK Karabaev, M Mustardy, S Morocz, D Dudits.

Activity of a chimeric promoter with the double CaMV enhancer element in protoplast-derived cells and transgenic plants in maize. Plant Mol Biol 21:415–428, 1993.

61. MV Golovkin, M Abraham, S Morocz, S Bottka, A Feher, D Dudits. Production of transgenic maize plants by direct DNA uptake into embyrogenic protoplasts. Plant Sci 90:41–52, 1993.
62. ACF Graves, SL Goldman. The transformation of *Zea mays* seedlings with *Agrobacterium tumefaciens*. Plant Mol Biol 7:43–50, 1986.
63. JH Gould, RH Smith. Transformation systems for corn. In D Wilkinson, ed. Proceedings of the Forty-Fourth Annual Corn and Sorghum Industry Research Conference, Washington, DC: American Trade Association, Inc., 1989, pp 1–10.
64. JC Sanford, TM Klein, ED Wolf, N Allen. Delivery of substances into cells and tissues using a particle bombardment process. Particulate Sci Technol 5:27–37, 1987.
65. D Pareddy, J Petolino, T Skokut, N Hopkins, M Miller, M Welter, K Smith, D Clayton, S Pescitelli, A Gould. Maize transformation via helium blasting. Maydica 42:143–154, 1997.
66. TM Spencer, JV O'Brien, WG Start, TR Adams, WJ Gordon-Kamm, PG Lemaux. Segregation of transgenes in maize. Plant Mol Biol 18:201–210, 1992.
67. DD Songstad, CL Armstrong, WL Petersen, B Hairston, MAW Hinchee. Production of transgenic maize plants and progeny by bombardment of Hi-II immature embryos. In Vitro Cell Dev Biol Plant 32:179–183, 1996.
68. EE Hood, DR Witcher, S Maddock, T Meyer, C Baszczynski, M Bailey, P Flynn, J Register, L Marshall, D Bond, E Kulisek, A Kusnadi, R Evangelista, Z Nikolov, C Wooge, RJ Mehigh, R Hernan, WK Kappel, D Ritland, CP Li, JA Howard. Commercial production of avidin from transgenic maize: Characterization of transformant, production, processing, extraction and purification. Mol Breed 3:291–306, 1997.
69. GY Zhong, D Peterson, DE Delaney, M Bailey, DR Witcher, JC Register III, D Bond, CP Li, L Marshall, E Kulisek, D Ritland, T Meyer, EE Hood, JA Howard. Commercial production of aprotinin in transgeneic maize seeds. Mol Breed 5:345–356, 1999.
70. BR Frame, Pr Drayton, SV Bagnall, CJ Lewnau, WP Bullock, HM Wilson, JM Dunwell, JA Thompson, K Wang. Production of fertile transgenic maize plants by silicon carbide whisker-mediated transformation. Plant J 6:941–948, 1994.
71. K D'Halluin, E Bonne, M Bossut, MD Beuckeleer, J Leemans. Transgenic maize plants by tissue electroporation. Plant Cell 4:1495–1505, 1992.
72. CM Laursen, RA Krzyzck, CE Flick, PC Anderson, TM Spencer. Production of fertile maize by electroporation of suspension culture cells. Plant Mol Biol 24:51–61.
73. SM Pescitelli, K Sukhapinda. Stable transformation via electroporation into maize Type II callus and regeneration of fertile plants. Plan Cell Rep 14:712–716, 1995.
74. G Bruening. Plant gene silencing regularized. Proc Natl Acad Sci USA 95:13349–13351, 1998.
75. J Sheng, V Citovsky. *Agrobacterium*-plant cell DNA transport: Have virulence proteins will travel. Plant Cell 8:1699–1710, 1996.
76. M-T Chan, H-H Chang, S-L Ho, W-F Tong, S-M Yu. *Agrobacterium*-mediated production of transgenic rice plants expressing a chimeric α-amylase promoter/β-glucuronidase gene. Plant Mol Biol 22:491–506, 1993.
77. Y Hiei, S Ohta, T Komari, T Kumashiro. Efficient transformation of rice (*Oryza sativa* L.) mediated by *Agrobacterium* and sequence analysis of the boundaries of the T-DNA. Plant J 6:271–282, 1994.
78. S Tingay, D McElroy, R Kalla, S Fieg, M Wang, S Thornton, R Brettell. *Agrobacterium tumefaciens*-mediated barley transformation. Plant J 11:1369–1376, 1997.
79. J Dong, W Teng, WG Bucholz, TC Hall. *Agrobacterium* mediated transformation of Japonica rice. Mol Breed 2:267–276, 1996.
80. M Uze, J Wunn, J Puonti-Kaerlas, I Potrykus, C Sautter. Plasmolysis of precultured immature embryos improves *Agrobacterium* mediated gene transfer into rice (*Oryza sativa* L.) Plant Sci 130:87–95, 1997.
81. Y Ishida, H Saito, S Ohta, Y Hiei, T Komari, T Kumashiro. High efficiency transformation of maize (*Zea mays* L.) mediated by *Agrobacterium tumefaciens*. Nat Biotechnol 14:745–750, 1996.
82. DA Walters, CS Vetsch, DE Potts, RC Lundquist. Transformation and inheritance of a hygromycin phosphotransferase gene in maize plants. Plant Mol Biol 18:189–200, 1992.
83. AR Howe, F Tamayo, S Brown, C Armstrong, M Fromm, J Hart, S Padgette, G Parker, R Horsch. De-

velopment of glyphosate as a selectable marker for the production of fertile transgenic corn plants. In Vitro Cell Dev Biol 28:12A, 1992.
84. CL Armstrong, GB Parker, JC Pershing, SM Brown, PR Sanders, DR Duncan, T Stone, DA Dean, DL Deboer, J Hart, AR Howe, FM Morrish, ME Pajeau, WL Petersen, BJ Reich, R Rodriguez, CG Santino, SJ Sato, W Schuler, SR Sims, S Stehling, L Tarochione, ME Fromm. Field evaluation of European corn borer control in progeny of 173 transgenic corn events expressing an insecticidal protein from *Bacillus thuringiensis*. Crop Sci 35:550–557.
85. RY Tsien. The green fluorescent protein. Annu Rev Biochem 67:509–544, 1998.
86. DW Owe, SL Medberry. Genome manipulation through site-specific recombination. Crit Rev Plant Sci 14:712–716, 1995.
87. JE Dixon, AC Shaikh, PD Sadowski. The Flp recombinase cleaves Holliday junctions in trans. Mol Microbiol 18:449–458, 1995.
88. LA Lyznik, KV Rao, TK Hodges. FLP-mediated recombination of FRT sites in the maize genome. Nucleic Acids Res 24:3784–3789, 1996.
89. SA Kempin, SJ Liljegren, LM Block, SD Rounsley, MF Yanofsky, E Lam. Targeted disruption in Arabidopsis. Nature 389:802–803, 1997.
90. HI Miller. Biotech offers (baby) food for thought. Scientist 13:13, 1999.
91. DA Russell, ME Fromm. Tissue-specific expression in transgenic maize of four endosperm promoters from maize and rice. Transgenic Res 6:157–168, 1997.
92. PH Abelson, PJ Hines. The plant revolution. Science 285:367–389, 1999.
93. B Mazur, E Krebbers, S Tingey. Gene discovery and product development for grain quality traits. Science 285:372–375, 1999.
94. SP Briggs. Plant genomics: More than food for thought. Proc Natl Acad Sci USA 95:1986–1988, 1998.
95. G Gaskell, MW Bauer, J Durant, NC Allum. Worlds apart? The reception of genetically modified foods in Europe and the U.S. Science 285:384–389, 1999.

# 44

# The Transformation of Onions and Related Alliums

**Colin C. Eady**
*New Zealand Institute for Crop & Food Research Ltd., Christchurch, New Zealand*

| | | |
|---|---|---|
| I. | INTRODUCTION | 655 |
| II. | CHARACTERISTICS OF *ALLIUMS* SUITABLE FOR MODIFICATION BY TRANSFORMATION | 656 |
| | A. Resistance Traits | 656 |
| | B. Quality Traits | 659 |
| | C. Breeding Attributes | 660 |
| III. | HISTORY OF *ALLIUM* TRANSFORMATION RESEARCH | 660 |
| | A. *Agrobacterium* Species–Mediated Transformation | 661 |
| | B. Microprojectile Bombardment of Garlic *Allium sativum* L. | 664 |
| IV. | CONCLUSION | 665 |
| | REFERENCES | 666 |

## I. INTRODUCTION

Onion (*Allium cepa* L.) and its close relatives, garlic (*A. sativum*), leek (*A. porrum*), and chives (*A. schoenoprasum*), are members of the Liliaceae family. They have been grown as vegetable crops throughout the world since 3000 B.C. Allium vegetables are the fourth largest group of commercially produced, nonleguminous vegetables after potatoes, cassava, and tomatoes. Their combined annual production around the world was $53 \times 10^6$ MT in 1998 (1).

Despite the huge commercial importance of this vegetable group, biotechnological advances in *Allium* species have lagged behind those in other commercially important vegetable species. For example, the first routine *Solanum* species transformations were reported in 1985 (2) yet successful *Allium* species transformation was only reported in 1998 (3). There are many reasons for the slow application of biotechnological techniques to *Allium* species. First, *Allium* species are biennial monocotyledons, which require specific day-length and temperature requirements for normal growth. This characteristic has produced a range of different cultivars adapted to specific environments. Biotechnological protocols that can be successfully applied to particular cultivars may not be transferable to others. Second, some *Allium* species, e.g., garlic and shallot, are poorly fertile and predominantly maintained by vegetative propagation, restricting the use of sexual reproduction in biotechnological techniques. Third, many cultivars are open-pollinated and maintain high levels of heterozygosity and thus variability. This limitation leads to difficulties in obtaining clearly defined responses to biotechnological procedures, such as in vitro culture and selection systems. Fourth, crop alliums have very large genomes (4), which have been reported to be more difficult to transform (5). In addition, many molecular biological techniques, such as Southern hybridization, genomic cloning, and polymease chain reaction (PCR), are more difficult to apply to larger genomes. Finally, allium breeding has traditionally been confined to small, specialized companies that lack the resources needed for technologically advanced research.

Despite these facts, a few research groups have persevered with allium research and have made considerable contributions to our knowledge about allium genetics. Research in allium interspecific hybridization (6–9), clonal propagation (10), and sulfur metabolism (11–13) and the nature of cytoplasmic male sterility (14) is particularly noteworthy, as are recent advances in our understanding of onion cytoplasmic and nuclear genomes (15). Research activity is set to gain momentum since the purchase of many small, specialized allium seed companies by multinationals. These companies are investing in techniques such as dihaploid production, transformation, and biochemical assessment of specific characteristics (e.g., pungency) to complement traditional breeding methods. The application of these techniques to *Allium* species should allow the modification of traits that otherwise would be extremely difficult or impossible to improve by traditional breeding techniques. In addition, the increasing use of $F_1$ hybrid varieties from male sterile lines and improved molecular marking of cultivars allow exclusive ownership of the parent material, a commercially important factor. Finally, as latecomers to the biotechnology from which many crops have benefited allium researchers can take advantage of studies performed on other species since the 1980s. This capacity reduces the risk that this technology will fail to deliver economic returns.

This chapter summarizes the history of *Allium* species transformation research and outlines current techniques and the results of their application.

## II. CHARACTERISTICS OF *ALLIUMS* SUITABLE FOR MODIFICATION BY TRANSFORMATION

Genes are being identified from all life-forms and indeed being created de novo in order to find deoxyribonucleic acid (DNA) sequences that can confer beneficial characteristics for medical, agricultural, and industrial use. This is a complex and difficult process, and the genes available today will no doubt be but a small fraction of those available in the future. Genes that may confer benefits in *Allium* species may be identified by looking at successes in other crop species. In addition, some genes involved in specific *Allium* species attributes have been isolated, e.g., allinase (16,17) and sucrose-sucrose fructosyltransferase (18), creating the potential for such attributes to be modified by manipulating these genes. To date genes with a proven commercial application have not been put into *Allium* species, apart from the *bar* gene encoding resistance to the herbicide phosphinothricin (Eady et al., unpublished). A summary of some, but by no means all, of the traits that could be potentially altered in *Allium* species by transformation is presented in the following discussion.

### A. Resistance Traits

#### 1. Herbicide Resistance

The creation of *Allium* sp. lines with resistance to herbicide is now possible. The insertion of the *bar* gene, which provides resistance to phosphinothricin, has produced onion tissue capable of regenerating in the presence of phosphinothricin, the active ingredient in the herbicide Basta (Eady et al., unpublished). *EPSP* synthase genes (*AroA*) (19) encoding tolerance to glyphosate (Roundup), the *bxn* gene encoding resistance to Bromoxynil (20), or mutant AHAS or CsrL-L genes from *Arabidopsis* species that code for resistance to sulfonylurea herbicides (21,22) could also be introduced into *Allium* species.

Herbicide resistance is a particularly useful trait to introduce into onions because their slow germination rate, upright narrow leaves, and slow growth make them particularly sensitive to early season weed competition. Such competition can decrease yields by 70% or more (23). At present the only effective management of weeds is by strategic use of expensive herbicide cocktails applied at specific times throughout the growing season or by manual hoeing, which is difficult and labor-intensive, especially at the seedling stage.

## 2. Disease Resistance

Rabinowitch and Brewster reviewed viral, bacterial, fungal, and insect pathogens of *Allium* species in 1990 (24). Present management of pests and diseases of *Allium* species requires the use of integrated pest management (IPM) programs (25). Although genetic transformation is unlikely to eliminate the need to implement an IPM program, the approach could be simplified by transforming cultivars so that they better withstand the effects of pests or diseases.

*a. Fungal Resistance.* The major fungal pathogens of onion roots and bulbs include white rot (*Sclerotium cepivorum*), pink root rot (*Pyrenochaeta terrestris*), fusarium basal rot (*Fusarium oxysporum*), and neck rot (*Botrytis allii*). Presently, such pathogens are controlled by fungicides in combination with particular curing and storage regimens (26).

Control of onion white rot by fungicides is becoming increasingly difficult and other disease control options are being investigated. The application of diallyl disulfide (DADS) in fallow years to stimulate sclerotial germination shows promise but is expensive and difficult to apply. A limited degree of natural protection to onion white rot has been identified in some cultivars of onion and garlic (27). Outside the crop species, but within the *Allium* species family, only one source of complete onion white rot resistance has been identified in an accession of *A. stipitatum* (Kik, personal communication). Biological control, using *Trichoderma* species, has also had some success in areas of low to medium onion white rot infestation (Stewart, personal communication). Antifungal resistance genes could also be used to combat fungal pathogens. Such genes have been identified in plants (28). They include pathogenesis-related genes PR-2 and PR-3 (chitinases and glucanases) (29–31), which act synergistically to prevent hyphal growth (32); PGIP genes encoding polygalacturonidase-inhibiting proteins, which inhibit enzymes released by the fungus that break down the plant cell wall (33); PR-5 genes encoding ribosome-inactivating proteins, which specifically act on fungal ribosomes (34); and genes encoding plant defensins, a class of small polypeptides that interfere with fungal cell wall extension (35). An antimicrobial gene from onion has also been used to confer resistance to *Botrytis cinerea* (36). In addition, oxalate oxidase genes have been transformed into plants and shown to inhibit fungal invasion by detoxifying oxalic acid, the toxin produced by the fungi (37).

Our research team has isolated an *Allium* sp. root alliinase that may be responsible for the production of sulfur compounds that are released into the soil and stimulate dormant fungal sclerotia to germinate. Experiments are currently underway to determine whether antisense technology can be used to reduce the expression of this gene in the root, thus reducing the release of sulfur compounds into the soil and consequently reducing sclerotial germination. Nonplant genes that may combat invasion by onion fungal pathogens include the antimicrobial genes outlined in Sec. II.A.2.b. Unfortunately, host-pathogen interactions are often very specific and a screening program would be necessary to identify which, if any, of these gene products has activity against specific fungal pathogens of *Allium* species but lacks activity against beneficial VA mycorrhiza. Gene discovery programs are advancing rapidly and identifying new genes for applications such as conferring fungal resistance. Testing their potential is difficult and likely to hinder their introduction to *Allium* species.

   b. *Bacterial Resistance.* Bacterial diseases of onion, such as leaf blight (*Xanthomonas* spp.), soft rot (*Erwinia* spp.), and sour skin/slippery skin (*Pseudomonas* spp.), cause severe damage. The unpredictable nature of these essentially facultative pathogens, which take advantage of poor environmental conditions, makes the development of spray-based control strategies difficult (38). Controlled atmosphere storage can reduce the effects of these diseases. However, in poor seasons the disease may take hold while the crop is still in the field. Antimicrobial genes, like those encoding for small channel-forming peptides such as magainins and cercopins (39), have been added to potato extract and shown to have a potent antibacterial effect (40). T4 lysozyme is another antibacterial gene that has been demonstrated to confer tolerance to bacterial pathogens (41). An antimicrobial gene from onion has also been shown to be active against gram-positive bacteria (42). However, this research is not complete and a greater understanding of temporal and spatial expression profiles and engineering of the structural gene to produce stable proteins with greater activity will enable advances in this technology to be made in the future. In addition, as with antifungal genes, it is likely that new genes will soon be discovered with many potential antibacterial properties. Again methods to assess these genes rapidly for activity in vivo will be required to check the efficiency of the technology in *Allium* species.

   c. *Viral Resistance.* *Allium* species suffer from many viral infections (43). The worst affected are the vegetatively propagated *Allium* species such as garlic and shallot. Under vegetative propagation there is no "cleansing" sexual round of propagation to eliminate non-seed-transmitted viruses. Consequences are gradual virus build-up and significant decrease in yield. Potyviruses (e.g., onion yellow dwarf virus, leek yellow stripe virus), carlaviruses (e.g., garlic latent virus, shallot latent virus (43), and garlic and shallot virus X (44) are the most devastating *Allium* species viruses. Although in vitro elimination is possible (45,46), in-built resistance would provide a simpler solution. The development and insertion of engineered viral protein genes (e.g., P1, P3 (47), or CI (48)) or gene sequences (e.g., Nib (49) or CP (50)) have proved to be effective measures to prevent viral overload (51). Since 1996 researchers have isolated and sequenced coat protein gene sequences from *Allium carla* (52) and potyvirus types (53,54). With this knowledge it should be relatively straightforward to engineer and express these sequences in *Allium* species to induce resistance.

   d. *Insect Resistance.* The orders Hemiptera, Thysanoptera, Lepidoptera, Diptera, and Coleoptera all contain *Allium* species pests (55). Sap sucking onion thrip (*Thrips tabacii*) is considered to be the major insect pest problem for *Allium* species, with damage levels on untreated crops reaching up to 55%. As well as causing physical damage, they spread viral diseases (55). Thrips are difficult to control by conventional means, although biological control, partial plant resistance, and chemical control measures do exist. Thrips resistant to the major chemical control group, the synthetic pyrethoids, have been reported (Stewart, personal communication), presenting a serious control problem. The transformation of *Allium* species with genes conferring resistance to thrips could reduce dependence on the limited existing control measures. Insertion of the tryptophan decarboxylase gene from *Catharanthus roseus* into tobacco reduced whitefly (*Bemisia tabacii*) emergence by 98.5% compared to that in nontransgenic controls (56). The insertion and expression of protease inhibitors also reduced insect feeding (57), as has the insertion of lectin genes, e.g., from snowdrop (58). By directing expression of these genes to the sap through phloem-specific promoters (59) it may be possible to target these insects without expressing the foreign genes in other parts of the plant. For the Lepidoptera, Diptera, and Coleoptera orders (and maybe the Thysanoptera) the use of specific insecticidal protein genes from *Bacillus thuringiensis* (*Bt* genes) may provide a control strategy. Specific forms of the gene are effective against certain insects (60). *Bt* genes have been engineered and introduced into plants to confer resistance (61) and are currently being used to improve the commercial production of cotton and corn, among other crops.

*e. Nematode Resistance.* Many types of nematode are capable of infecting *Allium* species. However, they are not considered a major pest and are at present well controlled by soil fumigants (62). Genetic engineering does offer the opportunity to control parasitic *Allium* species nematodes, such as root knot nematodes, which feed from giant transfer cells that are induced by the nematode to form in the plant. Two strategies have been developed to combat these types of nematode (63). The first relies on expression of a gene whose product is directed against the nematode or its secretions; the second relies on the specific expression of a phytotoxic product in the giant transfer cells so that the nematode has no structure upon which to feed.

## B. Quality Traits

### 1. Pungency

The secondary metabolite pathway leading to the production of onion pungency has been studied in great detail since 1992 (11,64). Unlike ~90% of flowering plant species, the *Allium* species do not accumulate significant pools of organic sulfur as cysteine or methionine. Instead a unique secondary pathway converts cysteine to several forms of *S*-alk(en)yl-cysteine sulfoxides (ACSOs) (65). The cleavage of these ACSOs by the enzyme alliinase, upon disruption of the cell, produces volatile flavor, odor, and lachrymatory compounds ("pungency"), as well as pyruvate and ammonia (66). Manipulation of the levels of these compounds may allow mild or pungent onions to be produced. Our research team has isolated some of the alliinase genes responsible for this process (67) and is currently regenerating transformed plants containing antisense versions of the allinase gene in an attempt to modify onion pungency. Other enzymes involved in the production of the ACSOs are being investigated (68). Identification and manipulation of these genes may allow the more subtle modification of the various components that make up onion pungency. For example, specific oxidases that can oxidize *S*-allyl-L-cysteine, a precursor of *S*-allyl-cysteine sulfoxide, have been detected in garlic (69). Upon lysis of *Allium* species cells the ACSOs react with alliinase to produce thiosulfinates and a subsequent cascade of additional organosulfur compounds. Work is currently being undertaken at the University of Wisconsin by Irwin Goldman's group to determine which thiosulfinates are derived from which ACSOs and identify those responsible for particular health benefits such as antiplatelet activity (70). This biochemical understanding, together with a knowledge of the genes responsible, may one day make it possible to improve the nutritional value of alliums further.

### 2. Carbohydrates

*Allium* species are among only 15% of flowering plants (71) that store the majority of their carbohydrate reserves as fructans (fructose and sucrose polymers). Many reasons for this have been proposed, including roles as an osmoregulant, cryoprotectant, and a source of stored carbohydrate that can be rapidly remobilized during the breaking of dormancy (72). The ability of *Allium* species to produce and store complex fructans could be useful. Fructans are being seen as an increasingly healthy component of the human diet (73) and are also used in products such as wash softeners, biodegradable plastics, and, in the inulin form, artificial sweeteners (74). Researchers have already demonstrated that it is possible to manipulate fructan products and levels by the introduction of specific sucrose-sucrose-fructosyltransferases (SSTs) (18) or fructan-fructan-fructosyltransferases (FFTs) isolated from onion (75). This work has led to the production of fructan-containing beets. SSTs or FFTs from onion could be useful in the further customization of carbohydrate content in such beets. The manipulation of fructan assimilation or degradation in *Allium* species via transformation may affect the health benefits of these vegetables as well as the storage characteristics, solids content, and sweetness of *Allium* species crops.

### 3. Other Quality Characteristics

Color, size, shape, number of skins, storage abilities, solid content, quercetin level, and sweetness are desirable traits in onion. The ability to engineer any of these traits precisely in alliums does not yet exist, although soon it may possible to alter fructan composition to enhance some of these traits (outlined earlier). Many genes involved in color regulatory pathways have been characterized and successfully introduced into other plants (76). The very existence of white, yellow, and red onions indicates that color pathways are present and that it should be possible to introduce anthocyanin regulatory genes specifically to modify *Allium* species color.

## C. Breeding Attributes

### 1. Male Sterility

The major onion cultivars now being released are hybrids, derived from crosses with male sterile germplasm. The principal source of male sterile S cytoplasm can be traced to a single plant identified in 1925 in Davis, California. Although other sources have been reported (e.g., T cytoplasm), they require complex fertility restoration (14). Considerable effort is underway to find new sources of cytoplasmic male sterility (CMS) in onion (14) and leek (77,78). The transfer of CMS from onion to leek by protoplast hybridization experiments has been achieved. However, the functionality of such somatic fusions has not been clearly demonstrated (79).

The ability to engineer sterility in *Allium* species would remove the limitation of essentially a single source of CMS, thus greatly enhancing the potential for new hybrid seed production. Engineered sterility, based on the *barnase/barstar* genes (80), is being used in the production of hybrid oilseed rape and could be applied to onion hybrid seed production.

### 2. Other Breeding Attributes

Other attributes sought by onion breeders include clonal seed production through apomixis and ability to manipulate flowering. This latter characteristic could help in the production of hybrid seed, which is often unreliable as a result of asynchronous flowering. At present these characteristics are beyond the scope of *Allium* species genetic engineers. However, it may one day be possible to manipulate them as our understanding of the physiological, biochemical, and genetic characteristics of these systems improves.

## III. HISTORY OF *ALLIUM* TRANSFORMATION RESEARCH

Like other monocotyledonous crop plants, such as rice, maize, wheat, barley, and asparagus, *Allium* species have been and still are among the most recalcitrant agriculturally important species to transform.

A crop species may be considered recalcitrant to transformation if one or more of the following restrictions apply: (a) specific DNA delivery techniques are required; (b) transgenic plants can only be regenerated from specific tissues or cultivars; (c) posttransformation selection has to follow precise protocols; or (d) the transformation frequency of initial explants is low.

The most common transformation methods applied to recalcitrant crops are *Agrobacterium tumefaciens*–mediated transformation and microprojectile bombardment (81–84), although many alternative DNA delivery systems are available (85).

The tissue used for transformation in recalcitrant monocotyledonous species is usually derived from embryo cell types (86). These cells are very plastic with respect to their regeneration/differentiation potential. In addition, their cell wall structure may be more amenable to *Agrobacterium* species attachment than that of cells from other monocotyledonous tissue whose wall structure may inhibit *Agrobacterium* species attachment (87).

Initial transformation of such embryogenic cultures is usually achieved by dedifferentiating the tissue with an initial culture on a dedifferentiating induction medium (88). However, this is not always necessary. Work in our laboratory (unpublished) demonstrates that direct secondary somatic embryo production occurs to produce chimeric secondary embryos. In both systems it is apparent that the initial transgenic tissue is still dependent upon nontransgenic tissue. Killing the nontransgenic tissue too early can prevent the transgenic tissue from surviving (unpublished observations). Thus, precise selection is often required in recalcitrant crops to promote the growth of transgenic sectors until they are large enough to survive unaided, at which point selection levels can be increased.

The general requirements outlined for transforming recalcitrant crops have been recognized in *Allium* species for some time (10). More specific research has targeted the various aspects of *Allium* species transformation: DNA delivery systems have been investigated (89–91) and in some cases optimized (90,92). Both *Agrobacterium* species–mediated and biolistic delivery of DNA have been shown to be efficient in *Allium* species tissues (90–94). Regulatory sequences required for a high level of constitutive expression in *Allium* species have been identified (90,91, 94,95). In all cases the CaMV35s promoter was shown to drive the highest levels of expression in *Allium* species tissues. Regeneration systems suitable for *Allium* species transformation experiments have been developed (10) and have been improved for immature embryos (96,97), mature embryos (98–100), and protoplasts (101). Levels of appropriate selectable agents for selecting transformants have only been determined in one study (102). Such experimentation is fraught with difficulty as optimized selectable levels can only be accurately determined once transgenic plants are available. Refinements based on these initial determinations are now possible and have been facilitated by use of the green fluorescent protein gene (*gfp*), a viable reporter gene that allows transgenic cells to be observed in situ during such selection experiments (103).

This work has led to the first routine *Agrobacterium* species–mediated transformation protocol for onion from immature (94) and mature embryos (Kik et al., personal communication). In addition, two less reliable microprojectile transformation protocols for onion (92) and garlic (93) have been developed. Protocols for an *Agrobacterium* species–mediated transformation system and a microprojectile transformation system are outlined in the discussion that follows.

## A. *Agrobacterium* Species–Mediated Transformation

### 1. Introduction

Two methods for transforming *Allium cepa* by using *Agrobacterium tumefaciens*–mediated transformation have been developed. The method of Eady and colleagues involves the use of immature embryos and was developed with the aid of the green fluorescent protein reporter gene *gfp* (103). It has been repeated over successive seasons on several cultivars using a variety of T-DNA constructs (unpublished results). Seeds are currently being produced from bulbs generated by this transformation process in order to demonstrate the stability of inheritance. The method of the *Allium* species group at CPRO-DLO (Wageningen, The Netherlands) (unpublished) has only recently been developed. It uses mature embryo-derived cultures for *Agrobacterium* species–

mediated transformation. Transformation via this system has been identified by using the histochemical reporter gene, *gus* (104). Transformants are presently regenerating so that transformation can be confirmed by Southern analysis.

### 2. Transformation of Immature Embyros [94]

*a. Bacterial Strain and Plasmids.* *Agrobacterium tumefaciens* strain LBA4404, containing the binary vector pBIN m-gfp-ER (103) or pCambia (105), was used. Cultures were grown overnight, with shaking, in 50 ml Luria broth (LB) medium containing appropriate selective agents. The following morning cultures were replenished with an equal volume of LB containing antibiotic and 100 µM acetosyringone and grown for a further 4 hours. Agrobacteria were isolated by 10-minute centrifugation at 4500 rpm and resuspended in an equal volume of P5 (90) containing 200 µM acetosyringone.

*b. Transformation Procedure.* Field-grown umbels of *Allium cepa* L. were used as a source of immature embryos and isolated (90). Batches of 40 embryos were isolated, cut into ~1-mm lengths, transferred into 0.8 ml of agrobacteria, vortexed for 30 seconds, and placed under vacuum (~25 mm Hg) for 30 minutes. Immature embryos were then blotted dry on filter paper before transfer to P5 medium (~40 embryos per plate). After 6 days cocultivation embryo pieces were transferred to P5 containing 200 mg/l timentin and an appropriate selection agent (12 mg/l geneticin or 5 mg/l Basta). Embryo pieces were cultured in the dark under the same conditions described for the production of secondary embryos (97). Cultures were transferred to fresh medium every 2 weeks. After three to four transfers, growing material (90) was transferred to P5 plus 25 mg/l geneticin or 5 mg/l Basta, respectively, and grown for a further three to four transfers. During this time pieces of putative transgenic tissue that reached ~3 mm$^2$ and were identified as fluorescing green under blue light (475 nm) excitation were transferred to regeneration medium (97) plus 20 mg/l of geneticin or 5 mg/l Basta, respectively. Shoot cultures were maintained for 12 weeks and developing shoots transferred to 1/2 Murashige and Skoog (MS) medium (106) plus an appropriate selection agent to induce rooting. Rooted plants were either transferred to 1/2 MS plus 120 g/l sucrose to induce bulbing or transferred to soil in the greenhouse (12 hours 12°–23°C day, 12 hours 4°–16°C night). Bulbing in greenhouse-grown plants was induced in plants naturally by increasing daylength. After 50% of the tops had fallen, bulbs were lifted and air-dried. Bulbs greater than 45 mm in diameter were cold-stored for 4 weeks to induce floral meristems prior to planting for seed production.

*c. Analyses of Transformants.* For GFP expression, tissue was examined by observation under a fluorescence microscope (excitation 475 nm, emission 510 nm) (103). Larger tissues with high levels of expression were observed by using fluorescence stereomicroscopes or hand-held flourescent lanterns.

Studies with the *gfp* reporter gene demonstrated that transient gene expression is first observed after 3 days of cocultivation (Fig. 1a). Variation between embryo pieces was large, but good initial transfer was recorded if 30% of immature embryo pieces had >20 GFP-positive cells. Over 90% of transiently expressing cells stopped fluorescing during the first 4 weeks of selection. However, up to 16% of initial tissue pieces produced stable expressing sectors (Fig. 1b). Genotype, condition of the embryo, size of the embryo, cocultivation conditions, and selection parameters all affected stable sector establishment.

Transgenic sectors were transferred to regeneration medium after 10–16 weeks. They responded in the same way as nontransgenic, embryo-derived cultures and produced shoots (Fig. 1c). Up to 2.7% of stable sectors produced shoot cultures from which plants were obtained (Table 1). Shoot cultures placed on rooting medium containing a selective agent produced actively growing roots that fluoresced (Fig. 1d). In differentiated structures most fluorescence was

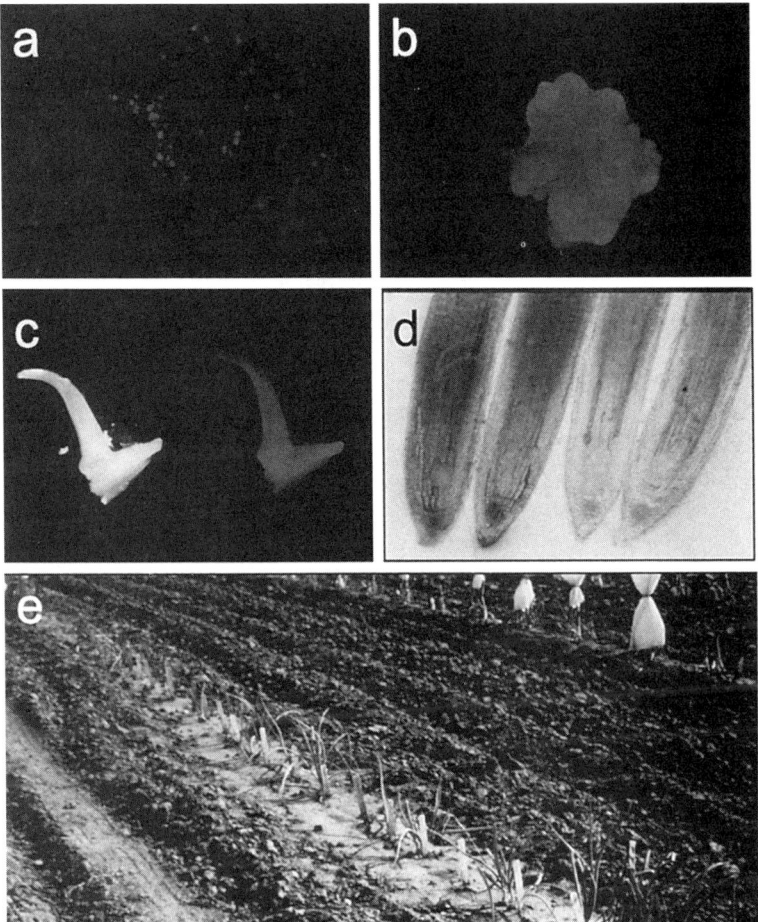

**Figure 1** (a) GFP expression in embryo tissue after 5 days of cocultivation (× 50). (b) Independent GFP positive tissue (× 5). (c) GFP positive onion shoot culture (× 5) under blue light (left) and white light (right). (d) Two GFP negative (left) and two GFP positive (right) roots (× 10). (e) Transgenic onion plant being grown in the field (× 0.3).

observed in root tips, whereas in shoots fluorescence was obscured by red autofluorescence from the chlorophyll and could only be seen in young leaves or stomatal guard cells.

Southern analysis was used to confirm transformation and transgene copy number of the regenerated onion plants (94) (Fig. 2). Approximately 80% of transgenic plants contained single-copy inserts, suggesting that integration of transgenes via *Agrobacterium* species–mediated transformation in *Allium* species is similar to that in other plant species. Chromosome counts in the primary transformants tested showed a diploid ($2n = 16$) chromosome complement (94).

Currently, these plants are being grown in the field (Fig. 1e) to maturity to assess the effect of transformation on phenotype and to generate seed to check the stability of the transgene in future generations. Genes encoding antisense alliinase and genes encoding resistance to the herbicide phosphinothricin have been transferred to onion embryos. Plants are in the process of being

**Table 1** Summary of Three Transformation Experiments

| Experiment | Number of embryos isolated | Embryos contaminated (%) | Stable GFP sectors after 8 wks | Independent plants | Positive Southern |
|---|---|---|---|---|---|
| 1 | ~360 | 40 | 15 (4.6) | 2 (0.6) | 1 of 1 tested |
| 2 | ~440 | 0 | 44 (10) | 12 (2.7) | 9 of 9 tested |
| 3 | ~520 | 60 | 11 (2.4) | 3 (0.7) | 2 of 2 tested |

Numbers in parentheses represent the percentage of transformants from uncontaminated embryos.

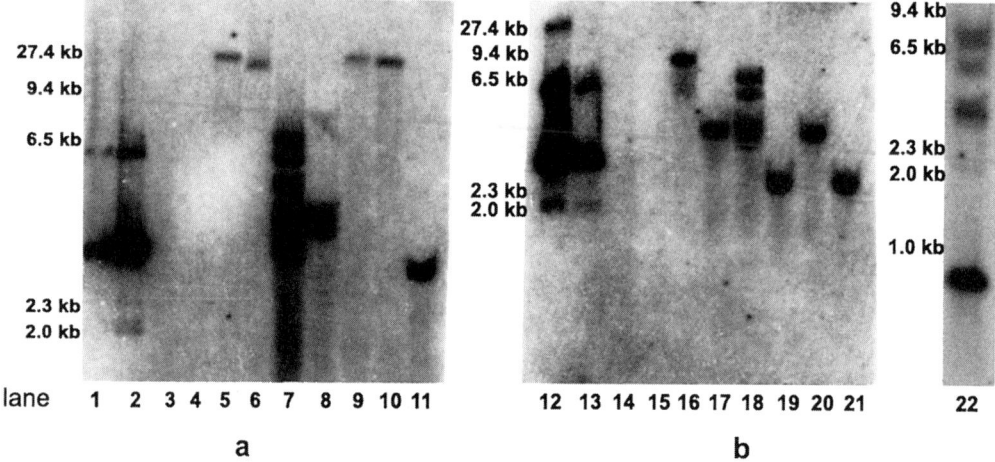

**Figure 2** (a) Southern analysis of *HindIII* digested DNA from primary transformants probed using the *gfp* gene: Bluescript plasmid containing the *gfp* gene (uncut) 1 copy number control (lane 1), 10 copy number control (lane 2), blank (lane 3), nontransgenic onion (lane 4), 7 transgenic onion plants (lanes 5–11), bluescript plasmid containing the *gfp* gene (uncut) 10 copy number control (lane 12), 1 copy number control (lane 13), blank (lane 14), nontransgenic onion (lane 15), 6 transgenic onion plants (lane 16–21). (b) Southern analysis of *EcoRI* digested primary transformants: transgenic onion (lane 22). Markers from lambda *HindIII* lanes (not shown).

regenerated from these embryos and will be analyzed for their altered pungency levels or their ability to grow in the presence of phosphinothricin.

## B. Microprojectile Bombardment of Garlic *Allium sativum* L.

### 1. Suspension Culture Preparation

Garlic suspension cultures of clone RAL27 (107) were collected by vacuum filtration onto filter paper and transferred on the paper to basal medium. After 1-day cultivation, cells were ready for bombardment.

## 2. Plasmid Deoxyribonucleic Acid

pBIN-121 (Clontech, Palo Alto, CA), containing the *gus* reporter gene and *nptII* selectable gene, was isolated by alkaline lysis and purified by using $CsCl_2$. Isolated plasmid was resuspended at 1 mg ml$^{-1}$ for use in bombardment.

## 3. Bombardment Conditions

The suspension cells were bombarded by the Bio-rad PDS 1000/He system (108). Seven milligrams of 1.6-μm-diameter gold particles in 100 μl of sterile water were vortexed for 2 minutes. During vortexing, 10 μl of plasmid DNA, 40 μl of spermidine (0.1 M), and 100 μl of $CaCl_2$ (2.5 M) were added and the mixture vortexed for a further 3 minutes. The gold was isolated by quick pulse centrifugation and resuspended in 120 μl of 100% ethanol. Twelve microliters of this were transferred to macrocarriers just prior to bombardment. For all bombardments, tissue was placed 9 cm from the rupture disk. The gap distance was 1 cm and the macrocarrier-to-stopping-plate distance was 6 mm. Rupture disks (1100 psi) were used and the vacuum pressure in the chamber was drawn to 26 inches Hg to reduce frictional drag.

## 4. Analyses of Transformants

Successful initial transfer was confirmed by GUS assays on sample tissue. Tissue was then transferred to medium containing the selecting agent, paromomycin sulfate, at 100 mg l$^{-1}$. Cells were maintained at 16-hour daylength at 26°C on this medium by fortnightly transfer to fresh medium. Putative transgenic leaf primordia were observed between 12 and 16 weeks. These were regenerated and transferred to the greenhouse after 12–13 months of culture. Putative transgenic shoots were analyzed for the presence of the *gus* or *nptII* genes by PCR and Southern analysis. PCR-amplified *gus* and *nptII* fragments were obtained from two of the five putative transformants. Southern analysis indicated that in these two lines the fragment for *nptII* was present in both, but that *gus* was only present in one. Both fragments were isolated by enzymes that cut within the T-DNA, so integration into the garlic genome could not be confirmed.

## IV. CONCLUSION

Transformation has the ability to produce incremental improvements in existing commercial cultivars with otherwise desirable characteristics. As gene discovery programs progress and our understanding of gene regulation, plant-microbe interactions, biotic and abiotic stresses, and plant biochemical characteristics advances, transformation will become an increasingly important tool, complementing conventional plant breeding of many crop species. However, for this to eventuate in recalcitrant systems, such as the *Allium* species, further advances in gene transfer technology are required.

As the knowledge to apply transformation technology to *Allium* species continues to progress, so too must our knowledge of the effect of transgenics on human health and the environment. It is easy to focus only on the potential benefits of transgenics to humankind. However, these benefits must be balanced against potential problems, and risk must be correctly assessed. Several potential problems (109–111) have been identified, and in some cases answers have been found (112). Care must be taken over decisions to release new transgenic crops while such questions remain unanswered. *Allium* crops are difficult to grow so it is extremely unlikely that transgenic alliums will ever become a weed problem. Crossing of transgenic *Allium* species with related wild species is also very improbable as few are readily cross-compatible without human intervention (113). The species are not wind-pollinated so the limited quantities of pollen shed will not contaminate the surrounding environment in the unlikely event that this should be a concern; in fact, the flowering stage is only required for seed production and not for crop production. This makes allium crops excellent candidates for field release studies and may assist allium plant breeders as they try to emulate genetic advances achieved in other crops.

## ACKNOWLEDGMENTS

Fig. 1 (a, d, e, f), Fig. 2, and the method described in Sec. III.A.2 are reproduced with permission from Springer-Verlag GmbH & Co. KG and have appeared in a journal article published by Springer Verlag: CC Eady, RJ Weld, CE Lister. *Agrobacterium tumefaciens*–mediated transformation and regeneration of onion (*Allium cepa* L.). *Plant Cell Reports.* (2000) 19:376–381. Section III.B.1 is used with the kind permission of J. M. Myers and P. W. Simon.

## REFERENCES

1. FAO statistics database, August 16, 1999. Available at http://apps.fao.org/lim500/nph-wrap.pl?Production. Crops.Primary&Domain=SUA&servlet=1
2. R Horsch, J Fry, N Hoffman, D Eichholtz, S Rogers, R Fraley. A simple and general method for transfering genes into plants. Science 227:1229–1231, 1985.
3. CC Eady, RJ Weld, CE Lister. Transformation of onion (*Allium cepa* L.). National Onion (and Other *Allium* Research) Conference, 1998, pp 91–95.
4. K Arumuganathan, ED Earle. Nuclear DNA content of some important plant species. Plant Mol Bio Rep 9:208–218, 1991.
5. P Elomaa, J Honkanen, R Puska, P Seppanen, Y Helariutta, M Mehto, M Kotilainen, L Nevalainen, T Teeri. *Agrobacterium*-mediated transfer of antisense chalcone synthase CDNA to *Gerbera hybrida* inhibits flower pigmentation. Biotechnology 11:508–509, 1993.
6. P Song, WH Kang, EB Peffley. Chromosome doubling of *Allium fistulosum* X a-*cepa* interspecific F-1 hybrids through colchicine treatment of regenerating callus. EUPHAA 93:257–262, 1997.
7. M Stevenson, SJ Armstrong, BV Ford-Lloyd, GH Jones. Comparative analysis of crossover exchanges and chiasmata in *Allium cepa* X *fistulosm* after genomic in situ hybridization (GISH). Chromosmone Res 6:567–575, 1998.
8. LI Khrustaleva, C Kik. Cytogenetical studies in the bridge cross *Allium cepa* X (a-*fistulosum* X a-*roylei*). Theor Appl Genet 96:8–14, 1998.
9. C Kik, AM Samoylov, WHJ Verbeek, LWD Vanraamsdonk. Mitochondrial DNA variation and crossability of leek (*Allium porrum*) and its wild relatives from the *Allium ampeloprasum* complex. Theor Appl Genet 94:465–471, 1997.
10. CC Eady. Towards the transformation of onions (*Allium cepa*) (review) NZ J Crop Hortic Sci 23:239–250, 1995.
11. E Block. The organosulfur chemistry of the genus *Allium*—implications for the organic chemistry of sulfur. Agnew Chem Int Ed Engl 31:1135–1178, 1992.
12. JE Lancaster, ML Shaw, WM Randle. Differential hydrolysis of alk(en)yl cysteine sulphoxides by alliinase in onion macerates—flavour implications. J Sci Food Agr 78:367–372, 1998.
13. WM Randle, JE Lancaster, ML Shaw, KH Sutton, RL Hay, ML Bussard. Quantifying onion flavor compounds responding to sulfur fertility-sulfur increases levels of alk(en)yl cysteine sulfoxides and biosynthetic intermediates. J Am Soc Hortic Sci 120:1075–1081, 1995.
14. MJ Havey. Male sterile cytoplasms used commercially to produce hybrid-onion seed. National Onion (and Other *Allium*) Research Conference, 1998, pp 35–39.
15. MJ Havey, DL Leite. Toward the identification of cytoplasmic male sterility in leek: Evaluation of organellar DNA diversity among cultivated accessions of *Allium ampeloprasum*. J Am Soc Hortic Sci 124:163–165, 1999.

16. EJM Van Damme, K Smeets, S Torrekens, F Van Leuven, WJ Peumans. Isolation and characterization of alliinase cDNA clones from garlic (*Allium sativum* L.) and related species. Eur J Biochem 209:751–757, 1992.
17. BJ Gilpin, DW Leung, JE Lancaster. Nucleotide sequence of a nuclear clone of alliinase (Accession No. L48614) from onion. Plant Physiol 110:336, 1996.
18. I Vijn, A Vandijken, M Luscher, A Bos, E Smeets, P Weisbeek, A Wiemken, S Smeekens. Cloning of sucrose-sucrose 1-fructosyltransferase from onion and synthesis of structurally defined fructan molecules from sucrose. Plant Physiol 117:1507–1513, 1998.
19. SR Padgette, KH Kolacz, X Delannay, DB Re, BJ Lavallee, CN Tinius, WK Rhodes, YI Otero, GF Barry, DA Eicholtz, VM Peschke, DL Nida, NB Taylor, GM Kishore. Development, identification, and characterization of a glyphosate-tolerant soybean line. Crop Sci 35:1451–1461, 1995.
20. CV Eberlein, MJ Guttieri, J Steffencampbell. Bromoxynil resistance in transgenic potato clones expressing the *Bxn* gene. Weed Sci 46:150–157, 1998.
21. JE Brandle, SG McHugh, L James, H Labbe, BL Miki. Instability of transgene expression in field grown tobacco carrying the *Csrl*-1 gene for sulfonylurea herbicide resistance. Biotechnology 13:994–998, 1995.
22. K Rajasekaran, JW Grula, RL Hudspeth, S Pofelis, DM Anderson. Herbicide-resistant acala and coker cottons transformed with a native gene encoding mutant forms of acetohydroxyacid synthase. Mol Breed 2:307–319, 1996.
23. B Raubin. Weed control In: HD Rabinowitch, JL Brewster, ed. Onions and Allied Crops. Vol. II. Agronomy, Biotic Interactions, Pathology and Crop Protection. Boca Raton, FL: CRC Press, 1990, pp 63–84.
24. HD Rabinowitch, JL Brewster. Agronomy, Biotic Interactions, Pathology, and Crop Protection. Boca Raton, FL: CRC Press, 1990.
25. R Voss, R Coulello, W Chaney, W Bentley, S Orlott, RM Davis, F Lammlen, C Bell, H Agamalian, D Cudney, B Fisher, B Westerdahl, U Kodira, ML Flint, B Ohlendort. Pest management guidelines: Onion and garlic from the University of California Statewide IPM Project. Proceedings of the 1998 National Onion (and other *Allium*) Research Conference, 1998, pp 361.
26. AR Entwhistle, HD Rabinowitch, JL Brewster. Onions and Allied Crops. Vol. II. Agronomy, Biotic Interactions, Pathology, and Crop Protection. Boca Raton, FL: CRC Press, 1990, pp 103–155.
27. F Crowe, MR McDonald. White rot information in conjunction with the 2nd International Symposium on Edible Alliaceae, November 1997. National Onion (and other *Allium* Research) Conference, 1998, pp 191–195.
28. D-J Yun, RA Bressan, PM Hasegawa. Plant antifungal proteins. Plant Breed Rev 14:39–87, 1997.
29. SA Masoud, Q Zhu, C Lamb, RA Dixon. Constitutive expression of an inducible beta-1,3-glucanase in alfalfa reduces disease severity caused by the oomycete pathogen *Phytophthora megasperma* F sp *medicaginis,* but does not reduce disease severity of chitin-containing fungi. Transgenic Res 5:313–323, 1996.
30. H Schickler, I Chet. Heterologous chitinase gene expression to improve plant defense against phytopathogenic fungi. J Ind Microbiotechnol 19:196–201, 1997.
31. E Jongedijk, H Tigelaar, JSC Vanroekel, SA Bresvloemans, I Dekker, PJM Vandenelzen, BJC Cornelissen, LS Melchers. Synergistic activity of chitinases and beta-1,3-glucanases enhances fungal resistance in transgenic tomato plants. EUPHAA 85:173–180, 1995.
32. R Marchant, MR Davey, JA Lucas, CJ Lamb, RA Dixon, JB Power. Expression of a chitinase transgene in rose (*Rosa hybrida* L.) reduces development of blackspot disease (*Diplocarpon rosae* Wolf). Mol Breed 4:187–194, 1998.
33. ALT powell, G D'Hallewin, BD Hall, H Stotz, JM Labavitch, AB Bennet. Glycoprotein inhibitors of fungal polygalacturonases: Expression of pear PGIP improves resistance in transgenic tomatoes (abstract). 4th International Congress of Plant Molecular Biology Amsterdam, The Netherlands, 1994, pp. 1904.
34. F Stripe, L Barbieri, LG Battelli, M Soria, DA Lappi. Ribosome-inactivating proteins from plants present status and future prospects. Biotechnology 10:405–412, 1992.
35. AD Conceicao, WF Broekaert. Plant defensins. In: SK Dattam S Muthukrishnan, ed. Pathogenesis related proteins in plants. 247–260, 1999.

36. YM Bi, BPA Cammue, PH Goodwin, S KrishnaRaj, PK Saxena. Resistance to *Botrytis cinerea* in scented geranium transformed with a gene encoding the antimicrobial protein Ace-AMP1. Plant Cell Rep 18:835–840, 1999.
37. Oxalic oxidase patent. International Publication Number WO 99/04012.
38. RB Maude. Storage diseases of onions. In: HD Rabinowitch, JL Brewster. Onions and Allied Crops. Vol. II. Agronomy, Biotic Interactions, Pathology, and Crop Protection. Boca Raton, FL: CRC Press, 1990, pp. 173–190.
39. B Bechinger. Structure and functions of channel-forming peptides: Magainins, cecropins, melittin and alamethicin. J Membr Biol 156:197–211, 1997.
40. ES Kristyanne, KS Kim, JM Stewart. Magainin 2 effects on the ultrastructure of five plant pathogens. Mycology 89:353–360, 1997.
41. J de Vries, K Harms, I Broer, G Kriete, A Mahn, K During, W Wackernagel. The bacteriolytic activity in transgenic potatoes expressing a chimeric T4 Iysozyme gene and the effect of T4 Iysozyme on soil- and phytopathogenic bacteria. Syst Appl Micro 22:280–286, 1999.
42. BPA Cammue, K Thevissen, M Hendriks, K Eggermont, IJ Goderis, P Proost, J Vandamme, RW Osborn, F Guerbette, JC Kader, WF Broekaert. A potent antimicrobial protein from onion seeds showing sequence homology to plant lipid transfer proteins. Plant Physiol 109:445–455, 1995.
43. DGA Walkey. Virus diseases: Onions and allied crops. In: HD Rabinowitch, JL Brewster, eds. Vol. II. Agronomy, Biotic Interactions, Pathology, and Crop Protection. Boca Raton, FL: CRC Press, 1990, pp 191–212.
44. SI Song, JT Song, CH Kim, JS Lee, YD Choi. Molecular characterization of the garlic virus X genome. J Gen Virol 79:155–159, 1998.
45. PJ Fletcher, JD Fletcher, SL Lewthwaite. In vitro elimination of onion yellow dwarf and shallot latent viruses in shallots (*Allium cepa* var. *ascalonicum* L.). N Z J Crop Hortic Sci 26:23–26, 1998.
46. U Robert, J Zel, M Ravnikar. Thermotherapy in virus elimination from garlic—influences on shoot multiplication from meristems and bulb formation in vitro. Sci Hortic 73:193–202, 1998.
47. M Moreno, JJ Bernal, I Jimenez, E Rodriguezcerezo. Resistance in plants transformed with the P1 or P3 gene of tobacco vein mottling potyvirus. J Gen Virol 79:2819–2827, 1998.
48. A Wittner, L Palkovics, E Balazs. *Nicotiana benthamiana* plants transformed with the plum pox virus helicase gene are resistant to virus infection. Virus Res 53:97–103, 1998.
49. HS Guo, JJ Lopez-Moya, JA Garcia. Mitotic stability of infection-induced resistance to plum pox potyvirus associated with transgene silencing and DNA methylation. Mol Plant Microbe Interact 12:103–111, 1999.
50. AF Hackland, EP Rybicki, JA Thomson. Coat protein mediated resistance in transgenic plants. Arch Virol 139:1–22, 1994.
51. F Revers, O Le Gall, T Candresse, AJ Maule. New advances in understanding the molecular biology of plant/potyvirus interactions. Mol Plant Microbe Interact 12:367–376, 1999.
52. T Tsuneyoshi, T Matsumi, TC Deng, I Sako, S Sumi. Differentiation of *Allium carlaviruses* isolated from different parts of the world based on the viral coat protein sequence. Arch Virol 143:1093–1107, 1998.
53. T Tsuneyoshi, T Matsumi, KT Natsuaki, S Sumi. Nucleotide sequence analysis of virus isolates indicates the presence of three potyvirus species in *Allium* plants. Arch Virol 143:97–113, 1998.
54. K Kobayashi, P Rabinowicz, F Bravoalmonacid, M Helguera, V Conci, H Lot, A Mentaberry. Coat protein gene sequences of garlic and onion isolates of the onion yellow dwarf potyvirus (Oydv). Arch Virol 141:2277–2287, 1996.
55. SK Soni, PR Ellis. Insect pests. In: HD Rabinowitch, JL Brewster, eds. Onions and Allied Crops. Vol. II. Agronomy, Biotic Interactions, Pathology, and Crop Protection. Boca Raton, FL: CRC Press, 1990, pp 213–272.
56. JC Thomas, D Adams, C Nessler, HJ Bohnert, JK Brown. Reduced reproduction of whitefly (*Bemisia tabaci*) on transgenic tobacco expressing tryptophan decarboxylase. Abstract No. 1924. 4th International Congress of Plant Molecular Biology, 1994.
57. VA Hilder, AMR Gatehouse, SE Sheerman, RFA Barker, D Boulter. A novel mechanism of insect resistance engineered into tobacco. Nature 330:160–163, 1987.

58. KV Rao, KS Rathore, TK Hodges, X Fu, E Stoger, D Sudhakar, S Williams, P Christou, M Bharathi, DP Bown, KS Powell, J Spence, AMR Gatehouse, JA Gatehouse. Expression of snowdrop lectin (GNA) in transgenic rice plants confers resistance to rice brown planthopper. Plant J 15:469–477, 1998.
59. E Stoger, S Williams, P Christou, RE Down, JA Gatehouse. Expression of the insecticidal lectin from snowdrop (*Galanthus nivalis* agglutinin; GNA) in transgenic wheat plants: Effects on predation by the grain aphid Sitobion avenae. Mol Breed 5:65–73, 1999.
60. E Schnepf, N Crickmore, J Vanrie, D Lereclus, J Baum, J Feitelson, DR Zeigler, DH Dean. *Bacillus thuringiensis* and its pesticidal crystal proteins (review). Microbiol Mol Biol Rev 62:775, 1998.
61. VA Hilder, D Boulter. Genetic engineering of crop plants for insect resistance—a critical review. Crop Protect 18:177–191, 1999.
62. CD Green. Nematode pests of *Allium* species. In: HD Rabinowitch, JL Brewster, eds. Onion and Allied Crops. Vol. II. Agronomy, Biotic Interactions, Pathology, and Crop Protection. Boca Raton, FL: CRC Press, 1990, pp 155–172.
63. Z Singh, S Sansavini. Genetic transformation and fruit crop improvement. Plant Breed Rev 16: 87–134, 1998.
64. JE Lancaster, ML Shaw, MD Pither, JP Farrant, JA McCallum. A review of the regulation of sulfur matebolism and its effect on quality in onions and garlic. National Onion (and Other *Allium* Research) Conference, 1998, pp 143–156.
65. JE Lancaster, MJ Boland. Flavour biochemistry. In: HD Rabinowitch, JL Brewster, eds. Onions and Allied Crops. Vol. III. Boca Raton, FL: CRC Press, 1990, pp 33–72.
66. SA Clark, ML Shaw, D Every, JE Lancaster. Physical characterization of Alliinase, the flavor generating enzyme in onions. J Food Biochem 22:91–103, 1998.
67. SA Clark. Molecular cloning of a cDNA encoding alliinase from onion (*Allium cepa* L.). PhD Dissertation, Christchurch, New Zealand, University of Canterbury, Christchurch, New Zealand, 1993.
68. JE Lancaster, ML Shaw. Characterization of purified γ glutamyl transpeptidase in onions: evidence for in vivo role as a peptidase. Phytochemistry 36: 1351–1358, 1994.
69. C Ohsumi, T Hayashi, K Sano. Formation of alliin in the culture tissues of *Allium sativum*. Oxidation of S-allyl-L-cysteine. Phytochemistry 33:107–111, 1993.
70. KS Orvis, CR Galamarini, IL Goldman, MJ Havey. A comparison of onion-induced platelet aggregation by plasma and whole blood aggregometry. Proceedings of the 1998 National Onion (and Other Allium) Research Conference, 1998, pp 178.
71. GAF Hendry, RK Wallace. The origin, distribution, and evolutionary significance of fructans. Sci Technol Fructans 119–139, 1993.
72. N Shiomi, S Onodera, H Sakai. Fructo-oligosaccharide content and fructosyltransferase activity during growth of onion bulbs. New Phytol 136:105–113, 1997.
73. MB Roberfroid, NM Delzenne. Dietary fructans (review). Annu Rev Nutr 18:117–143, 1998.
74. JW Yun. Fructo-oligosaccharides—occurrence, preparation, and application. Enzyme Microtechnol 19:107–117, 1996.
75. I Vijn, A Vandijken, N Sprenger, K Vandun, P Weisbeek, A Wiemken, S Smeekens. Fructan of the inulin neoseries is synthesized in transgenic chicory plants (*Cichorium intybus* L.) Harbouring onion (*Allium cepa* L.) fructan-fructan 6g-fructosyltransferase. Plant J 11:387–398, 1997.
76. Y Tanaka, S Tsuda, T Kusumi. Metabolic engineering to modify flower color (review). Plant Cell Physiol 39:1119–1126, 1998.
77. J Buiteveld, Y Suo, MMV Campagne, J Creemersmolenaar. Production and characterization of somatic hybrid onion (*Allium cepa* L.). Theor Appl Genet 96:765–775, 1998.
78. J Buiteveld, W Kassies, R Geels, MMV Campagne, E Jacobsen, J Creemersmolenaar. Biased chloroplast and mitochondrial transmission in somatic hybrids of *Allium ampeloprasum* L. and *Allium cepa* L. Plant Sci 131:219–228, 1998.
79. J Buiteveld. Regeneration and interspecific somatic hybridization in *Allium* for transfer of cytoplasmic male sterility to leek. Wageningen Agricultural University. 127. 1998.
80. C Mariani, MD Beuckeleer, J Truettner, J Leemans, RB Goldberg. Induction of male sterility in plants by a chimaeric ribonuclease gene. Nature 347:737–741, 1990.

81. AD Arencibia, ER Carmona, P Tellez, MT Chan, SM Yu, LE Trujillo, P Oramas. An efficient protocol for sugarcane (*Saccharum* spp. L.) transformation mediated by *Agrobacterium tumefaciens*. Transgenic Res 7:213–222, 1998.
82. M Cheng, JE Fry, SZ Pang, HP Zhou, CM Hironaka, DR Duncan, TW Conner, YC Wan. Genetic transformation of wheat mediated by *Agrobacterium tumefaciens*. Plant Physiol 115:971–980, 1997.
83. P Christou. Strategies for variety-independent genetic transformation of important cereals, legumes and woody species utilizing particle bombardment. EUPHAA 85:13–27, 1995.
84. P Christou. Rice transformation-bombardment. Plant Mol Biol 35:197–203, 1997.
85. DD Songstad, DA Somers, RJ Griesbach. Advances in alternative DNA delivery techniques. Plant Cell Tissue Org Cult 40:1–15, 1995.
86. G Hansen, JF Hubstenberger, GC Phillips. Regeneration of shoots from cell suspension-derived protoplasts of *Allium cepa*. Plant Cell Rep 15:8–11, 1995.
87. AT Mankarios, MA Hall, MC Jarvis, DR Threlfall, J Friend. Cell wall polysaccharides from onions. Phytochemistry 19:1731–1733, 1980.
88. Japan tobacco patent. WO 94/00977.
89. E Dommisse, MD Leung, WMM Shaw, LA Conner, J. Onion is a monocotyledonous host for *Agrobacterium*. Plant Sci 69:249–257, 1990.
90. CC Eady, CE Lister, YY Suo, D Schaper. Transient expression of uida constructs in in vitro onion (*Allium cepa* L.) Cultures following particle bombardment and *Agrobacterium*-mediated DNA delivery. Plant Cell Rep 15:958–962, 1996.
91. X Barandiaran, A Dipietro, J Martin. Biolistic transfer and expression of a uida reporter gene in different tissues of *Allium sativum* L. Plant Cell Rep 17:737–741, 1998.
92. CC Eady, CE Lister. Transformation of onion (*Allium cepa* L.) Alliums Australia 1997, Second International Symposium on Edible Alliaceae, 1997.
93. JM Myers, PW Simon. Microprojectile bombardment of garlic, *Allium sativum* L. 1998 National Onion (and Other Allium) Conference, 1998, pp 121–126.
94. CC Eady, RJ Weld, CE Lister. *Agrobacterium tumefaciens*-mediated transformation and regeneration of onion (*Allium capa* L.). Plant Cell Rep 19:376–381, 2000.
95. A Wilmink, BCE Vandeven, JJM Dons. Activity of constitutive promoters in various species from the Liliaceae. Plant Mol Biol 28:949–955, 1995.
96. HM Xue, H Araki, T Kanazawa, T Harada, T Yakuwa. Callus formation and plantlet regeneration through in vitro culture of immature embryo and seedling in Chinese chive (*Allium tuberosum* rottler) [Japanese]. J Jpn Soc Hortic Sci 66:353–358, 1997.
97. CC Eady, RC Butler, Y Suo. Somatic embryogenesis and plant regeneration from immature embryo cultures of onion (*Allium cepa* L.). Plant Cell Rep 18:111–116, 1998.
98. SJ Zheng, B Henken, E Sofiari, E Jacobsen, FA Krens, C Kik. Factors influencing induction, propagation and regeneration of mature zygotic embryo-derived callus from *Allium cepa*. Plant Cell Tissue Org Cult 53:99–105, 1998.
99. B Silvertand, A Vanrooyen, P Lavrijsen, AM Vanharten, E Jacobsen. Plant regeneration via organogenesis and somatic embryogenesis in callus cultures derived from mature zygotic embryos of leek (*Allium ampeloprasum* L). EUPHAA 91:261–270, 1996.
100. MM Saker. In vitro regeneration of onion through repetitive somatic embryogenesis. Biol Plant 40:499–506, 1998.
101. G Hansen, MS Wright. Recent advances in the transformation of plants (review). Trends Plant Sci 4:226–231, 1999.
102. CC Eady, CE Lister. A comparison of four selective agents for use with *Allium cepa* L. immature embryos and immature embryo derived cultures. Plant Cell Rep 18:117–121, 1998.
103. J Haseloff, KR Siemering, DC Prasher, S Hodge. Removal of a cryptic intron and subcellular localization of green fluorescent protein are required to mark transgenic arabidopsis plants brightly. Proc Natl Acad Sci USA 94:2122–2127, 1997.
104. RA Jefferson, TA Kavanagh, MW Bevan. GUS fusions: β-glucuronidase as a sensitive and versatile gene fusion marker in higher plants. EMBO J 6:3901–3907, 1987.
105. Y Hiei, S Ohta, T Komari, T Kumashiro. Efficient transformation of rice (*Oryza sativa* L.) mediated by *Agrobacterium* and sequence analysis of the boundaries of the T-DNA. Plant J 6:271–282, 1994.

106. TaS Murashige, F. A revised medium for rapid growth and bioassays with tobacco tissue cultures. Physiol Plant 15:473–497, 1962.
107. JM Myers, PW Simon. Continuous callus production and regeneration of garlic (*Allium sativum* L.) using root segments and shoot tip derived plantlets. Plant Cell Rep 17:726–730, 1998.
108. JC Sanford, TM Klein, ED Wolf. Delivery of substances into cells and tissues using a particle bombardment process. J Plant Sci Technol 5:27–37, 1987.
109. C Eady, D Twell, K Lindsey. Pollen viability and transgene expression following storage in honey. Transgenic Res 4:226–231, 1995.
110. EC Simpson, CE Norris, JR Law, JE Thomas, JB Sweet. Gene flow in genetically modified herbicide tolerant oilseed rape (*Brassica napus*) in the UK: Gene flow and agriculture: Relevance for transgenic crops 72:75–81, 1999.
111. BA Palevitz. Bt or not Bt: Transgenic corn vs. monarch butterflies. SCI 13:1–1999.
112. JE Wilkinson, K Lindsey, D Twell. Antisense-mediated suppression of transgene expression targeted specifically to pollen. J Exp Bot 49:1481–1490, 1998.
113. M Klaas. Applications and impact of molecular markers on evolutionary and diversity studies in the genus *Allium* (review). Plant Breed 117:297–308, 1998.

# 45
# Potato Transformation Produces Value-Added Traits

**Lawrence M. Kawchuk**
*Agriculture and Agri-Food Canada, Lethbridge, Alberta, Canada*

| | | |
|---|---|---|
| I. | INTRODUCTION | 673 |
| II. | AGRONOMIC TRAITS | 674 |
| III. | COMMERCIAL PRODUCTS | 675 |
| IV. | RESISTANCE | 676 |
| | A. Insects | 676 |
| | B. Fungi | 676 |
| | C. Bacteria | 677 |
| | D. Viroids | 677 |
| | E. Viruses | 677 |
| V. | CONCLUSIONS | 687 |
| | REFERENCES | 687 |

## I. INTRODUCTION

Potato (*Solanum tuberosum* ssp. *tuberosum* L.) is one of the most important food crops in the world. Production worldwide is approximately 293 million tons per annum and covers more than 18 million hectares. Consumption per capita in developing countries is rapidly increasing and has reached 14 kg per annum but is still far less than the European 86 kg or North American 63 kg, contributing to expectations of continued world expansion.

Potato is one of the most frequently genetically engineered crops since solanaceous plants have been relatively simple to transform. As a vegetatively propagated crop, potato provides many advantages in expressing transgenic traits and heterologous products. For example, transgenes are not diluted through meiosis, germplasm integrity is easily maintained, and the tuber provides an economical eukaryotic expression system. Various strategies have been developed to produce transgenic potato with improved agronomic performance, expressing economically important products, and exhibiting disease and pest resistance (Table 1).

**Table 1** Transgenic Potato: Improved Agronomic Performance, Economically Important Products, and Pest and Disease Resistance

| Trait | Gene | Plant host | Reference |
|---|---|---|---|
| *Agronomic* | | | |
| Tuber size and number | Invertase | Potato | (1) |
| | ADP-glucose pyrophosphorylase | Potato | (2) |
| | | Potato | (3) |
| Carbohydrate partitioning | Invertase inhibitor | Potato | (4) |
| | Starch binding protein | Potato | (5) |
| Nutritional value | Seed albumin | Potato | (6) |
| | Vitamin A | Rice | (7) |
| Physiological traits | Antifreeze | Potato | (8) |
| *Commercial product* | | | |
| Starch composition | Glycogen synthase | Potato | (9) |
| | Starch synthase | Potato | (10) |
| Vaccines | Enterotoxin | Potato, tobacco | (11) |
| | Cholera toxin B | Potato | (12) |
| Proteins | Insulin | Potato | (13) |
| | Monoclonal antibody | Tobacco | (14) |
| | Single-chain antibody | Potato | (15) |
| *Resistance* | | | |
| Insect | | | |
| Colorado potato beetle | *Bacillus thuringiensis* endotoxin cry III | Potato | (16) |
| Tuber moth | *Bacillus thuringiensis* endotoxin cry V | Potato | (17) |
| Aphid | Resistance gene *Mi* | Tomato | (18) |
| Hornworm | Proteinase inhibitor II | Tobacco | (19) |
| Fungi | | | |
| Late blight | Glucose oxidase | Potato | (20) |
| | Beta-1, 3-endoglucanase | Potato | (21) |
| Early blight | Endochitinase | Potato, tobacco | (22) |
| Black scurf | Endochitinase | Potato, tobacco | (22) |
| Gray mold | Endochitinase | Potato, tobacco | (22) |
| Bacteria | | | |
| Soft rot | Glucose oxidase | Potato | (20) |
| Blackleg | T4 lysozyme | Potato | (23) |
| Viroids | | | |
| Spindle tuber viroid | Ribozyme | Potato | (24) |
| | Ribonuclease | Potato | (25) |

## II. AGRONOMIC TRAITS

Several advances have been made in improving or modifying various agronomic characteristics associated with potato tubers. For example, tuber size was altered through modification of sucrose metabolism (1). Cytosolic localization of a yeast invertase gave rise to a reduction in tuber size and an increase in tuber number per plant, whereas apoplastic targeting led to an increase in tuber size and a decrease in tuber number per plant. Transgenic potato plants were also developed in which the expression of adenosine diphosphate– (ADP)-glucose pyrophosphorylase (AGPase) was inhibited by antisense ribonucleic acid (RNA) (2). This resulted in the reduction of starch formation in tubers, and up to 30% of the dry weight of the transgenic potato tubers was represented by sucrose and 8% by glucose. The process of tuber formation also changed, resulting in significantly more tubers both per plant and per stolon. A fivefold increase in activity of AGPase, achieved by transformation with the *Escherichia coli* AGPase gene glgC-16, resulted in a sucrose-to-starch conversion that increased roughly in proportion to the increase in AGPase activity (3). This conversion of sucrose could potentially reduce cold-temperature sweetening and the associated dark-colored processed products. However, the increased conversion into starch in the transformed tubers was accompanied by an increased rate of starch turnover.

Efforts to improve processing quality have focused on altering the tuber carbohydrate metabolism. Potato plants transformed with Nt-inhh complementary deoxyribonucleic acid (cDNA), encoding a putative vacuolar homolog of a tobacco cell wall invertase inhibitor, reduced cold-induced hexose accumulation by up to 75%, without any effect on potato tuber yield (4). Processing quality of tubers was greatly improved without changing starch quantity or quality. Another strategy involved a gene that was identified by the ability of its product to bind to potato starch granule (5). Reduction in the protein level of transgenic potatoes led to a reduction in the phosphate content of the starch. The reduced phosphate content in potato starch has some secondary effects on its degradability, as the respective plants displayed a starch excess in leaves and reduced in cold sweetening in tubers.

In an attempt to improve the nutritional value of potato, the seed albumin gene *AmA1* from *Amaranthus hypochondriacus* was successfully introduced and expressed in a tuber-specific and constitutive manner (6). There was a striking increase in the growth and production of tubers in transgenic plants and also of the total protein content with an increase in most essential amino acids. Expressed protein was localized in the cytoplasm as well as in the vacuole of transgenic tubers. Other improvements in potato nutritional value are expected, similar to the recombinant DNA technology recently used to allow biosynthesis of provitamin A in rice endosperm (7).

Many other potato agronomic improvements are predicted as genes are isolated and characterized. For example, expression of a synthetic antifreeze protein in potato plants reduced electrolyte leakage and increased tolerance to freezing (8). Other advances include the development of tubers resistant to bruising by reducing polyphenol oxidase and potato plants resistant to glyphosate herbicide.

## III. COMMERCIAL PRODUCTS

Starch is a major constituent of potato tubers. A chimeric gene containing the patatin promoter and transit-peptide region of the small-subunit carboxylase gene, directed expression of *E. coli* glycogen synthase (*glgA*) to potato tuber amyloplasts (9). Tubers from transgenic potato lines were found to have a novel potato starch with reduced phosphorus content and a very high degree of branching of the amylopectin. Carbohydrate partitioning was also modified in potato plants transformed to produce antisense RNA for a full-length granule-bound starch synthase (GBSS) cDNA that determines the presence of amylose in reserve starches (10). In those cases in which total suppression of GBSS activity was found, both GBSS protein and amylose were absent, giving rise to tubers containing amylose-free starch. Modified carbohydrate partitioning is becoming increasingly important in producing carbohydrates with commercial applications.

Alternatives to cell culture systems for production of recombinant proteins could make very safe vaccines at a lower cost. Genetically engineered plants facilitate expression of candidate vaccine antigens with the goal of using the edible plant organs for economical delivery or oral vaccines. Transgenic tobacco and potato plants were made with genes encoding binding subunit of *E. coli* heat-labile enterotoxin (LT-B) or an LT-B fusion protein with a microsomal retention sequence (11). The plants expressed the foreign peptides, both of which formed oligomers that bound the natural ligand. Mice immunized by gavage produced serum and gut mucosal anti-LT-B immunoglobulins that neutralized the enterotoxin in cell protection assays. Feeding mice fresh transgenic potato tubers also caused oral immunization. Similarly, transgenic potatoes engineered to synthesize a cholera toxin B subunit (CTB) pentamer with affinity for GMI ganglioside induced serum and intestinal CTB-specific antibodies in orally immunized mice (12). The cytopathic effect of cholera holotoxin (CT) on Vero cells was neutralized by serum from mice immunized with transgenic potato tissues. After intraileal injection with CT, the plant-immunized mice showed up to a 60% reduction in diarrheal fluid accumulation in the small intestine. These results demonstrate the ability of transgenic plants to generate protective immunity in mice against a bacterial enterotoxin.

Heterologous expression of important proteins in potato plants is another area expected to expand. Transgenic potato plants have been produced that synthesize human insulin, a major insulin-dependent diabetes mellitus autoantigen (13). To direct delivery of plant-synthesized insulin to the gut-associated lymphoid tissues, insulin was linked to the C terminus of CTB. Transgenic potato tubers produced 0.1% of total soluble protein as the pentameric CTB-insulin fusion. Nonobese diabetic mice fed transformed potato tuber tissues containing microgram amounts of the CTB-insulin fusion protein showed a substantial reduction in pancreatic islet inflammation, and a delay in the progression of clinical diabetes.

An example of the plants' ability to process heterologous molecules correctly is the production of functional antibodies (14). Furthermore, addition of the endoplasmic reticulum retention signal KDEL to the C terminus of single-chain antibody fragments (scFvs) resulted in significantly improved expression levels in transgenic potato (15). Addition of the KDEL motif is a simple and straightforward tool to stabilize in planta cytosolic expression of many scFvs.

## IV. RESISTANCE

### A. Insects

Progress has been made in developing resistance to pests that attack potato. Transgenic potato lines containing *Bacillus thuringiensis (Bt)* cryIIIA delta-endotoxin gene controlled first-instar Colorado potato beetle (CPB) larvae in bioassays, and no CPB larvae developed past the second instar on any of these plants (16). There was good correlation between insect control and the levels of delta-endotoxin RNA and protein. Transgenic potato cultivars resistant to CPB are commercially available in Canada and the United States. Resistance to the tuber moth (*Phthorimaea operculella*) in potato plants expressing the cryV *Bt* transgene has also been reported (17). Other examples of insect resistance that could be beneficial in potato include resistance against the aphid *Macrosiphum euphorbiae* observed in susceptible tomato transformed with the nematode resistance gene *Mi* that belongs to the nucleotide-binding, leucine-rich repeat family of resistance genes (18) and resistance to the tobacco hornworms (*Manduca sexta*) in transgenic tobacco plants expressing the potato proteinase inhibitor II gene (19).

### B. Fungi

Transgenic potato plants expressing a fungal gene encoding glucose oxidase, which generates $H_2O_2$ when glucose is oxidized, exhibited resistance to potato late blight caused by *Phytophthora infestans* (20). Similarly, potato plants that were transformed with the soybean beta-1,3-endoglucanase cDNA via *Agrobacterium tumefaciens* delivery (21) exhibited increased resistance to *P. infestans*. Disease resistance in transgenic potato and tobacco plants was also reported after the insertion of a gene encoding an antifungal endochitinase from the mycoparasitic fungus *Trichoderma harzianum* (22). Selected transgenic lines were highly tolerant or completely resistant to the foliar pathogens *Alternaria alternata, Alternaria solani,* and *Botrytis cinerea* and the soil-borne pathogen *Rhizoctonia solani.*

### C. Bacteria

Transgenic potato plants expressing a fungal gene encoding glucose oxidase exhibited strong resistance to a bacterial soft rot disease caused by *Erwinia carotovora* subsp. *carotovora* (20). This resistance to soft rot was apparently mediated by elevated levels of $H_2O_2$, because the resistance could be counteracted by exogenously added $H_2O_2$-degrading catalase. Control of *E. carotovora* was also observed in transgenic potato lines that synthesized bacteriophage T4 lysozyme (23).

### D. Viroids

A hammerhead ribozyme [R(–)] targeting the minus-strand RNA of potato spindle tuber viroid (PSTVd) and a mutated nonfunctional ribozyme [mR(–)] have been designed, cloned, and transcribed (24). Transgenic potato plants expressing the functional ribozyme were resistant to PSTVd. Transgenic potato lines and progeny expressing the yeast-derived double-stranded RNA-specific ribonuclease pac1 also suppressed PSTVd infection and accumulation (25). Double-stranded regions in PSTVd molecule and/or replicative intermediates may be targeted by pac1 gene product in the transgenic potato plant.

### E. Viruses

Potato is particularly subject to virus diseases, in part because vegetative propagation allows a virus several years to accumulate over successive generations. Strategies to control viruses need to be effective against viruses with diverse particle morphological characteristics, host ranges, vector specificities, and genome organizations. The potential targets for disruption of the virus infection process include entry, disassembly, translation, replication, encapsidation, movement, and transmission. Since 1986 (26), there have been more than 100 reports of plants with genetically engineered virus resistance. Several mechanisms that are not necessarily mutually exclusive have been proposed to explain the resistance observed with the various strategies. Many of these strategies have been or may be used to control viruses that infect potato.

#### 1. Coat Protein Gene

Genetically engineered plant virus resistance was first reported by Powell and coworkers (26). A cDNA clone of the tobacco mosaic tobamovirus (TMV) coat protein gene was constitutively expressed in *Nicotiana tabacum*, and plants that accumulated coat protein developed no disease symptoms or showed delayed symptom development when inoculated with TMV. The first report of genetically engineered plant virus resistance in potato described the use of the coat protein gene of potato potexvirus X (PVX) to delay symptom development and reduce virus accumulation (27). Genetic engineering of plant virus resistance using the coat protein gene is the most widely examined strategy and has been reported for plant viruses with single-stranded positive- and negative-sense RNA, double-stranded RNA, and DNA genomes (Table 2). Protection has been observed for plant viruses with differing particle morphological characteristics, genome organization, host range, and vector specificity. Resistance developed by expressing a coat protein gene has been reported for viruses from 15 groups.

The level of resistance conferred by coat protein genes is variable and depends on the virus

**Table 2** Genetically Engineered Virus Resistance Using Viral Coat Protein Genes

| Virus group | Virus | Abbreviation | Plant Host | Reference |
| --- | --- | --- | --- | --- |
| Alfamovirus | Alfalfa mosaic virus | AlMV | Tobacco | (42) |
| | | | Tobacco | (128) |
| | | | Tobacco, tomato | (57) |
| | | | Alfalfa | (129) |
| | | | Tobacco | (130) |
| Carlavirus | Potato virus S | PVS | *N. debneyii*[a] | (62) |
| | | | Potato | (41) |
| Cucumovirus | Cucumber mosaic virus | CMV | Tobacco | (50) |
| | | | Tobacco | (61) |
| | | | Tobacco | (131) |
| | | | Tobacco | (132) |
| | | | Cucumber | (133) |
| | | | Tobacco | (121) |
| | | | Squash | (122) |
| Geminivirus | Tomato yellow leaf curl virus | TYLCV | Tomato | (47) |
| | | | *N. benthamiana*[a] | (134) |
| Enamovirus | Pea enation mosaic virus | PEMV | Pea | (135) |
| Ilarvirus | Tobacco streak virus | TSV | Tobacco | (136) |
| Luteovirus | Potato leafroll virus | PLRV | Potato | (31) |
| | | | Potato | (32) |
| | | | Potato | (34) |
| | | | Potato | (35) |
| | | | Tobacco | (137) |
| | | | Potato | (36) |
| | | | Potato | (37) |
| | | | Potato | (38) |
| | | | Potato | (39) |
| | | | Potato | (33) |
| Nepovirus | Arabis mosaic virus | ArMV | Tobacco | (63) |
| | Grapevine chrome mosaic virus | GCMV | Tobacco | (138) |
| | Grapevine fanleaf virus | GFLV | Tobacco | (139) |
| | Tomato ringspot virus | TomRSV | Tobacco, *N. benthamiana*[a] | (140) |
| Pomovirus | Potato mop top virus | PMTV | *N. benthamiana*[a] | (141) |
| Potexvirus | Cymbidium mosaic virus | CyMV | *N. benthamiana*[a] | (142) |
| | Potato aucuba mosaic virus | PAMV | Tobacco | (143) |
| | Potato virus X | PVX | Tobacco | (44) |
| | | | Potato | (27) |
| | | | Potato | (28) |
| | | | Potato | (29) |
| | | | Potato | (30) |
| | | | Tobacco | (144) |
| Potyvirus | Bean yellow mosaic virus | BYMV | *N. benthamiana*[a] | (145) |
| | | | *N. benthamiana*[a] | (54) |
| | Lettuce mosaic virus | LMV | Tobacco | (66) |
| | Maize dwarf mosaic virus | MDMV | Corn | (67) |
| | Papaya ringspot virus | PRSV | Tobacco | (64) |
| | | | Papaya | (146) |
| | Plum pox virus | PPV | *N. benthamiana, N. clevelandii*[a] | (147) |
| | | | *N. benthamiana*[a] | (43) |

**Table 2** (*Continued*)

| Virus group | Virus | Abbreviation | Plant Host | Reference |
|---|---|---|---|---|
| Potyvirus (*continued*) | Potato virus Y | PVY | Potato | (29) |
| | | | Potato | (28) |
| | | | Tobacco | (40) |
| | | | Potato | (90) |
| | | | Tobacco | (124) |
| | Peanut stripe virus | PstV | *N. benthamiana*[a] | (148) |
| | Soybean mosaic virus | SMV | Tobacco | (59) |
| | Tobacco etch virus | TEV | Tobacco | (118) |
| | Watermelon mosaic virus | WMV | *N. benthamiana*[a] | (60) |
| | | WMV2 | Squash | (149) |
| | | | Squash | (122) |
| | Zucchini yellow mosaic virus | ZYMV | *N. benthamiana*[a] | (60) |
| | | | Melon, tobacco | (150) |
| | | | Squash | (149) |
| | | | Squash | (122) |
| Tenuivirus | Rice stripe virus | RSV | Rice | (45) |
| Tobamovirus | Pepper mild mottle virus | PMMV | Pepper | (151) |
| | Tobacco mosaic virus | TMV | Tobacco | (26) |
| | | | Tomato | (65) |
| | | | Tobacco | (48) |
| | | | Tobacco | (46) |
| | | | Tobacco | (152) |
| | | | Tobacco | (120) |
| | | | Tobacco | (153) |
| | | | Tobacco | (154) |
| | | | Tobacco | (56) |
| | | | Tobacco | (124) |
| | Tomato mosaic virus | ToMV | Tomato | (155) |
| Tobravirus | Tobacco rattle virus | TRV | *N. benthamiana*[a] | (128) |
| | | | Tobacco | (51) |
| | | | Tobacco | (52) |
| Tospovirus | Tomato spotted wilt virus | TSWV | Tobacco | (156) |
| | | | Tobacco | (157) |
| | | | Tobacco | (58) |
| | | | Tobacco | (158) |
| | | | Lettuce | (159) |
| | | | Tobacco | (123) |
| | | | *N. benthamiana*[a] | (160) |
| | Tomato chlorotic spot virus | TCSV | Tobacco | (123) |
| | Groundnut ringspot virus | GRSV | Tobacco | (123) |

[a]*Nicotiana* spp.

and host. Variability of resistance has also been observed within individual transformed lines since the random insertion of the coat protein gene into the genome can influence expression levels. In general, expression of a virus coat protein gene reduces symptom expression, slows virus movement, and reduces titers of closely related viruses. For example, reduced symptoms, movement and titers were observed in potato plants expressing the coat protein gene of PVX (27–30), potato leafroll luteovirus (PLRV) (31–39), potato potyvirus Y (PVY) (28,29,40), and potato carlavirus S (PVS) (41).

Resistance often depends on the level of coat protein expression, as observed with alfalfa mosaic alfamovirus (AlMV) (42), plum pox potyvirus (PPV) (43), PVX (27,44), rice stripe tenuivirus (RSV) (45), TMV (46), and tomato yellow leaf curl geminivirus (TYLCV) (47). Plants expressing higher levels of coat protein exhibited a greater level of protection. Resistance to TMV was attenuated by exposing transgenic plants to elevated temperatures that reduced coat protein but not viral RNA (48). Protoplasts from virus-resistant transgenic plants are also resistant, suggesting that protection occurs at the cellular level (49).

Experiments designed to determine the mechanism of the resistance suggest that the coat protein interferes with virus disassembly. Resistance to TMV was reduced by removing 5' subunits of the TMV particle as occurs during cotranslational disassembly (49) and there was a lack of resistance to AlMV (42), cucumber mosaic cucomovirus (CMV) (50), TYLCV (47), TMV (49), and tobacco rattle tobravirus (TRV) (51,52) when plants were inoculated with viral RNA. Partial transencapsidation of viruses has been observed with heterologous coat protein of transgenic plants (53,54), resistance was not reduced by the introduction of mutations into regions of the AlMV coat protein that are important in RNA interactions (55), and mutations in the TMV coat protein gene showed a strong correlation between the strength of subunit interactions and the level of resistance (56). Finally, increased virus concentrations were observed to reduce resistance conferred by the coat protein gene of TMV (26), AlMV (57), tomato spotted wilt tospovirus (TSWV) (58), soybean mosaic potyvirus (SMV) (59), and watermelon mosaic potyvirus (WMV) (60). The practical usefulness of coat protein–mediated resistance will depend on the level of inoculum during cultivation.

Interestingly, there are also several examples that suggest a mechanism independent of virus disassembly. Coat protein levels do not correlate with resistance levels to CMV (61), PLRV (32), PVY (29), and TSWV (58). Similar levels of resistance were observed in plants inoculated with virus or RNA of PVX (44) and PVS (41,62). Furthermore, resistance was not influenced by increasing inoculum levels of CMV (50), arabis mosaic virus (ArMV) (63), PLRV (32), or PVS (41,62).

Various levels of specificity have been reported for the resistance conferred by coat protein genes. In general, the highest resistance is to the virus from which the coat protein gene was derived. A correlation between the level of protection and the homology of the coat protein gene to that of the inoculated virus has been reported for carlaviruses (62), potyviruses (43,59,64), and tobamoviruses (64). However, exceptions include the lettuce mosaic potyvirus (LMV) coat protein gene, which confers complete protection to PVY but no protection against tobacco etch potyvirus (TEV) even though these viruses have considerable sequence homology (66), and maize dwarf mosaic potyvirus (MDMV) provided protection against the unrelated maize chlorotic mottle carmovirus (MCMV) (67).

## 2. Nonstructural Virus Genes

Several other virus genes have been used to engineer resistance, including the replicase, movement protein, and protease genes (Table 3). Unlike the expression of coat protein, the development of resistance with nonstructural genes usually requires the expression of an incompatible or defective protein that interferes with virus replication, movement, or maturation of virus proteins. This strategy has been used to engineer resistance to members of 12 plant virus groups and provides resistance ranging from specific resistance for the isolate from which the transgene was derived to resistance to unrelated viruses.

Transgenic plants expressing the putative 54-kDa protein encoded by the TMV replicase were completely resistant to TMV, even at inoculation concentrations 1000-fold greater than those that overcame resistance with the coat protein gene (68). This strategy has been applied to viruses from nine groups including potato virus X (69,70) and potato virus Y (71). Resistance lev-

**Table 3** Genetically Engineered Resistance Using Nonstructural Virus Genes

| Virus group | Virus | Abbreviation | Plant host | Reference |
|---|---|---|---|---|
| *Replicase* | | | | |
| Alfamovirus | Alfalfa mosaic virus | AlMV | Tobacco | (161) |
| Cucumovirus | Cucumber mosaic virus | CMV | Tobacco | (50) |
| | | | Tobacco | (162) |
| | | | Tobacco | (163) |
| | | | Tobacco | (164) |
| | | | Tobacco | (72) |
| | | | Tobacco | (73) |
| | | | Tobacco | (165) |
| Geminivirus | African cassava mosaic virus | ACMV | *N. benthamiana*[a] | (166) |
| | Tomato yellow leaf curl virus | TYLCV | *N. benthamiana*[a] | (134) |
| Luteovirus | Potato leafroll virus | PLRV | Potato | (75) |
| Potexvirus | Potato virus X | PVX | Tobacco | (69) |
| | | | Tobacco, *N. clevelandii*[a] | (70) |
| Potyvirus | Potato virus Y | PVY | Tobacco | (71) |
| | Plum pox virus | PPV | *N. benthamiana*[a] | (167) |
| | | | *N. benthamiana*[a] | (168) |
| | Tobacco etch virus | TEV | Tobacco | (89) |
| Sobemovirus | Rice yellow mottle virus | RYMV | Rice | (74) |
| Tobamovirus | Tobacco mosaic virus | TMV | Tobacco | (68) |
| | | | Tobacco | (169) |
| | | | Tobacco | (170) |
| | Pepper mild mottle virus | PMMV | *N. benthamiana*[a] | (171) |
| | | | *N. benthamiana*[a] | (172) |
| Tobravirus | Pea early browning virus | PEBV | *N. benthamiana*[a] | (173) |
| Tombusvirus | Cymbidium ringspot virus | CyRSV | *N. benthamiana*[a] | (174) |
| | | | *N. benthamiana*[a] | (175) |
| Tospovirus | Tomato spotted wilt virus | TSWV | Tobacco | (176) |
| *Movement* | | | | |
| Comovirus | Cowpea mosaic virus | CPMV | *N. benthamiana*[a] | (177) |
| Luteovirus | Potato leafroll virus | PLRV | Potato | (80) |
| Potexvirus | Potato virus X | PVX | Potato | (81) |
| | | | Tobacco | (178) |
| | White clover mosaic virus | WclMV | *N. benthamiana*[a] | (179) |
| Tobamovirus | Tobacco mosaic virus | TMV | Tobacco | (79) |
| | | | Tobacco | (78) |
| | | | Tobacco | (180) |
| | | | Tobacco | (178) |
| Tospovirus | Tomato spotted wilt virus | TSWV | Tobacco | (86) |
| *Protease* | | | | |
| Potyvirus | Potato virus Y | PVY | Tobacco | (83) |
| | Tobacco vein mottling virus | TVMV | Tobacco | (84) |

[a]*Nicotiana* spp.

els varied between transformed lines but were frequently higher than that observed using coat protein genes but extremely specific for the virus or isolate from which the replicase was derived. For example, potato plants transformed with the PVY$^O$ replicase gene were highly resistant to the homologous PVY$^O$ strain but not PVY$^N$ (71). Resistance obtained with a replicase gene has been reported to involve two independent mechanisms influencing virus replication and movement

(72,73). Transgenic rice encoding the RNA-dependent RNA polymerase of rice yellow mottle sobemovirus (RYMV) also suppressed virus multiplication (74). In the most resistant line, transcription analysis indicated that the resistance derives from an RNA-based mechanism associated with posttranscriptional gene silencing. Similar resistance was reported for plants expressing the PLRV replicase (75).

Cell-to-cell movement of viruses can involve a virus-encoded movement protein that interacts with the plasmodesmata to increase the size exclusion limit (76,77). Tobacco plants transformed with a defective TMV movement protein gene or with the movement protein gene from BMV, a virus that cannot move within tobacco, are resistant to TMV (78,79). Similarly, Tacke and coworkers (80) expressed modified movement proteins of PLRV in potato and observed broad-spectrum resistance to PLRV and unrelated PVX, and PVY. Resistance to PVX also was observed in potato plants expressing modified movement proteins of PVX (81). The broad-spectrum resistance induced by movement protein to viruses from five groups may be a consequence of defective or heterologous movement proteins competing with functional movement proteins for a limited number of target sites within the plasmodesmata.

Several plant viruses encode polyproteins that are processed by virus-encoded proteases. For example, the potyviruses produce a single polyprotein that is cleaved into mature virus proteins by three virus-encoded proteases (82). There are a limited number of examples of inhibiting polyprotein processing and all have involved potyviruses. When Vardi and associates (83) transformed tobacco with the PVY NIa protease, they observed that 2 of 50 plants were resistant to PVY. The authors suggested that changes during cloning may have produced dysfunctional protein. Similarly, resistance to TVMV in tobacco was conferred by genetically modified proteases (84).

## 3. Untranslatable Virus Sequences

Virus untranslatable nucleic acids have been exploited in engineering resistance to viruses from nine groups (Table 4). Antisense RNA and untranslatable sense RNA sequences derived from different regions of the viral genome have been used, including coat protein (32,34,40,44,50,85), movement protein (80,86), or replicase genes (87–89). Potato transformed with antisense RNA of the PLRV coat protein gene (32,34), a mutated sense RNA of the PLRV movement protein gene unable to produce detectable amounts of protein (80), or untranslatable sense and antisense constructs of the PVY coat protein gene (90) were resistant to the viruses from which the sequences were derived. The resistance is believed to involve an interaction of complementary virus and transgene sequences that disrupt translation, replication, stability, or access of host molecules. Resistance levels differ, depending on the sequence and virus being examined. High levels of resistance were observed with antisense RNA against specific plant viruses (32,87,91) but, surprisingly, other reports describe only limited protection against low levels of inoculum (44,50,85).

Sequences complementary to virus RNA have been developed to target sequence-specific ribozymes that catalyze biochemical restriction of single-stranded RNA at a specific sequence. Transgenic plants expressing a ribozyme specific for TMV or CMV RNA reduced virus titers (92,93). However, antisense RNA was reported to be as effective as the ribozyme (92).

## 4. Virus Replicated Sequences

Virus satellites and defective interfering molecules are parasites that rely on the associated helper virus for replication and have been reported to attenuate virus symptoms and multiplication (94–96). Transgenic plants expressing satellites or defective interfering sequences of viruses from six different groups have been developed (Table 5). Inoculation with the helper virus multiplies the transgene sequence, leading to a level of protection that is independent of inoculum levels and the

**Table 4** Genetically Engineered Virus Resistance Using Virus Untranslatable Nucleic Acid Sequences

| Virus group | Virus | Abbreviation | Plant host | Reference |
|---|---|---|---|---|
| *Antisense RNA* | | | | |
| Cucumovirus | Cucumber mosaic virus | CMV | Tobacco | (50) |
| | | | Tobacco | (85) |
| Geminivirus | Tomato golden mosaic virus | TGMV | Tobacco | (87) |
| | Tomato yellow leaf curl virus | TYLCV | *N. benthamiana*[a] | (181) |
| Luteovirus | Potato leafroll virus | PLRV | Potato | (32) |
| | | | Potato | (34) |
| | | | Potato | (182) |
| Nepovirus | Tomato ringspot virus | TomRSV | Tobacco, *N. benthamiana*[a] | (140) |
| Potexvirus | Potato virus X | PVX | Tobacco | (44) |
| | | | Tobacco | (88) |
| | Cymbidium mosaic virus | CyMV | *N. occidentalis*[a] | (183) |
| Potyvirus | Bean yellow mosaic virus | BYMV | *N. benthamiana*[a] | (145) |
| | Potato virus Y | PVY | Potato | (40) |
| | | | Tobacco | (90) |
| | Tobacco etch virus | TEV | Tobacco | (91) |
| | Zucchini yellow mosaic virus | ZYMV | Melon, tobacco | (150) |
| Tobamovirus | Tobacco mosaic virus | TMV | Tobacco | (184) |
| | | | Tobacco | (92) |
| Tospovirus | Tomato spotted wilt virus | TSWV | Tobacco | (185) |
| | | | Tobacco | (86) |
| *Sense RNA* | | | | |
| Luteovirus | Potato leafroll virus | PLRV | Potato | (80) |
| Potexvirus | Potato virus X | PVX | Tobacco | (88) |
| Potyvirus | Potato virus Y | PVY | Tobacco | (186) |
| | | | Tobacco | (40) |
| | | | Potato | (90) |
| | Tobacco etch virus | TEV | Tobacco | (91) |
| | | | Tobacco | (89) |
| | | | Tobacco | (119) |
| Tobamovirus | Tobacco mosaic virus | TMV | Tobacco | (187) |
| Tospovirus | Tomato spotted wilt virus | TSWV | Tobacco | (157) |
| | | | Tobacco | (158) |
| | | | Tobacco | (185) |
| | | | *N. benthamiana*[a] | (188) |
| | | | *N. benthamiana*[a] | (189) |
| | | | Tobacco | (86) |
| | | | *N. benthamiana*[a] | (167) |
| Tymovirus | Turnip yellow mosaic virus | TYMV | Rapeseed | (190) |
| *Ribozymes* | | | | |
| Cucumovirus | Cucumber mosaic virus | CMV | Tobacco | (93) |
| Tobamovirus | Tobacco mosaic virus | TMV | Tobacco | (92) |

[a]*Nicotiana* spp.

expression level of the transgene sequence (97,98). The virus-amplified transgene sequences are eventually encapsidated in helper virus coat protein and are transmitted along with the helper virus (97,99,100).

Full-length infectious virus clones of CMV or PVX expressed by transgenic plants conferred resistance to related viruses (101,102). Resistance may occur at several points during virus

**Table 5** Genetically Engineered Virus Resistance using Virus-Replicated Sequences

| Virus group | Virus | Abbreviation | Plant host | Reference |
|---|---|---|---|---|
| *Satellites* | | | | |
| Cucumovirus | Cucumber mosaic virus | CMV | Tobacco | (97) |
| | | | Tobacco | (98) |
| | | | Tobacco | (121) |
| | | | Tomato | (191) |
| Nepovirus | Tobacco ringspot virus | TobRV | Tobacco | (99) |
| Umbravirus | Groundnut rossette virus | GRV | *N. benthamiana*[a] | (192) |
| *Defective interfering sequences* | | | | |
| Geminivirus | African cassava mosaic virus | ACMV | *N. benthamiana*[a] | (100) |
| | Beet curly top virus | BCTV | *N. benthamiana*[a] | (193) |
| | | | *N. benthamiana*[a] | (194) |
| Tombusvirus | Cymbidium ringspot virus | CyRSV | *N. benthamiana*[a] | (195) |
| *Replicative virus clones* | | | | |
| Cucumovirus | Cucumber mosaic virus | CMV | Tobacco | (101) |
| Potexvirus | Potato virus X | PVX | Tobacco | (102) |

[a]*Nicotiana* spp.

multiplication, analogously to the original cross-protection phenomenon reported by McKinney (103). This strategy is a valuable tool in studying the mechanisms responsible for resistance. However, expression of a PLRV full-length clone was lethal in potato (104), and, as with satellites and defective interfering RNA, the potential application of replicating and complex constructs for developing resistance may be limited because of ecological concerns related to recombination.

## 5. Heterologous Genes

Heterologous proteins that confer plant virus resistance were identified (105–107) (Table 6). Resistance to PLRV, PVX, and PVY was reported in potato and tobacco expressing a pokeweed ribosome-inhibiting antiviral protein (105), and potato plants expressing a $2'$-$5'$ oligoadenylate synthetase, a component of the mammalian interferon-induced antiviral response, were resistant to PVX (107). Using the ACMV transactivatated promoter confined expression of the ribosome-inhibiting protein and the virus to cells involved in the primary infection (108). Expression of a double-stranded RNA yeast ribonuclease in plants provided resistance to ToMV, CMV, and PVY. The ribonuclease is thought to attack the double-stranded RNA that occurs during replication of many plant viruses (109). Expression of a rice cysteine proteinase inhibitor gene also induced resistance against two important potyviruses, tobacco etch (TEV) and potato virus Y (PVY), in transgenic potato plants (110).

*Nicotiana benthamiana* expressing single-chain plantibodies specific for the coat proteins of artichoke mottled crinkle tombusvirus (AMCV) or TMV exhibited virus-specific resistance (106,111). The expression of plantibodies to low-copy genes expressed early in the virus infection process such as the replicase may prove even more effective. Finally, isolation of the *N* gene that confers resistance to TMV facilitated the development of TMV-resistant transgenic tobacco and tomato plants (112–114). The N gene, a member of the Toll-IL-1 nucleotide binding–leucine-rich repeat (LRR) class of plant resistance genes, encodes two transcripts, N(S) and N(L), via alternative splicing (115). Only transgenic plants containing a cDNA-N(S)-bearing intron III and

**Table 6** Genetically Engineered Virus Resistance Using Heterologous Genes

| Virus group | Virus | Abbreviation | Plant host | Reference |
|---|---|---|---|---|
| *Antiviral* | | | | |
| Geminivirus | African cassava mosaic virus | ACMV | *N. benthamiana*[a] | (108) |
| Luteovirus | Potato leafroll virus | PLRV | Potato, tobacco | (105) |
| | | | Potato | (196) |
| Potexvirus | Potato virus | XPVX | Potato, tobacco | (105) |
| | | | Potato | (107) |
| | | | Potato | (196) |
| | | | Tobacco | (197) |
| Potyvirus | Potato virus Y | PVY | Potato, tobacco | (105) |
| | | | Potato | (196) |
| | | | Tobacco | (110) |
| | Tobacco etch virus | TEV | Tobacco | (110) |
| Tobamovirus | Tobacco mosaic virus | TMV | Tobacco | (198) |
| *Plantibody* | | | | |
| Furovirus | Beet necrotic yellow vein virus | BNYVV | *N. benthamiana*[a] | (199) |
| Tobamovirus | Tobacco mosaic virus | TMV | Tobacco | (111) |
| Tombusvirus | Artichoke mottled crinkle virus | AMCV | *N. benthamiana*[a] | (106) |
| *Resistance gene* | | | | |
| Tobamovirus | Tobacco mosaic virus | TMV | Tobacco | (112) |
| | | | Tobacco | (114) |
| | | | Tomato | (113) |
| | | | Tobacco | (115) |
| *Ribonuclease* | | | | |
| Tobamovirus | Tomato mosaic virus | ToMV | Tobacco | (109) |
| Cucomovirus | Cucumber mosaic virus | CMV | Tobacco | (109) |
| Potyvirus | Potato virus Y | PVY | Tobacco | (109) |

[a]*Nicotiana.* spp.

containing 3' N-genomic sequences, encoding both N(S) and N(L) transcripts, exhibit complete resistance to TMV. Heterologous proteins represent a diverse strategy with enormous potential for virus control that may be specific to a particular virus or provide resistance to several unrelated viruses.

## 6. Future Developments

Strategies for engineering virus resistance have increased rapidly since the initial report of transformation with the TMV coat protein in 1986. New strategies for engineering resistance to potato viruses will evolve from the characterization of pathogen- and host-derived genes, virus genome organization and expression, and the process of virus infection including disassembly, replication, encapsidation, movement, and transmission. Strategies are not necessarily independent of one another, and it is possible that resistance involves protein, nucleic acid, and other interactions. For example, homology-dependent silencing is a relatively recent mechanism that may also operate in many virus control strategies (116). This posttranscriptional mechanism causes suppression of transgene expression, possibly by methylation, and only requires sequence similarity between the transgene and virus to produce resistance (117). Experiments have confirmed that resistance to TEV (118), PVX (88), and PVY (119) involves homology-dependent silencing.

Gene expression can be localized in specific tissues (120,39), optimized to coincide with the virus infection (108), and adjusted to achieve the highest possible levels of resistance. In addition to potentially improving yield and quality, engineered virus resistance may reduce transmission by vectors (Table 7) and thereby reduce the use of pesticides. Furthermore, several strategies may be combined to increase the level and number of viruses that the host is resistant. Yie and coworkers (121) were able to increase resistance to CMV by expressing satellite RNA and the coat protein gene in tobacco. Barker and colleagues (36) transformed PLRV-resistant germplasm with the PLRV coat protein gene and reported the resistances were additive. Progress has also been made in combining the coat protein gene of more than one virus in a single transformed plant. Potato plants transformed with the coat protein genes of PVX and PVY were resistant to both viruses (29). Similarly, plants transformed with the coat protein genes of CMV, watermelon mosaic potyvirus (WMV2), and zucchini yellow mosaic potyvirus (ZYMV) (122) or tomato spotted wilt tospovirus (TSWV), tomato chlorotic spot tospovirus (TCSV), and groundnut ringspot virus (GRSV) (123) exhibited resistance to the viruses from which the sequences were derived. A sequence expressing TEV nuclear inclusion proteinase, TMV coat protein, and SMV coat protein from a single self-processing polypeptide was used to produce transgenic tobacco resistant to TEV, TMV, and PVY (124).

Several concerns have been expressed regarding the production of transgenic plants resistant to viruses. One concern is the possibility of recombination between a virus and virus-derived genes within the plant, creating a virus with new characteristics. Such recombinations have been reported to occur between viruses and transgenes of the cowpea chlorotic mottle bromovirus (CCMV) (125) and ACMV (126), but similar opportunities occur in mixed infections. Using the

Table 7  Influence of Genetic Engineering on Vector Transmission

| Virus group | Virus | Vector | Plant host | Reference |
|---|---|---|---|---|
| Cucumovirus | Cucumber mosaic virus | Aphid | Tobacco | (98) |
| | | | Tobacco | (61) |
| | | | Cucumber | (133) |
| | | | Tobacco | (164) |
| Geminivirus | Tomato yellow leaf curl virus | Whitefly | Tomato | (47) |
| Luteovirus | Potato leafroll virus | Aphid | Potato | (31) |
| | | | Potato | (33) |
| | | | Potato | (34) |
| | | | Potato | (35) |
| | | | Potato | (37) |
| | | | Potato | (38) |
| | | | Potato | (39) |
| | | | Potato | (33) |
| | | | Potato, tobacco | (105) |
| Pomovirus[a] | Potato mop top virus | Fungus | *N. benthamiana*[b] | (141) |
| Potexvirus | Potato virus X | Aphid | Potato | (29) |
| Potyvirus | Potato virus Y | Aphid | Potato | (29) |
| | Tobacco etch virus | Aphid | Tobacco | (91) |
| | | | Potato, tobacco | (105) |
| Tenuivirus | Rice stripe virus | Planthopper | Rice | (45) |
| Tospovirus | Tomato spotted wilt virus | Thrips | Tobacco | (157) |

[a]Previously classified as a furovirus.
[b]*Nicotiana* spp.

smallest possible fragment of the virus genome should reduce the possibility for recombination. Another concern is the potential for transformed potato plants to outcross with weedy species. However, many potato growing areas do not have weedy species that hybridize with potato, and where such species coexist, the incorporation of resistance in a weedy species could reduce virus reservoirs. Risk associated with a particular strategy depends on the gene, and each transformant must be evaluated individually. Overall, most examples to date have considerable merit, pose minimal risk, and represent a nucleic acid or protein equivalency in that the molecule occurs naturally. For additional details on the generalities of transgenic technologies against viruses, see Chapter 15.

## V. CONCLUSIONS

An increasing number of potato plants genetically engineered to express value-added traits are being developed and released for commercialization. Many strategies have been developed to improve agronomic performance by producing economically important products and exhibiting disease and pest resistance. Presently, the greatest impediments to further development of genetically engineered traits in crops such as potato, are business rather than science-related.

## REFERENCES

1. U Sonnewald, MR Hajirezaei, J Kossmann, A Heyer, RN Trethewey, L Willmitzer. Increased potato tuber size resulting from apoplastic expression of a yeast invertase. Nat Biotechnol 15:794–797, 1997.
2. B Muller-Rober, U Sonnewald, L Willmitzer. Inhibition of the ADP-glucose pyrophosphorylase in transgenic potatoes leads to sugar-storing tubers and influences tuber formation and expression of tuber storage protein genes. EMBO J 11:1229–1238, 1992.
3. LJ Sweetlove, MM Burrell, T ap Rees. Starch metabolism in tubers of transgenic potato (*Solanum tuberosum*) with increased ADPglucose pyrophosphorylase. Biochem J 320:493–498, 1996.
4. S Greiner, T Rausch, U Sonnewald, K Herbers. Ectopic expression of a tobacco invertase inhibitor homolog prevents cold-induced sweetening of potato tubers. Nat Biotechnol 17:708–711, 1999.
5. R Lorberth, G Ritte, L Willmitzer, J Kossmann. Inhibition of a starch-granule-bound protein leads to modified starch and repression of cold sweetening. Nat Biotechnol 16:473–477, 1998.
6. S Chakraborty, N Chakraborty, A Datta. Increased nutritive value of transgenic potato by expressing a nonallergenic seed albumin gene from *Amaranthus hypochondriacus*. Proc Natl Acad Sci USA 97: 3724–3729, 2000.
7. X Ye, S Al-Babili, A Kloti, J Zhang, P Lucca, P Beyer, I Potrykus. Engineering the provitamin A (beta-carotene) biosynthetic pathway into (carotenoid-free) rice endosperm. Science 287:303–305, 2000.
8. JG Wallis, H Wang, DJ Guerra. Expression of a synthetic antifreeze protein in potato reduces electrolyte release at freezing temperatures. Plant Mol Biol 35:323–330, 1997.
9. CK Shewmaker, CD Boyer, DP Wiesenborn, DB Thompson, MR Boersig, JV Oakes, DM Stalker. Expression of *Escherichia coli* glycogen synthase in the tubers of transgenic potatoes (*Solanum tuberosum*) results in a highly branched starch. Plant Physiol 104:1159–1166, 1994.
10. RG Visser, I Somhorst, GJ Kuipers, NJ Ruys, WJ Feenstra, E Jacobsen. Inhibition of the expression of the gene for granule-bound starch synthase in potato by antisense constructs. Mol Gen Genet 225:289–296, 1991.
11. TA Haq, HS Mason, JD Clements, CJ Arntzen. Oral immunization with a recombinant bacterial antigen produced in transgenic plants. Science 268:714–716, 1995.
12. T Arakawa, DK Chong, WH Langridge. Efficacy of a food plant-based oral cholera toxin B subunit vaccine. Nat Biotechnol 16:292–297, 1998.

13. T Arakawa, J Yu, DK Chong, J Hough, PC Engen, WH Langridge. A plant-based cholera toxin B subunit-insulin fusion protein protects against the development of autoimmune diabetes. Nat Biotechnol 16:934–938, 1998.
14. A Hiatt, R Cafferkey, K Bowdish. Production of antibodies in transgenic plants. Nature 342:76–78, 1989.
15. A Schouten, J Roosien, JM de Boer, A Wilmink, MN Rosso, D Bosch, WJ Stiekema, FJ Gommers, J Bakker, A Schots. Improving scFv antibody expression levels in the plant cytosol. FEBS Lett 4:235–241, 1997.
16. MJ Adang, MS Brody, G Cardineau, N Eagan, RT Roush, CK Shewmaker, A Jones, JV Oakes, KE McBride. The reconstruction and expression of a *Bacillus thuringiensis* cryIIIA gene in protoplasts and potato plants. Plant Mol Biol 21:1131–1145, 1993.
17. DS Douches, AL Westedt, K Zarka, EJ Grafius. Transformation of Cry V-Bt transgene combined with natural resistance mechanisms for resistance to tuber moth in potato (*Solanum tuberosum* L.). Hortscience 33:1053–1056, 1998.
18. M Rossi, FL Goggin, SB Milligan, I Kaloshian, DE Ullman, VM Williamson. The nematode resistance gene Mi of tomato confers resistance against the potato aphid. Proc Natl Acad Sci USA 95:9750–9754, 1998.
19. R Johnson, J Narvaez, G An, C Ryan. Expression of proteinase inhibitors I and II in transgenic tobacco plants: Effects on natural defense against *Manduca sexta* larvae. Proc Natl Acad Sci USA 86:9871–9875, 1989.
20. G Wu, BJ Shortt, EB Lawrence, EB Levine, KC Fitzsimmons, DM Shah. Disease resistance conferred by expression of a gene encoding $H_2O_2$-generating glucose oxidase in transgenic potato plants. Plant Cell 7:1357–1368, 1995.
21. M Borkowska, M Krzymowska, A Talarczyk, MF Awan, L Yakovleva, K Kleczkowski, B Wielgat. Transgenic potato plants expressing soybean beta-1,3-endoglucanase gene exhibit an increased resistance to *Phytophthora infestans*. Z Naturforsch 53:1012–1016, 1998.
22. M Lorito, SL Woo, I Garcia, G Colucci, GE Harman, JA Pintor-Toro, E Filippone, S Muccifora, CB Lawrence, A Zoina, S Tuzun, F Scala. Genes from mycoparasitic fungi as a source for improving plant resistance to fungal pathogens. Proc Natl Acad Sci USA 95:7860–7865, 1998.
23. K Düring, P Porsch, M Fladung, H Lörz. Transgenic potato plants resistant to the phytopathogenic bacterium *Erwinia carotovora*. Plant J 3:587–598, 1993.
24. X Yang, Y Yie, F Zhu, Y Liu, L Kang, X Wang, P Tien. Ribozyme-mediated high resistance against potato spindle tuber viroid in transgenic potatoes. Proc Natl Acad Sci USA 94:4861–4865, 1997.
25. T Sano, A Nagayama, T Ogawa, I Ishida, Y Okada. Transgenic potato expressing a double-stranded RNA-specific ribonuclease is resistant to potato spindle tuber viroid. Nat Biotechnol 15:1290–1294, 1997.
26. PA Powell, RS Nelson, B De, N Hoffmann, SG Rogers, RT Fraley, RN Beachy. Delay of disease development in transgenic plants that express the tobacco mosaic virus coat protein gene. Science 232:738–743, 1986.
27. A Hoekema, MJ Huisman, L Molendijk, PJM van den Elzen, BJC Cornelissen. The genetic engineering of two commercial potato cultivars for resistance to potato virus X. Biotechnoloy 7:273–278, 1989.
28. W Kaniewski, C Lawson, B Sammons, L Haley, J Hart, X Delannay, NE Tumer. Field resistance of transgenic Russet Burbank potato to effects of infection by potato virus X and potato virus Y. Biotechnology 8:750–754, 1990.
29. C Lawson, W Kaniewski, L Haley, R Rozman, C Newell, P Sanders, NE Tumer. Engineering resistance to mixed virus infection in a commercial potato cultivar: resistance to potato virus X and potato virus Y in transgenic Russet Burbank. Biotechnology 8:127–134, 1990.
30. E Jongedijk, AAJM de Schutter, T Stolte, PJM van den Elzen, BJC Cornelissen. Increased resistance to potato virus X and preservation of cultivar properties in transgenic potato under field conditions. Biotechnology 10:422–429, 1992.
31. LM Kawchuk, RR Martin, J McPherson. Resistance in transgenic potato expressing the potato leafroll virus coat protein gene. Mol Plant Microbe Interact 3:301–307, 1990.

32. LM Kawchuk, RR Martin, J McPherson. Sense and antisense RNA-mediated resistance to potato leafroll virus in Russet Burbank potato plants. Mol Plant Microbe Interact 4:247–253, 1991.
33. LM Kawchuk, DR Lynch, RR Martin, GC Kozub, B Farries. Field resistance to the potato leafroll luteovirus reduces tuber disease symptoms in transgenic and somaclonal variant potato plants. Can J Plant Pathol 19:260–266, 1997.
34. F van der Wilk, DP-L Willink, MJ Huisman, H Huttinga, R Goldbach. Expression of the potato leafroll luteovirus coat protein gene in transgenic potato plants inhibits viral infection. Plant Mol Biol 17:431–439, 1991.
35. H Barker, B Reavy, A Kumar, KD Webster, MA Mayo. Restricted virus multiplication in potatoes transformed with the coat protein gene of potato leafroll luteovirus: Similarities with a type of host gene-mediated resistance. Ann Appl Biol 120:55–64, 1992.
36. H Barker, KD Webster, CA Jolly, B Reavy, A Kumar, MA Mayo. Enhancement of resistance to potato leafroll virus multiplication in potato by combining the effects of host genes and transgenes. Mol Plant Microbe Interact 7:528–530, 1994.
37. GG Presting, OP Smith, CR Brown. Resistance to potato leafroll virus in potato plants transformed with the coat protein gene or with vector control constructs. Phytopathology 85:436–442, 1995.
38. PM Derrick, H Barker. Short and long distance spread of potato leafroll luteovirus: Effects of host genes and transgenes conferring resistance to virus accumulation in potato. J Gen Virol 78:243–251, 1997.
39. MW Graham, S Craig, PM Waterhouse. Expression patterns of vascular-specific promoters RolC and Sh in transgenic potatoes and their use in engineering PLRV-resistant plants. Plant Mol Biol 33:729–735, 1997.
40. HA Smith, SL Swaney, TD Parks, EA Wernsman, WG Dougherty. Transgenic plant virus resistance mediated by untranslatable sense RNAs: Expression, regulation, and fate of nonessential RNAs. Plant Cell 6:1441–1453, 1994.
41. DJ MacKenzie, JH Tremaine, J McPherson. Genetically engineered resistance to potato virus S in potato cultivar Russet Burbank. Mol Plant Microbe Interact 4:95–102, 1991.
42. LS Loesch-Fries, D Merlo, T Zinnen, L Burhop, K Hill, K Krahn, N Jarvis, S Nelson, E Halk. Expression of alfalfa mosaic virus RNA 4 in transgenic plants confers virus resistance. EMBO J 6:1845–1851, 1987.
43. M Ravelonandro, M Monsion, R Delbos, J Dunez. Variable resistance to plum pox virus and potato virus Y infection in transgenic *Nicotiana* plants expressing plum pox virus coat protein. Plant Sci 91:157–169, 1993.
44. C Hemenway, R-X Fang, WK Kaniewski, N-H Chua, NE Tumer. Analysis of the mechanism of protection in transgenic plants expressing the potato virus X coat protein or its antisense RNA. EMBO J 7:1273–1280, 1988.
45. T Hayakawa, Y Zhu, K Itoh, Y Kimura, T Izawa, K Shimamoto, S Toriyama. Genetically engineered rice resistant to rice stripe virus, an insect-transmitted virus. Proc Natl Acad Sci USA 89:9865–9869, 1992.
46. PA Powell, PR Sanders, N Tumer, RT Fraley, RN Beachy. Protection against tobacco mosaic virus infection in transgenic plants requires accumulation of coat protein rather than coat protein RNA sequences. Virology 175:124–130, 1990.
47. T Kunik, R Salomon, D Zamir, N Navot, M Zeidan, I Michelson, Y Gafni, H Czosnek. Transgenic tomato plants expressing the tomato yellow leaf curl virus capsid protein are resistant to the virus. Biotechnology 12:500–504, 1994.
48. A Nejidat, RN Beachy. Decreased levels of TMV coat protein in transgenic tobacco plants at elevated temperatures reduce resistance to TMV infection. Virology 173:531–538, 1989.
49. JC Register, RN Beachy. Resistance to TMV in transgenic plants results from interference with an early event in infection. Virology 166:524–532, 1988.
50. M Cuozzo, KM O'Connell, W Kaniewski, R-X Fang, N-H Chua, NE Tumer. Viral protection in transgenic tobacco plants expressing the cucumber mosaic virus coat protein or its antisense RNA. Biotechnology 6:549–557, 1988.

51. CMP van Dun, JF Bol. Transgenic tobacco plants accumulating tobacco rattle virus coat protein resist infection with tobacco rattle virus and pea early browning virus. Virology 167:649–652, 1988.
52. GC Angenent, JMW van Den Ouweland, JF Bol. Susceptibility to virus infection of transgenic tobacco plants expressing structural and nonstructural genes of tobacco rattle virus. Virology 175:191–198, 1990.
53. H Lecoq, M Ravelonandro, C Wipf-Scheibel, M Monsion, B Raccah, J Dunez. Aphid transmission of a non-aphid-transmissable strain of zucchini yellow mosaic potyvirus from transgenic plants expressing the capsid protein of plum pox potyvirus. Mol Plant Microbe Interact 6:403–406, 1993.
54. J Hammond, MM Dienelt. Encapsidation of potyviral RNA in various forms of transgene coat protein is not correlated with resistance in transgenic plants. Mol Plant Microbe Interact 10:1023–1027, 1997.
55. V Yusibov, LS Loesch-Fries. High-affinity RNA-binding domains of alfalfa mosaic virus coat protein are not required for coat protein-mediated resistance. Proc Natl Acad Sci USA 92:8980–8984, 1995.
56. M Bendahmane, JH Fitchen, G Zhang, RN Beachy. Studies of coat protein-mediated resistance to tobacco mosaic tobamovirus: correlation between assembly of mutant coat proteins and resistance. J Virol 71:7942–7950, 1997.
57. NE Tumer, KM O'Connell, RS Nelson, PR Sanders, RN Beachy, RT Fraley, DM Shah. Expression of alfalfa mosaic virus coat protein gene confers cross-protection in transgenic tobacco and tomato plants. EMBO J 6:1181–1188, 1987.
58. DJ MacKenzie, PJ Ellis. Resistance to tomato spotted wilt virus infection in transgenic tobacco expressing the viral nucleocapsid gene. Mol Plant Microbe Interact 5:34–40, 1992.
59. DM Stark, RN Beachy. Protection against potyvirus infection in transgenic plants: evidence for broad spectrum resistance. Biotechnology 7:1257–1262, 1989.
60. S Namba, K Ling, C Gonsalves, JL Slightom, D Gonsalves. Protection of transgenic plants expressing the coat protein gene of watermelon mosaic virus II or zucchini yellow mosaic virus against six potyviruses. Phytopathology 82:940–946, 1992.
61. HD Quemada, D Gonsalves, JL Slightom. Expression of coat protein gene from cucumber mosaic virus strain C in tobacco: protection against infections by CMV strains transmitted mechanically or by aphids. Phytopathology 81:794–802, 1991.
62. DJ MacKenzie, JH Tremaine. Transgenic *Nicotiana debneyii* expressing viral coat protein are resistant to potato virus S infection. J Gen Virol 71:2167–2170, 1990.
63. DJ Bertioli, JI Cooper, ML Edwards, WS Hawes. Arabis mosaic nepovirus coat protein in transgenic tobacco lessens disease severity and virus replication. Ann Appl Biol 120:47–54, 1992.
64. K Ling, S Namba, C Gonsalves, JL Slightom, D Gonsalves. Protection against detrimental effects of potyvirus infection in transgenic tobacco plants expressing the papaya ringspot virus coat protein gene. Biotechnology 9:752–758, 1991.
65. RS Nelson, SM McCormick, X Delannay, P Dube, J Layton, EJ Anderson, M Kaniewska, RK Proksch, RB Horsch, SG Rogers, RT Fraley, RN Beachy. Virus tolerance, plant growth, and field performance of transgenic tomato plants expressing coat protein from tobacco mosaic virus. Biotechnology 6:403–409, 1988.
66. S Dinant, F Blaise, C Kusiak, S Astier-Manifacier, J Albouy. Heterologous resistance to potato virus Y in transgenic tobacco plants expressing the coat protein gene of lettuce mosaic potyvirus. Phytopathology 83:818–824, 1993.
67. LE Murry, LG Elliott, SA Capitant, JA West, KK Hanson, L Scarafia, S Johnston, C DeLuca-Flaherty, S Nichols, D Cunanan, PS Dietrich, IJ Mettler, S Dewald, DA Warnick, C Rhodes, RM Sinibaldi, KJ Brunke. Transgenic corn plants expressing MDMV strain B coat protein are resistant to mixed infections of maize dwarf mosaic virus and maize chlorotic mottle virus. Biotechnology 11:1559–1564, 1993.
68. DB Golemboski, GP Lomonossoff, M Zaitlin. Plants transformed with a tobacco mosaic virus nonstructural gene sequence are resistant to the virus. Proc Natl Acad Sci USA 87:6311–6315, 1990.
69. CJ Braun, CL Hemenway. Expression of amino-terminal portions or full-length viral replicase genes in transgenic plants confers resistance to potato virus X infection. Plant Cell 4:735–744, 1992.
70. M Longstaff, G Brigneti, F Boccard, S Chapman, D Baulcombe. Extreme resistance to potato virus

X infection in plants expressing a modified component of the putative viral replicase. EMBO J 12:379–386, 1993.
71. P Audy, P Palukaitis, SA Slack, M Zaitlin. Replicase-mediated resistance to potato virus Y in transgenic tobacco plants. Mol Plant Microbe Interact 7:15–22, 1994.
72. K-H Hellwald, P Palukaitis. Viral RNA as a potential target for two independent mechanisms of replicase-mediated resistance against cucumber mosaic virus. Cell 83:937–946, 1995.
73. L Nguyen, WJ Lucas, B Ding, M Zaitlin. Viral RNA trafficking is inhibited in replicase-mediated resistant transgenic tobacco plants. Proc Natl Acad Sci USA 93:12643–12647, 1996.
74. YM Pinto, RA Kok, DC Baulcombe. Resistance to rice yellow mottle virus (RYMV) in cultivated African rice varieties containing RYMV transgenes. Nat Biotechnol 17:702–707, 1999.
75. W Kaniewski, C Lawson, J Loveless, P Thomas, T Mowry, G Reed, T Mitsky, J Zalewski, Y Muskopf. Expression of potato leafroll virus (PLRV) replicase genes in Russet Burbank potatoes provide field immunity to PLRV. In: M Mańka, ed. Environmental Biotic Factors in Integrated Plant Disease Control. Poznań: The Polish Phytopathological Society, 1995, pp 289–292.
76. CM Deom, M Lapidot, RN Beachy. Plant virus movement proteins. Cell 69:221–224, 1992.
77. S Wolf, CM Deom, RN Beachy, WJ Lucas. Movement protein of tobacco mosaic virus modifies plasmodesmatal size exclusion limit. Science 246:377–379, 1989.
78. M Lapidot, R Gafny, B Ding, S Wolf, WJ Lucas, RN Beachy. A dysfunctional movement protein of tobacco mosaic virus that partially modifies the plasmodesmata and limits virus spread in transgenic plants. Plant J 4:959–970, 1993.
79. SI Malyshenko, OA Kondakova, JV Nazarova, IB Kaplan, ME Taliansky, JG Atabekov. Reduction of tobacco mosaic virus accumulation in transgenic plants producing non-functional viral transport proteins. J Gen Virol 74:1149–1156, 1993.
80. E Tacke, F Salamini, W Rohde. Genetic engineering of potato for broad-spectrum protection against virus infection. Nat Biotechnol 14:1597–1601, 1996.
81. P Seppänen, R Puska, J Honkanen, LG Tyulkina, O Fedorkin, SY Morozov, JG Atabekov. Movement protein-derived resistance to triple gene block-containing plant viruses. J Gen Virol 78:1241–1246, 1997.
82. WG Dougherty, JC Carrington. Expression and function of potyviral gene products. Annu Rev Phytopathol 26:123–143, 1988.
83. E Vardi, I Sela, O Edelbaum, O Livneh, L Kuznetsova, Y Stram. Plants transformed with a cistron of a potato virus Y protease (NIa) are resistant to virus infection. Proc Natl Acad Sci USA 90:7513–7517, 1993.
84. IB Maiti, JF Murphy, JG Shaw, AG Hunt. Plants that express a potyvirus proteinase gene are resistant to virus infection. Proc Natl Acad Sci USA 90:6110–6114, 1993.
85. MA Rezaian, KGM Skene, JG Ellis. Anti-sense RNAs of cucumber mosaic virus in transgenic plants assessed for control of the virus. Plant Mol Biol 11:463–471, 1988.
86. M Prins, M Kikkert, C Ismayadi, W de Graauw, P de Haan, R Goldbach. Characterization of RNA-mediated resistance to tomato spotted wilt virus in transgenic tobacco plants expressing NS(M) gene sequences. Plant Mol Biol 33:235–243, 1997.
87. AG Day, ER Bejarano, KW Buck, M Burrell, CP Lichtenstein. Expression of an antisense viral gene in transgenic tobacco confers resistance to the DNA virus tomato golden mosaic virus. Proc Natl Acad Sci USA 88:6721–6725, 1991.
88. E Mueller, J Gilbert, G Davenport, G Brigneti, DC Baulcombe. Homology-dependent resistance: transgenic virus resistance in plants related to homology-dependent gene silencing. Plant J 7:1001–1013, 1995.
89. S Swaney, H Powers, J Goodwin, LS Rosales, WG Dougherty. RNA-mediated resistance with nonstructural genes from the tobacco etch virus genome. Mol Plant Microbe Interact 8:1004–1011, 1995.
90. HA Smith, H Powers, S Swaney, C Brown, WG Dougherty. Transgenic potato virus Y resistance in potato: evidence for an RNA-mediated cellular response. Phytopathology 85:864–870, 1995.
91. JA Lindbo, WG Dougherty. Untranslatable transcripts of the tobacco etch virus coat protein gene sequence can interfere with tobacco etch virus replication in transgenic plants and protoplasts. Virology 189:725–733, 1992.

92. R de Feyter, M Young, K Schroeder, ES Dennis, W Gerlach. A ribozyme gene and an antisense gene are equally effective in conferring resistance to tobacco mosaic virus on transgenic tobacco. Mol Gen Genet 250:329–338, 1996.
93. CS Kwon, WI Chung, KH Paek. Ribozyme mediated targeting of cucumber mosaic virus RNA 1 and 2 in transgenic tobacco plants. Mol Cells 7:326–334, 1997.
94. BI Hillman, JC Carrington, TJ Morris. A defective interfering RNA that contains a mosaic of a plant virus genome. Cell 51:427–433, 1987.
95. CW Collmer, SH Howell. Role of satellite RNA in the expression of symptoms caused by plant viruses. Annu Rev Phytopathol 30:419–442, 1992.
96. MJ Roossinck, D Sleat, P Palukaitis. Satellite RNAs of plant viruses: Structures and biological effects. Microbiol Rev 56:265–279, 1992.
97. BD Harrison, MA Mayo, DC Baulcombe. Virus resistance in transgenic plants that express cucumber mosaic virus satellite RNA. Nature 328:799–802, 1987.
98. M Jacquemond, J Amselem, M Tepfer. A gene coding for a monomeric form of cucumber mosaic virus satellite RNA confers tolerance to CMV. Mol Plant Microbe Interact 1:311–316, 1988.
99. WL Gerlach, D Llewellyn, J Haseloff. Construction of a plant disease resistance gene from the satellite RNA of tobacco ringspot virus. Nature 328:802–805, 1987.
100. J Stanley, T Frischmuth, S Ellwood. Defective viral DNA ameliorates symptoms of geminivirus infection in transgenic plants. Proc Natl Acad Sci USA 87:6291–6295, 1990.
101. M Suzuki, C Masuta, Y Takanami, S Kuwata. Resistance against cucumber mosaic virus in plants expressing the viral replicon. FEBS Lett 379:26–30, 1996.
102. SM Angell, DC Baulcombe. Consistent gene silencing in transgenic plants expressing a replicating potato virus X RNA. EMBO J 16:3675–3684, 1997.
103. HH McKinney. Mosaic diseases in the Canary Islands, West Africa, and Gibraltar. J Agric Res 39:557–578, 1929.
104. D Prüfer, J Schmitz, E Tacke, B Kull, W Rohde. In vivo expression of a full-length cDNA copy of potato leafroll virus (PLRV) in protoplasts and transgenic plants. Mol Gen Genet 253:609–614, 1997.
105. JK Lodge, WK Kaniewski, NE Tumer. Broad-spectrum virus resistance in transgenic plants expressing pokeweed antiviral protein. Proc Natl Acad Sci USA 90:7089–7093, 1993.
106. P Tavladoraki, E Benvenuto, S Trinca, D De Martinis, A Cattaneo, P Galeffi. Transgenic plants expressing a functional single-chain Fv antibody are specifically protected from virus attack. Nature 366:469–472, 1993.
107. E Truve, A Aaspollu, J Honkanen, R Puska, M Mehto, A Hassi, TH Teeri, M Kelve, P Seppänen, M Saarma. Transgenic potato plants expressing mammalian 2′-5′ oligoadenylate synthetase are protected from potato virus X infection under field conditions. Biotechnology 11:1048–1052, 1993.
108. Y Hong, K Saunders, MR Hartley, J Stanley. Resistance to geminivirus infection by virus-induced expression of dianthin in transgenic plants. Virology 220:119–127, 1996.
109. Y Watanabe, T Ogawa, H Takahashi, I Ishida, Y Takeuchi, M Yamamoto, Y Okada. Resistance against multiple plant viruses in plants mediated by a double stranded-RNA specific ribonuclease. FEBS Lett 372:165–168, 1995.
110. R Gutierrez-Campos, JA Torres-Acosta, LJ Saucedo-Arias, MA Gomez-Lim. The use of cysteine proteinase inhibitors to engineer resistance against potyviruses in transgenic tobacco plants. Nat Biotechnol 17:1223–1226, 1999.
111. A Voss, M Niersbach, R Hain, HJ Hirsch, YC Liao, F Kreuzaler, R Fischer. Reduced virus infectivity in N. tabacum secreting a TMV-specific full-size antibody. Mol Breed 1:39–50, 1995.
112. S Whitham, SP Dinesh-Kumar, D Choi, R Hehl, C Corr, B Baker. The product of the tobacco mosaic virus resistance gene N: Similarity to toll and the interleukin-1 receptor. Cell 78:1101–1115, 1994.
113. S Whitham, S McCormick, B Baker. The N gene of tobacco confers resistance to tobacco mosaic virus in transgenic tomato. Proc Natl Acad Sci USA 93:8776–8781, 1996.

114. SP Dinesh-Kumar, S Whitham, D Choi, R Hehl, C Corr, B Baker. Transposon tagging of tobacco mosaic virus resistance gene N: Its possible role in the TMV-N-mediated signal transduction pathway. Proc Natl Acad Sci USA 92:4175–4180, 1995.
115. SP Dinesh-Kumar, BJ Baker. Alternatively spliced N resistance gene transcripts: Their possible role in tobacco mosaic virus resistance. Proc Natl Acad Sci USA 97:1908–1913, 2000.
116. DC Baulcombe. Mechanisms of pathogen-derived resistance to viruses in transgenic plants. Plant Cell 8:1833–1844, 1996.
117. JJ English, E Mueller, DC Baulcombe. Suppression of virus accumulation in transgenic plants exhibiting silencing of nuclear genes. Plant Cell 8:179–188, 1996.
118. JA Lindbo, L Silva-Rosales, WM Proebsting, WG Dougherty. Induction of a highly specific antiviral state in transgenic plants: implications for regulation of gene expression and virus resistance. Plant Cell 5:1749–1759, 1993.
119. J Goodwin, K Chapman, S Swaney, TD Parks, EA Wernsman, WG Dougherty. Genetic and biochemical dissection of transgenic RNA-mediated virus resistance. Plant Cell 8:95–105, 1996.
120. U Reimann-Philipp, R Beachy. Coat protein-mediated resistance in transgenic tobacco expressing the tobacco mosaic virus coat protein from tissue-specific promoters. Mol Plant Microbe Interact 6:323–330, 1993.
121. Y Yie, F Zhao, SZ Zhao, YZ Liu, YL Liu, P Tien. High resistance to cucumber mosaic virus conferred by satellite RNA and coat protein in transgenic commercial tobacco cultivar G-140. Mol Plant Microbe Interact 5:460–465, 1992.
122. DM Tricoli, KJ Carney, PF Russell, JR McMaster, DW Groffl, KC Hadden, PT Himmel, JP Hubbard, ML Boeshore, HD Quemada. Field evaluation of transgenic squash containing single or multiple virus coat protein gene constructs for resistance to cucumber mosaic virus, watermelon mosaic virus 2, and zucchini yellow mosaic virus. Biotechnology 13:1458–1465, 1995.
123. M Prins, P de Haan, R Luyten, M van Veller, MQJM van Grinsven, R Goldbach. Broad resistance to tospoviruses in transgenic tobacco plants expressing three tospoviral nucleoprotein gene sequences. Mol Plant Microbe Interact 8:85–91, 1995.
124. JF Marcos, RN Beachy. Transgenic accumulation of two plant virus coat proteins on a single self-processing polypeptide. J Gen Virol 78:1771–1778, 1997.
125. AE Greene, RF Allison. Recombination between viral RNA and transgenic plant transcripts. Science 263:1423–1425, 1994.
126. T Frischmuth, J Stanley. Recombination between viral DNA and the transgenic coat protein gene of African cassava mosaic geminivirus. J Gen Virol 79:1265–1271, 1998.
127. Reference deleted.
128. CMP van Dun, JF Bol, L van Vloten-Doting. Expression of alfalfa mosaic virus and tobacco rattle virus coat protein genes in transgenic tobacco plants. Virology 159:299–305, 1987.
129. KK Hill, N Jarvis-Eagan, EL Halk, KJ Krahn, LW Liao, RS Mathewson, DJ Merlo, SE Nelson, KE Rashka, LS Loesch-Fries. The development of virus-resistant alfalfa Medicago sativa L. Biotechnology 9:373–377, 1991.
130. PEM Taschner, G van Marle, FT Brederode, NE Tumer, JF Bol. Plants transformed with a mutant alfalfa mosaic virus coat protein gene are resistant to the mutant but not to wild-type virus. Virology 203:269–276, 1994.
131. S Namba, K Ling, C Gonsalves, D Gonsalves, JL Slightom. Expression of the gene encoding the coat protein of cucumber mosaic virus (CMV) strain WL appears to provide protection to tobacco plants against infection by several different CMV strains. Gene 107:181–188, 1991.
132. M Nakajima, T Hayakawa, I Nakamura, M Suzuki. Protection against cucumber mosaic virus (CMV) strains O and Y and chrysanthemum mild mottle virus in transgenic tobacco plants expressing CMV-O coat protein. J Gen Virol 74:319–322, 1993.
133. D Gonsalves, P Chee, R Provvidenti, R Seem, JL Slightom. Comparison of coat protein-mediated and genetically-derived resistance in cucumbers to infection by cucumber mosaic virus under field conditions with natural challenge inoculations by vectors. Biotechnology 10:1562–1570, 1992.
134. E Noris, GP Accotto, R Tavazza, A Brunetti, S Crespi, M Tavazza. Resistance to tomato yellow leaf

curl geminivirus in *Nicotiana benthamiana* plants transformed with a truncated viral C1 gene. Virology 224:130–138, 1996.
135. GM Chowrira, TD Cavileer, SK Gupta, PF Lurquin, PH Berger. Coat protein-mediated resistance to pea enation mosaic virus in transgenic Pisum sativum L. Transgenic Res 7:265–271, 1998.
136. CM van Dun, B Overduin, L van Vloten-Doting, JF Bol. Transgenic tobacco expressing tobacco streak virus or mutated alfalfa mosaic virus coat protein does not cross-protect against alfalfa mosaic virus infection. Virology 164:383–389, 1988.
137. H Barker, B Reavy, KD Webster, CA Jolly, A Kumar, MA Mayo. Relationship between transcript production and virus resistance in transgenic tobacco expressing the potato leafroll virus coat protein gene. Plant Cell Rep 13:54–58, 1993.
138. V Brault, T Candresse, O le Gall, RP Delbos, M Lanneau, J Dunez. Genetically engineered resistance against grapevine chrome mosaic nepovirus. Plant Mol Biol 21:89–97, 1993.
139. N Bardonnet, F Hans, MA Serghini, L Pinck. Protection against virus infection in tobacco plants expressing the coat protein of grapevine fanleaf nepovirus. Plant Cell Rep 13:357–360, 1994.
140. LM Yepes, M Fuchs, JL Slightom, D Gonsalves. Sense and antisense coat protein gene constructs confer high levels of resistance to tomato ringspot nepovirus in transgenic *Nicotiana* species. Phytopathology 86:417–424, 1996.
141. B Reavy, M Arif, S Kashiwazaki, KD Webster, H Barker. Immunity to potato mop-top virus in *Nicotiana benthamiana* plants expressing the coat protein gene is effective against fungal inoculation of the virus. Mol Plant Microbe Interact 8:286–291, 1995.
142. T-F Chia, Y-S Chan, N-H Chua. Characterization of cymbidium mosaic virus coat protein gene and its expression in transgenic tobacco plants. Plant Mol Biol 18:1091–1099, 1992.
143. D Leclerc, MG AbouHaidar. Transgenic tobacco plants expressing a truncated form of the PAMV capsid protein (CP) gene show CP-mediated resistance to potato aucuba mosaic virus. Mol Plant Microbe Interact 8:58–65, 1995.
144. C Spillane, J Verchot, TA Kavanagh, DC Baulcombe. Concurrent suppression of virus replication and rescue of movement-defective virus in transgenic plants expressing the coat protein of potato virus X. Virology 236:76–84, 1997.
145. J Hammond, KK Kamo. Effective resistance to potyvirus infection conferred by expression of antisense RNA in transgenic plants. Mol Plant Microbe Interact 8:674–682, 1993.
146. MMM Fitch, RM Manshardt, D Gonsalves, JL Slightom, JC Sanford. Virus resistant papaya plants derived from tissues bombarded with the coat protein gene of papaya ringspot virus. Biotechnology 10:1466–1472, 1992.
147. F Regner, A da Câmara Machado, M Laimer da Câmara Machado, H Steinkellner, D Mattanovich, V Hanzer, H Weiss, H Katinger. Coat protein mediated resistance to plum pox virus in *Nicotiana clevelandii* and *N. benthamiana*. Plant Cell Rep 11:30–33, 1992.
148. BG Cassidy, RS Nelson. Differences in protection phenotypes in tobacco plants expressing coat protein genes from peanut stripe potyvirus with or without an engineered ATG. Mol Plant Microbe Interact 8:357–365, 1995.
149. M Fuchs, D Gonsalves. Resistance of transgenic hybrid squash ZW-20 expressing the coat protein genes of zucchini yellow mosaic virus and watermelon mosaic virus 2 to mixed infections by both potyviruses. Biotechnology 13:1466–1473, 1995.
150. G Fang, R Grumet. Genetic engineering of potyvirus resistance using constructs derived from the zucchini yellow mosaic virus coat protein gene. Mol Plant Microbe Interact 6:358–367, 1993.
151. PO Lim, JS Ryu, HJ Lee, U Lee, YS Park, JM Kwak, JK Choi, HG Nam. Resistance to tobamoviruses in transgenic tobacco plants expressing the coat protein gene of pepper mild mottle virus (Korean isolate). Mol Cells 7:313–319, 1997.
152. LA Wisniewski, PA Powell, RS Nelson, RN Beachy. Local and systemic spread of tobacco mosaic virus in transgenic tobacco. Plant Cell 2:559–567, 1990.
153. WG Clark, JH Fitchen, RN Beachy. Studies of coat protein-mediated resistance to TMV. I. The PM2 assembly defective mutant confers resistance to TMV. Virology 208:485–491, 1995.
154. WG Clark, J Fitchen, A Nejidat, CM Deom, RN Beachy. Studies of coat protein-mediated resistance to tobacco mosaic virus (TMV). II. Challenge by a mutant with altered virion surface does not overcome resistance conferred by TMV coat protein. J Gen Virol 76:2613–2617, 1995.

155. PR Sanders, B Sammons, W Kaniewski, L Haley, J Layton, BJ LaVallee, X Delannay, NE Tumer. Field resistance of transgenic tomatoes expressing the tobacco mosaic virus or tomato mosaic virus coat protein genes. Phytopathology 82:683–690, 1992.
156. JJL Gielen, P de Haan, AJ Kool, D Peters, MQJM van Grinsven, RW Goldbach. Engineered resistance to tomato spotted wilt virus, a negative-strand RNA virus. Biotechnology 9:1363–1367, 1991.
157. P de Haan, JJL Gielen, M Prins, IG Wijkamp, A van Schepen, D Peters, MQJM van Grinsven, R Goldbach. Characterization of RNA-mediated resistance to tomato spotted wilt virus in transgenic tobacco plants. Biotechnology 10:1133–1137, 1992.
158. S-Z Pang, P Nagpala, M Wang, JL Slightom, D Gonsalves. Resistance to heterologous isolates of tomato spotted wilt virus in transgenic tobacco expressing its nucleocapsid protein gene. Phytopathology 82:1223–1229, 1992.
159. S-Z Pang, F-J Jan, K Carney, J Stout, DM Tricoli, HD Quemada, D Gonsalves. Post-transcriptional transgene silencing and consequent tospovirus resistance in transgenic lettuce are affected by transgene dosage and plant development. Plant J 9:899–909, 1996.
160. AM Vaira, L Semeria, S Crespi, V Lisa, A Allavena, GP Accotto. Resistance to tospoviruses in *Nicotiana benthamiana* transformed with the N gene of tomato spotted wilt virus: correlation between transgene expression and protection in primary transformants. Mol Plant Microbe Interact 8:66–73, 1995.
161. FT Brederode, PEM Taschner, E Posthumus, JF Bol. Replicase-mediated resistance to alfalfa mosaic virus. Virology 207:467–474, 1995.
162. JM Anderson, P Palukaitis, M Zaitlin. A defective replicase gene induces resistance to cucumber mosaic virus in transgenic tobacco plants. Proc Natl Acad Sci USA 89:8759–8763, 1992.
163. JP Carr, A Gal-On, P Palukaitis, M Zaitlin. Replicase-mediated resistance to cucumber mosaic virus in transgenic plants involves suppression of both virus replication in the inoculated leaves and long-distance movement. Virology 199:439–447, 1994.
164. M Zaitlin, JM Anderson, KL Perry, L Zhang, P Palukaitis. Specificity of replicase-mediated resistance to cucumber mosaic virus. Virology 201:200–205, 1994.
165. WM Wintermantel, N Banerjee, JC Oliver, DJ Paolillo, M Zaitlin. Cucumber mosaic virus is restricted from entering minor veins in transgenic tobacco exhibiting replicase-mediated resistance. Virology 231:248–257, 1997.
166. Y Hong, J Stanley. Virus resistance in *Nicotiana benthamiana* conferred by African cassava mosaic virus replication associated protein (AC1) transgene. Mol Plant Microbe Interact 9:219–225, 1996.
167. HS Guo, MT Cervera, JA Garcia. Plum pox potyvirus resistance associated to transgene silencing that can be stabilized after different number of plant generations. Gene 206:263–272, 1998.
168. A Wittner, L Palkovics, E Balazs. *Nicotiana benthamiana* plants transformed with the plum pox virus helicase gene are resistant to virus infection. Virus Res 53:97–103, 1998.
169. JP Carr, M Zaitlin. Resistance in transgenic tobacco plants expressing a nonstructural gene sequence of tobacco mosaic virus is a consequence of markedly reduced virus replication. Mol Plant-Microbe Interact 4:579–585, 1991.
170. J Donson, CM Kearney, TH Turpen, IA Khan, G Kurath, AM Turpen, GE Jones, WO Dawson, DJ Lewandowski. Broad resistance to tobamoviruses is mediated by a modified tobacco mosaic virus replicase transgene. Mol Plant Microbe Interact 6:635–642, 1993.
171. F Tenllado, I Garcia-Luque, MT Serra, JR Diaz-Ruiz. *Nicotiana benthamiana* plants transformed with the 54-kDa region of the pepper mild mottle tobamovirus replicase gene exhibit two types of resistance responses against viral infection. Virology 211:170–183, 1995.
172. F Tenllado, I Garcia-Luque, MT Serra, JR Diaz-Ruiz. Resistance to pepper mild mottle tobamovirus conferred by the 54-kDa gene sequence in transgenic plants does not require expression of the wild-type 54-kDa protein. Virology 219:330–335, 1996.
173. SA MacFarlane, JW Davies. Plants transformed with a region of the 201-kilodalton replicase gene from pea early browning virus RNA1 are resistant to virus infection. Proc Natl Acad Sci USA 89:5829–5833, 1992.
174. L Rubino, R Lupo, M Russo. Resistance to cymbidium ringspot tombusvirus infection in transgenic *Nicotiana benthamiana* plants expressing a full-length viral replicase gene. Mol Plant Microbe Interact 6:729–734, 1993.

175. L Rubino, M Russo. Characterization of resistance to cymbidium ringspot virus in transgenic plants expressing a full-length viral replicase gene. Virology 212:240–243, 1995.
176. M Prins, R de Oliveira Resende, C Anker, A van Schepen, P de Haan, R Goldbach. Engineered RNA-mediated resistance to tomato spotted wilt virus is sequence specific. Mol Plant Microbe Interact 9:416–418, 1996.
177. T Sijen, J Wellink, J Hendriks, J Verver, A van Kammen. Replication of cowpea mosaic virus RNA1 or RNA2 is specifically blocked in transgenic *Nicotiana benthamiana* plants expressing the full-length replicase or movement protein genes. Mol Plant Microbe Interact 8:340–347, 1995.
178. X Ares, G Calamante, S Cabral, J Lodge, P Hemenway, RN Beachy, A Mentaberry. Transgenic plants expressing potato virus X ORF2 protein (p24) are resistant to tobacco mosaic virus and Ob tobamoviruses. J Virol 72:731–738, 1998.
179. DL Beck, CJ van Dolleweerd, TJ Lough, E Balmori, DM Voot, MT Andersen, IEW O'Brien, RLS Forster. Disruption of virus movement confers broad-spectrum resistance against systemic infection by plant viruses with a triple gene block. Proc Natl Acad Sci USA 91:10310–10314, 1994.
180. B Cooper, M Lapidot, JA Heick, JA Dodds, RN Beachy. A defective movement protein of TMV in transgenic plants confers resistance to multiple viruses whereas the functional analog increases susceptibility. Virology 206:307–313, 1995.
181. M Bendahmane, B Gronenborn. Engineering resistance against tomato yellow leaf curl virus (TYLCV) using antisense RNA. Plant Mol Biol 33:351–357, 1997.
182. A Palucha, W Zagorski, M Chrzanowska, D Hulanicka. An antisense coat protein gene confers immunity to potato leafroll virus in genetically-engineered potato. Eur J Plant Pathol 104:287–293, 1997.
183. SH Lim, MK Ko, SJ Lee, YJ La, BD Kim. Cymbidium mosaic virus coat protein gene in antisense confers resistance to transgenic *Nicotiana occidentalis*. Mol Cells 9:603–608, 1999.
184. PA Powell, DM Stark, PR Sanders, RN Beachy. Protection against tobacco mosaic virus in transgenic plants that express tobacco mosaic virus antisense RNA. Proc Natl Acad Sci USA 86:6949–6952, 1989.
185. S-Z Pang, JL Slightom, D Gonsalves. Different mechanisms protect transgenic tobacco against tomato spotted wilt and impatiens necrotic spot tospoviruses. Biotechnology 11:819–824, 1993.
186. RAA van der Vlugt, RK Ruiter, R Goldbach. Evidence for sense RNA-mediated protection to PVY$^N$ in tobacco plants transformed with the viral coat protein cistron. Plant Mol Biol 20:631–639, 1992.
187. A Nelson, DA Roth, JD Johnson. Tobacco mosaic virus infection of transgenic Nicotiana tabacum plants is inhibited by antisense constructs directed at the 5′ region of viral RNA. Gene 127:227–232, 1993.
188. S-Z Pang, JH Bock, C Gonsalves, JL Slightom, D Gonsalves. Resistance of transgenic *Nicotiana benthamiana* plants to tomato spotted wilt and impatiens necrotic spot tospoviruses: evidence of involvement of the N protein and N gene RNA in resistance. Phytopathology 84:243–249, 1994.
189. S-Z Pang, F-J Jan, D Gonsalves. Nontarget DNA sequences reduce the transgene length necessary for RNA-mediated tospovirus resistance in transgenic plants. Proc Natl Acad Sci USA 94:8261–8266, 1997.
190. B Zaccomer, F Cellier, J-C Boyer, A-L Haenni, M Tepfer. Transgenic plants that express genes including the 3′ untranslated region of the turnip yellow mosaic virus (TYMV) genome are partially protected against TYMV infection. Gene 136:87–94, 1993.
191. Y Saito, T Komari, C Masuta, Y Hayashi, T Kumashiro, Y Takanami. Cucumber mosaic virus-tolerant transgenic tomato plants expressing a satellite RNA. Theor Appl Genet 83:679–683, 1992.
192. ME Taliansky, EV Ryabov, DJ Robinson. Two distinct mechanisms of transgenic resistance mediated by groundnut rosette virus satellite RNA sequences. Mol Plant Microbe Interact 11:367–374, 1998.
193. T Frischmuth, J Stanley. Beet curly top virus symptom amelioration in *Nicotiana benthamiana* transformed with a naturally occurring viral subgenomic DNA. Virology 200:826–830, 1994.
194. DC Stenger. Strain-specific mobilization and amplification of a transgenic defective-interfering DNA of the geminivirus beet curly top virus. Virology 203:397–402, 1994.
195. A Kollar, T Dalmay, J Burgyan. Defective interfering RNA-mediated resistance against cymbidium ringspot tombusvirus in transgenic plants. Virology 193:313–318, 1993.

196. YH Moon, SK Song, KW Choi, JS Lee. Expression of a cDNA encoding *Phytolacca insularis* antiviral protein confers virus resistance on transgenic potato plants. Mol Cells 7:807–815, 1997.
197. NE Tumer, D-J Hwang, M Bonness. C-terminal deletion mutant of pokeweed antiviral protein inhibits viral infection but does not depurinate host ribosomes. Proc Natl Acad Sci USA 94:3866–3871, 1997.
198. H Takahashi, Z Chen, H Du, Y Liu, DF Klessig. Development of necrosis and activation of disease resistance in transgenic tobacco plants with severely reduced catalase levels. Plant J 11:993–1005, 1997.
199. LF Fecker, R Koenig, C Obermeier. *Nicotiana benthamiana* plants expressing beet necrotic yellow vein virus (BNYVV) coat protein-specific scFv are partially protected against the establishment of the virus in the early stages of infection and its pathogenic effects in the late stages of infection. Arch Virol 142:1857–1863, 1997.

# 46
# Transgenic Sweet Potato with Agronomically Important Genes

### Motoyasu Otani and Takiko Shimada
*Ishikawa Agricultural College, Nonoichi, Ishikawa, Japan*

| | | |
|---|---|---|
| I. | GENERAL | 699 |
| II. | *AGROBACTERIUM TUMEFACIENS*–MEDIATED TRANSFORMATION | 702 |
| | A. Target Tissues for Transformation | 702 |
| | B. Embryonic Callus Induction | 702 |
| | C. Factors Affecting Transformation Efficiency | 703 |
| | D. Plant Regeneration from Hygromycin Resistance Calli | 703 |
| | E. Histochemical GUS Assay of Transgenic Plants and Their Vegetatively Propagated Progenies | 706 |
| | F. Integration of Foreign DNA into the Genome of Transgenic Plants | 706 |
| | G. Inheritance of Introduced Genes | 706 |
| | H. Comparison with the Direct Gene Transfer | 706 |
| III. | *AGROBACTERIUM RHIZOGENES*–MEDIATED TRANSFORMATION | 709 |
| | A. Production of Transgenic Plants Mediated by *Agrobacterium rhizogenes* | 709 |
| | B. Morphological Characteristics of Ri-Transgenic Plants | 709 |
| | C. Production of Ri-Transgenic Plants Possessing *nptII* and *gusA* Genes | 710 |
| | D. Inheritance of Ri-Transformed Phenotypes | 711 |
| IV. | TRANSGENIC SWEET POTATO WITH AGRONOMICALLY IMPORTANT GENES | 711 |
| | A. A Coat Protein Gene of Sweet Potato Feathery Mottle Virus-S | 711 |
| | B. Fatty Acid Desaturase Gene | 711 |
| | C. Granule-Bound Starch Synthase Gene | 712 |
| | D. Genes for Insect Resistance | 712 |
| V. | PROTOCOLS | 712 |
| | A. Protocol for *Agrobacterium tumefaciens*–Mediated Transformation | 712 |
| | B. Protocol for *Agrobacterium rhizogenes*–Mediated Transformation | 713 |
| VI. | CONCLUDING REMARKS | 713 |
| | REFERENCES | 714 |

## I. GENERAL

Sweet potato [*Ipomoea batatas* (L.) Lam.] ranks seventh in annual production among the food crops in the world (1). The cultivation area of this plant species is localized mainly at low latitudes of South America, South Asia, and South Africa and spreads as far as some regions of Europe and the United States. The ancient civilization area of sweet potato was tropical America, and at the end of the 14th century it was transported to the Old World. In Japan, the sweet potato was introduced from China to Kagoshima in the Kyushu area at the beginning of the 17th century, by way of Ryukyu Islands, then rapidly spread to Honshu. A famous savant in the Edo era, Konyo Aoki (1698–1769), made an effort to propagate sweet potato. Today, it is called *Kara-imo* (Chinese potato) or *Ryukyu-imo* (Ryukyu potato) in Kyushu, and *Satsuma-imo* (Kagoshima potato) in Honshu (2).

Sweet potato is not only a good source of energy, supplying sugars and other carbohydrates, but also a nutritive food containing calcium, iron, and other minerals and vitamins, particularly vitamins A and C. It is also used for food processing as well as for starch and alcohol. Its productivity is very high: an average of 25 t/ha in Japan. Moreover, it does not require large amounts of fertilizers and other agricultural chemicals and is rather tolerant of environmental stresses. Therefore, the sweet potato is one of the most important crops for securing a stable food supply and has potential to alleviate some global environmental problems in the 21st century. Recently, the research on sweet potato in Japan has been designed to develop alternative uses for creating new ways to utilize the entire plant (3). New cultivars have been developed, for example, a cultivar to allow utilization of sweet potato tops (foliage and stem), as human food, not only livestock feed; cultivars created for high starch production, high-carotene-content cultivars, and a high-anthocyanin-content cultivar (4). New improved cultivars are awaited to sustain the production and use of this crop in the 21st century. However, the conventional breeding program based on sexual hybridization of sweet potato is limited by its sterility and cross-incompatibility. To overcome such limitations, novel approaches such as somatic hybridization and genetic transformation, especially the latter, must be incorporated into sweet potato breeding.

Biotechnology is not well developed for sweet potato because of the difficulty of establishing an efficient and reproducible system for plant regeneration from cultured tissues or cells. A number of reports have been published on the regeneration of sweet potato, using various tissues of various cultivars with varying levels of efficiency and reproducibility (5). However, there is a severe limitation of genotypes for regeneration, and efficiency is still low. In 1996 Otani and Shimada developed an efficient and variety-independent method for the production of embryogenic callus from meristem tissues using a medium containing altered plant growth regulators, picloram, dicamba, or 4-fluorophenoxyacetic acid (4FA) (6). Otani and coworkers have established an efficient method for the production of transgenic sweet potato plants via *Agrobacterium tumefaciens*–mediated transformation using embryogenic calli (7).

A few reports on the transformation of sweet potato have been published (Table 1). Dodds and colleagues (8) and Otani and associates (9) obtained transgenic sweet potato plants integrated with Ri transfer deoxyribonucleic acid (T-DNA) and/or *gusA* and *nptII* genes by using *Agrobacterium rhizogenes,* although those transgenic plants showed abnormal morphological characteristics due to integrated Ri T-DNA. By using a direct gene delivery system, Murata and coworkers (10) obtained transgenic plants regenerated from transformed protoplasts of a genotypically limited cultivar, and Parakash and Varadarajan (11) regenerated transgenic calli from bombarded leaves and petioles. On the other hand, transgenic plants have been produced by cocultivation of storage roots with *A. tumefaciens* (12), leaf disks (13), and embryogenic callus (7,14).

To date, the genes introduced into sweet potato have been marker and/or selectable marker genes and resistant genes for insect and virus diseases (7,10,12,13). We have introduced some agronomically important genes by using our transformation system. They are a fatty acid desaturase gene for low-temperature tolerance, granule bound starch synthase (*GBSS*) gene for the modification of starch structure, and a coat protein complementary DNA (cDNA) of a sweet po-

**Table 1**
Summary of Published Studies Concluded on Transformation of Sweet Potato

| Cultivar | Target tissue | Vector/method used | Genes | Observations/remarks | Reference (year) |
|---|---|---|---|---|---|
| Not stated | In vitro whole plant | *A. rhizogenes* | Synthetic sequence | Transgenic plants | Dodds et al. (8) (1991) |
| 'Jewel,' TIS-70357 | Leaves and petioles | Particle bombardment | *gusA nptII* | Transformed calli | Prakash and Varadarajan (11) (1992) |
| Five cultivars ('Chugoku 25,' etc.) | Leaves | *A. rhizogenes* 15334, etc. | *gusA, nptII* | Transgenic plants | Otani et al. (9) (1993) |
| 'Jewel' | Storage roots | *A. tumefaciens* LBA4404 | *gusA, nptII*, cowpea trypsin inhibitor, snowdrop lectin | Transgenic plants | Newell et al. (12) (1995) |
| 'White Star' | Embryogenic callus | *A. tumefaciens* EHA101 | *gusA, nptII* | Transgenic plants | Gama et al. (14) (1996) |
| 'Chikei 682-11' | Mesophyll protoplasts | Electroporation | *hpt*, SPFMV-S coat protein gene | Transgenic plants | Murata et al. (10) (1997) |
| 'Nanging' | Embryogenic callus | Particle bombardment | | | |
| 'Kokei 14' | Embryogenic callus | *A. tumefaciens* EHA 101 | *gusA, hpt* | Transgenic plants | Otani et al. (7) (1998) |
| 'Jewel' | Leaves | *A. tumefaciens* C58C1 | *cryIIIA, nptII* | Transgenic plants | Moran et al. (13) (1998) |
| 'Kokei 14,' 'Beniazuma' | Embryogenic callus | *A. tumefaciens* EHA 101 | *NtFAD3, hpt* | Transgenic plants | Otani et al. (in preparation) |
| 'Kokei 14' | Embryogenic callus | *A. tumefaciens* EHA 101 | *GBSS, hpt* | Transgenic plants | Kimura et al. (24) (1998) |

tato feathery mottle virus (SPFMV-S) for virus resistance. The present chapter describes a simple, efficient, and reproducible method for production of transgenic sweet potato plants using *Agrobacterium tumefaciens* as well as *A. rhizogenes*. We also describe the characteristics of transgenic plants and stability of the integrated genes.

## II. *AGROBACTERIUM TUMEFACIENS*–MEDIATED TRANSFORMATION

### A. Target Tissues for Transformation

We used embryogenic calli of sweet potato as target tissues for *A. tumefaciens*–mediated transformation. The embryogenic calli were yellow and friable and produced numerous embryoids when transferred onto LS medium (17) supplemented with abscisic acid (ABA) and $GA_3$; the embryoids developed into plantlets at high frequency when transferred onto the plant growth-regulator-free medium.

Newell and associates (12) obtained transgenic plants from disks of storage roots of sweet potato cv. 'Jewel' by *Agrobacterium* species–mediated transformation. The storage root may not be a good material for obtaining regenerated plants in sweet potato, because the culture responses varied with the genotypes, and moreover cultivars having the ability to regenerate plants from storage root disks are rare (15). For the same reason, leaf disks are not suited for transformation study, even though adventitious shoots regenerate at a high frequency in the limited genotypes (16). On the other hand, we obtained embryogenic calli from all 11 cultivars tested at relatively high frequencies by using altered plant growth regulators (6). The method of *A. tumefaciens*–mediated transformation using embryogenic callus might overcome the genotypic differences in genetic transformation of sweet potato.

### B. Embryogenic Callus Induction

Embryogenic calli were induced from shoot meristems of in vitro plants of sweet potato [*Ipomoea batatas* (L.) Lam.] cultivars 'Kokei 14' and 'Beniazuma,' the most important cultivars for table use in Japan, using LS medium containing 1 mg/l 4-fluorophenoxyacetic acid (4FA) or picloram, 3% (w/v) sucrose, and 0.32% (w/v) gellan gum, according to the method of Otani and Shimada (6). In our previous study using a medium supplemented with 1 mg/l picloram, more than 50% of the shoot apices formed embryogenic calli in all 11 genotypes cultured. In 'Kokei 14' and 'Beniazuma,' around 90% of the shoot meristems produced embryogenic calli (Fig. 1). These embryogenic calli were maintained at 26°C in the dark and proliferated by subculture on the same fresh medium every month.

### C. Factors Affecting Transformation Efficiency

We examined three factors, strains of *Agrobacterium*, inclusion of acetosyringone, and culture period of embryogenic calli, before bacterial inoculation, related to efficient transformation in sweet potato. Transient expression of GUS was examined with 1.5 g fresh weight of embryogenic calli per treatment after 3 days of cocultivation. The transient GUS expression varied with the bacterial strain. *A. tumefaciens* strain EHA101/plG121-Hm gave the highest GUS spots; the other bacterial strains gave no [for LBA4404/pTOK233 (18)] or very few spots (average six spots per gram fresh weight for R1000/pBl121). Gama and coworkers (14) also succeeded in obtaining transgenic sweet potato plants by using the same bacterial strain, EHA101. These results suggested that the supervirulent strain, EHA101, might be suited to the transformation of sweet potato.

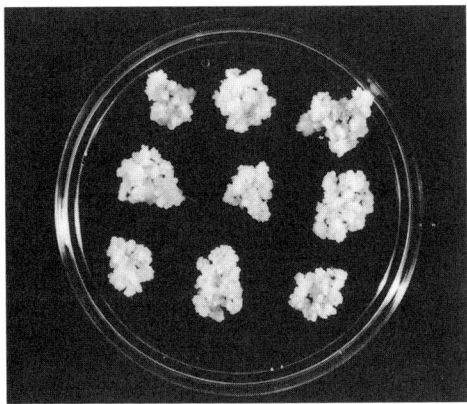

**Figure 1** Embryogenic calli from shoot apices of the sweet potato cultivar 'Kokei 14' on medium containing 1 mg/l 4FA.

The effect of acetosyringone on transient GUS expression was tested by using 14-day-old calli of the cultivar 'Kokei 14' and *A. tumefaciens* strain EHA101/plG121-Hm. As shown in Table 2, the addition of acetosyringone to both infection and cocultivation media clearly promoted the transient expression of GUS.

Embryogenic calli were infected with *A. tumefaciens* strain EHA101/plG121-Hm 3, 6, 10, 14, and 21 days after the beginning of subculture. The calli cultured for a short period (3 and 6 days) gave few GUS spots, and 14-day-old calli gave the largest number of GUS spots (Table 2). These findings clearly show that a 14-day culture period before bacteria infection was needed to obtain efficient expression of *gusA* gene in embryogenic callus 'Kokei 14.' Similar results were obtained in embryogenic callus of 'Beniazuma' (Fig. 2)

## D. Plant Regeneration from Hygromycin-Resistant Calli

The embryogenic calli infected with *A. tumefaciens* strain EHA101/plG121-Hm were cultured on hygromycin-containing media for 60 days. These calli produced several hygromycin-resistant calli, whereas uninfected embryogenic calli on the same media did not. The average number of

**Table 2** Effect of Actosyringone and Preculture Period on Trangient GUS Expression in Embryogenic Callus of Sweet Potato

| | No. of GUS spots/g fresh weight of embryogenic calli | |
|---|---|---|
| Culture period (days) | Acetosyringone (+) | Acetosyringone (−) |
| 3 | 514.7 ± 21.5 | — |
| 6 | 840.9 ± 242.2 | — |
| 10 | 1054.4 ± 255.6 | — |
| 14 | 1526.6 ± 266.7 | 382.2 ± 115.4 |
| 21 | 922.8 ± 32.4 | — |

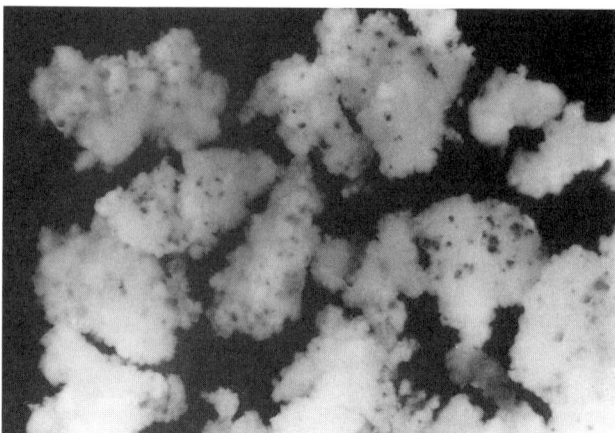

**Figure 2** Transient GUS expression in embryogenic callus of the sweet potato cultivar 'Beniazuma' after 3 days of cocultivation with *Agrobacterium tumefaciens* strain EHA101/plG121-Hm.

hygromycin-resistant calli produced per gram fresh weight of infected embryogenic calli was 10.7. These hygromycin-resistant calli produced numerous somatic embryos on the somatic embryo formation medium containing hygromycin (Fig. 3). The hygromycin-resistant plantlets were developed from these somatic embryos on the plant formation medium. An average of 53.1% of hygromycin-resistant calli regenerated plantlets. All of the regenerated plants grew further and rooted on the plant-growth-regulator-free LS medium supplemented with 25 mg/l hygromycin. Regenerated plants were transferred to pots containing a vermiculite and perlite mixture (3:1) and maintained at 26°C under a 16-hour photoperiod in a growth chamber for 14 days. Then, these regenerated plants were grown in a greenhouse. The transgenic plants grew normally (Fig. 4) and formed storage roots after 3 months (Fig. 5). Five transgenic plant independently regenerated from hygromycin-resistant calli were analyzed to determine phenotypic characteristics, such as the color of apical immature leaf, color of mature leaf, shape of mature leaf, stem color, number of storage roots per plant, fresh weight of storage roots, skin color of storage roots, and flesh color of storage roots. No morphological differences were observed between untransformed plants and transgenic plants (Table 3).

### E. Histochemical GUS Assay of Transgenic Plants and Their Vegetatively Propagated Progenies

Histochemical analysis of GUS activity was carried out on fully expanded leaves and storage roots of the plants regenerated from hygromycin-resistant calli. The tissues of regenerated plants were stained blue, indicating the expression of *gusA* gene, but not those of control plants. GUS expression was also observed in storage roots harvested after 3 months of cultivation in pots, whereas those of untransformed control plants did not show any GUS activity.

Leaves on the shoots freshly sprouting from harvested storage roots of transgenic plants also showed GUS activity (Fig. 6); suggesting that the *gusA* gene was transmitted to their vegetatively propagated progenies. We observed GUS activity in their fourth vegetatively propagated progenies. Since sweet potato is commonly propagated by using storage roots, this suggests the usefulness of genetically engineered sweet potato for the practical breeding of sweet potato.

**Figure 3** Numerous somatic embryos produced from hygromycin-resistant calli.

**Figure 4** Transgenic plants (left three pots) and an untransformed plant (right pot) established in soil and grown for 2 months in a greenhouse.

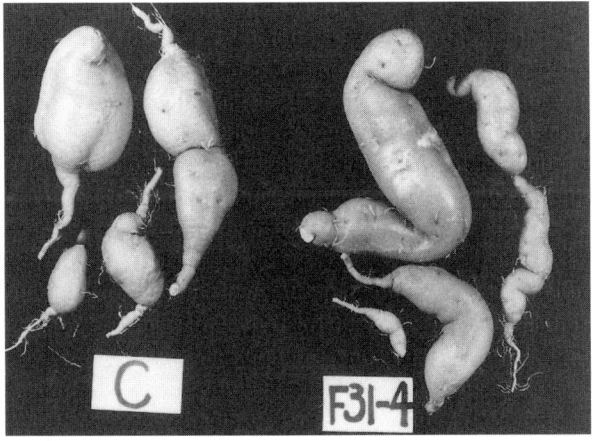

**Figure 5** Storage roots formed on untransformed (left) and transgenic plants (right).

**Table 3** Characterization of Independent Transgenic Sweet Potato Lines

| Characters | Control | Transgenic plant lines | | | | | |
| --- | --- | --- | --- | --- | --- | --- | --- |
| | | #1[a] | #5 | #6 | #7 | #11 | |
| Copy number of T-DNA | 0 | 3 | 3 | 2 | 2 | 1 | |
| Apical immature leaf color | Green | Green | Green | Green | Green | Green | |
| Mature leaf color | Green | Green | Green | Green | Green | Green | |
| Mature leaf shape | Heart shape with small lobe | Heart shape with small lobe | Heart shape with small lobe | Heart shape with small lobe | Heart shape with small lobe | Heart shape with small lobe | |
| Stem color | Green | Green | Green | Green | Green | Green | |
| Number of storage roots per plant | 5 | 4 | 7 | 5 | 5 | 5 | |
| Fresh weight of a storage root (g) ± SE | 26.1 ± 8.8 | 13.4 ± 2.9 | 24.2 ± 4.7 | 21.7 ± 7.7 | 25.7 ± 7.3 | 34.3 ± 6.8 | |
| Skin color of storage roots | Red | Red | Red | Red | Red | Red | |
| Fresh color of storage roots | Light yellow | Light yellow | Light yellow | Light yellow | Light yellow | Light yellow | |

[a] Plant line #1 was grown for 2 months in the greenhouse.

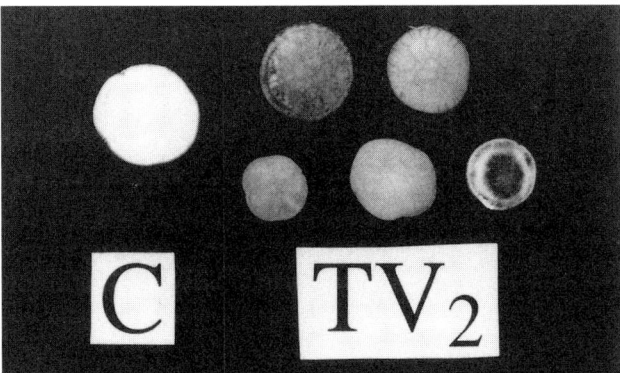

**Figure 6** Histochemical GUS assay in the second vegetatively propagated progenies of transgenic plants. Storage roots of five independent transgenic plants (TV$_2$) and untransformed plant (C).

## F. Integration of Foreign DNA into the Genome of Transgenic Plants

Genomic blot analysis was performed for transgenic plants regenerated from five independent hygromycin-resistant callus lines. A *SacI/XbaI* fragment of the gusA gene was used as a hybridization probe (Fig. 7a). Since there is only one *HindIII* site in the T-DNA region, hybridizing fragment(s) of different lengths indicated that the T-DNA was integrated at different location(s) in the sweet potato plant genome. The copy number was determined by the number of hybridizing fragments. Fig. 7b shows that the number and size of hybridizing fragments varied with the transgenic plant line. Transgenic plant lines possessed one to three copies of T-DNA.

## G. Inheritance of Introduced Genes

Since 'Kokei 14' is self-incompatible, one of the transgenic plants and a nontransformed 'Kokei 14' plant were crossed with 'Chikei 682-11,' which is compatible with 'Kokei14.' The F$_1$ seeds were germinated and the seedlings were investigated for GUS expression and hygromycin resistance.

The F$_1$ plants of a transgenic plant and 'Chikei 682-11' showed a segregation ratio of one GUS-positive and hygromycin-resistant to one GUS-negative and hygromycin-sensitive, indicating that T-DNA was inserted at a single locus. Harvested storage roots of F$_1$ plants and leaves on the shoot freshly sprouting from the roots also showed GUS activity, suggesting that the *gusA* gene was transmitted to their progenies. We also observed GUS activity in the vegetatively propagated progenies of these F$_1$ plants.

## H. Comparison with Direct Gene Transfer

In 1997 Murata and coworkers (10) obtained transformed sweet potato plants from both electroporated protoplasts and biolistically transformed suspension cultures. However, some problems remain, such as genotypic variation for transformation efficiency and inefficiency in the selection of the transformed cells. Moreover, transformation by direct gene transfer methods often results in complex integration (19–21). Production of transgenic plants having a low copy number (one to three) of integrated genes by the *A. tumefaciens*–mediated transformation method presented here has an advantage over this method.

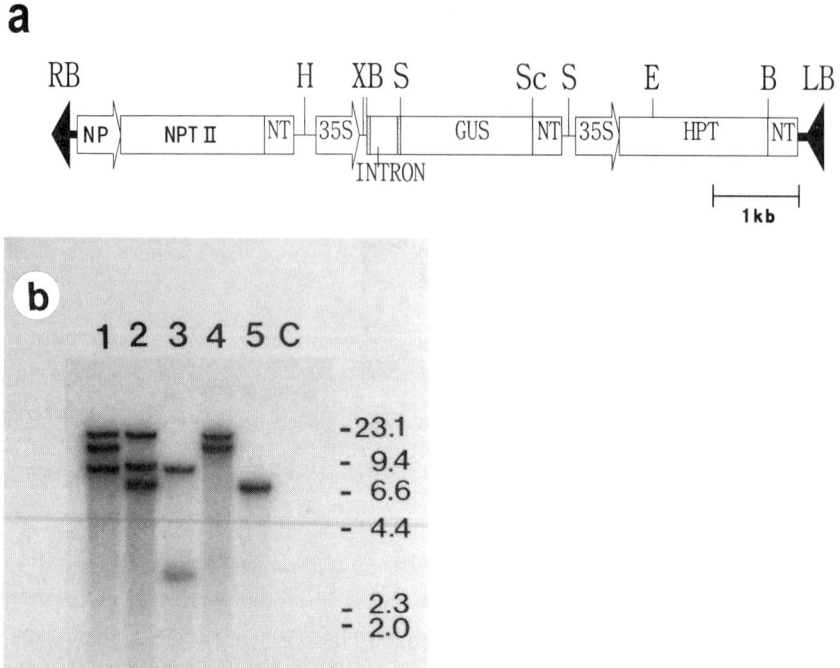

**Figure 7** Transformation vector and Southern blot analysis. (a) Schematic diagram of a part of the T-DNA region of transformation vector pIG121-Hm. RB, right border; LB, left border; NP, nopaline synthase promoter; NT, nopaline synthase terminator; 35S, 35S promoter of cauliflower mosaic virus; INTRON, the first intron of catalase gene of castor bean; NPTII, gene for neomycin phosphotransferase; GUS, gene for β-glucuronidase; HPT, gene for hygromycin phosphotransferase. Cutting sites of restriction enzymes are indicated: BamHI (B), EcoRI (E), HindIII (H), SalI (S), SacI (Sc), XbaI (X). (b) Southern blot analysis of five independent transgenic plants. DNA was digested with HindIII and allowed to hybridize to the gus probe: lanes 1–5, transgenic plants (No. 1, 5, 6, 7, and 11 shown in Table 4), which were regenerated from independent hygromycin-resistant calli; lane C, untransformed control plant.

## III. *AGROBACTERIUM RHIZOGENES*–MEDIATED TRANSFORMATION

### A. Production of Transgenic Plants Mediated by *Agrobacterium rhizogenes*

Hairy roots were produced on the wounded sites, for example, on the cut surface of excised petiole inoculated with *A. rhizogenes*. There were no significant differences in pathogenicity among the bacterial strains. On the other hand, genotypic differences in the frequency of hairy root formation were observed among the sweet potato cultivars when leaf disks were inoculated with *A. rhizogenes* strain A13 (Table 4). 'Chugoku 25,' 'Kanto 18,' 'Kokei 14,' and 'Shinya' produced hairy roots at a low frequency (23.5–27.8%), and the other cultivars produced hairy roots at a higher (51.4–88.6%) frequency.

Adventitious shoot formation was observed when hairy roots were cultured on 0.32% (w/v) gellan gum–solidified LS medium containing neither antibiotics nor PGR under continuous light. Regeneration of complete plants was obtained in 5, cultivars ('Chugoku 25,' 'Chugoku 35,' 'Hi-

**Table 4** Varietal Differences in Hairy Root Formation Induced by *A. rhizogenes* Strain A13

| Cultivar | No. of leaf disks inoculated | No. of leaf disks forming hairy roots (%) |
| --- | --- | --- |
| 'Bise' | 31 | 25 (80.6) |
| 'Chugoku 25' | 73 | 18 (24.7) |
| 'Chugoku 35' | 30 | 13 (43.3) |
| 'Hi-Starch' | 37 | 25 (67.6) |
| 'Kanto 18' | 34 | 9 (26.5) |
| 'Kanto 94' | 33 | 19 (57.6) |
| 'Kokei 14' | 36 | 10 (27.8) |
| 'Kyukei 17-3043' | 78 | 52 (66.7) |
| 'Naeshirazu' | 35 | 18 (51.4) |
| 'Norin 2' | 38 | 25 (65.8) |
| 'Okinawa 100' | 31 | 18 (58.1) |
| 'Shinya' | 34 | 8 (23.5) |
| 'Yamakawamurasaki' | 35 | 31 (88.6) |
| 'W51' | 40 | 30 (75.0) |
| *Ipomoea trichocarpa* (wild relative) | 37 | 34 (91.9) |

Starch,' 'Kyukei 17-3043,' and 'Yamakawamurasaki') of 10 cultivars tested. The percentages of hairy roots with shoots formed were 50% for 'Chugoku 25,' 100% for 'Chugoku 35,' 80% for 'Hi-Starch,' 40.9% for 'Kyukei 17-3043,' and 41.7% for 'Yamakawamurasaki.' These findings suggest that there are varietal differences in shoot regeneration from hairy roots. Shoot formation was not improved by the addition benzyladenine of (BA) (0.5 and 2 mg/l) to regeneration medium.

## B. Morphological Characteristics of Ri-Transgenic Plants

The adventitious shoots formed on the hairy roots grew and rooted on the 0.8% (w/v) agar-solidified LS medium. These plantlets were habituated and transplanted to the experimental field of Ishikawa Agricultural College. All of these plants survived.

All plants regenerated from the hairy roots showed Ri T-DNA-induced morphological characteristics. Aerial parts showed weak apical dominance and had short internodes (Fig. 8). The average stem length was about one-fifth of that in the normal plants. Stem length varied among the transgenic plants regenerated from each hairy root line. The leaves were wrinkled and smaller than normal. The shape of the flower was changed dramatically: hairy root–derived plants had small and star-shaped flowers. Pollen fertility of hairy root–derived plants was similar to that of normal plants: i.e., more than 90% of pollen grains were stained by 0.5% acetocarmine solution in both hairy root–derived and untransformed plants. The hairy root–derived plants had abundant roots with extensive branching and smaller storage roots.

Interestingly, the numbers of storage roots in the plants regenerated from the hairy root was equal to that of normal plants. The average fresh weight of the storage roots in hairy root–derived plants was 3.6 g, whereas that in normal plants was 39.4 g.

The storage roots of both hairy root–derived and normal plants were incubated at 28°C to allow adventitious shoots to sprout from them. The former formed few adventitious shoots and the shoots grew more slowly than the latter, whereas the storage roots of the former showed excessive rooting with intensive branching. Their vegetative progenies also exhibited similar morphological characteristics.

**Figure 8** Morphological abnormalities of regenerated plant of the cultivar 'Chugoku 25' transformed by *A. rhizogenes.* Aerial parts of transformed plant (left) and untransformed plant (right). C, untransformed plant; T, transformed plant.

Since dwarfness is one of the important characteristics in crop breeding, a short internode in transformed plants is a desirable character. However, the subterranean parts of the transformed plants showed abundant roots with extensive branching and smaller storage roots as described earlier. The transgenic tomato plants transformed with only the *rol B* gene of Ri-T-DNA were characterized by a reduction in both internode length and apical dominance; their root system was similar to that of untransformed plants (22). The subterranean parts of the transgenic sweet potato plant with the *rol B* gene or rol C gene should be compared with those of transgenic plants with the intact Ri plasmid T-DNA introduced.

### C. Production of Ri-Transgenic Plants Possessing *nptII* and *gusA* Genes

One plant each was regenerated from six hairy root clones of 'Yamakawamurasaki' induced by *A. rhizogenes* C8 possessing a binary vector, pBI121. All regenerated plants produced mikimopine. Shoots of these regenerated plants were transferred to LS medium containing 100 mg/l kanamycin. Three of the regenerated plants produced many roots days after the transfer, whereas the other plants did not form any roots. In our preliminary study using hairy roots transformed by wild-type *A. rhizogenes,* no root formation or root growth was observed on the medium containing 100 mg/l kanamycin. Thus, these three plants, which formed many roots on kanamycin-containing medium, were defined to be resistant to kanamycin. Root tips and leaf disks of the kanamycin-resistant plants were stained with 5-bromo-4-chloro-3-indolyl-glucuronide (X-gluc) for 16 hours and the expression of the *gusA* gene was confirmed. Integration of *nptII* and *gusA* genes was also confirmed by polymerase chain reaction (PCR).

In the present study, 50% (three of six) of hairy roots obtained were doubly transformed by Ri plasmid T-DNA and pBI121 T-DNA without any selection pressure.

We have reported here not only the successful transformation of Ri-T-DNA into sweet potato, but also the transformation of foreign genes mediated by *A. rhizogenes* C8 harboring a binary vector, pBI121. Since whole plants can be regenerated at a high frequency from hairy roots transformed by *A. rhizogenes,* the binary vector system of the Ri plasmid may be efficiently applied for genetic transformation in some genotypes of sweet potato.

## D. Inheritance of Ri-Transformed Phenotypes

Two transgenic 'Chugoku 25' plants (one transformed by *A. rhizogenes* strain A13, and the other by the strain 15834) and a nontransformed 'Chugoku 25' plant were crossed with nontransformed 'Yamakawamurasaki' plants.

The germination rate was 84.6–100%. There were no clear differences in germinability between the transformed and nontransformed $F_1$ seeds. Morphological phenotypes and opine synthesis of the $F_1$ seedings derived from the cross between the transgenic plants and nontransformed 'Yamakawamurasaki' segregated Ri-transformed phenotype: about 70% of these progeny demonstrated typical hairy root traits such as short internodes, short hypocotyl, and wrinkled leaf. The segregation data were consistent with the integration of one independent, dominant Ri T-DNA locus.

## IV. TRANSGENIC SWEET POTATO WITH AGRONOMICALLY IMPORTANT GENES

### A. A Coat Protein Gene of Sweet Potato Feathery Mottle Virus-S

Sweet potato feathery mottle virus severe strain (SPFMV-S) is the most serious virus disease in sweet potato in Japan. It causes decolored transverse bands of varying width on the surface of tubers, which reduce the commercial value. Farmers purchase the virus-free seedlings, but the problem of reinfection remains.

Murata and associates (10) introduced the coat protein gene of SPFMV-S into the protoplasts of sweet potato, cv. 'Chikei 682-11' to produce virus-resistant plants. The transgenic plants expressing the SPFMV-S CP gene had a lower tiler of virus than the nontransgenic plants after inoculation with SPFMV-S, indicating the suppression of virus proliferation in the transgenic plants. We also transformed the SPFMV-S CP gene into the commercially valuable sweet potato cultivar 'Kokei 14' by using the *A. tumefaciens*–mediated system described. Studies of the virus resistance in the transgenic plants are in progress.

### B. Fatty Acid Desaturase Gene

The degree of unsaturation of fatty acids is an important factor for metabolic adaptation of higher plants to temperature stress [reviewed by Sommerville and Browse (23)]. We introduced the microsomal ω-3 fatty acid desaturase gene isolated from tobacco *(NtFAD3)* that confers desaturation from linoleic acid (18:2) to linolenic acid (18:3) into sweet potato to improve cold tolerance (24). The fatty acid composition in roots was modified in the transgenic sweet potato plants: linoleic acid (18:2) content decreased from 53.6% in nontransgenic plants to 45.9% in transgenic plants and linolenic acid (18:3) content increased from 11.3% to 20.8%. The transgenic sweet potato plants expressing *NtFAD3* gene might have the potential not only to demonstrate resistance to chilling, but also to modify the nutritional quality and improve storage characteristics. Studies on the transgenic sweet potato plants are in progress.

### C. Granule-Bound Starch Synthase Gene

The reserve starch in sweet potato tubers might be increased or unique starch valuable for food or industrial purposes might be produced by genetic engineering of plants. The sweet potato starch is composed of about 16% amylose; the remainder is amylopectin. *GBSS* is a key enzyme in the formation of amylose in starch. We introduced *GBSS* cDNA from sweet potato under

CaMV 35S promoter into 'Kokei 14' by *A. tumefaciens*–mediated transformation (25). One transformed plant from 22 regenerated hygromycin-resistant plants had amylose-free starch in tubers, in which the endogeneou *GBSS* gene was cosuppressive with the introduced *GBSS* cDNA. Visser and associates (26) obtained amylose-free-starch potato by the introduction of antisense *GBSS* cDNA. We have obtained amylose-free sweet potato by sense *GBSS* cDNA. The results clearly demonstrate that in storage roots of sweet potato, starch composition is altered by either sense or antisense RNA.

### D. Genes for Insect Resistance

The sweet potato weevil (*Cylas formicarius* ssp.) is the major insect pest and the most important constraint in tropical regions (27). The genes for various types of protease inhibitors and some *Bacillus thuringiensis* genes to control coleopteran species have been demonstrated to confer insect resistance properties when introduced to plants (28). Newell and coworkers (12) successfully introduced the cowpea trypsin inhibitor gene (*CpTI*) and the GNA gene encoding a mannose-specific lectin from snowdrop (*Galanthus nivalis* L.) into the sweet potato cultivar 'Jewel' by using *A. tumefaciens*. Moran and associates (13) transformed sweet potato plants of the 'Jewel' cultivar with the *cryIIIA* delta-endotoxin gene from *Bacillus thuringiensis* subsp. *tenebrionis* (Btt), using *A. tumefaciens*. Biological tests with sweet potato weevil were developed under controlled conditions and field conditions. Although the level of the *CryIIIA* toxin in the transgenic plants was low, the transgenic plants showed resistance against sweet potato weevil and the damage in the tubers of some transgenic plants was lower than that in untransformed plants under both conditions.

## V. PROTOCOLS

### A. Protocol for *Agrobacterium tumefaciens*–Mediated Transformation

#### 1. Media

Media used in the following protocol are listed in Table 5.

#### 2. Induction of Embryogenic Callus

1. Place shoot meristems (0.5 mm in diameter) of in vitro plants onto LS medium (17) for embryogenic callus induction and culture at 26°C in the dark.
2. Maintain embryogenic calli by subculture on the same fresh medium every month. Use the calli subcultured at least three times for the transformation study.

#### 3. Transformation

1. Culture the *Agrobacterium* sp. for 2 to 3 days at 26°C on Luria broth (LB) (28) medium supplemented with 50 mg/l kanamycin, 50 mg/l hygromycin B, and 1.5% (w/v) agar. Then transfer the colony of bacteria to liquid LS medium and shake at 100 revolutions/minute for 30 minutes in the dark at 26°C.
2. Soak the embryogenic calli in a bacterial suspension for 2 minutes and blot dry with sterile filter paper to remove excess bacteria.
3. Then transfer the calli onto coculture medium and culture for 3 days in the dark at 26°C.
4. Wash the infected calli three times with sterile distilled water supplemented with 500

**Table 5** Media Used for Tissue Culture and Transformation of Sweet Potato

| Stage | Medium composition | Culture period |
|---|---|---|
| Embryogenic callus induction and proliferation | LS medium, 1 mg/l 4FA or picloram, 30 g/l sucrose, 3.2 g/l gellan gum, pH 5.8 | 14 days |
| Bacterial infection | LS medium, 1 mg/l 4FA or picloram, 10 mg/l actosyringone, 30 g/l sucrose, pH 5.8 | 2 minutes |
| Coculture | LS medium, 1 mg/l 4FA or picloram, 10 mg/l actosyringone, 30 g/l sucrose, 3.2 g/l gellan gum, pH 5.8 | 2 to 3 days |
| Selection | LS medium, 1 mg/l 4FA or picloram, 25 mg/l hygromycin B, 500 mg/l carbenicillin, 30 g/l sucrose, 3.2 g/l gellan gum, pH 5.8 | 50 to 60 days |
| Somatic embryo formation | LS medium, 4 mg/l ABA, 1 mg/l $GA_3$, 25 mg/l hygromycin B, 500 mg/l carbenicillin, 30 g/l sucrose, 3.2 g/l gellan gum, pH 5.8 | 14 to 21 days |
| Plant formation | LS medium, 0.05 mg/l ABA, 25 mg/l hygromycin B, 500 mg/l carbenicillin, 30 g/l sucrose, 3.2 g/l gellan gum, pH 5.8 | 14 to 21 days |
| Plant growth | LS medium, 30 g/l sucrose, 3.2 g/l gellan gum, pH 5.8 | |

mg/l carbenicillin; then transfer onto selection medium. Cultures were carried out at 26°C in the dark.

### 4. Selection and Regeneration of Transgenic Plants

1. After 2 week culture on the selection medium, wash the calli again as described, then transfer to fresh selection medium. Subculture the calli on fresh medium every 2 weeks.
2. After 60 days of culture on the selection medium, transfer the calli onto the somatic embryo formation medium and culture at 26°C under a 16-hour photoperiod 38 µmol $m^{-2}$ $s^{-1}$ from daylight fluorescent tubes.
3. After 21 days of culture on the somatic embryo formation medium, transfer the somatic embryos formed from hygromycin-resistant calli onto the plant formation medium for germination.

### B. Protocol for *Agrobacterium rhizogenes*–Mediated Transformation

1. Culture *Agrobacterium rhizogenes* for 16 hours at 27°C in a liquid YEB medium (30).
2. Immerse the apical one-third to one-fifth of fully expanded leaves of in vitro sweet potato plants into bacterial suspensions for 10 minutes.
3. Place the inoculated leaf disks on sterilized moist paper in a glass petri dish and incubate at 26°C in the dark for 3 to 5 days.
4. Transfer the leaf disks onto 1% (w/v) agar or 0.32% (w/v) gellan gum–solidified LS medium supplemented with 500 µg/ml Claforan (Hoechst) and incubate under the same conditions, after cocultivation.
5. Establish bacteria-free root lines after excision of single roots and propagate them on LS medium supplemented with 400 µg/ml Claforan and 0.32% (w/v) gellan gum with three subcultures.

6. Transfer the hairy roots (30 to 40 mm in length) onto 0.32% (w/v) gellan gum–solidified LS medium containing neither antibiotics nor plant growth regulators and culture at 26°C under continuous light at 38 $\mu$mol m$^{-2}$ s$^{-1}$ from daylight fluorescent tubes.

## VI. CONCLUDING REMARKS

A simple, efficient, and reproducible method for the transformation of commercial varieties of sweet potato mediated by *Agrobacterium tumefaciens* was established by using embryogenic calli as target tissues. *A. tumefaciens* strain EHA101/pIG121-Hm used in the present study contained a binary vector, genes for β-glucuronidase (*gus*A) and hygromycin resistance (*hpt*). About 10 hygromycin-resistant cell clusters were produced from 1 g fresh weight of the infected embryogenic calli. The hygromycin-resistant plantlets were regenerated from 53.1% of the hygromycin-resistant calli. Histochemical GUS assay and Southern hybridization analysis indicated that these plants were stably transformed. The copy number of introduced genes was one to three. Transgenic plants grew normally and formed storage roots after 3 months of cultivation in a greenhouse. By using the efficient *A. tumefaciens*–mediated transformation system, we introduced agronomically important genes, for example, a virus disease resistance–related gene, a cold tolerance–related gene, and a carbohydrate-modifying gene, into sweet potato.

Transformation was also accomplished by infection with *Agrobacterium rhizogenes*. Transgenic sweet potato possessing not only Ri-T-DNA but also *npt*II and *gus*A genes were obtained from hairy roots infected with *A. rhizogenes* containing the binary vector pBI121 in addition to the wild-type Ri plasmid. Leaf disks of in vitro plants were inoculated with different *A. rhizogenes* strains. Numerous hairy roots were induced on the leaf disks by both agropine-type and mikimopine-type strains. Whole plants transformed with Ri-T-DNA were regenerated from the hairy roots in five cultivars. These plants had wrinkled leaves, altered shape of flowers, reduced apical dominance, shortened internodes, small storage roots, and abundant, frequently branching roots that showed reduced geotropism.

Because of its high production yield of biomass, sweet potato is an attractive plant species as a target of "molecular farming." Recent developments of genetic engineering have allowed the production of various biomolecules such as carbohydrates, fatty acid, high-value pharmaceutical polypeptides, industrial enzymes, and biodegradable plastics in transgenic plants (31). Transgenic plants may become attractive and cost-effective alternatives to microbial and animal systems for the production of biomolecules.

The sweet potato is expected to be one of the most important crops globally in the 21st century, and its potential will be further developed by genetic engineering.

## REFERENCES

1. RK Jansson, KV Raman. Sweet potato pest management: A global overview. In RK Jansson, KV Raman, eds. Sweet potato pest management: A global perspective. Boulder, CO: Westview, 1991, pp 1–12.
2. J Kobayashi. Satsuma-imo no kita michi [The history of sweet potato]. Tokyo: Kokin-syoin, 1984.
3. O Yamakawa. Development of new cultivation and utilization system for sweetpotato toward the 21st century. Proceedings of International Workshop on Sweetpotato Production System Toward the 21st Century. Kyushu National Agricultural Experimental Station. Miyakonojo, Japan, pp 1–27, 1997.
4. K Komaki. Present status and future prospect of sweetpotato production and consumption in Japan. Proceedings of International Workshop on Sweetpotato Production System Toward the 21st Century. Kyushu National Agricultural Experimental Station. Miyakonojo, Japan, pp 149–158, 1997.
5. JM Lowe, WDO Hamilton, CA Newell. Genetic transformation in *Ipomoea batatas* (L.) Lam. (sweet

potato). In: YPS Bajaj ed. Biotechnology in Agriculture and Forestry. Vol 29. Plant Protoplasts and Genetic Engineering V. Berlin, Heidelberg, New York: Springer-Verlag, 1994, pp 308–320.
6. M Otani, T Shimada. Efficient embryogenic callus formation in sweet potato [*Ipomoea batatas* (L.) Lam.] Breed Sci 46:257–260, 1996.
7. M Otani, T Shimada, T Kimura, A Saito. Transgenic plant production from embryogenic callus of sweet potato [*Ipomoea batatas* (L.) Lam.] using *Agrobacterium tumefaciens*. Plant Biotechnol 15: 11–16, 1998.
8. JH Dodds, C Merzdorf, V Zambrano, C Siguenas, J Jaynes. Potential use of *Agrobacterium*-mediated gene transfer to confer insect resistance in sweet potato. In: RK Jansson, KV Raman eds. Sweet Potato Pest Management: A Global Perspective. Boulder, CO: Westview, 1991, pp 203–219.
9. M Otani, M Mii, T Handa, H Kamada, T Shimada. Transformation of sweet potato [*Ipomoea batatas* (L.) Lam.] by *Agrobacterium rhizogenes*. Plant Sci 94:151–159, 1993.
10. T Murata, Y Okada, H Fukuoka, A Saito, T Kimura, M Mori, M Nishiguchi, K Hanada, J Sakai, H Fukuoka. Transformation by direct gene transfer in sweetpotato [*Ipomoea batatas* L. (Lam)]. Proceedings of International Workshop on Sweetpotato Production System Toward the 21st Century. Kyushu National Agricultural Experimental Station. Miyakonojo, Japan, pp 159–180, 1997.
11. CS Prakash, U Varadarajan. Genetic transformation of sweet potato by particle bombardment. Plant Cell Rep 11:53–57, 1992.
12. CA Newell, JM Lowe, A Merryweather, LM Rooke, WDO Hamilton. Transformation of sweet potato [*Ipomoea batatas* (L.) Lam.] with *Agrobacterium tumefaciens* and regeneration of plants expressing cowpea trypsin inhibitor and snowdrop lectin. Plant Sci 107:215–227, 1995.
13. R Moran, R Garcia, A Lopez, Z Zasldua, J Mena, M Garcia, R Armas, D Somonte, J Rodriguez, M Gomez, E Primentel. Transgenic sweet potato plants carrying the delta-endotoxin gene from *Bacillus thuringiensis* var. *tenebrionis*. Plant Sci 139:175–184, 1998.
14. MICS Gama, RP Leite Jr, AR Cordeiro, DJ Cantliffe. Transgenic sweet potato plants obtained by *Agrobacterium tumefaciens*–mediated transformation. Plant Cell Tissue Org. Cult 46:237–244, 1996.
15. T Yamaguchi. Hormonal regulation of organ formation in cultured tissue derived from root tuber of sweet potato. Bull Univ Osaka Pref, Ser B, 30:54–88, 1978.
16. M Otani, M Mii, T Shimada. High frequency plant regeneration from leaf calli in sweet potato cv. Chugoku 25. Plant Tissue Cult Lett 13:23–27, 1996.
17. EM Linsmaier, F Skoog. Organic growth factor requirement of tobacco tissue culture. Physiol Plant 18:100–127, 1965.
18. Y Hiei, S Ohta, T Komari, T Kumashiro. Efficient transformation of rice (*Oryza sativa* L.) mediated by *Agrobacterium* and sequence analysis of the boundaries of the T-DNA. Plant J 6:271–282, 1994.
19. TM Klein, L Kornstein, JC Sanford, ME Fromm. Genetic transformation of maize cells by particle bombardment. Plant Physiol 91:440–444, 1989.
20. JC Register III, DJ Peterson, PJ Bell, WP Bullock, IJ Evans, B Frame, AJ Greenland, NS Higgs, I Jepson, S Jiao, JL Lewnau, JM Sillick, HM Wilson. Structure and function of selectable and non-selectable transgenes in maize after introduction by particle bombardment. Plant Mol Biol 25:951–961, 1994.
21. Y Wakita, M Otani, K Iba, T Shimada. Co-integration and co-segregation of an unlinked selectable marker gene and NtFAD3 gene in transgenic rice plants produced by particle bombardment. Genes Genet Syst 73:219–226, 1998.
22. AC van Altvorst, RJ Bino, AJ van Dijk, AM Lamers, WH Lindhour, F van der Mark, JJM Dons. Effect of the introduction of *Agrobacterium rhizongenes* rol genes on tomato plant and flower development. Plant Sci 83:77–85, 1992.
23. C Sommerville, J Browse. Plant lipids: Metabolism, mutants, and membranes. Science 252:80–87, 1991.
24. Y Wakita, M Otani, T Hamada, M Mori, K Iba, T Shimada. A tobacco microsomal ω-3 fatty acid desaturase gene increases the linolenic acid content in transgenic sweet potato (*Ipomoea batatas*). Plant Cell Rep 20:244–249, 2001.
25. T Kimura, M Otani, T Noda, O Ideta, T Shimada, A Saito. Decrease of amylose content in transgenic sweet potato. Breed Res 1:142, 1999.
26. RGF Visser, I Somhorst, GJ Kuipers, NJ Ruys, WJ Feenstra, E Jacobsen. Inhibition of the expression

of the gene for granule-bound starch synthase in potato by antisense constructs. Mol Gen Genet 225: 289–296, 1991.
27. DE Horton, PT Ewell. Sweet potato pest management: A social science perspective. In: RK Jansson, KV Raman, eds. Sweet Potato Pest Management: A Global Perspective. Boulder, CO: Westview, 1991, pp 407–427.
28. VA Hilder, AMR Gatehouse, SE Sheerman, F Barker, D Boulter. A novel mechanism of insect resistance engineered into tobacco. Nature 330:160–163, 1987.
29. J Sambrook, EF Fritsch, T Maniatis. Molecular cloning—a laboratory manual. 2nd ed. Cold Spring Harbor, NY: Cold Spring Harbor Laboratory Press, 1989.
30. G Vervliet, M Holsters, H Teuclay, M Van Montagu, J Schell. Characterization of different plaque forming and defective temperate phages in *Agrobacterium* strains. J Gen Virol 26:33–48, 1974.
31. OJM Goddijin, J Pen. Plants as bioreactors. Trends Biotechnol 13:379–387, 1995.

# 47
# Genetic Enrichment of Barley: Opportunities and Challenges

**Seedhabadee Ganeshan, Monica Båga, and Ravindra N. Chibbar**
*National Research Council, Saskatoon, Saskatchewan, Canada*

| | | |
|---|---|---|
| I. | INTRODUCTION | 717 |
| II. | BARLEY IMPROVEMENT | 718 |
| | A. Traditional Breeding | 718 |
| | B. Mutation Breeding | 718 |
| | C. Wide Hybridization | 719 |
| | D. Biotechnological Approaches | 720 |
| III. | GENETICALLY ENGINEERED BARLEY | 724 |
| | A. Herbicide Resistance | 724 |
| | B. Pest Resistance | 724 |
| | C. Improvement of Barley Quality | 725 |
| | D. Novel Uses | 727 |
| IV. | PROSPECTS AND CHALLENGES | 727 |
| | REFERENCES | 729 |

## I. INTRODUCTION

Barley is the world's fourth major cereal crop in acreage and production, surpassed by wheat, maize, and rice (1). Formerly the third founder cereal of Old World Neolithic agriculture after einkorn and emmer wheats (2–4), barley has gradually developed a diverse and unique niche of its own, with continued prominence in today's agriculture. Its domestication occurred in the Fertile Crescent about 10,000 years ago from brittle, two-rowed forms, bearing similarity to *Hordeum spontaneum* (4), which is considered to be the wild ancestor of cultivated barley, *Hordeum vulgare* L. (2,5). The primary center of diversity in Turkey, Jordan, Syria, Iraq, and Iran and a secondary center in Ethiopia are believed to harbor many wild species of barley (6), constituting, along with other wild species, about 30 species in the genus *Hordeum* (7).

The estimated world production of barley for the period 1999–2000 was around 128 million tons (8). The projected global production for the 2000–2001 period was about 133 million tons (8), and generally the production of barley hovers around these figures, depending on global marketing trends and environmental dicta. Significant increase in grain yields beyond the stagnant trend is likely to be attained with input from newer technologies (9).

NRCC No. 43788

Barley is mostly used as animal feed and as malt for the production of alcoholic beverages. In some parts of the world, barley is also used as human food. Extensive effort is being undertaken to improve barley quality for expanded consumption by humans. The main areas of focus include starch quality and protein composition and content. Besides food uses, there has also been interest in using specialized barley starches for industrial purposes. In this review available options for the genetic enrichment of barley are assessed, with particular emphasis on the biotechnological approaches that have opened up new vistas in this regard.

## II. BARLEY IMPROVEMENT

Because of barley's economic importance, considerable resources have been allocated for its improvement. Classical breeding approaches have been and will continue to be the foundation of all improvement initiatives. However, limitations such as lack of genetic variability, availability of desirable traits for introgression, and genetic complexity of some traits are often encountered in conventional breeding programs. To supplement these shortcomings, alternative resources have been deployed in breeding programs.

### A. Traditional Breeding

Barley breeding programs have been undertaken to improve varieties in terms of a number of characters, dictated by producer and consumer demands and environmental constraints. Although higher yields are conceivably the primary objective (10), other agronomic characters that indirectly affect grain yield and quality are also given due consideration. Some of the attributes, such as days to maturity, straw strength, plant height, disease resistance, malting quality, and protein content, are constantly being addressed. Among the breeding methods in practice are the bulk method, composite cross, pedigree, backcrossing, and single-seed descent. The choice of method depends on the objectives and resources available. The merits and demerits of these approaches can be found in plant breeding textbooks and are not discussed here. Suffice to mention that barley breeding has been a successful venture to date, leading to substantial yield increases. Lodging resistance, disease resistance, reduction in height, and increase in harvest index have been instrumental components contributing to yield increments (11). Newer malting varieties have additionally improved grain quality with respect to increases in plumpness, malt extract, $\alpha$-amylase activity, and diastatic power (12). With the advent of deoxyribonucleic acid (DNA) marker–assisted selection, further progress is being made toward incorporation of desirable traits, qualitative as well as quantitative. Although breeding efforts over the years have been highly successful, dwindling of the genetic diversity might stall progress in the development of more improved cultivars. Therefore, alternative genetic resources are being sought for subsequent incorporation into breeding programs.

### B. Mutation Breeding

Induced mutation is one method that has been proven to generate new sources of genetic variability. Mutation breeding involves the use of mutants either directly as new cultivars or via their incorporation into breeding programs to generate a valuable source of germplasm. Barley resistant specifically to powdery mildew was the first disease-resistant mutant to be generated by mutation breeding (13). Subsequently, mutant cultivars in many other crops have been released. As of December 2000, 2252 accessions were listed in the FAO/IAEA Mutant Varieties Database, of which 1072 were cereals, including 269 barley accessions (14).

Induced mutants of barley with superior agronomic and quality attributes have been generated and incorporated into many breeding programs. Although in the past few decades induced mutation was mainly focused on developing disease-resistant and high-grain-protein varieties (14), more favorable results were obtained with respect to lodging resistance (15,16), days to maturity (17–19), and grain protein content (20–23). Induced mutation for disease resistance is best exemplified by the X-ray-induced mutant of the barley cultivar 'Heine's Haisa,' which exhibits resistance to *Erysiphe graminis* f. sp. *hordei,* the causal agent of powdery mildew (24). Other mutants related to improved brewing and malting quality have also led to the release of new varieties (25). Mutant varieties with improved grain protein content are also available (25). More recently, mutational analysis was used to dissect the complex genetics of kernel rows in barley (26). It was shown that at least 12 gene loci (*hex-v* and 11 *int*) were involved in spike development of lateral florets, including size, awn development, fertility, and kernel development. Crosses involving a double mutant (*int/int*) with a *hex-v* mutant produced progenies exhibiting exceptional six-row type with prominent large spikes and thick culms.

It is evident that mutation breeding has played an important role in barley improvement and is likely to have more significant impact on the development of improved barley varieties. Physical and chemical mutagenic treatments currently available have been adequate but produce a large number of wasteful mutants, which consume enormous resources in terms of time, space, and labor. Thus, other means of generating genetic variation are being explored to allow a more targeted and/or controlled change.

## C. Wide Hybridization

The genetic erosion that results from the disappearance of exotic germplasm (e.g., land races) and stringent selection and use of a limited pool of germplasm for breeding purposes over the years has led to an interest in wild species and genera for wide crosses. Various barley improvement programs have been delving into the possibilities of identifying and introgressing genes from wild *Hordeum* species into cultivated barley. However, crossability of barley with other *Hordeum* species is very low (27). Specifically, chromosome pairing in interspecific hybrids is poor, except for those derived from *H. bulbosum* and *H. spontaneum* (28). Nonetheless, useful agronomic traits such as disease resistance and stress tolerance from many *Hordeum* species have been expressed in hybrids or backcross progenies (29). Attempts at generating intergeneric hybrids with *Hordeum* species have also progressed considerably and extensive analyses of the subject have been published elsewhere (29,30). It is interesting to note that intergeneric hybrids involving combinations of *Secale, Agropyron,* or *Elymus* species with *Hordeum* species outnumber the intergeneric hybrids made with *H. vulgare* (28). This is possibly a reflection of the closer phylogenetic relationship among wild genera within the tribe Triticeae than to domesticated barley. Irrespective of the crossing combinations, useful agronomic traits have been identified in *Hordeum, Elymus, Agropyron,* and *Secale* species, and the possibility of their hybridization with cultivated barley or other *Hordeum* species augurs well for the introgression of such traits (29).

The potential of alien gene introgression is likely to be fully realized as research continues to overcome some of the bottlenecks encountered with regard to crossability, hybrid sterilty, expression of introgressed genes, elimination of undesirable traits incorporated as a result of linkage drag, and selective elimination of chromosomes in hybrids. There are some strategies to overcome some of these barriers (for details refer to Ref. 31). For example, chromosome doubling can be used to circumvent hybrid sterility in $F_1$ hybrids. Hybrid embryos with undeveloped or poorly developed endosperms can be rescued by tissue culture techniques.

## D. Biotechnological Approaches

Biotechnology has unequivocally become an integral part of barley improvement programs. The access to additional gene pools has become a possibility as a result of the development of tissue culture and transgene technologies. For a long time, barley was considered to be recalcitrant to both technologies. However, major obstacles have now been overcome. DNA-based markers provided the breeder with another tool to select for desirable traits, thereby allowing more effective selection and also expediting the breeding process. The use of in vitro culture for barley improvement has been three-pronged, viz., induction of variation, doubled haploid production, and genetic transformation.

### 1. Tissue Culture and Somaclonal Variation

An essential requirement for the success of tissue culture is availability of an efficient regeneration system. Significant progress has been made over the years in barley tissue culture. Various sources of explants have been used for culture, including immature embryos, immature scutella, immature inflorescences, ovaries, ovules, anthers, microspores, apical meristems, mature embryos, and leaf bases. Immature tissues have been the preferred source of explants because of their responsiveness in culture. However, in recent years there has been interest in using meristems and mature embryos as explants. An obvious advantage of such explants is that they are readily available. Regardless of the source of the explant, plant regeneration can occur by either organogenesis or embryogenesis, via an intermediate callus induction phase.

It was thought that an in vitro selection system, as a novel source of variability, could mimic the type of selection employed by breeders and thus identify variant cell lines resistant to a particular stimulus. The concept of in vitro culture–induced variation (somaclonal variation) (32) further heightened interest in the possibility of inducing more genetic variability. The occurrence of chromosomal variability in callus and plants derived from ovarian tissue cultures of a *Hordeum vulgare* × *Hordeum jubatum* hybrid was already recognized as having potential for the generation of new chromosomal combinations and mutations and overcoming barriers inherent in conventional introgressive breeding techniques (33). A similar approach was explored with hybrid plants recovered by embryo rescue from a *Hordeum vulgare* × *Elymus canadensis* (Canada wild rye) cross (34). Canada wild rye exhibits winter-hardiness, drought tolerance, and resistance to barley yellow dwarf virus, but crosses with barley result in sterile $F_1$ hybrids. The tissue culture–derived regenerants showed evidence of chromosomal recombination. Many other studies have reported on the occurrence of aneuploids, polyploids, and chromosomal rearrangements in cultured cells of barley (e.g., 35–38). Several studies have attempted to capitalize on this tissue culture–induced variation. Heritable somaclonal variation was observed in *Hordeum spontaneum*, although chromosomal numbers and aberrations were not detected (39). Improved resistance to the net blotch pathogen was observed in barley plants regenerated from callus cultures subjected to selection on medium supplemented with the toxin produced by the pathogen (40). However, resistance was also observed in unselected controls, indicating that the observed resistance could be attributed to somaclonal variation. Resistance to *Helminthosporium sativum* was also observed in barley plants regenerated from callus cultures selected on partially purified toxins from the pathogen (41). Barley plants derived from callus cultures exhibited tolerance to the herbicide chlorsulfuron as a result of selection on media containing the herbicide (42). Increased tolerance to glyphosate was reported in some progenies of plants regenerated from callus subjected to glyphosate selection (43).

Unfortunately, somaclonal variation also impacts barley regenerants negatively, often leading to undesirable characteristics. In studies on malting quality of tissue culture–derived barley lines, it was found that most of the lines had poor malting quality (44). Besides undesirable characteristics in progenies of tissue culture–derived barley plants, genotype-dependent differences in the frequency and extent of the agronomic performances were observed (45). The evaluation of field performance of progenies from protoplast-derived barley plants did not show any abnormal phenotypes, but some agronomic characteristics were affected (46). Because of the adverse effects of somaclonal variation, it has been suggested that identification of useful regenerants from transformation experiments might be precluded (47). In a 1998 study on the performance of progenies of transgenic barley plants, somaclonal variation was deemed to be unforeseeably high, in the range of 14–60% (48). The report also claims that somaclonal variation was accentuated by the transformation procedure.

With conflicting reports on the occurrence of somaclonal variation in barley, there is clearly a need to revisit research on somaclonal variation with a new perspective. The high frequency and severity of somaclonal variation observed in some studies could be due to culture media components, which are constantly being altered to improve the responsiveness of cultured explants, i.e., to increase embryogenic callus induction and regeneration frequencies. Better callus induction and regeneration frequencies may not necessarily translate into production of normal plants. The low callus induction and regeneration frequencies obtained in earlier studies may not have been opportune for development of the undesirable variants. Most studies have indicated morphological uniformity among tissue culture–derived barley plants and infrequent occurrence of somaclonal variants (49–51). It has been suggested that either the barley genome could be tolerant to tissue culture environments or aberrant cell lines are at a competitive disadvantage during regeneration, thereby lowering the number of regenerated variants (52). Considering the few reports about useful somaclonal variants in barley and the inability to maximize culture-induced variation, it is not surprising that the euphoria surrounding somaclonal variation in barley has declined. The potential for creation of genetic variability does exist, especially with regard to culture of wide hybrids that would otherwise be infertile. In vitro culture in combination with mutagenic treatments could also be used to generate variants.

## 2. Doubled Haploid Production

Haploid breeding has become a significant component of barley breeding programs. Haploidy initiation followed by chromosome doubling allows homozygosity to be achieved in a single generation. Since improvement of inbreeding crops such as barley entails development of true breeding, homozygous lines by hybridization of two or more parental lines (53), followed by repeated selfing and selection (54), haploid breeding is an expedited approach to achieving similar results. At any stage of a breeding program, doubled haploids can be initiated, and the number of cycles of recombination (before homozygosity is attained) is at the discretion of the breeder (55,56).

There are several methods of haploid production in barley, including the bulbosum technique, haploid initiator gene method, ovary, anther, and microspore culture. Several reviews have been published on haploid production and use in barley breeding (e.g., 56–58,53). The bulbosum technique evolved from observations that wide hybrids obtained by embryo rescue from *H. vulgare* × *H. bulbosum* crosses were haploid in nature (59). Because of the possibility of producing haploids from a wide range of genotypes, this method has been favored for the production of doubled haploids in the past (60). The haploid initiator gene (*hap*), a result of induced mutation in barley, can also lead to the production of haploids (61). The advantage of using this approach is that in vitro culture can be obviated, but the inability to distinguish spontaneous doubled haploids from hybrid embryos and the imperative use of only genotypes carrying the *hap* gene have not led to the widespread use of this technique (53).

Haploid production via anther and microspore culture has become a method of choice because of the potentially large number of plants that can be generated. Requirements for androgenic culture are, however, critical for the success of efficient regeneration. Factors such as growth conditions of the donor plants and media components have been the subject of many studies, and significant improvements have been made in the responsiveness of anther and microspore culture. Although genetic instability and genotype dependency have been reported in plants derived from androgenic cultures, some progress toward circumventing these drawbacks has been made. One of the problems that still must be addressed is the observed distorted segregation ratios in doubled haploids derived from androgenic cultures (62–64). Nonetheless, androgenic doubled haploids are being used in plant breeding programs and have surpassed doubled haploid production by the bulbosum method (53).

The contribution of doubled haploid production in barley breeding has been fully recognized, as shown by the release of several varieties. Its implications for barley breeding are likely to expand further. In conjunction with mutagenesis, haploid production has been suggested to be useful in generating true-to-type mutants (65). By using such an approach several doubled haploid mutants of barley have been identified (66). However, the use of the haploid system is not restricted to the generation of doubled haploids for breeding programs. The prospects for using the haploid system for barley transformation also cannot be disregarded. Microspores are single-celled haploid entities that have been suggested to be attractive targets for transformation (67), and the practicality of such an approach has been demonstrated. For example, transient expression has been observed in electroporated microspores (68) and in those transformed by particle bombardment (69). Stable transformation of microspores (discussed later) has also been achieved.

## 3. Transgenic Plant Production

A suitable vehicle for DNA delivery; an appropriate target cell, tissue, or organ; and an efficient tissue culture regeneration system are pivotal to the success of transgene technology in barley. Although in planta transformation of barley plants has been attempted by macroinjection of DNA, results have not been encouraging enough to warrant further in-depth studies. For example, DNA from a two-row barley plant was injected into developing seeds of a six-row *waxy* barley (70). A few plants had normal starch composition and displayed two-row phenotype, but mendelian segregation was not observed in the progeny. Plasmids carrying the NPTII gene have also been injected into floral tillers (71), with inconclusive results. Application of plasmid DNA to stigmas (71) and inoculation of florets at anthesis with *Agrobacterium* sp. (72) are other approaches that have been tried to no avail. Several other attempts at DNA uptake by germinating seeds (73), by imbibing of embryos in DNA suspensions (74) and by electrophoretic transfection of seeds, have been equally unsuccessful. Moreover, such transformation systems are not specifically targeting meristematic cells, which would otherwise provide a degree of assurance as to the inheritance of integrated DNA in subsequent generations (52).

Microinjection of DNA and electroporation of cells have also been explored as potential DNA delivery systems. Although microinjection can be viewed as an elegant means of introducing DNA, it has the disadvantage of requiring highly specialized resources and skills. Besides being cumbersome, DNA delivery has to be precise, and the microinjection process should not inflict substantial injury to the cells. Since only a low number of cells can be microinjected because of the labor-intensive nature of the technique, there is a need for a highly efficient culture system for regeneration of treated cells. Potential target cells for microinjection are zygotic proembryos and microspore-derived proembryos (52). Multiple microinjections of proembryos have only produced putative primary chimeric transformants, a phenomenon ostensibly apparent when dealing with gene transfer to clumps of cells (75). Recent advances in improving efficiency of regeneration from barley microspore cultures could refurbish the microinjection technique.

Electroporation is a more subtle approach to transfer DNA to cells and tissues. Although initially protoplasts were investigated as the main targets for electroporation, plant cells and tissues have also been subjected to electroporation for the uptake of DNA. Many studies have used barley protoplasts for transient expression assays, using either electroporation (e.g., 76,77) or polyethylene glycol (PEG) (e.g., 78,79). Thus, the protoplast system seems to be an ideal target for DNA uptake by either electroporation or PEG-mediated transformation, except that some of the limitations of the techniques need to be addressed. Establishment of viable protoplasts requires specific culture conditions, is highly genotype-dependent, tends to lose morphogenic potential in long-term cultures, and is a laborious process (80). Somaclonal variation is another problem often encountered with protoplast regeneration. It has been suggested that use of barley protoplasts derived from primary callus from young scutella, microspores, or anthers may alleviate some of the problems associated with plant regeneration (81). Nonetheless, stable transformation of barley via PEG-mediated uptake of DNA by protoplasts has been reported (82).

Currently the method of choice for the transformation of barley is the microprojectile DNA delivery system. The physically harsh nature of this technique has not been a deterrent to the efficient regeneration of plants. Many reports on the use of this technique for barley transformation have been published (e.g., 83). Principles and parameters governing the microprojectile delivery system have been the subject of many studies and reviews (84) and are not discussed here. Improvements are constantly being made to the bombardment parameters in order to maximize efficiency of barley transformation. Most of the limitations encountered with protoplast transformation methods are nonexistent when using particle bombardment. Target cells or tissues are readily available and respond adequately upon culture, provided optimal culture conditions have been established in preliminary experiments. However, the haphazard nature of integration of the transgene in the genome has been an issue of concern with regard to multiple integration sites, rearrangements, truncations, gene stability, and unstable expression (85).

The use of *Agrobacterium* sp. for the transformation of barley has now been established (86–88). Considering the resources being allocated to research aimed at perfecting the process in several laboratories around the world, it is likely to become the method of choice. It is a cheaper alternative to particle bombardment, requiring simpler manipulative steps.

Thus far, DNA delivery systems have been discussed with ephemeral allusion to the target cells or tissues. An efficient transformation system is strongly dependent on the competency of the explant to be a recipient and the regeneration frequency of such an explant. It is therefore imperative that some time be devoted to carefully identifying the most appropriate target tissue. In barley, scutella from immature embryos have been the preferred material because of their responsiveness to in vitro regeneration. However, isolation of scutella is a time-consuming and labor-intensive process. Moreover, the scutella are not readily available and best response in culture is obtained from scutella ranging in size from 0.5 to 1.0 mm. Prolonged storage of the heads prior to isolation of scutella also reduces responsiveness. Because of these limitations with scutella, attempts have been made at identifying explants that are easily accessible and equally responsive. Regeneration from young leaf tissue–derived callus of barley was reported as early as 1988 (89). However, interest in leaf tissues has not been widespread, as evidenced by the limited reports of success. The use of barley leaf tissue as a potential explant was reviewed in a 1999 study, wherein all four genotypes under investigation regenerated plants (90). Further studies with a larger number of genotypes would confirm the suitability of this system for regeneration and transformation.

Mature embryos and meristems are appealing sources of explants because of the ease with which they can be obtained and manipulated. Callus induction from isolated apical meristems of barley was reported in 1975 (91). Subsequently, several reports have alluded to the competency of apical meristems together with the adjoining leaf tissues for somatic embryo induction and plant regeneration in barley (92–94). Most often mature embryos, seedlings, leaf bases, and meristems are mentioned as one explant, since their physical separation as individual explants is not mandatory to induce callus from and around the meristem tissues. Direct induction of multiple shoots from bombarded shoot meristematic cultures, without a callus induction phase, has also been achieved (95,96). Such a system may have the added advantage of reducing somaclonal variation. For mature embryos of barley, there is one successful report of callus induction and plant regeneration (35). No further studies using mature embryos as explants have been reported.

Other potential explants for transformation include immature ovaries, immature inflorescences, and anthers/microspores. Since the first report on the culture of unfertilized ovaries of barley (97), not much progress has been made in improving this culture system, in spite of its potential for the production of homozygous dihaploid plants. An improvement in regeneration frequencies was reported, provided ovaries were collected when pollen grains were at the trinucleate stage (98). Good regeneration was reported from immature inflorescences in the size range of 1.5–3.5 mm (99), but further progress in culturing immature inflorescences of barley has not been accomplished. The critical stages of ovaries and inflorescences to respond in culture may have been deterrent factors in exploring these explants for further use. Recent improvements in anther/microspore culture of barley offer another promising system for transformation. Transgenic plants from barley microspores have been reported (100,101), and research is likely to intensify, especially with *Agrobacterium* species–mediated transformation to make this system more efficient.

## III. GENETICALLY ENGINEERED BARLEY

Despite the limited successes in barley transformation as compared to that in many other cereal crops, several useful traits have already been introduced in barley. Although disease resistance is one of the important aspects of barley improvement, malting and brewing qualities are also becoming targets of investigation through manipulation of enzymes involved in these processes. Protein engineering for enhancing the feed and food value of barley is another area of focus. Although modification of starch composition could be a possible target, today there is no obvious use for barley starch with high amylose or high amylopectin content.

### A. Herbicide Resistance

Serious yield losses can occur in barley as a result of weed infestation. Although tillage practices and crop rotations can reduce weed populations to some extent, recourse to herbicide applications is sometimes necessary. However, the herbicide has to be chosen carefully. It has been reported that barley is sensitive to many herbicides at varying stages of growth (102). The incorporation of herbicide-resistant genes into barley by transformation may alleviate this situation, allowing effective control of weeds. Caution should nevertheless be exercised when using herbicides, and they should only be used in conjunction with an integrated weed management program. One of the first reports on the production of fertile transgenic barley by microprojectile bombardment employed a plasmid carrying the *bar* gene (103). Another study reported on the successful use of isolated barley microspores as target for biolistic transformation with a construct carrying the *bar* gene, wherein Basta-resistant plants were regenerated (100). Thus, besides serving as a selective system for regeneration from transformed cells in cultures, regenerated transformants are resistant to the herbicide Basta. However, Basta-resistant barley plants have not been commercially exploited so far.

### B. Pest Resistance

Barley, like other crops, is prone to a number of pests, such as viruses, bacteria, fungi, and insects. Yield as well as the quality of the grains for malting, feed, or food uses can be significantly affected without appropriate control measures. By far the best way to protect barley crops is by following sound agricultural practices such as crop rotation and by using resistant varieties. Only as a last resort should chemical control be undertaken. The use of pesticides is being discouraged mainly because of environmental concerns. Moreover, the costs associated with pesticides and their application may outweigh the returns. Although agricultural practices can be followed religiously, resistance or tolerance of varieties may not last, because of changes in pest populations, environmental conditions, and extensive use of a particular variety over a number of years, which could lead to a breakdown in resistance. Therefore, plant breeders are constantly breeding for resistance to pests. The limiting factors, however, are the lack of or inability to identify resistance genes in existing barley varieties and the difficulty in introgressing resistance genes from wild species, if any, into barley cultivars. The recently improved genetic engineering systems for barley are likely to play important roles in this respect. A number of candidate genes potentially capable of directly or indirectly contributing to enhanced pest resistance have been cloned (see Ref. 47 for a review). For example, coat protein–mediated resistance to barley yellow dwarf virus was demonstrated in transformed barley (104). When using complementary DNA (cDNA) of virus coat protein genes for transformation of barley, coat protein–mediated resistance to barley mild mosaic bymovirus and barley yellow mosaic bymovirus was observed (105). Enhanced resistance to *Botrytis cinerea* in excised leaves of barley was demonstrated by transformation with the stilbene synthase gene encoding synthesis of *trans*-resveratrol, a phenolic phytoalexin (106).

It is interesting to note that a number of genes related to insect resistance, bacterial disease resistance, and fungal disease resistance that have been cloned from barley have been transformed into tobacco or wheat. The barley trypsin inhibitor gene Cme was introduced into wheat immature embryos by particle bombardment, and regenerated plants were shown to reduce survival rate of *Angoumois* sp. grain moth significantly (107). In a study attempting to enhance fungal disease resistance in a quantitative manner, three barley genes encoding antifungal proteins were used for combinatorial expression in transgenic tobacco (108). The three barley genes encoded a class II chitinase, a class II β-1,3-glucanase, and a type I ribosome inactivating protein. Similarly, the precursor of small antibacterial proteins, hordothionins, present in barley endosperms, was expressed in tobacco to study expression, processing, sorting, and biological activity (109). In vitro toxicity assays with hordothionins extracted from transgenic tobacco showed toxicity similar to that of the barley hordothionins on the bacterial plant pathogen, *Clavibacter michiganensis* subsp. *michiganensis*. The recent advances in barley tissue culture and transformation might encourage use of such genes for barley transformation.

## C.  Improvement of Barley Quality

### 1.  Malting Barley

Malting barley quality is dependent on the inherent desirable traits that have been incorporated by breeders in response to specific demands by the malting industries. However, malting quality is also influenced by environmental conditions, which could adversely affect grain quality and result in rejection of the barley for malting. Visual inspection of the grains is the first assessment for malting potential; it focuses on plumpness, uniformity, and good color (110). If the visual inspection deems the barley to be good for malting, then more detailed protein and germination tests are performed to ascertain its malting potential (111). High protein content in barley grains is not desirable for malting, since it reduces the amount of extract produced from the malt. Malt destined for brewing should have high extract, an acceptable level of soluble protein, and the necessary starch degrading enzymes with optimal activity. The malting process is considered to proceed optimally if there is extensive endosperm degradation or modification during the germination stage of malting, with concurrent degradation of cell walls and protein leading to exposure of the starch granules (112).

Since barley is an important commodity for malting purposes, there is a need for production of high-quality grains. In this regard studies have been undertaken to render some of the biochemical modifications during malting more efficient. The cell wall polysaccharides can create a highly viscous solution during malting and brewing, and rapid degradation of these wall materials is desirable (113). The (1,3–1,4)-β-glucans are major cell wall polysaccharides in barley (114), and their complete degradation by (1,3–1,4)-β-glucanase during germination is essential for the production of good-quality malt. However, (1,3–1,4)-β-glucanase is inactivated at temperatures above 55°C (115). To overcome this limitation during the kilning process, which occurs from 40°C to 85°C, a codon-optimized thermostable (1,3–1,4)-β-glucanase gene from *Bacillus* sp. was engineered for optimal expression in transgenic barley (116). A subsequent study showed that the modified transgene driven by an α-amylase gene promoter displayed temporal and spatial expression patterns in scutella and aleurone layer during germination (117). As opposed to extensive degradation of cell wall materials and proteins during malting, the conversion of starch needs to be minimal, if the end use of the malt is brewing and distilling (111). The enzymes involved in this process are α-amylases, β-amylases, and limit dextrinase or pullulanase. A 1999 report on the increased activity of pullulanase in barley grains by coexpression of thioredoxin (118) could be useful in more efficiently hydrolyzing α-(1,6) linkages in amylopectin fractions of starch during brewing and distilling processes.

## 2. Nutritional Quality

There has been an increasing interest in improving the feed and food value of barley. Over 5% of the barley produced in the world is consumed as food (110); highest consumption occurs in North African countries at a rate of 65 kg/person/annum (119). The high dietary fiber content and its possible health benefits have led to improvement strategies that could popularize its consumption. However, the low content of lysine and threonine reduces its nutritional value for humans and monogastric animals. Therefore, attempts have been made to increase the levels of essential amino acids. High-lysine barley genotypes have been developed but have not been commercially acceptable because of the shrunken seed phenotype associated with elevated lysine content (120). Biotechnological approaches using transformation technology are also being explored. For example, the expression of mutant *Escherichia coli* genes encoding lysine feedback–insensitive forms of aspartate kinase and dihydrodipicolinate was studied in barley (121). Primary transformants carrying the modified aspartate kinase gene showed a 14-fold increase in free lysine and an 8-fold increase in free methionine. Mature seeds from transformants carrying the dihydrodipicolinate synthase gene showed a twofold increase in free lysine, arginine, and asparagine and a 50% decrease in free proline. These results indicate that lysine and methionine levels can potentially be enhanced by this approach.

As in malting barley, the high level of (1,3–1,4)-β-glucans can affect the feed value of barley, making it a low-metabolic-energy feed for poultry. In 2000, the codon-optimzed thermostable (1,3–1,4)-β-glucanase, as mentioned earlier but driven by the D hordein gene promoter, was transformed into barley by using *Agrobacterium* sp. (122). The transgenic lines were reported to have 40 times more recombinant enzyme than produced in the endosperm of the previously reported transgenic malt. Like (1,3–1,4)-β-glucans, arabinoxylan (another endosperm cell wall polysaccharide) has antinutritional value for monogastric animals. To hydrolyze the arabinoxylans, endo-1,4-β-xylanase of microbial origin has been added to the feed. To develop a more efficient system, a modified xylanase gene of fungal origin (123) driven by either a rice or a barley endosperm-specific promoter has been transformed via *Agrobacterium* sp. into barley (124). Although the estimated expression level was only 0.037% of total seed protein, there is scope for further improvement, possibly by optimized codon usage (124).

Barley for human consumption still requires extensive research and is yet to be in significant demand, in spite of increased usage as food (110). Barley has been reported to have several health benefits, such as reducing heart disease, colon cancer, and rate of absorption of sugar (beneficial to diabetics) (125), which are most likely due to the high level of β-glucans. Although β-glucan is not desirable in malting and feed barley, it is considered to be a neutraceutical for humans in this respect.

## D. Novel Uses

Barley is always thought of as a malting or feed crop and will continue to be important for such purposes. However, the potential for using barley for other end uses does exist. In the food industry it has found minor uses in wheat bread, soups, porridge, breakfast cereals, stews, bakery foods, and production of maltose syrups. Improvement of barley to cater to the primary niche markets, i.e., feed and malt, is an ongoing task and will not be surpassed by improvement for any other potential use. Novel uses of barley may arise from genetic engineering approaches, for example, in the production of enzymes, oral vaccines, antibodies, pharmaceuticals, and vitamins. The report on the large-scale production of xylanase in barley grains (124) is encouraging for exploring production of other industrial products in barley. Specialty starches for industrial purposes could also be produced in barley. For example, the production of industrial starch films is influenced by the amylose/amylopectin ratio among several other factors (126). High amylopectin content has been reported to decrease flexibility, burst resistance, and tensile strengths of films (127). Potato, corn, wheat, and rice starches have been studied for their bioplastics properties (128), and similar uses of barley starches as thermoplastic materials could be explored. In 2000, an antibody (scFv) against carcinoembryonic antigen was engineered and expressed in the leaves and seeds of rice and wheat, as an example for the production and storage pharmaceuticals in cereals (129). Such an approach could also be attempted with barley for similar or other types of antibodies or vaccines. Novel uses for barley are still in the exploratory phase and not extensively researched, possibly because of the priority of improvement in its primary uses as feed, malt, and food.

## IV. PROSPECTS AND CHALLENGES

From being recalcitrant to in vitro culture and transformation, barley can now be considered with conviction to be amenable to both. The number reports of success that have appeared are indeed encouraging, and this decade will most likely see a spurt of reports dealing with barley harboring desirable genes incorporated by direct gene transfer. Although particle bombardment will continue to be extensively used, the capability *Agrobacterium* sp.–mediated transformation systems provide, has given transgenic barley research further impetus.

However, there is a certain naïveté in harboring the notion that *Agrobacterium* sp.–mediated transformation will offer advantages such as more organized integration of a defined segment of DNA at low copy numbers, with higher, stable expression of transgenes as opposed to particle bombardment. Events such as DNA rearrangements, deletions, and multiple copy number and insertion sites are also observed with *Agrobacterium* sp. transformation (for a recent review see Ref. 130). Multiple copies of transgenes and integration sites have been linked to gene silencing, mostly believed to be a result of transgene hypermethylation. As early as 1992, it was recognized that DNA methylation had a significant role to play in the stability of foreign DNA in barley (131). The report suggested that the DNA to be transferred must have a certain methylation pattern for stability of the foreign DNA. Barley transformants generated by particle bombardment exhibited loss or variable expression levels of transgenes in some of the progeny (103). *Agrobacterium* sp.–mediated transformation of barley has also been reported to generate transformed lines with multiple insertions and transgene silencing (86). Furthermore, the transformation frequency obtained in this study was comparable to that obtained by particle bombardment.

As barley becomes more amenable to transformation, more reports are expected to allude to the issue of transgene stability and silencing. It is therefore imperative to identify ways to circumvent these problems. Since transgene silencing is often associated with multiple copies inserted at a single site or dispersed throughout the genome, it has been suggested that only transgenic plants carrying a single copy of the transgene be selected and advanced (132,133). However, introduced transgenes can also suppress expression of endogenous genes, if there is appreciable sequence homology. Part of this problem can be prevented by using different promoters for driving the expression of the transgenes (85). Stable integration and expression of transgenes have been shown to be improved by flanking trangenes with sequences of matrix attachment regions (MARs), thereby possibly shielding the transgene from influences of surrounding chromatin (134). Since barley has about 80% repeat DNA (135), this strategy employing MARs might prove to be useful in reducing transgene silencing, resulting in efficient expression of the transgenes.

*Agrobacterium* species–mediated transformation could also be used to deliver more specialized vectors carrying large inserts such as bacterial artificial chromosomes (BACs). Such an approach has been used for the transformation of tobacco, by the use of a binary bacterial artificial chromosome (BIBAC) vector (136). This approach is particularly appealing for transformation with regard to large fragments of DNA with several genes. Several recent reports have indicated clustering of genes affecting traits on specific chromosomal segments. For example, there are several reports pertaining to the existence of clusters of genes along chromosomal segments that are involved in disease resistance (*e.g.,* 137–139). Such clusters have been identified in barley (140,141) and could potentially be used for the transformation of barley by using the BIBAC system, when BAC libraries containing such clusters become available.

Ongoing research must continue to unravel further the biochemical events or pathways affecting the malting and feed quality attributes. Information thus obtained could be used to overexpress, down-regulate, or knock out specific genes for altered characteristics. Similarly, efforts need to be directed toward better understanding the importance of amylose and amylopectin in the malting process. Whereas waxy barley has been reported to have reduced cell wall modification (112), high-amylose barley has been reported to have other characteristics that adversely affect malting (142). However, many of the genes involved in the biosynthetic pathways of starch in cereals have now been cloned (143) and could be used for targeted modification of starch composition for malting. The characteristics of barley starch granules also determine the ease of breakdown of the starch. The large A-type and small B-type granules are more prone to degradation at different efficiencies during germination. However, because of their resistance to degradation by amylolytic enzymes during brewing, carryover of any remaining small granules into malt can affect brewing (111). Therefore, barley grains with higher concentrations of large A-type granules might be more desirable for brewing.

It is evident that opportunities for genetically engineering barley abound. Therefore, priorities need to be laid out as to the most important aspects of barley improvement for immediate consideration. Limitations due to availability of culture and/or transformation systems can now be considered insignificant. The general concept of having to establish systems capable of high transformation frequencies is a vagary to be given less importance. Resources should rather be directed toward production of useful transformation events. In other words, more consideration should be given to producing stable transformants and specialized constructs and maximizing transgene expression.

## REFERENCES

1. JM Poehlamn. Adaption and distribution. In: DC Rasmusson, ed. Barley. American Society of Agronomy, Inc., Crop Science Society of America, Inc., Soil Science Society of America, Inc., 1985.
2. JR Harlan, D Zohary. Distribution of wild wheats and barley. Science 153:1074–1080, 1966.
3. D Zohary. The origins of cultivated cereals and pulses. Chromosome Today 4:307–320, 1973.
4. D Zohary, M Hopf. Domestication of plants in the Old World. OxFord: Oxford University Press, 1988.
5. D Zohary. The progenitors of wheat and barley in relation to domestication and agriculture dispersal in the Old World. In: PJ Ucko, and GW Dimbleby, eds. The Domestication and Exploitation of Plants and Animals. London: Duckworth, 1969.
6. JM Poehlman. Breeding barley and oats. In: Breeding Field Crops, AVI, 1979.
7. R von Bothmer. The wild species of *Hordeum*: Relationships and potential use for improvement of cultivated barley. In: PR Shewry, ed. Barley: Genetics, Biochemistry, Molecular Biology and Biotechnology. Wallingford, England: CAB International, 1992, pp 3–18.
8. World Agricultural Production. Barley Area, Yield, and Production, Table 6, http://www.fas.usda.gov/wap/circular/2001/01-03/tables.html.
9. M Brophy. Global production and markets for barley production in the 21st century. In: G Scoles, B Rossnagel, eds. V International Oat Conference & VII International Barley Genetics Symposium: Proceedings, University Extension Press. University of Saskatchewan, Canada, 1996, pp 37–43.
10. MK Anderson, E Reinbergs. Barley breeding. In: DC Rasmusson, ed. Barley, Madison, WI: Agronomy series No. 26, ASA/CSSA/SSSA, Inc., Publishers, 1985, pp 232–268.
11. RH Busch, DD Stuthman. Self-pollinated crop breeding: Concepts and success. In: JP Gustafson, ed. Gene Manipulation in Plant Improvement II, 19th Stadler Genetics Symposium. New York: Plenum Press, 1990, pp 21–37.
12. RD Wych, DC Rasmusson. Genetic improvement in malting barley cultivars since 1920. Crop Sci 23:1037–1040, 1983.
13. R Freisleben, A Lein. Uber die Auffindung einer mehltauresistenten Mutante nach Röntgenbestrahlung einer anfalligen Linie von Sommergerste. Naturwissenschaften 30:608, 1942.
14. Mutant Varieties Database, Mutation Breeding Review, No. 12, FAO/IAEA Division of Nuclear Techniques in Food and Agriculture and FAO/IAEA Agriculture and Biotechnology Laboratory, Vienna, Austria, 2000.
15. Semidwarf Cereal Mutants and Their Use in Cross-Breeding. Vol. II. Vienna: IAEA, Vienna 1984.
16. Semidwarf Cereal Mutants and Their Use in Cross-Breeding. Vol. III. Vienna: IAEA, 1988.
17. W Gottschalk, G Wolff. Induced Mutations in Plant Breeding. Berlin: Springer-Verlag, 1983.
18. B Donini, T Kawai, A Micke. Spectrum of mutant characters utilized in developing improved cultivars. In: Selection in Mutation Breeding. Vienna: IAEA, 1984, pp 7–31.
19. CF Konzak. Role of induced mutations. In: PB Bose, SG Blixt, eds. Crop Breeding, a Contemporary Basis Oxford: Pergamon Press, 1984, pp 216–292.
20. A Micke. International research programmes for the genetic improvement of grain proteins. In: W Gottschalk, HP Müeller, eds. Seed Proteins: Biochemistry, Genetics, Nutritive Value. The Hague: Martinus Nijhoff/Dr W. Junk, 1983, pp 25–44.
21. Seed Protein Improvement in Cereals and Grain Legumes. Vienna: IAEA, 1979.
22. Cereal Grain Protein Improvement. Vienna: IAEA, 1984.
23. HP Müeller. Breeding for enhanced protein. In: PB Bose, SG Blixt, eds., Crop Breeding, a Contemporary Basis Oxford: Pergamon Press, 1984, pp 382–389.
24. A Micke. 50 years induction of mutations for improving disease resistance of crop plants. Mutat Breed Newsl 39:2–4, 1992.
25. Mutant Varieties—Databank: Cereals and Legumes. Vienna: IAEA, 1996.
26. U Lundqvist, A Lundqvist. Intermedium mutants of barley (*Hordeum vulgare* L.): Diversity, interactions and plant breeding value. J Appl Genet 39(1):85–96, 1998.
27. R von Bothmer, J Flink, N Jacobsen, M Kotimaki and T Landstrom. Interspecific hybridization with cultivated barley (*Hordeum vulgare* L.). Hereditas 99:219, 1983.

28. G Fedak. Wide hybridization for cereal improvement. In: M Maluszynski, ed. Current Options for Cereal Improvement. Dordrecht, The Netherlands: Kluwer Academic, 1989, pp 39–48.
29. G Fedak. Wide crosses in *Hordeum*. In: DC Rasmusson, ed. Barley. Madison, WI: Agronomy monograph No. 26, ASA-CSSA-SSSA, 1985, pp 155–186.
30. G Fedak. Intergeneric hybrids with *Hordeum*. In: PR Shewry, ed. Barley: Genetics, Biochemistry, Molecular Biology and Biotechnology. Wallingford, Oxford, CAB International, pp 1992, 45–70.
31. M Baum, ES Lagudah, R Appels. Wide crosses in cereals. Annu Rev Plant Physiol Plant Mol Biol 43:117–143, 1992.
32. PJ Larkin, WR Scowcroft. Somaclonal variation—a novel source of variability from cell cultures for plant improvement. Theor Appl Genet 60:197–214, 1981.
33. TJ Orton. Chromosomal variabilty in tissue cultures and regenerated plants of *Hordeum*. Theor Appl Genet 56:101–112, 1980.
34. LS Dahleen, LR Joppa. Hybridization and tissue culture of *Hordeum vulgare* × *Elymus canadensis*. Genome 35(6):1045–1049, 1992.
35. E Lupotto. Callus induction and plant regeneration from barley mature embryos. Ann Bot 54:523–529, 1984.
36. RJ Singh. Chromosomal variation in immature embryo derived calluses of barley (*Hordeum vulgare* L.). Theor Appl Genet 72:710–716, 1986.
37. AK Gaponenko, TF Petrova, A Iskakov, AA Sozinov. Cytogenetics of in vitro cultured somatic cells and regenerated plants of barley (*Hordeum vulgare* L.). Theor Appl Genet 75:905–911, 1988.
38. R Lührs, H Lörz. Initiation of morphogenic cell suspension and protoplast cultures of barley (*Hordeum vulgare* L.). Planta 175:71–81, 1988.
39. A Breiman, D Roten-Abarbanell, A Karp, H Shaskin. Heritable somaclonal variation in wild barley (*Hordeum spontaneum*). Theor Appl Genet 74:104–112, 1987.
40. R Hunold, H Hartleb, OS Afanasenko. Resistance against *Drechslera teres* (Sacc.) Shoem. in progenies of in vitro selected callus derived plants of barley (*Hordeum vulgare* L.). J Phytopathol 135:89–98, 1992.
41. PC Kole, HS Chawla. Variation of *Helminthosporium* resistance and biochemical and cytological characteristics in somaclonal generations of barley. Biol Planta 35(1):81–86, 1993.
42. AMR Baillie, BG Rossnagel, KK Kartha. In vitro selection for improved chlorsulfuron tolerance in barley (*Hordeum vulgare* L.). Euphytica 67:151–154, 1993.
43. M Concepcion-Escorial, H Sixto, JM Garcia-Baudin, MC Chueca. In vitro selection increases glyphosate tolerance in barley. Plant Cell Tissue Org Cult 46(3):179–186, 1996.
44. P Bregitzer, M Poulson, BL Jones. Malting quality of barley lines derived from tissue culture. Cereal Chem 72(5):433–435, 1995.
45. P Bregitzer, M Poulson. Agronomic performance of barley lines derived from tissue culture. Crop Sci 35(4):1144–1148, 1995.
46. M Kihara, S Takahashi, H Funatsuki, K Ito. Field performance of the progeny of protoplast-derived barley, *Hordeum vulgare* L. Breed Sci 48(1):1–4, 1998.
47. WR Bushnell, DA Somers, RW Giroux, LJ Szabo, RJ Zeyen. Genetic engineering of disease resistance in cereals. Can J Plant Pathol 20(2):137–149, 1998.
48. P Bregitzer, SE Halbert, PG Lemaux. Somaclonal variation in the progeny of transgenic barley. Theor Appl Genet 96(3/4):421–425, 1998.
49. A Karp, SH Steele, S Parmar, MGK Jones, PR Shewry. Relative stability among barley plants regenerated from cultured immature embryos. Genome 29:405–412, 1987.
50. RA Pickering. Plant regeneration and variants from calli derived from immature embryos of diploid barley (*Hordeum vulgare* L.) and *H. vulgare* L. × *H. bulbosum* L. crosses. Theor Appl Genet 78:105–112, 1989.
51. ML Ruíz, J Rueda, MI Peláez, FJ Espino, M Candela, AM Sendino, AM Vázquez. Somatic embryogenesis, plant regeneration and somaclonal variation in barley. Plant Cell Tissue Org Cult 28:97–101, 1992.
52. A Karp and PA Lazzeri. Regeneration, stability and transformation of barley. In: PR Shewry, ed. Barley: Genetics, Biochemistry, Molecular Biology and Biotechnology. Wallingford, Oxford: CAB International, 1992, pp 549–571.

53. BP Forster, W Powell. Haploidy in barley. In: SM Jain, SK Sopory, RE Veilleux, eds. In Vitro Haploid Production in Higher Plants. Vol 4. Dordrect, The Netherlands: Kluwer Academic, 1997, pp 99–115.
54. J Bingham. Winter wheat breeding methods and prospects. J. Agric Soc Engl 136:65–67, 1975.
55. JW Snape, E Simpson. The genetic expectations of doubled haploid lines derived from different filial generations. Theor Appl Genet 60:123–128, 1981.
56. TM Choo, E Reinbergs, KJ Kasha. Use of haploids in breeding barley. Plant Breed Rev 3:219–252, 1985.
57. KJ Kasha, A Ziauddin, U-H Cho. Haploids in cereal improvement: anther and microspore culture. In: JP Gustafson, ed. Gene Manipulation in Plant Improvement II. New York: Plenum Press, 1990, pp 213–235.
58. RA Pickering, P Devaux. Haploid production: approaches and use in plant breeding. In: PR Shewry, ed. Barley: Genetics, Biochemistry, Molecular Biology and Biotechnology. Wallingford, Oxford: CAB International, 1992, pp 519–547.
59. KJ Kasha, KN Kao. High frequency nhaploid production in barley (*Hordeum vulgare* L.). Nature 225:874–875, 1970.
60. KJ Kasha, E Reinbergs. Recent developments in the production and utilization of haploids in barley. In: MJC Asher, RP Ellis, AM Hayter, RNH Whitehouse, eds. Barley Genetics IV. Edinburgh: Edinburgh University Press, 1981, pp 655–665.
61. A Hagberg, G Hagberg. High frequency of spontaneous haploids in the progeny of an induced mutation in barley. Hereditas 93:341–343, 1980.
62. B Foroughi-Wehr, W Friedt. Rapid production of recombinant barley yellow mosaic virus resistant *Hordeum vulgare* lines by anther culture. Theor Appl Genet 67:377–387, 1984.
63. W Powell, EM Borrino, MJ Allison, DW Griffiths, MJC Asher, JM Dunwell. Genetical analysis of microspore-derived plants of barley (*Hordeum vulgare* L.). Theor Appl Genet 72:619–626, 1986.
64. DM Thompson, K Chalmers, R Waugh, BP Forster, WTB Thomas, PDS Caligari, W Powell. The inheritance of genetic markers in microspore-derived plants of barley (*Hordeum vulgare* L.). Theor Appl Genet 81:487–492, 1991.
65. M Maluszynski, BS Ahloowalia, B Sigurbjörnsson. Application of in vivo and in vitro mutation techniques for crop improvement. Euphytica 85:303–315, 1995.
66. di-U Umba, M Maluszynski, I Szarejko, J Zbieszczyk. High frequency of barley DH-mutants from $M_1$ after mutagenic treatment with MNH and sodium azide. Mutat Breed News Lett 38:8–9, 1991.
67. A Ziauddin, E Simion, KJ Kasha. Improved plant regeneration from shed microspore culture in barley (*Hordeum vulgare* L.) cv. Igri. Plant Cell Rep 9:69–72, 1990.
68. U Kuhlman, B Foroughi-Wehr, A Graner, G Wenzel. Improved culture system for microspores of barley to become a target for DNA uptake. Plant Breed 107:165–168, 1991.
69. QA Yao and KJ Kasha. Potential of biolistic transformation of barley microspores based on viability and transient beta-glucuronidase activity. Genome 40(5):639–643, 1997.
70. VN Soyfer. Hereditary variability of plants under the action of exogenous DNA. Theor Appl Genet 58:225–235, 1980.
71. RR Mendel, E Claus, R Hellmund, J SchulzeHH Steinbiss, A Tewes. Gene transfer to barley. Proceedings VIIth International Congress on Plant Cell and Tissue Culture, Amsterdam, The Netherlands: Kluwer Press, 1990, pp 73–38.
72. R Brettschneider, PA Lazzeri, P Langridge, S Hartke, R Gill, H Lörz. Cereal transformation with *Agrobacterium tumefaciens* via the pollen tube pathway. Abstracts of the VIIth International Congress of Plant Tissue and Cell Culture abstract A2-22. Amsterdam, The Netherlands: 1990, pp 49.
73. L Ledoux, R Huart. Fate of exogenous bacterial deoxyribonucleis acids in barley seedlings. J Mol Biol 43:243–262, 1969.
74. R Töpfer, B Gronenborn, J Schell, HH Steinbiss. Uptake and transient expression of chimaeric genes in seed-derived embryos. Plant Cell 1:133–139, 1989.
75. I Potrykus. Gene transfer to plants: Assessment of published approaches and results. Annu Rev Plant Physiol Plant Mol Biol 42:205–225, 1991.

76. TH Teeri, GH Patel, KA Aspegren, V Kauppinen. Chloroplast targeting of neomycin phophotransferase II with a pea transit peptide in electroporated barley mesophyll protoplasts. Plant Cell Rep 8:187–190, 1989.
77. M Salmenkallio, R Hannus, TH Teeri, V Kauppinen. Regulation of α-amylase promoter by gibberellic acid and abscisic acid in barley protoplasts transformed by electroporation. Plant Cell Rep 9:352–355, 1990.
78. B Junker, J Zimmy, R Lührs, H Lörz. Transient expression of chimaeric genes in dividing and nondividing cereal protoplasts after PEG-induced DNA uptake. Plant Cell Rep 6:329–332, 1987.
79. BT Lee, K Murdoch, J Topping, MGK Jones, M Kreis. Transient expression of foreign genes introduced into barley endosperm protoplasts by PEG-mediated transfer or into intact endosperm tissue by microprojectile bombardment. Plant Sci 78:237–246, 1991.
80. S Lütticke, A Gärtner, M Arndt, A Stöldt, H Lörz. Genetic transformation of barley—approaches towards molecular breeding. In: G Scoles, B Rossnagel eds. V International Oat Conference and VII International Barley Genetics Symposium. Proceedings, University Extension Press, University of Saskatchewan, Canada, 1996, pp 223–229.
81. A Stöldt, X-H Wang, H Lörz. Primary callus as source of totipotent barley (*Hordeum vulgare* L.) protoplasts. Plant Cell Rep 16(3/4):137–141, 1996.
82. PA Lazzeri, R Brettschneider, R Lührs, H Lörz. Stable transformation of barley via PEG-induced direct DNA uptake into protoplasts. Theor Appl Genet 81:437–444, 1991.
83. PG Lemaux, M-J Cho, S Zhang, P Bregitzer. Transgenic cereals: *Hordeum vulgare* L. (barley). In: IK Vasil, ed. Molecular Improvement of Cereal Crops. Dordrecht, The Netherlands: Kluwer Academic, 1999, pp 255–316.
84. TM Klein, T Jones. Methods of genetic transformation: the gene gun. In: IK Vasil ed. Molecular Improvement of Cereal Crops. Dordrecht, The Netherlands: Kluwer Academic, 1999, pp 21–42.
85. M Båga, RN Chibbar, KK Kartha. Expression and regulation of transgenes for selection of transformants and modification of traits in cereals. In: IK Vasil, ed. Molecular Improvement of Cereal Crops. Dordrecht, The Netherlands: Kluwer Academic, 1999, pp 83–131.
86. S Tingay, D McElroy, R Kalla, S Fieg, M Wang, S Thornton, R Brettell. *Agrobacterium tumefaciens*-mediated barley transformation. Plant J 11(6):1369–1376, 1997.
87. G Guo, F Maiwald, P Lorenzen, H-H Steinbiss. Factors influencing T-DNA transfer into wheat and barley cells by *Agrobacterium tumefaciens*. Cereal Res Commun 26(1):1998
88. H Wu, AC McCormac, MC Elliot, D-F Chen. *Agrobacterium*-mediated stable transformation of cell suspension cultures of barley (*Hordeum vulgare*). Plant Cell Tissue Org Cult 54:161–171, 1998.
89. BD Mohanty, PD Ghosh. Somatic embryogenesis and plant regeneration from leaf callus of *Hordeum vulgare*. Ann Bot 61:551–555, 1988.
90. TP Pasternak, VA Rudas, H Lörz, J Kumlehn. Embryogenic callus formation and plant regeneration from leaf base segments of barley (*Hordeum vulgare* L.). J Plant Physiol 155:371–375, 1999.
91. T-Y Cheng, HH Smith. Organogenesis from callus culture of *Hordeum vulgare*. Planta 123:307–310, 1975.
92. RC Weigel, KW Hughes. Long term regeneration by somatic embryogenesis in barley (*Hordeum vulgare* L.) tissue cultures derived from apical meristem explants. Plant Cell Tissue Org Cult 5:151–162, 1985.
93. Z Rengel, S Jelaska. Somatic embryogenesis and plant regeneration from seedling tissues of *Hordeum vulgare* L. J Plant Physiol 24:385–392, 1985.
94. Z Vitanova, V Vitanov, A Trifonova, D Savova, A Atanassov. Effect of 2,4-D precultivation on regeneration capacity of cultivated barley. Plant Cell Rep 14:437–441, 1995.
95. S Zhang, R Williams-Carrier, D Jackson, PG Lemaux. Expression of CDC2Zm and KNOTTED1 during in-vitro axillary shoot meristem proliferation and adventitious shoot meristem formation in maize (*Zea Mays* L.) and barley (*Hordeum vulgare* L.). Planta 204:542–549, 1998.
96. S Zhang, M-J Cho, T Koprek, R Yun, P Bregitzer, PG Lemaux. Genetic transformation of commercial cultivars of oat (*Avena sativa* L.) and barley (*Hordeum vulgare* L.) using in vitro shoot meristematic cultures derived from germinated seedlings. Plant Cell Rep 18:959–966, 1999.

97. LH San Noeum. Haploïdes d'*Hordeum vulgare* L. par culture in vitro d'ovaires non-fécondés. Ann l'Amélioration Plant 26:751–754, 1976.
98. AM Castillo, L Cistué. Production of gynogenic haploids of *Hordeum vulgare* L. Plant Cell Rep 12:139–143, 1993.
99. MR Thomas, KJ Scott. Plant regeneration by somatic embryogenesis from callus initiated from immature embryos and immature inflorescences of *Hordeum vulgare*. J Plant Physiol 12:159–169, 1985.
100. A Jähne, D Becker, R Brettschneider, H Lörz. Regeneration of transgenic, microspore-derived, fertile barley. Theor Appl Genet 89:525–533, 1994.
101. QA Yao, E Simion, JA Krochko, KJ Kasha. Transformation of isolated barley microspores and regeneration of transgenic plants. In Vitro Cell Dev Biol 32II:P-1172, 1996.
102. RJ Lallukka. Effects of dicamba/MCPA/mecoprop mixture on eight spring barley cultivars. Proceedings of the 1976 British Crop Protection Conference, British Crop Protection Council. Vol. 1. Londan, 1976, pp 143–150.
103. Y Wan, PG Lemaux. Generation of large numbers of independently transformed fertile barley plants. Plant Physiol 104:37–48, 1994.
104. PF McGrath, JR Vincent, CH Lei, WP Pawlowski, KA Torbert, W Gu, HFB Kaeppler, Y Wan, PG Lemaux, HR Rines. Coat protein-mediated resistance to isolates of barley yellow dwarf in oats and barley. Eur J Plant Pathol 103(8):695–710, 1997.
105. T Hagio. Studies on breeding of barley (*Hordeum vulgare* L.) and sorghum (*Sorghum bicolor* M.) using biotechnological and conventional methods. Bull Natl Inst Agrobiol Resources 13:23–96, 1999.
106. G Leckband, H Lörz. Transformation and expression of a stilbene synthase gene of *Vitis vinifera* L. in barley and wheatfor increased fungal resistance. Theor Appl Genet 96(8):1004–1012, 1998.
107. F Alpeter, I Diaz, H McAuslane, K Gaddour, P Carbonero, IK Vasil. Increased insect resistance in transgenic wheat stably expressing trypsin inhibitor Cme. Mol Breed 5(1):53–63, 1999.
108. G Jach, B Gornhardt, J Mundy, J Logemann, E Pinsdorf, R Leah, J Schell, C Mass. Enhanced quantitative resistance against fungal disease by combinatorial expression of different barley antifungal proteins in transgenic tobacco. Plant J 8(1):97–109, 1995.
109. DEA Florack, WG Dirkse, B Visser, F Heidekamp, WJ Stiekema. Expression of biologically active hordothionins in tobacco: Effects of pre- and pro-sequences at the amino and carboxyl termini of the hordothionin precursor on mature expression and sorting. Plant Mol Biol 24(1):83–86, 1994.
110. MJ Edney. Barley. In: RJ Henry, PS Kettlewell, eds. Cereal Grain Quality. London: Chapman & Hall, 1996, pp 113–151.
111. CW Bamforth, AHP Barclay. Malting technology and the uses of malt. In: AW MacGregor, RS Bhatty, eds. Barley: Chemistry and Technology. American Association of Cereal Chemists, Inc., St. Paul, MN: 1993, pp 297–354.
112. JS Swanston, RP Ellis, JR Stark. Effects on grain and malting quality of genes altering barley starch composition. J Cereal Sci 22:265–273, 1995.
113. RJ Stewart, TPJ Garrett, JN Varghese, PB Høj, GB Fincher. Protein engineering of enzymes for improved malting quality of barely. In: G Scoles, B Rossnagel, eds. V International Oat Conference and VII International Barley Genetics Symposium. Proceedings, University Extension Press, University of Saskatchewan, Canada, 1996, pp 7–19.
114. GB Fincher. Morphology and chemical composition of barley endosperm cell walls. J Inst Brew 81:116–122, 1975.
115. L Loi, PA Barton, GB Fincher. Survival of barley (1,3-1,4)-β-glucanase isoenzymes during kilning and mashing. J Cereal Sci 5:45–50, 1987.
116. LG Jensen, O Olsen, O Kops, N Wolf, KK Thomsen, D von Wettstein. Transgenic barley expressing a protein-engineered, thermostable (1,3-1,4)-β-glucanase during germination. Proc Natl Acad Sci USA 93:3487–3491, 1996.
117. LG Jensen, O Politz, O Olsen, KK Thomsen, D von Wettstein. Inheritance of a codon-optimized transgene expressing heat stable (1,3-1,4)-beta-glucanase in scutellum and aleurone of germinating barley. Hereditas 129(3):215–225, 1998.

118. M-J Cho, JH Wong, C Marx, W Jiang, PG Lemaux, BB Buchanan. Overexpression of thioredoxin leads to enhanced activity of starch debranching enzyme (pullulanase) in barley grains. Proc Natl Acad Sci USA 25:14641–14646, 1999.
119. RS Bhatty. Nonmalting uses of barley. In: AW MacGregor, RS Bhatty, eds. Barley Chemistry and Technology. St Paul, MN: American Association of Cereal Chemists, 1993, pp 355–417.
120. G Persson. Ideas and methods for genetic improvement of quality and quantity of barley. In: The use of nuclear techniques for cereal grain protein improvement, Vienna: STI/PUB/664, IAEA/FAO, 1984, pp 105–109.
121. H Brinch-Pedersen, G Galili, S Knudsen, PB Holm. Engineering of the aspartate family biosynthetic pathway in barley (*Hordeum vulgare* L.) by transformation with heterologous genes encoding feedback -insensitive aspartate kinase and dihydrodipicolinate synthase. Plant Mol Biol 32(4):611–620, 1996.
122. H Horvath, J Huang, O Wong, E Kohl, T Okita, CG Kannangara, D von Wettstein. The production of recombinant proteins in transgenic barley grains. Proc Natl Acad Sci USA 97(4):1914–1919, 2000.
123. GP Xue, SE Denman, D Glassop, JS Johnson, LM Dierens, KS Gobius, JH Aylward. Modification of a xylanase cDNA from an anaerobic rumen fungus *Neocallimastix patriciarum* for high-level expression in *Escherichia coli*. J Biotechnol 38:269–277, 1995.
124. M Patel, JS Johnson, RIS Brettell, J Jacobsen, G-P Xue. Transgenic barley expressing a fungal xylanase gene in the endosperm of the developing grains. Mol Breed 6:113–123, 2000.
125. DJA Jenkins, TMS Wolever, AR Leeds, MA Gassull, P Haisman, J Dilawari, DV Goff, GL Metz, KGMM Alberti. Dietary fibres, fibre analogues, and glucose tolerance: Importance of viscosity. Br Med J 1:1392–1394, 1978.
126. M Koskinen, T Suortti, K Autio, P Myllärinen, K Poutanen. Effect of pretreatment on the film forming properties of potato and barley starch dispersions. Ind Crops Prod 5:23–34, 1996.
127. IA Wolff, HA Davis, JE Cluskey, LJ Gundrum, CA Rist. Preparation of films from amylose. Ind Eng Chem 43:915–919, 1951.
128. JJG van Soest, SHD Hulleman, D de Wit, JFG Vliegenthart. Crystallinity in starch bioplastics. Ind Crops Prod 5:11–22, 1996.
129. E Stöger, C Vaquero, E Torres, M Sack, L Nicholson, J Drossard, S Williams, D Keen, Y Perrin, P Christou, R Fischer. Cereal crops as viable production and storage systems for pharmaceutical scFv antibodies. Plant Mol Biol 42:583–590, 2000.
130. M Stam, JNM Mol, JM Kooter. The silence of genes in transgenic plants. Ann Bot 79:3–12, 1997.
131. SW Rogers, JC Rogers. The importance of DNA methylation for stability of foreign DNA in barley. Plant Mol Biol 18:945–961, 1992.
132. J Finnegan, D McElroy. Transgene inactivation: plants fight back. Biotechnology 12:883–888, 1994.
133. T Demeke, P Hucl, M Bága, K Caswell, N Leung, RN Chibbar. Transgene inheritance and silencing in hexaploid spring wheat. Theor Appl Genet 99:947–953.
134. J-W Liu, LM Tabe. The influences of two plant nuclear matrix attachment regions (MARs) on gene expression in transgenic plants. Plant Cell Physiol 39(1):115–123, 1998.
135. G Moore. Cereal genome evolution: pastoral pursuits with 'Lego' genomes. Curr Opin Genet Dev 5:717–724, 1995.
136. CM Hamilton. A binary-BAC system for plant transformation with high-molecular-weight DNA. Gene 107–116, 1997.
137. T Pryor. The origin and structure of fungal disease resistance in plants. Trends Genet 3:157–161, 1987.
138. IR Crute, DAC Pink. Genetics and utilization of pathogen resistance in plants. Plant Cell 8:1747–1755, 1996.
139. RW Michelmore, BC Meyers. Clusters of resistance genes in plants evolve by divergent selection and a birth-and-death process. Genome Res 8:1113–1130, 1998.
140. JH Jørgensen. Genetics of powdery mildew resistance in barley. Crit Rev Plant Sci 13(1):97–119, 1994.

141. S Seah, K Sivasithamparam, A Karakousis, ES Lagudah. Cloning and characterisation of a family of disease resistance gene analogs from wheat and barley. Theor Appl Genet 97:937–945, 1998.
142. RP Ellis. The use of high amylose barley for the production of whisky malt. J Inst Brew 82:280–281, 1976.
143. M Båga, A Repellin, T Demeke, K Caswell, N Leung, EL Abdel-Aal, P Hucl, RN Chibbar. Wheat starch modification through biotechnology. Starch/Stärke 51:111–116, 1999.

# 48
# Transgenic Coffee

**Maria Filomena Carneiro**
*Instituto de Investigação Científica Tropical, Oeiras, Portugal*

**John I. Stiles**
*Integrated Coffee Technologies, Inc., Honolulu, Hawaii*

| | | |
|---|---|---|
| I. | INTRODUCTION | 737 |
| II. | HISTORICAL PERSPECTIVE | 738 |
| III. | MOLECULAR IMPROVEMENT OF COFFEE | 739 |
| | A. Insect Resistance | 739 |
| | B. Quality Traits | 740 |
| IV. | PROSPECTS | 742 |
| V. | CONCLUDING REMARKS | 744 |
| | REFERENCES | 745 |

## I. INTRODUCTION

Coffee is a major commercial crop and is second only to petroleum in value as a commodity in world trade. It is the major component of foreign exchange of a number of countries in Latin America, Africa, and Asia. Although not a primary source of food, coffee provides employment and social benefits for millions of people in developing countries and constitutes the major source of foreign exchange needed to buy goods and technology.

There are about 11.2 million ha of coffee worldwide and more than 6 million tons of coffee, worth approximately (U.S.) $10 billion, is grown annually. Brazil and Colombia are the first and second largest producers. Together these two countries are responsible for about 45% of the world's coffee production. Although there are many species of the genus *Coffea*, only two of them, *Coffea arabica* (Arabica) and *Coffea canephora* (Robusta), are widely cultivated. Other species, such as *C. liberica, C. dewevrei*, and *C. racemosa*, are only cultivated to satisfy local consumption. About 75% of the world's commercial coffee is Arabica. Essentially all coffee production in Latin America is Arabica. Some is also produced in African countries, such as Ethiopia and Kenya.

*C. arabica* is native to the highlands of southwestern Ethiopia and was introduced to the American continent from tropical Africa in the first decades of the 18th century. It is generally grown at altitudes between 1000 and 2000 meters. It is thought that all *C. arabica* plants are derived from a single plant collected by the Botanical Garden of Paris. Because of this derivation and because of its allotetraploid origin, *C. arabica* is characterized by low genetic diversity; thus, there is a lack of resistance to the main pests and diseases in most varieties of *C. arabica*. The majority of cultivars present homogeneous agronomic behavior characterized by high susceptibility to many pests and diseases and very low adaptability. The cup qualities of *C. arabica*, together with its autogamous and perennial characteristics, led to the development of homogeneous plantations all over the world.

*Coffea canephora* has a wide geographic distribution, extending from the western to the central tropical and subtropical regions of the African continent, from Guinea and Liberia to Sudan and the Uganda forest, with a high concentration of types in the Democratic Republic of Congo (old Republic of Congo Kinshasa). *C. canephora*, or Robusta, as it is commonly called, grows at low altitudes (about 850 m) and accounts for 80% of African coffee production. However, Robusta has also been cultivated in South America and Asian countries.

Conventional coffee breeding is a long process involving different methods, namely, selection from a wild population followed by hybridization and progeny evaluation, backcrossing, and interspecific crossing. It takes more than 30 years to obtain a new cultivar by using any of these methods. Because of the long breeding cycle and the lack of diverse germplasm, coffee is a good candidate for manipulation by genetic transformation. Furthermore, in vitro coffee tissue techniques including plant regeneration systems have been developed. Thus, the basic technology for manipulation of coffee plant cells at the cellular and molecular levels is available, making coffee a suitable crop for genetic transformation.

Several important agronomic traits could be considered for a coffee biotechnology program, namely, resistance to diseases and insects, resistance to herbicides, drought tolerance, cold/freezing tolerance, short maturation cycles, controlled ripening, and regulation of caffeine content. When multiple genes are responsible for the traits, as are the genes responsible for cold/freeze tolerance or drought resistance, the molecular approach is difficult to apply.

In this chapter we review the success already obtained in coffee transformation and the engineering of the first coffee plants carrying agronomic traits, which include *Bt* genes from *Bacillus thuringiensis* for resistance to coffee leaf miner and fruit-expressed 1-amino-cyclopropane-1-carboxylic acid (ACC) synthase and ACC oxidase genes that are responsible for the synthesis of ethylene, which governs fruit ripening and shortening of maturation cycles. We will review work on the gene encoding xanthosine-$N^7$-methyltransferase, one of the genes responsible for caffeine biosynthesis. Finally, we draw perspectives and discuss the scope of further studies and prospects for transgenic coffee production in the near future.

## II. HISTORICAL PERSPECTIVE

The technology of transgenic plant production can have an important and immediate impact on solving problems such as resistance to pests and diseases. In the future, as our understanding of more complex genetic systems increases and this technology is integrated with conventional breeding technology, biotechnology will have an increasingly important impact on the development of improved coffee cultivars. Using traditional breeding methods, it is possible to cross sexually compatible species and introduce sets of genes and chromosomes. Repetitive backcrossing to recover lines with the necessary commercial qualities is a long process in woody species such as coffee (1). Because of the time-consuming backcross process an alternative approach is desirable to solve some of the major problems of coffee production. Genetic engineering is an attractive and realistic option for solving this problem.

Theoretically, genetic engineering can be used to introduce any gene from any origin, effectively increasing the germplasm pool to include all organisms. However, one of the major limitations of genetic engineering is the lack of useful genes, since only a small number of useful genes have been characterized at the deoxyribonucleic acid (DNA) level in sufficient detail to impart precise and predictable phenotypes on whole plants.

Traits such as disease and insect resistance can be found in plant germplasm and in some cases have been shown to be controlled by a single locus. However, as yet, no classical plant disease resistance gene has been used to impart resistance to a heterologous host. However, transgenic plants with resistance to diseases, herbicides, and insect damage have been produced by using genes from both plant and nonplant sources (1).

Fraley and collaborators demonstrated that direct gene transfer was possible by using recombinant DNA technology (2). This ability to transfer genes among organisms without sexual crossing provides breeders with new opportunities to increase efficiency in developing new varieties and to add new and unique traits not present in the germplasm available for sexual crosses. The first reference to genetic transformation technology using the *Agrobacterium* species–mediated technique was published in 1983 (2–4). In the first experiments, protoplasts were used. Subsequently methods based on leaf and stem explants were developed, making it possible to link high-efficiency transformation with good regeneration capacity (5). After this breakthrough, many important traits were targeted for introduction into important crop plants, inducing resistance to insects, viruses, herbicide tolerance, and to the improvement of the seed storage proteins.

The first report on tissue culture of coffee was in 1970 (6) and in 1997 the first transgenic plants were reported by several research groups for a number of different characteristics, including resistance to coffee leaf miner, controlled ripening, and caffeine content regulation (7–11). It is clear that an enormous amount of effort was exerted in the 1990s in the area of coffee genetic transformation. The original coffee cell culture research focused on optimization of protocols for optimization of in vitro regeneration systems, both for plant production and for application of genetic engineering. A number of different approaches have been developed, including somatic embryogenesis, adventitious bud formation, cell suspension culture, and protoplast culture. Some of these regeneration systems have been successfully adapted to genetic engineering of coffee with reasonably high levels of transformation efficiency (12,13). *Agrobacterium* species–mediated transformation, direct gene transfer, and particle bombardment have all been adapted to coffee (14–20). With the development of transformation technology, research has begun on the introduction of specific agronomic traits, including resistance to insects such as coffee leaf miner and coffee berry borer, control of fruit ripening by using the ACC oxidase and ACC synthase genes, and regulation of caffeine content.

## III. MOLECULAR IMPROVEMENT OF COFFEE

### A. Insect Resistance

The most important insects that affect coffee are *Perileucoptera coffeella* (coffee leaf miner), *Hypothenemus hampei* (coffee berry borer), and some *Meloidogyne* species (nematodes). Coffee leaf miner is the most serious insect pathogen affecting coffee plants in Brazil. The crop losses can extend to 50%, and *Coffea arabica* varieties grown in Brazil are highly susceptible to this insect. Coffee berry borer causes great losses in certain countries where this insect is endemic. The control of both insects has been through cultural practices and insecticides, which greatly increase production costs and also lead to environmental pollution. Biological control programs using parasitoid wasps have been instituted to control the coffee berry borer in several countries, such as Colombia and Kenya. However, coffee berry borer is still a major problem in many coffee-growing regions. The use of resistant coffee varieties would be an ideal control method. However, there are no known resistance genes to the coffee berry borer, and even if resistance could be found, it would take many years to develop acceptable commercial varieties of coffee by conventional breeding. Genetic engineering can have a significant role in development of resistant varieties by allowing the introduction of effective genes for control of the insects and shortening the time necessary to obtain new resistant varieties.

As previously mentioned, the *Bt* toxin from *Bacillus thuringiensis* is effective in leaf miner control. *Bacillus thuringiensis* is a gram-positive bacterium that produces specific proteins toxic to some species of insects (21). The insecticidal activity of *Bacillus thuringiensis* is due to bodies of crystalline inclusions produced during the sporulation of the bacteria. There are several varieties of *Bacillus thuringiensis*, which differ in the structure of the *Bt* protein and the range of insecticidal activity. Most *Bt* toxins are effective against *Lepidopterae* species, but some are also toxic to Dipterae and Coleopteran and occasionally some other insects. The characterization of the activity of *B. thuringiensis* toxins against the coffee leaf miner and the development of a transgenic coffee plant expressing the selected toxin genes are valuable tools for assessing the impact of a transgenic approach on an obligate endocarp pest that feeds on perennial crops. Studies have now demonstrated that *Bt* toxins have activity against the coffee leaf miner (*Perileucoptera coffeella*). Immunocytochemical and biological assays revealed that the crystal proteins Cry IA (c) and Cry IB are toxic to *P. coffeella* and that the corresponding genes can be inserted into coffee by genetic transformation to yield coffee plants resistant to leaf miner (22,23). The results demonstrated that the coffee leaf miner is susceptible to at least two *Bacillus thuringiensis* δ endotoxins, Cry IA (c) and Cry IB (23). This work represented the first step in assessing the potential for insect resistance in a transgenic perennial tropical crop.

Transgenic plants of *C. canephora* (genotype 126) were obtained for the first time by cocultivating somatic embryos with *A. rhizogenes* (A4) and disarmed strains of *A. tumefaciens* (LBA4404) by using a modified binary plasmid, pBIN19. The transfer DNA (T-DNA) region of pBIN19 contains the GUS gene with the PIV2 intron, a gene conferring resistance to chlorsulfuron, a herbicide, and a wild-type or synthetic *Cry IA*(c) gene, which is active against the coffee leaf miner. The *CryIA* (c) gene is under the control of the EF1σ-A1 promoter (24), which was demonstrated to be expressed in coffee by transient expression studies (19). The herbicide-resistant selectable marker gene used was *csr*-1-1 (25), which confers resistance to chlorsulfuron. This gene was used under the control of the p70S promoter, which is a modification of the 35S CaMV promoter, in which the enhancer sequence was duplicated. The different constructs were introduced into *Agrobacterium* sp. LBA440, by triparental conjugation. Transgenic plantlets were regenerated through somatic embryogenesis. Stable integration of the T-DNA into coffee genome was verified by polymerase chain reaction (PCR) and Southern blot assays (7). Leroy and collaborators (8) have now obtained more than 100 transformed plants each of *C. canephora* and *C. arabica* by this protocol. Expression of the *cryIA*(c) gene was verified by Western blotting in the majority of the regenerated plants. Some of the transgenic plants have demonstrated resistance to leaf miner in bioassays. Field trials to test the effectiveness and genetic stability of these transgenic coffee plants are being carried out in French Guyana (8). This is the first report of transformation of *C. canephora* and *C. arabica* plants with an agronomic trait.

At the Centre de Coopération Internationale de Recherche Agronomique pour le Développement (CIRAD), tests to evaluate *Bt* genes for activity against the coffee berry borer are also under study. A large research program is also currently under way in Colombia to produce coffee berry borer–resistant transgenic plants of the variety 'Columbia' (20).

## B. Quality Traits

Quality traits are those that impart an improved plant product. In coffee examples of quality traits are uniformity of ripening, shorter maturation cycles, higher total soluble solids, larger bean size, increased bean density and texture, caffeine content, and improved aroma and taste (26,27). Significant progress has been made on two of these quality traits, control of ripening and control of caffeine content.

## 1. Control of Ripening

Coffee beverage quality is critically dependent on bean quality. Unripe beans give poor cup quality and if mixed with ripe beans result in a lower-quality product. Growth regulators and selective harvest can be used to alleviate the problem of nonuniform ripening but have the disadvantages of fruit fall and increased costs of production. The lack of uniform ripening also limits the effectiveness of mechanical harvesting and certain hand harvesting techniques that strip all of the fruit off the plants at one time. Ethylene controls many physiological and developmental processes in higher plants, including fruit ripening in climacteric plants. Coffee has now been shown to be a climacteric plant (9). The inhibition of ethylene synthesis has been achieved by the supression of ACC synthase and ACC oxidase genes by using antisense or cosuppression (28).

One approach to achieving uniformity of ripening is the inhibition of ethylene synthesis during the final stages of fruit maturation. The cloning of the genes involved in ethylene biosynthesis, ACC synthase and ACC oxidase, will make it possible to test this hypothesis. Neupane and collaborators reported the cloning of fruit-expressed ACC synthase and ACC oxidase complementary DNAs (cDNAs) (9). Reverse transcriptase PCR was used with degenerate primers specific for highly conserved regions of ACC synthases and ACC oxidases. The PCR products were then used as probes to screen a coffee fruit cDNA library, and full-length clones for each gene were obtained. Northern blotting experiments using these clones showed conclusively that coffee is a climacteric plant (9).

If ethylene production can be significantly reduced, fruit development should be blocked at the mature green stage. Synchronous ripening can then be induced by treatment of the plants with an ethylene-releasing chemical such as ethrel. Plants transformed with antisense genes to both ACC synthase and ACC oxidase have been produced and are awaiting field trials.

In other experiments, the ACC oxidase gene was tested in antisense orientation in order to obtain coffee transgenic plants exhibiting reduced ethylene production (LF Pereira, personal communication, 1999). Preliminary results of this study allowed the verification of the transient expression in embryogenic tissues. The regeneration of plants is still in progress.

## 2. Control of Caffeine Content

The coffee beans of *C. arabica* and *C. canephora* can contain more than 1% and 2% caffeine, respectively. Caffeine has been described as a stimulant that causes wakefulness, increases motor activity, enhances fatty acid metabolism, and has been commonly used in the treatment of migraine. However, there is a growing belief that the ingestion of large quantities of caffeine can have adverse effects on health (29). Consequently decaffeinated coffee consumption increased significantly in the early 1970s. The extraction of caffeine from the seed by using organic solvents was first reported in 1905 (30). Today, a number of sophisticated processes are in use, including treatment with hot water, supercritical carbon dioxide, and organic solvents such as dichloromethane and ethylacetate. Some consumers have health concerns about the decaffeination process, which have limited the decaffeination market. Also, even the best decaffeination treatments damage the beans and remove additional compounds, resulting in decreased cup quality. Species within the genus *Coffea* have reduced levels of or even no caffeine and *C. arabica* varieties such as 'Laurina' have reduced levels of caffeine. A breeding program to obtain plants with reduced levels of caffeine would take more than 30 years.

Although caffeine is an important secondary metabolite of coffee, tea, and a number of other beverages, little is currently known at the molecular level about the caffeine biosynthetic pathway. At this time only one gene of this pathway has been isolated. Moisyadi and collaborators reported the isolation of a cDNA encoding the first enzyme of the caffeine biosynthetic pathway xanthosine-$N^7$-methyltransferase (10,11). Xanthosine-$N^7$-methyltransferase was purified, and a partial amino acid sequence was obtained from fragments regenerated by partial tryptic digestion. On the basis of the amino acid sequence, degenerate polymerase chain reaction (PCR) primers were constructed and were used to amplify a segment of the gene, using cDNA synthetized from young leaf messenger ribonucleic acid (mRNA) as template. The PCR product was then used to screen a leaf cDNA library, and several clones containing the entire coding sequence were isolated. The sequence of xanthosine-$N^7$-methyltransferase deduced from the cDNA does not show significant homology to other known proteins but contains several consensus sequences common to $N$-methyltransferases (11). Proof of the identity of this cDNA has been obtained by heterologous expression in tobacco, which does not normally have xanthosine-$N^7$-methyltransferase activity (JI Stiles, unpublished). The xanthosine-$N^7$-methyltransferase cDNA has been inserted in antisense orientation into a transformation vector, and coffee plants have been transformed by the *Agrobacterium* sp.–mediated system. Plants have been recovered and are currently being analyzed for structure of the transgene, xanthosine-$N^7$-methyltransferase mRNA levels, xanthosine-$N^7$-methyltransferase enzyme activity, and caffeine levels (10). Preliminary results indicated that transgenic callus that produced only about 3% of the normal amount of caffeine could be recovered.

## IV. PROSPECTS

Biotechnology will play an important role in the future of the coffee industry. As a tropical crop, coffee is grown under conditions of high disease pressure and limited capital for purchase of pesticides. Since cup quality is a major factor in determining price and demand, especially for the higher-end gourmet market, breeding resistant cultivars is an especially arduous undertaking because both the agronomic and the quality factors must be managed. The ability to impart resistance by the introduction of single genes into existing highly accepted varieties through biotechnology will offer solutions much faster than traditional breeding program.

As is the case with most other crops, coffee biotechnology has focused much effort on disease and insect resistance. Like most tropical crops, coffee can be severely damaged by fungal and insect pathogens. Treatment with fungicides and insecticides is expensive, often beyond the means of small-scale farmers; ecologically damaging; and at times not highly effective. For example, coffee berry borer exists within the coffee fruit for a significant portion of its life cycle, making it difficult to control by chemical means. This relationship is somewhat analogous to that of maize with European corn borer. This insect is also difficult to eliminate with chemical treatments when it is in the maize plant. However, maize genetically engineered to express the *Bt* toxin is effective for control. Transgenic *Bt* maize in the United States has increased yields as much as 14% without the use of chemical pesticides (31). The losses reported for coffee berry borer range as high as 96%, depending on the location (32), and there is no known resistance gene. Effective biotechnology-based control using *Bt* or some other insecticidal protein would make a major contribution to the profitability of coffee production in many areas of the world. A secondary benefit is the reduction in fungal infections associated with coffee berry borer damage. Vega and Mercadier report that *H. hampei* can act as a vector for fungi such as the aflotoxin-producing *Aspergillus flavus* and *A. ochraceus* (33). Ochractoxin A, produced by *A. ochraceus* and occasionally a few other *Aspergillus* species, is a major concern in the European Union and is having serious economic consequences in several producing countries.

Broad-spectrum fungal resistance, such as that imparted by hydrolytic enzymes that attack the fungal cell wall, has great potential. Because of the growing environment of coffee, fungal diseases present severe problems. In addition to the major fungal diseases, coffee berry disease (*Colletotrichum kahawae*) and rust (*Hemileia vastatrix*), there are a number of less widespread diseases capable of inflicting significant losses caused by a variety of other fungi such as *Fusarium* and *Micena* species. Although coffee rust can be controlled by the use of resistant varieties, the resistance genes for most of these varieties are from the 'Timor Hybrid,' a natural *C. arabica* × *C. canephora* hybrid. However, many coffee buyers consider the cultivars developed from the 'Timor Hybrid' to be of inferior quality, presumably as a result of the introduction of genes from the lower-quality *C. canephora* along with the resistance genes. The introduction of specific fungal disease resistance genes into susceptible but high-quality *C. arabica* cultivars by using biotechnology would alleviate this problem.

Biotechnology can also play a role in ameliorating abiotic stresses. One of the most disruptive events in the coffee industry is a sudden freeze. Not only does it decrease income to producers through destruction of crops; coffee processors and roasters are also affected by the concomitant large price fluctuations that affect the commodity markets. It would be of great economic advantage to all segments of the industry if coffee had the ability to withstand the relatively brief and light freezes that are occasionally encountered in certain coffee growing regions. The mechanisms involved in cold acclimation and freezing tolerance are just beginning to be understood at a level necessary for intervention by biotechnology. Coffee may be an ideal candidate for such intervention since even a small increase in cold/freezing tolerance would make a major economic difference. Membrane damage is thought to be the primary site of freeze injury (34). Multiple changes occur in plants as a result of cold acclimation, including changes in the lipid composition, accumulation of sucrose and other sugars, and accumulation of certain proteins that help to stabilize the membranes such as late embryogenesis abundant (LEA) proteins and highly hydrophilic proteins such as the *Arabidopsis* COR15σ protein (34). Intervention to stabilize membranes by manipulating one or more of these factors may give coffee the ability to survive most low-temperature stresses found in the normal growing regions.

Although most quality traits are complex and not understood at a level to allow intervention by biotechnology, there are two potential areas for the application of biotechnology, control of fruit ripening and production of caffeine-free plants. One project currently under way is the production of caffeine-free coffee plants. As described, Moisyadi and collaborators isolated the first gene unique to the caffeine biosynthetic pathway. This gene, xanthosine-$N^7$-methyltransferase, is now being used in antisense to inhibit the caffeine biosynthetic pathway (10,11). The resulting coffee should have increased quality since only the caffeine will be removed and other changes resulting from the chemical decaffeination process will not have occurred. The caffeine-free coffee will also have added value since the cost of decaffeination will be eliminated. Biotechnology may also be used to control the fruit ripening process in coffee better, as described previously.

Most other coffee quality traits are too complex to be manipulated by the currently available biotechnologies. However, certain defects may be reduced or eliminated. For example, Robusta coffees have elevated levels of methylisoborneol (MIB). At this time it is not clear whether MIB is a product of the plant or of associated microbes. In either event, biotechnology could be utilized to eliminate MIB and the associated off flavor.

Much criticism has been targeted at the utilization of antibiotic resistance genes as selectable markers for genetic transformation of plants. Recently a number of new markers that give the plant cell the ability to utilize new carbon sources have been described. These marker genes select for transformed cells by giving them the ability to grow on a medium that cannot support the growth of untransformed plant cells. One such system uses the phosphate-6-mannose isomerase gene to give plant cells the ability to grow on mannose (35). This system is reportedly up to fivefold more efficient than traditional selectable markers that cause cell death.

New genetic transformation techniques that result in greatly increased transformation efficiencies are also being developed. Smith and collaborators, reported an electroporation-based pollen transformation system that yields up to 44% transgenic plants (36); a transformation system with such efficiency would not need a selectable marker system. This system could also be used to introduce more complex molecules such as those needed for site-specific mutagenesis. Two approaches show promise for introducing precise genetic modifications into specific plant genes without the introduction of transgenes. Kren and collaborators (37) reported site-specific alteration of the factor IX gene in rat liver hepatocytes at an efficiency of 40% by using a chimeric RNA/DNA oligonucleotide in the absence of any selectable markers. This chimeric oligonucleotide pairs with the homologous sequence in the genome and causes a mismatch repair or gene conversion event that produces the desired site-specific change. Although the size of alterations using this technology is relatively small, it could be used to insert a small deletion or nonsense mutation into a specific gene, replacing the antisense or cosuppression technologies currently utilized to silence plant genes. A second approach is the use of recombination proteins, such as the *Escherichia coli* RecA protein, to increase homologous recombination (38). Enhanced homologous recombination could be used to inactivate genes as described for the chimeric RNA/DNA but is more versatile in that larger alterations could also be produced. For example, an allele imparting resistance to a specific pathogen could be isolated from any species of *Coffea* and inserted at the homologous disease resistance locus in *C. arabica* without the introduction of other undesirable genes, producing a resistant cultivar with all of the other attributes of the original cultivar. As knowledge about the structure and function of disease genes increases, it may even be possible to design new alleles to increase disease resistance or overcome breakdowns in resistance that inevitably occur.

## V. CONCLUDING REMARKS

Advances in cellular, developmental, and molecular genetics, combined with traditional breeding, can target and achieve improvements in specific agronomic, processing, and consumer qualities. Although a major agricultural product, only now is coffee benefiting from the technological developments at the cellular and molecular levels. The first field trials of plants of *C. canephora* and *C. arabica* resistant to leaf miner, plants of *C. arabica* engineered for uniform ripening of fruits, and *C. arabica* with low levels of caffeine are just now beginning. From an economic perspective, caffeine-free coffee can have a considerable impact on reduction of production costs and maintenance of quality and flavor characteristics of the beverage. This can represent a large benefit in countries like the United States, where the decaffeination market represents about 22% of the 2.4 billion pounds of product consumed annually.

Although still preliminary, the results already obtained are an important first step in developing the genetic transformation mechanisms that will be required to utilize the potential of plant biotechnology fully. These initial results will give coffee biotechnologists and breeders the basic tools to develop plants resistant to the main coffee pests and diseases, plants resistant to herbicides, and plants with other specific characteristics, including cold/freeze and drought tolerance, and alteration of the amino acid content involved in coffee flavor such as cysteine and methionine.

The combination of conventional breeding methods and biotechnology has led to many new possibilities for crop improvement. In decade of the 1990s saw many important events in the development of coffee biotechnology and genetic transformation occurred, following the first successful transient expression results in the 1980s. As we move into the third millennium we anticipate the development of many new cultivars, with resistance to coffee leaf miner, coffee berry borer, and many other diseases. Improved varieties with uniformity of ripening, reduced maturation cycles, and caffeine-free beans will also become available.

Presently, the world is becoming more conscious of the environment and food safety issues. The distribution of coffee plants with resistance to pests, diseases, and herbicides will diminish the risk of environmental damage caused by continuing high-level of use of pesticides. Coffee plants engineered to resist diseases better will also increase food safety. Currently fungal contamination and resulting toxin production are serious concerns. The cautious use of transgenic plants will prevent environmental pollution and food safety problems and reduce the cost of coffee production, benefiting both producers and consumers.

## REFERENCES

1. AM Dandekar, GH McGranahan, DJ James. Transgenic woody plants. In: Shain-dow Kung, Ray Wu, eds. Transgenic Plants Present Status and Social and Economic Impacts. San Diego, California: Academic Press, 1993, pp 129–151.
2. RT Fraley, SG Rogers, RB Horsch, PR Sander, JS Flick, SP Adams, ML Bittner LA Brand, CL Fink, JS Fry, GR Galluppi, SB Goldberg, NL Hoffman, SC Woo. Expression of bacterial genes in plant cells. Proc Natl Acad Sci USA 80:4803–4807, 1983.
3. L Herrera-Estrela, M De Block, E Messens, J-P Hernalsteens, M Van Montagu, J Schell. Chimeric genes as dominant selectable markers in plant cells. Eur Mol Biol Org J 2:987–995, 1983.
4. M Bevan, R Flavell, M-D Chilton. A chimaeric antibiotic resistance gene as a selectable marker for plant cell transformation. Nature 304:184–187, 1983.
5. RB Horsch, J Fry, NL Hoffmann, D Eichholtz, SG Rogers, RT Fraley. A simple and general method for transferring genes into plants. Science 227:1229–1231, 1985.
6. G Staritsky. Embryoid formation in callus cultures of coffee. Acta Bot Neerl 19:509–514, 1970.
7. T Leroy, M Royer, M Paillard, M Berthouly, J Spiral, S Tessereau, T Legavre, I Altosaar. Introduction de génes d'intérêt agronomique dans l'espèce *Coffea canephora* Pierre par transformation avec *Agrobacterium* sp. Proceedings of 17th Colloquium of International Coffee Science Association (ASIC), Nairobi, 1997, pp 439–446.
8. T Leroy, AM Henry, R Philippe, M Royer, A Deshayes, R Frutos, D Duris, M Dufour, S Tessereau, I Jourdan, G Bossard, C Lacombe, C Fenouillet. Genetically modified coffee trees for resistance to coffee leaf miner: Analysis of gene expression, insect resistance and agronomic value. Proceedings of 18th Colloquium of International Coffee Science Association (ASIC), Helsinki, 1999, pp 332–338.
9. KR Neupane, S Moisyadi, JI Stiles. Cloning and characterisation of fruit-expressed ACC synthase and ACC oxidase genes from coffee. Proceedings of 18th Colloquium of International Coffee Science Association (ASIC), Helsinki, 1999, pp 322–326.
10. SM Moisyadi, KR Neupane, JI Stiles. Cloning and characterisation of a cDNA encoding xanthosine-N7-methyltransferase from coffee (*Coffea arabica*). Acta Hortic 461:367–377, 1998.
11. SM Moisyadi, KR Neupane, JI Stiles. Cloning and characterisation of xanthosine-N7methyltransferase, the first enzyme of the caffeine biosynthetic pathway. Proceedings of 18th Colloquium of International Coffee Science Association (ASIC), Helsinki, 1999, pp 327–331.
12. MF Carneiro. Coffee biotechnology and its application in genetic transformation. Euphytica 96:167–172, 1997.
13. MF Carneiro. Advances in coffee biotechnology. AgBiotechNet 1:1–7, 1999.
14. J Spiral, V Petiard. Protoplast culture and regeneration in coffee species. Proceedings of the 14th Colloqium of the International Coffee Science Association (ASIC), San Francisco, 1991, pp 383–391.
15. CR Barton, TL Adams, MA Zarowitz. Stable transformation of foreign DNA into *Coffea arabica* plants. Proceedings of the 14th Colloquium of the International Coffee Science Association (ASIC), San Francisco, 1991, pp 460–464.

16. CA Ocampo, LM Manzanera. Advances in genetic manipulation of the coffee plant. Proceedings of 14th Colloquium of International Coffee Science Association (ASIC), San Francisco, 1991, pp 378–382.
17. J Grèzes, B Thomasset, D Thomas. *Coffea arabica* protoplast culture: Transformation assays. Proceedings of 15th Colloquium of International Coffee Science Association (ASIC), Montpellier, 1993, pp 745–747.
18. J Spiral, C Thierry, M Paillard, V Petiard. Obtention de plantules de Coffea canephora Pierre (Robusta) transformés par *Agrobacterium rhizogenes*. Comptes Rendus Acad Sci 316 (Série III), 1–6, 1993.
19. J van Boxtel. Studies on genetic transformation of coffee by using electroporation and the biolistic method. PhD dissertation, University of Wageningen, 1994.
20. M de Peña. Development of a stable transformation procedures for the protoplasts of *Coffea arabica* cv Colombia. PhD dissertation, University of Pardue, 1995.
21. M Vaeck, A Reynaerts, H Hofte, S Jansens, M De Beuckeleer, C Dean, M Zabeau, M van Montagu, J Lemans. Transgenic plants protected from insects attack. Nature 327:33–37, 1987.
22. Guerreiro Filho, P Denolf, R Frutos M Peforoen, AB Eskes. L'identification de toxines de *Bacillus thuringiensis* actives contre *Perileucoptera coffeella*. Proceedings of 15th Colloquium of International Coffee Science Association (ASIC), Monpellier, 1993, pp 329–335.
23. Guerreiro Filho, P Denolf, M Peforoen, B Decazy, AB Eskes, R Frutos. Susceptibility of coffee leaf miner (*Perileucoptera* spp) to *Bacillus thuringiensis* δ-Endotoxins: A model for transgenic perennial crops resistant to endocarpic insects. Curr Microbiol 36:175–179, 1998.
24. C Curie, T Liboz, C Bardet, E Gander, C Médale, M Axelos, B Lescure. Cis and trans-acting elements involved in the activation of *Arabidopsis thaliana* A1 gene encoding the translation elongation factor EF-1 alpha. Nucleic Acids Res 1:1305–1310, 1991.
25. ACM Brasileiro, C Tourneur, JC Leplé, V Combes, L Jouanin. Expression of the mutant *Arabidopsis thaliana* acetolactate synthase gene confers chlorsulfuron resistance to the transgenic poplar plants. Transgenic Res 1:133–141, 1992.
26. MR Söndahl, WHT Loh. Coffee biotechnology. In: RJ Clarke, R Macrae, ed. Coffee Agronomy. Vol 4. London: Elsevier Applied Science, 1988, pp 235–262.
27. MR Söndahl, JA Lauritis. Coffee. In: Hammerschlag FA, Litz RE, eds. Biotechnology of Perennial Fruit Crops. New York: CAB International, 1992, pp 401–420.
28. JE Bourque. Antisense strategy for genetic manipulation in plants. Plant Sci 105:125–149, 1995.
29. MJ Rijo. Production of caffeine-deficient coffee plants using the RNA antisense technology. PhD dissertation, Vrije Universiteit Brussel, 1996.
30. G Bertrand. Sur les cafés sans caffeine. C R Acad Sci (Paris) 136:209–211, 1905.
31. L Gianessi. Agricultural Biotechnology: Insect Control Benefits. Washinton, DC: National Center for Food and Agricultural Policy, 1999.
32. BT Nyambo, DM Masaba. Integrated pest management in coffee: Needs, limitations and opportunities. Proceedings of 17th Colloquium of International Coffee Science Association (ASIC), Nairobi, 1997, pp 629–638.
33. EF Vega, G Mercadier. Fungi associated with the coffee berry borer *Hypotherimus Barufoei* (Ferrari) (Coleoptera: Seolytidae). Proceedings of 18th Colloquium of International Coffee Science Association (ASIC), Helsinki, 1999, pp 229–238.
34. MF Tomashow. Plant cold acclimation: Freezing tolerance genes and regulatory mechanisms. Annu Rev Plant Physiol Plant Mol Biol 50:571–599, 1999.
35. M Joersbo, I Donaldson, J Kreiberg, SG Petersin, J Brunsted, FT Okkels. Analysis of mannose selection used for transformation of sugar beet. Mol Breed 4:111–117, 1998.
36. CR Smith, JÁ Saunders, S Van Wert, J Cheng, BF Matthews. Expression of GUS and CAT activities using electrotransformed pollen. Plant Sci 104:49–58, 1994.
37. BT Kren, P Bandyopadhyay, CJ Steer. In vitro site-directed mutagenesis of the factor IX gene in rat liver and in isolated hepatocytes by chimeric RNA/DNA oligonucleotides. Nat Med 4:285–290, 1998.
38. S Pati, S Mirkin, B Feuerstein, D Zarling. Sequence-specific DNA targeting. Encyclopedia of Cancer. 3:1601–1625, 1997.

# 49
# Transgenic Linseed Flax

**Alan McHughen**
*University of Saskatchewan, Saskatoon, Saskatchewan, Canada*

|       |                                                       |      |
|-------|-------------------------------------------------------|------|
| I.    | INTRODUCTION                                          | 747  |
| II.   | TISSUE CULTURE IN VITRO                               | 748  |
| III.  | TRANSFORMATION TECHNOLOGY AND METHODOLOGY             | 749  |
|       | A. *Agrobacterium* Species–Mediated *Linum* Genetic Transformation | 749 |
|       | B. Selectable Markers                                 | 751  |
| IV.   | OTHER ATTEMPTS TO IMPROVE EFFICIENCY                  | 751  |
|       | A. Epidermal Stripping                                | 751  |
|       | B. Preculture                                         | 752  |
|       | C. Coculture Duration                                 | 752  |
|       | D. Chimeras                                           | 753  |
| V.    | OTHER IMPROVEMENTS                                    | 754  |
|       | A. Particle Bombardment                               | 754  |
| VI.   | APPLICATIONS                                          | 755  |
| VII.  | VALUE-ADDED TRAITS                                    | 756  |
| VIII. | POTENTIAL WEED CONTROL APPLICATIONS                   | 757  |
| IX.   | REGULATORY ISSUES                                     | 757  |
|       | REFERENCES                                            | 758  |

## I. INTRODUCTION

Linseed, flax, or flaxseed (*Linum usitatissimum* L.) is an ancient crop grown throughout the temperate regions of the world, but nowhere, except perhaps Canada, is it a major crop. The name *usitatissimum* means "most useful." The stems can be used for fiber, the seeds for oil, and the meal for feed (1). It is also the root of many common words. Some derivations are obvious (e.g., *flaxen*, as an adjective; *linen*, the textile derived from the bast fibers;) *linoleum*, the original flooring product made from both the fiber and the oil). Others are less apparent: *line*, from the long straight fibers; *lingerie*, made from the comfortable high-quality textile, even *linnet*, the European finch, which eats flaxseeds; and *lint*.

Most of the terms now relate to products from other sources: modern linoleum is made from vinyl, and most lint comes from cotton. However, flax (linseed) oil is still unsurpassed as an industrial oil in the production of paint, varnish, and linoleum and as a concrete sealant (2). The suitability of flax/linseed oil for these purposes is due to the high proportion of α-linolenic acid, resulting in a top-quality drying oil. Unfortunately, this same α-linolenic acid causes the oil to turn rancid quite rapidly, and thus the traditional oil cannot be used for commercial edible oil purposes (3). To overcome this limitation, mutations with reduced α-linolenic acid have been induced and identified in linseed cultivars; these have been developed into commercial cultivars for use in the human food oil market (3,4). There are renewed interest in and demand for linseed flax in fiber/textile as well as in industrial oil markets, largely due to the "natural" and "environmentally friendly" components of linseed flax production relative to those of the synthetic modern alternatives.

Unfortunately, *Linum usitatissimum* L. has limited genetic resources, as it has few close relatives amenable to crossing to introduce novel traits. Because of this limitation, gene transfer technology allows a much greater access to beneficial genetic traits than for many other crop species. For example, conventional breeding technology could not provide such traits as novel herbicide resistance to the crop because no interfertile germplasm with the desired herbicide-resistant trait exists. Similarly, the major commercial product of this crop, linseed oil, is a high-quality industrial drying oil with a stable but relatively small market. In order to increase market opportunities for flax producers, the characteristic oil might be modified by using gene transfer technology to yield different types of oil for different markets. Gene transfer technology for this (or any) species would be facilitated by reliable in vitro tissue culture and regeneration system, a reliable gene delivery/cell transformation system, and an efficient selection/regeneration system yielding fertile transgenic plants.

Linseed flax has a long history of manipulations in vitro, as it has been quite amenable to many tissue culture techniques, including regeneration from protoplasts as well as from hypocotyl, cotyledon, and leaf-derived callus. The ease with which plants can be regenerated from various flax seedling tissues, especially the hypocotyl and cotyledon, greatly facilitates its use in crop improvement programs requiring the production of fertile transgenic plants. The application of gene transfer technology also had an early start with linseed, as it was among the first crop species to be genetically transformed with disarmed *Agrobacterium* species, and one of the first crop species transformed with genes of potential agronomic value. Currently linseed is the target of "molecular" crop improvement programs in several countries, although it remains a relatively low priority in each of them.

## II. TISSUE CULTURE IN VITRO

Link and Eggars (5) reported that *Linum* species would regenerate shoot buds from decapitated flax seedlings, giving the impression that shoot regeneration in vitro might be simple. In spite of that early promise, in vitro studies exploiting the potential for linseed did not emerge until the 1970s. Then, a number of investigators studied linseed tissues in vitro to address a range of physiological and developmental questions. Ibrahim (6) provided a start in publishing satisfactory culture media, then Rybczynski (7) and Gamborg and Shyluk (8) studied factors affecting tissue growth and shoot regeneration in different cultivars. Murray and coworkers (9) contributed a seminal paper, suggesting that the shoot initials from regenerating hypocotyls in vitro might have a single-cell origin. Lane (10,11) also contributed to the growing body of knowledge of linseed physiological requirements for satisfactory tissue culture and shoot regeneration. Basic morphogenic studies with linseed still appear [e.g., Nataraja and Ravi (12)] but usually relate to specific cultivars or peculiar results. General improvements in the culture media and conditions continue to be reported, for example, by Bretagne and coworkers (13), who found thidiazuron to be generally beneficial to linseed regeneration.

Whereas shoot regeneration via organogenesis was simple in linseed stem or hypocotyl tissues in vitro, anther culture to produce haploid plants was, and still is, difficult. Sun (14) reported on preliminary experiments of linseed anther culture, but there were no subsequent reports of the degree of success. In spite of interest from a number of sources, linseed was recalcitrant to anther or microspore culture. Eventually, Nichterlein and Freidt (15) were able to report on plant regeneration from isolated microspores, to be followed by incorporation of androgenic techniques in a linseed breeding program (16). Later Canadian scientists were able to report anther culture success with their linseed cultivars (17).

Almost all reports of plant regeneration in linseed are based on a shoot organogenic process, in which shoots arise on the cultured tissues, then are excised and transferred to a different medium for rooting, prior to establishment as whole plantlets in soil. Although this method can be satisfactory, there are several advantages to also having an embryogenic system. Pretova and Williams (18) studied linseed embryos in vitro to try to develop an embryonic regeneration system, but the standard organogenic systems remain the preferred and most efficient means of obtaining regenerant linseed plants.

Others investigated protoplast culture. Barakat and Cocking (19) first reported linseed regeneration from protoplasts, but this work seems not to have been pursued further. Zhan and coworkers (20) reported protoplast work on the related species *L. marginale*, and Qing and Binding (21,22) used protoplasts of *L. usitatissimum* L. and *L. suffruticosum* L. in their attempts to obtain transgenic plants. Linseed has not been very amenable to protoplast manipulations, but various groups continue to develop the field. Roger and associates (23) tried immobilization of protoplasts to elicit a morphogenic response.

Finally, root tissues have caused problems, in that shoot regeneration from root tissue is rare, having been reported only once (24). It seems odd that a plant species so amenable to direct and indirect (i.e., via callus) shoot organogenesis, and subsequent rooting of regenerant shoots, would be so difficult in attempts to induce regeneration through other means, such as via embryogenesis, or from protoplasts, anther and microspore culture, or root tissue. Nevertheless, for transformation studies, the tissue culture and whole plant regeneration systems available are sufficient. For further information on the tissue culture aspects of *Linum* species, see the review by McHughen (25).

## III. TRANSFORMATION TECHNOLOGY AND METHODOLOGY

Linseed flax is quite amenable to in vitro culture and whole plant regeneration (usually via shoot organogenesis), with any of several different starting tissues. However, the most prolific regenerable tissue seems to be the hypocotyl; hence it has been a popular starting material for transformation studies.

### A. *Agrobacterium* Species–Mediated *Linum* Genetic Transformation

*Agrobacterium tumefaciens* is a naturally occurring genetic engineering agent. In the wild, it attaches to an exposed cell of a susceptible plant and transfers a piece of transfer deoxyribonucleic (T-DNA), into the host plant cell, where the T-DNA is incorporated into the cell's genome. The functional genes carried on the T-DNA are expressed by the infected cell, ultimately resulting in cell growth proliferation manifestation of crown gall disease. Modified strains of *Agrobacterium* sp. maintain the ability to transfer pieces of DNA but delete the disease-related genes; such strains are called *disarmed*. By using standard microbiological methods, one can introduce genes of interest into the modified (disarmed) *Agrobacterium* sp. strain, then mix the bacteria with susceptible plant tissue in vitro and let nature take its course.

*Agrobacterium tumefaciens* causes crown gall in *Linum* sp. (26), so this biological agent could potentially be used to transfer DNA to flax for crop improvement purposes. More direct evidence was provided when Hepburn and coworkers (27), using wild-type strains, documented the molecular process of flax genetic transformation in the lab. In theory, it would seem obvious and straightforward then to take similar but disarmed (nononcogenic) strains of *Agrobacterium* sp. containing a selectable marker (the most common of which is *npt*II, an antibiotic resistance marker) construct in the T-DNA and apply them to regenerable flax tissues in vitro. The flax cells successfully transformed would then be identified or selected (on the basis of expression of the introduced selection marker gene) and regenerated to whole plants. Flax is highly self-fertile, so the putative transformed regenerants/selectants could be allowed to self-pollinate; progeny could then be tested to confirm their transgenic nature and also to document segregation ratio.

In practice, it is not that simple. Although the disarmed *Agrobacterium* sp. seems quite efficient at transforming flax cells of regenerable (usually hypocotyl or cotyledon) tissues in vitro, the resulting regenerants tend to be nontransformed "escapes" (28). Flax was one of the first crop species to have active research on *Agrobacterium* sp.–based genetic transformation (29–31), but it was not until 1988 that the first transformed flax plants were confirmed by progeny analysis (32).

Since then, several specific protocols to generate transgenic linseed flax plants by using nononcogenic strains of *Agrobacterium* sp. have been published, and none of the protocols was particularly simple or efficient. Zhan and associates (33) avoided some of the problems associated with *Agrobacterium tumefaciens* by using the closely related *Agrobacterium rhizogenes* and regenerated transgenic linseed shoots. However, *A. rhizogenes* has its own problems (particularly troublesome are those related to regenerant phenotype and infertility) and has not been followed up as vector of choice for *Linum* sp. plant transformation, although it is still useful for the study of other aspects of genetic transformation such as development of root formation (34).

The problem with *Agrobacterium* sp.–mediated transformation in linseed flax seems to be that the cells susceptible to *Agrobacterium* sp. transformation tend not to be competent for shoot regeneration. When the bacteria encounter the hypocotyl in vitro, for example, it seems that either they preferentially transform cells not competent for regeneration or the successfully transformed cells lose their competence to regenerate shoots (28).

In either case, cellular transformation is not a problem, as *Agrobacterium* sp. strains carrying the *uid*-A (GUS) construct have shown the capacity to transform a large proportion of cells in the flax hypocotyl (35). A common feature of the successful protocols is either to stimulate these transformed cells to regain regenerable competence or to select the relatively few transformed cells maintaining shoot regenerability more effectively. However, none of the successful protocols is particularly efficient; in the best about 20% of *Agrobacterium* sp.–inoculated tissues give rise to a transgenic linseed flax plant (36–38). In any case, the trick to acquiring transgenic linseed flax plants is to combine cellular competence for transformation with competence for shoot regeneration; the competence can be either simultaneous or sequential: first transformation, then regeneration.

The undesirable limited gene pool in *Linum* sp. seems to have provided at least one advantage. Basic transformation techniques can be applied to many different genotypes. Techniques developed for transformation of Canadian linseed cultivars also seem to work for Eastern European flax cultivars (e.g., Ref. 39) and also for dual-purpose (fiber and oil) Egyptian cultivars (40). Chinese cultivars have also been transformed with *Agrobacterium* sp. (41), as have, as noted, French and Czech cultivars. With each of these, however, fine-tuning may be required to optimize the protocol for maximal efficiency.

## B.  Selectable Markers

One of the limitations of this method was relatively low efficiency, which was due mainly to a high incidence of escapes (wild-type nontransformed shoots appearing under selection conditions). Without selection at all, in the most successful inoculation procedure only about 2% of regenerant shoots are actually transformed (36). As this proportion is very low even under optimized conditions, and verification is extremely tedious and labor-intensive (because of the large number of shoots generated), an efficient selection system is desirable. Linseed flax hypocotyls placed on a regeneration medium readily produce shoots. Adding kanamycin to the medium prior to inoculation eliminates shoot formation entirely, indicating suitability of kanamycin as a selection agent for the *npt*II gene in transformation studies. Inoculating hypocotyls with *Agrobacterium* sp. carrying *npt*II in the T-DNA results in a reduction in the number of shoot regenerants, as one might expect in a perfect selection system. Unfortunately, most if not all of the shoots are escapes (28), as determined with subsequent assays for transformation. What appears to be occurring is that the *Agrobacterium* sp. does transform a certain number of cells in the hypocotyl [as determined by using the *uid*A gene followed by the X-Gluc histochemical assay on hypocotyls, as shown by Dong and McHughen (35)]. Those transformed cells express *npt*II, which detoxifies kanamycin passing through the cells, thus protecting neighboring, nontransformed cells from the effects of the selection agent. It seems it is the nearby protected, nontransformed cells that preferentially regenerate into shoots (42).

Another approach to overcome this limitation of the selection agent was to try other selectable marker genes and agents in linseed flax. Bretagne-Sagnard and Chupeau (38) tried spectinomycin as a selection agent and had good success at producing transgenic linseed flax of European cultivars.

The McHughen lab continued to investigate *npt*II as the selection gene of preference, trying to increase transgenic regenerant selection efficiency by testing varied concentrations of kanamycin and also other aminoglycosidic antibiotics such as G418 in the selection medium (36). Other selectable markers were tried also. One very effective one is *pat,* which confers resistance to phosphinothricin. Using this gene not only was very good for selection in vitro, but also provided excellent herbicide resistance in field trials with the transgenic linseed (43). However, because of other technical and nontechnical considerations, we returned to *npt*II as the selectable marker of choice.

## IV.  OTHER ATTEMPTS TO IMPROVE EFFICIENCY

### A.  Epidermal Stripping

Regeneration from linseed flax hypocotyl explants seems to result primarily from epidermal cells (9) whereas the cells transformed by *Agrobacterium* sp. tend to occur and proliferate around the cut ends of the hypocotyl explant. Unfortunately, the regenerated shoots usually arise from nontransformed epidermis tissue (28). To overcome this problem, Jordan and McHughen (32) used tweezers to peel the epidermis from along the hypocotyl, thus removing the cells less susceptible to *Agrobacterium* sp. transformation and simultaneously causing injury along the length of the hypocotyl, simulating greater *Agrobacterium* sp. virulence. An added benefit of this epidermal stripping was the reduction in occurrence of "escape" shoots, most of which arise from the non-transformed epidermal cells. The lower number of overall regenerant shoots made subsequent selection and analyses simpler and less labor-intensive, and the overall process became more efficient. Of course, it also removes the major source of regenerable cells, the reason for choosing the hypocotyl. In any case, this procedure led to the production of fully fertile transgenic flax plants that passed the foreign genes to their progeny in a normal mendelian manner (32).

## B. Preculture

Another modification to the basic procedure is the preculture of hypocotyls prior to *Agrobacterium* sp. inoculation. The idea behind this is to stimulate the transformation-competent cells to become regeneration-competent as well. As noted earlier, hypocotyl cells respond well to regeneration medium by generating shoots prolifically. Similarly, hypocotyl cells are quite amenable to transformation by *Agrobacterium* sp., but they tend not to be the same cells at the same time. By judicious timing, it should be possible to inoculate at a time when the hypocotyl cells are starting to respond to the regeneration stimuli in the medium yet are still competent and responsive to *Agrobacterium* sp. transformation.

McHughen and associates (44) found that when the explants were cultured for a period of 9 to 12 days on regeneration medium prior to removal of the epidermis, significantly more transformed shoots were regenerated when compared with either no preculture or a 3- to 7-day preculture period. They speculated that the cells become competent for regeneration during the preculture stage. Inoculation after this period results in transformation of cells already in regeneration mode, and those cells, now transformed, simply carry on with the shoot regeneration program. Cells transformed without a preculture period might be inhibited from subsequently regenerating shoots, possibly because of stress caused by the presence of the *Agrobacterium* sp., the selective agent (in this case, kanamycin), or other components of the medium. Whatever the reason, it is clear a preculture period increases transformant production.

Subsequent studies using the intron-containing *uid*A (GUS) gene construct (45) showed that longer preculture periods result in a reduced overall incidence of cellular transformation (35). This finding indicates that at least some transformation-competent cells lose that capacity during the preculture period. Presence of fewer transformed cells, yet greater numbers of transgenic shoots indicates that these conditions lead to preferential regeneration from transformed cells.

## C. Coculture Duration

Increasing the efficiency of recovery of transgenic flax plants from *Agrobacterium* sp. led to an investigation of coculture duration. In most plant transformation studies, the modified *Agrobacterium* sp. is mixed with the plant tissue for a few minutes, then plated on a standard culture medium (lacking antibiotics or other selection agents) for 2–3 days. After this short period, the plant tissues are transferred to a selection medium containing anti–*Agrobacterium* sp. agents to eliminate the vector organisms. If a coculture duration shorter than about 2 days were used, fewer transformation events ultimately resulted. If the period was increased to more than about 3 days, the *Agrobacterium* sp. overgrew and smothered the plant tissue (resulting in production of no transgenic plants). In order to try to increase coculture duration without *Agrobacterium* sp. overgrowth, Dong and McHughen (35) placed a layer of sterile filter paper on top of the coculture medium. The inoculated hypocotyls were placed on top of the paper and could remain there for several days without *Agrobacterium* sp. overgrowth. Using this setup, different coculture durations were then investigated, with the standard 2- to 3-day coculture as the control. Using the intron containing *uidA* gene as the scorable marker facilitated scoring of cellular transformation.

The results showed that the optimal coculture duration for cellular transformation (as determined by the transient expression histochemical assay) was 5–7 days, representing a 2- to 15-fold increase (depending on other culture factors) over the standard 2- to 3-day period.

## D. Chimeras

"Escape" regenerants have been a well-known part of plant transformation studies from the beginning. A less recognized phenomenon is that a regenerant plant, having passed all the requisite assays and tests for transformation, may give rise to seed progeny with fewer than expected or even no transmission of the transgene. Mendelian genetics dictates that a single locus insertion into a typical plant cell is hemizygous; self-pollination of the plant should result in a 3:1 positive-to-negative phenotypic segregation ratio in the progeny. Although most transgenic regenerants follow this pattern, some have a much higher proportion of positive progeny, indicating multiple insertion loci. It is easy to explain *no* positive progeny casually: either the plant was an escape, or the inserted construct was unstable. But how does one explain *fewer* than predicted positive progeny, say, 1 positive and 16 negative?

Chimeras are plants composed of cells of two (or more) distinct genotypes. It is not unusual, particularly with a shoot organogenic regeneration system (such as is common in linseed flax), to have more than one cell contribute to a shoot primordium. If the contributor cells are genetically distinct, say, one is transgenic, the other wild type, then the growing shoot will contain cells representing the two distinct genotypes. If the transgenic cell lineage contributes to any substantial proportion of the regenerant plant, it will score positive in the various tests for transformation. However, if it fails to contribute to the germ line, then the entirety of the gametes will be from the wild-type cells; none of the seed progeny will carry the transgenes. More commonly, the transgenic cell line contributes to only a portion of the gametes, yielding fewer seed progeny than would be expected from conventional mendelian genetics.

Chimeric regenerants can be easily detected by using the intron-containing *uid*A gene construct (45). As shoots arise on the hypocotyl (or whatever inoculum tissue), they can be excised for transfer to rooting or elongation medium. At this time, a small disk of shoot is excised and used for the histochemical X-Gluc assay. Since all cells in the shoot are derived from cell lines also represented in the disk, the disk serves as a full representative of the shoot. If the disk is fully negative in the X-Gluc assay, the associated shoot can be discarded as an escape. If the shoot is a chimera, only those transgenic cell lines stain indigo in the disk assay.

Using this marker, Dong and McHughen (46) showed that the incidence of chimeric shoots was much higher than previously thought or expected. Up to about 45% of regenerant linseed flax putative transformant shoots were actually chimeras, composed of both wild-type and transgenic cell lines. Most interesting are the chimeras composed of two or more transgenic cell lines. Chimeras are only relevant in regenerants themselves, as chimeras do not persist or appear in seed progeny. This means fully transformed (genetically solid) plants can in many cases be recovered from chimeras. One method is to test more seed progeny that ordinarily necessary under mendelian rules (if the germ line cells are derived from at least some transformed cell lines). Another method is internode propagation of the regenerant shoot and testing of several subregenerants. Finally, if researchers are desperate, genetically solid plants can be recovered from chimeras by culturing assay-positive leaf pieces from the chimeric plant on selective medium and regenerating shoots from any resulting callus (42,46).

Optimizing the conditions for regenerable cellular transformation (not necessarily the same as greatest cell transformation) and combining this with an efficient selection strategy led to an overall transformation efficiency of about 16% (i.e., 100 inoculated hypocotyls would yield about

16 transgenic regenerants) (36). Although this rate is not particularly efficient compared to, say, that for tobacco or some other species, it is quite workable for crop improvement programs.

The recovery of transgenic flax plants using *Agrobacterium* sp. as a vector system remains inefficient as a result of the relatively high incidence of chimeras and the relatively high proportion of escapes. The technical difficulties continue to be the assessment of the large number of regenerant shoots and the poor proportion of solid transgenics from among them.

## V. OTHER IMPROVEMENTS

Mylnarova and Pretova (37) developed an *Agrobacterium* sp. transformation system based on inoculation of hypocotyl-derived callus in an attempt to overcome some of these complicating factors. Ordinarily, callus is avoided, as the spontaneous mutations that occur in callus (somaclonal variation) might interfere with the agronomic performance of the resulting transgenic line. However, they encouraged callus growth to provide more cells for potential transformation and regeneration. They were indeed successful at increasing the efficiency to about 25% while reducing the incidence of escapes.

Taking different approach, Bretagne and colleagues (13) worked on improving the tissue culture and regeneration components of the process and found that thidiazuron improved the shoot regeneration response.

Another innovative attempt to increase linseed transformation efficiency involved using not *Agrobacterium tumefaciens* but its cousin, *Agrobacterium rhizogenes*. Zhan and associates (33, 34) used this bacterium and maintained an active program generating linseed transformants over several years. Unfortunately, they were not able to report an overall increase in transformation efficiency with recovery of transgenic plants.

### A. Particle Bombardment

An alternative to the *Agrobacterium* sp. gene transfer system is the particle gun, or biolistic method. Although *Agrobacterium* sp. can get the job done, it is neither efficient nor simple. The use of *Agrobacterium* sp. may entail potential other, nontechnical difficulties, such as intellectual property infringement or regulatory obstacles to commercial release of the final product. Although the particle gun certainly also may evoke intellectual property and regulatory issues, they might be simpler or less onerous. In any case, a particle gun transformation method might provide a functional alternative to *Agrobacterium* sp., especially where the obstacles to the latter prove prohibitive.

Gene transfer by particle bombardment can be a rapid and simple means for delivering functional DNA to intact cells and tissues (47,48). The concept is simple enough: the DNA of choice is affixed to microparticles that are shot, using a blast of gas or even gunpowder, into awaiting tissues in the trajectory flight path. At least some of the particles become embedded in the cell, thus delivering the DNA, which is then somehow inserted into the cell's genome. Although the likelihood of any one particle's succeeding is minuscule, the system seems to work because large numbers of particles can be shot into large numbers of cells. Transient expression of reporter genes such as β-glucuronidase (GUS) can be used to optimize factors affecting the success of particle bombardment–mediated DNA delivery. In practice, it is not usually as simple as the concept might indicate. However, transgenic plants of several species have been generated from particle bombardment, including soybean (49,50), tobacco (51,52), rice (53), and wheat (54). *Linum* species, would also seem to provide a suitable target, as they readily regenerate shoots in vitro. Linseed flax hypocotyls, for example, could be bombarded and cultured on a standard regeneration medium. As with the *Agrobacterium* sp. method, regenerating shoots could be excised and assayed for transient foreign gene expression early and quickly. Those shoots with no expression could be discarded, allowing the positive shoots to continue rooting, growth, and maturation.

Wijayanto and Wijayanto and McHughen documented a successful biolistic process to generate transgenic linseed flax (55,56). Testing and optimizing a number of variables, Wijayanto bombarded intact flax hypocotyl tissues with gold particles coated with plasmid DNA carrying the β-glucuronidase (GUS) (*uid*A) and neomycin phosphotransferase II (*npt*II) genes (55). Optimization of the parameters was facilitated by transient expression assays of the *uid*A gene. Among the critical variables (i.e., those providing significant differences in results) were hypocotyl preculture period, bombardment distances, degree of bombardment chamber vacuum, quantity of DNA that adhered to the microparticles, and size of the microparticles. In analyzing progeny from 10 transgenic regenerant lines, Wijayanto found that 7 lines segregated in a 3:1 ratio, suggesting a simple, single mendelian insertion. One line segregated according to a likely 15:1 ratio, suggesting two mendelian insertional loci. One was a chimera (only 2 of 66 progeny were positive; self-pollinated progeny from those 2 positive plants segregated in a normal 3:1 manner). One other line was either unstable or an escape, in that none of the progeny was transgenic (56).

It is difficult to compare the technical efficiency of the two methods (*Agrobacterium* sp. vs. biolistic) directly. The facile calculation would indicate the *Agrobacterium* sp. method is superior to the biolistic, in that many hypocotyls must be shot in order to recover one transgenic regenerant. However, many hypocotyls can be prepared and shot by the particle gun with the time and effort required to inoculate hypocotyls with the appropriate *Agrobacterium* sp. Furthermore, if the results from Wijayanto continue (and there's no indication that they wouldn't), the biolistic method appears to reduce the incidence of chimeric regenerants (to about 10%) and escapes (also to about 10%) dramatically. Both represent substantial improvements over the *Agrobacterium* sp. method, which yields up to 45% chimeric regenerants and many escapes.

## VI. APPLICATIONS

Like those of most plant species, early studies of genetic transformation in *Linum* sp. largely investigated basic processes and addressed fundamental questions of genetics, physiology, or other aspects of biology.

These studies need not involve regeneration of transformed plants but can use simple transformed cells or callus, as in that by Roberts and coworkers (57), who used transformed callus to study the effect of transposable elements. Others have used transformed *Linum* sp. to study other aspects, for example, the use of introduced transposable elements in flax to identify and isolate potentially useful genes such as the L6 gene for rust resistance (58,59). Cloning of a specific disease resistance gene represents a major advance in the understanding of phytopathological relationships with wide applicability to similar relationships in other crop species.

Most effort, however, has been and will likely continue to be invested in genetically transforming *Linum* sp. for a practical application. Not surprisingly, the first gene constructs available for transformation were developed to address problems in other crop species, and linseed flax happened to face similar issues. These, of course, were herbicide resistance genes. Linseed is a poor biological competitor, so weeds are invariably a concern to producers, and although conventional weed control is available, it is expensive and based on herbicides developed for larger and more lucrative crops.

Fortunately, several herbicide resistance genes were developed early, usually by the companies producing the herbicide. *Linum* sp. was among the first crop species to benefit from these constructs, as glyphosate (Roundup) resistance, sulfonylurea (e.g., Glean, Ally, Amber) resistance, and glufosinate (Liberty) resistance all were quickly inserted and field tested in commercial linseed flax genotypes (see Sec. VIII).

With respect to the other major agronomic problems that molecular genetics has been targeting, disease resistance and insect depredations, linseed flax is relatively well off in having fewer problems with these than other major crop species. Conventional breeders have done a superb job of overcoming the most common diseases in linseed [*Melampsora lini*, (rust)], and few insects seem to have linseed at the top of their menu, although certainly insects can and do occasionally cause problems for linseed flax farmers. Because of this and the issue of priorities, there does not seem to be much activity in transforming linseed with, for example, *Bacillus thuringiensis* (*Bt*) construct to confer resistance to lepidopteran insect pests. Although this trait would be useful in certain years, in certain locations, the problem is not important or common enough in itself to justify a higher cost to producers. Similarly, the value added by insertion of a disease resistance gene would probably not be sufficient to entice producers to pay a premium for the seed, at least not until the pathogens overcome and break down current levels of resistance.

Other agronomic characteristics for which linseed would be an early research candidate include mechanisms to deal with environmental stresses such as salt tolerance (e.g., 60) or cold tolerance (61), although neither of these has been reported to have been inserted into linseed to date.

## VII. VALUE-ADDED TRAITS

Whereas most of the applications of genetic transformation of *Linum* sp. have been in weed control, there are several other, nonagronomic breeding objectives being pursued and tested. Probably the most exciting is in oil profile modification. Linseed has a relatively simple oil profile of approximately 5% palmitic, 5% stearic, 25% oleic, 15% linoleic, and 50% linolenic (these figures vary according to genotype, location, year, etc.), and this profile is amenable to manipulation by a range of different mechanisms (4). One good example, although not transgenic, is edible oil linseed developed by mutation breeding (3,4), which has an oil profile similar to that of the high-quality sunflower or safflower oils. This mutation work shows that linseed can be genetically modified to produce a commercially viable vegetable oil for human consumption. It should be possible to use recombinant technology to achieve the same end by reducing the activity of $\delta$-15 desaturase; this would result in a dramatic reduction of linolenic acid, to provide an oil profile similar or identical to that of the mutants. At least two groups are actively working on this research (Scheffler, personal communication, 1998). The advantage here is overcoming the public reluctance to consuming mutated foods such as Linola.

Several recent developments in plant oil modification using gene transfer technology to produce new and useful fatty acid profiles have been reported (e.g., 63–65). Although these tend to be developed and tested on other oil crops, especially *Brassica* sp., there is no reason to suspect they would also not work in *Linum* sp.

If this is the case, it should be possible to generate linseed cultivars that produce one of many different types of oil by manipulating the fatty acid biosynthetic pathway. These might include industrial oils as well as those for mass and specialty (niche) edible markets, for example, change in the proportions of fatty acids to result in a cocoa butter–type oil (66). The "health" or functional food markets are also being targeted with products made from modified linseed cultivars.

Apart from direct crop value, the linseed plant is being investigated for use as a simple organic manufacturing plant. Linseed as a crop is fairly simple to grow, has no close relatives, is almost exclusively self-pollinating, does not readily outcross, and generates relatively large amounts of oil, both on an area basis and on a seed basis. Into this biological background one might place a gene coding for a high-value protein or protein by-product, then let the plant grow in the field and manufacture large quantities of the (ideally) valuable product (Moloney and Holbrook, personal communication, 1998).

## VIII. POTENTIAL WEED CONTROL APPLICATIONS

Flax is a poor biological competitor. Many common broadleaf and grassy plants successfully outcompete linseed, even when present in low populations, so good weed control is essential for maximal crop production. Although chemical herbicides are available for weed control, they are expensive and somewhat limited in their utility.

The earliest agronomic genes isolated and cloned confer resistance to herbicides. Linseed flax has been transformed with several of these genes, including those conferring resistance to glyphosate [the active ingredient in Roundup (Monsanto)], glufosinate [the active ingredient in Basta and Liberty (AgrEvo)], and sulfonylurea [the active ingredient in, e.g., Ally (DuPont)].

Transgenic linseed lines were generated and field-tested for each of these characteristics; however, only the sulfonylurea-resistant lines were considered for commercial release. The glyphosate-resistant lines, tested in collaboration with Monsanto, did not perform well enough in preliminary field trials to warrant continued evaluation (67). The glufosinate-resistant lines, tested in collaboration with AgrEvo, did perform well, and in fact one line was entered into the national cultivar registration trials, in which the agronomic performance was measured as being unaffected by the genetic transformation. The degree of herbicide resistance afforded by the activity of the gene was also considered commercially viable, but the line was not advanced for commercial release (43).

The gene conferring tolerance to sulfonylurea herbicides was isolated by Haughn and coworkers (68). The gene, from *Arabidopsis* sp., codes for a modified acetolactate synthase (ALS) protein that has reduced affinity for the herbicide. Sulfonylurea herbicides attack the ALS in susceptible plants. Providing linseed with an additional and alternate functional ALS may allow the plants to escape the damaging effects of the herbicide in soil residue trials. McHughen (69) transferred this gene to commercial linseed flax cultivars in an attempt to address this soil residue problem.

Transformed lines with the sulfonylurea tolerance gene have been tested in field trials since 1989. Initially about 30 different linseed lines were evaluated. Over the course of several years, most of the lines were eliminated as a result of either poor agronomic performance or poor herbicide tolerance (which was due to poor expression of the transgene). Eventually two transgenic lines remained. Both seemed to offer full herbicide resistance with no agronomic penalty (70–73). One of these lines was presented to and approved by Canada's national cultivar registration committee in February 1994 and issued its certificate of variety registration in 1996 (71). More recently, the same gene was introduced into Russian linseed cultivars (39) with the same ultimate objective.

## IX. REGULATORY ISSUES

Linseed flax is an international trade commodity. In addition to the usual regulatory oversight of any new cultivar, including variety registration; acquisition of intellectual property protection under plant breeders' rights, patents, or other systems; and the usual international phytosanitary inspections, transgenic linseed invokes various "genetically modified" organism regulations in each country of cultivation, transit, or destination. These regulations can be confused and onerous with little, if any coordination among nations. Even within nations, often confusion and ambiguity result from lack of clarify about which regulations exist and apply, and which agency administers them. Ordinarily, transgenic crops are scrutinized for variety registration by the appropriate agriculture department and for food safety by a health department; animal feed safety is often administered by another department, and, of course, environmental concerns are the responsibility of an environment department. To date, only one transgenic linseed has reached registered cultivar status, 'CDC Triffid' (71). In order to join other linseed cultivars in commercial production in Canada, 'CDC Triffid' needed, in addition to regular cultivar registration, approval from various government departments concerned with feed safety, human food safety, and environmental effects. In order for it to enter international trade, it was evaluated by the U.S. Food and Drug Administration (FDA) in the United States for human food and animal feed usage, and by the U.S. Department of Agriculture (USDA) for cultivation and marketing. Even with these approvals, 'CDC Triffid' must be segregated from conventional linseed while other international jurisdictions complete their respective evaluations. Currently, the only transgenic linseed in commercial production and international commerce is 'CDC Triffid.' That trade is limited to Canada and the United States.

## REFERENCES

1. J Janick, RW Schery, FW Woods, VW Ruttan. Plant Science: An Introduction to World Crops. San Francisco: WH Freeman, 1974, pp 518, 572–573.
2. E Kenaschuk. Flax breeding and genetics. In: JT Harapiak, ed. Oilseed and Pulse Crops in Western Canada. Calgary, Alberta: Western Canada Co-operative Fertilizers, Ltd. 1975, pp 203–221.
3. AG Green. Development of an edible-oil-flax genotype. Proceedings of the 51st Annual Flax Institute of the United States, Fargo ND, 1986.
4. GG Rowland, A McHughen, LV Gusta, RS Bhatty, SL McKenzie, DC Taylor. The application of chemical mutagenesis and biotechnology to the modification of linseed (*Linum usitatissimum* L.). Euphytica 85:317–321, 1995.
5. G Link, V Eggars. Mode, site and time of initiation of hypocotyledonary bud primordia in Linum usitatissimum L. Bot Gaz 107:441–454, 1946.
6. RK Ibrahim. Media for growth of flax tissue culture. Can J Bot 49:295–298, 1971.
7. J Rybczynski. Callus formation and organogenesis of mature cotyledons of *Linum usitatissimum* L. variety *szokijskij* [flax] in vitro culture. Genet-Pol, 16:161–166, 1975.
8. OL Gamborg, JP Shyluk. Tissue culture, protoplasts, and morphogenesis in flax. Bot Gaz 137:301–306, 1976.
9. BE Murray, RJ Handyside, WA Keller. In vitro regeneration of shoots on stem explants of haploid and diploid flax (*Linum usitatissimum*). Can J Genet Cytol 19:177–186, 1977.
10. WD Lane. Influence of growth regulators on root and shoot initiation from flax meristem-tips and hypocotyls in vitro. Physiol Plant 45:260–264, 1979.
11. WD Lane. Plant manipulation in vitro with hormones: *Linum-usitatissimum,* flax, *Malus,* apples. Comb-Proc-Int-Plant-Propagators-Soc. Boulder: The Society. 31:101–108, 1982.
12. K Nataraja, GM Ravi. Morphogenetic studies in vitro in a new variety of flax: *Linum usitatissimum.* L. cv. JLS-J-1. Beitr Biol Pflanz 60:199–206, 1985.
13. B Bretagne, M-C Chupeau, Y Chupeau, G Fouilloux. Improved flax regeneration from hypocotyls using thidiazuron as a cytokinin source. Plant Cell Rep 14:120–124, 1994.
14. HT Sun. Preliminary report on anther culture of flax [*Linum usitatissimum*]. K'o-Hsueh-T'ung-Pao-Kexue-tongbao 24:948–950, 1979.
15. K Nichterlein, W Friedt. Plant regeneration from isolated microspores of linseed (*Linum usitatissimum* L.). Plant Cell Rep 12:426–430, 1993.

16. W Friedt, C Bickert, H Shaub. In vitro breeding of high-linolenic, doubled-haploid lines of linseed (*Linum usitatissimum* L.) via androgenesis. Plant Breed 114:322–326, 1995.
17. YR Chen, EO Kenaschuk, JD Procunier. Plant regeneration from anther culture in Canadian cultivars of flax (*Linum usitatissimum* L.). Euphytica 102:183–189, 1998.
18. A Pretova, EG Williams. Direct somatic embryogenesis from immature zygotic embryos of flax (*Linum usitatissimum* L.). J Plant Physiol 126:155–161, 1986.
19. MN Barakat, EC Cocking. Plant regeneration from protoplast-derived tissues of *Linum usitatissimum* L. (Flax) [Isolation and culture]. Plant Cell Rep 2:314–317, 1983.
20. XC Zhan, DA Jones, A Kerr. In vitro plantlet formation in *Linum marginale* from cotyledons, hypocotyls, leaves, roots and protoplasts. Aust J Plant Physiol 16:315–320, 1989.
21. LH Qing, H Binding. *Agrobacterium tumefaciens*–mediated transformation in protoplasts of *Linum usitatissimum* L. and *L. suffruticosum* L. Physiol Plant 82:1, A34, 1991.
22. LH Qing, H Binding. Transformation in protoplast cultures of *Linum usitatissimum* and *L. suffruticosum* mediated with PEG and with *Agrobacterium tumefaciens*. J Plant Physiol 151:479–488, 1997.
23. D Roger, A David, H David. Immobilization of flax protoplasts in agarose and alginate beads: Correlation between ionically bound cell-wall proteins and morphogenetic response. Plant Physiol 112:1191–1199, 1996.
24. XC Zhan, DA Jones, A Kerr. Regeneration of shoots on root explants of flax. Ann Bot 63:297–299, 1989.
25. A McHughen. Flax (*Linum usitatissimum* L.): In vitro studies. In: YPS Bajaj, ed. Biotechnology in Agriculture and Forestry. Vol 10. Legumes and Oilseed Crops 1. Berlin, Heidelberg, New York: Springer, 1990, pp 502–514.
26. M De Cleene, J De Ley. The host range of infectious hairy root. Bot Rev 47:147–194, 1981.
27. AG Hepburn, LE Clarke, KS Blundy, J White. Nopaline T-plasmid, pTiT37, t-DNA insertions into a flax genome. J Mol Appl Genet 2:211–224, 1983.
28. MC Jordan, A McHughen. Transformed callus does not necessarily regenerate transformed shoots. Plant Cell Rep 7:285–287, 1988.
29. A McHughen, R Browne, D Kneeshaw, MC Jordan. Ti-mediated transformation and regeneration of flax plants in vitro. VI Intl Cong of Pl Tiss and Cell Cult. Minneapolis, 1986, p 130.
30. M Jordan, A McHughen. Flax improvement through genetic engineering: Paving the way. Proceedings of the 51st Annual Flax Institute of the U.S., Fargo, ND, 1986, pp 84–89.
31. N Basiran, P Armitage, RJ Scott, J Draper. Genetic transformation of flax (*Linum usitatissimum*) by *Agrobacterium tumefaciens*: Regeneration of transformed shoots via a callus phase. Plant Cell Rep 6:396–399, 1987.
32. MC Jordan, A McHughen. Glyphosate tolerant flax plants from *Agrobacterium* mediated gene transfer. Plant Cell Rep 8:281–284, 1988.
33. XC Zhan, DA Jones, A Kerr. Regeneration of flax plants transformed by *Agrobacterium rhizogenes*. Plant Mol Biol 11:551–559, 1988.
34. XC Zhan, DA Jones, A Kerr. The pTiC58 tzs gene promotes high efficiency root induction by agropine strain 1855 of *Agrobacterium rhizogenes*. Plant Mol Biol 14:785–792, 1990.
35. J-Z Dong, A McHughen. Patterns of transformation intensity on flax hypocotyls inoculated with *Agrobacterium tumefaciens*. Plant Cell Rep 10:555–560, 1991.
36. J-Z Dong, A McHughen. An improved procedure for the production of transgenic flax plants using *Agrobacterium tumefaciens*. Plant Sci 88:61–71, 1993.
37. L Mylnarova, A Pretova. High efficiency *Agrobacterium*-mediated gene transfer to flax. Plant Cell Rep 13:282–285, 1994.
38. B Bretagne-Sagnard, Y Chupeau. Selection of transgenic flax plants is facilitated by spectinomycin. Transgenic Res 5:131–137, 1996.
39. OF Chikrizova, AV Polyakov. Optimization of the conditions for producing transgenic plants of flax resistant to herbicides of the chlorsulfuron group. Sel'skokhozyaistvennaya Biol 3:117–120, 1996.
40. M Koronfel, A McHughen. An efficient regeneration procedure for flax (*Linum usitatissimum* L.): Advances in new methods of genetic engineering in linseed breeding: Bast Fibrous Plants Today and Tomorrow," Breeding, Molecular Biology and Biotechnology Beyond the 21st Century. St. Petersburg, Russia, September 28–30, 1998.

41. CX Yuan, YY Bai. An improved *Agrobacterium*-mediated transformation method for flax (*Linum usitatissimum* cv. Ningya No. II) cotyledon segments. Acta Phytophysiol Sin 19:387–390, 1993.
42. A McHughen, MC Jordan. Recovery of transgenic plants from "escape" shoots. Plant Cell Rep 7:611–614, 1989.
43. A McHughen, FA Holm. Development and preliminary field testing of a glufosinate-ammonium tolerant transgenic flax. Can J Plant Sci 75:117–120, 1995.
44. A McHughen, MC Jordan, G Feist. A preculture period prior to *Agrobacterium* inoculation increases production of transgenic plants. J Plant Physiol 135:245–248, 1989.
45. G Vancanneyt, R Schmidt, A O'Connor-Sanchez, L Willmitzer, M Rocha-Sosa. Construction of an intron-containing marker gene: Splicing of the intron in transgenic plants and its use in monitoring early events in *Agrobacterium* mediated plant transformation. Mol Gen Genet 220:245–250, 1990.
46. J-Z Dong, A McHughen. Transgenic flax plants from *Agrobacterium* mediated transformations: Incidence of chimeric regenerants and inheritance of transgenic plants. Plant Sci 91:139–148, 1993.
47. JC Sanford, TM Klein, ED Wolf, N Allen. Delivery of substances into cells and tissues using a particle bombardment process. Particulate Sci Technol 5:27–37, 1987.
48. TM Klein, ED Wolf, R Wu, JC Sanford. High-velocity microprojectiles for delivering nucleic acids into living cells. Nature 327:70–73, 1987.
49. DE McCabe, WF Swain, BJ Martinell, P Christou. Stable transformation of soybean (*Glycine max*) by particle acceleration. Biotechnology 6:923–926, 1988.
50. P Christou, WF Swain, NS Yang, D McCabe. Inheritance and expression of foreign genes in transgenic soybean plants. Proc Natl Acad Sci USA 86:7500–7504, 1989.
51. TM Klein, EC Harper, Z Svab, JC Sanford, ME Fromm, P Maliga. Stable genetic transformation of intact *Nicotiana* cells by the particle bombardment process. Proc Natl Acad Sci USA 85:8502–8505, 1988.
52. DT Tomes, AK Weissinger, M Ross, R Higgins, BJ Drummond, S Schaaf, J Malone-Schoneberg, M Staebell, P Flynn, J Anderson, J Howard. Transgenic tobacco plants and their progeny derived by microprojectile bombardment of tobacco leaves. Plant Mol Biol 14:261–268, 1990.
53. P Christou, TL Ford, M Kofron. Production of transgenic rice (*Oryza sativa* L.) plants from agronomically important indica and japonica varieties via electric discharge particle acceleration of exogenous DNA into immature zygotic embryos. Biotechnology 9:957–962, 1991.
54. V Vasil, AM Castillo, ME Fromm, IK Vasil. Herbicide resistant fertile transgenic wheat plants obtained by microprojectile bombardment of regenerable embryogenic callus. Biotechnology 10:667–674, 1992.
55. T Wijayanto. Gene transfer to flax (*Linum usitatissimum* L.) using particle bombardment. MSc Thesis, University of Saskatchewan, 1998.
56. T Wijayanto, A McHughen. Genetic transformation of *Linum* by particle bombardment. In Vitro Cell Dev Biol Plant 35:456–465, 1999.
57. M Roberts, A Kumar, R Scott, J Draper. Excision of the maize transposable element Ac in flax callus. Plant Cell Rep 9:406–409, 1990.
58. JG Ellis, EJ Finnegan, GJ Lawrence. Developing a transposon tagging system to isolate rust-resistance genes for flax. Theor Appl Genet 85:46–54, 1992.
59. EJ Finnegan, GJ Lawrence, ES Dennis, JG Ellis. Behavior of modified Ac elements in flax callus and regenerated plants. Plant Mol Biol 22:625–633, 1993.
60. LA Boyd, L Adam, LE Pelcher, A McHughen, R Hirji, G Selvaraj. Characteristics of an *Escherichia coli* gene encoding betaine aldehyde dehydrogenase (BADH): Structural similarity to mammalian ALDHs and a plant BADH. Gene 103:45–52, 1991.
61. KR Jaglo-Ottosen, SJ Gilmour, DG Zarka, O Schabenberger, MF Thomashow. Arabidopsis CBF1 overexpression induces COR genes and enhances freezing tolerance. Science 280:104–106, 1998.
62. Reference deleted.
63. SL MacKenzie. Current developments in the modification of oilseed composition. Outlook Agric 24:213–218, 1995.
64. K Dehesh, A Jones, DS Knutzon, TA Voelker. Production of high levels of 8:0 and 10:0 fatty acids in

transgenic canola by overexpression of CH Fat B2, a thioesterase cDNA from *Cuphea hookeriana*. Plant J 9:167–172, 1996.
65. RK Downey, DC Taylor. Diversification of canola/rapeseed fatty acid supply for the year 2000. Oleagineux Corps Gras Lipids 3:9–13, 1996.
66. G Thompson, R Jain, GG Rowland, D Taylor, SL MacKenzie, A McHughen. Fatty acid modifications in flax. Proc Agric Biotech International Conf. Saskatoon, Canada, 1996.
67. A McHughen, A Mitchell. Preliminary field test results of Roundup resistant transgenic flax. Proceedings of the 53rd Flax Institute of the United States, Fargo, ND, 1990.
68. G Haughn, J Smith, B Mazur, C Somerville. Transformation with a mutant *Arabidopsis* acetolactate synthase gene renders tobacco resistant to sulfonylurea herbicides. Mol Gen Genet 211:266–271, 1988.
69. A McHughen. *Agrobacterium* mediated transfer of chlorsulfuron resistance to commercial flax cultivars. Plant Cell Rep 8:445–449, 1989.
70. A McHughen, FA Holm. Herbicide resistant transgenic flax field test: Agronomic performance in normal and sulfonylurea-containing soils. Euphytica 55:49–56, 1991.
71. A McHughen, FA Holm. Transgenic flax with environmentally and agronomically sustainable attributes. Transgenic Res 4:3–11, 1995.
72. A McHughen, GG Rowland. The effect of T-DNA on the agronomic performance of transgenic flax plants. Euphytica 55:269–275, 1991.
73. A McHughen, GG Rowland, FA Holm, RS Bhatty, EO Kenaschuk. CDC Triffid. Can J Plant Sci 77:641–643, 1997.

# 50
# Antimicrobial Peptides from Macadamia Nuts: Potential Source of Novel Resistance in Transgenic Crops

**Kemal Kazan, John P. Marcus, Ken C. Goulter, and John M. Manners**
The University of Queensland, Brisbane, Queensland, Australia

| | | |
|---|---|---|
| I. | INTRODUCTION | 763 |
| II. | GENETIC ENGINEERING OF PLANTS FOR ENHANCED RESISTANCE TO DISEASES | 763 |
| III. | ANTIMICROBIAL ACTIVITY OF MACADAMIA ANTIMICROBIAL PEPTIDE 1 | 766 |
| IV. | CONSTITUTIVE EXPRESSION OF MACADAMIA ANTIMICROBIAL PEPTIDE 1 IN TRANSGENIC CANOLA | 767 |
| V. | CONCLUSION AND FUTURE PROSPECTS | 769 |
| | REFERENCES | 769 |

## I. INTRODUCTION

Plant diseases caused by fungal, viral, and bacterial pathogens are responsible for most heavy crop losses encountered in agriculture. Large stands of monoculture crops are especially susceptible to diseases that are difficult to control by conventional breeding or other means. Resistance genes currently present in the germplasm may be inadequate for certain diseases. In addition, new races of pathogens that overcome resistance genes may arise in a short period. Historically, heavy applications of chemical fungicides have often been required to minimize the crop losses. However, since chemical fungicides have potentially deleterious environmental and health consequences, it is desirable to enhance the plant's ability to combat diseases. Genetic engineering methods provide an attractive alternative to other more conventional approaches.

## II. GENETIC ENGINEERING OF PLANTS FOR ENHANCED RESISTANCE TO DISEASES

Genes that can be potentially deployed in transgenic plants for enhanced resistance can be broadly classified into three categories. Defense-related genes encode proteins that either (a) act directly against pathogens (e.g., antifungal proteins), (b) act against pathogenicity factors such as toxins, or (c) initiate and enhance the plant's natural defense responses (e.g., product of natural resistance and signaling genes).

Antifungal genes encode proteins that either are induced as part of the plant defense response or are simply present constitutively, acting as antifungal agents. These genes have been discovered by differential screening of messenger ribonucleic acids (mRNAs) in infected resistant tissues that are not expressed constitutively in uninoculated tissue (1) or by simple screening of proteins against fungal spore germination and hyphal growth (2). Some of the antifungal genes encode enzymes such as chitinase and β-1,3 glucanase, which inhibit the fungal growth by hydrolyzing chitin and glucan of the fungal membranes (3). Another group of defense-related proteins with antimicrobial properties are the cysteine-rich antimicrobial peptides (AMPs) commonly found in many plant species (4), which include plant defensins, lipid transfer proteins (LTPs), and thionins, which contain two, three, or four disulfide bridges for stabilization of short α helices (4). Representatives of these proteins have been isolated from the seeds of many plants. Other major groups of antimicrobial proteins are thaumatinlike proteins (TLPs) (5), ribosome inactivating proteins (RIPs) (6), and pathogenesis-related (PR) proteins (7). TLPs, which are peptides of 15–26 kDa, show homology to sweet protein isolated from the fruit of *Thaumatococcus danielle* (8). They include osmotin from tobacco (9), zeamatin from maize (10), and other pathogen-inducible TLPs from cereals (8). RIPs directly affect the protein synthesis machinery in sensitive plant, fungal, and animal cells (11). PR proteins accumulate in plants in response to infection (12). Although the function of most PR proteins is not known, some of them show toxicity to fungi or other pathogens (1).

Constitutive expression of antifungal proteins including glucanase (13), chitinase (14,15), pathogenesis-related protein 1 (16), osmotin (17), defensin (18), thionin (19), and a lipid transfer protein (20) in transgenic plants have been shown to be useful for enhancing resistance against pathogens.

Genes encoding proteins that are effective inhibitors of pathogenicity factors have also been proven useful. A well-known example of this strategy is the expression of a gene for detoxification of tablotoxin produced by the bacterium *Pseudomanas syringae* pv. *tabaci* (causative agent of the wildfire disease) in transgenic tobacco plants. Expression of this gene in tobacco resulted in a strong reduction in symptom formation by bacterial infection (21).

In contrast to the other two categories of defense genes, products of natural resistance (R) and defense signaling genes do not have direct antimicrobial activity against pathogens but function through activation of the plant's own defense system. Recognition of the pathogen-derived signals by the products of resistance genes usually activates downstream defense responses, which may include synthesis of phytoalexins, deposition of cell wall materials, and production of hydrolytic enzymes, reactive oxygen species, and antimicrobial proteins (22,23). However, it is not certain yet whether genes determining resistance in one species can also function in another species against different pathogens. This can only be determined empirically by transformation experiments.

In contrast to race-specific resistance determined by natural resistance genes, protection afforded by AMPs can be broader in spectrum. Therefore, AMPs may be more suitable for deployment especially against necrotrophic pathogens for which no gene-for-gene resistance mechanisms are available. However, regardless of their modes of action, large-scale exposure of a transgenic crop expressing defensive proteins to pathogenic fungi may render these proteins less effective. Therefore, novel forms of resistance should be continuously identified for potential deployment in transgenic crops. This task can be achieved by surveying the plant germplasm for novel plant compounds with antimicrobial activity. Once proteins that show in vitro antimicro-

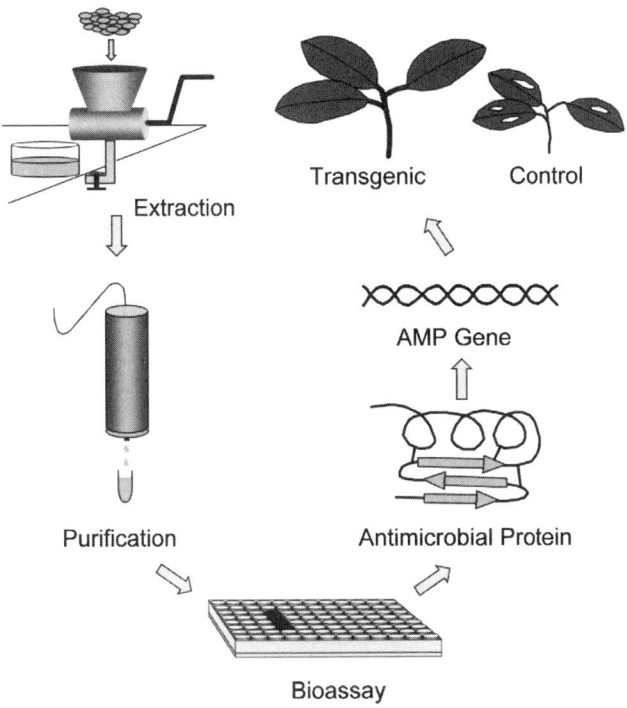

**Figure 1** Strategy for the isolation and application of antimicrobial peptides. Seed extracts with promising activities are purified and tested against pathogenic fungi in bioassays. Proteins with high antimicrobial activity are further characterized and their amino acid sequence determined. Corresponding antimicrobial peptide (AMP) genes are cloned and expressed constitutively in transgenic plants. Finally, transgenic plants expressing AMPs are then evaluated for enhanced resistance to the pathogens.

bial activity against important pathogens have been identified, genes encoding such proteins can be cloned and expressed in transgenic crop plants for enhanced protection (Fig. 1).

Native plants in Australian flora represent a relatively unexplored source of antimicrobial proteins (24,25). To identify such proteins with antimicrobial properties, we have screened crude seed extracts of approximately 250 accessions of Australian native plants. As a result of this survey, protein extracts from *Macadamia integrifolia* (Proteaceae) kernels were shown to exhibit promising antimicrobial activity against some of the economically important diseases. At least four peptides showing antimicrobial activity against a number of pathogens have been purified and their amino acid sequences determined and compared with the sequences of other proteins in the data bases. Two of the macadamia peptides were found to be similar to plant defensin–type AMPs previously identified in other plants. Another AMP showed similarity to portions of vicilin-type seed storage proteins (to be reported elsewhere). One of the peptides, called macadamia antimicrobial peptide 1 (MiAMP1), did not show similarity to any of the previously identified peptides (26). Complementary deoxyribonucleic acid (cDNA) clones corresponding to antimicrobial macadamia peptides have been isolated and fully sequenced to identify signal peptide sequences that mediate proper expression of the peptides.

AMPs from macadamia seeds have been routinely tested against plant and mammalian cells before being transferred to crop plants to ensure that there are no toxicity problems associated with these peptides. Although various other AMPs (e.g., cecropin and derivatives) isolated from sources other than plants (e.g., bacteria) have afforded a reasonable level of protection in transgenic plants (27), commercialization of such plants would be potentially more difficult than that of plants expressing proteins derived from edible plants. It is, therefore, envisaged that transferring antimicrobial genes from edible macadamia nuts to other crop species would be more easily accepted by consumers.

In this chapter, antimicrobial properties of MiAMP1 against important pathogens of canola and its expression in transgenic canola as a novel source of antimicrobial gene for enhanced disease protection are described.

## III. ANTIMICROBIAL ACTIVITY OF MACADAMIA ANTIMICROBIAL PEPTIDE 1

Purification of macadamia antimicrobial peptide 1 (MiAMP1) from the nut kernels of macadamia has been reported previously (26). Purified MiAMP1 was tested in vitro against a number of fungi. Antimicrobial activity assays were done in a medium described previously (2). Briefly, fungal spores or hyphal fragments were incubated in microtiter plates in the presence of varying concentrations of MiAMP1 peptide. A period of 24–48 hours of incubation was used for growth-inhibition measurements. The percentage growth inhibition was measured by following the absorbance at 600 nm and comparing experimental growth rates to those of the control. In these assays, purified MiAMP1 was active against a number of fungi, including *Leptosphaeria maculans* and *Sclerotinia sclerotiorum*, important pathogens of canola and other grains. Germination and growth of *L. maculans* and *S. sclerotiorum* spores were inhibited by low concentrations of MiAMP1. Fig. 2 shows the growth curves of *L. maculans* in the presence of various concentra-

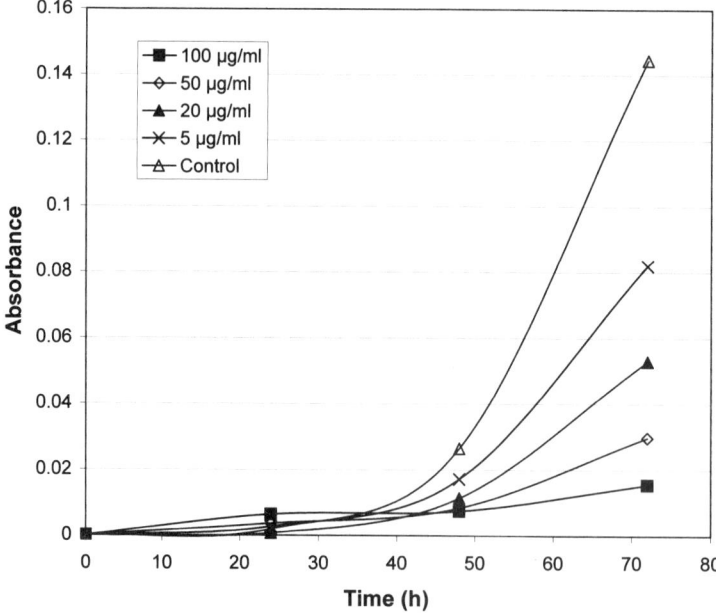

**Figure 2** In vitro inhibitory activity of macadamia antimicrobial peptide 1 (MiAMP1) on the growth of *Leptosphaeria maculans* (causative agent of black leg disease of canola). Fungal mycelial fragments are incubated either in the presence of various concentrations of MiAMP1 peptide or with water as control. Growth inhibition was evident after 40 hours of incubation and was highly correlated with the concentration of peptide used in the bioassay.

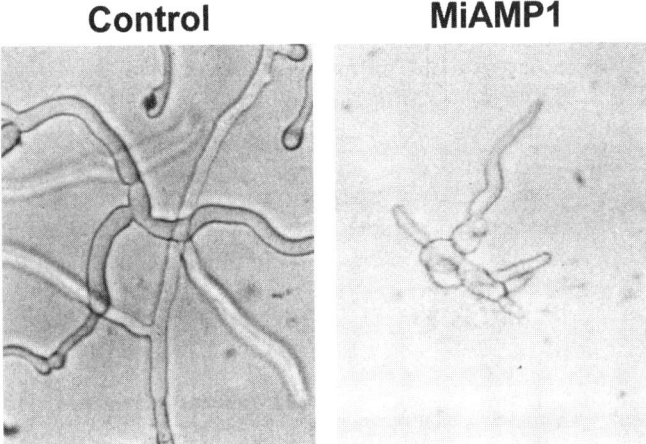

**Figure 3** Inhibition of *Sclerotina sclerotiorum* spores after 24 hours of incubation in the presence of 50 µg/ml MiAMP1 (right) and untreated control (left). Untreated germinating spores remain intact; MiAMP1 causes strong inhibition and distortion of the fungal hyphae.

tions of MiAMP1. Although low concentrations also inhibited the growth of *S. sclerotiorum*, the level of inhibition diminished rapidly after 24 hours. Figure 3 shows the clear inhibition of *S. sclerotiorum* spores after 24 hours of inhibition in the presence of 50 µg/ml MiAMP1, which caused marked morphological distortion of hyphal growth. Mammalian and plant cells treated in vitro with high concentrations of MiAMP1 (100 µM) did not show any loss of viability (26).

The nucleotide and amino acid sequence of MiAMP1 was previously reported (26). The amino acid sequence of MiAMP1 contains a 26-amino-acid signal peptide and a 76-amino-acid mature peptide, which contains six cysteine residues. The highly basic nature of MiAMP1 (pI 10.1) is thought to be important to antimicrobial activity, as positively charged residues often play roles in the interaction of peptides with negatively charged fungal membranes.

## IV. CONSTITUTIVE EXPRESSION OF MACADAMIA ANTIMICROBIAL PEPTIDE 1 IN TRANSGENIC CANOLA

Purified MiAMP1 was active against the most important pathogens of canola (*Brassica napus* L.), a major crop grown for its oil in Canada, India, China, and Europe. Crop losses due to these two pathogens may reach up to 95%, as fungicides are not very effective for eliminating the infections. In addition, breeding for resistance has not been effective because of a lack of resistant germplasm. Therefore, it is envisaged that canola could benefit greatly from the introduction of the MiAMP1 gene.

To express MiAMP1 peptides in transgenic plants, a binary vector harboring MiAMP1 cDNA (including the 26 aa signal peptide) has been constructed under the control of an enhanced double CaMV 35S RNA promoter. This construct also contained a second gene cassette conferring hygromycin resistance as a plant-selectable marker. The 35S-MiAMP1 construct was transferred into canola (*Brassica napus* L.) plants by using an *Agrobacterium tumefaciens*–mediated transformation technique. In these transformation experiments, double haploid lines derived from canola cultivars Westar and Oscar were used. Consequently, 10 Westar and 8 Oscar plants were generated and the transgenic status of the plants was confirmed by molecular analyses. These primary transgenic plants were selfed and their progeny were germinated on a medium containing hygromycin. Hygromycin-resistant progeny from these 18 lines were then further characterized by Northern and Western blot analysis. In these analyses 7 of 10 Westar and 6 of 8 Oscar lines showed a detectable level of MiAMP1 expression. MiAMP1 peptides were detected in the leaf tissue of transgenic plants by Western blot analyses (Fig. 4). These results were then confirmed by Northern blot analyses (Fig. 4). These plants are currently being evaluated for disease resistance against *L. maculans* and *S. sclerotiorum*. MiAMP1 peptides in canola leaves are possibly secreted into intercellular spaces, as previously demonstrated in tobacco transformed with the 35S-MiAMP1 construct (K. Kazan, unpublished). This could be potentially important for inhibiting the two canola pathogens, which invade intercellular spaces during infection. In fact, initial inoculation experiments undertaken with tobacco plants transformed with the 35S-MiAMP1 gene construct suggested a small but significant reduction (20–25%) in lesion development after infection with the tobacco pathogen *Cercospora nicotianae* (causative agent of frog eye disease) (K. Kazan, unpublished).

The use of macadamia antimicrobial peptides as a novel source of resistance in transgenic crop plants represents a potentially promising strategy for the management of plant diseases that are difficult to combat otherwise. However, for successful application of antimicrobial genes and corresponding peptides in engineering disease resistance in transgenic plants, AMPs should be evaluated for certain characteristics. First of all, the peptide should be highly toxic to the pathogen(s) of interest and exhibit minimal toxicity toward plant and mammalian cells. Second, the precursor peptide should be correctly processed and expressed in a suitable location of the plant cells where it is most likely to encounter the pathogen of interest. Expression of the peptide in plant cells should be at sufficiently high concentrations to cause growth inhibition of the pathogen. It would also be a benefit if the expressed peptide were resistant to fungal and plant proteases. MiAMP1 meets most of these criteria. Production of MiAMP1 peptides in canola did not cause any noticeable effect on the phenotype of these plants, confirming the results of our previous in vitro test with tobacco cell cultures, which indicated that MiAMP1 was not toxic to plant cells (26). It was also demonstrated that the signal peptide was correctly removed from the preprotein.

Recently, the three-dimensional (30) structure of MiAMP1 has been resolved by using nuclear magnetic resonance (NMR) spectroscopy techniques. On the basis of the 3D structure of the peptide certain amino acid residues that may be important in providing resistance have been identified. Using this 3D structure we have attempted to increase the activity of MiAMP1 by in vitro mutagenesis. Structural variants of MiAMP1 were generated, expressed in *Escherichia coli*, then purified. Some of these variants displayed greater antimicrobial activity than that of the native

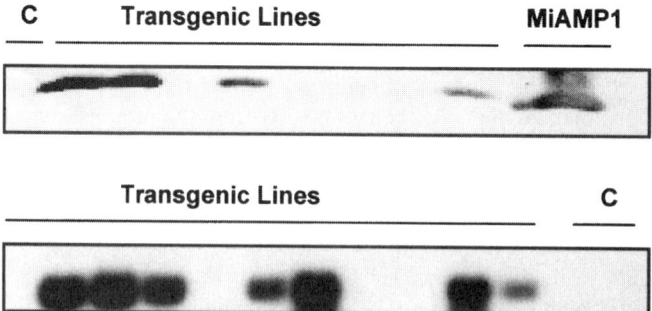

**Figure 4** Detection of MiAMP1 in the leaves of transgenic canola by Western (upper panel) and Northern blot (lower panel) analysis. C, protein or RNA samples from untransformed control plants; MiAMP1, MiAMP1 purified from macadamia nuts.

peptide (to be reported elsewhere). We are now in the process of introducing genes encoding these variants into canola plants.

## V. CONCLUSIONS AND FUTURE PROSPECTS

One of the major goals of plant improvement is to develop crops with resistance to diseases. Recently, it has become apparent that host plant resistance can be enhanced by application of gene transfer technologies. Expression of genes encoding novel antimicrobial proteins in macadamia represents a promising strategy to achieve this goal for important crop plants. Genetic engineering techniques can also be applied to the improvement of disease-resistant macadamia plants. This obviously requires development of a genetic transformation system for macadamia. As a first step toward development of such a system, the transformation techniques used in other tree species should be reviewed with respect to possible adaptation to macadamia transformation. If a genetic transformation system were available for macadamia, antimicrobial peptides from macadamia or other sources could be overexpressed in the seed tissue under the control of either constitutive or seed-specific promoters. Antimicrobial proteins might be useful for increasing the resistance of nuts to anthracnose (caused by *Colletotrichum* spp.) and to macadamia husk spot (caused by *Pseudocercospora* spp.). Expression of these peptides in other tissues may also be effective against *Phytophthora cinnamoni*, which produces a trunk cancer and may kill young macadamia seedlings.

The future strategies involving AMP genes from macadamia will be directed toward improvement of antifungal efficacy against specific pathogens, increasing the spectrum and durability of resistance. Site-directed mutagenesis of macadamia peptides based on 3D structure has already been shown to increase the activity of native MiAMP1, and similar structure-function analyses can be applied for further improvement of other macadamia peptides. Research on the mode of action of macadamia AMPs would be useful in determining the potential target molecules in the pathogen. This knowledge could then be used to design new AMPs and to enhance the effectiveness of chemical fungicides currently used to control fungal pathogens. Durability and spectrum of the AMP-mediated resistance could also be enhanced by combinatorial expression of AMP genes. As the modes of action of the AMPs may differ, this approach could lead to synergistic or additional enhancement of disease resistance in transgenic plants. To complement these approaches, suitable transgene expression strategies should be considered to maximize the level of AMP expression required for antimicrobial efficacy. It may be equally important to control and regulate spatial and temporal accumulation of AMPs in transgenic plants for optimal disease control.

## ACKNOWLEDGMENTS

This work is partially funded by the Grains Research and Development Corporation. We thank Ms. N. Willemsen for her assistance in Western and Northern blot analyses.

## REFERENCES

1. DJ Yun, RA Bressan, PM Hasegawa. Plant antifungal proteins. Plant Breed Rev 14:39–88, 1997.
2. WF Broekart, FRG Terras, BPA Cammue, J Vanderleyden. An automated quantitative assay for fungal growth-inhibition. FEMS Microbiol Lett 69:55–59, 1990.

3. T Boller. Hydrolytic enzymes in plant disease resistance. In: T Kosuge, EW Nester eds. Plant-Microbe Interactions: Molecular and Genetic Aspects. New York: McMillan, 1987, pp 384–413.
4. WF Broekart, BPA Cammue, MFC De Bolle, K Thevissen, GW De Samblanx, RW Osborn. Antimicrobial peptides from plants. Crit Rev Plant Sci 16:297–323, 1997.
5. AJ Vigers, S Wiedemann, WK Roberts, M Legrand, CP Selitrennikoff, B Fritig. Thaumatin-like pathogenesis-related proteins are antifungal. Plant Sci 83:155–161, 1992.
6. R Leah, H Tommerup, I Svendsen, J Mundy. Biochemical and molecular characterization of three barley seed proteins with antifungal properties J Biol Chem 266:1564–1573, 1991.
7. LC Van Loon. Induced resistance in plant and the role of pathogenesis-related proteins. Eur J Plant Pathol 103:753–765, 1997.
8. KC Lin, WR Bushnell, LJ Szabo, AG Smith. Isolation and expression of a host response gene family encoding thaumatin-like proteins in incompatible oat-stem rust fungus interactions. Mol Plant Microbe Interact 9:511–522, 1996.
9. A Stintzi, T Heitz, S Kauffmann, M Legrand, B Fritig. Identification of a basic pathogenesis-related thaumatin-like protein of virus infected tobacco as osmotin. Physiol Mol Plant Pathol 38:137–146, 1991.
10. WK Roberts, CP Selitrennikoff. Zeamatin, an antifungal protein from maize with membrane-permeablizing activity. J Gen Microbiol 136:1771–1778, 1990.
11. T Girbés, JM Ferreras, R Iglesias, L Citores, C de Torre, ML Carbajales, P Jiménez, FM de Benito, R Munoz. Recent advances in the uses and application of ribosome-inactivating proteins from plants. Cell Mol Biol 42:461–471, 1996.
12. HJM Linthorst. Pathogenesis-related proteins of plants. Crit Rev Plant Sci 10:123–150, 1991.
13. M Yoshikawa, M Tsuda, Y Takeuchi. Resistance to fungal diseases in transgenic tobacco expressing the phytoalexin elicitor-releasing factor, β-1,3-endoglucanase from soybean. Naturwissenschaften 80:417–420, 1993.
14. K Broglie, I Chet, M Holliday, R Cressman, P Biddle, S Knowlton, CJ Mauvais, R Broglie. Transgenic plants with enhanced resistance to the fungal pathogen *Rhizoctonia solani*. Science 254:1194–1197, 1991.
15. W Lin, CS Anuratha, K Datta, I Potrykus, S Muthukrishnan, SK Datta. Genetic engineering of rice for resistance to sheath blight. Biotechnology 13:686–691, 1995.
16. DRM Alexander, RM Goodman, M Gut-Rella, C Glascock, K Weymann, L Friedrich, D Maddox, P Ahl-Goy, T Luntz, E Ward, J Ryals. Increased tolerance to two oomycete pathogens in transgenic tobacco expressing pathogenesis-related protein 1a. Proc Natl Acad Sci USA 99:7327–7331, 1993.
17. D Liu, KG Raghothama, PM Hasegawa, RA Bressan. Osmotin overexpression in potato delays the development of disease symptomps. Proc Natl Acad Sci USA 91:1888–1892, 1994.
18. FRG Terras, K Eggermont, V Kovaleva, NV Raikhel, RW Osborn, A Kester, SB Rees, S Torrekens, F Van-Leuven, J Vanderleyen, BPA Cammue, WF Broekaert. Small cysteine-rich antifungal proteins from radish: Their role in host defense. Plant Cell 7:573–588, 1994.
19. MJ Carmona, A Molina, JA Fernandez, JJ Lopez-Fando, F Garcia-Olmedo. Expression of the α-thionin gene from barley in tobacco confers enhanced resistance to bacterial pathogens. Plant J 3:457–462, 1993.
20. A Molina, F Garcia-Olmedo. Enhanced tolerance to bacterial pathogens caused by the transgenic expression of barley lipid transfer protein LPT2. Plant J 12:669–675, 1997.
21. H Anzai, K Yonetyama, I Yamaguchi. Transgenic tobacco resistant to bacterial disease by the detoxification of a pathogen toxin. Mol Gen Genet 219:492–494, 1989.
22. VJ Higgins, L Huogen, T Xing, A Gelli, E Blumwald. The gene-for-gene concept and beyond: Interactions and signals. Can J Plant Pathol 20:150–157, 1998.
23. Y Yang, J Shah, DF Klessig. Signal perception and transduction in plant defense responses. Genes Dev 11:1621–1639, 1997.
24. SJ Harrison, JP Marcus, KC Goulter, JL Green, DJ Maclean, JM Manners. An antimicrobial peptide from the Australian native *Hardenbergia violacea* provides the first functionally characterised member of a subfamily of plant defensins. Aust J Plant Physiol 24:571–578, 1997.
25. DI Last, DJ Llewellyn. Antifungal proteins from seeds of Australian native plants and isolation of an antifungal peptide from *Atriplex nummularia*. NZ J Bot 35:385–394, 1997.

26. JP Marcus, KC Goulter, JL Green, SJ Harrison, JM Manners, J.M. Purification and characterisation of an antimicrobial peptide from *Macadamia integrifolia*. Eur J Biochem 244:743–749, 1997.
27. Y Huang, RO Nordeen, M Di, LD Owens, JH McBeath. Expression of an engineered cecropin gene cassette in transgenic tobacco plants confers resistance to *Pseudomonas syringae* pv. *tabaci*. Phytopathology 87:494–499, 1997.

# 51
# Transgenic Oilseed Brassicas

**Constantine E. Palmer**
*University of Manitoba, Winnipeg, Manitoba, Canada*

**Wilf A. Keller**
*National Research Council, Saskatoon, Saskatchewan, Canada*

|      |                                                              |     |
|------|--------------------------------------------------------------|-----|
| I.   | INTRODUCTION                                                 | 773 |
| II.  | BREEDING OBJECTIVES IN BRASSICA OILSEED IMPROVEMENT          | 774 |
| III. | THE PRODUCTION OF TRANSGENIC OILSEED BRASSICAS               | 775 |
|      | A. Plant Regeneration                                        | 775 |
|      | B. Gene Transfer and Plant Transformation                    | 775 |
|      | C. Selection and Characterization of Transgenics             | 776 |
| IV.  | TRANSGENIC TRAITS IN OILSEED BRASSICAS                       | 777 |
|      | A. Herbicide Tolerance                                       | 777 |
|      | B. Pathogen- and Insect-Tolerant Transgenic Oilseed Brassicas | 778 |
|      | C. Pollination Control Systems                               | 779 |
|      | D. Protein Modification in Transgenic Oilseed Rape           | 779 |
|      | E. Seed Oil Modification in Transgenic Oilseed Brassica      | 781 |
|      | F. Miscellaneous Transgenic Oilseed Brassica                 | 781 |
| V.   | TRANSGENE STABILITY                                          | 781 |
| VI.  | CONCLUSIONS AND FUTURE PROSPECTS                             | 783 |
|      | REFERENCES                                                   | 784 |

## I. INTRODUCTION

Oilseed brassicas are now established as a major source of vegetable oils for a variety of uses (1–3). This group is the third most important vegetable oil source after soybean and palm. This popularity can be attributed to the relatively wide adaptability of these species, which can be grown from subtropical to cool temperate regions. The crop is grown mainly for the oil, which is 40–45% in seed, but the meal is also valuable as a feed source, having about 40% protein on a dry weight basis.

Rapeseed, which includes *Brassica napus, B. rapa, B. juncea, B. carinata, B. nigra,* and *Sinapis alba,* is of Asian origin and contains high levels of erucic acid and glucosinolates. These are both undesirable in the oil and meal for domestic use and as feed. Through conventional breeding, both components were essentially removed or reduced to very low levels and the so-called double low, which refers to low erucic acid and low glucosinolates, characterizes canola (2,4,5). Canola is defined as containing less than 2% erucic acid in the seed oil and less than 30 µmole of aliphatic glucosinolates per gram of oil-free meal. As a consequence, there are several types of these oilseeds grown: high erucic acid rape (HEAR), low erucic acid rape (LEAR), and canola.

Among the brassica oilseeds, *Brassica napus* is the most widely grown, but others such as *B. rapa, B. juncea, B. carinata, B. nigra,* and *Sinapis alba* are also significant oilseeds. Some of these, e.g., *B. juncea,* are better adapted to water stress conditions compared to *B. napus* and *B. rapa,* and significant research efforts are aimed at the development of canola quality in these oilseeds (1). However, a significant amount of noncanola high-quality rapeseed is still grown, some for industrial purposes (6).

As with the improvement of any crop, improvements in brassica oilseeds necessitate the introgression of desirable genes from related species into these crops. The ease of such introgression depends on the relatedness of the species under consideration and the availability of the genes of interest. Where such genes are not available as natural variants in the population or from closely related species, mutagenesis is sometimes employed to increase the level of genetic variation (7). In addition, tissue culture–induced somaclonal variation may be a source of useful genetic variation (8). Plant tissue culture and biotechnology are playing increasingly important roles in the improvement of brassica oilseeds and techniques such as embryo rescue and the generation of doubled haploid plants through microspore culture are valuable adjuncts to brassica oilseed improvement programs. Even with these techniques, progress in brassica oilseed improvement will still be limited by the available genetic variation and that which can be induced by mutagenesis.

Advances in molecular biology, plant cell culture, and genetic engineering have enormously increased the range of genes available for transfer to plant cells. It is now possible to identify and isolate genes from any organism, transfer them to plant cells, and have them produce the desired gene products. As a consequence, the limitations to conventional plant breeding imposed by reduced genetic variability can be addressed through genetic engineering as there are no species or incompatibility barriers. Members of the Brassicaceae are well suited for genetic engineering and in vitro cell and tissue culture, and a variety of transgenic oilseed brassicas have been developed (9–12). The purpose of this review is to highlight recent developments in the production of transgenic oilseed brassicas. No attempt is made to cover the topic exhaustively as there are a number of excellent reviews to which the reader is referred (5,10,11,13,14).

## II. BREEDING OBJECTIVES IN BRASSICA OILSEED IMPROVEMENT

In any crop improvement program, the overall objective is to increase yield and adaptation to environmental conditions. To meet this objective, the genetic variation available must be exploited. By conventional breeding methods, the major limitation to crop improvement is likely to be the availability of germplasm from which desirable genes can be readily introgressed into the plant of interest. Wide hybridizations and mutagenesis are two techniques used to increase the range of germplasm. However, these are not without limitations.

In brassica oilseeds, seed mutagenesis was successful in the production of plants that had altered fatty acid composition in the seed oil (7,15). In these species, breeding objectives such as disease resistance, lodging resistance, increase in oil content, reduction in antinutritional components of the seed oil, and meal, and shatter resistance have been achieved to some degree by conventional breeding methods. However, some objectives such as the manipulation of seed oil fatty acid composition toward homogeneity may not be achievable by conventional means. As an example, the objective of superhigh erucic acid accumulation in rapeseed appears to be limited by the lack of erucic acid esterification of the SN-2 position of the glycerol molecule. This places a theoretical limit of 66% erucic acid on the seed oil. Some unrelated species have the capacity to esterify the SN-2 position with erucic acid quite readily. However, this enzyme system cannot be transferred to rapeseed by conventional means.

With the application of molecular biological techniques, such genes can be isolated, cloned, and transferred to rapeseed. With these techniques, genes from the entire biological world are potentially available for transfer into oilseed brassicas. As a consequence, almost any crop improvement objective is achievable by this extension in the range of available germplasm, coupled with plant biotechnology techniques. Notwithstanding this optimism, most transgenic traits are currently single genes and polygenic traits are more difficult to manipulate.

## III. THE PRODUCTION OF TRANSGENIC OILSEED BRASSICAS

The development of plant tissue culture and plant regeneration techniques applicable to brassicas along with transformation methods forms the basis for the production of these transgenic plants. The general aspects of plant regeneration have been addressed elsewhere in this volume and the treatment here is restricted to the brassicas.

### A. Plant Regeneration

Plant regeneration at high frequency can be achieved from a variety of explants, cotyledonary petioles, hypocotyl segments, roots, stem segments, thin cell layers, leaf disks, floral stalks, embryos, protoplast, and microspores (8,11,12,14,16–22). Frequency of regeneration is under genetic control as some genotypes are highly responsive and others are recalcitrant (23,24). Other factors such as hormone regimen, explant source, and age affect regeneration frequency (25,26).

Plant regeneration occurs mainly through organogenesis for most explants but somatic embryogenesis occurs in a number of cases (27,28). In isolated microspore culture regeneration occurs through direct embryogenesis with all developmental stages of embryogenesis recognizable. Mature embryos germinate into seedlings and entire plants are recovered. In some cases, especially in which the mode of plant regeneration is organogenesis, vitrification reduces the frequency of plant recovery. Manipulation of the medium conditions can reduce the occurrence of vitrified shoots and increase plant recovery (29).

### B. Gene Transfer and Plant Transformation

In all seed brassicas, there are a number of methods utilized for the transfer of genes into plant cells. Most of these methods require an efficient plant regeneration system from either single cells or complex tissues. There are a number of recent reviews and research articles on this topic on brassicas specifically and oilseeds in general, and the reader is directed to these reviews (8,14, 30–33) and other sections of this volume as only a few cases are referred to in this chapter.

The most widely used gene transfer technique in oilseed brassica is *Agrobacterium* species–mediated gene transfer. *Agrobacterium tumefaciens* is usually the vector of choice and is compatible with a number of brassica species and genotypes as well as explants (11,12,14,30,33). Even isolated microspores of oilseed brassicas have been transformed by using this vector (31). The efficiency of *Agrobacterium* species–mediated transformation is dependent on a number of factors (32) but can be as high as 50% (34).

*Agrobacterium rhizogenes* is sometimes used as a vector with relatively high efficiency of gene transfer (35–37), although induction of transformed roots and subsequent shoot regeneration from these roots are usually required.

The use of microinjection to deliver deoxyribonucleic acid (DNA) directly to the nucleus was reported to result in the recovery of transgenic plants from microinjected microspore-derived proembryos (38). This is the only reported case of transgenic plant recovery by DNA microinjection. However, this method is potentially useful because of the precision of DNA delivery and ability to select the targets for microinjection (39).

Particle bombardment is a widely used gene delivery method for species that are not readily amenable to *Agrobacterium* sp. infection. Its use has been successful in the regeneration of transgenic plants from cultured and isolated brassica microspores (40,41). The method is attractive for its high efficiency compared to that of *Agrobacterium* sp., but there is a tendency for multiple copy insertions and potential instability of the transgenics.

## C. Selection and Characterization of Transgenics

For most transformation methods, only a small number of cells are infected with the introduced DNA and the majority of the cell population do not contain the gene of interest. To give these transformed cells a selective advantage for growth and organ differentiation, it is necessary to screen them at an early stage after infection. To this end, some selectable marker is incorporated into the construct along with the gene of interest. Antibiotic resistance genes are commonly used, such as neomycin phosphotransferase (NPTII), which confers resistance to kanamycin and other aminoglycoside antibiotics.

Hygromycin phosphotransferase of *Escherichia coli* origin (hpt), which confers resistance to hygromycin, and aminoglycoside-3-adenyltransferase (aadA), the gene responsible for resistance to spectinomycin and streptomycin, are suitable selection markers for *Brassica napus* (42,43).

Herbicide-resistance genes such as bar gene, which confers resistance to both bialophos, and D+L-phosphinothricin, as well as the pat gene (phosphinothricin acetyltransferase), are frequently used as selectable markers (29). Among the selectable markers used in oilseed brassicas, kanamycin is the predominant marker. However, there is evidence that it limits regeneration in brassicas (44,45). To be effective as a selectable marker, the agent should reduce the incidences of escapes and be nondetrimental to transformed cell growth and morphogenesis, and the gene product should be expressed at a high level. Nontransformed tissue should have very low endogenous tolerance to the selective agent.

A positive selection employing the enzyme phosphomannose isomerase, which allows transformed cells to use mannose as a carbon source, has been applied successfully to brassica oilseed transformation (46).

### 1. Characterization of Transgenics

After in vitro selection and plant regeneration, it is essential that the introduced gene be stably integrated into the host genome and efficiently expressed. There are several methods commonly used to verify gene integration and expression.

1. Putative transformants can be assayed for a selectable marker enzyme system such as neomycin phosphotransferase or phosphinothricin acetyl transferase activity in the tissues.
2. The existence of GUS activity in putative transformed tissues is a commonly used assay. The *E. coli* enzyme β-glucuronidase (uidA) gene is incorporated into the plasmid construct used for transformation. The existence of transformed tissue can be determined by chromogenic assay for the enzyme by using the appropriate substrate. Though this method is frequently used, endogenous GUS activity has been reported in

brassica species (47), and results can be confused by the presence of *Agrobacterium* sp. in the plant tissues.
3. Polymerase chain reaction is used to amplify integrated sequences (48).
4. Well-established Southern and Northern hybridization techniques are used.

Once transformants are identified, stable inheritance can be determined by self-pollination.

Not only is the gene of interest expressed in the transgenic plant, the marker genes are as well. For example, with kanamycin resistance as a selectable marker, the tissues of the plant continue to be resistant to the antibiotic. For environmental reasons (49,50), it may be desirable to remove the marker genes. This has been achieved in brassicas by using a cotransformation procedure (51–53). The rationale is that with two different transfer DNAs (t-DNAs) in the *Agrobacterium* sp., insertion of both followed by segregation at meiosis will yield some plants that have the transgene but without the marker.

It may also be possible to excise the marker directly (54). In addition to the inheritance of selectable marker genes, there is now evidence that during transformation, DNA outside the T-DNA borders can be transferred and integrated into the plant genome (55). This finding has implications for the stability and authenticity of transgenics.

## IV. TRANSGENIC TRAITS IN OILSEED BRASSICAS

### A. Herbicide Tolerance

Herbicide-resistance genes were among the first to be engineered into plant cells and to be commercialized and allow the use of a number of postemergence, nonselective herbicides in weed control. This can be attributed to, among others, the enormous importance of weed control in crop production, existing knowledge of the physiological and biochemical characteristics of herbicide action, and single-gene control of the traits. In addition, herbicide-tolerant traits permit the use of the chemical as a selection agent during transformation. There are four classes of herbicides for which transgenic oilseed resistance has been obtained (Table 1). Glufosinate-ammonia resistance is conferred by the bar and pat genes, the former derived from *Streptomyces hydroscopicus* and the latter from *Streptomyces viridochromogens*. Resistant plants detoxify the herbicide by acetylation, whereas susceptible plants are killed by ammonia toxicity as a result of inhibition of the ammonia-metabolizing enzyme glutamine synthetase (29).

**Table 1** Herbicide-Tolerant Transgenic Oilseed Brassica

| Plant species | Resistant trait | Gene | Reference |
|---|---|---|---|
| *Brassica napus* | Basta, phosphinothricin | bar gene[a] | DeBlock et al., 1989 (29) |
| *Brassica napus* | Basta, phosphinothricin | Pat gene[b] | Oelck et al., 1991 (59) |
| *Brassica napus* | Glyphosate [(*N*-phosphonomethyl)] glycine | EPSPS[c] | Parker et al., 1991 (56) |
| *Brassica napus* | Acetohydroxy acid synthase | csr1-1[d] | Miki et al., 1990 (57) |
| *Brassica napus* | Acetohydroxy acid synthase | csr1-1[d] | Blackshaw et al., 1994 (60) |
| *Brassica napus* | Oxynil herbicides | Nitrilase[e] | Freyssinet et al., 1995 (58) |

[a]*Streptomyces hydroscopicus*.
[b]*S. viridochromogenes*.
[c]5-Enolpyruvylshikimate-3-phosphate synthase gene, CP4 *Agrobacterium* sp. strain.
[d]*Arabidopsis* sp. gene.
[e]*Klebsiella ozaenae*.

Glyphosate- (Round-Up) resistant canola was developed by transfer of a mutant gene of the enzyme 5-enolpyruvylshikimate-3-phosphate (EPSP) synthase of microbial origin to *B. napus* (56). *Brassica napus* plants tolerant to chlorsulfuron and related herbicides were obtained by transformation with a mutant gene for acetolactate synthase isolated from *Arabidopsis* sp. (57).

Transgenic brassica oilseed plants resistant to bromoxynil were also produced by insertion of a bacterial gene (*Klebsiella ozaenae*) that codes for a nitrilase enzyme to detoxify the herbicide (58). Those herbicides are photosynthetic inhibitors of most dicots.

Many of these transgenics have been developed commercially, mostly with *B. napus*, but the availability of efficient transformation methods and gene constructs will allow ready introduction of these genes into *B. rapa*, *B. nigra*, *B. carinata*, and *Sinapis alba*. These resistance traits have demonstrated good yield performance and are inherited in a mendelian fashion (59–62).

## B. Pathogen- and Insect-Tolerant Transgenic Oilseed Brassicas

Oilseed rape production is hampered by a number of pathogens and insect pests, and crop losses can be severe. Chemical and biological controls are not always effective and are sometimes undesirable. Pathogens such as *Leptosphaeria maculans* (Desm) ces. et De not., the causative agent of blackleg; *Sclerotinia sclerotiorum*; and *Pyrenopeziza brassicae* Raw. are not easily controlled by conventional means.

Knowledge of the biological characteristics of these fungi has allowed the use of gene transfer technique in the development of plants with high tolerance to these pathogens. Chitinases that hydrolyze chitin, a major component of fungal cell walls, have been engineered into oilseed rape and the transgenics have exhibited a high degree of pathogen tolerance (Table 2) (63–67). Transgenic rape plants expressing an oxalate oxidase gene derived from *Hordeum vulgare* roots showed enhanced resistance to *Sclerotinia* sp. The enzyme destroys the oxalate, which is required

**Table 2** Transgenic Pathogen- and Insect-Tolerant Oilseed Rape

| Plant species | Resistant trait | Gene | Reference |
| --- | --- | --- | --- |
| *Brassica napus* | Blackleg, *Sclerotinia* sp. | Chitinase | Grison et al., 1996 (63) |
| *Brassica napus* | Blackleg, *Sclerotinia* sp. | Chitinase | Grezes-Besset et al., 1995 (64) |
| *Brassica napus* | Blackleg | Pea defense gene DRR206 | Wang et al., 1999 (65) |
| *Brassica napus* | Fungal | 5'35S Bean chitinase | Broglie & Broglie, 1994 (66) |
| *Brassica napus* | Damping off | Bean endochitinase | Broglie et al., 1991 (67) |
| *Brassica napus* | Blackleg | PR peroxidase of *Stylosanthes humilis* | Kazan et al., 1998 (70) |
| *Brassica napus* | *Sclerotinia* sp. | Oxalate oxidase | Freyssinet et al., 1995 (69) |
| *Brassica napus* | *Sclerotinia* sp. | Oxalate oxidase | Thompson et al., 1995 (68) |
| *Brassica napus* | BWYV | Viral coat protein | Laucke et al., 1995 (71) |
| *Brassica napus* | CaMV | Viral coat protein | Hervé et al., 1993 (72) |
| *Brassica napus* | CaMV, TYMV, TMV | None | Spak et al., 1991 (73) |
| *Brassica napus* | TYMV | Viral noncoding region | Zaccomer et al., 1993 (37) |
| *Brassica napus* | Insect, *Plutella* sp., *Xylostella* sp. | Bt CryI Ac | Stewart et al., 1996 (80) |
| *Brassica napus* | Coleopteran insects | Proteinase inhibitor OG1, CII | Bonade-Bottino et al., 1998 (81) |
| *Brassica napus* | Coleopterans | Cysteine proteinase inhibitor | Girard et al., 1998 (82) |

for fungal infection (68,69). A peroxidase gene from *Stylosanthes humilis* conferred some measure of resistance to blackleg (70).

Several reports indicated that viral diseases of oilseed rape could be alleviated by expression of coat protein in transgenic plants (37,71–74).

The expression of insecticidal proteins such as those derived from *Bacillus thuringiensis (Bt)*, cholesterol oxidases, and proteinase inhibitors in transgenic plants is of increasing value in the management and control of plant insect pests (75–79). Some of these proteins, especially *Bt*-derived proteins and the proteinase inhibitors, have provided some degree of control of pests such as diamond back moth, *Plutella xylostella* L., and flea beetle, *Phyllotreta* spp., in transgenic oilseed rape (Table 2) (80–82).

## C. Pollination Control Systems

In oilseed brassica and other crop species, one objective is the development of hybrids that are likely to be superior to inbred lines in agronomic and other characteristics. For such development, a pollination control system is required. Although self-incompatibility and nuclear and cytoplasmic male sterility (CMS) exist as pollination control systems, these are not always stable and may not be available in the desired germplasm. A number of pollination control systems that use molecular biology and plant biotechnology methods have been reported (83–86).

In *B. napus,* male sterility was introduced by expression of ribonuclease (RNase) genes from *Aspergillus oryzae* and *Bacillus amyloliquefaciens* specifically in the tapetal cells of anthers, thus disrupting normal pollen development (84,87). These genes were linked to the bar gene, which confers herbicide resistance, allowing rogueing of the male fertile plants. Male fertility was restored by expressing an RNase inhibitor gene, barstar, in other plants (85). This represents a very elegant system of pollination control and is well established in oilseed brassica. Since CMS is associated with mitochondrial function, attempts have been made to transfer the mitochondrial gene to produce male sterile oilseed rape plants, as with the T-urf13 gene of CMS-T maize (86). In other transgenic oilseed brassica disruption of reproductive development has implications for pollination control (88–90).

## D. Protein Modification in Transgenic Oilseed Rape

There is considerable interest in the use of plants for the production of specialized proteins and other components of industrial, pharmaceutical, and nutritional interest. Seed storage protein modification for enhanced economic value as feed or food is also of interest. The production of plant-based recombinant enzymes for industrial uses is advantageous in low cost of production, stability, and potential for long-term storage. To exploit plants as a vehicle for such production it is necessary to manipulate the appropriate metabolic pathways and (or) the heterologous expression of desirable genes from other species, in plants.

In oilseed brassica, the oil-free meal is a useful source of animal feed and attempts have been made to modify the protein quality, especially the content of lysine and methionine (Table 3) (91–98). Stable increases in methionine content of *B. napus* were achieved by transfer of the methionine-rich 2S albumin protein gene from *Bertholettia excelsa* (Brazil nut) to rapeseed (96), and significant increases in lysine levels were reported by the expression of bacterial genes for dihydrodipicolinic acid synthase and aspartokinase in rapeseed (98). Other modifications to seed storage proteins have been noted in transgenic brassica (99–101). The oilbody proteins, oleosins, which may constitute 8–20% of the total seed protein (102) in oil seeds, are ideal targets for the production of fusion proteins as this simplifies both isolation and purification. There are a number of reports demonstrating the heterologous expression of oleosin genes in oilseed rape of both

**Table 3** Transgenic Oilseed Rape Engineered for Modified Protein Production

| Plant species | Modified trait | Gene | Reference |
|---|---|---|---|
| *Brassica napus* | Increased lysine, cysteine, methionine | Antisense cruciferin | Kohno-Murase et al., 1995 (91) |
| *Brassica napus* | Increased lysine, cysteine, methionine | Antisense cruciferin | Kohno-Murase et al., 1995 (92) |
| *Brassica napus* | Increased lysine, cysteine, methionine | Antisense napin | Kohno-Murase et al., 1994 (93) |
| *Brassica napus* | Storage protein | Chimeric 2S albumin | Denis et al., 1995 (94) |
| *Brassica napus* | Storage protein | Chimeric 2S albumin | Denis et al., 1995 (95) |
| *Brassica napus* | Increased methionine in storage protein | Brazil nut albumin | Altenbach et al., 1992 (96) |
| *Brassica napus* | Increased methionine in storage protein | 2S albumin | deClercq et al., 1990 (97) |
| *Brassica napus* | Increased lysine | Bacterial DHDPS and AK[a] | Falco et al., 1995 (98) |
| *Brassica napus* | 2S seed protein | napA promoter | Stålberg et al., 1998 (99) |
| *Brassica napus* | 2S seed protein | Modified napA promoter | Stålberg et al., 1996 (100) |
| *Brassica napus* | Oilbody protein | Soybean oleosin | Sarmiento et al., 1997 (103) |
| *Brassica napus* | Oilbody protein | *Arabidopsis* sp. oleosin promoter | Plant et al., 1994 (104) |
| *Brassica napus* | Oilbody protein | Oleosins | Hills et al., 1993 (105) |
| *Brassica napus* | Oilbody proteins | Maize oleosin | Lee et al., 1991 (106) |
| *Brassica carinata* | Oleosin | Oleosin-hirudin | Chaudhary et al., 1998 (107) |
| *Brassica napus* | Oilbody protein | β-Glucuronidase | Kühnel et al., 1996 (108) |
| *Brassica napus* | Oleosin | Hirudin | Parmenter et al., 1995 (109) |
| *Brassica napus* | Seed oilbodies | β-Glucuronidase | Rooijen & Moloney, 1995 (110) |
| *Brassica napus* | Seed protein | Phytase[b] | Verwoerd and Pen, 1996 (111) |
| *Brassica napus* | Root proliferation, peroxidase production | None[c] | Agostini et al., 1997 (112) |
| *Brassica napus* | 2S storage protein | Enkephalins | Vanderkerckhove et al., 1989 (101) |
| *Brassica napus* | Low glucosinolates | Tryptophan, decarboxylase | Chavadej et al., 1994 (113) |

[a]Genes derived from Corynebacterium dap A.
[b]Genes derived from *Aspergillus niger*.
[c]Transformation with *Agrobacterium rhizogenes*.

commercial and experimental interest (103–110). Significant production and recovery of the anticoagulant protein hirudin and the enzyme glucuronidase were obtained by oleosin fusion protein expression in oilseed rape (105–108). Increase in phytase occurred in seeds of *B. napus* expressing a fungal phytase gene (111), and levels of indoleglucosinolates were reduced by modification of tryptophan metabolism (112). High levels of phytase and glucosinolates are nutritionally undesirable in the seed meal, and with genetic manipulation significant reductions have been achieved. Even without specific targeting, protein modification can occur in transgenic oilseed rape, as in *B. napus* transformed with *Agrobacterium rhizogenes*, which showed increased levels of root peroxidase (113). Use of the appropriate promoters and gene modifications is required for high-level gene expression, as in some cases the targeted changes were not achieved (114).

### E. Seed Oil Modification in Transgenic Oilseed Brassica

Improvement in total seed oil yield increase and qualitative differences in the oils are important aspects of oilseed rape production. Although conventional breeding is useful in these improvements, the advent of plant molecular biology and advances in lipid biochemistry now form the bases for plant lipid manipulation so that the potential exists to synthesize almost any storage lipid in plant cells (5,115–120). Among the oilseed crops, brassica species are particularly amenable to genetic manipulation of the seed oil as they, in most cases, respond well to in vitro culture techniques, making regeneration of transgenic plants from transformed cells relatively easy. Also, very good germplasms have been accumulated over decades of conventional breeding. For these and other reasons, oilseed rape is regarded as a model crop for genetic manipulation (120), and it was the first transgenic crop with altered seed composition grown commercially in the United States (5). It would appear that metabolic engineering of oilseed rape toward the production of seed oils for industrial, nutritional, nutraceutical, and other uses is poised for significant advances as our understanding of the genetic mechanisms and regulatory control of fatty acid diversity in plants continue to increase. There are a number of excellent reviews of seed oil modification in crop plants (5,13,117,118,120–123) to which the reader is referred as experimental details are not repeated here. The purpose is to highlight in tabular form the current status of seed oil modification in transgenic oilseed rape (Table 4).

Even though the potential exists to produce oils of any fatty composition in oilseed rape, there are limitations yet to be resolved. In high-laurate transgenic rapeseed, very low levels of laurate were incorporated into the SN-2 position of the glycerol molecule. This places a theoretical limit on laurate oils at 66% (12,124). A similar situation exists with the quest for superhigh erucic acid rapeseed, because even when trierucin could be detected in the seed oil, the content of this fatty acid seldom exceeded 60% (125–127). An understanding of the basis for this limitation will be key to further advances in qualitative modification of seed oil composition.

### F. Miscellaneous Transgenic Oilseed Brassicas

The category of miscellaneous transgenic oilseed brassica includes transgenics that are primarily of experimental interest, although some are of potential commercial value (Table 5) (147–157). The use of antisense technology to reduce seed chlorophyll content is applicable to the removal of the green seed problem in canola (147). Some transgenic oilseed brassica are of value in the decontamination of polluted land sites, as genes for metal accumulation and chemical detoxification can be expressed in these species (148). The studies on promoters and gene expression in transgenic oilseed rape also have commercial implications (149,150). Other transgenics are useful in studies of plant hormone metabolism and the expression of specific enzymes (151–157).

## V. TRANSGENE STABILITY

The general assumption is that once the introduced gene is integrated into the genome, its expression and stability are assured. A body of evidence now indicates that this is not the case and transgenes can be silenced and even endogenous gene expression can be affected by the introduced gene (158–162). As a result of this phenomenon, the transgene phenotype may not be distinguishable from the nontransgenic. This condition presents some difficulty in monitoring the

**Table 4**  Oilseed Rape with Genetically Modified Seed Oil Composition

| Plant species | Modified trait | Gene | Reference |
|---|---|---|---|
| *Brassica napus* | Laurate accumulation | 12:0 acyl carrier thioester (BTE) | Voelker et al., 1992 (128) |
| *Brassica napus B. rapa* | Altered saturated fatty acids | Stearoyl-acyl carrier protein desaturase antisense | Knutzon et al., 1992 (129) |
| *Brassica napus* | Medium-chain fatty acid modification | Medium-chain hydrolase and acyl carrier Protein | Safford et al., 1993 (130) |
| *Brassica napus* | Medium-chain fatty acid modification | Malonyl CoA-ACP transacylase[a] (*Escherichia coli*) | Verwoert et al., 1994 (131) |
| *Brassica napus* | Increased palmatate | 16:0 ACP thioesterase[b] | Jones et al., 1995 (132) |
| *Brassica napus* | Erucic acid content | Erucoyl-CoA[c] acyltransferase | Wolter et al., 1995 (146) |
| *Brassica napus* | Polyunsaturates | Fatty acid desaturase and Acyl-ACP thioesterase | Hitz et al., 1995 (133) |
| *Brassica napus* | Erucic acid content | L-PAT acyltransferase[d] | Lassner et al., 1995 (121) |
| *Brassica napus* | eicosenoic acid content | SN-1 acylglycero-3-P[e] acyltransferase | Weier et al., 1998 (134) |
| *Brassica napus* | Erucic acid content | SN-2 acyltransferase gene (SLC1-1) | Zou et al., 1997 (126) |
| *Brassica napus* | 8:0 and 10:0 fatty acid levels | ch Fat B2 thioesterase | Dehesh et al., 1996 (135) |
| *Brassica napus* | Erucic acid content | LPAAT (*Limnanthes douglasii*) | Weier et al., 1997 (136) |
| *Brassica napus* | Stearate levels | Acyl-AcP thioesterase (Mangosteen) | Hawkins and Kridl, 1998 (137) |
| *Brassica napus* | Stearate levels | Modified acyl-acyl carrier protein thioesterase | Facciotti et al., 1999 (138) |
| *Brassica napus* | VLCFAS | β-Ketoacyl-CoA synthase (KCS) (*Brassica napus* | Han et al., 1998 (139) |
| *Brassica napus* | Fatty acid composition | L-PAAT (*Limnanthes douglasii*) | Wilmer et al., 1998 (140) |
| *Brassica napus* | Erucic acid content | LPAAT | Münster et al., 1998 (141) |
| *Brassica napus* | Saturated fatty acids | Stearoyl-acyl carrier protein desaturase (SAD) (flax) | Jain et al., 1998 (142) |
| *Brassica napus* | Unsaturated fatty acids | $\Delta^5$ desaturase (*Mortierella alpina*) | Knutzon et al., 1998 (143) |
| *Brassica napus* | Fatty acid composition | Homomeric acetyl CoA carbonylase (*Arabidopsis* sp.) | Roesler et al., 1997 (144) |
| *Brassica napus* | Fatty acid composition | *Escherichia coli* fabH 3-ketoacyl-acyl carrier protein synthase III | Verwoert et al., 1995 (145) |

[a] *Escherichia coli* gene.
[b] *Cuphea hookeriana* gene.
[c] *Limnanthes douglasii* gene.
[d] *L. douglasii* gene.
[e] *E. coli* gene.

**Table 5** Miscellaneous Transgenic Oilseed Brassica

| Plant species | Modified trait | Gene | Reference |
|---|---|---|---|
| *Brassica napus* | Seed chlorophyll content | cab Antisense | Morissette et al., 1998 (147) |
| *Brassica napus* | Cadmium resistance | Human metallothionein II | Misra and Gedama, 1990 (148) |
| *Brassica napus* | GUS expression tapetal cells | Oleosin-like promoter + GUS | Hong et al., 1997 (149) |
| *Brassica napus* | Pollen expression of GUS | PGU[b] promoter + GUS | Hong et al., 1997 (150) |
| *Brassica napus* | Kanamycin resistance | NPTII | deVries and Wackernagel, 1998 (151) |
| *Brassica napus* | Modified cytokinin levels | tzs | Roeckel et al., 1998 (152) |
| *Brassica napus* | Modified cytokinin levels | AT2S1-ipt | Roeckel et al., 1997 (153) |
| *Brassica napus* | Agravitropism | R1-plasmid (*Agrobacterium rhizogenes*) | Iversen et al., 1996 (154) |
| *Brassica napus* | Reduced gravitropism | Rt-plasmid *A. rhizogenes* | Legue et al., 1996 (155) |
| *Brassica napus* | GS overexpression | Cytosolic GS[a] (soybean) | Downs et al., 1994 (156) |
| *Brassica napus* | GUS | Mannopine synthase + GUS | Stefanov et al., 1994 (157) |

[a]Transformation with *Agrobacterium rhizogenes*.
[b]Polygalacturonase.

transgenics under growth conditions and for reliable production of gene products (163). Such silencing can be the result of transcriptional or posttranscriptional events (158,160,161) and can be influenced by environmental conditions such as heat stress (164). It is generally accepted that transgene inactivation is related to gene copy number, extensive DNA methylation of the promoter, and (or) transgene, and the homology between transgenes and endogenous genes (160). The latter, which is referred to as *cosuppression*, can lead to inactivation of both the transgene and the homologous endogenous gene.

Although transgene inactivation has been extensively studied in many species, there are very few reports for *brassica* species. Both physical loss of the transgene and loss of gene expression were observed in $F_1$ lines of transgenic *B. napus* (165). In a population of transgenic *B. napus*, some gene inactivation of the transgene for phosphinothricin (PPT) resistance was observed after selfing and backcrossing with nontransgenic rapeseed plants (166). Although transgene stability was not evaluated, differences in expression were noted for herbicide-resistant and marker traits in oilseed rape during two cycles of seed production (167). Homology-dependent transgene inactivation of an endogenous gene was reported in *B. napus*, in which silencing of the S-locus genes in self-incompatible plants caused a breakdown of self-incompatibility (168). This result was related to the level of DNA methylation of the endogenous gene.

## VI. CONCLUSIONS AND FUTURE PROSPECTS

There is now increased momentum in the drive to commercialize transgenic oilseed brassica, at least in North America. A significant portion of the cultivated brassica oilseed acreage in 1999 was seeded to transgenic oilseed rape (169) that had a number of genetically modified characteristics. Certain traits are single-gene-mediated and are relatively easy to manipulate. Consequently, herbicide-tolerant and insect- and disease-resistant transgenics will be the most prominent ones cultivated. Fatty acid and storage protein modifications in oilseed rape are now receiving extensive attention as a result of advances in lipid biochemistry, gene isolation, and manipulation, and cell culture techniques. With these advances, prospects are good for the production of a wide variety of plant-derived oils and other lipids for nutritional, nutraceutical, and industrial uses. These plants will also be useful as bioreactors for the production of medicinal and pharmaceutical proteins because of the advantages mentioned in the previous sections.

To date, no transgenic oilseed brassicas expressing polygenic traits such as those for flowering time, vernalization, and plant architecture have been uncovered. To improve plant performance, this point should be addressed, and the *Arabidopsis thaliana* model should prove useful in the manipulation of these traits once genes are mapped, identified, and assigned function in this species.

The production of oils of homogenous fatty acid composition is far from being assured in transgenic oilseed rape. There are a number of examples of plants with seed oil of close to homogenous fatty acid composition, and it may be fruitful to examine the biochemical and physiological bases for oil accumulation in such plants to identify any limitations in transgenic oilseed rape fatty acid biosynthesis and accumulation.

Compared to conventional breeding, the production of transgenic oilseed brassicas is quite rapid. Depending on the species and transformation method, plants can be obtained in 3–4 months. Since gene insertion is random, it is essential to produce a large number of independent transformants for transgene selection. Therefore, the efficiency of plant regeneration from a wide range of genotypes should be improved. It is noteworthy that of 29 agronomically important transgenic traits listed by Poulsen in 1996 (14) 19 were based on the *B. napus* cv. 'Westar.' This argues for a broadening of the germplasm used for the generation of transgenics.

To be acceptable, transgenics must have no negative attributes and be the same as the nontransgenic except for the engineered trait. There is evidence of changes in chemical composition of plant tissues as a result of unrelated genetic manipulation (170).

The stability of gene expression in transgenics is a fundamental issue for commercial acceptance. Gene silencing and inactivation occur frequently in transgenics and in some instances may not be evident for several generations (171). This phenomenon exists in transgenic oilseed rape but not to the extent observed in other species. Nevertheless, efforts should be made to improve transgene stability in oilseed brassicas to allow reliable and optimal gene expression. There is now evidence that such gene silencing may be reduced by gene manipulation (172,173). The wide adaptability of these oilseeds, coupled with efficient methods for gene transfer and plant recovery, make these species attractive for the production of both commercial and experimental transgenics, and the potential exists to diversify the production of a wide range of novel compounds such as industrially useful polymers by targeting transgenes to specific tissues and organs of oilseed rape (174–176).

## REFERENCES

1. G Rakow. Developments in the breeding of edible oil in other *Brassica* species. Proceedings of 9th International Rapeseed Congress, Cambridge, July 4–7, 1995, pp 401–406.
2. G Röbbelen. Rapeseed in a changing world: Plant production potential. In: DI McGregor, ed. Proceedings of 8th International Rapeseed Congress, Saskatoon, Canada, 1991, pp 29–38.
3. RA Carr, BE McDonald. Rapeseed in a changing world: Processing and utilization. In: DI McGregor, ed. Proceedings of 8th International Rapeseed Congress, Saskatoon, Canada, 1991, pp 39–56.
4. L Bush, V Gunter, T Montele, M Tachikawa, K Tanaka. Socializing nature: Technoscience and the transformation of rapeseed into canola. Crop Sci 34:607–614, 1994.
5. DJ Murphy. Biotechnology of oil crops. In: DJ Murphy ed. Designer oil crops: Breeding, Processing and Biotechnology. New York: VCH Publishers, 1994, pp 219–251.

6. W Friedt, W Lühs. Development in the breeding of rapeseed oil for industrial purposes. Proceedings of 9th International Rapeseed Congress, Cambridge, July 4–7, 1995, pp 401–406.
7. G Rakow. Selektion auf linol—und linolen—sauregehalt in Rapssamen nach mutagener. Dehandleurg. Z. Pflanzenzuchtg 69:62–82, 1973.
8. CE Palmer, WA Keller. In vitro culture of oilseeds. In. K Vasil, TA Thorpe, eds. Plant Cell and Tissue Culture. Dordrecht: Kluwer Academic, 1994, pp 413–455.
9. C Sjödin. Brassicaceae, a plant family well suited for modern biotechnology. Acta Agric Scand 42: 197–207, 1992.
10. RK Downey, WA Keller. Modifying oil and protein crop plants: New concepts and approaches. In: DR Buxton, R Shibles, RA Forsberg, BL Blad, RH Asay, GM Paulsen and RF Wilson, eds. Proceedings of International Crop Science Congress, Ames, IA, July 1994, pp 655–663.
11. PJ Dale, JA Irwin. The production and development of transgenic plants. Proceedings of 9th International Rapeseed Congress, Cambridge, July 9–11, 1995, pp 760–765.
12. O Damgaard, LA Jensen, OS Rasmussen. *Agrobacterium tumefaciens* mediated transformation of *Brassica napus* winter cultivars. Transgenic Res 6:279–288, 1997.
13. DJ Murphy. Production of novel oils in plants. Curr Opin Biotechnol 10:175–180, 1999.
14. GB Poulsen. Genetic transformation of *Brassica*. Plant Breed 115:209–225, 1996.
15. DL Auld, MK Keikkinen, DA Erickson, JL Sernyk, JE Ronoro. Rapeseed mutants with reduced levels of polyunsaturated fatty acids and increased levels of oleic acid. Crop Sci 32:657–661, 1992.
16. MM Moloney, JM Walker, KK Sharma. High frequency transformation of *Brassica napus* using *Agrobacterium vectors*. Plant Cell Rep 8:238–242, 1989.
17. JE Hackey, KK Sharma, MM Moloney. Efficient shoot regeneration of *Brassica campestris* is using cotyledon explants cultured in vitro. Plant Cell Rep 9:549–554, 1991.
18. PA Lazzeri, JM Dunwell. In vitro shoot regeneration from seedling segments of *Brassica aleracea* and *B. napus*. Ann Bot 54:341–350, 1984.
19. GS Khehra, RJ Mathias. The interaction of genotype, explant and media on the regeneration of shoots from complex explants of B. napus L. J Exp Bot 43:1413–1418, 1992.
20. V Kushinov, K Koivu, A Kanerva, E Pehu. *Agrobacterium tumefaciens*–mediated transformation of greenhouse grown *Brassica rapa* ssp. *oleifera*. Plant Cell Rep 18:773–777, 1999.
21. K Glimelius. High growth rate and regeneration capacity of hypocotyl protoplasts in some Brassicaceae. Physiol Plant 61:38–44, 1984.
22. R Lichter. Induction of haploid plants from isolated pollen of *Brassica napus*. Z Pflanzen Physiol 105:427–434, 1982.
23. SB Narasimhulu, VL Chopra. Species specific shoot regeneration response of cotyledonary explants of brassicas. Plant Cell Rep 7:104–106, 1988.
24. M Murata, JJ Orton. Callus initiation and regeneration in *Brassica* species. Plant Cell Tissue Org Cult 11:111–123, 1987.
25. J Ovesna, L Ptacek, Z Opatray. Factors influencing the regeneration capacity of oilseed rape and cauliflower in transformation experiments. Biol Plant 35:107–112, 1993.
26. T Takasaki, K Hatakeyama, K Ojima, M Wanatabe, K Toriyama, K Hinata. Effect of various factors (hormone combinations, genotypes and antibiotics) on shoot regeneration from cotyledon explants in *Brassica rapa* L. Plant Tissue Cult Lett 13(2):177–180, 1996.
27. AJ Graves, JK Hemphill, R Ram. Somatic embryogenesis in oilseed Brassica. Proceedings of the 8th International Rapeseed Congress, Saskatoon, Canada, July 9–11, 1991, pp 1801–1808.
28. EC Pua. Somatic embryogenesis and plant regeneration from hypocotyl protoplasts of *B. juncea* L. Czern and Cross. Plant Sci 68:231–238, 1990.
29. M DeBlock, D Debrouwer, P Tenning. Transformation of *Brassica napus* and *Brassica oleracea* using *Agrobacterium tumefaciens* and the expression of the bar and neogenes in the transgenic plants. Plant Physiol 91:694–701, 1989.
30. PJ Dale, JA Irwin. Transformation of oil crops. In: DJ Murphy, ed. Designer Oil Crops: Breeding, Processing and Biotechnology. New York, VCH, 1994, pp 195–218.
31. M Dormann, HM Wang, N Datla, AMR Ferrie, WA Keller, MM Oelck. Transformation of freshly isolated *Brassica* microspores and regeneration of fertile homozygous plants. In: D Murphy, ed. Proceedings of 9th International Rapeseed Congress. Vol. 3. Cambridge, 1995, 816–818.

32. T Takasaki, K Hatakeyama, K Ojima, M Watanabe, K Toriyama, K Hinata. Factors influencing *Agrobacterium*-mediated transformation of *Brassica rapa*. Breed Sci 47:127–134, 1997.
33. R Walden, R Wingender. Gene-transfer and plant regeneration techniques. Trends Biotechnol 13: 324–331, 1995.
34. V Babic, RS Datla, GJ Scoles, WA Keller. Development of an efficient *Agrobacterium*-mediated transformation system for *Brassica carinata*. Plant Cell Rep 17:183–188, 1997.
35. ME Boulter, E Croy, P Simpson, R Shields, RRD Croy, AH Shirsad. Transformation of *Brassica napus* L. (oilseed rape) using *Agrobacterium tumefaciens* and *Agrobacterium rhizogenes*: A comparison. Plant Sci 70:91–99, 1990.
36. O Damgaard, OS Rasmussen. Direct regeneration of transformed shoots in *Brassica napus* from hypocotyl injected with *Agrobacterium rhizogenes*. Plant Mol Biol 17:1–8, 1991.
37. B Zaccomer, F Cellier, JC Boyer, AL Haenni, M Tepfer. Transgenic plants that express genes including the 3' untranslated region of the turnip yellow mosaic virus (TYMV) genome are partially protected against TYMV infection. Gene 136:87–94, 1993.
38. G Neuhaus, G Spangenberg, OM Sheid, HG Schweiger. Transgenic rapeseed plants obtained by microinjection of DNA into microspore-derived embryoids. Theor Appl Genet 75:30–36, 1987.
39. E Jones-Villeneuve, B Huang, I Prudhomme, S Bird, R Kemble, J Hattori, B Miki. Assessment of microinjection for introducing DNA into uninuclear microspores of rapeseed. Plant Cell Tissue Org Cult 40:97–100, 1995.
40. JL Chen, WD Beversdorf. A combined use of microprojectile bombardment and DNA inbibition enhances transformation frequency of canola (*Brassica napus* L.). Theor Appl Genet 88:187–192, 1994.
41. H Fukuoka, T Ogawa, M Matsuoka, Y Ohkawa, H Yano. Direct gene delivery into isolated microspores of rapeseed (*Brassica napus* L.) and the production of fertile transgenic plants. Plant Cell Rep 17: 323–328, 1998.
42. P Van den Elzen, KY Lee, J Townsend, J Bedrock. A chimaeric hygromycin resistance gene as a selectable marker in plant cells. Plant Mol Biol 5:299–302, 1985.
43. M Schröder, C Dixelius, L Råhlen, K Glimelius. Transformation of *Brassica napus* by using the aadA gene as selectable marker and inheritance studies of the marker genes. Physiol Plant 92:37–46, 1994.
44. V Gupta, GL Sita, MS Shaila, V Jagannathan. Genetic transformation of *Brassica nigra* by *Agrobacterium* based vector and direct plasmid uptake. Plant Cell Rep 12:418–421, 1993.
45. JE Thomzik, R Hain. Transgenic *Brassica napus* plants obtained by cocultivation of protoplasts with *Agrobacterium tumefaciens*. Plant Cell Rep 9:233–236, 1990.
46. K Bojsen, I Donaldson, A Haldrup, M Joersboe, D Kreiberg, J Nielsen, FT Okkels, SG Petersen. Mannose and xylose based positive selection. PCT Int. Patent application, WO94/20627, 1994.
47. L Hodal, A Bochardt, JE Nielsen, O Mattson, FT Okkels. Detection, expression and specific elimination of endogenous β-glucuronidase activity in transgenic and nontransgenic plants. Plant Sci 87:115–122, 1992.
48. JD Hamill, S Rounsley, A Spencer, G Todd, MJC Rhodes. The use of polymerase chain reaction in plant transformation studies. Plant Cell Rep 10:221–224, 1991.
49. LJW Gilissen, PLJ Metz, WJ Steikema, P-J Nap. Biosafety of *E. coli* β-glucuronidase (GUS) in plants. Transgenic Res 7:157–163, 1998.
50. PLJ Metz, JP Nap. A transgene-centered approach to biosafety of transgenic plants: Overview of selection and reporter genes. Acta Bot Neerl 46:25–50, 1997.
51. M DeBlock, D Debrouwer. Two T-DNAs cotransformed into *Brassica napus* by a double *Agrobacterium tumefaciens* infection are mainly integrated at the same locus. Theor Appl Genet 82:257–263, 1991.
52. J Bade, B Damne. Selectable marker-free transgenic rapeseed using cotransformation and segregation. Proceedings 9th International Rapeseed Congress, Cambridge, July 4–7, 1995, pp 775–777.
53. M Daley, VC Knauf, KR Summerfeldt, JC Turner. Cotransformation with one *Agrobacterium tumefaciens* strain containing two binary plasmids as a method for producing marker-free transgenic plants. Plant Cell Rep 17:489–496, 1998.

54. SH Russell, JL Hoopes, JT Odell. Direct excision of a transgene from the plant genome. Mol Gen Genet 234:49–59, 1992.
55. SB Gelvin. The introduction and expression of transgenes in plants. Current Opin Biotechnol 9:227–232, 1998.
56. GB Parker, AH Mitchell, JL Hart, SR Padgette, MJ Fedele, GF Barry, DK Didier, DB Re, DA Eichholtz, GM Kishore, X Delannay. Development of canola genetically modified to tolerate Round-Up herbicide. Agron. Abs. 1991. Am. Soc. Agron. Annual Meeting Denver Co. p199.
57. BL Miki, H Labbe, J Hattori, T Ouellet, G Sunohara, PJ Charest, VN Iyers. Transformation of *Brassica napus* canola cultivar with *Arabidopsis* acetohydroxyacid synthase genes and analysis of herbicide resistance. Theor Appl Genet 80:449–458, 1990.
58. M Freyssinet, P Creange, M Renard, P McVetty, R Derose, G Freysinnet. Development of transgenic oilseed rape resistant to oxynil herbicides. Proceedings of 9th International Rapeseed Congress, Cambridge, July 4–7, 1995, pp 974–976.
59. MM Oelck, CV Phan, P Eckes, G. Donn, G Rakow and WA Keller. Field resistance of canola transformants (*Brassica napus* L.) to Ignite (phosphinotricin). Proceedings of 8th International Rapeseed Congress, Saskatoon, Canada, July 9–11, 1991, pp 292–297.
60. RE Blackshaw, D Kaneshiro, MM Moloney, WL Crosby. Growth, yield and quality of canola expressing resistance to acetolactate synthase inhibiting herbicides. Can J Plant Sci 74:745–751, 1994.
61. A Kumar, G Rakow, RK Downey. Genetic characterization of glufosinate-ammonium tolerant summer rape lines. Crop Sci 18:1489–1494, 1998.
62. RE Redmann, MQ Qi, M Belyk. Growth of transgenic and standard canola (*Brassica napus* L.) varieties in response to soil salinity. Can J Plant Sci 74:797–799, 1994.
63. R Grison, B Grezes-Besset, M Schneider, M Lucante, L Olsen, J-J Leguay, A Toppan. Field tolerance of fungal pathogens of *Brassica napus* constitutively expressing a chimeric chitinase gene. Nat Biotechnol 14:643–646, 1996.
64. B Grezes-Besset, R Grison, MJ Villege, C Nicolas, A Toppan. Field testing against four fungal pathogens of transgenic *Brassica napus* plants constitutively expressing chitinase gene. Proceedings 9th International Rapeseed Congress, Cambridge, July 9–11, 1995, pp 781–783.
65. Y Wang, G Nowak, D Calley, LA Hadwiger, B Fristensky. Constitutive expression of pea defense gene DRR206 confers resistance to blackleg (*Leptosphaeria maculans*) disease in transgenic canola (*Brassica napus*). Mol Plant Microbe Interact 12:410–418, 1999.
66. R Broglie, K Broglie. Chitinase gene expression in transgenic plants: A molecular approach to understanding plant defence responses. In: MW Bevan, BD Harrison, CJ Leaver, eds. The Production and Uses of Genetically Transformed Plants. Chapman and Hall, London, 1994, pp 77–82.
67. K Broglie, I Ohet, M Holliday, R Cressman, P Biddle, S Knowlton, CJ Mauvais, R Broglie. Transgenic plants with enhanced resistance to the fungal pathogen *Rhizoctonia solani*. Science 254:1194–1197, 1991.
68. C Thompson, JC Dunwell, CE Johnstone, V Lay, M Schmidt, H Watson, G. Nisbet. Degradation of oxalic acid by transgenic oilseed rape plants expressing oxalate oxidase. Euphytica 85:169–172, 1995.
69. M Freyssinet, B Dumas, A Sailland, R Pepin, G Freyssinet, K Pallett. Transgenic crops expressing oxalate oxidase as a way to increase resistance to oxalate-producing pathogens. Proceedings of 9th International Rapeseed Congress, Cambridge, July 4–7, 1995. pp 1278–1279.
70. K Kazan, KC Goulter, HM Way, JM Manners. Expression of a pathogenesis-related peroxidase of *Stylosanthes humilis* in transgenic tobacco and canola and its effect on disease development. Plant Sci 136:207–217, 1998.
71. G Laucke, E Maiss, J Schiemann. Transformation of cloned and partially modified coat protein and replicase sequences of beet Western yellows virus (BWYV) into *Nicotiana benthamiana* and *Brassica napus*. Proceedings of 9th International Rapeseed Congress, Cambridge, July 4–7, 1995, pp 787–789.
72. C Hervé, D Rouan, P Guerche, M-H Montane, P Yot. Molecular analysis of transgenic rapeseed plants obtained by direct transfer of two separate plasmids containing, respectively, the cauliflower mosaic virus coat protein gene and a selectable marker gene. Plant Sci 91:181–193, 1993.

73. J Spak, J Dusbadkova, D Kubelkova, J Necasek. Resistance of transformed and nontransformed oilseed rape CV HM-81 to the infection with cauliflower mosaic, turnip yellow mosaic, and turnip mosaic viruses. Biol Plant 33:234–239, 1991.
74. Reference deleted.
75. D Bosch, B Schipper, H Van der Kleif, RA deMaagd, WJ Stiekema. Recombinant *Bacillus thuringiensis* crystal proteins with new properties: Possibilities for resistance management. Biotechnology 12:915–918, 1994.
76. K Herbers, U Sonnewald. Production of new/modified proteins in transgenic plants. Curr Opin Biotechnol 10:163–168, 1999.
77. JJ Estruch, NB Carozzi, N Desai, NB Duek, GW Warren, MG Koziel. Transgenic plants: An emerging approach to pest control. Nat Biotechnol 15:137–141, 1997.
78. RT Roush. Managing pests and their resistance to *Bacillus thuringiensis:* Can transgenic crops be better than sprays? Biocontrol Sci Technol 4:501–516, 1994.
79. L Jouanin, M Bonade-Bottino, C Girard, G Morrot, M Giband. Transgenic plants for insect resistance. Plant Sci 131:1–11, 1997.
80. CN Stewart, MJ Adang, JN All, PL Raymer, S Ramachandran, WA Parrott. Insect control and dosage effects in transgenic canola containing a synthetic *Bacillus thuringiensis* crylAc gene. Plant Physiol 112:115–120, 1996.
81. M Bonade-Bottino, C Girard, M Le Metayer, AL Picard-Nizou, G Sandoz, J Lerin, MH Pham-Delegue, L Jouanin. Effects of transgenic oilseed rape expressing proteinase inhibitors on pests and beneficial insects. Acta Hort 459:235–239, 1998.
82. C Girard, M Le Metayer, B Zaccomer, L Bartlet, I Williams, M Bonade-Bottino, MH Pham-Delegue, L Jouanin. Growth stimulation of beetle larvae reared on a transgenic oilseed rape expressing a cysteine proteinase inhibitor. J Insect Physiol 44:263–270, 1998.
83. SF Fabijanski, PG Arnison. Antisense gene systems of pollination control for hybrid seed production. European patent application #89301053.8, 1989.
84. CM Mariani, M deBeukeleer, J Traettner, J Leemans, RB Goldberg. Induction of male sterility in plants by a chimaeric ribonuclease gene. Nature 347:737–741, 1990.
85. CM Mariani, V Gossele, M deBeukeleer, M deBlock, RB Goldberg, WD Greef, J Leemans. A chimaeric ribonuclease inhibitor gene restores fertility in a male sterile plant. Nature 357:384–387, 1992.
86. C Hartung, L Borchert, WO Abel, R Lührs. Transformation of *Brassica napus* with the CMS-associated mitochondrial gene T-urf13. Proceedings of 9th International Rapeseed Congress, Cambridge, July 4–7, 1995, pp 15–17.
87. M Denis, R DeLourme, JP Gouvret, C Mariani, M Renard. Expression of engineered nuclear male sterility in *Brassica napus.* Plant Physiol 101:1295–1304, 1993.
88. T Sato, MK Thorsness, KK Kandasamy, T Nisho, M Hirari, JB Nasrallah, ME Nasrallah. Activity of an S locus gene promoter in pistils and anthers of transgenic *Brassica.* Plant Cell 3:867–876, 1991.
89. MK Kandasamy, MK Thorsness, SJ Rundle, ML Golberg, JB Nasrallah, ME Nasrallah. Ablation of papillar cell function in *Brassica* flowers results in the loss of stigma receptivity to pollen. Plant Cell 5:263–275, 1993.
90. Y Sasaki, M Iwano, H Matsuda, G Suzuki, M Watanabe, A Isogai, K Toriyama. Localization of an SLG protein expressed under the regulation of a tapetum-specific promoter in anthers of transgenic *Brassica napus.* Sexual Plant Rep 11:245–250, 1998.
91. J Kohno-Murase, M Murase, H Ichikawa, J Imamura. Improvement in the quality of seed storage protein by transformation of *Brassica napus* with an antisense gene for cruciferin. Theor Appl Genet 91:627–631, 1995.
92. J Kohno-Murase, M Murase, H Ichikawa, J Imamura. Effects of an antisense cruciferin gene on seed storage protein in transgenic *Brassica napus* seeds. Proceedings of 9th International Rapeseed Congress, Cambridge, July 4–7, 1995, pp 772–774.
93. J Kohno-Murase, M Murase, H Ichikawa, J Imamura. Effects of an antisense napin gene on seed storage compounds in transgenic *Brassica napus* seeds. Plant Mol Biol 26:1115–1124, 1994.

94. M Denis, M Renard, E Krebbers. Isolation of homozygous transgenic *Brassica napus* lines carrying a seed specific chimeric 2S albumin gene and determination of linkage relationships. Mol Breed 1:143–153, 1995.
95. M Denis, A Van Veliet, F Leyns, E Krebbers, M Renard. Field evaluation of transgenic *Brassica napus* lines carrying a seed specific chimeric 2S albumin in gene. Plant Breed 114:97–107, 1995.
96. SB Altenbach, CC Kuo, LC Storaci, KW Pearson, C Wainwraight, A Georgescu, J Townsend. Accumulation of Brazil nut albumin in seeds of transgenic canola results in enhanced levels of seed methionine. Plant Mol Biol 18:235–246, 1992.
97. A deClercq, M Vandewiele, J Van Damme, P Guerche, M Van Montague, J Vanderkerckhove, E Krebbers. Stable accumulation of methionine contents in transgenic plants. Plant Physiol 94:970–979, 1990.
98. SC Falco, T Guida, M Locke, T Mauvais, C Sanders, RT Ward, P Webber. Transgenic canola and soybean seeds with increased lysine. Biotechnology 13:577–582, 1995.
99. K Stålberg, M Ellerström, S Sjodahl, I Ezcurra, P Wycliffe, L Rask. Heterologous and homologous transgenic expression directed by a 2S seed storage promoter of *Brassica napus*. Transgenic Res 7:165–172, 1998.
100. K Stålberg, M Ellerström, I Ezcurra, S Ablov, L Rask. Disruption of an overlapping E box/ABRE motif abolishes high transcription of the napA storage protein in transgenic *Brassica napus* seeds. Planta 199:515–519, 1996.
101. J Vanderkerckhove, J Van Damme, M Lijsebettens, J Botterman, M DeBlock, M Vendewiele, A deClercq, J Leemans, M Van Montague, E Krebbers. Enkephalins produced in transgenic plants using modified 2S seed storage proteins. Biotechnology 7:929–932, 1989.
102. DJ Murphy. Structure, function and biogenesis of storage lipid bodies and oleosins in plants. Prog Lipid Res 32:247–280, 1993.
103. C Sarmiento, JHE Ross, E Herman, DJ Murphy. Expression and subcellular targeting of a soybean oleosin in transgenic rapeseed: Implications for the mechanism of oilbody formation in seeds. Plant J 11:783–796, 1997.
104. AL Plant, GJH VanRooijen, CP Anderson, MM Moloney. Regulation of an *Arabidopsis* oleosin gene promoter in transgenic *Brassica napus*. Plant Mol Biol 25:193–205, 1994.
105. MJ Hills, MD Watson, DJ Murphy. Targeting of oleosins to the oilbodies of rape seed (*Brassica napus* L.). Planta 189:24–29, 1993.
106. WS Lee, JTC Tzen, JC Kridl, SE Radke, AHC Huang. Maize oleosin is correctly targeted to seed oilbodies in *Brassica napus* transformed with the maize oleosin gene. Proc Natl Acad Sci USA 88:6181–6185, 1991.
107. S Chaudhary, DL Parmenter, MM Moloney. Transgenic *Brassica carinata* as a vehicle for the production of recombinant proteins in seeds. Plant Cell Rep 17:195–200, 1998.
108. B Kühnel, LA Holbrook, MM Moloney, GJH VanRooijen. Oilbodies of transgenic *Brassica napus* as a source of immobilized β-glucuronidase. J Am Oil Chem Soc 73:1533–1538, 1996.
109. DL Parmenter, JG Boothe, GJH VanRooijen, EC Yeung, MM Moloney. Production of biologically active hirudin in plant seeds using oleosin partitioning. Plant Mol Biol 29:1167–1180, 1995.
110. GJH VanRooijen, MM Moloney. Plant seed oil bodies as carriers for foreign proteins. Biotechnology 13:72–77, 1995.
111. TC Verwoerd, J Pen. Phytase produced in transgenic plants for use as a novel feed additive. In: MRL Owen, J Pen, eds. Transgenic Plants: A Production System for Industrial and Pharmaceutical Proteins. John Wiley & Sons, 1996, pp 213–224.
112. E Agostini, SM DeForchetti, HA Tigier. Production of peroxidases by hairy roots of *Brassica napus*. Plant Cell Tissue Org Cult 47:177–182, 1997.
113. S Chavadej, N Brisson, JN McNeil, V DeLuca. Redirection of tryptophan leads to production of low indole glucosinolate canola. Proc Natl Acad Sci USA 91:2466–2170, 1994.
114. E Krebbers, P Rudelsheim, W DeGreef, J Vanderkerckhove. Laboratory and field performance of transgenic *Brassica* plants expressing chimeric 2S albumin genes. Proceedings of 8th International Rapeseed Congress, Saskatoon, Canada, July 9–11, 1991, pp 716–721.

115. JA Napier, LV Michaelson, AK Stobart. Plant desaturases: Harvesting the fat of the land. Curr Opin Plant Biol 2:123–127, 1999.
116. J Ohlrogge. Plant metabolic engineering: Are we ready for phase two? Curr Opin Plant Biol 2:121–122, 1999.
117. T Voelker. Transgenic manipulation of edible oilseeds. Rec Adv Phytochem 31:223–236, 1997.
118. M Lassner. Transgenic oilseed crops: A transition from basic research to product development. Lipid Technol 9:5–9, 1997.
119. EB Cahoon, Y Lindquist, G Schneider, J Shanklin. Redesign of soluble fatty acid desaturases from plants for altered substrate specificity and double bond position. Proc Natl Acad Sci USA 94:4872–4877, 1997.
120. DJ Murphy. Engineering oil production in rapeseed and other oil crops. Trends Biotechnol 14:206–213, 1996.
121. MW Lassner, CK Levering, HM Davies, DS Knutzon. Lysophosphatidic acid acyltransferase from meadow-foam mediates insertion of erucic acid at the SN-2 position of triacylglycerol in transgenic rapeseed oil. Plant Physiol 109:1389–1394, 1995.
122. AJ Del Vecchio. High-laurate canola. Inform 7:230–243, 1996.
123. A Merolli, J Lindemann, AJ Del Vecchio. Medium chain lipids: New sources, uses. Inform 8:597–603, 1997.
124. AJ Kinney. Genetic engineering of oilseeds for desired traits. JK Setlow, ed. Genetic Engineering. Vol. 19. New York: Plenum Press 1997, pp 149–166.
125. TA Voelker, TR Hayes, AM Cranmer, JC Turner, HM Davies. Genetic engineering of a quantitative trait: Metabolic and genetic parameters influencing the accumulation of laurate in rapeseed. Plant J 9:229–241, 1996.
126. J Zou, V Katavic, EM Giblin, DL Barton, SL MacKenzie, WA Keller, X Hu, DC Taylor. Modification of seed oil content and acyl composition in the Brassicaceae by expression of a yeast SN-2 acyltransferase gene. Plant Cell 9:909–923, 1997.
127. W Friedt, W Lühs. Development in the breeding of rapeseed oil for industrial purposes. Proceedings of 9th International Rapeseed Congress, Cambridge, July 4–7, 1995, pp 437–448.
128. TA Voelker, AC Worrel, L Anderson, J Bleibaum, C Fau, DJ Hawkins, SE Radke, HM Davies. Fatty acid biosynthesis redirected to medium chains in transgenic oilseed plants. Science 257:72–74, 1992.
129. DS Knutzon, GA Thompson, SE Radke, WB Johnson, VC Knauf, JC Kridl. Modification of *Brassica* seed oil by antisense expression of a stearoyl-acyl carrier protein desaturase gene. Proc Natl Acad Sci USA 89:2624–2628, 1992.
130. RM Safford, MT Moran, J deSilva, SJ Robinson, S Moscow, CD Jarman, AR Slabas, J deSilva. Regulated expression of the rat medium chain hydrolase gene in transgenic rapeseed. Transgenic Res 2:191–198, 1993.
131. IIGS Verwoert, HJJ Nijkamp, AR Stuitje, KH VanderLinden. Developmental specific expression and organelle targeting of the *Escherichia coli* fabD gene, encoding malonyl coenzyme A–acyl carrier protein transacylase in transgenic rape and tobacco seeds. Plant Mol Biol 26:189–202, 1994.
132. A Jones, H Mallor, TA Voelker. Palmitoyl–acyl carrier protein (ACP) thioesterase and the evolutionary origin of plant acyl ACP thioesterases. Plant Cell 7:359–371, 1995.
133. WD Hitz, CJ Mauvais, KG Ripp, RJ Reiter. The use of cloned rapeseed genes for the cytoplasmic fatty acid desaturases and the plastid acyl-ACP thioesterases to alter relative levels of polyunsaturated and saturated fatty acids in rapeseed oil. Proceedings of 9th International Rapeseed Congress, Cambridge, July 4–7, 1995, pp 470–472.
134. D Weier, W Lühs, J Dettendorfer, M Frentzen. SN-1 acylglycerol-3-phosphate acyltransferase of *E. coli* causes insertion of cis 11-eicosensic acid into the SN-2 position of transgenic rapeseed oil. Mol Breed 4:39–46, 1998.
135. K Dehesh, A Jones, DS Knutzon, TA Voelker. Production of high levels of 8:0 and 10:0 fatty acids in transgenic canola by overexpression of ch Fat B2, 9 thioesterase cDNA from *Cuphea hookeriana*. Plant J 9:167–172, 1996.

136. D Weier, C Hauke, A Eickelkamp, W Lühs, J Dettendorfer, E Schaffert, C Mollers, W Friedt, FP Wolter, M Frentzen. Trierucoylglycerol biosynthesis in transgenic plants of rapeseed (*Brassica napus* L.). Forschungsbeitrage Fett Lipid 99:160–165, 1997.
137. DJ Hawkins, JC Kridl. Characterization of acyl-AcP thioesterases of mangosteen (*Garcina mangostana*) seed and high levels of stearate production in transgenic canola. Plant J 13:743–752, 1998.
138. MT Facciotti, PB Bertain, L Yuan. Improved stearate phenotype in transgenic canola expressing a modified acyl-acyl carrier protein thioesterase. Nat Biotechnol 17:593–597, 1999.
139. J Han, W Lühs, K Sonntag, DS Borchardt, M Frentzen, FP Wolter. A *Brassica napus* cDNA restores the deficiency of canola fatty acid elongation at high level. In: J Sanchez, E Cerda-Olmeda, E Martinez-Force, eds. Advances in Plant Lipid Research. Universidad de Sevilla, 1998, pp 665–667.
140. JA Wilmer, M Brazier, M Craze, TR Slabas, T Barsby. Control of fatty acid composition in oilseed rape by introduction of a limnanthes acyl transferase (abstract b59). In: J Sanchez, E Cerda-Olmeda, E Martinez-Force, eds. Advances in Plant Lipid Research. Universidad de Sevilla. 1998.
141. AG Münster, W Lühs, DS Borchardt FP Wolter, M Frentzen. Experiments to optimize the channeling of eucric acid into the SN-2 position of transgenic rapeseed oil. In: J Sanchez, E Cerda-Olmeda, E Martinez-Force, eds. Advances in Plant Lipid Research, Universidad de Sevilla. 1998, pp 671–674.
142. RK Jain, GR Thompson, GG Rowland, A McHughen, SL MacKenzie, DC Taylor. Isolation and characterization of the two flax stearoyl-acyl carrier protein desaturase genes and promoters and their differential regulation in transgenic flax, tobacco and canola. In: J Sanchez, E Cerda-Olmeda, E Martinez-Force, eds. Advances in Plant Lipid Research. Universidad de Sevilla. 1998, pp 647–649.
143. DS Knutzon, JM Thurmond, Y-S Huang, S Chaudhary, EG Bobik Jr., GM Chan, SJ Kirchner, P Mukerji. Identification of $\Delta^5$ desaturase from *Mortierella alpina* by heterologous expression in bakers yeast and canola. J Biol Chem 273:29360–29366, 1998.
144. K Roesler, D Shintani, L Savage, S Boddupalli, J Ohlrogge. Targeting of the *Arabidopsis* homomeric acetyl-coenzyme A carboxylase to plastid of rapeseeds. Plant Physiol 113:75–81, 1997.
145. IIGS Verwoert, KH VanderLinden, MC Walsh, HJJ Nijkamp, AR Steitje. Modification of *Brassica napus* seed oil by expression of *E. coli* fabH gene encoding 3-ketoacyl-acyl carrier protein synthase III. Plant Mol Biol 27:875–886, 1995.
146. FP Wolter, C Hauke, A Eickelkamp, M Frentzen. Trierucin biosynthesis in transgenic rapeseed: Cloning and expression of cDNAs encoding an erucoyl-CoA specific acyltransferase. Proceedings of 9th International Rapeseed Congress, Cambridge, July 4–7, 1995, pp 473–475.
147. JCP Morissette, MN Kouschuh, AM Johnson-Flanagan, J Singh, L Robert. Reduction of chlorophyll accumulation in seed of transgenic *Brassica napus* using antisense technology. Acta Hortic 459:183–190, 1998.
148. S Misra, L Gedamu. Heavy metal tolerant transgenic *Brassica napus* L. and *Nicotiana tobacum* L. plants. Theor Appl Genet 78:161–168, 1990.
149. HP Hong, J Heross, JL Gerster, S Rigas, RSS Datla, P Hatzopoulos, G Scoles, W Keller, DJ Murphy, LS Robert. Promoter sequences from two different *Brassica napus* tapetal oleosin-like genes direct tapetal expression of β-glucuronidase in transgenic *Brassica* plants. Plant Mol Biol 34:549–555, 1997.
150. HP Hong, JL Gerster, RSS Datla, D Albani, G Scoles, W Keller, LS Robert. The promoter of a *Brassica napus* polygalacturonase gene directs pollen expression of β-glucuronidase in transgenic *Brassica* plants. Plant Cell Rep 16:373–378, 1997.
151. J deVries, W Wackernagel. Detection of NPT11 (kanamycin resistance) genes in genomes of transgenic plants by marker-rescue transformation. Mol Gen Genet 257:606–613, 1998.
152. P Roeckel, T Oancia, JR Drevet. Phenotypic alterations and component analysis of seed yield in transgenic *Brassica napus* plants expressing the tzs gene. Physiol Plant 102:243–249, 1998.
153. P Roeckel, T Oancia, JR Drevet. Effects of seed-specific expression of a cytokinin biosynthetic gene on canola and tobacco phenotypes. Transgenic Res 6:133–144, 1997.
154. TH Iverson, E Odegaard, T Beisvåg, A Johnson, O Rasmussen. The behaviour of normal and agravitropic transgenic roots of rapeseed (*Brassica napus* L.) under microgravity conditions. Biotechnol 47:137–154, 1996.

155. V Legue, D Driss-Ecol, R Maldiney, M Tepfer, G Perbal. The response to auxin of rapeseed (*Brassica napus* L.) roots displaying reduced gravitropism due to transformation by *Agrobacterium rhizogenes*. Planta 200:119–124, 1996.
156. CG Downs, MC Christey, KM Davies, GA King, JF Seelye, BK Sinclair, DG Stevenson. Hairy roots of *Brassica napus*. II. Glutamine synthetase overexpression alters ammonia assimilation and the response to phosphinothricin. Plant Cell Rep 14:41–46, 1994.
157. I Stefanov, S Fekete, L Bogne, J Pauk, A Feher, D Dudits. Differential activity of the mannopine synthase and the CaMV 35S promoters during development of transgenic rapeseed plants. Plant Sci 95:175–186, 1994.
158. MA Matzke, AJM Matzke. How and why do plants inactivate homologous (trans) genes? Plant Physiol 107:679–685, 1995.
159. J Finnegan, D McElroy. Transgene inactivation: Plants fight back. Biotechnology 12:883–888, 1994.
160. J Finnegan, D McElroy. Transgene stability. In: MR Lowen, J Pen, eds. Transgenic Plants: A Production System for Industrial and Pharmaceutical Proteins. John Wiley & Sons, New York, 1996, pp 169–186.
161. M Stam, JNM Mol, JM Kooter. The silencing of genes in transgenic plants. Ann Bot 79:3–12, 1997.
162. J-P Nap, L Mlynorova, WJ Steikema. From transgene expression to public acceptance of transgenic plants: A matter of predictability. Field Crop Res 45:5–10, 1996.
163. M DeNeve, S DeBuck, C DeWilde, H VanHoudt, I Strobbe, A Jacobs, M Van Montagu, A Depicker. Gene silencing results in instability of antibody production in transgenic plants. Mol Gen Genet 260:582–592, 1999.
164. I Broer. Stress inactivation of foreign genes in transgenic plants. Field Crop Res 45:19–25, 1996.
165. CS Jones, NE Owen, PJ Dale. Transgene stability-inheritance and expression in *Brassica napus*. Proceedings of 9th International Rapeseed Congress, Cambridge, July 4–7, 1995, pp 769–771.
166. PLJ Metz, E Jacobsen, WJ Steikema. Occasional loss of expression of phosphinothricin tolerance in sexual offsprings of transgenic oilseed rape (*Brassica napus* L.). Euphytica 98:189–196, 1997.
167. EJA Blakemore, SFE Turner, JC Reeves, RJ Cooke. Stability and uniformity of genetically manipulated oilseed rape. Proceedings of 9th International Rapeseed Congress, Cambridge, July 4–7, 1995, pp 1427–1429.
168. JA Conner, T Tantikanjana, JC Stein, MK Kandasamy, JB Nasrallah, ME Nasrallah. Transgene-induced silencing of S-locus genes and related genes in *Brassica*. Plant 11:809–823, 1997.
169. L Willmitzer. Plant biotechnology: Output traits—the second generation of plant biotechnology products is gaining momentum. Curr Opin Biotechnol 10:161–162, 1999.
170. SL Abidi, GR List, KA Rennick. Effect of genetic modification on the distribution of minor constituents in canola oil. J Am Oil Chem Soc 76:463–467, 1999.
171. SP Kumpatla, W Teng, WG Buchholz, TC Hall. Epigenetic transcriptional silencing and sazacytidmie-mediated reactivation of a complex transgene in rice. Plant Physiol 115:361–373, 1997.
172. KD Kasschau, JC Carrington. A counterdefensive strategy of plant viruses: Suppression of post transcriptional gene silencing. Cell 95:461–470, 1998.
173. P Vain, B Worland, A Kohli, JW Snape, P Christou, GC Allen, WF Thompson. Matrix attachment regions increase transgene expression levels and stability in transgenic rice plants and their progeny. Plant 18:233–242, 1999.
174. Y Poirier. Production of new polymeric compounds in plants. Curr Opin Biotechnol 10:181–185, 1999.
175. A Steinbüchel, B Füchtenbusch. Bacterial and other biological systems for polyester production. Trends Biotechnol 16:419–427, 1998.
176. C Hawrath, Y Poirier, C Somerville. Targeting the polyhydroxybutyrate biosynthesis pathway to the plastids of *Arabidopsis thaliana* results in high levels of polymer accumulation. Proc Natl Acad Sci USA 91:12760–12764, 1994.

# 52
## Studies on Genetic Engineering of Rice

**Honghong Zheng, Xiaotian Ming, Yi Li, Hongya Gu, and Zhangliang Chen**
*Peking University, Beijing, China*

| | | |
|---|---|---|
| I. | INTRODUCTION | 793 |
| II. | RICE TRANSFORMATION BY MICROPROJECTILE BOMBARDMENT | 794 |
| | A. Plant Material | 795 |
| | B. Plasmids | 795 |
| III. | TRANSFORMATION VIA PARTICLE BOMBARDMENT | 795 |
| | A. Plant Regeneration | 795 |
| | B. Molecular Analysis and Bioassay of Transgenic Rice Plants | 795 |
| IV. | RICE TRANSFORMATION BY THE *AGROBACTERIUM* SPECIES–MEDIATED APPROACH | 795 |
| | A. Plant Material | 796 |
| | B. *Agrobacterium* Species Strains | 796 |
| | C. Transformation | 796 |
| | D. Plant Culture Media | 797 |
| | E. Explants for Particle Bombardment | 797 |
| | F. Particle Bombardment | 797 |
| | G. Selection and Regeneration | 797 |
| | H. Polymerase Chain Reaction and Southern Blot Analysis Confirming the Integration of the S8 Gene in the Rice Genome | 798 |
| | I. Rice Dwarf Virus S8 Gene Expression Analysis | 798 |
| | J. Inheritance of the Introduced Rice Dwarf Virus S8 Gene in the Progeny | 798 |
| | K. Rice Dwarf Virus Resistance of Transgenic Rice Plants When Challenged | 798 |
| | REFERENCES | 798 |

## I. INTRODUCTION

Rice (*Oryzea sativa* L.) is one of the most important food resources for humankind, especially for people in developing countries. The progress in plant genetic engineering provides promising potential for the improvement of rice production and quality. Since the first transgenic rice was reported (1,2), remarkable progress was made in this field during the ensuing decade. Many contributions by various researchers have led to a routine transformation system (3) to obtain transgenic rice plants in many laboratories. Especially in the recent 3–4 years, a number of agronomically important genes have been introduced into rice and have proved to be effective. Engineered rice plants with modified *Bacillus Thuringiensis* (*Bt*) gene were obtained for insect resistance (4–7), and Cheng and coworkers (7) reported that *Agrobacterium* species–transformed rice plants expressing *cryIA(b)* and *cryA(c)* genes are highly toxic to striped stem borer and yellow stem borer. In order to obtain transgenic rice with resistance to fungal and bacterial diseases, chitinase gene *ChiI 1* was transferred into rice protoplast, and the regenerated rice plants showed resistance to sheath blight (8). Song and colleagues (9) cloned the rice blight resistance gene *Xa21*, and the transgenic rice plants with *Xa21* showed high levels of resistance against bacterial blight disease. For virus resistance, transgenic plants were produced with viral coat protein gene or part of another viral gene in order to add resistance against rice stripe virus (10), brome masaic virus (11), rice tungro virus (12), and rice dwarf virus (13). For stress tolerance, Xu and colleagues (14) transferred late embryogenesis abundant (LEA) protein gene from barley, *Hva 1*, into rice by particle bombardment; the transgenic rice plants obtained enhanced tolerance to drought and salinity. In addition, there are studies on rice quality improvements via transgenic approaches (15,16)

In recent years, two significant accomplishments have been achieved in the genetic transformation of cereal plants. First, biolistic technology (microprojectile bombardment) has been successfully developed, allowing routine production of transgenic plants in major cereals including rice. This system has solved many problems that exist in traditional methods of direct gene transfer, such as polyethylene glycol– (PEG)-mediated and electroporation gene transfer into protoplasts. Application of the direct transformation method to various cereal species was limited by some problems, for instance, the somaclonal variation derived from the long period of tissue culture procedure for protoplasts, the need for good protoplast culture, and the regeneration system, which is time-consuming, laborious, and highly genotype-dependent (17). One of the advantages of the biolistic method is that transgenic plants can be obtained after bombardment of various explants only if they are regenerable. In addition, its short period of tissue culture minimize the possibility of somaclonal variation and thus improves the fertility of transgenic plants. More importantly, it is considered to be variety-independent: transgenic plants of both the *Oryzea sativa japonica* and *O. sativa indica* rice varieties have been successfully obtained with relatively high transformation frequency.

The second significant accomplishment is that application of the *Agrobacterium* species–mediated method to rice transformation has become well established. The *Agrobacterium* species–mediated approach has significant advantages over the naked deoxyribonucleic acid (DNA) delivery methods, such as the introduction of fewer copies of genes into the plant genome, high coexpression of introduced genes, transformation of relatively large segments of DNA, and high fertility of transgenic plants (18,19). In the mid-1990s this method was successfully applied to many monocot crops that had been considered to be recalcitrant to *A. tumefaciens* infection for a long time. Transgenic rice, maize, and wheat plants mediated by *A. tumefaciens* were obtained subsequently with convincing and unambiguous data on transgene expression, gene segregation in the progeny, and DNA analysis (18–20). *Agrobacterium* species–mediated rice transformation has now been successfully applied to many cultivars of rice, e.g., *O. sativa japonica*, *indica*, and *javanica* (19–23).

## II. RICE TRANSFORMATION BY MICROPROJECTILE BOMBARDMENT

The microprojectile bombardment method is based on high-velocity bombardment of plant cells with DNA-coated microprojectiles (tungsten or gold) that can be accelerated by gunpowder explosion, pressurized helium, or electric discharge. Christou and coworkers (24) first reported the recovery of transgenic rice plants by a biolistic method in several commercially important *O.*

## Genetic Engineering on Rice

*sativa indica* and *japonica* cultivars. In the past few years, this bombardment-based methodology was further improved (25,26) and has been widely used in the production of transgenic rice plants. It seems to be the most versatile and successful approach (17).

There are many factors found to influence the successful transformation of rice by the biolistic method, including the choice of regenerable explants, the appropriate depth of penetration of particles, and the optimized selection and regeneration system. The following is a basic protocol for producing transgenic rice plants by particle bombardment under standard laboratory conditions.

### A. Plant Material

The explants used for bombardment can usually be rice immature embryos (12 days old) (24), embryogenic callus (26), or suspension cell aggregates induced from sterilized immature/mature embryos on Muroshige and Skoog (MS) (27) or N6 (28) medium supplemented with 2 mg/l 2,4-D (2,4-dichlorophenoxyacetic acid).

### B. Plasmids

The construct should contain the gene of interest and the selectable marker gene. For rice, *hpt* encoding for hygromycin resistance and *bar* encoding for herbicide resistance are often used.

## III. TRANSFORMATION VIA PARTICLE BOMBARDMENT

The explants were placed on the center of a 9-cm petri dish containing MS or N6 medium supplemented with 2 mg/l 2,4-D. Plasmid DNA was coated on microjectiles (tungsten or gold) according to the protocol provided by the manufacturer of the particle acceleration device. After bombardment, the explants were transferred to MS or N6 basal medium containing 2 mg/l 2,4-D for 2–7 days then transferred to the same medium with the selective agent, such as 30–50 mg/l hygromycin B (Sigma) or 3–8 mg/l bialaphos. To enhance the transformation efficiency the explants were cultured on osmotic treatment medium, as suggested by Vain and associates (29), for 4 hours before bombardment and for 16 hours after bombardment.

### A. Plant Regeneration

After 6–8 weeks of selection, the resistant calli were transferred onto regeneration medium [MS or N6 basal medium supplemented with 2–3 mg/l kinetin or 6-benzylaminopurine (BAP) and 0.5–1 mg/l naphthalene acetic acid (NAA)] to regenerate into plants. When the regenerated plantlets reached about 10 cm in height, they were transplanted into pots.

### B. Molecular Analysis and Bioassay of Transgenic Rice Plants

Putative transgenic rice plants were subjected to Southern blot analysis to confirm the presence and integration of the exogenous gene in the rice genome, Northern blot analysis for the transgene expression at ribonucleic acid (RNA) level, and Western blot analysis for protein expression.

## IV. RICE TRANSFORMATION BY THE *AGROBACTERIUM* SPECIES–MEDIATED APPROACH

There are many factors involved in *Agrobacterium* sp.–mediated gene transfer into rice. Genotype of plants, types and ages of tissues inoculated, kind of vectors, strains of *Agrobacterium* sp., selection marker genes and selective agents, and various conditions of tissue culture are of critical importance (30). During cocultivation, the addition of acetosyringone medium, a potent inducer of virulence genes, 2,4-D, acidic pH, and osmotic treatment has been reported to be important for the induction of Ti plasmid *vir* gene expression (31–33). Most reports have recommended the use of actively dividing, embryonic calli derived from rice mature seeds or immature embryos as competent cells for *A. tumefaciens* infection (34). The combination of vectors and *A. tumefaciens* strain is another important factor affecting the gene transfer: so-called ordinary strain (LBA4404) and superbinary vector (pTOK 233) combinations have been recommended by several scientists (19,21).

The following is a modified protocol for *Agrobacterium* sp.–mediated rice transformation according to the procedure provided by the Center for the Application of Molecular Biology to International Agriculture (*CAMBIA*) and Hiei and associates (19).

## A. Plant Material

Embryogenic calli derived from the scutellum of mature embryos were considered to be excellent material for rice transformation by *Agrobacterium* sp. (19), they can be obtained as described in the protocol for particle bombardment.

## B. *Agrobacterium* Species Strains

*Agrobacterium tumefaciens* strains LBA4404, EHA105, and AGL-1 were used. Plasmids were introduced into these strains by electroporation or triparental mating. *Agrobacterium tumefaciens* was grown on AB medium (35).

## C. Transformation

The embryogenic calli (actively growing, 1–2 mm in diameter) used for transformation were subcultured on 2N6 medium (19) at 26°C for 4 days. *Agrobacterium tumefaciens* was grown on AB medium with appropriate antibiotics and incubated at 28°C for 2 days. The bacterial cells were collected and resuspended in AAM medium (19) containing 100 µM acetosyringone (approximately 2.0 OD at 600 nm). The 4 day-incubated calli were immersed in the bacterial suspension for 30 minutes, then transferred to 2N6AS medium (19) and cocultivated in darkness for 3 days. Then the calli were rinsed thoroughly with water containing 0.25 g/l cefotaxime to remove the bacterial cells. The calli were blotted dry and cultured on 2N6 medium containing cefotaxime and selective agent.

The selection of transgenic callus and regeneration of transgenic plantlets were conducted as described in the protocol for particle bombardment.

In our lab, the National Laboratory of Protein Engineering and Plant Genetic Engineering (PRC), the biolistic method is frequently used for rice genetic transformation. The following is a detailed description of recovery of engineered rice plants expressing the rice dwarf virus (RDV) gene by the biolistic method.

Rice dwarf virus is the pathogen that causes rice dwarf disease, which causes great loss of rice productivity in Southeast Asia. However, conventional means of RDV control cannot solve the problem effectively. Transgenic virus resistance is an attractive alternate strategy.

Protection from viral infection can be gained by introduction of viral sequences into the genome of host plants. This pathogen-derived resistance (PDR) mediated by coat protein (CP), replicase, or parts of the viral genome in transgenic plants has been demonstrated for a number of viruses (36,37). We have developed or are developing a range of virus-resistance transgenes for RDV. In our lab, we have obtained all 12 segments (S1–S12) of the RDV genome (38). Some of the segments, such as S1, S2, S3, S6, S7, and S8, which encode replicase, coat proteins, core proteins, or nonstructural proteins, have been transferred into rice plants via particle bombardment. We hope that engineered rice plants with the RDV gene will provide an effective method for RDV resistance. At the same time, the expression of viral genes in vivo through transformation will be a good system for studying the functions of viral genes since this kind of double strand RNA virus cannot be transmitted mechanically.

In this experiment, outer coat protein gene (S8) of rice dwarf virus (RDV) was chosen as the target gene. We cloned S8 into plasmid vector for plant transformation. Using callus derived from mature rice embryos as the target tissue, we regenerated rice plants after particle bombardment with the plasmid containing the RDV S8 gene and a selection marker gene—neomycin phosphotransferase (NPTII). After integration, expression as well as stable inheritance of the S8 gene were confirmed by molecular analysis.

### D. Plant Culture Media

Modified N6 medium (28) was used for rice tissue culture with additional proline (500 mg/l) and enzymatic casein hydrolysate (300 mg/l). The pH was adjusted to 5.8 before autoclaving. The callus induction and maintenance medium contained 2 mg/l 2,4-D, and the regeneration medium contained 3 mg/l 6-benzylaminopurine (BAP) and 0.5 mg/l NAA in modified N6 medium. Modified N6 medium without hormone was used for plantlet growth. For osmotic treatment, sorbitol (0.4 M) was added 4 hours before and 12 to 16 hours after bombardment.

### E. Explants for Particle Bombardment

Cultivar 'Zhonghua 8' is one of the commercially important *O. sativa japonica* rice cultivars in China. Calli derived from mature rice embryos of this cultivar were used as the explants for particle bombardment. Mature rice seeds were dehusked and surface sterilized with 20% NaOCl for 20 minutes, followed by three to five washes with sterile water. The seeds were than placed on callus induction medium. The induced primary calli or short-term subcultured calli were used as target tissue for bombardment.

### F. Particle Bombardment

Bombardments were performed with a PDS 1000/He Biolistic Device (BioRad). Forty to sixty pieces of calli were placed in the center (2–3 cm in diameter) of each petri plate containing the osmotic treatment medium. The plasmid DNA was coated onto gold particles with an average size of 1.0 μm, following the procedures described in the device instructions. Bombardment parameters we used were the following: helium pressure, 1100 psi; gap distance, 1 cm; fly distance, 6 mm; target distance, 6 cm. The calli were transferred to callus maintenance medium after bombardment.

### G. Selection and Regeneration

After culture for 1 week on callus maintenance medium, the calli were transferred onto the same medium containing 50 mg/l G418 (Sigma). After 4 weeks, visible fresh G418-resistant calli appeared and were moved to fresh selection medium with 100 mg/l G418. After approximately another 3–4 weeks of selection with 100 mg/l G418, the proliferated calli were transferred onto the regeneration medium. Regeneration was performed under fluorescent illumination with a 16-hour photoperiod. Two weeks later, somatic embryos appeared; then shoots and roots developed. As soon as the shoots were 6–10 cm long, they were transplanted into soil in the greenhouse and grown to maturity. G418 was also added to the regeneration and plantlet growth medium at a concentration of 100 mg/l.

In our experiment, from a total of 110 G418-resistant cell clusters, 22 plant lines with G418-resistance were regenerated, and subsequent polymerase chain reaction (PCR) and Southern analysis have confirmed that 19 have RDV S8 gene present.

### H. Polymerase Chain Reaction and Southern Blot Analysis Confirming the Integration of the S8 Gene in the Rice Genome

Total DNA from the putatively transformed plants were screened by PCR using primers specific to the coding region of the RDV S8 gene to detect presence of S8. Further Southern analysis was carried out to confirm the presence and integration of the RDV S8 gene in the genome of transgenic rice plants.

### I. Rice Dwarf Virus S8 Gene Expression Analysis

Total protein were extracted from the transgenic plants and subjected to Western blotting using RDV antiserum. The S8 encoded protein was detected in transgenic rice plants.

Standard protocols for PCR and Southern and Western blotting we used have been described in our previous report (13).

### J. Inheritance of the Introduced Rice Dwarf Virus S8 Gene in the Progeny

Segregation of S8 gene in the progeny was analyzed. The results indicated that S8 has been stably transmitted into progeny. Some transgenic lines gave a good agreement with segregation ratios of 3:1, consistent with mendelian inheritance of a single dominant locus. But one transgenic line fit a 1:1 model, possibly as a result of lack of either male or female transmission. This phenomenon has also been mentioned in manipulations of other transgenic plants (39).

### K. Rice Dwarf Virus Resistance of Transgenic Rice Plants When Challenged

The progenies of the transgenic rice plants harboring the S8 gene, together with the transgenic rice plants harboring other gene segments of RDV (S7, S6), obtained by the same approach, were screened and selected. Experiments testing their resistance against RDV infection are ongoing now.

### REFERENCES

1. K Toriyama, Y Arimoto, H Uchimiya, K Hinata. Transgenic rice plants after direct gene transfer into protoplasts. Biotechnology 6:1072–1074, 1988.
2. HM Zhang, H Yang, EL Rech, TJ Golds, AS Davis, BJ Mulligan, EC Cocking, MR Davey. Transgenic rice plants produced by electroporation-mediated plasmid uptake into protoplasts. Plant Cell Rep 7: 379–384, 1988.
3. NM Ayers, WD Park. Genetic transformation of rice. Crit Rev Plant Sci 13:219–239, 1994.

4. H Fujimoto, K Itoh, M Yamamoto, J Kyozuka, K Shimamoto. Insect resistant rice generated by introduction of modified δ-endotoxin gene of *Bacillus thuringiensis*. Biotechnology 11:1151–1155, 1993.
5. J Wunn, A Kloti, PK Burkhardt, GC Ghosh-Biswas, K Launis, VA Iglesias, I Potrykus. Transgenic indica rice breeding line IR58 expressing a synthetic *cryIA(b)* gene from *Bacillus thuringiensis* provides effective insect pest control. Biotechnology 14:171–176, 1996.
6. C Wu, Y Fan, C Zhang, N Oliva, SK Datta. Transgenic fertile japonica rice plants expressing a modified *cryIA(b)* gene resistant to yellow stem borer. Plant Cell Rep 17:129–132, 1997.
7. X Cheng, R Sardana, H Kaplan, I Altosaaar. *Agrobacterium*-transformed rice plants expressing synthetic *cryIA(a)* and *cryIA(c)* genes are highly toxic to striped stem borer and yellow stem borer. Proc Natl Acad Sci USA 95:2767–2772, 1998.
8. W Lin, CS Anuratha, K Datta, I Potrykus, S Muthukrishnan, SK Datta. Genetic engineering of rice for resistance to sheath blight. Biotechnology 13:686–691, 1995.
9. WY Song, GL Wang, LL Chen, HS Kim, LY Pi, T Holsten, J Gardner, B Wang, WX Zhai, LH Zhu, C Fauquet, P Ronald. A receptor kinase–like protein encoded by the rice disease resistance gene, Xa21. Science 270:1804–1806, 1995.
10. T Hayakawa, Y Zhu, K Itoh, Y Kimura, T Izawa, K Shimamoto, S Toriyama. Genetically engineered rice resistant to rice stripe virus, an insect-transmitted virus. Proc Natl Acad Sci USA 89:9865–9869, 1992.
11. CC Huntley, TC Hall. Interference with brome mosaic virus replication in transgenic rice. Mol Plant Microbe Interact 9:164–170, 1996.
12. I Potrykus, PK Burkhardt, SK Datta, J Futterer, GC Ghosh-Biswas, A Kloti, G Spangenberg, J Wunn. Genetic engineering of indica rice in support of sustained production of affordable and high quality food in developing countries. Euphytica 85:441–449, 1995.
13. HH Zheng, Y Li, ZH Yu, W Li, MY Chen, XT Ming, R Casper, ZL Chen. Recovery of transgenic rice plants expressing the rice dwarf virus outer coat protein gene (S8). Theor Appl Genet 94:522–527, 1997.
14. D Xu, X Duan, B Wang, B Hong, T-H David Ho, R Wu. Expression of a late embryogenesis abundant protein gene, *HVA1*, from barley confers tolerance to water deficit and salt stress in transgenic rice. Plant Physiol 110:249–257, 1996.
15. Z Zheng, K Sumi, K Tanaka, M Murai. The bean seed storage protein β-phaseolin is synthesized, processed and accumulated in the vascular type II protein bodies of transgenic rice endosperm. Plant Physiol 101:777–786, 1995.
16. PK Burkhatdt, P Beyer, J Wunn, A Kloti, GA Amstrong, M Schledz, J von Lintig, I Potrykus. Transgenic rice (*Oryza sativa*) endosperm expressing daffodil (*Narcissus pseudonarcissus*) phytoene synthase accumulates phytoene, a key intermediate of provitamin A biosynthesis. Plant J 11:1071–1078, 1997.
17. P Christou. Rice transformation: Bombardment. Plant Mol Biol 35:197–203, 1997.
18. M Cheng, JE Fry, S Pang, H Zhou, CM Hironaka, DR Duncan, TW Conner and Y Wan. Genetic transformation of wheat mediated by *Agrobacterium tumefaciens*. Plant Physiol 115:971–980, 1997.
19. Y Hiei, S Ohta, T Komari, T Kumashiro. Efficient transformation of rice (*Oryza sativa* L.) mediated by *Agrobacterium* and sequence analysis of boundaries of the T-DNA. Plant J 6:271–282, 1994.
20. Y Ishida, H Saito, S Ohta, Y Hiei, T Komari, T Kumashiro. High efficiency transformation of maize (*Zea mays* L.) mediated by *Agrobacterium tumefaciens*. Nat Biotechnol 14:745–750, 1996.
21. J Zhang, RJ Xu, MC Elliott, and DF Chen. *Abrobacterium*-mediated transformation of Elite *indica* and *japonica* rice cultivars. Mol Biotechnol 8:223–231, 1997.
22. JJ Dong, WM Teng, WG Buchholz, TC Hall. *Agrobacterium*-mediated transformation of Javanica rice. Mol Breeding. 2:267–276, 1996.
23. RR Aldemita, TK Hodges. *Agrobacterium tumefaciens*–mediated transformation of *japonica* and *indica* rice varieties. Planta 199:612–617, 1996.
24. P Christou, TL Ford, M Kofron. Production of transgenic rice (*Oryza sativa* L.) plants from agronomically important *indica* and *japonica* varieties via electric discharge particle acceleration of exogenous DNA into immature zygotic embryos. Biotechnology 9:957–962, 1991.
25. J Cao, X Duan, D McElroy, R Wu. Regeneration of herbicide resistant transgenic rice plants follow-

ing microprojectile-mediated transformation of suspension culture cells. Plant Cell Rep 11:586–591, 1992.

26. L Li, R Qu, A de Kochko, C Fauquet, RN Beachy. An improved rice transformation system using the biolistic method. Plant Cell Rep 12:250–255, 1993
27. T Murashige, F Skoog. A revised medium for rapid growth and bioassays with tobacco tissue culture. Physiol Plant 15:473–497, 1962.
28. CC Chu, CC Wang, CS Sun, C Hsu, KC Yin, CY Chu, FY Bi. Establishment of an efficient medium for anther culture of rice through comparative experiments on nitrogen sources. Sci Sin 18:659–668, 1975.
29. P Vain, MD McMullen, JJ Finer. Osmotic treatment enhances particle bombardment–mediated transient and stable transformation of maize. Plant Cell Rep 12:84–88, 1993
30. S Toki. Rapid and Efficient *Agrobacterium*-mediated transformation in rice. Plant Mol Biol Rep 15:16–21, 1997.
31. MF Van Wordragen, HJM Dons. *Agrobacterium tumefaciens*–mediated transformation of recalcitrant crops. Plant Mol Biol Rep 10:12–36, 1992.
32. SCHJ Turk, LS Melchers, H den Dulk-Ras, AJG Regensburg-Tuink, PJJ Hooykaas. Environmental conditions differentially affect *vir* gene induction in different *Agrobacterium* strains: Role of the Vir A sensor protein. Plant Mol Biol 16:1051–1059, 1991.
33. S Usami, S Okamoto, I Takebe, Y Machida. Factor inducing *Agrobacterium tumefaciens vir* gene expression is present in monocotyledonous plants. Proc Natl Acad Sci USA 85:3748–3752, 1988.
34. Y Hiei, T Komari, T Kubo. Transformation of rice mediated by *Abrobcterium tumefaciens*. Plant Mol Biol 35:205–218, 1997.
35. MD Chilton, TC Currier, SK Farrand, AJ Bendich, MP Gordon, EW Nester. *Agrobacterium tumefaciens* DNA and PS8 bacteriophage DNA not detected in crown gall tumors. Proc Natl Acad Sci USA 71:3672–3676, 1974.
36. PP Abel, RS Nelson, B De, N Hoffmann, SG Rogers, RT Fraley, RN Beachy. Delay of disease development in transgenic plants that express the tobacco mosaic virus coat protein gene. Science 232:738–743, 1986.
37. TMA Wilson. Strategies to protect crop plants against viruses: Pathogen-derived resistance blossoms. Proc Natl Acad Sci USA 90:3134–3141, 1993.
38. Y Li, H Xu, MF Cheng, HH Zheng, ZJ Mao, FJ Zhang, XT Ming, L Qu, YF Liu, W Li, XL Zhao, NS Pan, ZL Chen. Functional genomic analysis of rice dwarf virus. Acta Sci Nat Univ Pek 34:332–341, 1998.
39. J Cooley, T Ford, P Christou. Molecular and genetic characterization of elite transgenic rice plants produced by electric discharge particle acceleration. Theor Appl Genet 90:97–104, 1995.

# 53
# Transgenic Sorghum with Improved Nutritional Quality

### Yohannes Tadesse and Michel Jacobs
*Free University of Brussels, Sint-Genesius-Rode, Belgium*

|      |                                                    |     |
|------|----------------------------------------------------|-----|
| I.   | INTRODUCTION                                       | 801 |
|      | A. Crop Distribution and Adaptation                | 801 |
|      | B. Basic Uses of Sorghum                           | 802 |
|      | C. Nutritional Value of the Crop                   | 802 |
|      | D. Insect Pest Resistance                          | 802 |
| II.  | IMPROVEMENT OF SORGHUM                             | 803 |
| III. | GENETIC MANIPULATION APPROACHES                    | 803 |
| IV.  | TRANSFORMATION METHODS AND SORGHUM MANIPULATION    | 804 |
| V.   | ENGINEERING THE ASPARTATE METABOLIC PATHWAY        | 806 |
| VI.  | MANIPULATION OF LYSINE CONTENT IN SORGHUM          | 807 |
| VII. | CONCLUSIONS                                        | 809 |
|      | REFERENCES                                         | 809 |

## I. INTRODUCTION

### A. Crop Distribution and Adaptation

Sorghum (*Sorghum bicolor* (L.) Moench) is an important staple food for about 500 million people in arid and semiarid regions of the world. It is the fifth important cereal crop in world production, following wheat, rice, maize, and barley. *Sorghum bicolor*, which belongs to the Gramineae family, originated in Africa and spread to the continents of Australia, America, and Asia. According to Doggett (1), this cultivated crop was domesticated in Africa, and the greatest variability of wild species occurred in the Ethiopia-Sudan region of East Africa.

Although when and how sorghum spread out from Africa are still matters of discussion, it has been shown to extend along the Nile to the Near East, and across India to Thailand and China. This crop was relatively recently introduced to the Americas, around 1850 in the United States; its cultivation in Central and South America occurred only in the 1950s (1). The genus *Sorghum*, as classified by Harlan and de Wet (2), includes three species: *S. bicolor*, an annual species, which includes all the cultivated sorghums, and two perennial species, *S. halepense* and *S. propinquum*. On the basis of morphological studies by Harlan and de Wet (2), the cultivated sorghum was divided into five races, bicolor, kafir, caudatum, durra, and guinea, on the basis of inflorescence types.

Sorghum adapts to wider environmental conditions but is primarily well suited to hot and dry conditions. In arid regions it performs better than maize, since it has an extensive and deeply penetrating root system. Because of its excellent drought resistance compared to that of most other crops, sorghum thus plays an essential role in regions where production systems of other crops cannot be maintained as a result of heat, drought, or poor soil condition (3). Sorghum can also stand a wide range of pH and soil textures and is relatively tolerant to salinity (4).

## B. Basic Uses of Sorghum

The basic importance of sorghum resides in its consumption as food by humans and as feed by animals, but the plant is also used as fuel and building material. Sorghum is consumed in a variety of ways, but the most important and widespread form of consumption is the flour form. Porridge (dough) is prepared under different names (*genfo, guma, mafo, nsima, sadza, tô, tuwo*, and *ugali*) and consumed in many sorghum producing countries (1). Fermented breads (*kisra* and *injera*) are consumed in parts of Africa, unfermented breads (such as *chapati, roti*, and *tortilla*) are consumed in various places of Asia and Central America. *Couscous* is another important sorghum food in the Sahel region of West Africa, where it is eaten in a mixture with baobab leaves and peanut butter. Furthermore, in many parts of Africa, malted sorghum is used for making traditional beers. In South Africa, the largest part of the total sorghum consumption is in the production of beer, most of which is home-brewed (5). Sorghum grains are also used for direct consumption as immature seeds and/or as boiled, roasted, or popped grains. Moreover, vegetative parts of the plant are fed to animals, the stem and foliage are used for green chop, hay, silage, and pasture.

## C. Nutritional Value of the Crop

Sorghum is the major food grain and an important protein source in the semiarid tropics. This crop, despite its wide adaptation and utilization, is known to have somewhat lower nutritional value than other cereal grains as a result of its characteristic low content of one of the essential amino acids, lysine. This low amount of lysine is due to the high content of lysine-poor prolamin proteins. Moreover, sorghum digestibility is low for livestock and also for humans (6). The fact that essential amino acids cannot be synthesized by humans and monogastric animals necessitates the consumption of a diet enriched with amino acids. In this respect sorghum amino acid composition limits its use as a sole food source. When sorghum protein is used as a livestock feed, it needs to be supplemented by a protein-rich feed containing a high level of lysine (1). Altogether the lysine content in cereals in general and sorghum in particular (ranging between 1% and 4%) fails to meet the level recommended (5.5%) by the World Health Organization (WHO) (7). This indicates the need for substantially elevating lysine content in sorghum by conventional as well biotechnological methods.

## D. Insect Pest Resistance

In addition to its low nutritional quality, sorghum is infested by a large number of insect species, of which aphids, head bugs, stem borers, and shoot fly are the most important pests, causing seedling death, stem destruction, falling of leaves, and grain shriveling. In 1992, the damage by these pests in Asia and Africa alone was estimated at a loss of about 9 million tons of grain (8). Although integrated pest management approaches had been attempted, problems related to socioeconomic conditions have become obstacles to the practical application of the concepts (9). On the other hand, screening of sorghum germplasm collections for resistance, though seemingly promising (10), has not yet yielded efficient resistance against the respective pests. As summarized by Nwanze and associates (9), genetic transformation could revolutionize the development of insect resistance in sorghum, assuming that feasible and reproducible gene delivery methods for this crop are developed and made routinely available.

## II. IMPROVEMENT OF SORGHUM

Up to the present time the multidisciplinary efforts to improve this crop have been focused on conventional breeding, involving such methods as germplasm screening, pedigree selection, backcrossing, hybridization, and mutation breeding. That approach has been successful in improving characteristics such as productivity, yield, stability, and resistance to biotic and abiotic stresses. Through germplasm enhancement, progress in developing resistance against *Striga*, a parasitic weed known for greatly affecting sorghum production in most parts of the sub-Saharan region, has been reported (11). Regarding nutritional properties, the discovery of high-lysine *opaque2* maize mutants (12) opened the way for similar attempts in other cereals. In 1970, a barley mutant with 30–40% more lysine than the wild type was identified. Similarly, Singh and Axtell (13) screened the world sorghum collection and came out with a high-lysine naturally occurring sorghum mutant containing a reduced level of prolamins. But these mutants suffered from some drawbacks such as dented endosperm, which was due to fewer and smaller protein bodies as a result of the decrease in prolamins, and a slow rate of dry matter accumulation during grain development, which resulted in reduced grain yield (14–17). Other high-lysine sorghum mutants with a 60% increase in total lysine were also induced by chemical mutagenesis (14,15). In further screening, sorghum lines with higher protein digestibility were found in populations that were developed from the latter type of mutants (18).

Thus, concerted efforts in classical breeding led to significant achievements in pest management, weed control, production, yield stability, and grain quality. The high-yielding varieties developed and released by conventional breeders have contributed a great deal to providing an adequate food supply for the ever-growing world population. But these traditional approaches cannot be efficient enough to meet the increasing demand for more food as a result of the highly expanding population size, especially in developing countries, where people have a more precarious food supply. Evidently sorghum is a crop of significance in these regions, which are dominated by subsistence agriculture. Therefore, if we have to overcome the challenges of malnutrition and famine and if we have to limit the destruction of natural habitats for agricultural land expansion, genetic engineering of crop plants has to be considered together with traditional approaches for crop improvement.

## III. GENETIC MANIPULATION APPROACHES

Plant genetic engineering has the potential to introduce totally novel traits into any crop species. Genetic modification of plants using in vitro culture and transformation techniques has opened the way to introducing foreign deoxyribonucleic acid (DNA) of agronomic interest into the plant genome. In a period of about 5 years from the first demonstration of successful production of transgenic plants (19), foreign gene transfer into a wide variety of plant species (from easy experimental model plants to difficult recalcitrant ones) were documented (for a review see Ref. 20). The development of techniques for stable integration of foreign genes into plants has led to new approaches to addressing agronomic problems. Manipulation of metabolic pathways (21); introduction of agronomically important genes such as insect, virus, and herbicide resistance (22,23); and hybrid seed production by engineering male sterile plants (24) are some of the achievements that witness the potential plant transformation technology holds for crop improvement. Although several transformation techniques have been developed, their application to cereals has encountered limitations as a result of host range specificity for natural vectors such as *Agrobacterium* and/or lack of efficient tissue culture systems.

## IV. TRANSFORMATION METHODS AND SORGHUM MANIPULATION

Progress in the development of gene transfer systems has conferred the ability to introduce foreign genes into plant cells and tissues and to regenerate fertile plants, providing the opportunity to modify and improve crop plants. Since the first transgenic plants were reported (19,25), derived vectors of the plant pathogen *Agrobacterium tumefaciens* have proved to be efficient and highly versatile vehicles for the introduction of genes into plant tissues and cells. Alternatively, direct DNA transfer systems based on the polyethylene glycol–(PEG)-mediated method (26), the high-voltage electrical impulse (electroporation) method (27), and techniques of microinjection (28) have been developed. Despite several advantages of *Agrobacterium*–mediated transformation, the problem of limited host range of this vector remains a problem, DNA transfer methods require the use of protoplasts, and unfortunately most cereals cannot be regenerated from single cells. Previously, monocotyledonous plants were known to be out of the natural host range of *Agrobacterium* (29). However, in the last few years genetic transformation mediated by *A. tumefaciens* infection of the major cereals has been attempted by some laboratories. As a result of these attempts, wheat tissues have been transformed by *Agrobacterium* (30), and transgenic barley plants have been obtained through *Agrobacterium*–mediated transformation (31). Transgenic rice plants have been produced by *A. tumefaciens* inoculation of immature embryos (32) and use of calli as starting explants (33). Maize and sugarcane, close relatives of sorghum, have also become amenable to this natural vector-mediated transformation. In addition to some reports giving evidence of *A. tumefaciens* infection of maize (34–36), an efficient method for producing transgenic maize plants from immature embryos cocultivated with *A. tumefaciens* has been reported (37). In 1998 an efficient protocol for *A. tumefaciens*–mediated sugarcane transformation was established (38). Compared to direct DNA transfer methods, *Agrobacterium*–mediated transformation offers some advantages, such as transfer of relatively large segments of DNA with little rearrangement, integration of low copy numbers of genes into the plant genome, and an economical and efficient procedure. However, in contrast to these benefits, no reports of the successful use of this method in sorghum have been published. Nevertheless, recent progress in the use of *Agrobacterium* for cereal transformation suggests that it is likely that sorghum may be within the natural host range. Zhao and coworkers recently reported the first successful use of *Agrobacterium* for production of stably transformed sorghum plants (71). As a whole, much attention has still to be focused on those cereals that are somewhat neglected, such as sorghum. They need continuous investigation and better understanding for the successful application of this transformation tool.

The first transgenic cereal plants were obtained by using direct DNA transfer into protoplasts (39,40). This method requires the establishment of homogeneous cell suspension cultures that can be used as a source of competent cells. But this process is time-consuming, difficult, and generally genotype-specific (20), though plant regeneration from protoplasts of sorghum has been described (41). The first stable sorghum transformants were obtained by bombardment of cell suspensions (42) and electroporation of protoplast cultures (43). However, plants were not regenerated from these transformed cells. Similarly our observation on sorghum in vitro culture is that although it is possible to initiate suspension cultures from embryogenic calli, the establishment of homogeneous cell suspension, the time required to do so, and the possibility of regenerating plants are highly influenced by genotype. Fortunately, these shortcomings are circumvented by the use of explants such as meristematic tissues that allow regeneration of plants through somatic embryogenesis.

In this regard, techniques based on particle bombardment have assumed an increasingly important role in extending recombinant DNA methodology to the genetic manipulation of recalcitrant species. Herbicide-resistant transgenic sorghum plants have been produced by this method (44–46). Additional efforts are required to further sorghum transformation technology utilizing traits of agricultural importance with the goal of generating nutritional improvement. At present, particle bombardment appears to be a method of choice for sorghum manipulation until alternative method (*Agrobacterium*–mediated) is developed.

The biolistic process known also as *microprojectile bombardment* or *particle acceleration* employs high-velocity metal particles to deliver biologically active DNA into plant cells or intact tissues (47). The acceleration impact is sufficient to penetrate plant cells, which can be regenerated into whole plants. The process is predominantly genotype-independent and imposes no biological limitations on DNA delivery. However, the settings for bombardment conditions are species- and tissue-specific and therefore require optimization in relation to the particular application (48,49). With this in mind, we first assessed the efficiency of different types of sorghum explants in combination with various physical parameters used in the biolistic treatment (Tadesse et al., submitted). Immature embryos, mature embryos, shoot tips, primary callus tissues and embryogenic calli were bombarded with vectors, including the reporter gene *uidA*, under the control of strong promoters. Transient GUS activity was observed 48 hours later. In general, immature embryos and shoot tips were superior to other tissues. In addition to the factors directly involved in the bombardment process, we have investigated different gene promoters by comparing their strength via transient expression of *uidA* gene in sorghum immature embryo and shoot tip tissues. We used different constructs: p35S-hpt, pBARGUS (50), pDB1 (51), and pAHC25 (52), each harboring the *uidA* gene, under the control of CaMV35S, maize Adh1, maize Ubi1, or rice Act1 promoter, respectively. Histochemical staining tests at 24, 48, and 72 hours after bombardment and fluorometric assays showed that the highest activity was achieved when ubiquitin promoter was used, followed by the actin promoter; Adh1 promoter was found to be about 1.5 and 3 times less active than the Act1 and Ubi1 promoters, respectively. The activity of the CaMV35S promoter was relatively weak, about 1.5 times lower than that of the Adh1 promoter. Similar promoter comparisons in rice (53) and in Liliaceae species (54) showed that the relative strength of promoters is a property of the promoters themselves but also varies significantly with the species in which they drive the introduced genes. Identification and use of strong promoters would facilitate high-level constitutive or specific expression of foreign genes in sorghum cells, and subsequently allow better growth of transformed cells on the respective selective medium. In genetic transformation of crop plants, the choice of selectable markers in monocots is narrower than that in dicots because of naturally existing endogenous resistance to some selective agents (55). Similarly, in our previous experiments on sorghum cell and tissue cultures, we observed the same trend of insensitivity to kanamycin selection. In this respect, we found geneticin (G-418) efficient in inhibiting tissue growth and induction of callus in sorghum immature embryos and shoot tips. When using geneticin as selective agent with *nptII* gene as selectable marker in sorghum transformation, the selection procedure has to be optimized to prevent potential protection of wild-type cells by transformants or unnecessary side effects resulting from stringency of selection. As part of establishing transformation conditions, the *nptII* gene under the control of the rice actin promoter was introduced into sorghum tissues by particle bombardment (56). Fertile transgenic sorghum plants resistant to geneticin were obtained and grown to maturity, giving seeds. The presence, stable integration, and inheritance of the foreign gene were analyzed at the molecular and the genetic

## V. ENGINEERING THE ASPARTATE METABOLIC PATHWAY

As do many bacterial species, higher plants synthesize the essential amino acid lysine from aspartate by a specific branch of the aspartate family pathway (57). In plants, this pathway is regulated by several end product feedback inhibition loops (see Fig. 1). Biochemical studies have shown that lysine feedback inhibits the activities of the two key enzymes of the pathway, aspartate kinase (AK) and dihydrodipicolinate synthase (DHDPS) (57). Analyses of mutant and transgenic plants containing enzymes with reduced sensitivity to feedback inhibition showed that lysine synthesis is regulated primarily by DHDPS (58–60), as this enzyme is highly sensitive to lysine. Earlier observation in microorganisms suggested that $S$-(2-aminoethyl)-L-cysteine (AEC), an analogue of lysine, appeared efficient in selecting mutants that overproduced lysine (61). By using *Nicotiana sylvestris* haploid protoplasts, AEC-resistant lysine overproducing mutant lines were obtained by Negrutiu and coworkers (62). Genetic analysis of the mutants showed that the resistance trait is due to a single dominant gene. There was a marked increase of free lysine in leaves and calli of such mutants. The identified mutation concerns the dihydrodipicolinate synthase gene (*dhdps*), which encodes the first regulatory enzyme of the lysine pathway (Fig. 1). The *dhdps-r1* mutation causes insensitivity of DHDPS enzyme to the feedback inhibition normally exerted by lysine and was identified as a substitution of two nucleotides, changing asparagine into isoleucine in a conserved region of the enzyme (63). Thus, the expression of a feedback-insensitive enzyme is associated with an increased production of free lysine. The expression of a feedback-insensitive DHDPS in plants, particularly in cereals such as sorghum, is of special interest from the nutritional point of view. The expression of this mutated gene together with the use of

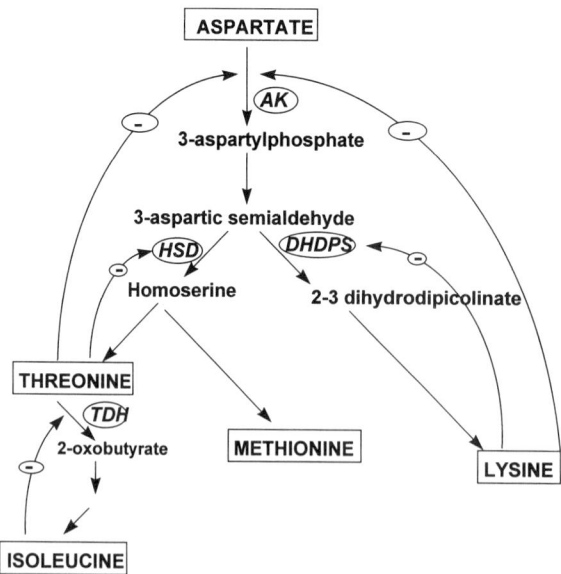

**Figure 1** The aspartate family biosynthetic pathway. Feedback inhibition by end-product amino acids is represented by curved arrows that have minus signs. AK, aspartate kinase; DHDPS, dihydrodipicolinate synthase; HSD, homoserine dehydrogenase; TDH, threonine dehydrogenase.

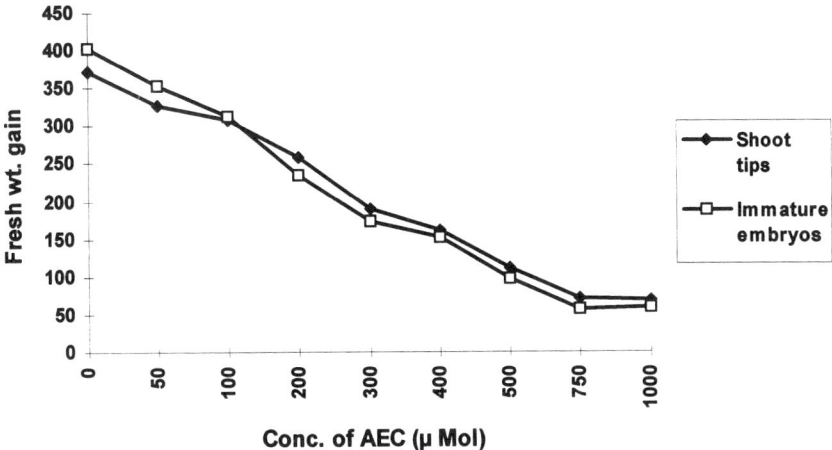

**Figure 2** Growth inhibition in shoot tips and immature embryos by $S$-(2-aminoethyl)-L-cysteine (AEC).

AEC as selective agent favors the growth of transformed cells and suppress the further development of the nontransformed cells. The use of AEC in selecting sorghum lines with increased lysine in the seeds was established by Vernaillen and coworkers (64) in our laboratory. The presence of 100 μM AEC in the germination medium was sufficient to inhibit elongation of roots after seed germination. On this basis, we determined the inhibiting concentration of AEC for callus induction in sorghum immature embryo and shoot tips (Fig. 2). We found that 300 μM of AEC inhibited callus formation in both types of tissues. In further transformation experiments this concentration was used in the first selection step of transformed cells and tissues.

## VI. MANIPULATION OF LYSINE CONTENT IN SORGHUM

Our goal was to express in sorghum the mutated form of the *dhdps* gene to evaluate a possible overproduction of lysine. Therefore, chimeric constructs harboring the *dhdps-r1* mutated allele under the control of constitutive or seed-specific promoters were developed. By setting up transformation experiments, the *dhdps-r1* allele under the control of the constitutive CaMV35S promoter was transferred through particle bombardment to immature embryos and shoot tips, which were then selected on progressively increasing concentrations of AEC. The expression of this mutated gene led to the synthesis of a feedback-insensitive DHDPS enzyme, which allowed growth and division of cell sectors on bombarded tissues under AEC selection. Total inhibition of callus initiation on nontransformed control tissues was indicative of the selection pressure exerted by the respective AEC concentration.

Stepwise selection of bombarded tissues has allowed the recovery of AEC-resistant embryogenic calli, which later gave rise to transgenic sorghum plants expressing the mutated *dhdps-r1* gene. Biochemical analyses of leaf extracts showed up to 5.5 times more DHDPS activity in transformants than in wild-type controls, although differences in the level of enzyme activity between different transformant lines were observed (Table 1). The differences in DHDPS activity level among individual lines could be due to differences in transgene copy number. Inhibition of DHDPS enzyme activity in the presence of increasing lysine concentrations was also tested (Table 2). Although the enzyme from the transgene did not show any significant inhibition, a clear drop in activity in the wild type was observed. The relatively higher specific activity of DHDPS

**Table 1** DHDPS Specific Activity in Leaf Extracts of Wild Type and Eight Transformed Sorghum Plants

| Samples | WT | 1 | 2 | 3 | 4 | 5 | 6 | 7 | 8 |
|---|---|---|---|---|---|---|---|---|---|
| DHDPS activity (unit/mg protein/min) | 505.2 | 096.8 | 2123.2 | 969.8 | 794.7 | 455.7 | 664.5 | 989.5 | 422.7 |

*Source*: Refs. 1–8. Enzyme assay was done according to Ghislain and coworkers (63). A unit of enzyme activity is defined as the amount of enzyme that causes an absorbance increase of 0.001/min at 550 nm. Data presented are means of three measurements.

**Table 2** Inhibition of DHDPS Activity by Lysine in Wild Type and Transformed Sorghum Plants

| | WT | 1 | 2 | 3 | 4 | 5 | 6 | 7 | 8 |
|---|---|---|---|---|---|---|---|---|---|
| Lysine concentration | | | | Enzyme activity (unit/mg protein/min) | | | | | |
| 0 µM | 497 | 2165 | 2201 | 1908 | 3211 | 1532 | 2550 | 2089 | 1410 |
| 25 µM | 260 | 1923 | 2045 | 1806 | 2807 | 1415 | 2343 | 1979 | 1360 |
| 50 µM | 147 | 1812 | 1921 | 1713 | 2981 | 1387 | 2121 | 1785 | 1324 |
| 75 µM | 130 | 1798 | 1818 | 1695 | 3056 | 1334 | 2113 | 1692 | 1298 |
| 100 µM | 111 | 1801 | 1796 | 1690 | 2666 | 1295 | 1943 | 1702 | 1267 |

*Source*: Refs 1–8. Enzyme assay was done according to Ghislain and coworkers (63). A unit of enzyme activity is defined as the amount of enzyme that causes an absorbance increase of 0.001/min at 550 nm. Data presented are means of three measurements.

and its insensitivity to lysine inhibition confirmed the expression of the mutated plant *dhdps-r1* gene under the control of the 35S promoter in transgenic sorghum plants. Previously the feedback-insensitive bacterial DHDPS enzyme coded by the *Escherichia coli dapA gene* was found to be about 20-fold less sensitive to lysine inhibition in transgenic tobacco plants than the wild type (65). The expression of the bacterial gene was accompanied by a substantial increase of free lysine in leaves. In sorghum, amino acid analysis results showed up to 3.5 times more free lysine in the leaves of transformed plants when compared to that in the wild-type line. Although constitutive in its expression, the CaMV35S promoter is known to be less strong in its expression in monocots than in dicots. In that context the use of *dhdps-rl* under another strong monocot promoter, such as maize ubiquitin, could allow better expression of the gene, resulting in much higher specific activity and leading to more lysine accumulation in transformed plants.

More interestingly, the expression of the *Corynebacterium dapA* gene alone or with a mutant *E. coli lysC* gene under seed-specific promoters has resulted in 100-fold or several hundred–fold increase in free lysine of canola and soybean seeds, respectively (66). This strategy is also being developed in sorghum transformation. The use of a mutated *dhdps-rl* gene under the transcriptional control of an endosperm-specific monocot promoter is under way. As the expression of this gene is expected to be limited at the seed level, another monocot promoter, such as rice actin, controlling the expression of the selectable marker *npt II* will be used together to allow the selection of transformants. On the other hand, a lysine breakdown as a result of lysine-ketoglutarate reductase (LKR) activity is known to prevent the accumulation of excess free lysine in tobacco seeds (67). To prevent this lysine catabolism, efforts have to be coordinated to altering the activity of the enzyme LKR via an antisense type of strategy. This could enable us to reach a rate of lysine synthesis that exceeds the rate of catabolism.

## VII. CONCLUSIONS

Despite the socioeconomic importance of sorghum, improving its characteristics has received relatively little attention, most of it in the form of classical breeding strategies. The fact that conventional approaches alone cannot satisfy the demand for adequate quantity and better quality of food for the ever increasing population of developing countries magnifies the need for new approaches based on plant biotechnology. As described by Bennetzen (68,69), the application of molecular techniques in DNA marker-assisted breeding and genetic engineering could be directed to the identification and use of agronomically important traits in sorghum. With this purpose we have obtained sorghum plants that express a mutated *dhdps* gene that codes for a lysine feedback-insensitive DHDPS enzyme. The expression of this gene under a strong seed-specific promoter, coupled with reduced catabolic activity of LKR, can bring about a significant increase of lysine in sorghum seed proteins. On the other hand, the elevation of lysine level in the whole plant is also advantageous where vegetative parts are used as feed for animals. Moreover, as sorghum is known to have wild and weedy relatives with which it can cross-pollinate (70), the introduction of *dhdps-rs* is not likely to have any negative effect if it escapes from transgenic plants. Furthermore, these achievements strengthen our expectation of the potential of biotechnology to produce transgenic sorghums that combine improved important traits.

## REFERENCES

1. H Doggett. Sorghum. 2nd ed. London: Longman Scientific & Technical, 1988, pp 428–453.
2. JR Harlan, JMJ de Wet. A simplified classification of cultivated sorghum. Crop Sci 12:172–176, 1972.
3. DT Rosenow, G Ejeta, LE Clark, ML Gilbert, RG Henzell, AK Borrell, RC Muchow. Breeding for pre- and post-flowering drought stress resistance in sorghum. Proceedings of International Conference on Genetic Improvement of Sorghum and Pearl Millet, Lubbock, TX, 1997, pp 400–411.
4. N Seetharama, AKS Huda, SM Virmani, JL Moneith. Sorghum in the semi-arid tropics: Agroclimatology, physiology and modelling. Proceedings of International Congress on Plant Physiology, New Delhi, 1990, pp 142–151.
5. SA Asante. Sorghum quality and utilization. Afr Crop Sci J 2:231–240, 1995.
6. RB Hamaker, JD Axtell. Nutritional quality of sorghum. Proceedings of International Conference on Genetic Improvement of Sorghum and Pearl Millet, Lubbock, TX, 1997, pp 531–538.
7. SWJ Bright, PR Shewry. Improvement of protein quality in cereals. CRC Crit Rev Plant Sci 1:49–93, 1982.
8. ICRISAT (International Crops Research Institute for the Semi-Arid Tropics). Medium Term Plan. Patancheru, India: ICRISAT, 1992.
9. KF Nwanze, N Seetharama, HC Sharma, JW Stenhouse. Biotechnology in pest management: Improving resistance in sorghum to insect pests. Afr Crop Sci J 3:209–215, 1995.
10. HC Sharma. Host plant resistance to insects in sorghum and its role in integrated pest management. Crop Protect 12:11–34, 1993.
11. G Ejeta. Development and enhancement of sorghum germplasm with sustained tolerance to drought, *Striga* and grain mold. Annual Report of Sorghum Millet Collaborative Research Support Program, 1996, pp 76–81.
12. ET Mertz, LS Bates, OE Nelson. Mutant gene that changes protein composition and increases lysine content of maize endosperm. Science 145:279–280, 1964.
13. R Singh, JD Axtell. High-lysine mutant gene (h1) that improves protein quality and biological value of grain sorghum. Crop Sci 13:535–539, 1973.

14. G Ejeta, JD Axtell. Protein and lysine levels in developing kernels of normal and high-lysine sorghum. Cereal Chem 64:137–139, 1987.
15. G Ejeta, JD Axtell. Dry-matter accumulation and carbohydrate composition in developing normal and high-lysine sorghum grain. J Agric Food Chem 35:981–985, 1987.
16. RP Johari, AB Dongre, SL Mehta. Protein, nucleic acids and enzyme levels during development in a high lysine sorghum grain. Phytochemistry 20:569–573, 1981.
17. RD Sullins, LW Rooney, DT Tosenow. Endosperm structure of high lysine sorghum. Crop Sci 15:599–600, 1975.
18. RB Hamaker. Chemical and physical aspects of food and nutritional quality of sorghum and millet. Annual Report of Sorghum Millet Collaborative Research Support Program, 1996, pp 105–119.
19. R Horsch, R Fraley, S Rogers, P Sanders, A Lioyd, N Hofman. Inheritance of functional foreign genes in plants. Science 223:496–498, 1984.
20. I Potrykus. Gene transfer to plants: Assessment of published approaches and results. Annu Rev Plant Physiol Plant Mol Biol 42:205–225, 1991.
21. R Müller, U Sonnewald, L Willmitzer. Inhibition of the ADP-glucose pyrophosphorylase in transgenic potatoes leads to sugar-storing tubers and influences tuber formation and expression of tuber storage protein genes. EMBO J 11:1229–1238, 1992.
22. MG Koziel, GL Beland, C Bowman, NB Carozzi, R Crenshaw, L Crossland, J Dawson, N Desai, M Hill, S Kadwell, K Launis, K Lewis, D Maddox, K McPherson, MR Meghji, E Merlin, R Rhodes, GW Warren, M Wright, SV Evola. Field performance of elite transgenic maize plants expressing an insecticidal protein derived from *Bacillus thuringiensis*. Biotechnology 11:194–200, 1993.
23. CJ Lamb, JA Ryals, ER Ward, RA Dixon. Emerging strategies for enhancing crop resistance to microbial pathogens. Biotechnology 10:1436–1445, 1992.
24. C Mariani, M De Beuckeleer, J Truettner, J Leemans, RB Goldberg. Induction of male sterility in plants by a chimeric ribonuclease gene. Nature 347:737–741, 1990.
25. M De Block, L Herrera-Esterella, M Van Montagu, J Schell, P Zambryski. Expression of foreign genes in regenerated plants and their progeny. EMBO J 3:1681–1689, 1984.
26. J Paszkowski, RD Shillito, M Saul, V Mandak, T Hohn, B Hohn, I Potrykus. Direct gene transfer to plants. EMBO J 3:2717–2722, 1984.
27. R Shillito, MWM Saul, J Paszkowski, M Muller, I Potrykus. High efficiency direct gene transfer to plant. Biotechnology 3:503–517, 1985.
28. G Neuhans, G Spangenberg. Plant transformation by microinjection techniques. Physiol Plant 79:213–217, 1990.
29. M De Cleene, J De Ley. The host range of crown gall. Bot Rev 42:389–466, 1976.
30. PA Mooney, PB Goodwin, ES Dennis, DJ Llewellyn. *Agrobacterium tumefaciens* gene transfer into wheat tissues. Plant Cell Tissue Org Cult 25:209–218, 1991.
31. S Tingay, D McElroy, R Kalla, S Fieg, M Wang, S Thornton, R Brettel. *Agrobacterium tumefaciens*-mediated barley transformation. Plant J 6:1369–1376, 1997.
32. MT Chan, HH Chang, SL Ho, WF Tong, SM Yu. *Agrobacterium* mediated production of transgenic rice plants expressing a chimeric α-amylase promoter/β-glucuronidase gene. Plant Mol Biol 22:491–506, 1993.
33. Y Hiei, S Ohta, T Komari, T Kumashiro. Efficient transformation of rice (*Oryza sativa* L.) mediated by *Agrobacterium* and sequence analysis of the boundaries of the T-DNA. Plant J 2:271–282, 1994.
34. N Grimsley, T Hohn, JW Davis, B Hohn. *Agrobacterium* mediated delivery of infectious maize streak virus into maize plants. Nature 325:177–179, 1987.
35. J Gould, M Devey, O Hasegawa, EC Ulian, G Peterson, RH Smith. Transformation of *Zea mays* L. using *Agrobacterium tumefaciens* and the shoot apex. Plant Physiol 95:426–434, 1991.
36. WH Shen, J Escudero, M Schläppi, C Ramos, B Hohn, Z Koukolikovä-Nicola. T-DNA transfer to maize cells: Histochemical investigation of β-glucuronidase activity in maize tissues. Proc Natl Acad Sci USA 90:1488–1492, 1993.
37. Y Ishida, H Saito, S Ohta, T Komari, T Kumashiro. Nat Biotechnol 14:745–750, 1996.
38. AD Arencibia, ER Carmona, P Téllez, MT Chan, SM Yu, LE Trjillo, P Oramas. An efficient protocol for sugarcane (*Saccharum* spp. L.) transformation mediated by *Agrobacterium tumefaciens*. Transgenic Res 7:213–222, 1998.

39. CA Rhodes, DA Pierce, IJ Mettler, D Mascarenhas, JJ Detmer. Genetically transformed maize plants from protoplasts. Science 240:204–207, 1988.
40. K Shimamoto, R Terada, T Izawa, H Fujimoto. Fertile transgenic rice plants regenerated from transformed protoplasts. Nature 338:274–276, 1989.
41. Z Wei, Z Xu. Regeneration of fertile plants from embryogenic suspension culture protoplasts of *Sorghum vulgare.* Plant Cell Rep 9:51–53, 1990.
42. T Hagio, AD Blowers, ED Earle. Stable transformation of sorghum cell cultures after bombardment with DNA-coated microprojectiles. Plant Cell Rep 10:260–264, 1991.
43. M Battraw, TC Hall. Stable transformation of *Sorghum bicolor* protoplasts with chimeric neomycin phosphotransferase II and β-glucuronidase genes. Theor Appl Genet 82:161–168, 1991.
44. AM Casas, AK Kononowicz, UB Zehr, DT Tomes, JD Axtell, LG Butler, RA Bressan, PM Hasegawa. Transgenic sorghum plants via microprojectile bombardment. Proc Natl Acad Sci USA 90:11212–11216, 1993.
45. AK Kononowicz, AM Casas, DT Tomes, RA Bressan, PM Hasegawa. New vistas are opened for sorghum improvement by genetic transformation. Afr Crop Sci J 3:171–180, 1995.
46. AM Casas, AK Kononowicz, TG Haan, L Zhang, DT Tomes, RA Bressan, PM Hasegawa. Transgenic sorghum plants obtained after microprojectile bombardment of immature inflorescences. In Vitro Cell Dev Biol 33:92–100, 1997.
47. JC Sanford. The biolistic process. Trends Biotechnol 6:299–302, 1988.
48. JC Sanford, FD Smith, JA Russell. Optimizing the biolistic process for different biological applications. Methods Enzymol 217:483–509, 1993.
49. JR Kikkert. The biolistic PDS-1000/He device. Plant Cell Tissue Org Cul 33:221–226, 1993.
50. ME Fromm, F Morrish, C Armstrong, R Williams, J Thomas, TM Klein. Inheritance and expression of chimeric genes in the progeny of transgenic maize plants. Biotechnology 8:833–839, 1990.
51. D McElroy, W Zhang, J Cao, R Wu. Isolation of an efficient actin promoter for use in rice transformation. Plant Cell 2:163–171, 1990.
52. AH Christensen, PH Quail. Ubiquitin promoter-based vectors for high level expression of selectable and/or screenable marker genes in monocotyledonous plants. Transgenic Res 5:213–218, 1996.
53. Z Li, MN Upadhyaya, S Meena, AJ Gibbs, PM Waterhouse. Comparison of promoters and selectable marker genes for use in Indica rice transformation. Mol Breed 3:1–14, 1997.
54. A Wilmink, BCE Van de Ven, JJM Dons. Activity of constitutive promoters in various species from the Liliaceae. Plant Mol Biol 28:949–955, 1995.
55. RM Hauptmann, V Vasil, P O-Akins, Z Abaeizadeh, SG Rogers, RT Fraley, RB Horsch, IK Vasil. Evaluation of selectable markers for obtaining stable transformants in the Gramineae. Plant Physiol 86:602–606, 1988.
56. Y Tadesse, L Sági, V Frankard, R Swennen, M Jacobs. Nutritional quality improvement of sorghum through genetic transformation. Proceedings of IX International Congress on Plant Tissue and Cell Culture, Jerusalem, Israel: Kluwer Ac. Publ, 1999, pp. 617–620.
57. JK Bryan. Synthesis of the aspartate family and branched-chain amino acids. In: BJ Miflin, ed. The Biochemistry of Plants. Vol 5. Amino Acids and Derivatives. New York: Academic Press, 1980, pp 403–452.
58. O Shaul, G Galili. Increased lysine synthesis in transgenic tobacco plants expressing a bacterial dihydrodipicolinate synthase in their chloroplasts. Plant J 2:203–209, 1992.
59. O Shaul, G Galili. Threonine overproduction in transgenic tobacco plants expressing a mutant desensitized aspartate kinase from *Escherichia coli.* Plant Physiol 100:1157–1163, 1992.
60. G Galili. Regulation of lysine and threonine synthesis. Plant Cell 7:899–906, 1995.
61. AL Demain. Overproduction of microbial metabolites and enzymes due to alterations of regulation. In: TK Ghose, A Fischer, eds. Advances in Biochemical Engineering. Vol. 1. New York, Berlin, Heidelberg: Springer, 1975, pp 113–142.
62. I Negrutiu, A Cattoir-Reynaerts, I Verbruggen, M Jacobs. Lysine overproducer mutants with an altered dihydrodipicolinate synthase from protoplast culture of *Nicotiana sylvestris* (Spegazzini and Comes). Theor Appl Genet 68:11–20, 1984.
63. M Ghislain, V Frankard, M Jacobs. A dinucleotide mutation in dihydrodipicolinate synthase of *Nicotiana sylvestris* leads to lysine overproduction. Plant J 5:733–743, 1995.

64. S Vernaillen, F Laureys, M Jacobs. A potential screening system for identifying sorghum ecotypes with increased lysine in the seeds. Plant Breed 111:295–305, 1993.
65. O Shaul, G Galili. Concerted regulation of lysine and threonine synthesis in tobacco plants expressing bacterial feedback-insensitive aspartate kinase and dihydrodipicolinate synthase. Plant Mol Biol 23:759–768, 1993.
66. SC Falco, T Guida, M Locke, J Mauvais, C Sanders, RT Ward, P Webber. Transgenic canola and soybean seeds with increased lysine. Biotechnology 13:577–582, 1995.
67. H Karchi, O Shaul, G Galili. Lysine synthesis and catabolism are coordinately regulated during tobacco seed development. Proc Natl Acad Sci USA 91:2577–2581, 1994.
68. JL Bennetzen. The potential of biotechnology for the improvement of sorghum and pearl millet. Proceedings of International conference on Genetic Improvement of Sorghum and Pearl Millet, Lubbock, TX, 1997, pp 13–20.
69. JL Bennetzen. Biotechnology for sorghum improvement. Afr Crop Sci J 2:161–170, 1995.
70. PE Arriola. Crop to weed gene flow in sorghum: Implications for transgenic release in Africa. Afr Crop Sci J 2:153–160, 1995.
71. Z Zhao, T Cai, L Tagliani, M Miller, N Wang, H Pang, M Rudert, S Schroeder, D Hondred, J Seltzer, D Pierce. *Agrobacterium*-mediated sorghum transformation. Plant Mol Biol 44:789–98, 2000.

# 54
# Sunflower Seed

**Günther Hahne**
*CNRS and Louis Pasteur University, Strasbourg, France*

| | | |
|---|---|---|
| I. | INTRODUCTION | 813 |
| | A. The Plant | 814 |
| | B. History of Sunflower Biotechnology Applicable to Genetic Engineering | 815 |
| II. | CULTURE AND REGENERATION SYSTEMS | 816 |
| | A. Regeneration from Cultured Tissues | 816 |
| | B. Regeneration from Isolated Protoplasts | 818 |
| | C. Properties of Regenerated Plants | 819 |
| | D. Problems and Solutions | 819 |
| III. | TRANSFORMATION AND TRANSGENIC PLANTS | 820 |
| | A. Direct Gene Transfer | 821 |
| | B. *Agrobacterium* Species–Mediated Gene Transfer | 822 |
| | C. Characterization of Transformants | 823 |
| | D. Introduced Characters | 824 |
| | E. Problems and Solutions | 827 |
| IV. | CONCLUSION AND PERSPECTIVES | 828 |
| | REFERENCES | 828 |

## I. INTRODUCTION

Most of the sunflower (*Helianthus annuus* L.) plants cultivated worldwide are grown for their seeds, which are rich in oil and protein. Although a small proportion of the production is directly consumed in the form of grains, e.g., as a snack or birdfeed, the largest part is processed to yield a highly appreciated oil with a balanced composition of fatty acids and good nutritious value. In several countries, sunflower oil is the exclusive source of vegetable oil; on a worldwide scale, it is among the four most important species cultivated as oil crops. This leading position has been made possible by a very successful application of conventional breeding. The introduction of high-performing $F_1$ hybrids, based on the cytoplasmic male sterility (CMS) system developed in 1969 by Leclercq (1), has promoted very fast development of this crop. The economic importance of sunflower has stimulated interest in the development of biotechnological approaches, which are hoped to extend the possibilities for further improvement of this crop. Simple conventional in vitro techniques are already applied with good results; the acceleration of the growth cycle by embryo culture, which allows the production of up to five generations per year, is a good example.

However, the potential impact of genetic engineering is estimated to exceed greatly that of conventional techniques. The transfer of isolated and well-characterized genes is expected to confer novel traits that would otherwise remain inaccessible on this important crop.

In vitro culture and transformation of sunflower have met with unexpected difficulties. Reliable and efficient protocols have been slow to be developed. To date, only a few publications relate information concerning the production and properties of transgenic sunflower plants. The purposes of the present review are to give an overview of the existing literature on this subject and to present some of the tissue culture systems that have been developed. The points that are most likely to be responsible for the difficulties encountered in the establishment of transformation systems are highlighted.

Any approach to transformation must take into account the specific biological properties of the plant studied. A brief description of some of the biological characteristics of sunflower will therefore facilitate the understanding of the choices made and difficulties encountered while developing transformation protocols for this recalcitrant species. Some background information on agronomy and breeding practices will be helpful for the identification of traits and characters to be introduced into sunflower by genetic engineering.

## A. The Plant

### 1. Biological Characteristics

The genus *Helianthus* (Asteraceae) comprises 67 species (2,3) which are either annual or perennial species. Some of these species, originating from arid to more temperate climates (4), are noxious weeds in their region of origin; others are appreciated for their ornamental value. Only *H. annuus* (the cultivated sunflower) and *H. tuberosus* (Jerusalem artichoke) are cultured on a large scale.

Sunflower seeds are large (6–10 mm) achenes that contain not only the embryo but also an air cavity, which often harbors fungal spores. Seedlings feature a long, sturdy hypocotyl and two large and fleshy cotyledons. Adult plants are tall (most commonly 1.5 to 3 m) and carry large leaves with an initially alternate, later spiral arrangement. Most commercial varieties and their female parents are characterized by a single stem bearing a single inflorescence, whereas ornamental varieties and male parent lines (pollinators) are most often branched for reasons of prolonged flowering time and thus more complete pollination. Floral induction, i.e., the transition from the vegetative to the reproductive state of the apical meristem, appears to be relatively independent of environmental stimuli, responding to developmental clues (5). Floral induction occurs early in sunflower development, approximately 20 days after germination. Anthesis occurs 8–12 weeks after germination, depending on genetic and environmental factors. As sunflower belongs to the tubuliflorae of the Asteraceae family, the inflorescence is radially symmetrical and composed of two types of flower. The outside rows are formed by sterile ray flowers, the fused petals of which are responsible for the attractive yellow ring. The incomplete disk flowers are fertile and responsible for the production of seeds. The opening of these flowers is asynchronous and proceeds in a centripetal fashion, as one to several rows of florets open per day. This process may occupy 5–10 days, depending on the size of the inflorescence and climatic conditions. The embryos develop quickly and may have reached their final size, but not their final volume and weight, by 12–15 days after pollination. The time to full maturity, however, is in the range of 2–3 months. For experimental purposes, immature embryos can be extracted as early as a few days after pollination. When germinated in vitro, they develop into normal plants that can be transferred to the greenhouse or field without difficulty. For more detailed information on sunflower anatomical and developmental characteristics the reader may refer to the authoritative publication edited by Carter (6).

## 2. Agronomy and Breeding

As is that of most *Helianthus* species, the natural habitat of sunflower is a warm and rather dry climate. In kernels of modern sunflower varieties, 50–60% or more of the dry weight is contributed by triglycerides (80% of the embryo). On average, sunflower oil is rich in unsaturated fatty acids, particularly linoleic acid (65%) and oleic acid (around 20%) (7). The composition of the oil is variable and depends on the genetic constitution but is also under environmental influence. The most important of the latter factors is the temperature. Whereas oil from plants grown in a temperate climate may contain 60–70% linoleic acid, this value may drop to 30% for plants grown in a Mediterranean or subtropical climate (7). Oil yield and oil composition remain the most important breeding goals. The production of specialty oils with reproducible composition is becoming increasingly desirable. Further applications of sunflower of more limited importance include its use as a fodder plant (either green or after silage), production of large grains with loose shells for human consumption, and cultivation for horticultural purposes, for which the most important traits are flower color and long flowering period (highly branched phenotypes).

Viral and bacterial diseases are of no particular concern, but a number of fungal pathogens are the cause of major yield losses. The most widespread fungal pathogens affecting sunflower include white rot (*Sclerotinia sclerotiorum*), gray mold (*Botrytis cinerea*), downy mildew (*Plasmopara halstedii*), stem canker (*Phomopsis helianthi*, formerly *Diaporthe*), *Alternaria helianthi*, phoma black stem (*Phoma oleracea* var. *helianthi-tuberosi*), as well as sunflower rust (*Puccinia helianthi*) (8). Genes conferring resistance to some of these diseases have been introgressed into cultivated sunflower from compatible *Helianthus* species by means of wide crosses, but other diseases still require heavy chemical treatments or are, to date, without effective remedy (*S. sclerotiorum*). Fungal pathogens continue to figure among the most important concerns of plant breeders. Broomrape (*Orobanche cumana* Wallr.) is a plant parasite that causes great crop losses in the region of the former USSR and areas bordering the Mediterranean Sea.

Of the hundreds of insect species that use sunflower for food or shelter, the small percentage of them that are the cause of serious yield losses still represent a large number (9). They may cause problems during seed storage, on the preestablishment level (planting through seedling), or on the postestablishment level (young plant through maturity). Insect pests are less perceived as a problem in Southern and Western Europe than in other areas where sunflower is grown as a field crop. Traditional pest management mainly relies on the use of chemical insecticides, although alternative strategies such as rotating crops, altering planting dates, and using pest-resistant cultivars may contribute to more profitable sunflower production.

Virtually all sunflower varieties grown today for oil production are $F_1$ hybrids based on the CMS system introduced by Leclercq in 1969 (1), and the breeding scheme is correspondingly complex. The selection and production of hybrid lines require the creation and maintenance of corresponding maintainer and restorer lines (7,8). This intricate system is likely to have consequences on the strategies to be chosen for the introduction of a foreign gene into a high-yielding commercial sunflower variety.

### B. History of Sunflower Biotechnology Applicable to Genetic Engineering

The use of sunflower for human consumption dates back to very ancient times, when it was used by, e.g., the American Indians. Sunflower became an important source of edible oil in the Russian empire at the beginning of the 20th century but gained its worldwide importance only in the 1980s (7). Consequently, the interest in biotechnological improvement of sunflower is also quite recent.

Modern researchers have little historical base to start from. Early literature on plant tissue culture does mention reports of in vitro culture of *Helianthus* spp., but they are sparse (e.g.,

10–13). The adaptation to sunflower of protocols that are successful for plant regeneration and transformation in other species has proved difficult. Long-term callus cultures have shown a strong tendency to lose their regeneration potential. Most reported of regeneration of plants from callus made use of immature embryos as donor material, and plant regeneration followed callus induction relatively fast (14–17). Induction of callus from other tissues or organs as well as long-term callus culture followed by plant regeneration have also been described (18–20). However, none of the callus-based regeneration protocols has given rise to a published reproducible transformation protocol, most probably because of the low efficiency of plant regeneration in all these protocols. Direct regeneration systems, i.e., systems in which shoots or somatic embryos are induced from cells originally contained in the explant and regenerated plants are formed without an intervening callus phase, have been established for several tissues. Immature embryos have been shown to be capable of direct induction of somatic embryos (21–23) or shoots (23). Large numbers of shoots can be induced on cotyledons of seedlings or mature seeds (24–29), or even on the primary leaves of such seedlings (19,30). Such epiphyllous shoot formation is, however, limited to rather particular conditions of in vitro culture. It cannot be induced on older leaves or leaves obtained from older donor plants.

The potential to transfer foreign genes to sunflower tissue by using *Agrobacterium tumefaciens* was recognized quite early. Crown gall tumors of sunflower have been the object of numerous studies (10,31–33), yet the combination of such infection with conventional regeneration protocols has been crowned with little success. The production of transgenic callus and subsequent plant regeneration was successful for *H. annuus* in a single instance (34) but suffered from low efficiency and poor reproducibility. A similar approach using the interspecific hybrid *H. annuus* × *H. tuberosus* yielded a transgenic plant that proved to be sterile (35). With these two notable exceptions, all reports on transgenic sunflowers published to date are based on *Agrobacterium* species–mediated gene transfer to embryonic axes of mature embryos with subsequent plant regeneration from the meristematic region (36–41).

## II. CULTURE AND REGENERATION SYSTEMS

### A. Regeneration from Cultured Tissues

In the following, we examine the regeneration systems that have been described in relation to transformation of sunflower and highlight their respective merits and shortcomings. Most known approaches to plant regeneration have been applied to sunflower and, in general, have given rise to regenerated plants in at least one instance. However, only a few approaches could be developed into efficient, reliable protocols that work in several laboratories, robust enough to be combined with gene transfer procedures. It is useful to distinguish indirect regeneration systems from direct ones that involve an undifferentiated callus phase.

An efficient system for the regeneration of true-to-type, fertile plants is the indispensable prerequisite for any transformation protocol. Any such transformation system must satisfy a number of conditions, including (a) gene transfer to a sufficient number of cells in the explant that will be or become involved in the morphogenic events leading to the regenerated plant; (b) an efficient selection agent capable of enriching the population of regenerating cells with those expressing the foreign genetic construct; and (c) the potential to maintain and multiply the obtained transformants in vitro or ex vitro, by vegetative multiplication or seed production.

#### 1. Somatic Embryogenesis

Vigorously growing callus cultures are easily obtained from most sunflower tissues. Such cultures can be maintained for prolonged periods and are convenient material for the induction of suspension cultures. However, sources of embryogenic callus are more limited. Embryogenic callus has been obtained from hypocotyl segments (19,42) and immature embryos (14). The regeneration potential of such cultures appears to be limited to a few selected genotypes, which are not freely available to the public. The feasibility of induction of somatic embryos in suspension cultures has been described (43). The more general experience with callus or suspension cultures of whatever origin is that they lose their regeneration potential very rapidly, often during the first subculture. It is not unusual to see somatic embryos develop on callus emerging from the explant. This observation is true for a wide range of media (unpublished). However, we have never been able to maintain this capacity for somatic embryo production once the callus was detached from the original explant, and we are not aware of any callus or suspension culture of *H. annuus* that has been maintained for a prolonged period and still has reasonably high embryogenic activity.

Essentially two types of explants have been used with success in the direct induction of somatic embryos in *H. annuus*, i.e., immature embryos (21–23,43) and epidermal thin layers from hypocotyl (44). Plants have also been regenerated via somatic embryogenesis from cotyledons in at least one instance (45). Direct embryogenesis in sunflower is distinguished by its rapidity. Embryonic features such as the presence of storage proteins may become detectable in developing somatic embryos as early as a few days after the explant is put in culture (46). Plantlets developed from such somatic embryos, separated from their explant 2 weeks after induction, may be transferred to the greenhouse 4–5 weeks after the beginning of the experiment.

The origin of somatic embryos has been the object of detailed study in only a few regeneration systems. Somatic embryos that develop directly on immature zygotic embryos have been shown to have a multicellular origin (47). Under the studied culture conditions, their induction is restricted to a particular region of the explant, i.e., the zone that will later become the crown. It appears that this particular location is not due to the presence of a predetermined cell type but rather to clues perceived in the functioning of the culture conditions (48,49). It is not known whether somatic embryos of unicellular origin occur in other culture systems of *H. annuus*. However, unicellular origin of both directly and indirectly forming somatic embryos has been demonstrated in callus cultures of the wild relative, *H. smithii* (50).

The induction of somatic embryos from cultured sunflower tissues is in general rather localized; thus, few cells are involved in the formation of a novel structure. Therefore, the probability is low that a foreign genetic construct, transferred by whatever method, will touch a cell that later will become incorporated in a somatic embryo. Since direct regeneration events are extremely fast [the induction phase may be in the order of only hours (48)], and since callus cultures lose their regeneration potential very rapidly, enrichment of transgenic cells by selection on a selective agent such as kanamycin is not practical. It is thus not surprising that no system for the transformation of sunflower based on either direct or indirect somatic embryogenesis of whatever origin has been proposed.

## 2. Caulogenesis

Depending on the exact combination of explant type and cultural conditions, normal, fertile sunflower plants can also be regenerated via adventitious shoot development. Shoots have been regenerated from callus in several instances (15,16,18,20). Such calli were induced on the same type of tissues as the embryogenic calli described. Furthermore, one report describes the regeneration of shoots from callus induced on cultured anthers (51). However, the limitations that restrict the use of calli for plant regeneration via somatic embryogenesis are also valid in the case of adventitious shoot induction: the morphogenic potential is extremely volatile, in addition to being highly genotype-specific (15). The reproducibility and efficiency of the pathway of indirect caulogenesis appear too limited to be applied to transgenic plant production.

The situation is different for direct caulogenesis. Several culture systems in which shoot induction and plant regeneration proceed rapidly and with high efficiency have been described. The explants that have been studied in most detail are cotyledons of mature seedlings (24–29,52–54) and immature embryos (14,17,46–49,55). In both cases, the morphogenic reaction may occur in a wide range of genotypes. The genotypic influence affects the efficiency of the system. The variation introduced by the genotype can be eliminated to a certain extent by the culture conditions of cotyledon explants (54). In this particular case, ethylene concentration was one of the key factors that determined the regeneration efficiency.

As is true of somatic embryogenesis, direct induction of adventitious shoots on explants of sunflower is extremely rapid (a few hours), and the time from the onset of the in vitro experiment to the harvest of seeds may be as short as 4 months. The induction time has been precisely studied in the case of immature embryos (48). The first characteristic metaphase plates can be observed as soon as 12 hours, and the morphogenic response is irreversibly determined approximately 24 hours after the explants are put in contact with the culture medium. Such a rapid response may be convenient for reasons of experimental design, but it leaves little possibility for selection of transformed cells. Furthermore, the conditions for optimal regeneration response and those for optimal gene transfer, respectively, can be quite different (40). In such a situation, the time window is too short for any experimental adjustment of the conditions between the two key steps, i.e., gene transfer and induction of the regeneration event.

## B. Regeneration from Isolated Protoplasts

Protoplasts can be obtained easily and in large number from diverse sunflower tissues including hypocotyl, cotyledons, and leaves [(56–59); for review see Ref. 60]. It is not difficult to obtain callus formation at high frequency from those protoplasts. However, the regeneration of plants from such calli is more delicate and proceeds at a rather low frequency: in the percentage range under the best conditions, and often several orders of magnitude below this value (61–66). For this reason, culture conditions for sunflower protoplasts and their effect on the regeneration capacity have been the subjects of numerous studies (67–72). These studies have demonstrated a peculiar type of behavior of sunflower protoplasts. They either form loose or compact calluslike colonies, as would be expected from experience with protoplast culture experiments in most other species, or they may develop into highly compact structures that, as a result of their suggestive appearance, have been termed *proembryoids* (59,67,69) although they were never shown to develop beyond an early stage. Their embryonic nature has been contested on grounds of histological appearance (73). Their occurrence is strictly linked to embedding in media solidified with agarose. This phenomenon of cytodifferentiation as a function of environmental conditions has triggered several detailed studies (74–76) which, however, had no immediate relevance to regeneration efficiency. Stimulated by the persistently low regeneration rate from sunflower, attempts have been made to increase this rate by electrostimulation (72) or to identify and select for protoplast subfamilies with increased regeneration potential (68,76–78). Although these studies have increased our understanding of the cellular responses that occur in the early phases of culture of sunflower protoplasts, regeneration frequencies are to date still too low to be exploited for direct gene transfer with subsequent plant regeneration. The production of transformed callus presents no major problem (79); however, the regeneration of plants from such callus has yet to be described.

From an analysis of the published results it becomes evident that the regeneration potential of sunflower protoplasts is extremely genotype-dependent. Frequencies of plant regeneration from sunflower protoplast cultures remain low in spite of rather intense efforts to improve culture conditions. Further work on media and culture conditions therefore appears not to be very promising unless novel fundamental approaches are found. However, no fundamental obstacle that would limit the use of protoplasts for sunflower transformation, other than the overall efficiency of this approach, has become evident.

## C. Properties of Regenerated Plants

Genotype, explant choice, and culture protocol have a decisive influence on the quality of the regenerating structures obtained. Both somatic embryos and adventitious shoots may be very vigorous and present no difficulty for root formation and subsequent transfer to the greenhouse (22, 23,26,27,45,63). These cases are most often encountered with direct regeneration systems. In those instances in which regeneration occurs from callus or cultured protoplasts, the regenerated shoots or embryos tend to be weak and generating root induction and development is difficult. One possibility for increasing the success rate of transfer to the greenhouse and the vigor of the plants obtained is to use the regenerated shoots or plantlets with defective root systems as scions and graft them onto sunflower rootstocks. The grafting can be performed in vitro, using in vitro grown sunflower seedlings, or performed directly in the greenhouse, using young sunflower plantlets (61).

Grafted or on their own roots, regenerated sunflower plants themselves (R0) generally present a peculiar phenotype. They are stunted, often not taller than 20–50 cm, although individual grafted specimens may almost reach the size of normal sunflowers. Even in genotypes characterized by a single head, multiple branching and production of several inflorescences are common phenomena. We have seen the size of the individual inflorescences vary between a few ray flowers surrounding a small number of disk flowers up to sunflower heads of respectable size with more than 200 seeds. However, the reasons for this altered phenotype appear to be a consequence of the in vitro culture on the physiological rather than the genetic level. In general, seed set presents no problems, even in the smallest inflorescences, and the subsequent generations consist of perfectly normal sunflower plants. Little published information is available concerning the genetic stability of regenerated sunflower plants. The most often used culture systems do not seem to cause a high rate of somaclonal variation or other (epi)genetic alterations. In various instances, no significant deviation from the parental phenotype has been detected over several generations (22,23). Karyotypic variations have not been detected in plants three generations after plant regeneration from immature embryos (23,80). Plants regenerated from protoplasts as well as their offspring have been followed over several generations with microsatellite markers. No sign of somaclonal variation could be detected by this technique, irrespective of the donor tissue used for protoplast isolation (hypocotyl, cotyledon: F. Charrière and G. Hahne, unpublished). However, plants with altered coumarin content were obtained after mercuric chloride treatment of immature embryo-derived callus (81), and Witrzens and coworkers (17) mention the possible occurrence of somaclonal variants among plants regenerated from a similar source. However, this phenomenon has not been studied in detail. From a practical point of view, lack of genetic stability is not perceived as a major obstacle to the application of biotechnology to *H. annuus*.

## D. Problems and Solutions

Most of the classic approaches to plant regeneration from in vitro cultured cells and tissues have been shown to be possible also with *H. annuus*. We can therefore conclude that no fundamental obstacle exists to applying such culture systems to the elaboration of transformation protocols. With this background, what are the reasons that few laboratories master the transfer of foreign genes into this species? Why is it that many researchers in this field still speak of a recalcitrant species? One possible answer that probably comes close to the truth is that the problem is not that there is a well-defined deficiency in one or several reactions but that the researcher is confronted with an accumulation of independent individual difficulties. The recurring problems are (a) low efficiency, (b) low reliability, and (c) low universality. Indirect culture systems, which could easily be adapted to gene transfer and selection of transformed cells, have regeneration efficiencies that are difficult to combine with the inherently low efficiency of gene transfer into a transformation system with acceptable overall output. Direct systems do not suffer the problem of low reliability and low efficiency to the same degree but are often too rapid to be submitted to a practically useful selection scheme. Selection therefore has to occur on the much more difficult shoot stage. Even more importantly, gene transfer does not necessarily target the few cells within the tissue that are competent for regeneration under the experimental conditions. Therefore, even if gene transfer and plant regeneration were each efficient processes, the combination of both would still be inefficient in many instances. Finally, there is a very pronounced genotype dependence in most sunflower in vitro culture systems. Although this is true for many species, this condition is aggravated by the difficulty of vegetatively propagating a sunflower clone with superior regeneration capacities. Multiplication by shoot cuttings can be successfully obtained, but because floral induction is rapid and more or less independent of environmental clues, such shoots invariably transit to the floral stage. This precocious flowering is a terminal stage that ends vegetative development, be it in planta or in vitro [e.g., Refs. (82–84)]. The consequence is that genotypes with superior regeneration capacity can only be maintained by genetic means. Breeding sunflower for this character has been shown to be feasible (19,34,52,85), but this approach is inconceivable for most focusing on basic research laboratories. Unfortunately, none of the tested public lines has proved extremely responsive for regeneration responses.

In view of the accumulation of difficulties that impede the simple adaptation of protocols proven in other species, it is not surprising that transgenic sunflower plants have been slow to come into existence. With one exception, all of the few publications describing transgenic sunflower plants are based on yet another culture system, i.e., regeneration from a tissue containing existing meristems. The most convenient tissue for this approach are the embryonic axes, which can be infected, under certain well-chosen conditions, by *Agrobacterium tumefaciens*; plants that bear transgenic seeds can be regenerated by this approach (39–41,86,87).

## III. TRANSFORMATION AND TRANSGENIC PLANTS

The capacity of *A. tumefaciens* to transform sunflower tissues was recognized quite early, and crown gall tumors obtained from sunflower plants have been the subject of numerous studies (e.g., Refs. 10,31–33). However, this transformation procedure has only more recently been adapted to nontumorous callus tissue, employing disarmed *Agrobacterium* sp. strains (34,88–91). Attempts to regenerate transgenic plants from such callus were unsuccessful because of the lack of regeneration systems from such material, with one exception (34), which involved a specially bred proprietary genotype. Confirmed transgenic plants were obtained in spite of the drastic loss of regenerative capacity of the callus under selective conditions. However, this initial report remained isolated, and a short note by Hartmann (92) indicates that difficulties with the regeneration and selection system made production of further transgenic plants by this system impractical. The leaf disk approach (93) that has become the standard transformation technique for many species is inapplicable to sunflower because of the lack of an appropriate regeneration system. It is not surprising, therefore, that a wide range of combinations of culture techniques and gene transfer methods had to be evaluated before transgenic sunflower plants could be generated in a reproducible albeit labor-intensive and comparatively inefficient manner.

## A. Direct Gene Transfer

Direct gene transfer, i.e., the introduction of foreign genetic material by physical methods and without the help of a biological vector, is based on three principal techniques: direct gene transfer to isolated protoplasts, microinjection, and bombardment with microparticles. Whereas the former technique requires an efficient regeneration system from isolated protoplasts, the latter two are theoretically applicable to intact plant organs such as meristems and embryos.

### 1. Protoplasts

Direct gene transfer to isolated protoplasts can be accomplished by either treatment with polyethylene glycol (PEG) or a short exposure to a strong electric field (electroporation) (94,95). Both approaches have yielded transgenic plants in an impressive number of species for review, see Ref. 96. Introduction of the foreign deoxyribonucleic acid (DNA) is followed by a short period (several hours) of expression that later decreases. Because of the high copy number that is generally introduced by these techniques this transient expression is usually quite strong compared to the expression of the stably integrated genetic material, which becomes apparent only later. Although transient expression may affect the majority of the protoplasts that survive the transformation procedure, stable integration of the introduced genetic construct is a rare event and can be detected in only a small fraction of the treated protoplast population.

Both PEG-mediated gene transfer and electroporation have been shown to be applicable to sunflower protoplasts for transient expression studies. The conditions for optimal gene transfer are comparable to those that have been found to be useful in other species (97). Callus resistant to selection on kanamycin-containing medium was obtained after PEG-mediated DNA uptake by sunflower protoplasts (79). The frequency of occurrence of stably transformed colonies was low in this study ($4 \times 10^{-6}$). No transgenic plants could be obtained since no regeneration protocol was available for the genotype used.

These two studies (79,97) have not been followed by other publications reporting on the use of sunflower protoplasts for direct gene transfer. It is clear from this work that no specific properties exist in sunflower protoplasts that would make them unsuitable to gene transfer experiments. The frequency of occurrence of stable transformants is low but not unusual for this type of experiment. However, the limitations for plant regeneration from sunflower protoplasts described, in particular the low efficiency of shoot induction and the rapid loss of the morphogenic potential, are the principal reasons why this approach to the production of transgenic sunflower plants was rapidly considered impractical.

### 2. Microinjection

Microinjection is a preferred technique for the transformation of animal cells, but the presence of a rigid cell wall complicates the application of this technique to plant cells. The production of transgenic rapeseed plants by this technique has been shown to be feasible by microinjection into microspore-derived embryoids (98). A subsequent step of secondary embryogenesis allowed nonchimeric transformants to be obtained. Although this result has been reported only once, it demonstrates that microinjection may be a valid alternative for the transformation of species when other approaches have failed. However, it requires the availability of an efficient regeneration system from a tissue or organ that is suitable for microinjection.

The sunflower ovule satisfies this criterion since it is possible to regenerate plants from young embryos, as has been demonstrated in an interspecific hybrid (99). Microinjection of such young embryos *in ovulo* was reported in one conference abstract (100). Technical details and a characterization of the plants obtained are not available since this initial report was not followed by further publications, but the approach has given rise to a patent application (Table 1). It is

**Table 1** Patent Applications with relevance Specific to sunflower transformation

| Title | Owner | Registration number |
|---|---|---|
| Plant transformation method using microprojectile bombardment and *Agrobacterium* sp. | Pioneer Hi-Bred Intern. | 92-09802 |
| Plant transformation using *Agrobacterium* sp. and microprojectile bombardment | Pioneer Hi-Bred Intern. | 92-09803 |
| Plant transformation by direct microinjection of exogenous DNA into fertilized ovule | South Dakota State University | 94-03332 |
| Production of entirely transformed transgenic plants | Biocem, France | 95-05874 |
| Recovery of transformed plants without selectable markers by nodal culture and enrichment of transgenic sectors | Pioneer Hi-Bred Intern. | 98-51806 |

*Source*: Derwent Biotechnology Abstracts (http://www.derwent.com/).

likely that the cumbersome and time-consuming isolation procedure of the starting material in combination with the inherent complexity of the microinjection technique are major obstacles to the development of a routine transformation system based on this approach.

### 3. Particle Gun

The bombardment of plant cells with microparticles that have been coated with DNA allows the introduction of this DNA into plant cells without prior removal of the cell wall. The transformation of species for which a transformation protocol had not been available has thus become possible (e.g., Ref. 101). The same technique also allows transient expression studies without the need to isolate protoplasts, thus maintaining a more natural cellular environment (e.g., Refs. 102, 103). However, prudence requires cautious interpretation of the results because transient expression often concerns dying cells (104).

The capacity bombarding intact tissues with inherent regeneration potential such as meristems (101,105) would make this approach to gene transfer appear ideal for species for which other efficient regeneration systems could not be established, such as sunflower. Yet the literature on this subject is very limited, although this method has been the subject of patent applications related to sunflower transformation (Table 1). Immature embryos have been bombarded successfully for transient expression, but no transgenic plants could be obtained (106). Bombardment of other sunflower tissues was inconclusive in this study. The only published account, of the use of the particle gun resulted in transgenic sunflower plants via an unusual modification of the technique. Naked particles (i.e., those having no foreign DNA) were used to introduce multiple wound sites and thus to increase the efficiency of infection with *Agrobacterium* sp. (37).

### B. *Agrobacterium* Species–Mediated Gene Transfer

*Agrobacterium tumefaciens* appears to be the most suitable vector for the transformation of sunflower: all transgenic sunflower plants for which details on the transformation protocol have been made available were obtained by using this agent. The difficulties that must be surmounted in adapting commonly used protocols for the infection of sunflower tissue to a suitable regeneration system reside essentially in the targeting of the infection to cells that retain their regeneration potential after treatment. A variety of regeneration systems have been tested for their compatibility with the *Agrobacterium* sp. transformation approach, including hypocotyl explants (34,90), seedling cotyledons, and immature embryos (40,107), and embryonic axes (36–39,86,108). However, all studies in which transgenic sunflower plants were obtained in a reproducible and reasonably efficient manner had made use of embryonic axes as the initial explant.

This technique was originally proposed by Schrammeijer and coworkers (36), who treated longitudinally cut apical parts of embryonic axis with an *Agrobacterium* sp. suspension. Shoots were regenerated from the persisting meristematic tissue. In some cases agrobacteria had been able to penetrate this tissue, and plants could be obtained with chimeric expression of the introduced marker gene (GUS). Although the efficiency was very low in this experiment (2 chimeric shoots were obtained from 1500 treated meristems) and the regenerated plants turned out to be sterile, this report demonstrated the compatibility of *Agrobacterium* species–mediated gene transfer with the regeneration from preexisting meristems. Subsequent work elaborated on this system. The introduction of multiple microwounds allowed Bidney and associates to increase the efficiency of the transformation system (37,109). Similar results were obtained by Knittel and colleagues (39), who also showed that in spite of the use of preexisting meristems, the overall efficiency remains a function of the genotype utilized. In fact, not all regenerated shoots originate in preexisting meristematic tissue, and one of the factors that determine the efficiency of regeneration and the proportion of transgenic shoots is the capacity for adventitious shoot induction in the vicinity of the original meristematic tissue (41). Transformation of the remaining meristematic tissue by the agrobacteria is possible but mainly leads to the development of chimeric shoots. Adventitious shoots tend to be more uniformly transformed (41). Other means of introducing wounds have since been devised, including vigorous shaking with glass beads (38) and partial digestion with pectolytic enzymes (108). The putatively transformed regenerated shoots tend to be weak, and their transfer to the greenhouse is a major difficulty in the process of producing transgenic offspring. Precocious induction of flowering and resulting poor root formation and vegetative growth are among the primary causes of this problem. Grafting the weak shoots at an early stage of their development may significantly increase the yield of transgenic offspring (M. Burrus, G. Hahne, C. Himber, J. Molinier, P. Rousselin, unpublished observation; for details, see http://ibmp.u-strasbg.fr).

It is difficult to compare the efficiencies obtained in these experiments because the basis of calculation is not standardized. However, it is fair to state that sunflower cannot yet be listed among the species for which transformation is efficient and problem-free. Although the production of regenerated shoots that express a visible marker gene such as *uid*A (GUS) at some point is usually in the range of 50–100%, the proportion of shoots that stably express the foreign construct in a tissue likely to give rise to transgenic offspring is much lower (< 10%). The overall efficiency from the explant (embryonic axis) to the confirmed transgenic seed-bearing plant is in the percentage range (fertile plants/explants). Although the transformation of embryonic axes by *A. tumefaciens* is conceptually simple, an easily applicable universal protocol is not yet available. Depending on the genotype, dissection technique, bacterial strain, selectable marker, in vitro culture, and greenhouse conditions, considerable fine-tuning is necessary before the technique is perfectly mastered by a laboratory.

As a result of the difficulties encountered by most laboratories in developing a practically useful transformation system for sunflower, many ingenious approaches have been explored. Some were novel developments that led to patents protecting the process. Some patent applications with relevance to sunflower transformation are listed in Table 1.

### C. Characterization of Transformants

Published information characterizing transgenic sunflower plants in detail is not available. Primary transformants that have integrated a genetic construct whose expression has no influence on plant development and morphological features are visually indistinguishable from untransformed regenerants. They tend to be small and weak plantlets, often highly branched and bearing small multiple inflorescences. Grafting on strong rootstocks often results in much more vigorous plants with significantly increased seed production. Depending on the origin of the regenerating shoots (preexisting meristematic tissue or adventitious buds), primary regenerants are more or less chimeric for the presence of the transgene. This is not a serious problem since vegetative multiplication is not feasible for sunflower, and passage to the following generation is an obligatory step for the maintenance of the transgenic state. The $F_1$ generation is of course heterogeneous but free of chimeras. The number of transgenic seeds obtained from a primary regenerant is highly variable and depends on many factors, including plant size, number and size of inflorescences, and degree of the chimeric state. As in untransformed regenerants, seed yield may vary from several to several hundred (G. Hahne, C. Himber, J. Molinier, unpublished observation).

Although numerous field trials with transgenic sunflowers have been conducted (Table 2), molecular and genetic data on the structure and inheritance of the transgenes have not been published.

## D. Introduced Characters

### 1. Genetic Markers

As in most transformation systems with low efficiency, the protocols used for sunflower transformation critically depend on an efficient selection scheme. The *npt*II gene, which codes for the enzyme neomycin phosphotransferase II (NPTII), is among the oldest and most widely used selection markers in plant transformation. This enzyme inactivates not only the frequently used kanamycin but also other compounds of the class of aminoglycoside antibiotics, such as gentamycin and paromomycin. Many of these compounds have been evaluated for their efficiency in selection for transgenic sunflower tissues. Depending on the tissue type and the origin of the regenerated shoots, different results have been obtained. In some cases, sunflower tissues, transgenic or not, were found to be highly resistant to aminoglycoside antibiotics, although differences were observed between individual compounds (90). In other instances, prolonged exposure of morphogenic callus to kanamycin resulted in a severe loss of its regeneration potential (34). In combination with the embryonic axis–based transformation system, selection on kanamycin seems to be the best compromise, although its use is treacherous. Because of the rapidity of the direct regeneration system, selection must be performed on the shoot level. At low and intermediate concentrations, this approach is much more prone to escapes than selection on the callus level. However, as a result of the chimeric nature of most primary transformants, selection pressure must remain on a nonlethal level because untransformed tissue at the base of the regenerants must survive in order to nourish the upper parts that potentially contain transformed portions. The selective agent, usually kanamycin, is therefore best employed in a concentration range in which untransformed, but not transformed, tissue is bleached. For the same reason, selective markers that invariably kill susceptible cells cannot be employed. Examples of such compounds are phosphinothricin (PPT) (Basta) and hygromycin. The ideal solution would consist of a nontoxic selection marker that allows visual identification of the transformants. Although the histological GUS assay yields visual information, it is unsuitable, for it requires fixation of the examined tissue. Two viable visual markers are available: The first, Luciferase, is an enzyme that produces chemiluminescence, which allows detection of transgenic plant material (110). This marker is highly specific and its detection relatively free of artefacts, but the level of the light emitted is very low. Consequently, the identification of transgenic plant material depends on the availability of sophisticated and costly equipment. The other marker is green fluorescent protein (GFP), which causes fluorescence characteristic of the transgenic plants and may be a useful selection tool (111). Its presence can be easily detected with a conventional fluorescence microscope equipped with a suitable filter set; however, a dissecting microscope with epifluorescence equipment may be required for the identification of individuals too large to fit under a microscope. The usefulness of GFP is limited by endogenous fluorescence of the plant tissue under study. This limitation is

**Table 2** Applications for Field Trials with Transgenic Sunflowers

| Character | Declared main trait | Company | Country | Date of application | Reference number |
|---|---|---|---|---|---|
| Seed quality | Altered seed storage protein | Pioneer | United States | 1991 | 91-067-01R |
| | Altered seed storage protein | Pioneer | United States | 1992 | 92-042-02R |
| | Altered seed storage protein | Pioneer | United States | 1993 | 93-033-02R |
| | Altered seed storage protein | Pioneer | United States | 1995 | 95-033-01R |
| | Albumin synthesis | Van der Have | NL | 1995 | B/NL/95/16 |
| | Asparagine synthesis | Van der Have | NL | 1995 | B/NL/95/16 |
| | Threonine synthesis | Van der Have | NL | 1995 | B/NL/95/16 |
| | Alteration of oil composition (high stearate content) | Rustica Prograin Génétique | F | 1997 | B/FR/97/11/29 |
| Fungal disease resistance | *Sclerotinia* sp. resistance | Van der Have | United States | 1995 | 95-033-01R |
| | Chalcone synthesis | Van der Have | NL | 1995 | B/NL/95/16 |
| | Chitinase synthesis | Van der Have | NL | 1995 | B/NL/95/16 |
| | Fungal resistance | Van der Have | NL | 1995 | B/NL/95/16 |
| | Glucanase synthesis | Van der Have | NL | 1995 | B/NL/95/16 |
| | Oxalate synthesis | Van der Have | NL | 1995 | B/NL/95/16 |
| | Oxalate oxidase synthesis | SES France Van der Have France | F | 1996 | B/FR/96/02/12 |
| | Oxalate oxidase synthesis | Pioneer Génétique France | F | 1997 | B/FR/97/05/29 |
| | Oxalate oxidase synthesis | Semillas Pioneer | E | 1998 | B/ES/98/14 |
| | *Sclerotinia* sp. resistance | Pioneer | United States | 1998 | 98-044-11N |
| | *Sclerotinia* sp. resistance | Pioneer | United States | 1998 | 98-201-05N |
| | *Sclerotinia* sp. resistance | Pioneer | United States | 1999 | 99-070-04N |
| Viral disease resistance | Virus resistance | Pioneer | United States | 1997 | 97-029-03R |

(*continues*)

**Table 2** (*Continued*)

| Character | Declared main trait | Company | Country | Date of application | Reference number |
|---|---|---|---|---|---|
| Insect resistance | *Bt*-derived insect resistance (cowpea trypsin inhibitor synthesis) | Van der Have | NL | 1995 | B/NL/95/16 |
| | Insect resistance | Van der Have | United States | 1996 | 96-071-02R |
| | Insect resistance | Pioneer | United States | 1999 | 99-070-05N |
| Herbicide resistance | Glufosinate tolerance | Van der Have | NL | 1994 | B/NL/94/18 |
| | Chlorsulfuron tolerance | Van der Have | NL | 1995 | B/NL/95/16 |
| | Glufosinate tolerance | Van der Have | NL | 1995 | B/NL/95/16 |
| Marker genes | Marker | Limagrain Genetics France | F | 1994 | B/FR/94/02/13 |
| | Marker | Van der Have | F | 1994 | B/FR/94/02/19 |
| Other traits | Visual marker (GUS) | Pioneer | United States | 1995 | 95-031-02R |
| | Visual marker (GUS) | Van der Have | United States | 1997 | 97-126-01R |
| | Drought tolerance | Van der Have Cubian | E | 1994 | B/ES/94/10 |
| | Levan sucrase synthesis | | | | |
| | Confidential business information | Van der Have | United States | 1994 | 94-025-01R |
| | Fructosyltransferase synthesis | Van der Have | NL | 1995 | B/NL/95/16 |
| | Levan sucrase synthesis | Van der Have | NL | 1995 | B/NL/95/16 |
| | Male sterility/fertility restoration | Van der Have | NL | 1995 | B/NL/95/16 |
| | Nitrate reductase synthesis | Van der Have | NL | 1995 | B/NL/95/16 |
| | Nitrite reductase synthesis | Van der Have | NL | 1995 | B/NL/95/16 |
| | Confidential business information | Van der Have | United States | 1996 | 96-071-04R |
| | Broomrape control | Semillas Pioneer | E | 1998 | B/ES/98/13 |

*Sources*: Robert-Koch-Institut (RKI) (http://www.rki.de/GENTEC/GENTEC.HTM), European Commission Joint Research Centre (jrc) (http://biotech.jrc.it/gmo.htm), USDA APHIS (http://www.aphis.usda.gov/biotech/), OECD (http://www.olis.oecd.org/biotrack.nsf).
NL = The Netherlands; F = France; E = Spain

highly variable between species and tissues. The diverse sunflower tissues we have studied were characterized by high endogenous background fluorescence (G. Hahne, J. Molinier, unpublished observations).

### 2. Genes of Interest

Although transformation of sunflower is still in its infancy, several laboratories have begun to transfer genes coding for characters of agronomic interest to this important oil crop. As can be expected of an important crop plant, most of the information relating to this subject is not in the public domain. However, the difficulties and constraints encountered during large-scale cultivation of sunflower as well as specific product-oriented requirements are reflected in the novel major traits of transgenic sunflower plants that have been submitted to field trials (Table 2). In addition to field evaluation of stability of expression and inheritance of the transgene, and general performance of the transgenic plants, evaluated with the help of marker genes, the target traits encountered in these field trials concern (a) characters related to plant cultivation, such as resistance to fungal and viral diseases, herbicides, insects, and adverse environmental conditions (drought), and (b) characters related to seed quality such as modified storage protein or fatty acid composition. To date and to the best of my knowledge, no application for the market introduction of a genetically modified sunflower has been made.

## E. Problems and Solutions

For the sunflower tissues studied so far, gene transfer does not appear to be subject to particular limitations or problems, whatever the technique used. Low efficiency of stable integration is not a characteristic peculiar to sunflower, it rather is the norm for most gene transfer systems. Why is it, then, that sunflower still figures on the list of the species most unamenable to genetic engineering? A major reason resides in the lack of a suitable in vitro culture system for efficient regeneration of the cells that have received and integrated the foreign DNA in their genome. In view of the low efficiency of gene transfer, the ideal culture system would allow efficient selection for and enrichment of the rare event before initiation of the morphogenic program. Such a culture system does not exist for sunflower. Furthermore, gene uptake cannot be efficiently directed to a particular cell population such as those cells that become involved in adventitious bud or somatic embryo formation. Conversely, no experimental means is available to direct the induction of the morphogenic program to those cells that are most likely to become involved in the uptake of the foreign DNA. In consequence, although DNA uptake and plant regeneration may each be within acceptable orders of magnitude for several culture/transformation systems, their combination causes overall efficiency to drop far below a practically useful level.

The challenge therefore consists of finding (a) a reasonably efficient culture system in which the regenerating cells are identical to those that receive the foreign DNA (b) which allows for a selection system with an acceptable level of enrichment of the transgenic shoots. The system that satisfies these requirements to an acceptable degree is *Agrobacterium* species–mediated gene transfer into the apical region of embryonic axes. It is obvious that many of the parameters that are decisive for the efficiency and localization of gene transfer, shoot development from preexisting meristematic tissue, and adventitious bud induction are highly sensitive to environmental and physiological clues. The practical result of this dependence is that all steps and parameters must be fine-tuned with high precision. Consequently, the time needed for the adaptation of the protocol to a new laboratory may be longer than is usually necessary for another species.

## IV. CONCLUSION AND PERSPECTIVES

Genetic engineering of sunflower has recently become a reality. Although sunflower transformation remains cumbersome, labor-intensive, and subject to unexpected difficulties, the technology is now available to a number of private and public laboratories, as shown by the fact that field trials have been conducted in several countries. However, substantial further efforts in fundamental research are needed to clarify the mechanisms regulating in vitro development of sunflower before further improvement of the transformation system can be envisaged on a rational basis. As in many other species with high economic impact, the flow of information between the laboratories involved in research and development of genetic engineering techniques (and the resulting intellectual cross-fertilization) is not stimulated by the inevitable considerations of intellectual property right protection.

The genes introduced vary according to the respective commercial or research interests but are not fundamentally different from the genes introduced to other species that respond to similar demands. Although simple, monogenic traits prevail at the moment, the engineering of more complicated biochemical pathways will become commonplace in the near future.

## REFERENCES

1. Leclercq P. Une stérilité mâle cytoplasmique chez le tournesol. Ann Amélior Plant 19:99–106, 1969.
2. Heiser C jr. Taxonomy of *Helianthus* and origin of domesticated sunflower. In: Carter JF, ed. Sunflower Science and Technology, Madison: American Society of Agronomy, Crop Science Society of America, Soil Science Society of America, 1978, pp 31–53.
3. Heiser CB. The Sunflower. Norman:University of Oklahoma Press, 1976.
4. Heiser CB jr, Smith DM, Clevenger SB. Martin WC jr. The North American Sunflowers (*Helianthus.*) Durham, NC; The Seeman Printery: 1969.
5. Steeves TA, Hicks MA, Naylor JM, Rennie P. Analytical studies on the shoot apex of *Helianthus annuus*. Can J Bot 47:1367–1375, 1969.
6. Carter JF. Madison, WI: Sunflower science and technology. American Society of Agronomy, Crop science Society of America, Soil Science Society of America, Inc., 1978.
7. Bonjean A. Le Tournesol. Paris: Les Editions de l'environnement, 1993.
8. Schuster WH. Die Züchtung der Sonnenblume. Berlin, Hamburg: Paul Parey Scientific Publishers, 1993.
9. Rogers CE. Insect pests and strategies for their management in cultivated sunflower. Field Crops Res. 30:301–332, 1992.
10. De Ropp RS. The isolation and behavior of bacteria-free crown-gall tissue from primary galls of *Helianthus annuus*. Phytopathology 37:201–206, 1946.
11. Hildebrandt AC, Riker AJ, Duggar BM. The influence of the composition of the medium on growth in vitro of excised tobacco and sunflower tissue cultures. Am J Bot 33:591–597, 1946.
12. Henderson JHM, Durrell ME, Bonner J. The culture of normal sunflower stem callus. Am J Bot 39:467–473, 1952.
13. Kandler O. Über eine physiologische Umstimmung von Sonnenblumenstengelgewebe durch Dauereinwirkung von β-Indolylessigsäure. Planta 40:346–349, 1952.
14. Wilcox McCann A, Cooley G, Van Dreser J. A system for routine plantlet regeneration of sunflower (*Helianthus annuus* L.) from immature embryo-derived callus. Plant Cell Tissue Org Cult 14:103–110, 1988.
15. Espinasse A, Lay C. Shoot regeneration of callus derived from globular to torpedo embryos from 59 sunflower genotypes. Crop Sci 29:201–205, 1989.
16. Espinasse A, Lay C, Volin J. Effects of growth regulator concentrations and explant size on shoot organogenesis from callus derived from zygotic embryos of sunflower (*Helianthus annuus* L.). Plant Cell Tissue Org Cult 17:171–181, 1989.

17. Witrzens B, Scowcroft WR, Downes RW, Larkin PJ. Tissue culture and plant regeneration from sunflower (*Helianthus annuus*) and interspecific hybrids (*H. tuberosus* × *H. annuus*). Plant Cell Tissue Org Cult 13:61–76, 1988.
18. Greco B, Tanzarella OA, Carrozzo G, Blanco A. Callus induction and shoot regeneration in sunflower (*Helianthus annuus* L.). Plant Sci Lett 36:73–77, 1984.
19. Paterson KE, Everett NP. Regeneration of *Helianthus annuus* inbred plants from callus. Plant Sci 42:125–132, 1985.
20. Lupi MC, Bennici A, Locci F, Gennai D. Plantlet formation from callus and shoot-tip culture of *Helianthus annuus* (L.). Plant Cell Tissue Org Cult 11:47–55, 1987.
21. Finer JJ. Direct somatic embryogenesis and plant regeneration from immature embryos of hybrid sunflower (*Helianthus annuus* L.) on a high sucrose-containing medium. Plant Cell Rep 6:372–374, 1987.
22. Freyssinet M, Freyssinet G. Fertile plant regeneration from sunflower (*Helianthus annuus* L.) immature embryos. Plant Sci 56:177–181, 1988.
23. Jeannin G, Hahne G. Donor plant growth conditions and regeneration of fertile plants from somatic embryos induced on immature zygotic embryos of sunflower (*Helianthus annuus* L.). Plant Breed 107:280–287, 1991.
24. Power CJ. Organogenesis from *Helianthus annuus* inbreds and hybrids from the cotyledons of zygotic embryos. Am J Bot 74:497–503, 1987.
25. Nataraja K, Ganapathi TR. In vitro plantlet regeneration from cotyledons of *Helianthus annuus* cv. Morden (sunflower). Indian J Exp Bot 27:777–779, 1989.
26. Knittel N, Escandon AS, Hahne G. Plant regeneration at high frequency from mature sunflower cotyledons. Plant Sci 73:219–226, 1991.
27. Chraibi B. KM, Latche A, Roustan J-P, Fallot J. Stimulation of shoot regeneration from cotyledons of *Helianthus annuus* by ethylene inhibitors, silver and cobalt. Plant Cell Rep 10:204–207, 1991.
28. Chraibi B. KM, Castelle J-C, Latche A, Roustan J-P, Fallot J. Enhancement of shoot regeneration potential by liquid medium culture from mature cotyledons of sunflower (*Helianthus annuus* L.). Plant Cell Rep 10:617–620, 1992.
29. Ceriani MF, Hopp HE, Hahne G, Escandón AS. Cotyledons: An explant for routine regeneration of sunflower plants. Plant Cell Physiol 33:157–164, 1992.
30. Konov A, Bronner R, Skryabin K, Hahne G. Formation of epiphyllous buds in sunflower (*Helianthus annuus* L.): Induction in vitro and cellular origin. Plant Sci. 135:77–86, 1998.
31. Matzke MA, Susani M, Binns AN, Lewis ED, Rubenstein I, Matzke AJM. Transcription of a zein gene introduced into sunflower using a Ti plasmid vector. EMBO J 3:1525–1531, 1984.
32. Ursic D. Eight DNA insertion events of *Agrobacterium tumefaciens* Ti-plasmids in isogenic sunflower genomes are all distinct. Biochem Biophys Res Comm 131:152–159, 1985.
33. Yao X, Jingfen J, Kuochang C. Transfer and expression of the T-DNA harboured by *Agrobacterium tumefaciens* in cultured explants of *Helianthus annuus*. Acta Bot Yunnanica 10:159–166, 1988.
34. Everett NP, Paterson-Robinson KE, Mascarenhas D. Genetic engineering of sunflower (*Helianthus annuus* L.). Biotechnology 5:1201–1204, 1987.
35. Pugliesi C, Biasini MG, Fambrini M, Baroncelli S. Genetic transformation by *Agrobacterium tumefaciens* in the interspecific hybrid *Helianthus annuus* × *Helianthus tuberosus*. Plant Sci 93:105–115, 1993.
36. Schrammeijer B, Sijmons PC, Van den Elzen PJM, Hoekema A. Meristem transformation of sunflower via *Agrobacterium*. Plant Cell Rep 9:55–60, 1990.
37. Malone-Schoneberg J, Scelonge CJ, Burrus M, Bidney DL. Stable transformation of sunflower using *Agrobacterium* and split embryonic axis explants. Plant Sci 103:199–207, 1994.
38. Grayburn WS, Vick BA. Transformation of sunflower (*Helianthus annuus* L.) following wounding with glass beads. Plant Cell Rep 14:285–289, 1995.
39. Knittel N, Gruber V, Hahne G, Lénée P. Transformation of sunflower (*Helianthus annuus* L.): a reliable protocol. Plant Cell Rep 14:81–86, 1994.
40. Laparra H, Burrus M, Hunold R, Damm B, Bravo-Angel A-M, Bronner R, Hahne G. Expression of foreign genes in sunflower (*Helianthus annuus* L.)—evaluation of three gene transfer methods. Euphytica 85:63–74, 1995.

41. Burrus M, Molinier J, Himber C, Hunold R, Bronner R, Rousselin P, Hahne G. *Agrobacterium*-mediated transformation of sunflower (*Helianthus annuus* L.) shoot apices: Transformation patterns. Mol Breed 2:329–338, 1996.
42. Paterson Robinson KA, Adams DO. The role of ethylene in the regeneration of *Helianthus annuus* (sunflower) plants from callus. Physiol Plant 71:151–156, 1987.
43. Prado E, Bervillé A. Induction of somatic embryo development by liquid culture in sunflower (*Helianthus annuus* L.). Plant Sci 67:73–82, 1990.
44. Pélissier B, Bouchefra O, Pépin R, Freyssinet G. Production of isolated somatic embryos from sunflower thin cell layers. Plant Cell Rep 9:47–50, 1990.
45. Fiore MC, Trabace T, Sunseri F. High frequency of plant regeneration in sunflower from cotyledons *via* somatic embryogenesis. Plant Cell Rep 16:295–298, 1997.
46. Jeannin G, Bronner R, Hahne G. Early cytological discrimination between organogenesis and somatic embryogenesis induced on immature zygotic embryos of sunflower (*Helianthus annuus* L.). Biotechnol. Biotechnol Eq 4:96–99, 1993.
47. Bronner R, Jeannin G, Hahne G. Early cellular events during organogenesis and somatic embryogenesis induced on immature zygotic embryos of sunflower (*Helianthus annuus*). Can J Bot 72:239–248, 1994.
48. Jeannin G, Charrière F, Bronner R, Hahne G. Is predetermined cellular competence required for alternative embryo of shoot induction on sunflower zygotic embryos? Bot Acta 111:280–286, 1998.
49. Jeannin G, Bronner R, Hahne G. Somatic embryogenesis and organogenesis induced on the immature zygotic embryo of sunflower (*Helianthus annuus* L.) cultivated in vitro: Role of the sugar. Plant Cell Rep 15:200–204, 1995.
50. Laparra H, Bronner R, Hahne G. Histological analysis of somatic embryogenesis induced in leaf explants of *Helianthus smithii* Heiser. Protoplasma 196:1–11, 1997.
51. Gürel A, Nichterlein K, Friedt W. Shoot regeneration from anther culture of sunflower (*Helianthus annuus*) and some interspecific hybrids as affected by genotype and culture procedure. Plant Breed 106:68–76, 1991.
52. Sarrafi A, Roustan JP, Fallot J, Alibert G. Genetic analysis of organogenesis in the cotyledons of zygotic embryos of sunflower (*Helianthus annuus* L.). Theor Appl Genet 92:225–229, 1996.
53. Chraibi B. KM, Castelle J-C, Latche A, Roustan J-P, Fallot J. Influence de l'acide 2-chloroéthylphosphonique et des polyamines sur la caulogenèse du tournesol (*Helianthus annuus* L.) à partir de cotylédons. Comptes Rend Acad Sci Paris 315:459–462, 1992.
54. Chraibi B. KM, Castelle J-C, Latche A, Roustan J-P, Fallot J. A genotype-independent system of regeneration from cotyledons of sunflower (*Helianthus annuus* L.): The role of ethylene. Plant Sci 86:215–221, 1992.
55. Charrière F, Hahne G. Induction of embryogenesis versus caulogenesis on in vitro cultured sunflower (*Helianthus annuus* L.) immature zygotic embryos: role of plant growth regulators. Plant Sci 137:63–71, 1998.
56. Bohorova N. Application of tissue and protoplasts culture in the genus *Helianthus* L. 12th International Sunflower Conference, Novisad, Yougoslavia, 1988, pp 300–304.
57. Guilley E, Hahne G. Callus formation from isolated sunflower (*Helianthus annuus*) mesophyll protoplasts. Plant Cell Rep 8:226–229, 1989.
58. Lenée P, Chupeau Y. Isolation and culture of sunflower protoplasts (*Helianthus annuus* L.): Factors influencing the viability of cell colonies derived from protoplasts. Plant Sci 43:69–75, 1986.
59. Moyne A-L, Thor V, Pélissier B, Bergounioux C, Freyssinet G, Gadal P. Callus and embryoid formation from protoplasts of *Helianthus annuus*. Plant Cell Rep 7:437–440, 1988.
60. Fischer C, Laparra H, Charrière F, Jung JL, Hahne G. Regeneration of plants from protoplasts of *Helianthus annuus* L. (sunflower). In: Bajaj YPS, ed. Plant Protoplasts and Genetic Engineering VII, 38. Berlin, New York: Springer Verlag, 1996, pp 48–63.
61. Fischer C, Klethi P, Hahne G. Protoplasts from cotyledon and hypocotyl of sunflower (*Helianthus annuus* L.): shoot regeneration and seed production. Plant Cell Rep 11:632–636, 1992.
62. Henn H-J, Wingender R, Schnabl H Plant regeneration from sunflower (*Helianthus annuus* L.) protoplasts and interspecific hybridization. 3rd European Symposium on Sunflower Biotechnology, Bad Münster a. Stein, Germany, 1995, pp 4.

63. Krasnyanski S, Menczel L. Somatic embryogenesis and plant regeneration from hypocotyl protoplasts of sunflower (*Helianthus annuus* L.). Plant Cell Rep 12:260–263, 1993.
64. Schmitz P, Schnabl H. Regeneration and evacuolation of protoplasts from mesophyll, hypocotyl and petioles from *Helianthus annuus* L. J Plant Physiol 135:223–227, 1989.
65. Trabace T, Vischi M, Fiore MC, Sunseri F, Vanadia S, Marchetti S, Olivieri AM. Plant regeneration from hypocotyl protoplast in sunflower (*Helianthus annuus* L.). J Genet Breed 49:51–54, 1995.
66. Wingender R, Henn H-J, Barth S, Voeste D, Machlab H, Schnabl H. A regeneration protocol for sunflower (*Helianthus annuus* L.) protoplasts. Plant Cell Rep 15:742–745, 1996.
67. Chanabé C, Burrus M, Alibert G. Factors affecting the improvement of colony formation from sunflower protoplasts. Plant Sci 64:125–132, 1989.
68. Chibbar RN, Shyluk J, Georges F, Constabel F. Biochemical aspects of sunflower protoplast culture. Plant Physiol 83:74, 1987.
69. Dupuis JM, Pean M, Chagvardieff P. Plant donor tissue and isolation procedure effect on early formation of embryoids from protoplasts of *Helianthus annuus* L. Plant Cell Tissue Org Cult 22:183–190, 1990.
70. Lenée P, Chupeau Y. Development of nitrogen assimilating enzymes during growth of cells derived from protoplasts of sunflower and tobacco. Plant Sci 59:109–117, 1989.
71. von Keller A, Coster HGL, Schnabl H, Mahaworasilpa TL. Influence of electrical treatment and cell fusion on cell proliferation capacity of sunflower protoplasts in very low density culture. Plant Sci 126:79–86, 1997.
72. Barth S, Voeste D, Wingender R, Schnabl H. Plantlet regeneration from electrostimulated protoplasts of sunflower (*Helianthus annuus* L.). Bot Acta 106:220–222, 1993.
73. Fischer C, Hahne G. Structural analysis of colonies derived from sunflower (*Helianthus annuus* L.) protoplasts cultured in liquid and semi-solid media. Protoplasma 169:130–138, (1992).
74. Barthou H, Brière C, Caumont C, Petitprez M, Kallerhoff J, Borin C, Souvré A, Alibert G. Effect of atmospheric pressure on sunflower (*Helianthus annuus* L.) protoplasts division. Plant Cell Rep 16:310–314, 1997.
75. Caumont C, Petitprez M, Woynaroski S, Barthou H, Brière C, Kallerhoff J, Borin C, Souvré A, Alibert G. Agarose embedding affects cell wall regeneration and microtubule organization in sunflower protoplasts. Physiol Plant 99:129–134, 1997.
76. Petitprez M, Brière C, Borin C, Kallerhoff J, Souvré A, Alibert G. Characterization of protoplasts from hypocotyls of *Helianthus annuus* in relation to their tissue origin. Plant Cell Tissue Org Cult 41:33–40, 1995.
77. von Keller A, Frey-Koonen N, Wingender R, Schnabl H. Ultrastructure of sunflower protoplast derived calluses differing in their regenerative potential. Plant Cell Tissue Org Cult 37:277–285, 1994.
78. Laparra H, Bronner R, Hahne G. Amyloplasts as a possible indicator of morphogenic potential in sunflower protoplasts. Plant Sci 122:183–192, 1997.
79. Moyne A-L, Tagu D, Thor V, Bergounioux C, Freyssinet G, Gadal P. Transformed calli obtained by direct gene transfer into sunflower protoplasts. Plant Cell Rep 8:97–100, 1989.
80. Jeannin G, Poirot M, Hahne G Régénération de plantes fertiles à partir d'embryons zygotiques immatures de tournesol. In: Doré C, ed. Cinquantenaire de la culture in vitro. Versailles, France: Ed. INRA, 1990, pp 275–276
81. Roseland CR, Espinasse A, Grosz TJ. Somaclonal variants of sunflower with modified coumarin expression under stress. Euphytica 54:183–190, 1991.
82. Henrickson CE. The flowering of sunflower explants in aseptic culture. Plant Physiol 29:536–538, 1954.
83. Paterson KE. Shoot tip culture of *Helianthus annuus:* Flowering and development of adventitious and multiple shoots. Am J Bot 71:925–931, 1984.
84. Trifi M, Mezghani S, Marrakchi M. Multiplication végétative du tournesol. 19:99–102, 1981.
85. Burrus M, Chanabé C, Alibert G, Bidney D. Regeneration of fertile plants from protoplasts of sunflower (*Helianthus annuus*). Plant Cell Rep 10:161–166, 1991.
86. Bidney DL, Scelonge CJ, Malone-Schonenberg JB. Transformed progeny can be recovered from chimeric plants regenerated from *Agrobacterium tumefaciens* treated embryonic axis of sunflower. In: Proceedings of the 13th International Sunflower Conference, Pisa, Italy, 1991, pp 1408–1412.

87. Scott Grayburn W, Vick BA. Transformation of sunflower (*Helianthus annuus* L.) following wounding with glass beads. Plant Cell Rep 14:285–289, 1995.
88. Everett N, Nutter B, Pierce D, Mettler I, Okubara P, Panganiban L, Johnson J, Lachmansingh R, Kostrikin A, Pomeroy L, Howard J. Control of kanamycin resistance in sunflower. J Cell Biochem Suppl 9c:253, 1985.
89. Escandón AS, Hahne G. Genotype and composition of culture medium are factors important in the selection for transformed sunflower (*Helianthus annuus*) callus. Physiol Plant 81:367–376, 1991.
90. Escandón AE, Hahne G. Sunflower transformation: a study of selectable markers. In: Sangwan RS, Sangwan-Norreel, BS, eds. The Impact of Biotechnology in Agriculture. Dordrecht: Kluwer Academic, 1990, pp 345–353.
91. Nutter R, Everett N, Pierce D. Factors affecting the level of kanamycin resistance in transformed sunflower cells. Plant Physiol 84:1185–1192, 1987.
92. Hartman CL. *Agrobacterium* transformation in sunflower. Fargo, ND: Sunflower Research Workshop 1991.
93. Horsch RB, Fry JE, Hoffmann NL, Eichholtz D, Rogers SG, Fraley RT. A simple and general method for transferring genes into plants. Science 227:1229–1231, 1985.
94. Jung J-L, Bouzoubaa S, Gilmer D, Hahne G. Visualisation of transgene expression at the single protoplast level. Plant Cell Rep 11:346–350, 1992.
95. Negrutiu I, Mouras A, Horth M, Jacobs M. Direct gene transfer to plants: Present developments and some future prospectives. Plant Physiol Biochem 25:493–503, 1987.
96. Potrykus I. Gene transfer to plants: Assessment of published approaches and results. Annu Rev Plant Physiol Mol Biol 42:205–225, 1991.
97. Kirches E, Frey N, Schnabl H. Transient gene expression in sunflower. Bot Acta 104:212–216, 1991.
98. Neuhaus G, Spangenberg G, Mittelsten-Scheid O, Schweiger HG. Transgenic rapeseed plants obtained by the microinjection of DNA into microspore-derived embryoids. Theor Appl Genet 75:30–36, 1987.
99. Espinasse A, Volin J, Dybing CD, Lay C. Embryo rescue through *in ovulo* culture in *Helianthus*. Crop Sci. 31:102–108, 1991.
100. Espinasse-Gellner A. A simple and direct technique of transformation in sunflower. Fargo, ND: Sunflower Research Workshop, National Sunflower Association, 1992, pp 50.
101. McCabe DE, Swain WF, Martinell BJ, Christou P. Stable transformation of soybean (*Glycine max*) by particle acceleration. Biotechnology 6:923–925, 1988.
102. Van der Leede-Plegt LM, Van de Ven BCE, Bino RJ, Van der Salm TPM, Van Tunen AJ. Introduction and differential use of various promoters in pollen grains of *Nicotiana glutinosa* and *Lilium longiflorum*. Plant Cell Rep 11:20–24, 1992.
103. Hamilton DA, Roy M, Rueda J, Sindhu RK, Sanford J, Mascarenhas JP. Dissection of a pollen-specific promoter from maize by transient transformation assays. Plant Mol Biol 18:211–218, 1992.
104. Hunold R, Bronner R, Hahne G. Early events in microprojectile bombardment: Cell viability and particle location. Plant J 5:593–604, 1994.
105. McCabe DE, Martinell BJ. Transformation of elite cotton cultivars via particle bombardment of meristems. Biotechnology 11:596–598, 1993.
106. Hunold R, Bronner R, Hahne G. Gus expression in sunflower following microprojectile bombardment. Biotechnol Biotechnol Eq 4:91–95, 1993.
107. Laparra H, Burrus M, Hunold R, Himber C, Damm B, Bravo-Angel AM, Knittel N, Bronner R, Hahne G. Approaches to the genetically engineered sunflower (*Helianthus annuus*). In: Caligari PDS, Hind DJN, eds. Compositae: Biology and Utilization. Vol. 2. Kew: Royal Botanic Gardens, 1996, pp 593–601.
108. Alibert B, Lucas O, Le Gall V, Kallerhoff J, Alibert G. Pectolytic enzyme treatment of sunflower explants prior to wounding and cocultivation with *Agrobacterium tumefaciens*, enhances efficiency of transient GUS expression. Physiol Plant (in press).
109. Bidney D, Scelonge C, Martich J, Burrus M, Sims L, Huffman G. Microprojectile bombardment of plant tissues increases transformation frequency by *Agrobacterium tumefaciens*. 18:301–313, 1992.

110. Barnes WM. Variable patterns of expression of luciferase in transgenic tobacco leaves. Proc Natl Acad Sci USA 87:9183–9187, 1990.
111. McCormac AC, Wu H, Bao M, Wang Y, Xu R, Elliott MC, Chen D-F. The use of visual marker genes as cell-specific reporters of *Agrobacterium* mediated T-DNA delivery to wheat (*Triticum aestivum* L.) and barley (*Hordeum vulgare* L.). Euphytica 99:17–25, 1998.

# 55
# Transformation of Wheat

**Mark C. Jordan**
*Agriculture and Agri-Food Canada, Winnipeg, Manitoba, Canada*

**Fredy Altpeter**
*Institut für Pflanzengenetik und Kulturpflanzenforschung Gatersleben, Gatersleben, Germany*

**Javed A. Qureshi**
*Novartis Agribusiness Biotechnology Research, Inc., Research Triangle Park, North Carolina*

| | | |
|---|---|---|
| I. | INTRODUCTION | 835 |
| II. | WHEAT TISSUE CULTURE | 836 |
| III. | TRANSFORMATION OF WHEAT | 837 |
| IV. | SELECTABLE MARKERS IN WHEAT TRANSFORMATION | 838 |
| | A. Herbicide Resistance | 838 |
| | B. Antibiotic Resistance | 839 |
| | C. Carbohydrate-Based Selection | 840 |
| V. | MOLECULAR IMPROVEMENT OF WHEAT | 841 |
| | A. Wheat Quality | 841 |
| | B. Crop Resistance Against Fungal Diseases and Insect Pests | 841 |
| | C. Male Sterility System | 842 |
| VI. | INHERITANCE AND EXPRESSION OF TRANSGENES IN WHEAT | 843 |
| VII. | FUTURE CHALLENGES | 844 |
| | REFERENCES | 844 |

## I. INTRODUCTION

Wheat is second to rice in world importance. It provides protein, minerals, and vitamins as well as complex carbohydrates. It is the main food staple of 35% of the world's population and provides almost 20% of their total food calories. Global demand for wheat will grow faster than that for any other major crop and is forecasted to reach 1.1 billion tons in the year 2020 (1), creating a challenge for molecular and conventional breeding. Wheat or its ancestral progenitor species was grown as a food crop as early as 8000 B.C. (2). During its early domestication important traits such as free threshing, nonshattering, and a move toward a more determinate growth habit were incorporated, and wheats changed from the emmer types to the common and durum wheats of today. In recent years different types of wheat have been bred for particular end uses: hard red for bread, durum for pasta, soft wheat for cakes and pastries, and white wheat (hard or soft) for noodles, bread, or pastries. A great deal of effort in wheat breeding has centered on improvement of quality characteristics for particular end uses. This, coupled with breeding for agronomic characteristics such as yield, resistance to lodging, pest resistance, and earliness, were the major focus of the world's wheat breeding programs in the 20th century. The introduction of semidwarf varieties that could provide high yields under optimal moisture and fertilizer regimens was the basis of the Green Revolution.

Harlan (3) proposed three categories of gene pools for cultivated plants: primary gene pool (GP-1), secondary gene pool (GP-2), and tertiary gene pool (GP-3). The GP-1 pool is the traditional biological species concept and for bread wheat corresponds to genotypes of *Triticum aestivum*. This forms the variation most commonly exploited in breeding programs. The secondary gene pool includes all biological species that will cross with the crop. Hybrids tend to be sterile, chromosomes pair poorly, but the gene pool can be utilized with effort. The tertiary gene pool is all species in which crosses can be made, but the hybrids are lethal and completely sterile and gene transfer either is not possible or requires such interventions as embryo culture, tissue culture, doubling of chromosome number, or use of bridging species. This pool represents genetic material that could be available for incorporation into a crop species if technological advances are capable of overcoming barriers.

With advances in plant transformation the addition of genetic material from GP-2 and GP-3 pools becomes much simpler. In fact, it is just as easy to transfer material from beyond the GP-3 pool as it is to transfer material from within GP-2, with the important distinction that the exact deoxyribonucleic acid (DNA) sequence desired must be known and the trait can only be conditioned by one or a very few genes at present.

Traditionally, the GP-2 and GP-3 pools have been used in wheat primarily to introduce disease resistance genes. As disease resistance genes are always in danger of becoming defeated by the pathogen, discovering new sources of resistance is an ongoing activity. Ultimately, the sources of resistance in the accessible gene pools will become fewer and novel resistance genes will become essential, especially with a demand for decreasing the amount of chemical pesticides used on food crops. Novel resistance strategies can also be used to augment traditional resistance breeding and could help to increase the durability of existing resistance genes.

## II. WHEAT TISSUE CULTURE

A prerequisite for gene transfer by the commonly used techniques of *Agrobacterium* species–mediated transformation or biolistics is a highly efficient method of producing undifferentiated callus tissue from a tissue explant and afterward triggering the regeneration of whole fertile plants. Wheat and monocots in general had been considered recalcitrant to such manipulations when compared to easily cultured species such as *Nicotiana tabacum* or *Daucus carota*. It was not until the 1980s that regeneration from callus culture could be routinely achieved. In wheat the explants that have been used to produce callus are those that are actively dividing or have only recently stopped dividing, which include nodes, leaf base, immature inflorescences, root tips, and scutellum (4). Regeneration from callus proceeds via either organogenesis or embryogenesis. Elucidation of factors that control the embryogenic pathway has resulted in efficient somatic embryogenesis from wheat explants, particularly scutellum and immature inflorescences. For scutellum regeneration both mature and immature embryos can be used, however, immature embryos have to date provided more efficient regeneration over a wider range of genotypes. For scutellum callus from immature embryos the important factors are stage of embryo development, composition of the nutrient medium, and phytohormones (5), genotype (6), and growth conditions of the donor plants (7,8). The effect of genotype is quite pronounced, and wheat genotypes vary considerably in the efficiency of embryogenic callus formation and shoot regeneration (9–16)

Similar parameters have been found to be important in embryogenesis from immature inflorescence explants. A low frequency of regeneration from such explants was reported in early studies (7,17,18). Sharma and coworkers (19) demonstrated that by choosing a suitable genotype and stage of development of the inflorescence a high degree of regeneration can be achieved.

Although the choice of genotype can be important, a genotype that has been considered poor can be improved by manipulating the other important factors. There are complex and significant interactions among stage of development, donor plant growth conditions, and media (20). Therefore, there is no reason why any wheat genotype would not be suitable for efficient regeneration and transformation provided the appropriate conditions are identified. It is a trade-off whether the advantages of the use of elite material justify optimizing donor plant growth conditions and tissue culture conditions for many genotypes. At the present time it seems that optimization of a single genotype that represents an elite adapted line for the target area is the best approach. This way a minimum of optimization has to be done while limiting the amount of linkage drag that could occur upon backcrossing from unadapted material.

## III. TRANSFORMATION OF WHEAT

Crop transformation has been accomplished by use of either *Agrobacterium tumefaciens* as a vector or direct DNA uptake. As a result of the perceived recalcitrance of monocot plants to *Agrobacterium* species, initial transformation studies focused on direct DNA uptake by protoplasts after treatment with polyethylene glycol (PEG) or electroporation. The isolation of regenerable protoplasts (21) from embryogenic cell suspensions (22) requires a long tissue culture period, reducing the chance of regenerating normal and fertile transgenic wheat (23). Alternative gene transfer systems, such as electroporation into intact cells (24), treatment of wheat cells with silicon carbide fibers (25), and biolistics (26) eliminated the need for protoplast isolation. It is the latter technique, that has provided the most success to date. Transient expression analysis with reporter genes (27) and stable expression in wheat callus (28) after biolistic gene transfer had to be combined with an efficient selection and regeneration system to produce the world's first transgenic wheat plants (29). This landmark is molecular wheat improvement motivated further refinements. In order to increase reproducibility and reduce the risk of somaclonal variation, instead of specific long-term callus (29), explants were used directly for biolistic gene transfer or after a short preculture period followed by a callus selection during several subcultures (30–33). These new protocols were a great improvement over the previous method and resulted in transgenic plants ready for transfer to soil 3–5 months after culture initiation compared to 12–15 months. All groups reported the production of 7 to 14 independent fully fertile transgenic wheat plants on the basis of marker gene activity assay. Further refinements were reported by Altpeter and coworkers (34). By optimizing the biolistic gene transfer parameters, reducing the period for callus initiation after bombardment to 2 weeks, and imposing selection during the regeneration period, transformed plants could be transferred to soil in as little as 2 months after culture initiation. The reported number of 80 independent transgenic wheat plants and the fact that this method could be successfully established in several labs (35,36) imply the high reproducibility of this procedure.

Immature inflorescences have been reported as an alternative target for biolistic wheat transformation (37,38). Although somewhat tedious and time-consuming in excision compared to immature embryos, donor plants can be harvested approximately 1 month earlier for immature inflorescences, reducing the risk of donor plant quality–associated problems and providing the possibility of extending efficient transformation to those genotypes that are still somewhat recalcitrant to plant regeneration from immature embryo-derived callus.

The first report of *Agrobacterium* species–mediated wheat transformation was published by Cheng and coworkers (35). As in the earlier reports of *Agrobacterium* sp.–mediated transformation of rice, maize, and barley (39–42) a key factor was the use of hypervirulent *Agrobacterium* sp. strains. *Agrobacterium* sp. methods may allow for less variation in copy number among transformed plants and make available to wheat *Agrobacterium* sp.–based technologies such as dual binary systems for the production of marker-free plants (43).

## IV. SELECTABLE MARKERS IN WHEAT TRANSFORMATION

Success in transformation experiments depends on effective selection of the few transformed cells from a mass of nontransformed tissue. Selection makes the experiments manageable and cuts down on labor, an important consideration for commercial exploitation of this technology. An optimal selection strategy should allow the recovery of normal and fertile transgenic plants at a high frequency, while reducing the workload associated with the handling of a large number of subcultures or nontransgenic plants that escaped the selection. However, reduction of the escape frequency in transformation experiments may be at the expense of transformation efficiency. Stringent selection throughout the selection cycle, including the regeneration step, would potentially eliminate the recovery of low-expressing phenotypes, leaving only high-expressing events. The expression levels of marker and useful genes do not necessarily correspond. Losing a population of weak-expressing transgenics through stringent selection might also be a disadvantage in answering basic questions about gene function, insertion, positional implications of insertion, and stability. Therefore, the application of fast screening methods to eliminate escaped plants rapidly seems more practical than aiming for an escape-free selection system. The selection of transformed cells in vitro is dependent upon many factors, including spatial and temporal expression of the introduced gene, level of expression, promoter elements used to drive the gene(s), chemical characteristics of the selection agent, composition of the medium containing the selective agent, and developmental stage of the tissue under selection.

### A. Herbicide Resistance

For wheat transformation, the most frequently used selectable marker is bar, which confers resistance to phosphinothricin (PPT), the active ingredient of both glufosinate (Sigma Chemical Company St. Louis, MO) and bialaphos (Meiji Seika Kasha Tokyo, Japan), commercially available as Basta and Herbiace, respectively. Glufosinate is the ammonium salt of PPT, whereas bialaphos is the tripeptide L-phosphinothricyl-L- alanine. Selection of transformed wheat tissue using Basta, PPT, or bialaphos (29,31–34,44) has been successful, but in general, this selection allows a considerable number of escapes. In plant cells, PPT targets glutamine synthatase (GS), slowing or blocking ammonia assimilation (45); thus nontransformed cells are adversely affected by ammonia accumulation and glutamine deprivation (46,47). Because of this mode of action of PPT the vigor and physiological state of the tissue under selection play an important role in the effectiveness of this selection agent. Some medium components, such as proline (47) and glutamine (48), also interfere with PPT action, reducing the efficiency of selection. Use of bialaphos instead of PPT, however, provides a higher level of toxicity for slowing the growth of nontransformed tissue during selection. Dennehey and colleagues (47) propose that "perhaps bialaphos, with its two alanine residues, is more readily transported to the plant cell than PPT." From a tissue culture standpoint, PPT or bialaphos selection can be made more effective by eliminating proline and glutamine from the selection medium if possible, by using very small tissue pieces for selection

# Transformation of Wheat

and by subculture of tissue under selection after a short interval, generally up to 2 weeks. Care must also be taken at the subculture to divide the selected tissue into very small segments. Larger tissue pieces allow cross-protection of nontransformed cells, hence more escapes.

Glyphosate (the active ingredient in the herbicide Roundup, Monsanto Company St. Louis, MO) resistance can be conferred upon plant cells by transforming them with either CP4 or GOX genes, and better yet with both. The CP4 gene was isolated from a glyphosate-tolerant strain (CP4) of *Agrobacterium* sp. (49,50), coding for a glyphosate-tolerant enolpyrovylshikimate-5-phosphate synthase (EPSPS). The GOX gene coding for a glyphosate oxidoreductase was cloned from *Achromobacter* sp. (49), which detoxifies glyphosate by converting it to aminomethyl phosphonic acid, which is nontoxic to plant cells. Glyphosate has emerged as a very potent selective agent for wheat transformation (51–53). In a wheat suspension culture–based study (52), glyphosate-resistant colonies were obtained from bombarded suspension culture cells in less than 3 weeks. Colonies were rapidly obtained from all the gene expression vectors used when the target genes were driven by 35S promoter, maize ubiquitin promoter, and rice actin promoter; however, more lines with higher tolerance to glyphosate were recovered from bombardments with genes driven by the rice actin promoter. In this study, irrespective of the promoter used, 21 of the 24 bombardments in four independent experiments yielded multiple independently transformed colonies, reflecting the effectiveness of glyphosate as a selection agent.

Escape-free multiple transformed wheat plants have also been obtained by using glyphosate as the selectable marker with a transformation frequency of 0.1–6% (51,53). The effectiveness of glyphosate as a selective agent can be attributed to its translocation and accumulation in fast-growing meristem tissues (51), properties that work against cross-protection of cells from glyphosate toxicity and, when applied during regeneration, kill all nontransformed meristems, which may otherwise be producing escape plants.

## B. Antibiotic Resistance

When aminoglycoside antibiotics such as geneticin and paromomycin were employed for selection, fewer escapes were reported, when compared to those noted with the use of bialaphos as a selection agent for wheat transformation. Witzrens and coworkers (44) reported an escape rate of over 90% in their experiments using bialaphos, compared to a 50% escape rate rate with geneticin. However, in this study (44) the use of different promoters to drive the individual selectable marker genes complicates the interpretation of the data. They further reported that the selection was even tighter when geneticin was replaced with paromomycin.

The effective use of aminoglycoside antibiotic geneticin for wheat transformation is very much dependent upon the developmental stage and size of the explants used for selection. Relatively older immature embryos (from longer preculture) and germinated somatic embryos are very tolerant to this antibiotic. The ability of 10- to 12-day postanthesis wheat embryos with 2–3 days of preculture to produce embryogenic callus is severely affected by geneticin even at a lower concentration (30 mg/l), meaning that effective selection can be done with this material, but if embryos of the same age are kept in culture for more than 6 days before exposing them to geneticin up to a concentration of 90 mg/l, the embryos show very little growth inhibition (48).

Hygromycin resistance conferred by the *hpt* gene is another selection marker successfully used in wheat transformation (54). In this study wheat embryogenic calli were selected with hygromycin after bombardment. Although escape frequency in these experiments remained high (over 60%), transformation frequency of up to 6% was achieved, whereas most other reports showed 0.1–1.5% when Basta, PPT, or bialaphos was the selective agent. It should be noted, however, that a direct comparison of transformation frequency of other reports employing immature embryos or isolated scutella and this report may not be valid since Ortiz and associates (54) bombarded immature embryo–derived calli. The high escape frequency might be due to the fact that the selective agent was not added to the regeneration medium. When hygromycin selection was applied during callus production and regeneration in barley (55), no escapes were found.

## C. Carbohydrate-Based Selection

Public acceptance of antibiotic- or herbicide-resistance-based selectable markers is low. An alternative is the use of enzymes that interfere with carbohydrate metabolism. Mannose-6-phosphate isomerase (MPI) encoded by the *manA* gene is an *Escherichia coli* glycolytic pathway enzyme that catalyzes the interconversion of mannose-6-phosphate and fructose-6-phosphate (56). MPI in plant cells makes mannose-6-phosphate available as a carbon source by converting it to fructose-6-phosphate. This ability of MPI has recently been exploited in plant transformation by using mannose as a selective marker. Unlike that employing antibiotics and herbicides, mannose-based selection creates a positive environment for transformant growth. This selection strategy has been shown to work for sugar beet transformation (57). Use of MPI for monocots like maize (58) and wheat (59) has the potential to be a key component of future transformation strategies (Table 1). Preliminary reports suggest that transformation frequencies up to 35% are possible (59).

Although the use of new selection strategies such as mannose bodes well for the future wheat transformation in many laboratories still needs optimization to obtain higher transformation efficiencies and extend the number of transformable genotypes. It also is desirable to reduce the effort required for donor plant production by using, for example, mature tissues or seed-derived tissues as gene transfer targets or even developing tissue culture–independent systems such as pollen transformation. Although wheat transformation has been established in a number of laboratories worldwide (60), only a few have highly efficient systems available today. This is a requirement for proceeding to the next phase—molecular wheat improvement by the introduction of useful transgenes—since a large number of transgenic lines has to be produced and screened to identify those that stably transmit the desired level of transgene expression over several generations along with the desired agronomic performance.

**Table 1** Selectable Markers and Promoters Used for Wheat Transformation

| Selectable marker | Function | Promoter | Refs. |
|---|---|---|---|
| *bar* | Basta resistance | CaMV 35S | 29 |
| | | | 33 |
| | | | 32 |
| | | Ubi1 | 31 |
| | | | 34 |
| | | | 44 |
| | | Act. 1 | 48 |
| *nptII* | Geneticin resistance | Act1 | 33 |
| | | Ubi1 | 44 |
| *CP4* | Glyphosate resistance | Act1 | 52 |
| | | | 53 |
| *GOX* | Glyphosate resistance | CaMV 35S | 51 |
| | | Act1 | 52 |
| | | | 53 |
| *hpt* | Hygromycin resistance | CaMV 35S | 55 |
| *pmi* | Mannose as a carbon source | Ubi | 60 |

## V. MOLECULAR IMPROVEMENT OF WHEAT

The importance of wheat is based on diverse uses, productivity, and genetic flexibility that allows adaptation to biotic and abiotic stresses in different climates. Molecular improvement of wheat is developing into an important tool to complement conventional plant breeding activities in maintaining and extending the value of this crop.

### A. Wheat Quality

Wheat gluten is unique among the cereals. The content and type of gluten proteins, especially the subunit composition and molecular mass of high-molecular-weight glutenin subunits (HMW-GSs), determine to a considerable extent the elasticity and extensibility of wheat dough through the formation of high-molecular-weight polymers stabilized by interchain disulfide bonds (61–63). Doughs that have high elasticity and reasonable extensibility are ideal for making bread, doughs that are highly extensible are good for cookies, and doughs with intermediate properties are used for making the flat breads of the Middle East and the Indian subcontinent and noodles in the Far East. The characteristic that the HMW-GSs determine wheat end use properties and yet represent a small percentage of the total seed protein made them a suitable target for genetic engineering (64). The introduction of additional HMW-GSs genes by conventional breeding is complicated by close linkage of these genes. Altpeter and coworkers (65) focused on the integration and expression of the natural HMW-GSs 1Ax1, known to be associated with good bread-making quality (62) in the cultivar 'Bobwhite,' which lacks this subunit in the wild type. Blechl and Anderson (66) introduced a hybrid Dy10:Dx5 HMW-GS gene. Both teams reported expression levels similar to or above those of native HMW-GSs. In contrast to the natural HMW-GSs, which assemble into extremely large elastic polymers, linked by noncovalent and intermolecular disulfide bonds, the hybrid Dy10:Dx5 HMW GSs showed a significant amount of monomeric protein. As a result of an unusual intramolecular disulfide bond, the ability to assemble into polymers was impeded in the hybrid Dy10:Dx5 HMW GSs (67). The expression level of transgenic HMW-GSs was consistent over several generations in controlled environmental conditions (64,66). For the value of this trait, reproducible processing characteristics in the flour need to be guaranteed under different environmental growth conditions. Therefore, quality tests of field-grown seeds expressing transgenic 1Ax1 HMW-GSs (64) are currently in progress (I. Vasil, personal communication). Meanwhile, data on significant improvement of dough strength in a small-scale mixograph analysis of greenhouse-grown T2 seeds of transgenic wheat lines expressing one or two additional natural HMW-GSs (1Ax1, Dx5) (68) provide a very optimistic perspective for the functional integrity of this trait.

### B. Crop Resistance Against Fungal Disease and Insect Pests

Fungal pathogens and insects cause substantial yield losses in wheat production around the world. The infection of the wheat spikes with certain fungal pathogens, e.g., fusarium, can also result in accumulation of mycotoxins, which even persist during food processing and therefore represent a serious risk for human and animal health. The use of agrochemicals to control fungal pathogens and pests is not always effective and economical and can be detrimental to the environment. Conventional plant breeding has produced fungus- and insect-resistant cultivars but can hardly keep pace with the rapid evolution of pathogenic microorganisms and insects. It would be desirable to pyramid several resistance genes to reduce genetic vulnerability. By overcoming crossing barriers, genetic engineering can use alternative sources of resistance genes to contribute to durable crop protection.

Disease resistance mechanisms include: the formation of physical barriers such as lignin, the production of pathogenesis-related proteins such as the hydrolytic enzymes chitinases and β-1,3-glucanases, plasma membrane–permeabilizing thaumatinlike proteins, ribosomal inhibitor proteins, and the induction of antimicrobial compounds called *phytoalexins* (for a review see Ref. 69). To date only a few attempts to tackle the problem of fungal pathogen attack in wheat through the introduction of transgenes have been reported (70–72).

Chitin is a major component of fungal cell walls and is predominant in hyphal tips. The introduction of a chitinase gene from rice under control of the constitutive 35S promoter resulted in instability of chitinase expression in the sexual progeny, and consequently plants were not challenged with the pathogen (70). Leckband and Lörz (71), reported the expression of the stilbene synthase gene from grape and the subsequent production of the phytoalexin resveratrol in wheat. The precursor of the phytoalexin stilbene exists in crop plants; therefore, the synthesis of stilbene requires only the expression of a stilbene synthase transgene. Data regarding the resistance of resveratrol-expressing wheat plants against major wheat pathogens have not been presented yet. Chen and colleagues (72) observed a significant delay in symptoms of wheat scab caused by *Fusarium graminearum* Schw. after constitutive expression of a thaumatinlike protein from rice in transgenic wheat.

Entomopathogenic proteins have been inserted into a variety of transgenic plants to create insect resistance. The major approach being followed is the expression of delta endotoxin from the bacterium *Bacillus thuringiensis* (*B*t endotoxin). With the wide application of this strategy (73) the risk of the development of a *B*t endotoxin–resistant insect population is increasing (74). An alternative or complement to the *B*t endotoxin in the defense against lepidopteran pests might be plant-derived or engineered enzyme inhibitors. These inhibitors interfere with the digestive process in insects. The success of this strategy (73) depends to a great extent on the affinity of the introduced enzyme inhibitor to the target insect proteases (75). Altpeter and coworkers (76) introduced the trypsin inhibitor CMe from barley into wheat. This inhibitor was chosen because its activity against lepidopteran proteases could be demonstrated in vitro; however, this trypsin inhibitor is sensitive to proteolytic inactivation by pepsin and has a very low affinity to human or porcine trypsin. Functional integrity of the transgenic inhibitor on ribonucleic acid (RNA), protein, and enzyme activity level could be demonstrated from extracts of transgenic wheat plants and their sexual progeny. A significant reduction in the survival rate of the Angoumois grain moth (*Sitotroga cerealella*), an important storage pest in many developing countries, was reported after rearing of neonate larvae on transgenic wheat seeds that expressed trypsin inhibitor CMe (76).

Aphids, which belong to the order Homoptera, are major wheat insect pests in Europe and North America. The content of proteases in the gut of sap-sucking insects such as aphids is, in contrast to that of lepidopteran pests, very low and therefore rules out a protease inhibitor strategy for their control. There is no report that *B*t endotoxin affects insects of the order Homoptera. However, expression of snowdrop lectin in transgenic potato inhibited the development and fecundity of aphids (77). The mechanism of lectin toxicity in insects is not clear, and its safe to use by human and animal consumers is currently under debate. Stoger and coworkers (78) reported the stable expression of the snowdrop lectin gene in transgenic wheat plants and their sexual progeny under a constitutive and phloem-specific promoter. Expression of snowdrop lectin in transgenic wheat at levels above 0.04% of the total extracted protein significantly decreased the fecundity, but not the survival of grain aphids (*Sitobion avenae* F.) (78).

## C. Male Sterility System

A wheat hybridization system combines the advantage of a trait protection system with hybrid vigor. In the past different cytoplasmic male sterile and nuclear male sterile systems were developed. However, a problem associated with these systems is creating a high degree of male sterility in all parts of the female parent under different environmental conditions (79). De Block and coworkers (80) reported the development of a nuclear male sterility system in wheat, which was due to the tapetum-specific expression of a barnase gene, which prevented normal pollen development. No self-fertilization occured under greenhouse conditions in otherwise normal plant phenotypes. Field evaluation is in progress (80).

## VI. INHERITANCE AND EXPRESSION OF TRANSGENES IN WHEAT

Once transgenic plants can be routinely produced, it is essential that the transgenes are expressed at the desired level, transmitted to the progeny, and stably expressed over many generations. As in other species transgene expression varies widely among individual transformed plants as a result of position effects. If a large number of individual plants can be routinely produced, then plants expressing the gene at the desired level can be identified. In wheat it is still difficult to produce large numbers of plants in many laboratories, where labor and funds are limited. Technologies that allow position effects to be reduced would be extremely valuable in decreasing the effort and cost involved in producing plants with adequate expression levels. The utilization of matrix attachment regions has been found to be useful in increasing transgene expression and/or reducing position effects in other species (81,82) and may be of value in wheat. As wheat matrix attachment regions were isolated in 1998 (83,84), they would be the ideal candidates for testing of their effect on transgene expression and stability.

Other problems with transgene stability could arise as a consequence of the use of direct gene transfer transformation schemes. Plants produced by using microprojectile bombardment have been shown to vary dramatically in copy number, stability, and heritability (85). Studies on transgene expression, stability, and heritability in wheat have been conducted in several laboratories. Srivastava and coworkers (86) examined six independent wheat lines over successive generations and found that five of the lines were stably transformed and exhibited mendelian segregation. The other line was unstable and first lost expression, then eliminated the transgenes in R3 plants. Even in the stable lines a decrease in GUS expression was observed: GUS expression was lost in R2 plants of all but one line. Rearrangements, amplifications, and methylation of the transgenes were also observed. Cannell and coworkers (87) observed six lines up to the R3 generation and found that five of the lines exhibited mendelian inheritance. Cannell and colleagues also observed loss of expression of the GUS gene over successive generations, which represents a form of progressive gene silencing. Loss of expression of GUS may be due to the prokaryotic origin of the gene; altering the codon usage and structure of such genes (for example, by the insertion of introns) to make them more plantlike may help to prevent induced silencing.

The studies described involved only a very few transgenic wheat lines. A 1998 study (36) investigated 70 independently derived transgenic wheat lines. Although chimerism and transgene elimination did occur, they were rare events. In addition, they compared transgene expression of the transgenic wheat with that of a similar population of transgenic rice. In contrast to the rice, in which there was no relationship between transgenic copy number and transgene expression or stability, in wheat multiple copy number lines tended toward higher levels of expression.

Although from these early results it appears that problems with stability may be isolated, further studies are necessary to investigate whether certain types of transgenes or insertion sites are more prone to elimination or silencing. All studies to date have focused on stability after selfing; however, it is possible that a greater degree of silencing can occur after crossing to nontransgenic plants, a condition that would have tremendous implications for a plant breeding program. Environmental effects on transgene expression and stability are also unknown. As wheat transformation moves from technology development into routine practice in cultivar development programs more information on the degree and importance of instability will become available.

## VII. FUTURE CHALLENGES

Now that wheat transformation has been firmly established in many laboratories it is clear that transgenic wheat will be in commercial production before 2005. Emphasis within many programs will shift from technology development to the introduction and evaluation of genes of potential economic importance. However, the transformation process itself must be improved before wheat will be amenable to many of the techniques available to less recalcitrant species. Generating the numbers required for transposon tagging populations, large studies on transgene stability and expression, or screening in a breeding program is time-consuming and labor-intensive. The number of cloned genes available for introduction into wheat is also increasing rapidly, and it is likely that extremely large numbers of genes will be cloned in the near future as a result of genomics programs. In addition to insertion of characterized genes into wheat for evaluation and commercial product development, there will be a requirement to insert unknown DNA sequences such as complementary DNA (cDNA) clones or Expressed Sequence Tags (ESTs), that have been identified as putatively associated with a certain trait through high-throughput analysis techniques such as microarrays. The demand for transgenic wheat lines will likely far exceed the capacity to produce them at the current rate of production unless large, well-funded laboratories dedicated solely to transformation are set up. The cost per transgenic line will be high as a result of the high cost of labor and the need for large numbers of embryo-donor plants to be grown under controlled environmental conditions year round.

In addition to optimizing the present methods to allow for more efficient production of transformed lines, research into strategies to reduce the numbers of independent transformed lines needed will also help to alleviate the problem. The use of explant material derived from mature seed would reduce the requirement for large amounts of growth space and reduce the cost as well as time involved. Increasing transgene stability and the ability to control expression levels will help to reduce the numbers required. The control of expression levels and stability could be accomplished by techniques for site-directed DNA insertion. These technologies are being developed and applied in model species now (88,89), and one can envision that they will be applied to wheat within 10 years.

Consumer acceptance of transgenic wheat will be a major issue in the near future. It is becoming apparent that the use of selectable markers such as antibiotic resistance or herbicide resistance will be unacceptable in several countries. The use of nonplant genes such as green fluorescent protein from jellyfish may also be less acceptable to many people than use of genes from plant sources. The development of non–tissue-culture-based transformation systems or new systems using plant genes will help to alleviate these concerns.

Although wheat transformation has come a long way since the first report of transgenic wheat in 1992 there is clearly much work to be done in the area, and with the number of laboratories conducting such research increasing, the next decade promises to be an exciting time.

## REFERENCES

1. WE Kronstad. Agricultural development and wheat breeding in the 20th century. In: H-J Braun, F Altay, WE Kronstad, SPS Beniwal, A McNab, eds. Wheat: Developments in Plant Breeding. Vol. 6. Prospects for Global Improvement. Proceedings of the 5th International Wheat Conference, Ankara, Turkey, Dordrecht: Kluwer Academic, 1998, pp 1–10.
2. JW Schmidt. Breeding and genetics. In: GE Inglett, ed. Wheat: Production and Utilization. Westport, CT: Avi, 1974, pp 8–30.
3. JR Harlan. Crops and Man. 2nd ed. Madison, WI: American Society of Agronomy, 1992, pp 106–110.
4. RJ Mathias. Factors affecting the establishment of callus cultures in wheat. In: YPS Bajaj, ed. Biotechnology in Agriculture and Forestry 13. New York: Springer-Verlag, 1990, pp 24–42.

5. JA Qureshi, KK Kartha, SR Abrams, L Steinhauer. Modulation of somatic embryogenesis in early and late-stage embryos of wheat (*Triticum aestivum* L.) under the influence of abscisic acid and its analogs. Plant Cell Rep 18:55–69, 1989.
6. KJ Scott, DG He, YM Yang. Somatic Embryogenesis in wheat. In: YPS Bajaj, ed. Biotechnology in Agriculture and Forestry 13. New York: Springer-Verlag, 1990, pp 46–63.
7. P Ozias-Akins, IK Vasil. Plant regeneration from cultured immature embryos and inflorescences of *Triticum aestivum* L. (wheat): Evidence for somatic embryogenesis. Protoplasma 110:95–105, 1982.
8. IK Vasil. Developing cell and tissue culture system for improvement of cereal and grass crops. J Plant Physiol 128:193–218, 1987.
9. SE Maddock, VA Lancaster, R Risiott, J Franklin. Plant regeneration from cultured immature embryos and inflorescenes of 25 cultivars of wheat (*Triticum aestivum*). J Exp Bot 34:915–926, 1983.
10. A Karp, SE Maddock. Chromosome variation in wheat plants regenerated from cultured immature embryos. Theor Appl Genet 67:249–255, 1984.
11. DG He, YM Yang, G Dahler, KJ Scott. A comparison of epiblast callus and scutellum callus induction in wheat: The effect of embryo age, genotype and medium. Plant Sci 57:225–233, 1988.
12. JG Carman, NE Jefferson, WF Campbell. Induction of embryogenic *Triticum aestivum* L. calli. I. Quantification of genotype and culture medium effects. Plant Cell Tissue Org Cult 12:83–95, 1988.
13. JP Fellers, AC Guenzi, CM Taliaferro. Factors affecting the establishment and maintenance of embryogenic callus and suspension cultures of wheat (*Triticum aestivum* L.). Plant Cell Rep 15:232–237, 1995.
14. RG Sears, EL Deckard. Tissue culture variability in wheat: Callus induction and plant regeneration. Crop Sci 22:546–550, 1982.
15. S Fennell, N Bohorova, M van Ginkel, J Crossa, D Hoisington. Plant regeneration from immature embryos of 48 elite CIMMYT bread wheats. Theor Appl Genet 92:163–169, 1996.
16. H Machii, H Mizuno, T Hirabayashi, H Li, T Hagio. Screening wheat genotypes for high callus induction and regeneration capability from anther and immature embryo cultures. Plant Cell Tissue Org Cult 53:67–74, 1998.
17. D Dudits, G Nemet, Z Haydu. Study of callus growth and organ formation in wheat (*Triticum aestivum*) tissue cultures. Can J Bot 53:957–963, 1975.
18. FA Redway, V Vasil, D Lu, IK Vasil. Identification of callus types for long-term maintenance and regeneration from commercial cultivars of wheat (*Triticum aestivum* L.) Theor Appl Genet 79:609–617, 1990.
19. VK Sharma, A Rao, A Varshney, SL Kothari. Comparison of developmental stages of inflorescence for high frequency plant regeneration in *Triticum aestivum* L. and *T. durum* Desf. Plant Cell Rep 15:227–231, 1995.
20. JG Carman, NE Jefferson, WF Campbell. Induction of embryogenic *Triticum aestivum* L. calli. I. Quantification of genotype and culture medium effects. Plant Cell Tissue Org Cult 12:83–95, 1988.
21. V Vasil, F Redway, IK Vasil. Regeneration of plants from embryogenic suspension culture protoplasts of wheat (*Triticum aestivum* L.). Biotechnology 8:429–434, 1990.
22. F Redway, V Vasil, IK Vasil. Characterization and regeneration of wheat (*Triticum aestivum* L.) embryogenic cell suspension cultures. Plant Cell Rep 8:714–717, 1990.
23. DG He, A Mouradov, YM Yang, E Mouradova, KJ Scott. Transformation of wheat (*Triticum aestivum* L.) through eletroporation of protoplasts. Plant Cell Rep 14:192–196, 1994.
24. A Kloti, VA Iglesias, J Wunn, PK Burkhardt, SK Datta, I Potrykus. Gene transfer be eletroporation into intact scutellum cells of wheat embryos. Plant Cell Rep 12:671–675, 1993.
25. O Serik, I Alnur, K Murat, M Tetsuo, I Masaki. Silicon carbide fiber-mediated DNA delivery into cells of wheat (*Triticum aestivum* L.) mature embryos. Plant Cell Rep 16:133–136, 1996.
26. JC Sanford, FD Smith, JA Russell. Optimizing the biolistic process for different biological applications. Methods Enzymol 217:483–509, 1992.
27. RN Chibbar, KK Kartha, N Leung, J Qureshi, K Caswell. Transient expression of marker genes in immature zygotic embryos of spring wheat (*Triticum aestivum*) through microprojectile bombardment. Genome 34:453–460, 1991.
28. V Vasil, SM Brown, D Re, ME Fromm, IK Vasil. Transformed callus lines from microprojectile bombardment of cell suspension cultures of wheat. Biotechnology 9:743–747, 1991.

29. V Vasil, AM Castillo, ME Fromm, IK Vasil. Herbicide resistant transgenic wheat plants obtained by microprojectile bombardment of regenerable embryogenic callus. Biotechnology 10:667–674, 1992.
30. V Vasil, V Srivastava, AM Castillo, ME Fromm, IK Vasil. Rapid production of transgenic wheat plants by direct bombardment of cultured immature embryos. Biotechnology 11:1553–1558, 1993.
31. JT Weeks, OD Anderson, AE Blechl. Rapid production of multiple independent lines of fertile transgenic wheat (*Triticum aestivum*). Plant Physiol 102:1077–1084, 1993.
32. D Becker, R Brettschneider, H Lorz. Fertile transgenic wheat from microprojectile bombardment of scutellar tissue. Plant J 5:299–307, 1994.
33. NS Nehra, RN Chibbar, N Leung, K Caswell, C Mallard, L Steinhauer, M Baga, KK Kartha. Self-fertile transgenic wheat plants regenerated from isolated scutellar tissues following microprojectile bombardment with two distinct gene constructs. Plant J 5:285–297, 1994.
34. F Altpeter, V Vasil, V Srivastava, E Stoger, IK Vasil. Accelerated production of transgenic wheat (*Triticum aestivum* L.) plants. Plant Cell Rep 16:12–17, 1996.
35. M Cheng, JE Fry, S Pang, H Zhou, CM Hironaka, DR Duncan, TW Conner, Y Wan. Genetic transformation of wheat mediatd by *Agrobacterium tumefaciens*. Plant Physiol 115:971–980, 1997.
36. E Stoger, S Williams, D Keen, P Christou. Molecular characteristics of transgenic wheat and their effect on transgene expression. Transgenic Res 7:1–9, 1998.
37. KL Caswell, NL Leung, RN Chibbar. Regeneration of fertile plants from immature inflorescences of four Canadian spring wheat cultivars. In: AE Slinkard, ed. Proceedings of the 9th International Wheat Genetics Symposium, Poster Presentations, University of Saskatchewan: University Extension Press, 1998, pp 166–168.
38. C Sparks, S Rasco-Gaunt, A Riley, P Barcelo, PA Lazzeri. Development of transformation systems for current wheat varieties. In: AE Slinkard, ed. Proceedings of the 9th International Wheat Genetics Symposium, Poster Presentations, University of Saskatchewan: University Extension Press, 1998, pp 212–214.
39. MT Chan, HH Chang, SL Ho, WF Tong, SM Yu. *Agrobacterium*-mediated production of transgenic rice plants expressing a chimeric α-amylase promoter/β-glucuronidase gene. Plant Mol Biol 22:491–506, 1993.
40. Y Hiei, S Ohta, T Komari, T Kumashiro. Efficient transformation of rice (*Oryza sativa* L.) mediated by *Agrobacterium* and sequence analysis of the boundaries of the T-DNA. Plant J 6:271–282, 1994.
41. Y Ishida, H Saito, S Ohta, Y Hiei, T Komari, T Kumashiro. High efficiency transformation of maize (*Zea mays* L.) mediated by *Agrobacterium tumefaciens*. Nat Biotechnol 14:745–750, 1996.
42. S Tingay, D McElroy, R Kalla, S Fieg, M Wang, S Thornton, R Brettell. *Agrobacterium tumefaciens*–mediated barley transformation. Plant J 11:1369–1376, 1997.
43. T Komari, T Hiei, Y Saito, N Murai, T Kumashiro. Vectors carrying two separate T-DNAs for co-transformation of higher plants mediated by *Agrobacterium tumefaciens* and segregation of transformants free from selection markers. Plant J 10:165–174, 1996.
44. B Witrzens, RIS Brettell, FR Murray, D McElroy, Z Li, ES Dennis. Comparison of three selectable marker genes for transformation of wheat by microprojectile bombardment. Aust J Plant Physiol 25:39–44, 1998
45. CJ Thompson, NR Movva, R Tizard, R Crameri, JE Davies, M Lauwereys, J Botterman. Characterization of the herbicide-resistance gene bar from Streptomyces hygroscopicul. EMBO J 6:2519–2523, 1987.
46. M DeBlock, J Botterman, M Vandewiele, J Dockx, C Thoen, V Gossele, NR Movva, C Thompson, M Van Montagu, J Leemans. Engineering herbicide resistance in plants by expression of a detoxifying enzyme. EMBO J 6:2513–2518, 1987.
47. BK Dennehey, WL Petersen, C Ford-Santino, M Pajeau, CL Armstrong. Comparison of selective agents for use with the selectable marker gene bar in maize transformation. Plant Cell Tissue Org Cult 36:1–7, 1994.
48. JA Qureshi, Z Basri, RA Burton, RR Singh, JE Kollmorgen, G Fincher. Production of fertile transgenic wheat using two different gene expression vectors. Cereals '96. North Melbourne, Australia: Royal Australian Chemical Institute, 1996, pp 422–424.
49. G Barry, G Kishore, S Padgette, M Taylor, K Kolacz, M Weidon, D Re, D Eichholtz, K Fincher, L Hal-

las. Inhibitors of amino acid biosynthesis: Strategies for imparting glyphosate tolerance to crop plants. BK Singh, HE Flores, JC Shannon, eds. Biosynthesis and Molecular Regulation of Amino Acids in Plants. Bethesda, MD: American Society of Plant Physiologists, 1992, pp 139–145.
50. GM Kishore, SR Padgette, RT Fraley. History of herbicide-tolerant crops, methods of development and current state of the art—emphasis on glyphosate tolerance. Weed Technol 6:626–634, 1992.
51. H Zhou, JW Arrowsmith, ME Fromm, CM Hironaka, ML Taylor, D Rodriguez, ME Pajeau, SM Brown, CG Santino, JE Fry. Glyphosate-tolerant CP4 and GOX genes as selectable marker in wheat transformation. Plant Cell Rep 15:159–163, 1995.
52. JA Qureshi, MC Jordan, RN Chibbar, KK Kartha, D Rodriguez, ME Fromm. Glyphosate tolerance: An efficient selection marker for stably transforming wheat cell cultures. Abstracts Conference on Value-Added Cereals Through Biotechnology. Plant Biotechnology Institute, National Research Council of Canada, Saskatoon, SK, Canada, 1995.
53. MC Jordan, JA Qureshi, RN Chibbar, KK Kartha, D. Rodrigue, ME Fromm. Genetic engineering of wheat for glyphosate tolerance. Abstracts Conference on Value-Added Cereals Through Biotechnology. Plant Biotechnology Institue, National Research Council of Canada, Saskatoon, SK, Canada: 1995.
54. JPA Ortiz, MI Reggiardo, RA Ravizzini, SG Altabe, GDL Cervigni, MA Spitteler, MM Morata, FE Elias, RH Vallejos. Hygromycin resistance as an efficient selectable marker for wheat transformation. Plant Cell Rep 15:877–881, 1996.
55. T Hagio, T Hirabayashi, H Machii, H Tomotsune. Production of fertile transgenic barley (*Hordeum vulgare* L.) plant using hygromycin-resistance marker. Plant Cell Rep 14:329–334, 1995.
56. JS Miles, JR Guest. Nucleotide sequence and transcriptional start point of the phosphomannose isomerase gene (manA) of *Escherichia coli*. Gene 32:41–48, 1984.
57. M Joersbo, I Donaldson, J Kreiberg, SG Petersen, J Brunstedt, FT Okkels Analysis of mannose selection used for transformation of sugar beet. Mol Breed 4:111–117, 1998.
58. G Hansen, MS Wright. Recent advances in the transformation of plants. Trends Plant Sci 4:226–231, 1999.
59. JR Reed, Y-F Chang, D McNamara, S Beer, P Miles. High frequency transformation of wheat with the selectable marker phosphomannose isomerase (PMI). Abstract In Vitro Biology Conference, New Orleans, 1999.
60. P Barcelo, S Rasco-Gaunt, C Sparks, M Cannell, S Salgueiro, L Rooke, GY He, C. Lamacchia, G De la Vina, PR Shewry, PA Lazzeri. Transformation of wheat: State of the technology and examples of application. In: AE Slinkard, ed. Proceedings of the 9th International Wheat Genetics Symposium, Keynote Addresses and Oral Presentations, University of Saskatchewan: University Extension Press, 1998, pp 143–147.
61. PR Shewry, BJ Miflin. Seed storage proteins of economically important cereals. In: Y Pomeranz, ed. Advances in Cereal Science and Technology. American Association of Cereal Chemists, St. Paul, MN, 1985, pp 1–83.
62. NG Halford, JM Field, H Blair, P Urwin, K Moore, L Robert. Analysis of HMW glutenin subunits encoded by chromosome 1A of bread wheat (*Triticum aestivum* L.) indicates quantitative effects on grain quality. Theor Appl Genet 83:373–378, 1992.
63. PR Shewry, NG Halford, AS Tatham. High molecular weight subunits of wheat glutenin. J Cereal Sci 15:105–120, 1992.
64. IK Vasil, OD Anderson. Genetic engineering of wheat gluten. Trends Plant Sci 2:292–297, 1997.
65. F Altpeter, V Vasil, V Srivastava, IK Vasil. Integration and expression of the high-molecular-weight glutenin subunit 1 Axl gene into wheat. Nat Biotechnol 14:1155–1159, 1996.
66. AE Blechl, OD Anderson. Expression of a novel high molecular weight glutenin subunit in transgenic wheat. Nat Biotechnol 14:875–879, 1996.
67. Y Shimoni, AE Blechl, OD Anderson, G Galili. A recombinant protein of two high molecular weight glutenins alters gluten polymer formation in transgenic wheat. J Biol Chem 272:15488–15495, 1997.
68. F Barro, L Rooke, F Bekes, P Gras, AS Tatham, R Fido, PA Lazzeri, PR Shewry, P Barcelo. Transformation of wheat with high molecular weight subunit genes results in improved functional properties. Nat Biotechnol 15:1295–1299, 1997.

69. RA Dixon, CJ Lamb, NL Paiva, S Masoud. Improvement of natural defense responses. Ann NY Acad Sci 792:126–139, 1996.
70. WP Chen, X Gu, GH Liang, S Muthukrishnan, PD Chen, DJ Liu, BS Gill. Introduction and constitutive expression of a rice chitinase gene in bread wheat using biolistic bombardment and the *bar* gene as a selectable marker. Theor Appl Genet 97:1296–1306, 1998.
71. G Leckband, H Lörz. Transformation and expression of a stilbene synthase gene of *vitis vinifera* L. in barley and wheat for increased fungal resistance. Theor Appl Genet 96:1004–1012, 1998.
72. WP Chen, PD Chen, DJ Liu, R Kynast, R Velazhahan, S Muthukrishnan, BS Gill. Development of wheat scab symptoms is delayed in transgenic wheat plants that constitutively express a rice thaumatin-like protein gene. Theor Appl Genet 99:755–760, 1999.
73. L Jouanin, M Bonade-Bottino, C Girard, G Morrot, M Giband. Transgenic plants for insect resistance. Plant Sci 13:1–11, 1998.
74. C Macilwain. Bollworms chew hole in gene-engineered cotton. Nature 382:289, 1996.
75. D Michaud. Avoiding protease-mediated resistance in herbivorous pests. Trends Biotechnol 15:4–6, 1997.
76. F Altpeter, I Diaz, H McAuslane, K Gaddour, P Carbonero, IK Vasil. Increased insect resistance in transgenic wheat stably expressing trypsin inhibitor CMe. Mol Breed 5:53–63, 1999.
77. RE Down, AMR Gatehouse, WDO Hamilton, JA Gatehouse. Snowdrop lectin inhibits development and decreases fecundity of the glasshouse potato aphid *(Aulacorthum solani)* when administered in vitro and via transgenic plants both in laboratory and glasshouse trials. J Insect Physiol 42:1035–1045, 1996.
78. E. Stoger, S Williams, P Christou, R Down, J Gatehouse. Expression of the insecticidal lectin from snowdrop (*Galanthus nivalis* agglutinin, GNA) in transgenic wheat plants: effects on predation by the grain aphid Sitobion avenae. Mol Breed 5:65–73, 1999.
79. AA Pickett. Hybrid wheat: Results and problems. G Röbbelen, WE Weber, eds. Advances in Plant Breeding, Supplements to Journal of Plant Breeding 15. Berlin: Paul Parey, 1993.
80. M De Block, D Debrouwer, T Moens. The development of a nuclear male sterility system in wheat: Expression of the barnase gene under the control of tapetum specific promoters. Theor Appl Genet 95:125–131, 1997.
81. F Schoffl, G Schroder, M Kliem, M Rieping. An SAR sequence containing 395 bp DNA fragment mediates enhanced, gene-dosage-correlated expression of a chimaeric heat shock gene in transgenic tobacco plants. Transgenic Res 2:93–100, 1993.
82. S Spiker, WF Thompson. Nuclear matrix attachment regions and transgene expression in plants. Plant Physiol 110:15–21, 1996.
83. MJ Christoffers, JP Gustafson. Identification and characterization of matrix attachment regions in wheat. AE Slinkard, ed. Proceedings of the 9th International Wheat Genetics Symposium, Poster Presentations, University of Saskatchewan: University Extension Press, 1998, p 1.
84. C. Rampitsch, MC Jordan, S Cloutier. A matrix attachment region is located upstream from the high-molecular-weight gluterin gene BX7 in wheat (*Triticum aestivum* L.). Genome 43:483–486, 2000.
85. WP Pawlowski, DA Somers. Transgene inheritance in plants genetically engineered by microprojectile bombardment. Mol Biotechnol 6:17–30, 1996.
86. V Srivastava, V Vasil, IK Vasil. Molecular characterization of the fate of transgenes in transformed wheat (*Triticum aestivum* L.). Theor Appl Genet 92:1031–1037, 1996.
87. ME Cannell, A Doherty, PA Lazzeri, P Barcelo. A population of wheat and tritordeum transformants showing a high degree of marker gene stability and heritability. Theor Appl Genet 99:772–784, 1999.
88. SA Kempin, SJ Liljegren, LM Block, SD Rounsley, MF Yanofsky. Targeted disruption in *Arabidopsis*. Nature 389:802–803, 1997.
89. AC Vergunst, LET Jansen, PJJ Hooykaas. Site-specific integration of *Agrobacterium* T-DNA in *Arabiopsis thaliana* mediated by Cre recombinase. Nucleic Acids Res 26:2729–2734, 1998.

# Index

Abiotic stresses, 743
ACC oxidase gene, 418, 425, 426, 738, 739, 741
Acetolactate synthase (ALS), 487, 757
Acetone cyanohydrin, 537
Acetosyringone effect, 489, 565, 581, 703
ACI gene, 533, 622
ACMD-causing geminivirus, 533
ACMV AC1 (Rep) gene, 532
  resistance, 540
*ACO* gene, 375, 377
*act 1D* gene, 362
ACT1 gene, 62, 630
  promoter, 630
*Actinidia* sp., 407–420
  *arguata,* 408–418
  *chinensis,* 407–418
  *deliciosa,* 407–418
  *eriantha,* 408–418
  *polygama,* 408–418
Acyl coenzyme A, 426
Acynaogenic cassava, 537
ADP glucose pyrophosphorylase 481, 674
ADPase genes, 536
Adventious organogenesis, 156
Advisory Committee on Novel Foods and Processes (ACNFP), 632
*ae* gene, 172
AEC (*see* S-(2-aminoethyl)-L-cysteine)
*Aequorea victoria*, 400

African cassava mosaic disease (ACMD), 532, 534, 535
African cassava mosaic virus (ACMV), 531
Agamous gene, 189
Agriculture
  and food crops 1–24,
  and food, politics of, 319
  and food, regulation of, 320
  innovations, 5
  low input sustainable, (LISA), 11
  prehistoric, 3
  technological changes, 5
*Agrobacterium rhizogenes*, 156, 350, 401, 451, 487, 504, 529, 580, 619, 621, 623–631, 700–709, 619, 740, 750, 754, 775, 780
*Agrobacterium tumefaciens*, 74, 156, 198, 329, 352, 362, 364, 376, 388–393, 400–402, 431, 451, 470, 480, 485, 489–490, 505, 511, 552, 555, 564, 580–583, 589–590, 604, 608, 618–623, 630, 631, 647–648, 652, 661, 700, 703, 740, 748, 750, 754, 767, 775, 796, 804, 816, 820, 822, 823, 837
  mediated delivery of DNA, 661
  oncogenic strains (A47, A281, A722 and C58), 480
  mediated transformation, 662, 480, 477, 563–565, 577, 580, 604–605, 607–610
*Agrobacterium* sp., 56, 71, 79, 385, 409, 451, 508, 509, 517, 535, 605, 620, 623–626,

849

[*Agrobacterium* sp.]
    629, 722–724, 726, 728, 739, 740, 749–752, 754, 755, 776, 777, 804, 820, 822, 823, 837–839
    mediated gene delivery, 619–624
    mediated transformation, 477–478, 489, 542, 631, 739, 750, 767, 775, 804, 805, 816, 822, 823, 827, 836, 838
*Agropyron*, 719
2S-albumin 480
    gene, 479
    protein gene, 779
*Aleurotachelus socialis,* 535
Allergen, in Brazil nut, 480
Allicin, 209
    alliinase, 659, 661
    genes, 659
    oxidase, 538
    *polygama,* 408
    synthase gene, 739, 741
Allium, 655–664
    *Agrobacterium*-mediated transformation, 664
    annual production, 655
    attributes, 661
    breeding attributes, 660
    carbohydrates, 659
    clonal propagation, 661
    clonal seed production, 660
    cytoplasmic male sterility (CMS), 660
    disease resistance, 657
    fungal pathogens, 658
    fungal resistance, 657
    genetics, 661
    insect resistance, 658
    interspecific hybridization, 661
    lachrymatory compounds, 659
    male sterility, 660, 661
    nematode, 659
        resistance, 659
    nutritional value, 659
    pests, 658
    production limitations, 655
    pungency, 659
    quality traits, 659
    root alliinase, 657
    sulfur metabolism, 659
    suspension cultures
    transformation research, 656, 660, 662
    vegetables, 655
    viral resistance, 658
    viruses of, 658
*Allium*
    *carla,* 658

[*Allium*]
    *cepa,* 652
    *porrum,* 655
    *sativum,* 209, 655, 664
    *solani,* 677
    *stripitatum,* 657
    *schoenoporasum,* 655
    spp. 50
Allotetraploid, 738
α-amylase, 718, 726
α-tubulin gene, 114
ALS (*see* Acetolactate synthase)
*Alternaria*
    *alternata,* 348, 424, 432, 677
    *brassicola,* 55
    *dauci,* 517
    *helianthi,* 815
    *radicini,* 515, 517
Alternaria rot, 424
Alternate oxidase, 426, 431
AMA1 gene, 675
*Amaranthus hypochondriacus*, 675
American
    Type Culture Collection (ATCC), 283
    Dietetic Association (ADA), 298
    Medical Association (AMA), 298
1-amino cyclopropane-1-carboxylate, 335, 336, 349, 375, 425, 427
    oxidase, 375, 377, 412, 417, 418, 426, 431, 439
    synthase, 417, 418, 426, 431, 439, 738
Aminoglycoside
    3-adenyltransferase, 776
    antibiotics, 751, 776, 824, 839
AMP (*see* Antimicrobial peptides)
*Amphipyra pyramidoies,* 331
*Amphorophorn agathonica*, 449
    *idaci,* 449
Amplified fragment length polymorphism (AFLP), 238, 241, 618
Amylase inhibitors, 256–257, 549
Amylopectin, 170–177, 481, 641–642, 712, 724–728
    modification, 173–175
Amylose, 170–177, 480, 642, 712, 724, 727, 728
    extender mutants, 172
Ancymidol, 469
Androgenesis, 43–50, 113, 590
    culture conditions, 449
    donor plant, 45
    influencing factors, 45
    media constituents, 47–49
    pollen grain development, 45–46

# Index

Androgenic culture, 721, 722
Angiosperm, 110, 112, 120
Angoumois grain moth, 842
Animal feed-grain, 637
    supplementation, 644
Anther culture, 469, 590, 721, 724, 749, 817
Anthocyanin, 71, 403
Anthracnose, 424, 428, 432, 438, 769
Antibiosis, 251–258
    amylase inhibitors, 256
    *Bacillus thuringiensis* toxins, 253–255
    chitinase, 257
    lectins, 257
    lipid disrupting proteins, 255
    and microbial proteins, 253
    *Photorhabdus luminescence* toxin, 255
    plant proteins, 256
    protease inhibitors, 256
    vegetative insecticidal protein, (VIP), 255
Antibiotic resistance, 743, 776, 839, 840, 844
Antiexinosis, 250, 258
Antifungal proteins, 428, 515
Antifungal resistance genes, 657. 764
Antigen, 675
Antimicrobial
    gene, 766, 658
    peptides, 763, 764, 768, 769
    proteins 365, 438, 764, 765, 769
Antioxidants, 208, 645
Antiplatelet activity, 659
*Antirrhinum majus,* 259, 596
Antisense genes, 426, 432, 456, 536
    sequence, 469
    starch gene, 530
    technology, 657
Antixenosis, 250, 258–265
    crossover strategies, 263–265
    and insect biochemistry, 261
    and multitrophic considerations, 262
    and pest management, 262–263
    phytochemical profiles, 259
    resistance, 258
    secondary metabolites, 260–261
    surface characteristics, 259
    tolerance, 262
    trichomes, 258–259
    wounding and oxidative stresses, 260
Apetala (AP) gene, 186–189
Aphid, 360, 449, 458, 616, 676
Apical meristem, 476
Apices, 490
*Apium graveolens,* 49

Apomixis, 118–120, 191, 527, 539, 660
    epigenetic engineering, 124–126
    and genomic imprinting, 118
    and seed quality, 120
    pseudogamous, 118–120
Apoplasm, 495, 497
Apples, 327–344
    carbohydrate metabolism, 332–334
    disease resistance, 332
    engineering features, 336
    ethylene biosynthesis, 334–326
    history, 328
    and insecticidal proteins, 329–331
    mealiness, 336
    other quality traits, 326–327
    and resistance strategies, 331–332
    texture, 336
    transformations, 328–329
    varieties, 328–329
1a-protein-encoding gene promoter, 620
Aprotinin, 649
*Apterix* spp. 408
*Arabidopsis thaliana*, 30, 32, 100, 110, 431, 621–622, 630
*Arabidopsis,* 37, 79, 186–188, 190, 192, 438, 458, 657, 743, 757, 784
Arabinoxylan, 726
Arabis mosaic virus, 455
Arachidonic acid, 210
*Arachis argyrosphila,* 331
    *cardenasii,* 241
    *hypogaea,* 77, 241
Arepas, 639
*argE* gene, 103, 104
*Arggyrotaenia citrona,* 331
    *arguta,* 408
*Argostis stollenefera,* 140
*Ascochyta rabiei,* 604
*Asparagus,* 465–471
    agronomic genes, 469
    anther culture system, 466
    biology, 465
    breeding goal, 466
    breeding, 466
    decline and replant syndrome, 466
    *densiflorus,* 466
    direct gene transformation, 480
    fruit, 465
    genetic transformation, 469, 480
    haploid embryos anther culture, 469
    homozygous *mm* female plant, 466
    *Mm* male plant, 466
    *officinalis* genome, 480

[Asparagus]
  *officinalis*, 145, 149, 465, 466, 470, 480
  origin, 465
  *plumosus*, 466
  production environment, 465
  productivity, 466
  protoplast, 470
  sexual modes, 466
  somatic embryogenesis, 466, 470
  supermale *MM* plant, 466
  world production, 465
Aspartate kinase gene, 726, 779, 806
Aspartate pathway, 806
*Aspergillus*
  *ochraceus*, 742
  *flavus*, 742
  *oryzae*, 779
  spp., 493
*Asteraceae* family, 613–617
*Asteraceae*, 492
*Asteromeria* spp., 115
*Atole*, 639
*Atropa* spp., 46
*Avena sativa*, 49, 116
*Avena sativa, sterilis*, 49
Average severity (AS), 575, 576
Avidin, 649
*Avocado*, 345–357
  *chinensis*, 408–418
  crop history, 346–347
  *delician*, 407
  embryogenic cultures, 351–352
  *erinatha*, 408
  genetic breeding, 347
  genetic manipulation, 351–353
  genetic manipulations, 351
  genetic transformation, 352–353
  genetics, 347
  history, 345
  *Phytophthora* root rot resistance, 350
  ripening control, 348–349
  root stock cultivars, 350–351
  salt tolerance, 351
  scion cultivars, 348
  somatic cell, 351–352
  somatic embryogenesis, 351
  sun blotch disease, 349–350
  taxonomy, 345–346
  world production, 345–347
Axillary meristem, 476
5-azacytidine, 631

$\beta$-amylase, 726
$\beta$-amylase inhibitor, 373
$\beta$-galactosidase, 426, 488
$\beta$-1, 3-endoglucanase, 103, 676
$\beta$-gluconidase, 514, 620, 649
$\beta$-glucuronidase, 740, 752, 754, 755, 776, 805, 823, 824, 843
$\beta$-carotene, 208, 503, 504, 647
$\beta$-1,3-glucanase, 538
$\beta$-1,3-glucanase gene, 621
*Bacillus* spp., 726
  *amyloliquefaciens*, 102, 497, 779
  *cereus*, 255
  coding, 330, 331, 392, 450, 454, 534
  *subtilis*, 497
  genes, 659
  *thuringienesis*, 243, 252, 253, 315, 392, 450, 454, 535, 534, 548, 590, 591, 649, 659, 676, 712, 738, 740, 742, 756, 779, 842
  var. *tenebrionsis*, 454, 712
Backcross, 440, 442, 445
Bacterial artificial chromosomes (BAC), 728
  library, 32, 241, 524, 530, 534, 540, 728
Bacterial blight resistance, 533
Bacterial resistance gene Xa21, 533
Bacterial resistance, 677
Bacterial rot, 619
Bacteriophage
  P1, 648
  T3, 427
*Bactrocera dorsalis*, 438
  *cucurbitae*, 438
Ballistic delivery of DNA into embryogenic cells, 647
Banana, 364–365
  pharmaceuticals, 366
  bracht virus, 362
  bunchy top virus (BBTV), 360
  constraints of production, 360–361
  embryogenic cell cultures, 362–363
  fruit quality control, 366
  genetic transformation, 361–365
  in vitro culture, 361
  microparticle bombardment, 362–363
  Moko disease, 361
  production and significance, 359–369
  resistance to pathogens, 365–366
  Sigatoka leaf disease, 360
  streak badnavirus (BBSV), 360
  world production, 360
*bar* gene encoding, 390, 403, 605, 615, 623, 629, 648, 656, 724, 776–779, 805, 838
Barley, 170, 717–735, 801, 803, 804, 838, 840, 842
  Heine's Haisa cultivar, 719
  history, 717

[Barley]
    malting, 725–726
    mutational breeding, 718
    novel uses, 727
    ribosome inactivating protein gene, 725
    *shx* mutants, 176
    somaclonal variation, 720
    traditional breeding, 710
    world production, 717
    yellow dwarf virus, 724
    yellow mosaic bymovirus, 724
*barnase barstar* genes, 102, 191, 244, 660
Barnase gene, 102, 366, 843
Basta, 202, 284, 364, 365, 399, 403, 535, 536, 662, 657, 724, 824, 839
Bean, 475–482
    *Agrobacterium* mediated, 477–478
    carbohydrate content, 480
    direct DNA transfer, 478
    gene transfer improvements, 479–482
    improved nutritional quality, 479–481
    manipulation of, 470
    organogenesis, 476
    particle bombardment, 478–479
        tissue culture, 476–477
        transformation, 469, 477–482
Beet, 485–502
    as a plant factory, 491
    biosafety, 489
    crop, 485
    engineered traits, 486
    fructans oligofructans, 492–498
    guard cell protoplasts, 490
    hexose content improvement, 488
    history, 485
    mannose, 490
    methods for engineering, 489
    necrotic yellow vein virus (BNYVV), 225, 487, 489
    particle bombardment, 490
    world production, 486
*Bemisia tabaci*, 532, 658
Benefin, 617
Bensulfide, 617
*Bertholettia excelsa*, 479, 779
*Beta*, 50, 51
    *patellaris*, 238, 242
    *procumbens*, 236, 488
    *vulgaris*, 50, 51, 236, 485–489
    *vulgaris* spp. *maritima*, 485, 489
    *vulgaris* spp. *vulgaris*, 485
    *webbiana*, 238, 242, 488
Betaine, 488
    dehydrogenase, 488

Bialaphos, 391, 423, 487, 490, 494, 623, 776, 838, 839
BIBAC (*see* Binary bacterial artificial chromosomes)
Binary
    bacterial artificial chromosomes (BIBAC) 728
    plasmid, 488, 580
Biodiversity of species, 13
Biolistic, 71, 387, 755, 794, 836, 837 (*see also* Particle bombardment)
Biological control, 487, 657
Biomass, 713
Bioplastic polymers, 537
Biotechnology, 22,
    and consumer concerns, 324
    and crop improvement, 384
    awareness and knowledge, 296–303
    complexity of issues, 323–324
    precautionary principle, 303
    protocol, 322
Bioreactors, 784
Biosafety, 489, 365
Blackberry, 449
Black headed fireworm, 393
Black vine weevil, 453
Blackleg, 778, 779
Bolting, 488
Bombardment (*see* Particle bombardment), 138
Borlug, Norman, 7, 10,
*Botrytis* spp., 456, 457, 589
    *allii*, 657
    *cineria*, 417, 418 456, 457, 517, 657, 677, 725
Bovine somatotropin, 295
Branching enzyme, 536
*Brassica*, 37, 48, 49, 115, 116, 209, *756*, 773–792
    antimicrobial protein, 767, 768
    *campestris*, spp. *pekinensis*, 77
    *carinata*, 30, 773, 774, 778
    glucosinolates, 209
    *juncea*, 30, 209, 773, 774
    *napus*, 30, 34, 37, 55, 56, 67 101, 104, 186, 261, 767, 774, 776, 778, 779, 780, 783, 784
    *nigra*, 30, 773, 774, 778
    *oleracea*, 30, 34, 101, 259
    *rapa*, 773, 774, 778
    spp. 30, 53, 356, 756, 773
    oils storage bodies, 210
    oilseed improvements, 773–784
    oilseed modification, 781
    oleosin, 210
    other transgenic oilseeds, 781
    production of trangenics oilseeds, 774, 775

[*Brassica*]
   plant regenerate, 775
   pathogen and insect tolerance, 778–779
   pollination control, 779
   protein modification, 779
   transgenic stability, 781–783
Brazil nut (*Bertholletia excelsa*), 469, 479, 480, 779
*Bremia lactucae,* 616, 621
Britain's Advisory Committee on Novel Foods and Processes, 632
Bromoxynil, 657
Broomrape, 815
Bt (*see Bacillus thuringiensis*)
Bt gene(s), 392, 535, 590, 738, 740
   *Cry*I gene, 590–594
   cry I B, 740
   *Cry* IA*(b-c)* gene, 243, 331, 535, 593, 610, 740, 794
   *Cry* IIIA gene, 590, 592, 676, 712
   *Cry* IIIB gene, 591, 592
   toxin, 243, 252, 258, 262, 263, 330–332, 393, 454, 534, 548, 558, 590, 649
   vip proteins, 255
Bu gene, 456, 459
Bud, 567, 568
Bulbosum technique, 721, 722
Bulking agent, 493
*bxn* gene, 657

Cabbage looper resistance, 615
Caffeine, 738, 739, 741, 742, 743, 744, 745
Calcium spermidine, 488
*Callaosobruchius* Beetle, 257
CaMV
   35S *gus*-intron reporter gene, 18, 621, 631
   *bar* gene, 629
   *gus*-gene, 631
   promoter, 539, 619, 631, 662, 580, 582
*Camellia japonica,* 49
*Canarhabdis elegans,* 111, 124, 243
Canola, 766–769, 773–792
   and antimicrobial proteins, 767
   apterous trait, 190
   high erucic acid rape, 774
   oilseed improvements, 773–784
   oilseed modification, 781
   other transgenic oilseeds, 781
   pathogen and insect tolerance, 778–779
   plant regenerate, 775
   pollination control, 779
   production of trangenics oilseeds, 774, 775
   protein modification, 779
   transgenic stability, 781–783

*Capsicum annuum,* 45, 564, 577
Carbenicillin, 555
Carbohydrate partitioning, 674
Carbohydrate, 470
*Carica* spp., 437–447
   *cauliflora, 438*
   *papaya,* 437
   *pubescens*, 438
   *quercifolia,* 438
   *stipulata*, 438
Carlavirus, 658
Carotenoids, 211
*Carotae*, 517
Carpel, 99
Carrots, 503–521
   disease tolerance, 504, 517
   explant sources, 504
   genetic engineering, 504–517
   genetic improvements
   history, 503–504
   molecular confirmation of transformants, 506–507
   tissue culture, 504, 507
   tolerance to fungi, 515–517
   transformation parameters, 506
      Agrobacterium strain, 504, 509, 511–515
      carrot cultivation, 508
      co-cultivation period, 508
      confirmation of, 506, 510–511
      explant age, 508
      regeneration, 510
   transformation procedures, 506
Cassava, 523–546
   African mosaic virus, 225, 531, 532
   agronomic cultivars, 536
   agronomic characteristics, 524
   bacterial blight (CBB), 533
   Biotechnology Network (CBN), 541–542
   biotechnology, 526, 534, 539–541
   common mosaic virus (CsCMV), 531
   cooking quality, 540
   cultivars, 537
   cyanogenic compounds, 526
   flour, 539
   gene, 541
   green mite (*Mononychellus tanajoa*), 534
   herbicide resistance, 535
   history, 523, 525
   postharvest deterioration, 526, 537, 538
   preharvest losses, 526
   processing, 537
   production, 523, 525
   seed propagation system, 539
   socioeconomic roles, 524

# Index

[Cassava]
   spp., 525–536
   starch products, 536
   transgenic technologies, 527, 539
   vein mosaic virus (CsVMV), 533, 539
   world production, 523–526
*Catharantus roseus,* 74, 658
Cauliflower, 547–559
   35S promotor, 479
   CAT, 583
   C-*cp,* 583
   coat protein gene, 562–566, 568–570, 576, 577
   coat protein, 564, 566, 568, 569–572, 576–577
   mosaic virus (CMV), 220, 562–577, 579
   O-*cp,* 561–568, 583
   regeneration, 554, 555
   resistance, 563–564, 567–568, 572–574, 577, 579, 583
   transformation 553
   Y, 583
   Y-c*p,* 580
   Coat Protein (CP) gene, 677, 711
   gene sequences, 658
   mediated protection, 564, 574
   mediated resistance, 621, 725
Caulogenesis, 530, 817, 818
CBB resistance, 533–534
*cbg3* promoter, 540
CDC Triffid lax, 758
Cdc 2ZM gene, 79
cDNA-AFLP differential display, 538
Cecropin, 765
Cell culture, 151–167
Cell suspension, 490, 804, 805
Cellulose binding protein, 236, 636
Centro Internacional de Agricultural Tropical (CIAT), 16, 527, 529, 534, 536, 539, 541
*cer* gene, 259
*Ceratitis capitata,* 438
*Cercospora nicotianae,* 768
Cereal(s)
  albinism, 52
  development, 120
  grain cooking quality, 120
  grains, 169
CGIAR centers, 23, 527, 541
Chalcone synthease, 102
*Chapati,* 802
Chemiluminescence, 824
Chenopodiaceae, 485, 489
*Chenopodium quinoa,* 572
Chicha morada (South America), 640

Chickpea, 608–610
  *Agrobacterium*-mediated transformation, 608, 610
  embryo co-cultivation, 608
  production statistics, 604
  selectable/screenable markers, 608, 610
  susceptibility to *Agrobacterium* strains, 608
  transformation by direct gene transfer, 610
  world production, 603–604
Chicory, 49, 493
*Chicoracearum* spp., 619
Chili pepper, 563–577
  and binary vectors, 564
  and molecular characterization, 567–572
  CMV resistance, 566, 572
  gene integration and expression, 567–571
  quality analysis of transgenic dry fruit, 577
  regeneration, 564–567
  T2, T3, and T4 progeny, 575–577
  transformation, 564
  transgenic plants, 565, 567–577
  virus infection protection, 572–573
*Chil 1* gene, 794
*Chilomina clarkei,* 534
Chimera, 753, 754, 755, 824
Chimeric, 362, 363, 364, 398, 399, 416, 440, 452, 744, 753, 755, 807, 821, 823, 824, 843
Chitinase, 257, 350, 384, 402, 428, 438, 440, 441, 457, 764, 778, 794, 842
  gene, 725
Chive, 655
*Chlamydomonas,* 171, 177
  *sta* mutants, 177
Chloramphenicol acetyl transferase (CAT), 490
  gene, 619
Chloroplast, 490
Chlorosulfuron, 490, 720, 740
Chlorpropham, 615
Cholera toxin B, 676
Cholesterol oxidase, 255, 779
Choline monoxygenase, 488
*Chroistoneura*
  *rosaceana,* 331
  *rosaceus,* 331
Chromosome
  doubling, 52
  rearrangements, 34
CIAT CBB gene-tagging project, 540
*Cicer*
  *arietinum,* 77, 603
  *echinospermum,* 115
*Cichorium ubtybus,* 49
CIRAD, 740

*Citrus*
  spp., 116
  *grandis*, 75
*Clavibacter michiganensis* subsp. *Michiganensis,* 725
*Clavicepes*
  *africana*, 17
  *purpurea*, 17
Climacteric, 335, 348, 349, 366, 424, 426, 452
Clonal selection, 397
CMS (*see* Cytoplasmic male sterility)
Co-cultivation, 565, 794
Co-culture, 581, 752, 753
Codex Alimentarius Commision (CODEX), 321
Codling moth, 329, 330, 332
Codon usage, 330
*Coffea*
  *arabica*, 737–741, 743, 744
  *canephora*, 737, 738, 740, 741, 743, 744
  *dewevrei*, 737
  *liberica*, 737
  *racemosa,* 737
Coffee 737–746
  berry borer, 739, 742, 745
  berry disease, 743
  biotechnology, 738, 739, 743,
  coffee content, 741
  history, 738–743
  leaf miner, 739, 740, 744, 745
  ripening, 741
  tissue culture, 739
  world production, 737–738
Cold/freeze tolerance, 738, 743, 744
*Coleoptera* spp., 534, 535
*Colletotrichum*
  *gloeosporioides* 348, 424, 428, 430, 438, 457
  *kahawae,* 743, 769
Colorado potato beetle, 256, 261, 591–596, 676
Commercial products, 675
Concanavalin A, 469
CONSTANS (CO) gene, 185–187, 428
Consumers attitude, 18, 302
Convention of Biological Diversity (CBD), 11
Corn, 727 (*see also* Maize)
  belt region, 640
  bread, 639
  chips, 639
*Corynebacterium* spp., 808
*Cosmopolites sordidus,* 361
Cosuppression, 432, 648, 741, 744, 783
Cotransformation 363
Cotyledon, 489, 564
  co-cultivation, 605
  node, 476

Coumarin, 819
*Couscus* (West Africa and Brazil), 639
Cowpea (*Vigna unguiculata*), 604–607
  *Agrobacterium*-mediated transformation, 604, 605
  chlorotic mottle virus (CCMV), 225, 686
  cotyledon co-cultivation, 605
  direct gene transfer, 605–606
  embryo co-cultivation, 605
  production statistics, 604
  selectable/screenable markers, 604, 606
  world production, 603–604
*cp* gene, 580, 648
CP4 gene, 839
Cranberry, 383–396
  blackheaded fireworms, 393
  explant age effects, 386
  explant orientation, 385–387
  growth regulators effects, 386
  history, 383
  in vitro cultures, 384–386
  regeneration, 384, 385
  screening systems, 391
  selection, 390
  transformants screening, 390–391
  transformation, 387–389
  transgenics, 391–394
  world production, 384
*cre* gene, 241, 648
Crop improvements, 35
Crop protection, 547
  seed market, 109
CsCMV coat protein (CP) gene, 531
CsCMV resistant strain, 532
Csr-1-1 gene, 740
CTAB method, 565
Cucumber, 579–584
  *Agrobacterium*, 580–581
  and protocol for transformation, 580–581
  CMV resistance, 583–584
  inoculation and co-cultivation, 581
  mosaic virus (CMV), 373, 564, 566, 577, 579, 583
  plants, 580
  role of acetosyringone, 581
  safety of transformed plants, 584
  selection and plant regeneration, 581
  transformation history of, 579–580
*Cucumis*
  *cantalupensis,* 371
  *conmon,* 371
  culture conditions, 161
  *dudaim,* 371
  *flexuosus,* 371

# Index

[*Cucumis*]
  *indorus*, 371
  *melo*, 371
  *momordica*, 371
  negative-positive selection, 161, 162
  *sativus*, 159
  var. *indoros*, 375
Cultivation costs,
Culture
  conditions, 161
    negative selection, 161
    positive selection, 161
Curly leaf gene, 122, 123, 192
*cut 1* gene, 259
*cvi* gene, 113
Cyanide, 537
Cyanogenesis, 537
Cyanogenic glycosides, 537
Cyclodextrins, 638
*Cydia pomonella*, 329
*Cylas formicarius* sp., 261, 712
*Cynara scolymus*, 495, 497
*Cyperaceae*, 492
Cyastatins, 242
Cytogenetics, 30
Cytokinins, 49, 76, 77, 375,
Cytoplasm, 495, 497
Cytoplasmic male sterility, 486, 504, 779, 813, 815, 842
Cytotoxic RNase, 191

$\delta$-15 desaturase, 756
2,4-D induction, 497
D+l-phosphothricin, 776
Daidzein, 211
*DapA* gene, 808
DAS-ELISA, 566, 591
*Datura* spp. 46
  *noxia*, 44
*Daucus carota* subsp. *sativa*, 146, 159, 503, 836
  disease resistance, 503, 504, 517
  fungal pathogens, 515
  genetic engineering, 504
  somatic embryogenesis, 504
  tissue culture, 504
*ddm* gene, 125, 126
*Def H9* gene, 595
*Def H9 iaam*, 596–597
*Def H9* promoter, 596
Defensin, 251, 428, 764, 765
*dek* mutation, 111
Delayed senescence, 621
Deoxyribonucleic acid methylation, 631
Dextrinase, 726

DHDPS (*see* Dihydrodipicolinate synthase)
Dihydrodipicolinate synthase, 726, 779, 806, 807, 808, 809
Diamond back moth larvae, 256, 779
*Dianthus caryophyllus*, 533
*Diaeretiella rapae*, 263
Dicamba, 700
Dichloromethane, 741
Diclofop-methyl, 617
Dihaploid plants, 724
Dihydroflavonid reductase, 403
Dimethylallyl pyrophosphate:AMP dimethylal-lyltransferase, 621
*Diosiorea* spp., 525
Direct
  embryogenesis, 398, 399
  DNA delivery/uptake, 137–141, 619
  gene transfer, 488, 605, 706, 739, 821
Disease, 590
  incidence (DI), 574–576
  resistance, 504, 35/15, 719, 738, 739, 742–745, 755, 756, 768, 769, 774, 783, 836, 842
Diuron, 615, 617
DNA
  amplification fingerprinting, 88
  amplified fragment length polymorphs (AFLP), 89, 90
  and embryogenic cells, 140
  and microparticle bombardment, 138–139
  delivery, 137–144, 145–149, 723
  electroporation, 139
  empirical evidence 147–148
  expressed sequence tags (EST), 91, 286, 844
  marker system, 86
  microsattelite loci, 89, 819
  simple sequence repeats (SSR), 88
  Southern hybridization, 619–620
  sequence characterization amplification sites (SCARS), 88, 241
  simple sequence length polymorphism (SSLP), 89
  theoretical background, 146
  tissue culture targets, 140
  whiskers, 139–140
Domestication of plants, 2, 109
Dominant resistance (*Dm*) genes, 616, 622
Double haploid production, 720, 721, 722, 767, 774
Downy mildew, 615, 815
  resistance, 616, 621, 622
*Drosophila* spp., 111–112, 122–124
Drought resistance, 738
  tolerance, 720, 744, 827
Dwarfness, 710

*dul* 1 mutants,
*dzr* –1 locus, 114

East African mosaic virus (EACMV), 532
EC countries, 12
*ech* gene, 609
Ecological risk, 459
Economics and
   transgenic plants, 311–325
   innovation, 313, 314
   scale and scope of, 314, 315
   agri-food industrial structures, 316–17
   trade and regional growth, 316
   winners and losers, 316–317
Edible oil, 815
EF1σ-A1 promoter, 740
Eggplant, 587–601
   and parthenocarpy, 593–598
   biotechnological approaches, 589
   breeding approaches, 589
   chimeric gene induced parthenocarpy, 594–597
   doubled haploid breeding, 589
   field performance of transgenic lines, 591
   genetic engineering, 590–593
   growth, 594
   greenhouse evaluation of parthenocarpic hybrids, 797–598
   history, 588
   insect resistance, 590
      *Cry* 1 transgenic lines
      *Cry* 3 transgenc plants
      Colorado potato beetle resistance, 590–591
   molecular markers, 589–590
   nutritional value, 600
   open field production, 593
   primary transformed plants, 896
   quality, 590, 594
   shoot and fruit borer resistance, 593
   somatic hybridization, 589
   world production, 588
EHA 101, 580, 703
Electroporation, 470, 478, 480, 485, 488, 490, 648, 722, 723, 744, 804, 821, 837
Electrostimulation, 818
Elongation factor, 644
Elongation, 565
*Elymus*, 719
   *canadiensis,* 720
Embryo co-cultivation, 605, 608–610,
Embryo, 110, 440, 719–723, 749, 775, 804, 807, 814–822, 836–844
Embryogenesis, 43–56, 70, 157, 361, 398, 399, 720, 749, 775, 816, 836
   and biochemical pathway, 56

[Embryogenesis]
   and double haploid plants, 52
   and gene transfer, 55–56
   and haploid plant production, 52–54
   and mutational breeding, 55
   application of, 54–56
   chromosome doubling, 52
   indirect, 398, 399
   plant regeneration, 52
   varietal development, 54–55
Embryogenic
   callus, 702
   cell suspension, 361, 362, 365
   cocultures, 352, 353
   culture maintenance, 351
   embryonic
      axis, 476
      callus, 258, 721, 817
*Emu* promoter, 362
Endo-1,4-β-xylanase, 726
Endochitinase, 332, 402, 677
Endosperm, 110–126
   shrunken, 112
   size, 125
Endotoxin, 676
Engineered viral protein genes, 658
Enhancer of *zeste* (Ez)-gene, 122
Enolpyruvylshikimate 3-phosphate synthase (ESPS) (*AroA*), 487, 648, 657, 778, 839
Enterotoxin, 675
Environmental stress, 334
Enzyme-linked immunosorbant assay (ELISA), 628, 644
Epidermal stripping, 751
Epigenetic, 111, 120, 124, 125, 819
   engineering, 125
Epimutations, 111
Epistasis, 120
*Eriodaphna* spp., 345, 350
ER-resident chaperonin (BiP), 644
Erucic acid, 773–775, 781
*Erwinia*
   *amylovora*, 332, 497
   *carotovora,* 361, 677
*Erinnyis ello*, 534
*ery* gene, 534
*Erysiphe cichoracearum*, 438, 619, 719
   *graminis* f. sp. *Hordei*, 719
   *heraclei*, 517
*Escherichia coli*, 103, 198, 210, 383, 441, 505, 550, 590, 591, 675, 726, 744, 768, 776, 808, 840
   RecA protein, 744
   enterotoxin, 675

Ethical issues, 119–121
Ethics, 20
Ethrel, 741
Ethylactate, 741
Ethylene, 250, 334–336, 348, 349, 366, 375, 377, 412–418, 425–427, 431, 439, 457, 555, 738, 741, 818
*Eucalyptus globulus,* 74, 156, 187
*Euchlaena mexicana,* 638
*Euphorbia pulcherrima,* 158
European corn borer, 332, 742
European Community-Union (ECEU), 12
Exochitinase, 332
Explants, 160
Expressed sequence tags (EST), 91, 286, 844
Expression, 564, 567–572, 577

Factor IX gene, 744
Factors related to maize kernel hardness, 641
FAO, 6,
Farming
   molecular, 713
   traditional, 23
Fatty acid
   acid elongase (FaLs), 37
   composition, 774, 783, 827
   desaturase, 210
   elongase, FALS, 37
   ω-3-fatty acid desaturase gene, 711
   fertilization independent endosperm (*fie*) gene, 112, 123, 192
   profile, 756 (*see also* Oil profile)
Fermented breads, 802
Ferridazon, 102
Fertile crescent, 2, 717
Fertilization independent seeds (*fis*), 112
*Festuca spp.,* 116
   *arundinacea*, 140
1-*fft* gene, 498
*fie* gene, 112, 123, 192
FIL gene, 123
*fis* gene, 123, 192
Field test, 564, 567, 574–576, 577
Field trial, 443, 445, 574–576, 590, 592
Fire blight 332
Flavor, 427, 458
FlavrSavr tomato, 307, 312, 450
Flax, 747–761
   CDC Triffid, 758
   L6 gene, 755
   regulatory issues, 757–758
   value added traits, 756
   weed control, 757
Floral organ identity (FOI) gene, 189–192

Floral homeotic (AGAMOUS) gene, 122
Flowering, 183–196
   and biotechnology, 187–188
   and floral meristem identity, 186–187
   and time, 185–187
   genes controlling, 184–185
   manipulation, 183, 660
   prevention, 187–188
   redesign of, 188–189
Flowers, 189–192
   and hybrid seed production, 191
   emasculation, 596
   new arrangements organs, 189–190
   on demand, 185–188
   organ identity genes, 189
   without petals, 190
Fluorometric assay, 628
Food
   and Agriculture Organization (FAO), 6,
   and Drug Administration (FDA), 21, 298, 303, 459
   crops and agriculture, 1–28
   grain, 638
   ingredients, 17
   marketing Institute, 295
   Quality Protection Act (USA), 450
   security, 11, 22, 524, 537
   type maize hybrids, 640
*Fragaria x ananassa,* 449
*Frankineliella occidentalis,* 589
*Freesia refracta,* 159
Friable embryogenic callus (FEC), 528, 530
   embryogenic callus, 497
Fructan, 485, 491–498, 654
   containing beets, 660
   fructan fructosyltransferse (1–FFT), 492, 494, 498, 660
   products, 660
Fructans, 659
Fructose, 333, 377
Fruitfly, 438
Fungal pathogens, 497
   resistance, 676, 742, 743, 827, 841
Fungicides, 657
*Fusarium*
   *oxysporum*, 360, 365, 373, 466, 589, 604, 657
   *graminearum*, 842
   *moniliforme*, 466
   resistance, 466
   *solani*, 17, 373
   spp., 17, 457, 589, 743, 841
Fusarium
   tolerance, 466
   basal rot, 657

[Fusarium]
  recessive gene, 616

g recessive gene, 616
G418, *see* Geneticin
*Galanthus nivalis*, 243, 257, 454
  *pallida,* 243
Gametes, 99
Gametocides, 100
Gametocytes, 100
Gametophyte indeterminate, 118
Garlic, 209, 655
  and shallot virus X, 658
  latent virus, 658
  microprojectile bombardment, 661, 665
  transformation protocol, 661
Geminivirus, 373
Gene(s)
  AFLP markers, 89–90
  amplification based markers, 87
  amplification fingerprints, 88
  analysis, 37
  and cytology, 30–31
  chips, 16, 159
  code, 621
  complex traits, 90
  DNA, 86
  Dose ratio (GDR), 116
  hybridization based
  inactivation, 783
  integration, 776
  marker genes, 85–98, 777, 823, 827, 837
  molecular marker systems, 85–90
  promoters, 629
  pyramiding, 263
  quantitative trait loci mapping, 91–92
  RAPD DNA, 87–88
  RFLP analysis, 86–87
  sequences, 658
  silencing, 376, 400, 631, 648, 727, 728, 784, 843
  simple sequence repeats, 88–89
  simple traits, 90
  single nucleotide polymorphism, 87
  single primer amplification reactions
  tagging, 90–91
  transfer, 20, 37, 748, 754, 769, 775, 804, 816, 818, 820–822, 827, 837, 840
Gene control
  in flowering time, 185
  for meristem identity, 186
General Agreement on Trade and Tariffs (GATT), 12–13, 280, 284

Genetic
  cosuppression, 432
  engineering for phytonutrients, 207–216, 591
  in valuable agronomic genes, 469
  map, 590
  marker based map, 86
  tagging, 90, 91
  transformation, 590
Genetically Modified organisms (GMO), 305–308, 312, 320, 323, 324
  Foods, 321
Geneticin (G418), 490, 648, 751, 839, 652
Genome(s), 29–41
  chromosomal rearrangement, 34
  comparative mapping, 35–36
  crop improvement gene analysis, 37
  dosage ratio in endosperm, 115–117
  dynamics, 35
  gene transfer, 37
  genomic evolution, 33–35
  instability, 116
  mapping, 31–32
  of plants, 29–38, 458
  plasticity, 20
  polyploidy, 34
  sequencing, 32–33
  structure, 33
  transposable elements, 35
Genomic(s), 14–15
  and human health factors, 212
  and integrated genetic transformation, 540
  and inter-ploidy crosses, 117
  comparative, 91
  functional, 15
  imprinting, 113–114, 118, 122–124
  use in potatoes and maize, 117–118
Genotype dependency, 176
Gentamycin, 824
*Gerbera* spp. 50
Germplasms, 13–14
  Elite, 115
GFP (*see* Green fluorescent protein)
*Gfp* expression, 652
*Gfp* reporter gene, 658
*Ginko biloba,* 49, 50
*gl*-1 gene, 59
*glgA* gene, 675
*Globdera* spp., 233
  *pallida,* 233
  *rostochiensis,* 233, 236, 238, 241
Glucan, 764
Glucanase, 725, 726, 764, 842
Glucose oxidase, 676

# Index

Glucosinolates, 209, 211, 773, 774, 780
Glucuronidase, 780
Glufosinate, 390, 392, 487, 624, 756, 757, 777, 838
   ammonium, 623
Gluten, 841
Glutenin subunits, 841
*Glycine max*, 49, 77, 146, 605
Glycogen synthase, 675
Glyphosate (Roundup), 487, 504, 535, 657, 720, 756, 757, 778, 839
   oxidase reductase (GOX), 487
   resistance, 648
Golden mosaic gemini virus (GMGV), 469
GNA gene, 257–265
GOX gene, 839
*Gpa* 2 locus, 241, 242
Graminan, 492
Granule-bound glucose synthetase (GBSS), 536, 712
Grape, 842
   *Agrobacterium*, 400–401
   Arabis mosaic virus (GAMV), 402
   Cabernet sauvingnon, 398
   Chardonnay, 398
   co-cultivation, 401–402
   Fanleaf virus (GFLV), 402
   gene delivery systems, 400–402
   mold resistance, 402
   Nepovirus, 402
   particle bombardment, 400
   proembryonal complexes, 398
   regeneration systems, 398–399
   seedless varieties, 402
   selectable markers, 399–400
   selection of transformed cultures, 401
   tissue culture, 401
   transgenic clones, 402–403
   vines, 397–405
   viruses A and B, 402
Gray mold, 456, 815
Green fluorescent protein (GFP), 55, 74, 163, 391, 365, 400, 452, 624, 648, 662, 824, 844
   reporter gene (*gfp*), 365, 658–662
GreenPeace, 300, 303
Green revolution, 7, 10, 22
*GroL* locus, 241
Grocery Manufacturers of America, 295
Guard cell protoplast, 490, 493–494, 498
Guma, 802
GUS (*see* β-Glucuronidase), 74, 362, 376, 391, 399, 415, 431, 452, 470, 477, 478, 580–582, 590, 605–608, 610, 631, 703–708, 750, 754

[GUS]
   activity, 631
   expression, 480, 540
   gene, 592, 630
   reporter gene, 362, 399, 652, 662
Gynogenesis, 43, 50, 113
   and developmental stage, 51
   culture conditions, 52
   donor plant, 50
   influencing factors, 50–52
   media constituents, 51

H1 locus, 241
*Ha1* and *Ha2* locus, 241
Hairy root, 709
*HAL-1* gene, 377
Haploid, 590
   breeding, 721
   initiator gene, (*Hap*) 721
   plant, 51, 749
Heat shock proteins (HSPs), 17, 53, 54, 68, 70
*Heliothis virescens,* 258
*Helianthus*
   *annuus* L., 78, 813–819
   *smithii*, 817
   spp., 51, 815
   *tuberosus*, 78, 492, 497–498, 814, 816
*Helicotylenchus multicictus,* 36
*Helicoverpa zia,* 260, 261
Helium shock wave, 488
*Helminthosporium sativum,* 720
*Hemileia vastatrix,* 743
Hemizygous seed progeny, 621
Hepatitis B antigen, 366
Herbicide, 487
   resistance, 365, 390, 391, 403, 459, 535, 536, 615, 617, 621–623, 649, 656, 661, 724, 738, 740, 744, 745, 748, 751, 755, 757, 776, 777, 783, 804, 805, 827, 838, 840, 844
   resistant biotype, 617
   resistant radicchio lettuce, 631
   resistant transgenics, 535
   tolerant, 486, 487
*Hero* gene, 238
*Heterodera*
   *aveanae*, 241
   *glycins*, 233, 241
   *schachtii*, 233, 236, 238, 487
   spp. 233
Heterosis, 99, 120–122
   additive effects, 121
   dominance effects, 121
   overdominance, 121

Hirudin, 780
*Hmg2* gene, 244
*Hominy*, 639
Homozygosity, 721
Homozygotes, 566, 567, 569–577
Homozygous
  line, 442
  progenies, 621
Homoptera, 842
Hop plants, 210
*Hordeum*
  *bulbosum*, 719, 721
  *jubatum*, 720
  *spontaneum*, 717–721
  *vulgare*, 116, 497, 717–720, 778
Hordothionins, 725
Hornworm, 534
*Hph* marker gene, 529, 580
*Hpt* (Hygromycin phosphotransferase) gene, 362, 452, 605, 648, 839
*Hs1* resistant genes, 480
*Hs1pro-1*, 236, 237, 242, 488
*Hs1web-1*, 488
*Hs1web-7*, 488
Human health, 207–216, 437
  and immunity, 366
  and insulin, 676
  and micronutrients, 210
  and mining of plant genomes, 212
  and nutrition, 208–209
  and phytonutrients, 207–208, 337, 402
  and plant micronutrients, 210
*Humulus lupulus*, 210
Hybrid sterility, 719
  vigour, 120
Hydroxynitrile lyase (HNL), 537
Hydroxyproline-rich glycoprotein, 538
Hygromycin, 580, 582, 605, 648, 703, 768, 776, 824, 839, 840
  resistant calli, 581
  phosphotransferase, 776, 805
*Hymenoptera* spp., 534
Hyperhydric, 352, 415, 430
Hypocotyl, 489, 554, 564, 565, 567, 577
*Hypothenemus hampei*, 739, 742

*Iaa*M gene, 596
IBA, 565
*Imperata cylindrica*, 535
Inbreeding depression, 122
*In planta*
  electroporation, 139, 605, 606
  transformation, 722

*In vitro*
  mutation, 430
  selection, 430
  inbred line, 445
Indirect DAS ELISA, 566, 575
Indole
  acetic acid (IAA), 565, 596
  acetamide, 596
  glucosinolates, 780
Inflorescences, 720
*Injera*, 640
Innovation, 311
  economics of, 313
  chain link model, 314
  first generation, 315
Insect(s)
  bioassay, 556, 591
  peritrophic matrix, 252, 257
  pest management, 262–265
  pests resistance, 534
  plant interaction, 249–263
  protected crops acceptance, 300
  resistance, 249–263, 547, 548, 558, 591, 649, 676, 738–742, 783, 802, 804, 827, 841, 842
Insecticidal
  (*Bt*) toxin, 534
  crystal protein 330, 331, 332
Insulin, 676
*int* gene, 250
Integrated pest management (IPM) programs, 263–265, 657
Intellectual property, 754, 757, 828
  freedom to operate (FTO), 319
  Paris Convention, 280
  principles, 280, 286
  politics of, 317
  protection, 263–265
  rights, 13, 18, 263–265, 657
  Trade related aspects (TRA), 318
International
  Cassava Biotechnology Research Unit (ICBRU), 540
  Food Information Council, 295–296
  Institute for Tropical Agriculture (IITA), 16, 527, 532, 541, 604
  International Laboratory for Tropical Agricultural Biotechnology (ILTAB), 529, 530, 533, 534, 536, 539
  ICARDA, 604
  ICRISAT, 604
  Plant Protection Convention (IPPC), 12, 321
  Office of Epizootics (IOE), 321

# Index

Interploidy crosses
   *Arabidopsis*, 117
   poato, 117
   maize, 118
Inulin, 492–9
Inventions, 260–262
Invertase, 458, 488, 674
   inhibitor, 675
2ip, 581
*Ipomoea batatas,* 699
*Ipt* code, 621
*Ipt* gene, 622
Isoflavones, 211
Isozyme analysis, 618
Isozymes, 85, 590

Jasmonic acid, 250
Jerusalem artichoke, 492–11, 498
Jumping genes, 35

*Kalanchoe* spp. 78
Kanamycin, 390, 399, 415, 443, 478, 490, 511–515, 555, 566–567, 569, 577, 580, 581, 604, 609, 620, 625, 751, 752, 776, 777, 805, 821, 824
   resistance, 470, 566–569, 572, 577, 591, 629, 648
   resistant calli, 581
   resistant plants, 619, 621, 628
   resistant shots, 621
   sulfate, 623–628
1-kestose, 492, 496
Kinetin, 49, 476, 504
Kisra, 802
Kiwi fruit, 407–420
   Agrobacterium mediated transformation, 409–415
   direct DNA transfer
   DNA transfer protocols, 409–415
   genetic improvement, 417–418
   genetic transformation of, 409–418
   history, 408
   progeny stability, 417
   regeneration systems, 408
   transgenic interaction, 417
Knotted gene (KN1), 79
Kokei 14, 702

*Lacanobia oleracea*, 257
*Lactuca*
   *perennis*, 619
   *deblis*, 619
   *indica*, 619
   *saligna,* 615–617, 619

[*Lactuca*]
   *sativa,* 613–619
   *serriola,* 613, 615–617, 619
   *tartarica,* 619
   *virosa,* 614, 615–616, 619, 623
Lachrymatory compounds, 659
*Lactuca* species, 614–616, 618
Late embryogenesis abundant proteins, 743
Laurate, 781
*LBA 4404-GUS* report gene, 477
*LC* gene, 259
Leaf aphid, 617, 619
   resistance, 615
Leaf disk, 489
LEAFY (LFY) gene, 186, 428
Leafy cotyledon 1 gene (*lec1*), 192
Lectin, 243, 257, 454, 842
   genes, 659
   toxcity, 842
*Leek,* 655
Leek yellow strip virus (LYSV), 658
Legumes, 197–204, 475–482, 603–610
Legume seed, 475
*Lemmi9* gene, 244
*Lens culinaris,* 603, 607
Lentil, 603–607
   Agrobacterium–mediated transformation, 607
   production statistics, 604
   screenable marker, 607
   susceptability to *A. tumefaciens*, 607
   transformation, 607
   world production, 603–604
*Lepidopterae*, 740
Lepidoptera, 253–256, 756, 842
*Lepitotera* spp., 534, 535
*Leptinotarsa decemlineata,* 589–593
*Leptosphaeria maculans,* 766, 768, 778
Ler *gene,* 113
Less Developing Countries (LDCs), 524, 526, 530, 536, 538–542
Lettuce, 613–632
   agronomically useful genes, 621
   biotechnology, 618
   breeding objectives, 615
   classification, 623
   conventional breeding, 614–615
   cultivation, 614
   disease resistance, 616
   downy mildew resistance, 616
   extended postharvest shelf life, 615
   genome, 631
   herbicide resistance, 617

[Lettuce]
   history of, 614
   inactivation of transgene expression, 629
   improved quality, 615
   increased postharvest shelf life, 618
   infectious yellow virus (LIYV), 629
LIYV resistance, 629
   insect resistant varieties, 615
   leaf aphid, 617
   male sterility, 615
   morphological types, 614
   mosaic virus (LMV), 615, 621, 629
   (LMV-CP) gene, 629
   coat protein, 621
   strain O (LMV-O), 621
   resistance, 615–616, 629
   nutritional content, 614
   origin, 613–614
   product quality, 617
   size and yield uniformity, 618
   transformation, 624
   world production, 614
*Leucinodes orbonalis*, 589–591
Leucoantheocynidine deoxygenase, 403
Levan, 498
   sucrase, 495
Liliaceae family, 465
*Liliaceae*, 492, 805
Linamarase cDNA clone, 540
Linamerin, 537
Linola, 756
Linoleic acid, 748, 756, 815
Linseed, 747–761
*Linum*
   *marginale,* 749
   spp., 756
   *suffruticosum*, 749
   *usitatissimum*, 747
Lipid transfer proteins, 764
*Liquidambar styraciflua*, 260
*Lolium* spp, 116, 140
   *multiflorum*, 140
   *perenne*, 140
Lotaustralin, 537
Low-calorie sweetener, 493
*Lox* gene, 648
*Lsc,* 497
*luc* marker gene, 104, 529
Luciferase, 488, 824
*Lycopersicon esculentum*, 45, 236
   *peruvianum*, 236
   *pimpinellifolium*, 116

*Lys*C gene, 808
Lysine, 726, 802, 803, 806–809
Lysin-ketoglutarate reductase (LKR), 808, 809
Lytic peptides, 332

Mac promoter, 629
Macadamia nuts, 763–771
   antimicrobial peptide 1 (MiAMP1), 765–769
   husk spot, 769
   nuts, 763
*Macadamia integrifolia,* 765
*Mahewu*, 639
MADS-box gene, 596
*mae, mag* genes, 241
*Mafo*, 802
Maize, 495, 637–638, 647, 742, 764, 779, 801, 802, 804, 808, 838–840
   Ac transposase, 621
   alkaline cooking processes, 641
   biochemical components, 641
   dent maize, 640
   double mutants, 643
   dry milling, 639
      fractions, 639
      milling industry, 640–641
   dry processing, 638
      products, 638
   endosperm protein, 644
   extruded corn snacks, 639
   fatty acid profiles, 646
   flint maize, 640
   food uses, 639, 640
   genetic engineering, 638, 647
   germ, 646
   gluten feed/meal, 640
   grain processing, 640
   grain types, 640
   herbicide-tolerant hybrids, 647
   high amylose *(ae)* mutant, 642
   history, 638
   improved lipid composition, 646
   insect resistant hybrids, 647
   kernel
      hardness, 641
      minerals, 646
      size, 118
   lipids, 645–646
   methionine-containing zeins, 645
   milling properties, 647
   modified traits, 649

# Index

[Maize]
  mutants, 643
    *ae* (amylose extender), 172, 643
    *bt* (brittle starch), 643
    double, 643
    *du* (dull starch), 172–178643
    *fl* (floury starch), 643
    *floury2*, 644
    high amylose *(ae)*, 642
    modified *opaque2*, 644
    *Opaque2*, 641, 644
    *sh* (shrunken starch), 643
    *su* (sugary starch), 172, 178, 643
    triple, 643
    *wx* (waxy starch), 172, 642–643
  nixtamol, 638
  non-traditional products, 638
  nutritional quality and value, 646, 647
  oil, 638, 640
  oil genotype, 645
  phytate, 464
  product development, 647
  products, 638
  protein, 641, 644
  protein bodies, 644
  quality protein (QPM), 641, 644
  starch, 640, 641
    alterations, 645
    derived products, 640
    modifications, 642
    processors, 642
  transformation, 647
  tripsicum hybrids, 119
  use, 638
  vitamins, 646
  wet
    milling industry, 640
    processing by-products, 638
    processing, 638
  world production, 637
Male sterile plants, 621, 804
  gene, 45
  restoration, 103
Malnourished, 10, 212, 524
Malthus Robert, 3, 10
Malting, 720, 724–728
*Malus foribunda*, 332
*Malus x domestica*, 328, 332
*Mamestra configurata*, 261
*Man*A gene, 840
Mendel, Gregory, 14
Mandioca, 524

*Manduca sexta*, 257, 258, 262, 676
*Mangifera indica,* 421–436
Mango, 421–436
  breeding, 424–429
  crop history, 421–423
  disease resistance, 428
  flavor, 427
  flowering, 428
  genetic manipulations, in vitro, 430–431
  genetic transformation, 423–432
  genetics and breeding, 423–427
  pigments, 427
  plant architecture, 428
  ripening, 424–425
  root stock cultivars, 428
  scion cultivars, 424–428
  selections, in vitro, 431
  somatic cell,
    embryogenesis, 429–430
    genetic manipulations, 429–430
  taxonomy, 422–423
  world production, 421–422
*Manihot esculenta* cultivars, 530, 532
  *flabelifolia*, 524
  *glaziovii*, 532
Manioc, 524
Mannitol, 48
Mannose-6-phosphate isomerase, 840
Marker assisted gene cloning, 236
*Masa*, 638
Matrix attachment regions (MARs), 728
McClintok, Barbara, 14, 35
*mea* gene, 122, 123–124
*mea* locus, 114, 122–124
*medea*-112
  gene, 122, 192
Mediterranean fruit fly, 438
Melon fly, 438
Mealybug, 534
  resistance, 540
*Melampsora lini,* 756
*Meloidogyne* spp., 233, 235, 236, 238, 243, 361, 589, 739
  *arenaria*, 238, 241
  *incognita*, 233, 236, 238, 243, 244
  *javanica*, 238
Melon, 371–381
  altered ripening, 377
  cantaloupe, 371
  casanba, 371
  explant factors, 375
  fly, 438

[Melon]
  fungal diseases of, 372–373
  growth media components, 375
  honeydew, 371
  improvements, 372
  introduced traits, 377
  muskmelon, 371
  regeneration, 373–374
  risk assessment of transgenic, 377–378
  somaclonal variation, 376
  source of explant, 373
  transformation, 376
  wintermelon, 371
Meristematic ring, 476
Methionine, 470, 779
Methionine-rich, 469
Methylisoborneol (MIB), 743
*Metopolophium dirhodum*, 261
*mi* gene, 236, 238, 242
*Micena*, 743
Microparticle bombardment, 128, 362
  tungsten, 138
Micropropagation, 70, 156, 443
*Microsiphum euporbiae*, 238, 676
Microinjection, 722, 775, 804, 821
Micronutrients, 208–212
  engineering in plants, 211–212
  and metabolites, 209
Microprojectile DNA delivery system, 723, 724
Microsatellite markers, 819
Microspore, 47, 720, 723, 775, 776, 821
  culture, 721, 724, 749, 774, 775
*Mingau*, 639
*mixta* gene, 259
*Mo* recessive gene, 616
Modification of physiological characteristics, 621
Moko disease 361
Molecular markers, 85–90, 239, 590, 600
Monelllin, 427
*Mononychellus tanajoa*, 534
*ms 10$^{53}$* gene, 45
MS medium, 564, 565, 567, 569
Multiple insertion loci, 753
Multitrophism, 262
*Munuca* (Brazil), 639
*Musa* spp. 76, 421
Mutagenesis, 774
Mutagenic treatments, 721
Mutant
  AHAS gene, 657
  CsrL-L genes, 657
Mutation breeding, 55, 718, 756, 803

*myb* gene, 259
Mycopathogens, 16
Mycorrhizal, 489
*Mycosphaerella*
  *fijiensis*, 360, 365
  *musicola*, 360
Mycotoxicosis, 16
Mycotoxins, 12
*Myzus persicae*, 261, 263

*nap*, 101
*Nasonovia ribis nigri*, 617, 619
Natural starch, 644
Neck rot, 657
Nematodes, 233–248, 255, 361, 365, 453, 487, 488
  cyst nematodes, 233–234
  disrupter of feeding cells, 244
  economic loss due to, 233
  life cycle and infection, 234–236
  marker assisted selection, 236–238
  root knot nematodes, 235–236
Nematodes resistance,
  artificial, 242–244
  crops, 241
  effector genes, 236, 242
  genes, 236–242
  in potato, 239–241
  in sugar beet, 238
  in tomato, 238–239
  natural, 236
*neo* gene, 362, 605, 609
Neomycin phosphotransferase II (NPTII), 530, 648, 722, 750, 751, 776, 805, 808, 824
  *npt*II gene, 353, 376, 390, 399, 416, 440, 452, 505, 506, 552, 580, 590, 591, 610, 619–620, 623, 629, 665, 722, 750, 751, 776, 805, 808, 824
  marker gene, 353, 529
  protein content, 623
*Nehpotettix cincitepes*, 260
*Nicotiana*
  *benthamiana*, 221, 225, 455, 456, 531, 532, 684
  *bigelorii*, 225
  *clevlandii*, 455
  *plumbaginifolia*, 628
  *sylvestris*, 806
  *tabacum*, 44, 146, 159, 219, 221, 455, 456, 677, 836
*Nilaparvata lugens*, 260
Nitrosamine, 618
*Nixtamal*, 638

Nonstructural virus genes, 680
Nopline, 470
Nopline synthesis, 480
North American Free Trade Agreement (NAFTA), 280, 284
Northern blot analysis, 592, 629
NOS, 580
   gene, 629
   promoter, 620
Nos.nptII.nos
   gene, 511, 512, 630–631
   terminator, 620
Npo virus, 455
NPT II (see Neomycin phosphotransferase II)
Nsima, 802
Nucellus, 351, 429
Nucleocapsid (N) protein gene, 621
Nutraceutical, 727, 784
Nutrition
   and crops, 10–14, 207–213
   quality improvement, 479
   value, 645, 675

Oat, 49
Ochractoxin A, 742
Oenethera spp., 116
   hooker, 116
Ogi, 639
Ogu, 101
Oidium caricae, 438
Oil profile, 756 (see also Fatty acids profile)
Oleic acid, 756, 815
Oleosin, 210, 779
Oligofructan, 485, 493, 495
Onion, 655–671
   Agrobacterium–mediated transformation, 661
   bacterial diseases, 658
   cytoplasmic genome, 661
   leaf blight (Xathomonas spp.), 658
   microprojectile transformation protocol, 661
   nuclear genome, 661
   soft rot (Erwinia spp.), 658
   sour/slippery skin (Pseudomonas spp.), 658
   transformation protocol, 661
   white rot control, 657
   yellow dwarf virus (OYDV), 658
Organ specific expression, 4/10
Organization for Economic Cooperation and Development (OECD), 320
Organogenesis, 70, 72, 156–160, 376, 476, 530
Organosulfur compounds, 659
Oriental fruitfly, 438

Orobanche cumana, 815
Orthosia hibisi, 331
Oryza sativa, 45, 50, 793–795
Oryzacystatin 1 gene, 242
Osmotin gene, 764
Ostrinia nubilalis, 331
Otioryhnchus sulcatus, 453
Over-expression, 644
Oxalate oxidase, 778
Oxycoccus spp., 384

P450 cytochrome, 537
pABD1, 619
pAHC25, 805
PAL-2 promoter, 540
Palm, 773
Palmitic acid, 756
Panama disease, 360
Pandemis prysuana, 331
Panicium spp., 119
   virgatum, 119
Pantalonia nigronervosa, 360
Papain, 437
Papaya, 437–447
   acreage in Hawaii, 438
   aluminium tolerance, 440
   annual production, 438
   Colombian, 438
   culture initiation, 440
   delayed ripening, 439
   desirable virus resistance, 439
   DNA extraction, 441
   extended shelf life, 442
   field evaluation of transgenic, 442–445
   fungal diseases, 438
   genetic engineering of, 438–445
   herbicide resistance, 440
   hermaphrodite plants, 443
   particle gun usage, 441
   plasmids, 440
   ringspot virus (PRSV), 223, 373, 438, 440, 441, 446,
   transformation, 439
      methods, 439
   virus resistance, 439
   world production, 438
Paramutation, 121
Parent of origin effect, 109–135
   and plant breeding, 114–115
   effects at gene level, 111–113
Paromomycin, 529, 824, 839
   resistant callus lines, 532
Parthenocarpy, 594

Particle bombardment, 400, 440, 478, 479, 488, 485, 490, 528, 532, 723, 727, 728, 739, 754, 755, 776, 804, 805, 807, 821, 822, 839
   based transformation, 535
*Paspalum notatum*, 119
PAT (*see* Phosphothricin acetyltransferase)
Patents, 279–291, 757
   Biotechnology Directive, 280, 284, 285
   Budapest Treaty, 283
   Canadian, tissue electroporation, 139
   claims define invention, 281
   Cooperation Treaty, 280
   deposit of organisms, 283
   disclosure, 283
   European Convention (EPC), 280, 285
   general principles, 280–281
   infringement, 284, 285–286
   interference, 284
   international framework, 280
   inventorship, 284–285
   microparticle bombardment, 138
   novelty, 281–282
   obviousness/inventive step, 282
   plants in U.S., 283, 286–287
   Plant Variety Protection Act, 283
   U.S. vs E. U., 283–290, 318
   utility/industrial applicability, 282
   whiskers, 139
Pathogen derived resistance (PDR), 218
   resistance genes (R-genes), 244
Pathogen related (PR) genes, 428, 620
Pathogenesis-related proteins, 350, 428, 430, 438, 515, 764
PBI121, 580
pCAMV CAT, 619
PCR, 565, 566, 568–572, 577
Pea, 197–205,
   enation mosaic virus (PEMV), 222
   improved agronomy, 202
   quality traits, 203
   regeneration, 201
   *rug 5* mutants, 176
   seed borne mosaic virus, 222
   transformation methods, 195–202
Pearl millet/*Pennisetum* hybrids, 119
Pectate lyase, 400
Pectolytic enzymes, 823
*Pennisetum* spp. 119
*Penax genseng*, 75
*Pentalonia nigronervosa*, 360
*Perileucoptera coffeella*, 739, 740
Perisperm, 110

Peroxidase, 779, 781
*Persea*
   *americana*, 345
   *borboria*, 350
   *cinerascens*, 350
   *pachypoda*, 350
   *schieneana*, 347, 350
Pest, 590
   management, 262–265
   resistance, 724, 745, 815
*pet E* gene, 629, 630
*PetE* promoter, 630–631
Petiole, 489
Petunia, 494, 631
*Petunia hybrida*, 621
Pharmaceutical, 727, 779, 784
Phaseolin, 480
*Phaseolus vulgaris*, 257, 475, 482
*Phenacoccus manihoti*, 534
Phenolics, 337, 393, 394
Phenotype
   hidden, 115
Phenylalanine ammonium lyase (PAL) gene, 538
*Ph 1* gene, 37
Phlein, 492
Phloem-specific promoters, 659
*Phoma* spp., 438
*Phomopsis helianthi*, 815
*Phomopsis oleracea* var. *helianthi-tuberosi*, 815
*Phomopsis* sp., 438
Phosphate-6–mannose isomerase gene, 743
Phosphinothricin (PPT), 487, 490, 648, 623, 751, 783, 824, 838, 839
   acetyl transferase (PAT), 487, 490, 494, 751, 777
   gene, 104, 390, 485, 776, 494
   resistance, 661
Phosphomannose isomerase, 776
Photosynthetic rate, 540
*Photorhabdus luminescense*, 255
*Phyllomorycter* spp. 331
*Phyllotreta* spp., 779
Phytase gene, 780
Phytate, 646
Phytoalexin, 725, 764, 842
Phytochrome, 428
Phytonutrients, 207–213
   chemicals, 259
   and health, 207–216
*Phytophthora*
   *cinnamomi*, 350, 769
   *fragariae*, 449, 458
   *infestens*, 676

[*Phytophthora*]
  *megasperma*, 350
  *palmivora*, 438
Phytophthora root rot, 350, 351, 438, 449, 457
*Phytorimaea operculella*, 676
*Picea sitchensis*, 51
Picloram, 700
Pigments, 336, 337, 427
Pink root rot, 657
*Pinus radiata*, 187, 188
*Pistalata* (Pi) gene, 190
*Pisum sativum*, 197–198
*P-Lec* gene, 264
*PL-mah* locus, 121
*Plamopara halstedii*, 815
Plant(s)
  biotechnology, 590
  breeder's rights, 757
  cell competence, 152
  chemical profiles, 259
  codon usage, 594
  competence, 152
  defenses to insects, 249–278
  deployment, 593
  designer, 189
  disease, 16
  domestication of, 2, 109
  growth regulators, 76
  micro- and macro- nutrients, 207–212
  morphological characteristics, 258, 259, 621
  number of species, 11
  oxidative stress, 260
  secondary metabolites, 260, 261
  systemic response, 250–252
  tolerance to insects, 250, 262
  totipotency, 152, 250, 528
  trichomes, 258, 657
  trichome development genes, 258
  Variety Protection Act (PVPA), 280–290, 318
  virus resistance, 217–222
  wax layers, 259
  wounding, 260
Plant regeneration, 151–167
  and markers, 158–159
  and methods, 161–163
  and selection agents, 161–162
  culture conditions, 161–163
  embryogenesis, 157
  explant, 160–161
  factors in, 151–159
  organogenesis, 156–157
  ovary culture, 721
  pathways, 155–157

[Plant regeneration]
  species and cultivars, 159–160
Plant variety protection, 287–290
  distinctiveness, 288
  novelty, 287–288
  stability, 288
  uniformity, 288
*Plantynota sultana*, 331
Plastocyanin promoter, 629
*Plodia interpunctella*, 256
Plum pox virus (PPV), 222
*Plutella xylostella*, 253, 556, 779
PluPoaceae, 492, 617
*Poa longifolia*, 119
  *pratensis*, 119
*Polenta*, 639
Political consequences and
  transgenic plants, 317–323
  intellectual property rights, 317–319
  agri-food policy, 319–320
  regulation, 320–321
  international problems, 321–323
  framework for managing complex issues, 17–23, 323–324
Pollen, 99–107, 489
  biotechnology, 99–107
  chemical hybridizing agents, 102
  control of development, 100–104
  development, 45, 100
  genetic transformation of pollen, 104
  grain, 45
  male sterile mutants, 101–102
  manipulation, 100–104
  transformation, 840
Polycomb (Pc-G) gene, 122–124, 192
Polyethylene glycol (PEG), 410, 415, 416, 478, 723, 804, 821, 837
  mediated DNA transfer, 490
  mediated transformation, 723
Polygalacturonase, 427, 457, 458
Polygalacturonase inhibiting protein, 457, 458
Polyhydroxybutyrate (PHB), 537
Polymerase chain reaction (PCR), 86–90, 353, 362–365, 376, 416, 441, 506, 510, 565, 591, 620, 710
*Polymixae betae*, 487
Polymorphism at single nucleotide, 87
Polyphenol oxidase, 158, 261, 337, 391, 403, 548
Polyploidy, 34, 376
Popcorn, 640
*Populus tremula*, 78
*Porrage*, 802

Posttranscriptional gene silencing, 621
Potato, 239, 495, 497, 673, 727, 842
   agronomic traits, 674
   antifreeze quality, 675
   bruise resistance, 675
   cholera toxin, 676
   consumption-North America, 673
   carlavirus (PCV), 679
   heterologous gene, 684
   human insulin in, 676
   insect resistance, 676
   leafroll virus (PLRV), 222, 679
   Naturemark, 301
   potyvirus (PPV), 679
   processing quality, 675
   spindle tuber viroids, 677
   transformation protocols, 712
   vaccine delivery, 675
   virus genes, 677, 680
   2X, 221, 677
   Y, 221, 564
   world production, 673
Potyviruses, 373, 658, 679
   sequence-mediated protection, 621, 622
Powdery mildew, 438, 619, 719
*Pozole*, 639
PPT (*see* Phosphinothricin)
*Pratylenchus penetrans*, 361, 453
Preculture, 752
*prf* gene, 242
PR-*gus*–transformed plants, 620
*Primula* spp., 116
Proembryoid, 818
Proembryonic
   masses, 157, 351, 352, 429–431, 504
   complexes, 157, 398
Proembryos, 776
Promoters, 362, 376, 391, 416, 418, 426, 440, 458, 805
Pronamide, 617
Propachlor, 617
Propham, 617
Propyzamide, 617
Protein
   bodies, 644
   content elevation, 539
   improved, 645
   marker system, 86
Proteinase
   cysteine, 242
   inhibitor II, 676
   inhibitors, 242, 256, 452–454, 549–551, 558, 659, 779
   serine, 256

Protoplast, 490, 493, 494, 498, 721, 723, 739, 748, 775, 804, 806, 818, 819, 821, 822, 837
   culture, 361, 618
*Prunus persica*, 88
*Pseudocercospora* spp., 769
Pseudogamous apomixis, 109
*Pseudomonas* spp., 427
   *cichorii*, 619
   *solanacerum*, 361
   *syringae*, 242, 260
   *syringae* pv. *savastonoi*, 596
   *syringae* pv. *tabaci*, 764
   *viridae flava*, 417
*Pseudoplusia includens*, 393
*Pseudoperonospora cubensis*, 373
Public perception, 293–304, 459
   and acceptance, 296
   awareness and knowledge, 296
   of technology, 294
*Puccina helianthi*, 815
Pulse crops, 603–666
Pullulanase, 726
Pungency, 661, 659
*Pyrenopesisa brasicae*, 778
Pyrethroids, 658
*Pyrinochaeta terrestris*, 657

Q-r locus, 113
Quantitative trait loci (QTL), 54, 55, 86, 90, 91, 112, 113, 186, 239, 241

*r*-locus, 113, 114
*Radopholus similis*, 361
Raffinose, 488, 495
Raisins, 403
*Ralstonia solanacearum*, 361, 589
Randomly amplified polymorphic DNAs (RAPDs), 15, 32, 87, 88, 238, 241, 431, 458, 540, 589–591,
Rape plant, 779
Rapeseed, 773, 774, 779, 781, 821
Raphanus, 37
Raspberry, 449–461
   red, 449–464
   black, 449–464
   aphid, 449
   bushy dwarf virus (RBDV), 456
Reactive oxygen species, 337
Recalcitrant monocotyledonous species, 137, 661
Reduced nitrate accumulation, 617
Regeneration, 564–565, 567, 570, 577, 580
Regulatory issues, 757
Regulatory sequence, 596
Resistance gene, 676

# Index

Resistance management, 331
Resistance to viruses, 217–231
  agricultural benefits, 223
  commercialization, 222–223
  engineering of, 218
  environmental risks, 223–226
  factors, 231
  field evaluation, 222
  mechanisms of, 219–221
  opposition to, 224
  potential impact, 223–224
  risk assessment, 224–226
Restriction fragment length polymorphism (RFLP), 15, 24, 31, 86–90, 238, 347, 534, 540, 590, 618
Retinol, 211
Retrotransposons, 35
Reverse transcriptase PCR, 596
*rhg-a* gene, 241
*Rhizoctinia solani*, 457, 515, 517, 677
Rhizomania, 486
*Rhizopus* spp., 438
  *stolonifera*, 438
*Rhopobota naevana*, 393
Ribonuclease, 677
  gene, 677, 779
  inhibitor gene (Barstar), 779
Ribosome inactivating proteins (RIPs), 350, 428, 457, 533, 764
Ribozyme, 677
Rice, 727, 754, 793–800, 801, 804, 805, 808, 835, 838, 839, 842, 843
  *Agrobacterium* mediated transformation, 795
  and culture media, 796
  and regeneration, 797–798
  explant, 797
  genetic engineering, 794–798
  historic yields, 4
  intensification of production, 4
  microprojectile bombardment, 795
  plant regeneration, 795
  *su* mutants, 177
  transformation, 794–797
  tungro spherical virus (RTSV), 222
  Yellow mottle virus (RYMV), 222
Rifamicin, 625,
Ripening, 335, 348, 349, 424, 425, 426, 427, 432, 457
Risk
  benefit-assessment, 18–21, 224, 377, 378
  environmental, 223
  gene flow, 224
  management, 294

[Risk]
  perception, 294–303,
  scientific facts, 236
RNA
  aberrant, 220
  turnover, 220
  silencing 220
  duplex, 220,
  satellite, 222
Rockefeller Foundation, 541–542
*rol* genes, 350, 412, 417, 710
*rol*ABC genes, 417, 621
Root hypersensitive response, 236
Root specific promoters, 538
Root, 565, 567–568
*Roti, chapati*, 639, 802
*R-r* locus, 121
RT-PCR, 566, 570, 577
*Rubus* spp. 449, 452
  *strigosus*, 459
Rust, 743, 755, 756, 815
*RX* gene, 241, 242
Rye, 170, 720

S-(2-aminoethyl)-L-cysteine (AEC), 806, 807
*Sac*B, 497
S-adenosyl methione (SAM), 335, 349, 425, 427
  decarboxylase, 336
  hydroxylase, 377, 427, 457, 458
*Sadza*, 802
Safety, 20–22, 489, 584
Safflower oil, 756
Safeners, 103
Salicylic acid, 250, 251
S-alk(en)yl-cycteine sulfoxides (ACSOs), 659
S-allyl L-cysteine, 659
Salt tolerance, 756
*Salvia sclarea*, 51
Sanitary and Phytosanitary Measures (SPS), 12–13
Sap sucking onion thrips (*Thrips tabacii*), 658
Scab, 332
*Sciara* spp., 113
*Scilla siberica*, 158
Sclerotinia stem rot, 190
*Sclerotinia*
  *sclerotiorum,* 766, 767, 768, 778, 815
  *rolfsii*, 515, 517
  *cepivorum*, 657
Screenable marker, 607, 622
*Secale*, 719
  *cereale*, 45
Secondary metabolite production, 260, 623

Seed
  albumin, 675
  giant size, 125
  hybrid production, 191,
  oil modification, 781
  quality traits, 120
  size manipulation, 125
  storage protein, 469
  without fertilization, 191
  world market for, 109
Seedless fruit, 594
Selectable markers, 399, 620, 751, 776, 777, 824, 839, 840
Selectable/screenable markers, 604–606
Selenium, 209
Self pruning (SP) gene, 188
Senescence-specific promoter, 621
Sesamine, 210
*Sesmum indicum,* 210
Sesquiterpene lactones, 622–623
6–*sft*, 497
Single primer amplification reaction (SPAR), 89
Shallot latent virus (SLV), 658
*Shiva*, 402
Shoot regeneration, 69–80
  abiotic factors, 76
  adventitious shoots, 71
  and carbon source, 77
  and growth regulators, 76–77
  and transgene technology, 70–79
  auxillary shoot formation, 71
  biotic factors, 75
  environmental factors, 77–78
  explant, 75–78
  factors involved, 69–80
  from callus, 74
  genotype development, 76
  role of culture media, 76
Shoot and fruit borer, 591
  resistance, 594
Shoot, 565, 567–568
  elongation, 385, 386
  regeneration, 69, 385
  formation, 70, 71
  production, 74
Sigatoka disease, 360
Signaling molecules, 250
Silicon carbide whiskers, 139, 648
*Sinapis alba*, 773, 774, 778
Single
  chain antibody, 676
  loci insertion, 753
  gene-mediated, 783

Sink strength, 488
Site-specific gene insertions, 648
*Sitobion avenae*, 842
*Sitotroga cerealella,* 842
Snowdrop lectin gene, 842
*Solanum*
  *acaule*, 118
  *berthaultii,* 258
  *commersonii,* 117
  *gourlayi*, 118
  *melongena* var. *insanum,* 589, 590
  *melongena*, 155, 163, 589
  transformation, 117, 655
  *tuberosum* ssp. *tuberosum*, 117, 673
Social issues of transgenic plants, 18–20
Somaclonal variation, 376, 619, 720, 721, 723, 724, 754, 774, 837
Somatic
  embryo production, 661, 723, 816, 817, 819, 827, 839
  embryo maturation, 352
  embryogenesis, 351, 376, 429, 431, 466, 477, 504, 527, 530, 739, 740, 775, 805, 816, 817, 836
  hybridization, 590, 615, 617–618
Sonication, 485, 490
Sorbitol 48, 332, 333, 334,
*Sorghum*
  *bicolor*, 261, 801
  *halepense*, 801
  *propinquum*, 801
Sorghum, 801–812
  and aspartate pathway, 806
  and lysine content, 807–808
  and nutritional value, 802
  improvements in, 803
  insect resistance, 802, 807
  mosaic virus (SMV), 222
  opaque mutant, 641, 645
  transformation methods, 804–806
  uses of, 802
South African mosaic virus (SACMV), 532
Southern analysis, 480, 631, 652
Southern blot analysis, 527, 592, 555, 565, 569–570, 577, 628
Soybean, 754, 773, 808
Soybean looper, 393
Sporamine, 550
*Sphaerotheca fuliginea*, 373
Special Programme on Biotechnology and Development Cooperation (DGIS/ BIOTECH), 541
Spectinomycin, 751, 776

# Index

*Spodoptera*
    *exigua*, 260, 261
    *littoralis*, 258
    *litura*, 552, 556
Spontaneous mutations, 754
Sporamine, 550, 551
Squash leaf curl virus (SLCV), 373
1-*sst* gene, 485, 492–498
Stab inoculation, 488
Starch, 169–182, 470, 641–644, 724, 700
    accumulation, 536
    amylopectin modification, 173–178, 742
    binding protein, 675
    biosynthesis, 170
    branching enzyme, 170, 481
    composition, 675
    debranching enzyme, 177, 481
    gelatinization, 172
    granules modification, 170–173
    granules, 170, 171, 728
    lipids, 646
    market size, 643
    modified structures, 169, 177
    retrogration, 172, 642
    synthase, 176, 481, 675, 711
Stearic acid, 756
Stem borers, 534
*Stemnophyllum* spp., 438
    *lycopersici*, 438
Stilbene synthase gene, 725, 842
Strawberries and raspberries, 449–464
    and genetic engineering, 449–459
    arable mosaic virus (ArMV), 455
    extended shelf life, 457
    *Fragaria* spp., 449–464
    fruit quality, 457–458
    fungal resistance, 456–457
    gene discovery, 458
    gene flow, 459
    genomics of, 458–459
    history of breeding, 449–450
    insect, nematode, mite resistance, 452–454
    latent ringspot virus (SLRV), 455
    mild yellow edge potex virus (SMYEV), 455
    promoters, 458
    regeneration methods, 450–451
    *Rubus* spp., 449–464
    transformation protocols, 451–452
    virus resistance, 454–456
*Streptomyces*
    *griseus*, 257
    *hygroscopicus*, 198, 623, 777
    *viridochromogens*, 777

Streptomycin, 776
Stress tolerance, 719
*Striga*, 803
*Stylosanthes humilis*, 779
*su*-2 gene, 172
Subtractive hybridization, 54
Sucrose, 485, 486, 488–489, 491–493, 495–514
    sucrose fructosyltransferase (1-SST), 485–498, 656–661
Sugar beet, 236–238, 485–502, 840
    agronomic traits, 486–487
    and *Agrobacterium*, 489
    and classical breeding, 486
    and electroporation/sonication, 490
    and engineering methods, 489
    biosafety issues, 489
    fructan production, 491–493
    gene of Jerusalem artichoke, 494
    genetic engineering, 485–502
    guard cells protoplasts, 490
    herbicide tolerance/resistance, 487
    industrial applications, fructans, 493
    nematode resistance, 487
    ogliofructans, 493–498
    particle bombardment, 490
    pest and disease resistance, 487
    quality improvement, 488
    short fructan synthesis, 492
    transgenic superior beets, 495
    virus resistance, 487
Sugarcane, 804
Sulfonylurea resistance, 617, 657
Sulfonylurea, 487, 756, 757
Sulfur-rich protein, 479
Sun blotch, 349
Sunflower, 756, 813–833
    biotechnology of, 815–816
    gene transfer and *Agrobacterium*, 822–823
    introduced traits, 824–827
    oil, 756, 813
    plant regeneration, 819
    protoplasts, 818–819
    seeds, 813–828
    tissue culture, 816
    transformation methods, 820–821
Superoxide dismutase, 211, 337
Susceptibility to *A. tumefaciens* strains, 607, 608
Suspension cultures, 817, 839
Sunflower, agronomy and breeding, 815
    somatic embryogenesis, 815
    gene markers, 824
Sweet corn, 640

Sweet potato, 699–716
  dwarfness, 710
  embryogenic callus, 702
  feathery mottle virus (SPFMV-S), 700, 711
  history, 700
  transformation efficacy, 702
  weevil, 712
  world production levels, 699
Swiss Institute of Technology (ETH), 529
*Synechocystis*, 212
Systematic acquired resistance, 259
Systemin, 250

T4 lysozyme, 677
Tablotoxin, 764
Tamales, 638, 639
Tapetum-specific promoter A9, 621, 630
Tapioca, 524
*tca-d* genes, 255
*T-cyt* gene, 621
T-DNA border, 478
T-DNA-mediated transformation, 648
TDZ (see Thidiazuron)
Technology
  and society interface, 18
  transfer, 265, 541
  use agreements, 317
Teosinte, 638
Terminal flower (*tfl*) mutants, 188
*Tetranychus cinnabornius*, 394, 589
Tesguino, 640
*Tetranychus urticae*, 453, 589
Tetraploidy, 376
Thaumatin, 373, 427
Thaumatin-like proteins, 764
*Thaumatococcus danielle*, 764
*Thermoanaerobacterium thermosulfurogenes*, 394
Thidiazuron, 749, 754
*Thiellaviopsos basicola*, 515
Thionin, 251, 764
Thioredoxin, 726
Thiosulfinates, 659
Threonine, 726
Thrips resistant, 658
*Thrips tabacci*, 658
Ti plasmid
  *vir* gene, 621, 796
  *virB* gene, 621
  *virC* gene, 621
Timenting, 162
Tissue culture, 504
  electroporation, 139

Tmr code, 622
To/tuwo/asido, 639, 802
Tobacco, 495, 497, 754, 764, 768
  coat protein (*cp*) gene, 564–566, 568–571, 574, 576, 577, 580
  coat protein, 454, 564, 566, 568–572, 577
  etch virus, 219
  mosaic virus (TMV), 218, 454, 564, 573–577
  TMV resistance, 564, 566–569, 572–574, 577
  Tob RB-7 gene, 244
  Rattle virus (TRV), 221
  pathogenesis-related protein gene promoter PR1a, 630
  spotted wilt virus (TSWV), 621
Tocopherols, 645
Tomato 45, 238, 455
  aspermy virus (TAV), 221
  black ringspot virus (TBRSV), 220
  bushy stunt virus (TBSV), 225
  spotted wilt virus (TSWV), 226
  wilt spot virus N (TWSVN), 564, 631
*Tortillas*, 637–639
*Tortillas* chips, 639, 802
*Towo*, 802
Trade related aspects of IPR (TRIPS), 280
Trademark, 290
Traditional breeding
  traits, 638
  procedure, 466
Transferring transgenic technologies, 542
Transformant, 566, 577
Transformation, 387, 563, 564, 566–568, 577
  by direct gene transfer, 605–608
  efficiency, 470
  of recalcitrant crops, 661
  selection mutants, 161–163
Transgene, 723, 742, 744, 753, 783, 784, 824, 827, 840, 842
  copy number, 630
  expression, 598, 728, 822, 843
  gain of function, 184
  inactivation, 629
  methylation, 222
  silencing, 629, 631
  stability, 728, 781, 784, 844
Transgenic, 638, 647
  AC1 callus lines, 533
  cucumber, 579
  maize research, 638
  maize, 647
  paradigm of, 17
  parthenocarpic hybrids, 597–599

# Index

[Transgene]
  plant lines, 536, 592
  plants-trade, 11
  resistance to ACMD, 533
  sweet potato, 673
  virus, 631
Transgenic plants, 305–309
  and agronomic benefits, 306
  and consumer attitude, 302
  and food security, 11
  and genomics, 14, 15
  and marketplace, 307–308
  and political consequence, 317–323
  and trade, 11–12
  consumer benefits, 306–307
  economic impact of, 313–317
Transient expression, 490
Transposon tagging, 622, 844
Transposons, 35
Trans-resveratrol, 725
*Trialeurodes vaporarium,* 589
Triazine-resistant weeds, 617
*Trichoderma harzanum,* 609, 677
Trichomes, 258, 657
  *cotyledon trichome* (COT1) gene, 258
  developmental genes, 258
  *globrous (gl)* gene, 258
  *transparent testa glabrous* (*ttg*) gene, 258
*Triapsaccum* spp., 119
*Trithorax* spp., 116
  *trx -G* gene, 122–124
Trierucin, 781
Trifluralin, 617
*Trimerotropis pallidipennis,* 261
Triploid, 486
*Triticum aestivum,* 35, 45, 50, 53, 836
Trypsin inhibitor, 550–556
  *Cme* gene, 725, 842
Tuber
  number, 674
  size, 674
*T-urf13* gene, 779
Type II embryogenic callus, 647

*ubi-1* gene, 362
*Ugali,* 639, 802
*Ugi,* 639
*uid* gene, 416, 417, 540
*UidA* gene, 605–610, 750, 752, 753, 755, 776, 805, 823
*uidA* marker gene, 529
UNICEF, 212
Untranslatable virus sequences, 682

Union for the Protection of New Varieties of Plants Convention (UPOV), 13, 280, 287–290, 318
USDA, 295, 303, 378, 385, 459
USEPA, 331

Vaccine, 366, 675
*Vaccinium*
  *macrocarpon,* 383
  *oxycocus,* 384
Vacuole, 492, 495, 497
Value-added traits, 756
*Venturia inaequalis,* 332
Vertical transmission of antibiotic resistance genes, 648
*Vicia faba,* 77
Vicilin, 765
*Vigna*
  *rhomboide,* 115
  *unguiculata,* 603
Vegetative insecticidal protein (VIP), 255
*Verticillium,* 457, 589
*Vf* gene, 332
Viral coat protein 377, 402, 442, 444, 450, 454–456
Viral resistance, 217–222, 440, 441, 563, 564, 577, 579, 677, 739, 804, 827
Viroid resistance, 677
Virus
  replicated sequences, 682
  resistant plants, 218–227, 531, 563–564, 577, 677, 739, 804, 827
  heterologous encapsidaion, 223
Vitamins,
  Required Daily Allowance (RDA) 209, 211
  A, 208, 437, 503, 647, 675
  B, 208, 359, 503, 647
  C, 208, 336, 359, 437, 503
  D, 208
  E, 337, 647
  K, 208
*Vitis*
  *vinifera,* 398–402
  *rupesteris,* 398–402
  *berlandieri,* 402
  spp., 397

Watermelon mosaic virus (WMV), 222, 373
Waxy
  allele, 222
  layers, 259
Weed beet, 489

Western blot, 479, 532, 555, 565–566, 572, 577, 591, 628
Wheat scab, 842
Wheat, 725, 727, 754, 801, 804, 835–848
   disease resistance, 842
   food crops, 835
   gene pools, 836
   grains supply, 7, 8
   harvest innovations, 5
   harvest, 5
   history, 836
   maize hybrids, 641
   mosaic virus (WMV), 579
   pest resistance, 841
   production, 6
   quality, 841
   tissue culture, 836
   transgenes, 844
Whiskers, 139, 648
*Win 6* promoter, 610
White rot, 657, 815
Whitefly, 532, 575, 616, 658
Wildfire disease, 764
Winter-hardiness, 720
World
   Bank, 10
   feeding, 7–9
   food demand, 1–2
   Intellectual Property Organization (WIPO), 280, 318
   Health Organization (WHO), 10, 802
   population, 523
   supply of grains, 7, 8
   Trade Organization (WTO), 12
World crop production, 6–9
   alliums, 655, 656
   apples, 328
   asparagus, 465, 466
   avocado, 345
   banana, 359, 360
   barley, 717
   carrot, 503
   cassava, 523–525
   chickpea, 603, 604

[World crop production]
   coffee, 737, 738
   cowpea, 603, 604
   cranberry, 383
   eggplants, 588
   food crops (1975–2000), 6–9
   kiwifruit, 407–408
   lentil, 603, 604
   lettuce, 613–614
   maize, 637, 638
   mango, 421, 422
   papaya, 437, 438
   potato, 673
   pulses, 603–604
   sorghum, 801, 802
   supply of grains, 7
   sweet potato, 699, 700
   wheat, 835, 836
Wound-inducible promoters, 538
Wounding, 388, 389, 402

*Xanthomonas* spp., 658
   *axonopodis* pv. *manihotis* (*Xam*), 533
   *oryzae* pv. *oryzae* (*Xoo*), 533
Xanthosine-N7-methyltransferase, 738, 742, 743
*Xiphinema diverrsicaudatum*, 455
X-GLUC, 751, 753
Xylanase, 726, 727

YACs, 32, 238
Yellow dwarf virus (YDV), 720
Yellow maize hybrids, 641
Yuca, 524

*Zea mays,* 35, 49, 53, 54, 71, 75, 110, 495, 637, 638, 647, 742, 764, 779, 801–808, 838–849
Zeamatin, 764
Zeatin, 159, 409, 413, 415, 470, 471
Zein, 641, 644
Zucchini yellow mosaic virus (ZYMV), 222, 377, 579, 686
*Zm* gene, 104
*Zobroties subfasciatus,* 257